Radiochemistry and Nuclear Chemistry

Radiochemistry and Nuclear Chemistry

By

GREGORY CHOPPIN

Professor (retired), Department of Chemistry, Florida State University, Tallahassee, FL, USA

JAN-OLOV LILJENZIN

Professor emeritus in Nuclear Chemistry, Department of Chemical and Biological Engineering, Chalmers University of Technology, Gothenburg, Sweden

JAN RYDBERG

Professor emeritus in Nuclear Chemistry, Department of Chemical and Biological Engineering, Chalmers University of Technology, Gothenburg, Sweden

CHRISTIAN EKBERG

Professor at Stena's Chair in Industrial Materials Recycling and Professor in Nuclear Chemistry, Department of Chemical and Biological Engineering, Chalmers University of Technology, Gothenburg, Sweden

Amsterdam • Boston • Heidelberg • London • New York • Oxford
Paris • San Diego • San Francisco • Sydney • Tokyo
Academic Press is an imprint of Elsevier

Academic Press is an imprint of Elsevier
The Boulevard, Langford Lane, Kidlington, Oxford OX5 1GB, UK
Radarweg 29, PO Box 211, 1000 AE Amsterdam, The Netherlands

Fourth edition 2013

Library of Congress Cataloging-in-Publication Data
Application submitted

British Library Cataloguing in Publication Data
A catalogue record for this book is available from the British Library

For information on all Academic Press publications
visit our web site at store.elsevier.com

ISBN: 978-0-12-405897-2

Photo courtesy: Cherenkov light from the TRIGA reactor at the University of California

CONTENTS

FOREWORD TO THE 4TH EDITION

Nuclear chemistry is one of the oldest nuclear sciences. It may be defined as the use of nuclear properties to study chemical phenomena, and the use of chemistry to study nuclear properties. However, the topics covered in this book are considerably wider.

The chemical behaviour of the heavy radioactive elements has lead to Nobel prices and international fame for many great scientists since the late 19th century. From the beginning the use of radiation was used for the good of humanity as treatment for different diseases, real or imagined. However, as will all new discoveries, many odd products such as e.g. radium soap hit the market and today we know that such use of radium is not healthy. However, the new knowledge also opened the possibility to use radioactive elements for the treatment of cancerous micro metastases by grafting them on antibodies seeking out the cancer. As the knowledge of nuclear chemistry advanced more elements were discovered. Today the periodic table is still expanding with new element as a result of the nuclear research. The powers of the nuclear forces have been used for both destruction and for building up peaceful power production for the good of society. In the future we expect that nuclear power will be made considerably more sustainable by a significantly better utilization of the uranium resource by the recycling of used nuclear fuel.

Seen in this context nuclear chemistry has still a significant part to play in the future development of a peaceful, sustainable society. Thus also the teaching on different levels will need to learn, not only the basic concepts of the atom, nuclear forces, detection and other basic facts but also look at which applications has been and will be used in the nuclear area.

The history of the present book dates back to 1964 when prof. Jan Rydberg first published a book in Swedish about nuclear chemistry to be used for teaching purposes. Later, this Swedish book was extended and translated to English together with prof. Gregory Choppin and the first edition of the current series was made. The present book is the 4th edition of the original Choppin–Rydberg book revised by Liljenzin and Ekberg. However, read and approved by Rydberg and Choppin. Essentially not much has been changed in the more basic chapters since few new facts have been discovered. However, the logic of the chapters have been made more closely to the first edition. The more applied chapters have been extensively edited and updated in order to mirror the latest discoveries in the still evolving areas such as production of new elements, new separation schemes for used nuclear fuel, and the discovery of new elementary particles.

This detailed update would not have been possible without consultation with experts in the different fields covered by the book. Especially we would like to express our gratitude to prof. Gunnar Skarnemark, prof. Henrik Ramebäck, prof. Henrik Nylén, prof. Janne Wallenius, and to all those who have helped us to improve the mauscript and to check the proofs.

We hope that his new edition will be an aid in the teaching of nuclear chemistry and also be a good companion after the education whenever some facts or equations have to be refreshed.

Gregory Choppin, Christian Ekberg, Jan-Olov Liljenzin, Jan Rydberg
Tallahassee, Florida, and Göteborg, Sweden, November 2012

Origin of Nuclear Science

Contents

1.1. RADIOACTIVE ELEMENTS

The science of the radioactive elements and radioactivity in general is rather young compared to its maturity. In 1895 W. Roentgen was working with the discharge of electricity in evacuated glass tubes. Incidentally the evacuated glass tubes were sealed by Bank of England sealing wax and had metal plates in each end. The metal plates were connected either to a battery or an induction coil. Through the flow of electrons through the tube a glow emerged from the negative plate and stretched to the positive plate. If a circular anode was sealed into the middle of the tube the glow (cathode rays) could be projected through the circle and into the other end of the tube. If the beam of cathode rays were energetic enough the glass would glow (fluorescence). These glass tubes were given different names depending on inventor, e.g. Hittorf tubes (after Johann Hittorf) or Crookes tubes (after William Crookes). Roentgens experiments were performed using a Hittorf tube.

During one experiment the cathode ray tube was covered in dark cardboard and the laboratory was dark. Then a screen having a surface coating of barium-platinum-cyanide started to glow. It continued even after moving it further away from the cathode ray tube. It was also noticed that when Roentgens hand partly obscured the screen the bones in the nad was visible on the screen. A new, long range, penetrating radiation was found. The name *X-ray* was given to this radiation. Learning about this, H. Becquerel, who had been interested in the fluorescent spectra of minerals, immediately decided to investigate the possibility that the fluorescence observed in some salts when exposed to sunlight also caused emission of X-rays. Crystals of potassium uranyl sulfate were placed on top of photographic plates, which had been wrapped in black paper, and the assembly was exposed to the sunlight. After development of some of the photographic plates, Becquerel concluded (erroneously) from the presence of black spots under the crystals

Radiochemistry and Nuclear Chemistry
ISBN 978-0-12-405897-2, http://dx.doi.org/10.1016/B978-0-12-405897-2.00001-X

that fluorescence in the crystals led to the emission of X-rays, which penetrated the wrapping paper. However, Becquerel soon found that the radiation causing the blackening was not "a transformation of solar energy" because it was found to occur even with assemblies that had not been exposed to light; the uranyl salt obviously produced radiation spontaneously. This radiation, which was first called uranium rays (or Becquerel rays) but later termed radioactive radiation (or simply *radioactivity*)[1], was similar to X-rays in that it ionized air, as observed through the discharge of electroscopes.

Marie Curie subsequently showed that all uranium and thorium compounds produced *ionizing radiation* independent of the chemical composition of the salts. This was convincing evidence that the radiation was a property of the element uranium or thorium. Moreover, she observed that some uranium minerals such as pitchblende produced more ionizing radiation than pure uranium compounds. She wrote: "this phenomenon leads to the assumption that these minerals contain elements which are more active than uranium". She and her husband, Pierre Curie, began a careful purification of pitchblende, measuring the amount of radiation in the solution and in the precipitate after each precipitation separation step. These first *radiochemical* investigations were highly successful: "while carrying out these operations, more active products are obtained. Finally, we obtained a substance whose activity was 400 times larger than that of uranium. We therefore believe that the substance that we have isolated from pitchblende is a hitherto unknown metal. If the existence of this metal can be affirmed, we suggest the name *polonium*." It was in the publication reporting the discovery of polonium in 1898 that the word radioactive was used for the first time. It may be noted that the same element was simultaneously and independently discovered by W. Marckwald who called it "radiotellurium".

In the same year the Curies, together with G. Bemont, isolated another radioactive substance for which they suggested the name *radium*. In order to prove that polonium and radium were in fact two new elements, large amounts of pitchblende were processed, and in 1902 M. Curie announced that she had been able to isolate about 0.1 g of pure radium chloride from more than one ton of pitchblende waste. The determination of the atomic weight of radium and the measurement of its emission spectrum provided the final proof that a new element had been isolated.

1.2. RADIOACTIVE DECAY

While investigating the radiochemical properties of uranium, W. Crookes and Becquerel made an important discovery. Precipitating a carbonate salt from a solution containing uranyl ions, they discovered that while the uranium remained in the supernatant liquid in the form of the soluble uranyl carbonate complex, the radioactivity originally associated

[1] The word radioactivity refers to the phenomenon *per se* as well as the intensity of the radiation observed.

with the uranium was now present in the precipitate, which contained no uranium. Moreover, the radioactivity of the precipitate slowly decreased with time, whereas the supernatant liquid showed a growth of radioactivity during the same period (Fig. 1.1). We know now that this measurement of radioactivity was concerned with only beta- and gamma-radiations, and not with the alpha-radiation which is emitted directly by uranium.

Similar results were obtained by E. Rutherford and F. Soddy when investigating the radioactivity of thorium. Later Rutherford and F. E. Dorn found that radioactive gases (*emanation*) could be separated from salts of uranium and thorium. After separation of the gas from the salt, the radioactivity of the gas decreased with time, while new radioactivity grew in the salt in a manner similar to that shown in Fig. 1.1. The rate of increase with time of the radioactivity in the salt was found to be completely independent of chemical processes, temperature, etc. Rutherford and Soddy concluded from these observations that radioactivity was due to changes within the atoms themselves. They proposed that, when radioactive decay occurred, the atoms of the original elements (e.g. of U or of Th) were transformed into atoms of new elements.

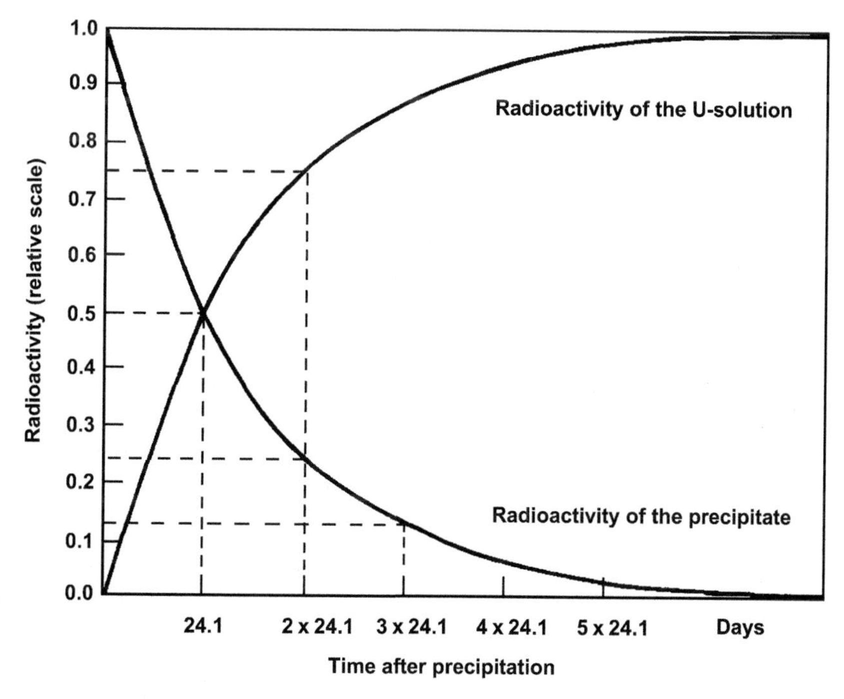

Figure 1.1 Measured change in radioactivity from carbonate precipitate and supernatant uranium solution, i.e. the separation of daughter element UX (Th) from parent radioelement uranium.

The radioactive elements were called *radioelements*. Lacking names for these radioelements, letters such as X, Y, Z, A, B, etc., were added to the symbol for the primary (i.e. parent) element. Thus, UX was produced from the radioactive decay of uranium, ThX from that of thorium, etc. These new radioelements (UX, ThX, etc.) had chemical properties that were different from the original elements, and could be separated from them through chemical processes such as precipitation, volatilization, electrolytic deposition, etc. The radioactive daughter elements decayed further to form still other elements, symbolized as UY, ThA, etc. A typical decay chain could be written: Ra → Rn → RaA → RaB → , etc.; see Fig. 1.2.

A careful study of the radiation emitted from these radioactive elements demonstrated that it consisted of three components which were given the designation alpha (α), beta (β), and gamma (γ). *Alpha-radiation* was shown to be identical to helium ions, whereas *beta-radiation* was identical to electrons. *Gamma-radiation* had the same electromagnetic nature as X-rays but was of higher energy. The rate of radioactive decay per unit weight was found to be fixed for any specific radioelement, no matter what its chemical or physical state was, though this rate differed greatly for different radioelements. The decay rate could be expressed in terms of a *half-life*, which is the time it takes for the radioactivity of a radioelement to decay to one-half of its original value. Half-lives for the different radioelements were found to vary from fractions of a second to millions of years; e.g. that of ThA is 0.1 of a second, of UX it is 24.1 days (Fig. 1.1), and of uranium, millions of years.

1.3. DISCOVERY OF ISOTOPES

By 1910 approximately 40 different chemical species had been identified through their chemical nature, the properties of their radiation, and their characteristic half-lives. The study of the generic relationships in the decay of the radioactive species showed that the radioelements could be divided into three distinct series. Two of these originated in uranium and the third in thorium. B. Boltwood found that all three of the series ended in the same element − lead.

A major difficulty obvious to scientists at that time involved the fact that while it was known from the Periodic Table (Appendix I) that there was space for only 11 elements between lead and uranium, approximately 40 radioelements were known in the decay series from uranium to lead. To add to the confusion was the fact that it was found that in many cases it was not possible to separate some of the radioelements from each other by normal chemical means. For example, the radioelement RaD was found to be chemically identical to lead. In a similar manner, spectrographic investigations of the radioelement ionium showed exactly the same spectral lines that had been found previously to be due to the element thorium.

In 1913 K. Fajans and Soddy independently provided the explanation for these seemingly contradictory conditions. They stated that by the radioactive α-decay a new

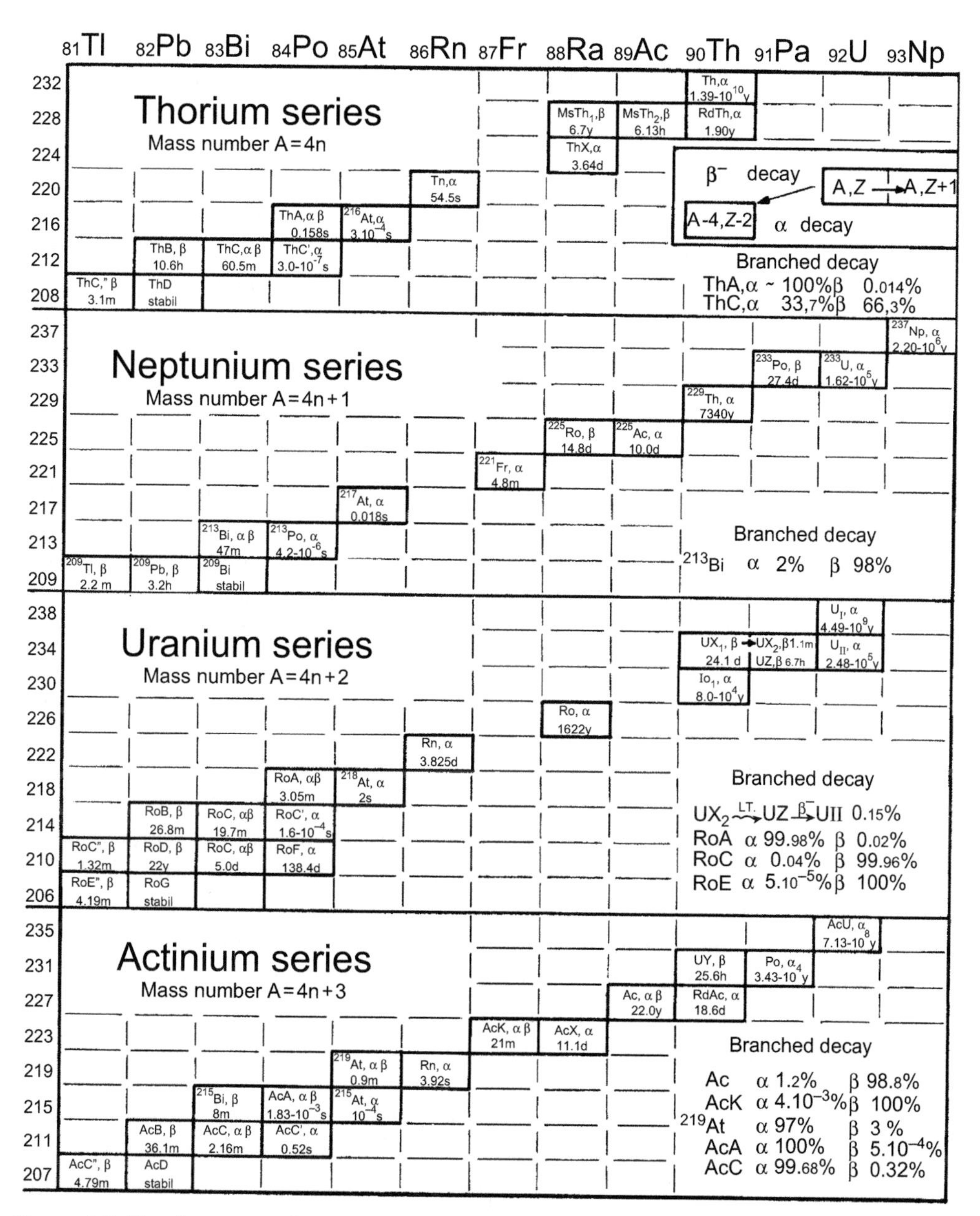

Figure 1.2 The three naturally occurring radioactive decay series and the man-made neptunium series. Although ^{239}Pu (which is the parent to the actinium series) and ^{244}Pu (which is the parent to the thorium series) have been discovered in nature, the decay series shown here begin with the most abundant long-lived nuclides.

element is produced two places to the left of the mother element in the periodic system and in α-decay a new element is produced one place to the right of the mother element (Fig. 1.2). The radioelements that fall in the same place in the periodic system are chemically identical. Soddy proposed the name *isotopes* to account for different radioactive species which have the same chemical identity.

Research by J. J. Thomson soon provided conclusive support for the existence of isotopes. If a beam of positively charged gaseous ions is allowed to pass through electric or magnetic fields, the ions follow hyperbolic paths which are dependent on the masses and charges of the gaseous ions (see Fig. 3.1 and associated text). When these ion beams strike photographic plates, a darkening results which is proportional to the number of ions which hit the plate. By using this technique with neon gas, Thomson found that neon consists of two types of atoms with different atomic masses. The mass numbers for these two isotopes were 20 and 22. Moreover, from the degree of darkening of the photographic plate, Thomson calculated that neon consisted to about 90% of atoms with mass number 20, and 10% of atoms with mass number 22.

Thus a chemical element may consist of several kinds of atoms with different masses but with the same chemical properties. The 40 radioelements were, in truth, not 40 different elements but were isotopes of the 11 different chemical elements from lead to uranium.

To specify a particular isotope of an element, the *atomic number* (i.e. order, number, or place in the Periodic Table of elements) is written as a subscript to the left of the chemical symbol and the mass number (i.e. the integer value nearest to the mass of the neutral atom, measured in atomic weight units) as a superscript to the left. Thus the isotope of uranium with mass number 238 is written as ${}^{238}_{92}\mathrm{U}$. Similarly, the isotope of protactinium with mass number 234 is designated ${}^{234}_{91}\mathrm{Pa}$. For an alpha-particle we use either the Greek letter α or ${}^{4}_{2}\mathrm{He}$. Similarly, the beta-particle is designated either by the Greek letter β or by the symbol ${}^{0}_{-1}\mathrm{e}$, and the electron antineutrino is designated as $\overline{\nu}$. However, the neutrino emitted is usually not written out in β-decay reactions like (1.2).

In radioactive decay both mass number and atomic number are conserved. Thus in the decay chain of the first two steps are written:

$$ {}^{238}_{92}\mathrm{U} \rightarrow {}^{234}_{90}\mathrm{Th} + {}^{4}_{2}\mathrm{He} \tag{1.1} $$

$$ {}^{234}_{90}\mathrm{Th} \rightarrow {}^{234}_{91}\mathrm{Pa} + {}^{0}_{-1}\mathrm{e} + \overline{\nu} \tag{1.2} $$

Frequently, in such a chain, the half-life ($t_{1/2}$) for the radioactive decay is shown either above or below the arrow. A shorter notation is commonly used:

$$ {}^{238}\mathrm{U} \xrightarrow[4.5\times10^{9}\mathrm{y}]{\alpha} {}^{234}\mathrm{Th} \xrightarrow[24\ \mathrm{d}]{\beta^-} {}^{234}\mathrm{Pa} \xrightarrow[1.1\ \mathrm{min}]{\beta^-} {}^{234}\mathrm{U} \xrightarrow[2.5\times10^{5}\mathrm{y}]{\alpha} {}^{230}\mathrm{Th},\ \text{etc.} \tag{1.3} $$

where the half-lives are given in years (y) and days (d)[2]. The three naturally occurring radioactive decay series, which are known as the *thorium series*, the *uranium series*, and the *actinium series*, are shown in Fig. 1.2. A fourth series, which originates in the synthetic element neptunium, is also shown. This series is not found naturally on earth since all of the radioactive species have decayed away long ago. Both the present symbolism of the isotope as well as the historical (i.e. "radioelement") symbolism are given in Fig. 1.2. Note that the *rule of Fajans and Soddy* is followed in each series so that α-decay causes a decrease in atomic number by two units and mass number by four, whereas β-decay produces no change in mass number but an increase in atomic number by one unit. Moreover, we see a pattern occurring frequently in these series where an α-decay step is followed by two β-decay steps. All known isotopes of elements $_{81}Tl$ to $_{92}U$ are given in Figure 13.4.

1.4. ATOMIC MODELS

Neither radioactive decay nor the discovery of isotopes provided information on the internal structure of atoms. Such information was obtained from scattering experiments in which a substance, such as a thin metal foil, was irradiated with a beam of ⟨-particles and the intensity (measured by counting scintillations from the scattered particles hitting a fluorescent screen) of the particles scattered at different angles measured (see Fig. 10.4). It was assumed that the deflection of the particles was caused by collisions with the atoms of the irradiated material. About one in 8000 of the á-particles was strongly deflected through angles greater than 90°. Consideration of these rare events led Rutherford in 1911 to the conclusion that the entire positive charge of an atom must be concentrated in a very small volume whose diameter is about 10^{-14} m. This small part of the atom he called the *nucleus*. The atomic electrons have much smaller mass and were assumed to surround the nucleus. The total atom with the external electrons had a radius of approximately 10^{-10} m in contrast to the much smaller radius calculated for the nucleus.

It was soon shown that the positive charge of the atomic nucleus was identical to the atomic number assigned to an element in the periodic system of Mendeleev. The conclusion, then, is that in a neutral atom the small, positively charged nucleus was surrounded by electrons whose number was equal to the total positive charge of the nucleus. In 1913 N. Bohr, using quantum mechanical concepts, proposed such a model of the atom which remains the basis of the modern atomic theory.

Our understanding of the nucleus has grown rapidly since Rutherford's scattering experiments. Some of the important steps in the history of nuclear science are listed in

[2] IUPAC recommends a for *annum*, instead of y, however y will be used throughout this text as it remains the commonly used term.

Table 1.1. Many of these discoveries and their practical consequences are discussed in the subsequent text.

1.5. NUCLEAR POWER

The peaceful use of the nuclear fission energy to produce electricity was first successfully employed in 1951 by the EBR-1 (Experimental Breeder Reactor) in the US. This was a fast breeder reactor using enriched uranium as fuel and cooled by a NaK eutectic melt. Originally it produced enough electricity to power four 200 W light bulbs and subsequently enough to power its own building. In 1955 the EBR-1 suffered a partial melt down and was finally deactivated in 1964. Admittedly the EBR-I produced the first electricity it was not really connected to the grid. This was achieved by the nearby plant, BORAX-III, which in 1955 made the nearby city of Arco to be the first city solely powered by nuclear energy.

Similar activities were performed in the Soviet Union where an existing graphite moderated channel-type plutonium production reactor was modified for heat and electricity generation by introducing a heat exchange system. It became the world's first thermal neutron nuclear powered electricity generator in 1954. The reactor was called AM-1 (Atom Mirny – peaceful atom) with a capacity of 30 MW_{th} or 5 MW_e. AM-1 produced electricity until 1959 and was used as a research facility and for production of radioisotopes until 2002. As seen above the general construction was close to the later infamous Chernobyl reactor RBMK (reaktor bolshoi moshchnosty kanalny – high power channel reactor).

The pressurised water reactor (PWR) history start with the use in naval propulsion. The Mark 1 prototype naval reactor became critical in 1953 which led to the first nuclear-powered submarine, *USS Nautilus*, was launched in 1954. This reactor led to the building in 1957 of the 60 MWe Shippingport demonstration PWR reactor in Pennsylvania. This reactor operated until 1982.

A boiling water reactor (BWR) prototype, Vallecitos, was operated from 1957 to 1963 but the first electricity producing one, Dresden-1 of 250 MWe was started in 1960 and was operating until 1978. There reactors also used enriched uranium fuel.

In Canada a different route was selected. Instead of going through the uranium enrichment process they created the CANDU reactor using natural uranium oxide fuel in a heavy water moderated and normal water cooled channel type reactor. The first one was started in 1962.

After these initial steps essentially four reactor systems are in general use: natural water moderated and cooled (PWR/VVER and BWR), heavy water moderated and natural water cooled (CANDU), and graphite moderated and water cooled (RBMK).

Table 1.1 Historical survey of nuclear science

Essential steps in the development of modern science	
~490-430 B.C.	**Empedocles** suggests that everything is made up of four elements: air, earth, water and fire. Every matter can be formed by transmutation between these. (This is principally correct if the four elements are interpreted as being the gaseous, solid and liquid states of matter, and fire interpreted as being energy.)
~460-370 B.C.	**Democritos** proposes that all matter consists of eternal, moving and indestructible atoms, qualitatively alike but differing in size, shape and mass.
1661	**Boyle** writes that the nature is made up of a limited number of substances (elements) which cannot be broken down into simpler ones.
1808	**Dalton**: All chemical compounds (molecules) are combinations of atoms in fixed proportions.
Important steps in the development of nuclear science	
1896	**Becquerel** discovers radiation from uranium (radioactivity). The intensity of the radiation is measured either through its ionization of air or through the scintillations observed when the radiation hits a fluorescent screen.
1896-1905	**Crookes, Becquerel, Rutherford, Soddy, Dorn, Boltwood** *et al.* Radioactive decay is found to be transformation of atoms leading to different radioelements which are genetically connected in radioactive decay series.
1898	**P. and M. Curie** discover polonium and radium; the first radiochemical methods.
1898-1902	**P. Curie, Debierne, Becquerel, Danilos** *et al.* discover that radiation affects chemical substances and causes biological damage.
1900	**Villard** and **Becquerel** propose that -radiation is of electromagnetic nature; finally proven in 1914 by **Rutherford** and **Andrade**.
1900	**Becquerel**: β-rays are identified as electrons.
1902	**M.** and **P. Curie** and **Debierne** isolate first macroscopic amounts of a radioactive element (radium)
1903	**Rutherford** shows α-radiation to be ionized helium atoms.
1905	**Einstein** formulates the law of equivalence between mass and energy.
1907	**Stenbeck** makes the first therapeutic treatment with radium and heals skin cancer.
1911	**Rutherford, Geiger,** and **Marsden** conclude from measurement of the scattering of α-radiation against thin foils that atoms contain a very small positive nucleus.
1912	**Hevesy** and **Paneth**, in the first application of radioactive trace elements, determine the solubility of $PbCrO_4$ using RaD.
1912	**Wilson** develops the cloud chamber, which makes tracks from nuclear particles visible.
1913	**Hess** discovers cosmic radiation.

(*Continued*)

Table 1.1 Historical survey of nuclear science—cont'd

1913	**Fajans** and **Soddy** explain the radioactive decay series by assuming the existence of isotopes. This is proven by **J. J. Thomson** through deflection of neon ions in electromagnetic fields. **Aston** separates the isotopes of neon by gas diffusion.
1913	**N. Bohr** shows that the atomic nucleus is surrounded by electrons in fixed orbitals.
1919	**Rutherford**: first nuclear transformation in the laboratory, $^{4}He + {}^{14}N \rightarrow {}^{17}O + {}^{1}H$.
1919	**Aston** constructs the first practical mass spectrometer and discovers that isotopic weights are not exactly integers.
1921	**Hahn** discovers nuclear isomers: $^{234m}Pa(UX_2) \xrightarrow[1.2\ min]{\gamma} {}^{234}Pa(UZ)$.
1924	**de Broglie** advances the hypothesis that all moving particles have wave properties.
1924	**Lacassagne** and **Lattes** use radioactive trace elements (Po) in biological research.
1925-1927	Important improvements of the Bohr atomic model: Pauli exclusion principle, Schrödinger wave mechanics, Heisenberg uncertainty relationship.
1928	**Geiger** and **Müller** construct the first GM tube for single nuclear particle measurements.
1931	**van de Graaff** develops an electrostatic high voltage generator for accelerating atomic ions to high energies.
1931	**Pauli** postulates a new particle, the neutrino, to be formed in β-decay.
1932	**Cockcroft** and **Walton** develop the high voltage multiplier and use it for the first nuclear transformation in the laboratory with accelerated particles (0.4 MeV $^{1}H + {}^{7}Li \rightarrow 2\ {}^{4}He$).
1932	**Lawrence** and **Livingston** build the first cyclotron.
1932	**Urey** discovers deuterium and obtains isotopic enrichment through evaporation of liquid hydrogen.
1932	**Chadwick** discovers the neutron.
1932	**Andersson** discovers the positron, e^{+} or β^{+}, through investigation of cosmic rays in a cloud chamber.
1933	**Urey** and **Rittenberg** show isotopic effects in chemical reactions.
1934	**Joliot** and **I. Curie** discover artificial radioactivity: $^{4}He + {}^{27}Al \rightarrow {}^{30}P + n$; $^{30}P \xrightarrow[2.5\ min]{\beta^{+}} {}^{30}Si$.
1935	**DeHevesy** develops neutron activation analysis.
1935	**Yukawa** predicts the existence of mesons.
1935	**Weizsäcker** derives the semiempirical mass formulae.
1937	**Neddermeyer** and **Andersson** discover μ-mesons in cosmic radiation using photographic plates.
1938	**Bethe** and **Weizsäcker** propose the first theory for energy production in stars through nuclear fusion: $3\ {}^{4}He \rightarrow {}^{12}C$.
1938	**Hahn** and **Strassman** discover fission products after irradiation of uranium with neutrons.

Table 1.1 Historical survey of nuclear science—cont'd

1938-1939	**Meitner** and **Frisch** interprets the discovery by **Hahn** and **Strassman** as fission of the U-atom by neutrons; this is almost immediately confirmed by several laboratories in Europe and the USA.
1938-1939	**F. Joliot, von Halban, Kowarski** and **F. Perrin** in France apply for patents for nuclear chain reacting energy producing devices and starts building a nuclear reactor; the work is interrupted by the war.
1940	**McMillan, Abelson, Seaborg, Kennedy,** and **Wahl** produce and identify the first transuranium elements, neptunium (Np), and plutonium (Pu), and with **Segré** discover that ^{239}Pu is fissionable.
1940	Scientists in many countries show that ^{235}U is fissioned by slow neutrons, but ^{232}Th and ^{238}U only by fast neutrons, and that each fission produces two or three new neutrons while large amounts of energy are released. The possibility of producing nuclear weapons and building nuclear power stations is considered in several countries.
1942	**Fermi** and co-workers build the first nuclear reactor (critical on December 2).
1944	First gram amounts of a synthetic element (Pu) produced at Oak Ridge, USA. Kilogram quantities produced in Hanford, USA, in 1945.
1944	**McMillan** and **Veksler** discover the synchrotron principle which makes it possible to build accelerators for energies > 1000 MeV.
1940-1945	**Oppenheimer** and co-workers develop a device to produce fast uncontrolled chain reactions releasing very large amounts of energy. First test at Alamogordo, New Mexico, USA, on July 16, 1945 produces an energy corresponding to 20,000 tons of TNT; this is followed by the use of atomic bombs on Hiroshima (Aug. 6, 1945) and on Nagasaki (Aug. 9, 1945).
1944-1947	Photo-multiplier scintillation detectors are developed.
1946	**Libby** develops the ^{14}C-method for age determination.
1946	First Soviet nuclear reactor starts.
1949	Soviet tests a nuclear bomb.
1950	**Mayer, Haxel, Jensen** and **Suess** suggest a nuclear shell model.
1951	The first breeder reactor, which also produces the first electric power, is developed by Argonne National Laboratory, USA, and built in Idaho.
1952	The United States test the first device for uncontrolled large scale fusion power (the hydrogen bomb).
1953-1955	**A. Bohr, Mottelson,** and **Nilsson** develop the unified model of the nucleus (single particle effects on collective motions).
1955	**Chamberlain, Segré, Wiegand,** and **Ypsilantis** produce antiprotons.
1955	First nuclear powered ship (submarine *Nautilus*).
1954-1956	A 5 MW_e nuclear power station starts at Obninsk, USSR, in 1954. First civilian nuclear power station (45 MW_e) starts at Calder Hall, England, in 1956.
1956	**Reines** and **Cowan** prove the existence of neutrinos.
1957	Fire in carbon dioxide cooled graphite reactor at Windscale, U.K.
1957	Explosion in nuclear waste storage facility at Kyshtym (Chelyabinsk), USSR, with contamination of large areas.

(*Continued*)

Table 1.1 Historical survey of nuclear science—cont'd

1959	First civilian ship reactor used in the ice-breaker *Lenin*, launched in the USSR.
~1960	**Hofstadter** et al.; protons and neutrons contain unevenly distributed internal charge.
~1960	**Lederman, Schwarz** and **Steinberger** discover the muon neutrino.
1961	A radionuclide (^{238}Pu) is used as power source in a satellite (Transit-4 A).
1961	Semiconductor detectors are developed.
1963	End of atmospheric testing of nuclear weapons (see below).
1965	**A. Penzias** and **R. W. Wilson** discover the 3 K cosmic microwave radiation background.
~1970	Theory of quarks developed (**Gell-Mann**); quarks proven in nuclear scattering experiments (**Friedman, Kendall and Taylor**).
1972	French scientists discover ancient natural nuclear reactor in Oklo, Gabon.
1979	Core melt-down in PWR reactor at the Three Mile Island plant near Harrisburg, USA; no environmental contamination.
1983	**Rubbia, van der Meer** & co-workers at CERN discover the W and Z weak interaction communicators.
1986	Explosion and fire in the Chernobyl-4 reactor unit at Pripyat, Ukraine, USSR, with contamination of large areas.
2011	Fukushima Daichi nuclear power station in Japan is hit, and badly damaged, by a very large tsumani.
2013	Higgs boson discovered at CERN
International events concerning radiation and nuclear energy	
1955	Formation of United Nations Scientific Committee on the Effects of Atomic Radiation (UNSCEAR).
1957	Formation of the International Atomic Energy Agency (IAEA), with headquarters in Vienna.
1963	Partial Test Ban Treaty bans nuclear tests in the atmosphere, in outer space, and under water.
1968	Treaty on the Non-Proliferation of Nuclear Weapons (NPT) is signed by the "three depository governments" (USSR, UK, and USA), all nuclear weapons countries (NWC), and 40 other signatory, non-nuclear weapons countries (NNWC).
1971	The IAEA takes the responsibility for a safeguards system for control of fissile material in non-nuclear weapons countries.
1991	140 states have signed the NPT agreement.
2011	Germany decides to abandon nuclear energy production within a few years.

1.6. LITERATURE

Historical reading and classics in nuclear chemistry:

E. Rutherford, J. Chadwick and C. D. Ellis, *Radiations from Radioactive Substances*, Cambridge University Press, Cambridge 1930 (reprinted 1951).
M. Curie, *Radioactivité*, Herrmann, Paris 1935.
O. Hahn, *Applied Radiochemistry*, Cornell University Press, 1936.
H. D. Smyth, *Atomic Energy for Military Purposes*, Princeton University Press, Princeton 1946.
R. T. Beyer, *Foundations of Nuclear Physics*, Dover Publ. Inc., New York 1949.
O. Hahn, *A Scientific Autobiography*, Charles Scribner's Sons, New York 1966.
Nobel Lectures, Chemistry and *Nobel Lectures, Physics*, Elsevier, Amsterdam 1966 and later.
S. Glasstone, *Source Book on Atomic Energy*, 3rd edn., van Nostrand, New York 1967.
A. Romer, *Radiochemistry and the Discovery of Isotopes*, Dover Publ. 1970.
G. T. Seaborg and W. D. Loveland, *Nuclear Chemistry*, Hutchinson, Stroudsberg 1982.
R. Rhodes, *The Making of the Atomic Bomb*, Simon & Schuster, 1986.
J. May, *The Greenpeace Book of the Nuclear Age*, Victor Gollancz Ltd., London 1989.
J. W. Behrens and A. D. Carlson (eds.), *50 Years with Nuclear Fission*, vol. 1 & 2, Am. Nucl. Soc., La Grange Park, Illinois, 1989.
G. T. Seaborg and W. D. Loveland, *The Elements Beyond Uranium*, John Wiley & Sons Inc., New York 1990.

Journals of importance to radiochemistry and nuclear chemistry:

Radiochimica Acta, R. Oldenbourg Verlag, Munich (Vol. 99, 2011).
Journal of Radioanalytical and Nuclear Chemistry, Akadémiai Kiadó, Budapest (Vol. 290, 2011).
Applied Radiation and Isotopes, Elsevier (Vol. 69, 2011).
Health Physics, Lippincott, Williams & Wilkins, Baltimore (Vol. 101, 2011).
International Atomic Energy Agency Bulletin, and other IAEA publications, Vienna.
Nuclear Engineering International, Global Trade Media (Vol. 49, 2011).
Nuclear Science and Engineering, Am. Nucl. Soc., La Grange Park, Illinois (Vol. 169, 2011)
Nuclear Technology, Am. Nucl. Soc., La Grange Park, Illinois (Vol. 176, 2011).

CHAPTER 2

Elementary Particles

Contents

This chapter gives the reader an insight in the world of "sub-electron" particles. It is in no way exhaustive but is intended to tease the interest of the reader into studying more literature dedicated to this specific subject. The basics of nuclear physics is of great importance to understand the mechanism governing the physics inside a nucleus.

The electron, proton and neutron are well known elementary particles. However, high energy nuclear collisions, such as those that occur in the atmosphere when it is hit by very high energy particles from space, are known to produce a wealth of other particles which once used to be called elementary particles. Such particles can now also be formed using large particle accelerators on earth, e.g. the one in CERN.

2.1. ELEMENTARY PARTICLES

The group of elementary particles began to be considerably expanded about 1947 when physicists discovered the first of the so-called "strange" particles in cloud chamber pictures of cosmic rays, see Figure 2.1. These new elementary particles were called strange because they lived almost a million million times[1] longer than scientists had any reason to expect at that time. The population of elementary particles has literally exploded since then, as physicists have built larger and larger particle accelerators, by which it is possible to impart sufficient kinetic energy to protons so that interaction with nuclides transform a large fraction of the kinetic energy into matter. The present limit is in the multi TeV range (Ch. 16). In 2011 it is thus possible to produce particles with a mass of up to ~3500 proton masses, and hundreds of new "strange" particles have been observed. Figure 2.2 is a typical picture of a reaction observed in a liquid hydrogen bubble chamber at an accelerator center.

[1] million million times = 10^{12} times

Radiochemistry and Nuclear Chemistry
ISBN 978-0-12-405897-2, http://dx.doi.org/10.1016/B978-0-12-405897-2.00002-1

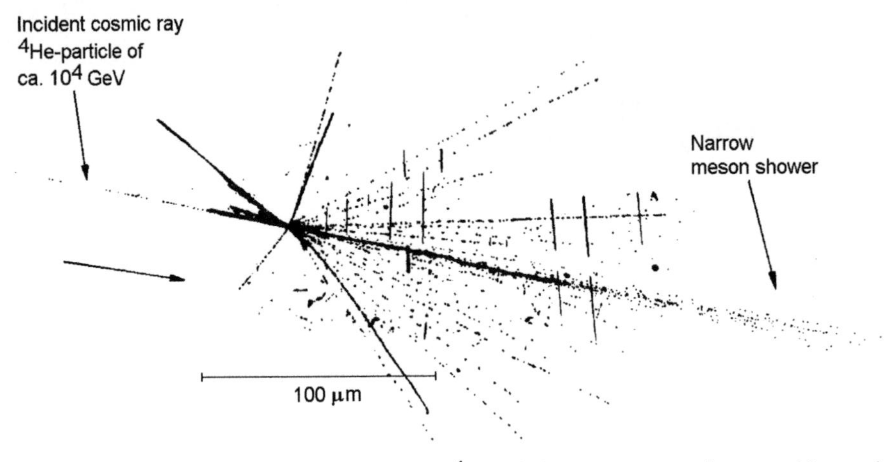

Figure 2.1 Secondary particles produced by a 10^4 GeV helium atom in a photographic emulsion.

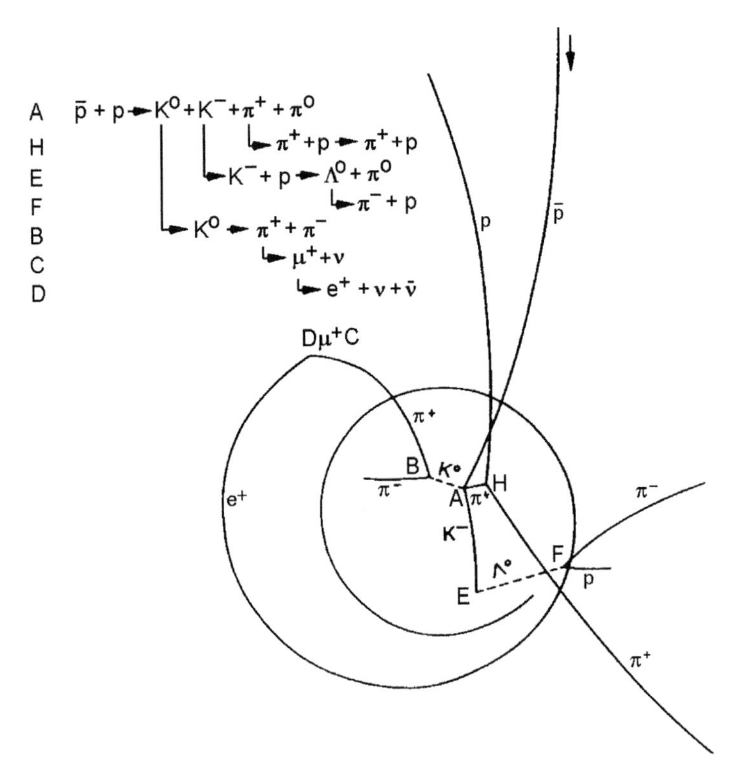

Figure 2.2 The reaction products of an annihilated antiproton as seen in the CERN liquid hydrogen bubble chamber. (*Annual report 1961*, CERN).

This has created a scientific area called *elementary particle physics*. It is quite different from nuclear physics, which is concerned with composite nuclei only. A principal objective of elementary particle physics has been to group the particles together according to their properties to obtain a meaningful pattern, which would describe all particles as parts of some few fundamental building blocks of nature. One step in this direction is to study how the elementary particles interact with each other, i.e. what kind of forces are involved.

Scientists have long doubted that all the particles produced with masses between the electron and the proton (loosely referred to as *mesons*, i.e. "intermediate"), and with masses greater than the proton (referred to as *baryons*, "heavy") really are "elementary". It was proposed that they have a substructure or constitute excited states of each other. Are they waves or particles since they serve as carriers of force? At this point it is important to understand what is meant by "particle" in nuclear physics.

2.2. FORCES OF NATURE

Considering what an immense and incredible diverse assembly the universe is – from the cosmos to man and microbes – it is remarkable that scientists have been able to discover with certainty only four basic forces, which govern the attraction and repulsion of all physical objects of nature. Let us consider these *forces of nature* in a qualitative way from the weakest to the strongest, see Table 2.1.

The first and weakest force of nature is that of *gravity*. The gravitational force, F_g, beween two masses, m_1 and m_2, at a distance r is given by eqn (2.1), where $\mathbf{G}$ is the universal constant of gravity ($\mathbf{G} = (6{,}6742 \pm 0{,}0007) \times 10^{-11}$ N m^2 kg^{-2}).

$$F_g = \mathbf{G} \times m_1 \times m_2/r^2 \tag{2.1}$$

F_g is the force that causes all objects to always attract one another and is responsible for the attraction of the planets to the sun in the solar system and of the solar system to the rest of the galaxy. It is also the force that holds us to the earth. It seems paradoxical that

Table 2.1 Forces of nature in order of strength and their exchange particles

Force	Relative strength	Range	Exchange particle	Spin	Rest mass
1. Gravitation	1	very long	graviton†	2	0
2. The weak force	10^{25}	very short	W^+, W^-	1	79.9 GeV
			Z^0	1	91.2 GeV
3. The electromagnetic force	10^{36}	long	photon	1	0
4. The nuclear force (the strong force)	10^{38}	10^{-14} m	gluon	1	0

†Not yet detected.

the weakest attraction of the four basic forces of nature is the force that is responsible for the assembly of the largest objects on the greatest scale. In modern physics it is believed that all forces are carried by "something" which either can be described as a wave or as a particle (see §2.3). The carrier of the gravitational force is the *graviton*. Experimenters have tried various ways to detect gravitational waves, but so far the results are negative.

The second weakest force is the one which is involved in the radioactive β-decay of atoms (see Ch. 5). It is known as the *weak interaction force*. This weak interaction force operates over extremely short distances and is the force that is involved in the interaction of very light particles known as *leptons* (electrons, muons, and neutrinos) with each other and as well as their interaction with mesons, baryons, and nuclei. The carrier of the weak interaction force is still a matter of considerable research; we will return to this point later.

The third weakest force of nature with which we are all relatively familiar is that of the *electromagnetic force*. The electromagnetic force, F_e, is expressed by Coulomb's law and is responsible for the attraction and repulsion between charged bodies, see eqn (2.2) where z_1 and z_2 are the charges, r is the distance and $\mathbf{k}_e$ is a constant ($\mathbf{k}_e = 1/(4\pi\varepsilon_0)$ in the SI-system).

$$F_e = -\mathbf{k}_e \times z_1 \times z_2/r^2 \tag{2.2}$$

Just as the gravitational force holds the planets in their orbits about the sun and explains the stability of the planetary systems, the electromagnetic force explains the attraction/repulsion between electrons in atoms, ionic bonds in molecules, molecules and ions in crystals. It is the force that holds the atomic world together. It is approximately 10^{36} times stronger than the gravitational force. If gravity is the force underlying the laws of astronomy, electromagnetism is the force underlying the laws of chemistry and biology. The carrier of the electromagnetic force is the *photon*.

The fourth major force in nature is the nuclear force. This force is also known as the *strong interaction force* and is the one responsible for holding nuclear particles together. Undoubtedly it is the strongest in nature but operates only over the very short distance of approximately 10^{-14} m and saturates rapidly, i.e. only a few nucleons are involved. Whereas electromagnetism binds electrons to nuclei in atoms with an energy corresponding to a few electron volts, the strong interaction force holds nucleons together in nuclei with energies corresponding to millions of electron volts. The carrier of the strong interaction force is recognized to be the *gluon*; we will return to this point in §2.6. One characteristic of leptons is that they seem to be quite immune to the strong interaction force.

The general comparison between the different forces is shown in Table 2.1. The strong nuclear force is approximately 10^2 times greater than the Coulombic force, while the weak interaction force is smaller than the strong attraction by a factor of approximately 10^{13}.

There is also a considerable difference in the reaction time of the different interactions. The strong interaction manifests itself in its ability to react in very short times. For example, for a particle, which passes an atomic nucleus of about 10^{-15} m in diameter with a velocity of approximately 10^8 m s^{-1} (i.e. with a kinetic energy of ~50 MeV for a proton and 0.03 MeV for an electron), the time of strong interaction is about 10^{-23} s. This is about the time of rotation of the atomic nucleus. The weak interaction force requires a much longer reaction time and explains why leptons such as electrons and photons do not react with atomic nuclei but do react with the electron cloud of the atom, which has a diameter on the order of 10^{-10} m. There is sufficient time in passing this larger diameter for the weak interaction force to be effective. It is expected that the electromagnetic force and the gravitational force are very fast compared to the weak nuclear force.

2.3. WAVES AND PARTICLES

It is daily experience that moving bodies have a kinetic energy, E_v, which involves a mass and a velocity, v. Less familiar is the concept of Planck (1900) in which light moves in wave packets of energy:

$$E = \mathbf{h}\nu \tag{2.3}$$

Here ν is the frequency of light with wavelength λ

$$\mathbf{c} = \nu\lambda \tag{2.4}$$

and **h** the Planck constant, 6.63×10^{-34} J s.

Einstein – in the theory of the photoelectric effect –, and Compton – in the theory of the scattering of photons (Ch. 6) – showed that photons have not only a discrete energy, but also a discrete momentum (p_v).

$$p_v = E_v/\mathbf{c} \tag{2.5}$$

Photons seem to collide with other particles as if they have a real mass and velocity as in the classical mechanical expression for momentum: $p = mv$. If we put $v = \mathbf{c}$ and equate with (2.5) we obtain a relativistic mass of the photon as

$$m_v = E_v/\mathbf{c}^2 \tag{2.6}$$

This is the mass-energy relation of Einstein.

There are many examples of mass properties of photons. To the two mentioned above we may add the solar pressure (i.e. photons from the sun which push atoms away from the sun and into space), which has played a significant part in the formation of our planetary system, and measurements showing that photons are attracted by large masses through the gravitational force. Thus we see the evidence for the statement in the beginning that all moving elementary particles must have relativistic mass, even if the rest mass is zero.

It is reasonable to assume, as de Broglie did in 1924, that since photons can behave as moving particles moving particles may show wave properties. From the previous equations, we can — by replacing **c** by v — derive that the wavelength of such *matter waves* is

$$\lambda = \mathbf{h}/mv \tag{2.7}$$

This relation is of importance in explaining nuclear reactions, and has led to practical consequences in the use of electron diffraction, and in the development of electron microscopy.

The wave and particle properties of matter complement each other (the *complementarity principle*; N. Bohr, 1928). Throughout this book we either use models based on wave properties or sometimes on particle properties, depending on which more directly explain the particular phenomenon under discussion.

2.4. FORMATION AND PROPERTIES OF SOME ELEMENTARY PARTICLES

The track formed by a moving particle in the magnetic field of a bubble chamber is characterized by its width, length and curvature. From a kinematic analyses of the tracks it is possible to determine the mass and charge of the particles involved, see Ch. 3, and Ch. 9. Further, as seen at points A-B and E-F in Figure 2.2, the interruption of a track can indicate formation of an uncharged particle, as they do not form visible tracks. From a knowledge of the tracks formed by known particles, "strange tracks" can be analyzed to identify new particles in bubble chamber pictures and to assign their properties. All these properties have to be quantized, so new quantum states have been introduced, like baryon number, statistics, symmetry, parity, hypercharge, isospin, strangeness, color, etc. in addition to spin. It is found that many of the particles observed are sensitive to only one or two of the forces of nature (§2.2), which serves as an additional aid in their classification. Nevertheless, the array of properties assigned to the hundreds of "elementary particles" which have been discovered resembles the situation when 40 radioelements were reported between uranium and lead, as described in chapter 1.

Before we proceed to the order which has evolved from this picture we must describe some of the concepts used to define these particles. It has been practical to divide the particles according to their masses: (i) *baryons* are the "heavy" ones (protons, neutrons, hyperons and nuclei), (ii) *leptons* are the "light" ones (the electron, the neutrino and the muon); (iii) *mesons* have "intermediate" masses; these include the π-meson, the K-meson, etc. The baryons and the mesons have also been considered as *hadrons*, "hard" or "strong" particles as they take part in the strong nuclear force. Such properties were used to develop a table of elementary particles, Table 2.2.

Table 2.2 Common "elementary particles" in nuclide reactions

Class	Name	Symbol	Rest mass (MeV)	Rest mass (u)	Life-time (s); decay mode
Baryons[†] (nucleons):	neutron	n	939.6	1.008665	889; $n \rightarrow p + e^- + \bar{\nu}_e$
S, A-yes, P+, s = ½	proton	p	938.3	1.007276	stable
Mesons: S, A-no, P−,	K-meson (kaon)	K^+, K^-	493.7	0.530009	$\approx 10^{-8}$; $K^+ \rightarrow \mu^+ + \nu_\mu, \pi^+ + \pi^0$, etc.[††]
s = 0	π-meson (pion)	π^+, π^0, π^-	139.6[†††]	0.149867	$\approx 10^{-8}$; $\pi^- \rightarrow \mu^- + \bar{\nu}_\mu$[††]
Leptons[††††]: W, A-no, P-no,	muon	μ^-	105.6	0.113366	2×10^6; $\mu^- \rightarrow e^- + \nu_e + \nu_\mu$
s = ½	electron stable	e^-	0.5110	0.0005486	
	neutrino	ν_e, ν_μ, ν_τ	0?	0?	stable
Photon: W, A-no, P-no, s = 1	photon, gamma	$h\nu$, γ	0	0	stable

S = Strong interaction; W = Weak interaction; A-yes = baryon number conserved; P± = conservation of parity and parity sign; P-no = parity not conserved; s = spin quantum number. All these particles have their anti-particles, except for the photon and mesons, which are their own anti-particles.

[†] Baryons include the nucleons (n and p), hyperons (also called "strange baryons", with $\mathbf{m}_0$ 1116 for Λ^0 to 1673 MeV for Ω^-, and lifetimes $\approx 10^{-10}$ s), and all nuclei.

[††] K^- and π^+ decay into corresponding anti-particles in a similar way.

[†††] π^0 rest mass is 135.0 MeV.

[††††] See also Table 10.2.

All the particles in Table 2.2 have *spin*[2]. Quantum mechanical calculations and experimental observations have shown that each particle has a fixed spin energy which is determined by the *spin quantum number s* ($s = \pm½$ for leptons and nucleons). Particles of non-integral spin are called *fermions* because they obey the statistical rules devised by Fermi and Dirac, which state that two such particles cannot exist in the same closed system (nucleus or electron shell) having all quantum numbers the same (referred to as *the Pauli principle*). Fermions can be created and destroyed only in conjunction with an anti-particle of the same class. For example if an electron is emitted in β-decay it must be accompanied by the creation of an anti-neutrino. Conversely, if a positron β^+ – which is an anti-electron – is emitted in the β-decay, it is accompanied by the creation of a neutrino.

Fermions are the *building blocks* of nature. There is another group of "particles", called *bosons*, to which the photon and mesons belong. The bosons are the *carriers of forces*. When two fermions interact they continually emit and absorb bosons. The bosons have an even spin (0, 1, etc), they do not obey the Pauli principle, and they do not require the formation of anti-particles in their reactions.

The well known Schrödinger wave equation is given by eqn (2.8). It requires that the mass of e.g. an electron is constant and independent of its velocity.

$$\mathrm{i}\hbar\frac{\partial\psi}{\partial t} = \widehat{\mathrm{H}}\psi \tag{2.8}$$

However, the well tested Lorenz equation (2.9) shows that this is not true in our universe.

$$m_v = \boldsymbol{m}_0/\left(1 - (v/\mathbf{c})^2\right)^{½} \tag{2.9}$$

Here m_v is the electron mass at velocity v, $\boldsymbol{m}_0$ is the rest mass, and $\mathbf{c}$ is the velocity of light. Hence there was a search for an alternative wave equation which would be valid in our universe with its limited speed of light. In 1928 Paul Dirac proposed his well known alternative (2.10) to the Schrödinger equation.

$$\left(\beta\boldsymbol{m}_0\mathbf{c}^2 + \sum_{k=1}^{3}\alpha_k p_k \mathbf{c}\right) \times \psi(x,t) = \mathrm{i}\hbar\frac{\partial\psi(x,t)}{\partial t} \tag{2.10}$$

where $\boldsymbol{m}_0$ is the rest mass of the electron, $\mathbf{c}$ is the speed of light, p is the momentum, understood to be an operator in the sense of the Schrödinger theory, x and t are the space and time coordinates and $\hbar$ is the reduced Planck constant, $\mathbf{h}$ divided by 2π, β and α_k are

[2] Spin is an intrinsic property of elementary particles, sometimes wrongly thought of as a rotation.

4×4 matrices, and ψ is the wave function which has four components. This equation is valid also when the speed of light is finite and it is thus relativistically correct.

The analytical solution of the Dirac wave equation for the energy a single electron is enclosed in a square root. Normally the negative sign is neglected for many such solutions. However, Dirac later recognized that also the solution with negative energy could have a physical meaning. The *hole theory* assumes that the negative energy levels of our universe are completely filled with all kind of particles at all possible quantum states. If sufficient energy is added ($>$ twice the rest mass of the particle) to a point in space it is possible to "lift" such a particle from the negative part of our universe into our normal space. Behind it is left a hole, which to us looks like the same particle but with opposite charge — an anti-particle. A similar theory is used to explain the electrical behavior of semiconductors, see Ch 9.

When the hole theory is generalized, it follows that all particles have their corresponding anti-particles, hence all the particles mentioned have their *antiparticles* (designated by a bar above the particle symbol), except the photon and the mesons, who are their own antiparticles. We may think about *antimatter* as consisting of antiprotons and antineutrons in an antinucleus surrounded by antielectrons (i.e. positrons). Superficially, there would be no way to distinguish such antimatter from our matter (sometimes called *koino matter*). It has been proposed that the universe is made up of matter and antimatter as a requirement of the principle of *symmetry*. In that case some galaxies, which perhaps can be observed, should be made up of antimatter. When such antimatter galaxies (or material expelled from them) collide with koino matter galaxies, both types of matter are annihilated and tremendous amounts of energy released.

Some of the problems with the Dirac hole theory, like a possibly unlimited amount of particles with negative energy, is addressed by the Quantum Field Theory (QFT). However, that theory also seems to have its problems like an infinite amount of vacuum energy. QFT will not be used here and the interested reader is referred to modern text books in particle physics.

In order to reach the goal of a comprehensive, yet simple, theory of the composition of all matter, the properties of the neutrinos and the quark theory must be considered.

2.5. THE NEUTRINO

The neutrino is a fermion that plays an essential role in the models of elementary particles and in the theory of the formation and development of the universe. The existence of the electron neutrino was predicted by Pauli in 1927 but it was not proven until 1956 when Reines and Cowan detected them in experiments at the Savannah River (USA) nuclear reactor. Since neutrinos are emitted in the β^- decays following fission, nuclear reactors are the most intense neutrino sources on earth. The detector in the discovery experiments consisted of a scintillating solution containing cadmium surrounded by photomultipliers to observe the scintillations which occurred as a consequence of the following reactions:

$$\overline{\nu} + {}^{1}\mathrm{H} \rightarrow \mathrm{n(fast)} + \mathrm{e}^{+} \tag{2.11}$$

$$\mathrm{e}^{+} + \mathrm{e}^{-} \rightarrow 2\gamma_1 \tag{2.12}$$

$$\mathrm{n(thermal)} + {}^{113}\mathrm{Cd} \rightarrow {}^{114}\mathrm{Cd} + \gamma_2 \tag{2.13}$$

The γ's emitted are of different energy; the γ_1 is 0.511 MeV, but γ_2 much higher. There is also a time lag between the γ's because of the time required for the fast neutrons to be slowed down to thermal energy. The detection system allowed a delay time to ascertain a relation between γ_1 and γ_2 (delayed coincidence arrangement). When the reactor was on, 0.2 cpm were observed, while it was practically zero a short time after the reactor had been turned off thereby demonstrating the formation of neutrinos during reactor operation.

Since the 1950s it has become clear that neutrinos exist as several types – called flavors. In β^- decay an "anti-neutrino" is formed, while a "neutrino" is emitted in β^+ decay. Both these neutrinos are now referred to as *electron neutrinos*, $\overline{\nu}_e$ and ν_e, respectively.

Other elemental particles are the pion and the muon. The pions formed in nuclear particle reactions are unstable and decay with a life-time of 3×10^{-8} s into a *muon* and a μ neutrino:

$$\pi^{\pm} \rightarrow \mu^{\pm} + \nu_{\mu} \tag{2.14}$$

The mass of the muon is 0.1135 u (105.7 MeV). The muon is also unstable and has a life-time of 2×10^{-6} s; it decays into an electron, an e neutrino and a μ anti-neutrino:

$$\mu^{\pm} \rightarrow \mathrm{e}^{\pm} + \nu_e + \overline{\nu}_{\mu} \tag{2.15}$$

A third type (flavor) of neutrino, the tau-neutrino, is also known. It also exists in normal and anti-particle forms. Tau-neutrinos are formed in the decay of tauons (a kind of very heavy leptons formed in high energy reactions) and designed as (ν_τ) and ($\overline{\nu}_\tau$).

In 1979 Reines, Sobel and Pasierb made new neutrino measurements with a detector containing heavy water so that the neutrinos would split the ^{2}H atom into two protons and a high energy electron, or convert it into two neutrons and a positron. The neutron will decay into a proton and an electron. Both reactions would only be sensitive to the ν_e; by measuring the neutron yield, the number of ν_e's could be calculated and compared to the calculated ν_e flux from the reactor. The two different decays could be followed by measuring the time delays between neutron capture and γ emission (see Ch. 9). The ratio of these two measurements was 0.43 ± 0.17.

In summary there are three flavors of neutrinos: (i) the electron neutrino, ν_e, which accompanies β decay, (ii) the muon neutrino, ν_μ, which accompanies pion decay, and (iii) the tau neutrino, ν_τ, which is only involved in very high energy nuclear reactions. The three flavors of neutrinos all have their anti particles. They have spin, no charge ("quark charge zero"), but possibly a small mass. They react very weakly with matter, the reaction cross section (cf. Ch. 11) being of the order of 10^{-43} cm^2, depending on neutrino energy.

Various attempts have been made to determine the neutrino rest mass, for example by measuring the decay of soft beta emitters like tritium, or double β decay as ^{82}Se $\rightarrow$ ^{82}Kr. The present published value for the mass of the electron neutrino is 0.07 eV or 1.25×10^{-37} kg. The three flavors of neutrinos are in a steady exchange, ν_e to ν_μ to ν_τ etc, i.e. they oscillate between the various flavors. Because speed is converted back and forth to neutrino rest mass the three flavors propagate with different speed. The total flux of neutrinos is large at the earth's surface, on the average 6.5×10^{10} s^{-1} m^{-2}.

An international collaboration built a neutrino detection station, Gallex, in a rock facility in the Gran Sasso mountain in the Appenines, Italy, which operated between 1991 and 1997. Some ^{71}Ga atoms in 30.3 tons of gallium metal was reacting with solar neutrinos to form ^{71}Ge ($t_{1/2}$ 11.4 d) which was converted to the gaseous hydride, GeH_4, and counted in a proportional detector.

In 1987 a large underground neutrino detector near Fairport, Ohio, in a few seconds registered a sudden burst of 8 events. Taking into account that the normal background rate is about 2 events per day, which is believed to be caused by neutrinos produced in the sun's fusion reactions, this was an exceptional occurrence not only because of the event rate but also because the source was located outside our solar system and was a bright new supernova, SN1987A, appearing in the Large Magellanic Cloud. This was a lucky observation because the previous "near by" supernova was observed in 1604 by Johannes Kepler. The neutrino observation preceded the optical confirmation and it has been calculated that about 10^{58} neutrinos were released in the explosion.

A very large neutrino detector, OPERA, is operating at the Gran Sasso facility replacing the earlier Gallex. It has been used to measure the speed of neutrinos by time of flight from a pulsed neutrino generating target at CERN.

One of the most significant effects of the neutrino mass relates to the mass of the Universe. According to the Big Bang Theory of the origin of the Universe (see Ch. 12) there should be as many neutrinos as there are photons in the microwave background radiation remaining from the Big Bang, or about 100 million times as many neutrinos as other particles. If these neutrinos had a mass >10 eV they would constitute the dominant mass in the Universe. This would mean that there would be enough mass in our Universe for gravitational attraction eventually to overcome the present expansion, and consequently we would have a closed, or possibly pulsating universe, instead of a universe which will continue to expand infinitely. Surprisingly, current observations of the most distant galaxies suggest an accelerating expansion of the Universe instead.

2.6. QUARKS AND THE STANDARD MODEL

All particles are considered to be possible states in which matter can condense. These states are related to the force that forms them. In this sense the solar system is a state of gravitational force, an atom is a state of electromagnetic force, and a nucleon is a state of the strong interaction force. A particle can represent a positive energy state of a system while its analog antiparticle represents the negative state of the same system. Some regular patterns have been found for the elementary particles which indicate that many of them in fact may only be exited states of the same particle, differing in quantum numbers such as spin (or "hyper charge"); in fact hundreds of such states are now known. For example, the neutron has a mass corresponding to 939 MeV and spin 1/2, and there is a baryon with mass 1688 MeV and spin 5/2 with all its other properties like those of the neutron: the heavier particle must be a highly excited state of the neutron.

Though many attempts have been made to unify all particles into one simple theory, this did not succeed until the quark theory was developed. To explain this we have to go back somewhat in time.

The spin of a charged particle leads to the formation of a *magnetic moment* directed along the axis of "rotation". It was discovered in the late 1930s that the magnetic moment of the proton spin ($M_p = 1.41 \times 10^{-26}$ J T^{-1}) is about 1/700 of the electron spin ($M_e = 9.27 \times 10^{-24}$ J T^{-1}), although theory predicts a ratio of 1/1836 ($= M_p/M_e$; see Ch. 6). Also, the neutron has a negative magnetic moment ($M_n = -0.97 \times 10^{-26}$ J T^{-1}). The only explanation scientists could offer for this deviation was that the proton is not an evenly charged rotating sphere, but contains some "internal electrical currents", and also the neutron must contain some internal charges, which balance each other to appear uncharged. Thus, it was doubtful that protons and neutrons were truly elementary.

Around 1960 Hofstadter and co-workers at the large Stanford Linear Accelerator Center (SLAC, Ch. 16) proved that both the proton and the neutron have an uneven internal nuclear charge density. This came from studies of the scattering of high energy electrons ($\sim$1 GeV) against protons and neutrons. It was suggested by Gell-Mann that this could mean that the proton and neutron were composed of smaller particles, with fractional charge and mass, which he called *quarks*. The intense search for such particles (leading to the discovery of many new "elementary" particles) culminated in the late 1970s in experiments in which still higher electron energies ($4 - 21$ GeV) were used and the energy and scattering angle of the electrons measured. These revealed that the nucleons had hard internal scattering centers with charge 1/3 that of the electron and mass 1/3 that of the nucleon. These particles, quarks, are held together by *gluons*, which are carriers of the nuclear force.

These results have led to the *Standard Model* of the building blocks of matter. According to this model all matter on Earth — and likely in the Universe — (and including our own bodies) consists to > 99% of quarks with associated gluons. The rest is electrons.

Elementary particles now come in only two kinds: quarks and leptons. There are only six quarks and six leptons, see Table 2.3. The leptons are the electron, e, the muon, μ, and the *tauon* (tau particle), τ, and their respective neutrinos. The quarks and leptons are grouped together in *three families* (or generations) of two quarks and two leptons each. This makes 12 elementary building blocks, or 24 if one includes their anti particles; Table 2.3 only refers to our matter (i.e. koino matter). The leptons and quarks all have different properties and names, sometimes also referred to as *colors*. The physical theory relating these particles to each other is therefore named *Quantum Chromo Dynamics* (QCD).

All matter in nature belongs to the first family, which consist of two leptons, the electron and electron-neutrino, and the up-quark and the down-quark. The proton is made up of 2 up-quarks and 1 down-quark, giving it a charge of +1 and mass 1, while the neutron is made up of 1 up- and 2 down-quarks giving it a charge of 0 and mass of 1:

$$n = u^{+2/3} + d^{-1/3} + d^{-1/3} \tag{2.16}$$

$$p^{+} = u^{+2/3} + u^{+2/3} + d^{-1/3} \tag{2.17}$$

where u and d represent the up and down quarks, respectively. The neutron decay (§5.4.5) can be written according to the quark model: i.e. a d-quark is transformed into a u-quark with the simultaneous emission of an electron and an anti-neutrino, see Figure 2.3.

In all reactions the lepton number must be conserved: the total number of leptons minus anti leptons on each side of a decay or reaction process must be the same. A similar law is valid for the quarks. In the reaction above several quantum numbers are obeyed: (i) the charge is the same on both side, (ii) the lepton number is zero on both sides (none = electron minus anti neutrino), (iii) the quark number is conserved. The elementary reactions in Figure 2.2 can all be described in terms of lepton and quark transformations.

All hadrons contain 3 quarks, while all mesons are made up of 2 quarks or anti quarks. The quarks move around in the nucleus, which makes it difficult to observe these minute particles: if an atom had the size of the earth, the size of the quark would be about half a cm. The quarks cannot appear free but must appear together in groups of two or three.

The second family in Table 2.3 contains the "heavy electron", the muon and the muon neutrino, and the charm and the strange quarks. The third family contains the tau particle, the tau electron, and the two quarks referred to as top (or truth) and bottom (or beauty). These quarks can only be produced in high energy particle reactions.

By combination of quarks and leptons, the true elementary particles of nature, it is possible to systematize all known particles. The success of this theory, founded on good

Table 2.3 Classification and properties of elementary particles according to the standard model

Basic nature	Family	Name and symbol		Forces involved	Charge	Rest mass
Basic building blocks of nature	I	Leptons:	electron, e	EM,W	±1	0.511 MeV
			e-neutrino, ν_e	W	0	<18 eV
		Quarks:	up	EM,W,S	+2/3	1/3 u
			down	EM,W,S	−1/3	1/3 u
Formed in Big-Bang, cosmic rays and high-energy accelerators	II	Leptons:	muon, μ	EM,W	±1	105.6 MeV
			μ-neutrino, ν_μ	W	0	0?
		Quarks:	charm	EM,W,S		
			strange	EM,W,S		
	III	Leptons:	tau, τ	EM,W	±1	
			τ-neutrino,	W	0	0?
		Quarks:	top (or truth)	EM,W,S		
			bottom (or beauty)	EM,W,S		
Carriers of force:		Photon, γ		EM	0	0
bosons; spin = 0, 1, ..		Pion, π		S	0,±1	137 MeV
Pauli principle not valid		Gluon		EM,W,S		
		W^+,W^-,Z^0		W		

EM = electromagnetic force, W = weak interaction, S = strong interaction.

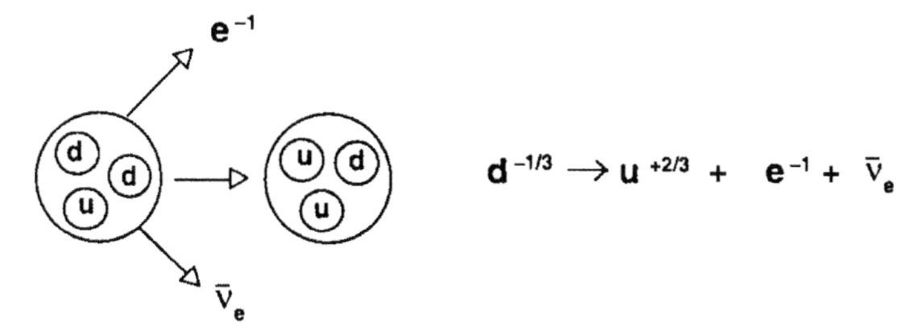

Figure 2.3 Neutron decay according to the quark model (simplified by omitting the intermediate vector boson, which decays to an electron and an antineutrino).

experimental evidence, has been so great that its name, the Standard Model of matter, is justified.

The force between two particles arises from the exchange of a "mediator" that carries the force at a finite speed: one of the particles emits the mediator, the other absorbs it. The mediator propagates through space and, briefly, is not lodged with either particle. These mediators have the same properties as the elementary particles — mass, electric charge, spin — so physicists often call them particles as well, even though their role in nature is quite different from that of the elementary particles. The mediators are the mortar that binds the particle building blocks together.

Three kinds of mediators — or exchange forces, as we have called them — are known in the nuclear world: photons which are involved in the electromagnetic force, gluons which are the mediators of the strong nuclear force, and the weak force mediators, which underlies the radioactive decay. We know now a great deal about the photons and the gluons, but little about the weak force. The weak force mediators are the W^+, W^- and Z^0 "particles". A high energy electron is supposed to be able to emit a Z^0, and then a positron can absorb the electron: the particles annihilate each other, leaving the Z^0 momentarily free. Afterwards the Z^0 must decay back into a pair of elementary particles, such as an electron and positron, or a quark and an antiquark. Z^0 particles are produced in high energy proton-antiproton colliders; around 1983 researchers at the CERN and SLAC laboratories were able to determine the Z^0 mass to 91.2 GeV. This is the heaviest known unit of matter today.

2.7. EXERCISES

2.1. Which type of mesons are released in high energy particle interactions, and why?

2.2. (a) What kinds of forces exist in nature? (b) How does the weak interaction manifest its properties?

2.3. (a) What are bosons and how do they differ from fermions? (b) Does the difference have any practical consequence?

2.4. What proof exists that the photon has matter properties?

2.5. How can the neutrino be detected?

2.8. LITERATURE

B. P. GREGORY, *Annual Report 1974,* CERN.

LANDOLT - BÖRNSTEIN, *Properties and Production Spectra of Elementary Particles*, Neue Serie I, **6**, Springer Verlag, 1972.

S. WEINBERG, Unified Theories of Elementary Particle Interaction, *Sci. Amer.*, **231** (1974) 50.

M. JACOB and P. LANDSHEFF, The Inner Structure of the Proton, *Sci. Amer.*, **242** (1980) 66.

K. S. KRANE, *Introductory Nuclear Physics*, J. Wiley & Sons, 1988.

J. R. REES, The Stanford Linear Accelerator, *Sci. Am.* Oct. 1989.

C. RUBBIA and M. JACOB, The Z^0, *Amer. Scient.*, **78** (1990) 502.

C. AMSLER et al. (Particle Data Group) "The Review of Particle Physics: Neutrino Mass, Mixing, and Flavor Change". Physics Letters **B 667**: 1 (2008).

CHAPTER 3

Nuclei, Isotopes and Isotope Separation

Contents

From a practical point of nuclear chemistry we accept the commonly used model that the nucleus is composed of only protons and neutrons. For the purposes of the scope of this book this definition is enough even if, as seen in Chapter 2, the world is far more complex than that. Although this is a simplification most observable patters of nuclear chemistry can be explained and the nomenclature can be kept simple. For example, the oxygen atom of mass number 16 has a nucleus which consists of 8 protons and 8 neutrons; since neutrons have no charge but are very similar to protons in mass, the net nuclear charge is +8. There are also 8 extranuclear electrons in the neutral atom of oxygen.

Radiochemistry and Nuclear Chemistry
ISBN 978-0-12-405897-2, http://dx.doi.org/10.1016/B978-0-12-405897-2.00003-3

3.1. SPECIES OF ATOMIC NUCLEI

Both the protons and the neutrons in the nucleus are commonly denominated *nucleons*. The *mass number* A is the total number of nucleons. Thus

$$A = N + Z \tag{3.1}$$

where Z is the number of protons, i.e. the *atomic number* and N is the number of neutrons. The elemental identity and the chemical properties are basically determined by the atomic number even if most chemical interactions take place via the interactions of the surrounding electron cloud.

As we have seen in Chapter 1, an element may be composed of atoms that, while having the same number of protons in the nuclei, have different mass numbers and, therefore, different numbers of neutrons. Neon, for example, has an atomic number of 10, which means that the number of protons in the nuclei of all neon atoms is 10; however, 90% of the neon atoms in nature have 10 neutrons present in their nuclei while 10% of the atoms have 12 neutrons. Such atoms of constant Z but different A are called *isotopes*. The heavy hydrogen isotopes ^{2}H and ^{3}H are used so often in nuclear science that they have been given special names and symbols, deuterium (D) and tritium (T), respectively.

The word isotope is often misused to designate any particular nuclear species, such as ^{16}O, ^{14}C, ^{12}C. It is correct to call ^{12}C and ^{14}C isotopes of carbon since they are nuclear species of the same element. However, ^{16}O and ^{12}C are not isotopic since they belong to different elemental species. The more general word *nuclide* is used to designate any specific nuclear species; e.g. ^{16}O, ^{14}C, and ^{12}C are nuclides. The term *radionuclide* should be used to designate any radioactive nuclide, although, somewhat illogically, *radioisotope* is a common term used for the same purpose.

In addition to being classified into isotopic groups, nuclides may also be divided into groupings with common mass numbers and common neutron numbers. *Isotopes* are nuclides with a common number of protons (Z), whereas *isobar* is the term used to designate nuclides with a common number of nucleons (A), i.e. the same mass number. Nuclei with the same number of neutrons (N) but different atomic numbers are termed *isotones* ^{40}K and ^{40}Ar are examples of isobars, while ^{3}H and ^{4}He are examples of isotones.

In some cases a nucleus may exist for some time in one or more excited states and it is differentiated on this basis. Such nuclei that necessarily have the same atomic number and mass number are called *isomers*. ^{60m}Co and ^{60g}Co are isomers; the ^{60m}Co nuclide exists in a high energy (*excited*) *state* and decays spontaneously by emission of a γ-ray with a half-life of 10.5 min to the lowest energy, *ground state*, designated by ^{60g}Co.

$$^{60m}\mathrm{Co} \xrightarrow[10.5\ \mathrm{min}]{\gamma} {}^{60g}\mathrm{Co} \tag{3.2}$$

The symbol m stands for *metastable*, while g (or no symbol) refers to the ground state. In order for a nuclide to be called *metastable* one usually demand that the excited state has a half-live of at least 10^{-9} s. This is about 1000 times longer than for a normal excited state to decay. There are a few nuclides having more than one metastable state. These are then designated as m_1, m_2, etc.

3.2. ATOMIC MASSES AND ATOMIC WEIGHTS

The *universal mass unit*, abbreviated u (sometimes amu for atomic mass unit), is defined as one-twelfth of the mass of the ^{12}C atom which has been defined to be exactly 12 u. The absolute mass of a ^{12}C atom is obtained by dividing the value 12 by the Avogadro number ($\mathbf{N}_A = 6.022137 \times 10^{23}$). The value for the mass of a ^{12}C atom, i.e. the nucleus plus the 6 extranuclear electrons, is thus $1.992\,648 \times 10^{-23}$ g. Atomic masses are expressed in units of u relative to the ^{12}C standard. This text uses M to indicate masses in units of u, and m in units of kilograms; $m = M/10^3\ \mathbf{N}_A$. One atomic mass unit corresponds to 931.5 MeV using the famous equation $E = m_0c^2$. Historically this has not always been the case. Up to as late as the 1960ies two different scales of atomic masses existed. The chemists used a scale based on the natural composition of oxygen and designated that to have a mass of 16 u while in the physics community the same mass was assigned to the most common oxygen isotope, ^{16}O. Since natural oxygen also contains the isotopes ^{17}O and ^{18}O this led to two different tables of atomic masses which naturally was highly inconvenient.

In nuclear science it has been found convenient to use the atomic masses rather than nuclear masses. The number of electrons are always balanced in a nuclear reaction, and the changes in the binding energy of the electrons in different atoms are insignificant within the degree of accuracy used in the mass calculations. Therefore the difference in atomic masses of reactants and products in a nuclear reaction gives the difference in the masses of the nuclei involved. In the next chapter, where the equivalence between mass and energy is discussed, it is shown that all nuclear reactions are accompanied by changes in nuclear masses.

The mass of the nucleus can be approximated by subtracting the sum of the masses of the electrons from the atomic mass. The mass of an electron is 0.000549 u which is equivalent to 9.1094×10^{-31} kg. Since the neutral carbon atom has 6 electrons, the approximate mass of the nucleus is $1.992\,648 \times 10^{-26} - 6 \times (9.1094 \times 10^{-31}) = 1.992\,101 \times 10^{-26}$ kg. This calculation has not included the difference in the mass of the 6 extra electrons attributable to the binding energy of these electrons. However, this binding energy has a mass equivalence which is smaller than the least significant figure in the calculation.

The mass of a neutron is 1.008 665 u while that of the hydrogen atom is 1.007 825 u. This shows the fact that the neutron decays to a proton and en electron. Since both neutrons and protons have almost unit atomic masses, the atomic mass of a nuclide should be close to the number of nucleons, i.e. the mass number. However, when the table of elements in the periodic system (Appendix I) is studied it becomes obvious that many *elements* have masses which are far removed from integral values. Chlorine, for example, has an atomic mass value of 35.453 u, while copper has one of 63.54 u. The variations in precision of the given atomic weights can be explained by the effect of the relative abundances of the stable isotopes of the elements contributing to produce the observed net mass.

If an element consists of n_1 atoms of isotope 1, n_2 atoms of isotope 2, etc., the *atomic fraction* x_1 for isotope 1 is defined as:

$$x_1 = n_1/(n_1 + n_2 + \dots) = n_1/\Sigma n_i \tag{3.3}$$

The *isotopic ratio* is the ratio between the atomic fractions (or *abundances*) of the isotopes. For isotopes 1 and 2, the isotopic ratio is

$$\zeta_1 = x_1/x_2 = n_1/n_2; \quad \zeta_2 = x_2/x_1 = n_2/n_1 \tag{3.4}$$

The *atomic mass of an element* (or *atomic weight*) M is defined as the average of the isotopic masses, i.e. M_i is weighted by the atomic fraction x_i of its isotope:

$$M = x_1 M_1 + x_2 M_2 + \dots = \Sigma x_i M_i \tag{3.5}$$

As an example, natural chlorine consists of two isotopes of which one has an abundance of 75.77% and an atomic mass of 34.9689 u and the second has an abundance of 24.23% and a mass of 36.9659 u. The resultant average atomic mass for the element is 35.453 u. The atomic mass of copper of 63.546 u can be attributed to the presence of an isotope in 69.17%

Table 3.1 Isotopic data for some elements

Element	*Z*	*N*	*A*	Atomic mass (u)	Abundance (%)	Atomic weight	Symbol
Hydrogen	1	0	1	1.007 825	99.985		^{1}H
	1	1	2	2.014 102	0.0155	1.00797	^{2}H, D
	1	2	3	3.016 049	0		^{3}H, T
Helium	2	1	3	3.016 030	0.0001		^{3}He
	2	2	4	4.002 603	100.00	4.0026	^{4}He
Lithium	3	3	6	6.015 121	7.42	6.939	^{6}Li
	3	4	7	7.016 003	92.58		^{7}Li
Beryllium	4	5	9	9.012 182	100.00	9.0122	^{9}Be
Boron	5	5	10	10.012 937	19.6	10.811	^{10}B
	5	6	11	11.009 305	80.4		^{11}B
Carbon	6	6	12	12.000 000	98.892	12.0112	^{12}C
	6	7	13	13.003 355	1.108		^{13}C
Nitrogen	7	7	14	14.003 074	99.635	14.007	^{14}N
	7	8	15	15.000 109	0.365		^{15}N
Oxygen	8	8	16	15.994 915	99.759		^{16}O
	8	9	17	16.999 131	0.037	15.999	^{17}O
	8	10	18	17.999 160	0.204		^{18}O
Chlorine	17	18	35	34.968 853	75.8	35.453	^{35}Cl
	17	20	37	36.965 903	24.2		^{37}Cl
Uranium	92	143	235	235.043 924	0.724	238.029	^{235}U
	92	146	238	238.050 785	99.266		^{238}U

abundance with a mass of 62.9296 u and of a second isotope of 30.83% abundance and 64.9278 u. Atomic masses and abundances of some isotopes are given in Table 3.1.

3.3. DETERMINATION OF ISOTOPIC MASSES AND ABUNDANCES

3.3.1. The mass spectrometer

The masses and relative abundances of different isotopes occurring naturally in an element can be determined with great exactness using the same technique J. J. Thomson employed to demonstrate the presence of isotopes in neon. The instrument used for this purpose is known as a mass spectrometer. The principles of the electromagnetic *mass spectrometer* are shown in Figs. 3.1 and 3.2.

Let us first consider the movement of an ion in electric and magnetic fields, as shown in Fig. 3.1. The ion of mass m is assumed to have a charge q (coulomb), which is an integer (z) multiple of the elementary charge $\mathbf{e}$ ($1.602\,177\times10^{-19}$ C): $q = z\mathbf{e}$. If it is accelerated from velocity zero to v (m s^{-1}) by an electric potential V (volts), it acquires a kinetic energy E_{kin} corresponding to joule (or Newton meter).

$$E_{\mathrm{kin}} = ½mv^2 = qV \tag{3.6}$$

If q is given in units of the elementary charge, the kinetic energy is in units of *electron volts* (eV). For transformation to other energy units, see Appendix IV.

Figure 3.1 shows the deviations of a positive ion in an electric field U (newton/coulomb) directed upwards, and a magnetic field B (tesla) directed into the page. The force F (newton) acting on the ion due to the electric field is

$$F_{\mathrm{e}} = qU \tag{3.7}$$

If the electric force would dominate, the ion would hit the screen at point P_1. For the magnetic field only, the force is

$$F_{\mathrm{m}} = qvB \tag{3.8}$$

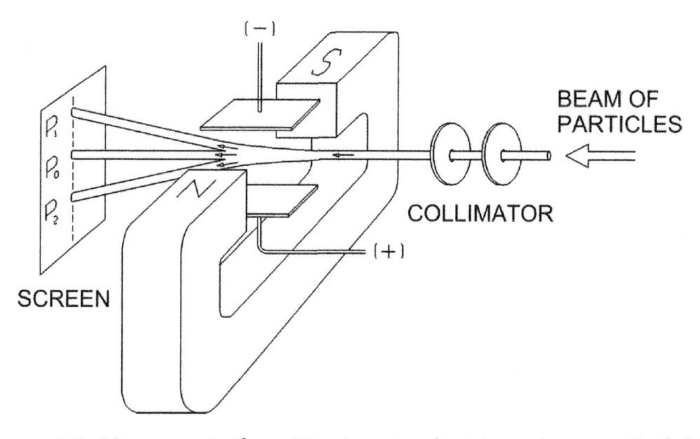

Figure 3.1 Movement of positive ions in electric and magnetic fields.

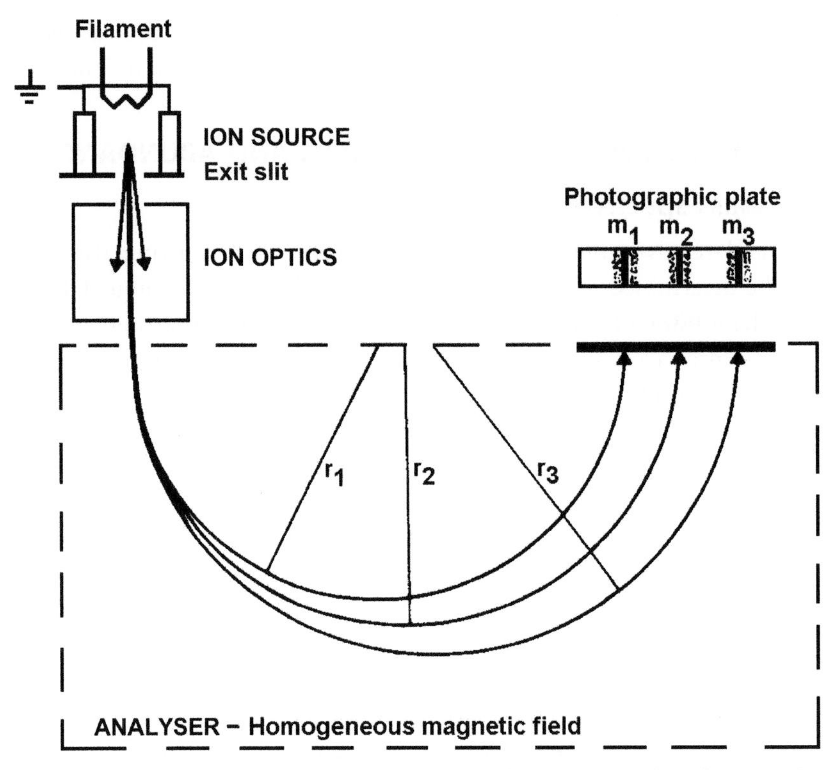

Figure 3.2 The principle of the mass spectro*graph* (*-meter*, if the photographic plate is replaced by an ion current meter).

and if this force would dominate the ion would hit the screen at point P_2. If the forces balance each other, i.e. when $F_e = F_m$, the ion hits the screen at point P_0 with a velocity given by

$$v = U/B \tag{3.9}$$

In either the fields (F_e or F_m) the deviation is counteracted by the centrifugal force F_c, where

$$F_c = mv^2/r \tag{3.10}$$

and r is the radius of curvature. In the magnetic field only, the balance between F_m and F_c leads to

$$q/m = v/Br \tag{3.11}$$

where q/m is denoted as the *specific charge* of the ion.

In the mass spectrometer gaseous ions are produced in an *ion source*, e.g. by electron bombardment of the gas produced after e.g. heating the substance in a furnace (Fig. 3.2) or by electric discharge, etc. If positive ions are to be investigated, the ion source is given a high positive potential relative to the exit slits. This results in the ions being accelerated away from the source and into the ion optic system. The purpose of the ion optic system is to produce ions of exact direction and of constant velocity, which is achieved through the use of electrostatic and magnetic fields as described; cf. (3.9).

The spectrometer commonly consists of a homogeneous magnetic field which forces the ions to move in circular paths according to (3.10). Combining (3.6) and (3.11) gives

$$m = q\,r^2\,B^2/2V \tag{3.12}$$

where V is the ion acceleration potential. Mass spectrometers are usually designed so that of the three variables V, B, or r, two are constant and the third variable, which allows ions of different q/m value to be focused on the detector. The minimum value of q/m is always $\mathbf{e}/m$ because singly charged ions of the atomic or molecular species are almost always present. In order to avoid collisions between ions and gaseous atoms, which would cause scattering of the ion beam, the system is evacuated. The detector is some charge collecting device (Ch. 9).

A common type of mass spectrometer (A. O. Nier 1940) uses a fixed magnetic field and a fixed radius of curvature for the ion beam. If the acceleration potential V is varied so that the masses m_1 and m_2 alternately are registered by the detector, producing ion currents I_1 and I_2, respectively, the abundance of each isotope can be calculated from the ratios $x_1 = I_1/(I_1 + I_2)$ and $x_2 = 1 - x_1$ when only two isotopes are present. The resolution of modern mass spectrometers can be extremely high, as indicated by the values in Table 3.1.

3.3.2. Applications

For several decades, mass spectrometers were used primarily to determine atomic masses and isotopic ratios. Now they are applied to a large variety of chemical problems and low resolution mass spectrometers are used for routine chemical analysis. For example, a modern mass spectrometer can easily distinguish between species such as $^{12}CH_4^+$ and $^{16}O^+$, having a mass difference of 0.03686 u. The resolution (R) of a mass spectrometer is defined as:

$$R = M/\Delta M \tag{3.13}$$

where M is the atomic mass and ΔM is the minimum peak separation. There are, however, two different methods to define the minimum peak separation:

The peak width definition: ΔM is given based on the width of the peak at a specific peak height, for example 5%. This definition is also used in e.g. gamma spectroscopy (Ch. 9)

The valley definition: ΔM is given by the closest spacing between two peaks of equal intensity where the intensity of the valley is less than a specific fraction of the peak height. Typical values are tenths of percent. For comparison, the value obtained from a 5% peak width is roughly equivalent to a 10% valley.

Some uses of mass spectrometry of interest to chemists involved in nuclear science are:

(a) *Molecular weight determination* can be made by mass spectrometry if gaseous ions can be produced with M/q values not exceeding about 400. This method is of great importance in radiation chemistry (Ch. 8) where often a large number of products are produced which may be quite difficult to identify by other means and in particular for the analysis of organic compounds.

(b) The *study of chemical reactions* directly in the gas phase by mass spectrometry is possible. Using an ion source in which molecules are bombarded by a stream of low energy ($\leq$100 eV) electrons, ionization and dissociation reactions can be studied, e.g.

$$C_8H_{18} + e^- \rightarrow C_8H_{18}^+ + 2e^-$$
$$\quad \hookrightarrow C_4H_9^+ + C_3H_6^+ + CH_3^{\bullet} + e^- \qquad (3.14)$$

This technique has practical application e.g. in the petroleum industry for determining the composition of distillation and cracking products.

(c) *Isotopic dilution* is a technique for determination of the number of atoms of an element (or isotope) in a composite sample (e.g. rock or biota), from which is difficult to recover the element reproducibly and determine it quantitatively. The technique is simple to use for radioactive nuclides, as described in §18.3.2, but more complicated and time consuming for stable nuclides. However, the high precision of mass spectrometry makes the latter alternative more accurate, and the technique is widely used for geologic dating.

Suppose we have a *sample* with *unknown* concentration of a certain element, consisting of the two isotopes 1 and 2, and want to determine the number of atoms, N_2, of isotope 2 in the sample (which, of course must be of known weight). The number of atoms of each isotope is N_1 and N_2, thus $N = N_1 + N_2$, in unit sample weight. By mass spectrometry we determine the isotope ratio $\zeta_N = N_1/N_2$.

In isotope dilution analysis the isotope ratio in a sample to be measured is altered by adding a known amount of a standard (spike) with a well known isotopic composition that differs from the isotopic composition in the sample. By measuring the isotopic composition of this mixture called blend, the amount of the analyte in the sample can be calculated. Of course, the isotopic composition of the sample has also to be known! Below, s refers to sample, sp to spike and B to blend. Moreover we consider only the two isotopes a and b. A mass balance for the two components 1 and 2 gives first for the number of moles of component a in the blend:

$$n_{1.B} = n_{1.s} + n_{1.sp} \qquad (3.15)$$

And for component b in the same way

$$n_{2.B} = n_{2.s} + n_{2.sp} \tag{3.16}$$

Combine these two equations (division), and we'll get

$$\frac{n_{2.B}}{n_{1.B}} = \xi_B = \frac{n_{2.s} + n_{2.sp}}{n_{1.s} + n_{1.sp}} \tag{3.17}$$

Where ξ_B is the isotope ratio (mole ratio) of component 1 and 2 in the blend. We know that the number of moles of isotope j in e.g. the blend is $n_{j.B} = f_{j.B} \cdot n_B$ ($n_B = n_{j.B} + n_{k.B}$) and that the abundance $f_{j.B} = \xi_{j.B}/\Sigma\xi_{j.B}$. From this we get the isotope ratio of 1 and 2 in the blend

$$\xi_B = \frac{\frac{\xi_{2.s}}{\Sigma\xi_s} n_s + \frac{\xi_{2.sp}}{\Sigma\xi_{sp}} n_{sp}}{\frac{1}{\Sigma\xi_s} n_s + \frac{1}{\Sigma\xi_{sp}} n_{sp}} \tag{3.18}$$

Note that here $\xi_{1.B} = 1$! After simplification we get

$$n_s = n_{sp} \frac{\xi_{sp} - \xi_B}{\xi_B - \xi_s} \cdot \frac{\Sigma\xi_s}{\Sigma\xi_{sp}} \tag{3.19}$$

But $n = C \cdot m$, so we get

$$C_s = C_{sp} \frac{m_{sp}}{m_s} \cdot \frac{\xi_{sp} - \xi_B}{\xi_B - \xi_s} \cdot \frac{\Sigma\xi_s}{\Sigma\xi_{sp}} \tag{3.20}$$

If the sample can be considered to have a natural isotopic composition of the analyte (or any other well known composition), we only have to measure the blend ratios. Using mass spectrometry we then get

$$C_s = C_{sp} \frac{m_{sp}}{m_s} \cdot \frac{\xi_{sp} - K \cdot r_B}{K \cdot r_B - \xi_s} \cdot \frac{\Sigma\xi_s}{\Sigma\xi_{sp}} \tag{3.21}$$

where r_i is an observed ratio and K a correction factor due to mass fraction within the mass spectrometer, i.e. $K = \xi_i/r_i$. Mass fractionation may depend on several processes depending on e.g. the ion source. The light isotope will deviate from the ion beam to a greater extent than the heavy one due to electrostatic repulsion. In ion sources where diffusion processes plays a role there will also be a mass fractionation since the light isotope will have a higher diffusivity compared to the heavy one.

Thus from 3 mass spectrometrically determined isotope ratios (ζ_s, ζ_{sp} and ζ_B) and known standard amount $P = P_1 + P_2$, the unknown number of atoms N_2 in the sample is determined. The precision of this technique is largest when sample and standard are added in such proportions that $N \sim P$. Examples of the use of this technique are given in §13.8 on geologic dating.

In the case of using a radioactive isotope for the isotopic dilution equation 3.20 simplifies to

$$W_x = W_0\left(\frac{S_0}{S_2} - 1\right) = W_0\left(\frac{A_0 W_2}{A_2 W_0} - 1\right) \tag{3.22}$$

where W_x is the weight of the unknown, W_0 is the weight of the spike added, A_0 its activity, and S_0 its specific activity ($S_0 = A_0/W_0$). Similarly W_2 is the weight of the analyte isolated, A_2 its activity, and S_2 its specific activity ($S_2 = A_2/W_2$). When only the ratio S_0/S_2 is used, the activities A_0 and A_2 can be replaced by the measured radioactivities R_0 and R_2 in case that the detection efficiency ψ is the same.

Because the variables in these equations often need multiple indexing (e.g. element, isotope and source) the amount of indexing can be reduced by using *italic* element symbols to refer to the specific radioactivity, or concentration, of that nuclide in the sample; e.g. ^{238}U means S_{238U} (radioactivity), or N_{238U} (atoms) per unit volume or weight of the *sample*. Extensive use of this formalism will be found in §18.3.2***ff***[1].

(d) *Analysis of gas purity* (e.g. in plants for separation of ^{235}U and ^{238}U) is done conveniently by mass spectrometry. Not only are the ratios of the uranium isotopes determined, but also the air and water vapour which leaks into the system can be measured. This produces such ions as O_2^+, N_2^+, CO_2^+, and HF^+, which can be measured easily because mass spectrometry detects the presence of impurities in parts per million (ppm) concentration.

(e) *Instrumental chemical analysis* can be done by using mass spectrometers having ion sources of the inductively coupled plasma type (ICP-MS systems), sometimes as on-line detectors for various chromatographic techniques. Due to the high resolution and sensitivity of this technique it is widely used in analysis of pollutants in the environment.

(f) *Measurement of products from nuclear reactions* can be made with special mass spectrometers on line to the irradiated target. With suitable arrangements both A and Z of the recoiling reaction products can be determined.

It needs to be noted here that although ppm or similar mass units are used in mass spectroscopy these units are not optimal for use in chemical analysis. Here there might be a need to assess whether there is enough of a specific ligand to reach equilibrium or to determine of a solution is super saturated with respect to an element. Thus for chemical use mass units are preferably converted to amount units like number of moles per volume.

[1] The Student making notes may have difficulties in distinguishing between italics and normal letters and is therefore recommended to adhere to the use of multiple indexing.

3.4. ISOTOPIC RATIOS IN NATURE

Mass spectrometric investigations of geologic material has shown that isotopic ratios vary slightly in nature with the largest variations observed for the heaviest and lightest elements. Hence the atomic weight of many elements is a little uncertain, which is indicated be giving fewer decimals than for their isotopes.

For the heaviest elements the reason for the variation in the ratio of isotopes in an element can be related directly to the consequence of naturally occurring radioactive decay. For example, in a thorium mineral the decay sequence terminates in the formation of an isotope of lead with mass number 208. By contrast, in a uranium ore, in which ^{238}U is the most abundant uranium isotope, the primary end product of the decay is an isotope of lead with mass number 206 (see Fig. 1.2). This means that thorium minerals and uranium minerals contain lead with different isotopic composition. In fact, one of the best confirmations of the existence of the radioactive decay series came from the proof that lead had different net atomic weights in different minerals.

Lead from different ores will contain slightly different isotopic abundances due to the fact that all long natural radioactive decay series end up in differet isotopes of lead An unusual use of this fact was made by Andrasko et al. to identify smears and fragments from lead bullets used in a homicide case, so that the suspect could be bound to the case, as the isotopic composition of lead bullets can be identified not only by manufacturer but also by manufacturing date.

The isotopic ratios for the lightest elements depend on in which natural material they are found. The ratio for $^{7}Li/^{6}Li$ varies from 12.20 to 12.48, while that of $^{12}C/^{13}C$ varies from 89.3 to 93.5. The isotopic ratio $^{18}O/^{16}O$ varies by almost 5% as shown in Fig. 3.3. However, since natural oxygen contains only 0.2% ^{18}O, even a 5% variation in the isotopic ratio has very little influence on the atomic weight of oxygen. The variation of natural isotopic ratios for boron and chlorine causes the uncertainties in their abundances shown in Table 3.

3.5. PHYSICOCHEMICAL DIFFERENCES FOR ISOTOPES[2]

Although the isotopic variations in the heaviest elements can be attributed to the consequences of radioactive decay, the variations observed in lighter elements are attributable to chemical behaviour. The rates and equilibria of chemical reactions depend on the masses of the atoms involved in the reactions as is explained in §3.6 and §3.7. As a consequence, isotopes may be expected to have somewhat different quantitative physicochemical values (the *isotope effect*). This effect is naturally more predominant for lighter elements than heavier since for the lighter elements the change in mass by adding a neutron

[2] §§3.5-3.7 outline the scientific basis for isotope effects and isotope separation. This is not essential for the Radiochemistry part of this book.

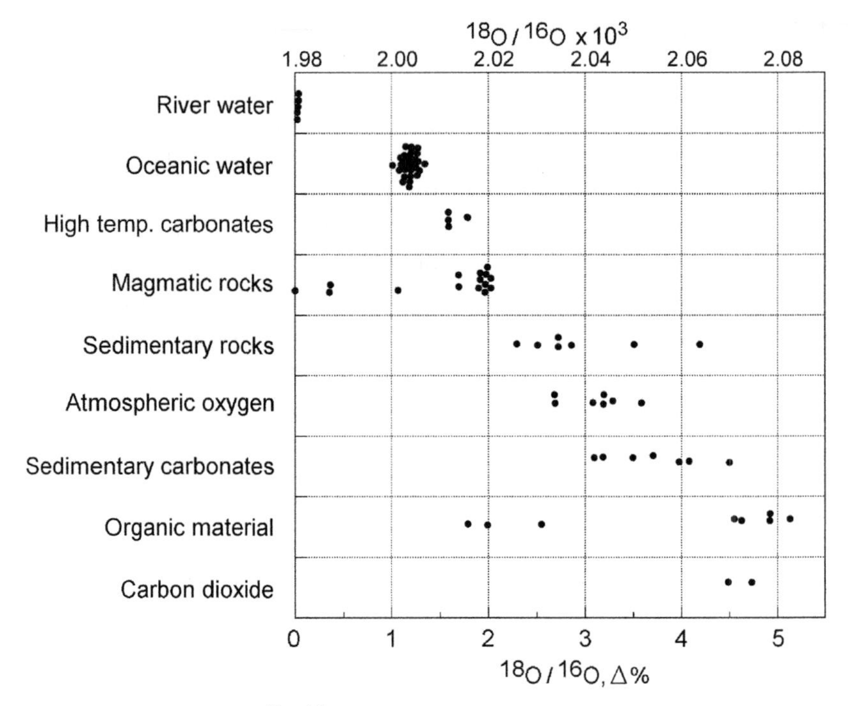

Figure 3.3 Observed $^{18}O/^{16}O$ isotope ratios. (According to Vinogradov.)

may be as high as 100% (for hydrogen) while the change in mass between e.g. ^{235}U and ^{238}U is only about 1%. As examples of the isotope effect, we may note that the freezing point of H_2O is 0°C (273.15 K), while that for heavy water D_2O is 3.82 degrees higher (276.97 K). The boiling point of D_2O is 1.43 K higher than that of H_2O. Similarly, while H_2 boils at 20.26 K, D_2 boils at 23.59 K. As a result of these differences in the boiling points, the vapor pressures at a particular temperature for H_2 and D_2 are not the same and distillation can be used to fractionate hydrogen isotopes. Other physical properties such as density, heat of vaporization, viscosity, surface tension, etc., differ similarly.

The optical emission spectra for isotopes are slightly different since electronic energy levels are dependent on the atomic masses. The light emitted when an electron falls from an outer orbit of main quantum number n_2 to an inner orbit of quantum number n_1 ($n_1 < n_2$) is given by

$$\bar{\upsilon} = \mathbf{R}_\infty Z^2 m_{red} \left(1/n_1^2 - 1/n_2^2\right)/\mathbf{m}_e \tag{3.23}$$

where $\bar{\upsilon}$ is the wave number ($1/\lambda$, m^{-1}), $\mathbf{R}_\infty$ is the Rydberg constant (1.097×10^7 m^{-1}), $\mathbf{m}_e$ is the electron (rest) mass, and m_{red} the reduced mass, according to

$$m_{red}^{-1} = m_e^{-1} + m_{nucl}^{-1} \tag{3.24}$$

where m_{nucl} is the nuclear mass. For the light hydrogen isotope, the H_α line (i.e. the transition energy between $n_1 = 1$ and $n_2 = 2$) occurs at 656.285 nm, while the corresponding line for the deuterium isotope of hydrogen occurs at 656.106 nm. This difference could be predicted from (3.23) and its observation experimentally in 1932 provided the first evidence for the existence of a heavy hydrogen isotope. This spectral difference has practical importance as it can be used for a spectroscopic analysis of the amount of heavy water in light water. Similar *isotopic line shifts* occur in the spectra of all elements, but are rarely as large as the shift of almost 0.2 nm observed for hydrogen. For the isotopes ^{235}U and ^{238}U, the isotopic shift is 0.025 nm.

3.6. ISOTOPE EFFECTS IN CHEMICAL EQUILIBRIUM

In a mixture of molecules AX and BX, with a common element X, an exchange of the atoms of the common element between the two molecules may occur. When the two compounds have different isotopes X and X*, we may have an *isotope exchange* according to

$$AX + BX^* = AX^* + BX \tag{3.25}$$

The equilibrium constant k in the reaction is given by

$$\Delta G^o = -\mathbf{R}T \ln k = -\mathbf{R}T \ln \{([AX^*][BX])/([AX][BX^*])\} \tag{3.26}$$

where ΔG^o is the (Gibb's) free energy and $\mathbf{R}$ is the molar gas constant. For values of fundamental constants see Appendix III.

3.6.1. The partition function

It has been shown that k deviates slightly from 1, giving rise to the isotopic effects observed in nature for the light elements. This deviation can be calculated by methods of statistical thermodynamics. Only the main features of this theory are given here. The equilibrium constant k can be written

$$k = (F_{AX^*}\ F_{BX})/(F_{AX}\ F_{BX^*}) \tag{3.27}$$

where F is the *grand partition function*, which for each molecule includes all possible energy states of the molecule and the fraction of molecules in each energy state under given external condition. The grand partition function is defined by

$$F = f_{tr} f_{rot} f_{vib} f_{el} f_{nsp} \tag{3.28}$$

where each term f_j refers to a particular energy form: translation, rotation, vibration, electron movement, and nuclear spin. The two latter will have no influences on the chemical isotope effect, and can therefore be omitted. It can be shown that each separate partition function f_j can be described by the expression

$$f_j = \Sigma g_{j,i} e^{-Ej,i/\mathbf{k}T} \tag{3.29}$$

where $E_{j,i}$ is the particular energy state i for the molecule's energy mode j; e.g. for j = vibration, there may be 20 different vibrational states (i.e. the maximum i-value is 20) populated. **k** is the Boltzmann constant; (3.29) is closely related to the Boltzmann distribution law (see next section).

The term $g_{j,i}$ is called the *degeneracy* and corrects for the fact that the same energy state in the molecule may be reached in several different ways. The summation has to be made over all energy states i.

3.6.2. Kinetic energy and temperature

The kinetic energy of one mole of an ideal gas at temperature T (K) is given by its translational energy, which according to the kinetic theory of gases is

$$E_{tr} = 3\mathbf{R}T/2 \ (\text{J mole}^{-1}) \tag{3.30}$$

Dividing by the Avogadro number $\mathbf{N_A}$ yields the average kinetic energy per molecule (or particle)

$$\overline{E}_{tr} = 3\mathbf{k}T/2 \ (\text{J particle}^{-1}) \tag{3.31}$$

where $\mathbf{k} = \mathbf{R}/\mathbf{N_A}$. From mechanics we know that the kinetic energy of a single particle of mass m and velocity v is

$$E_{kin} = ½mv^2 \tag{3.32}$$

Summing over a large number of particles, we can define an average kinetic energy

$$\overline{E}_{kin} = ½m\overline{v}^2 \tag{3.33}$$

where $\overline{v}^2$ is the mean square velocity. Because (3.31) is the average kinetic energy at temperature T, it must equal (3.33), *i.e.* $\overline{E}_{tr} = \overline{E}_{kin}$, thus

$$3\mathbf{k}T/2 = ½m\overline{v}^2 \tag{3.34}$$

Thus for a given temperature there is a corresponding *average particle velocity*. However, the individual particles are found to move around with slightly different velocities. J. C. Maxwell calculated the velocity distribution of the particles in a gas. For the simplest possible system it is a *Boltzmann distribution*. In a system of n_o particles the number of particles n_E that have kinetic energy > E is given by

$$n_E = n_o\left(2/\sqrt{\pi}\right)(\mathbf{k}T)^{-3/2} \int_E^{\infty} \sqrt{E}e^{-E/\mathbf{k}T}\mathrm{d}E \tag{3.35}$$

In Fig. 3.4 n_E/n_o is plotted as a function of E for three different T's. For the line at 290 K we have marked the energies $\mathbf{k}T$ and $3\mathbf{k}T/2$. While $3\mathbf{k}T/2$ (or rather $3\mathbf{R}T/2$)

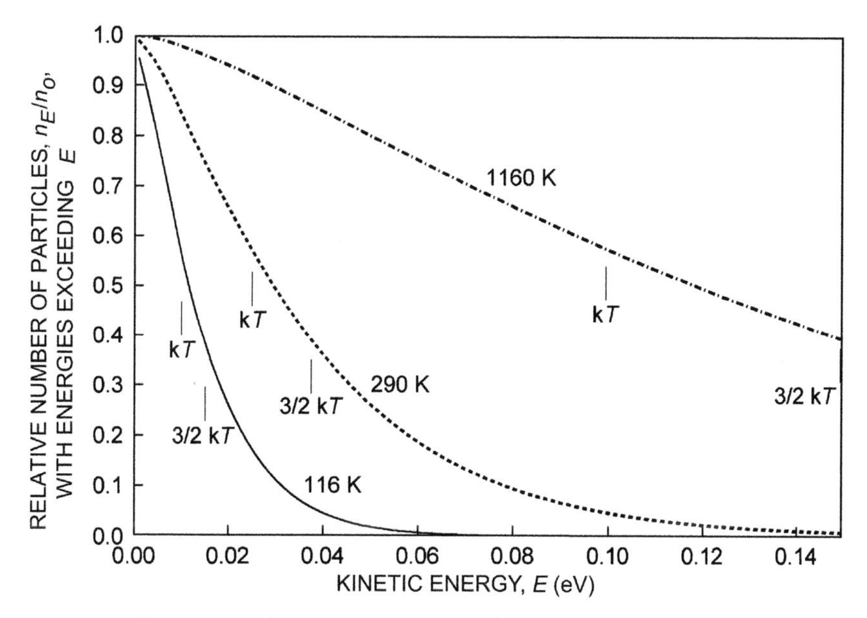

Figure 3.4 Relative number of particles as function of energy.

corresponds to the thermodynamic average translational energy, $\mathbf{k}T$ corresponds to the *most probable kinetic energy*: the area under the curve is divided in two equal halves by the $\mathbf{k}T$ line.

In chemistry, the thermodynamic energy (3.31) must be used, while in nuclear reactions the most probable energy E' must be used,

$$E' = \frac{1}{2} m(v')^2 = \mathbf{k}T \tag{3.36}$$

where v' is the *most probable velocity.*

Although the difference between $\overline{E}$ and E' is not large (*e.g.* at 17°C = 0.037 eV and $E' = 0.025$ eV), it increases with temperature. *The most probable velocity is the deciding factor whether a nuclear reaction takes place or not.* Using (3.35) we can calculate the fraction of particles at temperature T having a higher energy than $\mathbf{k}T$: $E > 2\mathbf{k}T$, 26%, $> 5\mathbf{k}T$; 1.86%, and $> 10\mathbf{k}T$, 0.029%. This high-energy tail is of importance for chemical reaction yields because it supplies the molecules with energies in excess of the activation energy. It is also of importance for nuclear reactions and in particular for thermonuclear processes.

3.6.3. The partial partition functions

So far there has been no clue to why isotopic molecules such as H_2O and D_2O, or $H^{35}Cl$ and $H^{37}Cl$, behave chemically in slightly different manner. This explanation comes from a study of the energy term E_j, which contains the atomic masses for all three molecular movements: translation, rotation, and vibration.

(a) *Translational energy.* The translational energy, as used in chemical thermodynamics involves molecular movements in all directions of space. The energy is given by the expression (3.31). A more rigorous treatment leads to the expression

$$f_{tr} = (2\pi M\mathbf{k}T)^{3/2} V_M \mathbf{h}^{-3} \tag{3.37}$$

for the translational partition function, where V_M is the molar volume and **h** the Planck constant. Notice that no quantum numbers appear in (3.37); the reason is that they are not known because of the very small ΔE's of such changes.

(b) *Rotational energy.* Taking the simplest case, a linear diatomic molecule with atomic masses m_1 and m_2 at a distance r apart, the rotational energy is given by

$$E_{rot} = I_{rot}\omega^2 \tag{3.38}$$

where I_{rot} is the rotational moment of inertia and ω the angular velocity of rotation (radians s^{-1}). I_{rot} is calculated from

$$I_{rot} = m_{red} r^2 \tag{3.39}$$

where the reduced mass $m_{red} = (m_1^{-1} + m_2^{-1})^{-1}$. Equation (3.38), derived from classical mechanics, has to be modified to take into account that only certain energy states are permitted

$$E_{rot} = \mathbf{h}^2 n_r(n_r + 1)/\left(8\pi^2 I_{rot}\right) \tag{3.40}$$

where n_r is the rotational quantum number. For transformation of energy in eV to the wave number $\bar{\nu}$ or wavelength λ of the corresponding photon energy, the relation

$$\Delta E\,(\text{eV}) = 1.23980 \times 10^{-4}\bar{\nu}\left(\text{cm}^{-1}\right) \tag{3.41}$$

is used, where $\bar{\nu} = 1/\lambda$. (For blue light of about 480 nm the following relations are obtained: 480 nm = 4.8×10^{-7} m = 0.48 μm = 20 833 cm^{-1} = 2.58 eV.) The rotational energies are normally in the range $0.001 - 0.1$ eV, i.e. the wavelength region $10^{-3} - 10^{-5}$ m. The partition function for the rotational energy is obtained by introducing (3.40) into (3.29). More complicated expressions are obtained for polyatomic and nonlinear molecules.

(c) *Vibrational energy.* For a diatomic molecule the vibrational energy is given by

$$E_{vib} = \mathbf{hc}\omega_v\left(n_v + ½\right) \tag{3.42}$$

where

$$\omega_v = ½(k'/m_{red})^{½}/(\pi\mathbf{c}) \tag{3.43}$$

is the zero point vibrational frequency (the molecule is still vibrating at absolute zero, when no other movements occur) and n_v the vibrational quantum number. k' is the *force constant* for the particular molecule.

3.6.4. The isotopic ratio

It is seen from this digression that the mass of the molecular atoms enters into the partition functions for all three modes of molecular movement. The largest energy changes are associated with the vibrational mode, and the isotopic differences are also most pronounced here. Neglecting quantization of translational and rotational movements, one can show that

$$F = f_{tr} f_{rot} f_{vib} = 2\left(2\pi M\mathbf{k}T/\mathbf{h}^3\right)^{3/2} V_M 8\pi^2 I_r \mathbf{k}T e^{-u/2}/\left\{\mathbf{h}^2 k_s\left(1-e^{-u}\right)\right\} \quad (3.44)$$

where k_s is a symmetry constant for rotation and $u = \mathbf{h}\mathbf{c}\omega/\mathbf{k}T$. This expression holds for all molecules in (3.25). Thus for the ratio one gets

$$F_{AX}/F_{AX^*} = \left(M^*/M\right)^{3/2} k_s I_r^*\left(1-e^{-u}\right)e^{-u^*/2}/\left\{k_s^* I_r\left(1-e^{-u^*}\right)e^{-u/2}\right\} \quad (3.45)$$

where the asterisk refers to molecule AX*. This relation shows the mass dependency of the equilibrium constant (3.26) (a similar relation holds for the BX-BX* combination). (3.45) contains factors all of which can be determined spectroscopically. Consequently, the equilibrium constants for isotope exchange reactions can be, and have been, calculated for many systems. We see here, as commented earlier that as for the physical effects of isotopic variations also the chemical one is highly dependent on the percentage change. Thus, also here we expect larger effects for lighter molecules.

3.6.5. Paleotemperatures and other applications

Eqn (3.45) contains the temperature in the exponent. Isotope exchange equilibria are thus temperature dependent. A practical use of this fact is the determination of paleotemperatures. In the isotope exchange reaction

$$C\,^{16}O_2(g) + H_2\,^{18}O\,(l) = C\,^{18}O\,^{16}O(g) + H_2\,^{16}O(l) \quad (3.46)$$

the oxygen isotopes in CO_2 exchange with the oxygen isotopes in H_2O. The value of the equilibrium constant (mole fractions) $k = 1.046$ (0°C) indicates that ^{18}O will be slightly enriched in the CO_2 by the reaction. Thus if carbon dioxide from combustion is bubbled through water, the emergent gas will be more enriched in the ^{18}O than the residual water. In this reaction, the *isotope effect* is said to be 4.6%. The following reaction occurs with carbonate ions in water:

$$Ca^{2+} + C\,^{16}O_3^{\;2-} + H_2\,^{18}O = CaC\,^{18}O\,^{16}O_2(s) + H_2\,^{16}O \quad (3.47)$$

In this reaction the isotope effect results in enrichment of ^{18}O in precipitated carbonate compared to dissolved carbonate (Fig. 3.5). The equilibrium constant for this reaction and, hence, the $^{18}O/^{16}O$ ratio, can be calculated to have a temperature dependence according to

$$T(°C) = 18 + 5.44 \times 10^6(\zeta_T - \zeta_{18}) \quad (3.48)$$

where ζ_{18} and ζ_T are the isotopic ratios (cf. (3.4)) at 18°C and at temperature T.

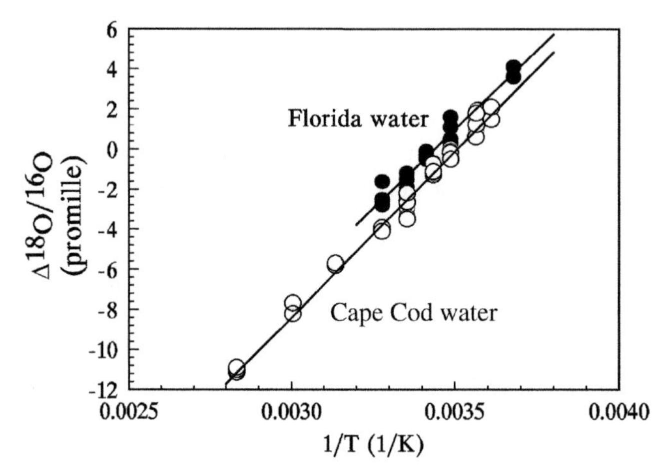

Figure 3.5 Variation of isotopic composition of $CaCO_3(s)$ with deposition temperature from water (○-Cape Cod; ●-Florida waters). (From McCrea.)

The isotopic ratios for sedimentary carbonate show a variation of 2.04 to 2.07×10^{-3} (Fig. 3.3). If it is assumed that these differences are due to the precipitation of the carbonate at different temperatures and that the isotopic composition of carbon in seawater was the same as today, one can use the isotopic ratios to obtain information on geologic temperature. In Fig. 3.6 the data for a shell from the Jura geologic period is shown. The oxygen ratio in the carbonate of the shell has been determined by starting from the center of the shell and measuring the isotopic ratio in each layer of annual growth.

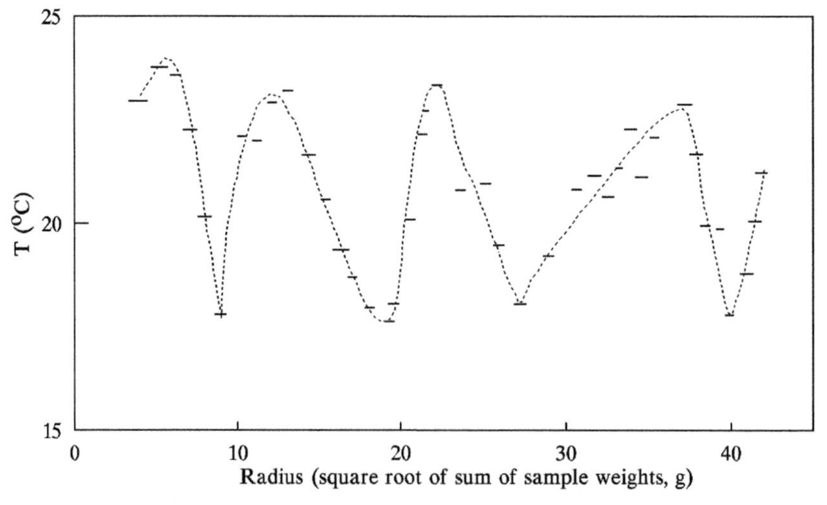

Figure 3.6 Temperature calculated from the $^{18}O/^{16}O$ ratio in carbonate of a shell from the Jura period as function of the distance from center of the shell. (According to Epstein.)

The result shows that the temperature at which the carbonate was formed varied during the life of the creature; further, the creature that inhabited the shell must have died during the spring of its fourth year since there was no further growth of the shell after the forth period of increasing temperature. Thus, by using this method it is possible to determine a small window of yearly temperature variations millions of years back in history.

Chemical isotope effects are particularly large for lighter elements in biological systems. The *chlorella algae* prefers deuterium over hydrogen, and tritium over deuterium. The enrichment factor depends on the conditions of growth; for deuterium to hydrogen an enrichment value of 1.6 − 3 has been found, while for tritium to hydrogen the enrichment factor is about 2.5. Bacteria behave similarly, e.g. *coli bacteria* showing an enrichment factor for deuterium of 3.9.

In as much as some of the hydrogen atoms are not exchanged readily due to the inertness of their chemical bonds, the isotopic fractionation which involves the easily exchangeable hydrogen atoms in these biological processes must have even larger enrichment factors for deuterium and tritium than their measured values would indicate.

The peculiarity of biological material to prefer certain isotopes has led to studies of how biological material behaves in an isotopic environment which differs substantially from that found in nature. Normally it is found that the organisms wither away and lose their ability to reproduce. Carp cannot survive a higher D_2O concentration than 30%, but, on the other hand, some organisms show a strong growth, and some microorganisms have been found to be able to live in pure D_2O or $H_2{}^{18}O$. It has been possible to raise mice with only ^{13}C in their organism (fed on ^{13}C algae). Exchanging natural $^{14}NH_3$ for $^{15}NH_3$ seems to have little effect on biological systems.

In all of these investigations it should be noted that even when we characterize an isotopic effect as large, it is still quite small by normal reaction criteria except for hydrogen isotopes. As pointed out previously, we can assume that for all but the very lightest elements in most chemical experiments that there is no isotope effect. This assumption forms a basis of the use of radioactive tracers to study chemical systems and diffusion at constant concentration.

3.7. ISOTOPE EFFECTS IN CHEMICAL KINETICS

The reason why higher organisms cannot survive when all light hydrogen atoms are replaced by deuterium is to be found not so much in a shift of chemical equilibria as in a shift in reaction rate leading to a fatal lowering of the metabolic rate when light isotopes are replaced by heavier.

In contrast to chemical equilibria, chemical reaction rates depend on the concentration of the reactants, the transition states and the activation energy but not on the product. The concentration of the transition states depends on the activation energy for its formation and the frequency for its decomposition into the products. These factors can be derived from the partition function which, as mentioned above, differ slightly for molecules of different isotopic composition. Let us consider the reaction

$$A + BC \rightarrow AB + C \tag{3.49}$$

The rate constant is given by the expression

$$-d[A]/dt = k[A][BC] \tag{3.50}$$

The reaction is assumed to take place over an intermediate compound ABC, usually denoted $ABC^{\#}$ where the # indicates a short-lived transition state. According to the transition state theory, derived by H. Eyring, J. Biegeleisen, and others, it is assumed that the intermediate complex undergoes internal vibrations, with such an energy E_v that the bond is broken along the vertical line in the complex AB|C, leading to the fragments AB and C. The rate of reaction is the rate at which the complex $ABC^{\#}$ decomposes into the products. It can therefore also be written

$$-d[A]/dt = \nu[ABC^{\#}] \tag{3.51}$$

where ν is the frequency at which the complex decomposes.

The reaction can be schematically depicted as in Figure 3.7, where indices 1 and 2 refer to two isotopic reactant molecules (e.g. H_2O and HDO), which must have different zero point energies with frequencies ν_1 and ν_2, respectively. For simplicity only the

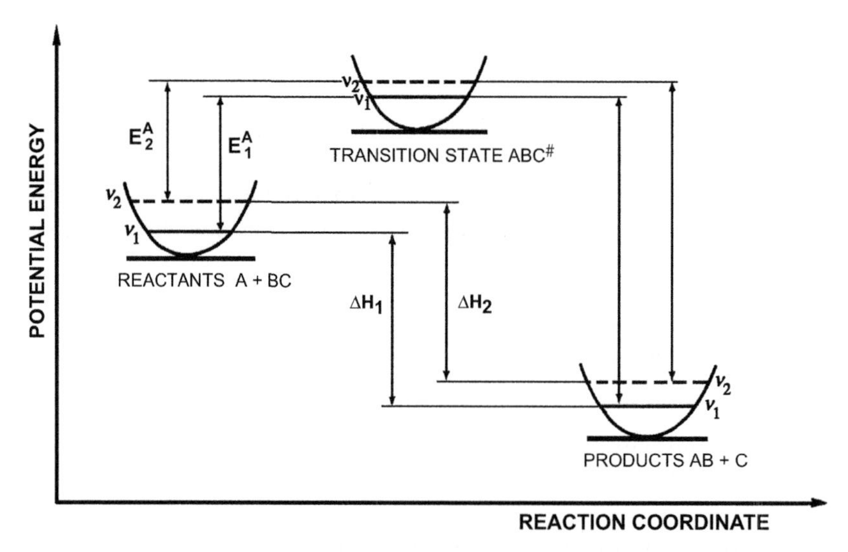

Figure 3.7 Schematic description of energies in a bimolecular reaction.

vibrational ground state is indicated; thus the energy change when going from reactants to products corresponds to the heat of reaction at absolute zero, ΔH_1 (0 K) and ΔH_2 (0 K) respectively. Because of the lower vibrational energy of the molecule indexed 1, this must contain the heavier isotope. In general, the difference in activation energy (E^A) is greater than the difference in heat of reaction ΔH for isotope molecules; thus, generally, isotope effects are larger in the kinetic than in the equilibrium effects.

When the molecule $ABC^{\#}$ decomposes into AB and C, the vibrational energy, given by the Planck relation

$$E_v = \mathbf{h}\nu \tag{3.52}$$

is changed into kinetic energy of the fragments, whose energy is (see §3.6.2)

$$E_{kin} = \mathbf{k}T \tag{3.53}$$

Because of the law of conservation of energy

$$E_v = E_{kin} \tag{3.54}$$

and

$$\mathbf{h}\nu = \mathbf{k}T \tag{3.55}$$

This development assumes that the vibrational energy is completely converted to fragment translational energy. This assumption is not always valid for polyatomic fragments, in which there is also contribution from vibration and rotation energies. Introducing (3.55) into (3.51) and equating (3.50) and (3.51) yields

$$k[A][BC] = \mathbf{k}T\mathbf{h}^{-1}[ABC^{\#}] \tag{3.56}$$

It is assumed that $ABC^{\#}$ is in dynamic equilibrium with the reactants A and BC. Thus

$$[ABC^{\#}]/([A][BC]) = k^{\#} \tag{3.57}$$

According to (3.27)

$$k^{\#} = F_{ABC^{\#}}/(F_A\,F_{BC}) \tag{3.58}$$

which with (3.56) yields

$$k = \mathbf{k}T\,F_{ABC^{\#}}/(\mathbf{h}\,F_A\,F_{BC}) \tag{3.59}$$

This expression must be multiplied by a factor κ, which is the probability that the complex will dissociate into products instead of back into the reactants as assumed in (3.57). The factor κ is called the *transmission coefficient*. The final rate expression thus becomes:

$$k = \kappa\,\mathbf{k}\,T\,F_{ABC^{\#}}/(\mathbf{h}\,F_A\,F_{BC}) \tag{3.60}$$

As is shown in §3.6, the grand partition functions F_i can be calculated from theory and spectroscopic data; because these functions are mass dependent k is mass dependent. In calculating the F_i's, all modes of energy must be included as well as the population of the different energy states.

For isotopes of the lighter elements, the activation energy term makes the main contribution to the reaction rate isotope effect, while for the heavier elements the vibrational frequency causing the decomposition into the products plays the larger role. Because the energy states usually are more separated for the isotopic molecules of the products and reactants than for the transition state, isotope effects are usually larger in reaction kinetics than in equilibria.

Studies of kinetic isotope effects are of considerable theoretical interest, particularly in organic chemistry. The practical applications are still meagre, but this will not necessarily be so in the future. An example is the decrease in metabolic rate for ^{13}C compounds, which has led to the suggestion of its use for treatment of certain diseases, as e.g. porphyria.

3.8. ISOTOPE SEPARATION PROCESSES

Many fields of fundamental science have found great advantage in using pure or enriched isotopes. The industrial use of nuclear power also requires the enrichment of particular isotopes, primarily of the uranium fuel but also separation of H, D and T in heavy water moderated reactors as well as nitrogen isotope separation for nitride fuels to be used in modern fast reactors (Ch. 21). The methods which have been developed to achieve isotopic fractionation may be divided into two groups.

(a) *Equilibrium processes* (§3.6). These processes consume little energy, but the size of the isotope effect in normal chemical equilibrium limits their use to the isotope fractionation of very light elements, usually atomic number less than about 10.

(b) *Rate processes* (§3.7). This includes processes which depend on such phenomena as ionic mobility, diffusion, electrolysis, electromagnetic separations, centrifugation, and kinetic processes of a chemical nature. While the isotopic effects in these processes are normally larger than for equilibrium processes, they require a large amount of energy and therefore have economic restrictions.

3.8.1. Multistage processes

Figure 3.8 shows a flow scheme for an isotopic fractionation process (or isotope *enrichment* process) based on an α-value near 1. Each *stage* consists of a number of *cells* coupled in parallel; in the Figure only one cell is shown for each separation stage, but in order to obtain a high product flow, the number of cells are usually high at the feed point and then decrease towards the product and waste stream ends. Each cell contains a physical arrangement, which leads to an isotope fractionation. Thus the atomic fraction

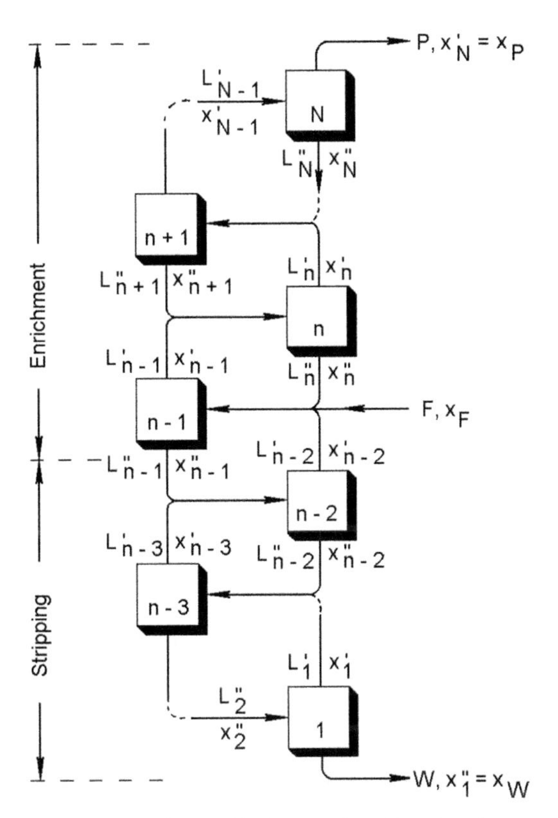

Figure 3.8 Flow arrangement for an ideal cascade with reflux for isotope separation.

for a particular isotope is different in the two outgoing streams from a cell; in the *product stream* the isotope is enriched (atomic fraction x'), while in the *waste stream* it is depleted (atomic fraction x''). The *separation factor* α is defined as the quotient between the isotopic ratios of the product and waste streams for a single step, thus ((3.3) and (3.4)):

$$\alpha = \zeta'/\zeta'' = [x'/(1 - x')]/[x''/(1 - x'')] \tag{3.61}$$

In most cases α has a value close to unity; $\alpha - 1$ is commonly called *enrichment factor.*

Since separation factors in general are small, it is necessary to use a multistage process to obtain a product with a high enrichment. The number of stages determines the degree of the enrichment of the product, while the number and size of cells in each stage determine the amount of product. This amount (P moles of composition x_P) is related to the amount of feed (F moles of composition x_F) and the amount of waste (W moles of composition x_W) by the equations

$$F = P + W \tag{3.62}$$

and

$$Fx_F = Px_P + Wx_W \tag{3.63}$$

From these equations we obtain

$$F = P(x_P - x_W)/(x_F - x_W) \tag{3.64}$$

$$W = P(x_P - x_F)/(x_F - x_W) \tag{3.65}$$

The number of stages required to separate feed into product and waste of specified composition is a minimum at *total reflux*, when $P = 0$. For this condition M. R. Fenske has derived a relation, which can be divided into one part for the enrichment:

$$N_P \ln \alpha = \ln[x_P(1 - x_F)/\{x_F(1 - x_P)\}] \tag{3.66}$$

and one for the stripping part of the cascade:

$$N_W \ln \alpha = \ln[x_F(1 - x_W)/\{x_W(1 - x_F)\}] \tag{3.67}$$

N_P and N_W are the minimum number of enrichment and stripping stages, respectively. In isotope separations α is often very close to one; ln α can then be replaced by $(\alpha - 1)$. In practice some product flow is desired, the fraction withdrawn at the enrichment stage being known as "*the cut*" P/F. The number of stages required to produce the composition x_P then increases. The most economic, and thus also the most common, type of cascade for processes with near unity α is the so-called *ideal cascade.* In this there is no mixing of streams of unequal concentrations, thus $x'_{n-1} = x''_{n+1}$ in Fig. 3.8. Although the number of stages required for a particular product is exactly twice the values given by (3.66) minus 1 and (3.67) minus 1,

$$N_{ideal} = 2N - 1 \tag{3.68}$$

the interstage flow becomes a minimum, thus minimizing the inventory and work required. Also for the ideal cascade, the minimum number of stages are obtained at total reflux.

The interstage flow rate L_i that must be maintained in order that there be any enrichment in the ideal cascade at the point, where the concentration x_i occurs, is given by

$$L_i = 2P(x_P - x_i)/[(\alpha - 1)x_i(1 - x_i)] \tag{3.69}$$

To obtain the flow rate in the stripping part, P is replaced by W, and $x_P - x_i$ by $x_i - x_W$.

For the enrichment of natural uranium (0.71% in ^{235}U) to 80% in ^{235}U, the minimum number of enrichment stages is found to be almost 3000 at an enrichment factor of 0.43%. The flow rate at the feed point becomes about 52 300 times the product flow P.

Isotope separation plants therefore become very large even for small outputs. Isotopic separation on a technical scale using cells with α near unity thus requires a large number of stages in order to obtain high enrichment and large amounts of materials in order to give a substantial amount of product. Small changes in α can have a large economic impact in isotopic separation.

3.8.2. Chemical exchange

As an example of an industrial isotope exchange process, let us consider the production of heavy water by the chemical reaction

$$H_2O(l) + HDS(g) = HDO(l) + H_2S(g) \quad \begin{matrix} k = 2.32(32°C) \\ k = 1.80(138°C) \end{matrix} \tag{3.70}$$

From the values of the equilibrium constants k we see that the enrichment of deuterium in water increases with decreasing temperature. Use is made of this property in the two-temperature H_2O-H_2S exchange process known as the G-S process (Girdler-Sulphide), which is used in many countries to produce heavy water. A typical plant consists of several units as shown in Fig. 3.9. Through the upper distillation tower natural water flows downward and meets hydrogen sulfide gas streaming upwards. As a result of the exchange between H_2O and H_2S, heavy hydrogen is enriched in the water. In the lower tower, which is operated at a higher temperature, the equilibrium conditions are such that deuterium is enriched in the hydrogen sulfide and moves with that gas to the upper tower. No catalyzer is required in order to achieve rapid equilibrium in this reaction. The product of the process is water which is enriched in deuterium from the top tower and water which is depleted in deuterium from the bottom tower. The hydrogen sulfide circulates through both towers with no net loss.

Plants capable to produce a total of more than 1200 tons annually exsist in Canada, India and the US. The largest exchange towers are 60 m high and have a diameter of 6 m. In 5 units (only one unit is indicated in Fig. 3.8) the D_2O concentration is raised from 0.014% to about 15%. The final concentration to 99.97% D_2O is then usually made by distillation of water. The 1990 price for pure D_2O was ~ US $400 per kg. It is important to recognize in tracer applications that commercially available D_2O always contains some tritium, which is co-enriched with deuterium; 2–7 kBq kg^{-1} D_2O.

As another example, the lithium isotope ^7Li (used as ^7LiOH for pH control in some nuclear power plants because of its small neutron cross-section) is produced in 99.99% purity by counter current chemical exchange of lithium between an aqueous solution of LiOH and lithium amalgam. A separation factor of 1.06 to 1.07 is reported thus requiring about 260 theoretical stages to reach the desired purity with about 0.3% of ^7Li remaining in the depleted stream. Reflux of lithium is obtained at one end by electrolytic

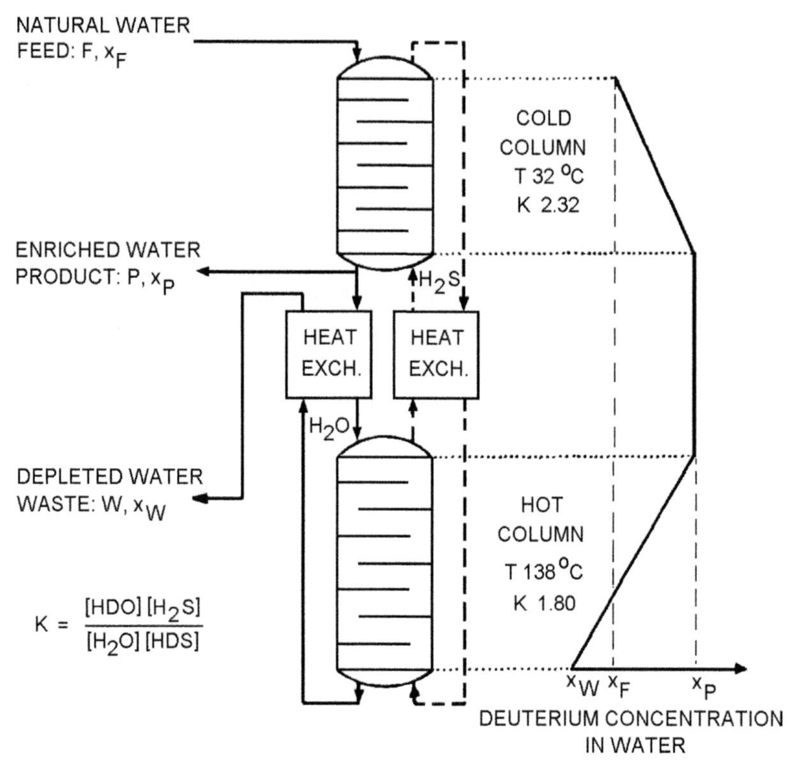

Figure 3.9 The water-hydrogen sulphide dual-temperature distillation process for enrichment of heavy water.

reduction of LiOH to Li(Hg) at a mercury cathode and at the other end by spontaneous oxidation of Li(Hg) on graphite by water producing hydrogen gas and LiOH.

3.8.3. Electrolysis

Electrolysis of water produces hydrogen gas at the cathode, which contains a lower proportion of deuterium than the original water. The isotope effect stems from the differences in the rates of dissociation of a proton (H^+) and a deuteron (D^+) from water, and the rates of neutralization of these hydrated ions, and thus has a kinetic basis. Depending on the physical conditions α-values between 3 and 10 are obtained. For $\alpha = 6$ it is necessary to electrolyze 2700 l natural water (deuterium content 0.014%) to produce 1 l water containing 10% deuterium, mainly as HDO. In a multistage process the hydrogen gas is either burnt to recover energy, or used in a chemical process, e.g. in ammonia synthesis by the Haber-Bosch process. Although this technique was once used industrially in Norway to produce ton amounts of pure D_2O, it is no longer considered economic except for final purification of D_2O.

3.8.4. Gaseous diffusion

In a gaseous sample the lighter molecules have a higher average velocity than the heavier molecules. In 1913 F. W. Aston in England showed that the lighter of the two neon isotopes ^{20}Ne, diffused through the walls of porous vessels somewhat faster than the heavier isotope, ^{22}Ne. In the gas the average kinetic energy of isotopic molecules must be the same, i.e. $1/2M_L v^2_L = 1/2M_H v^2_H$, where M_H and M_L are the masses of the molecules containing the heavy and light isotopes to be separated. The maximum theoretical separation factor in gaseous diffusion is given by

$$\alpha = v_L/v_H = (M_H/M_L)^{1/2} \tag{3.71}$$

The theory is more complicated, depending among other things upon the mean free path of the gaseous atoms, the pore length and diameter of the separating membrane, and the pressure difference over the membrane. If experimental conditions are carefully controlled, this theoretical maximum can be closely approached.

^{235}U is enriched through gaseous diffusion using the relatively volatile uranium compound UF_6. In addition to its volatility, UF_6 has the advantage that fluorine consists of only one isotope, ^{19}F. For the isotopic molecules $^{235}UF_6$ and $^{238}UF_6$, a value of 1.0043 is theoretically possible for α, cf. (3.71). The following conditions must be considered in the technical application of the separation.

(a) The cells are divided into two parts by a membrane which must have very small pores (e.g. 10–100 nm in diameter) in order to obtain isotopic separation. In order that large gas volumes can flow through the membrane, millions of pores are required for each square centimeter. Moreover, the membranes must have good mechanical stability to withstand the pressure difference across them.

(b) UF_6 sublimes at 64°C, which means that the separation process must be conducted at a temperature above this.

(c) UF_6 is highly corrosive and attacks most materials. The membrane must be inert to attack by UF_6. Water decomposes UF_6 according to the equation

$$UF_6(g) + 2H_2O(g) = UO_2F_2(s) + 4HF(g) \tag{3.72}$$

This reaction is quite undesirable as HF is highly corrosive and solid UO_2F_2 can plug the pores of the membranes. The tubing and cells of a plant are made principally of nickel and teflon to overcome the corrosion problems.

(d) In order to transport the large gas volumes and to keep the proper pressure drop across the membranes, a gaseous diffusion plant requires a large number of pumps. A large cooling capacity is needed to overcome the temperature rise caused by compression of the gas.

The work required to enrich uranium in ^{235}U increases rapidly with the mole fraction of ^{235}U in the product. Because of varying domestic prices on natural uranium, as well as varying content of ^{235}U in uranium obtained from used reactor fuel elements, so-called *toll enrichment* has been introduced. In this case, the purchaser himself provides the uranium feed into the separation plant and pays for the *separative work* required to make his desired product out of the uranium feed provided. Separative work is defined as

$$\text{Separative work} = W\ V(x_W) + P\ V(x_P) - F\ V(x_F) \tag{3.73}$$

where the *separation potential* $V(x_i)$ is defined by

$$V(x_i) = (2x_i - 1)\ln\{x_i/(1 - x_i)\} \tag{3.74}$$

As seen from (3.73), separative work has the dimension of mass, and can be thought of as the mass flow rate multiplied by the time required to yield a given quantity of product. The cost of isotope separation is obtained by assigning a value to one separative work mass unit (kgSW or SWU). A 1 GWe nuclear light water reactor station requires about 180×10^3 SWU in initial fueling and then $70{-}90 \times 10^3$ SWU for an annual reload.

In §3.8.1 the number of stages and the interstage flow relative to the product flow was given for enrichment of ^{235}U from its natural isotopic abundance of 0.71% to a value of 80%. With a waste flow in which the isotopic abundance of ^{235}U is 0.2%, (3.64) and (3.65) shows that for each mole of product obtained 156 moles of feed are necessary. In more recent designs the concentration of ^{235}U in the waste is increased to ~0.3% to minimize cost. Isotope separation through gaseous diffusion is a very energy-consuming process due to the compression work and cooling required. An annual production of 10 MSWU requires an installed capacity of ~2900 MW in present plants, or ~2500 kWh SWU^{-1}. Improved technology may reduce this somewhat. Gaseous diffusion plants are known to exist in Argentina, China, France, Russia and the United States. The combined capacity of these plants was about 40 MSWU/y at the end of 2000.

3.8.5. Electromagnetic isotope separation

During the Manhattan Project of the United States, electromagnetic separation was used to obtain pure ^{235}U. This process is identical in theory to that described for the mass spectrometer. The giant electromagnetic separators were called *calutrons* (California University Cyclotrons) and were after World-War II used at Oak Ridge National Laboratory to produce gram amounts of stable isotopes of most elements up to a purity of 99.9% or more. Large capacity electromagnetic separators have also been developed and operated in the former Soviet Union and a few other countries. Gram amounts of electromagnetically separated stable isotopes of most elements are commercially available on the world market in high isotopic purity. Electromagnetic separators are also used for on-line separation of nuclear reaction products (Ch. 17). A large number of such separators were also secretly built and operated by Irak for U-enrichment before the Gulf war.

3.8.6. Gas centrifugation

Though gas diffusion has been the dominating process for ^{235}U enrichment, its small separation factor, large energy consumption, and secrecy about the technique have encouraged interest in finding other and more advantageous processes. Most effort has been put into developing centrifugal processes because of the large separation factors achievable.

In a gas centrifuge, light molecules are enriched at the center and heavy molecules at the periphery (Fig. 3.10). It can be shown that the ratio of radial concentration to the axial concentration for isotopic molecules under equilibrium conditions and at low pressure (avoiding remixing due to the Brownian movement) is given approximately by

$$x_P / x_F \approx e^{\delta n} \tag{3.75}$$

for n centrifuges (stages) connected in series. Here

$$\delta = (M_H - M_L) 10^{-3} v_r^2 \, r^2 / 2\mathbf{R}T \tag{3.76}$$

where r is the centrifuge radius and v_r the speed of rotation. The separation factor $\alpha \approx 1 + \delta$.

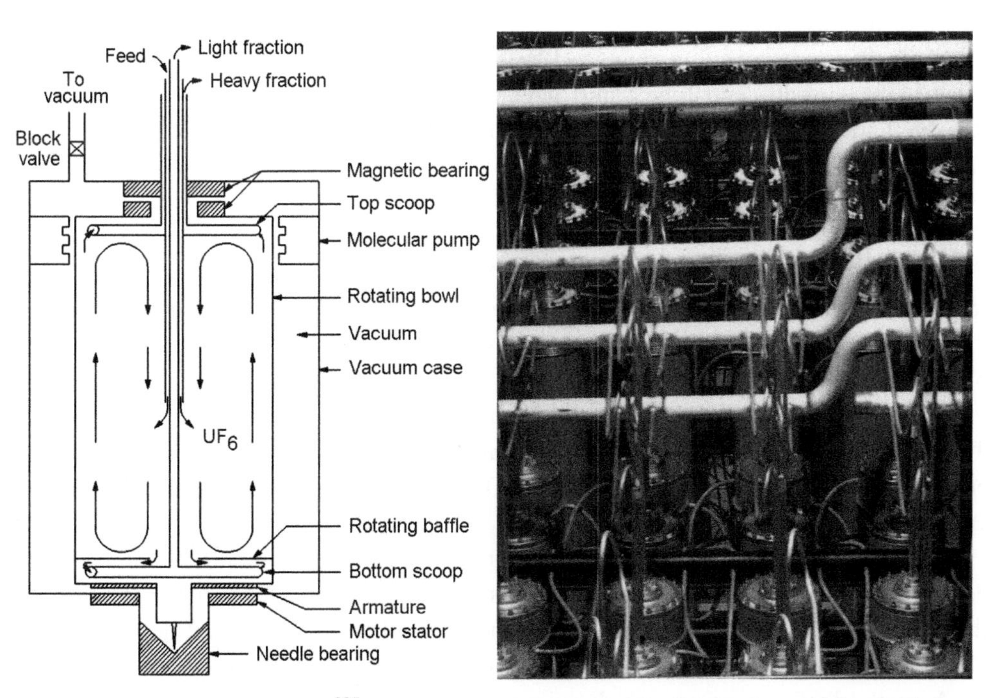

Figure 3.10 Gas centrifuge for $^{235}UF_6$ enrichment. Right shows detail of centrifuge cascade in operation at Almelo.

Estimated values for present centrifuges operating on UF_6 are: length 1–2 m; radius 0.2–0.4 m; rotational speeds of 50–80 × 10^3 rpm. The enrichment obtainable in one stage is limited by the material strength of the centrifuge bowl; the present materials limit $v_r\, r$ to < 800 m s^{-1}. Typical separation factors are 1.4–3.9 per stage, thus about 10 stages are required to enrich ^{235}U from 0.72% to 3% with a 0.2% tail. As compared, a diffusion plant would require ~1300 stages. The corresponding figures for production of 80% ^{235}U are ~45 and ~3600, respectively. Though rather few stages are required to upgrade natural uranium to reactor quality, a very large number of centrifuges are needed to produce large quantities of enriched material or highly enriched material.

Current centrifuge technology requires ~3% of the power consumed by a diffusion plant, or 50–100 kWh SWU^{-1}. This makes their environmental impact minimal, as compared to gas diffusion plants, which require substantial electric power installation and cooling towers with large water vapour effluent. Smaller plants of a few MSWU/y are economical, and their output can be readily multiplied by installation of parallel processing lines. The very large number of centrifuges required due to their small size (each having a capacity of = 15 SWU/y) and limited life-time does not lead to excessive construction costs due to continuous mass production on-site. On the whole, centrifuge separation now seems to have a lower enrichment cost than large scale diffusion plants. Centrifuge performance has been increased by a factor of ~25 since 1980 and is predicted to improve further.

Many plants are now operating, the largest known being near Ekaterinburg (Russia, 9 MSWU/y; Russian centrifuge plants have total capacity of ~19 MSWU/y), Almelo (the Netherlands, 1.7 MSWU/y), Capenhurst (United Kingdom, 1.1 MSWU/y) and Gronau (Germany, 1.1 MSWU/y).

3.8.7. Other methods of isotope separation

In theory all physicochemical procedures are capable of isotope separation. Some other methods which have been studied include distillation, solvent extraction, ion exchange, photoionization and photoexitation.

Tons of D_2O are purified annually in India by cryogenic *distillation* of hydrogen. Tenths of kilograms of pure ^{13}C and ^{15}N have been produced at the Los Alamos Scientific Laboratory through distillation of NO and CO at liquid air temperature. At the same time a fractionation between ^{16}O and ^{18}O occurs.

A *continuous ion-exchange* isotope separation process for uranium enrichment has been developed in Japan. Few details of this process have been disclosed. However, it is known that "reflux" is obtained by oxidation and reduction of U^{4+} and UO_2^{2+}. A demonstration plant with a capacity of 2 kSWU/y has been in operation at Hyuga.

A method of separation, involving passage of a mixture of UF_6 and helium or hydrogen at very high velocities through a *nozzle*, as seen in Fig. 3.11, has been

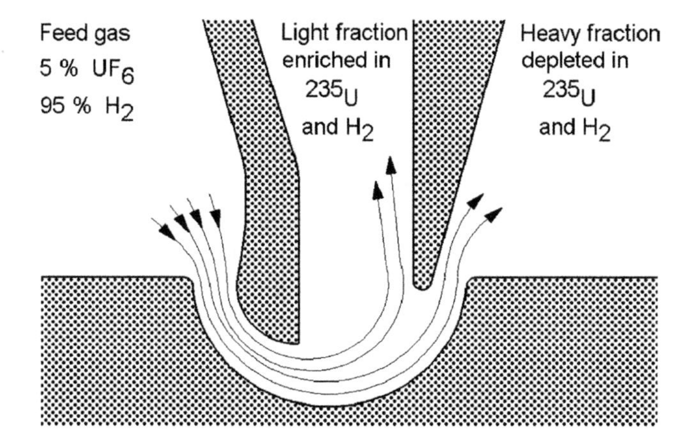

Figure 3.11 Section through a separation nozzle arrangement showing stream lines.

developed by E. W. Becker in Germany and in South Africa. The technique is sometimes referred to as "static" or "stationary-walled" gas centrifugation. The separation factor is typically 1.01 - 1.03 per stage, *i.e.* about three times better than in the gaseous diffusion process, and offers great possibilities for further improvements. Thus while the diffusion process requires about 1200 stages for a 3% ^{235}U enrichment (with 0.2% tails), the nozzle technique will require only about 500 stages. However, the power consumption is said to be larger, ~3000 kWh SWU^{-1} at a capacity of 10 MSWU/y, as compared to ~2500 kWh SWU^{-1} for a diffusion plant of the same size. The operating cost is said to be higher than for both the diffusion and the centrifuge methods. A plant based on the original German design is being built in Brazil (10 kSWU/y). A 0.3 MSWU/y separation plant at Valindaba, South Africa, used its own version of the nozzle technique (Helikon process), but is now shut down.

A number of *photoionization* and *photoexcitation* processes are being investigated for isotopic separation, especially of uranium. In one such process UF_6 is irradiated by a laser beam, producing selective vibrational excitation in the $^{235}UF_6$ molecule (cf. §3.5). By irradiation with ultraviolet light (possibly, but not necessarily, by laser) the excited molecule is caused to dissociate, leaving $^{238}UF_6$ undissociated. It is important that the ultraviolet pulse follows quite rapidly after the laser pulse, so that the vibrationally excited $^{235}UF_6$ molecule does not lose its excitation energy through collision with surrounding molecules. It is obvious that this necessitates gas phase reactions. The $^{235}UF_n^{6-n}$ ion formed through the dissociation ($n < 6$) is then collected by the action of electromagnetic fields. This technique is not limited to UF_6; pure uranium metal vapor and plutonium compounds have been separated into their isotopic constituents by two or three photon ionization with laser light (e.g. the US AVLIS process and the French SILVA process). Another alternative is to selectively excite $^{235}UF_6$ molecules by laser light in the presence of a reactive gas (the CRISLA process). The excited molecules then reacts preferentially with the gas forming molecules with lower vapor pressure than UF_6. Although research in these areas has indicated a large scale feasibility of several similar processes, no predictions can yet be made of their technological value. The energy consumption for the quantum processes discussed above are in the range 10–40 kWh SWU^{-1}. However, economic estimates indicate that their enrichment cost falls in the same price range as for centrifuge based plants. Hence, the interest in these methods has decreased and development of the ALVIS and SILVA processes has been terminated. Finally, it should be pointed out that the general concept of separative work is not valid for very high separation factors leading to very few units in a cascade (and always for mixtures of more than two isotopes).

3.9. EXERCISES

3.1. How many atoms of ^{235}U exist in 1 kg of uranium oxide, U_3O_8, made of natural uranium?

3.2. What is the atomic fraction of deuterium in water with the mole fraction of 0.81 for H_2O, 0.18 for HDO, and 0.01 for D_2O?

3.3. The translational energy of one mole of gas is given by 3/2**R***T*, which corresponds to an average thermal molecular velocity v (the root mean square velocity), while the most probable velocity $v' = \sqrt{0.67}\, v$.

(a) What is the most probable velocity of a helium atom at 800°C?

(b) What voltage would be required to accelerate an α-particle to the same velocity?

3.4. A Nier type mass spectrometer has a fixed radius of curvature of 5 cm and a magnetic field of 3000 G; 1 G $= 10^{-4}$ T. At what accelerating voltage will a Na^+ ion be brought to focus at the ion collector?

3.5. In a Dempster type (constant B and V) mass spectrograph utilizing 180° degree focusing, the ions $^{12}C^+$ and $^{11}BH^+$ are recorded simultaneously, the latter ion having a slightly larger orbit diameter. The separation between the lines recorded on the photographic plate is 0.0143 cm and the orbit diameter for the $^{12}C^+$ ion is 20 cm. What is the atomic mass of ^{11}B?

3.6. When the rotational quantum number n_r goes from 0 to 1 in $H^{35}Cl$, it is accompanied by the absorption of light with a wave number of 20.6 cm^{-1}. From this it is possible to calculate the interatomic distance between hydrogen and chlorine in the molecules. What is this distance?

3.7. In one mole of a gas at STP (standard temperature and pressure, i.e. 0°C and 1 atm) a small fraction of the molecules have a kinetic energy = 15**k***T*. (a) How many such molecules are there, and (b) what would their temperature be if they could be isolated?

3.8. In an investigation the isotope ratio $^{18}O/^{16}O$ was found to be 2.045×10^{-3} for fresh water and 2.127×10^{-3} for carbon dioxide in the atmosphere. Calculate the equilibrium constant (mole fractions!) for the reaction

$$H_2{}^{18}O(l) + CO_2(g) = H_2O(l) + C\,{}^{18}O^{16}O(g)$$

3.9. How many ideal stages in an ordinary cascade are required at an $\alpha = 6$ to produce water in which 10% of the hydrogen is deuterium?

3.10. In a distillation column with total reflux, ^{10}B is enriched through exchange distillation of $BF_3O(C_2H_5)_2$ from the natural value of 20 atom % to a product containing 95% ^{10}B. The packed column has a length of 5 m and a diameter of 3 cm. What is the approximate height of a theoretical stage if the enrichment factor is 0.026?

3.11. A gas centrifuge plant is set up in order to enrich UF_6 of natural isotopic composition in ^{235}U. The centrifuges, which each have a length of 100 cm and a diameter of 20 cm, rotate at 40000 rpm. The gas temperature is 70°C.

(a) Prove that the separation factor α in (3.61) can be approximated by e^{δ} according to (3.75) when the product flow is very small compared to the waste flow, and α is not far from 1.

(b) Using this approximation, what is the theoretical separation factor for one unit?

(c) Assuming that the enrichment factor obtained with the centrifuge is only 70% of the theoretical one, what number of units would be required in series in order to achieve UF_6 with 3% ^{235}U?

3.12. How much separative work is needed in order to produce 1 kg of uranium containing 4% ^{235}U from a feed of natural uranium and leaving 0.25% in the tail?

3.10. LITERATURE

J. M. McCrea, On the Isotopic Chemistry of Carbonates and a Paleotemperature Scale, *J. Chem. Phys.* **18** (1950) 849.

J. Biegeleisen, Isotopes, *Ann. Rev. Nucl. Sci.* **2** (1953) 221.

G. H. Clewlett, Chemical separation of stable isotopes, *Ann. Rev. Nucl. Sci.* **1** (1954) 293.

A. P. Vinogradov, *Bull. Acad. Sci. U.S.S.R.* **3** (1954) 3.

S. Epstein, quoted by A. P. Vinogradov.

T. F. Johns, Isotope separation by multistage methods, *Progr. Nucl. Phys.* (ed. O. R. Frisch) **6** (1957) 1.

J. Kistemacher, J. Biegeleisen, and A. O. C. Nier (eds.), *Proceedings of the International Symposium on Isotope Separation*, North-Holland, Amsterdam, 1958.

J. H. Beynon, *Mass Spectrometry and its application to Organic Chemistry*, Elsevier, Amsterdam, 1960.

L. Melander, *Isotope Effects on Reaction Rates*, Ronald Press, New York, 1960.

R. E. Weston Jr., Isotope effects in chemical reactions, *Ann. Rev. Nucl. Sci.* **11** (1961) 439.

H. London, *Separation of Isotopes*, G. Newnes, London, 1961.

A. E. Brodsky, *Isotopenchemie*, Akademie-Verlag, Berlin, 1961.

S. S. Roginski, *Theoretische Grundlagen der Isotopenchemie*, VEB Deutscher Verlag der Wissenschaften, Berlin, 1962.

IUPAC and IAEA, *Isotope Mass Effects in Chemistry and Biology*, Butterworths, London, 1962.

M. L. Smith (ed.), *Electromagnetically Enriched Isotopes and Mass Spectrometry*, Butterworths Sci. Publ., London, 1965.

F. A. White, *Mass Spectrometry in Science and Technology*, J. Wiley, 1968.

R. W. Levin, Conversion and enrichment in the nuclear fuel cycle, in D. M. Elliot and L E. S. Villani, *Uranium Enrichment*, Springer Verlag, 1979.

M. Benedict, T. Pigford, and H. W. Levi, *Nuclear Chemical Engineering*, 2nd ed., McGraw-Hill, New York, 1981.

FAO and IAEA, *Stable Isotopes in Plant Nutrition, Soil Fertility, and Environmental Studies*, IAEA, Vienna, 1991.

J. Andrasko, I. Kopp, Å. Åbrink and T. Skiöld, *J. Forensic Sci.* **38** (1993) 1161.

A calculator for enrichment of uranium isotope mixtures containing several uranium isotopes can be found on the internet at: http://www.wise-uranium.org/nfcjol.html

Nuclear Mass Stability

Contents

4.1. PATTERNS OF NUCLEAR STABILITY

There are today 255 different nuclei which have shown no evidence of radioactive decay and, hence, are said to be stable with respect to radioactive decay. Currently this means that if they have a half life it is greater than $\sim 10^{20}$ years, which is the longest half-life measured to this date. When these nuclei are compared for their constituent nucleons, we find that approximately 60% of them have both an even number of protons and an even number of neutrons (*even-even nuclei*). The remaining 40% are about equally divided between those that have an even number of protons and an odd number of neutrons (*even-odd nuclei*) and those with an odd number of protons and an even number of neutrons (*odd-even nuclei*). There are now only 4 stable nuclei known which have both an odd number of protons and odd number of neutrons (*odd-odd nuclei*); $^{2}_{1}$H, $^{6}_{3}$Li, $^{10}_{5}$B, and $^{14}_{7}$N, because recently $^{50}_{23}$Vwas found to be instable with a very long half-life. It is significant that the first stable odd-odd nuclei are abundant in the very light elements (the low abundance of $^{2}_{1}$H has a special explanation, see Ch. 12). The last nuclide is found in low isotopic abundance (0.25%) and has an extremely long half-life, 1.4×10^{17} y.

Considering this pattern for the stable nuclei, we can conclude that nuclear stability is favoured by even numbers of protons and neutrons. The validity of this statement can be confirmed further by considering for any particular element the number and types of stable isotopes; see Figure 4.1. Elements of even atomic number (i.e. even number

Radiochemistry and Nuclear Chemistry
ISBN 978-0-12-405897-2, http://dx.doi.org/10.1016/B978-0-12-405897-2.00004-5

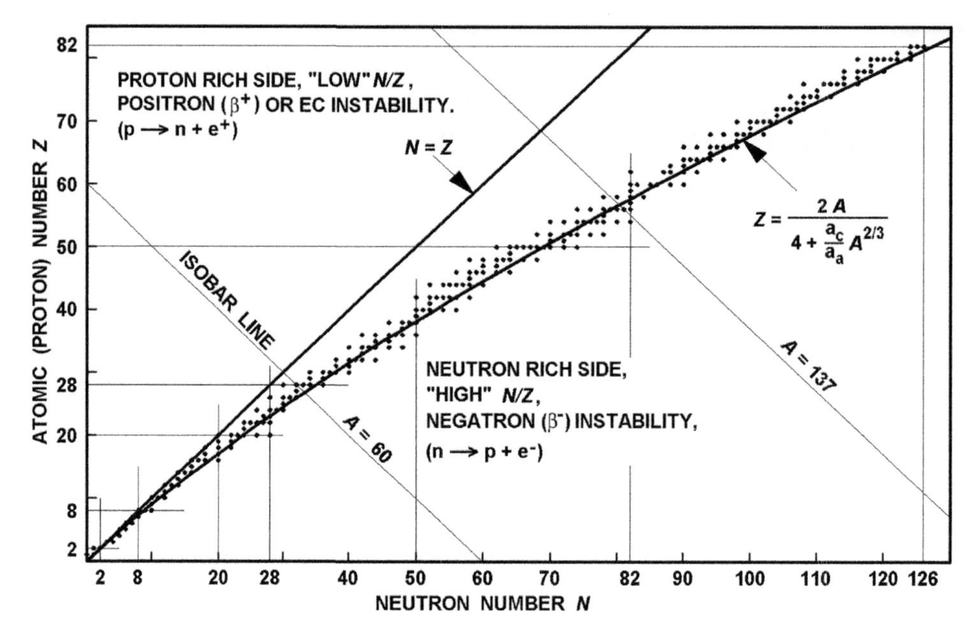

Figure 4.1 Chart of stable nuclides as a function of their proton (Z) and neutron (N) numbers. The numbers denoted 2, 8, etc., are discussed in Chapter 11.

of protons) are characterized by having a relatively sizable number of stable isotopes, usually 3 or more. For example, the element tin, atomic number 50, has 10 stable isotopes while cadmium ($Z = 48$) and tellurium ($Z = 52$) each have 8. By contrast silver ($Z = 47$) and antimony ($Z = 51$) each have only 2 stable isotopes, and rhodium ($Z = 45$), indium ($Z = 49$), and iodine ($Z = 53$) have only 1 stable isotope. Many other examples of the extra stabilization of even numbers of nucleons can be found from a detailed examination of Figure 4.1, or, easier, from nuclide charts, e.g. Appendix C. The guide lines of N and Z equal to 2, 8, 20, etc., have not been selected arbitrarily. These proton and neutron numbers represent unusually stable proton and neutron configurations, as will be discussed further in Chapter 6. The curved line through the experimental points is calculated based on the liquid drop model of the nucleus which is discussed later in this chapter.

Elements of odd Z have none, one or two stable isotopes, and their stable isotopes have an even number of neutrons, except for the 5 odd-odd nuclei mentioned above. This is in contrast to the range of stable isotopes of even Z, which includes nuclei of both even and odd N, although the former outnumber the latter. Tin ($Z = 50$), for example, has 7 stable even-even isotopes and only 3 even-odd ones.

The greater number of stable nuclei with even numbers of protons and neutrons is explained in terms of the energy stabilization gained by combination of like nucleons to form pairs, i.e. protons with protons and neutrons with neutrons, but not protons with neutrons. If a nucleus has, for example, an even number of protons, all these protons can exist in pairs. However, if the nucleus has an odd number of protons, at least one of these protons must exist in an unpaired state. The increase in stability resulting from complete pairing in elements of even Z is responsible for their ability to accommodate a greater range of neutron numbers as illustrated for the isotopes of germanium ($_{32}$Ge, 5 stable isotopes), relative to those of gallium ($_{31}$Ga, 2 stable isotopes), and arsenic ($_{33}$As, 1 stable isotope). The same pairing stabilization holds true for neutrons so that an even-even nuclide which has all its nucleons, both neutrons and protons, paired represents a quite stable situation. In the elements in which the atomic number is even, if the neutron number is uneven, there is still some stability conferred through the proton-proton pairing. For elements of odd atomic number, unless there is stability due to an even neutron number (neutron-neutron pairing), the nuclei are radioactive with rare exceptions. We should also note that the number of stable nuclear species is approximately the same for even-odd and odd-even cases. The pairing of protons with protons and neutrons with neutrons must thus confer approximately equal degrees of stability to the nucleus. This behaviour is rather similar to the electronic structure of the atoms where also paired electrons offer a greater stability. The similarities between the nucleons in the nucleus and the electrons in the atoms is further discussed in combination with the shell model for nuclear stability, Ch. 6.

4.2. NEUTRON TO PROTON RATIO

If a graph is made (Fig. 4.1)[1] of the relation of the number of neutrons to the number of protons in the known stable nuclei, we find that in the light elements stability is achieved when the number of neutrons and protons are approximately equal ($N = Z$). However, with increasing atomic number of the element (i.e. along the Z-line), the ratio of neutrons to protons, the N/Z ratio, for nuclear stability increases from unity to about 1.52 at lead. Thus pairing of the nucleons is not a sufficient criterion for stability: a certain ratio N/Z must also exist. However, even this does not suffice for stability, because at high Z-values, a new mode of radioactive decay, α-emission, appears. Above bismuth the nuclides are all unstable to radioactive decay by α-particle emission, while some are unstable also to β-decay.

If a nucleus has an N/Z ratio too high for stability, it is said to be *neutron-rich*. It will undergo radioactive decay in such a manner that the neutron to proton ratio decreases to approach more closely the stable value. In such a case the nucleus must decrease the value of N and increase the value of Z, which can be done by conversion of a neutron to a proton. When such a conversion occurs within a nucleus, β^- (or *negatron*) *emission* is the

[1] In graphs like Fig. 4.1, Z is commonly plotted as the abscissa; we have here reversed the axes to conform with the commercially available isotope and nuclide charts.

consequence, with creation and emission of a negative β-particle designated by β^- or ${}_{-1}^{0}e$ (together with an anti-neutrino, here omitted for simplicity, see Ch. 5). For example:

$$ {}_{49}^{116}\mathrm{In} \rightarrow {}_{50}^{116}\mathrm{Sn} \rightarrow {}_{-1}^{0}e^- \quad (4.1) $$

At extreme *N*/*Z* ratios beyond the so called neutron drip-line where the *N*/*Z* ratio is too high for stability, or for highly excited nuclei, *neutron emission* is an alternative to β^- decay.

If the *N*/*Z* ratio is too low for stability, then radioactive decay occurs in such a manner as to lower *Z* and increase *N* by conversion of a proton to neutron. This may be accomplished through *positron emission*, i.e. creation and emission of a positron (β^+ or ${}_{+1}^{0}e$), or by absorption by the nucleus of an orbital electron (*electron capture*, EC). Examples of these reactions are:

$$ {}_{51}^{116}\mathrm{Sb} \rightarrow {}_{50}^{116}\mathrm{Sn} + {}_{+1}^{0}e^+ \quad \text{and} \quad {}_{79}^{195}\mathrm{Au} + {}_{-1}^{0}e^- \overset{\mathrm{EC}}{\rightarrow} {}_{78}^{195}\mathrm{Pt} \quad (4.2) $$

Positron emission and electron capture are competing processes with the probability of the latter increasing as the atomic number increases. Beta decay is properly used to designate all three processes, β^-, β^+, and EC. (The term "beta decay" without any specification usually only refers to β^- emission.)

Thus in the early part of the Periodic Table, unstable neutron deficient nuclides decay by positron emission, but for the elements in the platinum region and beyond, decay occurs predominantly by electron capture. Both processes are seen in isotopes of the elements in the middle portion of the Periodic Table, see Figure 4.1 and Appendix C.

An alternative to positron decay (or EC) is *proton emission*, which, although rare, has been observed in about 40 nuclei very far off the stability line. These nuclei all have half-lives $\leq$1 min. For example: ^{115}Xe, $t_{1/2}$ (p) 18 s; proton/EC ratio, 3×10^{-3}.

We can understand why the *N*/*Z* ratio must increase with atomic number in order to have nuclear stability when we consider that the protons in the nucleus must experience a repulsive Coulomb force. The fact that stable nuclei exist means that there must be an attractive force tending to hold the neutrons and protons together. This attractive nuclear force must be sufficient in stable nuclei to overcome the disruptive Coulomb force. Conversely, in unstable nuclei there is a net imbalance between the attractive nuclear force and the disruptive Coulomb force. As the number of protons increases, the total repulsive Coulomb force must increase. Therefore, to provide sufficient attractive force for stability the number of neutrons increases more rapidly than that of the protons.

Neutrons and protons in nuclei are assumed to exist in separate *nucleon orbitals* just as electrons are in electron orbitals in atoms. If the number of neutrons is much larger than the number of protons, the neutron orbitals occupied extend to higher energies than the highest occupied proton orbital. As *N*/*Z* increases, a considerable energy difference can

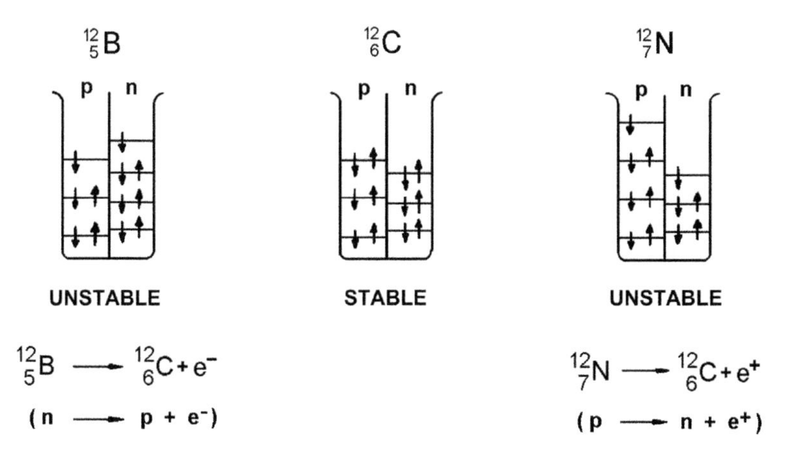

Figure 4.2 The separation and pairing of nucleons in assumed energy levels within the isobar $A = 12$. Half-life for the unstable ^{12}B is 0.02 s, and for ^{12}N 0.01 s.

develop between the last (highest energy) neutron orbital filled and the last proton orbital filled. The stability of the nucleus can be enhanced when an odd neutron in the highest neutron orbital is transformed into a proton fitting into a vacant lower energy proton orbital; see the example for $A = 12$ in Figure 4.2. In short, if the different "energy ladders" of the neutron and protons in the nucleus is investigated it becomes clear that the distance between the "steps" in the proton ladder is larger due to the repulsion force. Thus in order to achieve the lowest possible energy with a constant number of nucleons the "neuton ladder" will fill up faster thus making N/Z increase with the number of nucleons in the nucleus. These questions of nuclear forces and the energy levels of nucleons are discussed more extensively in Chapter 6.

4.3. MASS DEFECT

It was noted in Chapter 1 that the masses of nuclei (in u) are close to the mass number A. Using the mass of carbon-12 as the basis (^{1}C $\equiv$ 12 u), the hydrogen atom and the neutron do not have exactly unit masses. We would expect that the mass M_A of an atom with mass number A would be given by the number of protons (Z) times the mass of the hydrogen atom ($\mathbf{M}_\mathrm{H}$) plus the number of neutrons (N) times the mass of the neutron ($\mathbf{M}_n$), i.e.

$$M_A \approx Z\mathbf{M}_\mathrm{H} + N\mathbf{M}_\mathrm{n} \tag{4.3}$$

For deuterium with one neutron and one proton in the nucleus, we would then anticipate an atomic mass of

$$\mathbf{M}_\mathrm{H} + \mathbf{M}_\mathrm{n} = 1.007\,825 + 1.008\,665 = 2.016\,490 \text{ u} \tag{4.4}$$

When the mass of the deuterium atom is measured, it is found to be 2.014 102 u. The difference between the measured and calculated mass values, which in the case of deuterium equals −0.002 388 u, is called the *mass defect* (ΔM_A):

$$\Delta M_A = M_A - Z\mathbf{M}_{\mathrm{H}} - N\mathbf{M}_{\mathrm{n}} \tag{4.5}$$

From the Einstein equation, $E = m\mathbf{c}^2$, which is discussed further in Chapters 5 and 10, one can calculate that one atomic mass unit is equivalent to 931.5 MeV, where MeV is a million electron volts.

$$E = m\mathbf{c}^2 = 931.5\ \Delta M_A \tag{4.6}$$

where 931.5 is the mass energy content (in MeV) in 1u of mass. The relationship of energy and mass would indicate that in the formation of deuterium by the combination of a proton and neutron, the mass defect of 0.002 388 u would be observed as the liberation of an equivalent amount of energy, i.e. $931.5 \times 0.002\ 388 = 2.224$ MeV. Indeed, the emission of this amount of energy (in the form of γ-rays) is observed when a proton captures a low energy neutron to form H. As a matter of fact, in this particular case, the energy liberated in the formation of deuterium has been used in the reverse calculation to obtain the mass of the neutron since it is not possible to determine directly the mass of the free neutron. With the definition (4.5) all stable nuclei are found to have negative ΔM_A values; thus the term "defect".

In many nuclide (or isotope) tables the neutral atomic mass is not always given, but instead the *mass excess* (often, unfortunately, also called mass defect). We indicate this as δ_A and define it as the difference between the measured mass and the mass number of the particular atom:

$$\delta_A = M_A - A \tag{4.7}$$

Mass excess values are either given in u (or, more commonly, in micro mass units, μu) or in eV (usually keV). On of the advantages of using the mass excess is when calculating the energetics for nuclear reactions. In many cases the A will cancel out and thus it is possible to avoid comparing very large and very small numbers and thus increasing the accuracy of the calculations. However, with the computer power existing today this is rarely an issue. Table 4.1 contains a number of atomic masses, mass excess, and mass defect values, as well as some other information which is discussed in later sections.

When two elements form a compound in a chemical system, the amount of heat liberated is a measure of the stability of the compound. The greater this heat of formation (enthalpy, ΔH) the greater the stability of the compound. When carbon is combined with oxygen to form CO_2, it is found experimentally that 393 kJ of heat is evolved per mole of CO_2 formed. If we use the Einstein relationship, we can calculate that this would correspond to a total mass loss of 4.4×10^{-9} g for each mole of CO_2 formed (44 g).

Table 4.1 Atomic masses and binding energies

Element	Z	N	A	Atomic mass M_A (u)	Mass excess $M_A - A$ (μu)	Mass defect ΔM_A (μu)	Binding energy E_B (MeV)	E_B/A (MeV/A)
n	0	1	1	1.008665	8665	0	—	—
H	1	0	1	1.007825	7825	0	—	—
D	1	1	2	2.014102	14102	−2388	2.22	1.11
T	1	2	3	3.016049	16049	−9106	8.48	2.83
He	2	1	3	3.016029	16029	−8286	7.72	2.57
He	2	2	4	4.002603	2603	−30377	28.30	7.07
He	2	4	6	6.018886	18886	−31424	29.27	4.88
Li	3	3	6	6.015121	15121	−34348	32.00	5.33
Li	3	4	7	7.016003	16003	−42132	39.25	5.61
Be	4	3	7	7.016928	16928	−40367	37.60	5.37
Be	4	5	9	9.012182	12182	−62442	58.16	6.46
Be	4	6	10	10.013534	13534	−69755	64.98	6.50
B	5	5	10	10.012937	12937	−69513	64.75	6.48
B	5	6	11	11.009305	9305	−81809	76.20	6.93
C	6	6	12	12.000000	0	−98940	92.16	7.68
N	7	7	14	14.003074	3074	−112356	104.7	7.48
O	8	8	16	15.994915	−5085	−137005	127.6	7.98
F	9	10	19	18.998403	−1597	−158671	147.8	7.78
Ne	10	10	20	19.992436	−7564	−172464	160.6	8.03
Na	11	12	23	22.989768	−10232	−200287	186.6	8.11
Mg	12	12	24	23.985042	−14958	−212837	198.3	8.26
Al	13	14	27	26.981539	−18461	−241495	225.0	8.33
Si	14	14	28	27.976927	−23073	−253932	236.5	8.45
P	15	16	31	30.973762	−26238	−282252	262.9	8.48
K	19	20	39	38.963707	−36293	−358266	333.7	8.56
Co	27	32	59	58.933198	−66802	−555355	517.3	8.77
Zr	40	54	94	93.906315	−93685	−874591	814.7	8.67
Ce	58	82	140	139.905433	−94567	−1258941	1172.7	8.38
Ta	73	108	181	180.947993	−52007	−1559045	1452.2	8.02
Hg	80	119	199	198.968254	−31746	−1688872	1573.2	7.91
Th	90	142	232	232.038051	38051	−1896619	1766.7	7.62
U	92	143	235	235.043924	43924	−1915060	1783.9	7.59
U	92	144	236	236.045563	45563	−1922087	1790.4	7.59
U	92	146	238	238.050785	50785	−1934195	1801.7	7.57
Pu	94	146	240	240.053808	53808	−1946821	1813.5	7.56

Although chemists do not doubt that this mass loss actually occurs, at present there are no instruments of sufficient sensitivity to measure such small changes.

The energy changes in nuclear reactions are much larger. This can be seen if we use the relationship between electron volts and joules (or calories) in Appendix IV, and observe that nuclear reaction formulas and energies refer to single atoms (or molecules), while chemical reactions and equations refer to number of moles; we have:

$$1\ \text{eV/molecule} = 1.6022 \times 10^{-19} \times 6.0221 \times 10^{23} = 96.48\ \text{kJ mole}^{-1}$$
$$= 3.8268 \times 10^{-20} \times 6.0221 \times 10^{23} = 23.045\ \text{kcal mole}^{-1} \quad (4.8)$$

Thus, the formation of deuterium from a neutron and a hydrogen atom would lead to the liberation of 214.6×10^6 kJ (51.3×10^6 kcal) for each mole of deuterium atoms formed. By comparison, then, the nuclear reaction leading to the formation of deuterium is approximately half a million times more energetic than the chemical reaction leading to formation of CO_2.

It is not common practice to use mole quantities in considering nuclear reactions as the number of individual reactions under laboratory conditions is well below 6.022×10^{23}. Therefore, in nuclear science one uses the energy and mass changes involved in the reaction of individual particles and nuclei.

4.4. BINDING ENERGY

The energy liberated in the formation of CO_2 from the elements, the heat of formation, is a measure of the stability of the CO_2 molecule. The larger the heat of formation the more stable the molecule since the more energy is required to decompose the molecule into its component atoms. Similarly, the energy liberated in the formation of a nucleus from its component nucleons is a measure of the stability of that nucleus. This energy is known as the *binding energy* (E_B) and has the same significance in nuclear science as the heat of formation has in chemical thermodynamics. We have seen that the binding energy of deuterium is 2.22 MeV. The ${}^4_2\text{He}$ nucleus is composed of 2 neutrons and 2 protons. The measured mass of the ${}^4\text{He}$ atom is 4.002 603 u. The mass defect is:

$$\Delta M_{He} = M_{He} - 2\,\mathbf{M}_H - 2\,\mathbf{M}_n = 4.002603 - 2 \times 1.007825 - 2 \times 1.008665$$
$$= -0.030377\ \text{u} \quad (4.9)$$

According to eqn. (4.6) the binding energy, E_B, between the nucleons in a nucleus follows the simple relation[2]

$$E_B(\text{MeV}) = -931.5\ \Delta M_A(\text{u}) \quad (4.10)$$

Thus the binding energy for ${}^4\text{He}$ is 28.3 MeV. It is quite unlikely that 2 neutrons and 2 protons would ever collide simultaneously to form a ${}^4\text{He}$ nucleus; nevertheless, this calculation is useful because it indicates that to break ${}^4\text{He}$ into its basic component nucleons would require at least 28.3 MeV.

[2] The minus sign indicates that the loss of mass in formation of a nucleus releases energy.

A better indication of the relative stability of nuclei is obtained when the binding energy is divided by the total number of nucleons to give the average binding energy per nucleon, E_B/A. For ^{4}He the value of E_B/A is 28.3/4 or 7.1 MeV, whereas for ^{2}H it is 1.11 for the bond between the two nucleons. Clearly, the ^{4}He nucleus is considerably more stable than the ^{2}H nucleus. For most nuclei the values of E_B/A vary in the rather narrow range 5–8 MeV. To a first approximation, therefore, E_B/A is relatively constant which means that the total nuclear binding energy is roughly proportional to the total number of nucleons in the nucleus.

Figure 4.3 shows that the E_B/A values increase with increasing mass number up to a maximum around mass number 60 and then decrease. Therefore the nuclei with mass numbers in the region of 60, i.e. nickel, iron, etc., are the most stable. Also in this Figure we see that certain numbers of neutrons and protons form especially stable configurations – this effect is observed as small humps on the curve.

If two nuclides can be caused to react so as to form a new nucleus whose E_B/A value is larger than that of the reacting species, obviously a certain amount of binding energy would be released. The process which is called *fusion* is "exothermic" only for the nuclides of mass number below 60. As an example, we can choose the reaction

$$^{20}_{10}\text{Ne} + ^{20}_{10}\text{Ne} \rightarrow ^{40}_{20}\text{Ca} \tag{4.11}$$

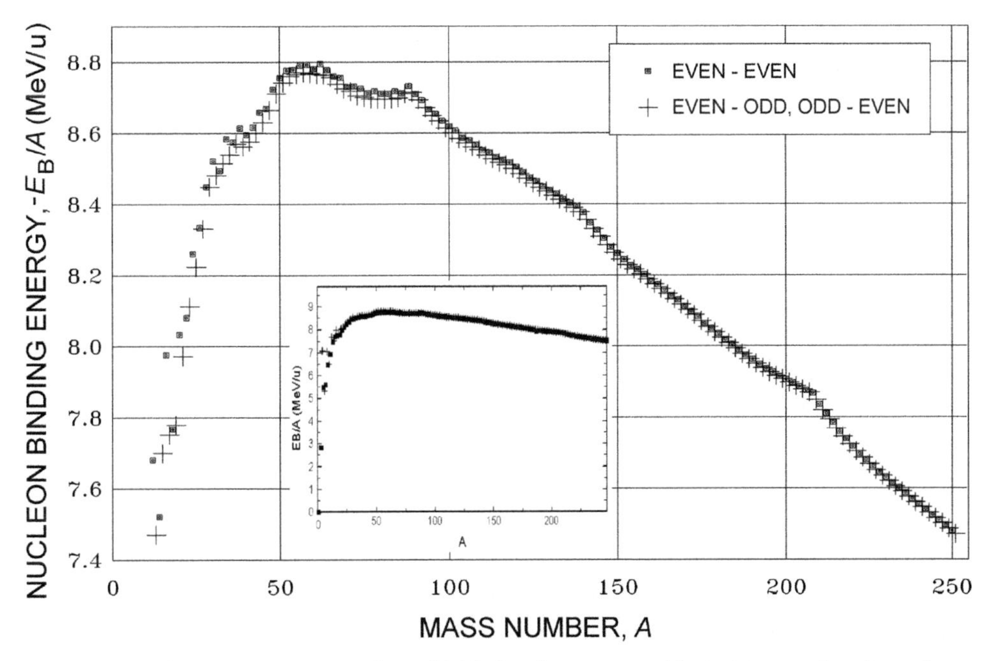

Figure 4.3 Binding energy per nucleon (E_B/A) for the most stable isobars as function of mass number (A).

From Figure 4.3 we estimate that E_B/A for neon is about 8.0 MeV and for calcium about 8.6 MeV. Therefore, in the 2 neon nuclei $2 \times 20 \times 8.0 = 320$ MeV are involved in the binding energy, while $40 \times 8.6 = 344$ MeV binding energy are involved in the calcium nucleus. When 2 neon nuclei react to form the calcium nucleus the difference in the total binding energy of reactants and products is released; the estimate gives $344-320 = 24$ MeV; a calculation using measured masses gives 20.75 MeV. It is also evident that the greater gain in fusion energy is in the combination of lighter elements where the derivative of the E_B/A curve is greater, e.g. $^2H + ^2H \rightarrow ^4He$ liberates 23.85 MeV per atom of ^4He produced.

Figure 4.3 also shows that a similar release of binding energy can be obtained if the elements with mass numbers greater than 60 are split into lighter nuclides with higher E_B/A values. Such a process, whereby a nucleus is split into two smaller nuclides, is known as *fission*. An example of such a fission process is the reaction

$$^{236}_{92}U \rightarrow ^{140}_{54}Xe + ^{93}_{38}Sr + 3n \tag{4.12}$$

The binding energy per nucleon for the uranium nucleus is 7.6 MeV, while those for the ^{140}Xe and ^{93}Sr are 8.4 and 8.7 MeV respectively. The amount of energy released in this fission reaction is approximately $140 \times 8.4 + 93 \times 8.7 - 236 \times 7.6 = 191.5$ MeV for each uranium fission.

4.5. NUCLEAR RADIUS

Rutherford showed by his scattering experiments that the nucleus occupies a very small portion of the total volume of the atom. Roughly, the radii of nuclei vary from 1/10 000 to 1/100 000 of the radii of atoms. While atomic sizes are of the order of 100 pm (10^{-10} m), the common unit of nuclear size is the femtometer (1 fm $= 10^{-15}$ m), sometimes referred to as 1 Fermi.

Experiments designed to study the size of nuclei indicate that the volumes of nuclei (V_n) are directly proportional to the total number of nucleons present, i.e.

$$V_n \propto A \tag{4.13}$$

Since for a sphere $V \propto r^3$, where r is the radius of the sphere, for a spherical nucleus $r^3 \propto A$, or $r \propto A^{1/3}$. Using r_o as the proportionality constant

$$r = r_o A^{1/3} \tag{4.14}$$

The implications of this is that the nucleus is composed of nucleons packed closely together with a constant density (about 0.2 nucleons fm^{-3}) from the center to the edge of the nucleus. This constant density model of the nucleus as been shown to be not completely correct, however. By bombarding nuclei with very high energy electrons or protons (up to $\geq$1 GeV) and measuring the scattering angle and particle energy, the

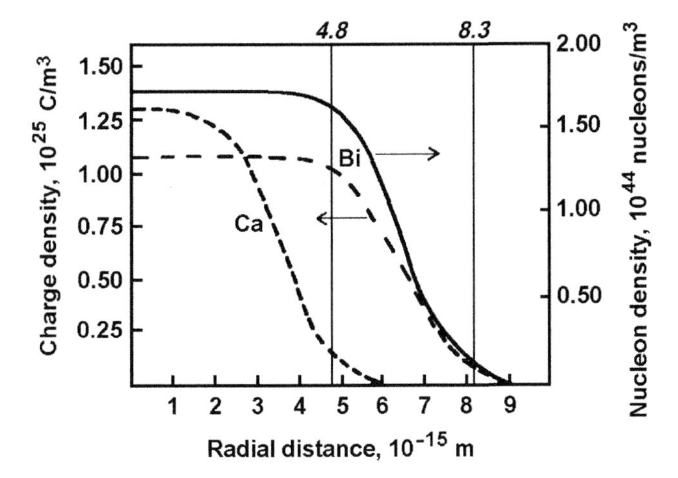

Figure 4.4 Experimentally measured charge and nuclear density values for ^{40}Ca and ^{209}Bi as a function of the nuclear radius.

charge and matter density near the surface of the irradiated nucleus can be studied. These experiments have led to the conclusion that nuclei do not possess a uniform charge or matter distribution out to a sharp boundary, but rather are fuzzy as indicated by the s-shaped curves in Figure 4.4. With an atomic number greater than 20 it has been found that a uniform charge and mass density exists over a short distance from the center of the nucleus, and this core is surrounded by a layer of decreasing density which seems to have a constant thickness of ~2.5 fm independent of mass number. In a bismuth nucleus, for example, the density remains relatively constant for approximately 5 fm then decreases steadily to one-tenth of that value in the next 2 fm (Fig. 4.4). This is not too surprising considering that as we get closer and closer to the edges of the nucleus not all the coordination sites of the nucleons are filled and thus the density will slowly decrease. It has also been found that not all nuclei are spherical, some being oblate and others prolate around the axis of rotation.

Despite the presence of this outer layer of decreasing density and the nonspherical symmetry, for most purposes it is adequate to assume a constant density nucleus with a sharp boundary. Therefore, use is made of the radius equation (4.14) in which the r_o value may be assumed to be 1.4 fm. Using this relationship, we can calculate the radius of ^{40}Ca to be $r = 1.4 \times 10^{-15} \times 40^{1/3} = 4.79$ fm, and for ^{209}Bi to be 8.31 fm. These values are indicated in Figure 4.4. For ^{80}Br a similar calculation yields 6.0 fm, while for ^{238}U the radius calculated is 8.7 fm. From these calculations we see that the radius does not change dramatically from relatively light nuclei to the heaviest.

4.6. SEMIEMPIRICAL MASS EQUATION

In preceding sections we have learned that the size as well as the total binding energy of nuclei are proportional to the mass number. These characteristics suggest an analogy between the nucleus and a drop of liquid. In such a drop the molecules interact with their immediate neighbours but not with other molecules more distant. Similarly, a particular nucleon in a nucleus is attracted by nuclear forces only to its adjacent neighbours. Moreover, the volume of the liquid drop is composed of the sum of the volumes of the molecules or atoms present since these are nearly incompressible. Again, as we learned above, this is similar to the behavior of nucleons in a nucleus. Based on the analogy of a nucleus to a droplet of liquid, it has been possible to derive a semiempirical mass equation containing various terms which are related to a nuclear droplet.

Let us consider what we have learned about the characteristics of the nuclear droplet.

(a) First, recalling that mass and energy are equivalent, if the total energy of the nucleus is directly proportional to the total number of nucleons there should be a term in the mass equation related to the mass number.

(b) Secondly, in the discussion of the neutron/proton ratios we learned that the number of neutrons could not become too large since the discrepancy in the energy levels of the neutron and proton play a role in determining the stability of the nucleus. This implies that the binding energy is reduced by a term which allows for variation in the ratio of the number of protons and neutrons.

(c) Since the protons throughout the nucleus experience a mutual repulsion which affects the stability of nucleus, we should expect in the mass equation another negative term reflecting the repulsive forces of the protons.

(d) Still another term is required to take into account that the surface nucleons, which are not completely surrounded by other nucleons, would not be totally saturated in their attraction. In a droplet of liquid this lack of saturation of forces on the surface gives rise to the effect of surface tension. Consequently, the negative term in the mass equation reflecting this unsaturation effect should be similar to a surface tension expression.

(e) Finally, we have seen that nuclei with an even number of protons and neutrons are more stable than nuclei with an odd number of either type of nucleon and that the least stable nuclei are those for odd numbers of both neutrons and protons. This odd-even effect must also be included in a mass equation.

Taking into account these various factors, we can write a semiempirical mass equation. However, it is often more useful to write the analogous equation for the mass defect or binding energy of the nucleus, recalling (4.10). Such an equation, first derived by C. F. von Weizsäcker in 1935, would have the form:

$$E_{\mathrm{B}}(\mathrm{MeV}) = a_{\mathrm{v}}A - a_{\mathrm{a}}(N-Z)^2/A - a_{\mathrm{c}}Z^2/A^{1/3} - a_{\mathrm{s}}\,A^{2/3} \pm a_{\delta}/A^{3/4} \qquad (4.15)$$

The first term in this equation takes into account the proportionality of the energy to the total number of nucleons (the volume energy); the second term, the variations in neutron and proton ratios (the asymmetry energy); the third term, the Coulomb forces of repulsion for protons (the Coulomb energy); the fourth, the surface tension effect (the surface energy). In the fifth term, which accounts for the odd-even effect, a positive sign is used for even proton-even neutron nuclei and a negative sign for odd proton-odd neutron nuclei. For nuclei of odd A (even-odd or odd-even) this term has the value of zero. Comparison of this equation with actual binding energies of nuclei yields a set of coefficients; e.g.

$$a_\text{v} = 15.5, \quad a_\text{a} = 23, \quad a_\text{c} = 0.72, \quad a_\text{s} = 16.8, \quad a_\delta = 34$$

With these coefficients the binding energy equations (4.5) and (4.10) give agreement within a few percent of the measured values for most nuclei of mass number greater than 40.

When the calculated binding energy is compared with the experimental binding energy, it is seen that for certain values of neutron and proton numbers, the disagreement is more serious. These numbers are related to the so-called "magic numbers", which we have indicated in Figure 4.1, whose recognition led to the development of the nuclear shell model described in a later chapter.

4.7. VALLEY OF β-STABILITY

If the semiempirical mass equation is written as a function of Z, remembering that $N = A - Z$, it reduces to a quadratic equation of the form

$$E_\text{B} = aZ^2 + bZ + c \pm d/A^{3/4} \tag{4.16}$$

where the terms a, b and c also contain A. This quadratic equation describes a parabola for constant values of A. Consequently, we would expect that for any family of isobars (i.e. constant A) the masses should fall upon a parabolic curve. Such a curve is shown in Figure 4.5. In returning to Figure 4.1, the isobar line with constant A but varying Z cuts

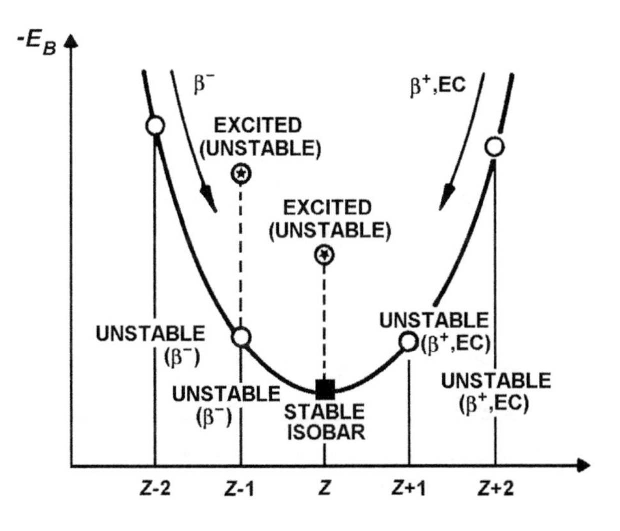

Figure 4.5 Isobar cut across the valley of stability showing schematically the position of different kinds of nuclei.

diagonally through the line of stable nuclei. We can picture this as a valley, where the most stable nuclei lie at the bottom of it (cf. Figs. 4.1 and 4.5), while unstable nuclei lie up the valley sides as shown in Figure 4.5. Any particular isobaric parabola can be considered as a cross-section of the valley of stability; Figure 4.5 would be seen by someone standing up to the right of Figure 4.1 and looking down the valley. The isobars located on the sides of the parabola (or slope of the valley) are unstable to radioactive decay to more stable nuclides lower on the parabola, though usually the most stable nucleus is not located exactly at the minimum of the parabola. Nuclides on the left hand side of the parabola (lower atomic numbers) are unstable to decay by β^- emission. Isobars to the right of the valley of stability are unstable to β^+ decay or electron capture. At the bottom of the valley the isobars are stable against β decay. The curved line in Figure 4.1 is calculated for maximum stability according to (4.15), and indicates the theoretical bottom of the valley. The minimum of the curve can be calculated from (4.15) to be

$$Z = 2A/[4 + (a_c/a_a)A^{2/3}] \tag{4.17}$$

and is shown in Figure 4.1. For small A values (4.17) reduces to $Z = A/2$ or $N = Z$; thus the bottom of the stability valley follows the $N = Z$ line as indicated in Figure 4.1 for the lighter nuclides.

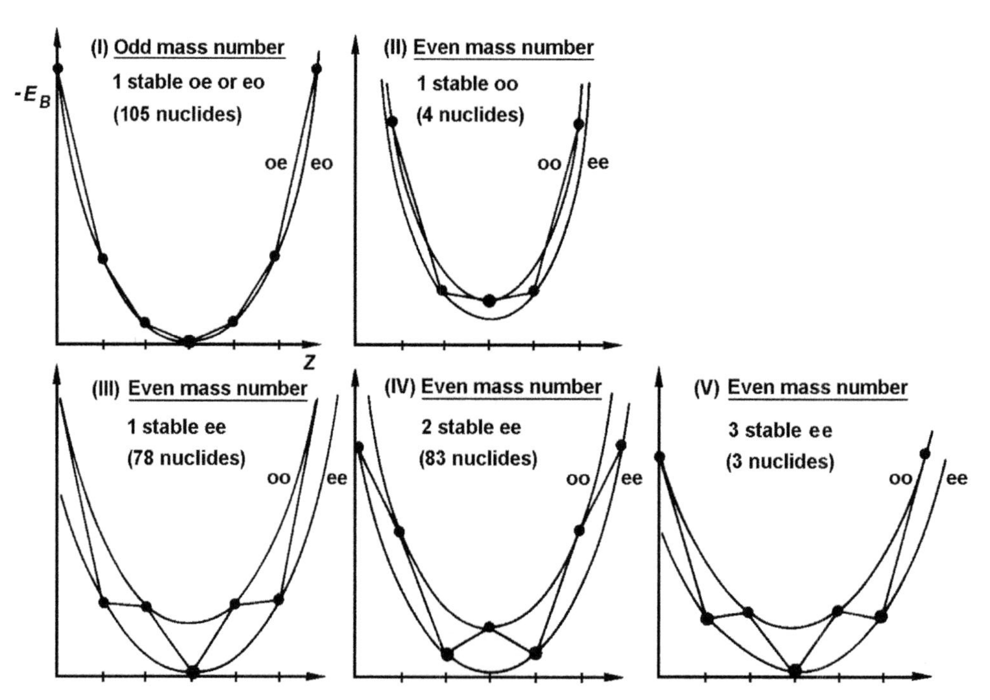

Figure 4.6 Isobar parabolas for odd mass numbers (I: odd-even or even-odd nuclides) and for even mass numbers (cases II - V). The stable nuclides are indicated by heavier dots.

A closer analysis of (4.16) makes us expect that the last term gives rise to three different isobaric parabola depending on whether the nuclei are odd-A (even-odd or odd-even), odd-odd, or even-even (Fig. 4.6). In the first case, in which the mass number is odd, we find a single parabola (I); whether all beta decay leads to changes from odd-even to even-odd, etc. For even mass numbers one finds a double parabola (II) – (V). When the individual nuclear properties are considered, the difference between the curves for the odd-odd and even-even nuclei may lead to alternatives with regard to the numbers of possible stable isobars: it is possible to find three stable isobars (case V) although two (case IV) are more common. Although the odd-odd curve always must lie above the even-even curve, still an odd-odd nucleus may become stable, as is shown for case II.

4.8. THE MISSING ELEMENTS: $_{43}$TC AND $_{61}$PM

Among the stable elements between $_1$H and $_{82}$Pb two elements are "missing": atomic number 43, named *technetium* (Tc), and atomic number 61, *promethium* (Pm). Though these elements can be produced through nuclear reactions and also have been found to exist in certain stars, they are not found on earth because their longest lived isotopes have much too short half-lives for them to have survived since the formation of our planet. This can be understood by considering the valley of β-stability. For pedagogic reasons we will first discuss promethium.

4.8.1. Promethium

The valley of β-stability for $Z = 61$ shows a minimum around mass number $A = 146$, for which the isotopes are either of the even-even or of the odd-odd type. Thus the binding energy curve should exhibit two isobar parabolas, as illustrated in Figure 4.7; the decay energy Q is released binding energy. ^{146}Pm has a 5.5 y half-life and decays either by electron capture (63%) to ^{148}Nd or by β^--emission (37%) to ^{146}Sm, who both are more stable (i.e. have a larger nucleon binding energy); the nuclear binding energy is given on the vertical scale in the Figure. The curves shown in Figure 4.7 differ from those calculated from eqn. 4.14 by about 1 unit in Z due to deviations from the simple liquid drop model in the lanthanide region, see §6.4.

The two adjacent isobars, mass numbers $A = 145$ and $A = 147$, are of the odd-even and even-odd types, thus only one isobaric β-decay curve exists for each of these. The decay scheme for $A = 145$ follows curve I in Figure 4.6 for which $^{145}_{60}$Nd is the stable isobar. ^{145}Pm is the longest lived promethium isotope ($t_{½}$ 17.7 y). For $A = 147$, the stable isobar is $^{147}_{62}$Sm; the half-life of ^{147}Pm is 2.62 y, which makes it the most convenient radioisotope of promethium for use in experiments.

Promethium is a fission product (Ch. 5 and 19) and can be chemically isolated in pure form. It exhibits typical lanthanide properties and is used in technology and medicine as a radiation source (Ch. 18).

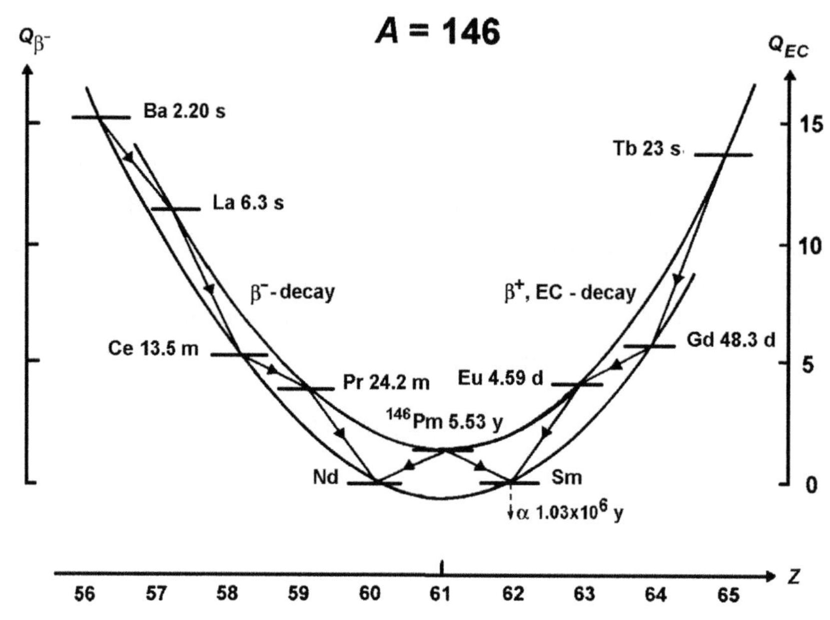

Figure 4.7 Decay scheme for $A = 146$, with isobar half-lives. Decay energy Q in MeV. ^{146}Nd and ^{146}Sm are β-stable.

4.8.2. Technetium

For technetium, $Z = 43$, the valley of β-stability has a minimum in the neighbourhood of $N = 55$ and thus, for $Z = 43$, A-values around 97 and 99 are most likely to be stable (recall that odd-odd nuclei are less stable than odd-even). If one considers all the isobars between $A = 95$ and 102 one finds that for each mass number in this range there is already at least one stable nuclide for the elements with $Z = 42$ (molybdenum) and $Z = 44$ (ruthenium). Since adjacent isobars cannot both be stable, this excludes the possibility of stable odd-even isotopes of technetium. The longest lived isotopes of technetium are those with $A = 97$ (2.6×10^6 y), $A = 98$ (4.2×10^6 y), and $A = 99$ (2.1×10^6 y). Figure 4.8 shows the decay scheme for $A = 99$, which is taken out of a standard Isotope Table; the vertical axis shows the relative binding energies (broken scale). The Figure illustrates the information normally presented in isotope tables, and will be further explained in subsequent chapters.

Hundreds of kilograms of ^{99}Tc and its mother ^{99}Mo are formed every year as fission products in nuclear reactors, and 10's of kg of Tc have been isolated and studied chemically. Its properties resembles those of its homologs in the Periodic Table - manganese and rhenium. Figure 4.8 shows decay schemes for mass number 99: the upper one from Shirley et al, 1986, the lower one from Dzhelepov et al, 1961; more detailed schemes appear in both references. The ones shown in Figure 4.8 were chosen

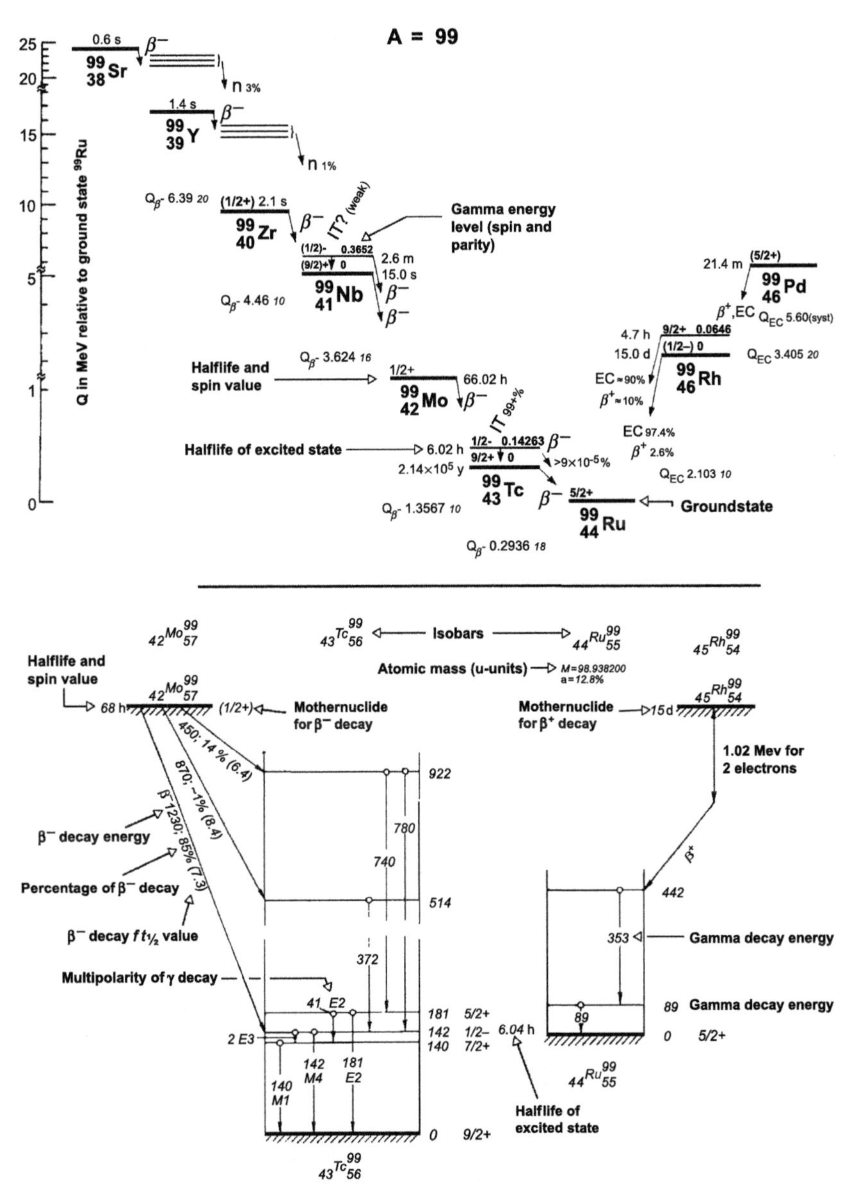

Figure 4.8 The $A = 99$ decay scheme. *(From E. Browne, R. B. Firestone and V. S. Shirley, Table of Radioactive Isotopes, and from B. S. Dzhelepov and L. K. Peker, Decay schemes of radioactive nuclei.)*

for pedagogic reasons, and, for this purpose, we have also inserted explanations, some of which will be dealt with later. Older references are often still useful for rapid survey, while the newest ones give the most recent information and refined numerical data.

The upper left part of Figure 4.8 shows a decay chain from fission of ^{235}U that ends in ^{99}Ru, the most stable isobar of $A = 99$. The lower diagram shows that the ^{99}Mo β^- decays all reaches the spin/parity ½-level (see § 3.6.1 and Ch. 11), designated ^{99m}Tc; this isomer decays with $t_{½}$ 6.02 h to long-lived ^{99}Tc, emitting a single γ of 0.142 MeV (> 99%, see upper diagram). The isomer ^{99m}Tc is a widely used radionuclide in nuclear diagnostics (§18.5), and can be conveniently "milked" from its mother ^{99}Mo, see §5.16.

4.9. OTHER MODES OF INSTABILITY

In this chapter we have stressed nuclear instability to beta decay. However, in §4.4 it was learned that very heavy nuclei are unstable to fission. There is also a possibility of instability to emission of α-particles in heavy elements and to neutron and proton emission.

Nuclei are unstable to various forms of decay as indicated in Figure 4.1. For example, making a vertical cut at $N = 100$, the instability from the top is first proton emission, then, α-emission (for $N = 60$ it would instead be positron emission or electron capture, as these two processes are about equally probable), and, after passing the stable nuclides (the isotones ^{170}Yb, ^{169}Tm and ^{168}Er), β^- emission and, finally, neutron emission. This is more clearly indicated in Appendix C, and for the heaviest nuclides (i.e. $Z \geq 81$) in Figures 5.1 and 14.1. For α-decay the Figure indicates that for $A > 150$ ($Z \geq 70$, $N > 80$) the nuclei are α-unstable, but in fact α-decay is commonly observed only above $A \approx 200$. This is due to the necessity for the α-particle to pass over or penetrate the Coulomb barrier (cf. §6.7.3). Although neutron and proton emissions are possible energetically, they are not commonly observed as the competing β-decay processes are much faster.

4.10. EXERCISES

4.1. Calculate the nucleon binding energy in ^{24}Mg from the atomic mass excess value in Table 4.1.

4.2. How many times larger is the nucleon binding energy in ^{24}Na than the electron binding energy when the ionization potential of the sodium atom is 5.14 V?

4.3. Assuming that in the fission of a uranium atom an energy amount of 200 MeV is released, how far would 1 g of ^{235}U drive a car which consumes 1 liter of gasoline (density 0.70 g cm^{-3}) for each 10 km? The combustion heat of octane is 5500 kJ mole^{-1}, and the combustion engine has an efficiency of 18%.

4.4. Estimate if fusion of deuterium into helium releases more or less energy per gram of material consumed than the fission of uranium.

4.5. When a neutron is captured in a nucleus, the mass number of the isotope increases one unit. In the following Table mass excess values are given for three important isotope pairs:

^{235}U	40 915 keV	^{236}U	42 441 keV
^{238}U	47 306	^{239}U	50 571
^{239}Pu	48 585	^{240}Pu	50 122

If the average nucleon binding energy in this region is 7.57 MeV one can calculate the difference between this average binding energy and the one really observed in the formation of ^{236}U, ^{239}U, and ^{240}Pu. Calculate this difference. Discuss the possible significance of the large differences observed for the ^{238}U/^{239}U pair as compared to the other pairs in terms of nuclear power.

4.6. With the semiempirical mass equation (4.15) estimate the binding energy per nucleon for ^{10}B, ^{27}Al, ^{59}Co, and ^{236}U. Compare the results with the observed values in Table 4.1.

4.7. With eqn. (4.17) determine the atomic number corresponding to maximum stability for $A = 10$, 27, 59, and 239. Compare these results with the data in the isotope chart, Appendix C.

4.11. LITERATURE

R. D. Evans, *The Atomic Nucleus*, McGraw-Hill, 1955.

B. S. Dzhelepov and L. K. Peker, *Decay schemes of radioactive nuclei*, Pergamon Press, 1961 (though grossly out-of-date, it is still useful as a source of simple decay schemes).

W. D. Myers and W. J. Swiatecki, Nuclear Masses and Deformations, *Nucl. Phys.*, **81** (1966) 1.

A. H. Wapstra and G. Audi, The 1983 Atomic Mass Evaluation, *Nucl. Phys.*, **A432** (1985) 1.

E. Browne and R. B. Firestone, V. S. Shirley (Ed.), *Table of Radioactive Isotopes*, J. Wiley & Sons, 1986.

K. S. Krane, *Introductory Nuclear Physics*, J. Wiley & Sons, 1988.

Nuclides and Isotopes, 14th Ed., General Electric Company, Nuclear Energy Operations, 175 Curtner Ave., M/C397, San Jose, CA 95125, USA, 1989.

R. B. Firestone, *Table of Isotopes*, 8th Ed., J. Wiley & Sons, 1996.

Unstable Nuclei and Radioactive Decay

Contents

Radiochemistry and Nuclear Chemistry
ISBN 978-0-12-405897-2, http://dx.doi.org/10.1016/B978-0-12-405897-2.00005-7

5.1. RADIOACTIVE DECAY

Radioactive decay is a spontaneous nuclear transformation that has been shown to be unaffected by pressure, temperature, chemical form, etc (except a few very special cases). This insensitivity to extranuclear conditions allows us to characterize radioactive nuclei by their decay period and their mode and energy of decay without regard to their physical or chemical condition.

The time dependence of radioactive decay is expressed in terms of the *half-life* ($t_{½}$), which is the time required for one-half of the radioactive atoms in a sample to undergo decay. In practice this is the time for the measured radioactive intensity (or simply, *radioactivity* of a sample) to decrease to one-half of its previous value (see Fig. 1.1). Half-lives vary from millions of years to fractions of seconds. While half-lives between a minute and a year are easily determined with fairly simple laboratory techniques, the determination of much shorter half-lives requires elaborate techniques with advanced instrumentation. The shortest half-life measurable today is about 10^{-18} s. Consequently, radioactive decay which occurs with a time period less than 10^{-18} s is considered to be instantaneous. At the other extreme, if the half-life of the radioactive decay exceeds 10^{15} y, the decay usually cannot be observed above the normal signal background present in the detectors. Therefore, nuclides which may have half-lives greater than 10^{15} y are normally considered to be stable to radioactive decay. However, a few unstable nuclides with extremely long half-lives, $\geq 10^{20}$ y, e.g. ^{82}Se, cf §5.4.4, have been identified. It should be realized that 10^{20} y is about 10 billion times larger than the estimated age of the universe.

Radioactive decay involves a transition from a definite quantum state of the original nuclide to a definite quantum state of the product nuclide. The energy difference between the two quantum levels involved in the transition corresponds to the *decay energy*. This decay energy appears in the form of electromagnetic radiation and as the kinetic energy of the products.

The mode of radioactive decay is dependent upon the particular nuclide involved. We have seen in Ch. 1 that radioactive decay can be characterized by α-, β-, and γ-radiation. *Alpha-decay* is the emission of helium nuclei. *Beta-decay* is the creation and emission of either electrons or positrons, or the process of electron capture. *Gamma-decay* is the emission of electromagnetic radiation where the transition occurs between energy levels of the same nucleus. An additional mode of radioactive decay is that of *internal conversion* in which a nucleus loses its energy by interaction of the nuclear field with that of the orbital electrons, causing expulsion of an electron and ionization of the atom instead of γ-ray emission. A mode of radioactive decay which is observed only in the heaviest nuclei is that of *spontaneous fission* in which the nucleus dissociates spontaneously into two roughly equal parts. This fission is accompanied by the emission of electromagnetic radiation and of neutrons. In the last decades also some unusual decay modes have been observed for nuclides very far from the stability line, namely *neutron emission* and *proton emission*. A few very rare decay modes like ^{12}C-emission have also been observed.

In the following, for convenience, we sometimes use an abbreviated form for decay reactions, as illustrated for the ^{238}U decay chain described in §1.3:

$$^{238}\mathrm{U}(\alpha)\,^{234}\mathrm{Th}(\beta^-)\,^{234}\mathrm{Pa}(\beta^-)\,^{234}\mathrm{U}(\alpha),\ \text{etc.,} \tag{5.1}$$

or, if half-lives and details are of importance:

$$^{238}\mathrm{U}(\alpha,\ 4.5\times10^9\ \mathrm{y})^{234}\mathrm{Th}(\beta^-,\ 24\ \mathrm{d})^{234\mathrm{m}}\,\mathrm{Pa}(\beta^-,\ \mathrm{IT},\ 1.17\ \mathrm{min})$$
$$^{234}\mathrm{Pa}(\beta^-,\ 6.7\ \mathrm{h})^{234}\mathrm{U}(\alpha,\ 2.5\times10^5\ \mathrm{y}),\ \text{etc.} \tag{5.2}$$

In the following chapter we discuss the energetics of the decay processes based on nuclear binding energy considerations and simple mechanics, then we consider the kinetics of the processes. In Ch. 6, where the internal properties of the nuclei are studied, the explanations of many of the phenomena discussed in this chapter are presented in terms of simple quantum mechanical rules.

5.2. CONSERVATION LAWS

In radioactive decay – as well as in other nuclear reactions – a number of *conservation laws* must be fulfilled. Such laws place stringent limitations on the events which may occur.

Consider the reaction

$$X_1 + X_2 \rightarrow X_3 + X_4 \tag{5.3}$$

where X represents any nuclear or elementary particle. In induced nuclear reactions X_1 may be the bombarding particle (e.g. a ^{4}He atom in a beam of α-particles) and X_2 the target atom (e.g. ^{14}N atoms), and X_3 and X_4 the products formed (e.g. ^{1}H and ^{17}O).

Sometimes only one product is formed, sometimes more than two. In radioactive decay several products are formed; reaction (5.3) is then better written $X_1 \rightarrow X_2 + X_3$. For generality, however, we discuss the conservation laws for the case (5.3).

For the general reaction (5.3):

(a) The *total energy* of the system must be constant, i.e.

$$E_1 + E_2 = E_3 + E_4 \tag{5.4}$$

where E includes all energy forms: mass energy (§4.3), kinetic energy, electrostatic energy, etc.

(b) The *linear momentum*

$$p = mv \tag{5.5}$$

must be conserved in the system, and thus

$$p_1 + p_2 = p_3 + p_4 \quad (5.6)$$

The connection between kinetic energy E_{kin} and linear momentum is given by the relation

$$E_{kin} = p^2/(2m) \quad (5.7)$$

(c) The *total charge* (protons and electrons) of the system must be constant, i.e.

$$Z_1 + Z_2 = Z_3 + Z_4 \quad (5.8)$$

where the charge is in electron units.

(d) The *mass number* (number of nucleons) in the system must be constant, i.e.

$$A_1 + A_2 = A_3 + A_4 \quad (5.9)$$

(e) The *total angular momentum* p_I of the system must be conserved, i.e.

$$(p_I)_1 + (p_I)_2 = (p_I)_3 + (p_I)_4 \quad (5.10)$$

Since there exist two types of angular momentum, one caused by orbital movement of the individual nucleons and the other due to the intrinsic spin of the nucleons (internal angular momentum), a more practical formulation of (5.10) is

$$\Delta I = I_3 + I_4 - I_1 - I_2 \quad (5.11)$$

where I is the (total) *nuclear spin quantum number.* The quantum rule is

$$\Delta I = 0, \pm 1, \pm 2, \pm 3, \ldots \quad (5.12)$$

i.e. the change of nuclear spin in a reaction must have an integral value.

The three first laws are general in classical physics; the last two refer particularly to nuclear reactions. In Ch. 6 and 11 other conservation laws are discussed for nuclear reactions, but these are less important in radioactive decay.

5.3. ALPHA DECAY

5.3.1. Detection

Alpha particles cause extensive ionization in matter, as discussed in more detail in Ch 8. If the particles are allowed to pass into a gas, the electrons released by the ionization can be collected on a positive electrode to produce a pulse or current. *Ionization chambers* and *proportional counters* are instruments of this kind, which permit the individual counting of each α-particle emitted by a sample. Alpha particles interacting with matter may also cause molecular excitation, which can result in fluorescence. This fluorescence – or *scintillation* – allowed the first observation of individual nuclear particles. The ionization in semiconductors caused by α-particles is now the most common means of detection, see §9.4.

5.3.2. Decay energy

Alpha decay is observed for the elements heavier than lead and for a few nuclei as light as the lanthanide elements. However, ^{8}Be can also undergo alpha decay, i.e. daughter and alpha particle are equal. Alpha decay can be written symbolically as

$$ {}^{A}_{Z}\mathrm{X} \rightarrow {}^{A-4}_{Z-2}\mathrm{X} + {}^{4}_{2}\mathrm{He} \tag{5.13} $$

We use X to indicate any element defined by its nuclear charge, Z and Z-2 in this equation. Examples are given in Ch. 1, and can be found e.g. in the natural radioactive decay series, see §13.5.2.

The decay energy can be calculated from the known atomic masses, because the binding energy released (spontaneous decay processes must be exoergic) corresponds to a disappearance of mass, cf. eqns. (4.5) and (4.10). This energy is also called the *Q-value of the reaction*

$$ Q\,(\mathrm{MeV}) = -931.5\,\Delta M\,(\mathrm{u}) \tag{5.14} $$

For α-decay we can define the Q-value as

$$ Q_{\alpha} = -931.5\left(M_{Z-2}^{A-4} + M_{\mathrm{He}} - M_{Z}^{A}\right) \tag{5.15} $$

We always write the products minus the reactants within the parenthesis. A decrease in total mass in α-decay means a release of energy. The minus sign before the constant 931.5 MeV/u, cf Ch. 3, is necessary to make Q positive for spontaneous decay.

Consider the following example. For the decay reaction $^{238}U \rightarrow {}^{234}Th + {}^{4}He$, the mass values for ^{238}U and ^{4}He are found in Table 3.1; for ^{234}Th it is 234.043 594 u. Thus we obtain $Q_{\alpha} = -931.5 \times (234.043\,594 + 4.002\,603 - 238.050\,7785) = 4.274$ MeV.

If the products are formed in their ground states, which is common for α-decay, the total decay energy is partitioned into the kinetic energies of the daughter nucleus (E_{Z-2}^{A-4}) and the helium nucleus (E_{α}):

$$ Q_{\alpha} = E_{Z-2}^{A-4} + E_{\alpha} \tag{5.16} $$

Because of conservation of energy (5.4) and momentum (5.6)

$$ E_{Z-2} = Q_{\alpha} M_{\alpha} / M_{Z}^{A} \tag{5.17} $$

and

$$ E_{\alpha} = Q_{\alpha}\, M_{Z-2}^{A-4} / M_{Z}^{A} \tag{5.18} $$

From these equations we can calculate the kinetic energy of the ^{234}Th daughter to be 0.072 MeV, while that of the α-particle is 4.202 MeV. Because of the large mass difference between the α-emitting nucleus and the helium atom, almost all of the energy is carried away with the α-particle.

Although the kinetic energy of the daughter nucleus is small in comparison with that of the α-particle, it is large (72 000 eV) in comparison with chemical binding energies (<5 eV). Thus the recoiling daughter easily breaks all chemical bonds by which it is bound to other atoms.

In 1904 it was observed by H. Brooks that measurements on ^{218}Po (RaA), obtained from radon, led to a contamination of the detection chamber by ^{214}Pb (RaB) and ^{214}Bi (RaC). This was explained by Rutherford as being due to daughter recoil in the α-decay of ^{218}Po in the sequence (written symbolically):

$$^{222}\mathrm{Rn}(\alpha, 3.8\ \mathrm{d})\ ^{218}\mathrm{Po}(\alpha, 3.05\ \mathrm{min})\ ^{214}\mathrm{Pb}(\beta^-, 27\ \mathrm{min})\ ^{214}\mathrm{Bi}(\beta^-, 20\ \mathrm{min})\ldots \quad (5.19)$$

This recoil led to ejection of ^{214}Pb into the wall of the instrument. The use of the *recoil* of the daughter to effect its separation was employed by O. Hahn beginning in 1909 and played a central role in elucidating the different natural radioactive decay chains.

The recoil may affect such chemical properties as the solubility or dissolution rate of compounds. For example the dissolution of uranium from uranium rich minerals is considerably higher than one would expect from laboratory solubility data because α and U-atom recoil have moved U-atoms away from their normal sites in the mineral.

Alpha-decay energies are most precisely measured in magnetic spectrometers. From (3.6) and (3.11) it is calculated that

$$E_\alpha = 2\mathbf{e}^2 B^2 r^2 / m_{\mathrm{He}} \quad (5.20)$$

From knowledge of the values of **e**, m_{He}, B, and r, E_α can be calculated. A more common technique is to use *semiconductor detectors* combined with pulse height analyzers ("α-spectrometers", Ch. 9).

5.4. BETA DECAY

5.4.1. Detection

Energetic electrons cause ionization and molecular excitation in matter, although the effect is weaker and more difficult to detect than for α-particles. As a result the effect must be amplified for counting individual β-particles. Ionization is used in *proportional* and *Geiger counters*. Scintillation counting can also be used with various detector systems (§9.5).

5.4.2. The β-decay process

The radioactive decay processes which are designated by the general name of β-decay include electron emission (β^- or ${}_{-1}^{0}e$), positron emission (β^+ or ${}_{+1}^{0}e$) and electron capture (EC). If we use the β-decay of ^{137}Cs as an example, we can write

$$^{137}\text{Cs} \rightarrow {}^{137m}\text{Ba} + \beta^- \tag{5.21}$$

This β-decay must occur between discrete quantum levels of the parent nuclide ^{137}Cs and the daughter nuclide ^{137m}Ba.

The quantum levels of nuclei are characterized by several quantum numbers, an important one being the *nuclear spin*. The spin value for the ^{137}Cs ground state level is 7/2, while that of ^{137m}Ba is 11/2. The electron emitted is an elementary particle of spin 1/2. In nuclear reactions the nuclear angular momentum must be conserved (5.10), which means that in radioactive decay processes the difference in total spin between reactant and products must be an integral value (5.12). Inspection of our example shows that this conservation of spin rule is violated if the reaction is complete as we have written it. The sum of the spin of the ^{137m}Ba and of the electron is 11/2 + 1/2 or 6, while that of the ^{137}Cs is 7/2. Therefore, the change in spin (ΔI) in the process would seem to be 5/2 spin units. In as much as this is a non-integral value, it violates the rule for conservation of angular momentum. Before accounting for this discrepancy let us consider another aspect of β-decay which seems unusual. Figure 5.1 shows the β-particle spectrum of

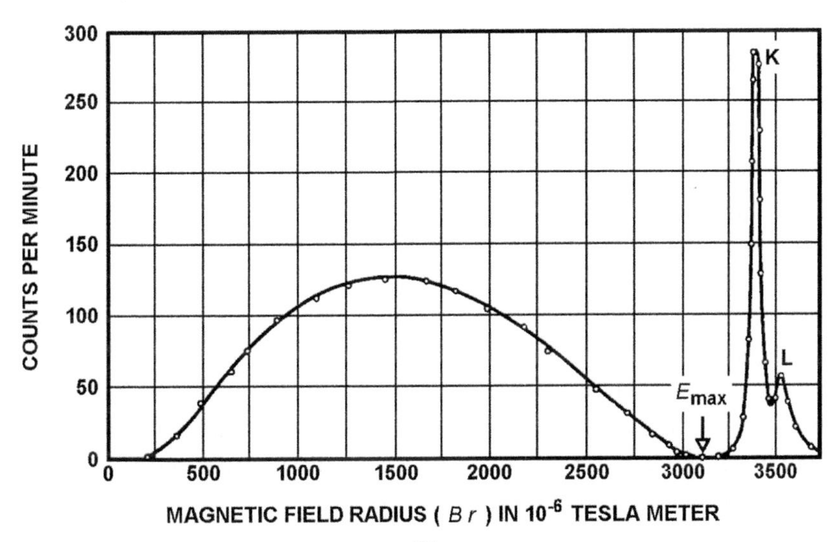

Figure 5.1 Spectrum for electrons emitted by ^{137}Cs as observed with a magnetic spectrometer. $Br \propto \sqrt{E}$, cf. Eqn (5.22). *(From W. Gentner, H. Maier-Leibnitz, and W. Bothe.)*

^{137}Cs as obtained by a magnetic spectrometer. The β-particle energy is calculated by the relation

$$E_\beta = \mathbf{e}^2 B^2 r^2/(2\, m_e) \tag{5.22}$$

where m_e is the electron relativistic mass, cf (2.9). The spectrum shows the number of β-particles as a function of *Br*, which is proportional to $\sqrt{E_\beta}$ through (5.22). We observe a continuous distribution of energies. This seems to disagree with our earlier statement that decay occurs by change of one nucleus in a definite energy state to another nucleus also in a definite energy state. The two sharp peaks designated K and L at the high energy end of the spectrum are not related to the beta spectrum itself and are discussed later in the chapter (§5.5.2).

5.4.3. The neutrino

This problem of "wrong" spin change and the continuous "non-quantized" spectrum led W. Pauli to the assumption that β-decay involves emission of still another particle which has been named the *neutrino* and given the symbol ν, cf. §2.5. The neutrino has a spin value of 1/2, an electric charge of 0, and a mass ≈ 0. It is therefore somewhat similar to the photon, which has neither mass, electric charge nor spin. However, while the photon readily interacts with matter, the neutrino does not. In fact the interaction is so unlikely that a neutrino has a very high probability of passing through the entire earth without reacting.

The spin attributed to the neutrino allows conservation of angular momentum; in our example, the total spin of the products would be 11/2 + 1/2 + 1/2 or 13/2, and when the spin of ^{137}Cs, 7/2, is subtracted from this the result is 6/2 = 3 which is an acceptable integral value. Thus the decay reaction above is incomplete and must be written

$$^{137}\mathrm{Cs} \rightarrow {}^{137m}\mathrm{Ba} + \beta^- + \overline{\nu} \tag{5.23}$$

Notice we have replaced ν by $\overline{\nu}$, which is the designation of the *antineutrino*. Antineutrinos are emitted in electron decay while "regular" neutrinos ν in positron decay. We can here consider the particles identical; cf. §2.3. Because of the extremely low probability of interaction or neutrinos with matter, they are unfortunately often omitted in writing β-decay reactions.

The neutrino theory also explains the energy spectrum in β-decay. However, this necessitates the introduction of another important nuclear concept, that of *relativistic mass* and *rest mass*. In 1901 S. G. Kaufmann showed in experiments that the mass of an electron m seemed to increase when its velocity v approached that of the speed of light $\mathbf{c}$. Later, it was found that this increase followed an expression

$$m = m^\circ \left(1 - v^2/\mathbf{c}^2\right)^{-1/2} \tag{5.24}$$

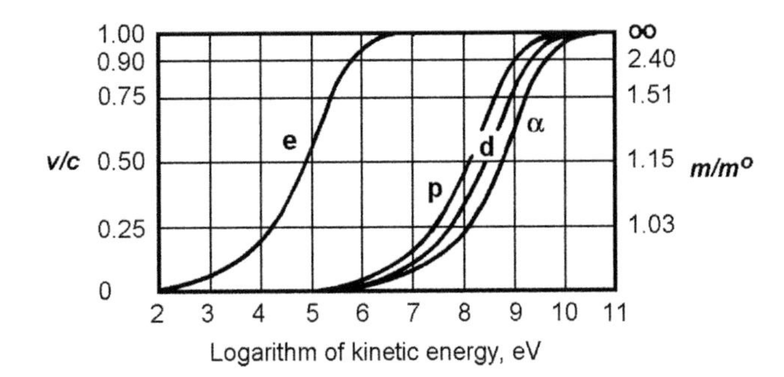

Figure 5.2 Relativistic masses m for some common nuclear particles, divided by their rest masses m^0, as a function of the kinetic energy of the particle.

based on H. Lorentz's studies of the relation between distance, speed of light, and time. m^o is the rest mass of the particle (at velocity $v = 0$), while m is referred to as the relativistic mass. This relation is valid for any moving object, macroscopic or microscopic, whether it is a "particle", a "wave-packet" or a space rocket. Figure 5.2 shows the quotient $v/\mathbf{c}$ as a function of the kinetic energy of the particle, E_{kin} (in some literature the quotient $v/\mathbf{c}$ is designated as β but for clarity compared to the radioactive decay we have not used β).

If the parentheses in (5.24) is expanded by means of the binomial theorem of algebra, it approximates to

$$m = m^o + \tfrac{1}{2}m^o v^2/\mathbf{c}^2 \tag{5.25}$$

The last term is approximately the kinetic energy of the particle (3.6) divided by $\mathbf{c}^2$, and thus

$$m \approx m^o + E_{\text{kin}}/\mathbf{c}^2 \tag{5.26}$$

The increase in mass, $\Delta m = m - m^o$, is due to the kinetic energy of the particle,

$$E_{\text{kin}} = \Delta m\mathbf{c}^2 \tag{5.27}$$

This was generalized by A. Einstein in the special theory of relativity, leading (after more detailed calculations) to the well known mass-energy relationship

$$E = m\mathbf{c}^2 \tag{5.28}$$

which we already have applied in the discussion of the nuclear binding energy (4.3).

When a neutrino is ejected from the nucleus it carries away energy of kinetic nature. Thus, according to (5.26) the neutrino has a relativistic mass > 0, and obviously also a momentum $p = mv$. Recoil studies of β-decay have proven this to be true.

In order to correctly apply (5.22) for the calculation of the β-decay energy, the relativistic electron mass must be used; as is seen from Figure 5.2, already at 0.1 MeV, the relativistic mass of the electron is 15% larger than the rest mass $\mathbf{m}_e^o$. (In the following the rest masses of the electron, neutron, etc., will be denoted simply as $\mathbf{m}_e$, $\mathbf{m}_n$, etc.; capital **M** if in universal mass units, u)

The energy released in β-decay is distributed between the neutrino, the electron, and the recoil of the daughter nucleus. This latter will be much smaller than the first two and can be neglected in a first approximation (§5.4.8). Therefore, the total β-decay energy can be considered to be distributed between the neutrino and the electron. For the decay $^{137}Cs \rightarrow {}^{137m}Ba$ it can be shown that the total decay energy Q_β is 0.514 MeV. This is also termed E_{max}. The neutrino energy spectrum is the complement of the β-particle energy spectrum – summing up to E_{max}. If the energy of the electron is 0.400 MeV, that of the neutrino is 0.114 MeV. If the electron energy is 0.114 MeV, the neutrino energy is 0.400 MeV.

In β^--decay the average value of the β^--particle energy is about 0.3 E_{max} due to attraction by the nuclear charge, and approximately given by the empirical equation (5.29).

$$E_{average} \approx 0.33 E_{max} \left(1 - \frac{\sqrt{Z}}{50}\right)\left(1 + \frac{\sqrt{E_{max}}}{4}\right) \quad \text{for } \beta^- \tag{5.29}$$

In positron emission, the average energy of the β^+-particle is higher and approximately 0.4 E_{max} as shown by the empirical equation (5.30).

$$E_{average} \approx 0.33 E_{max} \left(1 + \frac{\sqrt{E_{max}}}{4}\right) \quad \text{for } \beta^+ \tag{5.30}$$

With a few exceptions, equations (5.29) and (5.30) reproduce measured $E_{average}$ values to within 5%.

The assumption that the neutrino has a zero rest mass has been questioned by experimentalists and theorists. A number of experiments at the Kamiokande facility have established an upper limit of the rest mass of the electron neutrino as ~0.07 eV. The implications of a finite rest mass are broad as the nature of the neutrino and the theory of beta decay is involved.

5.4.4. Double beta decay

The rather unusual (and very slow) $\beta^-\beta^-$-decay mode is energetically possible for several even-even nuclei, and kinetically possible to observe in those cases where the separating odd-odd nucleus of higher energy prevents normal β^--decay, see Fig. 4.6 V. It has recently been observed for some such cases, e.g.

$$^{82}_{34}Se \rightarrow {}^{82}_{36}Kr + 2\beta^- + 2\overline{\nu} \tag{5.31}$$

The half-life of about 1.7 × 10^{20} y can be observed because small amounts of the noble gas Kr can be physically isolated from large amounts of ^{82}Se (9% natural abundance), and then measured. Studies of ββ-decay are of importance for evaluation of neutrino properties. However, neutrinoless ββ-decay may have been observed in the decay of $^{76}_{32}Ge$.

5.4.5. β^--Decay

This process can be written symbolically as follows:

$$^{A}_{Z}X \rightarrow {}^{A}_{Z+1}X + {}^{0}_{-1}\beta + \bar{\nu} \tag{5.32}$$

However, if we take the electrons into account, the neutral parent atom has Z orbital electrons, while the daughter atom, with a nuclear charge of $Z + 1$, must capture an electron from the surroundings, in order to become neutral:

$$^{A}_{Z+1}X^{+} + e^{-} \rightarrow {}^{A}_{Z+1}X \tag{5.33}$$

Moreover, since the beta particle emitted provides an electron to the surroundings, the total electron balance remains constant. As a result, in the calculation of the decay energy it is not necessary to include the mass of the emitted β-particle as the use of the mass of the neutral daughter atom includes the extra electron mass. The equation for calculating the Q-value β^- decay is thus:

$$Q_{\beta^-} = -931.5\,(M_{Z+1} - M_Z) \tag{5.34}$$

As an example we can take the decay of a free neutron in a vacuum; it transforms spontaneously with a half-life of 10.6 min to a proton through beta decay.

$$^{1}_{0}n \rightarrow {}^{1}_{1}H + {}^{0}_{-1}e^{-} \tag{5.35}$$

The Q-value for this reaction is

$$Q_{\beta^-} = -931.5\,(1.007\,825 - 1.008\,665) = 0.782\text{ MeV} \tag{5.36}$$

5.4.6. Positron decay

Positron decay can be written symbolically as

$$^{A}_{Z}X \rightarrow {}^{A}_{Z-1}X^{-} + {}^{0}_{+1}\beta + \nu \rightarrow {}^{A}_{Z-1}X + {}^{0}_{-1}e^{-} + {}^{0}_{+1}\beta + \nu \tag{5.37}$$

Here we must consider the net atomic charges. The daughter nucleus has an atomic number one less than the parent. This means that there will be one electron mass associated with the change in atomic number. Moreover, an electron mass must also be included for the positive electron emitted. When $^{22}_{11}Na$ decays to $^{22}_{10}Ne$, there are 11 electrons included in the ^{22}Na atomic mass but only 10 in the ^{22}Ne atomic mass. Consequently, an extra electron mass must be added on the product side in addition to the electron mass associated with the positron particle. The calculation of the Q-value must therefore include two electron masses beyond that of the neutral atoms of the parent and daughter

$$Q_{\beta^+} = -931.5\left(M_{Z-1}^{A} + 2\,\mathbf{M}_e - M_Z^A\right) \quad (5.38)$$

Each electron mass has an energy equivalent to 0.511 MeV, since 931.5 × 0.000 549 = 0.511.

Consider the calculation of the Q-value for the reaction

$$^{13}_{7}N \rightarrow {}^{13}_{6}C + \beta^+ \quad (5.39)$$

For this reaction we have

$$Q_{\beta^+} = -931.5\left(13.003\,355 - 13.005\,739\right) - 2 \times 0.511 = 1.20\ \text{MeV} \quad (5.40)$$

5.4.7. Electron capture

The EC decay process can be written symbolically

$$^{A}_{Z}X \overset{EC}{\rightarrow} {}_{Z-1}^{\ A}X + \nu \quad (5.41)$$

The captured electron comes from one of the inner orbitals of the atom. Depending on the electron shell from which the electron originates, the process is sometimes referred to as K-capture, L-capture, etc. The probability for the capture of an electron from the K-shell is several times greater than that for the capture of an electron from the L-shell, since the wave function of K-electrons is substantially larger at the nucleus than that of L-electrons. Similarly, the probability of capture of electrons in higher order shells decreases with the quantum number of the electron shell.

The calculation of the decay energy in electron capture follows the equation

$$Q_{EC} = -931.5\left(M_{Z-1}^{A} - M_Z^A\right) \quad (5.42)$$

Note that like the case of the β^- decay, it is not necessary to add or subtract electron masses in the calculation of the Q-value in EC. An example of EC is the decay of ^7Be to ^7Li for which it is possible to calculate that the Q-value is 0.861 MeV. This reaction is somewhat exceptional since for neutron deficient nuclei with values of Z below 30, positron emission is the normal mode of decay. Electron capture is the predominant mode of decay for neutron deficient nuclei whose atomic number is greater than 80. The two processes compete to differing degrees for the nuclei between atomic numbers 30 and 80. Electron capture is observed through the emission of electrons from secondary reactions occurring in the electron shell because of the elemental change (see §5.9).

5.4.8. Daughter recoil

If the β-particle and the neutrino are emitted with the same momentum but in opposite direction, the daughter nucleus experiences no recoil. On the other hand, if they are both emitted in the same direction, or if all the energy is carried away with one of the particles, the daughter experiences maximum recoil. The daughter therefore recoils with kinetic energies from zero up to a maximum value (when the β-particle is emitted with maximum energy). We can therefore write

$$Q_\beta = E_d + E_{max} \tag{5.43}$$

where E_d is the recoil energy of the daughter nucleus. From the laws of conservation of energy and momentum, and taking the relativistic mass changes of the electron into account, one finds that the daughter recoil energy is

$$E_d = \mathbf{m}_e E_{max}/m_d + E_{max}^2/\left(2\, m_d\, \mathbf{c}^2\right) \tag{5.44}$$

The recoil energy is usually ~100 eV, which still is sufficient for causing atomic rearrangements in surrounding molecules. In the decay of ^{14}C (to ^{14}N), E_{max} is 0.155 MeV, which gives $E_d = 7$ eV. However, by labelling ethane, $^{14}CH_3{}^{14}CH_3$, with ^{14}C in both C positions, it was found that $^{14}CH_3NH_2$ was formed in 50% of the cases when one of the ^{14}C atoms in ethane had decayed, although the C≡N bond strength is only 2.1 eV; most of the decays occur with less than the maximum recoil energy, which also can be averaged over the whole molecule. The small recoil also explains why decay reactions like

$$^{127}TeO_3{}^{2-} \rightarrow {}^{127}IO_3{}^{-} + \beta^- \tag{5.45}$$

and

$$^{52}MnO_4{}^{-} \rightarrow {}^{52}CrO_4{}^{2-} + \beta^+ \tag{5.46}$$

are possible, even when E_d is tens of electron volts. However, secondary effects tend to cause the chemical bond to break following radioactive decay (see §5.9).

5.5. GAMMA EMISSION AND INTERNAL CONVERSION

The α- and β-decay may leave the daughter nucleus in an excited state. This excitation energy is removed either by γ-ray emission or by a process called *internal conversion*.

5.5.1. Gamma emission

The α-emission spectrum of ^{212}Bi is shown in Figure 5.3. It is seen that the majority of the α-particles have an energy of 6.04 MeV, but a considerable fraction (~30%) of the α-particles have higher or lower energies. This can be understood if we assume that the

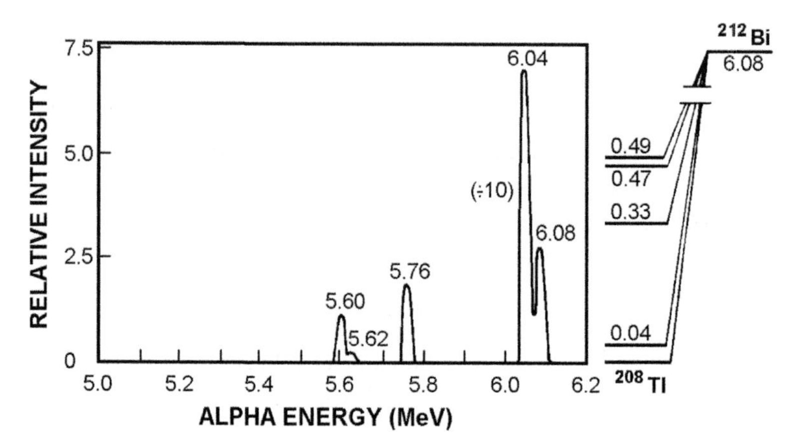

Figure 5.3 Alpha energy spectrum from $^{212}Bi \rightarrow {}^{208}Tl$. (According to E. B. Paul.)

decay of parent ^{212}Bi leads to excited levels of daughter ^{208}Tl. This idea is supported by measurements showing the emission of γ-rays of energies which exactly correspond to the difference between the highest α-energy 6.08 MeV, and the lower ones. For example, an ~0.32 MeV γ accounts for the 5.76 MeV α (6.08 − 5.76 = 0.32). The excited levels of ^{208}Tl are indicated in the insert in Figure 5.3.

Gamma rays produce very low density ionization in gases so they are not usually counted by ionization, proportional, or Geiger counters. However, the fluorescence produced in crystals such as sodium iodide make scintillation counting of γ-rays efficient. Gamma ray spectra can be measured with very high precision using semiconductor detectors, cf §9.4. The higher the density of the material the more efficient the semiconductor material is to absorb the energy. Figure 5.4 shows such a spectrum for the decay of various excited states of ^{197}Au. The $7/2^+$, etc. symbols are explained in §5.8.

In the great majority of cases the emission of the γ-ray occurs immediately after α- or β-decay, i.e. within $\leq 10^{-12}$ s, but in some instances the nucleus may remain in the higher energy state for a measurable length of time. The longer-lived exited nuclei are called *isomers,* for definition of life time of isomers cf Ch. 1. An example is ^{60m}Co, which decays with a half-life of 10.5 min to the ground state of ^{60}Co. The decay is referred to as *isomeric transition.*

The decay energy in γ-emission is distributed between the γ-ray quantum (E_γ) and the kinetic energy of the recoiling product nucleus (E_d). We can therefore write

$$Q_\gamma = E_d + E_\gamma \tag{5.47}$$

The distribution of energy between the γ-ray and the recoiling daughter, according to

$$E_d = E_\gamma^2 / \left(2\, m_d\, \mathbf{c}^2\right) \tag{5.48}$$

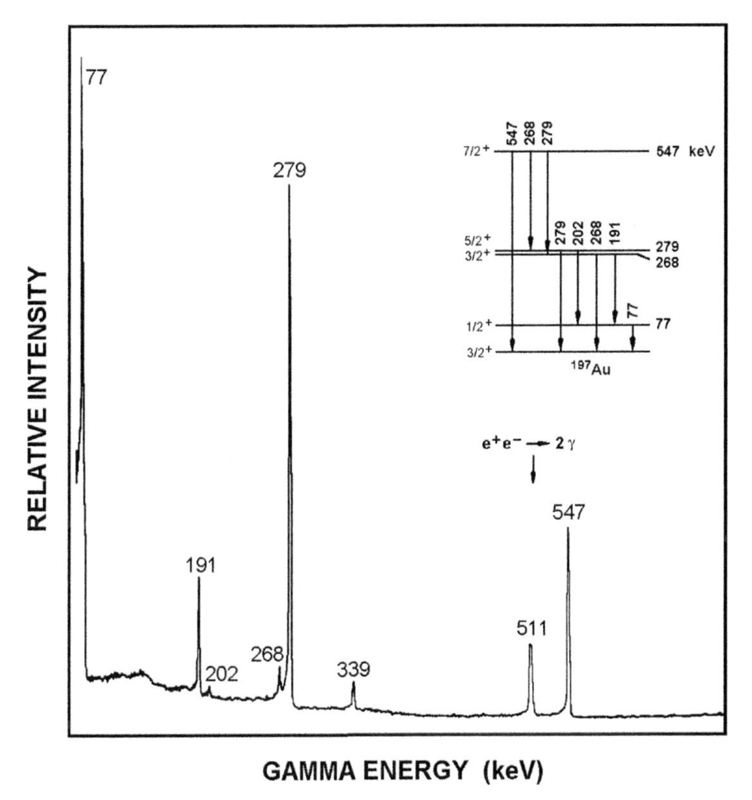

Figure 5.4 Gamma spectrum and decays scheme for ^{197}Au, produced through Coulomb excitation of gold by a 12 MeV ^{4}He beam. (According to M. G. Bowler.)

shows that $E_d < 0.1\%$ of E_γ. The amount of kinetic energy of the recoiling nuclide is therefore so trivial that it may be neglected when only the γ-ray energy is considered; cf. exercise 5.4. Even though in some cases the recoil energy from the gamma decay may be enough to break chemical bonds this is usually unimportant since the gamma follows a beta or alpha decay. Any bonds existing prior to the original decay are already been broken. However, for some longer lived meta stable nuclides, e.g. ^{99m}Tc new bonds could have been formed after the original decay and they may be broken by the gamma recoil.

5.5.2. Internal conversion

Gamma rays can interact with the orbital electrons of other atoms, so that the latter are expelled from that atom with a certain kinetic energy (see Ch. 7). A different process, called *internal conversion*, can occur *within* the atom undergoing radioactive decay. Because the wave function of an orbital electron may overlap that of the excited nucleus, the excitation energy of the nucleus can be transferred directly to the orbital electron (without the involvement of a γ-ray), which escapes from the atom with a certain kinetic energy E_e. No γ-ray is emitted in an internal conversion process; it is an alternate mode to γ-ray emission of de-excitation of nuclei.

Internal conversion can be represented symbolically as

$$^{Am}_{Z}X \rightarrow {}^{A}_{Z}X^{+} + {}^{0}_{-1}e^{-} \rightarrow {}^{A}_{Z}X \tag{5.49}$$

where we again have to consider the net atomic charge.

Part of the nuclear excitation energy is required to overcome the binding energy E_{be}, of the electron in its electronic orbital.[1] The remaining excitation energy is distributed between the recoiling daughter nucleus and the ejected electron E_e. The relationship is given by the equation

$$Q_\gamma - E_{be} = E_d + E_e \tag{5.50}$$

The ejected electron, known as the *conversion electron*, normally originates from an inner orbital, since their wave functions have greater overlap with the nucleus. It is to be noted that the conversion electrons are mono-energetic, i.e. there is no neutrino emission associated with the ejected electron.

In as much as the binding energies of the atomic orbitals are different, the values of E_e reflect the differences in the electronic binding energies. In Figure 5.1 two sharp peaks are observed just beyond E_{max}. The first peak, designated as K, is due to conversion electrons originating in the K atomic shell, while the peak labeled L is due to conversion electrons originating in the L atomic shell. Both of these groups of conversion electrons arise from the decay of ^{137m}Ba. Figure 5.5(f) shows schematically the decay process of $^{137}Cs \rightarrow {}^{137}Ba + \beta^-$ for the principal decay path; IT is an abbreviation for isomeric transition. The decay of ^{137m}Ba proceeds both by emission of a 0.66 MeV γ-ray and by the competitive process of internal conversion. The ratio between the number of conversion electrons and the number of γ-rays emitted in this competition is called the *conversion coefficient*. The amount of internal conversion is not indicated in simplified decay schemes like Figure 5.5.

If we denote the conversion coefficient as α_i, it is equal to the ratio of K-electrons ejected (which we may denote with I_{eK}) to that of gamma quanta emitted (I_γ):

$$\alpha_K = I_{eK}/I_\gamma \tag{5.51}$$

Usually $\alpha_K < 0.1$. Also $\alpha_K > \alpha_L > \alpha_M$, etc. For ^{137m}Ba the ratio of K-electrons to L-electrons emitted is about 5 while the value of α_K is 0.094.

It can be shown that the energy of the recoiling nucleus (E_d, (5.48)) is much smaller than the kinetic energy of the ejected electron E_e and may be ignored. The mathematical expression to use is (5.44).

[1] Electron-binding energies are tabulated in standard physics tables.

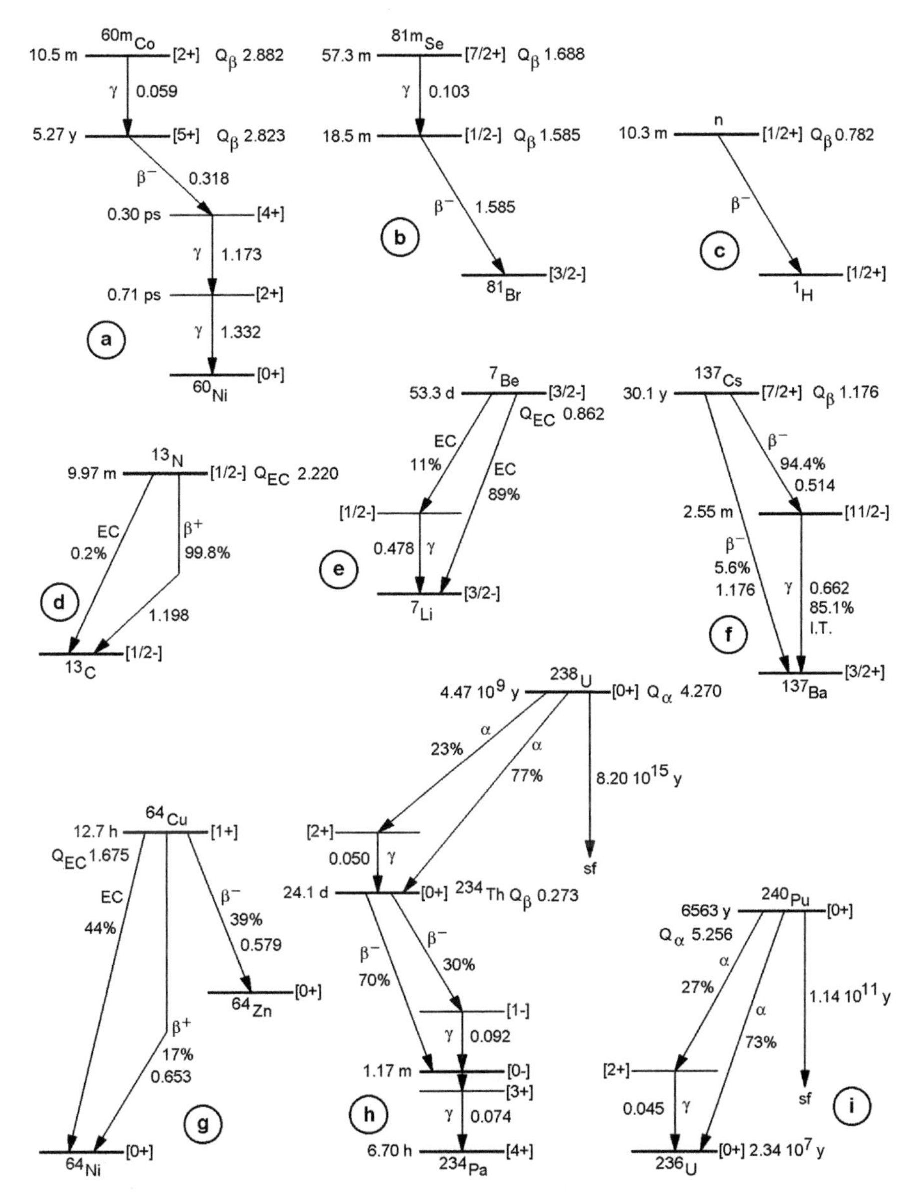

Figure 5.5 Examples of different decay schemes; energies are in MeV. The schemes are explained in the text.

A note on terminology. Consider Figure 5.5(f). Though the γ is emitted from an exited state of ^{137}Ba, in common language we nevertheless talks of "^{137}Cs-γ" (of 0.662 MeV), i.e. as if the γ was emitted from the ^{137}Cs nucleus. The quotation is to be interpreted as "^{137}Cs (decay through β^- followed by) γ(-emission)". A similar example is from the decay of ^{60}Co where two cascade gamma photons are emitted from the excited ^{60}Ni nucleus. These two characteristic gamma lines of 1173 keV and 1332 keV are always referred to as "cobalt gammas" whereas they are actually emitted by nickel.

5.6. SPONTANEOUS FISSION

As the nuclear charge increases to large values, nuclei become more unstable. This is reflected by decreasing half-lives for nuclei heavier than uranium. In 1940 K. Petrzak and G. Flerov found that ^{238}U in addition to α-decay also had a competing mode of radioactive decay termed *spontaneous fission*. In this mode two heavy fragments (*fission products*) are formed in addition to some neutrons. The reaction may be written

$$^{A}_{Z}\mathrm{X} \rightarrow \, ^{A1}_{Z1}\mathrm{X}_1 + \, ^{A2}_{Z2}\mathrm{X}_2 + \nu \mathrm{n} \tag{5.52}$$

where ν is the number of neutrons, usually 2–3. The half-life for spontaneous fission of ^{238}U is very long, about 8×10^{15} y. This means that about 70 fissions occur per second in 1 kg of ^{238}U, which can be compared with the simultaneous emission of 45×10^{9} α-particles.

With increasing Z, spontaneous fission becomes more common; i.e. the half-life for this decay process decreases. For $^{240}_{94}$Pu it is 1.2×10^{11} y; for $^{244}_{96}$Cm, 1.4×10^{7} y; for $^{252}_{98}$Cf, 66 y; and for $^{256}_{100}$Fm, 3×10^{-4} y. In fact, spontaneous fission becomes the dominating decay mode for the heaviest nuclei (see Fig. 4.3).

Spontaneous fission is in some ways similar to fission induced by bombardment with low energy neutrons (§11.7).

5.7. RARE MODES OF DECAY

Radioactive decay by *proton emission* is a very seldom observed decay mode for very neutron deficient nuclides because decay by β^{+} or EC normally has a very much shorter partial half-life (§5.14). Decay by p^{+} has been observed for ^{53m}Co (E_p 1.55 MeV, $t_{½}$ 0.25 s, ~1.5%). However, β^{+} decay sometimes leads to a proton-unstable excited state, which immediately ($<10^{-12}$ s) emits a proton. Several β^{+} emitters from ^{9}C to ^{41}Ti with $N = Z - 3$ have β^{+} delayed proton emission with half-lives in the range 10^{-3}–0.5 s. Also radioactive decay by simultaneous emission of two protons has been observed for a few proton rich nuclides, e.g. ^{16}Ne, $t_{½}$ ~10^{-20} s.

Among the very neutron rich nuclides, e.g. some fission products, β^{-} delayed neutron emission is observed. This decay mode is similar in nature to the β^{+} delayed p^{+} emission. Delayed n-emission is important for the safe operation of nuclear reactors, see Ch. 19.

Decay by emission of particles heavier than α, e.g. ^{12}C, ^{16}O, is energetically possible for some heavy α-emitters and has been observed in a few cases.

5.8. DECAY SCHEMES AND ISOTOPE CHARTS

Information on the mode of decay, the decay energy, and the half-life are included in the *nuclear decay scheme.* A number of simplified decay schemes are shown in Figure 5.5. Figure 4.8 explains a more detailed decay scheme, for $A = 99$.

Using Figure 4.8 as a guide, the decay schemes in Figure 5.5 are easily understood. Figure 5.5(c) shows the β^- decay of the neutron. The decay scheme of ^{137}Cs (Fig. 5.5(f)) differs somewhat from what we would expect from the curve in Figure 5.1. The reason is that in the electron spectrum of Figure 5.1, a small fraction of electrons (8%) emitted with an energy of 1.20 MeV could not be detected because of the insensitivity of the magnetic spectrometer used. It is common for nuclei to decay through different competing reactions, as in this case which involves different β-rays. If the higher energy β-decay had been as common as the lower one, we would have observed a mixed β-spectrum, as is indicated in Figure 5.6. Figure 5.5(a) shows the decay of ^{60}Co, and also its isomeric *precursor* ^{60m}Co. The β-decay is immediately followed by a cascade of two γ-rays. ^{64}Cu (Fig. 5.5(g)) decays through β^- (39%) and β^+ (17%) emission and electron capture (44%); this is referred to as *branched decay.* The vertical line in the angled arrow indicating the positron decay symbolizes the rest mass energy of the two electrons created, i.e. 1.02 MeV. Adding 1.02 MeV to 0.66 MeV gives 1.68 MeV, the Q-value for the decay from ^{64}Cu to ^{64}Ni. Figure 5.5(h) is a more complicated decay sequence for ^{238}U(α) ^{234}Th(β^-) ^{234}Pa. In the beginning of this chapter we pointed out that the decay of ^{238}U sometimes results in an excited state of the daughter ^{234}Th (in 23 out of

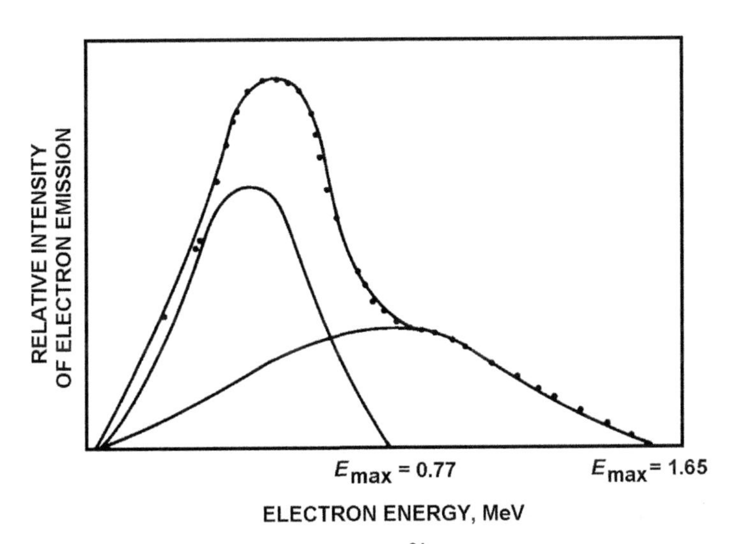

Figure 5.6 Beta spectrum for the positron decay of ^{84}Rb showing two positron groups, 0.77 MeV (11%) and 1.05 MeV (10%). (^{84}Rb also decays by β^- emission.) (Acc. to S. Eklund)

100 decays), although the excitation energy is comparatively small. Figure 5.5(i) shows how spontaneous fission competes with α-decay in ^{240}Pu. Instead of giving the percentage in the different decay branches, the half-life for that particular mode of decay may be given; conversion between half-lives and percentage is explained in §5.14.

"*Isotope charts*" can be considered as condensed nuclide tables. Figure 4.1 and Appendix C are such charts, though strongly abbreviated. Figure 5.7 shows the beginning of an ordinary chart (i.e. lower left corner of Figure 4.1 and App. C). The legend explains the information provided. Such nuclide charts are very useful for rapid scanning of ways to produce a certain nuclide and to follow its decay modes. Nuclide charts for $_{81}$Tl – $_{92}$U, and for $_{92}$U – $_{109}$Mt are shown in Figures 13.4 and 14.1, respectively.

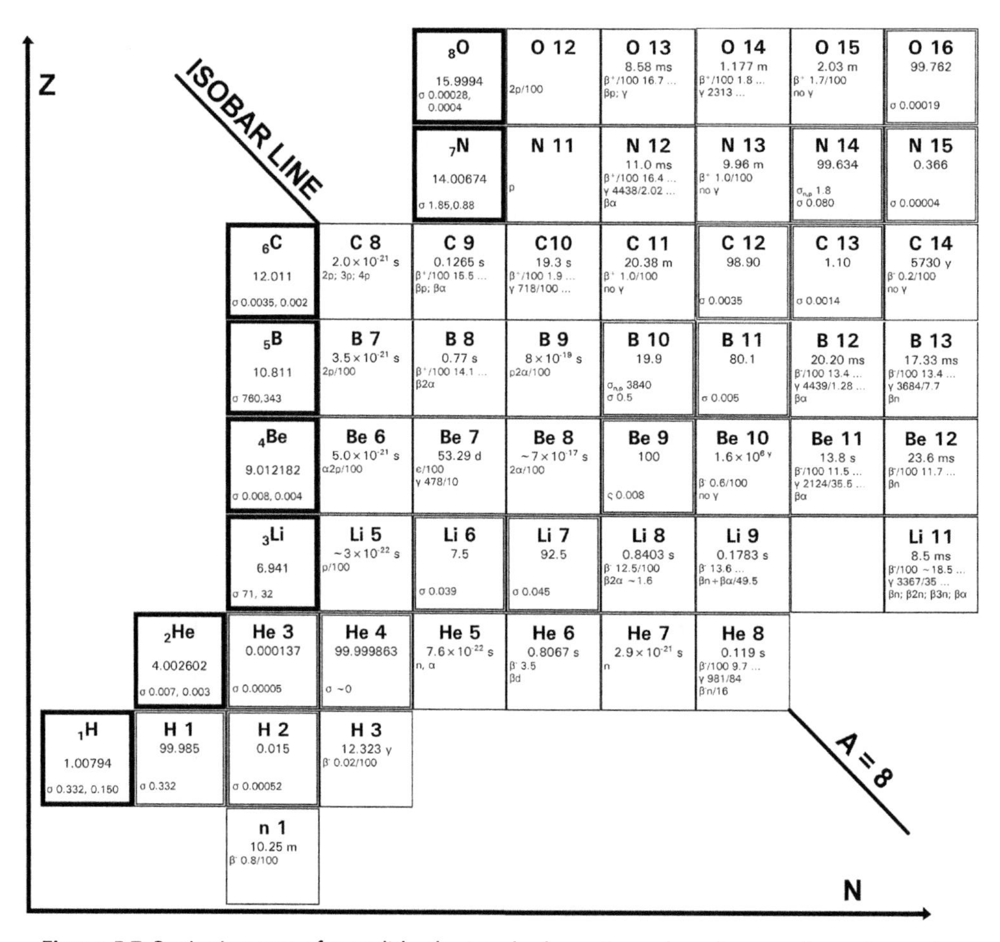

Figure 5.7 Beginning part of a nuclide chart and schematic nuclear decay and reaction paths.

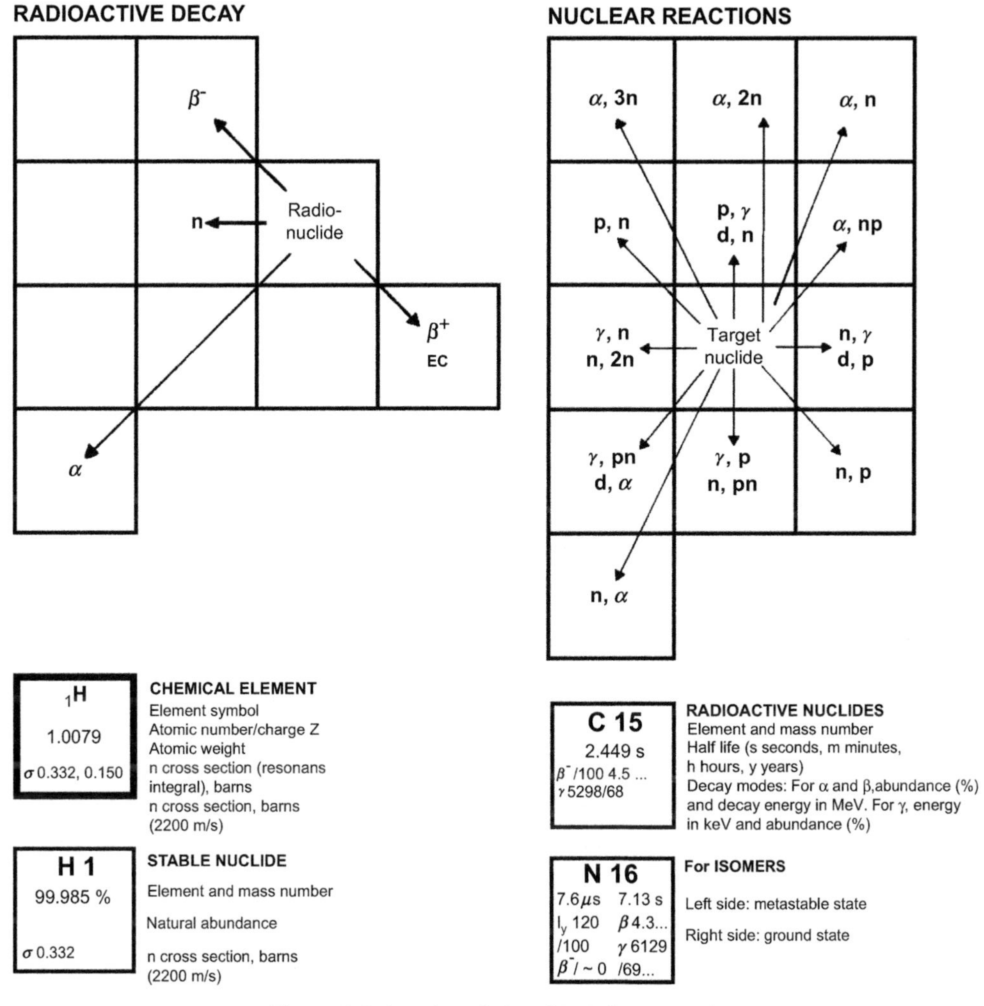

Figure 5.7 (*continued*). (see §5.20 for sources).

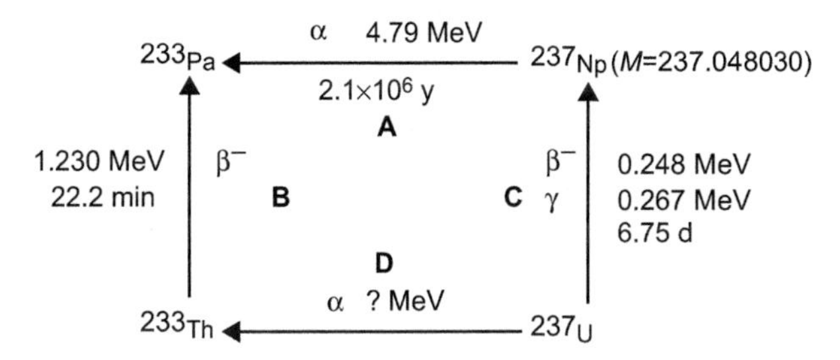

Figure 5.7.C (*continued*). Example of a closed cycle energy calculation.

5.9. SECONDARY PROCESSES IN THE ATOM

Once an electron is ejected from an atomic orbital due to internal conversion, electron capture, or some other process involved in radioactive decay, a vacancy is created in the electron shell which can be filled in several ways. Electrons from higher energy orbitals can occupy the vacancy. The difference in the binding energy of the two shells involved in the transition will be emitted from the atom as *X-rays*. This process is called *fluorescent radiation*.

If the difference in the binding energy for the transition is sufficient to exceed the binding energy of electrons in the L- or M-levels, emission of the energy as X-rays is not the predominant mode. Instead an *internal photoelectric* process can occur and the binding energy results in the emission of several low energy electrons which are called *Auger electrons*. Auger electrons are much lower in energy than the electron from the *nuclear internal conversion* process, since the difference in electronic binding energies is in the eV range compared to the energies in the nuclear conversion process which are in the MeV range. The atom may be left in a state of high ionization by Auger emission; positive charges of 10–20 have been observed. When such high charges are neutralized, the energy liberated is sufficient to break chemical bonds.

In isomeric decay the γ-energy is often so small that the daughter recoil is negligible. For example

$$^{80m}\mathrm{Br} \xrightarrow[4.4\ \mathrm{h}]{\gamma} {}^{80}\mathrm{Br} \qquad (5.53)$$

occur through the emission of a γ-ray of 0.049 MeV, giving the daughter a recoil energy of only 0.016 eV. Still the decay leads to the emission of ^{80}Br from ethyl bromide, when the parent compound is $C_2H_5\,{}^{80m}Br$, even though the bond strength is 2.3 eV. This is because the γ is highly converted and as the electron "hole" is filled, Auger emission occurs. Bromine ions from Br^+ to Br^{17+} have been observed through mass spectrometric analyses of the ethyl bromide gas phase.

5.10. CLOSED DECAY ENERGY CYCLES

The masses for many short-lived nuclei are unknown although their decay modes and energies have been determined. From this the nuclear masses may be calculated, and consequently Q-values of different unknown decay modes can be obtained. This can be done through the use of closed decay energy cycles.

Suppose we need to know if ^{237}U can decay to ^{233}Th through α-emission. Of course this is a simple calculation if the masses of ^{237}U and ^{233}Th are known, but let us assume they are not. We have data that ^{237}U decays through β-emission (E_{max} 0.248 MeV) followed by γ-decay (E_γ 0.267 MeV). ^{233}Th decays through β-emission (E_{max} 1.230 MeV) directly to ^{233}Pa. ^{237}Np undergoes α-decay to ^{233}Pa with $E_\alpha = 4.79$ MeV. We may construct a closed cycle including these decay energies as shown in Figure 5.7.C.

The Q-value for branch D is; $Q = -931.5\ (M_{233Th} + M_{He} - M_{237U})$. For branch A we can calculate (5.18); $Q = E_\alpha\ M_Z\ / M_{Z-2} = 4.79 \times 237/233 = -931.5\ (M_{233Pa} + M_{He} - M_{237Np})$. By introducing values for M_{He} and M_{237Np} we obtain $M_{233Pa} = 233.040\,108$. For branch B we calculate; $M_{233Th} = M_{233Pa} + 1.230/931.5 = 233.041\,428$. For branch C one obtains; $M_{237U} = M_{237Np} + (0.248 + 0.267)/931.5 = 237.048\,581$.

Thus all information is available for calculating branch D. The Q-value is found to be 4.23 MeV, and the $E_\alpha = 4.23 \times 233/237$ or 4.16 MeV. Although spontaneous α-decay is energetically possible, it has not been detected. The systematics of α-decay (§5.17) indicates an expected half-life of $>10^6$ y. Because the β-decay rate is much faster ($t_{½} = 6.75$ d), too few α's are emitted during the life-time of ^{237}U to be detected.

5.11. KINETICS OF SIMPLE RADIOACTIVE DECAY

Most radioactive isotopes which are found in the elements on earth must have existed for at least as long as the earth. The nonexistence in nature of elements with atomic numbers greater than 92 is explained by the fact that all the isotopes of these elements have life-times considerably shorter than the age of the earth.

Radioactive decay is a random process. Among the atoms in a sample undergoing decay it is not possible to identify which specific atom will be the next to decay. Let us make the following assumptions for a nucleus to undergo decay during a time interval Δt.

1. The probablity is independent on the history of the nucleus (i.e. equal for all t)
2. The probablity is independent on the surroundings of the nucleus
3. The probability is equal for all nuclei of a certain type and exitation energy

There are some cases where "2" is not valid but for the vast majority of cases the assumptions 1-3 are valid. Now, consider a selected nucleus and let s be the probability of decay during Δt. However, s increases proportionally to Δt according to 1. Let also p be the probability for not decaying during Δt. Then, $s + p = 1 \rightarrow p = 1 - s$ but $s = \lambda \Delta t$ and therefore $p = 1 - \lambda \Delta t$. Now, let q be the probability for not decaying during the time $t = \Sigma \Delta t$. (with $q = \Pi p$ and $p = t/\Delta t$). Since all p are equal for equal Δt one gets:

$$q = (1 - \lambda \Delta t)^{t/\Delta t} = \lim_{\Delta t \rightarrow 0} \left((1 - \lambda \Delta t)^{-1/(\lambda \Delta t)} \right)^{-\lambda t} = e^{-\lambda t} \tag{5.54}$$

For a large number of atoms (>100) probability may be approximated by frequency, why (according to 3 are all q equal): $N/N_0 = e^{-\lambda t} \rightarrow N = N_0 e^{-\lambda t}$, or $\ln(N/N_0) = -\lambda t$

Taking the derivative with respect to t yields:

$$dN/dt = -\lambda N \tag{5.55}$$

where λ is the proportionality constant known as the *decay constant.*

We denote the *decay rate* by A. It is a measure of the number of disintegrations per unit time:

$$A = -\mathrm{d}N/\mathrm{d}t \tag{5.56}$$

The decay rate is proportional to the number of radioactive atoms, N, present: $A \propto N$. If 10^5 atoms show a decay rate of 5 atoms per second then 10^6 atoms show a decay rate of 50 atoms per second. If the time of observation Δt during which ΔN atoms decay is very small compared to $t_{½}$ (usually <1%), one may simply write

$$A = \Delta N/\Delta t = \lambda N \tag{5.57}$$

If the number of nuclei present at some original time $t = 0$ is designated as N_0, (5.56) upon integration becomes the general equation for simple radioactive decay:

$$N = N_0\, \mathrm{e}^{-\lambda t} \tag{5.58}$$

In Figure 5.8 the ratio of the number of nuclei at any time t to the original number at time $t = 0$ (i.e. N/N_0) has been plotted on both a linear (left) and logarithmic (right) scale as a function of t. The linearity of the decay curve in the semi-logarithmic graph illustrates the exponential nature of radioactive decay. Since $A \propto N$, the equation can be rewritten as

$$A = A_0\, \mathrm{e}^{-\lambda t} \tag{5.59}$$

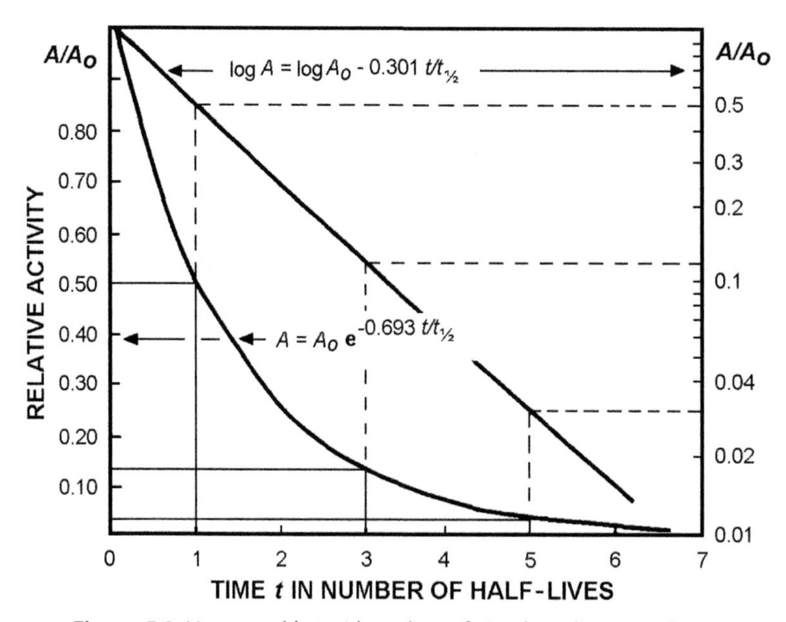

Figure 5.8 Linear and logarithm plots of simple radioactive decay.

Commonly, log A is plotted as a function of t since it is simpler to determine the disintegration rate than it is to determine the number of radioactive atoms in a sample. Instead of the decay constant λ, the average *lifetime* τ is sometimes used:

$$\tau = 1/\lambda \tag{5.60}$$

Even more common is the use of the *half-life*, $t_{½}$, which is the time needed to reduce the amount of radioactive material by a factor of 2. Thus

$$A/A_0 = N/N_0 = 1/2 = e^{-\lambda t_{½}} \tag{5.61}$$

and thus

$$t_{½} = \ln(2)/\lambda = 0.693/\lambda \tag{5.62}$$

$t_{½}$ is about 70% of the average lifetime τ.

The number of radioactive nuclei remaining at any time in a sample which at $t = 0$ had N_0 atoms can be calculated from the equation

$$N = N_0/2^n \tag{5.63}$$

where n is the number of half-lives which have passed. In radioactive work, 10 half-lives ($n = 10$) is usually considered as the useful lifetime for a radioactive species since $N = N_0/2^{10} = 10^{-3}\ N_0$; i.e. N, and hence A, is 0.001 of the original N_0 and A_0.

The decay rate is usually expressed as disintegrations per second (dps) or disintegrations per minute (dpm). In measuring radioactive decay, it is very rare that every disintegration is counted. However, a proportionality exists for any particular detection system between the absolute disintegration rate A and the observed decay rate:

$$R = \psi A \tag{5.64}$$

where R is the observed decay or *count rate* and ψ the proportionality constant, known as the *counting efficiency*, cf section 9.9. This counting efficiency depends on many factors including the detector type, the geometry of the counting arrangement, and the type and energy of the radioactive decay. ψ often has a value between 0.01 and 0.5. Equation (5.64) is only valid provided $\Delta t \ll t_{½}$ (in which case $\Delta N \ll N$), where Δt is the time of measurement; this is the normal situation; cf. (5.57).

Figure 5.9 shows the radioactivity of a ^{32}P sample measured every third day with a GM counter. It is seen that the activity decreases from about 8400 cpm at $t = 0$ to 4200 cpm in 14.3 days, and to about 2100 cpm in 2 × 14.3 days. The uncertainty in the measurements is about the size of the circles, i.e. about ± 110 cpm at $t = 0$ and about ± 65 cpm at 30 days. In plots of this kind, the count rate measured in the absence of the sample (the *background*) must be subtracted from that obtained with the sample present to yield the correct radioactivity for the sample alone. In Figure 5.9 the

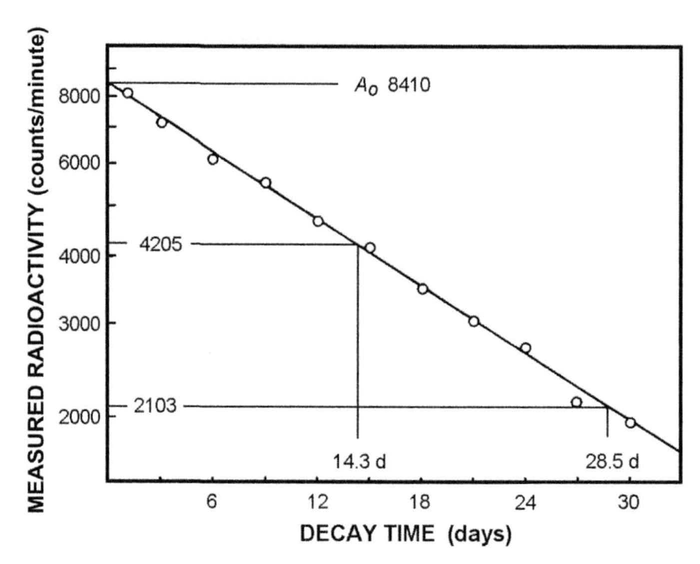

Figure 5.9 Semilogarithmic plot of the measured decay of ^{32}P using a GM counter with a background of 20 cpm.

background is so small (i.e. about 20 cpm) that it has very little influence on the decay curve.

The half-life is such a definitive characteristic of a radioactive species that knowledge of it plus the decay energy is often sufficient to allow identification of a nuclide. A radioactive sample, which exhibits a half-life of 4.5×10^9 y with α-decay energies of 4.8 MeV (77%) and 4.3 MeV (23%), is almost certainly ^{238}U as there is no other nuclide known with this exact set of properties.

With (5.57), (5.62), and (5.64) one obtains

$$R = \psi N \ln(2)/t_{1/2} \tag{5.65}$$

Knowing the counting efficiency ψ and the number of atoms N, the half-life can be calculated from measurement of R. For example, in a counting arrangement with $\psi = 0.515$ for α-particles, 159 cpm are observed from a ^{232}Th deposit of 1.27 mg (sample weight a). Thus $A = R/\psi = 309$ dpm, $N = a\,\mathbf{N_A}/M = 1.27 \times 10^{-3} \times 6.02 \times 10^{23}$ / $232.0 = 3.295 \times 10^{18}$ ^{232}Th atoms, and $t_{1/2} = 0.693 \times 3.295 \times 10^{18}/309 = 7.40 \times 10^{15}$ min $= 1.41 \times 10^{10}$ y.

It is quite obvious that it would not be possible to determine such a long half-life by following a decay curve like the one in Figure 5.9. Alternately, for short-lived nuclides such as ^{32}P one may use (5.65) with the known half-life and experimentally measured values of ψ and R to determine values of N.

5.12. MIXED DECAY

A radioactive sample may contain several different radioactive nuclides which are not genetically related. The decay of each nuclide follows the decay equations of the previous section. The detector measures a certain amount of the radioactivity of each species so that

$$R = R' + R'' \tag{5.66}$$

which with the introduction of (5.59) and (5.64) gives

$$R = \psi' A'_0 \, e^{-\lambda' t} + \psi'' A''_0 \, e^{-\lambda'' t} \tag{5.67}$$

Figure 5.10 shows the composite decay curve for the mixture of ^{71}Zn ($t_{½}$ 3.9 h) and ^{187}W ($t_{½}$ 23.8 h). If the half-lives of the species in the mixture differ sufficiently, as in this

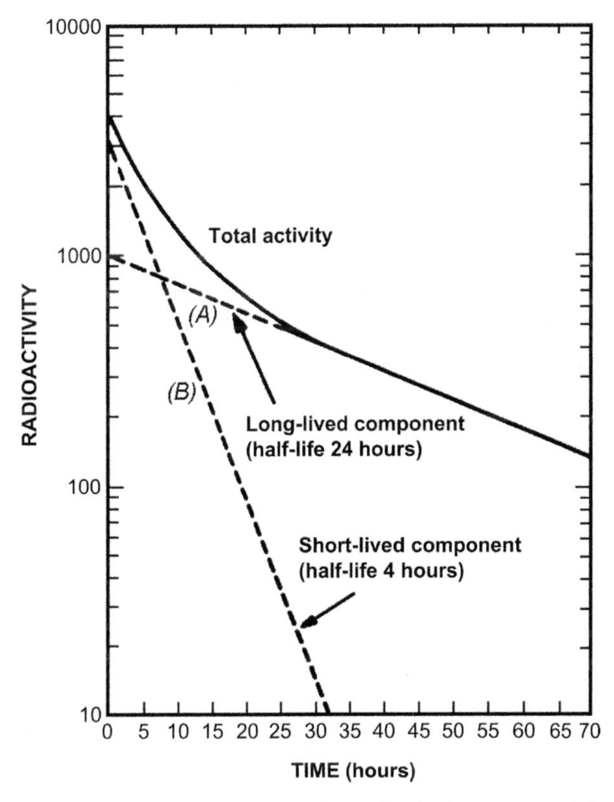

Figure 5.10 Decay diagram of a mixture of two independently decaying nuclides with half-lives of 4 and 24 h.

case, the decay curve can be resolved into the individual components. The long-lived nuclide (^{187}W, line A) can be observed to have linear decay at times long enough for the shorter-lived species to have died. The decay line related to this species can be extrapolated back to $t = 0$ and this line subtracted from the observed decay curve. The resulting curve should be the linear decay due to the shorter-lived species in the sample (line B). For more complex mixtures, this process may be repeated until the curve is completely resolved into linear components.

Sometimes it is possible in mixed decay to observe preferentially the decay of one species by proper choice of detection technique. For example, a proportional counter may be used at an operating voltage that allows detection of α-decay ($\psi_\alpha > 0$) but excludes detection of β-decay ($\psi_\beta = 0$). By contrast a typical Geiger counter can be used for β-decay but does not detect α-radiation since the α-particles do not penetrate the window of the Geiger tube. These problems are discussed more extensively in Ch. 9.

5.13. RADIOACTIVE DECAY UNITS

The SI unit for radioactivity is the *Becquerel (Bq)*, and the activity is given in reciprocal seconds, s^{-1}:

$$1\ \text{Becquerel}\ (\text{Bq}) = 1\ (\text{disintegration})\ \text{s}^{-1} \tag{5.68}$$

The *measured count rate* (R, in (5.64)) is given in *counts per second* (cps) or per minute (cpm); the abbreviations are in German Impulse pro minute (Ipm), and in French coups de minute (c/m).

An earlier unit, still in some use, is the *Curie unit* (abbreviated Ci) defined as:

$$1\ \text{Curie}\ (\text{Ci}) = 3.7 \times 10^{10}\text{s}^{-1}\ (\text{Bq}) \tag{5.69}$$

The Curie unit was originally defined as the number of decays per unit time and gram of ^{226}Ra, assuming its half-life to be 1580 y.

The *specific radioactivity* S is defined as the decay rate A per unit amount w of an element or compound,

$$S = A/w \tag{5.70}$$

The SI unit of specific radioactivity is Bq kg^{-1}. For practical purposes it is sometimes also defined in dpm g^{-1} or dpm mole^{-1}. *Activity concentration* (or "radioactive concentration") is given in Bq m^{-3} or Bq l^{-1}. With the half-life of 1599 ± 4 y the specific activity per gram of ^{226}Ra is 0.988 Ci or 3.65×10^{10} Bq or 2.19×10^{12} dpm. The specific activities of some of the longer-lived naturally occurring radioactive species are: ^{40}K, 31.0 kBq kg^{-1} of K; ^{232}Th, 4.06 MBq kg^{-1} of Th; ^{238}U, 12.3 MBq kg^{-1} of U.

5.14. BRANCHING DECAY

Several times in this chapter the possibility of competing modes of decay has been noted; see e.g. Figure 1.2. In such competition, termed *branching decay* (see e.g. Fig. 5.5(g)), the parent nuclide may decay to two or more different daughter nuclides: e.g.

$$ {}^{A}_{Z}X \begin{cases} \xrightarrow{\lambda_1} {}^{A_1}_{Z_1}X \\ \xrightarrow{\lambda_2} {}^{A_2}_{Z_2}X \end{cases} \tag{5.71} $$

where for each branching decay a *partial decay constant* can be determined. These constants are related to the total observed decay constant for the parent nuclide as

$$ \lambda_{tot} = \lambda_1 + \lambda_2 + \ldots \tag{5.72} $$

Each mode of decay in branching may be treated separately; the decay in an individual branch has a half-life based on the partial decay constant. Since only the total decay constant (the rate with which the mother nuclide, ${}^{A}_{Z}X$ in (5.71), decays) is observable directly, partial decay constants are obtained by multiplying the observed total decay constant by the fraction of parent decay corresponding to that branch. ^{64}Cu decays 43% by electron capture, 38% by negatron emission, and 19% by positron emission. The observed total decay constant is equal to 0.0541 h^{-1} based on the half-life of 12.8 h. The partial constants are:

$$ \lambda_{EC} = 0.43 \times 0.0541 = 0.0233 \text{ h}^{-1} \tag{5.73a} $$

$$ \lambda_{\beta^-} = 0.38 \times 0.0541 = 0.0206 \text{ h}^{-1} \tag{5.73b} $$

$$ \lambda_{\beta^+} = 0.19 \times 0.0541 = 0.0103 \text{ h}^{-1} \tag{5.73c} $$

These partial decay constants correspond to partial half-lives of 29.7 h for electron capture decay, 33.6 h for β^- decay, and 67.5 h for positron decay.

5.15. SUCCESSIVE RADIOACTIVE DECAY

There are many instances where a parent decays to a daughter which itself decays to a third species (i.e. a "grand-daughter"). The chains of radioactive decay in the naturally occurring heavy elements include as many as 10–12 successive steps (Fig. 1.2).

$$ X_1 \overset{\lambda_1}{\rightarrow} X_2 \overset{\lambda_2}{\rightarrow} X_3 \overset{\lambda_3}{\rightarrow} X_4 \ldots \tag{5.74} $$

The net rate of formation of the daughter atoms X_2 is the difference between the rate of formation of the daughter and her rate of decay, i.e.

$$dN_2/dt = N_1\lambda_1 - N_2\lambda_2 \quad (5.75)$$

where N_1 and N_2 are the number of parent and of daughter atoms, and λ_1 and λ_2, the decay constants of the parent and daughter, respectively. The solution of this equation is

$$N_2 = [\lambda_1/(\lambda_2 - \lambda_1)]\, N_{1,0}\left(e^{-\lambda_1 t} - e^{-\lambda_2 t}\right) + N_{2,0}\, e^{-\lambda_2 t} \quad (5.76)$$

where $N_{1,0}$ and $N_{2,0}$ are the amounts of parent and daughter respectively at time $t = 0$. The first term in this equation tells us how the number of daughter nuclei vary with time as a consequence of the formation and subsequent decay of the daughter nuclei, while the second term accounts for the decay of those daughter nuclei that were present at $t = 0$. It should be noted that the successive radioactive decay is a special case of the more general question of irradiation yield and decay, cf Ch. 17.

Let us illustrate this relationship by an example among the naturally occurring radioactive decay series. In an old uranium mineral all the products in the decay chain can be detected (see Fig. 1.2). Suppose now that we use a chemical separation to isolate two samples, one containing only uranium and one containing only thorium (relation (1.3) and Fig. 1.1). At the time of separation, which we designate as $t = 0$, there are $N_{1,0}$ atoms of ^{238}U and $N_{2,0}$ atoms of ^{234}Th. In the thorium fraction, which is free from uranium, $N_{1,0} = 0$ and, therefore, the thorium atoms decay according to the last term in (5.76). This sample gives a simple exponential decay curve with a half-life of 24.1 d, as shown by the precipitate curve in Figure 1.1. The uranium fraction at $t = 0$ is completely free of thorium; i.e. $N_{2,0} = 0$. However, after some time it is possible to detect the presence of ^{234}Th. The change in the number of ^{234}Th with time follows the first term of (5.78); in fact, in Figure 1.1 the measurements detect only the ^{234}Th nuclide (β-emitting), since the detection system used is not sensitive to the α's from ^{238}U ($\psi_\alpha = 0$). The time of observation is much smaller than the half-life of the ^{238}U decay, so there will be no observable change in the number of atoms of uranium during the time of observation, i.e. $N_1 = N_{1,0}$. Further, since $t_{1/2}$ for $^{238}U \gg t_{1/2}$ for ^{234}Th, i.e. $\lambda_1 \ll \lambda_2$, we can simplify (5.76) to

$$N_2 = (\lambda_1/\lambda_2)N_{1,0}\left(1 - e^{-\lambda_2 t}\right) \quad (5.77)$$

According to this equation, the number of ^{234}Th atoms, N_2, increases with time with the half-life of ^{234}Th. In other words, after a period of 24.1 d there is 50% of the maximum value of ^{234}Th, after 48.2 d there is 75% of the final maximum value, etc. This is illustrated by the change in the uranium fraction activity in Figure 1.1. Further, from (5.77) we can see that the maximum value of thorium ($t = \infty$) is given by

$$N_2\, \lambda_2 = N_{1,0}\, \lambda_1 \quad (5.78)$$

These equations, based on $\lambda_1 \ll \lambda_2$, show that the amount of daughter atoms becomes constant after some time. At that time the rate of decay of the daughter becomes equal to the rate of decay of the parent, i.e. $A_2 = A_1$, but the amounts of the parent and the daughter are not equal since N_2 is much smaller than N_1. This condition of $A_2 = A_1$ is known as *secular equilibrium*, which is a misnomer since this is a steady state and not a true equilibrium situation. It is also common to speak of *radioactive equilibrium* in referring to this steady state condition.

We can calculate that at secular equilibrium for each gram of ^{238}U there will be present 1.44×10^{-11} g of ^{234}Th and 4.9×10^{-16} g of ^{234}Pa. Since the specific radioactivity of ^{238}U is 746 000 dpm/g, the decay rate of 4.9×10^{-16} g ^{234}Pa is also 746 000 dpm.

When the time of observation is very short compared to the half-life of the parent nuclide, as in secular equilibrium, no change in the decay rate of the parent is observed for many daughter half-lives. Our example of ^{137}Cs, which decays via the isomeric state ^{137m}Ba to ^{137}Ba, presents another case of "secular equilibrium". If we have an "old" sample in which radioactive equilibrium has been reached (older than ~15 min, since the $t_{½}$ of the daughter is 2.6 min), and separate the cesium from the barium by precipitation and filtration of $BaSO_4$, the activity measured from the precipitate will follow curve (1) in Figure 5.11. In the filtrate solution the activity from ^{137}Cs, curve (2), is unchanged

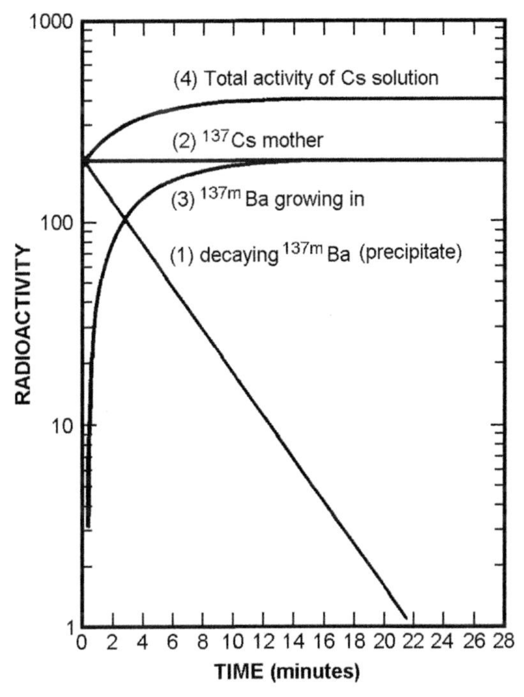

Figure 5.11 Case of radioactive equilibrium: successive decay chain ^{137}Cs($t_{½}$ 30 y) → ^{137m}Ba($t_{½}$ 2.6 min) → stable.

during our observation time. However, ^{137m}Ba grows into the solution, curve (3), so that the total activity of the solution, curves (2) plus (3), increases according to curve (4).

In many radioactive decay chains the half-life of the parent is longer than that of the daughter but it is short enough that a change in the disintegration rate of the parent is observable during the period of observation of the experiment. In such cases the system reaches the condition termed *transient equilibrium*. The length of time of observation of the activity of the sample may be the determining factor as to whether it appears to be transient or secular equilibrium. If a parent has a one month half-life and the observation of the change in decay rates of parent and daughter extends over an hour or even a few days, the data would follow the equation for secular equilibrium since the degree of change in the parent decay would be negligible. However, if the observation extends over a period of several weeks or months, then the change in the decay rate of the parent is significant and it would appear as transient equilibrium.

The case of transient equilibrium will be illustrated by an example, such as the decay chain

$$^{140}\mathrm{Ba}\left(\beta^-, 12.75\ \mathrm{d}\right)\ ^{140}\mathrm{La}\left(\beta^-,\ 1.678\ \mathrm{d}\right)\ ^{140}\mathrm{Ce}\left(\mathrm{stable}\right) \tag{5.79}$$

^{140}Ba is one of the most important fission products. If we isolate barium, lanthanum grows into the sample. Figure 5.12 shows the decay of ^{140}Ba in curve (1), which follows

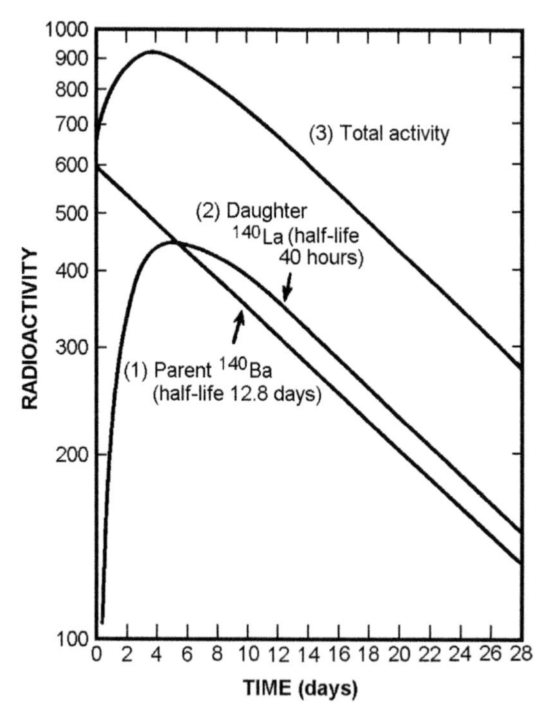

Figure 5.12 Case of transient equilibrium: successive decay chain ^{140}Ba($t_{1/2}$ 12.75 d) → ^{140}La($t_{1/2}$ 1.678 d) → stable.

the simple decay of (5.59). Curve (2) shows the activity of the daughter, for which the left half of eq. (5.76) is valid. Replacing decay constants by half-lives we can rewrite this equation as

$$A_2 = t_{½,1}/(t_{½,1} - t_{½,2})A_{1,0}\left(e^{-0.693t/t_{½,1}} - e^{-0.693t/t_{½,2}}\right) \tag{5.80}$$

At $t \ll t_{½,1}$ ($t \ll 12.8$ d) the first exponential term is very close to 1 and A_2 increases proportional to $(1 - e^{-0.693t/t_{½,2}})$; this is the increasing part of curve (2). At $t \ll t_{½,2}$ ($t \ll$ 40 h), the second exponential term becomes much smaller than the first one, and A_2 decreases proportional to $e^{-0.693t/t_{½,1}}$. For this part of the curve we may write

$$N_2 = N_1\lambda_1/(\lambda_2 - \lambda_1) \tag{5.81}$$

which is the relation valid for transient equilibrium. The total activity of the barium sample, curve (3), is the sum of curves (1) and (2).

If the parent is shorter-lived than the daughter, the daughter activity grows to some maximum value and then decays with its own characteristic half-life. This contrasts to the case of transient equilibrium where the daughter has an apparent decay given by the half-life of the parent. An example of this is shown in Figure 5.13 for the decay chain

$$^{218}\text{Po}\left(\alpha, 3\ \text{min}\right)\ ^{214}\text{Pb}\left(\beta^-, 27\ \text{min}\right)\ ^{214}\text{Bi} \tag{5.82}$$

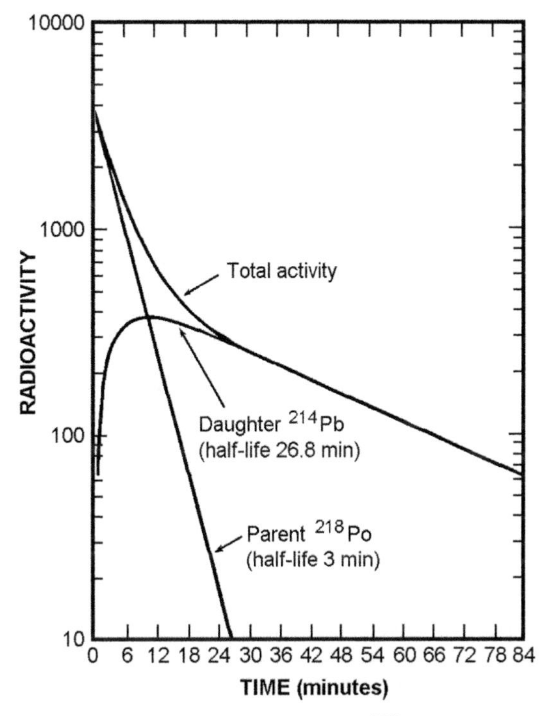

Figure 5.13 Case of no equilibrium: successive decay chain ^{218}Po($t_{½}$ 3 min) → ^{214}Pb($t_{½}$ 26.8 min) → stable.

The time necessary for obtaining the maximum daughter intensity in the non-equilibrium case of the shorter-lived parent is given by

$$t_{max} = (\lambda_2 - \lambda_1)^{-1}\ln(\lambda_2/\lambda_1) \quad (5.83)$$

5.16. RADIOISOTOPE GENERATORS

The growth of radioactive daughters frequently has practical significance. For example, in radiation therapy and diagnostic medicine it is preferable to use short-lived nuclides. In fact, it is preferable to conduct tracer experiments with short-lived nuclides as this eliminates the problem of disposal of residual radioactive waste after completion of the experiment. It is convenient to have a long-lived mother in storage from which a short-lived daughter can be removed as required for use in tracer work. A few examples of uses of such mother-daughter pairs are discussed; others are included in Table 5.1.

Such systems are called *radioisotope generators.* ^{222}Rn is sometimes used for the radiotherapeutic treatment of cancer. This product is isolated by separating it as a gas from the parent substance ^{226}Ra which is normally in the form of solid or a solution of $RaBr_2$.

Table 5.1 Some common radioactive milking pairs. The decay properties include decay energy (MeV), mode of decay and half-life

Mother nuclide	Decay properties	Daughter nuclide	Decay properties	Application
^{44}Ti	EC, γ; 47.3 y	^{44}Sc	1.5 β^+; 1.16 γ; 3.93 h	Teaching
^{68}Ge	EC; 270.8 d	^{68}Ga	1.9 β^+; 1.08 γ; 1.135 h	Medical
^{87}Y	EC; 3.35 d	^{87m}Sr	0.39 γ; 2.80 h	Medical & teaching
^{90}Sr	0.5 β^-; 28.5 y	^{90}Y	2.3 β^-; 2.671 d	Heat source†, Calibration source
^{99}Mo	β^-, γ; 65.9 h	^{99m}Tc	0.14 γ; 6.0 h	Medical
^{113}Sn	EC, γ; 115.1 d	^{113m}In	0.39 γ; 1.658 h	Medical
^{132}Te	β^-, γ; 78.2 h	^{132}I	2.1 β^-, γ; 2.28 h	Medical
^{137}Cs	β^-, γ; 30.0 y	^{137m}Ba	0.66 γ; 2.55 m	Gamma radiography, Radiation sterilization†
^{140}Ba	β^-, γ; 12.75 d	^{140}La	β^-, γ; 1.678 d	Lanthanum tracer
^{144}Ce	β^-, γ; 284.9 d	^{144}Pr	3.0 β^-; 17.28 m	Calibration source
^{210}Pb	β^-, γ; 22.3 y	^{210}Bi	1.2 β^-; 5.01 d	Calibration source
^{226}Ra	α; 1600 y	^{222}Rn	α; 3.825 d	Medical
^{238}U	α; 4.468×10^9 y	^{234}Th	β^-, γ; 24.1 d	Thorium tracer

†Main use of mother substance in large amounts.

^{222}Rn grows into the radium sample with a half-life of 3.8 d. After a 2-week period, following a separation of radon from radium, approximately 90% of the maximum amount of radon has grown back in the radium sample. Consequently, it is useful to separate ^{222}Rn each 2 weeks from the radium samples since further time provides very little additional radioactivity. The ^{222}Rn is an α emitter; the therapeutic value comes from the irradiation of the tissue by the γ-rays of the decay daughters ^{214}Pb and ^{214}Bi which reach radioactive equilibrium extremely rapidly with the ^{222}Rn.

^{99m}Tc is used for diagnostic purposes for liver, spleen, and thyroid scanning (§18.5). The ^{99}Mo parent, obtained by chemical separation from ^{235}U fission product mixture, is absorbed on a column of alumina and the daughter ^{99m}Tc removed by passage of saline solution at intervals governed by the equilibrium. The parent, when it is fixed in a semi-permanent sample as on an adsorbent column, is often known as a cow and the removal of the daughter activity from the radioisotope generator (the "cow") is termed *milking*.

Another commonly used radioisotope generator is ^{132}Te from which ^{132}I may be milked. In this case ^{132}Te is adsorbed as barium tellurite on an alumina column, and the ^{132}I removed by passage of 0.01 M ammonia through the column. The ^{132}I is used both diagnostically and therapeutically for thyroid cancer.

Many of these sources produce radionuclides with half-lives suitable for teaching purposes, e.g.

$$^{137m}\mathrm{Ba}(2.6\ \mathrm{min}),\ ^{144}\mathrm{Pr}(17.3\ \mathrm{min}),\ ^{44}\mathrm{Sc}(4.0\ \mathrm{h}),\ ^{99m}\mathrm{Tc}(6.0\ \mathrm{h})\ \text{and}\ ^{90}\mathrm{Y}(64\ \mathrm{h}). \quad (5.84)$$

5.17. DECAY ENERGY AND HALF-LIFE

It was observed early in both α- and β-decay that the longer the half-life the lower the decay energy. Although there are many exceptions to this observation H. Geiger and J. M. Nuttall formulated the law

$$\log \lambda_{\alpha} = a + b \log \hat{R}_{\mathrm{air}} \quad (5.85)$$

for the natural α-active nuclides. Here a and b are constants, and $\hat{R}_{\mathrm{air}}$ is the range of the α-particles in air which is directly proportional to the α-particle energy E_{α}. A similar relation was deduced by E. Fermi for the β-decay:

$$\log \lambda_{\beta} = a' + b' \log E \quad (5.86)$$

where a' is a constant related to the type of β decay and $b' \approx 5$.

Although these rules have been superseded by modern theory and the enormous amount of nuclear data now available, they may nevertheless be useful as rough guides in estimates of half-lives and decay energies. In §6.7 more valid but also more complicated relationships are discussed.

5.18. THE HEISENBERG UNCERTAINTY PRINCIPLE

In this chapter we have repeatedly stated that the nuclear decay energies are exact values as required by quantum mechanics. However, this is not exactly correct: the energy levels have a certain "spread". This was first stated by Heisenberg in 1927 and is of fundamental importance in all areas of nuclear physics.

The *uncertainty principle* states that it is impossible to measure simultaneously the exact position and the exact momentum of a particle. This follows from the wave properties of the particle. If for example, we attempt to measure the exact position of an electron by observing the light emitted when it hits a scintillating screen, this act interferes with the movement of the electron causing it to scatter, which introduces some uncertainty in its momentum. The size of this uncertainty can be calculated exactly and is related to the Planck constant. If Δx denotes the uncertainty in position and Δp the uncertainty in momentum along the x axis, then

$$\Delta x \, \Delta p \geq \mathbf{h}/(2\pi) \equiv \hbar \tag{5.87}$$

$\hbar$ is called "h-bar" (1.05×10^{-34} J s), and **h** is the *Planck constant.*

This principle holds for other *conjugate variables*, as angle θ and angular momentum p_θ

$$\Delta \theta \, \Delta p_\theta \geq \hbar \tag{5.88}$$

and time and energy

$$\tau \, \Delta E \geq \hbar \tag{5.89}$$

This latter equation relates the life-time τ of an elementary (or nuclear) particle to the uncertainty in its energy (ΔE). For excited nuclear states this can be taken as the width of the γ-peak at half-maximum intensity (the "FWHM value") (Fig. 5.14). For example, if $\Delta E_\gamma = 1.6$ keV, then $\tau \geq 1.05 \times 10^{-34}/1600 \times 1.60 \times 10^{-19}$ s $= 4.1 \times 10^{-19}$ s. This is a

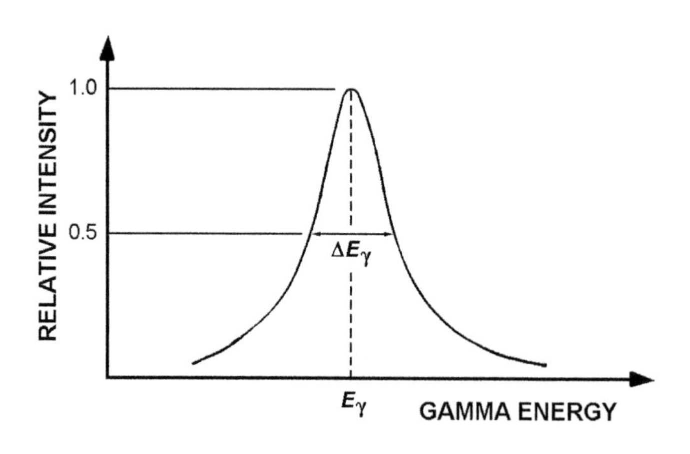

Figure 5.14 The energy half-width value ΔE_γ is the FWHM of the γ-peak (FWHM = Full Width at Half Maximum).

long time compared to that of a nuclear rotation, which is about 10^{-23} s. Consequently the wave mechanic properties of particles (and γ-rays) introduce a certain fundamental uncertainty in the particle energy. Several applications of this are described in later chapters.

5.19. EXERCISES

For some of the problems necessary nuclear data are given in the Tables or appendices.

5.1. ^{239}Pu emits α-particles of maximum 5.152 MeV. What is the recoil energy of the product formed?

5.2. Using a magnetic spectrometer the maximum energy of the electrons from ^{137}Cs was found in Figure 5.1 to correspond to 3.15×10^{-3} Tesla m. Calculate the energy (a) assuming that the electrons are non-relativistic, (b) with correction for relativistic mass increase.

5.3. ^{11}C decays through emission of positrons of a maximum energy of 1.0 MeV. Calculate the recoil energy of the daughter.

5.4. ^{16}N decays through β^- decay to ^{16}O with a half-life of 7.1 s. A number of very energetic γ's follow after the β-emission, the dominating one with an energy of 6.14 MeV. What is the ^{16}O recoil energy?

5.5. The binding energy of a K-electron in barium is 37 441 eV. Calculate from Figure 5.1 the internal conversion energy for ^{137m}Ba (Fig. 5.5).

5.6. From the specific activity of potassium (1850 dpm/g K) and the fact that it all originates in the rare isotope ^{40}K (0.0117%), calculate the half-life of ^{40}K.

5.7. One may assume that when ^{238}U was formed at the genesis an equal amount of ^{235}U was formed. Today the amount of ^{238}U is 138 times the amount of ^{235}U. How long a time ago did the genesis occur according to this assumption?

5.8. The interior of the earth is assumed to be built up of a solid core (radius 1371 km) followed by a molten core (radius 3471 km) and a molten mantle (radius 6354 km) covered by a 17 km thick crust. One assumes that 2% by weight of the molten mantle and crust is potassium; the average mantle density is assumed to be 6000 kg m^{-3} and that of the crust 3300 kg m^{-3}. What energy outflow will the radioactive decay of this element cause at the earth's surface? The decay scheme of ^{40}K is given in Eqn. (13.7). For the EC branch $Q = 1.505$ MeV, for the β^- branch 1.312 MeV. Each decay by the EC branch leads to emission of a 1.46 MeV γ. Compare this energy output to the solar energy influx to the earth of 3.2×10^{24} J y^{-1}.

5.9. A hospital has a 1.5 Ci source of ^{226}Ra in the form of a $RaBr_2$ solution. If the ^{222}Rn is pumped out each 48 h, what is (a) the radon activity (in Bq) at that moment, (b) the radon gas volume at STP?

5.10. (a) Prove the correctness of Eqn. (5.25) by using Newton's laws of motion. (b) Prove the correctness of Eqn. (5.48).

5.11. A recently prepared ^{212}Pb sample has the activity of 10^6 dpm. (a) What is the activity 2 h later? (b) How many lead atoms are left in the sample at this moment? $t_{½}$ 10.64 h.

5.12. A radioactive sample was measured at different time intervals:

Time (h)	Count rate (cpm)	Time (h)	Count rate (cpm)
0.3	11100	30	1015
5	5870	40	888
10	3240	50	826
15	2005	100	625
20	1440		

Determine the half-lives of the two nuclides (not genetically related) in the sample and their activities (in Bq) at time $t = 0$. The background of the detection device was 100 376 counts per 1 000 min; its counting efficiency was 17%.

5.13. The α-activity of a mixture of astatine isotopes was measured at different times after their separation, giving the following results:

t(min)	A(dpm)	t(min)	A(dpm)
12	756	121	256
17.2	725	140	215.5
23.1	638	161	178.5
30.0	600	184	150.7
37.7	545	211	127.3
47.5	494	243	101.9
59.5	435	276	84.9
73	380	308	68.2
87	341	340	55.0
102	288		

Calculate the half-lives of the isotopes and their activities at $t = 0$.

5.14. In the ion source of a mass spectrograph, UF_6 vapor is introduced which partly becomes ionized to UF_5^+. The ionic currents were measured at mass positions 333, 330, and 329. The ion current ratios were $I_{333}/I_{330} = 139$, and $I_{330}/I_{329} = 141.5$. What is the half-life of ^{234}U if that of ^{238}U is 4.5×10^9 y? Radioactive equilibrium is assumed to exist in the UF_6.

5.20. LITERATURE

W. F. LIBBY, Chemistry of energetic atoms produced in nuclear reactions, *J. Am. Chem. Soc.* **69** (1947) 2523.

W. GENTNER, H. MAIER-LEIBNITZ and H. BOTHE, *An Atlas of Typical Expansion Chamber Photographs*, Pergamon Press 1953.

R. D. EVANS, *The Atomic Nucleus*, McGraw-Hill, 1955.

E. B. PAUL, *Nuclear and Particle Physics*, North Holland 1969.

M. G. BOWLER, *Nuclear Physics*, Pergamon Press 1973.

K. H. LIESER, Chemische Gesichtspunkte fur die Entwicklung von Radionuklidgeneratoren, *Radiochim. Acta* **23** (2) (1976) 57.

E. BROWNE and R. B. FIRESTONE, V. S. SHIRLEY (Ed.), *Table of Radioactive Isotopes*, J. Wiley & Sons, 1986.

Nuclides and Isotopes, 14th Ed., General Electric Company, Nuclear Energy Operations, 175 Curtner Ave., M/C397, San Jose, CA 95125, USA, 1989.

M. S. ANTONY, *Chart of the Nuclides – Strasbourg 1992*, Jan-Claude Padrines, AGECOM, 1 route d'Epfig, Maison-Rouge, 67600 Sélestat, France, 1992.

Nuclear Structure

Contents

Throughout the ages and in every civilization, people have developed explanations of observed behaviors. These explanations are based on the principle of causality, i.e. every effect has a cause and the same cause produces always the same effect. We call these explanations *models*.

Scientists are professional model-builders. Observed phenomena are used to develop a model, which then is tested through new experiments. Actually, all things we claim that

Radiochemistry and Nuclear Chemistry
ISBN 978-0-12-405897-2, http://dx.doi.org/10.1016/B978-0-12-405897-2.00006-9

we know today are in reality models of some unknown reality. This is continuously noted when all of a sudden a new discovery seem to falsify an older model e.g. the current discussion of the possibility that neutrinos actually travel faster than the speed of light. If these experiments are true we have to remodel much of that which we today consider being the truth. Indirect model building is familiar to every chemist: although we cannot see the atoms and molecules which we add into a reaction vessel, we certainly have some idea about what is going to happen. It is indeed our enjoyment in developing models which causes us to experiment in science. We also want to be able to make quantitative predictions based on our models which we therefore formulate in mathematical terms. To allow tractable calculations most models involves simplifications of the "real world". Of course, since man is fallible some models may turn out to be wrong but as new data accumulate, wrong or naive models are replaced by better ones.

We have already shown how one model for the nuclear structure, the liquid drop model, has helped us to explain a number of nuclear properties, the most important being the shape of the stability valley. But the liquid drop model fails to explain other important properties. In this chapter we shall try to arrive at a nuclear model which takes into account the quantum mechanical properties of the nucleus based mainly on the interactions and properties of neutrons and protons. Naturally it is also possible to go even deeper and discuss the nuclear structure through the interactions of quarks or using the string theory but that is beyond the scope of this book and we instead refer to textbooks in nuclear physics.

6.1. REQUIREMENTS OF A NUCLEAR MODEL

Investigation of light emitted by excited atoms (J. Rydberg 1895) led N. Bohr to suggest the quantized model for the atom, which became the foundation for explaining the chemical properties of the elements and justifying their ordering in the periodic system. From studies of molecular spectra and from theoretical quantum and wave mechanical calculations, we are able to interpret many of the most intricate details of chemical bonding.

In a similar manner, patterns of nuclear stability, results of nuclear reactions and spectroscopy of radiation emitted by nuclei have yielded information which helps us develop a picture of nuclear structure. But the situation is more complicated for the nucleus than for the atom. In the nucleus there are two kinds of particles, protons and neutrons, packed close together, and there are two kinds of forces - the electrostatic force and the short range strong nuclear force. This more complex situation has caused slow progress in developing a satisfactory model, and no single nuclear model has been able to explain all the nuclear phenomena.

6.1.1. Some general nuclear properties

Let us begin with a summary of what we know about the nucleus, and see where that leads us.

In Chapter 4 we observed that the average binding energy per nucleon is almost constant for the stable nuclei (Fig. 4.3) and that the radius is proportional to the cube root of the mass number. We have interpreted this as reflecting fairly uniform distribution of charge and mass throughout the volume of the nucleus. Other experimental evidence supports this interpretation (Fig. 4.4). This information was used to develop the liquid drop model, which successfully explains the valley of stability (Fig. 4.1). This overall view also supports the assumption of a strong, short range nuclear force.

A more detailed consideration of Figures 4.1 and 4.3 indicates that certain mass numbers seem to be more stable, i.e. nuclei with *Z*- or *N*-values of 2, 8, 20, 28, 50, and 82 (see also Table 4.1). There is other evidence for the uniqueness of those numbers. For example, if either the probability of capturing a neutron (the neutron capture cross-section) or the energy required to release a neutron is plotted for different elements, it is found that maxima occur at these same neutron numbers, just as maxima occur for the electron ionization energy of the elements He, Ne, Ar, Kr, etc. (i.e. at electron numbers of 2, 8, 16, 32, etc.). The nuclear *N*- or *Z*-values of 2, 8, 20, 28, 50, and 82 are called "magic numbers".

It seems logical that these magic numbers indicate some kind of regular substructure in the nucleus similar to the orbitals for the atomic electrons. Moreover, since the same magic numbers are found for the neutrons and for the protons, we would further assume

Table 6.1 Summary of the properties of the atomic constituents (independent movements in a central potential field).

Line property	Electron	Proton	Neutron
1 Mass (u)	0.000 5486	1.007 276	1.008 665
2 Charge (**e** units)	−1	+1	0
3 Spin quantum number s	1/2	1/2	1/2
4 Spin dipole moment	$\mu_e = 1$ B.m.	μ_p=2.793 n.m.	μ_n= 1.913 n.m.
5 Orbital quantum number	$0 \leq l \leq n-1$		$0 \leq l$
6 Permitted orbital field projections	$-\Sigma l \ldots + \Sigma l$		$-\Sigma l \ldots + \Sigma l$
7 Total angular quantum number	$(j = l \pm s)$		$j = l \pm s$
8 Total angular momentum	$(p_j = \Sigma\sqrt{(j(j+1))})$		$p_j = \Sigma\sqrt{(j(j+1))}$
9 Particle symbolism	nl^i		nl_j

Notes

2. The electron unit charge: $\mathbf{e} = 1.602 \times 10^{-19}$ C.
3. The spin angular momentum has the magnitude $p = \Sigma\sqrt{(s(s+1))}$, with the permitted projections on an external field axis: $\pm s\Sigma$ (for e, p, and n =½Σ).
7. l and s couple only in the one-electron system; see §11.2.4.
8. Permitted projections on an external field axis: Σm_j, where $j \leq m_j \leq j$.
9. n is the principal quantum number, l the azimuthal (orbital, radial) quantum number, i is the number of electrons in the particular n,l state, and j is the total angular momentum quantum number. For $l = 0$, 1, 2, etc., the symbols s, p, d, f, g, etc., are used.

that the neutrons and the protons build their substructure independently of each other, but in the same way. Another fact that must be indicative of the nuclear substructure is the stability for nuclei with even proton or even neutron numbers. Since we know that the individual nucleons have spin, we could postulate that nucleons in the nucleus must pair off with opposed spins.

6.1.2. Quantized energy levels

The nucleus would thus seem to consist of independent substructures of neutrons and protons, with each type of nucleon paired off as far as possible. Further, the nucleons are obviously grouped together in the magic numbers. From the decay of radioactive nuclei we know that the total decay energy (Q-value) of any particular nuclide has a definite value. Moreover, γ-emission from any particular nucleus involves discrete, definite values. These facts resemble the quantized emission of electromagnetic radiation (X-ray, UV, visible light, etc.) from atoms. We may conclude a similar explanation for the nucleus: decay of radioactive nuclei, whether α, β, or γ, involves a transition between discrete quantized energy levels.

6.1.3. The nuclear potential well

In our development of a model of the nucleus we have so far not considered the nuclear binding energy. Let us imagine the situation wherein a neutron of low kinetic energy approaches a nucleus (Fig. 6.1). Since the neutron is uncharged it is not affected by the Coulomb field of the nucleus and approaches the nucleus with no interaction, until it is close enough to experience the strong nuclear force F_n, which is always attractive, i.e. F_n is a positive quantity. At the point r_n the neutron experiences the strong attraction to the nucleus and is absorbed. The surface of the nucleus is assumed to extend from r_n to r_s since this distance represents the radius of the nucleus over which the nuclear force is constant. When the neutron is absorbed the nucleus is not organized optimally and the excess energy is released and emitted in the form of a γ-quantum. The energy of the γ-ray can be calculated from the known masses of the reactants and product nuclides: $E_\gamma = -931.5(M_{A+1} - M_A - M_n)$ The energy released is the neutron *binding energy* part of the total binding energy in the nucleus E_B. The total energy of the nucleus has thus decreased as is indicated in Figure 6.1(B); it is common to refer to this decrease as a *potential well*. The nucleons can be considered to occupy different levels in such a potential well. The exact shape of the well is uncertain (parabolic, square, etc.) and depends on the mathematical form assumed for the interaction between the incoming particle and the nucleus.

Protons experience the same strong, short range nuclear force interaction as they contact the nucleus. However, they also experience a long range repulsive interaction due to the Coulomb force between the positive incoming protons and the positive charge of the nucleus. This repulsion prevents the potential well from being as deep for

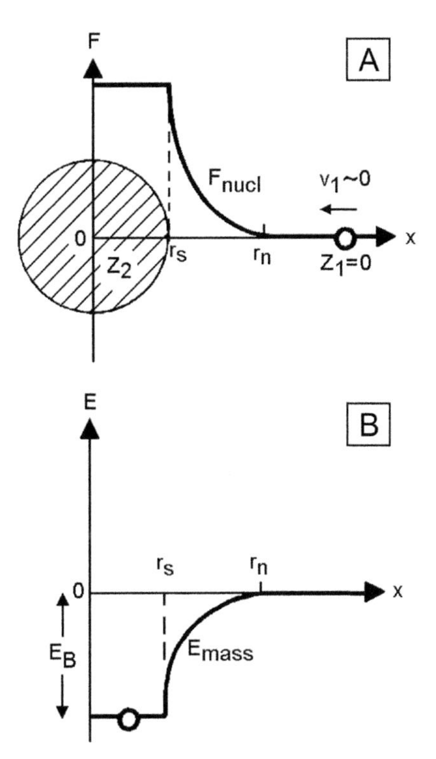

Figure 6.1 The approach along the x-axis of a neutron (charge $Z_1 = 0$, velocity $v_1 \approx 0$) towards a nucleus (charge Z_2; Fig. A) leading to the neutron being trapped at a certain energy level, E_B (Fig. B). F_{nucl} is the nuclear force extending from the surface of the "solid core" at r_s to the distance r_0. E_B is the nuclear binding energy in the potential well.

protons as for neutrons. Figure 6.2 shows the energy levels of the protons and of the neutrons in the nucleus of $^{116}_{50}Sn$.

6.2. ROTATIONAL ENERGY AND ANGULAR MOMENTUM

It is an intriguing fact that nucleons in nuclei, electrons in atoms, as well as large cosmic objects such as solar systems and even galaxies, are more dominated by rotational than by linear motion, although in our daily life the latter seems to be a more common phenomenon. In rotation there is a balance between two forces: the centrifugal force of inertia, which tries to move a body away from a center point, and an attractive force (gravitational, electrostatic, etc.), which opposes the separation.

In the preceding section we assumed that the nucleus existed in some kind of a potential well. One may further assume that a nucleon moves around in this well in a way not too different from the way the electron moves around the atomic nucleus, i.e. with an oscillation between kinetic and potential energy. With some hypothesis about the shape

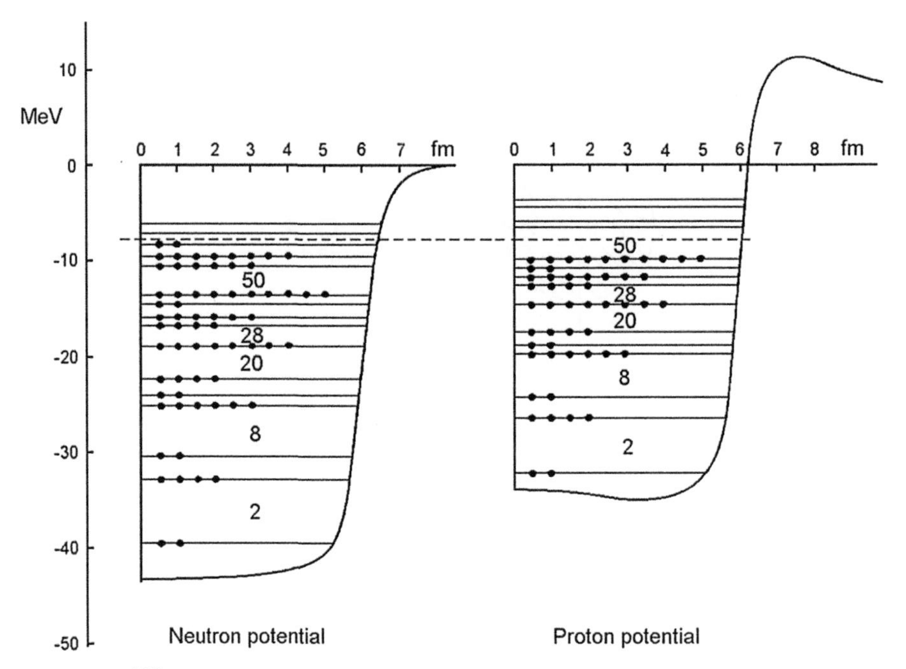

Figure 6.2 The ^{116}Sn neutron and proton shell structure in the potential well. (According to S. G. Nilsson.)

of the nuclear potential well we can apply the *Schrödinger wave equation* to the nucleus. Without being concerned at this point with the consequences of assuming different shapes we can conclude that the solution of the wave equation allows only certain energy states. These energy states are defined by two quantum numbers: the *principal quantum number n*, which is related to the total energy of the system, and the *azimuthal* (or *radial*) *quantum number l*, which is related to the rotational movement of the nucleus.

6.2.1. Rotational (mechanical) energy

If a mass m circles in an orbit of radius r at a constant angular velocity ω (rad s^{-1}), the tangential velocity at this radius is

$$v_r = \omega r \tag{6.1}$$

The kinetic energy in linear motion is $E_{kin} = \frac{1}{2}mv^2$, and we therefore obtain for the (kinetic) rotational energy

$$E_{rot} = \frac{1}{2}\, m\omega^2 r^2 \tag{6.2}$$

The angular velocity is often expressed as the frequency of rotation, $\upsilon_r = \omega/2\pi$ s^{-1}. Equation (6.2) can also be written

$$E_{rot} = \frac{1}{2}\, I_{rot}\omega^2 \tag{6.3}$$

Where

$$I_{rot} = \Sigma m_i r_i^2 \tag{6.4}$$

is the rotational *moment of inertia*. We consider as the rotating body a system of *i* particles of masses m_i, each individual particle at a distance r_i from the axis of rotation; then (6.3) is valid for any rotating body. In nuclear science we primarily have to consider two kinds of rotation: the *intrinsic rotation* (or *spin*) of a body around its own axis (e.g. the rotation of the earth every 24 h), and the *orbital rotation* of an object around a central point (e.g. rotation of the earth around the sun every 365 days). Eqn (6.3) is valid for both cases, but (6.2) only for the orbital rotation of a particle of small dimensions compared to the orbital radius, in which case $I_{rot} = mr^2$. For a spherical homogeneous spinning body (e.g. the earth's intrinsic rotation) of external radius r_{ex}, (6.3) must be used, where $I_{rot} = 2mr_{ex}^2/5$.

6.2.2. Angular momentum

Like linear motion, rotation is associated with a momentum, called the *angular momentum*. For the orbital rotation the *orbital angular momentum* (p_l) is

$$p_l = mv_r r \tag{6.5}$$

while for spin the spin angular momentum (p_s) is

$$p_s = \omega I_{rot} \tag{6.6}$$

Angular momentum is a *vector quantity*, which means that it has always a certain orientation in space, depending on the direction of rotation. For the rotation indicated in Figure 6.3(a), the vector can only point upwards. It would not help to turn the picture

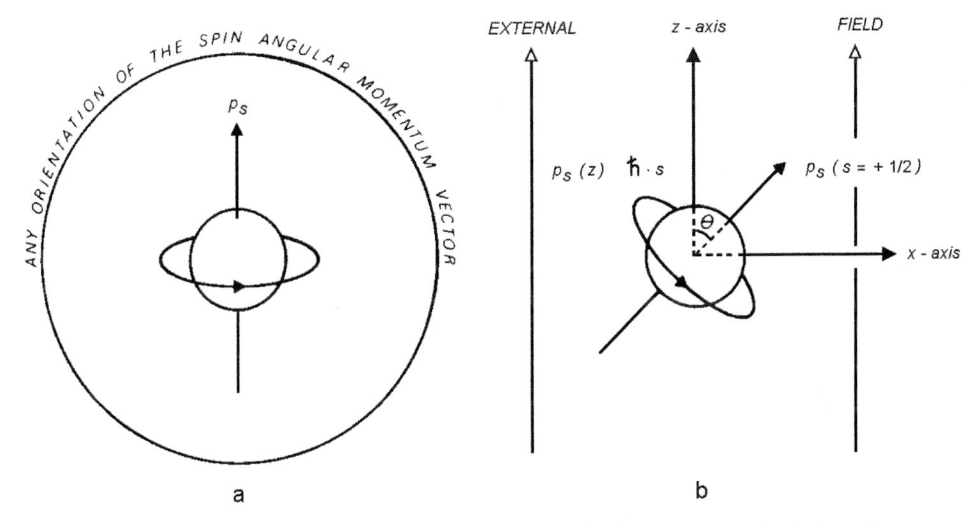

Figure 6.3 A particle spinning in (a) field-free space and (b) in an external field along the *z*-axis.

upside down because the coupling between rotational direction and its vector remains the same, as students south of the equator will agree.

Quantum mechanics prescribes that the spin angular momentum of electrons, protons, and neutrons must have the magnitude

$$p_s = \hbar[s(s+1)]^{1/2} \tag{6.7}$$

here s is the *spin quantum number.* For a *single* particle (electron or nucleon) the spin s is always ½. In addition to spin, the three atomic particles can have orbital movements. Again quantum mechanics prescribes that the magnitude of the orbital angular momentum of these particles

$$p_l = \hbar[l(l+1)]^{1/2} \tag{6.8}$$

We shall refer to l as the *orbital (angular momentum) quantum number.* Only certain values are permitted for l, related to the main quantum number n:

for electrons: $0 \leq l < n-1$
for nucleons: $0 \leq l$

For nucleons but not for electrons l may (and often does) exceed n.

6.2.3. Coupling of spin and orbital angular moments

A rotating charge gives rise to a *magnetic moment* μ_s. The rotating electron and proton can therefore be considered as tiny magnets. Because of the internal charge distribution of the neutron it also acts as a small magnet. In the absence of any external magnetic field these magnets point in any direction in space (Fig. 6.3(a)), but in the presence of an external field they are oriented in certain directions determined by quantum mechanical rules. This is indicated by the angle θ in Figure 6.3(b), when we have the spinning particle in the center of a coordinate system. The quantum mechanical rule is that the only values allowed for the projections of spin angular momentum $p_s(z)$ on the field axes are:

$$p_s(z) = \hbar m_s \tag{6.9}$$

For composite systems, like an electron in an atom or a nucleon in a nucleus, the *(magnetic) spin quantum number* m_s may have two values, +½ or −½, because the spin vector has two possible orientations (up or down) with regard to the orbital angular momentum.

The orbital movement of an electron in an atom, or of a proton in the nucleus, gives rise to another magnetic moment (μ_l) which also interacts with external fields. Again quantum mechanics prescribes how the orbital plane may be oriented in relation to such a field (Fig. 6.4(a)). The orbital angular momentum vector l can assume only such directions that its projection on the field axes, $p_l(z)$, has the values

$$p_l(z) = \hbar m_l \tag{6.10}$$

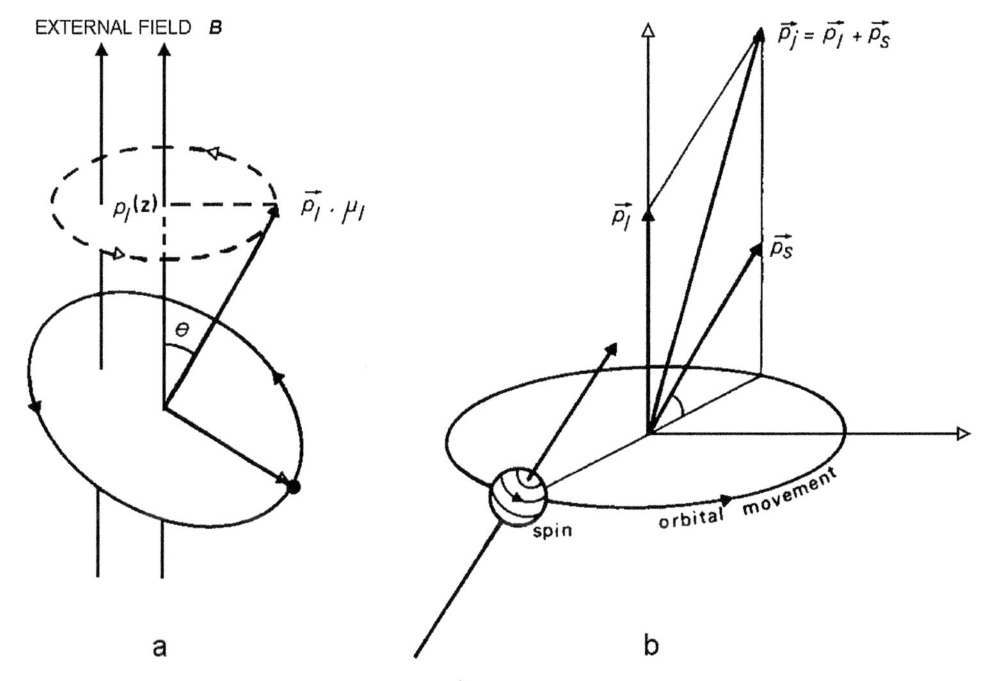

Figure 6.4 (a) Angular momentum caused by an orbiting particle and permitted value projected on the external field axis. The vector p_l precesses around the field axis as indicated by the dashed circle (ellipse in the drawing). (b) The l–s coupling of orbital angular momentum and spin leads to a resultant angular momentum p_j.

m_l is referred to as the *magnetic orbital quantum numbers;* m_l can have all integer values between $-l$ and $+l$.

For a single-particle, nucleon or electron, the orbital and spin angular moments add vectorially to form a *resultant* vector (Fig. 6.4(b)),

$$\vec{p}_j = \vec{p}_l + \vec{p}_s \tag{6.11}$$

$\vec{p}_j$ will orient itself towards an external field so that only the projections

$$p_j(z) = \hbar m_j \tag{6.12}$$

are obtained on the field axes, m_j is the *total magnetic angular momentum quantum number*; it can have all integer values between $-j$ and $+j$. The magnitude of p_j is

$$p_j = \hbar[j(j+1)]^{1/2} \tag{6.13}$$

where

$$j = l \pm s \tag{6.14}$$

Here j is the *total* (*resultant*) *quantum number of the particle.*

6.2.4. Magnetic moments

a. Single particles. Dirac showed in 1928 that the spin magnetic moment of the electron is:

$$\mu_s(\text{electron}) = \mathbf{\hbar e}/2\mathbf{m}_e = \mathbf{B}_e \tag{6.15}$$

where $\mathbf{B}_e = 9.273 \times 10^{-24}\ \mathrm{JT}^{-1}$ (joule per tesla). This value is referred to as one *Bohr magneton* (B.m.). For the proton one would expect the spin magnetic moment μ_s to be

$$\mathbf{B}_e \cdot \mathbf{m}_e/\mathbf{m}_p = \mathbf{B}_n \tag{6.16}$$

$\mathbf{B}_n$ (n for nucleon) has the value $5.051 \times 10^{-27}\ \mathrm{JT}^{-1}$ and is referred to as one *nuclear magneton* (n.m.). However, measurements show that

$$\mu_s(\text{proton}) = 14.1 \times 10^{-27}\ \mathrm{JT}^{-1} = 2.793\ \text{n.m.} \tag{6.17}$$

The reason for the higher value is found in the uneven charge distribution within the proton.

Recall that the neutron also has an uneven charge distribution. As pointed out earlier, this gives rise to a neutron spin magnetic moment

$$\mu_s(\text{neutron}) = -1.913\ \text{n.m.} \tag{6.18}$$

The magnetic moment μ_l, caused by a charge q in *circular orbit*, can be calculated from classical physics: the current caused in the orbit, qv_r, times the area encircled, πr^2. Therefore

$$\mu_l = qv_r r/2 \tag{6.19}$$

Dividing (6.19) by p_l according to (6.5), gives

$$\gamma = \mu_l/p_l = q/2m \tag{6.20}$$

This ratio γ is called the *gyromagnetic ratio.* If γ and p_l are known, μ_l may be calculated. Equation (6.20) is valid for electrons ($\mathbf{e}$, $\mathbf{m}_e$), but for protons only the left part. Because of the quantization of $p_l(z)$, (6.12), the component of the orbital magnetic moment in the field direction (Fig. 6.4(a)) is also quantized:

$$\mu_l(\text{electron}) = m_l \mathbf{B}_e \tag{6.21}$$

Such a simple approach is not possible for a nucleon in a nucleus, because no nucleon moves completely independent of the other nucleons, nor is the orbital path always circular.

b. Atoms and nuclei. In an atom with many electrons, the spin and angular moments of the electrons couple vectorially and separately to form of resultant quantum numbers (using conventional symbolism)

$$S = \Sigma \vec{s}_i \tag{6.22}$$

$$L = \Sigma \vec{l}_i \tag{6.23}$$

which couple to form the total angular momentum (or internal) quantum number of the atom

$$\vec{J} = \vec{L} + \vec{S} \quad (6.24)$$

(see Table 6.2). Equations (6.22) to (6.24) are referred to as *Russell-Saunders coupling*. For the atom as a whole, the magnetic moment is

$$\mu(\text{atom}) = g_j \, \mathbf{e} \, p_j / 2 \, \mathbf{m}_\text{e} = g_j \, \mathbf{B}_\text{e} \, m_j \quad (6.25)$$

where g_j is the Landé factor and $+J \leq m_j \leq -J$. The Landé factor accounts for the effect of mutual screening of the electrons.

The *nuclear magnetic moment* depends on the spin and angular moments of the neutrons and protons. For the nucleus it is given by

$$\mu(\text{nucleus}) = g_I \, \mathbf{e} \, p_I / 2 \, \mathbf{m}_\text{p} \quad (6.26)$$

where g_I is the nuclear *g*-factor and p_I, the magnitude of the nuclear spin angular moment. Because this moment can have only the projections $\hbar m_I$ on the axes of a magnetic field where m_I is the nuclear magnetic angular momentum quantum number $(-I \leq m_I \leq I)$, we may write this equation

$$\mu(\text{nucleus}) = g_I (\mathbf{e} \, \hbar / 2 \, m_p) m_I = g_I \, \mathbf{B}_\text{n} \, m_I \quad (6.27)$$

I is the *total nuclear spin*. We shall see in the next section how I can be determined.

In Table 6.1 we have summarized the most important properties of the atomic constituents, and in Table 6.2 their modes of interaction. In Table 6.3 some spin and magnetic moments are given for stable and radioactive nuclei.

6.2.5. Precession

Before going into the details of nuclear structure, there is one more property of the nucleon which must be considered. Both types of angular momenta, p_s and p_l, as well as their corresponding magnetic moments, are vector quantities. Quantum mechanics forbids $\vec{p}$ (and consequently μ) to be exactly parallel with an external field. At the same time the external field tries to pull the vector so that the plane of rotation becomes perpendicular to the field lines. The potential magnetic energy is

$$E_\text{magn} = \vec{B} \cdot \vec{\mu} = B \cdot \mu \cos \theta \quad (6.28)$$

(see Figs. 6.3 and 6.4) where B is the magnetic field strength. However, because of the inertia of the particle this does not occur until the particle has moved somewhat along its orbit, with the consequence that the vector starts to rotate around the field axes as shown in Figure 6.4(a). The angular momentum vector therefore precesses around the field axes, like a gyroscope axis. We may rewrite (6.20) as

$$\gamma = \vec{\mu} / \vec{p}_\text{rot} \quad (6.29)$$

Table 6.2 Summary of atomic and nuclear properties associated with particle interactions (q.n. = quantum number)

Line property	Electron-electron interaction	Nucleon-nucleon interaction
10 Spin-spin (*s-s*) coupling, q.n.	$S = \Sigma s_i$ (strong)	For each nucleon $j = l \pm s$
11 Total spin magnetic moment	$\mu_s = 2(S(S+1))^{1/2}$ B.m.	and $I = \Sigma j_i$
12 Orbit-orbit (*l-l*) coupling, q.n.	$L = \Sigma l_i$ (strong)	For ee nuclei: groundstate $l = 0$
13 Total orbital magnetic moment	$\mu_L = 2(L(L+1))^{1/2}$ B.m.	For eo, oe nuclei: groundstate $l = j$
14 Spin-orbit coupling, q.n.	$J = S + L$ (weak)	For oo nuclei: groundstate varies
15 Resulting angular momentum	$p_J = \hbar(J(J+1))^{1/2}$ B.m.	$p_I = \hbar(I(I+1))^{1/2}$ n.m.
16 Resulting magnetic moment	$\mu_J = g_J(J(J+1))^{1/2}$ B.m.	$\mu_I = g_I(I(I+1))^{1/2}$ n.m.
17 Transition selection rules	$\Delta L \pm 1$, $\Delta J = 0$ or ± 1	$\Delta_I =$ integer, $\Delta_j > 0$
18 Multipole moment	Electric dipole	Electric or magnetic multipole
Electron-nucleon interactions		
19 Grand atomic angular mom., q.n.	$F = J + I = S + L + I$	
20 Grand atomic angular mom.	$p_F = \hbar(F(F+1))$; field projections $0, \ldots, \hbar F$	
21 External field: none	Hyperfine spectrum (hfs): J levels split due to nuclear spin I	
22 External field: weak ($\leq 10^{-2}$ T)	Hfs F-levels split into $2F+1$ levels (Zeeman effect)	
23 External field: average (~ 10 T)	Electron-nucleon q.n. decouple, producing $(2J+1)(2I+1)$ levels	
24 External field: very strong	Electron spin-orbit decouple, producing separate S- and L-levels	

Table 6.3 Spin I, parity (+ even, − odd), nuclear magnetic moment (μ n.m.) and quadrupole moment $\hat{Q}$ (10^{-28} m^2) for some nuclides ^{A}X

Stable nuclides				Radioactive nuclides				
^{A}X	I (±)	μ_I	$\hat{Q}$	^{A}X	I (±)	μ_I	$\hat{Q}$	$t_{1/2}$
^{1}H	1/2 +	+2.793		^{3}H	1/2 +	+2.979		12.33 y
^{2}H	1 +	+0.857	+0.003	^{14}C	0 +			5730 y
^{10}B	3 +	+1.801	+0.085	^{24}Na	4 +	+1.690		14.959 h
^{11}B	3/2	+2.689	+0.041	^{32}P	1 +	−0.252		14.262 d
^{12}C	0 +			^{36}Cl	2 +	+1.285	−0.018	3.01×10^5 y
^{13}C	1/2	+0.702		^{45}Ca	7/2	−1.327	+0.046	163.8 d
^{14}N	1 +	+0.404	+0.019	^{55}Fe	3/2			2.73 y
^{16}O	0 +			^{60}Co	5 +	+3.799	+0.44	5.271 y
^{17}O	5/2 +	−1.894	−0.026	^{64}Cu	1 +	−0.217		12.70 h
^{19}F	1/2 +	+2.629		^{95}Zr	5/2 +			64.02 d
^{23}Na	3/2 +	+2.218	+0.101	^{131}I	7/2 +	+2.742	−0.40	8.0207 d
^{31}P	1/2 +	+1.132		^{137}Cs	7/2 +	+2.841	+0.051	30.0 y
^{33}S	3/2 +	+0.644	−0.076	^{140}La	3	+0.730	+0.094	1.678 d
^{39}K	3/2 +	+0.391	+0.049	^{198}Au	2	+0.593	+0.68	2.6952 d
^{59}Co	7/2	+4.627	+0.404	^{232}Th	0 +			1.405×10^{10} y (α)
^{87}Sr	9/2 +	−1.094	+0.335	^{235}U	7/2	−0.38	+4.55	7.038×10^8 y (α)
^{141}Pr	5/2 +	+4.275	−0.059	^{238}U	0 +		13.9	4.468×10^9 y (α)
^{197}Au	3/2 +	+0.146	+0.547	^{239}Pu	1/2 +	+0.203		2.411×10^4 y (α)

This form makes it more obvious why γ is referred to as the gyromagnetic ratio. Equation (6.29) is valid both for angular momentum and spin, but of course the value is different for electrons and protons. The angular velocity ω_γ of the precession is found to be

$$\omega_\gamma = \mu \vec{B} / \vec{p}_{\rm rot} \tag{6.30}$$

By replacing angular velocity with frequency ($\omega = 2\pi\upsilon$)

$$\upsilon_\gamma = \gamma \cdot \vec{B} / 2\pi \vec{p}_{\rm rot} \tag{6.31}$$

or

$$\upsilon_\gamma = \gamma \cdot \vec{B} / 2\pi \tag{6.32}$$

where υ_γ is the Larmor precession frequency.

6.3. THE SINGLE-PARTICLE SHELL MODEL

6.3.1. Quantum number rules

Let us assume that a nucleon moves around freely in the nuclear potential well, which is spherically symmetric, and that the energy of the nucleon varies between potential and kinetic energy like a harmonic oscillator, i.e. the potential walls (see Figs. 6.1 and 6.2) are parabolic. For these conditions the solution of the Schrödinger equation yields:

$$E(\text{nucleon}) = \hbar\left(2\,U_o/mr^2\right)^{1/2}[2(n-1)+l] \tag{6.33}$$

where U_o is the potential at radius $r = 0$ and m is the nucleon mass. We have defined n and l previously. The square root, which has the dimension s^{-1}, is sometimes referred to as the oscillator frequency (ω in Table 6.6). The following rules are valid for nucleons in the nuclear potential well:

(a) l can have all positive integer values beginning with 0, independent of n;
(b) the energy of the l state increases with increasing n as given by (6.33);
(c) the nucleons enter the level with the lowest total energy according to (6.33) independent of whether n or l is the larger;
(d) there are independent sets of levels for protons and for neutrons;
(e) the Pauli principle is valid, i.e. the system cannot contain two particles with all quantum numbers being the same;
(f) the spin quantum numbers must be taken into account (not included in (6.33)).

Following these rules, the nucleons vary greatly in energy and orbital motion. However these rules do not exclude the existence of two nucleons with the same energy (so-called degenerate states), provided the quantum numbers differ

6.3.2. Nuclei without nucleon spin-orbit coupling

If we calculate the sequence of energy levels on the assumption that n is constant, we obtain the pattern in Table 6.4. This sequence correspond exactly to the electronic states in atoms but it does not agree with the observed magic numbers 2, 8, 20, 28, 50, etc., for nuclei.

Applying the rules of the previous section a new set of nucleon numbers is obtained: 2, 8, 20, 40, 70, etc. (Table 6.5, and left column Table 6.6). This level scheme allows a large amount of degeneracy. For example, for $n = 1$ and $l = 3$ (1f-state) we find from (6.33) that $[2(n-1)+l] = 3$, which value is obtained also for $n = 2$ and $l = 1$ (2p-state).

Table 6.4 Energy levels derived on basis of permitted values for the azimuthal quantum number l and spin quantum number s

l	State	Possible quantum values	Number of states (including spin)	Accumulated nucleon number
0	s	0	$1 \times 2 = 2$	2
1	p	−1,0,+1	$3 \times 2 = 6$	8
2	d	−2,−1,0,+1,+2	$5 \times 2 = 10$	18
3	f	−3,−2,−1,0,+1,+2,+3	$7 \times 2 = 14$	32
4	g	−4,−3,−2,−1,0,+1,+2,+3,+4	$9 \times 2 = 18$	50
l	-	$-l, \ldots, +l$	$2(2l+1)$	—

Table 6.5 Energy levels according to equation (6.25)

Levels	Number of nucleons	Accumulated nucleons
1s	2	2
1p	6	8
1d and 2s	10 + 2	20
1f and 2p	14 + 6	40
1g, 2d, and 3s	18 + 10 + 2	70
1h, 2f, and 3p	22 + 14 + 6	112

Since the f-state can have 14 and the p-state 6 nucleons, the degenerate level of both states can contain 20 nucleons. However, the numbers still do not correspond to the experimental magic numbers.

A further refinement is possible if we assume that the nuclear potential well has straight walls, i.e. the potential energy $U(r)$ is $-U_o$ at $r < r_n$ while it is infinite at $r \geq r_n$, where r_n is the nuclear radius. This assumption, when introduced (see Fig. 6.2 and figure in Table 6.6) in the Schrödinger-equation, leads to a splitting of the degenerate levels, so that the lowest energy is obtained for the state with lowest main quantum number n. For our example of the 1f and 2p-states the 1f orbitals are lower in energy than the 2p. This refinement yields the middle row of levels in Table 6.6 but still does not lead to the correct magic numbers.

6.3.3. Nuclear level scheme with nucleon spin-orbit coupling

In multielectron atoms, the Russell-Saunders coupling is present in light atoms. However, in the heaviest atoms of many electrons and in highly charged nuclei, the j–j (spin-orbit) coupling better describes the systems. Haxel, Jensen, Suess, and Goeppert-Mayer in 1949 suggested that the nucleons always experience a strong spin-orbit coupling according to (6.12)

$$j = l \pm s \tag{6.34}$$

and that the total spin of the nucleus I is the sum of the nucleon spins

$$I = \Sigma j \tag{6.35}$$

In this way, all l levels are split into two levels with quantum values $l + ½$ and $l - ½$, of which the former has the lowest energy value (the opposite of the electron case). This yields the row of levels on the right in Table 6.6. Because of the energy splitting the new levels group together so they fit exactly with the experimental magic numbers.

As an example, consider the level designation $1i_{11/2}$. This has the following interpretation: the principal quantum number is 1; i indicates that the orbital quantum

Table 6.6 Nuclear energy levels obtained for the single-particle model by solving the schrödinger-equation (a) for a harmonic oscillator and rounded square-well potentials (b) without and (c) with spin-orbit coupling. The $n\hbar\omega$ figures to the left are energies for model a, according to eqn (11.25). numbers in ( ) indicate orbital capacities and those in [ ] give cumulative capacity up to the given line

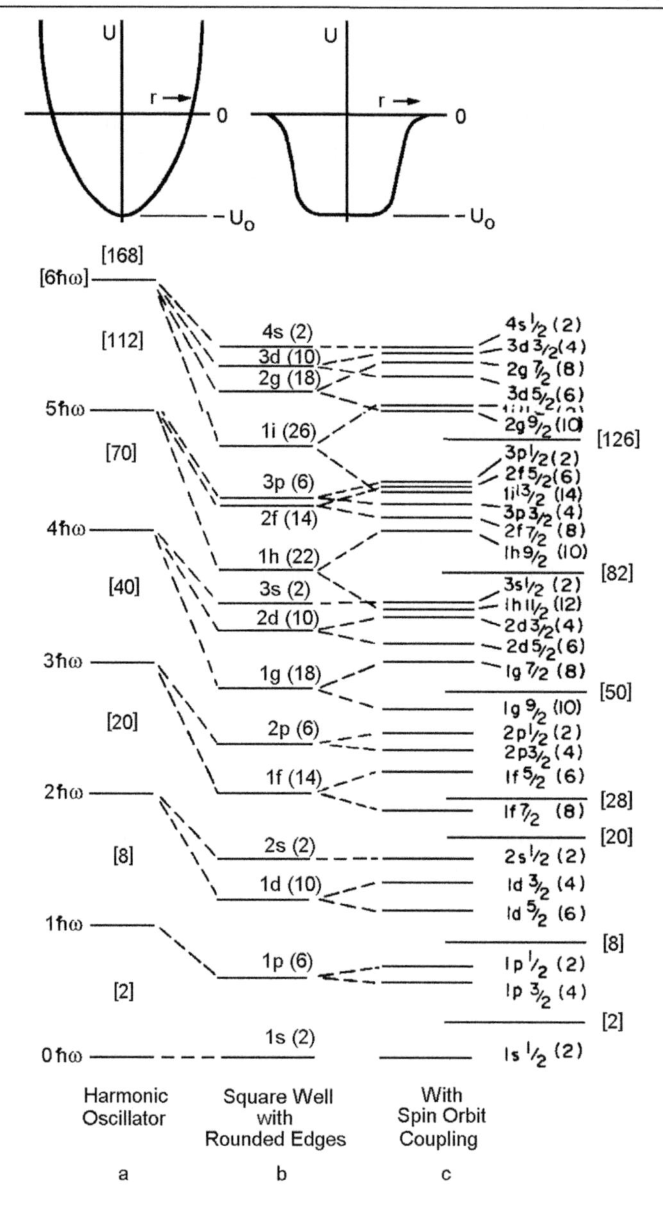

From Gordon and Coryell.

number l is 6; the angular momentum quantum number j is 11/2 ($j = l - 1/2$). The number of permitted nucleons in each level is $2j + 1$, thus 12 for $j = 11/2$.

The magic numbers correspond to sets of energy levels of similar energy just as in the atom the K, L, M, etc., shells represent orbitals of similar energy. The N-electronic shell contains the 4s, 3d, 4p sets of orbitals while the 4th nuclear "shell" contains the $1f_{5/2}$, $2p_{3/2}$, $2p_{1/2}$, and $1g_{9/2}$ sets of orbitals. Remember that there are separate sets of orbitals for protons and for neutrons.

6.3.4. The nuclear spin

The nuclear spin I is obtained from (6.34) and (6.35). Since j is always a half-integer, nuclides with odd number of nucleons (odd A) must have odd spin values (odd I), while those with even A must have even I. The nucleons always pair, so that even numbers of protons produce no net spin. The same is true for even numbers of neutrons. For nuclei of even numbers for both N and Z, the total nuclear spin I is always equal to zero. Some ground state nuclear spins are given in Table 6.3.

For odd A the nuclear spin is wholly determined by the single unpaired nucleon (single particle model). Let us take the nucleus $^{13}_{6}C$ as an example. It contains 6 protons and 7 neutrons. From Table 6.6 we conclude that there are 2 protons in the $1s_{1/2}$ level and 4 protons in the $1p_{3/2}$ level. A similar result is obtained for the first 6 neutrons but the 7th neutron must enter the $1p_{1/2}$ level. The value of I for $^{13}_{6}C$ is thus ½. Another example is $^{51}_{23}V$, which has 23 protons and 28 neutrons. Since $N = 28$, the neutrons do not contribute to I. From Table 6.6 we see that we can accommodate 20 protons in the orbitals $1s^2$, $1p^6$, $1d^{10}$, $2s^2$. The next 3 protons must go into the $1f_{7/2}$ level ($1\ f^3_{7/2}$, where, however, 2 of them are paired. Therefore, the single unpaired proton has $j = 7/2$, leading us to predict $I = 7/2$, which also is the measured nuclear spin value.

When a nucleus is excited, either through interaction with other particles or in a decay process, for nuclei with N or Z values near magic numbers the paired nucleons seem not to be perturbed by the excitation (if it is not too large). As a result, we can associate the excitation with any unpaired nucleons. Let us choose the decay of ^{47}Ca to ^{47}Sc (see Fig. 6.5). ^{47}Ca has 27 neutrons and in the ground state the last 7 neutrons must occupy the $1f_{7/2}$ level. The ground state of (the unstable) ^{47}Sc has 21 protons, the unpaired proton can only be accommodated in the $1f_{7/2}$ level. Thus both of the ground states have $I = 7/2$. The next higher energy states for ^{47}Sc involve the levels $1f_{5/2}$ and $2p_{3/2}$. In the Figure it is seen that these levels are observed although their order is reversed from that in Table 6.6. This reflects some limitation of the single-particle model, and is explained in §6.5.

For odd-odd nuclei the nuclear spin is given by

$$I(\text{odd-odd}) = j_p + j_n = (l_p \pm ½) + (l_n \pm ½) \qquad (6.36)$$

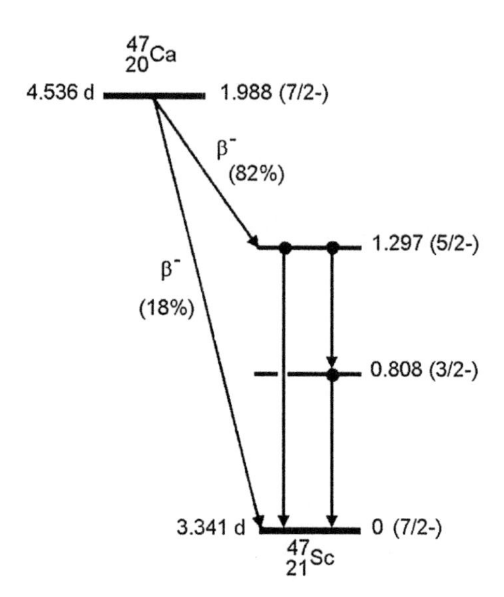

Figure 6.5 Part of the decay scheme for $A = 47$; the data are experimental values.

According to the rules formulated by Brennan and Bernstein (useful in the range 20 $< A <$ 120 for ground states and low-lying long-lived isomeric states), if the odd particles are both particles (or both holes) in their respective unfilled subshells then[1]

$$\begin{aligned} &\text{if } j_p + j_n + l_p + l_n \text{ is even,} \quad \text{then } I = |j_p - j_n| \\ &\text{if } j_p + j_n + l_p + l_n \text{ is odd,} \quad \text{then } I = |j_p \pm j_n| \end{aligned} \tag{6.37}$$

but if we have both particles and holes, $\quad$ then $I = j_p + j_n - 1$

The use of these rules can be illustrated by the case of Cu which has its odd proton in the $1f_{5/2}$ orbital and its odd neutron in the $2p_{3/2}$ orbital. Thus for the proton, $j = 5/2$, $l = 3$; for the neutron, $j = 3/2$, $l = 1$. Because $j_p + j_n + l_p + l_n$ is even, we use $I = |5/2 - 3/2| = 1$, which is the observed spin value. The lightest nuclei are exceptions to this rule since they often exhibit LS coupling according to (6.18). ^{10}B is an example; it has 5 protons and 5 neutrons, the fifth nucleon being in the $1p_{3/2}$ state. Thus $I = j_1 + j_2 = 3$, which is observed.

We mentioned in §6.2.4 that theoretical calculations of the nuclear magnetic moment, (6.21), are usually not satisfactory. The value of μ(nucleus) can have a number of values depending on m_I, with a maximum value of I. The parity[2] of the nucleus follows the rules: (i) when both particles are in states either of even parity or of odd parity,

[1] Vertical bars used here denote absolute value.

[2] Parity is the behaviour of wave functions when all coordinate signs are reversed. Even parity — no effect, else odd parity.

they combine to a system of even parity. (ii) one particle in a state of even parity and one in a state of odd parity combine to a system of odd parity.

6.4. DEFORMED NUCLEI

6.4.1. Deformation index

Both the liquid-drop model and the single-particle model assume that the mass and charge of the nucleus are spherically symmetric. This is true only for nuclei close to the magic numbers; other nuclei have distorted shapes. The most common assumption about the distortion of the nuclide shape is that it is ellipsoidal, i.e. a latitudinal cross-section of the nucleus is an ellipse. Figure 6.6 shows the oblate (flying-saucer-like) and prolate (egg-shaped) ellipsoidally distorted nuclei; the prolate shape is the more common. Deviation from the spherical shape is given by

$$\beta = 2(a - c)/(a + c) \tag{6.38}$$

where a and c are the elliptical axes as shown on in Figure 6.6(c). For prolate shape $\beta > 0$, and for oblate shape $\beta < 0$. The maximum deformation observed is about $\beta = \pm 0.6$.

The deformation is related to the nuclear shell structure. Nuclei with magic numbers are spherical and have sharp boundary surfaces (they are "hard"). As the values of N and Z depart from the magic numbers the nucleus increases its deformation.

6.4.2. Electric multipoles

In a spherical nucleus we assume the charge distribution to be spherical and the nucleus acts as a monopole. In the deformed nuclei, the nuclear charge has a non-spherical distribution. The potential at a point, $\bar{x}$, $\bar{y}$, $\bar{z}$ (Fig. 6.6) will be found to vary depending on the charge distribution and mode of rotation of the nucleus. The nuclear charge may be distributed to form a dipole, a quadrupole, etc. Nuclei are therefore divided into

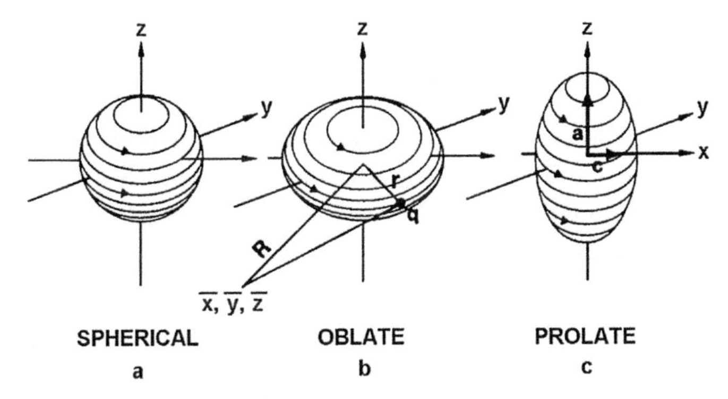

Figure 6.6 Three nuclidic shapes: (a) spherical nucleus, (b) oblate (extended at the equator), and (c) prolate (extended at the poles).

different classes depending on their electrical moments: monopoles, dipoles, quadrupoles, octupoles etc.

It has been found that nuclei with spin $I = 0$ have no multipole moment which corresponds to the reasoning made above regarding spherical shape and magic numbers. According to theory, nuclei with $I = ½$ can have a dipole moment, but this has not yet been shown experimentally. Nuclei with $I = 1$ have quadrupole moments; they are fairly common. The quadrupole moment $\hat{Q}$ can be calculated for spheroidal (i.e. deformation not too far from a sphere) nuclei in terms of the electron charge, **e**, by

$$\hat{Q} = (2/5)Z\left(a^2 - c^2\right) \quad (6.39)$$

$\hat{Q}$ is usually referred to as the internal quadrupole moment, i.e. the expected value for a rotation around the z-axis. However, quantum mechanics makes this impossible and gives for the maximum observable quadrupole moment

$$\hat{Q}_{obs} = \hat{Q}(I - 1/2)/(I + 1) \quad (6.40)$$

Thus $\hat{Q}_{obs} = 0$ for $I \leq 1/2$. $\hat{Q}_{obs}$ is usually given in area (m^2). Most commonly $10^{-28} m^2$ is used as unit and referred to as one barn, $\hat{Q}_{obs}$ is > 0 for the more common prolate shape, and < 0 for oblate. Some measured values are given in Table 6.3.

The rotation of nuclei with electric multipoles gives rise to formation of magnetic multipoles. Nuclei can therefore also be divided according to the magnetic moments in the same way as according to their electrical moments.

6.4.3. The collective nuclear model

In 1953 A. Bohr and Mottelsen suggested that the nucleus be regarded as a highly compressed liquid, undergoing quantized rotations and vibrations. Four discrete collective motions can be visualized. In Figure 6.6 we can imagine that the nucleus rotates around the y-axis as well as around the z-axis. In addition it may oscillate between prolate to oblate forms (so-called *irrotation*) as well as *vibrate*, for example, along the x-axis. Each mode of such collective nuclear movement has its own quantized energy. In addition, the movements may be coupled (cf. coupling of vibration and rotation in a molecule).

The model allows the calculation of rotational and vibrational levels as shown in Figure 6.7. If a ^{238}U nucleus is excited above its ground state through interaction with a high energy heavy ion (coulomb excitation), we have to distinguish between three types of excitation: (a) nuclear excitation, in which the quantum number j is changed to raise the nucleus to a higher energy level (according to Table 6.6); (b) vibrational excitation, in which case j is unchanged, but the nucleus is raised to a higher vibrational level, characterized by a particular vibrational quantum number (indicated in the figure); (c) rotational excitation, also characterized by a particular rotational quantum number. The Figure shows that the rotational levels are more closely spaced and thus transitions

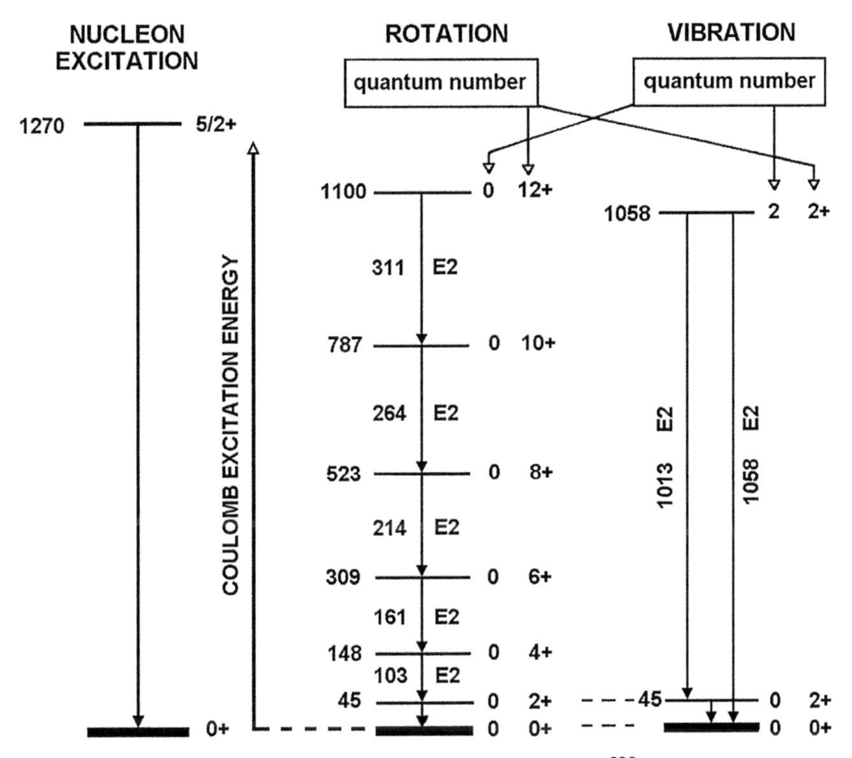

Figure 6.7 Nucleonic, rotational, and vibrational levels observed in ^{238}U; energy in keV. *(From E. K. Hyde.)*

between rotational levels involve lower energies than de-excitation from excited nuclear or vibrational states.

In case of even-even nuclei, the rotational energy levels can be often calculated from the simple expression

$$E_{rot} = \left(\hbar^2/2I_{rot}\right)n_r(n_r + 1) \tag{6.41}$$

where I_{rot} is the moment of inertia and n_r the rotational quantum number; this equation is identical to (3.39). The validity of this equation depends on whether the different modes of motion can be treated independently or not, which they can for strongly deformed nuclei like ^{238}U.

6.5. THE UNIFIED MODEL OF DEFORMED NUCLEI

The collective model gives a good description for even-even nuclei but cannot account for some of the discrepancy between observed spins and the spin values expected from the single-particle shell model. The latter was developed on the assumption of a nucleon

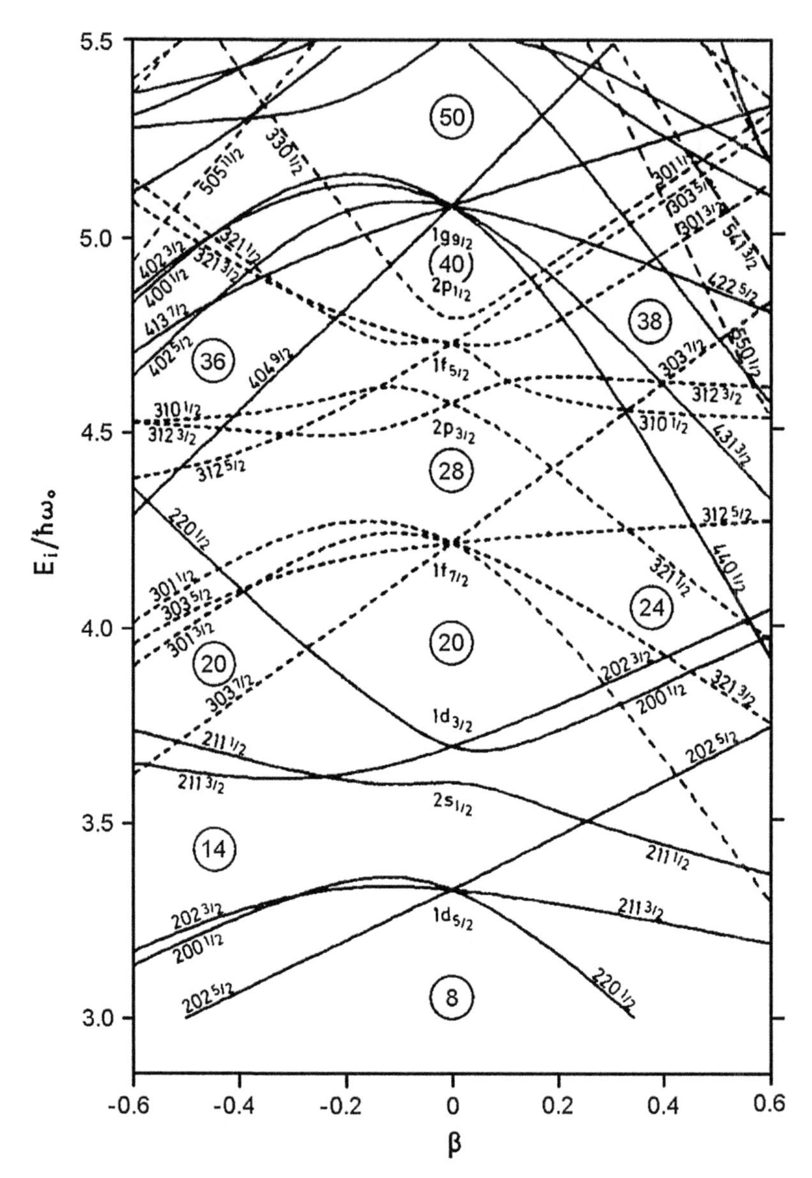

Figure 6.8 Nilsson diagram of single-particle energy levels for deformed nuclei. The energy scale is in units of $\hbar\omega_0$ ($\hbar\omega_0$ is approximately $41A^{-1/3}$ MeV). The figures along the center give the neutron or proton numbers and l–j values for the single nucleon. The figure combinations at the end of the lines are approximate quantum numbers; the first figure is the principal quantum number n. Full-drawn lines are for even parity, broken lines for odd. *(From Larsson, Leander, Ragnarsson and Alenius.)*

moving freely in a symmetrical potential well, a situation which is valid only for nuclei near closed shells. The angular momentum of an odd-A deformed nucleus is due both to the rotational angular momentum of the deformed core and to the angular momentum of the odd nucleon.

Consequently the energy levels for such a nucleus are different from those of the symmetric shell model.

This situation was taken into account by S. G. Nilsson, who calculated energy levels for odd nuclei as a function of the nuclear deformation β. Figure 6.8 shows how the energies of the Nilsson levels vary with the deformation β of the potential well. Each shell model level of angular momentum j splits into $j + ½$ levels (called *Nilsson levels* or *states*). Each level may contain up to two nucleons and form the ground state of a rotational band. In addition the undeformed levels ($\beta = 0$) appear in somewhat different order than for the symmetric shell model (Table 6.6). This leads to a reversal in order for some of the levels, e.g. $1f_{5/2}$ and $2p_{3/2}$ (this explains the observed level order for ^{47}Sc, Fig. 6.5). The Nilsson levels are quite different in all characteristics from the shell model states, and their prediction of energies, angular momenta, quantum numbers, and other properties agrees better with experimental data for deformed nuclei than those of any other model.

As an example we may choose $^{23}_{11}Na$, which has a quadrupole moment of 0.101 barn. Assuming the nuclear radius to be 1.1 $A^{1/3} = (a+c)/2$ (fm) we can use (6.38) and (6.39) to calculate a deformation index $\beta = 0.12$. (The value 1.1 for the constant r_o in (4.14) gives better agreement in nuclei where the inertia of the nucleus is involved.) From Figure 6.8 the 11th proton must enter the 3/2 level rather than the $1d_{5/2}$ level as the symmetric shell model indicates in Table 6.6. The experimental spin of 3/2 confirms the Nilsson prediction. Similarly, for the deformed $^{19}_{9}F$ and $^{19}_{10}Ne$, we expect from Table 6.8 the odd nucleon to give spin of ½ and not 5/2, as would be obtained from Table 6.6. Again, experiment agrees with the prediction of ½.

6.6. INTERACTION BETWEEN THE NUCLEAR SPIN AND THE ELECTRON STRUCTURE

We have already seen how the spin and orbital angular momentum of the electrons and of the nucleus produce magnetic fields that interact with each other. The field produced by the electrons is much larger than that of the nucleus, and consequently the nuclear spin is oriented in relation to the field produced by the electron shell. By contrast the effects of the nuclear spin on the electron structure is so small that it usually neglected. Nuclear physics has provided us with instruments of such extreme sophistication and resolution that there are many ways of measuring with great accuracy the interaction between the nucleus and the electrons. The result has been research tools of utmost importance, most prominent being the *nuclear magnetic resonance* (nmr) techniques. The separate disciplines of chemistry, atomic physics, nuclear physics, and solid state physics approach each other closely in such techniques, and an understanding of the theory and experimental methods requires knowledge of all these subjects.

In this section only a few important aspects of the interaction between nuclear spin and electronic structure are reviewed. The methods described are usually not considered to fall within the framework of nuclear chemistry, but in all scientific fields it is important to be able to reach the border and look at developments and techniques used on the other side. Such information is often the seed to further scientific development.

6.6.1. Hyperfine spectra

In §6.3.4 we mentioned that the electrons in the atomic shell have Russell-Saunders coupling, (6.24), $\vec{J} = \vec{L} + \vec{S}$, where $\vec{J}$ is referred to as the *internal quantum number.* The magnetic field created by the electrons interacts with that caused by the nuclear spin to yield the grand atomic angular momentum vector

$$\vec{F} = \vec{J} + \vec{I} \tag{6.42}$$

The magnitude of this momentum is

$$p_F = \hbar[F(F+1)]^{1/2}, \tag{6.43}$$

but only projections $p_F(z) = 0, \ldots \hbar F$ are permitted on an external (to the atom) field axis. This leads to a large number of possible energy levels, although they are limited by certain selection rules.

The nuclear spin can orient itself in relation to $\vec{J}$ in

$$2I + 1 \ \text{ directions if } I \leq J$$

$$2J + 1 \ \text{ directions if } J \leq I$$

Consider ^{23}Na as an example (Fig. 6.9);[3] it has a nuclear spin $I = 3/2$. The yellow sodium line of 589.6 nm is caused by de-excitation of its electronically excited p state to the ground state $2s_{1/2}$.

The difference between the two p-states is very small, only 0.0022 eV (or about 0.6 nm), and can only be observed with high resolution (*fine spectra*). To each of these three levels the nuclear spin I has to be added, yielding the quantum number F according to (6.31). It is easy to determine that the level number rule holds, e.g. for $I = J$ we must have $2I + 1 = 4$ possible levels. These levels can be observed in optical spectrometers only at extremely high resolution (*hyperfine spectrum*, hfs). The energy separation between the hfs lines depends on the nuclear magnetic dipole μ_I, the spin value I, and the strength

[3] The letters S and P stand for $\vec{L} = 0$ ($\Sigma l_i = 0$) and 1, respectively. The superscript 2 refers to the number of possible S-values; thus $2S + 1 = 2$, i.e. $S = \Sigma s_i = 2$. The subscript gives the $\vec{J}$-value.

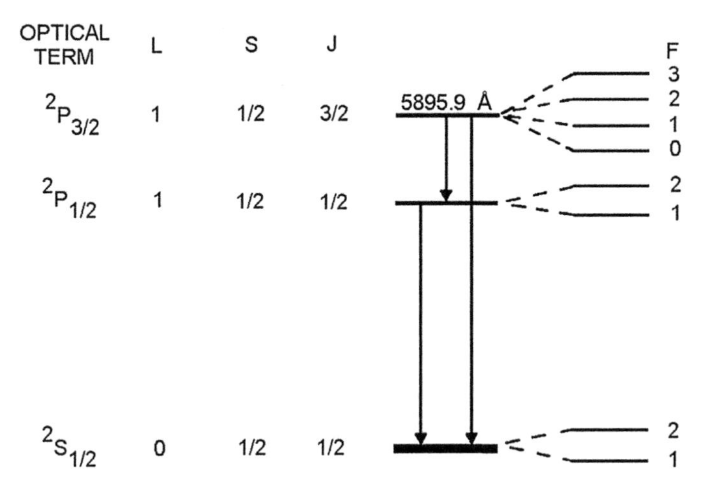

Figure 6.9 The development of hyperfine lines in an optical spectrum of sodium due to the nuclear spin I, which enters through $\vec{F} = \vec{J} + \vec{I}$.

of the magnetic field produced at the nucleus, as discussed in §6.2.4 (see also §6.6.3). The energy separation is very small, on the order of 10^{-5} eV, corresponding to a wavelength difference of about 1/1000 of a nm. Even if there is great uncertainty in the energy determination of the levels, simply counting the number of hyperfine lines for a certain electronic $\vec{J}$-level gives the value of the nuclear spin, because $-I \leq m_I \leq I$ (number of m_I values $= 2I + 1$).

At very high resolutions it may be possible to determine μ_I from hfs, and from this to calculate I. The hyperfine splitting and shifting of optical lines have yielded important information not only about nuclear spins and magnetic moments, but also about the electric charge distribution and radius of the nucleus.

6.6.2. Atomic beams

The hyperfine spectra are obtained from light sources in the absence of any external (to the atom) magnetic field. If the source is placed between the poles of a magnet, whose strength B is progressively increased, the following sequence of changes takes place (see Table 6.2, electron-nucleon interaction).

Suppose B is increased slowly to 10^{-2} T. Each hyperfine level F is found to split into $2F + 1$ levels, since the quantum mechanical rule of permitted projections on an external field vector comes into operation. These permitted projections can vary from zero to a maximum value of $\hbar F$; for $F = 3$ seven new lines are obtained. Further increases in B to 10 T leads to a decoupling of F into its components J and I (Fig. 6.10). For each projection of J on the field vector there are $2I + 1$ lines $(-I...0...+I)$, giving

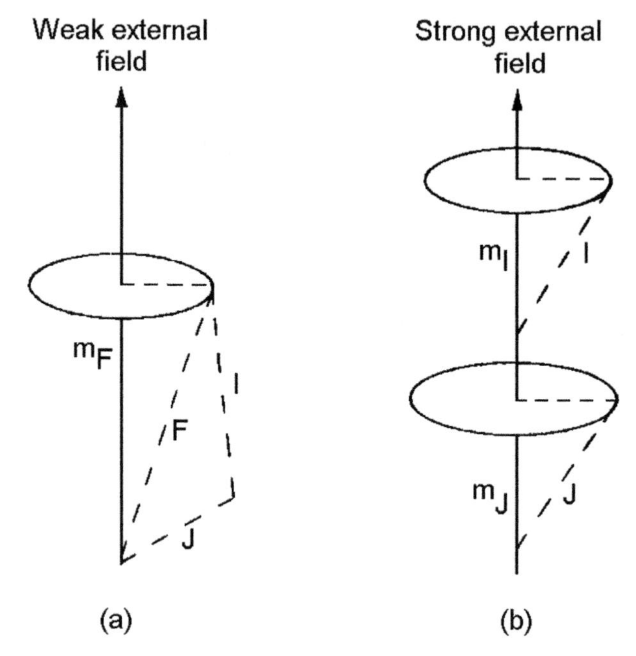

Figure 6.10 (a) Coupling of electronic and nuclear angular moments J and I in a very weak magnetic field. Their sum F forms the quantum number m_F along the direction of the magnetic field, the maximum magnitude of which is $p_F(z) = \hbar F$. In a strong magnetic field the electronic and nuclear angular moments decouple, so that I and J act independently, giving projections m_j and m_i on the field vector.

altogether $(2J + 1)(2I + 1)$ lines. The splitting of the spectral lines in a weak magnetic field is called the *Zeeman effect*.

The decoupling of the angular momentum of the atom into its electronic and nuclear components is used in the *atomic beam apparatus* to determine nuclear magnetic moments and spin values. A beam of atoms, produced in an oven, is allowed to enter a tube along which a series of magnets (usually three) have been placed. The magnetic field splits the atomic beam into several component beams, each containing only atoms which have the same values for all the quantum numbers. Between the magnets there is a small coil connected to a high frequency oscillator, which produces a weak oscillating magnetic field. When this oscillator is tuned to an energy **h***υ*, which exactly matches the energy difference between two quantum states of the atom, energy may be absorbed producing a transition from one of the states to another. Subsequently the atoms are deflected differently by the magnetic field than they were before the energy absorption. By a combination of homogeneous and heterogeneous magnetic fields the atomic beam apparatus allows only those atoms which have absorbed the energy quantum **h***υ* to reach the detector. From the properties of the magnetic fields and the geometric dimensions and frequency of the instrument, the magnetic moment of the atom in different quantum states can be determined. This allows calculation of the magnetic moment and spin of the nucleus. This technique is of interest to the nuclear chemist because spin values can be obtained for short-lived nuclei in submicroscopic amounts. In contrast the hfs technique requires macroscopic amounts of atoms.

6.6.3. Nuclear magnetic resonance[4]

When an atom is placed in an external magnetic field (strength B) so that J and I decouple (cf. (6.42) and (6.43)), the nuclear magnetic moment vector μ_I must precess around the field direction, with the components in the direction of the field restricted to

$$\mu_I = g_I\,\mathbf{B}_\mathrm{n}\,m_I \tag{6.44}$$

In the external field the states with different m_I have slightly different energies. The potential magnetic energy of the nucleus is

$$E_\mathrm{magn} = -\overrightarrow{\mu}_\mathrm{I}\overrightarrow{B} = -g_\mathrm{I}\,\mathbf{B}_\mathrm{n}\,m_I\,B \tag{6.45}$$

The energy spacing between two adjacent levels is:

$$\Delta E = g_I\,\mathbf{B}_\mathrm{n}\,B \tag{6.46}$$

because $\Delta m_I = \pm 1$. For example, consider the case of the nucleus ^{19}F for which $g_I = 5.256$ in a field of 1 T, $\Delta E = 5.256 \times 5.0505 \times 10^{-27} \times 1 = 2.653 \times 10^{-26}$ J. The frequency of electromagnetic radiation corresponding to this energy is 4.0×10^7 s^{-1} or 40 MHz. This lies in the short wave length region, $\lambda = 7.5$ m. This frequency is also the same as that of the Larmor precession, as given by (6.32).

These relations can be used to calculate the nuclear magnetic moment, if I is known, or vice versa. Figure 6.11 shows the results of an experiment in which a sample has been placed in a variable magnetic field containing two coils, one connected to a radio transmitter operating at 5 MHz and the other to an amplifier. The sample is a glass tube (B, O, Na, Al, Si atoms) containing a piece of copper alloy (Cu, Al) in water (H, D, O). By varying the magnetic field a number of resonances are observed. For example, for ^{23}Na one has a resonance at 0.443 T. With (6.46) we can calculate $g_I = 1.48$, and, consequently, the magnetic moment of ^{23}Na is 2.218 (Table 6.3). Further, for ^{23}Na, $m_I = 3/2$ and $I = 3/2$.

The magnetic field experienced by the nucleus is not exactly equal to the external field because of the shielding effect of the electron shell, even if I and J are decoupled. Although this shielding of the nucleus is very small, about 10^{-5} B, it can still easily be detected with modern equipment. The shielding effect depends on the electronic structure.

The structural information that can be provided by this method is very detailed, and a new and deeper insight in chemical bonding and molecular structure is provided. The

[4] Today usually called Magnetic Resonance when used in medicine, etc.

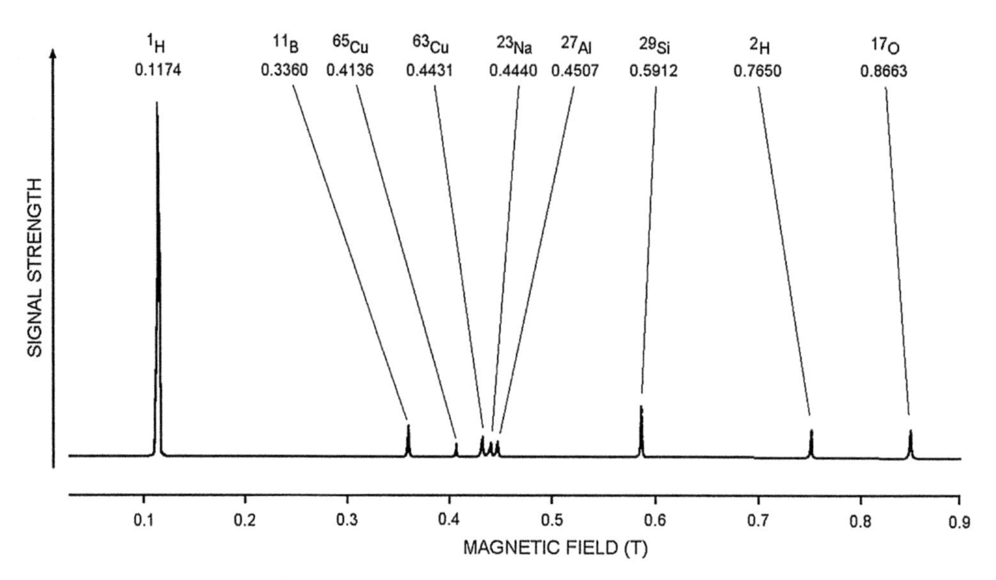

Figure 6.11 Nmr spectrum for a glass vessel containing water and some copper. The magnetic field is given in tesla (T).

nmr technique has therefore become a central tool for the investigation of chemical structures in solids, liquids, and the gaseous state. Tomography by nmr is also a valuable tool for medical imaging. To avoid the unease of the patients caused by the word "nuclear", nmr in hospitals is now usually called magnetic resonance, mr.

6.7. RADIOACTIVE DECAY AND NUCLEAR STRUCTURE

In the preceding section we have described three methods of determining nuclear spin - one optical and two magnetic. The nuclear spin plays a central role in forming the nuclear energy states. It is therefore to be expected that it also should be of importance in nuclear reactions and in radioactive decay. Let us consider some rules for the lifetimes of unstable nuclei, for their permitted modes of decay, and for the role of nuclear spin. Knowing these rules, it is, for example, possible from a decay scheme to predict the spin states of levels which have not been measured.

6.7.1. Gamma-decay

Photons are emitted in the transition of a nucleus from a higher energy state (level) to a lower

$$E_\gamma = \mathbf{h}\upsilon = E_f - E_i \tag{6.47}$$

where f and i refer to the final and initial states. Because the photon has spin 1, de-excitation through γ-emission is always accompanied by a spin change $\Delta I = 1$. This

Table 6.7 Classification of radiation emitted in γ-decay and lifetime calculations of the excited states for given γ-energies

Type of radiation	Name of transition	Spin change $\Delta I = \Delta l$	Parity change	Average lifetime (τ) in seconds for γ-energy level		
				1MeV	0.2 MeV	0.05MeV
E1	Electric dipole	1	Yes	4×10^{-16}	5×10^{-14}	3×10^{-12}
M1	Magnetic dipole	1	No	3×10^{-14}	4×10^{-12}	2×10^{-10}
E2	Electric quadrupole	2	No	2×10^{-11}	6×10^{-8}	6×10^{-5}
M2	Magnetic quadrupole	2	Yes	2×10^{-9}	5×10^{-6}	5×10^{-3}
E3	Electric octupole	3	Yes	2×10^{-4}	2	70 h
M3	Magnetic octupole	3	No	2×10^{-2}	180	200 d

(From Blatt and Weisskopf.)

leads to a change in the charge distribution of the nucleus and hence also to a change in its magnetic properties. Depending on the type of change occurring through the γ-emission, the radiation is classified as electric or magnetic (§6.4) according to the scheme in the left part of Table 6.7. Although parity need not be conserved in γ-decay, the parity change associated with the different types of γ-transitions is listed in that table.

Based on the single-particle model, Blatt and Weisskopf have calculated probable lifetimes for excited states assuming a model nucleus with a radius of 6 fm. For 2^L-multipole transitions of electric (E) or magnetic (M) type they derived the following equations

$$\lambda_{EL} = 4.4 \times 10^{21}\left\{(L+1)/\left[L((2L+1)!!)^2\right]\{3/(L+3)\}^2(E_\gamma/197)^{(2L+1)}r^{2L}\right. \tag{6.48}$$

$$\lambda_{ML} = 1.9 \times 10^{21}\left\{(L+1)/\left[L((2L+1)!!)^2\right]\{3/(L+3)\}^2(E_\gamma/197)^{(2L+1)}r^{2L-2}\right. \tag{6.49}$$

where L is the angular momentum carried away by the photon ($= \Delta l$), λ is the decay constant in s^{-1}, E_γ is the γ-ray energy in MeV and r is the nuclear radius in fm.

As seen, for decay involving electric multipole transitions the average lifetime τ is proportional to $r^{-2\Delta l}$, and for decay involving magnetic multipoles, to $r^{-2(\Delta l-1)}$; in the decay the nuclear spin quantum number s does not change. Calculated values are included in Table 6.7.

The decay of ^{24m}Na ($I = 1+$) to ^{24}Na ($I = 4+$) provides a useful example. The transition involves $\Delta l = 3$, no (i.e. there is no parity change), so it is designated as M3. The γ-energy is 0.473 MeV and the nuclear radius of ^{24}Na is about 3.7 fm. In order to compare this energy with those given in Table 6.7, a radius correction from the assumed 6 fm to the observed 3.7 fm must be made; accordingly $E_{corr} = E_{obs}(r_i/6)^{-2(\Delta l-1)}$ for

M-type radiation. This gives a hypothetical energy of $0.473(3.7/6)^{-4} = 3$ MeV. According to Table 6.7 the lifetime should be >0.02 s; the observed value is 0.035 s. The agreement between experiment and calculation is sometimes no better than a factor of 100.

The Blatt-Weisskopf relationship between energy and lifetime is only applicable for excited nucleonic states, not for rotational states. In the decay of these states the rotational quantum number always changes by two units and the lifetime of the states is proportional to $E_\gamma^{-5}\ \hat{Q}^{-2}$. The lifetimes are so short that no rotationally excited isomers have been observed as yet.

6.7.2. Beta-decay

Beta-decay theory is quite complicated and involves the weak nuclear interaction force, which is less understood than the strong interaction. The theory for β-decay derived by Fermi in 1934 leads to the expression

$$\lambda = G|M|^2 f \tag{6.50}$$

for the decay constant, λ. G is a constant, $|M|$ is the nuclear matrix element describing the change in the wave function during the β-transformation (i.e. of a proton into a neutron for positron emission, or the reverse for negatron emission), f is a function of E_{max} and Z. $|M|$ depends on the wave functions before and after the transformation and gives the "order" of decay. Since $t_{½} = \ln 2/\lambda$

$$\ln 2 = G|M|^2 f t_{½} \tag{6.51}$$

we see that the product $ft_{½}$ should be constant for a decay related to a certain $|M|$. The ft value (omitting index ½) is often referred to as the comparative β half-life, and nomograms for its calculation are given in nuclear data tables and decay schemes. The lower the ft value the higher is the probability for decay, and the shorter is the half-life. Gamow and Teller have given selection rules for β-decay which are useful for estimating decay energy, half-life, or spin in a certain decay process, if two of these properties are known. These rules are summarized in Table 6.8.

Table 6.8 Gamow Teller Selection Rules for β-Decay

Transition type	ΔI	Parity change	Log *ft*
Super allowed	0	No	3
Allowed	$0, \pm1^{(a)}$	No	4-6
First forbidden	$0, \pm1, \pm2$	Yes	6-9
Second forbidden	$\pm2, \pm3^{(b)}$	No	10-13

(a) Not $0 \rightarrow 0$
(b) Also $0 \rightarrow 0$.

For example, the decay of ^{24}Na occurs 99% through β-emission (with an $E_{max} = 1.4$ MeV) to an excited state of ^{24}Mg (Fig. 4.7). The log *ft* value of the transition is 6.1. The ground state of ^{24}Mg is 0+; the excited state has positive parity. Thus the selection rules indicate an allowed transition for which the only spin changes permitted are 0 and ±1. The ground state of ^{24}Na is 4+, and the observed excited state (at the 4.12 MeV level) is 4+, in agreement with the rule.

6.7.3. Alpha-decay theory

If we calculate the Q-value for the α-decay reaction (5.13) from the mass formulae, (4.15), we find that $Q > 0$ for all nuclei with $A > 150$, which means that we would expect all elements heavier than the rare earths to be unstable with respect to α-decay. However, the accuracy of (4.15) decreases when we move away from the region of β stability. Hence, we should not be surprised that nuclides, with a high Z/A ratio, far from stability have been found to decay by α-emission down to $A = 106$. For nuclei nearer β-stability, decay by emission of α-particles is observed for some isotopes of rare earths and heavier elements, but it occurs frequently only for $A = 210$, i.e. nuclides heavier than ^{209}Bi.

In §4.17 we mentioned the discovery by Geiger and Nuttall that the lower the α-energy the longer was the half-life of the α-decay; doubling the decay energy may reduce the half-life by a factor of 10^{20}. Alpha-decay has been observed with energies from slightly greater than 1.8 MeV (e.g. ^{144}Nd, $E_\alpha = 1.83$ MeV, $t_{½} = 2.1 \times 10^{15}$ y) to about 10 MeV (e.g. ^{262}Ns, $E_\alpha = 10.38$ MeV, $t_{½} = 4.7$ ms). The isotopes of the actinide elements typically have α-energies between 4 and 10 MeV.

The observed stability against α-decay for most nuclei in the range $102 \leq A \leq 210$ having a positive Q_α can be explained by assuming the α-particle exists as a (preformed) entity inside the nucleus but with insufficient kinetic energy to overcome the "internal Coulomb barrier". This barrier is assumed to be of the same type, although of somewhat different shape, as the external Coulomb barrier, which is discussed in some detail in §10.3.

Assume that the average kinetic energy is at the level marked E_α in Figure 6.12. If the particles in the nucleus have a Boltzmann energy distribution, an α-particle could in

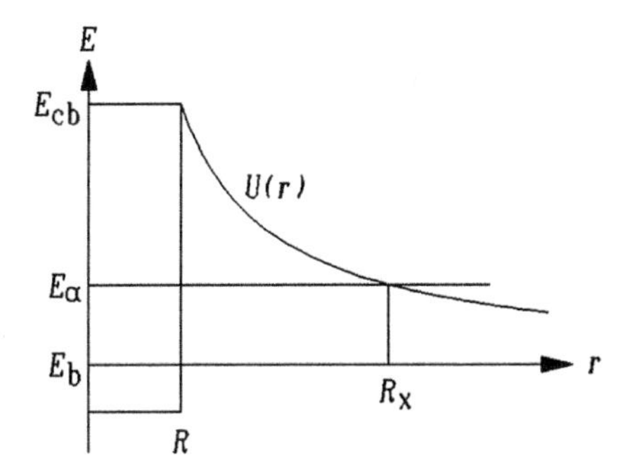

Figure 6.12 Alpha-penetration through the potential wall.

principle form in the nucleus and perhaps acquire sufficient kinetic energy through collisions to overcome the barrier (E_{cb}). It would then be emitted with an energy E_{cb}, which for an element like uranium is 26 MeV. However, the observed α-energy is only 4.2 MeV. This reasoning is very similar to chemical reactions having to pass an activation energy to proceed and that reaction component energy follow a Boltzmann distribution in energies. Thus, a chemical reaction rate can be determined from the activation energy and the half life of an alfa decay can be estimated from the alpha energy, i.e. essentially from the value of the α-particle wave function outside the coulomb barrier.

This contradiction was explained by Gamow, and independently by Gurney and Condon, in 1928, by using a quantum mechanical model, which retained the feature of the "one-body model" with a preformed α-particle inside the nuclear potential well of even-even nuclei.

The time independent solution to the Schrödinger wave equation for an α-particle inside the nuclear potential well is a wave function which has a small, but non-zero, value even outside the potential well. The probability, p, to find the α-particle outside the potential well is the square of the wave function and is found to be

$$p = \exp\left[-\left\{4\pi\sqrt{2\mu}/\mathbf{h}\right\} \int_{R}^{R_x} \sqrt{U(r) - Q_\alpha} \mathrm{d}r \right] \quad (6.52)$$

where index α refer to the α-particle, μ is the reduced mass of α-particle and residual nucleus, $\mu = M_\alpha M_1/(M_\alpha + M_1)$, R and R_x are inner and outer integration limits respectively (where the potential energy of the barrier is equal to the energy of the emitted particle, cf. Fig 6.12), r is the distance from the center of the nucleus, $U(r)$ is the potential energy of the α-particle, and Q_α is the total α-decay energy. Index 1 refer to the nucleus remaining after emission of the α-particle.

The decay constant can be regarded as the product of p and the frequency, f, by which the α-particle hits the barrier from inside. If we assume that the deBroglie wavelength, $\mathbf{h}/\mu v$, for an α-particle of velocity v inside the nucleus is approximately equal to the nuclear radius, R, we obtain

$$\mathbf{h}/\mu v \approx R \quad (6.53)$$

The frequency f can then be estimated from v if we assume that the α-particle bounces back and forth inside the potential well with constant velocity.

$$f = v/2R \approx \mathbf{h}/2\mu R^2 \quad (6.54)$$

Combining (6.40) with (6.38) we obtain the following expression for the decay constant

$$\lambda \approx \left[\mathbf{h}/2\mu R^2\right] \exp\left\{ -\left[4\pi\sqrt{2\mu}/\mathbf{h}\right] \int_{R}^{R_x} \sqrt{U(r) - Q_\alpha} \mathrm{d}r \right\} \quad (6.55)$$

The same relation also holds — with appropriate substitutions — for any charged particle trying to enter the nucleus from outside the potential barrier (cf. §10.3), where, however, only one impact is possible, i.e. $f = 1$.

For some simple mathematical forms of the potential energy $U(r)$ it is possible to find analytical solutions to the integral in (6.55). The simplest form of $U(r)$, a square well nuclear potential according to Figure 6.12, yields the following expression after integration and some algebra[5]

$$\lambda \approx \left[\mathbf{h}/\left(2\mu R^2\right)\right]\exp\left\{-\left[\sqrt{2\mu}\mathrm{e}^2 Z_1 Z_\alpha/\left(\varepsilon_0\mathbf{h}\sqrt{Q_\alpha}\right)\right]\left[\operatorname{asccos}(\mathrm{u}) - \mathrm{u}\sqrt{1-u^2}\right]\right\} \tag{6.56}$$

where

$$u = \left(E_\alpha/E_{\mathrm{cb}}\right)^{1/2} = \left[4\pi\varepsilon_0 Q_\alpha R/\left(Z_1\ Z_\alpha\ \mathbf{e}^2\right)\right]^{1/2} \tag{6.57}$$

The radius of the decaying nucleus can be estimated from

$$R = r_0 A_1^{1/3} + r_\alpha \approx 1.30\, A_1^{1/3} + 1.20\ (\mathrm{fm}) \tag{6.58}$$

where A_1 is the mass number of the nucleus after emission of the α-particle and r_α is the effective radius of the α-particle. Equation (6.56) can not only be used to compute decay constants, but also to estimate the nuclear radius for even-even α-emitters from measured half-lives and decay energies. The calculated decay constant is very sensitive to Q_α: a 1 MeV increase in Q_α increases λ (and decreases $t_{1/2}$) by a factor of about 10^5. It is also very sensitive to the nuclear radius: a 10% increase in R (or the Coulomb radius r_c; see Fig. 12.4) which means a corresponding decrease of the Coulomb barrier height, increases λ by a factor of 150.

Decay constants and partial α half-lives computed from (6.56) are normally within a factor 4 of the measured values for even-even nuclei. The half-lives of even-odd, odd-even, and odd-odd nuclides are often longer than predicted by equations like (6.56), even after inclusion of more elaborate nuclear potentials and angular momentum effects in the theory. The ratio between the observed and predicted half-lives is called *hindrance factor* and ranges from one to ~3000.

Assuming the Coulomb barrier to be much larger than Q_α, i.e. u small, designating the resulting exponential term in (6.56) by e^{-2G} and taking the logarithm of the resulting expression we obtain

$$\log\lambda = \text{constant} - 2G \tag{6.59}$$

which is very similar to the empirical Geiger-Nuttall law.

[5] It is important to recognize the difference between the classical CGS system where the force F between point charges in vacuum is $F = z_1\ z_2\ \mathbf{e}^2/r^2$ and the SI system where $F = z_1\ z_2\ \mathbf{e}^2/(4\pi\varepsilon_0\ r^2)$.

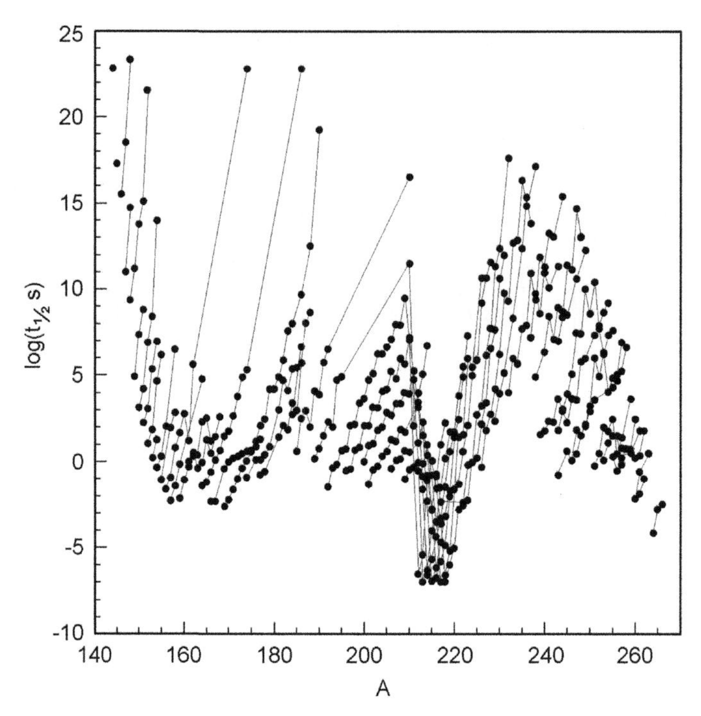

Figure 6.13 The variation of α-half-life with mass number. Lines connect isotopes.

Figure 6.13 shows the systematic change of α-decay half-life with nuclear charge and mass for heavy nuclei, lines connect data for constant Z. Odd-even effects and influence from magic numbers are visible. Such diagrams have historically played an important role in the synthesis and identification of isotopes of the heaviest elements.

The α-decay theory was the first successful (quantum mechanical) explanation of radioactive decay, and as such played a major role in further development of nuclear theories and models. Although its simplicity causes it to fail for non-spherical nuclei as well as those near closed shells, such effects can be taken into account in more advanced Nilsson-type calculations.

6.7.4. Spontaneous fission

In §5.4 we found that fission of heavy nuclei like ^{236}U is exoergic and in §5.6 that it is a common decay mode for the heaviest nuclei. From the semiempirical mass equation (4.15) it can be found that fission of nuclei with $A \geq 100$ have positive Q-values. Why is the decay by spontaneous fission only observed for nuclei with $A \geq 230$?

The breakup of a large even-even nucleus into two positively charged fragments of roughly equal mass and charge can be treated in a way similar to that of α-decay.

Assuming separation into two spherical fragments, A_1 Z_1 and A_2 Z_2, in point contact the Coulomb energy can be calculated to be

$$E_{cb} = 0.96\ Z_1\ Z_2/\left(A_1^{1/3} + A_2^{1/3}\right) \quad \text{(MeV)} \tag{6.60}$$

where $r_0 = 1.5$ fm has been assumed. In a real case it is necessary to consider that the newly formed fragments have non-spherical shapes and the value obtained from (6.46) is thus very approximate but sufficient for our discussion. The spontaneous fission of a nucleus is obviously hindered by a Coulomb barrier, the *fission barrier*, and the process should be treated as a barrier penetration problem for $Q < E_{cb}$. When $Q = E_{cb}$, breakup of the nucleus will occur within a few nuclear vibrations, $\sim 10^{-22}$ s.

The critical condition $Q = E_{cb}$ for fission of an even-even nucleus into two equal fragments with an unchanged charge to mass ratio can be estimated by equating E_{cb} from (6.46) with the Q-value computed from (3.8). Neglecting the pairing term this results in

$$\left(Z^2/A\right)_{crit} = 37.89 \tag{6.61}$$

Because asymmetric fission is more common than symmetric and the emerging fission fragments have non-spherical form, the numerical value derived above is not very accurate. However, the concept of a critical value of Z^2/A is important. A more sophisticated treatment results in the equation

$$\left(Z^2/A\right)_{crit} = 50.883\left[1 - 1.7826\,(N - Z)^2/A^2\right] \tag{6.62}$$

We can then define a *fissionability parameter*, x,

$$x = \left(Z^2/A\right)/\left(Z^2/A\right)_{crit} \tag{6.63}$$

as a measure of how prone to fission a nucleus is.

Figure 6.14 shows fission barrier height as function of Z^2/A. Calculations by Myers and Swiatecki using a refined liquid drop model predict that the barrier height should pass through a maximum around $Z^2/A \sim 16$. The penetrability of the barrier increases roughly exponentially with its height. Hence, low values of Z^2/A (but above the value corresponding to the maximum barrier height) implies extremely long half-lives. As an example, ^{238}U has $Z^2/A = 35.56$, $E_{cb} = 5.8$ and a partial half-life for spontaneous fission, $t_{½,SF}$, of $\sim 10^{16}$ y. By comparison we can estimate that a nucleus with $A = 100$ and $Z = 44$ (^{100}Ru), $Z^2/A = 19.36$, has a fission barrier tens of MeV high and a practically infinite half-life with regard to spontaneous fission.

So far we have neglected the effects of pairing, nuclear shell structure and nuclear deformation on the fission process. Odd-even, even-odd and odd-odd nuclei exhibit

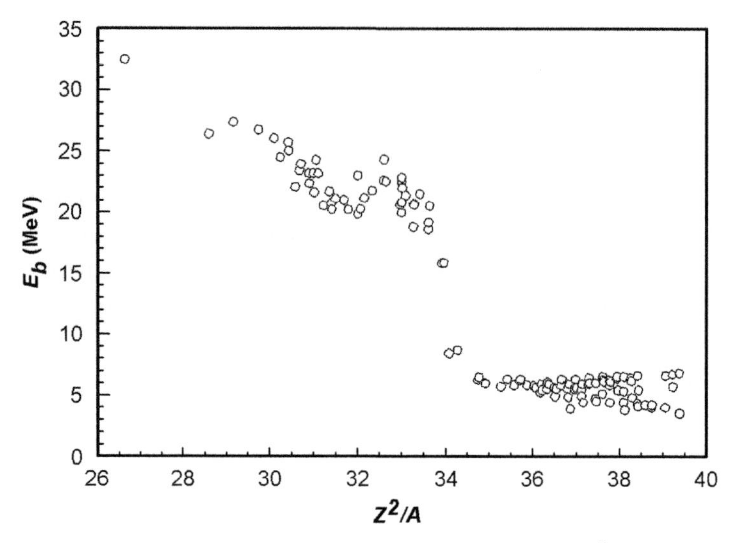

Figure 6.14 Fission barrier height as function of Z^2/A.

large hindrance factors, HF, for spontaneous fission somewhat similar to the phenomenon observed in α-decay. The presence of one or two odd nucleons leads to spin and parity values which must be conserved during the deformations leading to fission and thus constrains the possible shapes and energy levels. This contributes to the occurrence

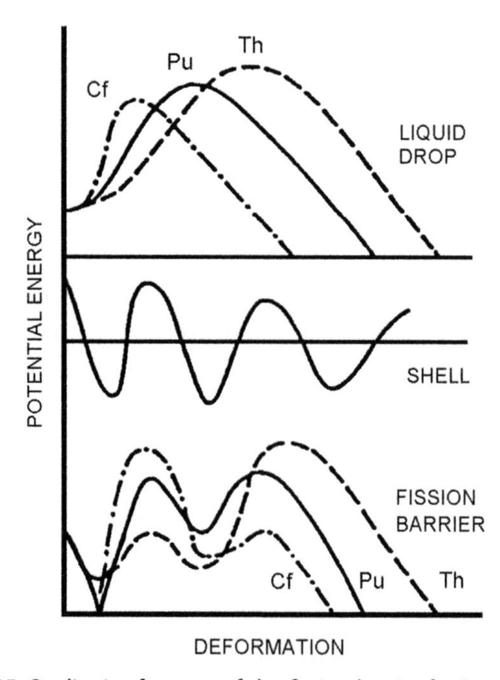

Figure 6.15 Qualitative features of the fission barrier for Pu. *(From Britt.)*

of hindrance factors. Hence, o-e, e-o and o-o nuclei have normally much longer spontaneous fission half-lives than their neighboring e-e nuclei.

The ground state energy, E, of a nucleus can be regarded as a sum of the liquid drop model energy (including deformation), E_{LDM}, the pairing correction, δ_{P}, and the shell correction, δ_{S}, to that energy.

$$E = E_{\mathrm{LDM}} + \delta_{\mathrm{P}} + \delta_{\mathrm{S}} \tag{6.64}$$

The use of (6.64), with deformation dependent shell corrections, leads to fission barriers with two maxima, see Figure 6.15. The occurrence of a secondary minimum is consistent with the observation of spontaneous fission isomers. Shell and pairing effects also give rise to long spontaneous fission and α-decay half-lives for nuclides around magic N or Z numbers. Detailed calculations based on advanced theories of nuclear structure lead to the prediction of an area of spheroidal nuclei with increased nuclear stability (in their ground state) around $Z = 114$ and $N = 184$, the so called *superheavy elements*, see Ch. 14.

A number of very rare decay modes have also been observed where clusters of nucleons heavier than α-particles are emitted, see §5.7. Some of these are difficult to distinguish from highly asymmetrical fission.

6.8. EXERCISES

6.1. The quantum numbers $s = ½$ and $l = 2$ are assigned to a particle. (a) If spin and orbital movements are independent, how many space orientations (and thus measured spectral lines if no degeneration of energy states occur) are possible in an external field of such a strength that both movements are affected? (b) How many lines would be observed if spin and orbital movements are coupled?

6.2. In the hydrogen atom the K-electron radius is assumed to be 0.529×10^{-10} m (the Bohr radius). (a) Calculate the orbital velocity of the electron assuming its mass to be $\mathbf{m}_{\mathrm{e}}$. (b) How much larger is its real mass because of the velocity? Does this affect the calculations in (a)?

6.3. A beam of protons pass through a homogeneous magnetic field of 0.5 T. In the beam there is a small high frequency coil which can act on the main field so that the proton spin flips into the opposite direction. At which frequency would this occur?

6.4. Calculate the nuclear Landé factor for ^{11}B.

6.5. In Table 6.5 the first degenerate levels have been given. Using the same assumptions, what states will be contained in the next level and how many nucleons will it contain?

6.6. How deep is the nuclear well for ^{116}Sn if the binding energy of the last nucleon is 9 MeV?

6.7. Calculate the spins and nuclear g factors for (a) ^{45}Ca, (b) ^{60}Co, and (c) ^{141}Pr, using data in Table 6.3.

6.8. The observed quadrupole moment of ^{59}Co is 0.40 barn. (a) What is the deformation value β? (b) What spin value is expected from the Nilsson diagram?

6.9. Which neutron and proton states account for the spin value I of ^{14}N?

6.10. A γ-line at 0.146 MeV is assigned to a $+4 \rightarrow +0$ rotational level change in ^{238}Pu. (a) What should the energy of the +2 and +6 rotational levels be? Compare with the measured values of 0.044 and 0.304 MeV. (b) If ^{238}Pu is considered to be a homogenous sphere, what will its apparent radius be? Compare with that obtained using (4.14).

6.11. A ^{239}Pu compound is placed in a test tube in a 40 MHz nmr machine. At what field strength does resonance occur with the nuclear spin? Is the measurement possible? Relevant data appear in Table 6.3.

6.12. Using the Gamow theory the probability for tunneling of an α-particle in the decay of ^{238}U is $1{:}10^{38}$, and the α-particle hits the walls about 10^{21} times per second. What average lifetime can be predicted for ^{238}U from this information?

6.13. Calculate the half-life for α-decay of ^{147}Sm assuming that Q_α is 2.314 MeV. Compare the result with the measured half-life and compute the hindrance factor.

6.9. LITERATURE

I. PERLMAN, A. GHIORSO, and G. T. SEABORG, Systematics of alpha-radioactivity, *Phys. Rev.* 77 (1950) 26.
I. PERLMAN and J. O. RASMUSSEN, Alpha radioactivity, *Handbuch der Physik* 42, Springer-Verlag, 1957.
W. J. MOORE, *Physical Chemistry*, 3rd edn., Prentice-Hall, I. 1962.
W. D. MYERS and W. J. SWIATECKI, Nuclear Masses and Deformations, *Nuclear Physics* **81** (1966) 1.
E. K. HYDE, Nuclear models, *Chemistry* 40 (1967) 12.
H. C. BRITT, in N. M. EDELSTEIN (Ed.), *Actinides in Perspective*, Pergamon, 1982, p. 245.
K. S. Krane, *Introductory Nuclear Physics*, Wiley, 1988.
G. T. SEABORG and W. D. LOVELAND, *The Elements Beyond Uranium*, Wiley-Interscience, 1990.
D. N. POENARU (Ed.), *Handbook of Decay Modes*, CRC Press, 1993.

CHAPTER 7

Absorption of Nuclear Radiation

Contents

Our understanding of the nature of nuclear particles is based on their mode of interaction with matter. If the interaction of the various nuclear particles with various material is known it is possible to use this knowledge for e.g. proper utilization and construction of detection and measuring devices for radiation, the design of radiation shielding, the medical and biological applications of radiation, radiochemical synthesis, etc.

The term *nuclear radiation* is used to include all elementary particles, both uncharged (e.g. photons) and charged (e.g. electrons), having energies in excess of approximately 100 eV. This is valid whether the particles have been produced through nuclear reactions (spontaneous or induced) or have acquired their energy in electrostatic accelerators.

Radiochemistry and Nuclear Chemistry
ISBN 978-0-12-405897-2, http://dx.doi.org/10.1016/B978-0-12-405897-2.00007-0

100 eV is a high energy in comparison to ionization energies (usually <15 eV) and to the energies involved in chemical bonds (normally 1−5 eV). Therefore, nuclear radiation can cause several ionizations in its passage through matter; this is reflected in the common name *ionizing radiation*. Neutrons of energies <100 eV are included because their absorption (capture) by nuclei results in emission of nuclear radiation with energies much greater than 100 eV.

The passage of such high energy radiation through matter results in the transfer of energy to the atoms and molecules of the absorber material. This transfer of energy continues until the impinging particle of the radiation has reached the same average kinetic energy as the atoms comprising the material; i.e. until *thermal equilibrium* is obtained.

In considering the absorption of nuclear radiation it is appropriate to view the overall process from two aspects: (1) from the impinging particles point of view, i.e.processes occurring to the nuclear particles themselves as their energies are reduced to the thermal equilibrium value; such *absorption processes* are the principal consideration of this chapter; (2) from the absorbing materials point of view, i.e. processes in the absorbing material due to the effect of the transfer of energy. This transfer results initially in excitation and ionization which cause physical and chemical changes. The study of these effects is the domain of *radiation chemistry* and is considered in Chapter 8.

7.1. SURVEY OF ABSORPTION PROCESSES

The reduction in the intensity of a beam of ionizing particles can be caused either by reaction with the nuclei of the absorbing material (nuclear reactions) or with the atomic electrons (electron collision). In Table 7.1 the most important processes involved in the absorption of nuclear radiation in matter are listed along with the probability for each process. Comparison shows that the probability of interactions with electrons is considerably greater than that of a nuclear reaction; the only exception to this is the case of neutron absorption. In fact the principal mode of interaction between the particle and the atoms of the absorbing material involves the electromagnetic fields of the particle and the atomic electrons. Since neutrons are neutral particles which do not with electrons, in order for them to transfer energy to an atom it is necessary that they experience a direct collision with a nucleus. Consequently for all particles except neutrons, nuclear reactions can be neglected in considering the processes involved in the reduction of the intensity of the particle beam.

As nuclear radiation passes the atoms of an absorber, it can transfer some of its energy to the atomic electrons. If the amount of energy transferred is sufficient, ionization of the atom results. The positive ion and the electron thus formed are known as an *ion pair*. Frequently the electrons from this primary ionization have sufficiently high kinetic energy to cause *secondary ionization* in other atoms. The number of electrons produced in

The reaction cross-sections (σ) given are only the order of magnitude at about 1 MeV in $Z \approx 20$

	Reacting particles and fields	Type of reaction	σ (barns)	Name of process
1	*Protons and heavier ions* react with			
1a	orbital electrons	Particle energy loss through atomic excitation and ionization	$\leq 10^5$	Ionization, (atomic) excitation
1b	atomic nucleus (1b–1d)	Particle elastically scattered	≤ 10	Nuclear scattering
1c	atomic nucleus	Particle inelastically scattered	<1	Nuclear (coulomb) excitation
1d	atomic nucleus (1b–1d)	Particle captured, formation of compound nucleus ($E_p > E_{cb}$(min))[c]	≤ 0.1	Nuclear transmutation
2	*Electrons* (e^-, β^-, β^+) react with			
2a	orbital electrons (2a–2b)	Particle energy loss through atomic excitation and ionization	$>10^2$	Ionization, (atomic) excitation
2b	orbital electrons (2a–2b)	Slow β^+ annihilated, 2–3 photons formed	(100%)	Positron annihilation
2c	electric field of nucleus	Particle scattered with energy loss, continuous emission of hν (E_e »1 MeV)	>1	Bremsstrahlung
3	*Photons* (γ) react with			
3a	field of orbital electrons	γ scattered without energy loss	≤ 0.01	Coherent scattering
3b	free (outer) electrons	γ scattered with energy loss, ionization	$\leq 10^{(a)}$ (3b–3d)	Compton effect
3c	bound (inner) electrons	γ completely absorbed, one electron knocked out	$\leq 10^{(a)}$	Photo effect
3d	field of nuclear force	γ annihilated, formation of positron-negatron pair ($E > 1.02$ MeV)	$\leq 10^{(a)}$ (3b–3d)	Pair formation
3e	atomic nucleus (3e–3g)	γ scattered without energy loss	$\leq 10^{\ 3}$ (3e–3g)	Mössbauer effect
3f	atomic nucleus	γ scattered with energy loss	$\leq 10^{\ 3}$ (3e–3g)	Nuclear excitation
3g	atomic nucleus (3e–3g)	γ absorbed by nucleus, nuclear transmutation ($E > 5$ MeV)[b]	$\leq 10^{\ 3}$ (3e–3g)	Nuclear photo effect
4	*Neutrons* react with			
4a	atomic nucleus (4a–4b)	n scattered with energy loss	≤ 10	Neutron moderation
4b	atomic nucleus (4a–4b)	n captured, nuclear transformation	$\leq 10^4$	Neutron capture

[a]See Fig. 7.17; σ increases strongly with decreasing energy.
[b]Threshold energy for Be(γ,α)^{4}He 1.6 MeV, D(γ,n)H 2.2 MeV.
[c]E_{cb}(min) is the Coulomb barrier energy, eqn. (10.15).

secondary ionization is often larger than that of the primary ionization but the average kinetic energies of the secondary electrons are lower than those of the primary electrons. In many interactions the initial radiation transfers insufficient energy for ionizations; instead an electron is raised to a higher, excited energy level of the atom. These *excited atoms* rapidly return to lower energy states by emission of electromagnetic radiation such as X-rays, visible light, etc. For neutrons the absorption process involving the capture of the neutron imparts sufficient recoil energy to cause ionization and excitation.

7.2. ABSORPTION CURVES

In order to measure the absorption of nuclear radiation, the experiments must be performed in such a manner as to eliminate as many of the interfering factors as possible. Usually a well-collimated beam is used. This is illustrated in Figure 7.1 for a point radioactive source. The relation between the disintegration rate A and the count rate R is given by Eqn (7.1):

$$R = \psi A \tag{7.1}$$

The *counting efficiency* ψ includes a number of factors:

$$\psi = \psi_{\text{sample}} \psi_{\text{abs}} \psi_{\text{det}} \psi_{\text{geom}} \tag{7.2}$$

If conditions were ideal, there would be no self-absorption or scattering in the sample (in which case $\psi_{\text{sample}} = 1$), no absorption of radiation between the sample and the detector window ($\psi_{\text{abs}} = 1$), and the detector would have a 100% efficiency (sensitivity) to a "count" for each particle reaching its window ($\psi_{\text{det}} = 1$).

The geometric efficiency ψ_{geom}, is 1 for 4π-geometry, i.e. for a spherical detector subtending a 360° solid angle about the sample. Although such detectors exist (Ch. 9), more commonly the sample is counted outside the detector at some distance r, as indicated in Figure 7.1. If the detector window offers an area of S_{det} perpendicular to the radiation, the geometrical efficiency is approximated by (for small ψ_{geom})

$$\psi_{\text{geom}} \approx S_{\text{det}} / \left(4\pi r^2\right) \tag{7.3}$$

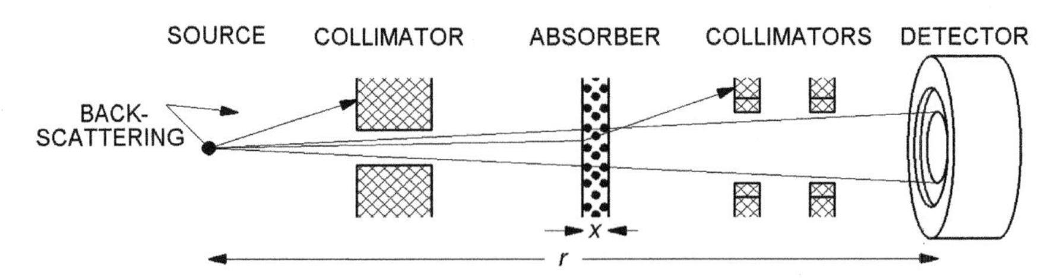

Figure 7.1 Geometrical arrangement for measuring absorption curves.

If a detector with a circular window of radius s is at a distance r from a source of radius d, the geometrical efficiency is given by

$$\psi_{geom} = \tfrac{1}{2}\left[1 - \left(1 + s^2/r^2\right)^{-\frac{1}{2}}\right]k \quad (7.4)$$

When the sample is a point source, $k = 1$, else k can be read from the series of curves in Figure 7.2.

The activity measured is proportional to the particle flux φ reaching the detector

$$R = k_{det}\,\varphi \quad (7.5)$$

where $k_{det} = S_{det}\,\psi_{det}$ and

$$\varphi = \psi_{abs}\,\varphi_0 \quad (7.6)$$

and

$$\varphi_0 = \psi_{sample}\,nA/\left(4\pi r^2\right) \quad (7.7)$$

where φ_0 is the flux of particles (particles $m^{-2}\,s^{-1}$) which reach the detector from the source when $\psi_{abs} = 1$, and n is the number of particles emitted per decay ($n > 1$ only for γ following α- and β-decays). Thus if every β-decay yields 2γ (in cascade) and γ is the measured radiation, then $n_\gamma = 2$, and $\varphi_0 = n_\gamma\,A_\beta/(4\pi\,r^2)$ γ-quanta $m^{-2}\,s^{-1}$. In branched decay n will not be an integer. Equation (7.7) is the so called $1/r^2$ - *law* since the measured flux varies as the inverse of the square of the distance to the source.

These equations are valid as long as the conditions at the source and at the detector, as well as r, are kept constant. When an absorber is inserted between the source and detector (Fig. 7.1), ψ_{abs} depends on the absorber thickness x (m). For zero absorber thickness, $\psi_{abs} = 1$ in accurate measurements. There is a small absorption due to the air between the sample and detector unless the measurement is done in a vacuum.

Absorption curves relate the variation of either R or φ to the thickness of the absorbing material. In Figure 7.3 the relative transmission φ/φ_0 is plotted as a function of absorber thickness for different kinds of radiation. For charged particles, i.e. electrons, protons, and heavier ions, φ/φ_0 reaches zero at a certain x-value (x_{max}); this is referred to as the maximum range of the particles. The range can be expressed by either the *average range* ($x = C_1$ for heavy ions and C_3 for electrons) or the *maximum* (or extrapolated) *range* (C_2 and C_4, respectively, in Fig. 7.3). The loss of energy involves collisions with atomic electrons, and the energy loss per collision and the number or collisions varies from one ionizing particle to the next, resulting in a slight *straggling* in the range. The average range is the meaningful one.

Figure 7.4 shows an absorption curve for radiation from ^{32}P. The radioactivity R has been measured as a function of aluminum absorber thickness in *linear density* (kg m^{-2} in SI-units, sometimes also in mg cm^{-2}). The low activity "tail" (C_4 in Fig. 7.4) is the

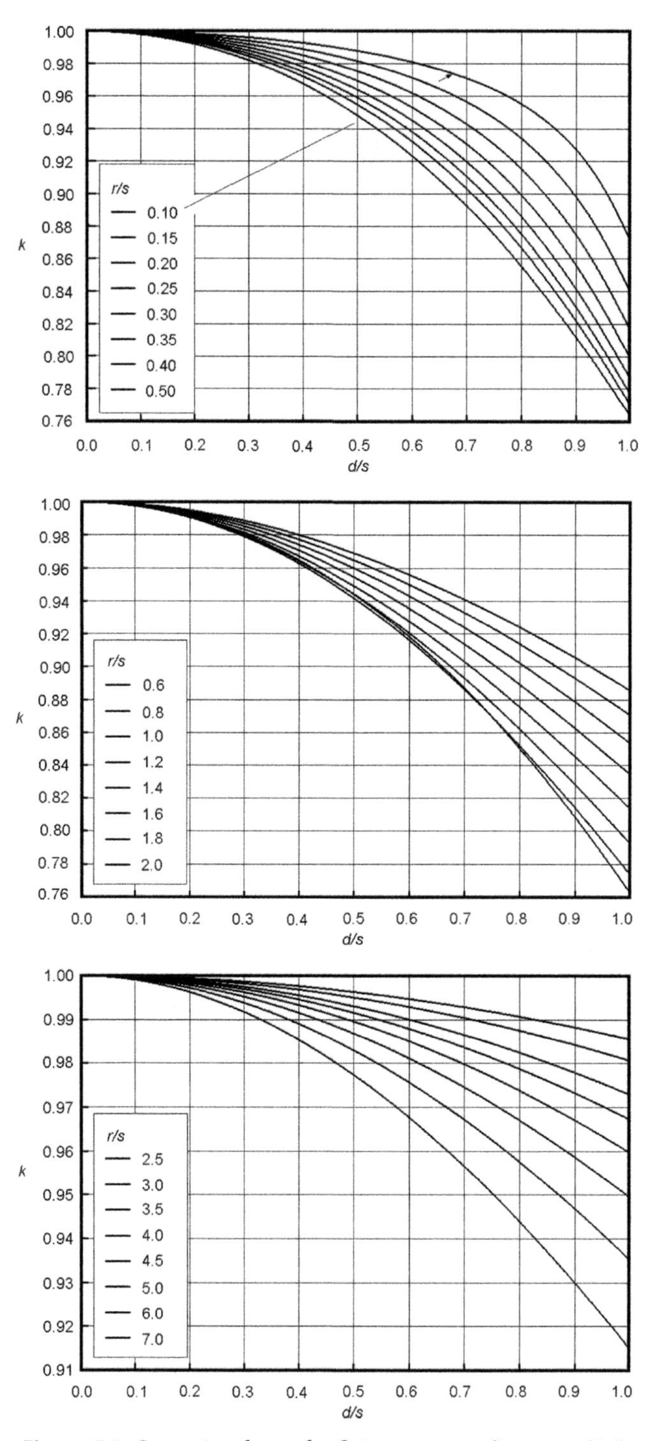

Figure 7.2 Correction factor for finite source radius, eqn. (7.4).

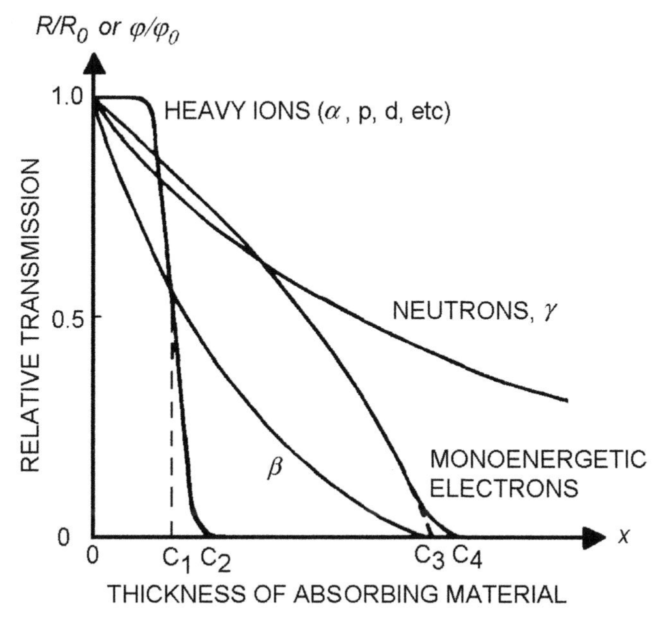

Figure 7.3 Curves showing relative transmission φ/φ_0 (or R/R_0) as function of absorber thickness x. C_1 and C_3 are average, C_2 and C_4 maximum range.

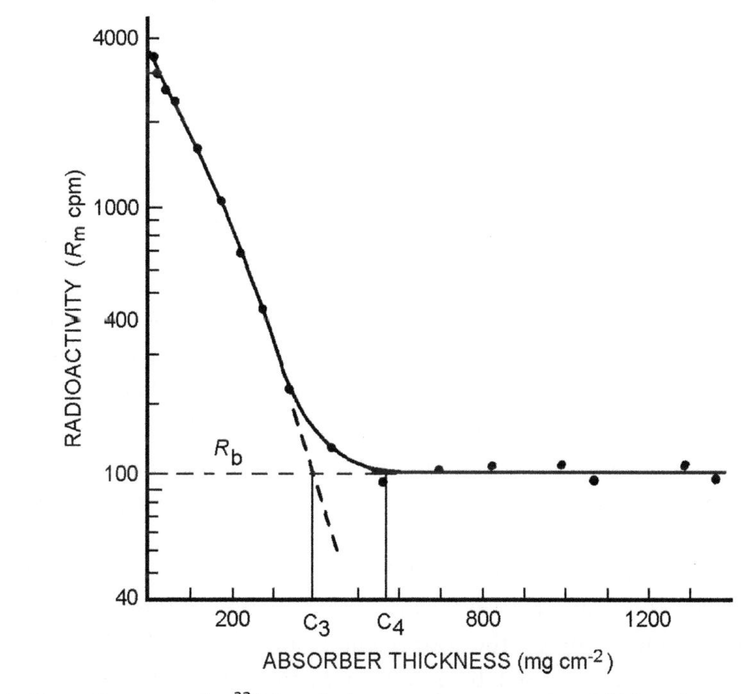

Figure 7.4 Absorption curve for ^{32}P β-radiation showing extrapolated (C_4) and average (C_3) ranges. The dashed curve is obtained after subtraction of background.

background activity R_b, which has to be subtracted from the measured value R_m to obtain the true value for the radiation (e.g. from ^{32}P): $R = R_m - R_b$. The extrapolation of R to a value equal to R_b (i.e. C_3) gives the range.

Whereas it is possible to specify maximum ranges for charged particles, this is not possible for neutral particles such as neutrons and γ-quanta. If the absorber is not too thick, these particles undergo only one collision, or at the most a few, before they are absorbed. As a result the absorption curve has an exponential form.

$$\varphi = \varphi_0 e^{-\mu x} \tag{7.8}$$

where μ is the *total attenuation coefficient*. Thus for n and γ we have

$$\psi_{abs}(x) = e^{-\mu x} \tag{7.9}$$

The reduction in intensity of a beam can occur by two mechanisms. One involves the deflection or scattering of the particles from the direct line of path between the source and the detector and is described by the *scattering coefficient* μ_s. The second mode of reduction is the complete transfer of the projectile energy to the absorbing material (the particles are "captured") and is designated by the *(energy) absorption coefficient* μ_a. The (total) attenuation coefficient in Eqn (7.8) is the sum of both these mechanisms.

$$\mu = \mu_s + \mu_a \tag{7.10}$$

Both μ_s and μ_a can be measured independently. The (total) attenuation coefficient is of primary interest in radiation shielding, while the (energy) absorption coefficient is important in considering the effect of radiation on matter.

7.3. ABSORPTION OF PROTONS AND HEAVIER IONS

The mode of interaction of protons and heavier charged particles with the atoms of the absorbing material can be illustrated by considering the absorption of α-particles. With rare exception, α-particles emitted by radioactive nuclides have energies between 4 and 9 MeV. Since the α-particles are so much heavier than electrons, they are deflected very slightly when their Coulomb fields interact with atoms or molecules to form ion pairs. As a result, α-particles travel in a straight line as they pass through matter, which explains the straight paths observed for α-particles in cloud chamber photographs (Fig. 7.5). This is in contrast to the very curved or irregular paths of the secondary electrons emitted in the formation of the ion pair. For a 5 MeV α-particle the maximum energy of the secondary electrons is 2.7 keV. However only a small fraction of the secondary electrons actually receive this much energy; the average energy of the secondary electrons is closer to 100 eV. The ionization caused by more energetic secondary electrons is usually referred to as *δ-tracks* (cf. §8.2).

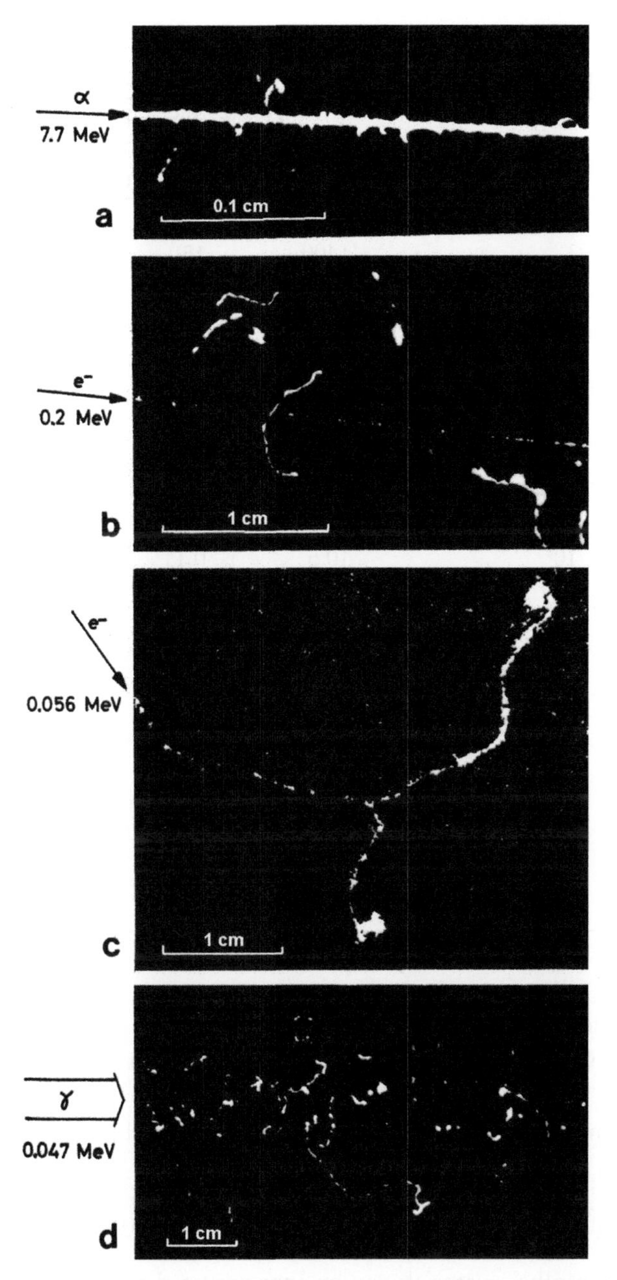

Figure 7.5 Cloud chamber tracks of α, β, (e^-), and γ-rays at 1 bar in air ((a), (b), and (c)) and in methane (d). *(From W. Gentner, H. Maier-Leibnitz, and H. Bothe.)*

Table 7.2 Range in water, and average linear energy transfer (LET) values for different radiation

Upper Half Refers to Monoenergetic (Accelerated) Particles. For β-decay $E_{abs} = 1/3\ E_{max}$

Radiation	Energy (MeV)	Maximum range		Average LET value in water (keV/μm)
		cm air	mm water	
Electron	1	405	4.1	0.24
	3	1400	15	0.20
	10	4200	52	0.19
Proton	1	2.3	0.023	43
	3	14	0.014	21
	10	115	1.2	8.3
Deuteron	1	1.7	—	—
	3	8.8	0.088	34
	10	68	0.72	14
Helium	1	0.57	0.0053	190
	3	1.7	0.017	180
	10	10.5	0.11	92
Fiss. fragment	100	2.5	0.025	3300
^{226}Ra (α)	E_α 4.80	3.3	0.033	145
^{210}Po (α)	E_α 5.30	3.8	0.039	136
^{222}Rn (α)	E_α 5.49	4.0	0.041	134
^{3}T (β)	E_{max} 0.018	0.65	0.0055	1.1
^{35}S (β)	E_{max} 0.167	31	0.32	0.17
^{90}Sr (β)	E_{max} 0.544	185	1.8	0.10
^{32}P (β)	E_{max} 1.71	770	7.9	0.07
^{90}Y (β)	E_{max} 2.25	1020	11	0.07
^{137}Cs (γ)	E_γ 0.66	$x_{½} = 8.1$ cm H_2O		0.39
^{60}Co (γ)	E_γ 1.20−1.30	$x_{½} = 11.1$ cm H_2O		0.27

In solids and liquids the total path length for α-particles from radioactive decay is quite short, only some micrometers. However, in gases at standard temperature and pressure the paths are several centimeters long (Table 7.2). The range in air, $\hat{R}_{air}$, for α-particles with an initial energy E_α MeV can be calculated by the empirical equation ($\rho_{air} = 1.293$ kg m^{-3}; 1 mg/cm^2 = 0.01 kg/m^2):

$$\hat{R}_{air} = 0.31\ E_\alpha^{3/2}(\text{cm}) = 0.40\ E_\alpha^{3/2}\left(\text{mg cm}^{-2}\right) = 0.004\ E_\alpha^{3/2}\left(\text{kg m}^{-2}\right) \quad (7.11)$$

The range $\hat{R}_z$ in other materials can be approximated roughly by

$$\hat{R}_z = 0.173\ E_\alpha^{3/2} A_z^{1/3}\left(\text{mg cm}^{-2}\right) = 0.00173\ E_\alpha^{3/2} A_z^{1/3}\left(\text{kg m}^{-2}\right) \quad (7.12)$$

where A_z is the atomic weight of the absorber. Figure 7.6 shows the range of various charged particles in an aluminum absorber. The range of a 5 MeV α is 6 mg cm^{-2}; thus

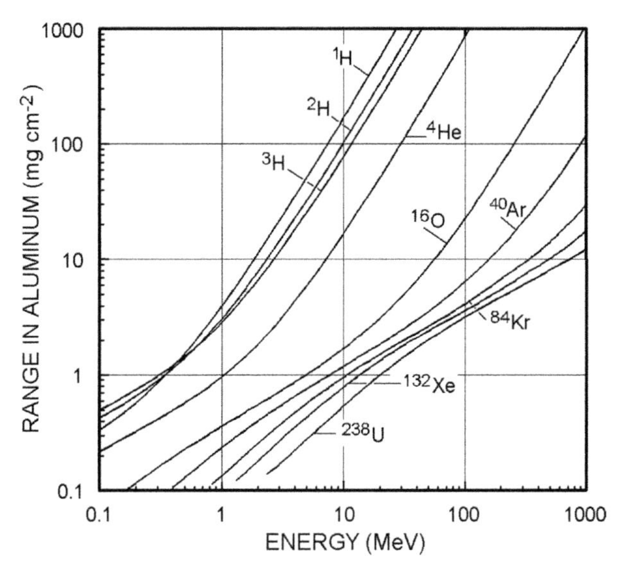

Figure 7.6 Range of some energetic particles in an aluminum absorber.

$\hat{R}_{Al} = 6 \times 10^{-3}/\rho_{Al}$ cm $= 0.002$ mm. Alpha-particles from radioactive decay are easily stopped even by a sheet of paper.

When the absorber consists of a composite material, containing the weight fractions w_1, w_2, w_3, etc of elements 1, 2, 3, etc with ranges $\hat{R}_1$, $\hat{R}_2$, $\hat{R}_3$, etc, the range $\hat{R}_{comp}$ in the absorber is obtained from the relation

$$1/\hat{R}_{comp} = w_1/\hat{R}_1 + w_2/\hat{R}_2 + w_3/\hat{R}_3 + \dots \tag{7.13}$$

The number of ion pairs formed per millimeter of range for α-particles, protons, and electrons are shown in Figure 7.7(a). The larger *specific ionization* of the α-particles compared to the protons is related to the fact that the former are doubly charged. In general the specific ionization increases with the ionic charge of the particle for the same kinetic energy. Fission fragments that initially have very large energies also have very large ionic charges leading to quite high specific ionization in their absorption in matter; their range is 2–3 cm in air and 2–3 mg cm^{-2} in aluminum.

A quantum-mechanical and relativistic analysis of the interaction between a fast moving positive ion of atomic number z and the electrons in the absorber leads to the following expression for the energy loss per unit distance, dE/dx (J/m), travelled in an absorber

$$\begin{aligned} -\,dE/dx = & \left\{\left(4\pi(\gamma z\mathbf{e})^2\mathbf{e}^2 NZ\right)/\left(\mathbf{m}_e v^2\right)\right\}\left[\ln\left\{\left(2\mathbf{m}_e v^2\right)/I\right\}\right. \\ & \left. -\ln\left(1-(v/\mathbf{c})^2\right) - (v/\mathbf{c})^2\right] \end{aligned} \tag{7.14}$$

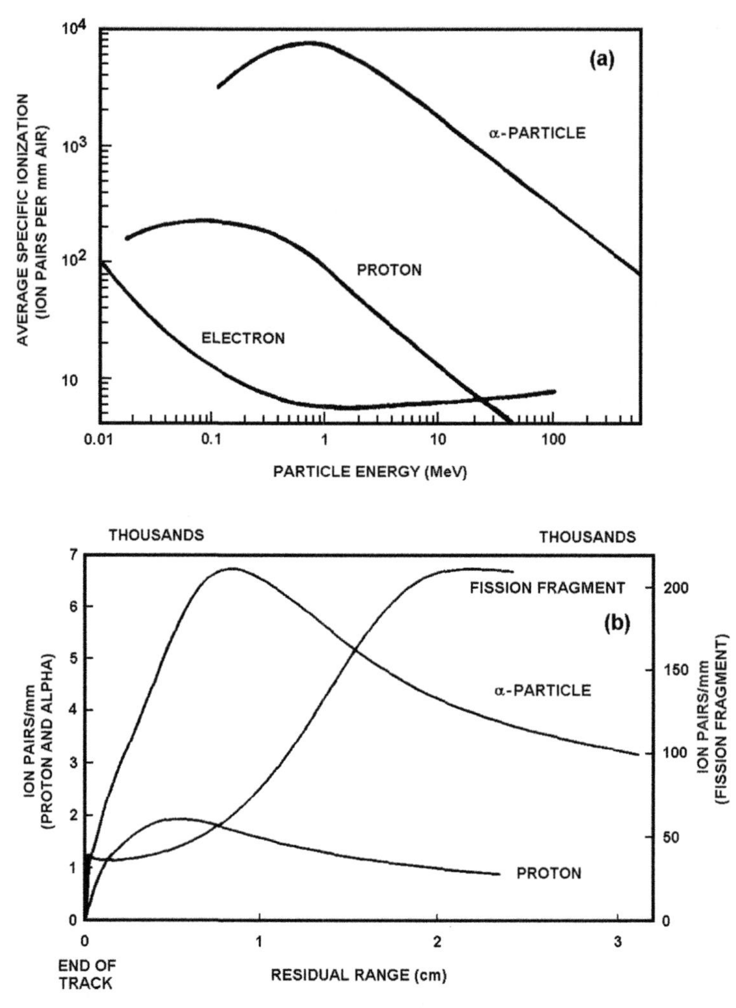

Figure 7.7 Number of ion pairs formed per mm air at STP as function of particle energy. (a) average specific ionization at maximum particle energy, (b) same for residual range. *(From H. A. C. McKay.)*

where ($\gamma z\mathbf{e}$) is the charge of the ion moving at velocity v through an absorber containing N atoms of atomic number Z per volume unit and having an effective ionization potential I. For a completely stripped ion, $\gamma = 1$. The range, $\hat{R}$, of an ion may be calculated by integrating the energy loss expression

$$\hat{R} = \int_{E_0}^{0} (\mathrm{d}E/\mathrm{d}x)^{-1}\mathrm{d}E \tag{7.15}$$

from the initial energy E_0 to zero.

Charged particles decrease in velocity as they lose their energy in traversing an absorber. As a result they spend progressively longer times in the vicinity of any particular atom, which results in an increase in the probability of interaction with that atom. Consequently there is a steady increase in the number of ion pairs formed along the path (or track) of the particle rather than a constant density of ion pairs. As mentioned above, near the end of the range for heavy charged particles a maximum is observed for the number of ion pairs formed per unit path length (the *Bragg peak*) (Fig. 7.7b). At a distance just beyond the Bragg peak maximum the kinetic energy of the particles is comparable to those of the orbital electrons of the absorber. As a result, the particle can acquire electrons, $\gamma < 1$ in Eqn. (7.14), finally becoming uncharged, $\gamma = 0$, and thereby losing its ability to cause further ionization. When scaled by the maximum ionization, the Bragg peak is most pronounced for protons.

7.4. ABSORPTION OF ELECTRONS

Absorption of high energy electron beams occurs through interaction with the orbital electrons and the electromagnetic field at the atom. The processes are summarized in Table 7.1 and Figure 7.8. In order to distinguish between electrons from accelerators and those from β-decay we refer to the latter as *β-particles*.

7.4.1. Ionization

Beta-particles lose their energy primarily by the same mechanism as α-particles (Fig. 7.7a and b); however, there are several important differences. Since the masses of the β-particles and of the orbital electrons are the same at non-relativistic velocities, the β-particles can lose a large fraction of their energy in a single collision. The β-particle undergoes a wide angle deflection in such collisions and consequently β-particles are scattered out of the beam path all along the length. The secondary electrons ionized from the atom have such high energies that they cause extensive secondary ionization which provides 70–80% of the total ionization in β-absorption processes (Fig. 7.5b). Approximately half of the total energy of the β-particle is lost by ionization and half by excitation. The track formed is further discussed in §8.2.

The specific ionization from a β-particle is much lower than that from a heavy ion as can be seen in Figure 7.7a. This is due to the fact that for the same initial energy, β-particles have much greater velocity than have α-particles or protons because their mass is very much smaller than the mass of the heavy particles. This greater velocity results in a correspondingly lower ionization due to the considerable shorter time available for interaction with the electrons of a specific atom and gives a much longer range to β-particles. The erratic path observed for β-particles in Figure 7.5(c) is a result of the large energy transfer and the resulting large deflections involved in the encounters with the orbital electrons. However, at very high energies the β-particles have straight

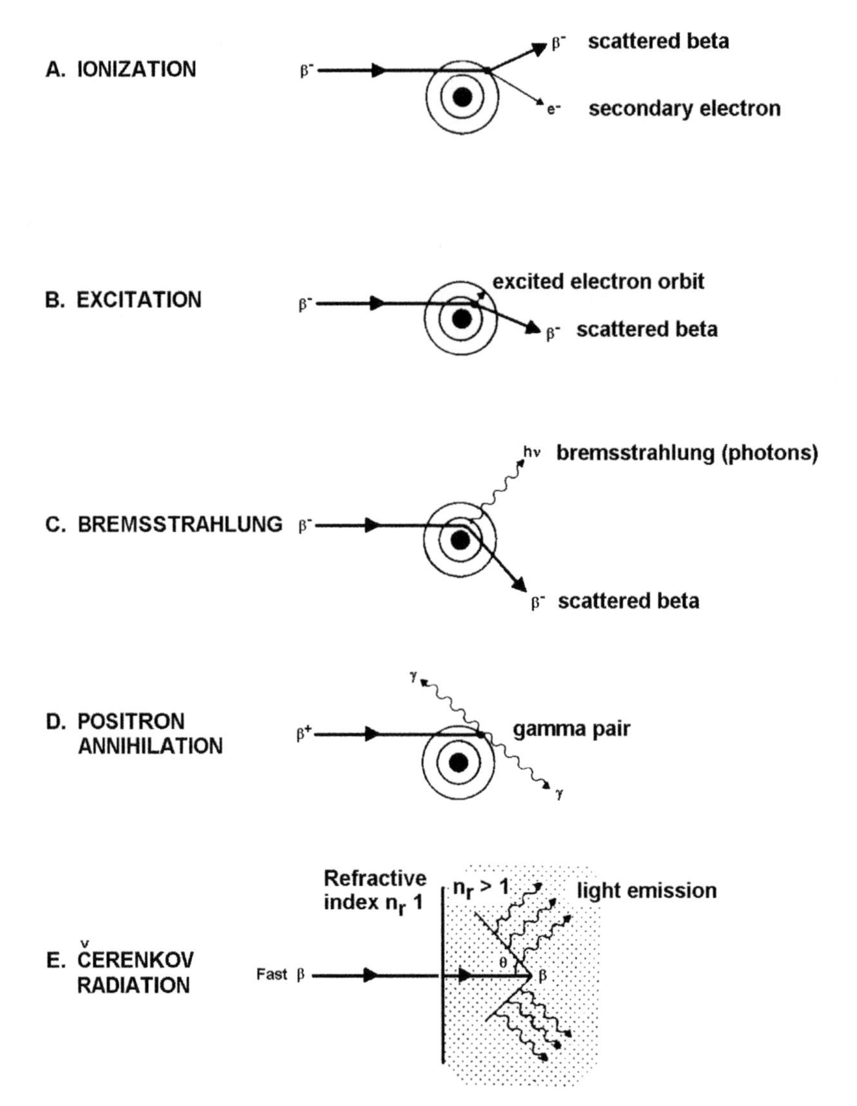

Figure 7.8 Schematic description of the five processes accounting for β-particle absorption.

paths as a result of the fact that very energetic β-particles have a momentum considerably in excess of that of the orbital electron.

7.4.2. Bremsstrahlung

As a β-particle approaches an atomic nucleus, it is attracted by the positive field of the nucleus and deflected from its path. The deflection results in an acceleration that, according to classical electrodynamics, leads to emission of electromagnetic radiation (Fig. 7.8c), cf. synchrotron radiation Ch. 16. Therefore the encounter with the positive

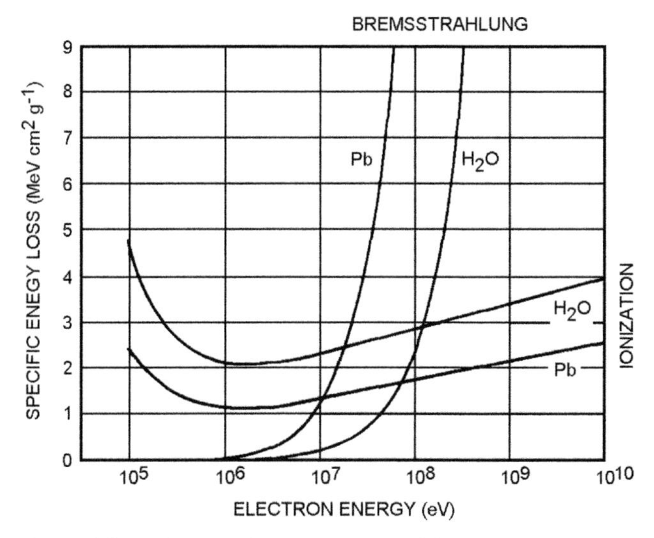

Figure 7.9 Energy loss of fast electrons by ionization and bremsstrahlung. *(From Gentner, Maier-Leibnitz, and Bothe.)*

charge of the nuclear field decreases the energy of the β-particle by an amount exactly equal to the amount of electromagnetic radiation emitted. This radiation is known as *bremsstrahlung* (braking radiation). The loss of energy by emission of bremsstrahlung radiation increases with the β energy and with the atomic number (positive charge) of the absorber material (Fig. 7.9). In aluminum approximately 1% of the energy of a 1 MeV electron is lost by bremsstrahlung radiation and 99% by ionization whereas in lead the loss by bremsstrahlung is about 10%. For electrons of greater than 10 MeV energy, bremsstrahlung emission is the predominant mode of energy loss in lead. However, for the energies in radioactive decay, bremsstrahlung can usually be neglected - particularly for absorption processes in material of low atomic weight. The ratio of *specific energy loss* (dE/dx) through bremsstrahlung to that through collision (i.e. all other processes) is approximately:

$$(\mathrm{d}E/\mathrm{d}x)_{\mathrm{brems}}/(\mathrm{d}E/\mathrm{d}x)_{\mathrm{coll}} \approx E_{\mathrm{e}}Z/800 \qquad (7.16)$$

where E_e is kinetic energy of the electron (MeV) and Z the atomic number of the absorber atoms.

Figure 7.10 shows the bremsstrahlung spectrum obtained in aluminum for β-particles emitted by ^{147}Pm. In this case a very small fraction of the β-energy is converted into radiation. The bremsstrahlung spectrum is always of much lower energy than that of the β-spectrum. Bremsstrahlung sources of a wide variety of energies are commercially available. Recently special electron accelerators have been designed for the sole purpose of producing bremsstrahlung radiation to be used for (analytical) X-ray excitation and for medical irradiation purposes; see Ch. 16.

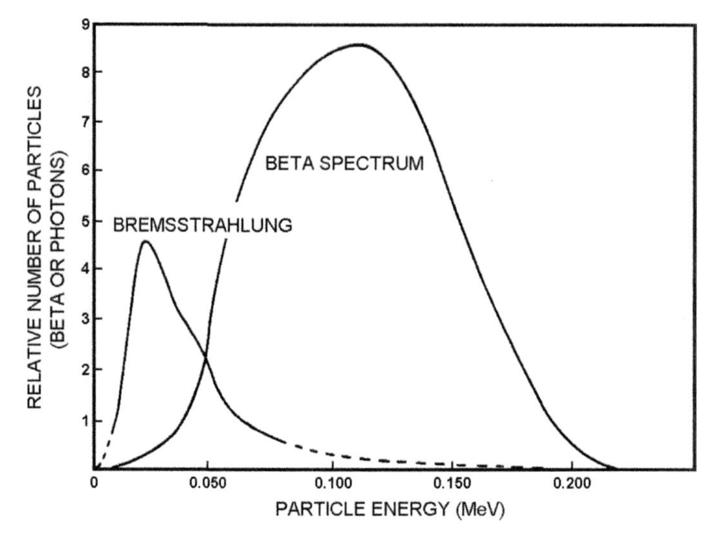

Figure 7.10 Beta-spectrum (right curve) and bremsstrahlung spectrum in aluminum for ^{147}Pm. The ordinate of the bremsstrahlung spectrum is enlarged about 100 times.

7.4.3. Čerenkov radiation

The velocity of light in matter c depends on the refractive index n_r

$$c = \mathbf{c}\, n_r^{-1} \tag{7.17}$$

In water $n_r = 1.33$, in plexiglass 1.5. β-particles with energies >0.6 MeV move faster than light in water. When the particle velocity (v_p) > c, electromagnetic radiation is emitted coherently in a cone whose axis is the direction of the moving particle, Fig. 7.8(e). This is called Čerenkov radiation. The angle of the cone θ is obtained from

$$\sin \theta = c/v_p \tag{7.18}$$

Čerenkov radiation is continuous. Around the visible spectrum, the relative intensity per unit frequency is approximately proportional to the frequency. Thus, higher frequencies are more intense in Čerenkov radiation. This radiation is the source of the bluish light observed in highly radioactive solutions (Fig. 7.11) and around used reactor fuel elements submerged in water. The radiation can be used for detecting β-particles and for measuring high particle energies (from θ). For a fast electron the energy loss through radiation is $\leq$0.1% of the energy loss through other processes. Čerenkov detectors are described in Ch. 9.

7.4.4. Positron annihilation

Positrons interact with matter through ionization, excitation, emission of bremsstrahlung, and Čerenkov radiation in the same manner as negative electrons. As the kinetic

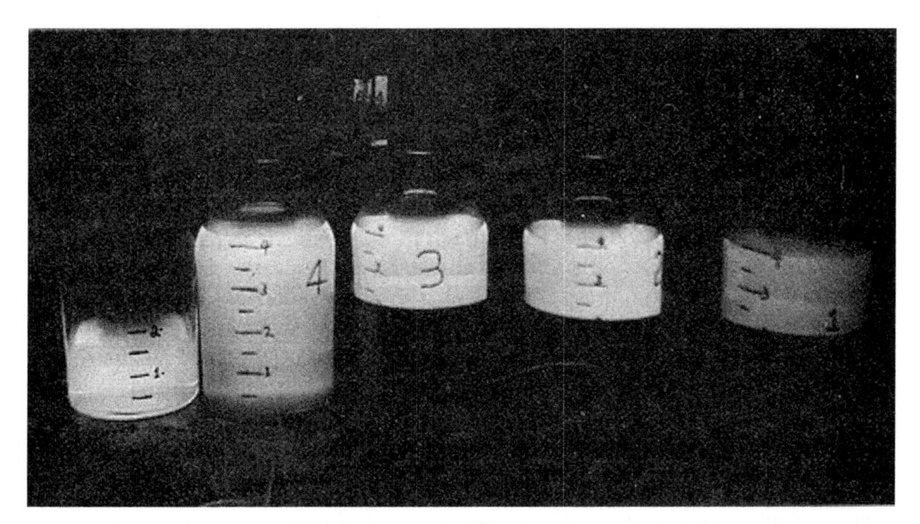

Figure 7.11 Bottles containing highly radioactive ^{90}Sr solutions glow in the dark due to Cerenkov radiation from daughter ^{90}Y (E_{max} 2.3 MeV).

energy of the positron decreases in the absorber, there is an increase in probability of direct interaction between the positron and an electron (Fig. 7.8(d)) in which both the positron and electron are annihilated. The energy of the two electron masses is converted into electromagnetic radiation. This process, known as *positron annihilation*, is a characteristic means of identification of positron emission. Since an electron mass is equivalent to 0.51 MeV, and the kinetic energy of the particles of annihilation is essentially zero, the total energy for the annihilation process is 1.02 MeV. In order to conserve momentum the photons must be emitted with equal energy and in exactly opposite direction in case of only two photons (the dominating case). These photons of 0.51 MeV each are referred to as *annihilation radiation*. The presence of γ-rays at 0.51 MeV in the electromagnetic spectrum of a radionuclide is strong evidence for the presence of positron emission by that nuclide. In rare cases three γ-rays of correspondingly lower energy are omitted.

7.4.5. Absorption curves and scattering of β-particles

An absorption curve for β-particles has a quite different shape than it has for α-particles (cf. Fig. 7.3). The continuous spectrum of energies in radioactive β-decay plus the extensive wide angle scattering of the β-particles by the absorber atoms account for the fact that range for β-particles continuously decrease. Even for a beam of initially mono-energetic electrons, the continuous removal of electrons from the beam path by wide angle deflection results in a plot showing a continuous decrease in the numbers of electrons with distance, with approximately 95% of the original β-particles stopped in the first half of the range. It is more common to speak of the

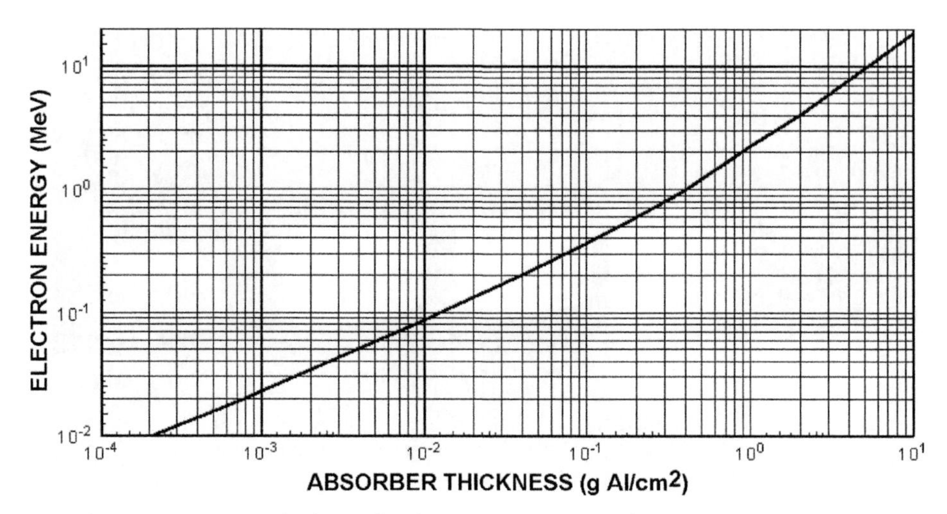

Figure 7.12 Empirical relation for the maximum range of β-particles in aluminum.

absorber thickness necessary to stop 50% of the particles than to speak of the range itself. This *half-thickness value* is much easier to ascertain experimentally than is an apparent range. It should be remembered that the energy deposited at complete β-absorption is $E_{abs} \approx 0.3\, E_{max}$ (Ch. 5).

The absorption curve for β-particles formed in radioactive decay can be described with fair approximation by Eqn (7.7). This is due to the continuous energy spectrum resulting in an exponential relationship for the range curve. In the E_{max} range 0.7–3 MeV the range in aluminum closely follows the relation (*Feather's rule*), with E_{max} in MeV:

$$\left(\text{g Al cm}^{-2}\right) = 0.543\ E_{max} - 0.160 \qquad (7.19)$$

This is the range C_3 in Figure 7.4.

Figure 7.12 shows an empirical relationship between the maximum energy of β-particles and the extrapolated range in aluminum. Due to their relatively few interactions per travelled distance, compared to α-radiation, β-radiation has a much longer range. For example, the range of an α-particle of 5 MeV is 3.6 cm in air while that of a β-particle of 5 MeV is over 17 m. A comparison of the range in air and water for electrons and heavy particles is given in Table 7.2. Figure 7.13 is useful for a rapid estimate of the absorber thickness needed to protect against β-particles.

An additional complication in the experimental measurements of absorption curves for β-particles is found in the fact that a certain fraction of β-particles which are not originally emitted in the direction of the detector may be deflected to the detector by the large angle scattering. This process is known as *backscattering*, since the *backing* (or *support*) for radioactive samples may cause scattering of a certain fraction of the particles through as much as 180°. The fraction of back-scattered β-radiation depends on the geometry of

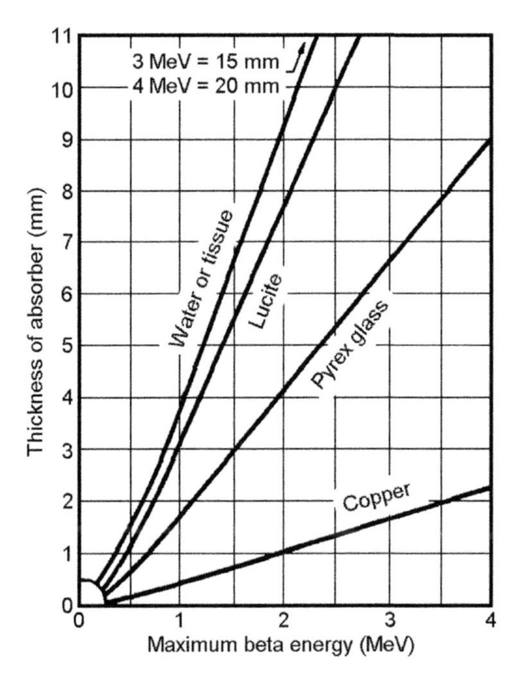

Figure 7.13 Thickness of various materials needed to completely stop β-particles.

the measuring system, the energy of the β-particles, and the thickness and electron density of the backing material. In Figure 7.14 the percent backscattering as a function of the atomic number of the backing material is shown for four β-energies (E_{max}); the radioactive sample is considered infinitely thin (i.e. no self-absorption). From the curve for beta rays from ^{32}P on platinum ($Z = 78$) we see that about 40% of the measured radiation is due to back-scattered radiation ($0.8/(1.0 + 0.8) = 0.4$). Backscattering increases with the thickness of the backing material up to a saturation value which is

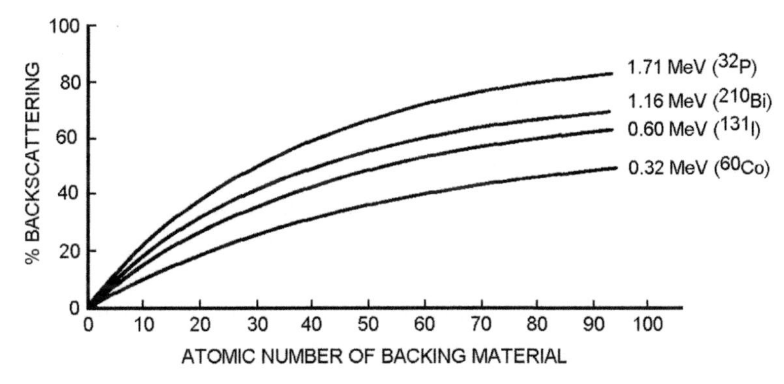

Figure 7.14 Backscattering of β-particles of different energy as function of the atomic number of thick backing materials.

reached when the thickness of the backing is about one-fifth of the extrapolated range of the β-particles in that material.

7.5. ABSORPTION OF γ-RADIATION

The absence of charge and rest mass for γ-rays results in little interaction with the absorbing atoms and in long ranges. The number of ion pairs produced in a given path length by γ-rays is only 1−10% of that produced by β-particles of the same energy (Fig. 7.7); e.g. a 1 MeV γ-ray produces only about one ion pair per centimeter of air. As a consequence of this low specific ionization of γ-rays, the ionization is almost completely secondary in nature resulting from the action of a few high energy primary ion pairs.

7.5.1. Attenuation coefficient

Unlike heavy particles and electrons which lose their energy as a result of many collisions, γ-rays are completely stopped in one or, at most, a few interactions. For thin absorbers the attenuation of γ-rays follows Eqn (7.8), where φ is the number of photons $m^{-2}\ s^{-1}$. The proportionality factor μ is called the *(total) attenuation coefficient*. When it has the dimension of m^{-1} and the thickness x is expressed in meters, μ is referred to as the *linear* attenuation coefficient. The attenuation coefficient can be expressed in other ways:

$$\mu_{\mathrm{m}} = \mu/\rho = \sigma_{\mathrm{a}}\mathbf{N}_{\mathrm{A}}/M = \sigma_{\mathrm{e}}Z\,\mathbf{N}_{\mathrm{A}}/M \qquad (7.20)$$

where ρ is the density, M the average atomic weight, and Z the average atomic number of the absorber. $\mathbf{N}_{\mathrm{A}}\rho/M$ can be replaced by N_{v}, by which we can define a *macroscopic absorption cross-section* Σ (cf. §11.1). Σ^{-1} is the mean free path or *relaxation length* of the radiation in the absorbing material. μ_{m} (in $cm^2\ g^{-1}$ when x is in centimeters) is the *mass attenuation coefficient*; σ_{a} is the probability of reaction between a γ-ray and the electron cloud of the absorber atom (*atomic reaction cross-section*, m^2 atom^{-1}); σ_{e} is the probability of the reaction of a γ-ray with a single electron of the absorber (*electron reaction cross-section*, m^2 electron^{-1}). σ_{a} and σ_{e} are analogous to the nuclear reaction cross-sections discussed later (cf. Ch. 11). In Table 7.1 only the equivalent nuclear and atomic reaction cross-sections are given.

Since a γ-ray may be removed from the beam in the first few Ångströms of its entrance into the absorber or may travel several centimeters with no interaction at all and then be removed, it is not possible to apply the range concept to γ-rays in the way that it is applied to charged particles. However, it is experimentally easy to measure the thickness of absorber necessary to remove half of the initial γ-rays (half thickness value) from a beam, or reduce it to 1/10, 1/100, etc. Figure 7.15 shows the required thickness of concrete and lead necessary to reduce γ-rays of different energies by factors of 10 (shielding thicknesses are further discussed in §7.7). The half-thickness value is

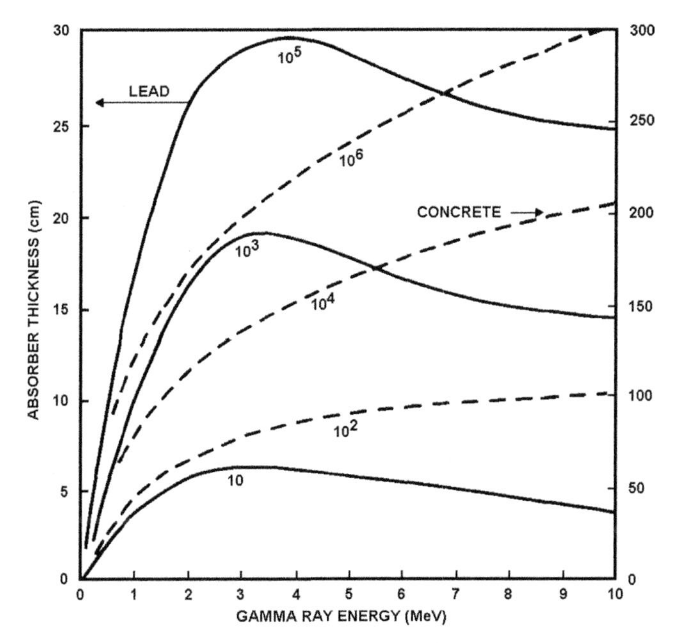

Figure 7.15 Thickness of lead and concrete reducing γ-ray flux by different powers of 10, as function of γ-ray energy. The curves are for thick absorbers and include build-up factors.

$$x_{1/2} = \ln(2)/\mu \tag{7.21}$$

and the 1/10 value

$$x_{1/10} = \ln(10)/\mu \tag{7.22}$$

7.5.2. Partial absorption processes

Gamma-ray absorption occurs as illustrated in Figure 7.16 by four different processes: coherent scattering, photoelectric effect, Compton effect, and pair production. For each of these processes, a partial coefficient can be expressed:

$$\mu = \mu_{\text{coh}} + \mu_{\text{phot}} + \mu_{\text{Comp}} + \mu_{\text{pair}} \tag{7.23}$$

Comparing with Eqn (7.10), μ_{phot} and μ_{pair} are absorption processes, while μ_{coh} is only a scattering process; μ_{Comp} contributes to both the absorption and the scattering terms. In Figure 7.17 the total attenuation, absorption, and the partial coefficients are given for water, aluminum, and lead as a function of the γ-ray energy. The corresponding linear coefficients are obtained by multiplying with ρ (for aluminum 2.7 g/cm^3, for lead 11.3 g/cm^3). It should be noted that the aluminum curves also can be used for absorption in concrete.

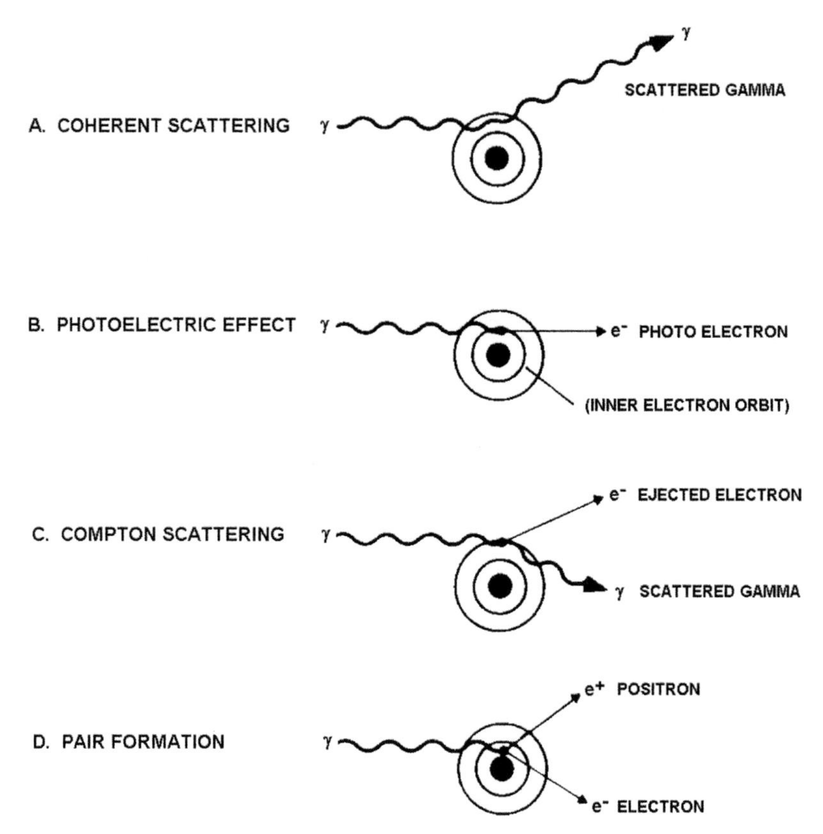

Figure 7.16 Schematic description of the four main processes accounting for γ-ray interaction and absorption.

In *coherent scattering* (also called Bragg or Rayleigh scattering, denoted σ_r in Fig. 7.17) the γ-ray is absorbed and immediately re-emitted from the atomic nucleus with unchanged energy but in a different direction. Coherently scattered radiation can give interference patterns, so the process is used for structural analysis of absorbing material in the same way as X-rays are. The probability for coherent scattering increases with the square of atomic number of the absorber and decreases with γ-ray energy. In lead, coherent scattering amounts to about 20% of the total attenuation for γ-energies of 0.1 MeV but decreases in importance for higher energy γ-rays.

In absorption of γ-rays by the *photoelectric effect* (denoted τ in Fig. 7.17 and Fig. 7.19) the photons are absorbed completely by the atom electrons. This absorption results in excitation of the atom above the binding energy of some of its orbital electrons with the result that an electron is ejected and an ion pair formed. The energy of the emitted photoelectron, E_e, is the difference between the energy of the γ-ray, E_γ, and the binding energy, E_{be}, for that electron in the atom.

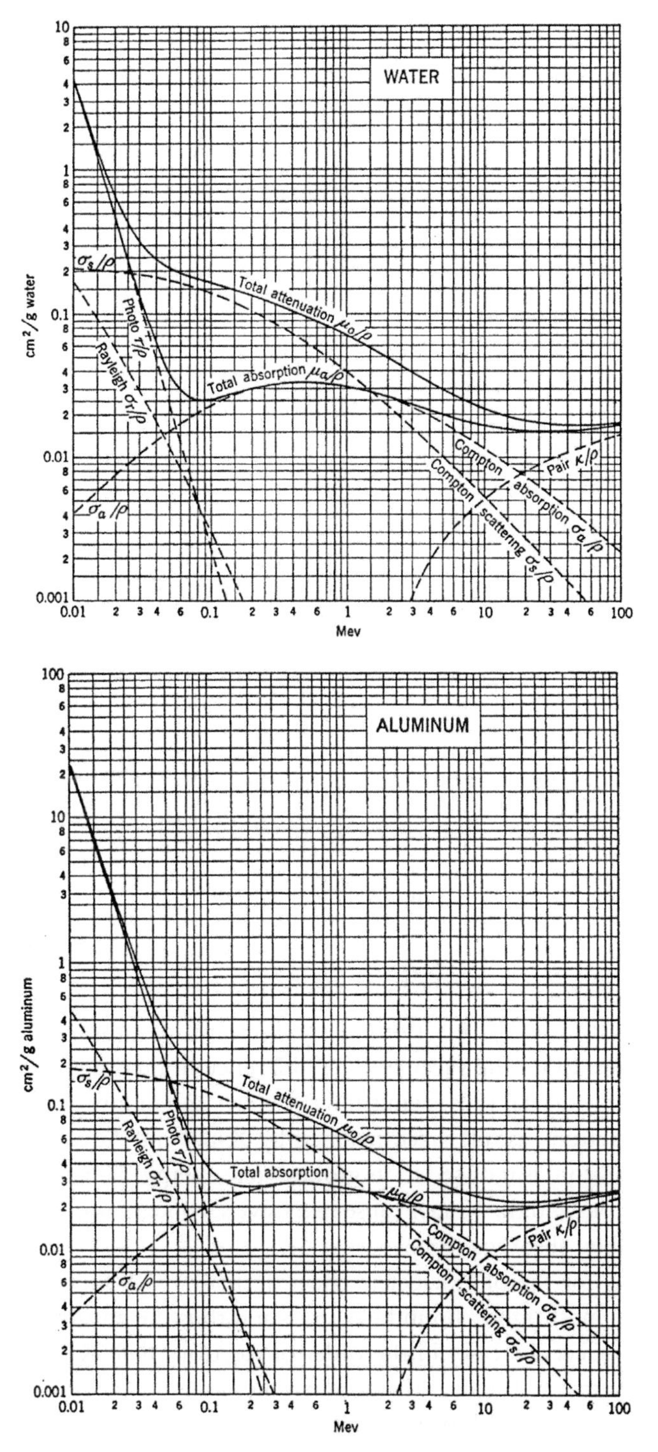

Figure 7.17 Total and partial mass absorption and attenuation coefficients for γ-rays in water, aluminum and lead. *(From R. D. Evans.)*

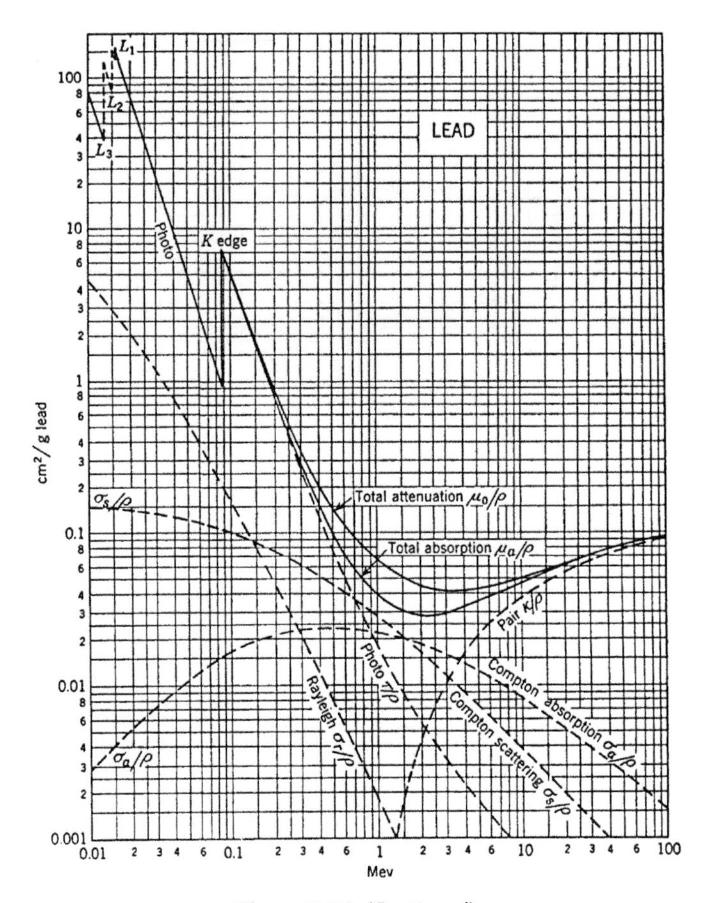

Figure 7.17 *(Continued)*

$$E_e = E_\gamma - E_{be} \tag{7.24}$$

If the photoelectron originates from an inner electronic orbital, an electron from a higher orbital moves to fill the vacancy. The difference in binding energy of the higher and the lower energy orbital causes emission of X-rays and of low energy *Auger electrons*. The process of electron cascade, accompanied by X-ray and Auger electron emission, continues until the atom is reduced to its ground state energy. The photoelectron as well as the Auger electrons and the X-rays cause extensive secondary ionization by interacting with the surrounding absorber atoms.

The probability for the photoelectric effect decreases with increasing γ-ray energy. It is largest for the most tightly bound electrons and thus the absorption coefficient for the photoelectric effect increases in the order of electron shells K > L > M >, etc. In Figure 7.17 we see that in lead it is the dominating mode of absorption up to about 0.7 MeV. Discontinuities observed in the graph of μ vs E_γ are related to the differences in binding energies of the electrons in the different shells as the increasing γ-ray energy allows more tightly bound electrons to be emitted. These discontinuities coincide with the K, L, etc., edges observed in X-ray absorption.

Gamma-rays of higher energy, rather than interacting with the field of the whole atom as in the photoelectric effect, interact with the field of one electron directly. This mode of interaction is called the *Compton effect* after its discoverer, A. H. Compton. In the Compton effect an electron is ejected from an atom while the γ-ray is deflected with a lower energy. The energy of the scattered γ-ray, E'_{γ}, is expressed by the equation

$$E'_{\gamma} = E_{\gamma} - E_{e} \quad (7.25)$$

where E_e is the kinetic energy of the Compton electron. The probability for Compton scattering increases with target Z and decreases with E_{γ}. Since the Compton interaction occurs only with the most weakly bound electrons and high energy γ-rays, the binding energy of the electron is negligible compared to E_{γ}. The Compton electrons and scattered γ-rays have angles and energies which can be calculated from the relationships between the conservation of energy and momentum, correcting for the relativistic mass of the electrons at these kinetic energies. The scattered γ-ray may still have sufficient energy to interact further by the Compton effect, the photoelectric effect or pair production. Again, emission of X-rays and Auger electrons usually accompanies Compton interaction and extensive secondary ionization follows. Since the Compton electron can have a spread of energies, the scattered γ-rays exhibit a broad spectrum. The Compton electrons, as in the case of photoelectrons, are eventually stopped by the processes described for β-particles.

Figure 7.17 shows the division of energy between the scattered Compton γ and the Compton electron as a function of γ-ray energy. Only the energy of the electron is deposited in the absorber as the scattered γ-ray has a high probability of escape. Thus Compton electrons contribute to the *(energy) absorption coefficient* μ_a while the Compton γ contributes to the *total attenuation coefficient* μ through the scattering coefficient μ_s in Eqn (7.10).

The fourth mode of interaction for γ-rays with an absorber involves conversion in the Coulomb field of the nucleus of a γ-ray into an electron and a positron (Fig. 7.18). This process is termed *pair production* since a pair of electrons, one positive and one negative, is produced. The process can be considered as the inverse phenomenon of positron annihilation. Since the rest mass of an electron corresponds to 0.51 MeV, the γ-ray must have a minimum value of 1.02 MeV to interact by pair production. As the energy of the γ-ray is increased beyond this value, the probability of pair production increases (see Figs. 7.17 and 7.19, where μ_{pair} is denoted κ). The excess energy (above the 1.02 MeV) appears as the kinetic energy of the electron pair,

$$E_{\gamma} = 1.02 + E_{e^-} + E_{e^+} \quad (7.26)$$

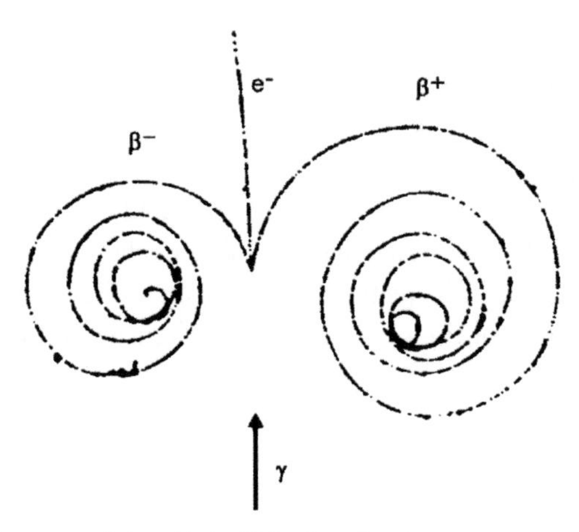

Figure 7.18 Tracks of electron pair in a H_2 bubble chamber in a strong magnetic field perpendicular to the plane of the tracks. *(Courtesy Lawrence Radiation Laboratory.)*

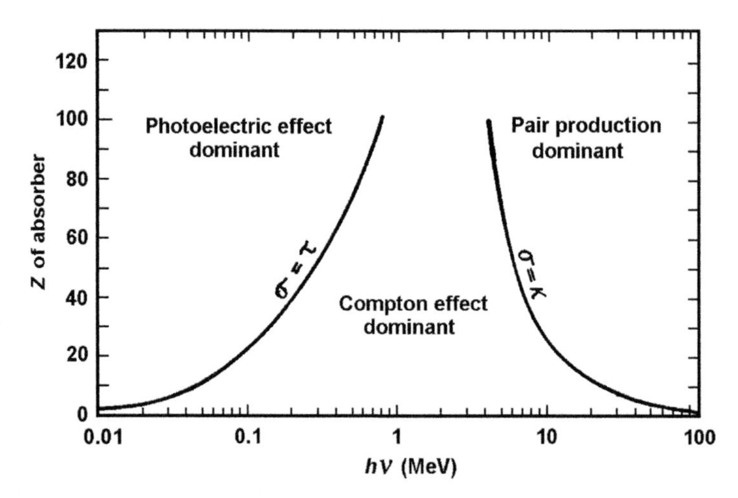

Figure 7.19 Dominance regions for the three γ-ray absorption processes, c.f. Fig. 7.17

where all energies are in MeV. The pair of electrons are absorbed as described in §7.4. The annihilation of positrons normally produce pairs of 0.51 MeV gammas, which are absorbed by the processes described previously[1].

Figure 7.19 summarizes the domains of interaction of the main γ-ray absorption processes as a function of γ-ray energy and absorber Z-value.

[1] Annihilation of positrons can in a very few cases also produce three gamma rays.

7.6. ABSORPTION OF NEUTRONS

A beam of collimated neutrons is attenuated in a thin absorber through scattering and absorption processes in a similar manner to the attenuation of γ-rays; these processes are described in previous sections. In a thick absorber the neutrons are slowed from incident energy at the absorber face to thermal energies if the absorber is thick enough. The ultimate fate of the neutron is capture in an absorber nucleus. Because of the spread in neutron energy and the energy dependency of the capture cross-sections, no simple relation can be given for the attenuation of the neutron beam (cf. next section).

7.7. RADIATION SHIELDING

The absorption properties of nuclear radiation in different materials must be known in order to design shielding to avoid unwanted radiation effects on the surroundings by nuclear radiation sources.

For charged particles the shielding is usually slightly thicker than that required for the maximum range of the particles in the material. Absorption thicknesses of 0.2 mm are adequate to completely absorb the particles from α-decay. By contrast 15 mm of materials of low Z such as water, plastic, etc., are required for absorption of β-radiation with energies up to 3 MeV. Radiation shielding constructed from materials of higher atomic number require correspondingly thinner thicknesses. The data in Table 7.2 and the curves in Figs. 7.6 and 7.13 provide information on the thickness of absorber material required for the energy of various types of radiation.

Since γ-rays and neutrons have no definite range but exhibit a logarithmic relation between thickness and intensity, only a partial reduction of the radiation can be obtained. By combining Eqns (7.7) and (7.8), Eqn (7.27) is obtained

$$\varphi = \left(n\, A/\left(4\pi\, r^2\right)\right)\mathrm{e}^{-\mu x} \tag{7.27}$$

It is seen that the intensity from a point source of radiation can be decreased by increasing either the distance from the source r or the thickness x of the absorber. Alternately, an absorber with a higher absorption coefficient, μ, can be chosen to reduce the thickness required.

Equation (7.27) is valid only for point sources with ideal geometry, i.e. no back or multiple scattering, etc. For γ-radiation the thicker the absorber the higher is the percentage of radiation which may be scattered backwards through secondary (mainly Compton) scattering. The effect of geometry and absorber thickness can be taken into account by including a constant B in the absorption equation:

$$\varphi = B\,\varphi_0\,\mathrm{e}^{-\mu x} \tag{7.28}$$

The "*dose build-up*" *factor* B not only takes into account multiple Compton and Rayleigh scattering but also includes correction for positron formation at high γ-energies

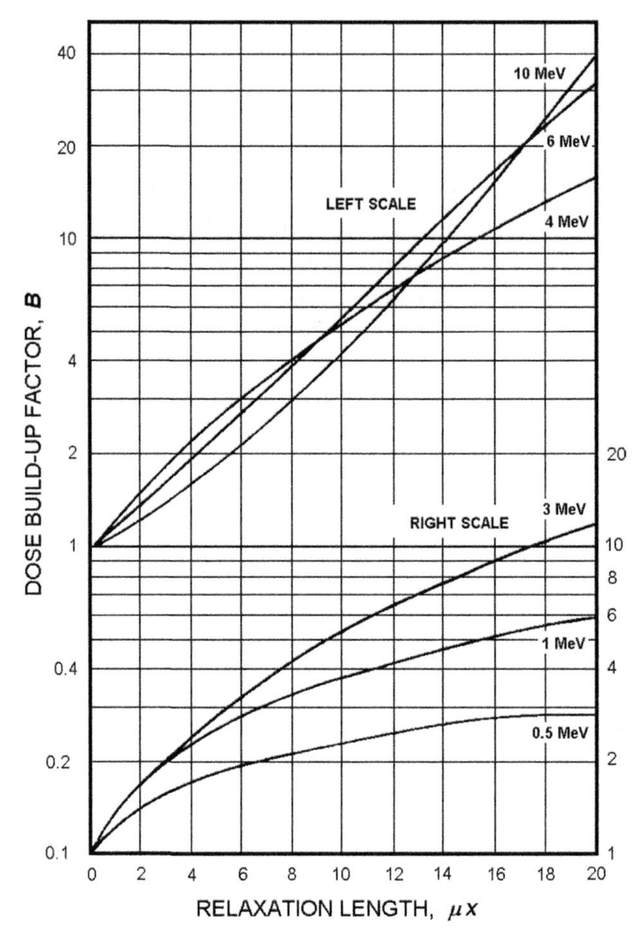

Figure 7.20 Dose build-up factors in lead for a point isotropic source. Upper curves have scale to the left, lower curves to the right. *(From Radiological Health Handbook.)*

and subsequent annihilation. Since for thick radiation shielding and high γ-energies the factor B may reach several powers of 10, it is quite important to be considered in designing biological shielding for radiation. Calculation of B is difficult and empirical data are most commonly used. Figure 7.20 gives B-values for a lead shield. The thickness of the shielding is given in relaxation lengths μx. This value is obtained from diagrams like Figure 7.17; e.g. for a 3 MeV γ, μ_m is found to be 0.046 $cm^2\,g^{-1}$. If the absorber is 0.1 m thick the linear density is $10\rho_{Pb} = 113\ g\,cm^{-2}$ and the relaxation length becomes 5.2. With Figure 7.20 this gives (for the 3 MeV γ-line) a dose build-up factor B of 3. Thus the lead shield transmits three times more radiation than is expected by the simple Eqn (7.26). The flux reduction values for concrete and lead shielding in Figure 7.15 have been adjusted to take the dose build-up into consideration.

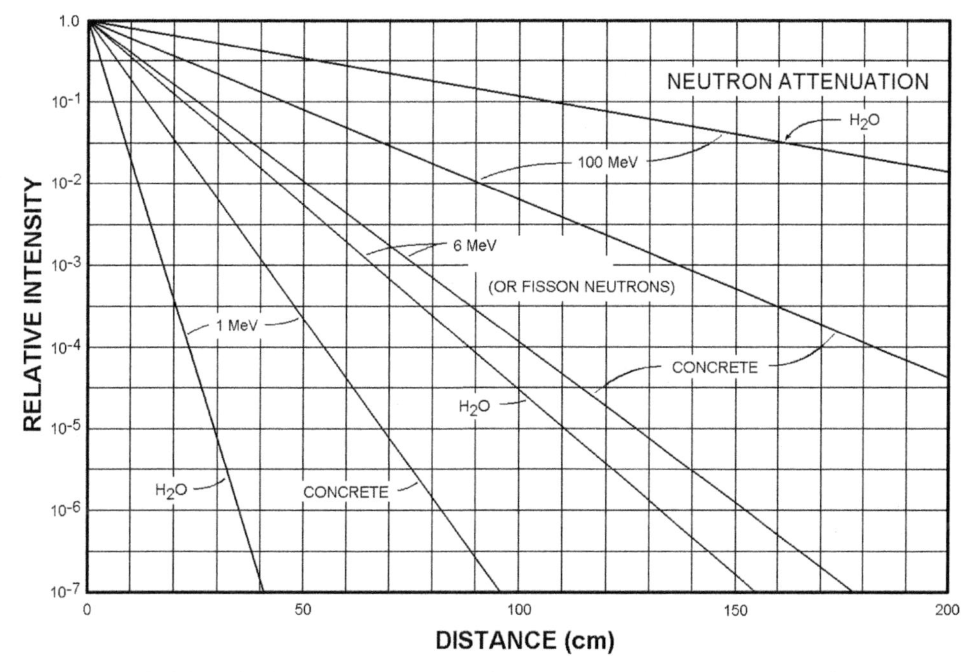

Figure 7.21 Shielding thickness necessary to reduce a neutron beam in water and concrete. The 6 MeV lines can be used for fission neutrons. *(From Nuclear Data Tables.)*

Equation (7.27) is not directly applicable for neutrons. For an estimate of required shielding we can use diagrams like that in Figure 7.21, which shows the attenuation of neutrons of three different energies in concrete and water, the most common neutron-shielding materials. It is necessary also to take into account the γ-rays emitted in neutron capture, which increases the shielding thickness required. (Please note that the order of absorption change as the energy of the neutron is increased. This is due to the energy dependence of the dominating cross sections, especially for the heavier atoms present in concrete.)

7.8. ANALYTICAL APPLICATIONS OF RADIATION ABSORPTION

In previous sections of this chapter we have shown how different particles emitted in nuclear reactions are stopped in matter without causing nuclear reactions (nuclear reactions are treated in Ch. 10 and 11). If the property of the particle is well defined (e.g. mass, charge, energy, etc) its interaction with the absorbing material can be well predicted, provided the composition of the absorbing material also is well defined. Conversely, the composition of the absorbing material (a "sample") can be determined from studying the absorption process (the "irradiation"). For example, atoms of the absorber (sample) may be knocked out and can be collected for analyses. When electrons are knocked out from the atomic shells of the absorber (sample) atoms, either the energy of these electrons may be analyzed, or the energy of the electromagnetic radiation emitted when the shells are refilled. These energies are characteristic of the sample and can thus be used for its identification. There are many analytical applications based on these principles, the most important ones will described in this section.

7.8.1. SIMS (secondary ion mass spectrometry)

When heavy ions of energy largely exceeding the chemical binding energies, but with energies much lower than needed to cause nuclear reactions, hit a surface then atoms of this surface are *sputtered* out. These atoms, or actually ions, can be introduced in a mass spectrometer to determine the exact masses and mass/charge ratios, from which the element is identified. This is the bases for the SIMS analytical method for studying surfaces, particulary semiconductor surfaces. By bombarding with O^{2+} or Cs^{+} of ≤ 10 keV most surface elements can be detected (in fortunate cases down to the ppb range) with a resolution of the order 100−200 Å.

7.8.2. PIXE (proton or particle induced X-ray emission)

In the PIXE technique high energy protons (or heavier ions) are used to irradiate a thin sample (say 0.1 mg/cm^2). The probability for expulsion of an electron followed by the emission of a K-X-ray decreases with increasing Z of the sample; e.g. for 5 MeV protons the reaction cross section is ca. 10^{-25} m^2 for $Z = 10$ to 20 and about 10^{-28} m^2 (1 *barn*) for $Z = 50$; cf. Table 7.1. It can be shown that the reaction probability has a maximum when the incident particle has a velocity equal to the Bohr orbital velocity of the electron. Though the sensitivity decreases with Z elements up to Pb can be determined. The technique has primarily been used to determine trace elements in environmental and biological samples, see Figure 7.22. It has also been used by several Mars landers to analyse the composition of various rocks and minerals on the surface of Mars.

7.8.3. ESCA (Electron spectrometry for chemical analysis)

High resolution β-spectroscopy can be used to determine chemical properties. A sample is irradiated with mono-energetic photons of $E_{\mathbf{h}\nu}$, leading to the emission from the sample surface of photoelectrons. The relevant equation is

$$E_{\mathbf{h}\nu} = E_{\mathrm{be}}(\mathrm{X}, \mathrm{Y}) + E_{\mathrm{e}} \tag{7.29}$$

where E_{e} is the kinetic energy of the emitted electrons, which can be determined very accurately (presently to about 0.01 eV) in the magnetic spectrograph; *photoelectron spectroscopy*. This sensitivity is much greater than chemical binding energies, E_{be}(X,Y), where X refers to an atom in compound Y. The probability for ejection of photoelectrons increases with decreasing photon energy (Fig. 7.16 and 7.19) and therefore low energy X-rays are used as a source.

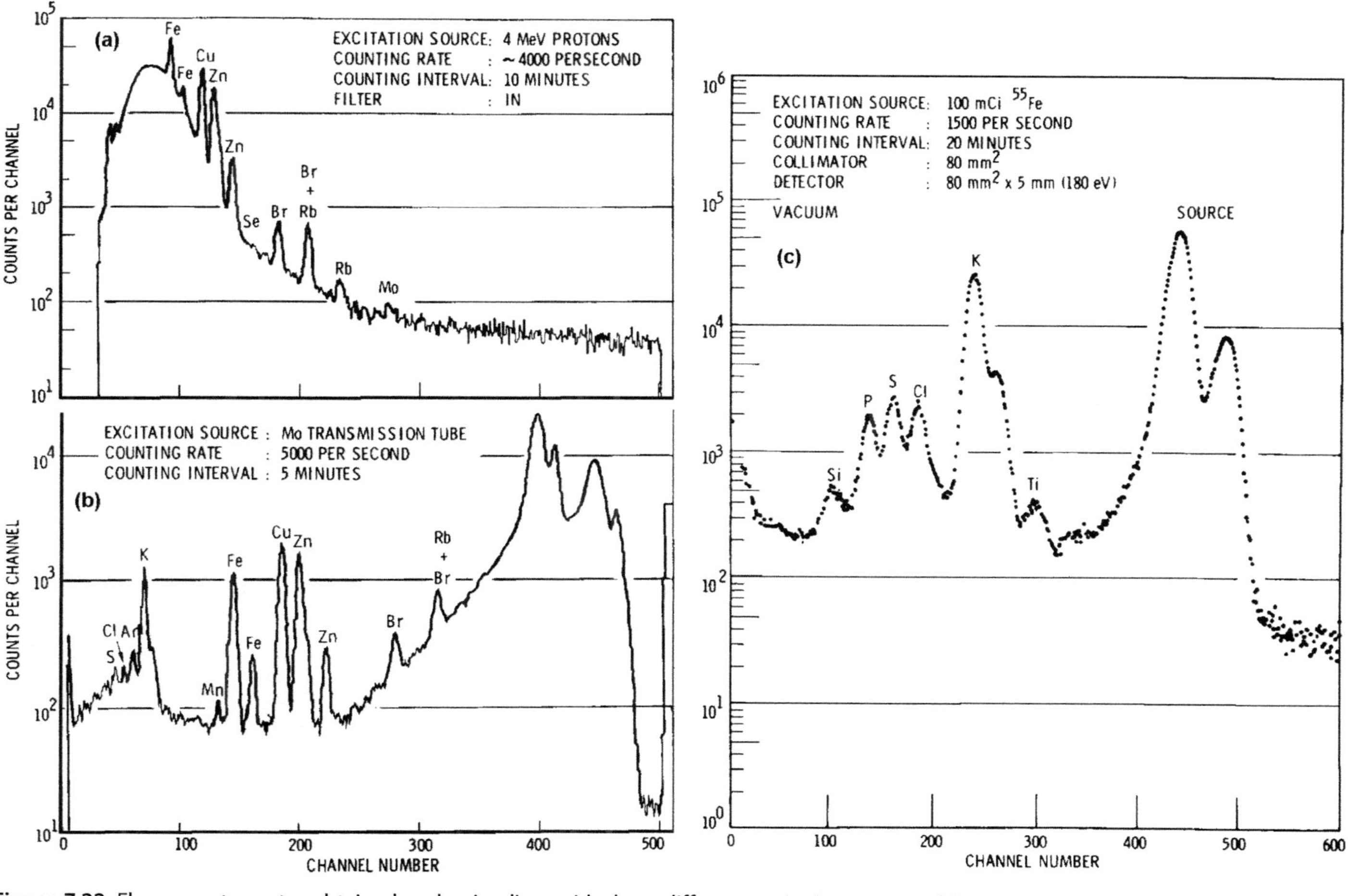

Figure 7.22 Fluorescent spectra obtained on bovine liver with three different excitation sources: (a) 4 MeV protons (PIXE), (b) (conventional) molybdenum transmission tube, and (c) radioactive ^{55}Fe.

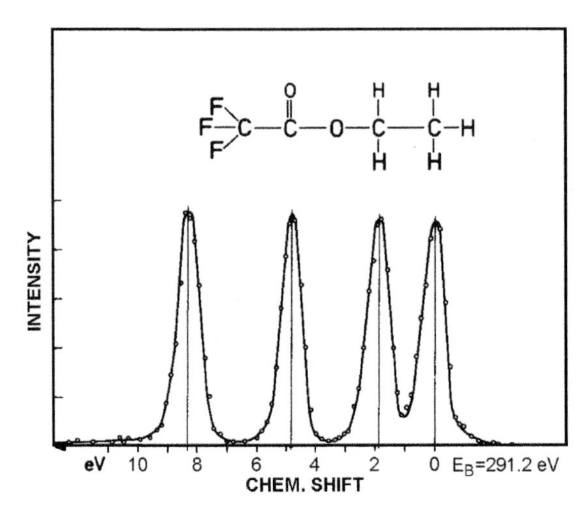

Figure 7.23 ESCA spectrum of trifluoroacetate.

Although it is the outermost (or most weakly bound) electrons which form the valence orbitals of a compound, this does not leave the inner orbitals unaffected. An outer electron (which we may refer to as e_L) of an atom X_1 which takes part in bond formation with another atom X_2 decreases its potential, which makes the inner electrons (which we may call e_K) more strongly bound to X_1. Thus $E_{be}(e_K)$ increases by an amount depending on $E_{be}(e_L)$. Although this is a somewhat simplified picture, it leads to the practical consequence that the binding energy of e_K, which may be in the 100–1000 eV range, depends on the chemical bond even if its orbital is not involved directly in the bond formation. Figure 7.23 shows the ESCA spectrum of trifluoroacetate. Since the largest chemical shift, relative to elementary carbon, is obtained for the most electronegative atoms, the peaks refer to the carbon atoms in the same order as shown in the structure.

7.8.4. XFS (X-ray fluorescence analysis)

If a sample containing atoms of a particular element (e.g. Ag) is irradiated with photons of energy high enough to excite an inner electron orbital (e.g. the K_α orbital in Ag at 22.1 keV), X-rays are emitted in the de-excitation. If the photon source is an X-ray tube with a target made of some element (Ag in our example), the probability is very high that the K_α X-ray emitted from the source would be absorbed by the sample atoms and re-emitted (*fluorescence*). (This is an "electron shell resonance absorption" corresponding to the Mössbauer effect, §7.8.5, although the width of the X-ray line is so large that recoil effects can be neglected.) The spectrum of the scattered X-radiation (or, more correctly, photon radiation emitted from the sample) is referred to as the *X-ray fluorescence spectrum*. It contains lower energy radiation including K_α radiation

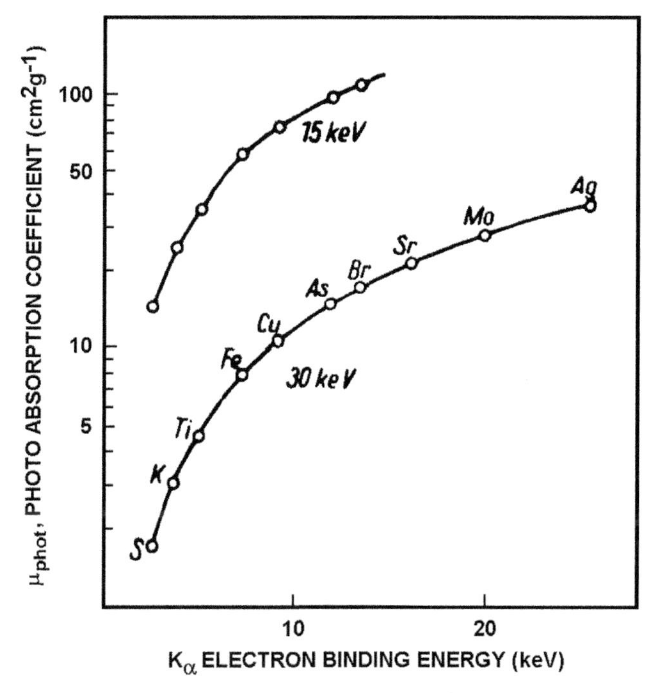

Figure 7.24 Photoelectron absorption coefficients at K edges for 10 and 30 keV γ-rays as function of absorber material. The photo effect is the dominating absorption mode.

emitted by atoms of lower atomic number. The height of these other peaks is lower because of a lower reaction cross-section (Fig. 7.24). Figure 7.22 illustrates an XFS analysis of a biological sample.

X-ray fluorescence analysis using vacuum tube sources have become a well-established analytical technique in the last decade. Nuclear interest stems partly from the possibility of using radioactive sources for stimulating X-ray fluorescence in a sample. These sources can be classified depending on the mode of production of the X-rays:

(i) γ-ray sources: decay between closely located nuclear energy levels, e.g. a 59.5 keV γ emitted in the α-decay of ^{241}Am: also "broad spectrum" γ-sources like ^{125}I are used.

(ii) X-ray sources: (a) radiation emitted in rearrangement of electron orbitals following α- or β-decay (primary X-rays), e.g. 11.6—21.7 keV uranium L-X-rays from ^{238}U formed in the α-decay of ^{242}Pu, or 41.3—47.3 keV europium K-X-rays from β-decay of ^{153}Gd; (b) irradiation of a target with α-, β-, or γ-radiation leading to ionization and excitation of the target atoms and its de-excitation by X-ray emission.

(iii) Bremsstrahlung sources, e.g. T in titanium, or ^{147}Pm in aluminum (Fig. 7.10).

An important advantage of radioactive X-ray fluorescence sources is that very small instruments requiring no (X-ray) high voltage or current can be designed for field applications. Such instruments are used in geological investigations (bore holes, mineral samples, etc.), in-line mineral analysis (Zn, Cu, and other ores in flotation cycles, Ca in cement raw-material, etc.), on-line analysis of surface depositions (Zn on iron sheets, Ag in photographic emulsions, etc.). Figure 7.22 shows fluorescent spectra obtained on bovine liver with three different excitation sources: (a) 4 MeV protons (PIXE), (b) (conventional) molybdenum transmission tube, and (c) radioactive ^{55}Fe.

7.8.5. Mössbauer effect

According to the wave model of the atom, electrons in the innermost orbitals have a finite probability of existence within the nucleus. These electrons interact with the nuclear charge distribution, and thereby affect the nuclear energy levels (cf. §11.3.3). The extent of the effect depends on the exact properties of the electron orbitals involved, which vary with different chemical compounds. Therefore a γ-ray emitted from an isomeric state of an atom bound in one chemical compound may have a slightly different energy than from the same atom bound in another compound. This difference, referred to as the *isomer (energy) shift*, is extremely small, only about 10^{-10} of the energy of the emitted γ. Nevertheless, it can be measured by a technique developed by R. Mössbauer. The fundamental physics involved and technique used is well illustrated by Mössbauer's original experiment. Mössbauer placed an ^{191}Os source about a half-meter from a γ-ray detector A as shown in Figure 7.25. An iridium foil absorber was placed between the source and detector so that some of the photons of 129 keV energy from the ^{191}Os were absorbed by the iridium atoms in the foil, exciting these atoms from the ground state

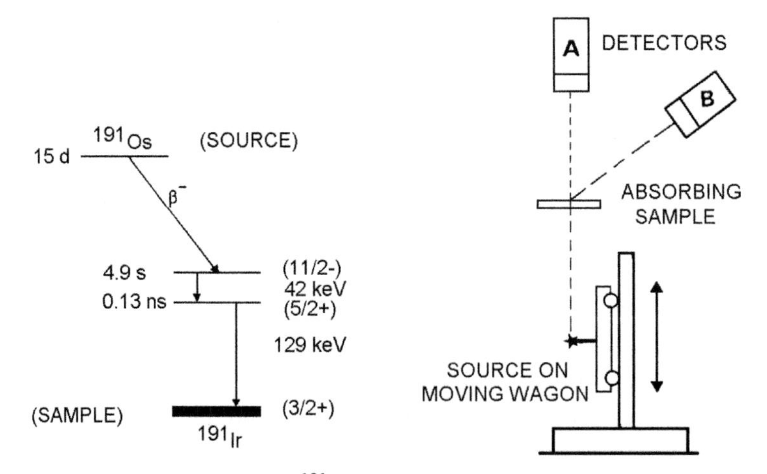

Figure 7.25 Decay scheme of ^{191}Os and principle of a Mössbauer experiment.

(3/2+) to the 5/2+ state. Because of the short half-life of the latter state it immediately decayed, re-emitting the γ-ray. The emission was isotropic, i.e., occurs in all directions. The result was a reduction in intensity measured by detector A but an increase in the count rate in detector B.

The conditions for such a *nuclear resonance absorption* are very stringent. Using the Heisenberg relationship, Eqn (5.89), it is possible to estimate the half-value width of the 129 keV peak to be 5×10^{-6} eV. Eqn (5.48) can also be used to calculate the iridium atom recoil energy. It is then found to be 46×10^{-3} eV. Thus the γ-ray leaves the source with an energy of ($129 \times 10^3 - 46 \times 10^{-3}$ eV). Also, in order for the 129 keV γ-ray to be absorbed in ^{191}Ir, it must arrive with an excess energy of 46×10^{-3} eV to provide for the conservation of momentum of the absorbing atom. Thus there is a deficit of $2 \times 46 \times 10^{-3}$ eV, which is very different from the value of the very narrow energy width of the γ-ray. Consequently, no resonance absorption can take place.

The limitation posed by the recoil phenomenon can be circumvented. If the source and absorber atoms are fixed in a crystal, the recoil energy may be insufficient to cause bond breakage. The energy is absorbed as an atomic vibration in the crystal, provided the quantization of the vibrational states agree exactly with the recoil energy. If not, which is often the case, the absorber atom stays rigid in the lattice, and the recoil energy is taken up by the whole crystal. In this case it is necessary to use the mass of the crystal in Eqn (4.34) rather than the mass of a single atom. Under these circumstances the recoil energy becomes infinitesimally small for the emitting as well as the absorbing atom; this is called *recoilless absorption*. The probability for recoilless absorption is improved if the source and absorber are cooled to low temperatures.

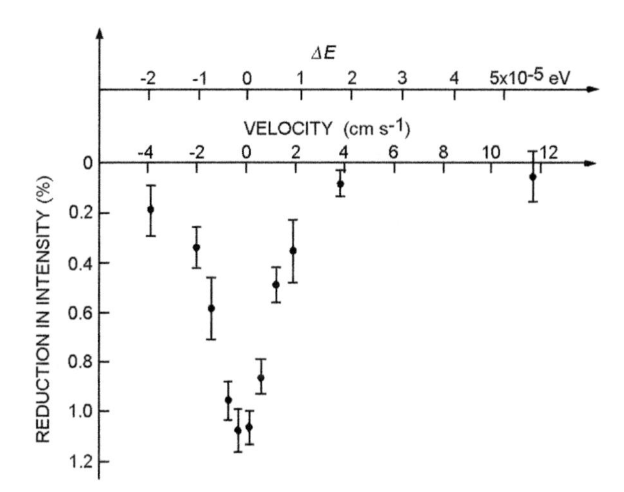

Figure 7.26 Mössbauer spectrum of ^{191}Ir metal. *(From R. Mössbauer.)*

The data of Figure 7.26 were obtained by recoilless absorption in osmium metal containing ^{191}Os (source) and Ir metal, both cooled in cryostats. By slowly moving the source (with velocity v) towards or away from the absorber (see Fig. 7.25), some kinetic energy ΔE_γ is added or subtracted from the source energy E_γ as "detected" by the absorber (Doppler effect). The energy and velocity relationship is given by the Doppler equation

$$\Delta E_\gamma / E_\gamma = v/\mathbf{c} \tag{7.30}$$

The velocity is shown in the Figure, where a value of v of 1 cm s^{-1} corresponds to 4.3×10^{-6} eV. The half-value of the γ-peak is found to be about 20×10^{-6} eV, i.e. a factor 4 times higher than calculated by the Heisenberg relationship. This is due to Doppler broadening of the peak as a consequence of some small atomic vibrations. Although the Mössbauer method can be used for measurements of γ-line widths, the results are subject to considerable errors.

One of the most striking uses of the extreme energy resolution obtainable by the Mössbauer effect was achieved by R. V. Pound and G. A. Rebka, who measured the emission of photons in the direction towards the earth's center, and in the opposite direction from the earth's center. They found that the photon increased its energy by one part in 10^{16} per meter when falling in the earth's gravitational field. This can be taken as a proof that the photon of $E_{\mathbf{h}\nu} > 0$ does have a mass, cf. §2.3.

When a "Mössbauer pair" (like ^{191}Os/Ir, or ^{57}Co/Fe, ^{119m}Sn/Sn, ^{169}Er/Tm, etc.) have source and absorber in different chemical states, the nuclear energy levels differ for the two Mössbauer atoms by some amount ΔE_γ. By using the same technique as in Figure 7.25, resonance absorption can be brought about by moving the source with a velocity corresponding to ΔE_γ. In this manner, a characteristic *Mössbauer spectrum* of the compound (relative to a reference compound) is obtained; the location of the peaks (i.e. the absorption maxima) with respect to a non-moving source (the *isomer shift*) is usually given in mm s^{-1}.

Figure 7.27 shows the isomer shifts obtained for a number of actinide compounds. The positions of the isomer shifts show the effect of valence states due to different population of the 5f orbitals. The different shifts for compounds of the same valence state is a measure of the variation in the covalency of the bonding. The compounds on the left are metallic. The shifts reflects the contributions of conduction electrons to the electron density at the nucleus of neptunium.

Mössbauer spectroscopy is limited to the availability of suitable sources. About 70 Mössbauer pairs are now available. The technique provides a useful method for studying chemical compounds in the solid state, especially compounds which are nontransparent to light and chemically or radioactively unstable. Mössbauer spectroscopy has also been used by several Mars landers, e.g. Spirit and Opportunity, to analyse rocks and minerals on the surface of Mars looking for signs left from ancient surface water.

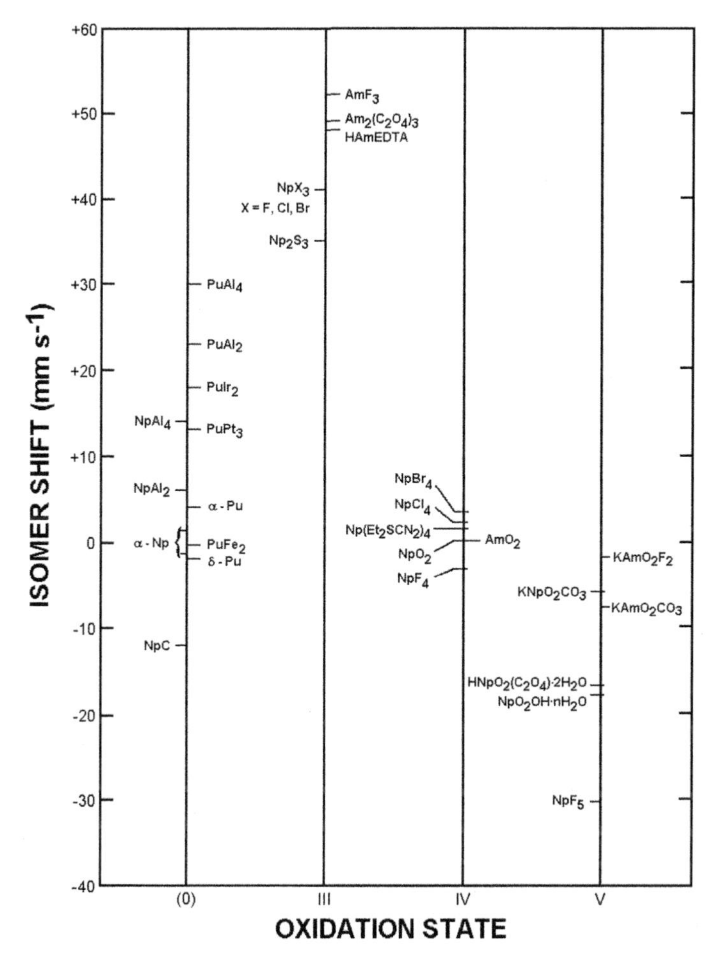

Figure 7.27 Isomer shifts for some actinide metals and actinide compounds.

7.9. TECHNICAL APPLICATIONS OF RADIATION SOURCES

Nuclear radiation absorption methods have many technical applications. These methods are not to be confused with radioisotope tracer methods, although radioisotopes may be used as radiation sources. In the tracer method the chemical properties of the radionuclide are important while in the applications discussed in this section only the type and energy of radiation emitted are important.

As a *source* of radiation in such technical applications, either accelerators or radiation from radionuclides can be used. Interchangeable radionuclides have the advantage over accelerators as radiation sources in that they can cover a larger energy range from high energy γ-rays to low energy β-rays in a much simpler way. This makes it possible to select the type and energy of radiation which have the most advantageous properties for a particular use. An additional advantage of radionuclides is that the sources can usually be made much smaller than X-ray sources, enabling them to be used in places where larger equipment is inconvenient or impossible to place. The fact that radionuclides require neither electric power nor cooling also renders them more suitable for field applications. Further, their independence from effects of temperature, pressure, and many other factors results in higher reliability compared to X-ray generators. Finally, and, perhaps as important as any of the other factors, they are in general much less expensive than accelerators.

To counterbalance these advantages is the disadvantage of the inability to turn off the radiation from radionuclides. This often requires that the radiation source be well shielded, adding to its weight and cost. An additional drawback to the use of radionuclides is that they have to be replaced after a few half-lives. The seriousness of this disadvantage depends upon the lifetime of the particular nuclide and is unimportant in cases where long-lived sources can be used.

The extensive use of radionuclides in industry is illustrated by Table 7.3, which summarizes the various studies undertaken by I.C.I., UK, in "a typical year".

7.9.1. Radionuclide gauges

Radionuclide gauges are measurement systems consisting of two parts, a radioactive source and a detector, fixed in some geometry to each other. They are used mainly for control in

Table 7.3 Radioisotope based studies undertaken annually by a large chemical company

Technique[(a)]	Number of applications
Level and interface measurements:	
γ-ray absorption	210
Neutron backscatter	480
γ-ray backscatter (storage cavities)	71
Blockage detection and deposition:	
γ-ray absorption	132
Neutron backscatter	129
Entrainment and voidage:	
γ-ray absorption	86
Thickness and corrosion measurements	15
Distillation column scans	108
Flow measurements:	
Pulse velocity	483
Dilution techniques	84
Leak detection	90
Residence-time studies	21
Carry-over studies (tracer)	6

[(a)]Less commonly used techniques have not been included.

industrial processes but can also be applied for specific analyses. The gauges come in two types. In one type the radiation source and the detector are on opposite sides of the technical arrangement to be measured; these are known as transmission or absorption instruments. In the second type, known as reflection or back-scattering instruments, the radiation source and the detector are on the same side. The instruments are also classified with respect to the kind of radiation involved. For example, γ-transmission, β-reflection, secondary X-ray instruments, etc. These radioisotope gauges are used for measurements of thicknesses, densities, etc. In Figure 7.28 a number of types of application for transmission instruments are illustrated. Illustrations a–c show measurements of level control. Type a can be used only for liquids while b and c can be used for all kinds of material. The latter instruments are uniquely suitable for application to large storage containers for grain, wood chips, oil, sand, cement, etc., and for material under extreme conditions such as molten glass and metal, explosives, etc. Level gauges are also used in the control of the filling or packages, cans, etc. in industry as illustrated in d.

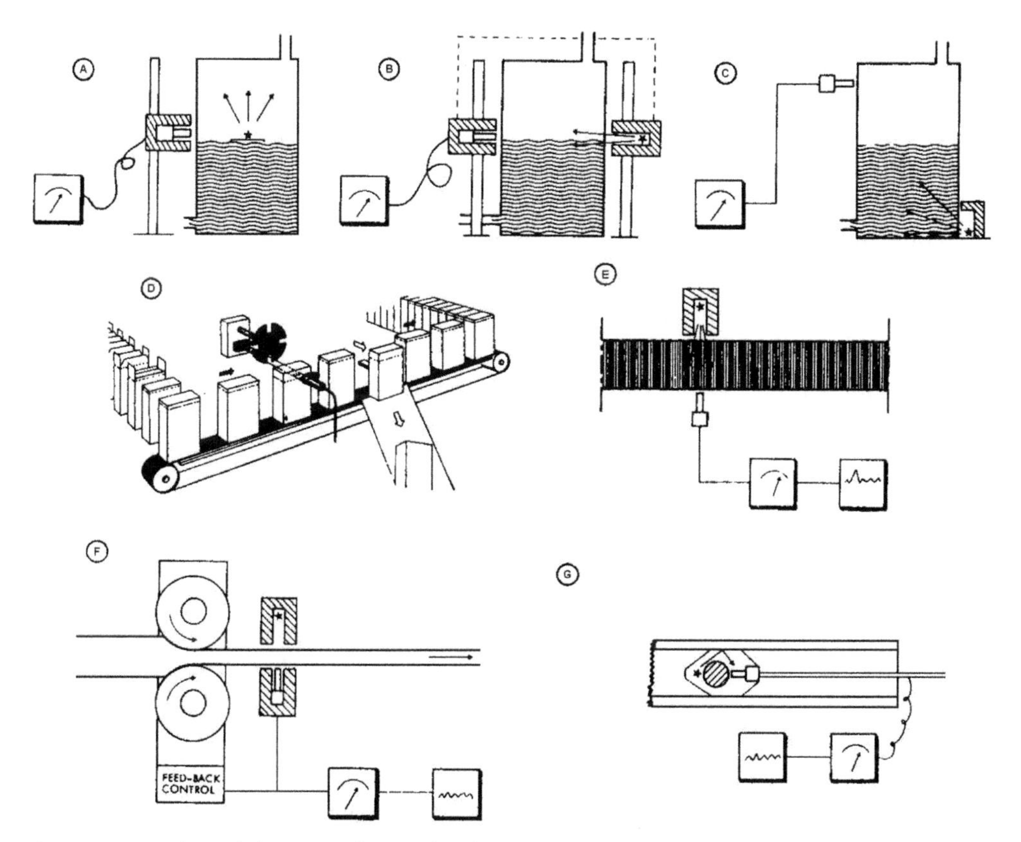

Figure 7.28 Radionuclide gauges for: A-C level measurements in tanks, D control of package filling, E flow density measurements, F thickness control, and G wall material in pipes and bore holes.

Let us consider a somewhat unusual application of this type of instrument. In the manufacture of titanium it is important to keep liquid $TiCl_4$ at a particular high temperature and pressure in a vessel. $TiCl_4$ has a triple point (i.e. the pressure and temperature conditions at which all three phases - solid, liquid, and vapor - of a substance are at equilibrium) in the neighbourhood of the particular technical conditions. By using a γ-density gauge it is possible to detect when the triple point is exceeded because of the disappearance of the vapor-liquid interface. This allows a simple method of control of the process conditions in the vessel.

The use of radioisotope gauges in density measurements is dependent upon Eqn. (7.27) in which r and x are constant while the absorption coefficient is density dependent (i.e. dependent on the average atomic composition of the absorber). A practical arrangement is illustrated in Figure 7.28(e), where the density of a medium in a pipeline is measured. This medium may be a mixture of gas and liquid such as water and water liquids with different composition and different amounts of dissolved substances as, for example, oil, salts or acids in water, process solutions in general or sludges, and emulsions such as fruit juices, latex emulsions, etc. From the variation in density the concentration and composition may be determined. Such density gauges are also used for control in filling of soft drink bottles and cans and submerged in rivers and lakes for measuring the depth of the bottom silt, etc. Density gauges are used in the production and the fabrication of such diverse products as automobile tires and cigarette packages.

Thickness gauges are the most common type of instrument using the absorption technique. In this case x in Eqn. (7.27) is varied. Measurements can be carried out on all kinds of materials with thicknesses of $\leq 100 \text{ g cm}^{-2}$ and is independent of the temperature and of whether the material is stationary or in motion. Figure 7.28(f) illustrates the application of a thickness gauge in a rolling mill where material of constant thickness is produced by using the signal from the detector for control purposes. Thickness gauges are used in the fabrication of glass, metal, paper, plastic, rubber, candy bars, etc. They have been used for measuring the thickness of snow in polar regions, icing on airplane wings, and other applications in which it is necessary to use remote operation.

In order to measure very thin layers of material such as coatings of paint, wax, and plastic films on papers or other material, two thickness gauges are used with a differential coupling so that one detector measures the uncovered and the other the covered or treated portion of the material. Thickness gauges also are used in industry to measure the degree of wear in industrial machinery. For surface measurements of thicknesses $\leq 0.8 \text{ g cm}^{-2}$ most thickness gauges use radiation sources with β-emitters while for thicknesses of $0.8-5 \text{ g cm}^{-2}$ bremsstrahlung radiation sources are most suitable. For even thicker materials γ-emitters are used.

Use of reflection gauges depends on the fact that the intensity of the scattered radiation under conditions of constant geometry depends on the thickness of the scattering material and its electron density (if β- or γ-sources are used). If neutrons are used the mass number of the scattering material is of prime importance. The electron density of the scattering material varies with the particular element and the chemical composition. Frequently, it is possible to determine the thickness and the nature of a surface layer by means of β-scattering. Reflection gauges have been applied to on-line analysis of tin-covered iron plates, metal coatings on plastics, paint layers, and to measuring the protective coating inside pipelines, Fig. 7.28(g). In some instances γ-radiation sources are preferred over β-emitters in measurements of material with greater wall thicknesses, particularly when transmission measurements are not feasible. Steel thickness from 1 to 20 cm has been measured with 5% accuracy using backscattering from ^{60}Co or ^{137}Cs sources of ≈20 μCi intensity (≈750 kBq).

Scattering and reflection are dependent on the electron density of the absorber, which is approximately proportional to the value of Z/A. Backscattering of β-particles from organic compounds is therefore very dependent on the hydrogen concentration ($Z/A = 1$) but fairly independent of the concentration of C, N, and O ($Z/A = 0.5$). This has led to the development of sensitive instruments for hydrogen analysis for various organic and water-containing materials. In an instrument using 10 mCi ^{90}Sr (370 MBq), the hydrogen concentration in a 10 ml sample can be determined in 20 min with 0.03% accuracy. This is not only superior to other conventional analytical methods, it also has the advantage of being a non-destructive technique.

Neutrons are slowed down most effectively by light elements (§10.5). As a consequence, neutron scattering can be used for the analyses of light elements, particularly hydrogen. In one type of instrument the radiation source consists of ^{252}Cf which produces fast neutrons (from spontaneous fission), while the detector is sensitive only to slow neutrons. This system is used for studies of ground water and analysis of bore holes in wells, Fig. 7.28(g). These analyses are usually combined with density determinations using a γ-source, thereby making it possible to identify strata of water, oil, coal, etc.

Some properties and uses of commercial radionuclide gauges are listed in Table 7.4.

7.9.2. Radiography

Radiography is a photographic technique in which nuclear radiation is used instead of light. Medical examination and nondestructive industrial testing using X-rays generated by high-vacuum tubes are the most important areas. A number of suitable sources of radioactive nuclides for producing radiograms are given in Table 7.4.

Beta-radiography is only suitable for thin objects and not widely applied. On the other hand, γ-radiography is a common nondestructive test technique in which normally ^{137}Cs or ^{60}Co has been used. γ-radiography has advantages in field use and in detection of sensitive objects. The radiation source can be ^{60}Co, which is normally kept in a portable radiation shielding of 30−60 kg of lead and situated at the end of a rod so that it can be pushed out of the shielding for use. The photographic film is located in a cassette surrounded by amplifying screens.

Table 7.4 Some commercially available radionuclide gauges and γ-sources for radiography (France)

Radiation	Source	$t_{½}$	Application
α	U or ^{226}Ra	Long	Thickness control in manufacturing paper, aluminum; 60 g/m^2
Soft β	^{147}Pm(0.2 MeV)	2.6 y	Thickness control; ≤400 g/m^2
β, soft γ	^{204}Tl(0.8 MeV)	3.8 y	Thickness: 1.10 mm steel, 3–50 mm glass; 8–100 kg/m^2
Hard β	^{144}Ce(3 MeV)	0.78 y	Thickness 1 mm steel; ≤10kg/m^2
X	^{109}Cd(88 keV)	1.24 y	Detection of S-content in hydrocarbons
n,γ	RaBe, ^{137}Cs	30 y	Moisture-density meter for civil engineering and agriculture
γ	^{60}Co(1.3 MeV)	5.3 y	4 MBq source for backscatter on ≤20 mm steel, 0.4–40 GBq for remote level indication
Soft γ	^{192}Ir(0.3 MeV)	74 d	400 GBq, 26 kg: ≤40 mm steel radiography
Medium γ	^{137}Cs(0.7 MeV)	30 y	400 GBq, 45 kg: ≤70 mm steel pipeline inspection
Hard γ	^{60}Co(1.3 MeV)	5.3 y	10 TBq, 900 kg: ≤180 mm steel radiography

The exposure At (GBq hours) required for an optical density ($\hat{D} = \log$ (incident light/transmitted light)) ~ 2 at an absorber (object) thickness x (cm) using a typical industrial X-ray film and a ^{60}Co γ-ray source positioned at a distance of 1 m from the film can be estimated by the expressions

$$\log(At) = 1.068 + 0.135\,x \quad \text{for an iron absorber} \tag{7.31}$$

$$\log(At) = 1.068 + 0.040\,x \quad \text{for an concrete absorber} \tag{7.32}$$

Exercise 7.14 is an example of the use of these expressions.

Gamma-radiography has been used for determining the number of reinforced iron bars in concrete construction, cavities in various kinds of castings (as explosives, plastics or metals), cracks or other defects in turbine blades in airplane parts, detonators in unexploded bombs, welded joints in pressure vessels, distillation towers and pipes, corrosion inside pipes and furnaces, and medical field X-rays, to mention only a few applications. Gamma-radiography is used throughout the world for product control leading to improved working safety and economy.

Because γ-absorption occurs through interaction with the electrons, objects of high atomic numbers show the strongest absorption. By using neutrons instead of γ-rays, the opposite effect is achieved, i.e. low Z objects are most effective in removing neutrons from a beam. This is used in *neutron radiography* in which both reactor neutrons and neutrons from ^{252}Cf sources are applied. Because of the higher neutron flux from the reactor than from ^{252}Cf sources of normal size (i.e. $\leq$1 mg) the exposure time at the reactor is much shorter. On the other hand, the small size of the ^{252}Cf source offers other conveniences.

7.9.3. Radionuclide power generators

The absorption of radiation leads to an increase in the temperature of the absorber. An example of this is the absorption of the kinetic energy of fission products in nuclear reactor fuel elements which is a main source of the heat production in reactors. The absorption of decay energy of radioactive nuclides in appropriate absorbing material can be used in a similar - albeit more modest - way as an energy source.

Figure 7.29 shows the principles of two different radioisotope power generators, the larger (to the left) is of SNAP-7 type and produces ~60 W, the smaller one produces ~10 mW. The radiation source for the larger generator consists of 15 rods (A) clad with hastelloy and containing approximately 7 kg of $SrTiO_3$ which has approximately 225 000 Ci of ^{90}Sr. This heat source is surrounded by 120 pairs of lead telluride thermoelements (B) and a radiation shield of 8 cm of depleted uranium (C). The whole arrangement is surrounded by a steel cover with cooling fins. The weight of this generator is 2.3 t with dimensions of 0.85 m in length and 0.55 m in diameter. It is estimated that the lifetime of such an energy source is at least 5 y, although the half-life of ^{90}Sr (30 y) promises a longer period. Radionuclide generators in unmanned lighthouses, navigation buoys, automatic weather stations, etc., in sizes up to about 100 W, have been in use in a number of countries, e.g. Japan, Sweden, the UK, the USA, etc. Since no moving parts are involved, these generators need a minimum of service. Their reliability makes them valuable in remote areas like the Arctic regions where several such generators have been installed. ^{90}Sr is the preferred radionuclide for terrestrial uses. However, at present the fear of a possible release of radioactivity have largely led to their recovery, or replacement by other power sources. The power density of ^{90}Sr is relatively high, 0.93 W g^{-1}, as compared to some other possible radionuclides as ^{137}Cs (0.26 W g^{-1}) and ^{238}Pu (0.55 W g^{-1}), but it is lower than ^{244}Cm (2.8 W g^{-1}) and ^{245}Cm (121 W g^{-1}). Recently ^{60}Co has come into use; the heating mainly comes from absorption of the γ's in a uranium shielding.

^{238}Pu has been used as an energy source in space. Several satellites with radioisotope generators of 25 W have been placed in space. The Apollo project employed a generator "SNAP-27" containing ^{238}Pu with a total weight of 14 kg and producing 50 W power. The Viking landers on the planet Mars used ^{238}Pu as the main energy source — a number of so called RTG-generators. RTG-generators were also used to power the Pioneer vehicles which gave us the first close pictures of the planet Jupiter and in the Voyager missions to the outer planets and beyond (at present they are all leaving the solar system, but some still keep radio contact with Earth) where the ^{238}Pu power source produced

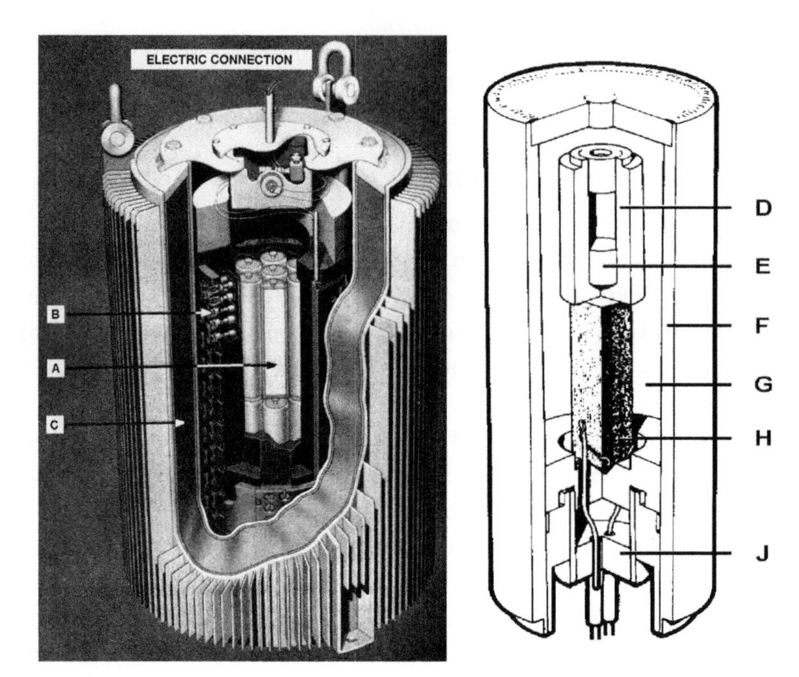

Figure 7.29 A - $^{90}SrTiO_3$; B - thermocouples; C - radiation shield (depleted U); D - capsule; E - ^{238}Pu; F - secondary containment; G - thermal insulation; H - thermopile; J - insulator.

450 W. One will reach the star Sirius in about 300 000 years! In 2012, the most recent RTG use as a power source in space vehicles is on the New Horizons mission to Pluto and Charon.

7.10. EXERCISES

7.1. In §7.3 two equations are given for calculating the range of α-particles in air and in other material from the particle energy as well as a curve for the range in aluminum. How different are the values from the equations and from the curve for a 5 MeV α?

7.2. What is the minimum energy that an α-particle must have to be detected by a GM tube having a mica window of 1.5 mg cm^{-2} (the density of mica is approximately equal to aluminum)?

7.3. For an irradiation experiment it is necessary to extract a beam of deuterons from an accelerator. The projectile energy is 22 MeV D^+. For this purpose the beam is deflected and permitted to pass through a thin titanium foil (density 4.5 g cm^{-3}). Assuming that $\hat{R}_1 \rho_1 M_1^{-1/2} = \hat{R}_2 \rho_2 M_2^{-1/2}$ (Bragg-Kleman rule), what is the maximum thickness of the foil? Give the answer in millimeters.

7.4. Make a rough estimate of the range in air for a 1 MeV α, 1 MeV H^+ and 1 MeV e^- using the plot in Fig. 7.7. The energy to form an ion pair in air is 14.6 eV but assume that twice as much energy is lost through excitation.

7.5. What is the range of a 6.3 MeV α-particle in (a) aluminum, (b) nickel, (c) platinum?

7.6. What is the γ-ray flux from a 3.7 GBq ^{60}Co source at a distance of 3 m? Assume $\psi_{sample} = 1$.

7.7. What is the maximum range in millimeters of β-particles from T, ^{14}C, ^{32}P and ^{90}Sr in a photographic emulsion if its absorption efficiency is assumed to be the same as aluminum? The density of the emulsion is assumed to be 1.5 g cm^{-3}.

7.8. The E_{max} of ^{32}P β-particles is 1.71 MeV. To what electron velocity does this correspond?

7.9. In a laboratory an irradiation area must be designed for γ-radiography using a 3.7×10^{11} Bq ^{60}Co source. For this purpose a cubic building is erected with an interior side length of 2 m. The desired flux reduction is 10^6. How thick must the wall be and how much will the shielding material cost (i.e. not including labor costs) if it is made of (a) concrete? (b) lead? Assume lead blocks cost \$1.50 per kg and concrete \$40 per m^3.

7.10. An experiment is done with ^{60m}Co which emits 0.05860 MeV γ. The detector used is a NaI crystal. What photo peaks will be observed if the electron binding energies in sodium are K 1072 and L 63 eV, and in iodine K 33 170 and L 4800 eV?

7.11. A human body may be considered as consisting of water. Radiation from ^{137}Cs is absorbed by a 15 cm thick body. How much is the γ-ray flux reduced by the body, and how much of the beam energy (β plus γ) is absorbed?

7.12. For iron the mass attenuation coefficients are: at 0.5 MeV γ, 0.083; at 1.0 MeV, 0.059; at 1.5 MeV, 0.047 cm^2 g^{-1}. Calculate the corresponding one-tenth values.

7.13. An absorption curve of a sample emitting β- and γ-rays was taken with aluminum absorber using a gas-flow proportional counter. The data obtained were:

Absorber thickness (g cm^{-2})	Activity (counts min^{-1})	Absorber thickness (g cm^{-2})	Activity (counts min^{-1})
0	5800	0.700	101
0.070	3500	0.800	100
0.130	2200	1.00	98
0.200	1300	2.00	92
0.300	600	4.00	80
0.400	280	7.00	65
0.500	120	10.00	53
0.600	103	14.00	40

(a) Estimate the maximum energy of the β-spectrum. (b) Find the energy of the γ-ray.

7.14. A 40-story high modern business building is supported by 0.9 m thick pillars of reinforced concrete. The insurance company must check that the number of iron bars are as many as required, and therefore they want to investigate the pillars by γ-radiography. What exposure times are required for (a) a small 200 GBq ^{60}Co source, (b) for a large 150 TBq source? Use the same film data as in 7.9.2.

7.15. A swimming-pool reactor produces a flux of 3×10^{16} thermal neutrons $m^{-2}\,s^{-1}$ at 1 m from the reactor center. Assuming a parallel beam of neutrons diffusing up to the surface of the pool where the neutron flux is measured to be 10^8 n $m^{-2}\,s^{-1}$, calculate the thickness (x m) of the water layer required. For thermal neutrons the flux is reduced exponentially with the exponent $x\,L^{-1}$, where L is the diffusion length (2.75 cm in H_2O).

7.16. In a sample of 10.4 TBq of old fission products, the average γ-ray energy is 0.5 MeV and on the average 0.4 γ's are emitted per β-decay. (a) What is the lead shielding required to reduce the γ flux to 10^2 γ $cm^{-2}\,s^{-1}$ at 1.5 m from the source assuming only exponential attenuation? (b) What is the relaxation length? (c) What is the build-up factor?

7.11. LITERATURE

B. T. Price, C. C. Horton, and K. T. Spinney, *Radiation Shielding*, Pergamon Press, Oxford, 1957.
S. Flügge (ed.), *Handbuch der Physik*, Band 34, 1958, and 38/2, 1959, Springer-Verlag.
R. L. Mössbauer, Recoilless nuclear resonance absorption, *Ann. Rev. Nucl. Sci.* **12** (1962) 1.
J. C. Rockley, *An Introduction to Industrial Radiology*, Butterworths, London, 1964.
C. S. Fadley, S. B. M. Hagström, J. M. Hollander, M. P. Klein, and D. A. Shirley, Chemical bonding information from photoelectron spectroscopy, *Science* **157** (1967) 1571.
D. A. Shirley, Chemical tools from nuclear physics, *Science* **161** (1968) 745.
IAEA, *Nuclear Well Logging in Hydrology*, Tech. Report 126, Vienna, 1971.
IAEA, *Commercial Portable Gauges for Radiometric Determination of the Density and Moisture Content of Building Materials*, Tech. Report 130, Vienna, 1971.
G. M. Bancroft, *Mössbauer Spectroscopy*, J. Wiley, 1973.
J. A. Cooper, Comparison of particle and photon excited X-ray fluorescence applied to trace element measurements on environmental samples, *Nucl. Instr. Methods* **106** (1973) 525.
H. W. Thummel, Stand und Entwicklungstendenzen auf dem Gebiet der Isotopen und Strahlenanalytik. Physikalische Analysenverfahren mit Radionukliden, *Isotopenpraxis* **11** (1975)1, 41, 87, 117, 172.
C. H. Wang, D. H. Willis, and W. D. Loveland, *Radiotracer Methodology in the Biological, Environmental and Physical Sciences*, Prentice Hall, 1975.
W. D. Ehmann, and D. E. Vance, *Radiochemistry and Nuclear Methods of Analysis*, Wiley Interscience, 1991.
G. Furlan, P. Cassola Guida, and C. Tuniz (Ed.), *New Paths in the Use of Nuclear Techniques for Art and Archeology*, World Scientific Pub., Singapore, 1986.
M. J. Rycroft (ed.), *The Cambridge Encyclopedia of Space*, Cambridge University Press, 1990.

Radiation Effects on Matter

Contents

Soon after the discoveries of X-rays and radioactivity it was learned that radiation could cause changes in matter. In 1901 P. Curie found that when a radium source was placed on his skin, wounds were produced that were difficult to heal. In 1902 skin cancer was shown to be caused by the radioactivity from radium but 5 years later it was learnt that radium therapy could be used to heal the disease. Large radiation doses were found to kill fungi and microorganisms and produce mutations in plants.

Glass ampoules containing milligrams of radium darkened within a few months and became severely cracked, allowing the leakage of radon gas. In the early years of the investigation of radioactivity, emphasis was on radium and its decay products. Among the radiation effects observed were the fluorescence induced in different salts and the changes in their crystallographic form resulting in colour changes from delocalised charges in the crystals. Metals were found to lose their elasticity and become brittle due to delocalised atoms breaking the inherent structure. Radiation was also found to have a profound effect on the chemical composition of solutions and gases. Water, ammonia and simple organic substances decomposed into more elementary constituents and also combined

Radiochemistry and Nuclear Chemistry
ISBN 978-0-12-405897-2, http://dx.doi.org/10.1016/B978-0-12-405897-2.00008-2

into more complex polymeric products. Radiation decomposition (*radiolysis*) of water caused evolution of hydrogen and oxygen gas and formation of hydrogen peroxide. Conversely, it was shown that water could be synthesized through irradiation of a mixture of H_2 and O_2. In 1911 S. Lind found that 1 g of radium exposed to air resulted in the production of 0.7 g of ozone per hour. By relating this *radiation yield* to the number of ions produced by the amount of radiation, Lind initiated the quantitative treatment of radiation-induced changes.

8.1. ENERGY TRANSFER

The chemical effects of radiation depend on the composition of matter and the amount of energy deposited by the radiation. In this section we consider only the energy transfer. For this purpose it is practical to divide high energy radiation into (1) charged particles (e^-, e^+, α, etc) and (2) uncharged particles (n) and electromagnetic radiation (γ). The latter produce recoil atomic ions, products of nuclear reactions and electrons as charged secondary ionizing particles. The terms direct and indirect ionizing radiation are often used for (1) and (2) respectively.

The amount of energy imparted to matter in a given volume is

$$E_{imp} = E_{in} + \Sigma Q - E_{out} \tag{8.1}$$

where E_{in} is the energy (excluding mass energy) of the radiation entering the volume, E_{out} is the energy of the radiation leaving the volume, ΣQ is the sum of all Q-values for nuclear transformation that have occurred in the volume. For a beam of charged particles $E_{in} = E_{kin}$; for γ-rays it is E_γ. If no nuclear transformations occur, $\Sigma Q = 0$. For neutrons which are captured and for radionuclides which decay in the absorber, $\Sigma Q > 0$; in the case of radionuclides already present in the absorber, $E_{in} = 0$.

8.1.1. Charged particles

We learned in the previous chapter that the energy of charged particles is absorbed mainly through ionization and atomic excitation. For positrons the annihilation process (at $E_{kin} \approx 0$) must be considered. For electrons of high kinetic energy bremsstrahlung must be taken into account. However, in the following text we simplify by neglecting annihilation and bremsstrahlung processes. The bremsstrahlung correction can be made with the aid of Figure 7.9 which gives the *average specific energy loss* of electrons through ionization and bremsstrahlung.

It has been found that the average energy, w, for the formation of an ion pair in gaseous material by charged particles is between 25 and 40 eV. For the same absorbing material it is fairly independent of the type of radiation and the energy. Table 8.1 lists values of w in some gases. The ionization potentials j of the gases are lower than the

Table 8.1 Ion pair formation energies for charged particles. All values in eV

Absorber	w	j	w-j
He(g)	43	24.5	18.5
H_2(g)	36	15.6	20.4
O_2(g)	31.5	12.5	19
Air	34	15	19
H_2O(g or l)	38	13	25
Ar(g) 5 MeV α	26.4		
Ar(g) 340 MeV p^+	25.5		
Ar(g) 10 keV e^-	26.4		
Ar(g) 1 MeV e^-	25.5		
Ar(g) average	26	15.7	10.3

typical w-values and thus the rest of the energy, $w-j$, appears as excitation energy. Since the excitation energies per atom are $\leq$5 eV, several excited atoms are formed for each ion pair formed. While it is easy to measure w in a gas, it is more difficult to obtain reliable values for liquids and solids. They also differ more widely; e.g. w is 1300 eV per ion pair in hexane (for high energy electrons) while it is about 5 eV per ion pair in inorganic solids.

The *specific energy loss* of a particle in matter is called the *stopping power* $\widehat{S}$,

$$\widehat{S} = \mathrm{d}E_{\mathrm{loss}}/\mathrm{d}x \ (\mathrm{J/m}) \tag{8.2}$$

where x is the distance traversed by the particle. To a good first approximation the stopping power of a material is determined by its atomic composition and is almost independent of the chemical binding of the atoms. Stopping power is a function of the particle velocity and changes as the particle is slowed down.

The *specific ionization* J is the number of ion pairs produced per unit path length

$$J = \mathrm{d}N_{\mathrm{j}}/\mathrm{d}x \ (\text{ion pairs/m}) \tag{8.3}$$

The value of J depends on the particle and its energy as seen from Figure 7.7. The relation between $\widehat{S}$ and J is

$$\widehat{S} = wJ \ (\mathrm{J/m}) \tag{8.4}$$

The *mass stopping power,* $\hat{S}/\rho$, is commonly expressed in units of MeV/g cm^2.

Another important concept is the *linear energy transfer* (abbreviated as *LET*) of charged particles. It is defined as the energy absorbed in matter per unit path length travelled by a charged particle

$$LET = \mathrm{d}E_{\mathrm{abs}}/\mathrm{d}x \tag{8.5}$$

Values of *LET* in water are given for various particles and energies in Table 7.2. For the same energy and the same absorbing material, the *LET* values increase in the order:

high energy electrons (also approximatively γ-rays
< β-particles (also approximately soft X-rays)
< protons
< deuterons
< heavy ions (ions of N, O, etc.)
< fission fragments

The relationship between *LET*, which refers to the energy absorbed in matter, and the stopping power, which refers to the energy loss of the particle, is

$$dE_{loss}/dx = dE_{abs}/dx + E_x \quad (8.6)$$

The difference E_x in these two energy terms is related to the energy loss by electromagnetic radiation (mainly bremsstrahlung).

8.1.2. Uncharged radiation

When neutrons or photons having the incident particle energy E_{in} are absorbed, a certain fraction of energy E_{tr} is transferred into kinetic energy of charged particles when traversing the distance dx. We can define an *energy transfer coefficient* as

$$\mu_{tr} = E_{in}^{-1} dE_{tr}/dx \left(m^{-1}\right) \quad (8.7)$$

If we neglect the bremsstrahlung associated with the absorption of the secondary charged particles formed in the initial absorption processes, E_{tr} is the energy absorbed (designated as E_{abs}), and we can write Eqn (8.7) as

$$\mu_a = E_{in}^{-1} dE_{abs}/dx \left(m^{-1}\right) \quad (8.8)$$

8.2. RADIATION TRACKS

The energy lost when a high energy charged particle is slowed in matter gives rise to a trail (or, more commonly, track) of ionized and excited molecules along the path of the particle; Figure 8.1 schematically depicts a track. Cloud chamber photographs of such tracks are shown in Figure 7.5. A photon imparts a large fraction of its energy to a single electron which subsequently ionizes and excites many other molecules along its path. The absorption of any type of ionizing radiation by matter thus leads to the formation of tracks of ionized and excited species. Whereas these species generally are the same in a particular sample of matter regardless of the type of radiation, the tracks may be sparsely or densely populated by these species. Expressions such as linear energy transfer (*LET*) and stopping power ($\hat{S}$) are based on the implicit assumption of a continuous slowing down process and thus gives quantitative information on the average energy loss but only qualitative information on the densities of reactive species.

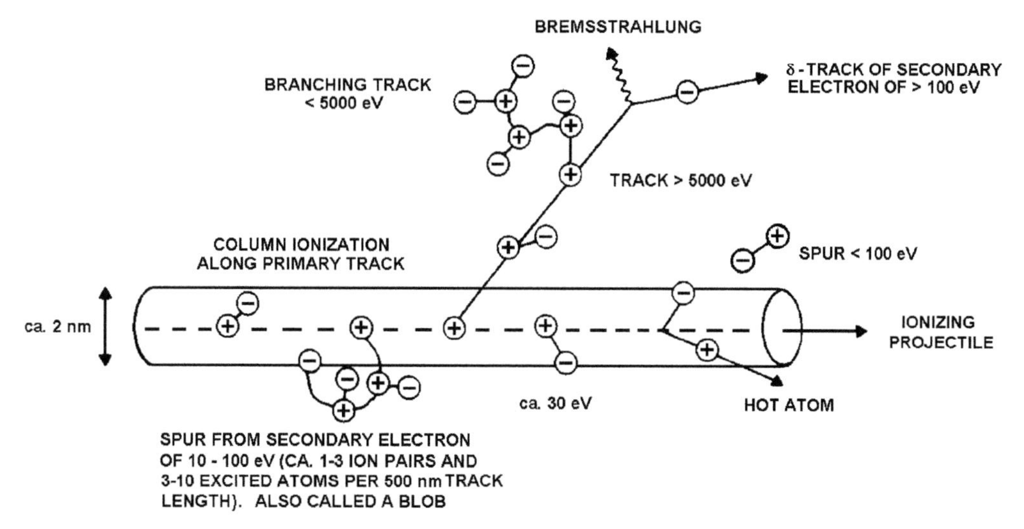

Figure 8.1 Track formed by energetic ionizing particle in condensed matter. Distances between ion pairs along track are: ~1000 nm γ, 500-100 nm fast electron, ~1 nm slow electron and α.

Most of the radiation-matter interactions are involving the electrons rather then the nucleus itself. As seen previously these interactions with the electrons may occur in several ways. Thus, electrons liberated in the ionization process will have greatly varying kinetic energy. If the energy of these secondary electrons is relatively low (<100 eV) their range in liquids and solids is short and the ionizations and excitations caused by these electrons take place close to the primary ionizations leading to the formation of small *spurs* containing ionized and excited species. Secondary electrons with high kinetic energy form tracks of their own branching from the primary track. Such electrons are called *δ-rays*. For high energy electrons, spurs are formed at well separated intervals along the track whereas for densely ionizing radiation such as α particles, protons and recoil atomic ions the spurs overlap and form *columns* of excited and ionized species. Differences in chemical and biological effects caused by different radiations reflects the varying track structures.

The tracks of X-ray and γ-radiation result in tracks of fast electrons. The energy of these fast electron is consumed by the formation of spurs (6–100 eV), blobs (100–500 eV) and short tracks (500–5000 eV), see Figure 8.1. For a primary electron in the range 10^4–10^7 eV the distribution of energy of secondary electrons is approximately 40% <3.4 eV, 20% 3.4–6.8 eV, 18% 6.8–13.5 eV and 12% 13.5–27.1 eV. The radius of the tracks core at low electron velocities is of the order of 1 nm. Thus it is possible to calculate in e.g. a solution how many molecules will be affected by the tracks based in the concentration on a certain molecule.

The energy transferred to the electrons by an energetic ion depends on the mass, the charge (Z) and velocity (v) of the ion, cf. Eqn (7.14): the probability of interaction is proportional to the ratio Z^2/mv^2. Heavy ions produce track structures similar to those of fast electrons. However, the spurs are quite close to each other. For the case of α-particles, the original distance between spurs is of the order of fractions of a nanometer. Immediately at their formation they comprise a continuous cylindrical column. The column consists of a dense core surrounded by a more diffuse shell of tracks of high energy δ-electrons.

8.3. RADIATION DOSE AND RADIATION YIELD

The oldest radiation unit still in use is the *roentgen* (R). It applies only to photons and is defined as the *exposure*. The exposure is the energy flux of the unperturbed photon radiation when hitting matter. 1 R is the exposure that in air produces ion pairs with total charge per unit mass of 2.58×10^{-4} C/kg. This corresponds to 1.61×10^{15} ion pairs per kg air or 8.8×10^{-3} J/kg absorbed energy at an average ion pair formation energy of 34 eV. The roentgen has been in use for more than 60 years and is still not uncommon in the medical profession in some countries.

The *absorbed dose* (*D*) is the amount of radiation energy absorbed per unit mass:

$$D = \mathrm{d}E_{\mathrm{abs}}/\mathrm{d}m \tag{8.9}$$

According to Eqn (8.1), in the absense of nuclear transformations, $E_{\mathrm{abs}} = E_{\mathrm{in}} - E_{\mathrm{out}}$. The SI unit is the *Gray* (Gy)

$$1\ \mathrm{Gy} = 1\ \mathrm{J/kg} \tag{8.10}$$

An old, more or less obsolete, unit is the *rad* (for radiation absorbed dose)

$$1\ \mathrm{Gy} = 100\ \mathrm{rad} \tag{8.11}$$

The rad was proposed in 1918 as "that quantity of X rays which when absorbed will cause the destruction of the malignant mammalian cells in question..." Formally it was defined in 1953 as the dose causing 100 ergs of energy to be absorbed by one gram of matter. An erg is the unit of energy and mechanical work in the old centimetre-gram-second (CGS) system of units. The name comes from the greek word for work, ergon. One erg is the amount of work done by a force of one dyne exerted for a distance of one centimeter. The dyne also originates from a greek word, namely dynalis meaning force or power. The dyne is defined as "the force required to accelerate a mass of one gram at a rate of one centimeter per second squared" Although these units may still be found in the literature their use is generally discouraged in education for other pursposes than for relating old measures to units in the SI system. In some countries, however, they are still used as an industrial standard.

The *dose rate* is the absorbed dose per unit time. The SI unit is Gray per second (Gy/s).

The *specific γ-ray dose rate, $\dot{D}$,* is a practical measure for estimation of the radiation hazard to people from γ-emitting radionuclides:

$$\dot{D} = A\, r^{-2} \Sigma\, n_i\, k_i\, B_i\, \mathrm{e}^{-\mu_i x} \tag{8.12}$$

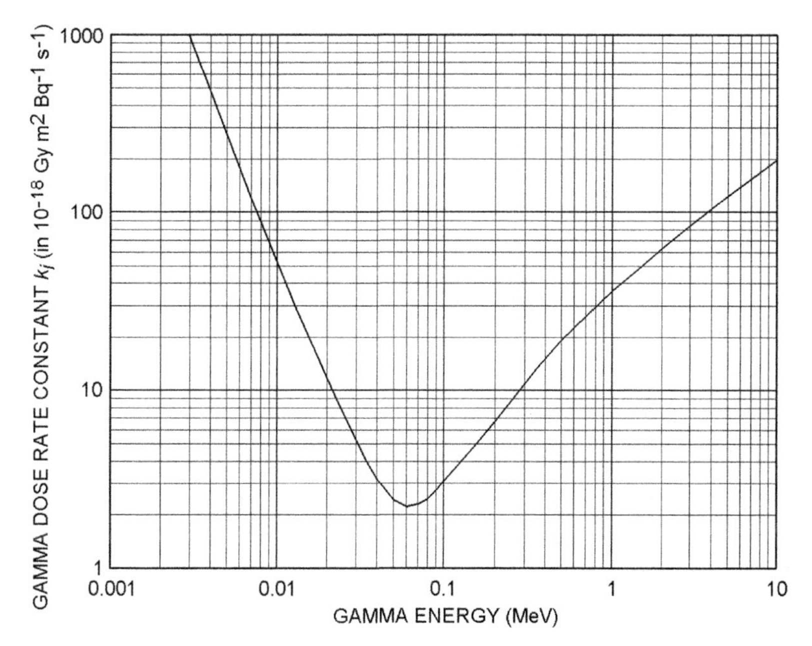

Figure 8.2 Dose rate constant for a monoenergetic point isotropic γ-source as a function of γ-energy.

where $A\,r^{-2}\,\Sigma\,n_i\,k_i = \dot{D}_0$ is the relative dose rate (Gy/s) without any radiation shielding, and the same for a radiation shield with build-up factor B_i and attenuation factor $e^{-\mu_i x}$ (see eqns. 7.8 and 7.27–7.28); n_i is the fraction of all decays (given by the decay scheme of the source nuclide) yielding a γ-ray of energy E_i corresponding to the source constant k_i according to Figure 8.2, and A is the source strength (Bq). E.g. ^{24}Ne decays to 8% with two γ's (0.878 and 0.472 MeV) and to 92% with one γ (0.472 MeV). Thus the sum in Eqn (8.12) contains two terms, one for 0.878 MeV γ ($n_1 = 0.08$) and one for 0.472 MeV γ ($n_2 = 1$). The k_i-values are taken from Figure 8.2. The relaxation length $\mu_i\,x$ is computed for each energy E_i and the corresponding attenuation factors of the radiation shield $e^{-\mu_i x}$(x m thick) are calculated using e.g. Figure 7.17. Build-up factors B_i can be estimated from Figure 7.20. It should be noted that Eqn (8.12) does not account for scattering around the shield ("sky-shine"), which has to be estimated separately; see text books on radiation shielding.

Closely related to the absorbed dose is the *kerma* (K), which is the kinetic energy released per unit mass by uncharged particles and electromagnetic radiation (n and γ):

$$K = dE_{tr}/dm = \varphi\,E\,\mu_{tr}\,\rho^{-1} \tag{8.13}$$

where E is the radiation energy, φ is the particle fluence (particles/m^2) and μ_{tr}/ρ is the mass energy transfer coefficient. The SI unit of the kerma is J/kg.

In the following we refer to all absorbed radiation energy as the radiation dose independent of whether the incident radiation is charged or uncharged particles or photons.

Radiation chemical yield is described in terms of *G-values*. Originally G(x) was defined as the number of molecules of x transformed per 100 eV absorbed energy (a practical notation as most reactions have G-values of <10). In the SI system the symbol G(x) is the same but the unit is mol/J. The conversion factor between the two units is 1 mol/J = 9.649×10^6 molecules per 100 eV absorbed. Today the symbol *g*(x) usually refer to the number per 100 eV and G(X) to mol/J.

8.4. METALS

Metals consist of a solid lattice of atoms whose valence electrons cannot be considered to belong to any particular atom, but rather to a partially filled energy band (the conduction band) established by the total lattice network. Interaction of radiation with the metal can cause *excitation* of bound electrons in the atoms to the conduction band. This, however, due to the excellent conducting properties of metals do not lead to prevailing electron deficiencies resulting in e.g. colour changes.

While irradiation by γ-rays and electrons has little influence on metallic properties, heavy particles cause serious damage through their collision with atoms in the metal lattice network. This results in *displacements* of the atoms from their lattice positions. The number of displacements (n_{disp}) depends on the amount of energy transferred in the collision event (E_{tr}) to the recoiling (target) atom, and the energy required for moving this atom from its lattice position. This so-called *displacement energy* averaged over all directions in the crystal lattice (E_{disp}) is 10–30 eV for most metallic materials. According to the Kinchin-Pease rule,

$$n_{disp} \leq \tfrac{1}{2} E_{tr}/E_{disp} \qquad (8.14)$$

The maximum energy transferred can be calculated assuming purely elastic collisions between hard spheres (§10.1). Thus for a 1.5 MeV fission neutron, E_{tr}(max) is 425 keV in C, 104 keV in Fe, and 25 keV in U. With $E_{disp} \approx 25$ eV, up to 8500, 2080 and 500 displacements, respectively, occur in these metals due to the absorption of a fission neutron (neglecting nuclear reactions). In practice the numbers are somewhat smaller, especially at the higher E_{tr} (where it may be about one third of the calculated value). The reason for this is the recombination of defects at the end of the collision cascade. Atomic displacements cause many changes in the properties of metals. Usually electrical resistance, volume, hardness and tensile strength increase, while density and ductility decrease.

The microcrystalline properties of metals are particularly influenced by irradiation. Although low-alloy steel in modern reactor tanks are rather radiation resistant (provided they are free of Cu, P, Ni, Mn and S impurities). Significant radiation damage in the reactor vessels due to impurities of Ni and Mn has been observed in both BWR and VVER reactors. Stainless steel (e.g. of the 18% Cr, 8% Ni type) has been found to become brittle upon irradiation due to the formation of microscopic helium bubbles, probably due to n,α reactions in ^{54}Fe and impurities of light elements (N, B, etc.). This behaviour is accentuated for metallic uranium in reactors because of the formation of fission products, some of which are gases. As a result of this radiation effect it is not possible to use uranium metal in modern power reactors, where high radiation doses are accumulated in a very short time. The fuel elements for power reactors are therefore made of non-metallic uranium compounds. However, in some early fast reactors, e.g. EBR II, metallic uranium fuel was used successfully by increasing the gap between the fuel and the can and filling it with sodium.

The displaced atoms may return to their original lattice positions through diffusion if they are not trapped in energy wells requiring some activation energy for release. Such energy can be provided by heating or by irradiation with electrons or γ-rays (these do not cause new displacements). This "healing" of particle radiation damage is commonly referred to as annealing. The *thermal annealing rate* increases with temperature as does *radiation annealing* with radiation dose. Doses in the 10 kGy range are usually required for appreciable effect.

8.5. INORGANIC NON-METALLIC COMPOUNDS

The time for a high energy particle to pass by an atom is $\leq 10^{-16}$ s. In this time the atom may become excited and/or ionized, but it does not change position (the *Franck-Condon principle*) provided there is no direct collision. The excited atoms are de-excited through the emission of *fluorescence radiation*, usually within 10^{-8} s. The ionization can result in simple trapping of the electrons and production of "electron holes" in the lattice, especially at impurity sites. The local excess (or deficiency) of charge produced in this way leads to electronic states with absorption bands in the visible and ultraviolet regions of the spectrum. For example, irradiation of LiCl results in a change of the colour of the compound from white to yellow. Similarly, LiF becomes black, KCl blue, etc. The irradiation of ionic crystals also leads to changes in other physical properties much as conductivity, hardness, strength, etc. Frequently, heating (sometimes room temperature is enough) returns the properties and colour to the normal state (or close to it) accompanied by the emission of light; this forms the basis for a radiation dose measurement technique named "thermoluminescence dosimetry" (§8.10).

Following a collision between a heavy particle (n, p, etc.) and an absorber atom in a crystalline material the recoiling ion produces lattice vacancies and, upon stopping, may occupy a non-equilibrium interstitial position (Fig. 8.3). The localized dissipation of energy can result in lattice oscillations, terminating in some reorientation of the local

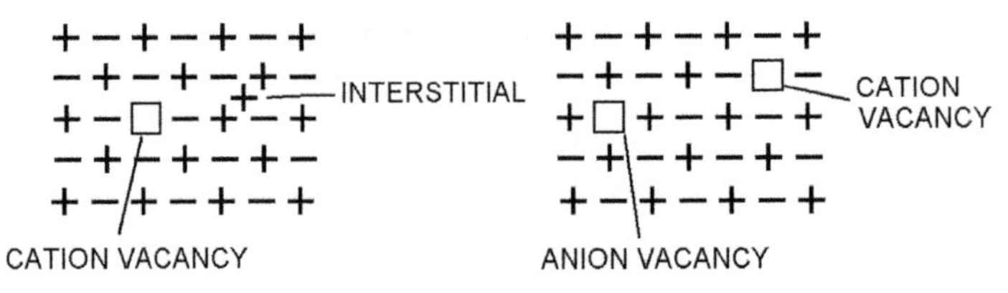

Figure 8.3 Irradiated NaCl type crystal showing negative and positive ion vacancies.

regions in the crystal lattice. These crystal defects increase the energy content of the crystal. Semiconductors, where the concentration of charge carriers is very small, have their conductivity reduced by introduction of lattice defects during irradiation. The production of interstitial atoms makes the graphite moderator in nuclear reactors stronger, harder, and more brittle. Since these dislocated atoms are more energetic than the atoms in the lattice, the dislocations lead to an energy storage in the material (the *Wigner effect*), which can become quite significant. For reactor graphite at 30°C this frequently reaches values as high as 2000 kJ/kg for fluencies of 2×10^{21} n/cm^2. At room temperature the interstitial atoms return to their normal positions very slowly (annealing), but this rate is quite temperature dependent. If the elimination of the interstitial atoms occurs too rapidly, the release of the Wigner energy can cause the material to heat to the ignition point. This was the origin of a fire which occurred in a graphite moderated reactor in England in 1957, in which considerable amounts of radioactive fission products were released into the environment.

Inorganic substances exposed to high fluencies of neutron and γ-radiation in nuclear reactors are found to experience decomposition. Thus:

$$KNO_3 \sim\sim\sim\sim\rightarrow KNO_2 + \tfrac{1}{2}O_2 \quad (8.15)$$

The notation $\sim\sim\sim\sim\rightarrow$ symbolizes a radiation induced reaction. At high fluxes the oxygen pressure in the KNO_3 causes the crystal to shatter. However, some crystals are remarkably stable (although they may become coloured), e.g. Li_2SO_4, K_2SO_4, $KCrO_4$, and $CaCO_3$.

Theory has not been developed sufficiently to allow quantitative calculation of the radiation sensitivity of compounds. Usually covalent binary compounds are highly radiation resistant. Two examples are UO_2 and UC, whose insensitivity to radiation has led to their use as reactor fuels. In the fuel elements of commercial water cooled reactors UO_2 is used in the form of small, sintered pellets about 1 cm^3 in volume. Because of the build-up of pressure from fission gases, these pellets crack at high fluxes ($\geq 10^{22}$ n/cm^2); Figure 8.4. Another binary compound, CO_2, which is used as a coolant in some older reactors, decomposes by irradiation to graphite and polymeric species.

In mixtures of inorganic compounds many unexpected and even undesirable reactions may occur. For example, radiolysis of liquid air (often used in radiation

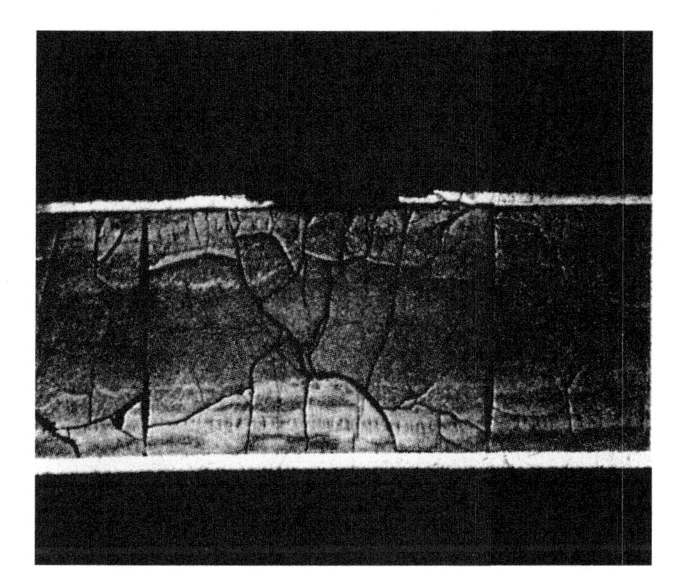

Figure 8.4 Zircaloy canned UO_2 fuel after exposure to very high power (high dose and heat) in a test reactor. The pellets have fractured and the canning has been penetrated. (SKI report)

research) yields ozone, while radiation of humid air yields HNO_3. One of the first observations of radiation-induced changes was the darkening of glass. Glass often contains iron, manganese, and other metals that can exist in several oxidation states with different colours. As a result of the irradiation, the oxidation state can change, resulting in change in colour. Dislocations as well as trapped electrons also contribute to the colour changes in glass. In chemical and metallurgical work with highly active substances it is desirable to observe the experiment through a thick glass window, which provides protection from the radiation. In order to avoid the colouring of the glass, a small amount (1−2%) of an element which can act as an electron trap is added, e.g. CeO_2, which acts by the reaction $Ce^{4+} + e^- \rightarrow Ce^{3+}$. After an exposure of 10^4 Gy the transmission to light of ordinary glass had been reduced to 44%, while for a CeO_2 protected glass it was still 89%.

Glass is very resistant to radiation damage because it is a non-crystalline "solid" liquid. Therefore, in such a material, it is not possible to speak of dislocations: the random structure of the glass allows it to include foreign species throughout the sample. This is why glass has been intensely studied, including full scale tests, as the matrix ("container") for high active waste (HAW) consisting of fission products and actinides.

8.6. WATER

The consequences of ionization and excitation depend on the physical state and the molecular composition of the irradiated material. In this section we introduce chemical phenomena into the description using the radiolysis of water as an example.

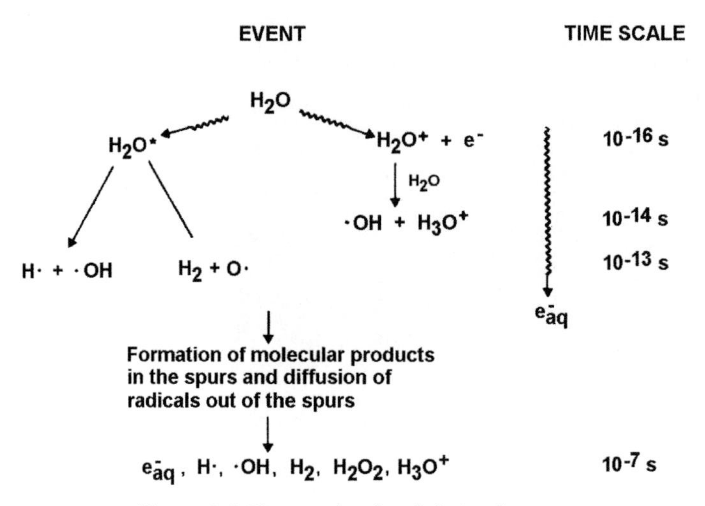

Figure 8.5 Time scale of radiolysis of water.

The time scale for the sequence of events on the radiolysis of water is shown in Figure 8.5. The ionization event occurs on the time scale of an electronic transition ($<10^{-16}$ s). The positive ion H_2O^+ formed reacts with water within 10^{-14} s, forming an $\cdot OH$ radical and H_3O^+. The electron, if liberated with sufficient kinetic energy, can ionize further water molecules before its energy falls below the ionization threshold of water (12.61 eV). The electron thereafter loses the rest of its energy by causing vibrational and rotational excitation of the water molecules and, finally, becomes solvated. The solvated electron is one or the most powerful reducing agents.

The solvation process has been shown to occur within 10^{-12} s. The excited states dissociate within 10^{-14}–10^{-13} s, i.e. on the same time scale as a molecular vibration to form O·, H·, ·OH and H_2. The physical and physicochemical (pre-thermal) processes are thus completed within 10^{-12} s leaving the species in thermal equilibrium with the water.

The radiolysis products are clustered in spurs (Fig. 8.1); i.e., they are inhomogeneously distributed in the water and proceed to diffuse out of the spur volume. During this "spur diffusion" process, recombination reactions take place leading to the formation of molecular or secondary radical products. The spur expansion is complete within 10^{-7} s, at which time the radiolysis products are those shown in Figure 8.5. Spur reactions are listed in Table 8.2. *G*-values for the radiolysis products in water irradiated with different types of radiation are given in Table 8.3. It is seen that, as the *LET* of the radiation increases, the *G*-values of radical products decrease while the molecular *G*-values increase, which evidences the formation of tracks. The higher *G*-values for water consumption and for molecular product formation in irradiated water vapour, as compared to liquid water, show that spur formation and spur reactions are of minor importance in the

Table 8.2 Spur reactions in water

$e_{aq}^- + e_{aq}^- \rightarrow H_2 + 2OH^-$
$e_{aq}^- + \cdot OH \rightarrow OH^-$
$e_{aq}^- + H_3O^+ \rightarrow H\cdot + H_2O$
$e_{aq}^- + H\cdot \rightarrow H_2 + OH^-$
$H\cdot + H\cdot \rightarrow H_2$
$\cdot OH + \cdot OH \rightarrow H_2O_2$
$\cdot OH + H\cdot \rightarrow H_2O$
$H_3O^+ + OH^- \rightarrow H_2O$

Table 8.3 Product yields (μmol/J) in irradiated neutral water

Radiation	G(-H_2O)	G(H_2)	G(H_2O_2)	G(e_{aq}^-)	G(H·)	G(·OH)	G(·HO_2)
Liquid water:							
γ and fast electrons	0.43	0.047	0.073	0.28	0.062	0.28	0.0027
12 MeV He^{2+}	0.294	0.115	0.112	0.0044	0.028	0.056	0.007
Water vapor:							
γ,electrons	0.85	0.05	0	G(e^-) = 0.31	0.75	0.85	

vapour phase. This is to be expected in a less dense medium where the average distance between separate spurs will be larger. Also the diffusivities will be larger in the vapour phase.

In the absence of solutes, reactions between radical and molecular species occur in the bulk water. In pure water irradiated with γ or X-rays, water is reformed via the reactions

$$H_2 + \cdot OH \rightarrow H\cdot + H_2O \tag{8.16}$$

$$H\cdot + H_2O_2 \rightarrow H_2O + \cdot OH \tag{8.17}$$

$$\cdot OH + H\cdot \rightarrow H_2O \tag{8.18}$$

and no net decomposition of water is observed. The yields depend on the *LET*-value of the radiation, as illustrated in Figure 8.6 for acidic and neutral water. Irradiation with higher *LET* radiation, which have higher *G*-values for molecular products than for radical products, causes net decomposition of water. The reason for this is that with a high LET more radiolysis products are formed close to each other and may thus react forming longer lived molecular species. This will also have the effect that e.g. alpha radiation is more prone to cause oxidation but gamma radiation is more prone to cause a reductive environment.

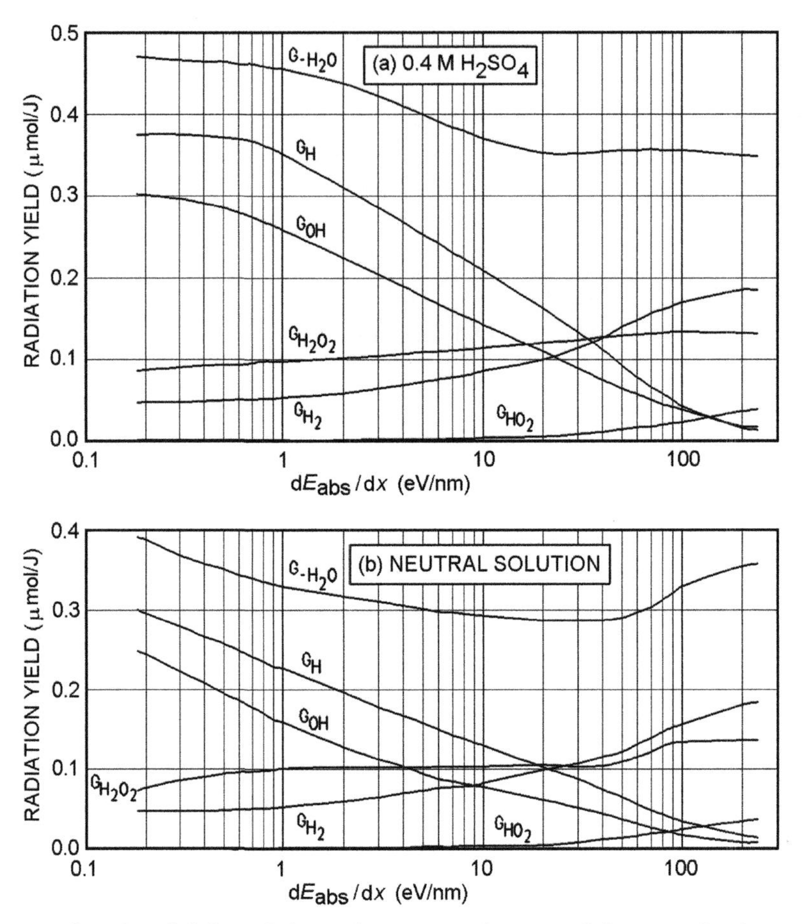

Figure 8.6 *G*-values (μmol/J) for radiolysis of water as a function of the *LET* value (eV/nm) of the system: (a) 0.4 M H_2SO_4, and (b) neutral solution. *(From A. O. Allen.)*

In nuclear reactors water used as a coolant or moderator should be as pure as possible to minimize dissociation during the time in the reactor. The formation and accumulation of an explosive gas mixture of H_2 and O_2 must be carefully avoided in all reactors in order to prevent accidents. Moreover the decomposition products of water can increase the corrosion of fuel elements, structural material, etc. Many reactors use N_2 as a protective gas. In this case the radiolysis can lead to the formation of HNO_3 unless suppressed by an excess of H_2 which preferentially yields NH_3. The pH of the water may be regulated by the H_2(g) pressure. The problem here is obvious though. Reactor water chemistry is monitored by taking samples and then measuring composition etc when cooler. By then the most reactive radiolysis products have already reacted and thus no real information of the corrosivity close to, or in, the reactor core can be obtained.

8.7. AQUEOUS SOLUTIONS

In irradiated dilute aqueous solutions at concentrations <0.1 mol/l, practically all the energy absorbed is deposited in the water molecules. Hence, the observed chemical changes are the result of the reactions between the solutes and the products of the water radiolysis. With increasing solute concentrations, the direct radiolysis of the solute gradually becomes important and the solute may also interfere with the spur reactions. The use of high concentrations of solutes which react selectively with the radical products — so-called *scavengers* — have provided experimental support for the existence of spurs.

In dilute solutions the chemical changes of a specific solute primarily reflect its reactivity towards e^-_{aq}, H· and ·OH. The hydrated electron e^-_{aq} is a strongly reducing species ($E_0 = -2.9$ V) whereas the hydrogen atom is a less powerful reductant ($E_0 = -2.3$ V). The H atom can be considered as a weak acid with a pKa of 9.6

$$e^-_{aq} + H^+ \rightleftharpoons H\cdot \tag{8.19}$$

The hydroxyl radical ·OH is a strong oxidant ($E_0 = 2.7$ V in acidic and 1.8 V in basic solution). In strongly alkaline solution ·OH dissociates into its anionic form O^- which has nucleophilic properties

$$\cdot OH \rightleftharpoons \cdot O^- + H^+ (pKa = 11.9) \tag{8.20}$$

In oxygenated solutions the perhydroxyl radical $\cdot HO_2$ is formed in reactions between e^-_{aq}, ·H and O_2

$$e^-_{aq} + O_2 \rightarrow \cdot O_2^- \tag{8.21}$$

$$H\cdot + O_2 \rightarrow \cdot HO_2 \tag{8.22}$$

$$\cdot HO_2 \rightleftharpoons \cdot O_2^- + H^+ \ (pKa = 4.7) \tag{8.23}$$

In saline waters (0.1 M Cl^-, pH 4—10) the dominant radiolysis reactions are

$$\cdot OH + Cl^- \rightleftharpoons HClO^- \tag{8.24}$$

$$HClO^- + H^+ \rightleftharpoons Cl\cdot + H_2O \tag{8.25}$$

$$Cl^- + Cl\cdot \rightleftharpoons Cl_2^- \tag{8.26}$$

These equilibria are pH-dependent, and the reactions may have important consequences for storage of nuclear waste in salt media.

The oxidation of Fe^{2+} to Fe^{3+} in acidic oxygenated solution is the basis for the *Fricke dosimeter*, §8.10. The reaction scheme is:

$$e^-_{aq} + H^+ \rightarrow H\cdot \tag{8.27}$$

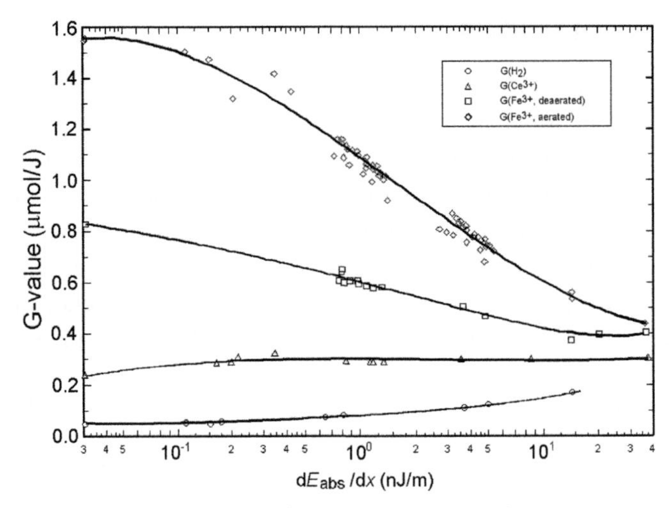

Figure 8.7 Radiation yield for oxidation of Fe^{2+} and reduction of Ce^{4+} in slightly acid (H_2SO_4) solution as a function of the radiation *LET* value. *(From A. O. Allen.)*

$$H\cdot + O_2 \rightarrow \cdot HO_2 \tag{8.28}$$

$$\cdot HO_2 + Fe^{2+} \rightarrow Fe^{3+} + HO_2^- \tag{8.29}$$

$$HO_2^- + H^+ \rightarrow H_2O_2 \tag{8.30}$$

$$H_2O_2 + Fe^{2+} \rightarrow Fe^{3+} + OH^- + \cdot OH \tag{8.31}$$

$$\cdot OH + Fe^{2+} \rightarrow Fe^{3+} + OH^- \tag{8.32}$$

The yield of Fe^{3+} is given by the equation

$$G\left(Fe^{3+}\right) = 2G(H_2O_2) + 3\left[G\left(e_{aq}^-\right) + G(\cdot H) + G(\cdot HO_2)\right] + G(\cdot OH) \tag{8.33}$$

Figure 8.7 shows radiation yields for oxidation of Fe^{2+} in acidic solution as a function of the *LET* value (see also §8.10).

As pointed out earlier, when the concentration of the aqueous solution is greater than about 0.1 M, the solutes may undergo direct radiolysis. The products of the radiolysis of the solute can react with water itself or with the radiolytic products of water. Irradiation of solutions containing sulfate, sulfite, or sulfide ions with *fast neutrons* yields radioactive phosphorus, through the $^{32}S(n,p)^{32}P$ reaction, almost exclusively as orthophosphate. However, depending on the redox conditions of the solution, reduced species of phosphorus also appear in minor amounts. Somewhat in contrast to this, *slow neutron* irradiation of solutions of $NaHPO_4$ yields phosphorus in many different oxidation states;

thus P(+1), P(+3) and P(+5) appear in species such as hypophosphite, phosphite, and o-phosphate. These species are all rather stable in aqueous solutions, and have been identified through paper electrophoretic analysis. In general, it may be assumed that slow neutron irradiation of solutions of oxyanions changes the central atom to another (usually a more reduced) valence state through n,y-reactions. For example, while manganese is in the Mn(+7) state in MnO_4^-, neutron capture leads to the formation of $^{56}Mn^{2+}$ species. However, if the product valence state is unstable at ambient solution conditions, it may be immediately oxidized to a more stable higher valence state.

The *self radiolysis* of a solution may change the chemical equilibria of the solution components. For example, the α-decay of plutonium decomposes water; in a solution containing 1 mole of ^{239}Pu, ca. 0.01 mole of H_2O_2 is produced per day. This hydrogen peroxide can react with the plutonium to form a precipitate of plutonium peroxide. To avoid this precipitation, nitrite ions are added to the solution to react with the hydroxyl radicals formed by the radiolysis and to eliminate the H_2O_2.

8.8. ORGANIC COMPOUNDS

As discussed above, the chemical consequences of radiolysis depends on the physical state and the molecular composition of the irradiated material. Two properties, the dielectric constant and the electron mobility, are of great importance for the fate of the ion pairs (the radical cations and the electrons) formed on ionization.

The distance r_c (the *Onsager radius*) at which the potential energy of an ion pair corresponds to the thermal energy $\mathbf{k}T$ is, according to Onsager,

$$r_c = \mathbf{e}^2/4\pi\,\boldsymbol{\varepsilon}_0\,\boldsymbol{\varepsilon}_r\,\mathbf{k}\,T \tag{8.34}$$

where $\boldsymbol{\varepsilon}_0$ and $\boldsymbol{\varepsilon}_r$ is the permittivity of free space and in the medium at distance r_c, respectively (the *Onsager equation*). The probability of an electron escaping its positive ion to become a free ion is equal to $e^{-r_c/r}$ (r is the distance travelled by the electron before it becomes thermalized) depends on the electron mobility in the medium. Most organic liquids have a lower dielectric constant than that of water and the free ion yields are, therefore, generally lower, see Table 8.4.

Table 8.4 Onsager radii, electron mobilities and free ion yields

Liquid	r_c (nm)	Mobility (cm^2/Vs)	G(free ions) (μmol/J)
Neopentane	32	55	0.09–0.11
Cyclohexane	28	0.35	0.016–0.02
Benzene	25	–	0.005–0.008
Methanol	2.3	–	0.2
Water	0.7	–	0.28

Electrons escaping their positive ions are gradually thermalized and solvated whereas electrons that recombine with their geminate (i.e. original ion pair) radical cations form excited molecules. The radical cations may react with solvent molecules. The types of reactions to be expected are e.g. proton, hydrogen atom and hydride ion transfer. Excited molecules are thus formed directly and by geminate ion recombination

$$A \rightsquigarrow A^* \quad (8.35)$$

$$A \rightsquigarrow A^+ + e^- \quad (8.36)$$

$$A^+ + e^- \rightarrow A^{**} \rightarrow A^* \quad (8.37)$$

The highly excited (electronically and/or vibrationally) molecule A^{**} formed by charge recombination may (within 10^{-11} s) lose part of its energy rapidly through collisions with its neighbouring molecules. The excited molecules can return to their ground state by several processes. *Singlet states* (antiparallel spins of valence electrons) de-excite within 10^{-8} s by *fluorescence*, whereas the *phosphorescence* of *triplet states* (parallel spins of the outer electrons) requires 10^{-5} to 10 s. Alternatively the excited molecules can undergo unimolecular isomerization reactions or dissociate into highly reactive radicals.

Organic molecules in general have more atoms than water and, therefore, the formation of a higher variety of products is to be expected. Both molecules smaller than the original and polymeric products are formed.

The effects of radiation on organic molecules have been shown to be strongly dependent on the molecular structure. Since the excitation energy is rapidly spread out over the whole molecule, one would expect the weakest bond to rupture, producing two radicals, provided the excitation energy exceeds the bond energy. Typical bond energies in eV are C−C $\leq$3.9, C=C (aliph) 6.4, C=C (arom) 8.4, C≡C $\leq$10, C−H 3.5−4.5, C−O 3.7 and C=O 7.7. Although C−C bonds are weaker than C−H bonds, C−H rupture predominates. For reasons as yet unknown, the C−H bond rupture is neither random nor localized to the weakest bond. Nevertheless, it is expected that compounds with unsaturated bonds are more radiation resistant than those with saturated bonds. Thus, upon irradiation with γ-rays from a ^{60}Co source, the hydrogen yield from cyclohexane is 150 times greater than that from benzene. This has been interpreted to be a result of the greater stability of the excited states of aromatic systems.

The presence of π-electrons diminishes the probability of a localization of excitation energy at a specific bond. As a result, the excitation energy is spread over the whole carbon ring and de-excitation is more likely to occur through processes such as collisional transfer rather than by dissociation. The organic compounds which are most radiation resistant contain aromatic rings (polyphenyls) and condensed ring systems (naphthalene, etc.). Their insensitivity to radiolysis has led to studies of the use of such

Table 8.5 Effects of γ-radiation on organic compounds of technical interest

Compound	Observed change at (kGy)	Useless at (kGy)	Compound	25% Reduction of desired property (kGy)
Olefins	5	10	Teflon	0.1
Silicones	5	50	Cellulose acetate	2
Mineral oils	10	100	Polyethylene	9
Alkyl aromatics	100	500	Polyvinylchloride	10
Polyphenyls	500	5000	Polystyrene	400
			Neoprene, silicon rubber	0.6
			Natural rubber	2.5

aromatic liquid hydrocarbons as cooling media in nuclear reactors. Radiation sensitivity of these compounds increases with increasing size of the aliphatic side chains but never reaches a *G*-value as high as that for a pure aliphatic compound. The primary radiolytic products of aromatic compounds are polymers. *G*-values for a large number of systems have been tabulated and can be found in the literature.

In Table 8.5 effects of γ-radiation on some organic compounds of technical interest are given. For oils the effects are mainly observed as changes in viscosity and acidity while for plastics they are associated with formation or rupture of the cross-linking. For elastomers like rubber there are changes in elasticity. For polyethylene, the following effects were observed: at 10–100 kGy the tensile strength increased; between 100 and 2000 kGy the irradiated substance became rubber and jelly-like; at 2000–5000 kGy it became hard; at >5000 kGy it became glassy but with high elasticity.

Solutes in low concentration in organic solvents react primarily with the radicals formed from the solvent. High concentration of solutes, e.g. of alkyl halides, may interfere with the charge recombination process through reactions such as

$$e^- + CH_3Cl \rightarrow \cdot CH_3 + Cl^- \quad (8.38)$$

Direct energy transfer from excited solvent molecules (A^*) to solute molecules (B) to form excited solute molecules may also take place:

$$A^* + B \rightarrow B^* \quad (8.39)$$

The reprocessing of used reactor fuel elements involves solvent extraction processes with organic solvents. In these processes the solvents are subjected to high radiation fields with subsequent decomposition of the organic solvent. The design of chemical reprocessing systems must take into account any interference by the radiolytic products (Ch. 21).

Labelled compounds experience self-radiolysis induced by the radioactive decay. The extent of such radiation effects depends on the half-life, the decay energy, the specific activity of the sample, and the *G*-value for decomposition. The presence of other substances can considerably affect the amount of damage. Aromatic compounds such as benzene (as a solvent) can serve as a protective medium to minimize radiation self-decomposition, whereas water or oxygen enhance it.

Radiation doses of 10^5 Gy can induce decomposition effects of the order of 1%. Samples whose specific activity exceed 40 GBq (1 Ci) per mol for ^{14}C or about 400 GBq (10 Ci) per mol for ^3H will receive a dose of this magnitude in a period of a year. Samples may be stored in benzene solution in vacuo or in deep freeze to minimize self-radiation effects and should be re-purified before use if the decomposition products are likely to affect the experiment.

The instability towards irradiation is also a problem for e.g. processes for recycling of nuclear waste by solvent extraction. Then organic molecules are used as extractant and they, together with the organic diluent decompose with time and cleaning stages have to be included. The most famous of these degradation routes is the one relating to the commonly used extraction molecule tri-butyl phosphate (TBP) which form dibutyl phosphoric acid (HDBP) with a yield of approximately $G_{HDBP} = 0.07-0.12$ μmol/J and to a lesser extent bis(1,1,3,3-tetramethylbutyl)phosphinic acid (HMBP). Both these decomposition products are extractants in their own way and disturb the extraction process by increasing the extraction of some unwanted fission and corrosion products.

8.9. EXPERIMENTAL METHODS

Radiation chemistry is characterized by the very fast generation of reactive species followed by extensive competition between recombination reactions and reactions with solutes. A complete description of a radiation chemical process requires information about the final products and the transient species.

The final products can be analyzed with standard chemical methods and much information has been gained through the use of selectively reacting scavengers.

Information about structures and identities of primary ionic species has been obtained from mass spectrometry, photoelectron and vacuum ultraviolet spectroscopy. Electron spin resonance techniques have been used extensively in the study of free radicals. Low temperature, matrixes and "spin traps" have been employed to stabilize the short lived radicals.

Typical irradiation facilities are in the case of gamma radiolysis based on either ^{60}Co or ^{137}Cs. Samples introduced in these, often intense gamma fields (up to tenths of kGy/h), will not become radioactive and the radiolysis products can be studied in normal chemical laboratories. In the case of alpha radiolysis the situation is more complicated. Alpha particles have so short range that external irradiation into a sample vial is almost impossible. Therefore it is common practice to dissolve a alpha radioactive sample into the liquid of interest. The problem is that there will be residual alpha activity present in the sample preventing analysis of degradation product in a non-radioactive laboratory and thus highly sophisticated equipment has to be dedicated and contaminated.

Accelerators producing high energy He beams can also be used because the high energy α-particles will penetrate through the vial walls. Then there will be no residual activity in the sample. Recently a new method based on "in vitro" radiation of solution using ^{211}At has been developed at Chalmers. This method will also produce an, in principle, non radioactive solution with radiolysis products.

Much insight into radiation processes and radiation induced radical reactions has been gained by means of the *pulse radiolysis* method which is based on irradiation of samples with a short pulse of ionizing radiation. The radiation source is generally an electron accelerator or, to a much lesser extent, a positive particle accelerator. Techniques used to follow the transient behaviour of the radiolytically generated short-lived species are optical absorption spectroscopy, esr, conductivity and polarography. The time resolution is generally in the femtosecond to microsecond range.

8.10. DOSE MEASUREMENTS

The amount of radiation energy absorbed in a substance is measured with *dose meters* (or *dosimeters*). These may react via a variety of processes involving (a) the heat evolved in a calorimeter. (b) the number of ions formed in a gas, (c) the chemical changes in a liquid or in a photographic emulsion, and (d) the excitation of atoms in a glass or crystal. The first two ones are *primary meters* in the sense that they can be used to accurately calculate the exposure or dose absorbed from a radiation source. They can be used to calibrate the *secondary meters*.

In 1925 C. D. Ellis and W. A. Woorter, using RaE (^{210}Bi) in *calorimetric measurements*, obtained the first proof that the maximum energy and the average energy of β-radiations were different. A precision of about 1% can be obtained in a calorimeter for an energy production rate of $\sim 10^{-6}$ J/s which corresponds to approximately 0.7 MeV average energy for a sample of 40 MBq (1 mCi). If the average energy of an α- or β-emitting nuclide is known, calorimetric measurement of the energy production rate can be used to calculate the specific activity. This technique is not suitable for γ-sources.

A more sensitive and general instrument for the measurement of ionizing radiation is the *condenser ion chamber.* This is a detector which has a small gas-filled volume between two charged electrodes. When radiation ionizes the gas between the two electrodes, the cations travel to the cathode and the electrons to the anode, thus preventing recombination of the ion pairs. Measurement of the amount of discharge provides a determination of the ionization and consequently of the dose delivered to the instrument. This type of instrument is described in more detail in Ch. 9. The flexibility and accuracy of this dosimeter have led to it being widely employed for the exact measurement of γ-dose rates. The most common version of this type of instrument is the pen dosimeter (Fig. 8.8), which can be made to provide either a direct reading of the absorbed dose or

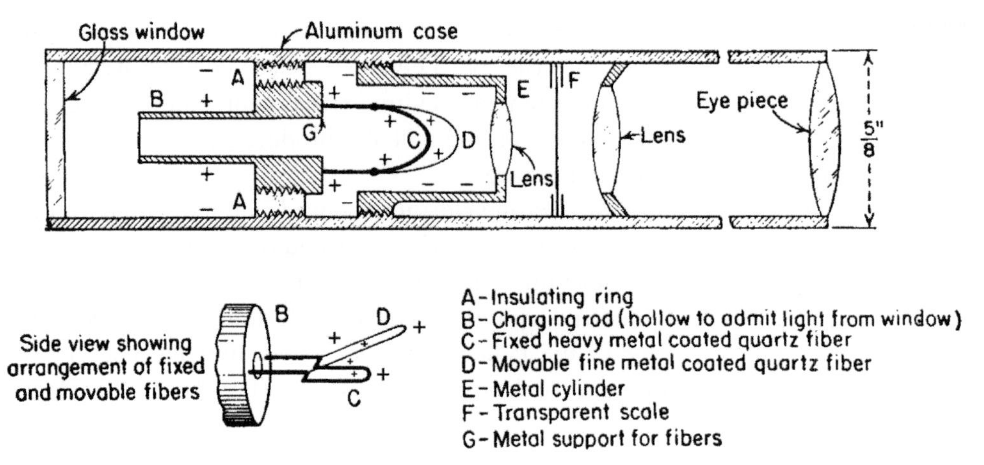

Figure 8.8 Pen dosimeter with direct readout of the dose. *(From Lapp and Andrews, 1948.)*

indirectly via an auxiliary reading instrument. Instruments with ranges from 0.0002 to 10 Gy (full scale) are available commercially.

There are numerous *chemical dosimeters* based on the radiolysis of chemical compounds, both organic and inorganic. An illustrative example is the $CHCl_3$ dosimeter. This is a two- phase aqueous-organic system. Radiation produces HCl which changes the pH of the almost neutral aqueous phase as shown by the colour change of a pH indicator. This dosimeter is suitable only for rather high doses, 10^2-10^5 Gy.

The most common chemical dosimeter is the *Fricke dosimeter* (§8.7), which consists of an aqueous solution of approximately the following composition: 0.001 M $Fe(NH_4)_2(SO_4)_2$, 0.001 M NaCl, and 0.4 M H_2SO_4. The amount of Fe^{3+} formed through irradiation is determined spectrophotometrically and the dose absorbed in Gy calculated by the equation:

$$D(\mathrm{Gy}) = A/\{\varepsilon x \rho \, \mathrm{G}(\mathrm{Fe}^{3+})\} \quad (\mathrm{J/kg}) \tag{8.40}$$

where A is the change in absorbance, $G(Fe^{3+})$ is the yield of Fe^{3+} in mol/J (cf. eqn. 8.33), ε is the molar extinction coefficient (217.4 m^2/mol at 304 nm), x is the length of the cell (in m), and ρ is the density of the solution (1024 kg/m^3 at 15–25°C). The *G*-value depends somewhat on the *LET* value of the radiation as seen in Figure 8.7. The Fricke dosimeter is independent of dose rate up to dose values of about 2×10^6 Gy/s and can be used in the range of 1–500 Gy. In a common modification, the solution also contains NaSCN, leading to the formation of the intensely red complex ion $Fe(SCN)_6^{3-}$ upon irradiation.

Photographic emulsions are sensitized by ionizing radiation resulting in darkening upon development. This is used in the film dosimeter for measurement of β-, γ-, or n-doses. In order to differentiate between various types of radiation, the film can be surrounded by filters or transfer screens. Although any type of film may be used, special nuclear emulsions have been designed. The dose received is directly proportional to the optical density of the exposed film. Film dosimeters are useful in the same range as the pen dosimeter, and both are used for personnel measurements. While the pen dosimeter can be read directly, the film dosimeter requires development.

The *glass dosimeter* is made of phosphate glass containing 5–10% of $(AgPO_3)_n$ polymer. The small piece of glass, a few cubic centimeters in size, is protected from light by means of a coating. When radiation strikes the glass, trapped electrons are produced which can be released by irradiation with ultraviolet light after removal of the protective coating. This results in the emission of fluorescent radiation which can be measured photometrically. The amount of fluorescent radiation is proportional to the dose received for doses up to 10 Gy.

The *thermoluminescent dosimeter* (*TLD*) covers the range 10^{-6}–10 Gy. The detector consists of a crystalline powder of CaF_2, LiF, or similar compound, either pure or incorporated in a plastic material like teflon. The irradiation leads to ionization and trapping of the electrons in imperfections in the crystal lattice. Upon heating, recombination occurs with light emission, which is measured photometrically. The electrons are trapped at different energy levels, and slow heating releases the electrons in order of increasing energy of the trapping levels. Consequently it is possible to take a dose reading on the lowest-energy trapped electrons and still retain a memory of the dose through the electrons left in more energetic traps. The dose can be read at a later time by releasing the remaining electrons at a higher temperature. The TLD can be designed like the pen dosimeter in which it is surrounded by screens to differentiate between different kinds of radiation. By using a lead filter the dosimeter can be made energy independent in the range 0.02–20 MeV for X-rays and γ-radiation.

This is also the basis for *thermoluminescence dating*. When geologic minerals like quartz, feldspars, etc as well as ceramic materials like fired clay are exposed to high energy cosmic radiation during the ages, imperfections are produced in the crystal lattice. When heated these crystals produce light in proportion to the dose received. If the dose rate is known the age of the exposed material can be calculated. There are numerous examples of the use of this technique e.g. for authentication of old porcelain or ceramics. Recently British Museum was forced to remove a large number of Mexican, Greek and Roman sculptures and vases because TLD-dating showed that they had been manufactured during the 19th century. Also samples of the enamel covering human teeth can be used to determine the accumulated dose by thermoluminescense. However, irradiation of the teeth by UV-light, often used by dentists, will erase some — or all — of the accumulated energy.

A modification is the *thermocurrent dosimeter*. In this case the detector may be a thin crystal of synthetic sapphire (ϕ 10 mm, thickness $\leq$1 mm) between thin metal electrodes. When the crystal is heated after having received a certain dose, the trapped electrons are released and cause a current to pass between the electrodes. The peak of the integrated current is a measure of absorbed dose. The effect is referred to as RITAC (radiation induced thermally activated current) and the technique is named TC (thermocurrent) dosimetry. Because it is instrumentally easier to measure an electric current than light, the TC dosimeter may replace the TLD with time.

Doses can also be calculated from the product of the dose rate and the time of exposure. The most common *dose rate meter* is the ionization chamber. Because of the close connection between this instrument and pulse type ionization counters, which measure individual nuclear particles entering the detector, the discussion of ionization chambers is deferred to Ch. 9, which deals in more detail with radiation measurement techniques.

If the number, energy, and type of nuclear particles being absorbed in a material can be measured or estimated, the absorbed dose can be calculated as described previously. Such calculations are very important, particularly in the field of radiation protection.

8.11. LARGE-SCALE NON-BIOLOGICAL APPLICATIONS

In this section we deal only with the more important non-biological applications, as the latter are treated in Ch. 15, which specifically deals with biological effects of radiation.

Ionizing radiation produces ionized and excited atoms and molecules in all materials. Excited molecules formed directly or by recombination reactions between electrons and cations decompose in the vast majority of systems to highly reactive free radicals. The reactive species formed on radiolysis are precursors of further reactions, such as reduction, oxidation, polymerization, cross linking and so on. It should therefore be possible to apply radiation chemical methods to industrial processes and, consequently, extensive applied research and development of radiation chemistry has been carried out during the past three decades.

A large proportion of the radiation induced reactions can be brought about by thermal, photochemical or chemical initiation, but the advantages of radiation initiation are generally claimed to be:

- No contamination by catalyst and catalyst residue.
- Temperature independence.
- Easy control of radiation intensity and hence rate of induced reactions.
- High speed treatment capability.
- Ionizing radiation offers the advantage of greater penetrating power compared to initiation by UV-light.

8.11.1. Radiation sources

The radiation sources normally used can be divided into two groups, those employing radioactive isotopes (e.g. ^{60}Co) and those employing electron beam accelerators (EBA).

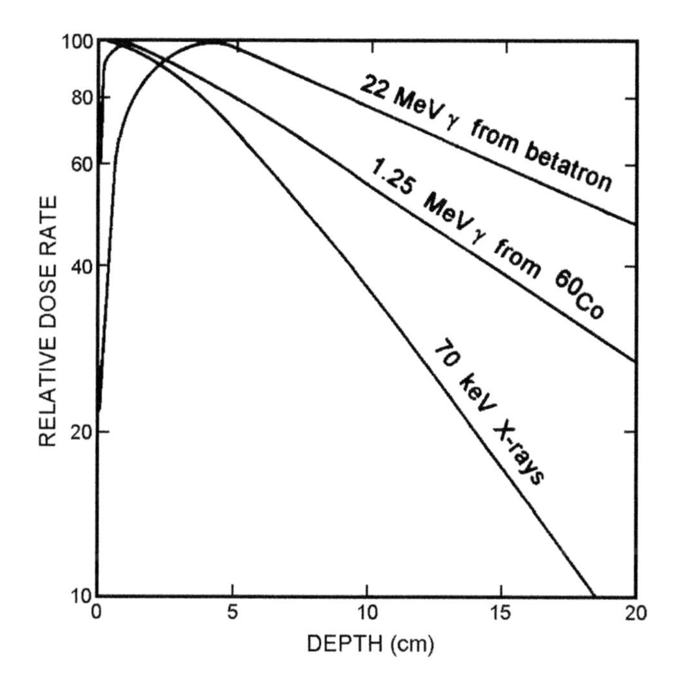

Figure 8.9 Depth-dose curves for electromagnetic radiation in water. Distance between source and water surface is 0.8 m; area of beam at water surface 0.01 m^2. *(From Spinks and Woods.)*

Gamma radiation, having a large penetration range (Fig. 8.9), is advantageously used for initiation of reactions in solids and liquids, food irradiation and sterilization of medical products. The useful penetration depth of electrons is approximately given by $0.4.E/\rho$ (cm), where E is the electron energy (in MeV) and ρ the density of the irradiated material (in g/cm^3). For EBA radiation, dose rate and penetration depth are easily controlled by varying the beam current and acceleration voltage. Very high radiation intensity can be provided and hence be applied to high speed processes. Radiation absorption is discussed in detail in Ch. 7, and particle accelerators in Ch. 16

8.11.2. Process criteria

The radiation dose required to completely convert 1 kg material to product is

$$D = (M\ G)^{-1}\ \mathrm{J\ kg^{-1}} \tag{8.41}$$

where D is the radiation dose (Gy), G is the radiation yield (mol J^{-1}), and M is the molar mass (kg mol^{-1}).

Ionizing radiation is an expensive form of energy, whether the radiation source is ^{60}Co or an accelerator, and generally at least one of the following criteria are fulfilled in the established radiation processes:

- A small amount of chemical change causes a marked change in physical properties (e.g. for polymers or biological systems).

- The radiation induced reaction has a high yield (e.g. polymerization, chain reaction).
- The radiation has a specific effect or process technical advantages which not easily can be obtained by other methods (e.g. staining of glass).

About 80% of ongoing radiation processes can be ascribed to the first category. Some examples of radiation processing are briefly described in the remainder of this chapter.

8.11.3. Radiation induced synthesis

A great number of reaction types have been investigated, the majority involve radical initiated organic chain reactions. A limited number of radiation induced synthesis has, however, been developed to the pilot plant or industrial scale. Examples of reported industrial synthesis are sulfoxidation and sulfochlorination of hydrocarbons for detergent production and polymerization of ethene.

The bromoethane process, which from 1963 and a number of years onwards, was used by the Dow Chemical Company to produce 500 tons bromoethane per year is of especial interest. In the process a 67 GBq ^{60}Co-source was utilized to irradiate hydrogen bromide-ethane mixtures and initiate the reaction sequence

Initiation
$$HBr \rightsquigarrow H\cdot + Br\cdot \qquad (8.42)$$

$$C_2H_4 \rightsquigarrow C_2H_2 + 2\,H\cdot \qquad (8.43)$$

$$H\cdot + HBr \rightarrow H_2 + Br\cdot \qquad (8.44)$$

Propagation
$$Br\cdot + C_2H_4 \rightarrow \cdot CH_2CH_2Br \qquad (8.45)$$

$$\cdot CH_2CH_2Br + HBr \rightarrow CH_3CH_2Br + Br\cdot \qquad (8.46)$$

Termination
$$2\,Br\cdot \rightarrow Br_2 \qquad (8.47)$$

$$2\,\cdot CH_2CH_2Br \rightarrow BrCH_2CH_2CH_2CH_2Br \qquad (8.48)$$

The chemical yield was, depending on reaction conditions, in the range 0.001–0.01 mol/J.

8.11.4. Industrial radiation processing

Today a large number of ^{60}Co and ^{137}Cs based irradiation sources and electron beam accelerators (EBA) are in use. In the following some of the more important uses will be summarized.

Sterilization: There has been a steadily increasing demand for single use disposable medical supplies (e.g. injection syringes, injection needles, surgical gloves) as well as surgical implants, contact lens solution and laboratory animal feed. Radiation sterilization doses in the range 25–50 kGy have been adapted in many countries. Both ^{60}Co and high energy electron accelerators are used as radiation sources. Radiation sterilization has the advantage of avoiding the use of toxic chemicals (e.g. ethylene oxide) and high temperature. A further advantage is that the material can be sterilized in a continuous process after final packaging.

Wire and cable: Thermoplastics and elastomers are widely used as electrical insulating material due to their physical properties and processability. Cross linking is an effective means for improving e.g. the thermal resistance and tensile strength. EBA-irradiation ($\approx$50 kGy) affords a rapid, well controlled cross linking and is used by several major producers of thin wires and cables.

Shrinkable film and tubing: Cross linked semi-crystalline thermoplastics display rubberlike properties at temperatures above their melting points. On deformation followed by fast cooling the polymer maintains its deformed shape. The polymer returns to its original shape when reheated. This memory effect is applied in the production of heat shrinkable films and tubing. Radiation doses of the order of 40–100 kGy are used in the production of heat shrinkable products.

Curing of surface coatings and inks: Irradiation is used to cure printing inks and varnishes and to bond coatings to surfaces of paper, fabrics, polymer films and steel. The absorbed doses are in the range 10–100 kGy and use of accelerators with low energy and high beam current affords selective energy deposition in the surface layer and high product throughput. Industrial radiation curing lines are operating in the car, wood panel and steel industries. Due to the low electron energy, no heavy shielding is required and the accelerators can easily be placed in normal industrial environment.

Potential applications: Immobilization or trapping of bioactive material such as enzymes, antigens, antibodies and drugs in polymer matrixes by radiation processes have received much attention.

Radiation treatment: To remove organics and metallic pollutants from waste water and SO_X and NO_X from flue gases emitted from coal power stations and industrial plants have been studied in great detail. The cleaning of flue gases has been developed to the pilot plant stage.

8.12. TECHNICAL USES OF SMALL DOSE-RATES

The ability of nuclear radiation to ionize gases is used in several applications. Static electricity can be eliminated through the installation of an α- or β-source in microbalances. Similarly, but on a much larger scale, elimination of static electricity is used in the paper, textile and paint industries. An ionization instrument for the analysis of gas has been developed in which the gas passes through a small chamber where it is irradiated by a small radioactive source. For a constant source of radiation, the ions produced in the gas depend on the flow velocity of the gas and on its temperature, pressure and atomic composition. The dependence of the ionization on the atomic composition is a consequence of the different ionization potentials of the different types of atoms of the gas and the different probabilities for electron capture and collision. The ion current is collected on an electrode and measured. This current is a function of the gas pressure and velocity since the higher the pressure, the more ions form, while at higher velocity, the fewer ions are collected as more ions are removed by the gas prior to collection. Such ionization instruments are used in gas chromatographs and other instruments as well as in smoke detection systems (the normal radiation source is ^{241}Am, usually $\leq$40 kBq), where secondary electrons condense on smoke particles, leading to lower mobility for the electrons and a decreased ion current.

Both α- and β-emitters are used in luminescent paint. The fluorescent material is usually ZnS. T and ^{14}C are preferred sources since their β-energies are low, but ^{85}Kr, ^{90}Sr, and ^{147}Pm are also used. The amount or radioactivity varies, depending on the need (watches, aircraft instruments, etc.) but it is usually $<$400 MBq ($<$10 mCi), although larger light panels may require $>$50 GBq (several Curies). For such high activities only T or ^{85}Kr are acceptable because of their relatively low radiotoxicity.

8.13. EXERCISES

8.1. How many ion pairs are produced in 10 m of air of STP by one (a) 5 MeV α-particle, (b) 1 MeV β-particle, and (c) 1 MeV γ-quantum (μ_m(air) ≈ μ_m(water))?

8.2. Estimate the fraction of energy lost through bremsstrahlung for a β-emission of $E_{max} = 2.3$ MeV, when absorbed in aluminum. The effective β-energy must be taken into account.

8.3. A freshly prepared small source of ^{24}Ne had a measured decay rate of 1 GBq 1 s after its preparation. The source is shielded by 10 cm of Pb. ^{24}Ne emits γ-rays; 8% with 0.878 MeV and 100% with 0.472 MeV. Estimate the total integrated dose received at 2 m distance during its life-time after preparation. Neglect build-up factors.

8.4. An acidic aqueous solution is irradiated by α-particles from dissolved ^{239}Pu at a concentration of 0.03 M. The plutonium is originally in its hexavalent state, but is reduced to the tetravalent state by the reaction Pu(VI) + 2H· → Pu(IV) + $2H^+$. How much of the plutonium can be reduced in one week?

8.5. An acid solution of fresh fission products contains 0.8 g l^{-1} cerium as Ce^{4+}. The γ-flux in the solution corresponds to 520 GBq l^{-1} of an average energy of 0.7 MeV. If half of the γ-flux is absorbed in the solution, what fraction of cerium is reduced to Ce^{3+} in 24 h? Assume the same *G*-value as in Figure 8.7.

8.6. Estimate the average *LET* value in water for β-particles from ^{90}Sr and ^{90}Y, and in Al for T.

8.7. A ^{60}Co irradiation source is calibrated by the Fricke dosimeter for which the *G*-value is assumed to be 1.62 μmol/J. Before the irradiation the optical density *D* of the solution at 305 nm was 0.049 in a 1 cm cuvette. After exactly 2 h the *D* had changed to 0.213. Calculate the dose rate when the molar extinction of Fe^{3+} is 217.5 m^2 mol^{-1}.

8.8. A direct reading condenser chamber (pen dosimeter) is charged from a battery pack so that full scale (100) is obtained at 20 V. When completely discharged, the accumulated dose is 5.00 mGy. The gas volume is 4 cm^{-3} air (STP). What is the capacitance of the condenser chamber?

8.14. LITERATURE

S. C. LIND, *Radiation Chemistry of Gases*, Reinhold, 1961.

A. O. ALLEN, *The Radiation Chemistry of Water and Aqueous Solutions*, D. van Nostrand, 1961.

M. HAISSINSKY and M. MAGAT, *Radiolytic Yields*, Pergamon Press, Oxford, 1963.

R. O. BOLT and J. G. CARROLL, *Radiation Effects on Organic Materials*, Academic Press, 1963.

B. T. KELLY, *Irradiation Damage to Solids*, Pergamon Press, Oxford, 1966.

J. L. DYE, The solvated electron, *Sci. Am.*, Febr. 1967.

J. R. CAMERON, N. SUNTHARALINGHAM and G. N. KENNEY, *Thermoluminescent Dosimetry*, University of Wisconsin Press, 1968.

E. J. HART (Ed.), *Radiation Chemistry*, Advances in Chemistry Series 81. Am. Chem. Soc., Washington DC 1968.

E. J. HENLEY and E. R. JOHNSON, *The Chemistry and Physics of High Energy Reactions*, University Press, Washington DC 1969.

W. SCHNABEL and J. WENDENBURG, *Einführung in die Strahlenchemie*, Verlag-Chemie, 1969.

N. E. HOLM and R. J. BERRY (Eds.), *Manual on Radiation Dosimetry*, Marcel Dekker, 1970.

J. H. O'DONNELL and D. F. SANGSTER, *Principles of Radiation Chemistry*, Elsevier, 1970.

A. R. DENARO and G. G. JAYSON, *Fundamentals of Radiation Chemistry*, Butterworths, 1972.

G. E. ADAMS, E. M. FIELDEN, and B. D. MICHAEL (Eds.), *Fast Processes in Radiation Chemistry and Biology*, J. Wiley, 1975.

T. CAIRNS, *Archeological Dating by Thermoluminescence*, *Anal. Chem.* **48** (1976) 267A.

G. V. BUXTON and R. M. SELLERS, The radiation chemistry of metal ions in aqueous solutions, *Coord. Chem. Rev.* **22** (1977) 195.

J. FOELDIAK (Ed.), *Radiation Chemistry of Hydrocarbons*, Elsevier, 1981.

FARHATAZIZ and M. A. J. RORGERS (Eds.), *Radiation Chemistry, Principles and Applications*, VCH Press Publ. Co, Weinheim 1987.

G. FREEMAN (Ed.), *Kinetics of Nonhomogeneous Processes*, J. Wiley, 1987.

J. BEDNÁR, *Theoretical Foundations of Radiation Chemistry*, Kluwer Academic Publishers, 1989.

J. R. GREENING, *Fundamentals of Radiation Dosimetry*, Adam Hilger, 1989.

V. K. MILINCHUK and V.I. PUPIKOV (Eds.), *Organic Radiation Chemistry Handbook*, Ellis Horwood/J. Wiley, 1989.

J. W. T. SPINKS and R. I. WOODS, *An Introduction to Radiation Chemistry*, 3rd edn., J. Wiley, 1990.

Y. TABATA (Ed.), *Pulse Radiolysis*, CRC Press Inc., 1991.

D. W. CLEGG and A. A. COLLYER (Eds.), *Irradiation Effects on Polymers*, Elsevier, 1991.

The Journal of Radiation Physics and Chemistry (Elsevier) regularly contains survey papers of various fields of radiation chemistry.

Detection and Measurement Techniques

Contents

Radiochemistry and Nuclear Chemistry
ISBN 978-0-12-405897-2, http://dx.doi.org/10.1016/B978-0-12-405897-2.00009-4

The ionization and/or excitation of atoms and molecules when the energies of nuclear particles are absorbed in matter is the basis for the detection of individual particles. Macroscopic collective effects, such as chemical changes and heat evolution, can also be used. The most important of the latter have been described before because of their importance for dose measurements (e.g. the blackening of photographic films and other chemical reactions, excitation of crystals (thermoluminescence), and heat evolved in calorimeters; Ch. 8). Although animals have no known senses for detection of nuclear radiation, it has been found that subletal but large radiation fields can affect animals in various ways such as disturbing the sleep of dogs or causing ants to follow a new pathway to avoid a hidden radiation source. Apollo astronauts observed scintillations in their eyes when their space ship crossed very intense showers of high energy cosmic rays. People who have been involved in criticality accidents experiencing high intensities of n and γ have noted a fluorescence in their eyes and felt a heat shock in their body.

However, we are not physiologically aware of the normal radiation fields of our environment. In such low fields we must entirely rely on instruments.

In this chapter we consider only the common techniques used for detection and quantitative measurement of *individual* nuclear particles. We also discuss the problem of proper preparation of the sample to be measured as well as consideration of the statistics of the counting of nuclear particles necessary to ensure proper *precision* (i.e. how well a value is determined) and *accuracy* (i.e. agreement between measured and true value).

9.1. TRACK MEASUREMENTS

The most striking evidence for the existence of atoms comes from the observation of tracks formed by nuclear particles in cloud chambers, in solids and in photographic emulsions. The tracks reveal individual nuclear reactions and radioactive decay processes. From a detailed study of such tracks, the mass, charge and energy of the particle can be determined.

The tracks formed can be directly observed by the naked eye in cloud and bubble chambers, but the tracks remain only for a short time before they fade. For a permanent record we must use photography. On the other hand, in solid state nuclear track detectors (SSNTD), of which the photographic emulsion is the most common variant, the tracks have a much longer lifetime during which they can be made permanent and visible by a suitable chemical treatment. Because of the much higher density of the absorber, the tracks are also much shorter and often therefore not visible for the naked eye. Thus the microscope is an essential tool for studying tracks in solids.

9.1.1. Cloud and bubble chambers

The principle of a *cloud chamber* is shown in Figure 9.1. A volume of saturated vapour contained in a vessel is made supersaturated through a sudden adiabatic expansion. When ionizing radiation passes through such a supersaturated vapour the ionization produced

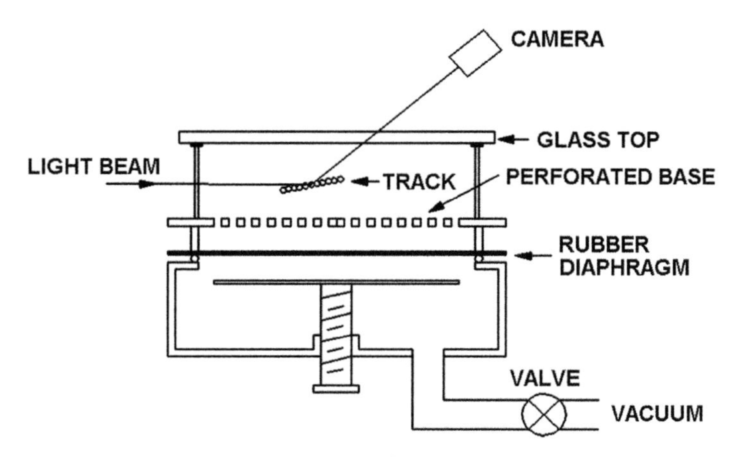

Figure 9.1 Principle of a cloud chamber.

in the vapour serves as condensation nuclei. As a result small droplets of liquid can be observed along the path of the radiation. These condensation tracks have a lifetime of less than a second and can be photographed through the chamber window. The density of the condensation depends on the ionization power of the projectile as well as on the nature of the vapour, which is often an alcohol or water. Cloud chamber photographs are shown in Figure 7.5.

In a similar fashion *bubble chambers* operate with superheated liquids in which gas bubbles are produced upon the passage of ionizing radiation. The most commonly used liquid in bubble chambers is hydrogen, and, as a consequence, the chamber must be operated at low temperatures (23 K for H_2). Since the liquid medium in a bubble chamber is much denser than the vapour medium in a cloud chamber, the former are more suitable for studies of reactions of more energetic projectiles. The high energy p-p reaction shown in Figure 2.2 has been recorded in a large hydrogen bubble chamber at CERN; see also Figure 7.18.

9.1.2. Solid state nuclear track detectors (SSNTD)

The main types of SSNTD (or DTD, for dielectric track detector) are photographic emulsions, crystals, glasses, and plastics. Because the density of these materials is much higher than for the gases often used in the previous detector group (section 9.1.1), nuclear particles can spend all their kinetic energy in these detectors, allowing identification of the particle. Since the SSNTD retains the particle path, they can be used to record reactions over a long time period. These advantages have made SSNTD especially valuable in the fields of cosmic ray physics, radiochemistry, and earth sciences.

Nuclear emulsions are similar to optical photographic emulsions. They contain AgBr crystals embedded in gelatin to which small amounts of sensitizing agents have been added. The AgBr content is as much as four times (i.e. 80% AgBr) greater than in optical film. Also the crystals are much smaller (developed grain 0.1–0.6 μm) and well separated. The emulsions come in thicknesses from a few μm up to 1 mm. Nuclear radiation passing through the emulsion causes ionization and excitation which activates the AgBr crystals, producing a latent image of the particle path. Upon development the activated crystals serve as centers for further reduction to silver metal, leading to visible grains. It is assumed that at least 3 silver atoms must be activated to produce a visible grain, while about 30 atoms are needed for normal blackening. Each activated grain seems to require an average absorption of about 2.5 eV.

While the memory effect of the properly developed film is almost infinite, this is not the case for the latent image which slowly fades, depending on the number of originally activated silver atoms, the film type and external conditions like temperature, humidity, etc. When stored under ambient conditions, about 80% of the latent image disappears in half a year.

The developed grains form an interrupted track along the original path of the energetic particle (Fig. 9.2(a)). The specific energy loss of the particle, dE/dx (i.e. the stopping power of the absorber), depends on the mass, charge, and velocity of the particle, and on the composition of the absorber. From the track length, grain density (i.e. grains per track length), and gap length between the grains, the particle and its energy can be determined (cf. sections 7.3 and 8.1). For a given particle, the range is proportional to the energy as shown in Table 9.1. The range decreases with increasing mass of the particle and density of the absorber. The grain density depends on the specific ionization of the particle which does not vary linearly with the particle energy (or velocity), as seen from Figure 7.7; thus the grain density changes along the track.

Other solid material may be used as SSNTD instead of AgBr emulsions: *plastics* (cellulose nitrate and polycarbonate films), *glass*, *crystals*, etc. In order to make the tracks visible in the microscope the surface of the SSNTD must be polished and etched, usually with alkali.

Because of the natural radiation background, every SSNTD has a memory of past nuclear events, which must be erased as far as possible before a new exposure. In nuclear emulsions an α-radiation background of 20–60 tracks cm^{-2} per day is normal. The technique of *background eradication* prior to exposure may consist of treating an emulsion with chromic acid, H_2O_2-vapour or heating (annealing) a glass plate. Because this technique more easily removes weak images, it may also be used after exposure, e.g. to remove fainter α-tracks from heavier fission tracks.

Let us consider some examples of uses of SSNTD. Tracks obtained under various conditions are shown in Figure 9.2.

(i) As mentioned SSNTD has been used in cosmic ray experiments at high altitudes and in space journeys where the memory effect and simple construction make them especially suitable. Many elementary particles have been discovered by this

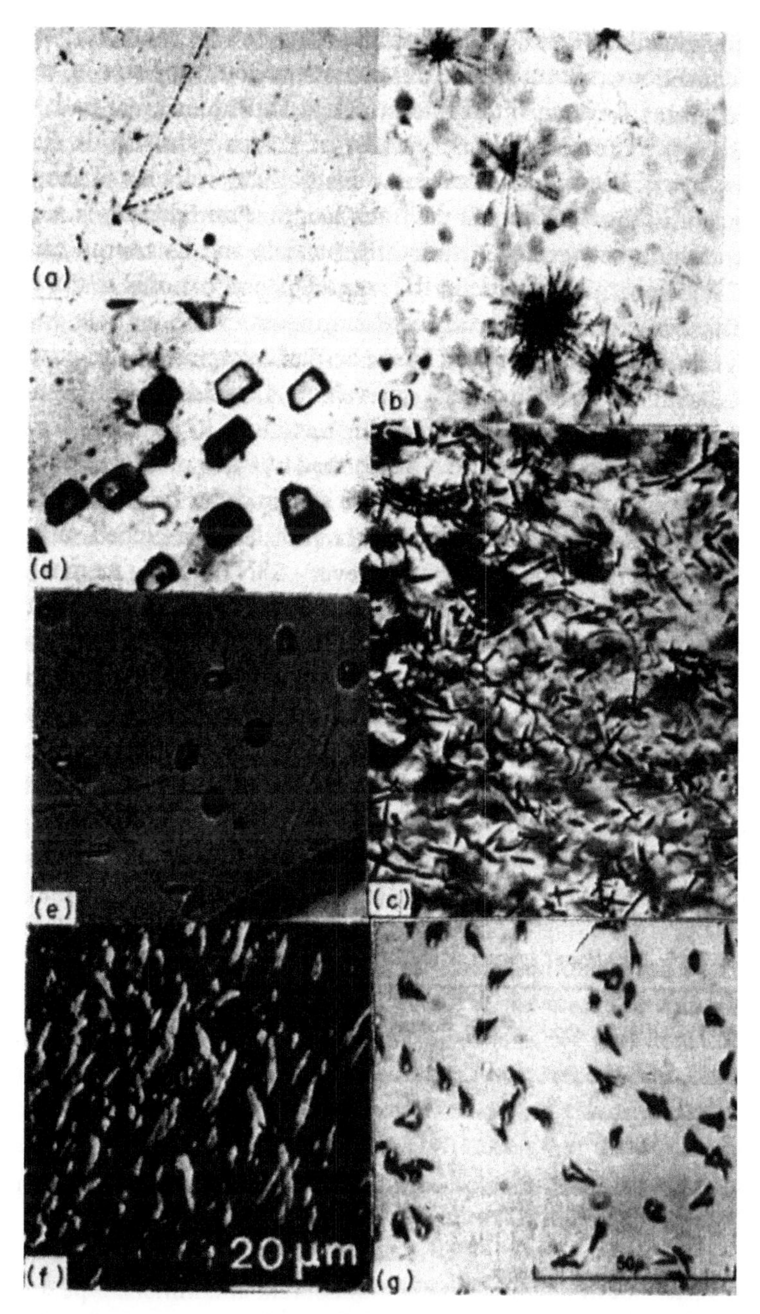

Figure 9.2 Tracks of highly ionizing particles in solid absorbers. (a) Alpha-tracks originating in the same point in a nuclear emulsion. (Acc. to P. Cuer.) (b) Autoradiography of a lung showing deposits of inhaled plutonium. (Acc. to B. A. Muggenburg.) (c) Fission tracks in neutron irradiated apatite containing some evenly distributed uranium. (Acc. to E. I. Hamilton.) (d) Neutron-induced fission tracks in muscovite mineral. (Acc. to E. I. Hamilton.) (e) Neutron-induced fission tracks in volcanic glass recovered from deep sea sediments. (Acc. to J. D. Macdogall.) (f) Fission tracks in mineral from the Oklo mine. (Acc. to J. C. Dran *et al.*) (g) Fission tracks from ^{252}Cf in Lexan polycarbonate. *(From Fleischer, Price, and Walker).*

Table 9.1 Range of energetic high-ionizing particles in various solids

Particle	Energy (MeV)	Absorber (Density)	Range (μm)
H	10	Ilford C2 (3.8)[(a)]	540
^{3}H	10	Ilford C2 (3.8)[(a)]	230
^{4}He	10	Ilford C2 (3.8)[(a)]	57
^{4}He (^{214}Bi)	7.7	Eastman NTA (3.6)	38
		Mica (3.1)	36
		Glass (2.5)	41
		Water (1.0)	60
^{4}He (^{238}U)	4.2	Mica (3.1)	13
^{4}He (U-series)		Pitchblende (7.0)	23[(b)]
		Carnotite (4.1)	32[(b)]
^{235}U Light fiss. fragm.	~150	Eastman NTC (~3.4)	14 } ~25
^{235}U Heavy fiss. fragm.		Eastman NTC (~3.4)	10.5 }
^{238}U Both fiss. fragm.	~160	Leopoldite (~4)	20

[(a)]Density of AgBr 6.47; of gelatin 1.31.
[(b)]Range of the predominating α-particles.

technique, notably the π- and μ-mesons. Figure 13.2 shows tracks of high energy cosmic ray particles, probably iron atoms, which have been stopped in Apollo astronaut helmets.

(ii) Nuclear reactions can be studied by SSNTD. The target material may either be regular atoms of the detector (H, O, Ag, Si, etc.) or material introduced into the matrix, e.g. thin threads of target metals or uranium atoms. The former have been used in high energy physics for hadron-induced reactions, and the latter for studying fission processes. From experiments with uranium the frequency of spontaneous fission of ^{238}U has been determined, and also the rate of ternary fission and emission of long range α's.

(iii) When emulsions are dipped into solution, some of the dissolved material is soaked up or absorbed in the emulsion. For example, if the solution contained samarium, some α-tracks of its spontaneous decay (decay rate 127 Bq g^{-1}) appear in the emulsion (cf. Fig. 9.2(a)). Since ^{147}Sm has an isotopic abundance of 15%, its half-life is calculated to be 1.06×10^{11} y. The lower limit of detection is about 500 tracks cm^{2} d^{-1}, so quite low decay rates can be accurately measured, making this a valuable technique for determination of long half-lives.

(iv) ^{222}Rn is released through the earth's surface from uranium minerals. The amount released varies not only with the uranium content and mineral type, but also with the time of the day; variation from 1–75 Bq l^{-1} has been registered during a 24 h period. To avoid this variation, cups containing a piece of plastic TD are placed upside down in shallow (0.5–1.0 m) holes for about 3 weeks, after which the SSNTD are etched and α-tracks from radon counted. Mineral bodies hundreds of meters underground can be mapped with this technique in great detail in a reasonably short time. The US Geological Survey uses a similar technique to predict earthquakes; it has been observed that just before earthquakes the radon concentration first increases, then suddenly decreases, the minimum being observed about one week before the earthquake appears.

(v) The average radon concentration in houses can be measured by hanging a plastic film inside the house over a time period of some weeks. The film is returned to a laboratory where it is etched and the number of α-tracks per unit area counted.

(vi) Fission fragments make dense tracks in all solid material. The tracks are short and thick: in a crystal material like zircon (a common mineral, $ZrSiO_4$) they may not be more than 10^{-2} μm (10 nm) in diameter, and 10–20 μm in length. They are therefore not visible even in the best optical microscopes. Using scanning electron microscopy, it has been found that the hole formed retains the crystal structure or regains it (Fig. 9.2(d)). On the other hand, if the track is formed in glass, a gas bubble appears instead of a track (Fig. 9.2(e)); these slightly elongated bubbles can be distinguished from other completely spherical bubbles formed by other processes. To make the tracks visible, the specimen is embedded in a resin, then one surface is ground and carefully polished after which it is dipped in an acid, e.g. HF. Because of defects in the crystal structure along the fission track, the track and its close surroundings are attacked by the acid, and the diameter of the track increases a hundredfold to a micron or so. The tracks are then visible under a microscope with a magnification of 500–1000×.

This procedure has been used as an analytical tool for determination of uranium and plutonium in geological and environmental samples. In this technique, the sample (either a ground and polished surface of a mineral, or a dust sample on tape) is firmly pressed against a photographic film, and the package is irradiated by slow neutrons. From the fission track count of the developed film the uranium or plutonium content can be calculated. Thus a Swedish shale was found to contain 4 ± 1 ppm U, and a bottom sediment in a Nagasaki water reservoir 0.44 ± 0.04 Bq ^{239}Pu per kg sediment. In the latter case, the ratio between the number of fission tracks $N_{f\,t}$ and α-tracks $N_{\alpha\,t}$ from ^{239}Pu is

$$N_{f\,t}/N_{\alpha t} = \sigma_f \varphi / \lambda_\alpha \quad (9.1)$$

where σ_f is the fission cross section, φ the n-flux, and λ_α the decay constant of ^{239}Pu. This technique is very useful for routine measurements of fissionable material in very low concentrations and for the safeguards program. Figure 9.2(c) shows fission tracks in a uranium-containing mineral which has been exposed to neutrons.

Fission track counting is also important for dating of geological samples (Ch. 13) and for estimation of the maximum temperature experienced by sedimentary rocks. Temperature history information can be obtained from a combination of age (estimated by other radioactive methods, cf. Ch. 13), uranium content of crystals of several minerals in the rock, their fission track count and track-length distribution caused by thermal annealing. Zircon, titanite and apatite are examples of three such minerals in order of increasing annealing temperature. The temperature history of a rock formation is important in oil prospecting operations as the maximum temperature seems to be a useful indicator on whether to expect oil, natural gas or nothing. For too low temperatures neither oil nor gas is expected, for intermediate temperatures oil may be present and for high temperatures only natural gas.

9.2. GENERAL PROPERTIES OF DETECTORS

The principle for detection of nuclear radiation is the same as discussed previously for both the interaction of radiation with matter and radiation production namely: excitation, ionisation, deflection, and atom recoil. In the case of detection only the first two of the interactions are primarily used even if, naturally, the others also occur. A nuclear particle entering a detector produces excitation and ionization, both of which can be used for detection. When the excitation is followed by fluorescent de-excitation (section 7.8.4 and Ch. 8) the light emitted can be registered by light-sensitive devices, e.g. the photo-multiplier tube (PMT) which transforms the light into an electric current. Scintillation and Čerenkov detectors are based on light emission. A similar current is generated when production of *charge carriers* (i.e. ions, electrons and holes) takes place between the charged electrodes of a detector. Detectors based on the production of charge carriers are either gas-filled (ion chambers, proportional and Geiger-Müller tubes in which charge carriers are produced by ionization of a gas) or solid, usually semiconductor crystals. In the latter case electrons and holes are produced in pairs (section 9.4).

An ionizing particle or photon will produce a collectable charge ΔQ in the detector

$$\Delta Q = 1.60 \times 10^{-19} E_{\text{loss}} \, \eta \, w^{-1} \quad (9.2)$$

where E_{loss} is the total energy lost by the particle to the detector, η is the charge collection efficiency, w is the energy required for the formation of a pair of charge carriers in the detector medium, while the constant is the charge (Coulomb) of a single charge carrier (the pairs must be regarded as singles because they move in opposite direction in the electric field due to their opposite charges).

Because of the short duration of the absorption process for a single particle (10^{-4}–10^{-9} s) the short current, i, is referred to as a *pulse* of charge ΔQ

$$i = \Delta Q / \Delta t \quad (9.3)$$

If this current passes through a resistor R it will produce a voltage pulse (cf. Eqn (9.2))

$$\Delta V = R \, \Delta Q / \Delta t \quad (9.4)$$

The pulse is usually referred to as the *signal* (for the preamplifier). While the signal from semiconductor detectors is used as a charge pulse, most other detectors immediately convert the current to a voltage drop over a resistor. In either case, the output from the preamplifier is usually a voltage pulse. Preamplifiers are often integrated with the detector.

Pulse counting *per se* does not distinguish between different nuclear particles (α, β, γ, etc.) or between particles of different energy. Such distinction is obtained by choosing detectors of unique (or exceptionally high) sensitivity to the particles of interest. Energy analysis, if desired, is achieved by the accompanying electronic circuitry because the pulse charge or voltage (ΔQ or ΔV) is proportional to the energy of the absorbed particle.

9.2.1. Pulse generation

Figure 9.3 will be used to describe the formation of a voltage pulse. A detector is connected between points A and B. The detector has an *internal* resistance (because of the limited charge carrier mobility in the detector) and capacitance (because of mechanical construction), indicated by R_i and C_i. Figure 9.3(a) does *not* show the physical design of the detector, but only its electrical equivalents; this will make it easier to understand its function. When a particle enters the detector it produces charge carriers (this is symbolized by the closing of switch S), and the collection of these at the electrodes gives a current which, together with a small current from the bias supply through R_e, flows through R_i to ground. R_e is the resistance between the detector anode and the positive terminal of the bias voltage supply (voltage $+ V_0$); the other terminal is grounded. We shall concentrate our interest on the potential V_p at point P which is connected via the comparatively large capacitance C_e to the output. In general $R_e > R_i$ (under conducting

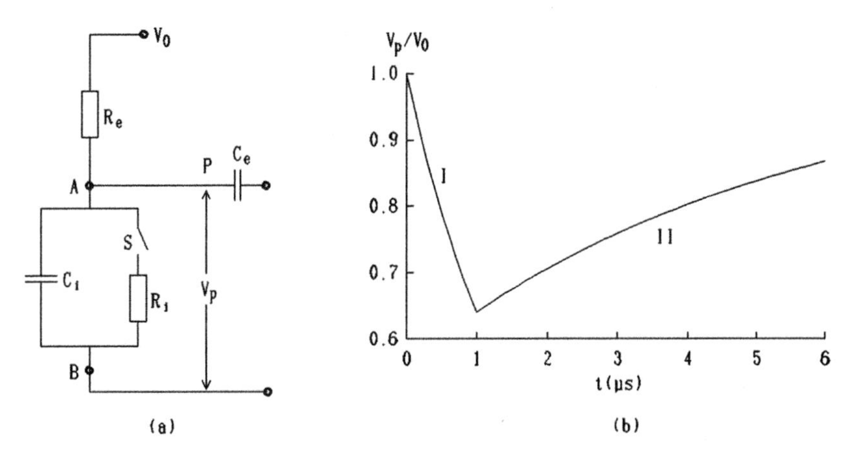

Figure 9.3 (a) Equivalent circuit (b) pulse shape in point P.

conditions), and $C_e \gg C_i$; for illustrative purposes we will assume $R_e = 50$ kΩ, $R_i = 10$ kΩ and $C_e+C_i=100$ pF.

When S is open (no ionization in the detector), the potential at point P must be $V_p = V_0$, i.e. the potential of the bias voltage. At time $t = 0$, S is closed (production of charge carriers has occurred in the detector because of interaction with a nuclear particle), and the charge of $C_i + C_e$ flows through R_i together with a small current from the bias supply through R_e. The potential in P decreases according to

$$V_p/V_0 = a\left(1 - e^{-bt}\right) + e^{-bt} \tag{9.5}$$

where $a = R_i/(R_i + R_e)$, $b = (a\,R_e\,C)^{-1}$, and $C = C_i + C_e \approx C_e$. For $R_e = 50$ kΩ, $R_i =$ 10 kΩ and $C = 100$ pF, $a = 1/6$ and $b = 12\,000$ s^{-1}. $1/b$ is referred to as the *time constant* of the system (the decay time of the charge of the system); in the time $1/b$ the voltage has dropped to $V_0/\mathbf{e}$. Assuming the charge collection ends after 1 μs this is equivalent to opening switch S at that time. The voltage in P has now fallen to its minimum value as indicated in Figure 9.3(b).

When S is opened charge starts building up on the capacitors by current flowing from the bias voltage through resistor R_e. The voltage in P begins to increase again according to

$$V_p/V_0 = 1 - (1 - V_m/V_0)e^{-t/(R_e C)} \tag{9.6}$$

where V_m is the voltage in P when the switch is opened again, t is now the time after opening of S and $C = C_e + C_i$. If we assume $R_e = 50$ kΩ the voltage build-up will follow the climbing curve to the right of 1 μs in Figure 9.3(b). In practice, the current flow stops more gradually and the potential in P follows a smoother curve.

The voltage drop at P results in a negative voltage pulse after the capacitor C_e. In Figure 9.3(b) the voltage change at point P is more than 40% of the bias voltage; this is highly exaggerated, since the change is normally a very small fraction of the bias voltage. This is due to the fact that the current is carried only by the charge carriers formed in the detector. If the number of primary charge carriers formed is n_i, the charge transport ΔQ is

$$\Delta Q = -1.60 \times 10^{-19} n_i a \text{ (Coulomb)} \tag{9.7}$$

where a is the multiplication factor (1 in solid state detectors, $\gg 1$ in gas-filled detectors). When $a < 1$ it has the same meaning as the collection efficiency η in (9.2). The maximum voltage drop is then

$$\Delta V \approx \Delta Q/C \tag{9.8}$$

If 10^5 charge carrier pairs have formed in the detector and $C = 100$ pF, then $\Delta V = 0.16$ mV. In practice the effective voltage drop ΔV_{eff} is somewhat less than the calculated ΔV.

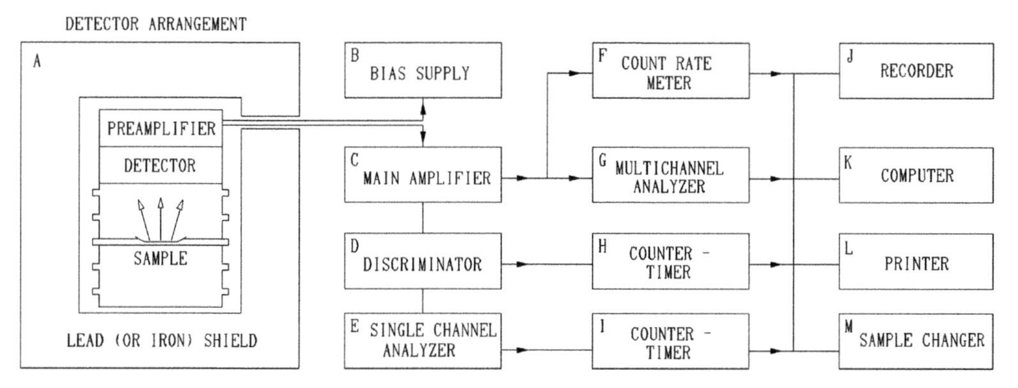

Figure 9.4 The most common pulse-type measuring units and combinations.

9.2.2. Basic counting systems

The block diagram of Figure 9.4 indicates the most common components in a measuring circuit. The detector (in its shield A) is connected to a stabilized bias voltage power supply B which furnishes the potential difference necessary for the detector to operate. The magnitude and type of the signal from the detector varies, depending on the type of detector, for the type of circuit illustrated in Figure 9.3 from a few hundred microvolts to several volts. In most cases it is necessary or preferable to have a *preamplifier* connected directly to the detector, cf. section 9.8.1. The preamplifier often increases the pulse size to the 0.1–10 V preferred for the auxiliary system. This initial amplifier also supplies the power needed to drive the load presented by the connecting cable and input impedance of the *main amplifier* C. The total gain in the system may vary from 10 for a Geiger-Müller counter to 10^4 for some solid state detectors.

A limit is set on the voltage gain of an amplifier by the presence of electronic noise at the amplifier input due to thermal noise in the detector and preamplifier input stage, microphonics, and components which cause small, random voltage changes. A part of this noise can be reduced by cooling the input circuitry. The noise is amplified with the signal and can mask small pulses from the detector. Some of the noise may be eliminated by proper design and operation of the electronic circuitry, but a small inherent noise is always present (Fig. 9.5(a) and section 9.8.1). The *signal to noise ratio* is a ratio of output pulse (for a given input signal) to the noise level at the output. Noise pulses can be rejected by a *discriminator* D which serves as a filter to allow only pulses of a certain minimum size to be passed on to the rest of the system; in Figure 9.5(a) the discriminator rejects all pulses below the dashed line. The signal pulse, after amplification, is sufficiently large to operate the electronic *counter* H in Figure 9.4 causing registration of the pulses. In some cases the number of pulses counted are often shown on light-emitting diode (LED), gas-plasma, or liquid crystal displays.

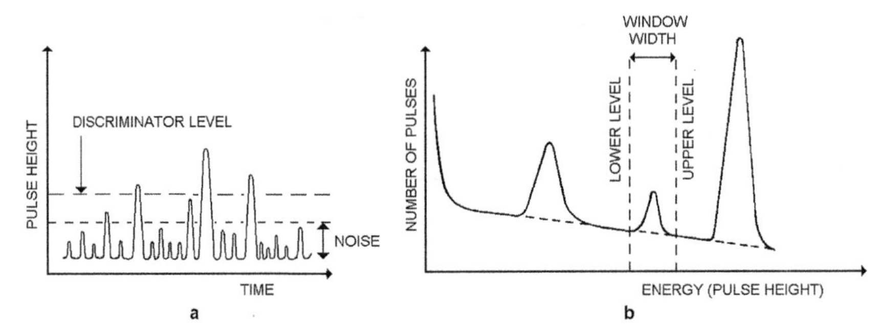

Figure 9.5 Idealized pulse spectrum obtained (a) from a shaping amplifier, and (b) from a multichannel analyzer using a solid state detector. The figure shows the effect of different discriminator settings.

9.2.3. Pulse shape and dead time

From Figure 9.3(b) it is obvious that it takes some time to restore proper measuring conditions. Such conditions may be assumed to have occurred when the potential V_p has returned above a certain level. During the first part of this time the detector is unable to collect a new charge because working conditions have not been sufficiently restored. This time, which in the Figure is something like 10 μs, any new events would not produce a pulse crossing the discriminator level. This interval is properly called *dead time*, see Figure 9.6. Somewhat later the initial operating conditions are still not

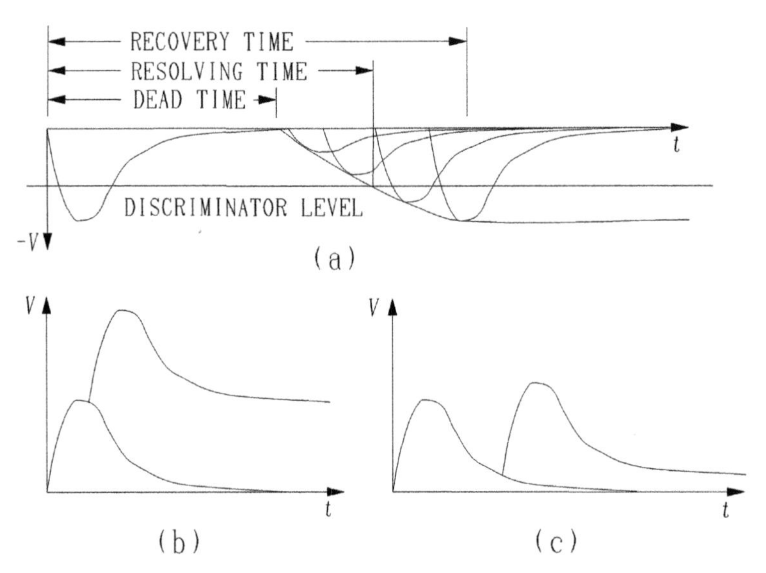

Figure 9.6 Effects of short interval between events registered in detectors. (a) G.M. tube, (b) peak pile-up and (c) tail pile-up for fast detectors.

fully restored but the detector is now able to produce a pulse of larger magnitude which triggers the discriminator. Still later, after the *recovery time*, the initial conditions are restored. If the second event occur within a short time, *peak pile-up* will occur. At a somewhat later time the new pulse overlaps with the tail of the pulse from the previous event causing so called *tail pile-up*. Pulse pile-up may make two or more closely spaced events look like a single more energetic event, Figure 9.6. The time needed to separate two events is referred to as *resolving time* (for simplicity dead time and resolving time will be used as synonyms in this text). Thus the detector and measuring circuit needs a certain time to register each individual event separately with correct magnitude. In many cases the measuring circuitry is much faster than the detector and the dead time is a function of the detector only. Since radioactive decay is a statistical random process and not one evenly spaced in time, see section 9.11, even for relatively low count rates a certain percentage of events will occur within the resolving time of the system. In order to obtain the true count rate it is necessary to know the correction that must be made for this *random coincidence loss*. In systems using a MCA for pulse height analysis, the MCAs pulse conversion time is usually determining the system dead time and not the detector.

Two different models exist for the dead time, t_d, of counting systems depending on system behavior after a pulse, see Figure 9.7. In a *nonparalyzable* system the pulses following the first within the dead time are lost, but the system is ready to accept another event immediately after the dead time has expired. If N_{obs} is the counted number of pulses during the measurement time, t_{meas}, and R_{real} the real number of particles which hit the detector per unit time, then the total lost time, t_{lost}, is

$$t_{lost} = N_{obs} \times t_d \tag{9.9}$$

and

$$N_{obs} = R_{real} \times \left(t_{meas} - N_{obs} \times t_d\right) \tag{9.10}$$

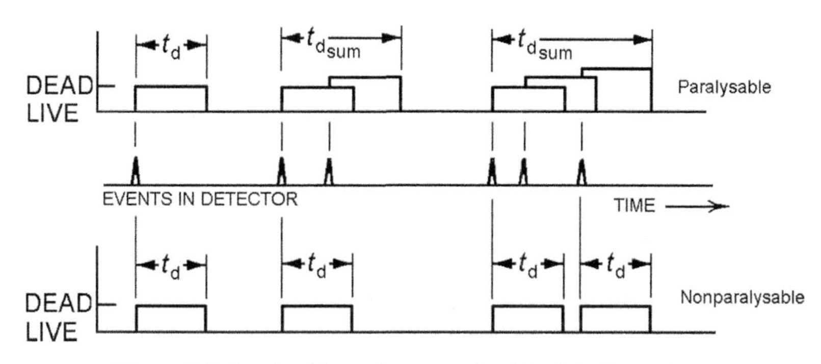

Figure 9.7 Paralysable and nonparalysable detector systems

but

$$R_{obs} = N_{obs}/t_{meas=} = R_{real} \times (1 - R_{obs} \times t_d) \tag{9.11}$$

thus

$$R_{real} = R_{obs}/(1 - R_{obs} \times t_d) \tag{9.12}$$

In a *paralyzable* system each event starts a new dead time period whether or not it generates an output signal. The dead time for the paralysable detector is a bit more tricky to estimate. However, assume that $P_1(t) \times dt$ is the probability that the next event will happen during the short interval dt after a delay of t,. Further more, assume that $P(0)$ is the probability that no event will occur between 0 and t and $R_{real} \times dt$ is the probability of an event occurring during dt. Then

$$P_1(t) \times dt = P(0) \times R_{real} \times dt \tag{9.13}$$

Poisson statistics gives that

$$P(0) = e^{-R_{real} \times t} \tag{9.14}$$

and hence

$$P_1(t) \times dt = R_{real} \times e^{-R_{real} \times t} \times dt \tag{9.15}$$

Assume now that R_{obs} is the count rate of time intervals betwen real events spaced more than t_d apart. The probability for intervals longer than t_d, $P_2(\Delta t > t_d)$, are given by integration of:

$$P_2(\Delta t > t_d) = \int_{t_d}^{\infty} P_1(r) \times dt = e^{-R_{real} \times t_d} \tag{9.16}$$

The rate of such events is then given by multiplying with R_{real} yielding:

$$R_{obs} = R_{real}\, e^{-R_{real} t_d} \tag{9.17}$$

At very low count rates it can be shown the result is independent of the type of system, i.e. $R_{obs} \approx R_{real}\ (1 - R_{real}\ t_d)$. However, the behavior of these two system types at high count rates are different. A nonparalyzable system shows an asymptotic approach to a maximum count rate with increasing source strength whereas the count rate on a paralyzable system passes through a maximum and then decreases again due to the exponential term in Eqn (9.17). Hence each reading on a paralyzable system corresponds to one of two values, one low and one high. Dangerous mistakes can occur by misinterpreting the reading from a paralyzable dose rate meter.

Most pulse height analyzer systems (MCAs) contain a dead-time metering circuitry. This is especially needed for MCA:s with varying dead-time (see section 9.8.5).

The simplest technique for measuring the resolving time t_r of a nonparalyzable counting system uses a method of matched samples. Two samples of similar counting rates are counted separately and then together. The combined sources should give about 20% fractional dead time, $R_{obs}\, t_r$. From the difference between the measured count rate of the pair together and the sum of their individual rates, the resolving time is calculated using the equations:

$$x = R_a R_b - R_{ab} R_0 \tag{9.18}$$

$$y = R_a R_b (R_{ab} + R_0) - R_{ab} R_0 (R_a + R_b) \tag{9.19}$$

$$z = y (R_a + R_b - R_{ab} - R_0)/x^2 \tag{9.20}$$

$$t_r = x\left[1 - (1 - z)^{1/2}\right]/y \tag{9.21}$$

where R_a, R_b, and R_{ab}, are the measured count rate of samples a and b separately, a and b together, respectively. R_0 is the background count rate for the system. The correction for the dead-time can then be made according to Eqn (9.12).

A more accurate technique is based on the use of a short-lived radionuclide, e.g. ^{99m}Tc or ^{116m}In. The count rate is then measured a number of times during at least one half-life with the source left untouched in position all the time. When the background count rate can be neglected (which is usually the case) combining Eqn (9.12) with the equation for radioactive decay gives after some algebra

$$R_{obs}\, e^{\lambda t} = R_0 - R_0\, t_d\, R_{obs} \tag{9.22}$$

where t is the time of measurement, λ the decay constant and R_0 the initial count rate. A plot of $R_{obs}\, e^{\lambda t}$ as function of R_{obs} should yield a straight line with slope $-R_0\, t_d$ and intercept R_0 on the vertical axis. The dead time is then obtained as the absolute value of slope divided by intercept. If the plotted data deviates strongly from a straight line it indicates that the system investigated is paralyzable.

A peculiar property of nuclear decay is that events do not occur evenly distributed in time but as bursts of various amount of particles within short time intervals. The reason for this is found in the independence of each individual decay and is a side effect of the law of radioactive decay as derived in section 5.11. Accordingly the most probable time interval between any two decays is in fact zero and independent of the actual average decay rate! This leads to the requirement that nuclear pulse handling equipment must be able to separate events very close in time, i.e. have an extremely high band width.

9.3. GAS COUNTERS

All gas-filled counters are in principle ion chambers (with the exception of the less common gas scintillation counters). The ionization produced in an ion chamber by a single nuclear particle produces too low a charge pulse to be easily detectable except for α- particles. However, an ion chamber can be designed so that the number of ion pairs formed in each event is multiplied greatly.

Consider an ion chamber with a hollow cylindrical cathode and a thin central wire as an anode (Fig. 9.9(a) shows an old type GM-tube). The ion pairs formed in the gas by the passage of the ionizing radiation are separated from each other by their attraction to the electrodes. The small, very mobile electrons are rapidly collected on the anode, which is maintained at a high positive potential above ground, $\geq$1000 V. Most of the voltage pulse which appears on the anode arises by induction by the positive ions as they move away from the immediate region of the anode. This step is responsible for the rise time of the pulse, curve I in Figure 9.3(b). The potential decrease is only momentary as the anode is rapidly recharged by the power supply. The time necessary to restore the original potential is a function of the decrease in electric field near the anode due to the build-up of a layer of slow-moving positive ions, and of the time constant circuitry (curve II in Fig. 9.3(b)).

The electric field strength at a distance x from the anode is proportional to $1/x$. If the applied voltage of a cylindrical chamber of 1 cm radius is 1000 V, the potential in the immediate vicinity of a center wire of 0.0025 cm diameter is approximately 7×10^4 V cm^{-1}. As the primary electrons reach the vicinity of this high field and increase their kinetic energy, they cause secondary ionization which increases the pulse detected at the wire anode. The collected charge is given by Eqn (9.7), where a, *the gas multiplication factor*, is $\gg$1.

This gas multiplication factor varies with the applied voltage as illustrated in Figure 9.8. In region II of the curve a flat plateau over a relatively wide voltage range is observed. Prior to the attainment of the threshold voltage for the plateau, the ions would not have sufficient *drift velocity* to prevent elimination of some ion pairs by recombination. Throughout the plateau range the drift velocity is sufficient to make recombination negligible and since secondary ionization is not present, $a = 1$. Ionization chambers operate in this voltage region. In region III the electrons from the primary ion pairs receive sufficient acceleration to produce additional ionization and the process of gas multiplication ($a > 1$) increases the number of collected charges. This is the region of *proportional counter operation* as the pulses are proportional in size to the energy deposited in the detector by the passage of the initial radiation. Region IV is the one used for *Geiger-Müller counter* (or GM counter) operation. In this region the gas multiplication is very high ($\geq 10^6$) and the pulse size is completely independent of the initial ionization. Beyond region IV continuous discharge in the detector occurs.

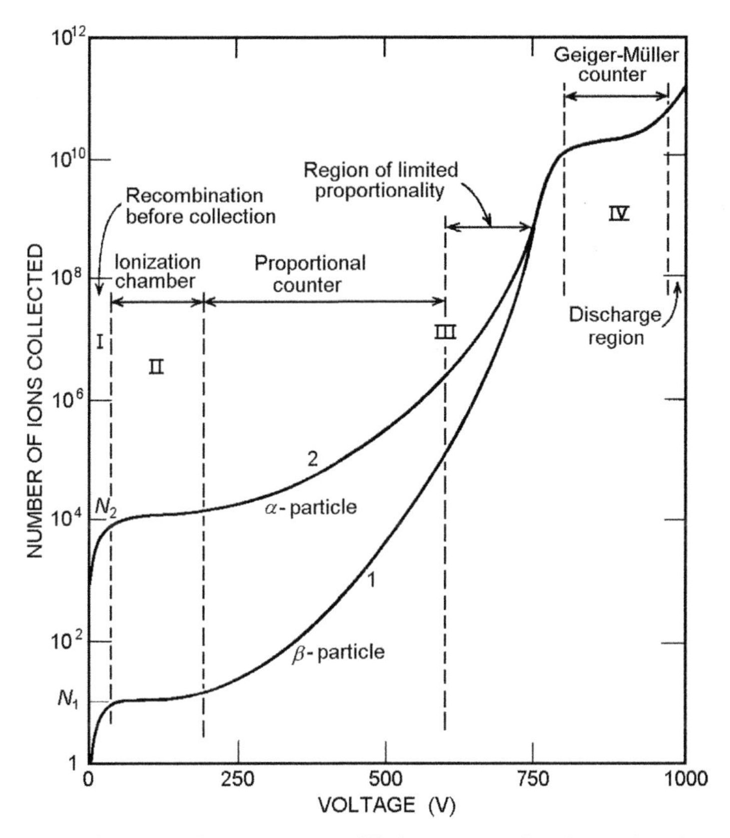

Figure 9.8 Number of ion pairs formed in a gas-filled ionization chamber with a thin wire anode as function of anode voltage. (Acc. to C. G. Montgomery and D. D. Montgomery.)

9.3.1. Ion chambers

The ion chamber is a gas-filled space between two electrodes. In Figure 9.10(a) the electrodes are two parallel plates, but another common geometrical arrangement uses the cathode as a hollow cylinder and the anode as a thin wire in its center, e.g. the GM tube in Figure 9.9(a). In other chambers the chamber walls serve as the cathode with a thin wire loop as anode as illustrated in Figure 9.9(b). The chamber may be designed for recording radiation reaching it from the outside, or it may be used for measuring radioactive samples placed within it. Some chambers have additional electrodes, usually a thin central grid to improve measuring conditions. The anode is kept at a positive potential 100–1000 V above the cathode.

Ions and electrons formed in the gas by nuclear radiation move towards the electrodes where they are discharged. If the gas is pure argon, only Ar^+ and e^- are formed. The electrons move rapidly towards the anode, and then through the electrical circuitry, as shown in Figure 9.10(b), over the resistor R towards the cathode where they neutralize

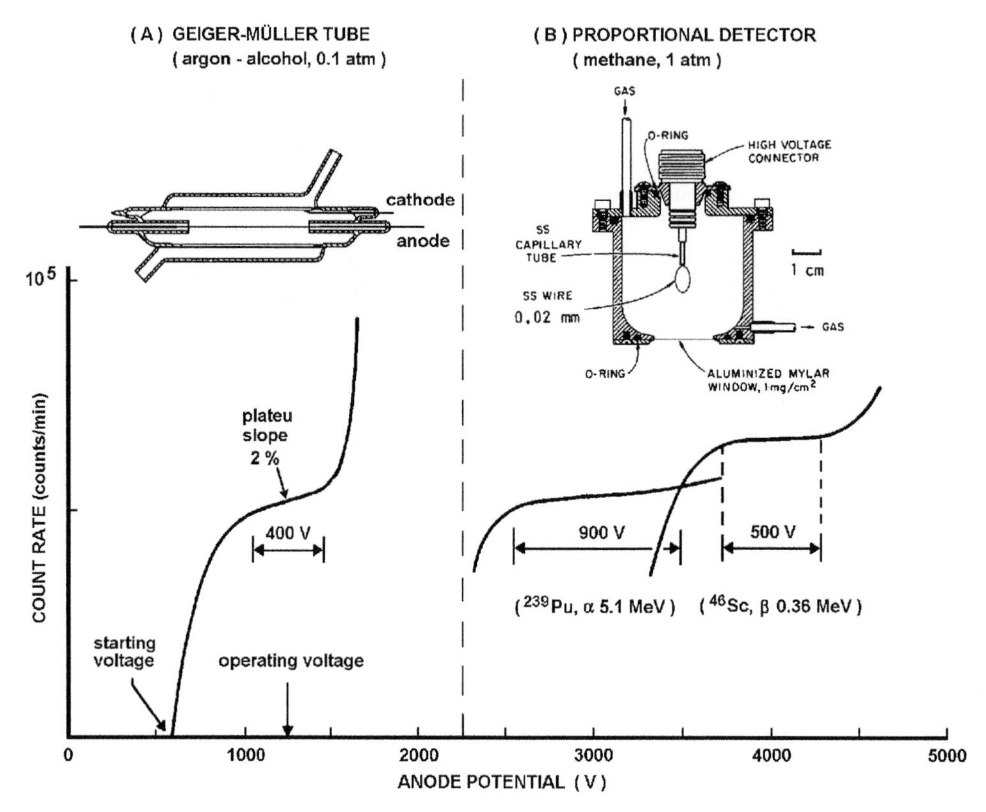

Figure 9.9 Characteristics for GM and proportional counter tubes. The GM tube (a) is designed for flowing liquids. The proportional tube (b) uses a flowing counting gas.

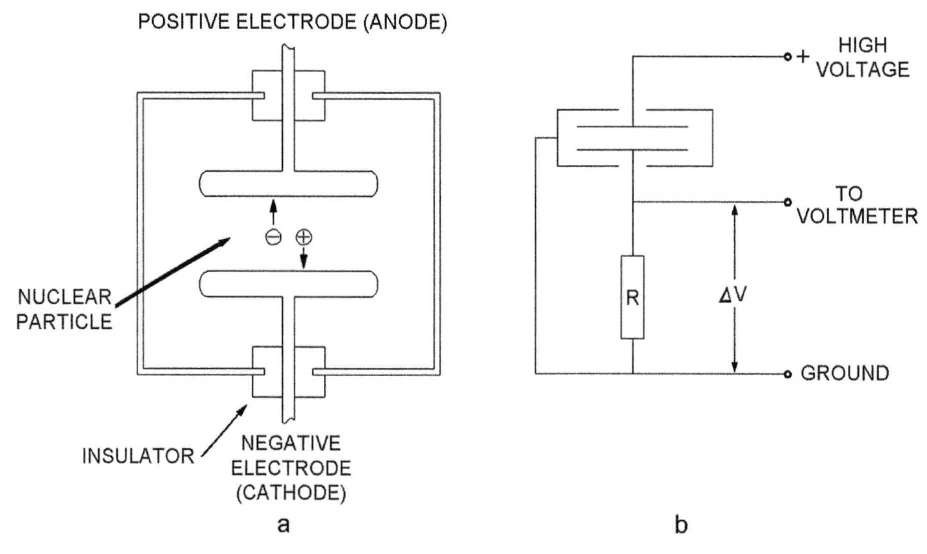

Figure 9.10 (a) A parallel plate ionization chamber and (b) its measuring circuitry.

the argon ions: $Ar^+ + e^- \rightarrow Ar$. The gas is therefore not used up. The current i through the chamber and through the resistor R causes a voltage drop V,

$$V = R\,i \tag{9.23}$$

which can be continuously recorded by a sensitive voltmeter. If the current or voltage drop, ΔV, is measured as a function of the voltage V applied over the electrodes, it is found that the current (or ΔV) increases with V up to a saturation value. The reason for this is that at low voltages some of the positive ions formed initially by the radiation recombine with the electrons, reducing the collected charge. With higher voltages the cations and electrons separate more rapidly with less recombination, and at the saturation value essentially no recombination occurs. The ionization chamber will always be operated at saturation voltage.

Suppose a 100 kBq α-sample is placed within an argon-filled chamber of sufficient size that all the 5 MeV α's emitted are stopped in the gas volume. The saturation ion current will be found to be:

$$i = 1.60 \times 10^{-19} A\, E_{\text{loss}}\, \eta\, \psi_{\text{geom}}\, w^{-1} \tag{9.24}$$

where A is the radioactivity of the sample in Bq, E_{loss} is the total energy lost per particle (5×10^6 eV) to the detector, η is the collection efficiency, ψ_{geom} is the geometric efficiency (for a thin solid sample we shall assume it to be half of a full sphere, i.e. 0.50), w is the energy required for the formation of an ion pair in the gas (for argon 26 eV), while the constant is the charge (Coulomb) of a single ion (the ion pairs must be regarded as single charges): thus $i = 1.54 \times 10^{-9}$ A. With a resistor of $10^9\ \Omega$, the voltage drop ΔV is 1.5 V.

Two types of ion chamber are common: (i) Simple, portable, rugged instruments with resistors $\leq 10^{13}\ \Omega$. With these, radiation intensities of $\geq 10^3\ \beta\ s^{-1}$ and of $\geq 10^5\ \gamma\ s^{-1}$ can be measured. They are usually calibrated in dose rate (e.g. Gy h^{-1}) and used for radiation protection measurements under field conditions (*dose rate meters*). (ii) Advanced, very sensitive instruments, with very high resistors ($\sim 10^{15}\ \Omega$) or special circuitry as in the vibrating reed electrometer. The chamber must always be designed with extreme care to avoid leaking currents from the anode over the chamber casing to the cathode. One way to minimize this is to ground the casing, as shown in Figure 9.10(b). In the best of these instruments, currents as low as 10^{-18} A can be measured, corresponding to less than 1 $\alpha\ h^{-1}$, 10 $\beta\ min^{-1}$, or 10 $\gamma\ s^{-1}$. These instruments are best suited for measurement of radioactive gases, like tritium or radon in nature.

The *beta-current neutron detector* is a solid state ion chamber which is used in nuclear reactor technology. It consists of an emitter in which a nuclear reaction occurs, leading to the emission of primary β^- particles (e.g. through the reaction ($^{103}Rh(n,\gamma)^{104}Rh(\beta^-$, 4.2 s) ^{104}Pd) or secondary electrons (e.g. through absorption of the prompt γ's emitted in the neutron capture). These electrons represent a current and are collected by an anode. The radioactive decay type detectors have a response time depending on the product half-life, which the capture-γ detectors lack. These detectors have a limited lifetime; for the $^{59}Co(n,\gamma)^{60}Co$ it amounts to 0.1% per month at 10^{13} n cm^{-2}s^{-1}. The lifetime depends on the $\sigma_{n\gamma}$ value (37 b for ^{59}Co, 146 b for ^{103}Rh).

9.3.2. Proportional counters

Values of a in Eqn (9.5) of 10^3–10^5 are commonly achieved in proportional counter operation. If $a = 10^3$, essentially all the gas multiplication occurs within 10 mean free path lengths from the wire for the electron in the gas (2^{10}=1024). At 1 atm the mean free path length is approximately 10^{-6} m, which means that the gas multiplication occurs within 0.01 mm of the wire.

The gas multiplication factor varies with the applied voltage but for a given voltage a is constant so the detector pulse output is directly proportional to the primary ionization. As a result it is possible to use a proportional counter to distinguish between α- and β-particles and between identical particles of different energies in as much as different amounts of primary ionization are produced in these cases.

The output pulse in proportional counter operation is not dependent on the collection of the positive ions by the cathode. Consequently the rate of detection depends on the time necessary for the primary electrons to drift into the region of high field strength near the anode wire. As a result, proportional counters have a much shorter resolving time than ion chambers which depend on the slow-moving positive ions. In fact the detector tube in a proportional counter can amplify a new pulse before the positive ion cloud of the previous pulse has moved very far if the new ionization occurs at a different location on the center wire. Time intervals necessary to enable the counter to measure two distinct pulses can be as low as 0.2–0.5 μs. Frequently the associated measuring equipment is a greater determinant of the resolving time than the detector itself. If, however, a proportional counter is being used for the measurement of particle energies, any residual positive ion cloud must have time to drift an appreciable distance before a new pulse is generated. In this case the resolving time is closer to 100 μs.

Counting gases consists usually of one of the noble gases mixed with a small amount of polyatomic gas. The latter makes the gas multiplication factor less dependent on applied voltage, and increases the speed of electron collection. Typical counting gas mixtures are 90% Ar + 10% CH_4 ("P-gas") and 96% He + 4% i-C_4H_{10} ("Q-gas"). Many other pure gases or combinations are possible, but molecules which readily attach electrons must be avoided. Figure 9.9(b) shows a proportional counter tube using pure CH_4 and the tube characteristics (i.e. count rate versus voltage) for ^{239}Pu α-particles and for ^{46}Sc β-particles entering through the thin aluminized mylar window.

The gas mixture and electrodes may be separated by a thin window from the radioactive sample, or the counter may be operated windowless. For windowless operation, after insertion of the sample the chamber must be flushed with gas to eliminate all oxygen and water vapour as these molecules absorb electrons readily to form negative ions and, by

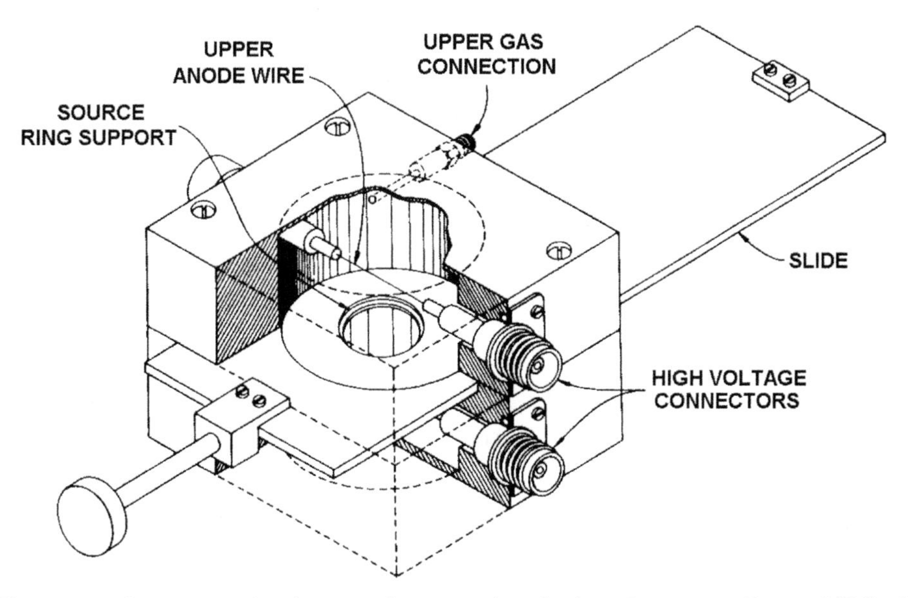

Figure 9.11 A 4π proportional counter for measuring absolute decay rates. (Acc. to O'Kelley.)

so doing reduce the pulse size. For α- and β-particles whose ranges do not exceed the dimensions of the chamber, windowless counters are often referred to as 2π counters since a solid angle of 2π is subtended above the sample. In these cases, with proper care, the measured count rate is very close to 50% of the true disintegration rate. Such windowless proportional counters are very useful for measuring low energy radiation such as the β-emissions of ^{14}C and ^{3}H, and for absolute counting. Figure 9.11 shows a 4π proportional counter in which the sample is inserted in the middle between the two half-domes (cathodes) giving near 100% geometric efficiency.

Proportional counters can be used also for *neutron detection* by using a gas containing nuclei that capture neutrons and in the nuclear reaction produce charged particles, e.g. ^{10}B or ^{3}He. In BF_3-counter the gas is BF_3, usually enriched in ^{10}B. With neutrons the reaction

$$^{10}B + n \rightarrow {}^{7}Li + {}^{4}He \quad (Q = 2.78 \text{ MeV}) \tag{9.25}$$

occurs (cf. Figure 11.4). The ionization of the two products produces a heavy pulse, which is easy to discriminate against an intense γ-background. Some properties of a BF_3 neutron detector are given in Table 9.2. From Figure 11.4 it is clear that the BF_3 counter has a higher efficiency for thermal than for fast neutrons.

Another technique for neutron detection uses a *fission chamber.* One design contains a stack of alternate anodes and cathodes, one of the electrodes being covered by a thin layer of uranium enriched in ^{235}U. The fission fragments produce large ionization even

Table 9.2 Properties of some representative counter tubes

	Geiger Müller counters		Proportional counters	
Purpose	β,γ	α,β	n	α,β
Wall thickness	0.1mm glass	1.5 mg/cm^2 mica	1 mm steel	0.3 mg/cm^2 foil
Filling gas	Ne+Ar+halogen	Ar+organic	^{10}B enriched BF_3	Pure CH_4
Operating voltage	700 V	1250 V	2200 V	α 1900 V, β 2600 V
Plateau length	>250 V	~300 V	>300 V	α > 800 V, β~400V
Plateau slope	8%/100 V	<4%/100 V	<2%/100 V	<2%/100 V
Lifetime	>3 × 10^{9}counts	3 × 10^{8}counts		Infinite
Dead time	140 s	300 s		3 s
Background count rate	20–30 cpm	80 cpm	1–2 cpm	α 0.1 cpm β 20–25 cpm
Background shielding	50 cm Pb	10 cm Pb		50 cm Pb
Special features	Insensitive to overvoltage		Insensitive to γ	

though the gas multiplication is quite low. This detector is more sensitive to fast neutrons than the BF_3 counter, and can be used for fast neutron fluxes up to ~10^{10} n m^{-2} s^{-1} with a background of a few cps.

9.3.3. Geiger-Müller counters

In region IV (Fig. 9.8), the proportionality between the primary ionization and the output pulse disappears and the latter becomes the same size for all initial ionization whether it be a 6 MeV α-particle or a 50 keV X-ray. Geiger-Müller (GM) counters which operate in this region have high sensitivity to all different kinds of radiation and the large size of the output pulse (from a tenth of a volt to one volt, compared to the several tenths of a millivolt output of ionization chambers) requires much less external amplification. This considerably reduces the complexity of the auxiliary electronic equipment. The detector tubes for GM counters are quite simple and allow a great deal of flexibility in design. Figure 9.9(a) shows a GM tube with a jacket for flowing liquids; Table 9.2 gives the properties of some other typical GM tubes. In general GM counters are limited to handling lower count rates than proportional counters.

As in the case of proportional counters, the primary electrons from the ionizing radiation cause secondary ionization near the center anode wire in GM detectors. This initial avalanche ends when the very mobile electrons are all collected by the anode. However, the neutralization of the electrons at the wire produces photons, which react with the gas leading to the emission of photoelectrons. These trigger further avalanches and an overall avalanche spreads along the complete length of the center wire and continues until the build-up of the positive ion sheath progresses to a point sufficient to reduce the field strength sufficiently to prevent further ionization. This build-up takes place because the heavy positive ions have such a slow rate of movement that they are essentially stationary during the time interval of the electron avalanches. The time required to reach this point in the process is of the order of a few microseconds. Since the dead-time is of the order of 100 μs due to the slow movement of the heavy cations to the cathode, it is necessary to make dead-time corrections in GM counting at much lower count rates than in the case of proportional counting. Typically, corrections are significant for count rates exceeding 10 000 cpm.

When the positive ion sheath reaches near the cathode it may induce new avalanches as it collides with atoms in the gas and produces UV-light or emission of electrons. To avoid a recurring pulsing, which would render the counter useless, it is necessary to prevent further avalanches at this point by a process of quenching. This is usually accomplished by the addition of a small amount of an organic compound such as ethyl alcohol or ethyl formate or a halogen to the counting gas. Since the ionization potential of the organic molecule is lower than that of argon, the usual counting gas, when the positive argon sheath moves to the cathode and encounters organic molecules the following reaction occurs:

$$Ar^{+} + C_2H_5OH \rightarrow Ar + C_2H_5OH^{+} \tag{9.26}$$

In this reaction the charge of the argon ion is transferred to the organic molecule which gains an electron upon striking the cathode. The energy acquired in the neutralization causes dissociation into uncharged fragments rather than producing photon or electron emission. In as much as the quenching gas is dissociated in the process of counting, such GM tubes have limited lifetimes which usually amount to approximately 10^9 discharges.

Chlorine and bromine have strong absorption bands below about 251 nm for the photons emitted, leading to dissociation. In the recombination the halogen molecule returns to its ground state via a series of low energy excited states. This makes halogens useful as quenchers in GM tubes. Halogen-filled GM tubes are popular because of their infinite lifetime and low operating voltage. Another advantage is that they will not be damaged by wrong polarity or excess voltage, as is the case with the organic quenched tubes.

GM tubes are available in a wide variety of shapes and sizes. Tubes have been used successfully which varied from approximately 1 mm to several centimeters in diameter and from 1 cm to almost a meter in length. The cathode can be made by coating the inside of a glass cylinder with a conducting material such as metal or graphite while the anode may be a tungsten wire mounted coaxially. The "end-window" GM tube has a thin mica window and the center wire is terminated with a glass bead. The cathode is always at ground potential while the anode is at a high positive potential.

When the radiation intensity is measured as a function of the electrode potential it is found for GM counters (as for proportional counters) that over a certain voltage interval there is little, if any, change in the measured count rate. This is known as the *plateau region* (Fig. 9.9). This is the voltage range in which the detectors are operated since they are relatively insensitive to small voltage changes in this region. For organic quenched GM tubes the plateau commonly occurs between 1200 and 1500 V and its slope should not exceed 5% per 100 V. Halogen-quenched tubes operate at a lower voltage but have higher plateau slopes (cf. Table 9.2).

9.4. SEMICONDUCTOR DETECTORS

The semiconductor detector is similar to an ordinary semiconductor diode composed of p-type and n-type semiconductor material. This detector has become dominant for nuclear spectroscopy (i.e. determination of the energy of nuclear radiation) but it is not so often used for simple measurement of count rates.

Semiconductors are materials like silicon (resistivity ~10^3 Ωm) and germanium (0.6 Ωm) with resistivities between metals (e.g. copper, 10^{-8} Ωm) and insulators (e.g. quartz, 10^{12} Ωm). A crystal of pure silicon placed between two electrodes is almost nonconducting. The electrons in the material are almost all *valence electrons*, bound to specific silicon atoms with an energy of 1.115 eV at room temperature (0.75 eV for Ge at 80 K). If 1.115 eV is given to an electron in silicon, it moves to a band of overlapping energy levels which are not associated with specific atoms. The electron moves readily through the crystal in this "conduction band", i.e. the crystal conducts electricity. At a certain temperature some electrons, according to the Maxwell energy distribution, always have the necessary energy to be in the conduction band. These electrons provide a very small conductivity for pure silicon; this is referred to as the *intrinsic conductivity.* For diamond, the gap between the valence and conducting band is 7 eV, which is so large that essentially no electrons are found in the conduction band at room temperature, and thus diamond is an insulator.

The energy needed for transferring valence electrons to the conduction band can be supplied by nuclear radiation. The average energy needed to produce an electron-hole pair in silicon at room temperature is not 1.115 eV but 3.62 eV because some energy is lost as crystal excitation (at 80 K the exitation energy is 3.72 eV in Si and 2.95 eV in Ge). The electron removed from the valence band leaves a vacancy or "hole". The ionization is said to give an *electron-hole pair.* Just as the electrons move towards the anode, the holes move towards the anode.

Si has 4 valence electrons while P has 5 and In 3. If we introduce a very small amount of phosphorus into silicon, the phosphorus atoms substitute for silicon in the crystal lattice. Each such phosphorus has an excess of 1 electron. These electrons are not free but are very weakly bound, such that only 0.04 eV is needed to transfer them into the conduction bands. Because phosphorus donates extra electrons to the system, it is referred to as a *donor material*. Silicon which contains small amounts of donor material (usually referred to as "impurity") is called *n-type silicon* since it has excess negative charge and conduction is by electrons.

If instead, indium is the impurity in the silicon crystal structure, the opposite effect is produced. Such material contains a number of energy levels only 0.06 eV above the valence band; the result is holes in the valence bands. Such material is referred to as *acceptor material*. Silicon with acceptor material is called *p-type silicon*, since the holes are considered to be positively charged. Conduction is in this case by movement of holes. The addition of controlled amounts of impurity atoms thus provide *charge carriers* (as the electrons and holes collectively are called, c.f. section 9.2) and produces the desired properties in semiconductor materials.

The most interesting effect comes from the combination of two types of silicon, one n-type and another p-type (Fig. 9.12(b)). The contact surface is referred to as a *p-n junction*. At such a junction some positive holes move to the n-type material, and vice versa. As a result a "depletion" layer a few microns thick is established at the junction where all the holes are filled with electrons and the layer is depleted of charge carriers. A p-n junction can be produced in a single piece of silicon by doping it with the proper impurity from either side of the crystal and by other techniques like ion implantation.

If a voltage is applied over the junction by connecting the negative terminal to the p-type region and the positive terminal to the n-type region, the junction is said to be *reversely biased*. With such reverse bias, the barrier height and depletion layer thickness increase. As a result the crystal opposes any current as the resistivity becomes very high. In the reverse direction the semiconductor represents a high resistance shunted by a capacitive component (Fig. 9.3(a)) due to the dielectric of the barrier layer (the p-n junction diode).

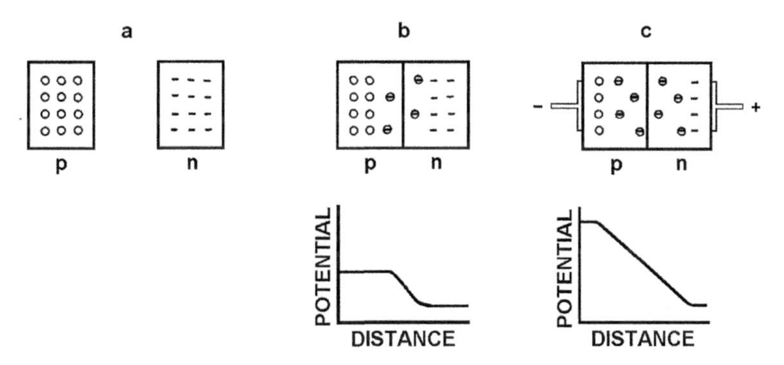

Figure 9.12 Formation and properties of a p-n junction in a semiconductor.

A number or variations of this basic design exist. The point of importance for semiconductors as nuclear detectors is that a depleted layer with a high space charge is formed. A nuclear particle entering this volume forms electron-hole pairs, which are rapidly and efficiently collected at the electrodes due to the high potential gradient. By this a charge is transported through the crystal while the original conditions are restored.

There are some formal similarities between the function of a semiconductor detector and a parallel plate ion chamber. In comparison with ionization chambers the semiconductor detector (i) requires only 3.62 eV (Si, 300 K) or 2.95 eV (Ge, 80 K) for an electron-hole pair (as compared to about 15−35 eV for an ion pair in a gas), (ii) collects the charge much faster (no slow positive ions), (iii) has a much higher stopping power, (iv) but does not normally have the property of charge multiplication. The charge through the semiconductor detector is given by Eqn (9.2) where η (the collection efficiency) is usually very close to 1, and w is 3.62 eV for silicon at 300 K and 2.95 eV for germanium at 80 K. It is seen that the signal is directly proportional to the energy absorbed in the detector as long as η is constant. Because w is much smaller in semiconductor detectors than in gas and scintillation detectors, more primary electrons are released in each absorption event, which gives better "statistics" (see section 9.11) and higher energy resolution. This makes semiconductor detectors useful for nuclear spectroscopy, although for α- and β-spectroscopy the precision is higher with magnetic deflecting devices (cf. section 3.3).

9.4.1. Surface barrier detectors

The surface barrier detector is a p-n type silicon diode wafer characterized by a rather thin depletion layer (Fig. 9.13(a)). It is made of n-type silicon on which one surface has been etched prior to coating with a thin layer of gold (typically ~40 μg/cm^2) and the other surface coated with a thin layer of aluminum (typically ~40 μg/cm^2) to provide electrical contact. This results in a window-layer which is equivalent to ~800 Å of Si. Depending on the applied voltage, the detector can be partially depleted (inactive entrance layer), totally depleted (no inactive layer), or over-depleted (higher applied potential than required for total depletion). Surface barrier detectors are used mainly for α- and β-spectroscopy and for dE/dx and E measurements for high energy particles, although the efficiency is limited by the sensitive surface diameter (≤10 cm) and the energy range by depleted layer thickness (≤5 mm).

The radiation sensitive depleted layer is available in various thicknesses, ≤ 5 mm, enough to stop electrons of ≤2.2 MeV, p of ≤32 MeV, and α of ≤120 MeV. A typical silicon surface barrier detector for α-spectroscopy has a sensitive area of 300 mm^2, 300 μm depletion depth, 20 keV FWHM (full width at half maximum) and operates at 100 V reverse bias. The resolving time is about 10^{-8} s. Special "rugged" detectors are

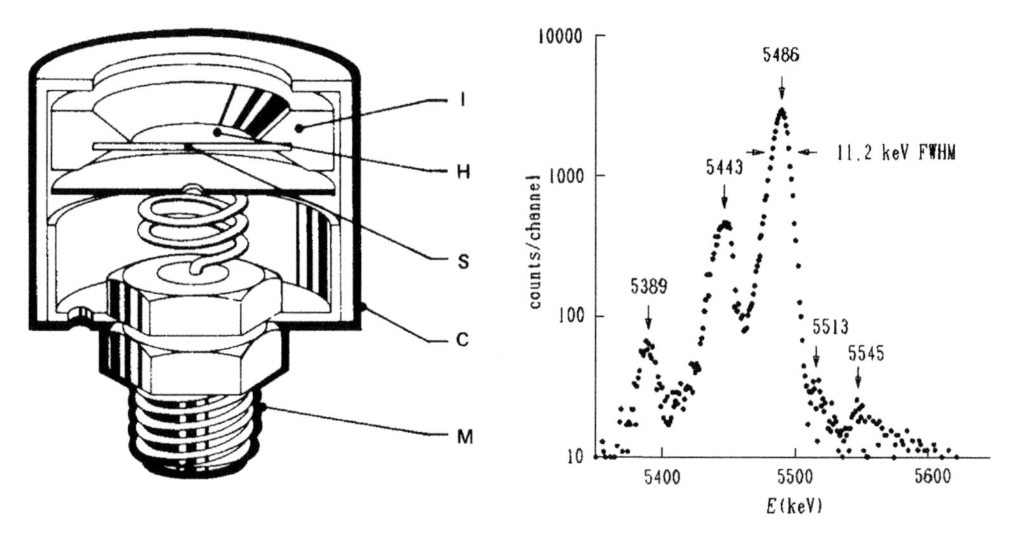

Figure 9.13 (a) Surface barrier detector; S detector wafer, H gold surface layer, M connector. (b) α-spectrum of ^{241}Am measured with a high resolution detector (Acc. to ORTEC.)

available which have an acid resistant SiO_2 surface layer permitting cleaning and contact with liquids. Detailed information for detector selection is available from various detector manufacturers.

When used for α- or β-spectroscopy, a vacuum is applied between the detector and the radiation source. In the absence of a vacuum for α-radiation the energy loss is about 1 keV per 0.001 atm per cm distance between source and detector. The absorption in the detector window for a 6 MeV α is less than 6 keV. A resolution of about 12 keV FWHM can be obtained for a 6 MeV α, Figure 9.13.

In totally depleted silicon surface barrier detectors the sensitive region extends through the whole thickness of the silicon, which may be in the form of a very thin slice (e.g. 20 μm) within two very thin conducting surface layers. A particle passing through such a detector loses a small fraction of its energy $\mathrm{d}E/\mathrm{d}x$ and may then be completely stopped in a second (much thicker depleted layer) detector to lose the remainder of its energy, which may essentially be its original total E_{kin}. Particle mass A and charge Z can be determined from $\mathrm{d}E/\mathrm{d}x$ and E, e.g. with the aid of the proportionality

$$E\,\mathrm{d}E/\mathrm{d}x \propto A\,Z^2 \tag{9.27}$$

Figure 9.14 shows a hypothetical distribution when the recorded intensities of ^1H, ^2H, and ^3H are plotted against $E\,\mathrm{d}E/\mathrm{d}x$.

As an alternative to X-ray film large plates containing a matrix of surface barrier detectors on their surface are in use. The individual detectors are pre-charged and can after exposure be individually addressed for electric readout of remaining charge in a

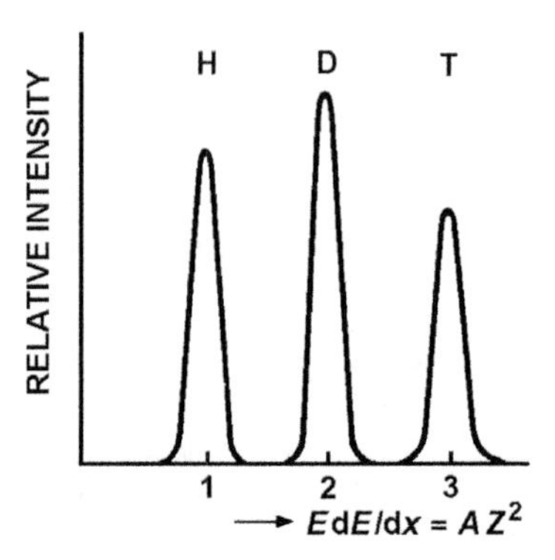

Figure 9.14 Plot of relative particle intensities for H, D, and T versus *EdE/dx* in a two-detector telescope, where the first detector records d*E*/d*x* and the second *E*.

similar way as optical MOS-sensors in digital cameras. Such plates are mainly used for X-ray imaging but also for autoradiography.

High energy particles not only cause ionization in the detector crystal but may displace some detector atoms from the crystal lattice. Radiation damage decreases with applied bias and increases with the particle mass. Such radiation damage to the crystals limits the lifetime of the detectors. The threshold dose (in particles/cm^2) is about 10^8 for fission fragments, 10^9 for α, 10^{10} for p^+, 10^{12} for fast neutrons, and 10^{13} for e^-. Radiation damage can usually be removed if the detector can be annealed at 200°C.

9.4.2. Lithium-drifted detectors

The probability of γ-interaction is so small in the small depletion depth of the surface barrier detectors that they are not very useful for γ-spectroscopy. Large depleted volumes can be created by drifting lithium atoms into a silicon or germanium crystal. Lithium does not occupy a crystal site in the crystal, but is small enough to go into interstitial sites. The ease of ionization of Li to Li^+ makes it a donor impurity. The lithium is drifted from one side of the crystal using an electric field. Its concentration at the "entrance" side becomes high and then decreases towards the other end of the crystal. The amount of lithium in the crystal adjusts itself during the drifting process so that the lithium atoms compensate the impurities. This is, however, not a stable situation when the drifting operation is stopped. Hence, the final state is "frozen" by cooling of the drifted crystal to liquid nitrogen temperature. Accidental heat-up will destroy the lithium compensation and the detector must be redrifted.

When a potential is applied over such a crystal, with the positive terminal at the high lithium side, three volumes are created, one of p-type, a middle "intrinsic" one, and an n-type one (p-i-n detectors). In the intrinsic volume the lithium donor electrons neutralize any original impurities, which are of acceptor p-type. The intrinsic volume

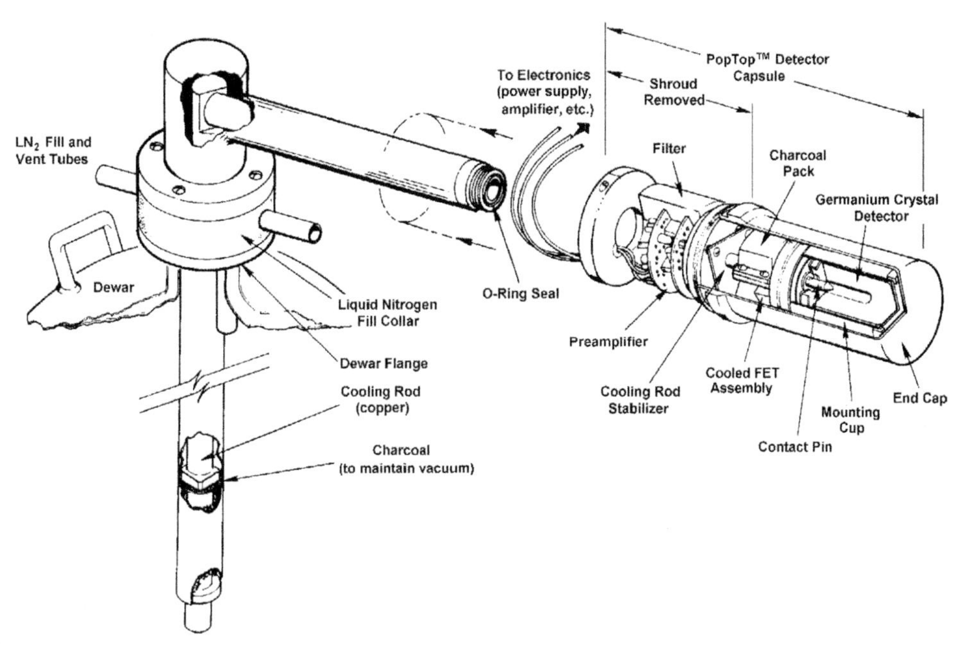

Figure 9.15 Cut-away view of a Ge-detector showing Dewar, cold-finger, preamplifier and germanium crystal. (Acc. to ORTEC.)

becomes depleted and thereby sensitive to nuclear radiation, and detectors with depleted volumes up to more than 100 cm^3 are commercially available. Figure 9.15 shows the arrangement of the Dewar vessel with liquid N_2, cold-finger, detector, and preamplifier.

Lithium-drifted detectors are made either from silicon (Si(Li) detectors) or from germanium (Ge(Li) detectors). The latter has a higher atomic number and density than silicon and is therefore preferable for γ-spectrometry. For 60 keV X-rays, the efficiency of a Si(Li) detector may be 5%, while for a comparable Ge(Li) detector it may be 100%. At lower energy the Si(Li) detector is preferable, especially if the measurements are carried out in a high γ-background. Si(Li) detectors are of particular importance in X-ray fluorescence analysis (cf. section 7.8.4).

Both types give excellent resolution, for a good Si(Li) X-ray detector about 160 eV at 5.9 keV and for a good Ge(Li) detector about 1.75 keV FWHM for 1.33 MeV γ. Energy resolution of Ge detectors varies with energy below 1.5 MeV according to the empirical eqn. FWHM $\approx (n^2 + kE_\gamma)^{1/2}$, where FWHM is the full peak width at half maximum, n the noise line width, $k = 2$, and E the γ-energy, all in eV units. This resolution is far superior to that of the scintillation detector, as is seen from Figure 9.16. On the other hand, the detection efficiency is usually smaller. However, an efficiency up to about 200% of that obtained with a $3' \times 3'$ NaI(Tl) scintillation detector for 1.33 MeV γ-rays (which is the reference for Ge-detector efficiency) has been obtained in the best designs, but typical values are 10–100%.

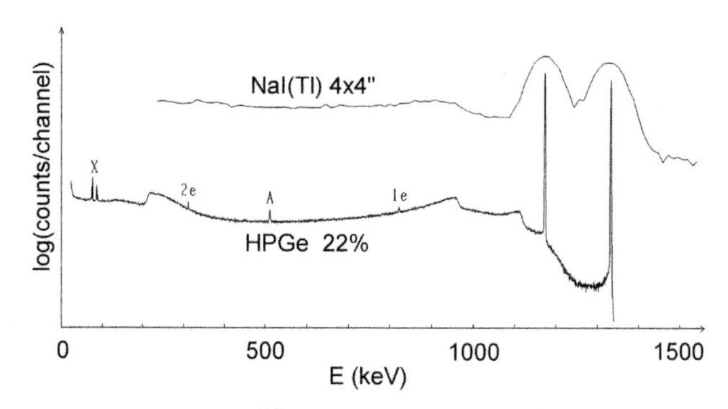

Figure 9.16 Energy spectrum of ^{60}Co obtained with scintillation and HPGe detectors.

9.4.3. Intrinsic detectors

One reason for drifting lithium into silicon and germanium is the necessity to compensate some p-type (acceptor) impurities normally present in pure materials. Detectors without lithium drifting are called *intrinsic detectors*. It is possible to increase the purity of germanium to 1:10^{13} (compared to earlier 1:10^{11}) which makes lithium-drifting unnecessary. Such material is called high purity germanium or HPGe. Detectors made of HPGe make uninterrupted cooling less important; it is not required when storing the crystals, but should be used when measuring in order to improve resolution and prevent crystal overheating. Depending on the type of dominating impurity, HPGe crystals can be of either n- or p-type. HPGe-detectors of p-type have usually a lower energy cutoff, beginning at about 100 keV, caused by the detector capsule wall and insensitive entrance layer. Commercially available p-type detectors have relative efficiencies of 10–100%. On the other hand n-type HPGe-detectors can be made with a very thin entrance layer and, when fitted with a Be-window, have a low energy cutoff beginning at about 6 keV. Relative efficiencies for n-type HPGe detectors are in general somewhat less than for p-type detectors, i.e. 10–80%. Intrinsic germanium detectors are to a large extent replacing the lithium-drifted germanium detectors because of their greater handling simplicity at no higher cost or loss of resolution.

9.5. SCINTILLATION DETECTORS

In 1908 Rutherford and Geiger established the reliability of a method of counting α-particles by observing visually the flashes of luminescence[1] produced in a thin layer of ZnS by the α-particles. Since the development of reliable *photomultiplier* tubes (PMT) in 1946, scintillating counting techniques have played an important role in nuclear science. A scintillation detector consists of a *scintillator*, or *phosphor* optically coupled to a PMT

[1] *Luminescence* includes both fluorescence and phosphorescence (sections 7.5 and 7.8).

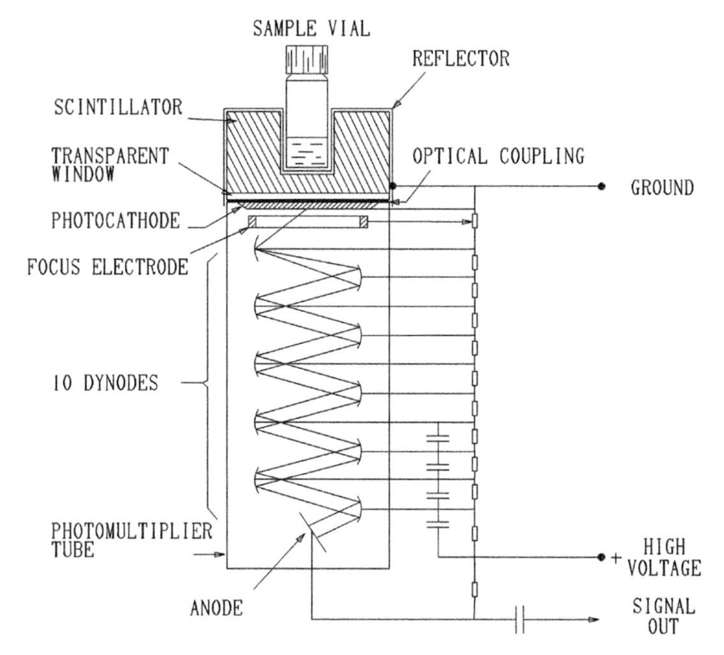

Figure 9.17 Well-type scintillation detector with photomultiplier circuitry.

which produces a pulse of electric current when light is transmitted to the tube from the scintillator (see Fig. 9.17). The scintillating material can be an inorganic crystal or an organic solid, liquid, or gas. Detectors with sandwiched scintillators of different kind, phoswich-detectors, can, with suitable electronics, use the difference in scintillator properties (light output and decay time), thickness and position in the sandwich to differentiate between radiation types and radiation energies permitting e.g. simultaneous separate counting of α- and β,γ-radiation.

In organic substances the absorbtion of energy raises the organic molecule to one of the vibrational levels of an excited electronic state (see sections 8.5 and 8.8). Through lattice vibrations some of the excitation energy is dissipated as heat and the molecule decays to lower vibrational levels of the exited electronic state. After approximately 10^{-8} s, a time sufficient for many molecular vibrations, the molecule may return to the ground electronic state with emission of light photons. Since the energy which excites the molecule is in general larger than that emitted in any single step in the decay back to the ground state, reabsorption of these emitted photons is unlikely, and the crystal is consequently transparent to the emitted photon. This transparency is necessary if the scintillations are to escape the scintillator and reach the PMT. Aromatic hydrocarbons such as anthracene and stilbene which have resonance structures are excellent scintillators. Liquid and solid solutions of such organic substances as p-terphenyl are also used as scintillators. In these systems the energy absorbed through the interaction of radiation with the solvent molecules is transferred rapidly by the latter to the solute which undergoes excitation and fluorescence as described above. The exact mechanism of the transfer of energy from solvent to solute is not fully understood.

It is necessary to have small amounts of impurities in inorganic crystals to have luminescence. In ionic crystals in the ground state all the electrons lie in a lower valence band of energy. Excitation promotes the electrons into a higher conduction band of energy.

If impurities are present they can create energy levels between the valence and conduction bands, as described in section 9.4. Following excitation to the conduction band through absorption of energy an electron may move through the conduction band until it reaches an impurity site. At this point it can "decay" to one of the impurity electron levels. The de-excitation from this level back to the valence band may occur through phosphorescent photon emission. Again, since this photon would have an energy smaller than the difference between the valence and conduction bands, these crystals are transparent to their own radiation.

To be useful as a scintillator a substance must possess certain properties. First, there must be a reasonable probability of absorption of the incident energy. The high density in solid and liquid scintillators meets this condition. Following absorption, emission of luminescence radiation must occur with a high efficiency and — as mentioned — the scintillator must be transparent to its own radiations. Finally, these radiations must have a wavelength that falls within the spectral region to which the PMT is sensitive. Since this is not always the case, particularly with liquid scintillators, "wave-length shifters" are added (e.g. diphenyl-oxazolbenzene (POPOP) to solutions of p-terphenyl in xylene). Further, "quenching" substances which absorb the light emitted from the scintillator should be absent. This is a particular problem in liquid scintillation counting.

Table 9.3 lists the properties of some common scintillators. The data indicate that the greater density of inorganic crystals makes them preferable for γ-ray counting. The resolving time is shorter for the organic systems whether liquid or solid. When large detector volumes are necessary a liquid solution system is the simplest and most economical.

The scintillator must be coupled optically to the PMT so that there is a high efficiency of transfer of the light photons to the PMT photo cathode. Since PMTs are sensitive to light in the visible wavelength region, both scintillator and PMT must be protected from visible light. Figure 9.17 shows a typical combination of a "well-type" crystal phosphor and PMT. The light sensitive photo cathode of the PMT is a semi-transparent layer of a material such as Cs_3Sb which emits electrons when struck by visible light. The emitted photoelectrons are accelerated through a series of 10 to 14 electrodes (*dynodes*) between which a constant voltage difference is maintained. When the photoelectrons strike the nearest dynode, secondary electrons are emitted as the dynodes are

Table 9.3 Properties of some common phosphors

Material	Density ($g\ cm^{-3}$)	Wavelength of maximum emission (nm)	Decay constant for emission (μs)	Relative pulse height
Gaseous				
Xe	n.a.	325	small	9[b]
Kr	n.a.	318	small	5[b]
Inorganic				
NaI(Tl)[a]	3.67	410	0.23	100[b]
CsI(Na)	4.51	420	0.63	85[b]
CsI(Tl)	4.51	565	1.0	45[b]
ZnS(Ag)	4.09	450	0.20	130[b]
CaF_2(Eu)	3.19	435	0.9	50[b]
Organic				
Anthracene	1.25	440	0.032	100[d]
Stilbene	1.16	410	0.006	60[d]
Plastic phosphors	1.06	350–450	0.003–0.005	28–48[d]
Liquid phosphors	0.86	355–450	0.002–0.008	27–49[d]

[a](Tl), (Na), (Ag), and (Eu) indicate small amounts of these elements added as activators.
[b]For α; NaI(Tl) = 100.
[c]For γ; NaI(Tl) = 100.
[d]For β; Anthracene = 100.

also covered with Cs_3Sb. Consequently, there is a multiplication of electrons at each dynode stage and at the last dynode the number of original electrons have been increased by about a factor of 10^6 over a total voltage drop in the photo tube of 1000–2000 V. The electrical signal is normally generated from a voltage change between ground and the anode caused by a resistor between anode and bias supply.

9.5.1. Gas scintillator detectors

Several high purity gases are useful scintillators, notably N_2, He, Ar, Kr and Xe. Except for N_2, much of the emitted light lies in the UV range. Hence, PMTs sensitive to UV must be used or a wave-length shifting gas like N_2 added. The scintillations produced are of very short duration, a few ns or less, which puts them among the fastest of radiation detectors. Gas scintillators have easily variable size, shape and stopping power. The latter by changing the gas pressure. They are often unusually linear over a wide range of particle energy and dE/dx. However, the light yield is, at best, an order of magnitude lower than that of NaI(Tl). This has mostly limited their use to counting of α and other energetic multiply charged particles. Gas-flow detectors are discussed in section 9.8.

9.5.2. Liquid scintillator detectors

Liquid scintillators have a wide use for routine measurement of β-emitters, particularly low-energy ones like ^{3}H and ^{14}C, in liquid samples, especially in biochemistry and in hospitals. Since these isotopes are very important in biochemical applications, most of the development of liquid scintillation technique has been focused on them.

The sample is directly dissolved in the liquid scintillator solution (*scintillator cocktail, a mixture of scintillating molecules and solvent)* and the light output measured by PMTs. Normally two PMTs are used in order to eliminate much of their internal noise by only accepting coincident pulses from both tubes. Liquid scintillation counting offers several advantages when measuring low-energy β-emitters compared to most other detectors. Problems like attenuation by the detector window, self-absorption and backscattering are avoided. However, the introduction of the sample into the scintillator medium often reduces the light output considerably, see Table 9.4 and Fig. 9.18. This effect is called *quenching* and depends on phenomena such as chemical reactions that absorb some of the deposited energy (*chemical quenching*) and changes in optical properties (*colour quenching*). A reduction in light output reduces the efficiency, especially at low β-energies. For low quenched samples the efficiency may approach 100% since the β-particles almost always have to encounter the scintillator. The energy resolution is sufficiently good to differentiate between e.g. ^{3}H ($E_{\beta,max}$ 18 keV) and ^{14}C ($E_{\beta,max}$ 160 keV).

Measurement of α-emitters is also feasible. In this case several MeV of energy is deposited in the scintillation cocktail, usually yielding near 100% detection efficiency. The high amount of energy deposited also reduces sensitivity to quenching. The energy resolution for α-particles is at best 5–10% and thus far inferior to that of surface barrier detectors. By means of special electronics the difference in light-pulse decay time

Table 9.4 Quenching ratios for organic and aqueous solutions relative to water. Each sample contains 0.2 ml sample, 4 ml scintillator cocktail (Beckman ready solv CP) and 5 μl ^{147}Pm solution

Sample solution	Concentration	Quenching ratio (cpm/cpm_{aq})
1,2-Dichloroethane	100%	0.96
Carbon tetrachloride	100%	0.06
n-Hexane	100%	1.01
Acetylacetone	100%	0.16
Acetylacetone in n-Hexane	0.1 M	0.97
MIBK	100%	0.82
MIBK in n-Hexane	0.1 M	1.01
Benzoylacetone in n-Hexane	0.1 M	0.90
HTTA in n-Hexane	0.1 M	0.49
Nitric acid	1 M	0.87
Sulfuric acid	1 M	0.89
Hydrochloric acid	1 M	0.93

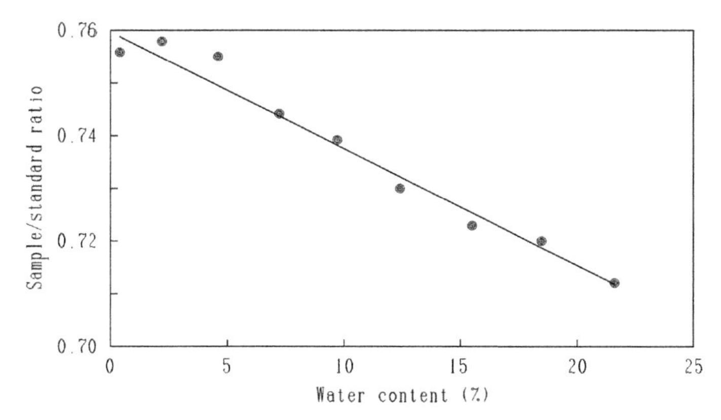

Figure 9.18 Quenching effect of water. The sample/standards ratio is the count rate ratio between 4 ml scintillator cocktail (Beckman Ready-Solv HP) + water and Packard reference standard.

between scintillations caused by α (long decay time) and by β,γ (short decay time) can be used to measure α-emitters in samples with a high β,γ-background. Liquid-flow detectors are discussed in section 9.8.

In principle the set up for a liquid scintillator can also be used for detection of high energy electrons by their Čherenkov radiation. This method can be used if no other interfering radiation types are in the sample. In this case no scintillation cocktail is needed but rather the Čherenkov photons are detected directly by the photomultiplication equipment.

9.5.3. Solid scintillator detectors

ZnS(Ag) is a traditional phosphor for α-detection while anthracene and stilbene can be used for β-particle detection. For γ-rays, sodium iodide with a small amount of thallium impurity, NaI(Tl), is the most common phosphor. CsI(Tl) is another often used scintillator because it can be formed to special shapes, e.g. thin sheets, much easier than NaI(Tl). Plastics with incorporated organic scintillators are often used in nuclear physics experiments because they produce short light pulses and can be made in various shapes and sizes.

Detectors with scintillation crystals are used commonly for routine radioactivity measurements, particularly of γ-emitters, because of their reliability. As compared to GM tubes they have the advantage of shorter resolution time and higher γ-efficiency, although they require a more stable high voltage supply. Particularly the well-type crystal shown in Figure 9.17 is popular because of the high counting efficiency for samples introduced into the well ($\psi = 0.9$). For counting very large liquid volumes (e.g. environmental, water samples) a specially designed sample vessel is used which fits over and around the cylindrical detector arrangement (Marinelli beaker).

9.6. ČERENKOV DETECTORS

The Čerenkov effect described in section 7.4.3 can be used for detection of high energy β-radiation because the velocity of the nuclear particle must exceed the ratio **c**/*n*, where *n* is the refractive index of the absorber.

$$E_{\beta\text{threshold}} = 0.511\left[-1 + \left\{1 + \left(n^2 - 1\right)^{-1}\right\}^{1/2}\right]\ (\text{MeV}) \qquad (9.28)$$

The β-threshold energy in lucite ($n = 1.5$) is 0.17 MeV, so lucite and similar plastics are often used as particle absorbers in Čerenkov detectors. In order to detect the light emitted, PMTs are placed in the direction of the emitted light. There are many similarities between scintillation and Čerenkov detectors; however, the light pulse from the Čerenkov detector is faster, $\sim 10^{-10}$ s, but smaller than from scintillation detectors. The advantage of the Čerenkov detector is that aqueous or organic (uncoloured) solutions can be used without the need to add a scintillator, that soft β-emitters and γ-emitters give little or no signal, and that the angle of emitted light reveals the velocity of the absorbed particles, cf. Eqn (7.18). At energies <10 MeV, only β-particles are detected.

9.7. MICROCHANNEL PLATE DETECTORS

A microchannel plate detector (MCP) is a planar component produced by cutting packages of sintered microtubes at an angle to their axis and used for detection and amplification of particles and radiation. The MCP consists of an insulating thin plate penetrated by very many holes and sandwished between two thin conducting surfaces which are also penetrated by the holes. In use, a high voltage is applied between the conducting surfaces. The MCP is a relative to the PMT-tube, as both intensify single particles or photons by the multiplication of electrons via secondary emission. However, a microchannel plate detector has many separate channels in parallel, each of which works like a small windowless PMT. Hence, it can also provide some spatial resolution. In general two types of MCP:s are in use. In one type the channels are straight and at an angle of about 8° relative to the surface. In another type two plates are sandwished with their hole-angles in a zig-zag arrangement, which yields a higher amplification. The channels have a diameter in the range 5–10 μm and are spaced about 15 μm apart. MCP-devices are used as detectors in mass spectrometers and as amplifiers in light-intensifying devices.

By using a very thin carbon foil positioned at an angle above the entrance side of the MCP, electrons released when charged particles transit through the carbon foil with very small energy loss can be collected, amplified and detected. This gives an electric signal from any passing particle or bunch of particles which can be used for timing purposes.

9.8. ELECTRONICS FOR PULSE COUNTING

A wide variety of counting systems has been developed for various purposes. Equipment is often built as NIM (Nuclear Instrument Module standard) or CAMAC (Computer Automated Measurement And Control standard) modules, which fit into standard bins (or crates) containing power supplies and some inter-module connections. Cable connectors are also largely standardized. This facilitates combinations of bias supplies, amplifiers, discriminators, SCAs, ADCs, counters, and other circuitry to fit any need as well as their connection to computers. Several types of counting systems are also designed as plug-in cards to AT-type PC:s. Some non-standard units, e.g. portable instruments, have their own power supply, main amplifier, counter or rate-meter, etc. Today, it is common that even though a detector system comprises all the components above they are not always separate units but integrated in the detector which in turn is connected to a computer using e.g. a USB cable.

9.8.1. Preamplifiers

The purpose of a preamplifier may be twofold. First of all it should increase the energy in the detector signal to such a level that it can drive a reasonable length of low impedance coaxial cable properly terminated at the other end with only a small loss in pulse amplitude. When the detector produces a very weak signal it may be advantageous to amplify that signal to a level where external noise becomes negligible. However, in some applications the detector signal is already large enough, the detector capacitance constant (e.g. a PMT) and *voltage sensitive preamplifiers* with unit gain are used, Figure 9.19(a).

A 0.5 MeV γ absorbed in a germanium crystal will only produce a charge of 2.7×10^{-14} C, Eqn (9.2). Moreover, the detector capacitance may change with operating conditions which make voltage sensitive preamplifiers less suitable and charge sensitive preamplifiers are preferred. The weak charge signal has to be integrated and converted to a voltage pulse by the preamplifier without adding too much noise. Figure 9.19(b) shows a typical input stage for a preamplifier used with germanium γ-ray detectors.

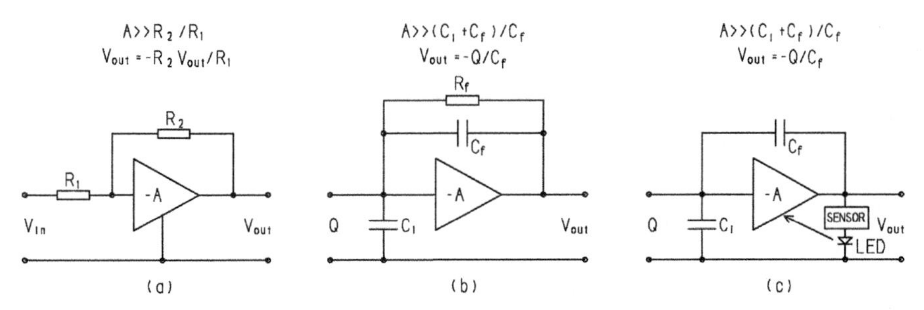

Figure 9.19 Voltage sensitive (a), charge sensitive resistive feedback (b), and pulsed optical feedback (c) preamplifiers. A is amplifier gain.

The function of this stage is to integrate the total charge of the pulse through the detector converting it to a voltage signal with an amplitude that is proportional to the energy deposited in the detector. Often the input FET (Field Effect Transistor) of the preamplifier and the feedback resistor (R_f) are cooled to reduce their contribution to system noise. The charge accumulated on the FET is removed through a resistor – capacitor network, R_f and C_f in Figure 9.19(b), which is part of the amplifier feedback loop. This arrangement is called a *resistor feedback preamplifier.* The feedback resistor adds its inherent Johnson noise to system noise; typically 300 eV in FWHM with a resistor at room temperature. The contribution to the total noise level is, at least for a cooled resistor and FET, negligible in normal γ-spectrometry. However, when measuring low energy X-rays with a Si(Li) detector it may be important to reduce the noise further.

In the *pulsed optical feedback preamplifier,* Figure 9.19(c), the resistor R_f is omitted and the input FET is permitted to charge-up in steps by each new pulse. The output signal is in the form of a staircase function and this is transformed to voltage pulses by differentiating circuits in the main amplifier. When the accumulated charge becomes too large (i.e. the output voltage is near its highest possible value) the sensor circuit triggers and the FET is irradiated with a short light pulse from the built-in light emitting diode thereby causing a short circuit in the FET, removing the accumulated charge. The reset operation produces a large voltage swing at the preamplifier output which may easily overload the main amplifier and make it distort pulses rapidly following the reset pulse. In order to avoid making measurements during and immediately after the big reset pulse a blocking signal is often available from the preamplifier during the reset pulse. A typical pulsed optical feedback preamplifier can, with a Si(Li) X-ray detector, give a FWHM of <150 eV at 5.9 keV. The good resolution is important in low energy applications, e.g. X-ray measurements. A minor drawback of pulsed optical feedback is the extra dead time introduced by blocking measurements during each reset operation.

The output from a voltage sensitive or resistor feedback preamplifier is a tail pulse with a rather long decay time. Hence, some pulse pile-up is unavoidable, except at very low count rates. Pile-up will cause the average level of this signal to increase with pulse rate, which may approach the limit of linear operation of the preamplifier.

9.8.2. Amplifiers

The purpose of an amplifier to amplify a voltage pulse in a linear fashion and to shape the pulse so that the event can be analyzed easily and correctly in a short time. A linear amplifier accepts tail pulses as input, usually of either polarity, and produces a shaped and amplified pulse with standard polarity and amplitude span (NIM standard is positive polarity and 0 to 10 V amplitude). On most commercial linear amplifiers, the time constants for the various pulse shaping circuits are adjustable to fit various detector and count-rate requirements.

Biased amplifiers shifts the zero of the amplified pulse down while still producing only positive output pulses. This effect is important in e.g. α-spectrometry as the lower energy range is normally of no interest and it is better to use the available output amplitude span to enlarge the interesting high energy region than to cover the whole energy range.

Amplifiers with a logarithmic amplitude response have use in liquid scintillation counting for compression of the higher energy part of the β-spectrum.

9.8.3. Single channel analyzers

The pulse obtained from many detectors is directly proportional to the energy absorbed in the detector. If all of the energy of the nuclear particle is absorbed in the detector, which is possible for all kinds of ionizing radiation, and only one particle or photon interacts at a time with the detector, the energy of the particle or photon can be determined from the size of the generated pulse. In single-channel analyzers (SCA) only one channel exists which serves as a "window" to accept only pulses of a certain size corresponding to a limited range of energy as indicated in Figure 9.5(b). This window can be moved in steps through the entire energy range, thereby obtaining a measure of the count rate of particles having different energies. In the Figure the window position and width are set so as to cover only the middle peak. By narrowing the window, and moving it from zero to maximum pulse amplitude, the whole particle energy spectrum is obtained. Spectrometry represents one of the principal techniques whereby the energy of ionizing radiation is measured. The single channel spectrometer is a system for energy determinations, although multichannel systems are simpler to use, more accurate and faster.

The main use of SCAs is to select events within a given energy interval. Thereby the background count rate is very much reduced which increases sensitivity. One of the most important routine uses of SCAs is in simultaneous ^{3}H and ^{14}C counting using liquid scintillators. Two SCAs are normally used in, one set to cover pulses corresponding to the very soft β^{-} from ^{3}H and the other set to cover the more energetic pulses caused by the β^{-} from ^{14}C.

9.8.4. Counters and rate meters

Ultimately the amplified pulses must often be accumulated over a fixed time interval and the resulting number displayed. The device used is based on a simple electronic register which is incremented by one each time a pulse arrives at its input. Such devices are often called *counters* or *scalers*. Counters can usually be operated in two modes, *preset time* or *preset count*. Counting is started and stopped by manual push buttons or by logic signals from other units, e.g. the logic stop signal from a timer in the preset time mode. In the preset count mode, the scaler stops at a given count and produces a logic output signal which can be used by other units, e.g. to stop a timer.

Multi-channel scalers are devices where the counts are accumulated in memory cells similar to a MCA, cf. section 9.8.5. In operation counts are accumulated in a current memory cell and the cell is changed to the next higher cell after a preset time interval. These devices are often in the shape of a circuit board for use in a PC. Start, stop, channel change, reset, and readout are then performed through software running on the PC.

In some applications a continuous display or record of the count rate is desired, e.g. in a survey meter. This is achieved by use of a *count rate meter.* Count rate meters can be analog (based on a diode pump) or digital (based on a recycling scaler-timer-display unit or a multi-channel scaler with preset dwell time for the channels). In either case, the value obtained follows the input rate with a certain time-lag, determined by the *time constant* used. A high time constant gives a smooth reading as it averages input pulse rate fluctuations over a longer time interval, but a very sluggish response to true input rate changes. Correspondingly a short time constant gives a rapid response, but also a very fluctuating reading.

9.8.5. Multichannel analyzers

Multichannel analyzers (MCA) consisting of an analog to digital converter (ADC), controller and storage unit which may have 16 000 channels or more (i.e. the energy scale is split up into that number of steps). In this case the pulses are sorted immediately into the various channels which record the counts as they occur rather than scan over an energy range in steps. In many designs the MCA is interfaced to a computer to provide display, recording and analysis of the energy spectrum. MCAs are also made as circuit boards that fit inside a personal computer and have suitable software by which the PC can emulate a stand-alone dedicated MCA. The ADC unit is normally based on one of two principles.

In a Wilkinson ADC, the beginning of an input pulse starts a pulser (the clock) and a ramp voltage. When the input signal culminates, its amplitude is "frozen" by a sample-and-hold gate. The pulser runs until the ramp voltage crosses the frozen input signal level. By accumulating the pulser signals in a counter the final count is proportional to the amplitude of the input pulse. This count can then be used as a digital address, *channel*, telling the attached digital circuitry where to increment the stored counts. The dead time of this ADC type is roughly proportional to the amplitude of the input pulse and to the highest address permitted. A typical conversion time with a 100 MHz clock may be $1 + 0.01n$ μs, for channel n. Wilkinson ADCs are mostly used in MCAs with a relatively small number of channels.

The successive approximation ADC also locks the maximum amplitude in a sample-and-hold circuit and then uses an ultra fast digital voltmeter of successive approximation type to measure the amplitude. A typical conversion time is ~25 μs at 4000 channels full scale. The voltmeter reading is then used as address in the same way as for the Wilkinson ADC. This type of ADC has usually a fixed dead time, independent of the pulse amplitude and only moderately dependent on the maximum address (each doubling of the total number of channels adds the same increment to the conversion time). Successive approximation ADCs are preferred for MCAs with very many channels as they then, on the average, become faster than Wilkinson ADCs. On the other hand the linearity of a Wilkinson ADC is usually slightly better than for a successive approximation ADC.

MCAs are probably the most versatile instruments for nuclear particle detection because of their usefulness both for qualitative identification and quantitative determination of radioactive nuclei. Practically all α- and γ-spectra reproduced in this book have been obtained through this technique.

In case correlated events registered by several ADCs are measured the addresses from the ADCs and time is usually recorded digitally on a magnetic storage medium in real time. The desired information, e.g. $\gamma-\gamma$ coincidences, is then obtained by reading and sorting the stored data after the experiment.

9.8.6. γ-spectrometry

As we learned in section 7.5, the capture of a γ-ray in an absorber such as a NaI(Tl)- or Ge-crystal occur by any of three processes – photoelectric effect, Compton effect, and pair production. In energy analysis of γ-rays it is desirable to capture the total energy and to minimize the loss of energy by escape of the scattered γ-rays from Compton interaction. This increases the number of events contributing to the photo peak which corresponds to the total γ-ray energy. Also, if lower energy γ-rays are present their photopeaks may be obscured by the Compton distribution from higher energy γ-rays. With anthracene crystals little resolution is seen in γ-spectra as the low atomic number of the absorber makes the principal interaction the Compton effect. However, in NaI(Tl), CsI(Tl), Si(Li), Ge(Li) and HPGe crystals the photoelectric effect is the more important. Increasing the size of the crystal increases the probability of photon capture and, therefore, the probability of capture of the scattered γ-ray in Compton events is increased. As a result, the increase in crystal size results in more capture of the total incident γ-ray energy which appears under the photo peaks.

For an ideal infinitely small detector a γ-spectrum from ^{137}Cs would have the shape shown in Figure 9.20(a). For comparison theoretical γ-spectra for finite size and infinitely large detectors are shown in the same Figure. The counts located between the compton edge and photo peak for the medium size detector are due to multiple Compton events. As can be seen detector size has a big influence of the Compton continuum and photo peak. A measured spectrum is shown in Figure 9.20(b). The energy of the Compton edge, E_{CE} (keV), caused by a γ-line of energy E_γ (keV), is given by

$$E_{CE} = E_\gamma/\left[1 + 511/\left(2E_\gamma\right)\right] \tag{9.29}$$

where 511 is the energy in keV equivalent of one electron mass.

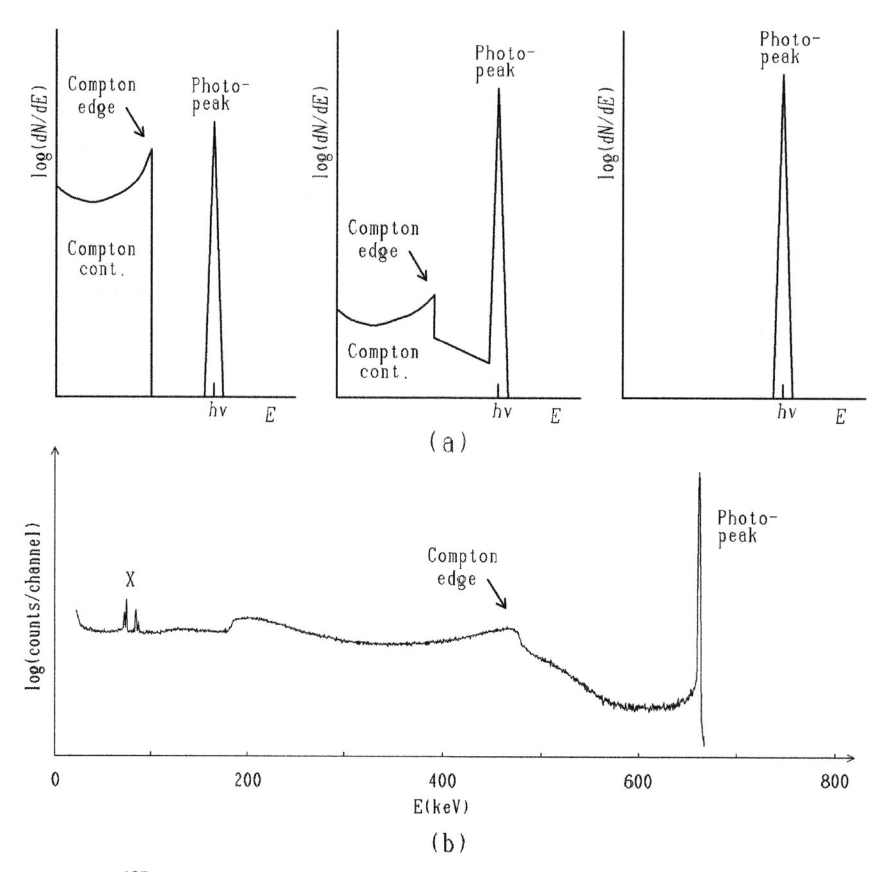

Figure 9.20 (a) ^{137}Cs γ-spectrum as calculated for "ideal" very small, medium and infinite size γ-detectors (b) the spectrum really recorded with a 22% HPGe detector.

The broadening of the photo peak has many causes such as inhomogeneities in the crystals and variations in charge, or light, collection and noise in the preamplifier. However, for scintillation detectors the main cause is found in the PMT where non-uniformity in the photocathode, fluctuations in the high voltage imposed on each dynode, and statistical variations in the small number of photoelectrons formed at the photocathode are all contributing factors.

The resolution is the determining factor in the ability of the system to differentiate between photopeaks of γ-rays of similar energy. Other features of the spectrum in Figure 9.20(b) are the backscatter peak and the X-ray peaks. The broad backscatter peak, located between the X-rays and the compton edge, arises by the absorption in the crystal of scattered photons resulting from γ-ray absorption via Compton interactions in the material surrounding the crystal. Obviously the magnitude of this peak is dependent on the distance of this material from the crystal and on the nature and amount of the material. The X-ray peaks (X in Figures 9.16 and 9.20(b)) are due to the absorption of the X-rays emitted in the electronic rearrangement following the nuclear disintegration, following internal conversion or after excitation of materials near the detector. Bursts of very low energy electrons are generated thermally in a semiconductor or emitted spontaneously from the photocathode in a PMT and are the cause of a noise peak at very low discriminator settings. This often limits the photon energies that can be studied to a minimum of some keV.

For γ-energies above 1.02 MeV, pair production in the detector followed by annihilation of the positron leads to the generation of pairs of 0.511 MeV photons. The probability increases rapidly with energy above 1.02 MeV. One or both of these photons may escape from the detector giving a deposited energy corresponding to the initial γ-energy less one or two times 0.511 MeV. Positrons are also generated outside the detector and their annihilation radiation reaches the detector. This gives rise to three smaller photo peaks called *escape peak*, *annihilation peak* and *double escape peak* at the corresponding energies, $E_\gamma - 0.511$ MeV, 0.511 MeV and $E_\gamma - 2 \times 0.511$ MeV, marked 1e, A and 2e respectively in Figure 9.16.

A complicating factor in gamma spectroscopy is the possibility of summing of two or more different gammas into one peak, so called coincident summing. This effect can originate from two ways: random coincidence and true coincidence summing (TCS). Random coincidence is the result of two separate gamma photons from different decays hitting the detector at the same time, i.e. within the resolution time of the detector. The energy of these gammas will then be summed and registered as a single event. This effect is dependent on the actual activity of the sample. The higher the activity the more probable that two or more decays occur simultaneously. In the second case, true coincedence summation, the origin is from a complex decay scheme, i.e. for each original decay there are several gammas emitted more or less simultaneously. This is not activity dependent and can cause severe calibration problems if not taken into account. This phenomenon is sometimes also called cascade summing. In more detail one may say that coincident γ-rays emitted in rapid cascade during one decay may hit the detector at the same time and register as one event with the combined energy. This gives two effects, firstly the generation of a *sum peak* containing one count per coincidence and secondly corresponding loss of one count in each of the peaks corresponding to the coincident photons. Summing is strongly dependent on detector size and source to detector distance.

The photo peaks from a well behaved detector are almost Gaussians. Thus the FWHM value determines the peak shape. The energy scale is calibrated against standard isotopes and is normally almost linear. In most applications only the area below the photo peak but above the underlying continuum is used as a measure of activity. The total efficiency (based on photo peak area) varies usually with energy in the way illustrated by Figure 9.21.

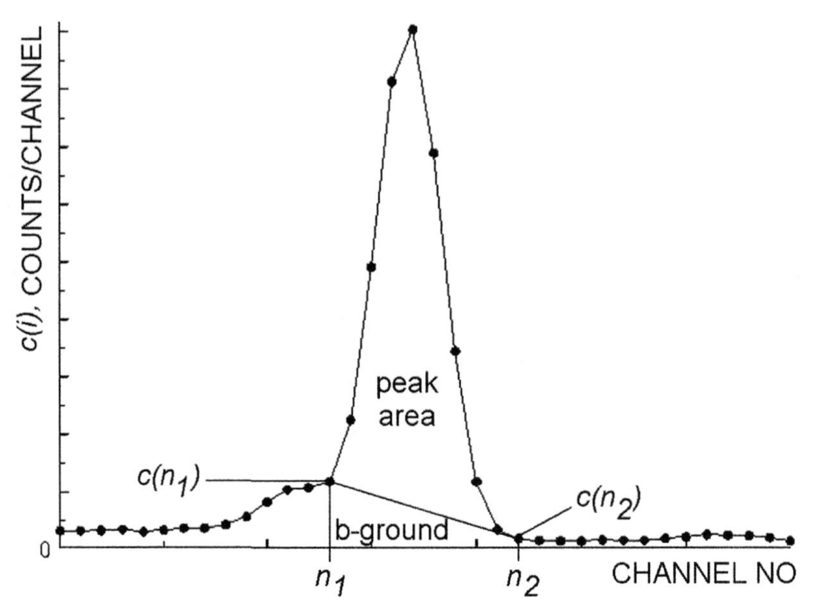

Figure 9.21 Typical efficiency curves for HPGe detectors. GEM and GMX are detectors of n- and p-type with Be-window respectively. (Acc. to ORTEC.)

The determination of a net peak area in α, γ, and X-ray spectroscopy requires a calculation and subtraction of the background in the peak region. Figure 9.22 shows a typical peak which begins in channel n_1 and ends in channel n_2. If no other information about the shape of the background is available it is usually assumed to be a linear function

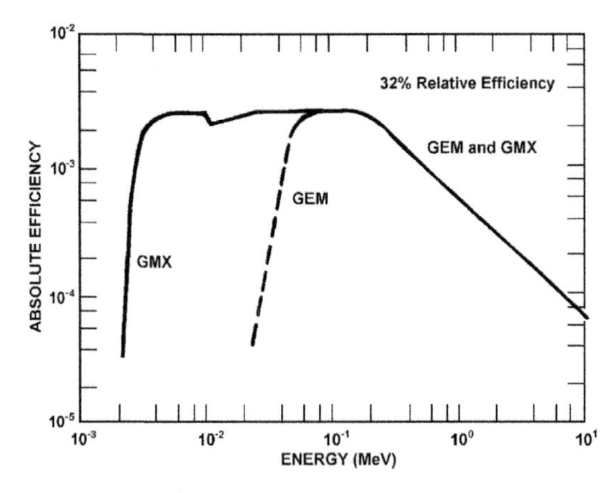

Figure 9.22 A typical peak from a HPGe detector and notations used in eqns. (9.30) and (9.31) for calculation of the net peak area and its standard deviation.

of the channel number between the two limiting channels, n_1 and n_2. The net peak area count, N_{peak}, is then calculated from

$$N_{peak} = \left[\sum_{i=n_1}^{n_2} c(i)\right] - 0.5 \times (n_2 - n_1) \times [c(n_1) + c(n_2)] \tag{9.30}$$

where $c(i)$ is the count in channel i. The uncertainty in N_{peak}, $s_{N_{peak}}$, can be estimated from

$$s_{N_{peak}} = \sqrt{\left[\sum_{i=n_1}^{n_2} c(i)\right] + 0.5 \times (n_2 - n_1) \times [c(n_1) + c(n_2)]} \tag{9.31}$$

9.9. SPECIAL COUNTING SYSTEMS

For *low intensity measurements*, in which a level of radioactivity comparable to the normal background radiation is to be measured, special electronic circuits incorporating two detectors are used. The detectors are coupled so that a signal registers only when both detectors are activated at the same time (*coincidence circuit*) or, alternatively, when only one but not the other is activated at the same time (*anticoincidence circuit*). With such arrangements the normal β, γ background of a detector may be decreased by more than a factor of 100. In the most advanced coincidence techniques both detectors are energy sensitive as well, providing information on the type of radiation being measured. For example, β-γ coincidence measurements are used for absolute determination of radioactivity for samples in which β-decay is immediately followed by γ-emission.

Detectors which are *direction sensitive* have been developed primarily for use in medical diagnosis with radioactive isotopes. The simplest version involves a scintillation detector surrounded by a lead shield with a small hole (collimator) through which radiation reaches the detector. More complex systems with many detector/collimator pairs are in common use (γ-cameras); cf. section 18.5. Scanning instruments (Fig. 18.13) have been developed which permit the measurement of radioactivity as a function of several coordinates as illustrated in Figure 18.14. Such instruments make it easy to detect the accumulation of a radioactive tracer in a particular organ of the body.

Whole-body counters were originally developed for investigation of poisoning by radioactive substances such as radium. They are now used diagnostically and consist of a large scintillation, or semiconductor, detector with the whole system, including the patient, placed in a heavily shielded room. The sensitivity is sufficient to measure natural radioactivity in the human body from such nuclides as ^{40}K.

In nuclear installations like uranium mines, nuclear reactors, reprocessing plants, etc., it is necessary to continuously monitor gas and liquid effluents. Figure 9.23 shows an

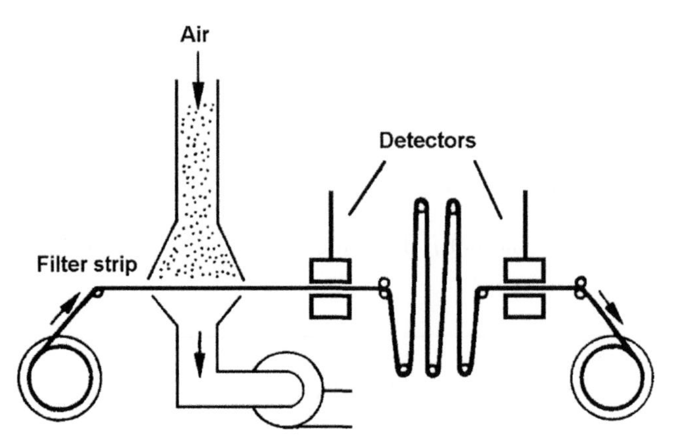

Figure 9.23 Monitor for radioactive aerosols with prompt and delayed measurements. *(From H. Kiefer and R. Maushardt.)*

arrangement for monitoring of *radioactive aerosols* (e.g. Tc, Ru, actinides in air). Two detectors are used, so that some activity (e.g. mother or daughter activity) is allowed to decay between the two detectors. The delay time is adjusted by varying the length of the paper strip between the detector and the rate of movement of the paper strip. The detector may be energy sensitive or simple GM, proportional or scintillation devices. In monitoring of water the detectors (e.g. GM tubes) may dip into the streaming water, or the water may flow around the detector as in Figure 9.9(a).

Radioactive gas flows can be measured by using capillaries or packed columns of 2–20 ml volume containing scintillating material, either for monitoring purposes or, commonly, combined with a gas chromatograph for obtaining "radio-chromatograms" of ^{3}H or ^{14}C labeled organic substances. These detectors, which are commercially available, are viewed by PMTs connected to PC based analysis and display systems.

Radioactive liquid flows can be monitored in several ways depending on the decay mode, energy and liquid film thickness. Liquid flow GM-counters (Fig. 9.9.a), or other types of flow cells combined with solid state detectors, are used at high β- and γ-energies. Glass scintillators embedded in tubings (alternatively tubings containing scintillators) are used for low-energy β-emitters (e.g. ^{3}H, ^{14}C, ^{35}S). The flow cells have typically volumes from 0.01 to 5 ml. By using peak analysis (i.e. analyzing each scintillation intensity versus time, so-called "time-resolved technique") it is possible to measure α-emitters in high β-γ fluxes and even to obtain crude α-spectra, as well as to discriminate different β-emitters and to reduce background. A weakness of this technique is the memory effect in the detectors caused by radionuclide sorption. Therefore, the commercial systems have easily exchangeable flow cells. This demand is circumvented in the LISOL-system, in which the liquid flow is premixed with acid and scintillator solution before counting; the use of that system is described in section 18.4.3.

9.10. ABSOLUTE DISINTEGRATION RATES

The determination of absolute disintegration rates is of great importance in all areas of nuclear chemistry, tracer work, age calculation, etc. Numerous methods have been employed, many using techniques described above, as track counting, liquid scintillation measurements, 4π proportional counters, etc. If the nuclei decay through β-γ emission, the absolute rate may be obtained by two detectors placed close to each side of a thin sample, one detector β-sensitive and the other γ-sensitive.

When only a single detector in a conventional counting set-up is available (e.g. detector arrangement in Fig. 9.4), absolute counting rates can be obtained for unknown samples by comparison with known standards.

When standards are not available it is possible to obtain an approximate estimation of the absolute disintegration rate from a knowledge of the various factors that influence the counting efficiency. The detection efficiency ψ is defined as a ratio between the count rate and the absolute disintegration rate Eqn (5.64). This detection efficiency, which was discussed briefly in section 7.2, is the product of all the factors which influence the measured count rate and may be expressed as

$$\psi = \psi_{det}\psi_{res}\psi_{geom}\psi_{back}\psi_{self}\psi_{abs} \tag{9.32}$$

where

ψ_{det} = counting efficiency of detector,
ψ_{res} = resolving time correction (see section 9.3),
ψ_{geom} = geometry factor (see section 7.2),
ψ_{back} = backscattering factor,
ψ_{self} = self-absorption factor ($\psi_{sample} = \psi_{back}\psi_{self}$), and
ψ_{abs} = absorption factor (see section 7.2).

The efficiency of the detector is a measure of the number of counts registered compared to the number of particles that enter the sensitive volume of the detector. This efficiency is approximately 100% for α- and high-energy β-particles in most detectors, but often substantially lower for γ-rays. In as much as it is quite difficult to apply simple geometric considerations to the solid angle subtended by a detector for a source which is not concentrated at a point the factors ψ_{det} and ψ_{geom} may be determined experimentally by using a very thin standard source of approximately the same area as the unknown. The factor ψ_{geom} can be calculated for circular samples and detector windows, see section 7.2

It was noted in section 7.4 that β-rays undergo large angle deflections. As a result, β-particles from the sample which may start in a direction away from detector can be deflected by several scattering events back into the detector. Such backscattering is dependent upon the atomic number of the material upon which the sample is supported (cf. Fig. 7.14). ψ_{back} increases with backing material thickness up to a saturation thickness beyond which it is constant. Counting is usually done with either an essentially weightless backing ($\psi_{back} = 1$) or with a backing sufficiently thick as to have saturation. The saturation thickness corresponds to approximately 20% of the range of the β-particles in the backing material. Scattering can also occur from the walls of the sample holder, but this is usually less important than the backscattering from the sample backing material.

If the sample is thick, the count rate may be increased by internal backscattering, but the decrease of the count rate due to self-absorption within the sample is a greater factor. This sample self-absorption is inversely proportional to the β-ray energy and directly proportional to the thickness of the sample. For β^- from ^{32}P (1.72 MeV) sample thicknesses of 15 mg cm^{-2} show little self-absorption, while for β^- from ^{14}C (0.15 MeV), the absorption for sample thicknesses as small as 1 mg cm^{-2} is significant. The absorption factor can be determined by counting a series of samples of different thicknesses but with the same total disintegration rate, and then extrapolating that to zero sample weight.

The absorption factor ψ_{abs} is related to the absorption of the particles after they leave the sample by any covering over the sample, by the air between the sample and the detector, by the absorption in the detector window, etc. Again, this factor is usually determined by experimental comparison with a sample of known absolute disintegration rate, a *standard*.

9.11. SAMPLE PREPARATION

From the discussion of the factors that enter into the counting efficiency it is obvious that the preparation of the counting sample must be done with care and must be reproducible if several samples are to be compared. Counting of α- and β-emitters *in solution* is best achieved by means of liquid scintillation counting. Because in this technique the emitters are included in the detection system itself the efficiency is very high and reproducible.

The advantage of using *solid samples* for counting is that the samples can be made very robust and small, allowing the use of either very simple counting systems (e.g. GM-counters), or the use of commercial very efficient high-capacity (>1000 samples/hour) automatic counting systems. Alpha-emitters can only be counted efficiently if the sample is very thin so that the self-absorbtion is eliminated. For α-spectrometry the surface density should be <0.1 mg cm^{-2}. Preferably, α's should be counted by surface barrier detectors, windowless proportional counters or internal ion chambers. Counting of solid samples of β-emitters may or may not be a problem depending on the energy of the β-emission. Again, care must be taken with uniform thickness of sample, backscattering, etc. The use of energy sensitive detectors makes possible a reliable measurement of one particular radioactive nuclide in the presence of other radiation of secondary importance.

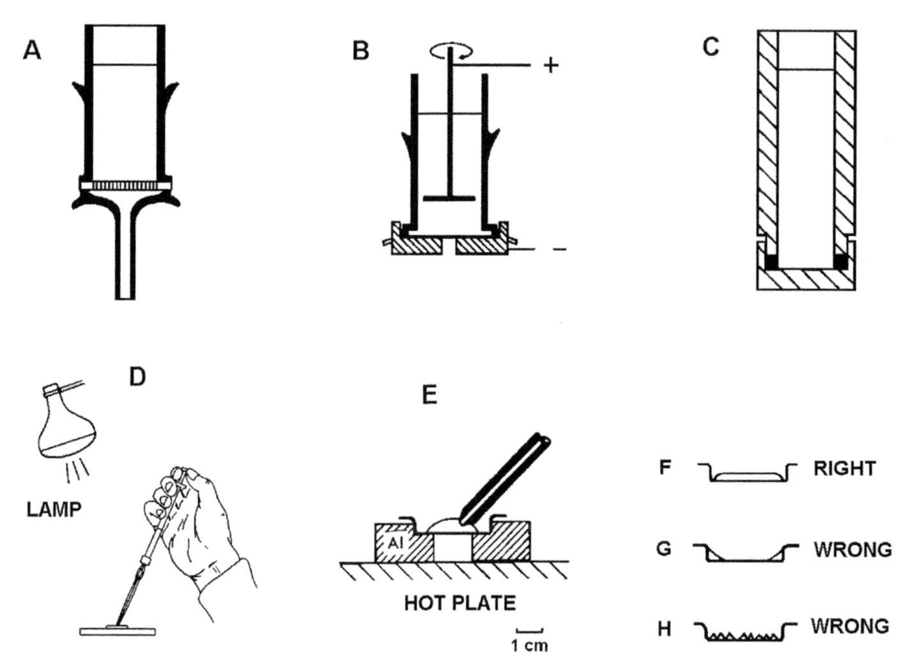

Figure 9.24 The preparation of thin, uniform samples of known surface density. A, filtering arrangement; B, electrolytic plating system; C, centrifugation vessel; D E, pipetting of a solution into an evaporation tray (glass, stainless steel, platinum), yielding a correct even deposit F or uneven samples G H. The aluminum ring E should have a temperature only slightly exceeding the boiling point of the liquid.

Solid samples can be prepared by a variety of techniques such as precipitation, evaporation, and electrolysis (Fig. 9.24). When the precipitation technique is used the radioactive material must always be precipitated for comparative counting with the same amount of carrier and all samples must have the same surface density. The precipitate is filtered on a filter plate or filter paper of known reproducible geometry A. If filtration is not feasible the precipitate may be centrifuged in special vials C, or the precipitate, after centrifugation and decanting, may be slurried with ether or alcohol, and the slurry transferred by pipette to a counting disc of fixed geometry; when the organic liquid evaporates, it hopefully leaves a uniform deposit of the precipitate slurry D, F. Problems are plentiful in obtaining a deposit of uniform thickness by evaporation of a solution. However, an arrangement such as that shown in Figure 9.24(E) has been found to be suitable; slow and even evaporation of 0.1−1.0 ml samples result in an even deposit if the amount of solid material is small. More even deposits can be obtained by electrodeposition of samples from solution B. This method can be used also for nonaqueous solutions provided that the organic solvents contain traces of water and a potential of several hundred volts per centimeter is used. To speed-up sample preparation, different types of semi-automatic sample preparation systems are commercially available. They are particularly useful in bio-medical tracer research, where often large numbers of samples are produced.

9.12. STATISTICS OF COUNTING AND ASSOCIATED ERROR

Even if the experimental design and execution are perfect so that the determinant error is eliminated in experiments involving radioactivity there is always a random error due to the statistical nature of radioactive decay. Each radioactive atom has a certain probability of decay within any one time interval. Consequently, since this probability allows unlikely processes to occur occasionally and likely processes not to occur in any particular time interval, the number of decays may be more or less than the number in another similar interval for the same sample. It is necessary, when counting a sample, to be able to calculate the probability that the recorded count rate is within certain limits of the true (or average) count rate.

The binomial distribution law correctly expresses this probability, but it is common practice to use either the Poisson distribution or the normal Gaussian distribution functions since both approximate the first but are much simpler to use. If the average number of counts is high (above 100) the Gaussian function may be used with no appreciable error. The probability for observing a measured value of total count N is

$$P(N) = (2\pi\overline{N})^{-1/2}\, e^{-1/2(\overline{N}-N)^2/\overline{N}} \tag{9.33}$$

The standard deviation σ is given in such cases by

$$\sigma = \sqrt{N} \tag{9.34}$$

In these equations $P(N)$ is the probability of the occurrence of the value N while $\overline{N}$ is the arithmetic mean of all the measured values and N is the measured value. Figure 9.25 shows the Poisson and Gaussian distributions for $\overline{N} = 20$ counts. The standard deviation ("statistical error"), according to the theory of errors, indicates a 68% probability that the measured value is within $\pm\sigma$ of the average "true" value $\overline{N}$. For 100 measured counts the value 100 ± 10 indicates that there is a 68% probability that the "true" value will be in the interval between 90 and 110 counts. If the error limit is listed as 2σ the probability that the "true" count will be between these limits is 95.5%; for 3σ it is 99.7%. Figure 9.26 shows the relationship between K, the number of standard deviations, and the probability that the true figure lies outside the limits expressed by K. For example, at $K = 1$ (i.e. the random error is $\pm 1\sigma$) the Figure indicates that the probability of the true value being outside $N \pm \sigma$ is 0.32 (or 32%). This agrees with the observation that the probability is 68%, i.e. the "true" value is within the limits of $\pm 1\sigma$. Figure 9.25 can be used to establish a "rule of thumb" for rejection of unlikely data. If any measurement

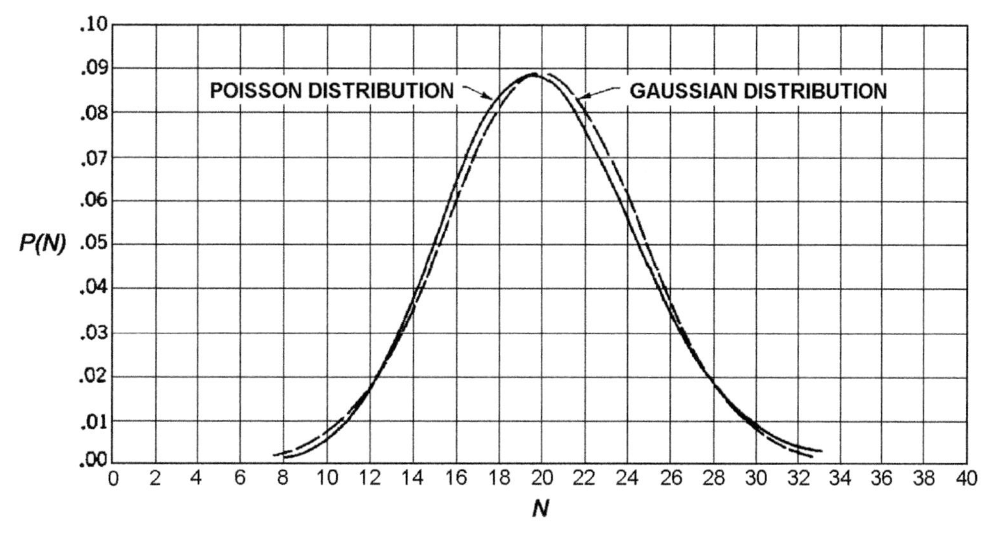

Figure 9.25 Poisson (smooth) and Gaussian (dashed line) distributions for $\overline{N} = 20$.

differs from the average value by more than five times the probable error it may be rejected as the probability is less than one in a thousand that this is a true random error. The probable error is the 50% probability which corresponds to 0.67σ.

From the relationship of σ and N it follows that the greater the number of collected counts the smaller the uncertainty. For high accuracy it is obviously necessary to obtain a large number of counts either by using samples of high radioactivity or by using long counting times.

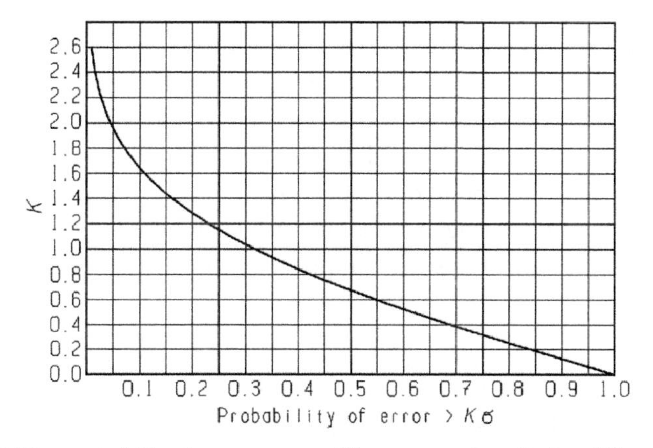

Figure 9.26 The probability that an error will be greater than $K\sigma$ for different K-values.

In order to obtain the value of the radioactivity of the sample, corrections must be made for background activity. If our measurements give $N \pm \sigma$ counts for the sample and $N_0 \pm \sigma_0$ for the background count, the correct value becomes

$$N_{\text{corr}} = (N - N_0) \pm \left(\sigma^2 + \sigma_0^2\right)^{1/2} \tag{9.35}$$

If the sample was counted for a time Δt and the background for a time Δt_0, the measured rate of radioactive decay is

$$R = N/\Delta t - N_0/\Delta t_0 \pm \left[(\sigma/\Delta t)^2 + (\sigma_0/\Delta t_0)^2\right]^{1/2} \tag{9.36}$$

It is extremely important in dealing with radioactivity to keep in mind at all times the statistical nature of the count rate. Every measured count has an uncertainty and the agreement between two counts can only be assessed in terms of the probability reflected in terms of σ.

The statistical nature of radioactive decay also leads to an uneven distribution of decays in time which is important when handling dead-time corrections and discussing required system time resolution. Let us first assume that a decay has occurred at time $t = 0$. What is then the differential probability that the next decay will take place within a short time interval, dt, after a time interval t has passed? Two independent processes must then occur in series. No decay may take place within the time interval from 0 to t, probability $P(0)$, followed by a decay within the time interval dt, probability rdt. The total probability is then given by the product of the individual probabilities of the two processes, i.e. if the combined probability is denoted $P(t)$dt then

$$P(t)\mathrm{d}t = P(0) \times r\,\mathrm{d}t \tag{9.37}$$

By applying the Poisson distribution and noting that r is equal to λN_0 we obtain

$$P(t)\mathrm{d}t = \lambda N_0\,\mathrm{e}^{-\lambda N_0 t}\,\mathrm{d}t = A_0\,\mathrm{e}^{-A_0 t}\,\mathrm{d}t \tag{9.38}$$

Equation (9.38) implies that the decays are not evenly distributed i time but often occur in clusters. The time interval between pairs of decays has indeed its maximum probability at zero.

In experimental work with radionuclides many other errors occur in addition to the statistical uncertainty in the count rate. Such errors may originate in the weighing or volumetric measurements, pH determination, etc. Such errors must also be considered in presenting the final results. For such composite errors, the *law of error propagation* must be applied:

$$\sigma_{\mathrm{F}} = \left[(\sigma_{\mathrm{A}}\,\mathrm{d}F/\mathrm{d}A)^2 + (\sigma_{\mathrm{B}}\,\mathrm{d}F/\mathrm{d}B)^2\right]^{1/2} \tag{9.39}$$

where σ_{F} is one standard deviation in F, which is a function of the uncorrected variables A, B, ..., with the standard errors σ_{A}, σ_{B}, etc. Commonly the unbiased estimation of the standard deviation of σ is expressed as s. For n events, x_i, having the average $\overline{x}$, s is estimated according to:

$$s = \left\{ [1/(n-1)] \times \sum_{i=1}^{n} (x_i - \overline{x})^2 \right\}^{1/2} \quad (9.40)$$

For the product $A \times B$ and ratio A/B, one obtains

$$F = (A \times B)(1 \pm s) \quad (9.41)$$

and

$$F = (A/B)(1 \pm s) \quad (9.42)$$

where

$$s = \left[(\sigma_A/A)^2 + (\sigma_B/B)^2 \right]^{1/2} \quad (9.43)$$

For the function A^x and $\log A$ the following relations are valid

$$F = A^x \pm x \times A^{x-1} \, \sigma_A \quad (9.44)$$

$$F = \log A \pm [\sigma_A/(2.303A)]^{1/2} \quad (9.45)$$

A useful technique for checking that the error in the measurements has a Gaussian distribution is the so-called "χ-square" test. The quantity χ^2 is calculated from

$$\chi^2 = \left[\sum_{i=1}^{M} (\overline{F} - F_i)^2 \right] \Big/ \left[\overline{F}(k-1) \right] \quad (9.46)$$

where M is the number of measurements (e.g. points on a curve) for which the function F is (believed to be) valid. χ^2 would have a value 0.5−1.0 when the Gaussian fit exceeds 50%. k is the degree of freedom, i.e. M minus the number of determined variables. This relation is generally valid; for simple counting systems F is replaced by N, the number of counts in a given time interval, and $k = M$.

9.12.1. One-atom-at-a-time measurements

The difficulty to obtain good measuring statistics makes measurement of *e.g.* distribution coefficients (D) difficult with a small amount of atoms. Figure 9.27 shows the uncertainty of distribution coefficients based on 20 and 70 decay events, respectively, which may correspond to all atoms present given a short half-life, a sufficiently large Δt, and a detection efficiency of ~100%. Not surprisingly the uncertainty has a minimum at D = 1. Thus, if performing experiments where the heaviest elements are compared to lighter homologues it is favorable if the heavy elements are measured under conditions where D

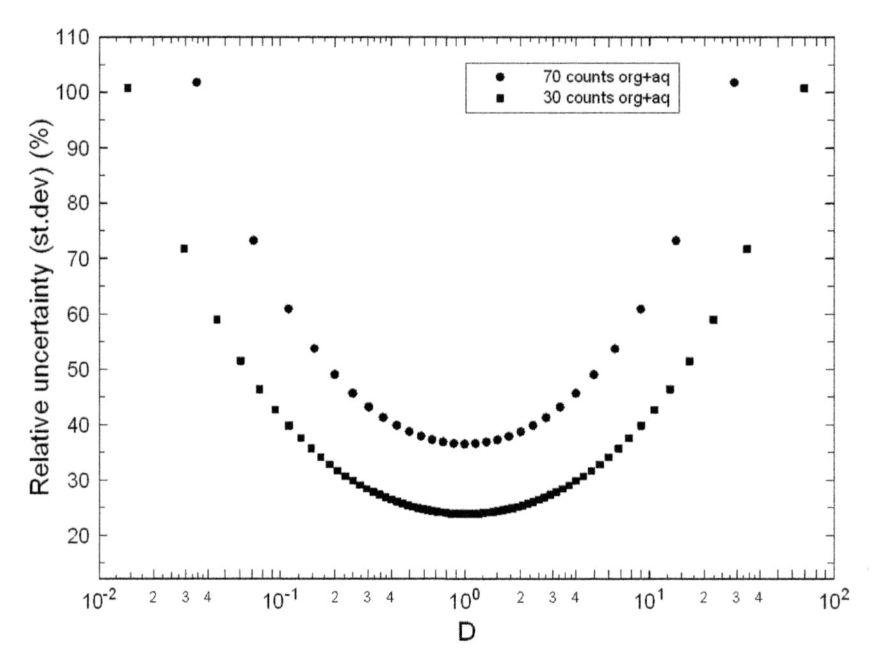

Figure 9.27 The relative size of one standard deviation in the D-value as a function of the D-value and a given total number of counts.

is not too far from 1 and the homologues, which can be produced in larger amounts, at a higher or lower D.

However, if fewer atoms than ~20 are detected other approaches are needed. One method is to calculate the probability for an atom to be present in the organic phase at a certain D-value. Then a random number is picked from some distribution, preferable a uniform distribution, thus assigning this particular atom to either the organic or aqueous phase. This may be repeated for the number of atoms desired by the aid of a simple computer program. If a large number of such calculations are performed they will approach the value of the uncertainty given above. Alternatively one may use the students T statistics to modify the multiple of the standard deviation needed for a given confidence interval.

When the total number of atoms used in an experiment is extremely small the usual fundamental thermodynamic descriptions no longer hold. Thus, the phase concept disappears when the medium contains only one or a few radioactive atoms. For reasons that are easily understood the law of mass action is also not applicable. If you, for example, only have one atom it is impossible for that atom to be in equilibrium with itself.

More specifically, in solvent extraction the distribution ratio between two phases looses its significance: a single atom can only exist in one of the two phases at the moment of separation. An average distribution can, however, be obtained by performing a large number of identical experiments. The number of decay events in each phase is recorded. The decay statistics are replaced by the statistical distribution of experimental results.

Studies using extremely small amounts of any shortlived radionuclide involve a number of statistical considerations. The binomial law has to be used since the conditions for the Poisson distribution, *i.e.* that $N > 100$, and that the number of observed decays, $\overline{N}_{\Delta t}$, in the measuring time, Δt, is much smaller than N, are seldom fulfilled in this case. The probability, $P(1)$, of exactly one event happening is according to the binomial distribution:

$$P(1) = \left(\frac{n!}{(n-1)!}\right)p(1-p)^{(n-1)} \tag{9.47}$$

where p is the probability of decay, and n is the total number of atoms.

For a radioactive decay, the probability for exactly one decay during the time Δt in a sample containing 1 atom is thus:

$$p(\Delta t) = 1 - e^{-\lambda \Delta t} \tag{9.49}$$

However, when the half-life is long compared to Δt it is possible to simplify this expression by a truncated Taylor series expansion to:

$$p(\Delta t) = \lambda \Delta t \tag{9.50}$$

In the general case involving N atoms, the following expressions for the expected average of decayed atoms, $\overline{N}_{\Delta t}$, and its variance, σ, are obtained:

$$\overline{N}_{\Delta t} = N\left(1 - e^{-\lambda \Delta t}\right) \tag{9.51}$$

and

$$\sigma = \sqrt{\overline{N}_{\Delta t}(1 - e^{-\lambda \Delta t})} \tag{9.52}$$

9.13. EXERCISES

9.1. A detector has a 1 cm^2 efficient area perpendicular to a γ-particle flux produced by a source 7 m away. The sensitivity for the 0.73 MeV γ-radiation is 8.2%. (a) What must the source strength be for the detector to register 1000 cpm? (b) What fraction of the radiation is absorbed in the air space between source and detector?

9.2. A ~100 MeV fission fragment is stopped in a plastic plate with density ~1 and an average atomic spacing of 0.25 nm. Estimate (a) the range in the plate, and (b) the ionization density (ion pairs μm^{-1}). If the ionization along the track is spread out perpendicular from the track so that 1 in 10 atoms are ionized (c) what would be the diameter of the track? From the track dimensions (d) calculate the average energy deposition to each atom within the "cylinder", and, using the relation $E = 3\mathbf{k}T/2$, (e) estimate the average temperature within the track volume. In lack of basic data for the plastic material, use data for water.

9.3. Plutonium in a urine sample is soaked into a photographic emulsion so that the emulsion increases its volume by 20%. The 12 μm thick emulsion is dried to original thickness and then left in darkness for 24 h. After development, α-tracks are counted and an average of 2356 tracks cm^{-2} found. If the plutonium consists of 67% ^{239}Pu and 33% ^{240}Pu, what was the plutonium concentration in the urine?

9.4. A ^{244}Cm sample is measured in an ion chamber (Fig. 9.11). The voltage drop over a 3×10^{13} Ω resistor is measured to be 0.47 V. What is the activity of the sample if all α's emitted in the chamber (2π geometry) are stopped in the gas?

9.5. In a proportional counter filled with methane of 1 atm the gas multiplication is 2×10^4. What is the maximum pulse size for a 5.4 MeV α, if the ion-pair formation energy is assumed to be 30 eV? The capacitance of the circuit is 100 pF.

9.6. In a GM counter, sample A gave 12630, B 15480, and A + B together 25147 cpm. (a) What is the resolving time of the counter? (b) With the same counter, the distribution of radioactive samarium between an organic phase and water was measured according to $D_m = R_{org}/R_{aq}$. The measured R_{org} is 37160 cpm, and that of R_{aq} is 2965. (b) What is the measured D_m? (c) Using corrections for resolving time, what is the true D-value?

9.7. Assume that 10^9 alcohol molecules are dissociated per discharge in a GM tube of 100 cm^3 filled with 90% Ar and 10% ethyl alcohol vapour at a pressure of 100 mmHg (25°C). What is the lifetime of the tube in terms of total counts assuming this coincides with the dissociation of 95% of the alcohol molecules?

9.8. A 1 mm tick surface barrier detector of 10 mm diameter has a resistivity of 7000 Ω cm and a capacitance of 50 pF at 300 V reverse bias. Calculate the resolving time (time constant).

9.9. A plastic scintillation detector was to be calibrated for absolute measurements of β-radiation. For this purpose a 2.13×10^{-5} M $^{204}TlCl_3$ solution was available with a specific activity of 13.93 μCi ml^{-1}; ^{204}Tl emits β-particles with E_{max} 0.77 MeV. Of this solution 0.1 ml is evaporated over an area of exactly 0.1 cm^2 on a platinum foil. The sample is counted in an evacuated vessel at a distance of 15.3 cm from the detector, which has a sensitive area of 1.72 cm^2. The detector registers 2052 cpm with a background of 6 cpm. What is (a) the surface weight of the sample, (b) the backscattering factor, and (c) the detector efficiency for the particular β's?

9.10. A sample counted for 15 min gave 9000 total counts. A 30 min background measurement registered 1200 counts. Calculate (a) the count rate or the sample alone, with its standard deviation, and (b) with its probable error.

9.11. A certain sample has a true average counting rate of 100 cpm. What is the probability that 80 counts would be obtained in a 1 min recording?

9.14. LITERATURE

H. YAGODA, *Radioactive Measurements with Nuclear Emulsions*, Wiley, 1949.

G. B. COOK and J. F. DUNCAN, *Modern Radiochemical Practice*, Oxford University Press, 1952.

S. FLÜGGE and E. CREUTZ (Eds.), *Instrumentelle Hilfsmittel der Kernphysik II. Handbuch der Physik*, XLV, Springer-Verlag, 1958.

G. D. O'KELLEY, *Detection and Measurement of Nuclear Radiation*, NAS-NS 3105, Washington DC, 1962.

E. SCHRAM and R. LOMBAERT, *Organic Scintillation Detectors*, Elsevier, 1963.

W. H. BARKAS, *Nuclear Research Emulsions*, Academic Press, 1963.

W. J. PRICE, *Nuclear Radiation Detection*, McGraw-Hill, 1964.

P. C. STEVENSON, *Processing of Counting Data*, NAS-NS 3109, Washington DC, 1966.

W. B. MANN and S. B. GARFINKEL, *Radioactivity and its Measurement*, van Nostrand, 1966.

Nuclear spectroscopy instrumentation, *Nucl. Instr. Meth.* **43** (1966) 1.

G. BERTOLINI and A. COCHE (Eds.), *Semiconductor Detectors*, North-Holland, 1968.

R. J. BROUNS, *Absolute Measurement of Alpha Emissions and Spontaneous Fission*, NAS-NS 3112, Washington DC, 1968.

J. M. A. LENIHAN and S. J. THOMSON, *Advances in Activation Analysis*, Academic Press, 1969.

K. BÄCHMANN, *Messung Radioaktiver Nuklide*, Verlag-Chemie GmbH, 1970.

C. E. CROUTHAMEL, F. ADAMS, and R. DAMS, *Applied Gamma Ray Spectrometry*, Pergamon Press, Oxford, 1970.

O. C. ALLKOFER, *Teilchen-Detectoren*, Thiemig, 1971.

P. QUITTNER, *Gamma Ray Spectroscopy*, Adam Hilger, London, 1972.

H. KIEFER and R. MAUSHART, *Radiation Protection Measurement*, Pergamon Press, Oxford, 1972.

R. A. FAIRES and B. H. PARKS, *Radioisotope Laboratory Techniques*, Butterworth, 1973.

J. A. COOPER, Comparison of particle and photon excited X-ray fluorescence applied to trace element measurements or environmental samples, *Nucl. Instr. Meth.* **106** (1973) 525.

J. KRUGERS, *Instrumentation in Applied Nuclear Chemistry*, Plenum Press, 1973.

Users' Guide for Radioactivity Standards, NAS-NS 3115, Washington DC, 1974.

D. L. HORROCKS, *Application of Liquid Scintillation Counting*, Academic Press, 1974.

P. J. OUSEPH, *Introduction to Nuclear Radiation Detection*, Plenum Press, 1975.

R. L. FLEISCHER, P. B. PRICE, and R. M. WALKER, *Nuclear Tracks in Solids*, University of California Press, 1975.

S. DEFILIPPIS, Activity analysis in liquid scintillation counting, *Radioactivity and Radiochemistry* **1**, 4 (1990) 22.

G. F. KNOLL, *Radiation Detection and Measurement*, 4nd Ed., John Wiley & Sons, 2010.

CHAPTER 10

Energetics of Nuclear Reactions

Contents

Reactions between an atomic nucleus and another particle are called nuclear reactions. In some such reactions, the nucleus is unchanged (*elastic scattering*), in others the original nucleus is excited to a higher energy state (*inelastic scattering*); in a third class new nuclei are formed (*nuclear transmutations*). Spontaneous nuclear transformations, which are involved in the radioactive decay of unstable nuclei, have been discussed in Chapter 5. In this chapter the emphasis is on the mass and energy relationships when a projectile interacts with a nucleus.

10.1. CONSERVATION LAWS IN NUCLEAR REACTIONS

We have previously discussed a number of laws governing nuclear processes. A summary of these laws provides us with a picture of the mechanics of nuclear interactions occurring below ~100 MeV/A. At higher energies the mass number may not be conserved.

Let us begin by considering the simple process shown in Figure 10.1, in which a *projectile*, indicated by subscript 1, collides with a *target* atom, subscript 2, forming an intermediate system, subscript i. The intermediate system splits into the products, which are designated by subscripts 3 and 4. We call this a *nuclear reaction*. The velocity of the projectile v_1 is greater than zero, but the velocity of the target atom v_2 is made to be zero by using the target nucleus as the origin of the system of reference.

Radiochemistry and Nuclear Chemistry
ISBN 978-0-12-405897-2, http://dx.doi.org/10.1016/B978-0-12-405897-2.00010-0

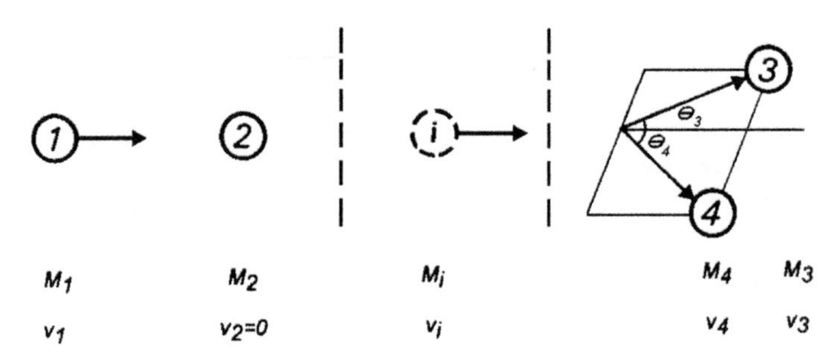

Figure 10.1 Schematic picture of a nuclear reaction.

All the *conservation laws* derived in §5.2 are applicable to nuclear reactions. These are:

(a) the conservation of total energy:	$\Delta E = 0$	(5.4)
(b) the conservation of linear momentum:	$\Delta p = 0$	(5.6)
(c) the conservation of total charge:	$\Delta Z = 0$	(5.8)
(d) the conservation of mass number:	$\Delta A = 0$	(5.9)
(e) the conservation of spin:	$\Delta I = 0$	(5.11)

"Conservation of total energy" means that the total energy of the products must equal the total energy of the reactants, i.e. $E_{products} - E_{reactants} = \Delta E = 0$. For (5.6) it should be remembered that linear momentum is a vector property; thus

$$p_1 + 0 = p_3 \cos \theta_3 + p_4 \cos \theta_4 \tag{10.1}$$

$$p_3 \sin \theta_3 = p_4 \sin \theta_4 \tag{10.2}$$

where θ_3 and θ_4 are the respective emission angles, see Fig. 10.1. As pointed out earlier, some of these conservation laws (e.g. (d) and (e)) are not always obeyed in high energy reactions in which new elementary particles may be formed. The kinetic energy equation $E_{kin} = ½mv^2$ (3.6) was deduced by Newton in 1687 assuming that the mass of the particle was independent of its velocity. We have seen that this is not true at high particle velocities (cf. Fig. 5.2). For example the relativistic mass increase of a bombarding proton is about 1% at a kinetic energy of 10 MeV. In all dynamic relations involving moving particles, the relativistic mass must be used. This is particularly important in Ch. 16, where we discuss the effect of the acceleration of charged particles to high energies in nuclear particle accelerators.

10.2. THE MASS ENERGY

As for radioactive decay, the energy of a nuclear reaction is given by its Q-value (4.10, 5.14):

$$Q\,(\text{MeV}) = -931.5\,\Delta\, M^0(\text{u}) \tag{10.3}$$

where

$$\Delta M^0 = M_3^0 + M_4^0 - M_1^0 - M_2^0 \tag{10.4}$$

The reaction energy may thus be computed from the rest masses of the reactants and products. If mass disappears in the reaction ($\Delta M^0 < 0$), energy is released: the reaction is said to be *exoergic*, and Q is positive. For $Q < 0$ the reaction is *endoergic* and $\Delta M^0 > 0$. (For comparison, in chemistry a negative value of ΔH is associated with exothermic reactions.) Tables of Q-values, especially for light projectiles, can be found in several literature sources.

The relativistic mass equation (5.26)

$$E_{\mathrm{kin}} = \left(m - m^0\right)\mathbf{c}^2 \tag{10.5}$$

can be separated into two terms if we define

$$E^0_{\mathrm{mass}} \equiv m^0\mathbf{c}^2 \tag{10.6}$$

and

$$E_{\mathrm{tot}} \equiv m\mathbf{c}^2 \tag{10.7}$$

Then

$$E_{\mathrm{tot}} = E_{\mathrm{kin}} + E^0_{\mathrm{mass}} \tag{10.8}$$

We will call E^0_{mass} the *mass energy*, which is independent of the kinetic energy of the particle. E^0_{mass} is *potential* energy, and can in principle be converted into any other energy form because (10.6) is a form of Einstein's mass-energy relation. E^0_{mass} is closely related to the nuclear binding energy (cf. §4.4).

In the previous section we pointed out that for $Q > 0$ the reaction is exoergic and, as a consequence of that, mass has to disappear ($\Delta M^0 < 0$). If the total energy of the system is constant (conservation rule (a)), it becomes obvious from (10.8) that when E^0_{mass} decreases E_{kin} must increase. Thus the products of the exoergic nuclear reaction have a higher kinetic energy than the reactants.

10.3. THE COULOMB BARRIER

Equation (10.8) is a special case of a more general equation applicable to all systems of particles:

$$E_{\mathrm{tot}} = E_{\mathrm{kin}} + E_{\mathrm{pot}} \tag{10.9}$$

where E_{kin} = translational, rotational, vibrational, etc., energy and E_{pot} = mass energy, gravitational, electrostatic energy, surface energy, chemical binding energy, etc. This is the total energy referred to in (5.2).

In nuclear reactions we would include the mass energy for the atomic masses in their ground state E^0_{mass}, the excitation energy of the nucleus above its ground state E_{exc}, the absorption or emission of photons in the reaction E_υ, and – in reactions between charged particles – the electrostatic potential (Coulomb) energy E_{coul}. Since we are concerned only with reactions induced by neutral or positively charged particles, the Coulomb energy is zero or positive (i.e. repulsive). If charged, the incoming projectile must possess sufficient kinetic energy to overcome possible repulsion. During the intermediate stage, this energy appears as partly as recoil energy of the center of mass and partly as potential energy. In the second step of the reaction the repulsion of charged products results in greater kinetic energy. Thus, generally,

$$E_{tot} = E_{kin} + E_{coul} + E^0_{mass} + E_{exc} + E_\upsilon \qquad (10.10)$$

In nuclear reactions the total energy must be conserved, although the distribution of this energy in the different forms of (10.10) usually changes during the course of the reaction. For example, any decrease in the mass energy term must be balanced by a complimentary increase in one or more of the other energy terms. An example of this occurs in the explosion of nuclear weapons where a fraction of the rest mass is transformed to other forms of energy. The opposite reaction, the transformation of kinetic energy to mass, occurs in the production of elementary particles and in high energy acceleration of particles in cyclotrons, synchrotrons, etc.

Let us, as an example of the use of (10.10), consider a reaction of a positively charged particle (M_1, Z_1, v_1) with a target atom (M_2, Z_2, $v_2 \approx 0$). We make two simplifying assumptions: the target nucleus is in the center of the coordinate system (i.e. $v_2 = 0$) and relativistic mass corrections can be neglected.

Because both projectile and target have positive charge they must repel each other according to the Coulomb law:

$$F_{coul} = k\,\mathbf{e}\,Z_1\,\mathbf{e}\,Z_2/x^2 \qquad (10.11)$$

where k is 8.99×10^9 (N m^2 C^{-2}). This force is shown as a function of the distance between the particles in Figure 10.2a. At a distance greater than r_n only the Coulomb repulsive force, F_{coul}, is in operation; however, for distances less than r_n both the attractive nuclear force F_{nucl} and the repulsive Coulomb force F_{coul} act upon the system. The total force is given by $F_{tot} = F_{coul} + F_{nucl}$. This is shown by a solid line in the Figure. At some particular distance designated r_c, the forces balance each other and at shorter distances ($x < r_c$), the attractive nuclear force dominates. The distance r_c is known as the Coulomb radius; to be more exact, r_c is often considered to be the sum of the projectile and target radii.

In Figure 10.2b the variation in the value of the different forms of (10.10) is indicated as a function of the distance x between the two particles. At long distances from the target

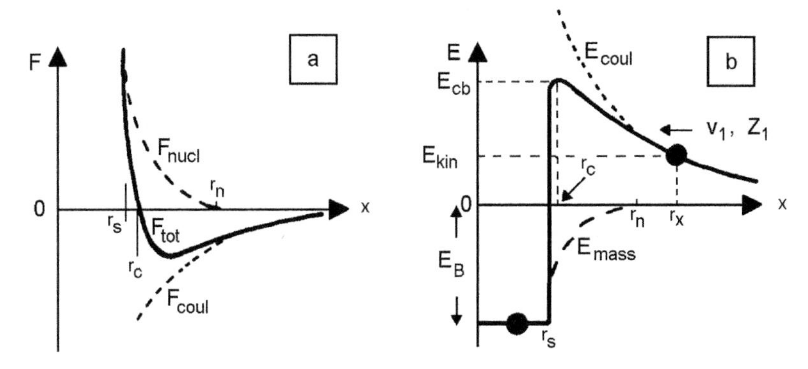

Figure 10.2 Forces (a) and energy (b) conditions when a charged projectile (Z_1, v_1) reacts with a target nucleus. In (a) the coulomb repulsion force is directed away from the nucleus, hence it is shown as negative, and the nuclear force is directed towards the nucleus and shown as positive.

nucleus the kinetic energy of the projectile is decreased due to the Coulomb repulsion. For such distances the nuclear force can be neglected and

$$E_{kin}^{0} = E_{kin} + E_{coul} \tag{10.12}$$

Where E_{kin}^{0} is the original kinetic energy of the projectile and E_{coul} is the electrostatic (Coulomb) potential energy. The initial projectile energy is thus divided into remaining kinetic enegy, E_{kin} and potential coulomb enegy, E_{coul}, which at the distance from the center x can be written as

$$E_{coul} = k\, Z_1\, Z_2\, \mathbf{e}^2/x \tag{10.13}$$

As the projectile approaches the target nucleus, the Coulomb repulsion causes the potential energy to increase as the kinetic energy of the particle decreases. If this decrease in kinetic energy of the particle is such that the kinetic energy reaches a value of zero at any distance greater than r_c, the particle is reflected away from the nucleus before it is close enough to experience the attractive nuclear force. The projectile is thus hindered by a Coulomb potential barrier from causing nuclear reaction. A necessary condition for charged projectiles to cause nuclear reactions is that E_{in} exceeds the Coulomb barrier height, E_{cb}, defined as:

$$E_{cb} = k\, Z_1\, Z_2\, \mathbf{e}^2/r_c \tag{10.14}$$

where r_c is the *Coulomb radius*. This equation is useful for determining the Coulomb radius ("distance of closest approach") of nuclei; for conservation of momentum see (10.15 and 10.16).

We learned in Chapter 4 that experimentally, the nuclear radius is given by $r = r_o\, A^{1/3} \times 10^{-15}$ m. Thus far we have discussed three different kinds of nuclear radius constants — r_s, r_c, and r_n. In addition we have treated the target and the projectile as points in space.

Experiments have indicated that the values of r_o are approximately 1.1 for the radius r_s of constant nuclear density, 1.3 for the Coulomb radius r_c, and 1.4 for the nuclear radius r_n, which includes surface effects.

To calculate the value of the energy of the Coulomb barrier we can use a model in which the target nuclei and the projectile are just touching so that r_c is taken as the sum of the radii of the projectile and the target nucleus. Moreover, we must now consider that the center of mass is not immobile in the collision but has a certain kinetic energy determined by the conservation of momentum. With this model the Coulomb barrier energy is given by the equation

$$E_{cb(min)} = 1.109\,(A_1 + A_2)\left[Z_1 Z_2 / \left\{A_2\left({A_1}^{1/3} + {A_2}^{1/3}\right)\right\}\right] \text{ (MeV)} \tag{10.15}$$

$E_{cb(min)}$ is the minimum energy that a projectile of mass A_1 and charge Z_1 must have in order to overcome the Coulomb barrier of a target nucleus of mass A_2 and charge Z_2 in a central collision. For the reaction between an α-particle and the nucleus ^{14}N, $E_{cb(min)}$ has a value of 4.99 MeV; it is obvious that Rutherford's α-particles of 7.68 MeV from the decay of ^{210}Po had sufficient kinetic energy to cause reaction.

In Figure 10.3 the values of $E_{cb(min)}$ for protons, α-, and ^{16}O-particles are shown as a function of the atomic number of the target nucleus. The A-values of the target are of the most stable isotope of the element Z. The initial decrease in the curves occur because in that range the projectile is heavier than the target nuclei.

In most cases the projectile-target collision is not head-on. Consider a projectile that would pass the target nucleus with a shortest distance, x, between their centers if no forces acted between them. This distance is called the *impact parameter.* The repulsive Coulomb force will cause the projectile to pass at a larger distance. By considering conservation of momentum and energy, and assuming that the distance of closest approach, d, is outside the range of the nuclear force one obtains

$$d = r + \left[r^2 + x^2\right]^{1/2} \tag{10.16}$$

In this equation we have introduced the *collision radius*, r, defined as

$$r = \mathrm{k}\; Z_1\; Z_2\; \mathbf{e}^2 / \left[m_{\mathrm{red}}\left(v_1^0\right)^2\right] \tag{10.17}$$

where $\mathrm{k} = 8.99 \times 10^9$ N m^2 C^{-2}, $m_{red}^{-1} = m_1^{-1} + m_2^{-1}$ and v_1^0 is the initial projectile velocity in the laboratory coordinate system.

A technique called *Coulomb excitation* is used to induce rotationally excited states in nuclei; an example of this was shown in Figure 6.7. In order to impart high energy to the nucleus without causing nuclear transformation, heavy ions with kinetic energy below that required for passing over the Coulomb barrier ($E_1 < E_{cb(min)}$) may be used as projectiles.

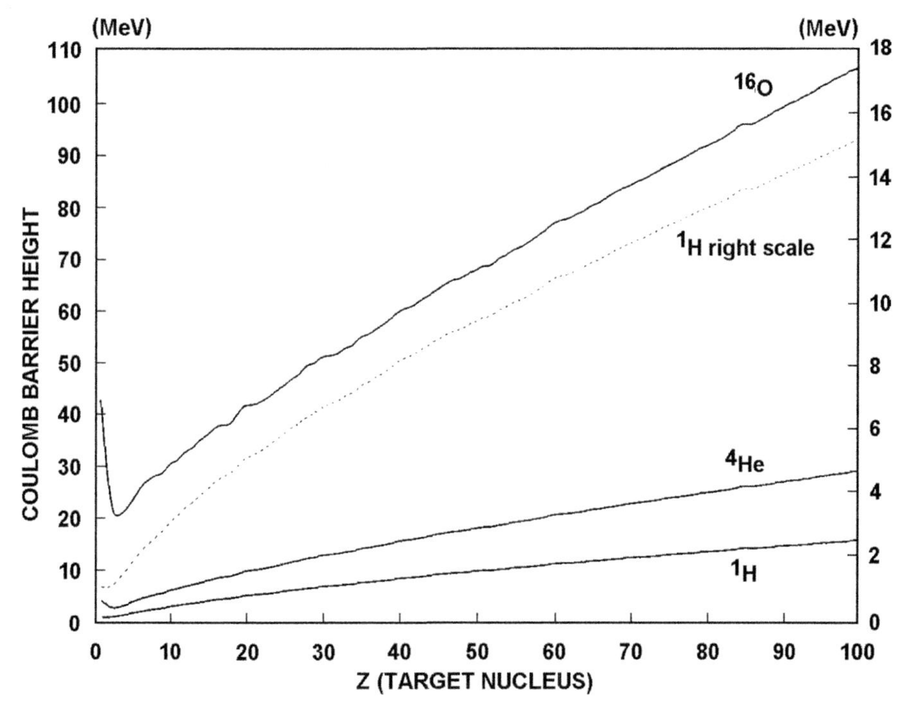

Figure 10.3 The Coulomb barrier height, E_{cb}(min) according to (10.15) for reactions between a target element Z (of most common A) and projectiles ^{1}H, ^{4}He, and ^{16}O.

10.4. RUTHERFORD SCATTERING

If a collimated beam of particles (Z_1, A_1, $E_{\text{kin},1}$) strikes a foil (Z_2, A_2) so that most of the particles pass through the foil without any reduction in energy, it is found that many particles are scattered away from their incident direction (Fig. 10.4). We can neglect multiple scattering and nuclear transformations, since they are many times less than the number of scattering events. Solid angles are measured in steradians, where 1 steradian corresponds to $1/(4\pi)$:th of the surface area of an enclosing sphere such that all particles will cross its surface area.

Geiger and Marsden found that scattering follows the relation

$$\mathrm{d}\sigma/\mathrm{d}\Omega = \left[k\, Z_1\, Z_2\, \mathbf{e}^2/\left\{2m_{\text{red}}\left(v_1^0\right)^2\right\}\right]^2/\sin^4(\theta/2) \qquad (10.18)$$

Where $\mathrm{d}\sigma/\mathrm{d}\Omega$ is the differential cross-section (reaction cross-section in barns or m^2 per steradian), θ is the scattering angle (Fig. 10.4) and v_1^0 incident particle velocity. The differential cross-section is a measure of the probability σ of a scattering event per unit solid angle Ω.

$$\mathrm{d}\sigma/\mathrm{d}\Omega = n/(n_\text{o}\ \mathrm{N_v}\ x\ \Delta\Omega) \qquad (10.19)$$

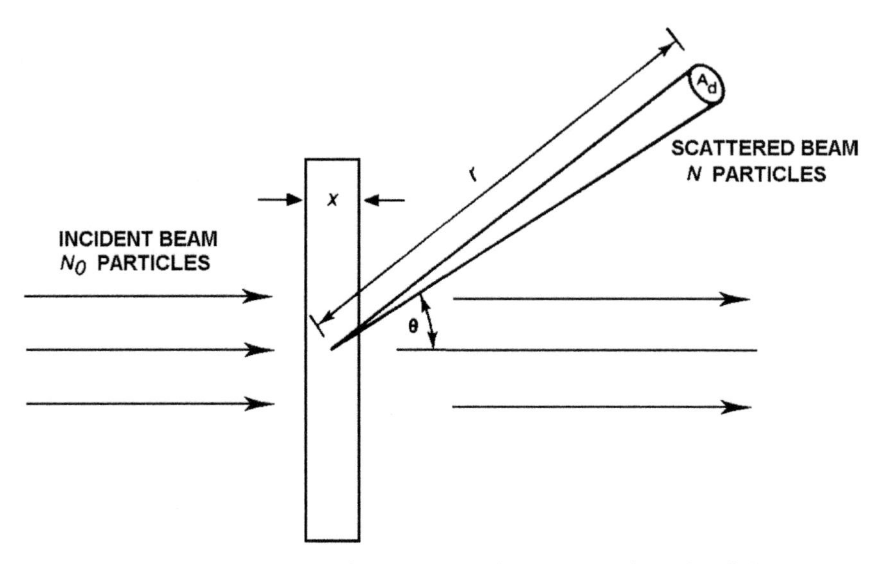

Figure 10.4 Geometry for measuring the scattering by a thin foil.

where n is the number of projectiles scattered into the detector which subtends a solid angle

$$\Delta\Omega = A_d r^{-2} \tag{10.20}$$

with respect to the center of the target; A_d is the detector area and r is the distance between target and detector; n_o is the number of projectiles hitting the target, which is x thick and contains N_v scattering atoms per unit volume ($N_v = \rho \mathbf{N}_A/M$, where ρ is the density).

By bombarding foils made of different metals with α-particles from radioactive elements, (10.18) can be used to prove that Z_2 is identical to the atomic number of the metal. This implies that the atom has a central core with a positive charge of that magnitude. Such experiments showed that some particles scattered almost directly back (i.e. $\theta > 90°$), a fact which baffled Rutherford: "It was quite the most incredible event that ever happened to me in my life. It was almost as incredible as if you fired a fifteen inch shell at a piece or tissue paper and it came back and hit you." From the scattering equations he concluded that the scattering center, the "nucleus", had a diameter which was 1000 times smaller than the atom. This same scattering technique was used 55 years later in the US lunar explorer Surveyor 5 (1967) to determine the composition of the lunar surface for the first time.

10.5. ELASTIC SCATTERING

In elastic scattering energy is exchanged between the projectile and the target nucleus but the value of Q is zero:

$$X_1(v_1) + X_2(v_2 \approx 0) \rightarrow X_1(v_1') + X_2(v_2' > 0) \quad Q = 0 \tag{10.21}$$

From the conservation of momentum and energy one finds that for a central collision:

$$v_1' = v_1(M_1 - M_2)/(M_1 + M_2) \tag{10.22}$$

$$v_2' = 2v_1 M_1/(M_1 + M_2) \tag{10.23}$$

v_1 is the projectile velocity, and v_1' and v_2' are the velocities of the scattered species.

An important example of elastic scattering reaction involves the slowing down of neutrons in nuclear reactors from high kinetic energies, which they possess when emitted in nuclear fission, to very low energies at which they have much higher reaction probabilities with the nuclear fuel. The neutrons are slowed to energies comparable to those of a neutron gas at the temperature of the material in which they are moving and, hence, they are known as *thermal neutrons*. The most probable kinetic energy for particles at thermal equilibrium at the temperature T is given by the relation $E_{kin} = \mathbf{k}T$. At 25°C the most probable energy for thermal neutrons is 0.026 eV, while the average energy is approximately 0.040 eV. In nuclear reactions it is necessary to consider the most probable kinetic energy rather than the average energy (cf. §3.6.2).

Equation (10.22) shows that for the case of head-on (or central) collisions between energetic neutrons and protons at thermal energies ($v_1 \approx 0$), the velocity of the neutron is reduced in a single step to thermal energy as a result of the approximately equal masses of the neutron and proton. This effect can be seen in a head-on collision between equally heavy billiard balls. If the target atom is ^{12}C, the velocity of a neutron decreases approximately 15% in a single collision, while the energy of the neutron decreases 28%. If the target atoms are as heavy as uranium, the decrease in velocity and in kinetic energy is on the order of about 1% per collision. Fast neutrons lose roughly their energy in a similar way. However, for ^{235}U fast neutrons lose most of their energy by elastic scattering. (This is important for fast reactors)

The slowing down of energetic neutrons to low kinetic energies is called *moderation*. Equation (10.22) shows that the lightest atoms are the best *moderators* for fast neutrons – ($M_1 - M_2$) smallest – and heavier atoms are poorer moderators because that difference is now rather large. Conversely, heavier atoms will be better *reflectors* for energetic neutrons since they will scatter them back with little loss in energy. For thermal neutrons, however, heavy atoms are not good reflectors because of their greater tendency to absorb the neutrons. Noncentral collisions ($\theta > 0$) are more common than central ones, so the average decrease in neutron energy per collision is less than the values calculated for central collisions (see further Ch. 19).

In a point neutron source, surrounded by a moderator, the thermal neutron flux (particles per unit area and unit time) as a function of the distance from the source is found initially to increase and then to decrease; see Fig. 10.5. The distance from the source at which the flux is at maximum is given in Table 10.1. This is the optimum position for thermal neutron induced reactions.

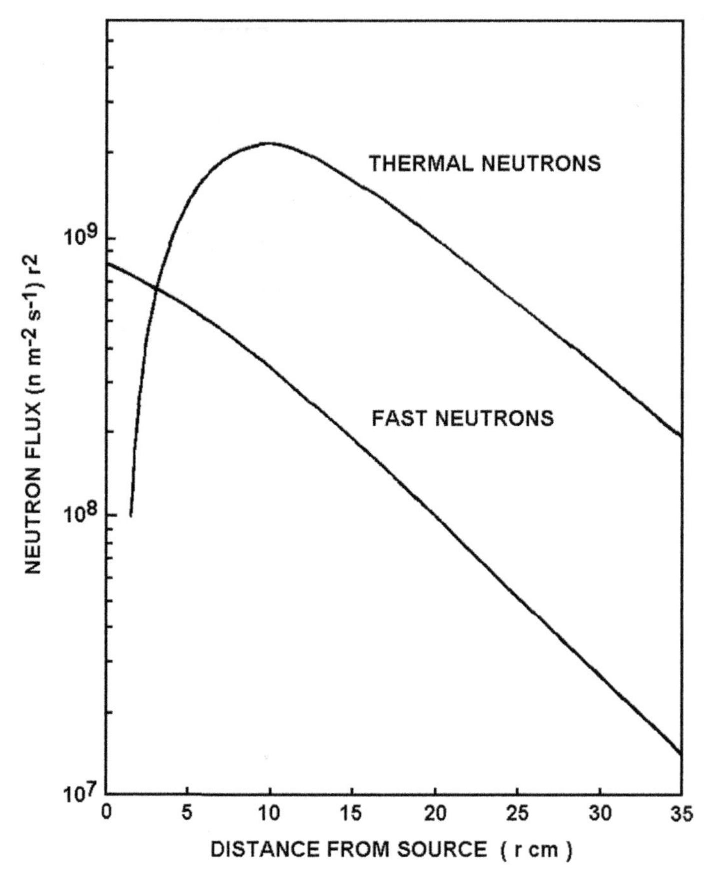

Figure 10.5 The thermalization in water of fast neutrons from a RaBe point neutron source (intensity 10^6 n s $^{-1}$).

Table 10.1 Distance from neutron source for maximum thermal flux

Neutron source	Neutron energy (at source)	Moderator	
		H_2O	D_2O
RaBe	Average 4 MeV	10 cm	
TD reaction	14 MeV	15 cm	
Fission	Average 2 MeV	7 cm	21 cm

10.6. INELASTIC SCATTERING

If energy is transferred not only to change the velocity of the projectile and the target but also the energy states, excitation, it is called inelastic scattering. Here a part of the kinetic energy of the projectile is transferred to the target nucleus as excitation energy without changing the values of A or Z of either target or projectile. In case the projectile is a heavy ion it may also become excited. However, the collision of the projectile and target forming the products does result in a value of Q different than zero. For light projectiles the process can be written:

$$X_1 + X_2 \left\langle \begin{array}{ll} a & \rightarrow X_1 + X_2 + \gamma \\ & \qquad\qquad Q<0 \\ b & \rightarrow X_1 + X_2^* \end{array} \right. \qquad \begin{array}{r} (10.24a) \\ \\ (10.24b) \end{array}$$

The reaction path *a* indicates that the energy Q is emitted as a γ-ray. In reaction path *b* the Q is retained as excitation energy of the target nuclide. The latter exists in an excited state and may transform to the ground state quite rapidly or may exist for a measurable time as an isomer. An example of inelastic scattering reaction (10.24b) is the formation of an isomer of silver by the irradiation of ^{107}Ag with neutrons

$$^{107}\mathrm{Ag(n, n')}^{107m}\mathrm{Ag} \rightarrow {}^{107}\mathrm{Ag} + \gamma \qquad (10.25)$$

the half-life of ^{107m}Ag is 44 s. The energy relationships in inelastic scattering are the same as for other nuclear reactions.

10.7. DISSECTING A NUCLEAR REACTION

As an example of a nuclear transmutation reaction, let us consider the following:

$$^{14}_{7}\mathrm{N} + {}^{4}_{2}\mathrm{He} \rightarrow [^{18}_{9}\mathrm{F}]^* \rightarrow {}^{17}_{8}\mathrm{O} + {}^{1}_{1}\mathrm{H} \qquad (10.26)$$

For pedagogic reasons we treat this reaction as if an intermediate *compound nucleus* is formed. The compound nucleus $^{18}_{9}$F is in square brackets to indicate its transitory nature and marked with an asterisk to indicate that it is excited. Induced nuclear reactions are often written in an abbreviated manner indicating first the target and then, in parentheses, the projectile and the smaller product, followed by the major product outside the parentheses. In the case of the sample reaction, we would write ^{14}N(α,p)^{17}O The abbreviations used for ^{4}He, ^{1}H, ^{2}H (=D), ^{3}H (=T), etc., are α, p, d, t, etc. The reactions may be classified by the particles in parentheses; the sample reaction is called an (α,p) type.

The sample reaction is of historical interest since this is actually the reaction studied by Rutherford in 1919 when he produced the first induced nuclear transformation in the laboratory. A cloud chamber photograph of the reaction is shown in Figure 10.6.

From the atomic rest masses we can calculate that the change in mass for this reaction is

$$\begin{aligned} \Delta M^0 &= M_3^0 + M_4^0 - M_1^0 - M_2^0 \\ &= (16.999\,131 + 1.007\,825 - 14.003\,074 - 4.002\,603) = 0.001\,279(\mathrm{u}) \end{aligned} \qquad (10.27)$$

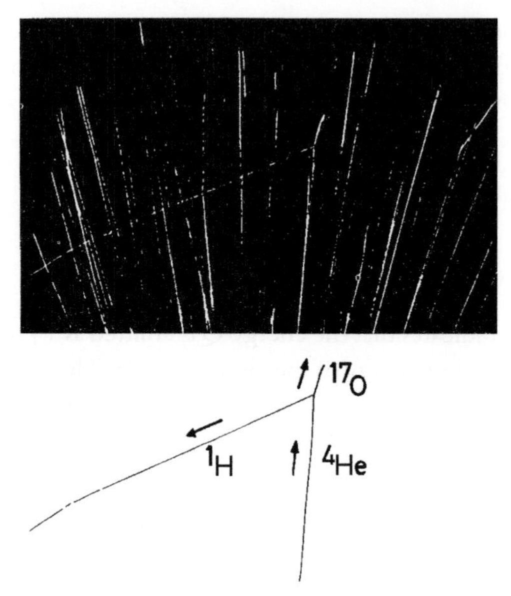

Figure 10.6 Cloud chamber photograph of the Rutherford reaction $^{4}He + {}^{14}N \rightarrow {}^{17}O + {}^{1}H$. *(From Blackett and Lees.)*

This increase in mass corresponds to a Q-value of −1.19 MeV (10.3). The energy required for this endoergic reaction can only be obtained through the kinetic energy of the projectile.

In the collision between the projectile and the target nucleus, which we have assumed to be stationary, the compound nucleus formed always acquires a certain kinetic energy, a recoil. The recoil energy can be calculated from the conservation laws to be

$$E_{\mathrm{kin,c}} = E_{\mathrm{kin,1}} M_1/(M_1 + M_2) \approx E_{\mathrm{kin,1}} A_1/A_c \tag{10.28}$$

The mass numbers can be substituted for the atomic masses since we are carrying the calculation of the energy to three significant figures only. In order for the reaction to occur, the kinetic energy of the projectile $E_{\mathrm{kin,1}}$ must exceed the kinetic energy of the compound nucleus $E_{\mathrm{kin,c}}$ by the value of Q:

$$E_{\mathrm{kin,1}} = E_{\mathrm{kin,c}} - Q \tag{10.29}$$

The minimum projectile energy necessary for a reaction to occur is called the *threshold energy* E_{tr}, which according to (10.28) and (10.29) is

$$E_{\mathrm{tr}} = -Q(M_1 + M_2)/M_2 \approx -QA_c/A_2 \quad (\text{if } Q < 0) \tag{10.30}$$

For our sample reaction (10.26), the threshold energy is 1.19 × 18/14 = 1.53 MeV.

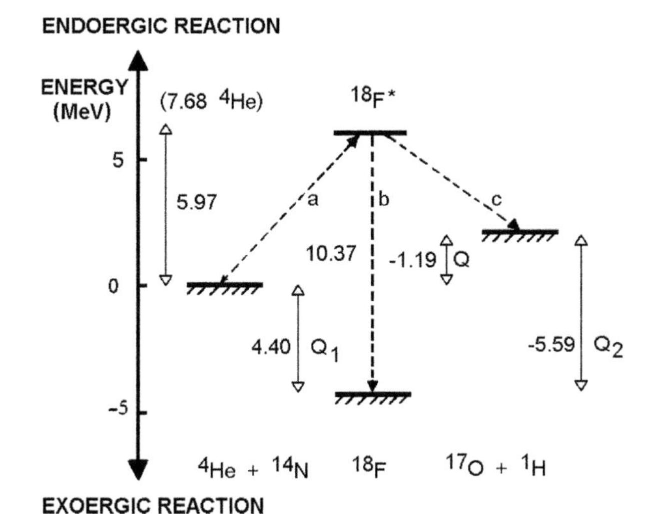

Figure 10.7 Energy diagram of the transmutation $^4He + ^{14}N \rightarrow ^{18}F^* \rightarrow ^{17}O + ^1H$ caused by bombarding ^{14}N with 7.68 MeV α-particles from ^{214}Po.

If the projectile has a higher kinetic energy than the threshold energy, the products would have a correspondingly higher combined kinetic energy, assuming no transformation into other forms of energy. Rutherford used α-particles from the radioactive decay of ^{214}Po, which have kinetic energies of 7.68 MeV. The products, therefore, had considerable kinetic energy, which is shown by the thick tracks in the cloud chamber picture of Figure 10.6.

In Figure 10.7 the energy pattern for our reaction is summarized. The Q-value for the first step of the reaction ($^4He + ^{14}N \rightarrow ^{18}F$) is +4.40 MeV ($Q_1$). The kinetic energy of the ^4He is 7.68 MeV, so conservation of momentum requires that the ^{18}F nucleus have $7.68 \times 4/(4 + 14) = 1.71$ MeV of kinetic energy. Thus, the internal excitation energy of $[^{18}F]^*$ is $4.40 + 7.68 - 1.71 = 10.37$ MeV. For the second step, $^{18}F \rightarrow ^{17}O + ^1H$, the Q_2-value is -5.59 MeV. This shows that the total excitation and kinetic energy of the products is $10.37 + 1.71 - 5.59 = 6.49$ MeV.

10.8. THE COMPOUND NUCLEUS MODEL

We have used a model which assumes the formation of an excited intermediary nucleus in the previous section to discuss some aspects of nuclear reactions. Let us now study this model a little closer.

If $E^0_{kin} > E_{cb}$ the attractive nuclear force dominates and the particle is absorbed by the target nucleus. Assuming $Q > 0$, the E^0_{mass} decreases. This means that E_{exc} increases, and the system is transformed into an excited compound nucleus. Moreover, since the projectile had a certain kinetic energy, the excitation of the compound nucleus is

$$E_{exc} = Q + E^0_{kin} \tag{10.29}$$

This equation is based on the assumption that the compound nucleus is the center of reference. It should be noted that the height of the Coulomb barrier does not influence the excitation energy of the nucleus in any way other than that a projectile must have a kinetic energy greater than that value before the reaction can occur. In the reaction all E_{kin} and E_{coul} are transferred into E_{kin}, E_{exc} and E^0_{mass} for the compound nucleus. The excitation of the compound nucleus can be removed either through the emission of γ-rays, neutrons, or by the decay of the compound nucleus into different product nuclei. In the former case, $E_{kin} = E_{\upsilon}$. In the latter case we again have to bring in (10.10).

If the reaction passes through formation of a compound nucleus, having a lifetime of many nuclear vibrations, we should expect that the amounts and types of reaction products were independent of the projectile – target combination used to produce a specific compound nucleus and only depend on its excitation energy. Essentially the compound nuclear does not have any memory of how it was formed and its decay is controlled by its own features only. This has indeed been found to be the case for some kinds of nuclear reactions, see Ch. 11.

10.9. RADIOACTIVE NEUTRON SOURCES

Neutrons were discovered through the reaction between α-particles, emitted by radioactive substances, and light elements: ^{4}He (from Po) + ^{9}Be → ^{12}C + n (Chadwick 1932). The coulomb barrier of light elements is sufficiently small to be penetrated by α-particles emitted by radioactive nuclides; in the reaction used by Chadwick the coulomb barrier is only ~3.5 MeV, eqn. (10.15) and Fig. 10.3. All early neutron research was conducted with sources of this kind, the most popular being the RaBe mixture. A number of radioactive n-sources are listed in Table 10.2.

Table 10.2 Radioactive neutron sources

Material	Half-life	Neutron yield (n s^{-1} Bq^{-1})	γ-dose rate[a] (mSv h^{-1} m^2 Bq^{-1})
^{226}Ra + Be	1600 y	3.5×10^{-4}	850
^{239}Pu + Be	24110 y	2.4×10^{-4}	4
^{239}Pu + ^{18}O	24110 y	7.8×10^{-6}	[b]
^{241}Am + Be	433 y	6.8×10^{-5}	2.5
^{210}Po + Be	138 d	6.8×10^{-5}	<0.3[c]
^{124}Sb + Be	60 d	$\sim5\times10^{-6}$	1000
^{252}Cf	2.6 y	5×10^{12} n s^{-1} g^{-1}	[d]

[a]Dose rate at 1 m from source, cf. eqn. (7.9).
[b]Variable source of Pu^{16}O$_2$ + H$_2$ ^{18}O; by transferring water through heating into the porous oxide, the reaction ^{18}O(α,n)^{21}Ne is initiated; by cooling the water evaporates and is returned to storage whereby the n output decreases to < 0.01 of maximum output.
[c]The half-life can be increased to 22 y by including the precursor nuclide ^{210}Pb.
[d]γ-dose rate increases with time due to formation of γ-emitting daughters.

The nuclide $^{252}_{98}$Cf emits neutrons through spontaneous fission in ~3% of all decays, the rest being α-decays. All the other neutron sources listed involve a radioactive nuclide whose decay causes a nuclear reaction in a secondary substance which produces neutrons. For example, $^{124}_{51}$Sb produces neutrons in beryllium powder or metal as a result of the initial emission of γ-rays, in which case there is no coulomb barrier to penetrate. Radium, polonium, plutonium, and americium produce neutrons by nuclear reactions induced in beryllium by the α-particles from their radioactive decay. For the neutrons produced either by spontaneous fission in californium or by the α-particle reaction with beryllium, the energy is between 0 and 10 MeV, while for the neutrons emitted in the γ,n-reaction involving $^{124}_{51}$Sb, the energy is 0.02 MeV. Neutron sources are commercially available.

Neutron multipliers have also come into use. These usually consist of a tank containing highly enriched ^{235}U. The uranium releases 100 n for each trigger neutron sent in; the common source is ^{252}Cf. These units are used for neutron radiography (§7.9.2), activation analysis (§18.3.3), etc.

Although the neutron flux from radioactive sources is comparatively small, they are still quite useful for specific purposes, due to their extreme simplicity, reliability (they cannot be shut off), and small size. The γ-radiation which is also present is, however, a disadvantage. From this point of view, ^{252}Cf is the most preferable neutron source, but, on the other hand, it has a relatively short half-life compared with some of the other sources. ^{252}Cf is further discussed in Chapter 14.

10.10. EXERCISES

10.1. (a) What kinetic energy must be given to a helium atom in order to increase its mass by 1%? (b) What are the mean velocity and the mean kinetic energy of a helium atom at STP?

10.2. Calculate the distance of closest approach for 5 MeV α-particles to a gold target.

10.3. In a Rutherford scattering experiment ^{2}H atoms of 150 keV are used to bombard a thin ^{58}Ni foil having a surface density of 67×10^{-6} g cm^{-2}. The detector subtends a solid angle of 1.12×10^{-4} sr and detects 4816 deuterons out of a total of 1.88×10^{12} incident on target. Calculate (a) the differential cross-section (in barns). (b) What is the distance between target and the solid state detector, which has a surface area of 0.2 cm^{2}?

10.4. In Rutherford scattering on a silver foil using α-particles from a thin-walled radon tube, the following data were observed: $d\sigma/d\Omega = 22(\theta\ 150°)$, 47(105°), 320(60°), 5260(30°), 105 400(15°) barns per steradian. Calculate the energy of the incident α-particles.

10.5. Alpha-particles from ^{218}Po (E_α 6.0 MeV) are used to bombard a gold foil. (a) How close to the gold nucleus can these particles reach? (b) What is the nuclear radius of gold according to the radius-mass relation ($r_o = 1.3$ fm)?

10.6. What is the Q-value for the reactions: (a) $^{11}B(d,\alpha)^{9}Be$; (b) $^{7}Li(p,n)^{7}Be$?

10.7. What is the maximum velocity that a deuteron of 2 MeV can impart to a ^{16}O atom?

10.8. Calculate the mass of an electron accelerated through a potential of 2×10^{8} V.

10.9. ^{12}C atoms are used to irradiate ^{239}Pu to produce an isotope of berkelium. What is the Coulomb barrier height?

10.10. Measurements made on the products of the reaction $^{7}Li(d,\alpha)^{5}He$ have led to an isotopic mass of 5.0122 for the hypothetical nuclide ^{5}He. Show that this nuclear configuration cannot be stable by considering the reaction $^{5}He \rightarrow {}^{4}He + n$.

10.11. In an experiment one hopes to produce the long-lived (2.6 y) ^{22}Na through a d,2n-reaction on neon. What is (a) the Q-value, (b) the threshold energy, (c) the Coulomb barrier height, and (d) the minimum deuteron energy for the reaction? The mass excesses (in keV) are -5185 for ^{22}Na and -8027 for ^{22}Ne.

10.11. LITERATURE

K. S. Krane, *Introductory Nuclear Physics*, J. Wiley & Sons, 1988.

Californium-252 Progress, US ERDA, Savannah River Operations Office, PO Box A, Aiken, South Carolina 29801.

W. N. Cottingham, D. A. Greenwood, *An introduction to nuclear physics*, Cambridge University Press, 2001, ISBN 0521657334.

CHAPTER 11

Mechanisms and Models of Nuclear Reactions

Contents

The variety and complexity of nuclear reactions make this a fascinating area of research quite apart from the practical value of understanding fusion and fission. From studies of such properties as the relative amounts of formation of various competing products, the variation of the yields of these with bombarding energy, the directional characteristics and kinetic energies of the products, etc., we may formulate models of nuclear reaction mechanisms. Such models lead to systematics for nuclear reactions and make possible predictions of reactions not yet investigated. As such, the nuclear reactions can be simulated in fashons very similar to those used for thermodynamic modelling of chemical reactions where the yield of different species are governed by stability constants which in turn may be differently affected by, e.g. changes in temperature etc.

11.1. THE REACTION CROSS-SECTION

The probability for a nuclear reaction to occur is expressed in terms of the *reaction cross-section*. The geometric cross-section that a nucleus presents to a beam of particles is πr^2. If

Radiochemistry and Nuclear Chemistry
ISBN 978-0-12-405897-2, http://dx.doi.org/10.1016/B978-0-12-405897-2.00011-2

we use 6×10^{-15} m as an average value for the nuclear radius, the value of πr^2 becomes $3.14\ (6 \times 10^{-15})^2 \approx 10^{-28}\ \text{m}^2$. This average geometric cross-section of nuclei is reflected in the unit of reaction probability which is the *barn*, b, where $1\ \text{b} = 10^{-28}\ \text{m}^2$. The name barn has a historic background – cross sections of $\sim 10^{-28}\ \text{m}^2$ were said to be "big as a barn".

Consider the bombardment of a target containing N_v atoms per m^3 by a homogeneous flux φ_o of particles (Fig. 11.1). The flux is expressed in units of particles $\text{m}^{-2}\ \text{s}^{-1}$. The target atoms N_v refer only to the atoms of the species involved in the nuclear reaction. If a Li-Al alloy is bombarded to induce reactions with the lithium, N_v is the number of lithium atoms per m^3 in the alloy, not the total of lithium and aluminum atoms. The change in the flux, $d\varphi = \varphi - \varphi_o$, may be infinitesimal as the particles pass through a thin section of target thickness dx. This change depends on the number of target atoms per unit area (i.e. $N_v dx$), the flux ($\varphi_o \approx \varphi$), and the reaction cross-section σ.

$$-\, d\varphi = \varphi\ \sigma\ N_v dx \tag{11.1}$$

The negative sign indicates that the flux decreases upon passing through the target due to reaction of the particles with the target atoms: thus $-d\varphi$ is the number of reactions. Integration gives:

$$\varphi = \varphi_o e^{-\sigma N_v x} \tag{11.2}$$

where φ_o is the projectile flux striking the target surface. For targets which have a surface area of S (m^2) exposed to the beam, for the irradiation time t, the total number of nuclear reactions ΔN is:

$$\Delta N = (\varphi_o - \varphi) S\, t \equiv \varphi_o S\, t \left(1 - e^{-\sigma N_v x}\right) \tag{11.3}$$

For a thin target in which the flux is not decreased appreciably upon passage through the target, i.e. $\sigma\ N_v\ x \ll 1$ and hence $e^{-\sigma N_v x} \approx 1 - \sigma\ N_v\ x$, (11.3) can be reduced to:

$$\Delta N = \varphi_o S\ t\ \sigma\ N_v x = \varphi_o \sigma\ t\ N_v V = \varphi_o \sigma\ t\ N_t \quad \text{(thin target)} \tag{11.4}$$

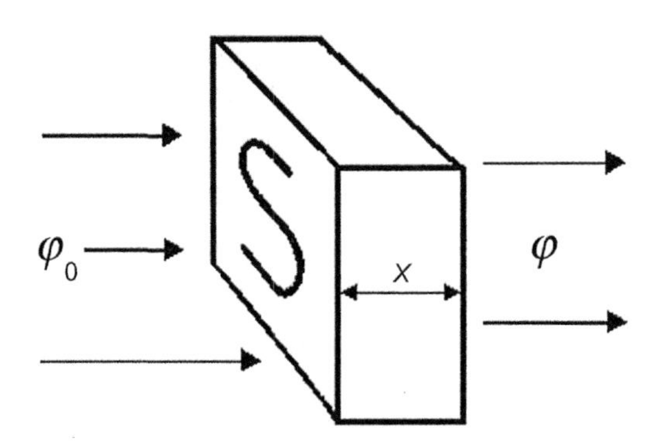

Figure 11.1 Reduction of particle flux by absorption in a target.

where $V = S\,x$ is the target volume, and $N_t = N_v\,V$ is the number of target atoms. Notice that as a result of the product $S\,x$, which equals the volume of the target, the relationship on the right of (11.4) is independent of the geometry of the target and involves only the total number of atoms in it.

Equation (11.4) can be used only when particle fluxes are homogeneous over the whole irradiated sample. In nuclear reactors, where the area of the sample is much smaller than the area of the flux, it is convenient to express the flux in terms of neutrons $m^{-2}\,s^{-1}$ and the target in terms of total number of atoms, as above.

By contrast, in an accelerator the target surface is often larger than the cross-section of the ion beam, and (11.4) cannot be used without modification. For accelerators the beam intensity (i.e. particle current) I_o (particles s^{-1}) is given by (16.3); 1 A (ampere) correspond to 6.24×10^{18} times the charge of the particle per second. Hence we have ½ × 6.24×10^{18} particles per second for a 1 A He^{2+} beam. In (11.4) $\varphi_o\,S$ must be substituted by $6.24 \times 10^{18}\,i\,z^{-1}$, where i (A) is the electric current and z is the number of electron charges.

In (11.4) $N_v\,x$ has the dimensions of atoms m^{-2}. This is a useful quantity in many calculations and for a pure elemental target is equal to

$$N_v x = 1000\,\mathbf{N}_{\mathbf{A}} \rho\, x\, y_i / M \tag{11.5}$$

where ρ is the density of the target (kg m^{-3}), M the atomic weight (in gram/mole) of the element, x the target thickness in m, and y_i the isotopic fraction of reactive atoms of kind i in the target.

As an example of the use of these equations, consider the irradiation of a gold foil by thermal neutrons. Assume the foil is 0.3 mm thick with an area of 5 cm^2 and the flux is 10^{17} n $m^{-2}\,s^{-1}$. The density of gold is 19.3 g cm^{-3} while the cross-section for the capture of thermal neutrons by ^{197}Au is 99 b. Transferring these units into SI units and introducing them into (11.3) for an irradiation time of 10 min yields a value of 4.6×10^{15} for the number of ^{198}Au nuclei formed. If the thin target equation is used (11.4), the value of 5.0×10^{15} nuclei of ^{198}Au is obtained. If the same gold foil is bombarded in a cyclotron with a beam of protons of 1 μA when the cross-section is 1 b and $t = 10$ min, the number of reactions is 6.6×10^{12}.

Frequently, the irradiated target consists of more than one nuclide which can capture bombarding particles to undergo reaction. The *macroscopic* cross-section, which refers to the total decrease in the bombarding particle flux, reflects the absorption of particles by the different nuclides in proportion to their abundance in the target as well as to their individual reaction cross-sections. Assuming that the target as a whole contains N_v atoms m^{-3} with individual abundances y_1, y_2, etc., for nuclides 1, 2, etc., the individual cross-sections are σ_1, σ_2, etc. The macroscopic cross-section $\hat{\sum}(m^{-1})$ is

$$\hat{\sum} = N_v \sum_1^n y_i \sigma_i \tag{11.6}$$

For a target which is x m thick one obtains

$$\varphi = \varphi_o e^{-\hat{\sum} x} \tag{11.7}$$

The value $\hat{\sum}^{-1}$ is the average distance a projectile travels between successive collisions with the target atoms (the *mean free path*).

11.2. PARTIAL REACTION CROSS-SECTIONS

The irradiation of a target may lead to the formation of a number of different products. For example, the irradiation of ^{63}Cu with protons can produce the nuclides ^{62}Zn, ^{63}Zn, ^{62}Cu, all of which are radioactive:

$$^{63}Cu + {}^{1}H \rightarrow \begin{cases} (p,n)\ {}^{63}Zn + n & (11.8a) \\ (p,2n)\ {}^{62}Zn + 2n & (11.8b) \\ (p,pn)\ {}^{62}Cu + {}^{1}H + n & (11.8c) \end{cases}$$

The formation probability of each product corresponds to a *partial reaction cross-section*. The total reaction cross-section σ_{tot} is the sum of all the partial cross-sections and measures the probability that the projectile causes a nuclear reaction independent of the products formed. Thus, the decrease in intensity of the particle flux is proportional to σ_{tot}. The amount of an individual product formed is proportional to σ_i, where σ_i corresponds to the partial reaction cross-section for the formation of the i:th product.

Some partial cross-sections have their own names such as the *scattering cross-section* for elastic and inelastic scattering (σ_{scat}), the *activation cross-section* (σ_{act}) for the formation of radioactive products, the *fission cross-section* (σ_f) for fission processes, and *adsorption* or *capture cross-sections* (σ_{abs} or σ_{capt}) for the absorption or capture of particles. If all of these processes take place, one obtains (with caution to avoid overlapping reactions)

$$\sigma_{tot} = \sigma_{scat} + \sigma_{act} + \sigma_f, \text{ etc} \tag{11.9}$$

In the irradiation of ^{235}U with thermal neutrons, σ_{scat} is about 10 b, σ_{act} (for forming ^{236}U) is approximately 107 b, and σ_f is 582 b.

The reaction cross-section depends on the projectile energy as shown in Figure 11.2. The curves obtained for the partial reaction cross-section as a function of projectile energy are known as *excitation functions* or *excitation curves*.

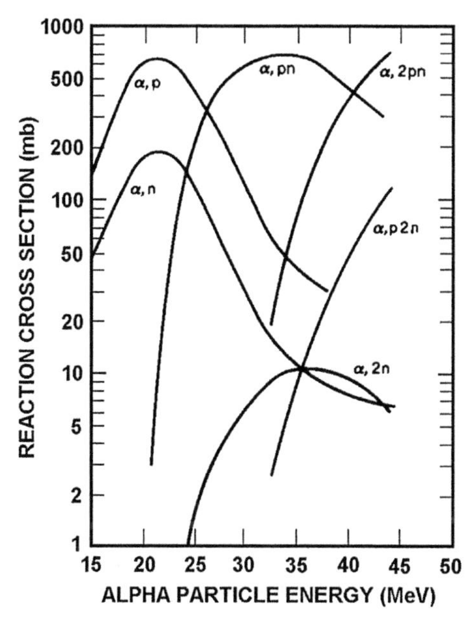

Figure 11.2 Excitation functions for reactions between ^{4}He ions and ^{54}Fe target nuclei. The kinetic energy of the projectile is in the laboratory system. (From Houck and Miller.)

11.3. RESONANCE AND TUNNELING

Experimentally, it is found that nuclear reactions sometimes occur at energies less than that required by the Coulomb barrier. This behavior is related to the wave mechanical nature of the particles involved in a nuclear reaction.

As a projectile approaches a target nucleus in a nuclear reaction, the probability that there will be overlap and hence interaction in their wave functions increases. This concept was used in §6.7.3 to explain the emission of α-particles with energies less than that required to overcome the Coulomb barrier. Such tunneling may also occur for projectiles approaching the nucleus from the outside. An example is provided by the reaction of protons with lithium (Fig. 11.3).

$$ {}^7_3\mathrm{Li} + {}^1_1\mathrm{H} \rightarrow {}^8_4\mathrm{Be}^* \rightarrow \begin{cases} 2\,{}^4_2\mathrm{He} & Q = 17.4\ \mathrm{MeV} \quad (11.10a) \\ {}^8_4\mathrm{Be} + \gamma & Q = 17.2\ \mathrm{MeV} \quad (11.10b) \\ {}^7_4\mathrm{Be} + \mathrm{n} & Q = -1.64\ \mathrm{MeV} \quad (11.10c) \end{cases} $$

For this reaction the value of $E_{cb(min)}$ is 1.3 MeV. However, due to tunneling the reactions begin to occur at lower proton energies. At an energy of 0.15 MeV about 0.1% of the protons penetrate the Coulomb barrier, at 0.3 MeV about 1%, and at 0.6 MeV about 20%.

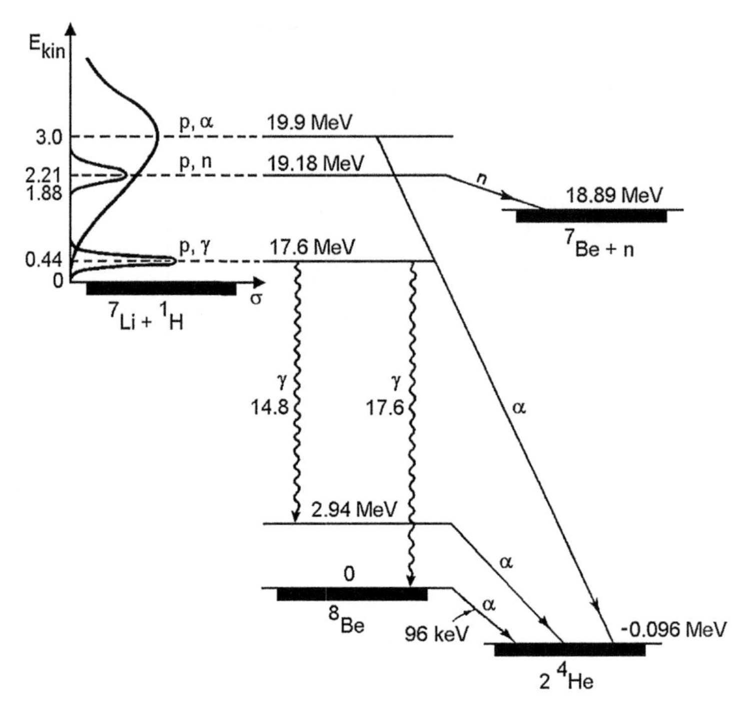

Figure 11.3 Yield curves for the reaction between protons and ^{7}Li, leading to different excited levels in ^{8}Be, followed by decay to stable end products.

The reaction cross-section is closely related to the excited energy states of the compound nucleus. Four such levels are shown for ^{8}Be* in Figure 11.3. To the left of the figure the (p,γ), (p,n) and (p,α) partial cross-sections (excitation functions) are shown as function of the proton kinetic energy. The maximum cross-section for reaction (b) occurs at a proton kinetic energy of 0.44 MeV, which, together with the Q-value, 17.2 MeV, of the reaction ^{7}Li + ^{1}H → ^{8}Be, leads to an excitation energy of 17.6 MeV, which exactly matches an excited level of the same energy in ^{8}Be*. At an excitation energy of 19.18 MeV another energy level is reached in the compound nucleus leading to its decay into ^{7}Be + n, reaction (c). The excitation energy is achieved from the release of nuclear binding energy (17.25 MeV) and from the proton kinetic energy. The amount needed is 19.18 − 17.25 = 1.93 MeV. In order to conserve momentum, the proton must have (8/7) × 1.93 = 2.21 MeV in kinetic energy. The increase in cross-section when the total excitation energy matches an excited energy level of the compound nucleus is known as a *resonance.*

This particular reaction is of interest for several reasons. It was the first nuclear reaction that was produced in a laboratory by means of artificially accelerated particles (Cockcroft and Walton 1932; cf. §16.3). Reaction (b) is still used for the production of γ-radiation (17 MeV), while reaction (c) is used as a source of mono-energetic neutrons. The energy of the neutrons from reaction (c) is a function of the proton energy and the angle between the neutron and the incident proton beam. A necessary requirement, however, is that the threshold energy ($1.64 \times (8/7) = 1.88$ MeV) must be exceeded, the Q-value for reaction (c) being -1.64 MeV.

11.4. NEUTRON CAPTURE AND SCATTERING

Unlike charged particles, no Coulomb barrier hinders neutrons from reaching the target nucleus. This leads to generally higher reaction cross-sections for neutrons, particularly at very low energies. Moreover, since neutrons can be produced in very high fluxes in nuclear reactors, neutron-induced processes are among the more important and well studied nuclear reactions.

We have seen that the geometric cross-section of a target nucleus is in the order of 1 b, or 10^{-28} m^2. Experimentally, the cross-sections for capture of energetic ("fast") neutrons ($E_n \geq 1$ MeV) are often close to 1 b. However, for neutrons whose kinetic energy is in the 1−100 eV region, some nuclei show very large cross-sections — as high as 10^5 b. Such values can be explained as being due to neutron capture where the compound nucleus is excited exactly to one of its discrete energy levels (resonance capture). This does not mean that the nucleus is larger than its calculated geometric cross-section but that the interaction probability is very large in such cases — greater than the calculation of πr^2 would indicate. This can be understood as a case where sometimes reaction occur between the tails of the overlapping wave functions even when the overlap is small.

For low energy ("slow") neutrons ($E_n \ll 1$ MeV) the cross-section is also larger than the πr^2 value, and decreases as the velocity increases; this relation, $\sigma \propto v^{-1}$, is shown for boron in Figure 11.4. The relationship between the cross-section and the neutron velocity can be understood in wave mechanical terms since the wavelength associated with the neutron increases with a decrease in velocity. According to the matter-wave hypothesis (§2.3) the wavelength associated with a moving particle is $\lambda = \hbar/(mv)$ which can be written:

$$\lambda = \hbar(2mE_{\text{kin}})^{-1/2} = 0.286 \times 10^{-8}(mE_{\text{kin}})^{-1/2} \tag{11.11}$$

While the wavelength of a slow neutron is about 0.1 nm, that of a fast neutron is less than 1/1000 of that. Since the reaction probability increases with increasing particle wavelength (lager wave function overlap), a slow-moving particle has a higher probability of reaction than a faster one of the same kind.

In effect, the wave properties make neutrons appear much larger than their geometric size and increases the probability of interaction with the nucleus. From (11.9) it follows that

$$\sigma_{\text{capt}} \propto \lambda_n \propto E_n^{-1/2} \propto v_n^{-1} \tag{11.12}$$

where λ_n, E_n, and v_n are the wavelength, kinetic energy, and velocity, respectively, of the neutron. This relation is known as the $1/v$ law. This law is valid only where no resonance

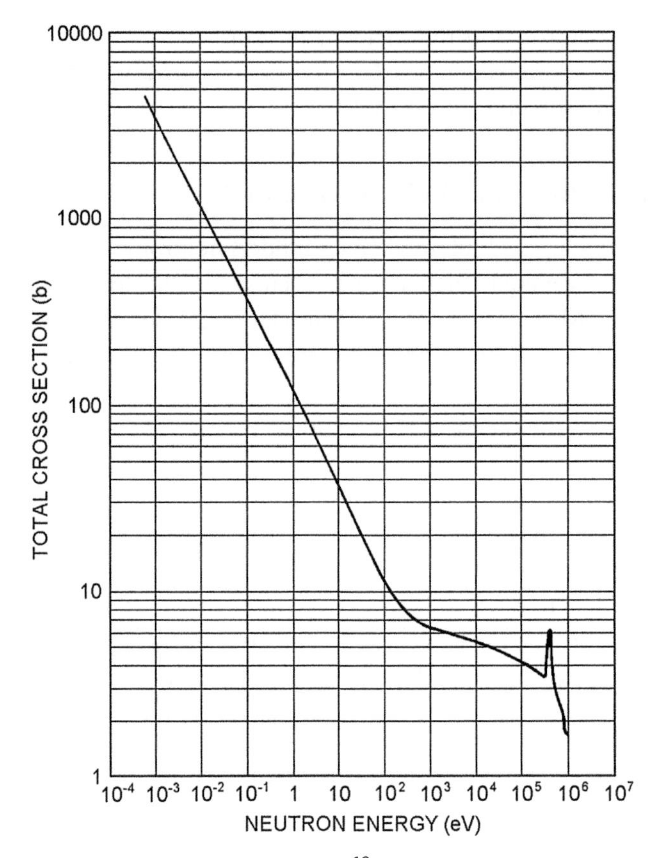

Figure 11.4 The total reaction cross-section of ^{10}B for neutrons of various kinetic energies.

absorption occurs. In Figure 11.4 $\sigma_{tot} = \sigma_{capt} + \sigma_{scat}$, but σ_{scat} is approximately constant. Since σ_{capt} decreases as E_n increases, at higher energies ($\geq$ 500 eV) σ_{scat} dominates over σ_{capt}, except for the resonance at 0.3 MeV. The capture or neutrons in ^{10}B leads to the formation of ^7Li and ^4He.

11.5. NEUTRON DIFFRACTION

From (11.12) we calculated that for neutrons the wavelength at thermal energies is on the order of 0.1 nm, i.e. of the same order of magnitude as the distance between atomic planes in a crystal. Thermal neutrons can, therefore, be scattered by crystals in the same manner as X-rays. For studies of crystal structures by neutron diffraction, a beam of mono-energetic neutrons can be obtained from the spectrum of neutron energies in a reactor by the use of monochromators.[1]

[1] Another method is mechanical time-of-flight based selectors.

As described in §11.4 neutrons interact with the nucleons in the nucleus. On the other hand photons mainly interact with the electrons surrounding the nucleus (which number is proportional to the atom number) The probability of scattering of neutrons without energy change (*coherent scattering*) is approximately proportional to the area of the nucleus. As a consequence, coherent scattering of neutrons is less dependent on the atomic number than the scattering of X-rays which is proportional to the electron density (i.e. $\propto Z^2$). Thus it is clear that for a compound consisting of both heavy and light atoms, the position of the lighter atoms can be more easily determined using neutron diffraction, while X-ray diffraction is better for locating the heavier atoms. Neutron diffraction is, therefore, valuable for complementing the information obtained on the position of heavy atoms by X-ray diffraction. Neutron diffraction is particularly valuable in the location of hydrogen atoms in organic and biological materials.

11.6. MODELS FOR NUCLEAR REACTIONS

No single model is successful in explaining all the aspects of the various types of nuclear reactions. Let us consider three models which have been proposed for explaining the results of nuclear reaction studies.

11.6.1. The optical model

In the process of elastic scattering the direction of the particles is changed but none of the kinetic energy is converted to nuclear excitation energy. This would indicate that the reaction is independent of the internal structure of the nucleus and behaves much like the scattering of light from a crystal ball. Consequently, a model has been developed based on the mathematical techniques used in optics. Light shining on a transparent crystal ball is transmitted with some scattering and reflection but no absorption. Light shining on a black crystal ball is all absorbed and there is no transmission or scattering. In nuclear reactions the incoming particles are scattered in elastic scattering and are absorbed in induced transmutations. Therefore, if the nucleus is to act as a crystal ball it can be neither totally transparent nor totally black. The optical model of the nucleus is also known as the *cloudy crystal ball model*, indicating that nuclei both scatter and absorb the incoming particles.

The nucleus is described as a potential well containing neutrons and protons. The equation for the nuclear potential includes terms for absorption and scattering. This potential can be used to calculate the probability for scattering of incident particles and the angular distribution of the scattering. The model is in excellent agreement with experiments for scattering. Unfortunately, this model does not allow us to obtain much information about the consequences of the absorption of the particles which lead to inelastic scattering and transmutation.

11.6.2. Liquid-drop model

As the excitation energy of an excited nucleus increases, the energy levels get closer together. Eventually, a continuum is reached where the density of nuclear levels is so great that it is no longer possible to identify individual levels. (This is similar to the case for electronic energy levels of atoms.) When the excited nucleus emits a proton or neutron while in the continuum energy, the resultant nucleus may be still sufficiently energetic that it remains in the continuum region.

N. Bohr offered a mechanism to explain nuclear reactions in nuclei which are excited into the continuum region. When a bombarding particle is absorbed by a nucleus, the kinetic energy of the bombarding particle plus the binding energy released by its capture provides the excitation energy of the compound nucleus. In this model, the *compound nucleus* becomes uniformly excited in a manner somewhat analogous to the warming of a small glass of water upon addition of a spoonful of boiling water. As the nucleons move about and collide in the nucleus, their individual kinetic energies vary with each collision just as those of molecules in a liquid change in molecular collisions. As this process continues, there is an increase in the probability that at least one nucleon will gain kinetic energy in excess of its binding energy (assuming the total excitation energy to be larger than the binding energy). That nucleon is then evaporated (i.e. leaves the nucleus) analogously to the evaporation of molecules from liquid surfaces.

The evaporation of the nucleon decreases the excitation energy of the residual nucleus by an amount corresponding to the binding energy plus the kinetic energy of the released nucleon. The evaporation process continues until the residual excitation energy is less than the binding energy of a nucleon. The excitation energy remaining at this point is removed from the nucleus by emission of γ-rays.

Assume that the compound nucleus $^{188}_{76}Os$ is formed with a total excitation energy of 20 MeV. If the average binding energy of a neutron is 6 MeV and if each neutron leaves with 3 MeV of kinetic energy, evaporation of a neutron de-excites the nucleus by 9 MeV. Therefore evaporation of two neutrons would leave the residual ^{186}Os nucleus with an excitation energy of only 2 MeV. Since this is below the binding energy of a neutron, further evaporation is not possible and γ-ray emission removes the final 2 MeV. If the ^{188}Os compound nucleus was formed by α-bombardment of ^{184}W, the reaction is represented as

$$^{184}_{74}W + \alpha \rightarrow \left[^{188}_{76}Os^*\right] \rightarrow {}^{186}_{76}Os + 2n \qquad (11.13)$$

The "evaporation" of neutrons is also used in the terminology for the production of superheavy elemens using "cold fusion" (few neutrons "evaporate") and "hot fusion" (many neutrons "evaporate"), cf Ch 11.

11.6.3. Lifetime of the compound nucleus

An important assumption of the compound nucleus theory is that the nucleon system is held together long enough for the energy to be shared by all nucleons. Furthermore, it is assumed that the time it takes for the accumulation on one nucleon of enough energy to allow evaporation is even longer by nuclear standards. This time is of the order of 10^{-14} s as compared to a time of 10^{-20} s required for a nucleon to cross the nuclear diameter once. Since the time is so long and there are so many inter-nucleon collisions, the nucleus retains no pattern ("no memory") of its mode of formation, and the mode of decay should therefore be independent of the mode of formation and only depend on the amount of excitation energy, the *nuclear temperature* (usually expressed in MeV). For example, ^{150}Dy ($t_{½} = 7.2$ min) can be formed in the following two ways:

$$\left.\begin{array}{l} {}^{12}_{6}\mathrm{C} + {}^{144}_{60}\mathrm{Nd} \\ \\ {}^{20}_{10}\mathrm{Ne} + {}^{136}_{56}\mathrm{Ba} \end{array}\right\rangle \rightarrow [{}^{156}_{66}\mathrm{Dy}^{*}] \rightarrow {}^{150}_{66}\mathrm{Dy} + 6\mathrm{n} \qquad (11.14)$$

both form the excited ^{156}Dy*. To a first approximation, the probability for formation of ^{150}Dy is dependent only on the excitation energy of the compound nucleus ^{156}Dy* but

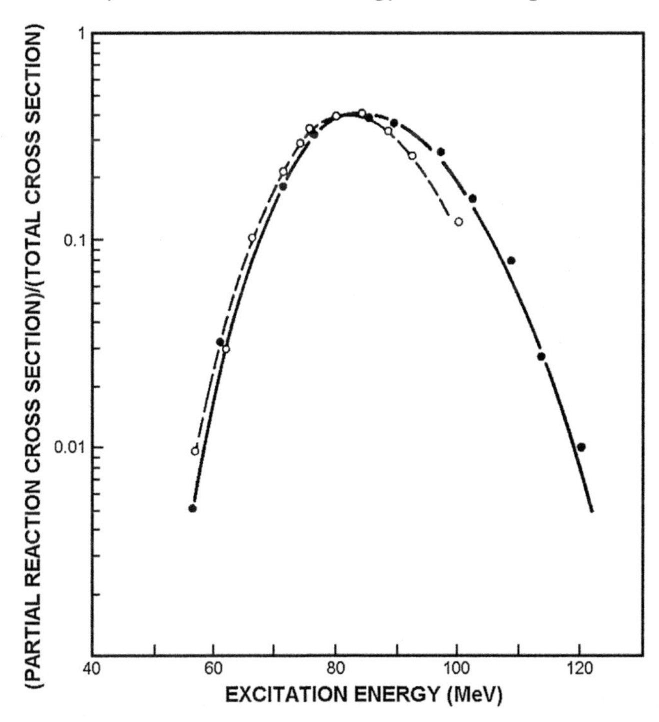

Figure 11.5 Excitation functions for the formation of ^{150}Dy, through bombardment of either ^{144}Nd with ^{12}C (○) or ^{136} Ba with ^{20}Ne (•). (From Lefort)

not on the manner in which this compound nucleus is formed (Fig. 11.5). This assumption of no memory of the mode of formation is not valid if very different amounts of angular momenta are involved in different modes of formation or the collision energy is very high. For example, for proton induced reactions, the excitation energy is essentially all available for internal (nucleon) excitation. By contrast, as heavier bombarding particles are used, the average angular momentum of the compound system increases and the excitation energy is divided between the rotation and internal (nucleon) excitation. The modes of subsequent decay of the compound nucleus is affected by the amount of excitation energy that was involved in the angular momentum of the compound nucleus.

In general, neutron emission is favoured over proton emission for two reasons. First, since there are usually more neutrons than protons in the nucleus, a neutron is likely to accumulate the necessary evaporation energy before a proton does. Second, a neutron can depart from the nucleus with a lower kinetic energy — the average neutron kinetic energy is 2–3 MeV. On the other hand, evaporating protons must penetrate the Coulomb barrier, so they often need about 5 MeV above their binding energy. It takes, as an average, a longer time for this amount (~12 MeV *vs.* ~9 MeV for neutrons) to be concentrated on one nucleon.

Obviously, such a simple picture ignores a large number of complicating effects that can, in particular cases, reverse the order of these cross-sections. Nevertheless, despite its simplicity, the compound nucleus theory has been of great value in explaining many aspects of medium energy nuclear reactions (i.e. $\leq$ 10 MeV per nucleon of the bombarding particle).

11.6.4. Direct interaction model

The compound nucleus theory assumes that the bombarding projectile interacts with the nucleus as a whole. The nucleus is excited uniformly and evaporation of low energy nucleons follows. This model fails to explain some of the phenomena observed as the kinetic energy of the bombarding particle increases. One such observation is the occurrence of high energy neutrons and protons among the emitted particles. Another is the large cross-sections for reactions such as $X_1(p, pxn)X_2$ at energies where 6 or 7 nucleons are evaporated in order to de-excite the nucleus. At still higher energies compound nucleus formation is too slow to occur at all and the target nucleus (and also projectile in case of heavy ions) splits rapidly into several fragments.

Figure 11.6 shows the cross-section for production of nuclides of $A = 20 - 200$ when ^{208}Pb is bombarded with protons of energies 40, 480, and 3000 MeV. Such reactions, yielding a large number of products at high projectile energies, have been extensively studied by nuclear chemists, partly because the mixture of products required separation by radiochemical techniques. Although some ambiguity exists in its use, the term *spallation* is often used for reactions in which a number of particles are emitted as a result of a direct interaction. At very high energies, not only is a broad range of products

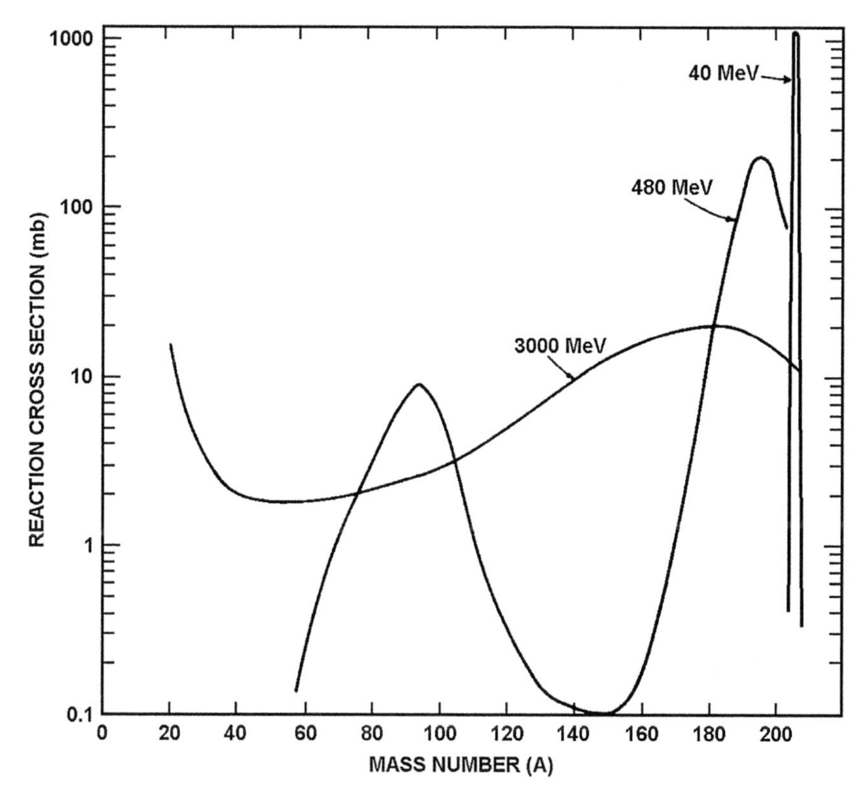

Figure 11.6 Mass yield curve obtained by bombardment of lead with high energy protons. (From Miller and Hudis.)

formed but the probability for the formation of these products is, within an order of magnitude, similar for every mass number, except around the projectile and target masses where strong peaks occur (Fig. 11.7). Further, studies at bombarding energies above 100 MeV/u show that high energy protons, neutrons, and heavier particles are emitted from the nucleus in a forward direction. Compound nucleus evaporation would be expected to be isotropic (i.e. show no directional preference in the center-of-mass system).

Serber has suggested a mechanism that satisfactorily accounts for many features of nuclear reactions at bombardment energies above 50 MeV for protons, deuterons, and α-particles. At such energies the relative speed between projectile and target nuclei is so high (near **c**) that the time available for distribution of energy between all nucleons is too short and we can initially consider projectile and target nuclei as consisting of fairly isolated nucleons. The concept of a common nuclear temperature is no longer valid. He proposed that high energy reactions occur in two stages.

(i) During the first stage the nucleons in the incoming particle undergoes direct collision with individual target nucleons. In these collisions the struck nucleon

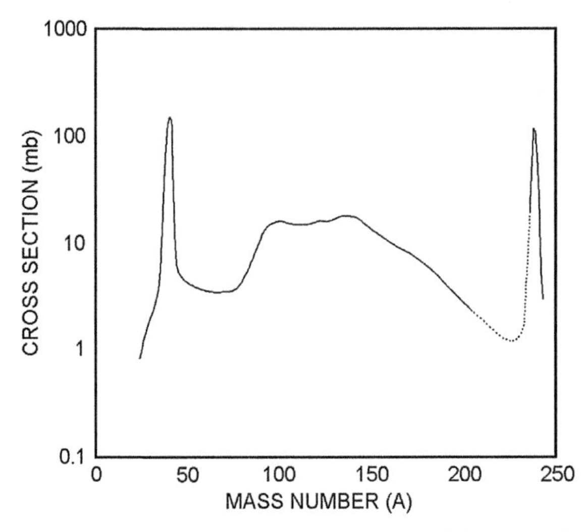

Figure 11.7 Mass yield curve obtained by bombardment of ^{238}U with ^{40}Ar-ions at 7.2 MeV/u. (From Kratz, Liljenzin, Norris, and Seaborg.)

often receives energy much in excess of its binding energy. Consequently, after each collision both the nucleon belonging initially to the bombarding particle and the struck nucleon have some probability of escaping the nucleus since their kinetic energies are greater than their binding energies. If both particles escape, the nucleus is usually left with only a small amount of excitation energy. This explains the high cross-section for (p,pn) reactions. In support of this explanation, both the emitted proton and neutron have large kinetic energies. Either one or both of the original pair may collide with other nucleons in the nucleus rather than escape. During this initial stage, known as the *knock-on-cascade process*, the total number of direct collisions may be one or many. After a period lasting about 10^{-19} s, some of the struck nucleons have left the nucleus.

(ii) In the remaining nucleus the residual excitation energy is uniformly distributed. The reaction then enters its second and slower stage, during which the residual excitation energy is lost by nucleon evaporation. This stage resembles the compound nucleus process very closely.

In Figure 11.6 at 40 MeV only the second, evaporation, stage is observed as seen by the narrow mass distribution curve. The curves for 480 and 3000 MeV reflect the increased importance of the first, direct interaction stage which leads to a broad spectrum of product mass numbers.

The experimental data have been successfully reproduced using a calculation technique known as the Monte Carlo method and assuming a Fermi gas model for the nucleus. This model treats the nucleons like molecules of a very cold ideal gas in a potential well. The nucleons do not follow the Pauli exclusion principle and fill all vacant orbitals.

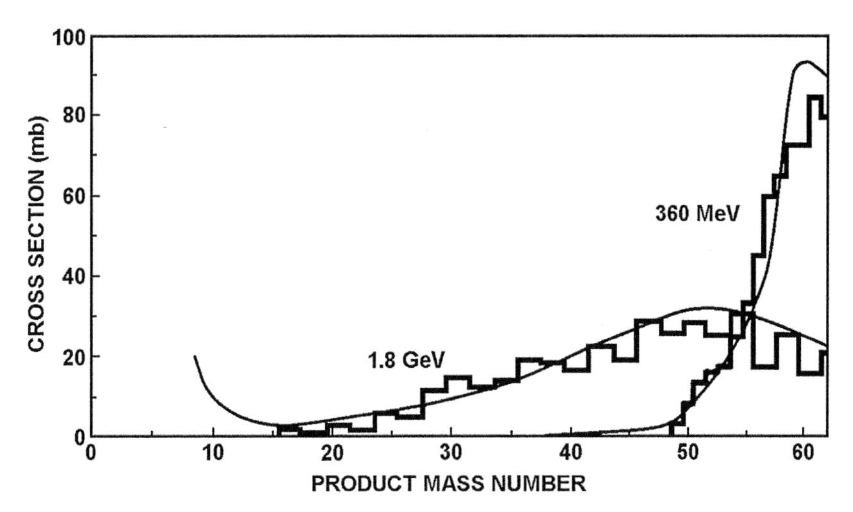

Figure 11.8 Comparison of mass yield curves for proton interaction with copper as predicted by Monte Carlo calculations (histogram) with experimental results (curve). (From G. Friedlander.)

In a *Monte Carlo calculation*, the history of each incident nucleon is studied in all its collisions in the nucleus. Each collision with a target nucleon is characterized by probability distributions for occurrence, for energy, for angular distribution, and for pion formation. The outcome of each process is obtained from a series of properly distributed random numbers and scores are accumulated for each process. The calculations are repeated for different impact points. It is the use of random numbers that is the basis of the Monte Carlo technique. Modern computers allow so many random number calculations that a useful pattern of events emerges from the cumulative scores.

Figure 11.8 shows the quite satisfactory agreement for two bombardment energies between experimental yields of A in proton bombardment of copper (solid lines) and a Monte Carlo calculation (histograms).

11.7. NUCLEAR FISSION

Nuclear fission can be either spontaneous or induced by particle (e.g. neutron) bombardment. It is fairly unique among nuclear reactions since the nucleus divides into a few heavy parts with the release of a large amount of energy, about 200 MeV per fission. It is probably no overstatement to say that fission is the most important nuclear process among man made ones, both for its potential to destroy civilization through the use of weapons and its potential through reactors to supply abundant power for all people.

An aspect of fission which has been studied extensively is the distribution of mass, charge and kinetic energy among the fragments formed in fission. No matter how nuclei are made to undergo fission, fragments of various masses are formed which result in production of chemical elements as light as zinc (atomic number 30) and as heavy as gadolinium (atomic number 64), with half-lives from fractions of seconds to millions of years. Approximately 400 different nuclides have been identified as products in the fission of ^{235}U by neutrons. Study of these fission products has required extensive radiochemical work and continues as a very active field of research – for example, in the measurement of very short-lived products. Although fission is an extremely complicated process, and still challenges theorists, satisfactory models for most of the fission phenomena have been developed.

11.7.1. Mass and charge distribution

If fission was symmetric, i.e. if two fragments of equal mass (and charge) were formed, the thermal neutron fission of $^{235}_{92}U$ would lead to the production of two $^{118}_{46}Pd$ nuclides,

$$^{235}_{92}U + n \rightarrow 2\,^{118}_{46}Pd \tag{11.15}$$

However, a plot showing the amount of different masses formed is a curve with two maxima – one near mass number 97 (close to the n-shell $N = 50$) and a second near mass number 137 (close to the n-shell $N = 82$) (Fig. 11.9). These two masses are formed together in the most probable split which is asymmetric ($A_1 \neq A_2$). As Figure 11.9 shows, symmetric fission ($A_1 = A_2$) is rare in fission of ^{235}U by thermal neutrons – the yield for $A = 115$ is only 0.01% compared to 6% for $A_1 = 97$ (or $A_2 = 137$). Since two fission products are formed in each fission event, mass yield curves like that in Figure 11.9 must total 200% on a number basis.

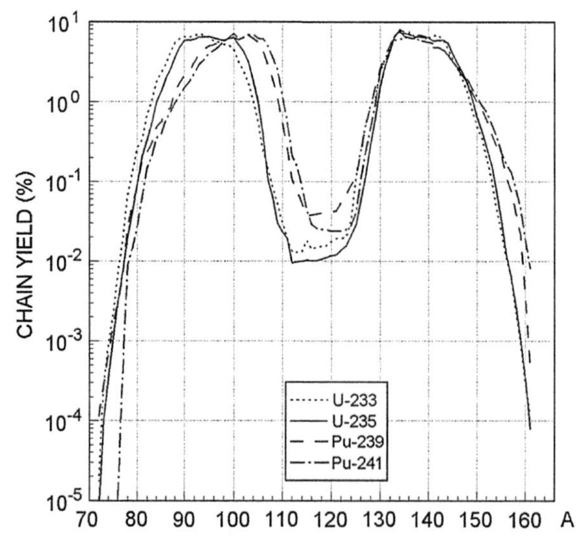

Figure 11.9 Chain yield curves for fission of ^{233}U, ^{235}U, ^{239}Pu and ^{241}Pu with thermal neutrons.

From the mass yield curve in Figure 11.9 we learn that complementary fission products (i.e. the two products A_1 and A_2 with identical yield values symmetrically located around the minimum) add up to about 234, not 236. Direct neutron measurements reveal that on the average 2.5 neutrons are emitted in fission of ^{235}U by thermal neutrons. This number increases as the Z of the target and as the bombarding energy increases.

Because the N/Z ratio for $^{236}_{92}$U(the fissioning nucleus) is 1.57, while the ratio necessary for stability is 1.2 − 1.4 in the elements produced in fission, fission fragments always have a too large N/Z ratio. This is partially compensated by the emission of several neutrons in the act of fission, *prompt neutrons*. However, the number of neutrons emitted is not sufficient to lower the N/Z ratios to stable values. To achieve further lowering, the fission fragments, after neutron emission, undergo a series of radioactive decay steps in which β^--particles are emitted. Since the β-decay occurs with no change in A, successive β-decay steps follow the isobar parabola of the stability valley (see Fig. 4.5; cf. also the nuclide chart, Appendix C). For $A = 137$, $^{137}_{52}$Te is the first nuclide measured. The chain sequence is

$$^{137}\mathrm{Te} \xrightarrow{3.5\ \mathrm{s}} {}^{137}\mathrm{I} \xrightarrow{24.5\ \mathrm{s}} {}^{137}\mathrm{Xe} \xrightarrow{3.82\ \mathrm{min}} {}^{137}\mathrm{Cs}$$
$$\xrightarrow{30.0\ \mathrm{y}} {}^{137m}\mathrm{Ba} \xrightarrow{2.6\ \mathrm{m}} {}^{137}\mathrm{Ba(stable)} \qquad (11.16)$$

Thermal fission of ^{235}U leads to a yield of 6.183% for the $A = 137$ chain. In a small number of cases the decay chains passes nuclides which emit a neutron after β^--decay, so called *delayed neutrons* (e.g. 0.016% of all neutrons from thermal fission of ^{235}U are delayed neutrons). Neutrons are emitted in 2% of the ^{137}Te β-decays and in 6.4% of the ^{137}I β-decays. The existence of delayed neutrons is important for nuclear reactor control, see Ch. 19.

In addition to measuring the variation of mass yield, the variation of fission yield in isobaric mass chains as a function of the proton number has been studied. In Figure 11.10, "individual" yield data are presented for the $A = 93$ chain. In general, the charge distribution yields follow a Gaussian curve with the maximum displaced several

~3×10^{-6}%		4×10^{-3}%		0.67%		2.96%		2.66%		0.074%		3×10^{-4}%		~2×10^{-8}%
⇓		⇓		⇓		⇓		⇓		⇓		⇓		⇓
$^{93}_{34}$Se	? →	$^{93}_{35}$Br	? →	$^{93}_{36}$Kr	1.3 s →	$^{93}_{37}$Rb	5.8 s →	$^{93}_{38}$Sr	7.45 min →	$^{93}_{39}$Y	10.1 h →	$^{93}_{40}$Zr	1.5 My →	$^{93}_{41}$Nb (stable)
↓ β⁻ n?		↓ β⁻ n		↓ β⁻ n 1.9%		↓ β⁻ n 1.4%								
$^{92}_{35}$Br		$^{92}_{36}$Kr		$^{92}_{37}$Rb		$^{92}_{38}$Sr		Total chain yield = 6.375%						

Figure 11.10 Fission product decay chain of $A = 93$. The independent yields in the upper row refer to nuclides believed to be formed directly in fission of ^{235}U by thermal neutrons.

units below the value of Z for stable nuclides with the same A. For $A = 93$ the yield is largest for $Z = 37$ and 38 (*most probable charge*, Z_p) compared to the stable value of $Z = 41$.

$$y(A,Z) = y(A)s^{-1}(2\pi)^{-1/2}e^{-(Z-Z_p)^2/(2s^2)} \tag{11.17}$$

where $y(A,Z)$ is the initial yield of the fission fragment with mass A and atomic number Z, $y(A)$ the total yield of mass A (usually given in % of all fissions) and s the width parameter for the charge distribution at mass A. Tables of $y(A)$, Z_p and s for fission of several nuclides by thermal and energetic neutrons are available in the literature. Values of $y(A)$, the *chain yield*, for fission of ^{235}U by thermal neutrons are usually also given in nuclide charts. However, it should be noted that several of the individual yields given in the data-bases or nuclide charts are not measured but calculated from a fitted model.

Fission of heavy elements other than uranium can be made to occur by particle bombardment, particularly if we use high energy neutrons or high energy charged particles such as protons. The mass distribution curve for this type of fission is interesting. At low bombarding energies it is asymmetric for many heavy elements as it is with low energy neutrons. However, as the energy of bombardment is increased the valley between the peaks of the curve becomes more shallow and, at high energies, a single-humped symmetric curve is obtained (Fig. 11.11). Thus, the most probable mode of mass split changes from asymmetric at low energies to symmetric at high energies. As the

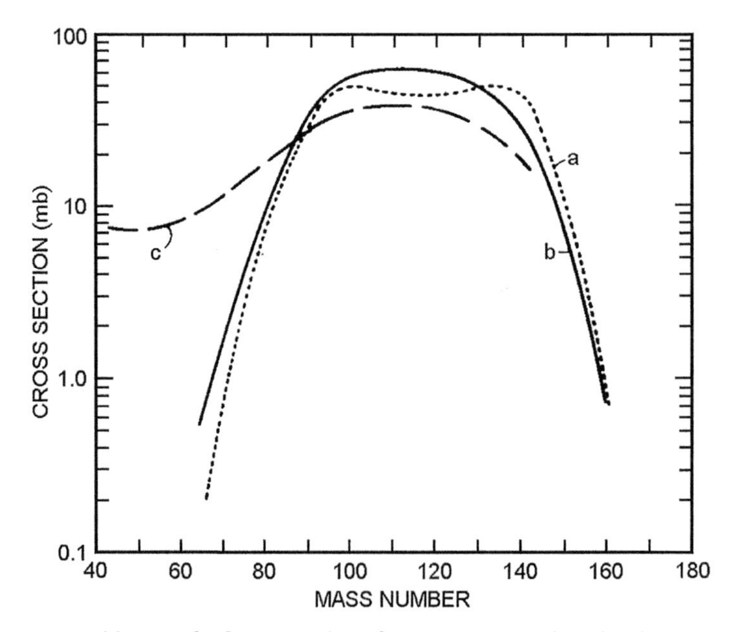

Figure 11.11 Mass yield curves for fission products from uranium irradiated with protons: (a) 100 MeV, (b) 170 MeV, and (c) 2.9 GeV. (From Friedlander.)

energy is increased, the fission yield curve $y(A)$ or cross-section curve $\sigma_{fiss}(A)$ becomes indistinguishable from yield curves of the type in Figure 11.6 ascribed to direct interaction mechanisms.

In contrast to the above, fission of nuclei in the radium region results in mass yield curves with three peaks symmetric around $\sim A/2$. Furthermore, low energy or spontaneous fission of some heavy actinide isotopes, e.g. ^{259}Fm, produces symmetric mass yield curves.

The light mass peak of double-humped fission yield curves shifts towards heavier masses when heavier nuclides undergo fission, but the position of the heavier mass peak remains almost constant. The Z_p-value increases somewhat with increasing charge of the fissioning nucleus. As an example the average mass of the light and heavy mass yield peaks in thermal fission of ^{235}U, ^{239}Pu and ^{241}Pu are 96.57, 100.34, 102.58 and 139.43, 139.66, 139.42 u, respectively, c.f. Fig. 11.9.

11.7.2. Energy of fission

From the curve of the binding energy per nucleon (see Fig. 4.3), we calculated in §4.4 that about 200 MeV would be released in the fission of a heavy element. In this section we consider how this fission energy is partitioned.

The neutrons emitted have an average kinetic energy of $\sim$2 MeV. For the average of 2.5 neutrons emitted in fission of ^{235}U by thermal neutrons, about 5 MeV of the fission energy is required. The emission of γ-rays in the act of fission, *prompt* γ-rays, accounts for another 6–8 MeV. The largest part of the fission energy is observed as the kinetic energy of the fission products. We can estimate this by calculating the Coulombic repulsion energy of a probable fission product pair, $^{93}_{37}$Rb and $^{143}_{55}$Cs. The model used is two touching spherical nuclei with a charge center distance $d = 12.7$ fm (Fig. 11.12a). From (10.16) we calculate that the repulsive Coulomb energy (and, hence, the kinetic energy of separation) is 175 MeV. If the model is modified slightly to include a neck between the two nuclei, increasing the distance between the charge centers to 13.5 fm, the kinetic energy would be 165 MeV, as experiment requires. Such an elongated shape with a small neck at the time of separation in fission (the "scission" shape) is supported by several types of evidence.

The remaining $\sim$23 MeV of fission energy is retained in the fission product nuclei as internal excitation and mass energy. This energy is released in a sequence of β-decay steps in which the N/Z values are adjusted to stability. Table 11.1 summarizes the distribution of fission energy.

11.7.3. Fragment kinetic energies

In fission induced by thermal neutrons (or spontaneous fission) the fissioning nucleus has a low kinetic energy. Thus the kinetic energies of the fragments are very near the same in the laboratory coordinate system and in the center-of-mass system. During fission, the

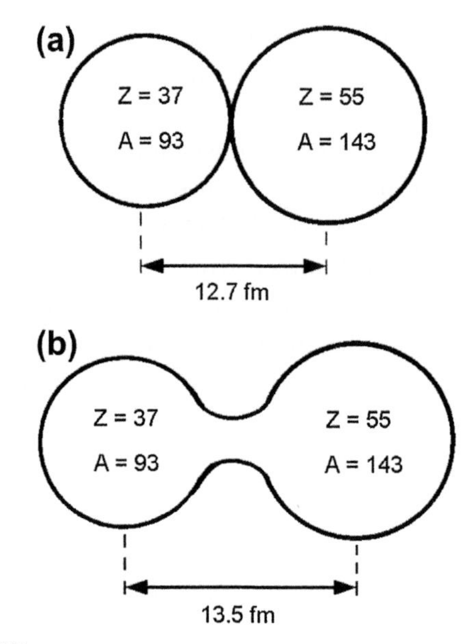

Figure 11.12 Fission of ^{236}U showing the distance between the effective centers of charge. (a) touching spheroids, (b) a scissioning dumbbell. Nuclear radii are calculated by $r = A^{1/3}$.

center of mass must remain stationary and the total impulse must be zero. Assuming $A \propto m$, the fragment kinetic energy, E_1, of a fragment with mass A_1 is given by

$$E_1 = [A_2/(A_1 + A_2)]E_{kin} \tag{11.18}$$

where E_{kin} is the total kinetic energy of the two fission fragments and $A_1 + A_2$ is the mass of the fissioning nucleus. Hence the lighter fragment initially carries a larger part of the kinetic energy than the heavier fragment. The initial distribution of kinetic energy is distorted by the emission of energetic neutrons shortly after fragment separation. Fission

Table 11.1 Data for energy distribution in thermal fission of ^{235}U in MeV

Prompt energy		*176.5 ± 5.5*
of which the	kinetic energy of fission products	164.6 ± 4.5
	kinetic energy of 2.5 prompt neutrons	4.9 ± 0.5
	γ-energy (prompt)	7.0 ± 0.5
Delayed energy from fission product decay		23.5 ± 5.0
of which the	kinetic energy of β's	6.5 ± 1.5
	neutrino radiation	10.5 ± 2.0
	γ-energy	6.5 ± 1.5
Total		200.0 ± 6

fragments are often used when calibrating the energy scale of heavy ion surface barrier detectors. The number-averaged fragment energies in spontaneous fission of ^{252}Cf are 105.71 MeV for light fragments and 80.01 MeV for heavy fragments. Rocket engines based on direct fission fragment emission have been comtemplated by NASA and seems today to be the only viable means to reach our nearest stars in less than a few 100 years.

11.7.4. Fission models

The analogy between nuclei and liquid droplets was found to be useful in deriving the semi-empirical mass formula (§4.6). Bohr and Wheeler explained fission just months after its discovery by using the same model. The surface tension of a liquid causes a droplet to assume a spherical shape, but if energy is supplied in some fashion this shape is distorted. If the attractive surface tension force is greater than the distorting force, the drop oscillates between spherical and elongated shapes. If, however, the distorting force becomes larger than the attractive force, the drop elongates past a threshold point and splits (fission).

In §4.6 we described how the repulsive forces between the protons in the nucleus could be expressed by a term a_c proportional to $Z^2/A^{1/3}$, and the surface tension attraction by another term a_s proportional to $A^{2/3}$. The repulsive Coulomb force tends to distort the nucleus in the same way a distorting force does a droplet, while the surface tension tries to bring it in to a spherical form. The ratio between the two opposing energies should measure the instability to fission of the nucleus. As shown in §6.7.4, the liquid drop model predicts that the probability of fission should increase with increasing Z^2/A. Of all naturally occurring nuclides only ^{235}U can be fissioned by thermal neutrons, while ^{238}U fission requires energetic neutrons ($\geq$ 2 MeV). With increasing Z ($>$ 92) the fission probability with thermal neutrons increases and the half-life of radioactive decay by spontaneous fission decreases. Both of these processes are more probable for even Z-elements than for odd Z-elements. The half-life for spontaneous fission decay is given in Figure 11.13 as a function of the fissionability parameter, x.

In the semi-empirical mass equation a spherical shape is assumed. If N and Z are kept constant and the potential energy of the nuclear liquid drop is calculated as a function of deformation from spherical to prolate, the curve in Figure 11.14a is obtained. The nucleus exists normally in the ground state level of the potential well. In order to undergo fission it must be excited above the fission barrier which is about 5–6 MeV. As the diagram shows, this means excitation of the nucleus into the continuum level region if the nucleus retains the shape associated with the potential well. However, if the nucleus deforms, some excitation energy goes into deformation energy. At the top of the barrier, the nucleus is highly deformed and has relatively little internal excitation energy. It exists in well-defined vibrational levels, and fission occurs from such a level. This is known as the "saddle point" (the top of the barrier) of fission.

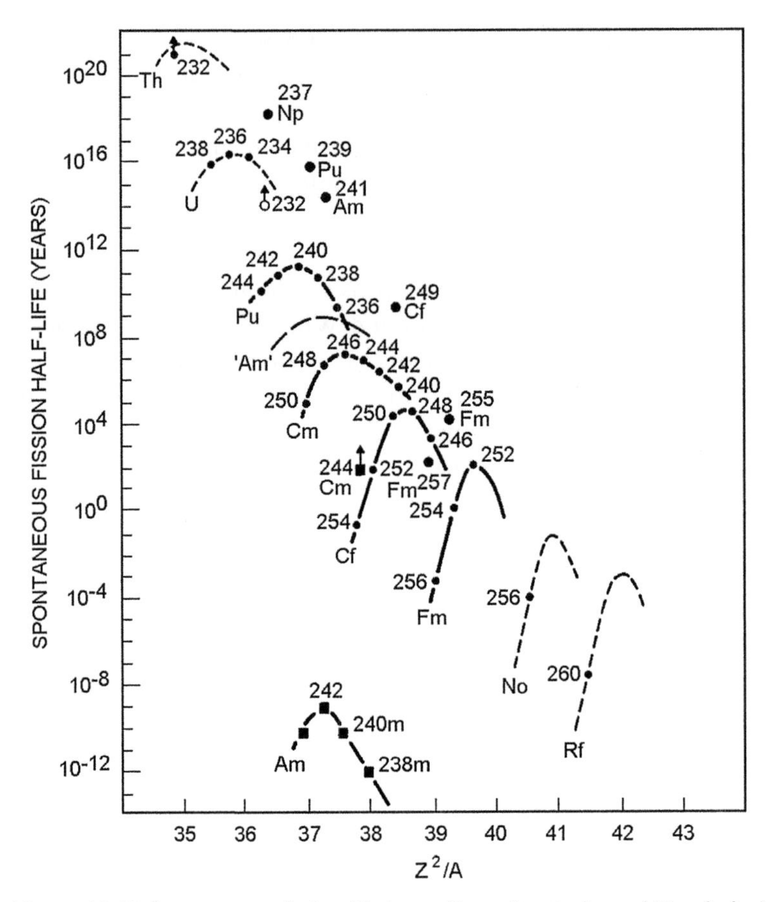

Figure 11.13 Spontaneous fission lifetimes. (From Strutinsky and Bjørnholm.)

It has long been recognized that the liquid-drop model semi-empirical mass equation cannot calculate the correct masses in the vicinity of neutron and proton magic numbers. More recently it was realized that it is less successful also for very deformed nuclei midway between closed nucleon shells. Introduction of magic numbers and deformations in the liquid drop model improved its predictions for deformed nuclei and of fission barrier heights. However, an additional complication with the liquid-drop model arose when isomers were discovered which decayed by spontaneous fission. Between uranium and californium a number of nuclides were found to decay by spontaneous fission with half-lives of 10^{-2} to 10^{-9} s, millions of times slower than prompt fission which occurs within 10^{-14} s but millions of times faster than normal spontaneous fission (Fig. 11.13). For example, ^{242}Cm has a ground state half-life to spontaneous fission of 10^6 years, while an isomeric state of ^{242}Cm has been found to fission with $t_{½}$ of 10^{-7} s.

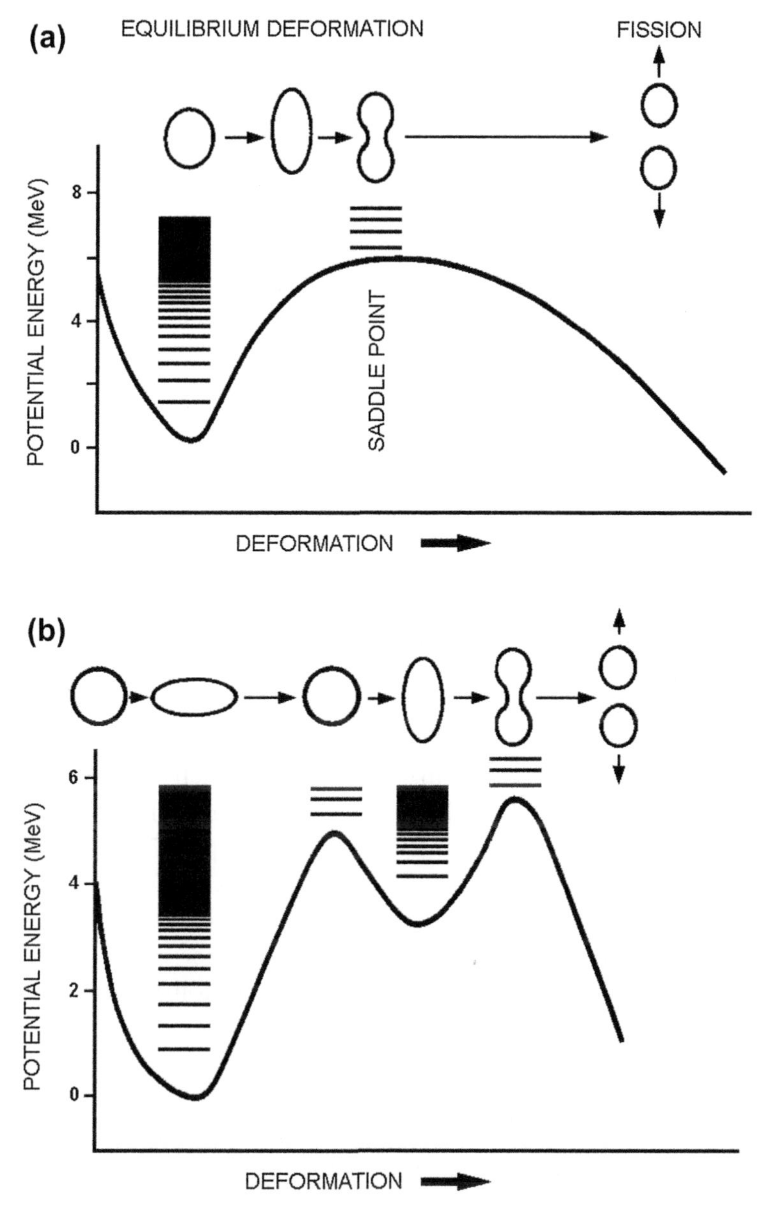

Figure 11.14 (a) The liquid-drop model potential energy curve. (b) Same, but modified by shell corrections.

Strutinsky developed an extension of the liquid drop model which satisfactorily explains the fission isomers and asymmetric fission. For such short half-lives the barrier must be only 2–3 MeV. Noting the manner in which the shell model levels vary with deformation (§6.5, the "Nilsson levels"), Strutinsky added shell corrections to the basic liquid-drop model and obtained the "double-well" potential energy curve in Figure 11.14b. In the first well the nucleus is a spheroid with the major axis about 25% larger than the minor. In the second well, the deformation is much larger, the axis ratio being about 1.8. A nucleus in the second well is metastable (i.e. in isomeric state) as it is unstable to γ-decay to the first well or to fission. Fission from the second well is hindered by a 2–3 MeV barrier, while from the first well the barrier is 5–6 MeV, accounting for the difference in half-lives.

The single-well curve in Figure 11.14a predicts symmetric fission whereas the double-well curve (Fig. 11.14b) leads to the correct prediction of asymmetric fission and a thin neck. Incorporation of shell effects in the fission model also leads to the prediction that the half-lives of very heavy nuclides ($Z \geq 106$) must be longer than the simple liquid-drop model would indicate. This led to a search for "super heavy" elements with $Z = 110 - 118$. The major difficulty is the release of reaction energy which causes immediate fission of the compound nucleus. Hence, strongly endothermic reactions have been tried. However, these have extremely small reaction yields. The heaviest man made elements are discussed in Ch. 14.

11.8. PHOTONUCLEAR REACTIONS

If a photon transfers sufficient energy to a nucleus to excite it to a higher state, three possibilities for de-excitation exist: (a) the same energy is immediately re-emitted isotropically, (b) a long-lived isomer may be formed which decays through emission of one or more γ-rays, and (c) the nucleus disintegrates. The first process (a) is referred to as the Mössbauer effect and is discussed in chapter 7. The second process (b) of nuclear de-excitation has also been discussed earlier. The third process (c) is referred to as *photonuclear disintegration.* The energy transferred to the nucleus must be sufficient to excite it above the dissociation energy for a proton, neutron, or other particle. A large energy transfer can also induce fission of heavy nuclei, *photo fission.*

The simplest photodisintegration process is that of the deuteron, whose binding energy is 2.23 MeV. If the γ-ray energy absorbed exceeds this value, a neutron and a proton are formed. This is a common reaction in nuclear reactors using heavy water as moderator, because the fission γ-ray energy is often several MeV. However, γ-rays of such energy rarely occur in the radioactive decay of nuclides (i.e. with half-lives of hours or longer). The cross-section for photodisintegration of ^2H has a maximum value of 2.4 mb at 4.3 MeV E_γ.

The energy necessary for photodisintegration of a nucleus is calculated from known nuclear masses. It is obviously easier to remove one particle than several from a nucleus. As a result we find that E_γ must be $\geq$ 5 MeV for photodisintegration of heavier nuclei.

For $10 \leq E_\gamma \leq 40$ MeV the photon wavelength is comparable to the nuclear size. It is therefore easily absorbed, which causes collective nuclear vibrational motions (so-called dipole vibrations, because the neutrons and protons are assumed to vibrate in separate groups). De-excitation occurs through γ-emission. This is known as the "giant resonance" region, because the total cross-section for heavier nuclides goes up to hundreds of millibarns. For higher E_γ, nucleons may be expelled, the main reactions being (γ,n), (γ,2n) and (γ,np) in order of descending importance. As the γ-energy increases and the wavelength decreases to nucleon dimensions, interaction with nuclear groups (e.g. deuterons) or single nucleons takes place. Below 550 MeV one pion plus a nucleon may be emitted in the de-excitation following the photon absorption. At higher energies, several pions may be emitted. Very little is known about the details of these processes.

11.9. EXERCISES

11.1. A 0.01 mm thick gold foil, 1 cm^2 in area, is irradiated with thermal neutrons. The (n,γ) cross-section is 99 b. What is the transformation rate at a n-flux of 10^{19} n m^{-2} s^{-1}?

11.2. Assume the irradiation time of the gold foil in the previous problem is one week. What percentage of the original gold atoms in the target have undergone transformation?

11.3. A water-cooled copper foil (0.1 mm thick) is irradiated by the internal beam of a sector focused cyclotron with 1.2 mA H^+ ions of 24 MeV for 90 min. The reaction ^{63}Cu(p,pn)^{62}Cu occurs with a probability of 0.086 b. Copper consists to 69% of ^{63}Cu. The proton beam has a cross-section of only 15 mm^2. (a) How many ^{62}Cu atoms have been formed? (b) What fraction of the projectiles have reacted to form ^{62}Cu? (c) What cooling effect is required (kW) at the target?

11.4. Calculate the macroscopic cross-section for reaction of natural uranium with thermal neutrons. See Figures 14.1 and 19.5.

11.5. (a) Estimate the yield (% of fissions) of ^{142}La in thermal fission of ^{235}U given a chain yield of 5.839% for A = 142, most probable charge = 55.86 and a width parameter of 0.56. (b) Is an appreciable amount of ^{142}Nd formed directly in thermal fission of ^{235}U?

11.6. The total kinetic energy of the fragments from thermal fission of ^{239}Pu is 177.7 MeV and the average fragment masses are 100.34 and 139.66 u respectively. What are the kinetic energies of the average light and heavy mass fragments?

11.7. Calculate the kinetic energy of the ^{4}He ion formed through thermal neutron capture in ^{10}B.

11.8. What is the minimum photon energy required for the reaction ^{11}B(γ,n)^{10}B?

11.9. A 2 MeV neutron collides elastically with an iron atom (^{56}Fe). What is the average temperature (corresponding to the maximum velocity) which can be ascribed to the iron nucleus after the collision?

11.10. LITERATURE

P. M. ENDT and M. DEMEUR, *Nuclear Reactions*, North-Holland, 1959.

V. GOLDANSKII and A. M. BALDWIN, *Kinetics of Nuclear Reactions*, Pergamon Press, Oxford, 1961.

Physics and Chemistry of Fission, IAEA, Vienna, 1965, 1969, and 1973.

V. S. FRASER and J. C. D. MILTON, Nuclear fission, *Ann. Rev. Nucl. Sci.,* **16** (1966) 379.

M. LEFORT, *Nuclear Chemistry*, van Nostrand, 1968.

R. VANDENBOSCH and J. R. HUIZINGA, *Nuclear Fission*, Academic Press, 1973.

K. WOLFSBERG, *Estimated Values of Fractional Yields from Low Energy Fission and a Compilation of Measured Fractional Yields*, Report LA-5553-MS, UC-34c, Los Alamos Scientific Lab., May 1974.

R. VANDENBOSCH, Spontaneously fissioning isomers, *Ann. Rev. Nucl. Sci.*, **27** (1977) 1.

Nuclear Structure Physics, ed. S. J. HALL and J. M. IRVINE, Proc. 18th Scottish U. Sum. Sch. in Physics, Edinburgh, 1977.

G. F. CHAPLINE, W. M. HOWARD, and B. G. SCHNITZLER, *Fission Fragment Rockets – A New Frontier*, in *50 Years with Nuclear Fission*, ISBN 0-89448-144-4, (1989) p. 601.

D. C. HOFFMAN, T. M. HAMILTON and M. R. LANE, *Spontaneous Fission*, Report LBL-33001, UC-413, Lawrence Berkeley Lab., Oct. 1992.

Link to some sources of fission yield data (2012): http://www-nds.iaea.org/indg_eval_fpy.html

CHAPTER 12

The Origin of the Universe and Nucleosynthesis

Contents

Today it is assumed that our universe consists of ~70% dark energy, ~25% dark matter, and ~5% of radiation and visible matter, see Table 12.1. Hydrogen, deuterium and most of the helium atoms in the normal matter part of the universe are believed to have been created 13.75 ± 0.11 billion years ago in a primary formation process referred to as the *Big Bang*, while all other elements have been formed – and are still being formed – in nuclear reactions in the stars and today also in our nuclear reactors. Most of these reaction processes can only be understood in an astrophysical context, as briefly outlined in this chapter, which also describes how nuclear science has provided much understanding about the universe, our solar system and our planets. The fusion reactions, which created the lightest elements, may in the future also provide us with a large source of energy in controlled thermonuclear reactors, see §19.17. It follows from the understanding of the processes involved that both fission and fusion used for power producing purposes should be named *nuclear power.*

Radiochemistry and Nuclear Chemistry
ISBN 978-0-12-405897-2, http://dx.doi.org/10.1016/B978-0-12-405897-2.00012-4

Table 12.1 Composition of the universe

Constituent	% of Universe mass
Dark energy	70
Dark matter	25
Free H and He	4
Stars	0.5
Neutrinos	0.3
Elements heavier than He	0.03

12.1. OBSERVATIONS FROM SPACE PROBE EARTH

12.1.1. Our place in the universe

The stars we directly can see with our eyes all belong to our Galaxy, the *Milky Way*, which is a barred spiral galaxy, about 30 kpc (kpc = kiloparsec) across and about 1 kpc thick, See Fig.12.1. The parsec (or parallax second) is the common astronomical unit of distance; 1 parsec is 3.1×10^{16} m, or 3.26 light years, ly, which corresponds to a one second change in direction angle when the earth moves between two opposite positions around the sun. Thus light travels across our galaxy in about 100 000 y. The Milky Way contains some 200×10^{9} stars, interstellar dust and gas (~200 pc thick) which spreads out to a diameter of about 50 kpc (hot gas atoms and single stars, the *halo*). It has two long spiral arms and a short one beginning from the ends of a central bar. Our sun is located in a spur from the outer edge of the short spiral arm, about 8.5 kpc from the galactic center as shown in Figures 12.1 and 12.2. The dust limits the sight at visual wave-lengths towards the center to only a few kpc; without this dust the Galaxy center would shine equally bright as our sun. Infrared- and radio-telescopes have made it possible to penetrate much of the dust and show stars circling very rapidly around the center. The stars in the bar seems at the same time to rotate like a cylinder. The other stars in our Galaxy move tangentially around its center with angular velocities also increasing closer to the center. Both observations indicate the existence of an extremely heavy central object, called Sagittarius (Sgr) A^{*}.

The Milky Way belongs to the *Local Group*, a cluster of some 20 galaxies which include, i.a. the Large Magellanic Cloud, our nearest galaxy, 50 kpc away and the Andromeda galaxy, 650 kpc away. The Local Group is part of the larger *Virgo super-cluster.* The universe contains some 10^{10} galaxies, but galaxies, stars and gases only fill a fraction of space, <5%, the rest appears void of matter. On a grand scale it looks like the galaxies and empty space form a bubbly structure, see Figure 12.3.

In the 1930s Hubble discovered that galaxies on the whole are equally distributed in all directions of space as observed from the earth. Thus space — on a large scale — seem to be isotropic. This idea of uniformity of the universe is called *the cosmological principle.*

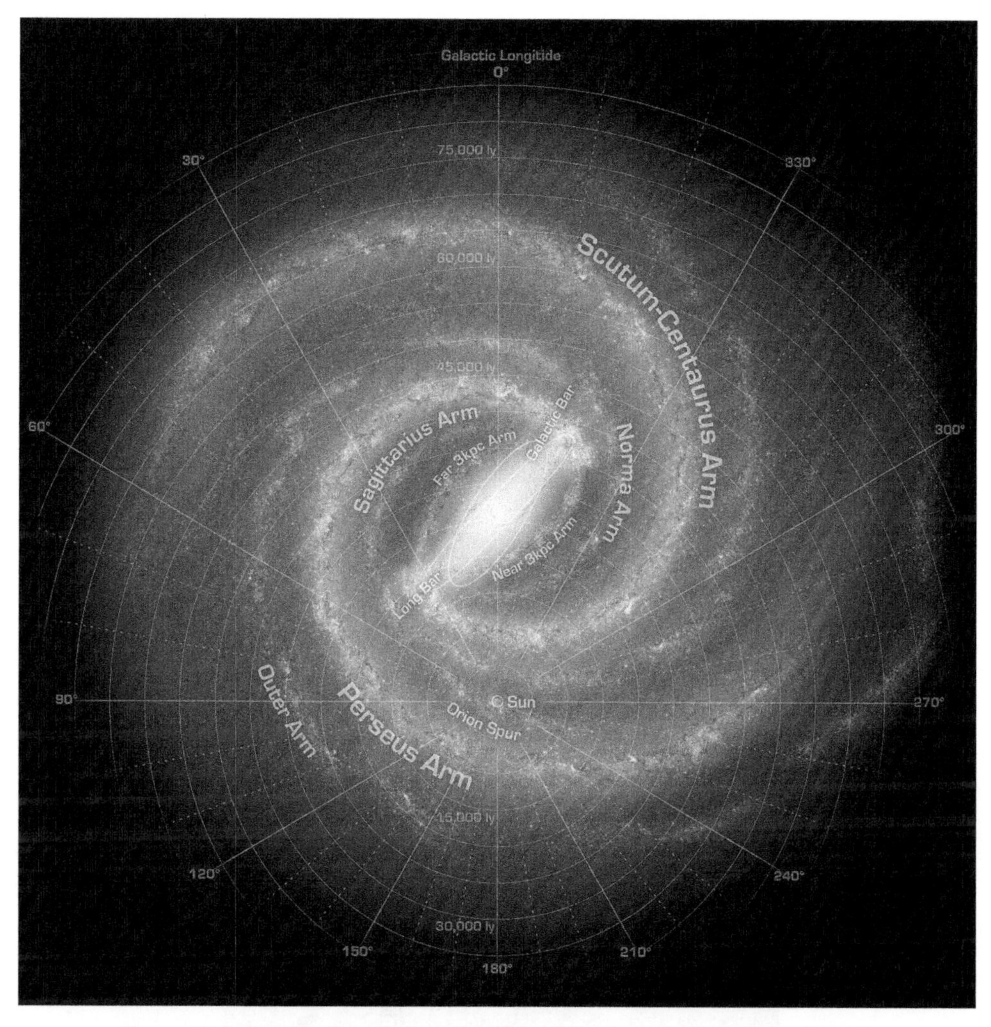

Figure 12.1 Sketch of our galaxy, the Milky Way, as seen from above (NASA).

This information has been deduced from *celestial mechanics* (movements of bodies according to Newton's fundamental laws) and from spectroscopic analysis of light and other kinds of radiation. It has been found that the mass of our sun is 1.99×10^{30} kg ($= 1$ solar mass, $\mathbf{M}_{\odot}$). The mass of the Milky Way is $>2 \times 10^{11}$ $\mathbf{M}_{\odot}$. About 10% of the mass is interstellar gas, and 0.1% is dust (typically particles with diameter 0.01–0.1 μm). The interstellar gas density varies considerably in our galaxy; in our part of space it varies from about 10^9 (in dark clouds) to 10^5 atoms m^{-3} (on the average ~1 atom cm^{-3}). Though it contains mainly H and He, also large rather complex molecules containing H, C (up to C_{15} molecules), N and O (including amino acids) have been discovered.

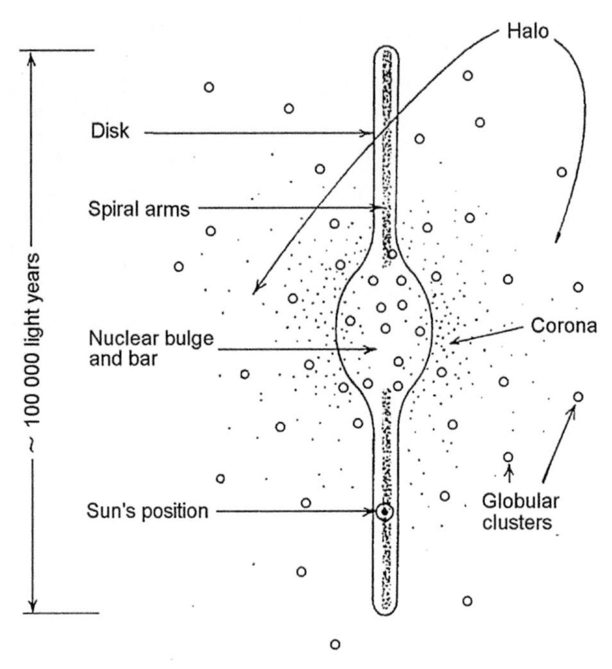

Figure 12.2 Sketch of the Milky Way as seen from the side (edge) and showing the halo.

Figure 12.3 Sketch of the general structure of our universe showing its bubbly structure. Each object plotted is a galaxy.

12.1.2. Dark matter

Astronomical models of the universe indicate that it will expand forever if the observed galaxies alone should account for the total mass of the universe. Moreover, the visible mass within galaxies also seems to be too small to hold them together. Based on the average density of the observable universe, 90% of the mass needed for a slowing down and ultimately contracting universe is missing. Most cosmologists believe that the mass of the observed galaxies is $\leq$5% of the mass of the universe, the main part consisting of "*dark matter*"; this includes interstellar and intergalactic matter, brown dwarfs, *neutron stars*, "*black holes*" and other little known sources of radiation, like quasars, whose masses are unknown. Measurements of the cosmic background electromagnetic radiation indicate that the universe now is infinite and flat.

From mechanics and Newton's gravitational law one can calculate the velocity needed for a body, m_x, to escape the gravitational pull of a larger mass, m, where $m_x \ll m$. For example, if m is the earth's mass (5.94×10^{24} kg), a rocket (mass m_x) must have a velocity of about 11 km/s to escape from the earth's surface (the *escape velocity*, v_e). Conversely, for a given velocity, v_e, one can calculate the mass and size of the large body needed to hamper such an escape. A body with one solar mass, $\mathbf{M}_\odot$, but a radius of only 3 km, requires an escape velocity $>3 \times 10^{10}$ m/s. Thus not even light will escape such a body, which therefore is termed a *black hole*. Though we cannot observe the black holes directly, some secondary effects can be observed.

Astronomical observations of star movements support the existence of black holes. For example, from movements of stars close to our galactic center, it is believed that a black hole is located at SgrA* in the center of the Milky Way, with a mass $>3 \times 10^6\ \mathbf{M}_\odot$. The radius[1] of such a hole would be of the same size as that of our sun. The density of matter in the hole would be several million times the density of our sun (average value for the sun is about 1400 kg/m^3). Obviously matter cannot be in the same atomic state (i.e. nuclei surrounded by electrons) as we know on earth. Instead we must assume that the electron shells are partly crushed; we refer to this as *degenerate matter*, because the electron quantum rules, Tables 6.1–6.2, cannot be upheld. For completely crushed atoms, matter will mainly consist of compact nuclei. For example, for ^{40}Ca the average nuclear density is $\sim 7.3 \times 10^{16}$ kg/m^3 (r $\sim$ 6 fm, cf. Fig. 4.4).

Even if black holes are given a considerable portion of the missing mass, this will not be enough. However, a recent discovery may perhaps provide the "needed" mass: Detailed analysis of the variation in luminosity (a factor of about 2.5) for some 10 million double stars in the Large Magellanic Cloud, gives support for the existence of nearby "*gravitational microlenses*", which are believed to be unborn stars (so-called *brown dwarfs*, §12.3) of sizes $\sim 10\ \mathbf{M}_\odot$. When such a dark object passes the line of sight to a distant star

[1] The radius at which light can no longer escape from the central gravity field.

it acts as a focusing lens for the light, thereby temporarily increasing that star's apparent luminosity. As these gravitational microlenses seem to be especially abundant in the halo of our galaxy (and presumably in halos of other galaxies), they — together with neutron stars and black holes — could correspond to a part of the dark-matter needed to hold galaxies and clusters of galaxies together by gravity.

12.1.3. Dark energy

Measurements of the distances and red-shifts for many distant galaxies, using type 1a supernovas as standard candles, indicate that the expansion of the universe is not decreasing as expected but expanding with increasing velocity. In practice this *red shift* is a Doppler effect shifting to longer wave lengths as a light source moves away from the observer, see eqn 12.2. The opposite is called *blue shift* and is the corresponding shift to shorter wave lengths as the source is moving close to the observer. Measurements and calculations indicate the there was a shift from retardation to expansion about 7.9×10^9 years ago.

The accelerating expansion of remote objects away from us is thought to be due to a weak repulsive force that penetrates the whole universe. Nothing is known about this force, but when it acts accelerating upon the universe an enormous energy is needed to overcome gravity and increase speed — the so called dark energy. When converted to mass by the Einstein equation it corresponds to about 70% of the total mass of the universe.

12.1.4. Light, energy flux and the hubble law

Spectral analysis of the light received from astronomical objects have provided us with information of their (surface) *temperature* (from their continuous spectrum) and outer *chemical composition* (from identification of spectral line frequencies), while bolometric measurements have given their *luminosity* (energy flux density, F Wm^{-2}). In 1911 Hertzsprung and Russell discovered that if the luminosity and colour (or temperature) of stars in different galaxies were compared with similar type of stars in the Milky Way, the stars become distributed according to a certain pattern, the so-called *Zero Age Main Sequence* (ZAMS) of stars, see Fig. 12.4. It is believed that most stars in their evolution follow the diagram beginning at the lower right side, along the main sequence into the red giant phase, then to the left and down, decreasing in size and temperature to end as blue or white dwarfs. *Hertzsprung-Russell (HR-)diagrams*, like the one in Figure 12.4, are valid for stars of about 0.7–70 $\mathbf{M}_{\odot}$: from such diagrams conclusion can be drawn about the *size* (or *mass*) and *relative age* of the star, as it is assumed that stars of a given mass follow the same sequence as they age. The apparent luminosity, F, which we observe with our telescopes, is related to the *absolute luminosity*, L^*, i.e. the total *energy flux* in all directions from a star, by the relation

$$F = L^*/4\pi d^2 \tag{12.1}$$

where d is the *distance* from the star. The historical classification of stars into *brightness classes* is now usually replaced by their relative (or apparent) *magnitude*, m^*, defined as

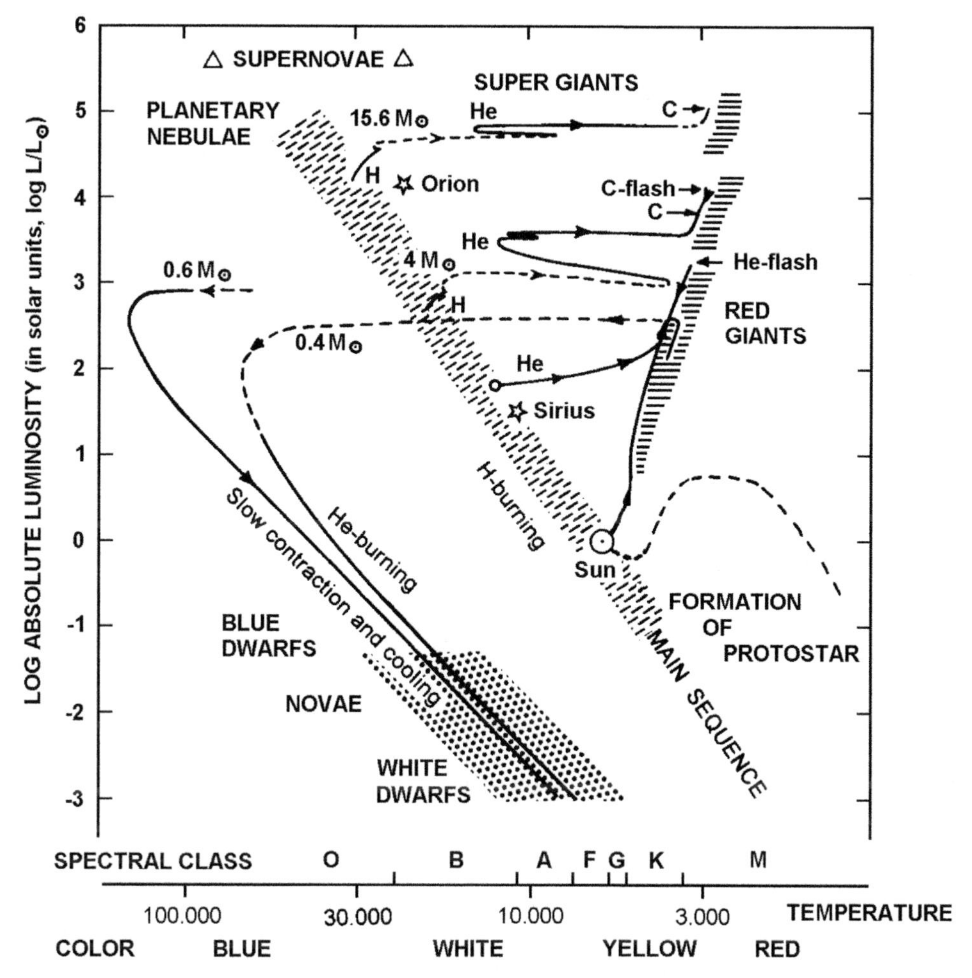

Figure 12.4 HR-diagram showing surface temperature, spectral class and color for stars born on the main sequence as a function of their absolute luminosity relative to the solar mass $M_{\odot}$.

$m^* = -2.5 \log(F/F_0)$, where F_0 is a reference flux density of light. Historically, the star Vega has been defined as having magnitude zero, but Vega is now known to vary its intensity somewhat with time and a better standard is needed.

Hubble discovered that all galaxies, except for those in the Virgo, show a *spectral redshift* (i.e. increased distance between known frequency lines). This is assumed to be a Doppler effect due to that the objects move away from us (compare the lowering of the pitch from the horn of a train moving away from us). The red shift z is

$$z = (\lambda - \lambda_o)/\lambda_o = \mathbf{H}\, d/\mathbf{c} \tag{12.2}$$

where **H** is the *Hubble constant*, λ and λ_0 are the shifted and non-shifted wave length of a specific spectral line. For *velocities* $v \ll \mathbf{c}$, the relation becomes $z = v/\mathbf{c}$, hence

$$v_r = \mathbf{H}\, d \tag{12.3}$$

which is the common expression of the *Hubble law* and v_r is the *radial velocity*. If the red shift is plotted against the apparent magnitude of the brightest star in a large number of galaxies it is seen to increase with decreasing luminosity, which is interpreted so as that more distant (faintest) galaxies move away faster from us than the closer ones. Except for the galaxies in the Local Group, all galaxies recede from us with velocities up to 20 000 km/s; hence it is concluded that *the universe expands*. This conclusion and the underlying calculations was first published in French in a small Belgian scientific journal by a catholic priest – George Lemaître. At a later date Edvin Hubble draw the same conclusion from the same type of calculations using the same data and published a paper in English – and became famous.

12.1.5. Determination of ages

If the universe in general is expanding, the galaxies were once much closer to each other. If the rate of expansion had been unchanged, the inverse of the Hubble constant, $\mathbf{H}^{-1}$, would represent the *age of the universe*.

In (12.3) v_r is the radial velocity of a galaxy at distance d from us. But velocity is just distance divided by time; i.e. $v_r = d/t_o$, where t_o is the time the movement (expansion) has gone on, assuming a constant speed. Thus,

$$d/t_o = \mathbf{H}\, d \tag{12.4}$$

or

$$t_o = 1/\mathbf{H} \tag{12.5}$$

The numerical value of t_o is only an approximate estimate of the age of the universe, because we have all reasons to believe that the movement first slowed down due to gravitational pull and then began to accelerate again. According to the best present estimate a **H**-value of 70.4 ± 1.4 km s^{-1} pc^{-1}, after some corrections, corresponds to an age of 13.7 ± 0.13 Gy (giga years, 10^9 years, or *eons*). Cosmologists also give the age in the "scale factor" $(1 + z)$-values, i.e. red-shift values; e.g. we would observe a z-value of 10 for an object about one billion years old from the formation of the universe.

The *lifetime of a solar system* can be estimated from its composition and energy production. The average surface temperature of our sun is 5780 K while that of its center is about 1.5×10^7 K. The energy production rate is 3.76×10^{26} J s^{-1} $(= L)$ and is assumed to have been relatively constant since the formation of the sun about 5 eons ago. The energy production is so immense that gravitation alone cannot account for it, leading to the conclusion that the main solar energy source is not gravitational but nuclear. This is strongly supported by the cosmic abundance of the elements and measurements of the neutrino flux from the sun.

The high temperature allows hydrogen to fuse into helium by the formal reaction

$$4^1\mathrm{H} \rightarrow {}^4\mathrm{He} + |2\beta^+ + 2e^-| + 2\nu_e \quad Q = 24.7 + 2 \times 1.02 = 26.7\ \mathrm{MeV} \qquad (12.6)$$

(neutrinos take away 0.06 MeV). The particles within || annihilate and disappear. H-burning is assumed to be the main source of nuclear energy in a star of mass m. In 5 eons $\sim 9 \times 10^{28}$ kg ^{1}H was consumed, which is $\sim$5% of the suns current content of ^{1}H. It is possible to calculate the time it takes to consume all available hydrogen; this is referred to as *the nuclear time scale*. From (12.6) one finds that about 0.70% of the hydrogen mass is converted into energy. As only about 10% of the hydrogen can be consumed before other more rapid evolutionary mechanisms set in (see below), the *nuclear lifetime*, t_n, is

$$t_n = 0.0070 \times 0.10 \times m\,\mathbf{c}^2/L \qquad (12.7)$$

which, for our sun ($m = \mathbf{M}_\odot$), gives a value of $\sim 10^{10}$ years, i.e. 10 eons.

The *age of matter* can be determined from its content of radioactive (and sometimes also stable) isotopes, as described in Chapter 13. From analyzes of lunar and meteoritic samples the age of the solar system has been calculated to be about 4.7 eons. The spectra of the light from some stars in our galaxy show the existence of relatively short lived elements like technetium (longest lived isotope has a half-life of 4.1×10^6 y). Because these elements cannot be formed in the main sequence stars, as we shall see later, they must have been accumulated "recently" from interstellar matter, indicating that these stars are rather young. From HR-diagrams, stars in our galaxy appears to have ages from <0.1 to >6 eons. Thus the evidence is that the formation of stars has gone on for a long period of time and is still occurring.

12.1.6. Elemental composition

Figure 12.5 shows the *relative abundance* of the elements of the universe and of the earth. The abundances are approximate, as a consequence of the difficulties in their assessment and limitations of experimental techniques. The abundances in the universe (based on spectral measurements on stars and interstellar matter) are used as a refinement of data obtained for the solar system. Stellar light is divided in spectral classes depending on the surface temperature of the star, see Fig. 12.4. The various classes (Harvard Spectral Classification) show lines of the elements as listed below in approximately decreasing intensity:

O	He^+, N^{2+} and Si^{3+}; weak H lines
B	He, $Si^{+,2+}$, O^+, Mg; more H than in Class O
A	Strong H; Mg^+, Si^+, Fe^+, Ti^+, Ca^+ and some other metals
F	H weaker than Class A; also Ca^+, Fe^+, Cr^+ and other metals
G	Lines of ionized metals; less H; also CH and hydrocarbons
K	Lines of neutral metals; also CH present
M	Lines of neutral metals; also molecular bands from C_2, CN, CH, TiO, ZrO, etc

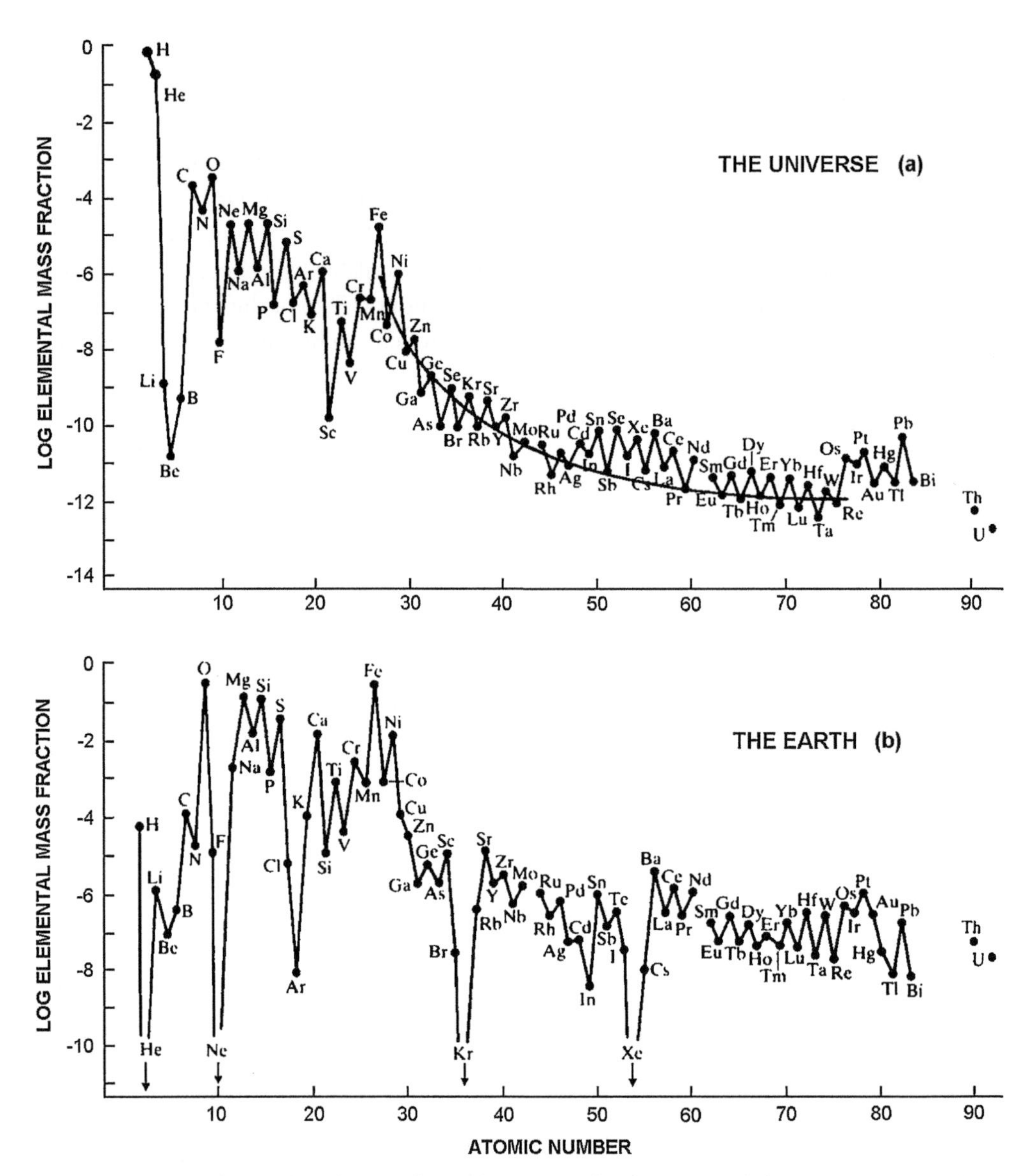

Figure 12.5 Relative abundances of the elements in the universe and the earth. *(From Cox.)*

Some stars are labelled chemically peculiar (CP-stars) because the spectrum of their light is different from most stars in the main sequence, e.g. they have very strong lines of rare earths or heavy elements. Strong lines are assumed to reflect that these elements have been moved up from the core by heavy turbulence. The Hubble Space Telescope (HST), which was placed in orbit around the earth in 1990, has allowed detailed spectroscopic analysis of CP-stars. For example in X-Lupi, high concentrations of heavy metals like

Hg, Ru and Pt have been discovered; and 100% of the mercury is ^{204}Hg, while the solar abundance of that isotope is only 7%.

The abundance data in Figure 12.5 show some regularities:

(i) H and He are the dominating elements in the universe (>99 %),

(ii) the abundances show a rapid, exponential decrease with increasing Z,

(iii) there is a pronounced abundance peak $Z=26$ (Fe), and minor peaks at 6 and 8 (C and O),

(iv) elements of even atomic number are more abundant than their neighbours of odd atomic number.

Further, a study of isotopic abundances shows:

(a) for the lightest elements, isotopes of masses which are multiples of 4 show the highest abundances, indicating their formation through fusion of He-atoms;

(b) for isobars lighter than $A \approx 70$ (corresponding to $Z \approx 28$) the most proton rich (or most neutron deficient) isobar is the most abundant, while the opposite goes for isobars with $A > 70$, indicating two different modes of synthesis for elements lighter than and heavier than $Z \approx 28$;

(c) elements that have N and Z values close to the magic numbers (Ch. 6) are more abundant then neighbouring ones, further supporting the idea of elemental formation through nuclear processes.

From these and additional observations Burbidge, Burbidge, Fowler, and Hoyle ("B^{2}FH") and others have developed a model for elemental formation in the universe, which is the basis for much of the subsequent discussion.

12.1.7. Microwave radiation and the big bang

In 1965 it was discovered that low energy microwave radiation (at 7.35 cm uncorrected) reaches us from all directions in space (about 400 photons cm^{-3}). This is referred to as the *cosmic background radiation* whose wavelength corresponds to radiation from a black body of temperature 2.725 K (about 0.0003 eV). Thermodynamic calculations show that this is the temperature reached after adiabatic expansion of a very hot cloud for some 10 billion years. The existence of such a background radiation was predicted by Gamow decades before in a cosmological hypothesis referred to as the *Big Bang model.* Recent measurements have shown that the intensity of this radiation varies slightly ("ripples") in different directions in space.

The Big Bang hypothesis requires an instantaneous beginning of our universe at a point to which all energy (in the Planck-Einstein matter-radiation-energy concept, Ch 10) is concentrated. Ordinary nuclear reactions cannot model this beginning and we must turn to particle physics for information (Ch 2).

In the following we describe the formation of the universe and the elements according to the so-called *Standard Model of Stellar Evolution*, which is based on the models originally developed by Bethe and Weizsäcker in the 1930's for the reactions in the sun, and the Big Bang hypothesis for the formation of the universe as originally suggested in 1948 by Gamow, Alpher and Herman, and later developed by the B^2FH-group, Weinberg and others.

12.2. IN THE BEGINNING OF TIME

Around "time zero" the Universe consisted of an immensely dense, hot sphere of photons, quarks and leptons, and their antiparticles, in thermal equilibrium, particles being created by photons and photons created by annihilation of particles. The temperature must have been $\geq 10^{13}$ K, but no light was emitted, because of the enormous gravitational force pulled the photons back. The system was supposed to be in a unique state with no repulsion forces. However, just as a bottle of supercritical (overheated) water can explode by a phase transition, so did the Universe, space came into being and time began. The Universe expanded violently in all directions, and as age and size grew, density and temperature fell.

A one hundreds of a second later all the quarks were gone, and the Universe consisted of an approximately equal number of electrons, positrons, neutrinos and photons, and a small amount of protons and neutrons; the ratio of protons to photons is assumed to have been about 10^{-9}. The temperature was about 10^{11} K and the density so high, about 4×10^6 kg m^{-3}, that even the unreactive neutrinos were hindered to escape.

The conditions can be partly understood by considering the relations

$$E\,(\mathrm{MeV}) = m\,\mathbf{c}^2 = 931.5\ \Delta M\,(\mathrm{u}) \qquad (4.6 \text{ and } 12.8)$$

which gives the energy required to create a particle of mass ΔM, and

$$E\,(\mathrm{MeV}) = \mathbf{k}\ T = 8.61 \times 10^{-11}\,T \qquad (3.53 \text{ and } 12.9)$$

which gives the average kinetic energy of a particle at temperature T (K, Kelvin). As the photon energy of E (eV) corresponds to the wavelength λ (m) according to (3.52)

$$E\,(\mathrm{eV}) = \boldsymbol{h}\,\nu = \boldsymbol{h}\ \mathbf{c}/\lambda = 1.240 \times 10^{-6}/\lambda\,(\mathrm{m}) \qquad (12.10)$$

one can estimate that the creation of a proton or a neutron (rest mass 940 MeV) out of radiation requires a temperature of 1.1×10^{13} K, corresponding to a photon wave length of about 10^{-15} m, i.e. the size of a nucleon. At these temperatures nucleons are formed out of radiation, but are also disrupted by photons, leading to an equilibrium with about an equal number of protons and neutrons. At temperatures below the threshold formation energy, no nucleons are formed. However, it should be remembered that particles and radiation are distributed over a range of energies, according to the Boltzmann (cf. Fig. 3.4) and Planck distribution laws. Thus some formation (and disruption) of nucleons occurs even at lower temperatures.

In about 0.1 s the temperature is assumed to have decreased to 3×10^{10} K (corresponding to 2.6 MeV). Now, the equilibrium between protons, neutrons, electrons and neutrinos can be written

$$p + \nu \rightleftharpoons n + e^{+} \quad Q = -1.80 \text{ MeV} \tag{12.11}$$

$$n + \nu \rightleftharpoons p + e^{-} \quad Q = 0.78 \text{ MeV} \tag{12.12}$$

The mass of the neutron exceeds that of the proton by a small margin of 0.0013884 u, corresponding to 1.293 MeV; thus reaction (12.11) requires energy, while reaction (12.12) releases energy. The formation of protons was therefore favoured over neutrons, leading to 38% neutrons and 62% protons.

As temperature and density further decreased the neutrinos began to behave like free particles, and below 10^{10} K they ceased to play any active role in the formation sequence (matter became transparent to the neutrinos). The temperature corresponded then to ~1 MeV, i.e. about the threshold energy for formation of positron/electron pairs. Consequently they began to annihilate each other, leaving, for some reason, a small excess of electrons. Though neutrons and protons may react at this temperature, the thermal energies were still high enough to destroy any heavier nuclides eventually formed.

After 14 s the temperature had decreased to 3×10^{9} K (0.27 MeV) and 3 min later to about 10^{9} K (<0.1 MeV). Now, with the number of electrons, protons and neutrons about equal (though the universe mostly consisted of photons and neutrinos), some protons and neutrons reacted to form stable nuclides like deuterium and helium:

$$^{1}\text{H} + {}^{1}\text{H} \rightarrow {}^{2}\text{H} + \left| e^{-} + \beta^{+} \right| + \nu_{e} \quad Q = 0.42 + 1.02 = 1.44 \text{ MeV} \tag{12.13}$$

This reaction is exothermic. However, for the ^{2}H to be stable the temperature must decrease below the Q-value, i.e. to about 10^{10} K, and, in reality, to the much lower value of about 10^{9} K, because of the high photon flux which may dissociate ^{2}H into 2 ^{1}H. Two deuterium atoms then fuse, probably in several steps as discussed below, to form He:

$$^{2}\text{H} + {}^{2}\text{H} \rightarrow {}^{4}\text{He} \quad Q = 23.85 \text{ MeV} \tag{12.14}$$

^{4}He is an extremely stable nucleus, not easily destroyed, as compared to nuclides with masses >4, whose binding energies (per nucleon) are only a few MeV (Fig. 4.3). As the universe expanded, the probability for particle collisions decreased, while the kinetic energy available for the fusion reactions was reduced. Therefore, the nucleon build-up in practice stopped with ^{4}He, leading to an average universal composition of 73% hydrogen and 27% helium. A very small amount of deuterium atoms was still left, as well as a minute fraction of heavier atoms, formed by the effects of the "Boltzmann tail" and "quantum tunnelling" (see §12.4 and Fig. 12.6). The remaining free neutrons (half-life 10.4 min) now decayed to protons.

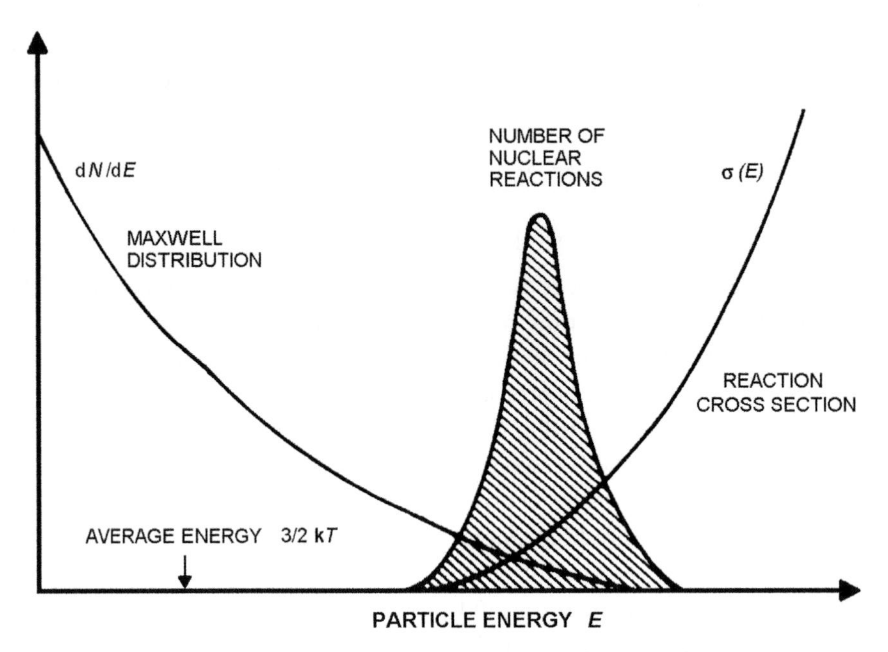

Figure 12.6 Fusion reactions occurring in the shaded area, as a function of particle energy, showing Maxwell particle distribution and reaction cross section.

The situation about 35 minutes after time zero was then the following: Temperature was 3×10^8 K, density about 10^{-4} kg m^{-3}. The Universe consisted of 69% photons, 31% neutrinos, and a fraction of 10^{-9} of particles consisting of 72–78% hydrogen and of 28–22% helium, and an equivalent number of free electrons, all rapidly expanding in all directions of space.

It was still too hot for the electrons to join the hydrogen and helium ions to form neutral atoms. This occurred not until about 500 000 years later, when temperature had dropped to a few 1000 K. The disappearance of free electrons broke the thermal contact between radiation and matter, and radiation continued then to expand freely. An outside spectator would have observed this as a hugh flash and a rapidly expanding fireball. In the adiabatic expansion the radiation cooled further to the cosmic background radiation level of ~2.7 K measured today.

The recent observation that the cosmic background radiation shows ripples in intensity in various directions of space indicates a slightly uneven ejection of matter into space, allowing gravitational forces to act, condensing the denser cloud parts into even more dense regions, or "islands", which by time separated from each other, leaving seemingly empty space in between. Within these clouds, or proto-galaxies, local higher densities lead to the formation of stars, as will be explained in next section.

12.3. STAR IGNITION

The condensation of matter releases gravitational energy which is transformed into kinetic energy of the particles in the gas cloud, increasing temperature and pressure. As the density increases, radiation transport through the cloud becomes more difficult, leading to more rapid heating of the gas.

At a critical density, the *Chandrasekhar limit*, the hydrogen degenerates into H^+ ions and non-localized ("free") electrons (this is the *plasma state*). The ionization energy of H (13.6 eV) corresponds to an average kinetic particle energy of 160 000 K, eqn. (3.31), but because of the particle energy distribution (Fig. 3.4), degeneration begins already at lower temperature. Because the elimination of the electron shell leads to a pressure reduction, the gravitational forces cause an increased rate of condensation, with a further rise in temperature until eventually all the gas is ionized.

Because the temperature decreases from the center towards the shell, the light emitted from the center may be absorbed in outer layers. Thus the protostar may be invisible, but as it gradually heats up it begins to emit light. There are many such faint objects in our galaxy.

Depending on the mass of the contracting cloud, the star evolution follows different paths.

(i) For a cloud with $m < 0.08\ \mathbf{M}_\odot$, temperature and pressure never reach high enough values for hydrogen ignition and such "stars" will contract to planet-like *brown dwarfs*.

(ii) For masses $>0.08\ \mathbf{M}_\odot$ fusion reactions begin when the core reaches $\geq 4 \times 10^6$ K. In Figure 12.4 we have indicated the birth of a star like our sun ($1\ \mathbf{M}_\odot$) which appears initially as a red object in the sky, indicating a low surface temperature, but the luminosity slowly increases over a condensation time of $10^5 - 10^6$ y.

The Coulomb barrier for interaction of protons with each other is 1.1 MeV, eqn. (10.15). In the sun, density and temperature increases strongly from the surface towards the core. At a core temperature of 1.4×10^7 K, the average kinetic energy of a proton is only 1.8 keV (the most probable energy is 1.2 keV, § 3.6.2, while the fraction of protons with energies ≥ 1.1 MeV is only $\sim 10^{-398}$, eqn. (3.35). Formation of a ^{2}He nucleus from two protons is endothermic and will not occur by tunnelling due to the lack of a driving force. On the other hand the reaction forming ^{2}H is exothermic. Hence, quantum mechanical tunnelling can allow two protons to interact to form a ^{2}H nucleus according to (12.13) during the brief encounter of a scattering event even at energies $\ll 1.1$ MeV. The reaction cross section is extremely small due to the involvement of the weak force and a change in spin for one of the nucleons ($\sim 10^{-47}$ barn at 1 MeV collision energy). Once deuterium has been formed it may fuse into helium. Thus the star has begun its "childhood" by burning hydrogen into helium: it has entered the Zero Age Main Sequence of stars. Stars with smaller masses enter lower on the Main Sequence, and evolve very slowly.

(iii) Stars with 0.08 $\mathbf{M}_{\odot} < m < 0.26$ $\mathbf{M}_{\odot}$ are completely convective, leading to hydrogen burning to helium, until all hydrogen is exhausted, after which the star contracts to a *white dwarf*.

Other stars, with $m > 0.26$ $\mathbf{M}_{\odot}$, evolve differently depending on their masses, as shown in Figure 12.4 and described in the following sections.

12.4. FUSION PROCESSES IN STARS

The elemental composition of our sun is about 73% hydrogen, 25% helium, and 2% carbon, nitrogen, oxygen, and other elements distributed as shown in Figure 12.5. In all, approximately 70 elements have been detected in the solar spectrum and there are reasons to believe that all the elements to uranium are present in our sun. Let us now consider the reactions for the formation of all these elements and the energy producing nuclear processes in our sun and other stars.

The present understanding of processes in the interior of stars is the result of combined efforts from many scientific disciplines such as hydrodynamics, plasma physics, nuclear physics, nuclear chemistry and not least astrophysics. To understand what is going on in the inaccessible interior of a star we must make a model of the star which explains the known data: mass, diameter, luminosity, surface temperature and surface composition. The development of such a model normally starts with an assumption of how elemental composition varies with distance from the center. By solving the differential equations for pressure, mass, temperature, luminosity and nuclear reactions from the surface (where these parameters are known) to the star's center and adjusting the elemental composition model until zero mass and zero luminosity is obtained at the center one arrives at a model for the star's interior. The model developed then allows us to extrapolate the star's evolution backwards and forwards in time with some confidence. Figure 12.7 shows results from such modelling of the sun.

12.4.1. Hydrogen burning to helium

Helium can be formed from hydrogen in several ways, the least likely one is (12.5), which would require that 4 protons come together simultaneously. Similarily to normal chemical rate reactions a reaction of this type involving a multi body reaction it is likely that the reaction is a multi-step process which one rate determining step. The possible multi-step processes are (not showing γ's)

$$^{1}H + {}^{1}H \rightarrow {}^{2}H + |e^{-} + e^{+}| + \nu_{e} \quad Q = 0.42 + 1.02 = 1.44 \text{ MeV} \qquad (12.13)$$

$$^{1}H + {}^{2}H \rightarrow {}^{3}He \quad Q = 5.49 \text{ MeV} \qquad (12.15)$$

$$^{1}H + {}^{3}H \rightarrow {}^{4}He \quad Q = 19.81 \text{ MeV} \qquad (12.16)$$

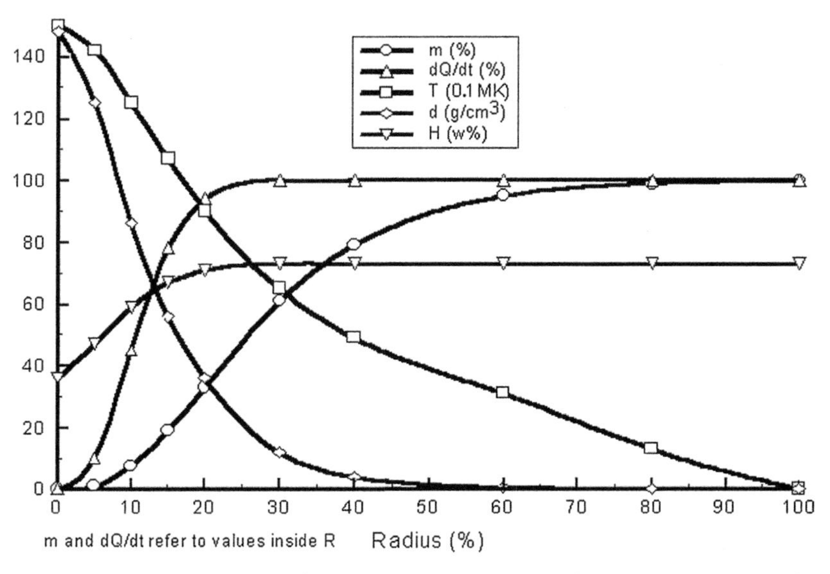

Figure 12.7 Calculated composition of our sun (mass, energy production, temperature, density and hydrogen content) as function of the radius.

$$^{2}\text{H} + \text{n} \rightarrow {}^{3}\text{H} \qquad Q = 6.25\ \text{MeV} \qquad (12.17)$$

$$^{2}\text{H} + {}^{2}\text{H} \rightarrow {}^{3}\text{H}\ (1.01\ \text{MeV}) + {}^{1}\text{H}\ (3.02\ \text{MeV}) \qquad Q = 4.03\ \text{MeV} \qquad (12.18)$$

$$^{2}\text{H} + {}^{2}\text{H} \rightarrow {}^{3}\text{He}\ (0.82\ \text{MeV}) + \text{n}\ (2.45\ \text{MeV}) \qquad Q = 3.27\ \text{MeV} \qquad (12.19)$$

$$^{2}\text{H} + {}^{2}\text{H} \rightarrow {}^{4}\text{He} \qquad Q = 23.85\ \text{MeV} \qquad (12.14)$$

$$^{2}\text{H} + {}^{3}\text{H} \rightarrow {}^{4}\text{He}\ (3.52\ \text{MeV}) + \text{n}\ (14.06\ \text{MeV}) \qquad Q = 17.58\ \text{MeV} \qquad (12.20)$$

$$^{2}\text{H} + {}^{3}\text{He} \rightarrow {}^{4}\text{He}\ (3.67\ \text{MeV}) + {}^{1}\text{H}\ (14.67\ \text{MeV}) \qquad Q = 18.35\ \text{MeV} \qquad (12.21)$$

$$^{3}\text{H} + {}^{3}\text{H} \rightarrow {}^{4}\text{He}\ (3.78\ \text{MeV}) + 2\text{n}\ (7.55\ \text{MeV}) \qquad Q = 11.33\ \text{MeV} \qquad (12.22)$$

$$^{3}\text{He} + 2\text{n} \rightarrow {}^{4}\text{He}\ (4.27\ \text{MeV}) + \text{n}\ (17.09\ \text{MeV}) \qquad Q = 21.36\ \text{MeV} \qquad (12.23)$$

$$^{3}\text{He} + {}^{3}\text{He} \rightarrow {}^{4}\text{He}\ (4.29\ \text{MeV}) + 2\,{}^{1}\text{H}(8.57\ \text{MeV}) \qquad Q = 12.86\ \text{MeV} \qquad (12.24)$$

All reactions are exothermic. Figure 12.8 shows how reaction cross sections increase with temperature. As the number of particles of a particular energy decreases with energy

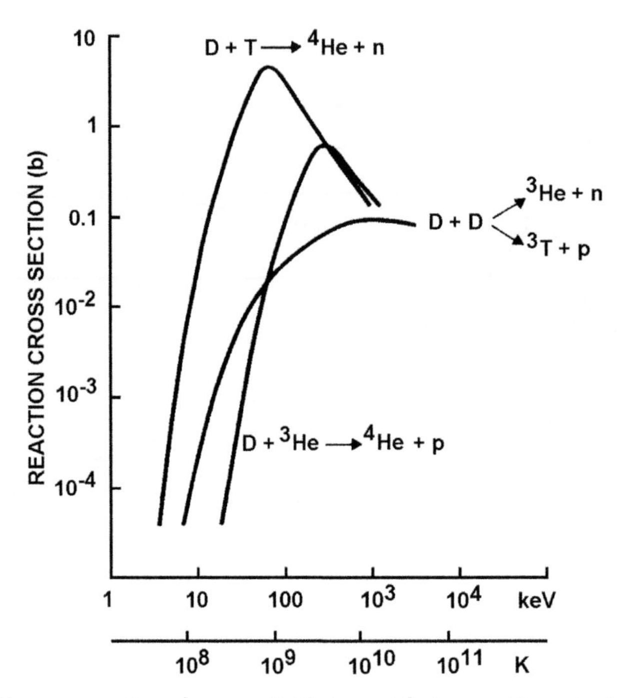

Figure 12.8 Reaction cross sections for some light element fusion reactions at solar conditions. *(From Bowler.)*

(Fig. 12.6) this results in a maximum number of reactions at a certain energy, E_{max}. The fastest hydrogen fusion reaction is (12.20), which, however, requires that tritium has been synthesized in an earlier step, e.g. by (12.17) or (12.18).

(i) For stars with $m \leq 1.5\ \mathbf{M}_\odot$ and $T \geq 2 \times 10^7$ K, the main reaction sequence is (12.13) followed by (12.15) and (12.24). This is referred to as the *proton-proton chain*, summarized in Table 12.2. Reaction times in Tables 12.2 and 12.3 are from Gamow and refer to solar conditions.

In 9% of the pp-chain, reaction (12.21) is replaced by

$$^3He + {}^4He \rightarrow {}^7Be \qquad Q = 1.59\ \text{MeV} \tag{12.25}$$

$$^7Be + e^- \rightarrow {}^7Li + \nu_e(4\%) \qquad Q = 1.37\ \text{MeV} \tag{12.26}$$

Table 12.2 The proton-proton chain for ^{4}He formation; about 90% of the solar energy production

Reaction steps	Q (MeV)	Mean reaction time	Eqn.
$^1H + {}^1H \rightarrow {}^2H + \mid e^- + \beta^+ \mid + \nu$ (2%)	1.44 MeV	1.4×10^{10} y	(12.13)
$^2H + {}^1H \rightarrow {}^3He + \gamma$	5.49 MeV	6 s	(12.15)
$^3He + {}^3He \rightarrow {}^4He + 2\,{}^1H$	12.86 MeV	9×10^5 y	(12.24)
Overall $4\,{}^1H \rightarrow {}^4He + \mid 2e^- + 2\beta^+ \mid + 2\nu$	26.7 MeV	1.4×10^{10} y	(12.6)

Table 12.3 The CNO cycle for helium production; about 10% of the solar energy

Reaction steps	Q (MeV)	Mean reaction time	eqn.
$^{12}C + {}^{1}H \quad {}^{13}N + \gamma$ ↓	1.94	10^{7} y	(12.41)
$^{13}C + \vert e^{-} + \beta^{+} \vert + \nu$	1.20	7 min	(12.42)
$^{13}C + {}^{1}H \rightarrow {}^{14}N + \gamma$	7.55	3×10^{6} y	(12.43)
$^{14}N + {}^{1}H \rightarrow {}^{15}O + \gamma$ ↓	7.29	3×10^{8} y	(12.44)
$^{15}N + \vert e^{-} + \beta^{+} \vert + \nu$	1.74	2 min	(12.45)
$^{15}N + {}^{1}H \rightarrow {}^{12}C + {}^{4}He$	4.86	10^{5}y	(12.46)
Overall $4\,{}^{1}H \rightarrow {}^{4}He + \vert 2e^{-} + 2\beta^{+} \vert + 2\,\nu$	26.7	3×10^{8} y	(12.6)

$$^{7}Li + {}^{1}H \rightarrow 2{}^{4}He \qquad Q = 17.34 \text{ MeV} \qquad (12.27)$$

The neutrino takes 4% of the decay energy. To a very small amount (<1%) also the reaction sequence

$$^{7}Be + {}^{1}H \rightarrow {}^{8}B; \quad {}^{8}B + e^{-} \rightarrow {}^{8}Be \rightarrow 2{}^{4}He \quad Q = 18.21 \text{ MeV} \qquad (12.28)$$

occurs. Thus isotopes of Li, B and Be are formed as intermediates. ^{7}Be decays by electron capture (see Fig. 5.7) while ^{7}Li is stable. ^{8}B is very shortlived, decaying to ^{8}Be which immediately decays into 2 He. The number of p-p-fusion reactions in our sun amounts to 1.8×10^{38} s^{-1}. The solar neutrino emission comes to ~15% from reaction (12.26) (E_{ν} 0.38 – 0.86 MeV) and to ~85% from (12.13) ($E_{\nu} \leq 0.4$ MeV).

(ii) For stars with $m \geq 1.4\ \mathbf{M}_{\odot}$ Bethe and Weizsäcker in the 1930s deduced the so-called "*CNO-*" or *carbon cycle*, Table 12.3. In such stars temperature and pressure reach higher values, and the consumption of hydrogen is faster. A star of 20 $\mathbf{M}_{\odot}$ burns its hydrogen through the CNO-cycle in some 10 My, compared to the suns pp-cycle, which burns hydrogen at a slower rate for about 10 000 My.

The CNO-cycle requires the presence of some ^{12}C, which acts as a catalyst. In a hydrogen burning star some small amounts of ^{12}C is always produced through reaction (12.29).

In our sun ~6% of the hydrogen originally in the core has now been burnt. Since helium has a greater mass than hydrogen, it accumulates in the core, while most of the hydrogen burning moves to a layer around the He. As the hydrogen fuel is consumed, the temperature of the core decreases somewhat. However, as the amount of helium is increased and the hydrogen depleted, the core contracts through the increased gravitational attraction of the helium, and the temperature of the core rises. This heats the outer layers of hydrogen and results in an expansion of the outer mantle of the star, which in turn results in a cooler surface, so that the star irradiates more red light. The star at this period of its life is referred to as a *Red Giant*, see Figure 12.4. We can expect that our sun in about five eons from now will pass through a red giant stage at which time its diameter should expand sufficiently to engulf the inner planets of the solar system.

12.4.2. Helium-burning to iron

(i) For stars with 0.26 $\mathbf{M}_\odot < m < 1.5\ \mathbf{M}_\odot$, the contracting helium core degenerates and warms. When a density of $\sim 10^7$ kg m^{-3}, and a temperature of $\sim 1.5 \times 10^8$ K are reached, He begins to fuse. Even though ^{8}Be has an extremely short lifetime, there is always a small equilibrium amount present. It has been calculated that in some red giants one ^{8}Be nucleus is in equilibrium with $\sim 10^9$ nuclei of ^{4}He. This amount is sufficient to allow some capture of a third helium nucleus to form ^{12}C. The reaction is

$$^{8}\text{Be} + {}^{4}\text{He} \rightarrow {}^{12}\text{C} \tag{12.29}$$

also referred to as the "3α-process" as 3 He form a C atom. This produces a sudden heating at the center, and in fact the core can be considered to detonate, the *helium flash* (Fig. 12.4), though this detonation is dampened by the large outer masses. In such a star helium burning goes on in the core and hydrogen burning at the outer shell. As the central helium is exhausted, helium burning proceeds in an outer shell, while carbon collects in the center. For our sun the He-burning period is expected to last for only 60 My after which $\sim$50% of the core will be C and O. The fusion reactions will then halt. The outer envelope, which through turbulent mixing includes some core material, expands and the star loses some of its mass. The expanding envelope forms a *planetary nebula* with the star in the center termed a white dwarf, see Figure 12.4.

(ii) Stars of masses $>1.5\ \mathbf{M}_\odot$ burn hydrogen through the CNO cycle. This requires that the core be convective. The main sequence phase ends as the hydrogen in the core is exhausted, and shell burning begins. The helium core remains convective and non-degenerate, and helium burning begins without perturbations, leading to the formation of mainly carbon and oxygen.

(iia) Stars with 3 $\mathbf{M}_\odot < m < 15\ \mathbf{M}_\odot$ burn hydrogen and helium in outer shells. For stars of masses $>3.5\ \mathbf{M}_\odot$ helium burning becomes the important energy source. Once ^{12}C has been formed, further reactions with helium can explain the formation of oxygen, neon and higher elements according to

$$^{12}\text{C}\,(^{4}\text{He},\gamma)\,^{16}\text{O}(^{4}\text{He},\gamma)\,^{20}\text{Ne}\,(^{4}\text{He},\gamma)\,^{24}\text{Mg}\,(^{4}\text{He},\gamma)\,^{28}\text{Si} \tag{12.30}$$

These reactions occur with increasing yields in stars of increasing mass.

(iib) Carbon fusion can occur in stars $>7.5\ \mathbf{M}_\odot$ and at core temperatures $\geq 8 \times 10^8$ K:

$$
\begin{aligned}
{}^{12}\mathrm{C} + {}^{12}\mathrm{C} &\rightarrow {}^{24}\mathrm{Mg} \\
&\rightarrow {}^{23}\mathrm{Na} + {}^{1}\mathrm{H} \\
&\rightarrow {}^{20}\mathrm{Ne} + {}^{4}\mathrm{He}
\end{aligned}
\tag{12.31}
$$

This occurs suddenly, and is observed as a *carbon flash*, Figure 12.4. The star then either continues to burn carbon, or explodes (a *supernova*) with destruction of most of the star.

(iic) In the very heavy stars, $m > 15\ \mathbf{M}_\odot$, the He-burning only lasts for a few My. The carbon core formed remains convective, and carbon burns to oxygen and magnesium. Further fusion synthesis occurs in several zones, leading to the production of elements up to ^{40}Ca, ^{44}Ti, ^{48}Cr, ^{52}Fe and ^{56}Ni, partly by He-capture partly by direct fusion of heavier nuclides. The heaviest elements may be formed in reactions like

$$
{}^{28}\mathrm{Si} + {}^{28}\mathrm{Si} \rightarrow {}^{56}\mathrm{Ni}\left(\beta^+, 6.1\ \mathrm{d}\right) {}^{56}\mathrm{Co}\left(\beta^+, 77.3\ \mathrm{d}\right) {}^{56}\mathrm{Fe}\left(\mathrm{stable}\right) \tag{12.32}
$$

From Figure 4.3 it is seen that formation of elements higher than those of A around 60 through fusion processes are exoergic (i.e. requires energy). When fusion processes end the star consists of an iron core surrounded by shells with silicon, oxygen, carbon, helium and hydrogen. Material is continually lost to space, propelled by strong solar winds. Massive stars of 20 $\mathbf{M}_\odot$ lose as much as 1/100 000 of their mass every year. The outer layer may have a temperature of only a few million degrees, while the center may be 10^9 K. In the lover density outer layer we might expect to find Li, Be, and B. However, these elements have been consumed in reactions with lighter nuclides.

The last steps of production of heavy elements (up to Fe/Ni) occurs rather rapidly in a few thousand years. When the nuclear fuel for fusion is exhausted the star collapses and results in a *supernova*. Figure 12.9 illustrates the composition of a 20 $\mathbf{M}_\odot$ star just before a supernova explosion.

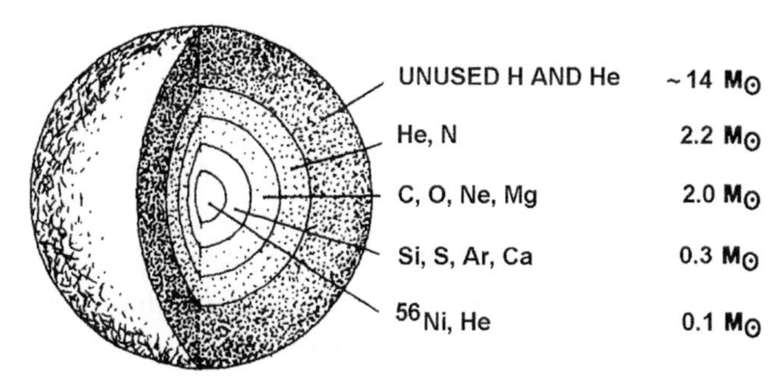

Figure 12.9 Shell structure of a 20 M$_\odot$ star just before supernova explosion.

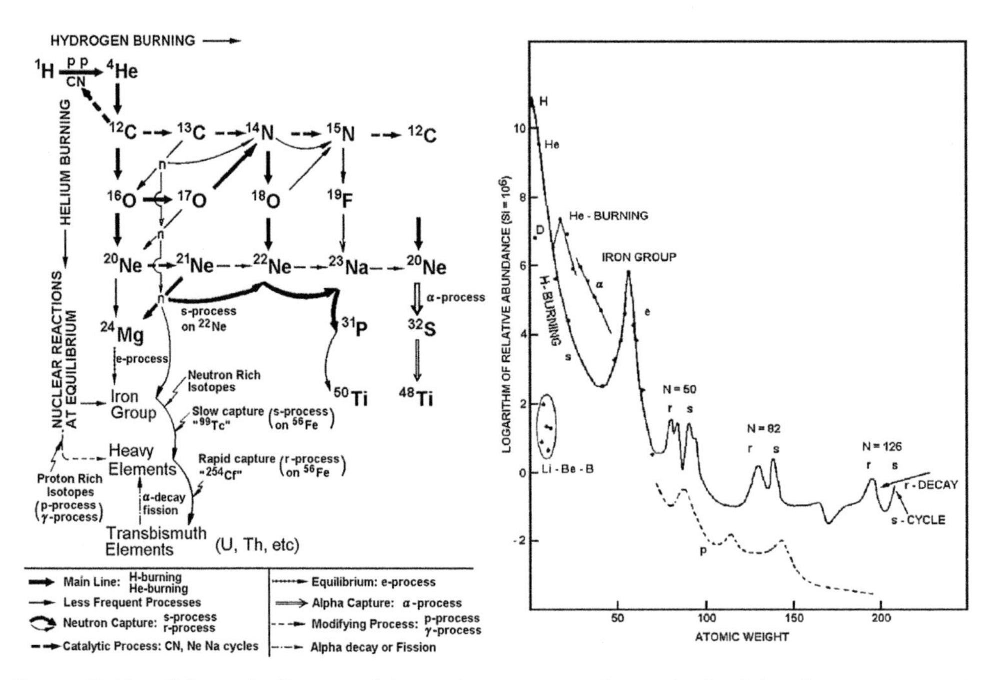

Figure 12.10 a. Schematic diagram of the nuclear processes for synthesis of the elements in stars. b. Calculated atomic abundances. *(From Burbidge, Burbidge, Fowler and Hoyle.)*

The sequence of element formation is summarized in Figure 12.10a (from the famous paper by Burbidge, Burbidge, Fowler and Hoyle). Figure 12.10b is from same paper and shows the calculated elemental composition, which is not too different from that in Figure 12.5a.

The short half-life of ^{8}Be plays a significant role. If ^{8}Be was much more stable, the helium-consuming chain would have proceeded much more rapidly. In fact after the helium ignition the energy production rate would have increased enormously and the red giant would have exploded as a supernova. However, the burning would not have gone much further than to ^{12}C. On the other hand, was ^{8}Be less stable, the fusion synthesis would never have been able to bridge the mass 8, and no elements higher than Be would have been formed.

The Q-value for the reaction $^{4}\mathrm{He} + {}^{12}\mathrm{C} \rightarrow {}^{16}\mathrm{O}$ is 7.16 MeV. The excited levels of ^{16}O near this value are 7.12 and 8.87 MeV. The 8.87 MeV level is not useful for ^{16}O production since the Boltzmann distribution provides too small a fraction of particles with the needed kinetic energy. However, the resonance width of the 7.12 MeV level makes the reaction possible for the high energy tail of the Boltzmann distribution. Had it been easier to make ^{16}O, we would have had less carbon and more oxygen in the universe. This probably would have been a hindrance in the development of life, since it is believed that life as we know it must start in a reducing (low oxygen containing) atmosphere.

Our galaxy has stars much larger and much smaller than our sun, and of widely different ages. Thus the processes we have described are occurring at present in the Milky Way and in other galaxies as well. Moreover, the stars emit considerable amounts of matter into space. As a result, interstellar gas, out of which new stars are formed, contains atoms heavier than helium, although hydrogen is the most abundant element.

12.5. NEUTRON CAPTURE PROCESSES: FROM IRON TO URANIUM

Figure 4.3 shows that the maximum nucleon binding energy occurs at $A \approx 60$, i.e. around iron, which we may consider to be the most thermodynamically stable element in the universe. At lower values of A, fusion of lighter elements releases energy while the exothermic reactions to form heavier elements ($A > 60$) involve neutron capture.

12.5.1. Slow neutron capture

Through hydrogen and helium burning, neutrons are formed. The most important reaction is believed to be

$$^{22}\mathrm{Ne} + {}^{4}\mathrm{He} \rightarrow {}^{25}\mathrm{Mg} + \mathrm{n} \tag{12.33}$$

As the heavier elements form in the star, the neutron production increases considerably since such reactions become more prevalent as heavier elements are involved in the reactions. In reactions like (12.31) and (12.32) energetic γ-rays are emitted, which decompose D into H and n (the binding energy is 2.22 MeV). The mode of production of the elements changes from that of helium capture to that of neutron capture, so that the elements from iron to bismuth can be formed by a slow process of neutron capture (n,γ-reactions), interrupted by β-decay whenever it is faster than the next capture step.

Such a process is known as the *slow*, or *s-process*. While the reaction probability for the capture of neutrons increases with the atomic number of the element, the relative amount of the elements in the star will decrease with increasing atomic number, because of the successive higher order of reaction (§17.3). The result is the observed flattening of the abundance curve for $A > 100$, see Figures 12.5a and 12.10b.

The formation of ^{104}Pd from ^{100}Ru can serve as an example of the steps in the s-process of element formation

$$^{100}_{44}\mathrm{Ru}\,(n,\gamma)\;^{101}_{44}\mathrm{Ru}\,(n,\gamma)\;^{102}_{44}\mathrm{Ru}\,(n,\gamma)\;^{103}_{44}\mathrm{Ru}$$

$$\beta \downarrow 39\ \mathrm{d}$$

$$^{103}_{45}\mathrm{Rh}\ (\text{stable})\,(n,\gamma)\;^{104}_{45}\mathrm{Rh} \tag{12.34}$$

$$\beta \downarrow 42\ \mathrm{s}$$

$$^{104}_{46}\mathrm{Pd}\ (\text{stable})$$

Note that ^{100}Ru to ^{102}Ru are stable.

The discovery of the element promethium (for which the longest-lived isotope has a half-life or only 18 y) in a star (HR 465) in the Andromeda constellation shows that an s-process must be occurring. A possible reaction path is

$$^{146}\mathrm{Nd}\left(n,\gamma\right)^{147}\mathrm{Nd}\left(\beta^{-},11\mathrm{d}\right)^{147}\mathrm{Pm}\left(\beta^{-},2.6\ \mathrm{y}\right) \tag{12.35}$$

The s-process is believed to be extensive in Red Giant stars of mass 3–8 $\mathbf{M}_{\odot}$ and to last for about 10 My, a short period in the total lifetime of a star.

As the star proceeds through the s-process stage we can expect that the fusion reactions decrease while the gravitational contraction of the star continues. To conserve angular momentum the rotational velocity increases, resulting in the ejection of some of the outer mantle into space, thereby exposing the inner, hotter core. Thus in the radio source Cassiopeia A one can see fast moving "knots" of O, S and Ar. The turbulence is likely to bring up heavier elements from the core, as observed in the CP-stars (§12.1.6). For example, the light from the planetary nebula nucleus of FG Sagittae suddenly showed strong barium lines between 1967 and 1972. The turbulence may become so violent that hydrogen from outer layers is mixed in with deeper layers of higher temperature, leading to instantaneous hydrogen burning and a very rapid rise in energy production. This is observed as a sudden light increase from the star (*nova*). After the nova stage (or stages, as several such may occur) the star would continue to cool until it becomes a white dwarf. The density of such a body would be very large, about $10^{8}\ \mathrm{kg\,m^{-3}}$.

12.5.2. Supernova explosions

The s-process cannot explain the formation of the elements heavier than bismuth as the trans-bismuth elements have a number of short-lived isotopes which prevent the formation of thorium and uranium in the amounts observed in nature. The heaviest elements are believed to be formed in supernova explosions.

In a star in which heavier elements are accumulated in its center, the energy production occurs in the layer surrounding the core, see Figure 12.9. For stars of an original mass >7.5 $\mathbf{M}_\odot$ (or >3.5 $\mathbf{M}_\odot$ at the end of the helium-burning period) the energy loss through photon and, especially, through neutrino emission is very large. This has several consequences:

(i) the emission of energy into space cools the core and the giant begins to contract,

(ii) the atoms in the core can no longer resist the tremendous pressure and their electron shells collapse (forming degenerate matter),

(iii) the contraction and pressure in the core increase to nuclear density (ca. 10^{17} kg/m^3), and the released gravitational energy increases temperature and pressure also in the outer layers.

Under the development of these conditions, the elements in the core disintegrate (especially iron) releasing helium and neutrons; e.g.

$$^{56}\mathrm{Fe} + \gamma \rightarrow 13\,^{4}\mathrm{He} + 4\mathrm{n} \tag{12.36}$$

The helium immediately fuses and the intense heat developed spreads as a heat shock, which passes to the cooler outer shells of hydrogen and helium, initiating new thermonuclear reactions in the mantle. As a result the whole star explodes as a supernova. While the outer layers expand into space, the core contracts to a *black hole.* Other mechanisms producing supernovae are also known.

A supernovae can be seen by the naked eye as a new star if it occurs within our galaxy. The light intensity slowly decreases and the star may be too weak to see after a year. A supernova discovered by Kepler in 1604 could be seen even in day time and its expanding gas cloud is still easily observed by telescope. The next "nearby" supernova (SN1987A) was seen, on February 24, 1987, in the Large Magellanic Cloud. Before the light arrived to us, a high energy (20–40 MeV) neutrino emission (a total of 19 electron anti-neutrinos was measured in 10 s, compared to a normal 2 per day) was observed in Japanese and American detectors (cf. § 2.5), proving that a large neutrino burst ($\sim 10^{58}$ neutrinos) accompanied the supernova explosion. The neutrinos had an energy corresponding to 5×10^{10} K, and the total gravitational energy released in the process was estimated to 3×10^{46} J.

SN1987A was of a composition like that in Figure 12.9. At the high temperature of the explosion, most of the lighter atoms fused into Fe/Ni, which, thus is not only consumed but also produced "further out". The ^{56}Ni decays according to

$$^{56}\mathrm{Ni}\,(\mathrm{EC}, 6.1\ \mathrm{d})\,^{56}\mathrm{Co}\,(\mathrm{EC}, 77.3\ \mathrm{d})\,^{56}\mathrm{Fe}\,(\mathrm{stable{:}}\ 91.7\%) \tag{12.37}$$

The light from the supernova could be related to 0.85 and 1.24 MeV γ-lines from the decay of ^{56}Co; Figure 12.11 shows the decay curve. One could actually calculate that in the first moments after the explosion about 0.1 $\mathbf{M}_\odot$ of ^{56}Ni was formed. Most of the Fe in the Universe is likely produced by this kind of process.

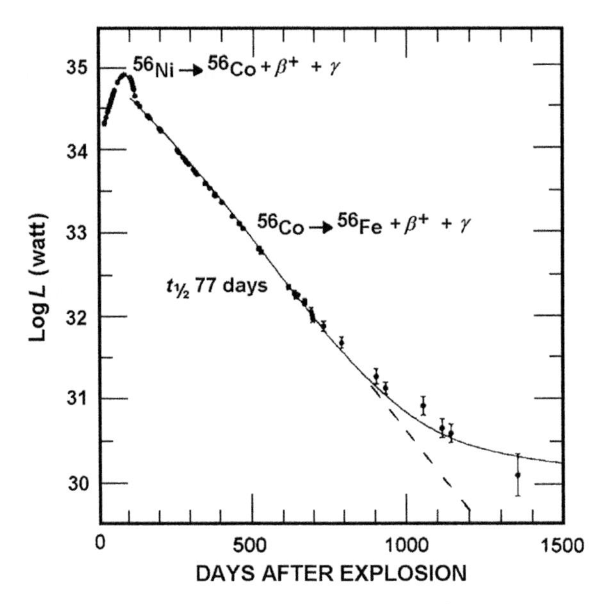

Figure 12.11 Luminosity of SN1987A as function of time. *(From Fransson.)*

The mass of heavy elements spewed out into space enriches it for the formation of later generation stars. The transition elements, which are only formed in such stellar explosions, are a necessity for the existence of life on earth.

12.5.3. The rapid (r-) process

The supernova stage is very short-lived with extremely intense neutron production. It provides a method whereby the barrier of the short-lived isotopes between polonium and francium is overcome and the heaviest elements synthesized. This mode of element formation is known as the *rapid* or *r-process*; see Figure 12.12.

The n-capture in the r-process has been suggested to go up to Z about 100 and $N \leq 184$. In the intense neutron field a considerable amount of (mainly fast) fission of the newly synthesized heavy elements probably also occurs. This partly explains the peaks at $N = 50$ and 82 in Figure 12.10b, which also correspond to maximum yields at $A = 95$ and 140 in thermal fission. Some stars are unique in that they have an unusually high abundance of fission products; spectral lines from heavy actinides, like americium and curium, have also been observed in such stars.

The explosion of thermonuclear hydrogen bombs containing uranium resulted in the formation of elements 99 (Es) and 100 (Fm) (§14.2.9). These elements were synthesized in the extremely short time of the explosion by the intense neutron flux bombarding the uranium (shaded area in Figure 14.5). The explosion of the hydrogen

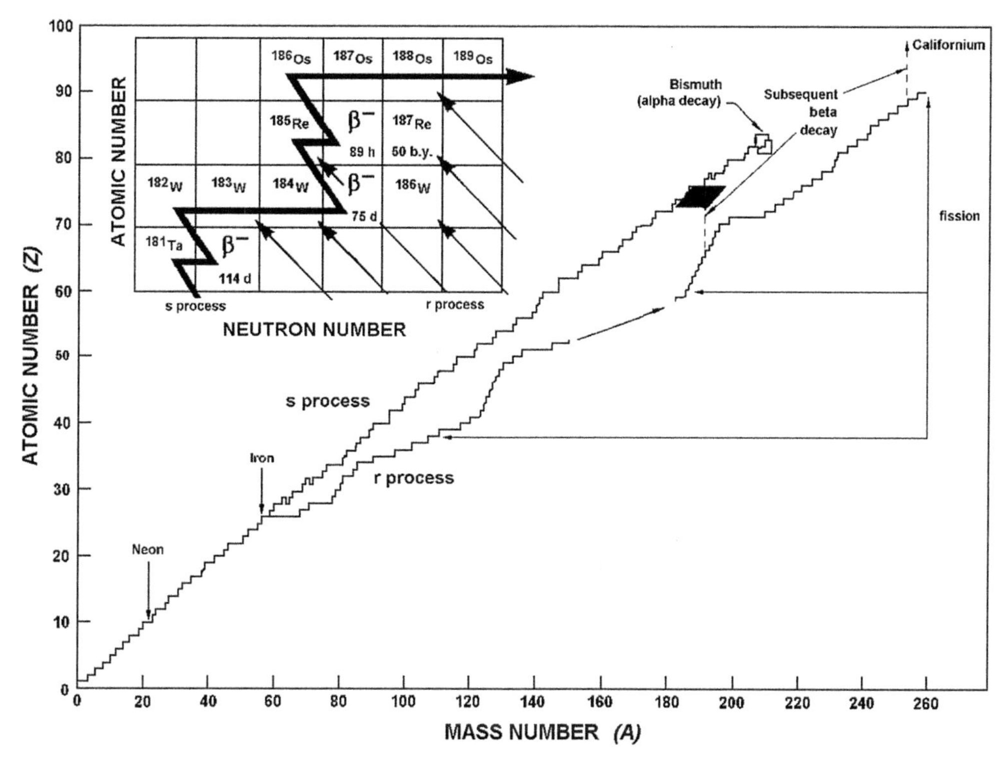

Figure 12.12 Part of isotope chart showing the formation of W, Re and Os isotopes by the s- and the r-processes.

Table 12.4 Comparison of conditions for n-capture processes

Process	Flux (n m^{-2} s^{-1})	×	time	=	exposure (n m^{-2})
s-process	$\sim 10^{18}$		~ 1000 y		$\sim 3 \times 10^{29}$
r-process	$>10^{29}$		$1 - 100$ s		$>10^{29}$
nuclear explosion	$>10^{33}$		<1 μs		$\sim 10^{27}$
nuclear reactor	$\sim 10^{16}$		~ 1 y		$\sim 10^{23}$

bomb duplicated in a very small way what is believed to be the process of the formation of the heaviest elements in supernovae. The neutron fluxes and exposures in the s- and r-processes as compared to those in a nuclear explosion and a reactor are given in Table 12.4.

The intensity of the neutron flux as well as the very short time preclude β-decay as a competitor to neutron capture in the r-process. This results in a different isotopic distribution of the elements for the r-process compared to that formed in the s-process. The following reaction sequence illustrates the r-process in which β-decay can occur only after the explosion has terminated and the intense neutron fluxes decreased (compare with the sequence (12.34)):

$$\begin{array}{ccccccccccccc} ^{100}\text{Ru} & (n,\gamma) & ^{101}\text{Ru} & (n,\gamma) & ^{102}\text{Ru} & (n,\gamma) & ^{103}\text{Ru} & (n,\gamma) & ^{104}\text{Ru} & (n,\gamma) & ^{105}\text{Ru} & (n,\gamma) & ^{106}\text{Ru} \\ & & & & & & \beta\downarrow 39\text{ d} & & & & \beta\downarrow 4.4\text{ h} & & \beta\downarrow 368\text{ d} \\ & & & & & & ^{103}\text{Rh} & & & & ^{105}\text{Rh} & & ^{106}\text{Rh} \\ & & & & & & \text{(stable)} & & & & & & \\ & & & & & & & & & & \beta\downarrow 36\text{ h} & & \beta\downarrow 30\text{ s} \\ & & & & & & & & & & ^{105}\text{Pd} & & ^{106}\text{Pd} \\ & & & & & & & & & & \text{(stable)} & & \text{(stable)} \end{array} \quad (12.38)$$

After completion of the r-process, ^{103}Ru, ^{105}Ru, ^{106}Ru undergo β-decay to isotopes of Rh and Pd. In this r-sequence ^{104}Ru – ^{106}Ru are formed, but in the s-sequence beginning with ^{100}Ru, the heaviest ruthenium isotope has $A = 103$.

In a supernova explosion a large mass of material is ejected into interstellar space. This contributes to the higher abundance of heavy elements in cosmic rays as compared with the cosmic abundance. In fact, even uranium has been observed in cosmic rays and in our sun. Since our sun is undergoing the simplest type of hydrogen-burning cycle, it is not possible for the heavier elements (~2%) to have been synthesized by the sun. Consequently, their presence indicates that the sun has been formed as a second (or later) generation star from material that included matter ejected by an earlier supernova, or has accumulated matter from such a star.

The carbon cycle stars are likely to be second generation stars because ^{12}C is needed in the core for the carbon cycle to start. The same star may pass through several novae explosions whereby it loses large amounts of the lighter elements from the outer mantle in each explosion. The chemical composition of a star thus not only indicates its age but also tells us to which generation of stars it belongs.

12.5.4. Neutron stars

A supernova explosion leaves a residue at the center of the expanding gas cloud generated by the explosion. This residue is believed to contract into a neutron star. Assuming a nucleon radius of 0.85 fm, the density becomes about 2×10^{18} (cf. also Figure 4.4 for densities of other nuclear matter). When the central (remaining) mass of the supernova has contracted to a density of $>10^{17}$ kg m^{-3} matter will consist of relatively closely packed nuclei and free electrons. It now becomes favorable for the protons to capture electrons

$$p + e^- \rightarrow n + \nu \quad (12.39)$$

The neutrinos are emitted and only neutrons are left.

The neutron star has a considerable energy reservoir in its rotation, typically one rotation per second. Particles ejected from this rapidly rotating object would be caught in the rotating magnetic field and accelerated to relativistic tangential velocities. They emit high intensity bremsstrahlung, which appear as pulses of radiation with the frequency of the rotation. Such radiation pulses have been observed from cosmic sources known as *pulsars*. The discovery of a pulsar at the center of the expanding Crab nebula, which is the remnants of a supernova explosion observed by the Chinese in the year 1054 has provided strong evidence for the model outlined above. Pulsars have been suggested to be the source of the high energy cosmic ray particles observed in our atmosphere.

12.6. AGE OF THE GALAXY

Estimates of the age of our galaxy comes mainly from theoretical calculations. However, radiochemical investigations can yield information about several important steps. Thus the time since the final solidification of the earth's crust is obtained from the dating of geologic samples (Ch. 13). Nuclear clocks like U → Pb give only the age of minerals; the oldest known minerals on earth are 3.7×10^9 y.

Some glassy materials in meteorites and from the moon are assumed to be original condensed matter which never has undergone melting. The oldest ages seem all to converge at 4.6×10^9 y. Thus we conclude that this was the time when our planetary system began to form (the basis for the estimate of the age of our sun in §12.1.5).

Further information about this event has been obtained by studying tracks which nuclear decay processes leave in certain minerals (§9.1.2). Fission tracks can only persist in minerals that have not been heated because heating above 600°C erases the tracks. The fact that ^{244}Pu fission tracks have been found in iron meteorites and in lunar samples shows that ^{244}Pu existed when the planetary system formed. Because of the short half-life of ^{244}Pu (8×10^7 y) it can be concluded that such mineral samples must have formed within a few hundred million years after the nuclide ^{244}Pu itself was formed. This is probably also the time for planetary formation. The existence of primordial plutonium indicates that an r-process preceeded the formation of the planets.

Apparently, our solar system was showered by debris from a recent but distant supernova explosion which may have disturbed the "peaceful" gas cloud in our part of the universe and initiated the condensation of the solar system. At this time newly formed elements stopped being added to the solar system from the galaxy. Sufficient amounts of ^{129}I and ^{244}Pu remained after the formation of solid materials to produce characteristic isotope anomalies and fission track excesses in meteoritic and lunar materials. The spontaneous fission decay of ^{244}Pu yields ^{129}I, which decays as

$$^{129}\text{I} \xrightarrow[1.6 \times 10^7 \text{ y}]{\beta^-} {}^{129}\text{Xe (stable)} \qquad (12.40)$$

The fact that meteorites containing about 100 times excess fission tracks also contain ^{129}Xe is evidence both for the existence of ^{244}Pu at the beginning of the collapse of the solar nebula, and for the rapid condensation of the meteorites from freshly produced new elements. This time has been calculated to about 80 My from the isotopic concentrations. By similar analysis we can go further back in the history of the universe and calculate the time between the formation of the elements in the galaxy and the condensation of the solar system; the result yields a time of 9.9×10^9 y (or 9.9 eons).

There are also some radiochemical evidence about the age of our Galaxy. Consider Figure 12.12, which shows a part of a nuclide chart; the zigzag arrow shows nuclides formed in the s-process, the lower right arrows nuclides formed by the r-process. ^{186}Os can only form in the s-process and ^{187}Re only in the r-process. It is likely that our solar system, and ^{187}Re, was formed from a supernova in the r-process. ^{187}Re decays ($t_{1/2}$ 46×10^9 y) to ^{187}Os, which also is formed from slow n-capture in ^{186}Os. The amount of ^{187}Os formed in the s-process can be calculated from estimated flux values and known cross sections (at 30 keV; 3.6×10^8 K). Thus the amount of ^{187}Os formed by decay of ^{187}Re can be calculated, and the time since the r-process determined. This rhenium-osmium clock gives us an age of 10.8 ± 2.2 eons from formation of the Galaxy until our solar system formed. If this is combined with the age of the solar system of 4.7 eons, the age of the galaxy becomes 15.5 ± 2.3 eons. Adding 1.5 eons for the time between the Big Bang and the formation of the Galaxy, the universe should have an age of 17 ± 3 eons.

This value should be compared with the most recent value of t_o (12.5), which without a correction for any change in the rate of expansion, gives 13.18 ± 0.11 eons. The disagreement between the uncorrected value, 13.2 ± 0.1 eons, and the value, 17 ± 3 eons, above indicates either that the Universe does not continue to expand at an unchanged rate, or if it does, that

(i) the expansion of the universe accelerates slowly (as indicated by resent measurements), or
(ii) Hubble's law is not valid, or
(iii) the cosmological (Einstein) constant is not unity, or
(iv) something is wrong with the Big Bang hypothesis, or
(v) the radiometric data are too uncertain.

12.7. THE EVOLUTION OF THE PLANETS AND THE EARTH

The planets are assumed to have formed out of the same cloud as, and simultaneously with, the sun. The sun contains 99.85% of the mass of the solar system, the planet Jupiter 0.1%, and all other planets 0.04%. The originally spherical cloud was disturbed and caused to rotate through some external action (e.g. supernova explosion or passage of some celestial body), the rotation leading to the formation of a disk-like cloud, and later to cloudy rings similar to those one can see around many stars and some of the planets.

The cloud condensed successively into dust particles, which through collisions formed larger particles until rather large clumps of sizes similar to the present asteroids were obtained (so-called *planetesimals*, with diameters of several km; compare the sizes of the craters on the moon). By the time these clumps drifted together (a process referred to as *accretion*) to form the protoplanets, which collected more dust and gas from the surrounding cloud. When the H-burning started in the sun, a strong solar wind was produced, which blew away the gas and dust left in the cloud, and the planet formation was in principle finished. Some cosmogonists think this whole process took place in only a few million years.

The temperature gradient in the primordial elliptical cloud is given in Figure 12.13 with condensation temperatures for important elements and compounds. The volatile elements are almost completely absent in the innermost ("terrestrial") planets (Mercury, Venus, Earth, Mars). Here rock-forming minerals were produced in high-temperature chemical processes (Al_2O_3, SiO_2, $CaTiO_4$, $Ca_2Al_2SiO_7$, etc) and condensed out together with little volatile metals (Fe, Ni, etc). At lower temperatures further out in the solar system, H_2O was formed and condensed out as ice; e.g. some moons of Saturn are almost pure ice. At still lower temperatures NH_3 and CH_4 would liquefy. Jupiter and Saturn contain 99% H_2, 0.1% He, the rest being mainly H_2O (ice), CH_4 and NH_3; wide storms of NH_3-crystals have been observed on Saturn; Pluto has a thin atmosphere of only CH_4. Owing to the high pressures and low temperatures at the giant planets,

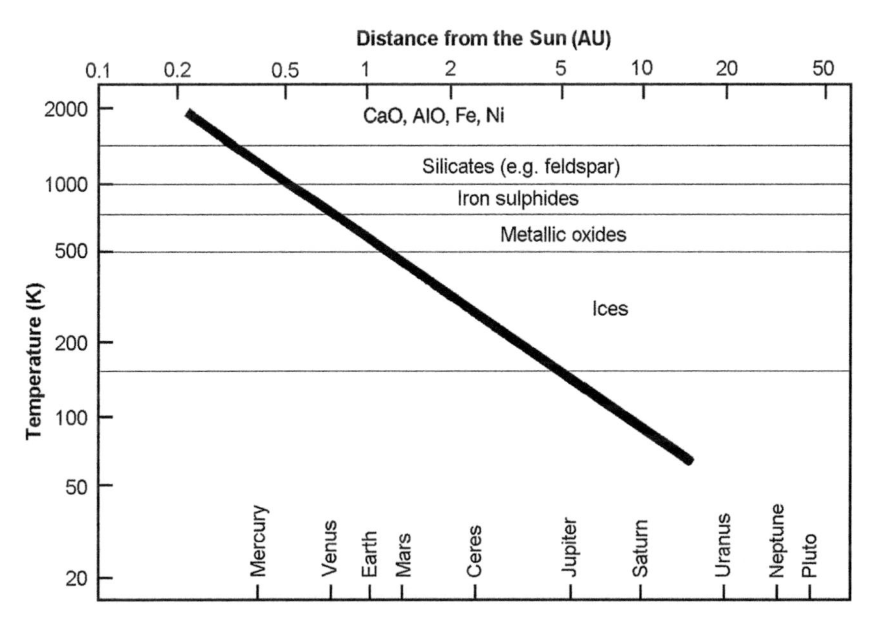

Figure 12.13 Temperature distribution in the solar system during planet formation. *(From Karttunen et al.)*

hydrogen is either in the liquid or solid (metallic) state. However, if the planets have migrated much over time to their present positions, this simple explanation does not hold.

From assumptions of the elemental composition and temperature-pressure conditions of the solar system one can in principle calculate the average mineral composition of the Earth using thermodynamic data of known compounds. Several such calculations have been made, which more or less well account for the present composition of the crust, magma, etc. There are, however, some other observations which need comments.

According to Urey, Suess, and others, the accretion of cosmic particles during the formation process released gravitational energy; also the higher concentration of radioactive nuclides in the primordial cloud (Fig. 12.14) contributed to heat and melt the solid material of our Earth-to-be. The Earth's temperature was 40–50% higher than today about 3.3 Gy ago. Not until a solid crust was formed could the radioactive clock begin to tick. This did not occur until about 3.7 eons ago, as this is the age of the oldest minerals found on Earth. Thus, it should have taken about 1 eon for the Earth to solidify. As soon as the temperature fell below 100°C water could begin to condense, possibly leading to an abrupt global temperature reduction, due to cloud formation and high heat of vaporization.

All planets were originally formed with some atmosphere, partly as a remnant of the primordial cloud, partly due to outgassing of the interior through volcanos and to impacts of comets. The atoms/molecules in the atmosphere have an average velocity, which must

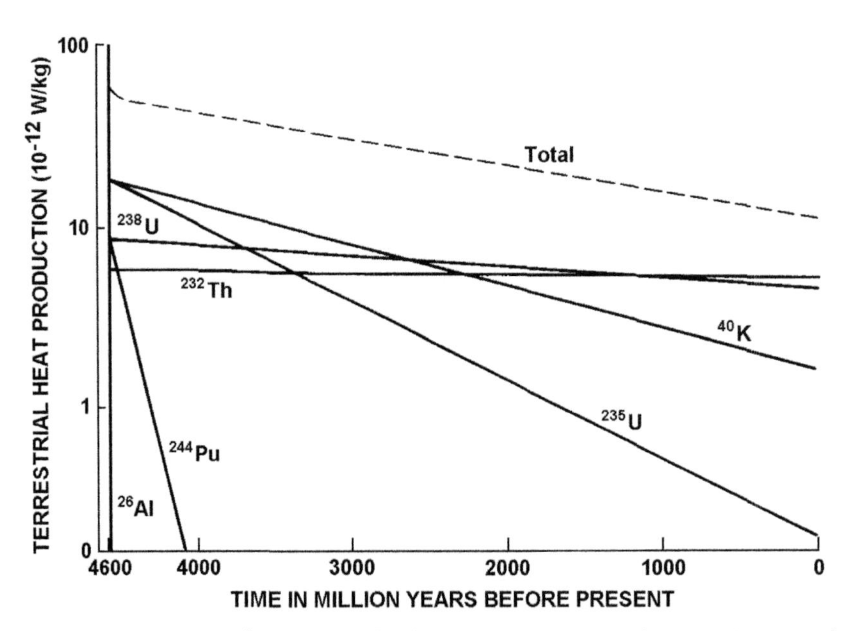

Figure 12.14 Heat production from decay of radioactive species in earth. *(From Brown and Musset, acc. to Cox.)*

be less than the escape velocity from the planet if the gases should not evaporate into space. The escape velocity depends of temperature and mass of the planet (cf. §12.1.2); for example, the moon has too small a mass to keep any atmosphere. The escape velocity of H_2 and He from the Earth is too large for these elements to be retained. The original atmosphere of the Earth is therefore believed to have been mainly H_2O, CO_2, N_2, CH_4 and NH_3. This is referred to as a *reducing atmosphere*, which probably is typical for all new-born Earth-sized planets. The UV-radiation from the sun decomposes these molecules (though the fragments may recombine), releasing H_2, which escapes. The O_2 released is very reactive and combines with many elements to form oxides; most of the red surface of Mars is FeO. The common assumption is that very little oxygen was originally present as free O_2. However, rather rapidly, processes led to a shift of the atmosphere from reducing towards *oxidizing conditions*. Life appears to have begun before this transition as evidenced by fossile bacteria-like organisms in very old rocks formed in a reducing atmosphere.

12.8. EXERCISES

12.1. (a) What is the most probable kinetic energy of a hydrogen atom at the interior of the sun ($T = 1.5 \times 10^7$ K)? (b) What fraction of the particles would have energies in excess of 100 keV?

12.2. What amount (kg) of hydrogen is consumed per second by the sun in the fusion reaction 12.6 in Table 12.2? The solar constant (energy flux at earth's orbit) is 1.353 kW m^{-2} and the earth's average distance to the center of the sun is 149 599 000 km.

12.3. Assuming the sun generates its energy to 2/3 by the proton-chain and the rest by the CNO cycle, what would the expected solar neutrino flux be at the center of the earth?

12.4. In a water power station with a fall height of 20 m and a water flow of 500 $m^3 s^{-1}$, the electric power output at 100% efficiency can be calculated. If the heavy water in the fall could be extracted and converted in a fusion reactor according to (12.14) with a 25% efficiency, which of the two power sources would yield more energy?

12.5. The supernova SN1987A was at a distance of about 51.47 kPc from us and had about 10^{46} J total energy release as estimated from the neutrino flux hitting Earth. How much mass in kg was then converted to energy in the explosion?

12.9. LITERATURE

G. GAMOW and C. L. CRITCHFIELD, *The theory of atomic nucleus and nuclear energy sources*, The Clarendon Press, 1949.

E. M. BURBIDGE, G. R. BURBIDGE, W. A. FOWLER and F. HOYLE, Synthesis of the elements in stars, *Rev. Mod. Phys.*, **29** (1957) 547.

H. Craig, S. L. Miller and G. J . Wassenburg (Eds.), *Isotopic and Cosmic Chemistry*, North-Holland, 1963.
Abundances of the elements in the universe, Landolt-Börnstein Physical Tables, New Series VI/1, 1965.
D. D. Clayton, *Principles of Stellar Evolution and Nucleosynthesis*, McGraw-Hill, 1968.
V. Trimble, The origin and abundances of the chemical elements, *Rev. Mod. Phys.*, **47** (1975) 877.
J. E. Ross and L. H. Aller, The chemical composition of the sun, *Science*, **191** (1976) 1223.
S. Weinberg, *The First Three Minutes*, Basic Books Inc., New York 1977.
J. Rydberg and G. Choppin, Elemental evolution and isotopic composition, *J. Chem. Ed.*, **54** (1977) 742.
The Cambridge Encyclopedia of Astronomy, Trewin Coppelstone Publ. Ltd, London 1977.
Landolt-Börnstein, New Series VI, Astronomy and Astrophysics Vol. 2a-c, Springer Verlag, 1981–1982.
P. K. Kuroda, *The Origin of the Chemical Elements*, Springer Verlag, 1982.
B. Mason and C. B. Moore, *Principles of Geochemistry*, J. Wiley & Sons, 1982.
H. D. Holland, *The chemical evolution of the atmosphere and the oceans*, Princeton Univ. Press, 1984.
H. Karttunen, P. Kröger, H. Oja, M. Poutanen and K. J. Donner, *Fundamental Astronomy*, Springer Verlag, 1987.
G. O. Abell, D. Morrison and S. C. Wolff, *Exploration of the Universe*, Saunders College Publishing, 5th Ed., 1987.
P. A. Cox, *The Elements*, Oxford Univ. Press, 1989.
M. Roos, *Expansion of the Universe - Standard Big Bang Model*, arXiv:0802.2005v1 [astro-ph] 14 Feb 2008.

CHAPTER 13

Cosmic Radiation and Radioelements in Nature

Contents

The early workers in nuclear science found that their measuring equipment had a constant background level of radiation which could be eliminated only partially even with the aid of thick shielding walls of iron and lead. It was assumed initially that this radiation had its origin in naturally radioactive elements present in the materials in the laboratory. However, in 1911 Hess carried measuring equipment into the atmosphere with the aid of balloons and learned that some background radiation increased with altitude. Obviously, at least a component of the laboratory background radiation had its

Radiochemistry and Nuclear Chemistry
ISBN 978-0-12-405897-2, http://dx.doi.org/10.1016/B978-0-12-405897-2.00013-6

origin in some extra-terrestrial source. Test equipment carried outside of the earth's atmosphere by rockets has given us data which provide a fairly accurate picture of the composition of the radiation that comes to the earth from space.

The investigation of cosmic radiation has had a profound influence on nuclear science. When Chadwick in 1932 discovered the neutron, the picture of matter seemed complete: all matter appeared to be composed of four fundamental particles: protons, neutrons, electrons, and photons. The existence of the electron neutrino was predicted by Pauli in 1927 but it was not proven until 1956. However, through studies of the cosmic radiation Anderson discovered the *positron* (the first *antiparticle*) in the same year. Five years later Anderson and Neddermeyer discovered another new particle with a mass about one-tenth of a proton or about 200 times heavier than the electron. This particle is the *muon*, designated by μ. Since that time a large number of subatomic particles have been discovered.

13.1. PRIMARY COSMIC RADIATION

A rather small fraction of the cosmic radiation consists of electromagnetic radiation and electrons. The former vary in energy from a small percentage of γ-rays to a considerable intensity of X-rays, to visible light and to radiation in the radiofrequency region. The types and intensities of this radiation have been of great importance to development of models of the formation and composition of the universe.

The major part of the cosmic radiation is nuclear particles with very high energy: approximately 70% protons, 20% α-particles, 0.7% lithium, beryllium, and boron ions, 1.7% carbon, nitrogen, and oxygen ions, the residual 0.6% ions of $Z > 10$. These ions are bare nuclei prior to interaction since their kinetic energies exceed the binding energies of all of the orbital electrons.

The cosmic particle radiation can be divided by energy into two major groups (Fig. 13.1). One group has energies mainly below 1 GeV and consists primarily of protons. This group originates mainly from the sun. Its intensity varies in relation to solar eruptions since at the time of such an eruption a large amount of solar material, primarily hydrogen, is ejected into space.

The second group has energies up to about 3×10^{11} GeV, although the intensity of the particles decreases with increasing energy, following the relation $N(E) \propto E^{-1.6}$, where $N(E)$ is the number of particles with energies in excess of E. Thus particles of 10^3 GeV have an intensity of about 10^{11} higher than particles of 10^{10} GeV. Within this high energy group the particles at the lower end of the energy spectrum are assumed to originate from sources within our galaxy (the Milky Way), while the particles of the higher energy end are assumed to come from sources outside of our galaxy. Different hypotheses, which for the most part are untested, suggest that the particles come from astronomical radio sources, exploding super novae, or colliding galaxies, etc. At least a

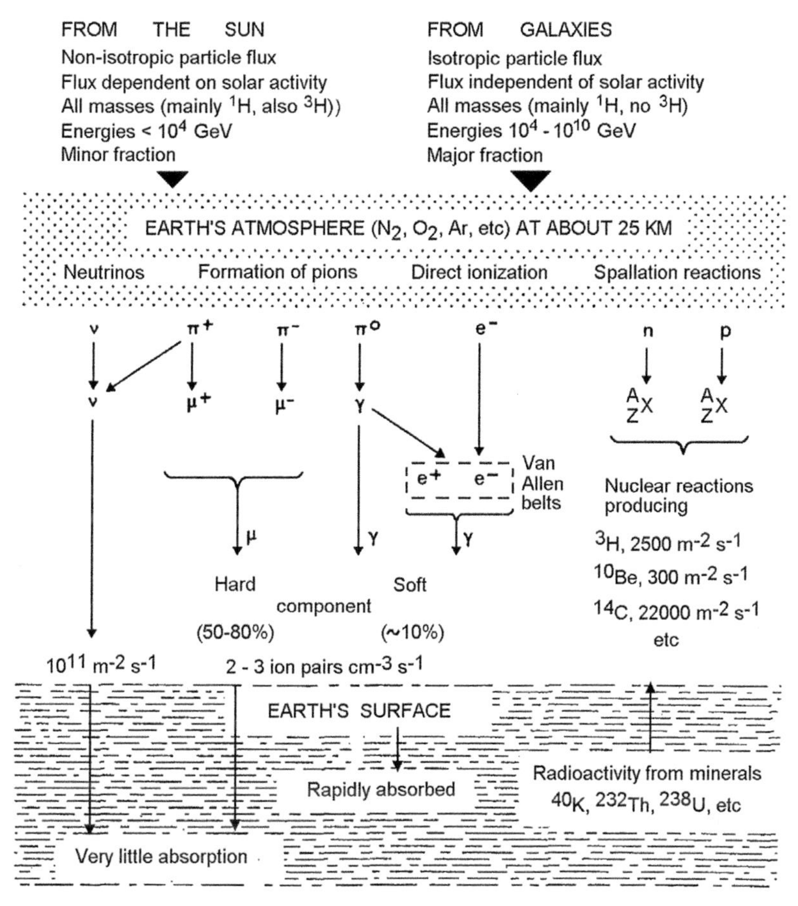

Figure 13.1 Cosmic radiation consists of atoms and photons which react with the atmosphere leading to the formation of numerous secondary particles, some (but not all) detectable at the Earth's surface.

portion of this radiation is a residue of the processes involved in the original formation of the universe. It is assumed that the high energy particles obtain their tremendous kinetic energies through acceleration in the magnetic field of galactic objects (synchrotron acceleration, §16.6).

When the primary cosmic particles enter the earth's atmosphere, they collide with the matter of the atmosphere and are annihilated. In this annihilation process a large number of new particles are formed whose total kinetic energy is less than that of the original primary radiation but whose total rest mass is larger than that of the primary particle. A 10^4 GeV cosmic helium ion may produce a shower of 50−100 new highly ionizing particles, cf. Fig. 2.1. The main reaction products are particles, which are known as pions, designated π.

Figure 13.2 Holes in helmet of Apollo 12 astronaut caused by high energy cosmic rays. The holes have been made visible in a microscope by etching *(From New Scientist, April 22, 1973).*

Figure 13.2 shows the effect of high energy cosmic rays hitting the helmets of Apollo 12 astronauts. It is probable that the cosmic ray intensity will put a limit to how long man can endure in outer space: it has been calculated that in a journey to the planet Mars about 0.1% of the cerebral cortex will be destroyed. The annihilation process occurs to such an extent that below an altitude of approximately 25 km above the earth the number of primary cosmic particles has been reduced to quite a small fraction of the original intensity.

The discovery of π-mesons (or *pions*) was reported by Powell and Occhialini in 1948 after they had analyzed tracks in photographic emulsions placed for some months on a mountain top to get a high yield of cosmic ray interactions. Pions are produced in large amounts in all high energy (≥400 MeV) nuclear reactions. In 1935 Yukawa suggested that the nucleons in a nuclide were held together through the exchange of a hypothetical particle, which we now recognize as the pion, just as hydrogen atoms in H_2 are held together through the exchange of an electron:

$$p \leftrightarrow n + \pi^+; \quad n \leftrightarrow p + \pi^-; \quad p \overset{\pi^\circ}{\leftrightarrow} p; \quad n \overset{\pi^\circ}{\leftrightarrow} n; \tag{13.1}$$

Pions are the particles of *strong interaction,* cf. §2.2. Since the pion rest mass is 0.147 u, the energy to produce a pion is at least 137 MeV (in practice, in order to conserve momentum it exceeds 400 MeV, cf. §4.6 and 5.7). The pions are unstable in free form.

13.2. SECONDARY REACTIONS IN THE EARTH'S ATMOSPHERE

Few of the pions formed in the annihilation process reach the earth's surface. They undergo radioactive decay (life-time about 10^{-6} s) to muons and muon neutrinos (which will in a very short time undergo flavour oscillations) with a probability of about 99.99%. The second most probable fate of the pion is by decaying into an electron and a electron neutrino with a probability of about 0.01%. This latter decay was discovered in CERN in 1958. If none of the previous decay modes are realised they collide with other particles in the atmosphere and are annihilated. The muon is also an unstable subatomic particle with a mean lifetime for stationary muons of 2.2 μs. This comparatively long decay life time (the second longest known) is due to being mediated by the weak nuclear force. The only longer lifetime for an unstable subatomic particle is that for the free neutron which also decays via the weak force. All muons decay to three particles: an electron plus two neutrinos of different types.

The collision reactions of the pions result in the formation of a large number of other particles such as electrons, neutrons, protons, and photons. Some of the electrons so formed are captured in a thick zone around the earth known as the inner van Allen belt.

The main part (50–80%) of the cosmic radiation which reaches the earth's surface consists of high energy muons. Muons have much less tendency to react with atomic nuclei than pions and, therefore, can penetrate the atmosphere and the solid material of the earth relatively easily. The remaining part, which is the lower energy component of the cosmic radiation that strikes the earth, consists of photons, electrons, and positrons. At sea level this part of the cosmic radiation gives rise to approximately 2–3 ion pairs s^{-1} cm^{-3} of air. It is this component of the cosmic radiation that gives rise to the cosmic ray portion of the natural background that is measured by nuclear detection devices in laboratories.[1] Some of the cosmic radiation interacts to make atmospheric radioactivity which is principally the nuclides ^{3}H and ^{14}C.

13.3. RADIOELEMENTS IN NATURE

Practically all natural materials contain radioactive nuclides, though usually in such low concentrations that they can be detected only by very sensitive analysis. This is true for water (rain water, rivers, lakes, sea), rocks and soil, and all living matter, as well as for structures based on natural raw materials except where great care has resulted in use of a non-radioactive material. The radionuclides in our environment can be divided into (i) those formed from cosmic radiation, *cosmogenic,* (ii) those with lifetimes comparable to the age of the earth together with those that are part of the natural decay chains beginning with thorium and uranium, *primordial,* and (iv) those introduced into nature by man using modern techniques, *anthropogenic.*

[1] The remainder of the natural background comes from naturally occurring radioactive elements in the laboratory materials and surrounding building.

13.4. COSMOGENIC RADIONUCLIDES

13.4.1. Survey

Cosmic irradiation of the atmosphere produces neutrons and protons which react with N_2, O_2, Ar, etc. resulting in the production of radioactive nuclides, some of which are listed in Tables 13.1 and 13.2. These nuclides are produced at constant rates and brought to the earth surface by rain water. Though they are formed in extremely low concentrations, the global inventory is by no means small (§§13.4.2 and 13.4.3). Equilibrium is assumed to be established between the production rate and the mean residence time of these radionuclides in terrestrial reservoirs (the atmosphere, the sea, lakes, soil, plants, etc) leading to constant specific radioactivity of the elements in each reservoir. If a reservoir is closed from the environment, its specific radioactivity decreases. This can be used to determine exposure times of meteorites to cosmic radiation (and the constancy of

Table 13.1 Long-lived cosmogenic radionuclides appearing in meteorites and rain water

Nuclide	Half-life (years)	Decay mode & particle energy (MeV)	Atmospheric production rate (atoms m^{-2} s^{-1})
^{3}H	12.32	β^- 0.0186	2500
^{10}Be	1.52×10^6	β^- 0.555	300
^{14}C	5715	β^- 0.1565	17 000–25 000
^{22}Na	2.605	β^+ 0.545	0.5
^{26}Al	7.1×10^5	β^+ 1.16	1.2
^{32}Si	160	β^- 0.213	1.6
^{35}S	0.239 (87.2 d)	β^- 0.167	14
^{36}Cl	3.01×10^5	β^- 0.709	60
^{39}Ar	268	β^- 0.565	56
^{53}Mn	3.7×10^6	EC (0.596)	
^{81}Kr	2.2×10^5	EC (0.28)	

Values within parenthesis after EC are decay energies.

Table 13.2 Short-lived cosmogenic radionuclides appearing in rain water

	Decay mode	
Nuclide	Half-life	Particle energy (MeV)
^{7}Be	53.28 d	EC (0.862)
^{24}Na	14.96 h	1.389
^{28}Mg	21.0 h	0.459
^{32}P	14.28 d	1.710
^{33}P	25.3 d	0.249
^{39}Cl	55.6 min	1.91

Value within parenthesis after EC is decay energy
Production rates (atoms m^2 s^1): ^{7}Be 81, ^{39}Cl 16

the cosmic radiation field, using ^{81}Kr), dating marine sediments (using ^{10}Be, ^{26}Al), groundwater (^{36}Cl), glacial ice (^{10}Be), dead biological materials (^{14}C), etc. The shorter-lived cosmogenic radionuclides have been used as natural tracers for atmospheric mixing and precipitation processes (e.g. ^{39}Cl or ^{38}S). Only T and ^{14}C are of sufficient importance to deserve further discussion.

13.4.2. Tritium

Satellite measurements have shown that the earth receives some of the tritium ejected from the sun. Much larger amounts are formed in the atmosphere through nuclear reactions; e.g., between fast neutrons and nitrogen atoms

$$n(\text{fast}) + {}^{14}N \rightarrow {}^{12}C + {}^{3}H \qquad (13.2)$$

The yield for this reaction is about 2500 atoms tritium per second per square meter of the earth's surface; the global inventory is therefore about 1.3×10^{18} Bq. Tritium has a half-life of 12.33 y, decaying by weak β^- emission to ^3He. It is rapidly incorporated in water, entering the global hydrological cycle. The average residence time in the atmosphere is about 2 y which is a small fraction of the half-life, as once the tritiated water reaches the lower troposphere, it rains out in 5–20 days. If we define 1 TU (Tritium Unit) as 1 tritium atom per 10^{18} hydrogen atoms, 1 TU corresponds to 118 Bq/m^3. Before the advent of nuclear energy, surface waters contained 2–8 TU (an average value of 3.5 TU is commonly used). The tritium content in water now commonly is of the order 20–40 TU. Rainwater contains between 4 and 25 TU, lower at the equatorial zone and increasing with latitude.

Tritium is also a product in the nuclear energy cycle, some of which is released to the atmosphere and some to the hydro sphere. The emissions differ between reactor types (usually in the order HWR > PWR > BWR, see Ch. 19 and 20) and is a function of the energy production. Assuming the annual releases to be 40 TBq/GW_e (Giga Watt electricity) from an average power plant and 600 TBq/GW_e from a typical reprocessing plant, the annual global injection of tritium in the environment was estimated to ~10 PBq in 1992. Though this is a small fraction compared to the natural production of ~150 PBq/y, it causes local increases.

The hydrogen bomb tests conducted in the atmosphere during the decade of the 1950's and early 1960's injected large amounts of tritium into the geosphere; 2.6×10^{20} Bq up to the end of the tests in 1962. This considerably exceeds the natural production inventory.

Before 1952 (first hydrogen bomb tests) the tritium content could be used to date water (i.e. determine when it became isolated from contact with the atmosphere). This was very useful e.g. for determining ice ages. However, due to the much larger content of anthropogenic tritium presently, this is no longer a useful technique for such dating.

Tritium in concentrations as low as 1 TU can be measured in low background proportional counters, and, after isotope enrichment (e.g. by electrolysis of alkaline water, by which tritium is enriched in the remainder), down to 0.01 TU. For very low concentrations mass spectrometry is preferred.

13.4.3. 14-Carbon

^{14}C is produced in the atmosphere by a variety of reactions, the most important being between thermalized neutrons from cosmic radiation and nitrogen atoms:

$$n(\text{slow}) + {}^{14}N \rightarrow {}^{14}C + {}^{1}H \tag{13.3}$$

If we compare this reaction with the one forming tritium by the interaction of fast neutrons with ^{14}N we observe that the results are intuitive. The higher the energy of the incoming projectile, the larger the fraction of the compound nucleus is ejected. The production of ^{14}C is approximately 22 000 atoms per s and m^2 of the earth's surface; the global annual production rate is ~1 PBq, and global inventory ~8500 PBq (corresponding to ~75 tons). Of this amount ~140 PBq remain in the atmosphere while the rest is incorporated in terrestrial material. All living material (incl. body tissue) has a ^{14}C concentration of ~227 Bq/kg. The half-life of ^{14}C is 5715 y; it decays by soft β^- emission (E_{max} 158 keV).

^{14}C is also formed by reaction (13.3) in nuclear tests. From these 220 PBq is assumed to have been injected into the atmosphere up to 1990. This ^{14}C comes to equilibrium with other atmospheric carbon (CO_2) in 1–2 years. Figure 13.3 shows the measured atmospheric concentration of ^{14}C as function of time after a nuclear weapons test in the Arctic. Some ^{14}C, about 18 TBq/GW_e per year, is also released from nuclear power plants (mainly from HWR, see Ch. 19–22). The global atmospheric value is <300 TBq/y.

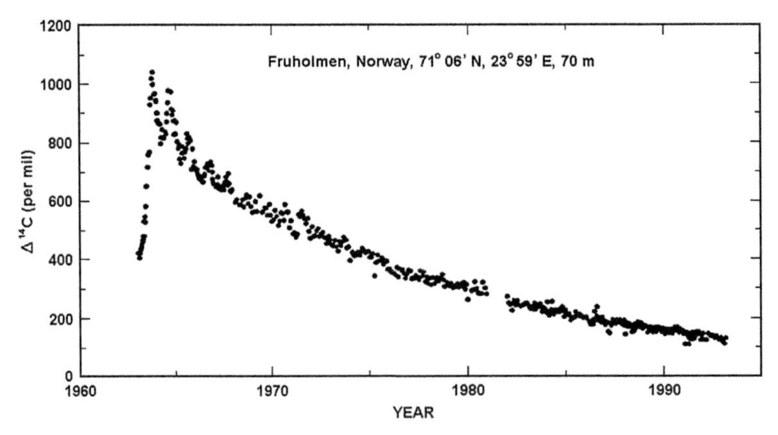

Figure 13.3 Atmospheric concentration of ^{14}C measured at Furuholmen after Soviet nuclear weapons test *(Data from Univ. of Oslo)*.

The combustion of fossil fuel adds CO_2, which is almost free of ^{14}C, to the atmosphere, thus reducing the specific activity (the dilution was about 3% for the period 1900–1970). Taking all anthropogenic sources into account, a global average specific activity of modern carbon is now 13.56 ± 0.07 dpm/g C. In §13.8.1 we discuss dating by the "C14-method", for which such a figure is important.

13.5. PRIMORDIAL RADIONUCLIDES

Some radionuclides have halflives so long that some of what was formed in the aftermath of the Big Bang and later supernova explosions still remains on the Earth, and some of their daughters.

13.5.1. Very long-lived nuclides lighter than lead[2]

As the detection technique for radioactivity has been refined, a number of long-lived radionuclides have been discovered in nature. The lightest have been mentioned in §13.4.1. The heavier ones, not belonging to the natural radioactive decay series of uranium and thorium, are listed in Table 13.3. ^{50}V is the nuclide of lowest elemental specific activity (~0.0001 Bq/g) while the highest are ^{87}Rb and ^{187}Re (each ~900 Bq/g). As our ability to make reliable measurements of low activities increases, the number of elements between potassium and lead with radioactive isotopes in nature can be expected to increase.

Table 13.3 Primordial radionuclides for Z < 82 (Pb)

Nuclide	Isotopic abundance %	Decay mode and particle energy MeV	Half-life (years)
^{40}K	0.0117	β^- EC 1.31	1.26×10^9
^{50}V	0.250	β^- EC (0.601)	$>1.4 \times 10^{17}$
^{87}Rb	27.83	β^- 0.273	4.88×10^{10}
^{115}In	95.72	β^- 1.0	4.4×10^{14}
^{123}Te	0.905	β^- EC (0.052)	1.3×10^{13}
^{138}La	0.092	β^- EC	1.06×10^{11}
^{144}Nd	23.80	α	2.1×10^{15}
^{147}Sm	15.0	β^- 2.23	1.06×10^{11}
^{148}Sm	11.3	β^- 1.96	7×10^{15}
^{176}Lu	2.59	β^-(1.188)	3.8×10^{10}
^{174}Hf	0.162	α	2×10^{15}
^{187}Re	62.60	β^- 0.0025	4.2×10^{10}
^{190}Pt	0.012	α	6.5×10^{11}

Values within parenthesis refer to decay energies.

[2] The heavier elements are discussed in §13.5.2 and in Ch. 14.
[3] The student can easily trace the decay series in Fig. 13.4 with a transparent marker.

Because of the long half-lives of these nuclides they must have been formed at the time of (or possibly even before) the formation of the solar system and of the earth. When the earth's crust solidified, these radionuclides became trapped in rocks. As they decayed, decay products accumulated in the closed rock environment. By measuring the amount of parent and daughter nuclides, it is possible with the half-life to calculate how long this environment (e.g. a rock formation) has existed. This is the bases for *nuclear dating* (also called "radioactive clocks"), and almost all of the nuclides in Tables 13.1 and 13.2 can be used for this purpose. In §13.8 we discuss dating methods for the K—Ar, Rb—Sr, and Re—Os systems.

A careful look at these naturally occurring long-lived nuclei reveals that some of them appear in *short* decay series, e.g. $^{152}Gd \rightarrow {}^{148}Sm \rightarrow {}^{144}Nd \rightarrow {}^{140}Ce$ and $^{190}Pt \rightarrow {}^{186}Os \rightarrow {}^{182}W$. The heavy element series beginning with U and Th isotopes are therefore referred to as *long* decay series.

13.5.2. Elements in the natural radioactive decay series

In Chapter 1 we briefly discussed the existence of four long series of genetically related radioactive nuclides beginning with Th, U or Np and ending with Pb or Bi. In Figure 13.4 we present all known isotopes of elements $_{81}Tl$ to $_{92}U$. The chemistry and other data for Th and U are given in detail in Ch 14. Some of the nuclides occur naturally in the long decay series shown in Figure 1.2. Others are produced through nuclear reactions according to the schemes in Figure 5.7 and principles described in Ch. 10 & 17. The first series in Figure 1.2 is known as the *thorium decay series*, and consists of a group of radionuclides related through decay in which all the mass numbers are evenly divisible by four (the $4n$ series). It has its natural origin in ^{232}Th which occurs with 100% isotopic abundance. Natural thorium has a specific activity S of 4.06 MBq/kg, as its half-life through α-decay is 1.41×10^{10} y. The terminal nuclide in this decay series is the stable species ^{208}Pb (also known as ThD). The transformation from the original parent to the final product requires 6 α- and 4 β-decays. The longest-lived intermediate is 5.76 y ^{228}Ra.

The *uranium decay series* consist of a group of nuclides that, when their mass number is divided by 4, have a remainder of 2 (the $4n + 2$ series). The parent of this series is ^{238}U with a natural abundance of 99.3%; it undergoes α-decay with a half-life of 4.46×10^9 y. The stable end product of the uranium series is ^{206}Pb, which is reached after 8 α- and 6 β-decay steps.

The specific activity of ^{238}U is 12.44 MBq/kg ^{238}U. However, because *natural uranium* consists of 3 isotopes, ^{238}U, ^{235}U and ^{234}U, whose isotopic abundances are 99.2745%, 0.7200% and 0.0055%, respectively, the specific activity of natural uranium is 25.4 MBq/kg.

The uranium decay series provides the most important isotopes of elements radium, radon, and polonium, which can be isolated in the processing of uranium minerals. Each ton of uranium is associated with 0.340 g of ^{226}Ra. Freshly isolated ^{226}Ra reaches

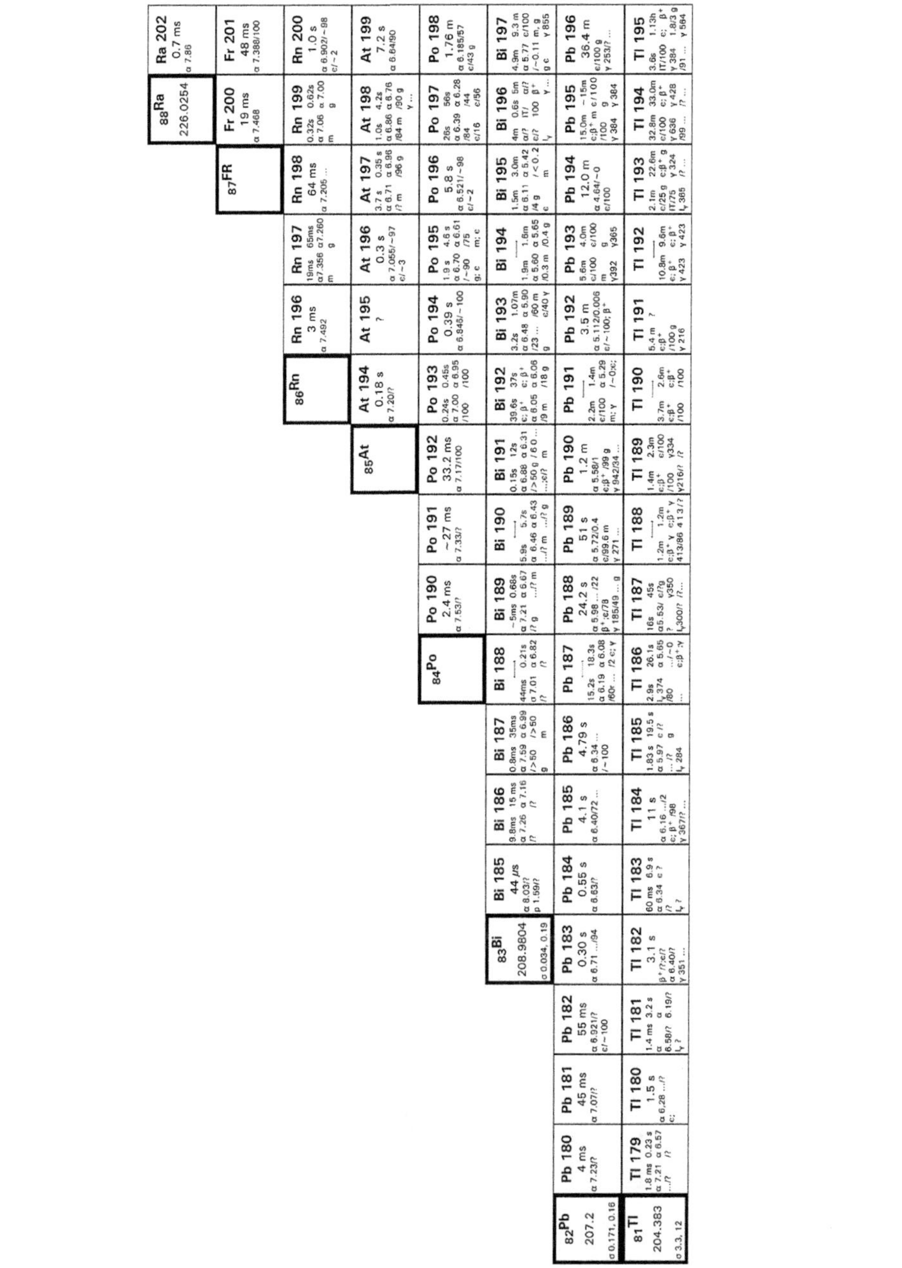

																$_{88}$Ra 226.0254	**Ra 202** 0.7 ms α 7.86
															$_{87}$FR	**Fr 200** 19 ms α 7.468	**Fr 201** 48 ms α 7.388/100
												$_{86}$Rn	**Rn 196** 3 ms α 7.492	**Rn 197** 19ms 65ms α7.356 α7.260 m g	**Rn 198** 64 ms α 7.205 ...	**Rn 199** 0.32s 0.62s α 7.06 α 7.00 m g	**Rn 200** 1.0 s α 6.902/~98 ε/~2
											$_{85}$At	**At 194** 0.18 s α 7.20/?	**At 195** ?	**At 196** 0.3 s α 7.055/~97 ε/~3	**At 197** 3.7 s 0.35 s α 6.71 α 6.96 /? m /96 g	**At 198** 1.0s 4.2s α 6.86 α 6.76 /84 m /90 g γ ...	**At 199** 7.2 s α 6.64/90
								$_{84}$Po	**Po 190** 2.4 ms α 7.53/?	**Po 191** ~27 ms α 7.33/?	**Po 192** 33.2 ms α 7.17/100	**Po 193** 0.24s 0.45s α 7.00 α 6.95 /100 /100	**Po 194** 0.39 s α 6.846/~100	**Po 195** 1.9 s 4.6 s α 6.70 α 6.61 /~90 /75 g; ε m; ε	**Po 196** 5.8 s α 6.521/~98 ε/~2	**Po 197** 26s 56s α 6.39 α 6.28 /84 /44 ε/16 ε/56	**Po 198** 1.76 m α 6.185/57 ε/43 g
				$_{83}$Bi 208.9804 σ 0.034, 0.19	**Bi 185** 44 μs α 8.03/? p 1.59/?	**Bi 186** 9.8ms 15 ms α 7.26 α 7.16 /? /?	**Bi 187** 0.8ms 35ms α 7.59 α 6.99 />50 />50 g m	**Bi 188** ⟶ 44ms 0.21s α 7.01 α 6.82 /? /?	**Bi 189** ~5ms 0.68s α 7.21 α 6.67 /? g .../? m	**Bi 190** ⟶ 5.9s 5.7s α 6.46 α 6.43 .../? m .../? g	**Bi 191** 0.15s 12s α 6.88 α 6.31 />50 g /60 ... ...;ε/? m	**Bi 192** 39.6s 37s ε; β⁺ ε; β⁺ α 6.05 α 6.06 /9 m /18 g	**Bi 193** 3.2s 1.07m α 6.48 α 5.90 /23 ... /60 m g ε/40 γ	**Bi 194** ⟶ 1.9m 1.6m α 5.60 α 5.65 /0.3 m /0.4 g	**Bi 195** 1.5m 3.0m α 6.11 α 5.42 /4 g /<0.2 ε m	**Bi 196** 4m 0.6s 5m α/? IT/ α/? ε/? 100 β⁺ I$_γ$ γ ...	**Bi 197** 4.9m 9.3 m α 5.77 ε/100 /~0.11 m, g g ε γ 855
$_{82}$Pb 207.2 σ 0.171, 0.16	**Pb 180** 4 ms α 7.23/?	**Pb 181** 45 ms α 7.07/?	**Pb 182** 55 ms α 6.921/? ε/~100	**Pb 183** 0.30 s α 6.71 .../94	**Pb 184** 0.55 s α 6.63/?	**Pb 185** 4.1 s α 6.40/72 ...	**Pb 186** 4.79 s α 6.34 ... /~100	**Pb 187** ⟶ 15.2s 18.3s α 6.19 α 6.08 /60r ... /2 ε; γ	**Pb 188** 24.2 s α 5.98 ... /22 β⁺;ε/78 γ 185/49 ... g	**Pb 189** 51 s α 5.72/0.4 ε/99.6 m γ 271 ...	**Pb 190** 1.2 m α 5.58/1 ε;β⁺ /99 g γ 942/34 ...	**Pb 191** ⟶ 2.2m 1.4m ε/100 α 5.29 m; γ /~0;ε;	**Pb 192** 3.5 m α 5.112/0.006 ε/~100; β⁺	**Pb 193** 5.6m 4.0m ε/100 ε/100 m g γ392 γ365	**Pb 194** 12.0 m α 4.64/~0 ε/100	**Pb 195** 15.0m ~15m ε;β⁺ m ε/100 /100 g γ 384 γ 384	**Pb 196** 36.4 m ε/100 g γ 253/? ...
$_{81}$Tl 204.383 σ 3.3, 12	**Tl 179** 1.8 ms 0.23 s α 7.21 α 6.57 .../? /?	**Tl 180** 1.5 s α 6,28 .../? ε;	**Tl 181** 1.4 ms 3.2 s α α 6.58/? 6.19/? I$_γ$?	**Tl 182** 3.1 s β⁺/?;ε/? α 6.40/? γ 351 ...	**Tl 183** 60 ms 6.9 s α 6.34 ε ? /? I$_γ$?	**Tl 184** 11 s α 6.16 .../2 ε; β⁺ /98 γ 367/? ...	**Tl 185** 1.83 s 19.5 s α 5.97 ε /? ... /? g I$_γ$ 284	**Tl 186** 2.9s 26.1s I$_γ$ 374 α 5.65 /80 .../~0 ... ε;β⁺;γ	**Tl 187** 16s 45s α5.53/ ε/?g ? γ350 I$_γ$300/? /?...	**Tl 188** ⟶ 1.2m 1.2m ε;β⁺ γ ε;β⁺ γ 413/86 413/?	**Tl 189** 1.4m 2.3m ε;β⁺ ε/100 /100 γ334 γ216/? /?	**Tl 190** ⟶ 3.7m 2.6m ε;β⁺ ε;β⁺ /100 /100	**Tl 191** 5.4 m ? ε;β⁺ /100 g γ 216	**Tl 192** ⟶ 10.8m 9.6m ε; β⁺ ε; β⁺ γ 423 γ 423	**Tl 193** 2.1m 22.6m ε/25 g ε;β⁺ g IT/75 γ 324 I$_γ$ 365 /? ...	**Tl 194** 32.8m 33.0m ε/100 ε; β⁺ γ 636 γ 428 /99 ... /? ...	**Tl 195** 3.6s 1.13h IT/100 ε; β⁺ γ 384 1.8/3 g /91 ... γ 564

Figure 13.4 Chart of all known isotopes of elements $_{81}$Tl to $_{92}$U. The legend to the chart is given in Fig. 5.7.

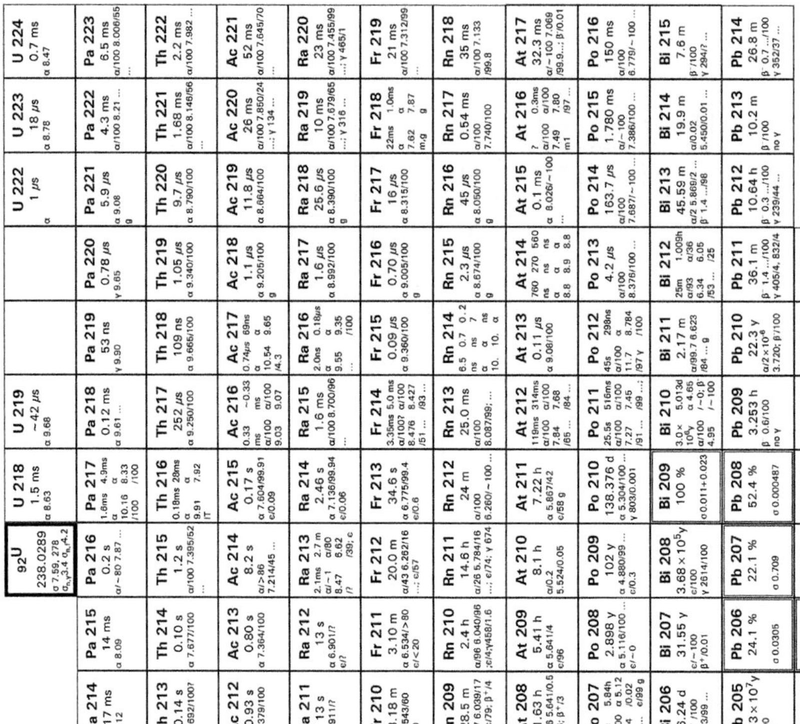

										$_{92}$U 238.0289 σ 7.59, 278 σn,γ 3.4 σn,f 4.2	U 218 1.5 ms α 8.63	U 219 ~42 μs α 9.68			U 222 1 μs α	U 223 18 μs α 8.78	U 224 0.7 ms α 8.47
					$_{91}$Pa 231.0359	Pa 212 5.1 ms α 8.27	Pa 213 5.3 ms α 8.24	Pa 214 17 ms α 8.12	Pa 215 14 ms α 8.09	Pa 216 0.2 s α/~80 7.87 ...	Pa 217 1.6ms 4.9ms α α 10.16 8.33 /100 /100	Pa 218 0.12 ms α 9.61 ...	Pa 219 53 ns γ 9.90	Pa 220 0.78 μs γ 9.65	Pa 221 5.9 μs α 9.08 g	Pa 222 4.3 ms α/100 8.21 ...	Pa 223 6.5 ms α/100 8.006/55 ...
					$_{90}$Th 232.0381 σ 7.40, 85	Th 211 37 ms α 7.79	Th 212 30 ms α 7.80	Th 213 0.14 s α 7.692/100?	Th 214 0.10 s α 7.677/100	Th 215 1.2 s α/100 7.395/52 ...	Th 216 0.18ms 28ms α α 9.91 7.92 IT	Th 217 252 μs α 9.250/100	Th 218 109 ns α 9.665/100	Th 219 1.05 μs α 9.340/100	Th 220 9.7 μs α 8.790/100	Th 221 1.68 ms α/100 8.146/56 ...	Th 222 2.2 ms α/100 7.982 ...
		$_{89}$Ac 227.0278	Ac 207 22 ms α 7.712	Ac 208 25ms 95ms α 7.76 α 7.57 m2 g	Ac 209 90 ms α 7.59/100	Ac 210 0.35 s α 7.462/100	Ac 211 0.25 s α 7.481/? ε/?	Ac 212 0.93 s α 7.379/100	Ac 213 0.80 s α 7.364/100	Ac 214 8.2 s α/>86 7.214/45 ...	Ac 215 0.17 s α 7.604/99.91 ε/0.09	Ac 216 0.33 ~0.33 ms ms α/100 α/100 9.03 9.07	Ac 217 0.74μs 69ns α α 10.54 9.65 /4.3	Ac 218 1.1 μs α 9.205/100 g	Ac 219 11.8 μs α 8.664/100	Ac 220 26 ms α/100 7.850/24 ...; γ 134 ...	Ac 221 52 ms α/100 7.645/70 ...
Ra 203 33ms ~1ms α 7.62 α 7.58 m g	Ra 204 59 ms α 7.484	Ra 205 0.17s 0.21s α 7.37 α 7.34 m g	Ra 206 0.24 s α 7.270/100	Ra 207 55ms 1.3s α 7.32 α 7.13 /15 m /90 g IT? ε/10	Ra 208 1.3 s α 7.133/~95 ε/~5	Ra 209 4.6 s α 7.010/100	Ra 210 3.7 s α 7.019/100	Ra 211 13 s α 6.911/? ε/?	Ra 212 13 s α 6.901/? ε/?	Ra 213 2.1ms 2.7 m α/~1 α/80 8.47 6.62 /? /39; ε	Ra 214 2.46 s α 7.136/99.94 ε/0.06	Ra 215 1.6 ms α/100 8.700/96 ...	Ra 216 2.0ns 0.18μs α α 9.55 9.35 ... /100	Ra 217 1.6 μs α 8.992/100	Ra 218 25.6 μs α 8.390/100 g	Ra 219 10 ms α/100 7.679/65 ...; γ 316 ...	Ra 220 23 ms α/100 7.455/99 ...; γ 465/1
Fr 202 0.34s 0.34s α 7.24 7.24 m g	Fr 203 0.55 s α 7.132/100	Fr 204 ~1 2.6s 1.7s s α α α 6.9 7.0 7.0 ... ... g	Fr 205 3.9 s α 6.916/100	Fr 206 0.7s 16s 16s α α α 6.9 6.7 6.8 g	Fr 207 14.8 s α 6.766/93 ε/7	Fr 208 58.6 s α 6.636/100 ε/23	Fr 209 50.0 s α 6.646/100 ε/11	Fr 210 3.18 m α 6.543/60 ε/40	Fr 211 3.10 m α 6.534/>80 ε/<20	Fr 212 20.0 m α/43 6.262/16 ...; ε/57	Fr 213 34.6 s α 6.775/99.4 ε/0.6	Fr 214 3.35ms 5.0 ms α/100? α/100 8.476 8.427 /51 ... /93 ...	Fr 215 0.09 μs α 9.360/100	Fr 216 0.70 μs α 9.005/100 g	Fr 217 16 μs α 8.315/100	Fr 218 22ms 1.0ms α α 7.62 7.87 m,g g	Fr 219 21 ms α/100 7.312/99 ...
Rn 201 3.8 s 7.0 s α 6.77 α 6.72 /~90 /~80 ε m ε g	Rn 202 9.85 s α 6.641/85 ε/~15 g	Rn 203 28s 45s α 6.55 α 6.50 /100 /66 m ε g	Rn 204 1.24 m α 6.417/68 ε/32	Rn 205 2.83 m α/23 6.263/23 ...; ε/77 g γ 265/? ...	Rn 206 5.67 m α 6.260/68 ε/32 γ 498/? ...	Rn 207 9.3 m α/23 6.126/23 ...; ε;β+/77 g	Rn 208 24.4 m α/60 6.138/60 ...; ε/40; γ 427	Rn 209 28.5 m α/17 6.039/17 ...; ε/79; β+/4	Rn 210 2.4 h α/96 6.040/96 ;ε/4;γ458/1.6	Rn 211 14.6 h α/26 5.784/16 ...; ε/74; γ 674	Rn 212 24 m α/100 6.260/~100 ...	Rn 213 25.0 ms α/100 8.087/99; ...	Rn 214 6.5 0.7 0.2 ns ns 7 α α ns 10. 10. α	Rn 215 2.3 μs α 8.674/100 g	Rn 216 45 μs α 8.050/100 g	Rn 217 0.54 ms α/100 7.740/100	Rn 218 35 ms α/100 7.133 /99.8
At 200 3.5s 47s 43s α α α 6.5 6.4 6.4 4 1 6	At 201 1.5 m α 6.344/71 ε/29	At 202 0.5s 3.0 3.1 α m α m α 6.2 6.1 6.2 8 4 3	At 203 7.4 m α 6.088/31 g ε; β+ /69 g,m	At 204 9.2 m α 5.951/4 ε/68;β+/29;	At 205 26.2 m α 5.902/10 ε;β+ /90	At 206 29.4 m α 5.703/1; ε/82 β+ 3.1.../17	At 207 1.8 h α 5.759/10 ε;β+ /90	At 208 1.63 h α/0.6 5.641/0.5 ε/96; β+/3	At 209 5.41 h α 5.641/4 ε/96	At 210 8.1 h α/0.2 5.524/0.05	At 211 7.22 h α 5.867/42 ε/58 g	At 212 119ms 314ms α/100 α/100 7.84 7.68 /65 ... /84 ...	At 213 0.11 μs α 9.08/100	At 214 760 270 560 ns ns ns α α α 8.8 8.9 8.8	At 215 0.1 ms α 8.026/~100 ...	At 216 ? 0.3ms α/100 α/100 7.49 7.80 m1 /97 ...	At 217 32.3 ms α/~100 7.069 /99.9...; β-/0.01
Po 199 4.2m 5.2m α 6.06 α 5.95 /39 m /12 g ε/61 g ε/88 m	Po 200 11.5 m α 5.863/15 ε/85 m	Po 201 8.9 m 15.3 m α 5.79 α 5.68 /3;ε/57 /2 g IT/40 ε/98 m	Po 202 44.7 m α 5.588/2 ε/98	Po 203 45s 36m IT α5.384 Iγ /0.1 641/51 ε/93;	Po 204 3.53 h α 5.377/0.7 ε/99.3	Po 205 1.66 h α 5.22/0.1 g ε/98; β+/2	Po 206 8.8 d α 5.223/5 ε/95 g	Po 207 2.8s 5.84h IT/100 α 5.12 γ 814 /0.02 /99 ... ε/99 g	Po 208 2.898 y α 5.116/100 ... ε/~0	Po 209 102 y α 4.880/99 ... ε/0.3	Po 210 138.376 d α 5.304/100 ... γ 803/0.001	Po 211 25.5s 516ms α/100 α/100 7.27 7.45 /91 ... /99 ...	Po 212 45s 298ns α/100 α 11.7 8.784 /97 γ /100	Po 213 4.2 μs α/100 8.376/100 ...	Po 214 163.7 μs α/100 7.687/~100 ...	Po 215 1.780 ms α/~100 7.386/100 ...	Po 216 150 ms α/100 6.779/~100 ...
Bi 198 7.7 11. 10. s 6 3 IT m m Iγ ε; ε;	Bi 199 24.7m 27m α ε/100 5.484 γ 424/? /<2.8 m; g	Bi 200 31m 36.4m ε;β+ ε/89 /100 β+/11 γ 1027 γ 1027	Bi 201 59.1m 1.8h α 5.24 ε/100 /~0; ε; γ 629 g IT; /? ...	Bi 202 1.72 h ε; β+; g γ 579/7 ...	Bi 203 11.76 h ε/~100 g; m β+ 1.4 .../0.2	Bi 204 11.22 h ε/100 γ 899/98 ...	Bi 205 15.31 d ε/~100 β+/0.1	Bi 206 6.24 d ε;β+ /100 γ 803/99 ...	Bi 207 31.55 y ε/~100 β+/0.01	Bi 208 3.68 × 10^5 y ε/100 γ 2614/100	Bi 209 100 % σ 0.011+0.023	Bi 210 3.0× 5.013d 10^{6}y α 4.65 α/100 /~0; β- 4.95 /~100	Bi 211 2.17 m α/99.7 6.623 /84 ... g	Bi 212 25m 1.009h α/93 α/36 6.34 6.05 /53 ... /25	Bi 213 45.59 m α/2 5.869/2 ... β- 1.4 .../98	Bi 214 19.9 m α/0.02 5.450/0.01 ...	Bi 215 7.6 m β-/100 γ 294/? ...
Pb 197 43m ~8m ε/81 ε;β+ IT/19 /100 γ 386 γ 386	Pb 198 2.4 h ε/100 g γ 290/36 ...	Pb 199 12.2m 1.5h IT/93 ε/99 ε/7 β+/1 Iγ 424 γ 367	Pb 200 21.5 h ε/100 γ 148/38 ...	Pb 201 1.02m 9.33h IT/100 ε/~10 Iγ 629 0 /54 β+	Pb 202 3.62h 5.3× ε/10 10^{4}y IT/90 ε/100 Iγ 961 no γ	Pb 203 6.3s 2.169d IT/100 ε/100 γ 825 γ 279 /72 ... /80 ...	Pb 204 1.4 % 1.12 h IT/100	Pb 205 1.53 × 10^7 y ε/100 no γ	Pb 206 24.1 % σ 0.0305	Pb 207 22.1 % σ 0.709	Pb 208 52.4 % σ 0.000487	Pb 209 3.253 h β- 0.6/100 no γ	Pb 210 22.3 y α/2×10^{-6} 3.720; β-/100	Pb 211 36.1 m β- 1.4 .../100 γ 405/4, 832/4	Pb 212 10.64 h β- 0.3 .../100 γ 239/44 ...	Pb 213 10.2 m β-/100 no γ	Pb 214 26.8 m β- 0.7 .../100 γ 352/37 ...
Tl 196 1.41h 1.84h ε/96 ε; IT/5 β+/14 γ 426 γ 426	Tl 197 2.84 h ε; β+/2 g γ 426 /13 ...	Tl 198 1.87h 5.3h ε/54 ε; β+/1 IT/46 γ 412 γ 412 /83 ...	Tl 199 7.42 h ε/100 g γ 455/12,	Tl 200 1.088 d ε/~100 β+/0.4	Tl 201 3.05 d ε/100 γ 167/9 ...	Tl 202 12.23 d ε/100 γ 440/91 ...	Tl 203 29.524 % σ 11.0	Tl 204 3.78 y ε/2.6 β- 0.8/97.5	Tl 205 70.476 % σ 0.11	Tl 206 3.76 m 4.20 m IT/100 β-1.5... γ 687 /100; γ /100 ... 803/~	Tl 207 1.33s 4.77m IT/100 β-/100 γ 1000 γ 897 /87 ... /0.2 ...	Tl 208 3.053 m β- 1.8 .../100 γ 2614/99.8 ...	Tl 209 2.16 m β- 1.8 .../100 γ 1566/98 ...	Tl 210 1.30 m β- 1.9 .../100 γ 798/99 ...			

Figure 13.4 *(continued)*.

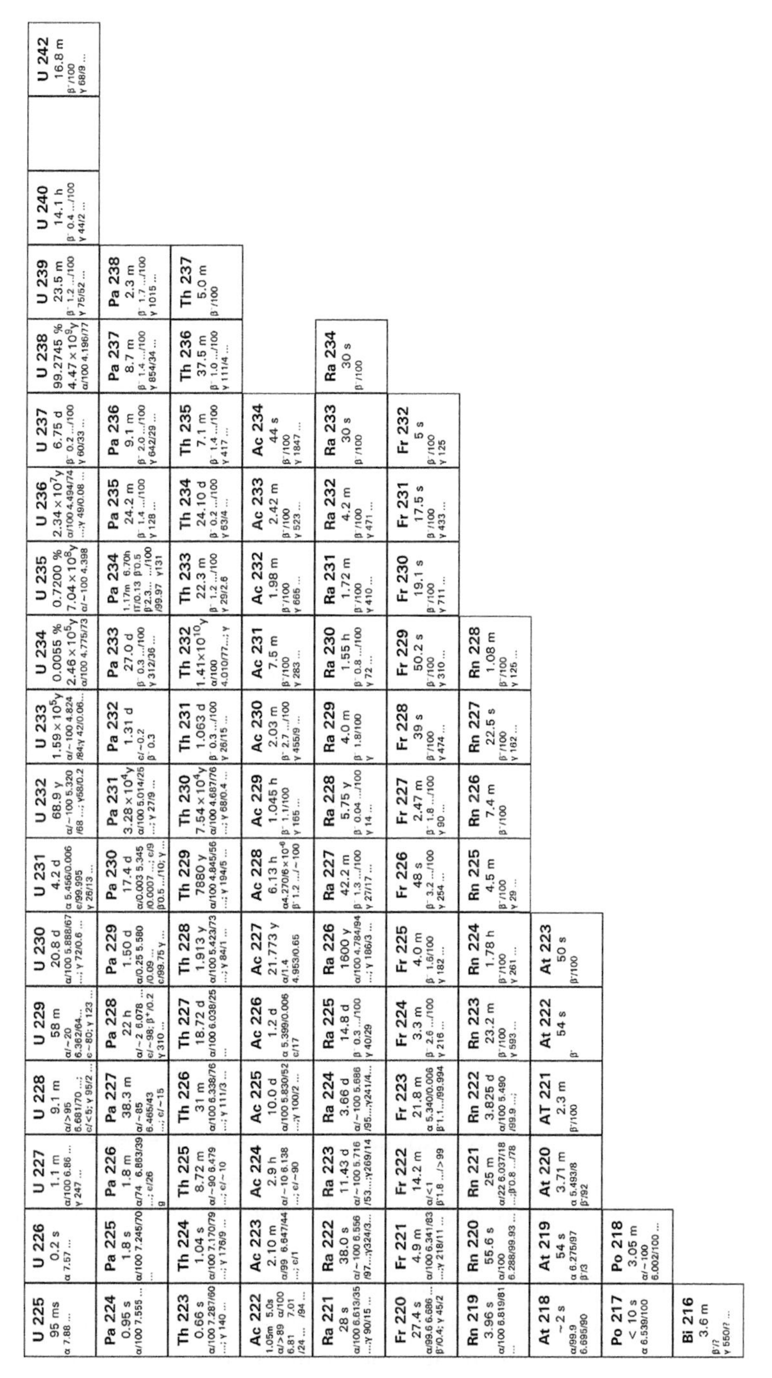

Figure 13.4 (*continued*).

radioactive equilibrium with its decay products to ^{210}Pb in about two weeks (see Fig. 1.2). Many of these products emit energetic γ-rays, which resulted in the use of Ra as a γ-source in medical treatment of cancer (radiation therapy). However, the medical importance of radium has diminished greatly since the introduction of other radiation sources, and presently the largest use of radium is as small neutron sources (see Table 10.2).

Although the chemistry of radium is relatively simple (like barium), the fact that it produces a radioactive gas (radon) complicates its handling. The decay of radon produces "airborne" radioactive atoms of At, Po, Bi, and Pb. Since uranium is a common element in rocks it is also a common constituent of building materials. Such material emits Rn, as discussed further in §13.6. Work with radium compounds should be carried out within enclosures to avoid exposure to Rn and its daughters. In recent years the use of ^{223}Ra has become increasingly common when treating bone cancer. The use is both for easing the pain but also for direct treatment. Radium is an alkali metal with the homologues Ca, Sr and Ba and substitutes readily with these, e.g. with the Ca in bone.

The *actinium decay series* consists of a group of nuclides whose mass number divided by 4 leaves a remainder of 3 (the $4n + 3$ series). This series begins with the uranium isotope ^{235}U, which has a half-life of 7.04×10^8 y and a specific activity of 8×10^4 MBq/kg. The stable end product of the series is ^{207}Pb, which is formed after 7 α- and 4 β-decays. The actinium series includes the most important isotopes of the elements protactinium, actinium, francium, and astatine. Inasmuch as ^{235}U is a component of natural uranium, these elements can be isolated in the processing of uranium minerals. The longest-lived protactinium isotope, ^{231}Pa ($t_{1/2}$ 3.28×10^4 y) has been isolated on the 100 g scale, and is the main isotope for the study of protactinium chemistry. ^{227}Ac ($t_{1/2}$ 21.8 y) is the longest-lived actinium isotope.

A fourth long radioactive decay series, the *neptunium series* (Fig. 1.2), is composed of nuclides having mass numbers which divided by 4 have a remainder of 1 (the $4n + 1$ series). The name comes from the longest lived $A = 4n + 1$ nuclide heavier than Bi, ^{237}Np, which is considered as the parent species; it has a half-life of 2.14×10^6 y. In as much as this half-life is considerably shorter than the age of the earth, primordial ^{237}Np no longer exists on earth, and, therefore, the neptunium series is not found as a natural occurrence. Np found in nature on Earth has now human origin. However, Np has been discovered in the spectrum of some stars. All known Np-isotopes are presented in Figure 14.1.

Very small amounts of ^{237}Np, as well as of ^{239}Pu, have been discovered on earth; the half-lives of ^{239}Pu (in the $4n + 3$ series) is 2.411×10^4 y. Both isotopes are too short-lived to have survived the 4 eons since the solar system was formed. However, they are always found in minerals containing uranium and thorium and it is believed that the neutrons produced in these minerals through (α,n) and (γ,n) reactions with U and Th as well as by spontaneous fission of ^{238}U form the neptunium and plutonium through n-capture and β-decay processes. The n-production rate in the uranium mineral pitchblende (containing ~50% U) is about 50 n/kg s. The typical value for the ^{239}Pu/^{238}U ratio in minerals is 3×10^{-12}.

The end product of the neptunium series is ^{209}Bi, which is the only stable isotope of bismuth. Seven α- and four β-decays are required in the sequence from the parent ^{237}Np to ^{209}Bi. An important nuclide in the neptunium decay series is the uranium isotope ^{233}U, which has a half-life of 1.59×10^5 y (the most stable intermediate) and, like ^{235}U, is fissionable by slow neutrons.

The long-lived plutonium isotope ^{244}Pu (belonging to the $4n$ series; see also Fig. 14.1), which decays through α-emission and spontaneous fission (0.13%) with a total half-life of 8.26×10^7 y, was discovered in rare earth minerals in 1971. If this is a survival of primeval ^{244}Pu, only 10^{-15} % of the original can remain. An alternate possibility is that this ^{244}Pu is a contaminant from cosmic dust (e.g. from a supernova explosion in more recent times than the age of the solar system).

13.6. RADIUM AND RADON IN THE ENVIRONMENT

In uranium ore radioactive equilibrium is established between the mother ^{238}U and daughters in the decay chain (see Fig. 1.2, $4n + 2$ series) at *the rate of the daughter half-life* and at the *level of the mother decay rate.* The chain passes ^{226}Ra and ^{222}Rn and daughters down to ^{206}Pb at a rate corresponding to the original amount of ^{238}U, as long as the host material is undisturbed. The equilibrium between ^{226}Ra and ^{210}Pb is established within some weeks. Since ^{222}Rn is a gaseous intermediate, its daughters are likely to be formed in air ("radon daughters"). Radon diffuses out of thorium and uranium minerals, and adds radioactivity to the ground water and to the atmosphere both by its own presence and that of its daughters. Since Ra and Rn are among the most radio-toxic substances existing, causing bone and lung cancer at relatively low concentrations (the risk levels are discussed in Ch. 15), special attention must be devoted to their appearance in nature.

Common Rn-concentrations in ground water are 5−300 kBq/m^3, but in areas of U-rich granite values $\gg$1 MBq ^{222}Rn/m^3 occur. Tap water usually contains of the order of 1 kBq/m^3. In many places water from hot mineral wells is considered beneficial to health both for bathing and for drinking ("spas" or hot springs). The water may be warm due to radiogenic heating at the source (minerals rich in U or Th) and have a high content of dissolved radium and radon. Thus, in famous spas in Europe the ^{222}Rn concentration for "therapeutic inhalation" may be 1 MBq/m^3 air (Baden-Baden). From the Joachimsthal U-mine in Bohemia, which contains a number of hot wells (29°C), water containing 10−15 MBq ^{222}Rn/m^3 is pumped to spas, where it is used as medical treatment against rheumatism (a 30 min bath a day).

The average exhalation rate of radon from the ground is 5−50 mBq/m^{2}s, leading to a near ground level radon concentration of 1−10 Bq/m^3, but varies widely with ground conditions. In soil over Swedish uranium bearing shale ($\approx$300 ppm U) the ^{222}Rn concentration can exceed 1 MBq/m^3, though on the average the concentration in Scandinavian air (the main geology consists of granitic rocks) is only 3 Bq/m^3.

Representative values are for the US 0.1–10 Bq/m^3, UK and Germany ~3, and 10 (average) for France. The concentration in air above ground depends also on temperature and wind conditions.

Many ores contain small amounts of uranium. During processing, uranium and/or its daughters may enter the product, causing a radioactive contamination problem. For example, when apatite is used to produce phosphoric acid, the gypsum by-product contains all the radium originally present, producing a γ-ray and inhalation hazard from Rn-daughters, making it unsuitable for building material.

Radon concentrations in indoor air may be quite high, depending on site and building material. The ^{226}Ra content e.g. in German building materials varies from >500 (gypsum) to 60 (brick) Bq/kg; slag used in Poland contains <800, Italian tuff ≈280, concrete in Hungary ≈13, and white bricks in the UK only ≈4 Bq/kg. The indoor concentration of radon also depends on the way in which the house is built and used (poorly ventilated, etc). In the US it varies between <1 to >1000 Bq/m^3. Authorities no longer recommend very tight houses, as suggested in the 1970's to reduce heating costs. At levels <70 Bq/m^3 the Rn-hazard is considered negligible. In Sweden (pop. 8 M people) 50% of the houses have 70–200 Bq/m^3, while 40 000 houses have been classified as "Rn-houses", i.e. their indoor concentration of Rn plus daughters exceed 400 Bq/m^3. If the ground is the main source of radon, ventilation of the basement may be sufficient to declassify a "Rn-house".

Even coal contains small amounts of uranium, 4 to 300 kBq/ton, a typical value being 20 kBq/ton. When coal is burnt, the more volatile U-daughter products are released into the atmosphere, while the less volatile ones are deposited in the dust filters. A 1 GW$_e$ coal-fired power plant typically releases 60 GBq ^{222}Rn and 5 GBq ^{210}Pb+Po annually, while ~3 MBq daughter products are obtained per ton of fly ash.

13.7. DISEQUILIBRIUM

Uranium ores at surface level are usually identified from the penetrating γ-ray emissions of the daughter products. However, it has been observed that some peat, which absorb uranium from local drainage, contain very little of the daughter products because they have been formed so recently that radioactive equilibrium has not been established.

The decay products of uranium passes over 10 elements (Fig. 1.2), all with very different chemical properties. These elements are transported by groundwater (*migrate*), the solute composition of which varies with the surrounding rock/soil minerals. The different elements migrate at different rates due to their different chemistries, dissolving in some areas and precipitating in others. If the mother and daughter in a radioactive chain migrate at different velocities during a time which is short in relation to the daughter half-life, the radioactive equilibrium is disturbed; this is referred to as *disequilibrium*. Such disequilibrium can be used to measure the age of the sample.

Let us consider the essential steps of the uranium decay series:

$$
\begin{array}{cccccccccccc}
 & \alpha+2\beta & & \alpha & & \alpha & & 4\alpha+2\beta & & 2\beta+\alpha & \\
^{238}\text{U} & \dashrightarrow & ^{234}\text{U} & \dashrightarrow & ^{230}\text{Th} & \dashrightarrow & ^{226}\text{Ra} & \dashrightarrow & ^{210}\text{Pb} & \dashrightarrow & ^{206}\text{Pb} \\
4.5\times10^{9}\text{ y} & & 2.5\times10^{5}\text{ y} & & 7.5\times10^{4}\text{ y} & & 1.6\times10^{3}\text{ y} & & 22\text{ y} & & \\
|\Leftarrow \approx 10^{6}\text{ y} \Rightarrow| & & \Leftarrow \approx 3\times10^{5}\text{ y} \Rightarrow| & & \Leftarrow \approx 10^{4}\text{ y} \Rightarrow| & & \Leftarrow \approx 100\text{ y} \Rightarrow| & & \Leftarrow \approx 100\text{ y} \Rightarrow| & &
\end{array}
\qquad (13.4)
$$

In this decay series the nuclide pairs which are suitable for determination of ages are associated with the time periods given between the double arrows of the two connected isotopes. For example, the ^{238}U-decay to ^{234}U passes over the short-lived intermediates ^{234}Th ($t_{½}$ 24.1 d) and ^{234}Pa ($t_{½}$ 1.17 m). The Th-isotope is long-lived enough to follow its own chemistry in a dynamic system. In strongly acidic solutions it forms Th^{4+} ions, while U forms UO_2^{2+} ions; the behavior of these two ions with regard to complex formation (e.g. by carbonates, hydroxyl or humic acid) and solubility is drastically different in neutral waters, leading to different migration rates for the two elements. For example, ^{238}U may migrate away, while ^{234}Th is precipitated or sorbed. As a result, when ^{234}Th decays (via the rapid equilibrium with ^{234}Pa) to ^{234}U, the latter is free from ^{238}U. From the deviation from the original activity ratio of $^{238}U/^{234}U = 1.0$, (^{238}U refers to the concentration of that isotope, see §3.3.2) the time since ^{238}U and ^{234}Th separated (i.e. the age of the sample) is deduced.

An example of such a system is the sedimentary uranium deposit in Tono, Japan, studied by Nohara et al. The ground water moves through the area with a velocity of 0.001 to 1 m/y. The specific activity ratio, measured by α- and βγ-spectrometry, of $^{234}U/^{238}U$ is plotted against $^{230}Th/^{234}U$ (a), and $^{226}Ra/^{230}Th$ against $^{210}Pb/^{226}Ra$ (b) for a large number of rock samples in Figure 13.5. If radioactive equilibrium existed all ratios

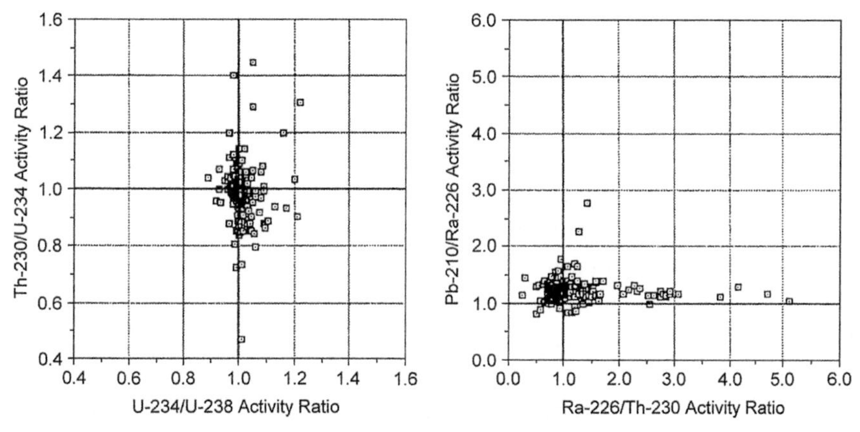

Figure 13.5 Activity ratios $^{234}U/^{238}U$ vs $^{230}Th/^{234}U$, and $^{226}Ra/^{230}Th$ vs $^{210}Pb/^{226}Ra$ for rock samples from the Tono uranium deposit. *(From Nohara et al.).*

should be 1.0. The observation of deviations from 0.5 to 1.5 in (a) and up to 5 in (b) indicate that U, Th, Ra and Pb have migrated at different velocities in recent times. A detailed analysis yields the age of the U deposit and the migration rates of the daughter elements: the U and Th has not migrated during the past several hundred thousand years (this is probably also the age of the U deposit); Ra has migrated a few meters in the last 10 000 y.

13.8. AGE DETERMINATION FROM RADIOACTIVE DECAY

Prior to the discovery of radioactivity, geologists could obtain only poor estimates of the time scale of the evolution of the earth. The oldest geologic materials were assumed to be some 10 million years old, and it was believed that this represented the age of the earth. However, with the discovery of radioactivity early in this century, geologists developed more objective methods for such age determination ("nuclear clocks"). In 1907 B. B. Boltwood obtained a value of 2.2×10^9 y for the age uranium and thorium minerals, assuming that all U and Th ultimately decayed to lead. Considering how few isotopes in the chains had been discovered at the time, the calculation was surprisingly good. Nuclear clocks have provided primary data on the age and evolution of the earth (*nuclear geo-chronology*) as well as the formation of elements and of the universe (*cosmo-chronology*).

The cosmogenic radionuclides with relatively short half-lives can be used to date materials of more recent origin; e.g., ^{3}H for water movements in the geosphere, and ^{14}C for organic material of archaeologic interest. Practically all of the primordial radionuclides (Table 13.3) can be used for dating geologic materials: ^{40}K/^{40}Ar for igneous (plutonic) rocks (i.e. rocks which have solidified from a rather homogenous melt); ^{87}Rb/^{87}Sr for metamorphic and sedimentary rocks; ^{147}Sm/^{143}Nd for rock-forming silicate, phosphate and carbonate minerals; ^{187}Re/^{187}Os for sulfides and metallic material like iron meteorites, etc.

13.8.1. Dating by the ^{14}C method

It is reasonable to assume that the production of ^{14}C in the atmosphere has been rather constant for at least a million years, which means that equilibrium exists between the rates of formation and decay of the ^{14}C in the atmosphere. Moreover, the half-life of ^{14}C is sufficient to allow equilibrium between the ^{14}C in the atmosphere, the oceans (including precipitations to ocean bottoms), and exchangeable carbon in natural materials. Thus from measurement of the specific radioactivity of carbon, it should be possible to determine when the sample became isolated from its natural environmental compartment.

The discovery that all living organic material has a certain specific radioactivity due to ^{14}C led W. Libby to a new method for determination of the age of biological material. This method, which has been of great importance in dating archeological, geological, etc, materials, is based on the assumption (i) that cosmogenic ^{14}C has been produced at a constant rate, (ii) that the amount of anthropologic ^{14}C is negligible compared to the cosmogenic, (iii) that after the organism incorporating the biological material died, no exchange occurs between the carbon atoms of the material and those of the surroundings. In such material the number of ^{14}C atoms decreases with time according to the half-life of ^{14}C. The equation is (for notation, see §5.13)

$$^{14}C\,(\mathrm{Bq/g}) = {}^{14}C_o\,\mathrm{e}^{-0.693t/5568} \tag{13.5a}$$

or

$$t\,(\mathrm{y}) = (\log{}^{14}C_o - \log{}^{14}C)5568/0.301 \tag{13.5b}$$

where $^{14}C_o$ is the initial ^{14}C activity of a standard ($^{14}C_o \approx 14$ dpm/g, c.f. §13.4.3) at time of death of the plant, etc ($t = 0$). The reference time for ^{14}C ages is AD 1950, which is indicated with the letters bp or BP (for "before present"). The half-life of 5568 y is a standard reference value introduced by Libby. For example, if the specific activity of a sample is measured to be 0.1 dpm/g, then (13.5b) gives a value of 39 700 years as the time since the material ceased to exchange its carbon. Only with extreme care and very sophisticated equipment can a specimen very old be determined with reliability, but shorter times can be measured more accurately since the specific activities are larger. ^{14}C ages are used for dating specimen of 300 to 50 000 y, with an uncertainty of 10 − 100 y. ^{14}C-determinations were originally made by transferring the carbon into carbon dioxide, which was measured in an internal GM-counter (Ch. 9). Later, transformation into methane was preferred, and the CH_4 counted in an internal proportional counter. The most sensitive technique today is to introduce the sample into the ion source of a tandem-van-der-Graaff accelerator and "count" the amount of ^{14}C-ions relative to ^{12}C-ions in the beam by mass-spectrometry.

There are many cautions that must be observed in the use of carbon dating. In addition to those mentioned above, there is the possibility of isotopic effects in metabolic processes. These could cause ^{14}C to be slightly depleted relative to ^{12}C due to chemical reactions in the biological material. Many plants prefer ^{12}C over ^{13}C and ^{14}C. To take this into account to get the correct "solar age" instead of a slightly misleading "^{14}C age", the isotopic depletion can be determined from measured deviations in the $^{12}C/^{13}C$ ratio. A "$\delta^{13}C$ correction" is introduced according to

$$^{14}C_{\mathrm{corr}} = {}^{14}C\{1 - 2(\delta^{13}C + 25)/1000\} \tag{13.6}$$

The value of $\delta^{13}C$ differs for various substances; e.g. −35 to −20 for terrestrial organic matter, −8 to −7 for atmospheric CO_2, etc. Figure 13.6 illustrates the use of various $\delta^{13}C$ values for biological matter. These corrections lead to considerable adjustments of conventional ^{14}C ages; e.g. for the period 7 000 to 2 000 years ago,

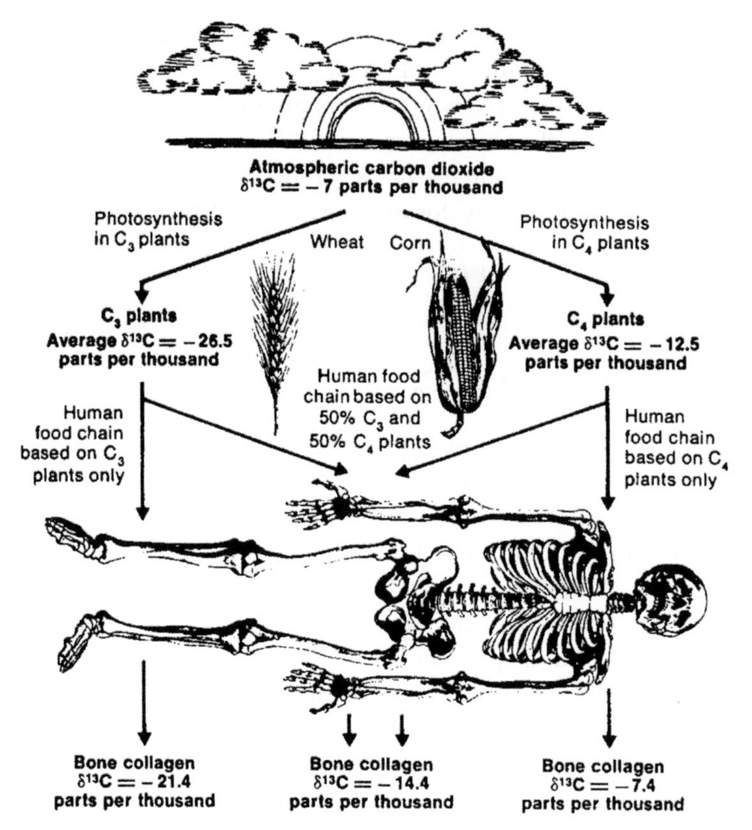

Figure 13.6 Ratio of $^{13}C/^{12}C$ in human bone depends on diet and affects the ^{13}C correction factor. *(From American Scientist,* ***70*** *(1982) 602).*

corrections increase ages by up to 1 000 y. The $\delta^{13}C$ can also be used to infer the relative size of the human contribution to atmospheric CO_2 by combustion of organics.

Another necessary correction has to be made for changes in the cosmic ray irradiation, which, *a priori* was assumed to be constant, but is not over long geological periods. By counting the number of annular rings on old trees such as the Sequoia gigantea, which can be almost 4 000 years old, and determining the ^{14}C content of each ring, the variation of cosmic radiation through thousands of years can be studied accurately, see Figure 13.7. Obviously there is a periodic cycle of some 10 000 y, which coincides with the variation in warm periods and small ice ages.

A large number of important age determinations have been made with the ^{14}C dating method. It was believed that a large ice cap had covered parts of the North American continent until about 35 000 years ago. Dating of wood and peat by the ^{14}C method has shown that the ice must have lasted until about 11 000 years ago. Moreover, several hundred pairs of sandals found in a cave in Oregon have been shown to be about 9000

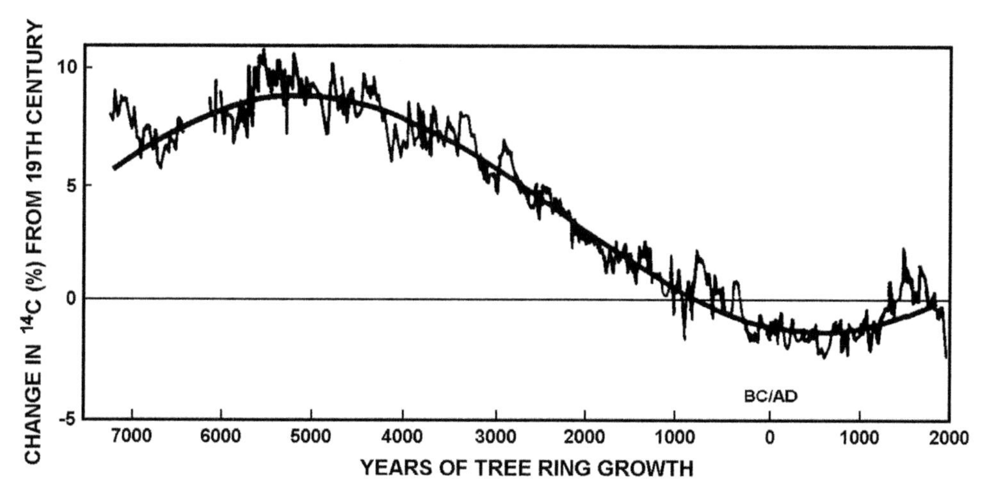

Figure 13.7 Long term variation in atmospheric ^{14}C activity as determined from tree rings. *(From H. E. Suess, La Jolla Radiocarbon Lab).*

years old, indicating that tribes with significant cultures had developed soon after the withdrawal of the ice cap from North America. By analyzing inorganic carbon in bone apatite, it has been determined that hunters in Arizona killed mammoths there 11 300 years ago. Another example of the use of ^{14}C dating, which has attracted widespread attention, involves the Dead Sea scrolls. There was considerable controversy about their authenticity until ^{14}C dating showed their age to be slightly more than 1900 years.

Some researchers claim that ^{14}C-ages lag behind with as much as 3500 years for times close to 20 000 y, when compared to U–Th dating (§13.8.4). Such adjustments play a large role in the debate about cultural migrations. It is interesting to note that adjusted ^{14}C-dates now have led to the conclusion that e.g. the stone monuments in Carnac, Brittany, are believed to be >6000 years old, i.e. older than the cultures in Egypt and Babylon. Similarly, grave mounds near the ancient edge of the inland ice in southern Norway and in Newfoundland have been found to be >7000 years old.

13.8.2. Dating by K-Ar method

Potassium is the eighth most abundant element in the earth's crust and occurs in many important rock-forming minerals. The radioactive isotope ^{40}K is present only to 0.0117% in natural potassium. It has a branched decay as follows:

$$^{40}\mathrm{K}\left(t_{1/2}1.28 \times 10^{9}\,\mathrm{y}\right)\begin{cases}\xrightarrow{\mathrm{EC},\ 10.7\%}\ ^{40}\mathrm{Ar}\ \ (99.6\%\ \text{of natural Ar})\\ \xrightarrow[\beta^-,\ 89.3\%]{}\ ^{40}\mathrm{Ca}\ \ (96.8\%\ \text{of natural Ca})\end{cases} \tag{13.7}$$

$\lambda_{EC} = 0.578 \times 10^{-10}$ y^{-1}, $\lambda_{\beta} = 4.837 \times 10^{-10}$ y^{-1}, E$_{max}$ 1.32 MeV. Because the ^{40}K half-life is of the same magnitude as the age of the earth, measurement of the $^{40}K/Ar$ ratio can be used to determine ages of the oldest K-containing minerals (for notation, see §3.3.2). From the decay scheme one can derive the equation

$$t = \lambda^{-1} \ln[\{^{40}Ar/(0.107\, ^{40}K)\} + 1] \tag{13.8}$$

where λ is the total decay constant ($= \lambda_{EC} + \lambda_{\beta}$; see (5.72)), ^{40}Ar represents the concentration of ^{40}Ar-atoms in the sample (i.e. number of radiogenic atoms of ^{40}Ar per unit sample weight), and ^{40}K is the present atomic abundance of ^{40}K-atoms. This equation assumes that all ^{40}Ar in the mineral is radiogenic. Any amount of non-radiogenic ^{40}Ar present initially (e.g. dissolved in the magma from which the mineral formed) is denoted as $^{40}Ar_i$. Thus the present amount of ^{40}Ar is the sum of the radiogenic amount and that originally present:

$$^{40}Ar = {}^{40}Ar_i + (\lambda_{EC}/\lambda)\{^{40}K(e^{\lambda t} - 1)\} \tag{13.9}$$

The problem is that both t and $^{40}Ar_i$ are unknown. This can be solved by dividing equation (13.9) by the number of ^{36}Ar atoms in the sample, yielding

$$^{40}Ar/^{36}Ar = (^{40}Ar/^{36}Ar)_i + (\lambda_{EC}/\lambda)(^{40}K/^{36}Ar)\{e^{\lambda t} - 1\} \tag{13.10}$$

Because ^{36}Ar is a stable non-radiogenic isotope, its amount in the sample should not change by time. Thus the measured $^{40}Ar/^{36}Ar$ ratio is the sum of the original and the radiogenic contributions from ^{40}Ar. Mineral samples of the same origin (so-called *cogenetic*) should have the same initial $^{40}Ar/^{36}Ar$ ratio and be of the same age (same t), though the $^{40}K/^{36}Ar$ may vary. Thus for cogenetic samples a plot of $^{40}Ar/^{36}Ar$ versus $^{40}K/^{36}Ar$ should yield a straight line, from the slope of which t is calculated. This line is referred to as an *isochron*. From the isochron for the rock sample from Tanzania in Figure 13.8, its age is calculated to be $(2.04 \pm 0.02) \times 10^6$ y, which is an interesting result as remains of early humans has been found in the same tuff matrix.

13.8.3. Dating by Rb–Sr method

Some uncertainty is associated with the use of the $^{40}K/^{40}Ar$ ratio because of the possibility of the loss of gaseous argon from minerals. An alternative method is based on ^{87}Rb/Sr system:

$$^{87}\text{Rb} \xrightarrow[4.8 \times 10^{10}\,\text{y}]{\beta^-} {}^{87}\text{Sr (7.00\% of natural Sr)} \tag{13.11}$$

The measurement of the $^{87}Rb/^{87}Sr$ ratio by isotope dilution and mass spectrometry is one of the most reliable methods for geologic age determinations. Meteorite values as

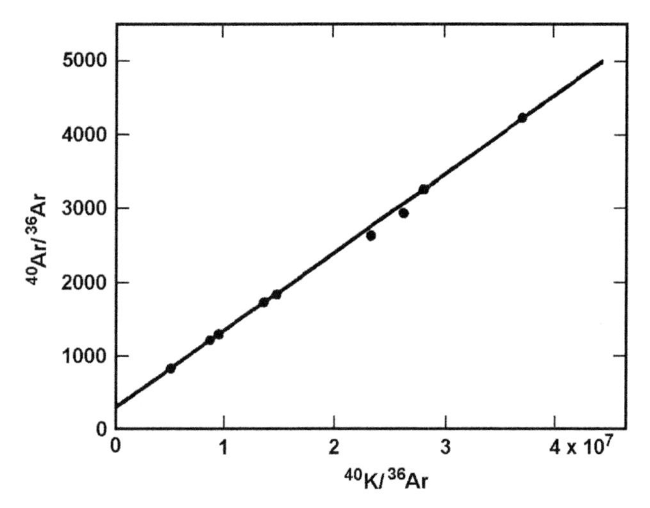

Figure 13.8 K-Ar isochron of tuff from Olduvai Gorge, Tanzania, yielding a slope corresponding to an age of 2.04 ± 0.02 My. *(From Faure).*

high as 4.7×10^9 y have been obtained, indicating that this is the age of the solar system. For the Rb–Sr clock a relation of type (13.11) is valid although corrections must be made for any non-radiogenic ^{87}Sr present; it is believed that the primordial $^{87}Sr/^{86}Sr$ ratio was 0.70.

13.8.4. Dating by the Re–Os method

The measurement of ^{187}Re, ^{186}Os, and ^{187}Os has been used to measure the age of meteorites and this yields some data about the age of our Milky Way galaxy.

$$^{187}\text{Re} \xrightarrow[4.6 \times 10^{10}\,\text{y}]{\beta^-} {}^{187}\text{Os} \tag{13.12}$$

The insert in Figure 12.9 shows a part of the nuclide chart where the zigzag arrow shows nuclides formed in the s-process and the lower right arrows show nuclides formed in the r-process (c.f. §12.5.3). ^{186}Os can only form in the s-process and ^{187}Re only in the r-process. However, ^{187}Os is also formed by slow n-capture in ^{186}Os. The amount of ^{187}Os formed in the s-process can be calculated from estimated flux values and known cross sections (at 30 keV; 3.6×10^8 K). Thus the amount of ^{187}Os formed by decay of ^{187}Re can be calculated. This Re–Os clock gives us an age of 10.8 ± 2.2 eons from the formation of our Galaxy until our solar system formed. When combined with the age of the solar system, 4.7 eons, it yields an age of 15.5 ± 2.3 eons for our Galaxy, which seems to be too long when compared with the current best estimate of the time since the Big Bang, 13.75 ± 0.11 eons.

13.8.5. Dating based on ^{238}U decay

In the uranium decay series 8 α-particles are emitted in the decay from ^{238}U to ^{206}Pb. Thus for every 8 helium atoms found in a uranium mineral, one atom of ^{238}U must have decayed to ^{206}Pb. Designating the number of original uranium atoms in the sample at time 0 as $^{238}U_o$, the number which has decayed with time t would be $^{238}U_o - {}^{238}U$, where the last number refers to the uranium atoms present now. Then $^{238}U_o - {}^{238}U = He/8$, where He is the number of helium atoms made. From this it relatively easy to derive an equation from which t, i.e. the age of the mineral, can be calculated once the values of ^{238}U and He are known. For accurate calculations, it is necessary to correct for the formation of helium from the decay of the ^{235}U and ^{232}Th also present in the mineral (see Fig. 13.4). Further, if the mineral has lost any helium through diffusion or other processes during its existence, the helium content would be abnormally low, leading to erroneously small values of t. This method, therefore, can only give lower limits of the ages of minerals.

Another common method of dating U-minerals is by considering its content of lead isotopes. Lead has four stable isotopes of which three are end products of radioactive decay series. The fourth lead isotope, ^{204}Pb, is found in lead minerals in about 1.4% isotopic abundance and has no radio-genetic origin. At the time of formation of the earth, all the ^{204}Pb in nature must have been mixed with unknown amounts of the other lead isotopes. If a lead-containing mineral lacks ^{204}Pb, it can be assumed that presence of the other lead isotopes together with uranium and/or thorium must be due to their formation in the decay series. In such ^{204}Pb-free minerals if it is possible to determine the amount of the parent nuclide ^{238}U and of the end product ^{206}Pb, the age of the mineral can be obtained from the general equation

$$t = \lambda^{-1} \ln\left(1 + N_d/N_p\right) \tag{13.13}$$

where N_p is the number of parent atoms (e.g. ^{238}U) and N_d the number of radiogenic daughter atoms (e.g. ^{206}Pb), and λ is the decay constant of the parent. The atomic ratios between a number of isotopic pairs as a function of time is shown in Figure 13.9.

This method is more reliable than the helium method since there is very little possibility of any of the lead that has been formed by radioactive decay having diffused or been leached from the mineral during its geologic age. Mineral samples from the earth have yielded values as great as 3×10^9 y by the lead content method. Unfortunately, there is a problem with this method also. The decay series all pass through isotopes of the inert gas radon, and if some of the radon is lost from the mineral the ^{206}Pb content leads to an erroneously low age. However, if relation (13.13) is used for different decay series (i.e. the ^{238}U and the ^{235}U series) and the same t is obtained, the data are said to be *concordant*. Such data increases the confidence in the measured age.

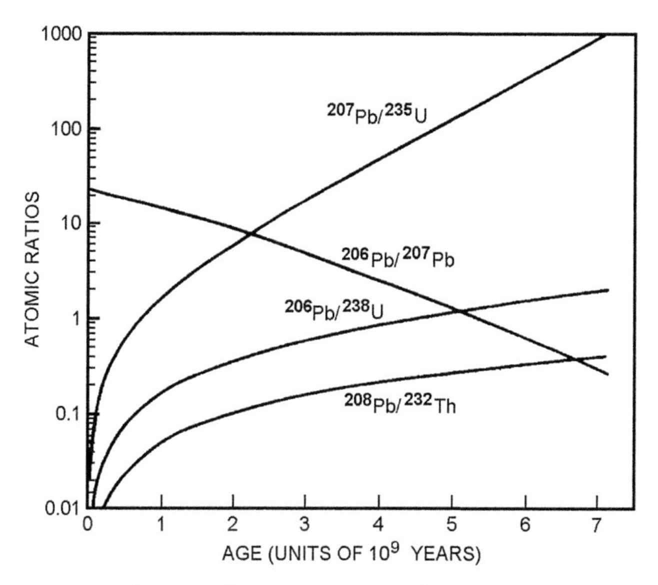

Figure 13.9 Atomic ratios at the time of measurement as a function of the age of the minerals. *(From E. K. Hyde).*

Since the lifetimes of the uranium isotopes ^{238}U and ^{235}U are different, the isotopic ratio between their end products ^{206}Pb and ^{207}Pb can also be used for age determination. One can derive the relationship

$$^{207}Pb/^{206}Pb = (1/138)\left(e^{\lambda_{235}t} - 1\right)/\left(e^{\lambda_{238}t} - 1\right) \tag{13.14}$$

where the factor 1/138 is the present isotopic abundance ratio of the uranium isotopes. This method has given values of 2.6×10^9 y for uranium and thorium minerals. When applied to stony meteorites, a somewhat longer age of $(4.55 \pm 0.07) \times 10^9$ y is obtained.

When lead is extracted from the ore it is in secular equilibrium with its precursors radium and uranium. The radium and most of its descendants are removed during processing while the ^{210}Pb accompanies the other lead isotopes. The separation of radium from lead is not always complete. Because of the long half-life of ^{226}Ra (1600 y) as compared to the short- lived ^{210}Pb (22.3 y), the determination of the excess ^{210}Pb over the equilibrium amount received from ^{226}Ra decay provides a scale for the time since the manufacture of the lead. This was used to verify forgeries of paintings claimed to be made by the Dutch artist Vermeer van Delft (1632—1675). The paintings were in fact made by H van Meegeren (1889—1948), in the 1940's and were so excellent in style that every authority accepted them as authentic. When threatened by the death penalty for selling such national treasures to the Germans during World War II, van Meegeren confessed the forgery, and his story was verified by dating the lead in the "lead white", $PbSO_4$, in the fake "Vermeers".

13.9. NATURAL RADIOACTIVITY OF THE OCEANS

The total amounts of ^{238}U and ^{232}Th in the ocean are 4.3×10^{12} kg (53 EBq) and 6.9×10^{10} kg (0.3 EBq), respectively. The oceans contain a much higher concentration of uranium than would be expected from its abundance in rocks. It also occurs at much larger concentration than thorium, even though the latter element is more abundant in rocks. This is explained by the absence of easily soluble thorium complexes, in contrast to the strong $UO_2(CO_3)_3^{4-}$ complex. ^{238}U decays to ^{234}Th ($t_{½}$ 24.1 d), which precipitates out or absorbs to colloids before it decays to ^{234}U, causing a slight disequilibrium between the two U-isotopes. This is used to determine the mean residence time of Th in the oceans: ~200 y; by contrast, the residence time of U in the ocean is 5×10^5 y. The decay passes the long-lived ^{230}Th ($t_{½}$ 75 400 y), which deposits in sediments before radioactive equilibrium is reached with ^{226}Ra (t 1600 y); consequently the specific activity of ^{226}Ra is less than expected from the U-content (39 Bq/m^3), only 1–10 Bq/m^3. It should be noted that the ^{222}Rn concentration in surface waters is in disequilibrium with ^{226}Ra because Rn is vented by heat, turbulence, etc.

As the total content of ^{40}K in the oceans is 7.4×10^{13} kg, corresponding to 1.94×10^4 EBq, ^{40}K is the largest source of radioactivity in the oceans. Minor activities come from ^3H as HTO and from ^{14}C as dissolved CO_2 or HCO_3^-.

13.10. ANTHROPOGENIC RADIOACTIVITY IN NATURE

In the analysis of a sample for its content of natural radioactivity it is necessary today to consider the possibility that the sample has become contaminated by "non-natural" radioactivities, i.e. radionuclides added to by human activities (so-called *anthropogenic sources*). Nuclear weapons tests, nuclear satellites burnt-up in the atmosphere, and nuclear power accidents may release large amounts of activities, see Table 13.4. The nuclear power industry is permitted by health authorities to continually release small, controlled amounts of specified radionuclides into the atmosphere and into open waters, Tables 19.7 and 21.10. Effluents may also come from large radiotracer users (especially

Table 13.4 Events leading to large injections of radionuclides into the atmosphere

Source	Country	Time	Radioactivity (Bq)	Important nuclides
Hiroshima & Nagasaki	Japan	1945	4×10^{16}	Fiss.prod., Actinides
Atmospheric weapons tests	USA USSR	-1963	2×10^{20}	Fiss. prod., Actinides
Windscale	UK	1957	1×10^{15}	^{131}I
Chelyabinsk (Kysthym)	USSR	1957	8×10^{16}	Fiss. prod. ^{90}Sr, ^{137}Cs
Harrisburg	USA	1979	1×10^{12}	Noble gases, ^{131}I
Chernobyl	USSR	1986	2×10^{18}	^{137}Cs
Fukushima	Japan	2010	2×10^{18}	^{131}I, ^{137}Cs

hospitals). In most cases these releases are well known and the area concerned ("contaminated") identified. From a global standpoint, compared to natural radioactivity, these releases are minor, but they do add to the our "natural" radiation background (also called eco radiation), as discussed in Chapters 15 and 22.

Usually one distinguishes between "near field" and "far field" effects of radioactivity releases. Near field effects are observed close to the release source, as for example the nuclear power plant or nuclear waste storage facility. The dissolution of nuclear waste by rain or ground water is a typical near field problem. As the source is known, it can be controlled and its environment monitored. If the radioactivity exceeds permitted levels, access to the contaminated area can be restricted. Far field effects involve the behavior of radionuclides which have spread out of such a restricted area, caused either by nuclear power accidents and weapons tests or by leakage from nuclear power plants.

In this paragraph we briefly describe some of the largest anthropogenic sources causing far field effects, i.e. nuclear weapons tests and nuclear power plant accidents. The cause of the releases is discussed in Chapter 19. Chapter 22 discusses both near and far field effects in further detail, particularly with regard to chemical properties: liquid releases from nuclear power plants, dissolution of solidified nuclear waste and of fall-out particles, migration in the environment, and possible consequences.

13.10.1. Nuclear weapons

Nuclear weapons were tested up to 1963, the year of the test ban, with total releases up to 2×10^{20} Bq fission products, as well as some lesser amounts of Pu isotopes; cf. Table 22.2. Most of the debris injected into the troposphere had a mean residence time of ~30 d, causing fall-out mostly in the neighborhood of the test area. Some of the debris passed through the tropopause and entered the stratosphere where it was carried by winds around the globe at approximately the latitude of the release. The residence time in the stratosphere is 3–24 months depending on latitude (greatest at the equator). The average surface deposition is 2500 Bq/m^2 at 40–50°N, decreasing towards north an south (<800 at >70°N and at <20°N). The global integrated deposition density for ^{241}Pu is 440 Bq/m^2 and the air concentration is 0.8 mBq/m^3.

Because the atmospheric tests stopped in 1963, the fission products have now decayed for ~50 y, leaving only ^{90}Sr, ^{137}Cs and Pu "active". Comparing with global natural radioactivity the contributions of these isotopes to our environment is negligible today, except for T and ^{14}C.

13.10.2. Nuclear power plant accidents

In 1957 a fire developed in one of the gas cooled graphite moderated rectors at *Windscale*, UK. The main radionuclides released were (in TBq) ^{131}I 700, ^{137}Cs 20, ^{89}Sr 3 and ^{90}Sr

0.3. The largest air concentration close to the plant was 20 kBq/m^3. The Cs activity deposited was ≤ 4 kBq/m^2. The ground activity is now back to normal.

Also in 1957 an explosion occurred in a nuclear waste storage facility (probably due to reactions between organic matter and nitrate) at Kyshtym south of Sverdlovsk, former USSR, leading to the contamination of approximately 1600 km^2 of land by 8×10^{16} Bq fission products, causing local contaminations exceeding 10^{10} Bq/m^2 (2×10^8 Bq/m^2 ^{90}Sr and ^{137}Cs). The area is still uninhabitable, see also §21.10.

In 1979 a partial core melt-down occurred in a reactor at Three Mile Island power station (TMI), *Harrisburg*, Pennsylvania. Although the fission products released from the core were almost completely contained within the building, the Xe and Kr and some iodine (~1 TBq ^{131}I) were released. No deposition occurred on ground outside the building.

The explosion and fire in one of the power reactors at *Chernobyl*, then Soviet Union – now Ukraine, in 1986, was a more severe accident (§§20.1.2 and 22.3). For several days large amounts of fission products and actinides were ejected and spread over large areas of the former USSR and Europe, see Figure 13.10. Almost 20% of the

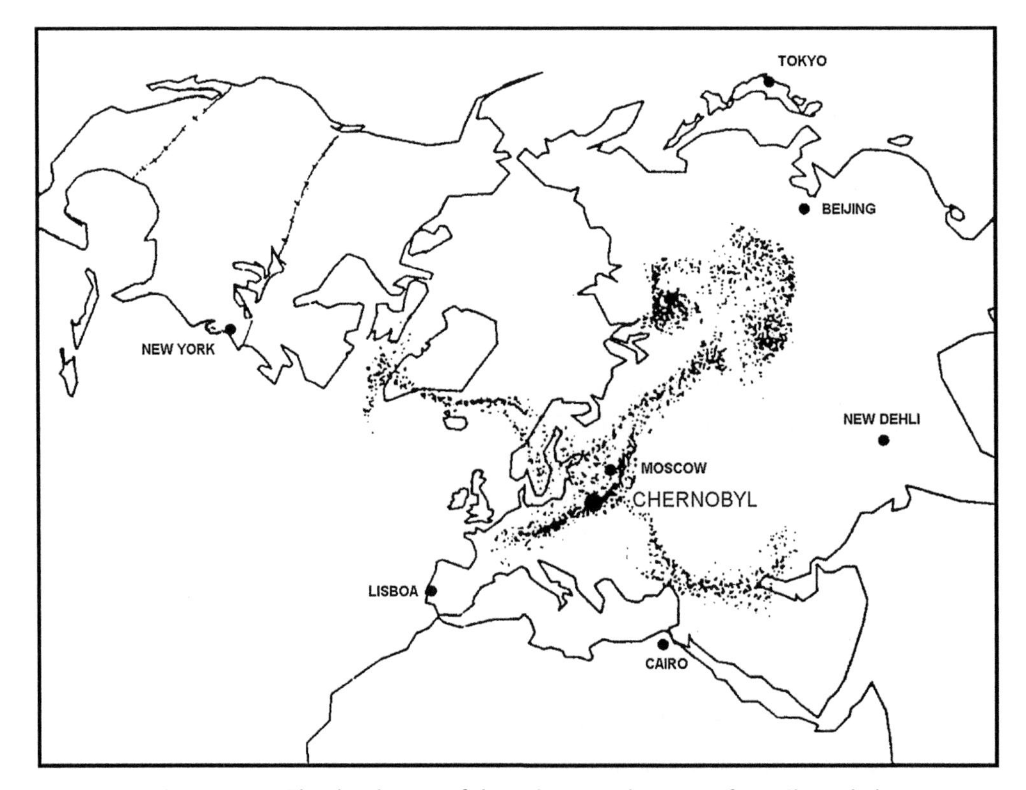

Figure 13.10 The distribution of the airborne radioactivity from Chernobyl.

fission products came down over Scandinavia, causing a deposition >120 kBq $^{137}Cs/m^2$ over the city Gävle in Sweden (170 km north of Stockholm). The plume spread down over Central Europe, causing very uneven deposition due to wind pattern and local rains: e.g., Munich, Germany, received ~25 kBq ^{137}Cs and 0.2 kBq ^{90}Sr per m^2, while Mainz (400 km away) received 180 Bq ^{137}Cs and ~0.001 Bq ^{90}Sr per m^2, see also §22.3.

On the 11th of March 2011 an earthquake of the magnitude 9.0 on the Richter scale occurred outside the eastern coast of Japan. All affected nuclear reactors shut down as planned as a response to the earthquake. It should be noted that some of the reactors only had a design basis of 8.2 but handled the larger magnitude anyway. However, as a result of the earthquake a tsunami was also formed. As it reached the Japanese coast the devastation was huge resulting in more than 20 000 casualties. Some of the nuclear reactors were also affected and most severely where the ones at the Fukushima Daiichi plant. Even if the reactors handled the unusually strong earthquake the protection against a tsunami of this size was not good enough. The wall designed to protect the reactor from tsunamis was about 7 meters tall but the wave was more than 10 meters high as it hit the coast. The water masses affected the power supply and also the possibility to run the on site emergency diesel generators. Thus all six reactors suffered a total station blackout when the battery power was lost after a few hours. With a total loss of power the auxiliary cooling systems stopped and cooling water was lost to the reactor cores which were heated up due to the remaining decay heat in the fuel. However, the chain reactions had stopped before this point. As the fuel heated up the zircaloy cladding underwent a highly exothermic oxidation reaction according to (19.44), where T is in °C.

$$\mathrm{Zr} + 2\mathrm{H_2O} \rightarrow \mathrm{ZrO_2} + 2\mathrm{H_2} \quad \Delta H = -6.67 + 2.57 \times 10^{-4}\ \mathrm{T}\ (\mathrm{MJ/kg\ Zr}) \quad (19.44)$$

The hydrogen produced would normally have been converted to water by the installed electrically heated recombiners or vented by the fans to the stack. However, nothing of this worked after the total power loss. Thus hydrogen escaped up into the secondary reactor building where it accumulated and later exploded. The explosion did not affect the primary containment. However, the spent fuel storage on unit 4 was damaged and the stored fuel could no longer be cooled. That increased the risk for damage not only to the fuel inside the reactor cores but also to the spent fuel in the storage pond, which was not inside an effective containment. This increased the risk for another large release of radioactive material.

The release of radioactive material from the Fukushima accident was modest when compared to Chernobyl, ~5200 PBq total from Chernobyl and ~900 PBq total into the air from Fukushima. The release at Fukushima contained little, or no, corium and consisted mainly of the volatile fission products iodine and cesium (probably in the form of CsOH, CsI, and I_2). However, significant land masses were contaminated. In most areas the contamination levels were comparable to those seen in many areas (e.g close to Gävle) in Sweden after the Chernobyl accident (about 120 kBq/m^2) where no evacuation or cleaning up measures were deemed necessary. However in the plume north of the plant in Fukushima there were areas with about 3 MBq/m^2.

An interesting observation from the Fukushima accident is the release pattern and the effect of weather at the time of the release. Most of the released activity ended up in the ocean due to the prevailing wind direction. However, over time the wind changed direction and the release was then spred over land. In retrospect it is now possible to see where it was raining at the time since CsOH and CsI are both water soluble and were washed out by the rain while some the elemental iodine (or iodine oxides) was transported through the precipitation zones.

13.10.3. Releases from nuclear power plants

All nuclear power plants are carefully controlled by the national radiation protection boards, and strict limits for releases are set. Usually it is quite easy for the power plants to meet these requirements. These releases are small compared to the natural radioactivities and mostly negligible in the far field, see §§20.4, 21.8, and 21.11.

13.10.4. Other anthropogenic additions

In 1975 the *London Dumping Convention* set limits to dumping of nuclear waste in the Oceans to places far of shipping trades and fishing areas and at depths >4000 m. Earlier dumping had been common practice by the nuclear powers even in narrow (the British Channel) and shallow (up to 50 m) waters to get rid of large volumes of low level long-lived waste. Officially reported dumpings were made as early as in 1946 and continued in the Pacific and the North Atlantic until 1982 and in the Barents and Kara seas even later. East of the British Islands the U.K. has dumped 665 TBq, France 134 TBq and other countries together some 35 GBq. In total some 45 PBq have been dumped at 46 different sites, mostly at depths of 1400–6500 m. Measurements reveal that in some cases radioactive nuclides are leaking out from the containers, causing bottom contamination at the site; however, the radionuclides are rapidly diluted as they are carried away by bottom currents.

In the 1990ies it was disclosed that the former USSR made huge dumps of nuclear waste in the waters east of Novaya Zemlya: some 10 000 containers and 13 nuclear reactors (8 still containing fuel), mainly from nuclear submarines. In some places the waste is at very shallow waters (some 30 m). The total activity is estimated to >60 PBq. The waste will ultimately leak out and spread with westerly currents towards the Barents Sea and the North Atlantic. However, the releases are expected to be rapidly diluted to harmless concentrations.

13.11. EXERCISES

13.1. What proof exists that some cosmic rays do not come from the sun?

13.2. (a) What is the primary cosmic radiation hitting earth's atmosphere? (b) Does it penetrate to the earth's surface?

13.3. What background from cosmic radiation is expected for an unshielded 100 ml ion chamber which exhibits an area of 100 cm^2 perpendicular to the direction of the cosmic radiation?

13.4. Which type of mesons are released in high energy particle interactions, and why?

13.5. Cosmic-ray irradiation of the atmosphere yields 0.036 ^{10}Be atoms $cm^{-2}\ s^{-1}$. If this ^{10}Be is rapidly carried down into seawater, which is assumed to have a volume of $1.4 \times 10^{18}\ m^3$, what will the equilibrium radioactivity of ^{10}Be in 1 m^3 seawater be? The earth's total surface area is $510 \times 10^6\ km^2$.

13.6. In Greenland ice the ^{10}Be radioactivity has been measured to be 0.0184 dpm m^{-3}. How old is this ice if it was formed out of water in equilibrium with cosmic-ray ^{10}Be (see previous question)?

13.7. The CO_2 in the atmosphere is in exchange with carbon in living organisms, humus, dissolved organic compounds, and carbonate in the sea water, the latter being the main reservoir (88%) of all exchangeable carbon. The amount of such exchangeable carbon is estimated to be 7.9 g cm^{-2} of the earth. When cosmic-ray produced ^{14}C is mixed into this exchangeable carbon, what will the specific activity become?

13.8. 0.11 cm^3 helium gas at NTP was isolated from 100 g of uranium mineral containing 5 ppm uranium. How old is the mineral?

13.9. A mineral was found to contain 39.1 g K and 87.2×10^{-6} liter Ar at NTP. How old is the mineral?

13.10. A uranium mineral was found to contain the lead isotopes ^{204}Pb, ^{206}Pb, and ^{207}Pb in the ratio 1:1087:388, as determined with a mass spectrometer. Estimate the age of the mineral.

13.11. The heat flow from the earth's crust is 0.060 W m^{-2}. The mean thickness of the crust is 17 km and the earth's radius is 6371 km. The average concentration of uranium, thorium, and potassium in granite is estimated to be 4 ppm (by weight) 18 ppm, and 3.6%, respectively. Assuming that 7% by volume of the crust is made up of granite (feldspar + quartz, density 2.6 g cm^{-3}), what will the heat flow at the earth's surface be from each of these elements? Assume β-heat as 1/3 E_{max}; for α-decay assume $E_\alpha = Q_\alpha$. Each ^{40}K decay by the EC-branch emits one 1.46 MeV γ. Discuss the results.

13.12. A 1 GW_e nuclear power station uses annually about 30 t uranium enriched to 3% ^{235}U. (a) How much natural uranium consumed each year to keep it running? Assume waste stream from isotope separation plant to contain 0.3% ^{235}U.

(b) How much low grade ore (assume 0.06% uranium) must be mined at steady state, if the uranium recovery efficiency in the process is 70%?

13.13. The assumed uranium resources in Japan are 4 kt, in Argentina 12 kt and in France 48 kt U_3O_8. How many 1 GW_e reactor years can these uranium amounts sustain in each country at the uranium consumption rate (a) of the previous exercise?

13.12. LITERATURE

Cosmic radiation

S. FLÜGGE (ed.), Kosmische Strahlung, *Handbuch der Physik,* Band XLV 1/1, Springer-Verlag, Berlin, 1961.
G. BURBIDGE, The origin of cosmic rays, *Sci. Am.* **215**, Aug. 1966.
A . WEBSTER, The cosmic background radiation, *Sci. Am.* **231**, Aug. 1974.
D. D. PETERMAN and E. V. BENTON, High LET particle radiation inside Skylab command module, *Health Physics* **29** (1975) 125.

Radioelements in nature and dating

E. K. HYDE, *Natural Radioactivity of the Heavy Elements: A Comprehensive Review,* Univ. of California Radiation Laboratory, Report UCRL-10612.
W. F. LIBBY, *Radiocarbon Dating,* University of Chicago Press, 1956.
C. E. JUNGE, *Air Chemistry and Radioactivity,* Academic Press, 1963.
H. CRAIG, S. L. MILLER, and G. J. WASSERBURG (Eds.), *Isotopic and Cosmic Chemistry,* North-Holland, 1963.
E. I. HAMILTON and L. H. AHRENS, *Applied Geochronology,* Academic Press, 1965.
B. KEISCH, Dating works of art through their natural radioactivity: improvements and applications, *Science* **160** (1968) 413.
D. YORK and R. M. FARQUHAR, *The Earth's Age and Geochronology,* Pergamon Press, 1972.
P. K. KURODA, Fossil nuclear reactor and plutonium-244 in the early history of the solar system. In IAEA, *The Oklo-phenomenon*, Vienna, 1976.
Ionizing Radiation: *Sources and Biological Effects*, United Nations 1982.
P. S. ZURER, Archeological Chemistry, *Chem. & Eng. News Febr.* **21**, 1983, p. 21.
G. FAURE, *Principles of Isotope Geology,* 2nd Ed. J Wiley & Sons 1986.
M. EISENBUD, *Environmental Radioactivity,* 3rd Ed., Academic Press 1987.
M. A. GEYH and H. SCHLEICHER, *Absolute Age Determination*, Springer-Verlag 1990.
G. BRIMHALL, The Genesis of Ores, *Sci. Am.*, May 1991, p. 48
Fuel Review 1991, Nucl. Eng. Int. Spec. Publ. 1991.
L. MOBERG (Ed.), *The Chernobyl Fallout in Sweden*, The Swedish Radiation Protection Institute, Box 60024, S-10401 Stockholm 1991.
T. NOHARA, Y. OCHIAI, T. SEO AND H. YOSHIDA, PNC, Gify, Japan, Private communication.
Nuclear Geophysics (Int. J. Radiation Appl. and Instrumentation Part E), published by Pergamon Press *World Nuclear Industry Handbook*, Nucl. Eng. Int. Spec. Publ. 2000.

CHAPTER 14

The Actinide and Transactinide Elements

Contents

Radiochemistry and Nuclear Chemistry
ISBN 978-0-12-405897-2, http://dx.doi.org/10.1016/B978-0-12-405897-2.00014-8

14.1. EARLY "TRANSURANIUM" ELEMENTS

After the discovery of the neutron by Chadwick in 1932, the group led by Fermi in Rome began bombarding different elements with neutrons to study the radioactivity induced through (n,γ) reactions. The decay characteristics of the radioactivity in neutron-irradiated uranium was interpreted to indicate that some of the products were probably transuranium elements. Similar conclusions were reached by other research groups; for example in 1937 Meitner, Hahn and Strassman in Berlin proposed the following reaction/decay series (omitting mass numbers):

$$_{92}\mathrm{U} + \mathrm{n} \rightarrow {}_{92}\mathrm{U} \xrightarrow[10\ \mathrm{s}]{\beta} {}_{93}\mathrm{EkaRe} \xrightarrow[2.2\ \mathrm{m}]{\beta} {}_{94}\mathrm{EkaOs} \xrightarrow[59\ \mathrm{m}]{\beta} \mathrm{etc} \tag{14.1}$$

However, many anomalies existed in the chemical properties of these "transuranium elements" as pointed out by Ida Noddack, Irene Curie, Savitch and others.

In 1939 Hahn and Strassman conducted a series of extremely careful chemical investigations which showed that these "transuranium elements" were in fact isotopes of elements in the middle of the Periodic Table, such as Sr, La, Ba etc. The suggestion that their observations were due to a fission of uranium was first made in a scientific paper by Ida Noddack, but ignored. About two years later Meitner and Frisch postulated that the uranium atoms were caused to split by thermal neutrons into two approximately equal fragments.

Further investigation showed that it was the isotope ^{235}U which had undergone fission, and that large amounts of energy and about 2.5 neutrons were released in the process. It occurred to a number of scientists that, if at least one of these neutrons could be captured by other uranium atoms to cause further fission, a chain reaction should be the consequence. This led to the development of nuclear energy and nuclear weapons, as described in Chapter 19.

To the present time 21 transuranium elements have been synthesized; the isotopes known are summarized in Figure 14.1; except elements above 114, which are still uncertain.

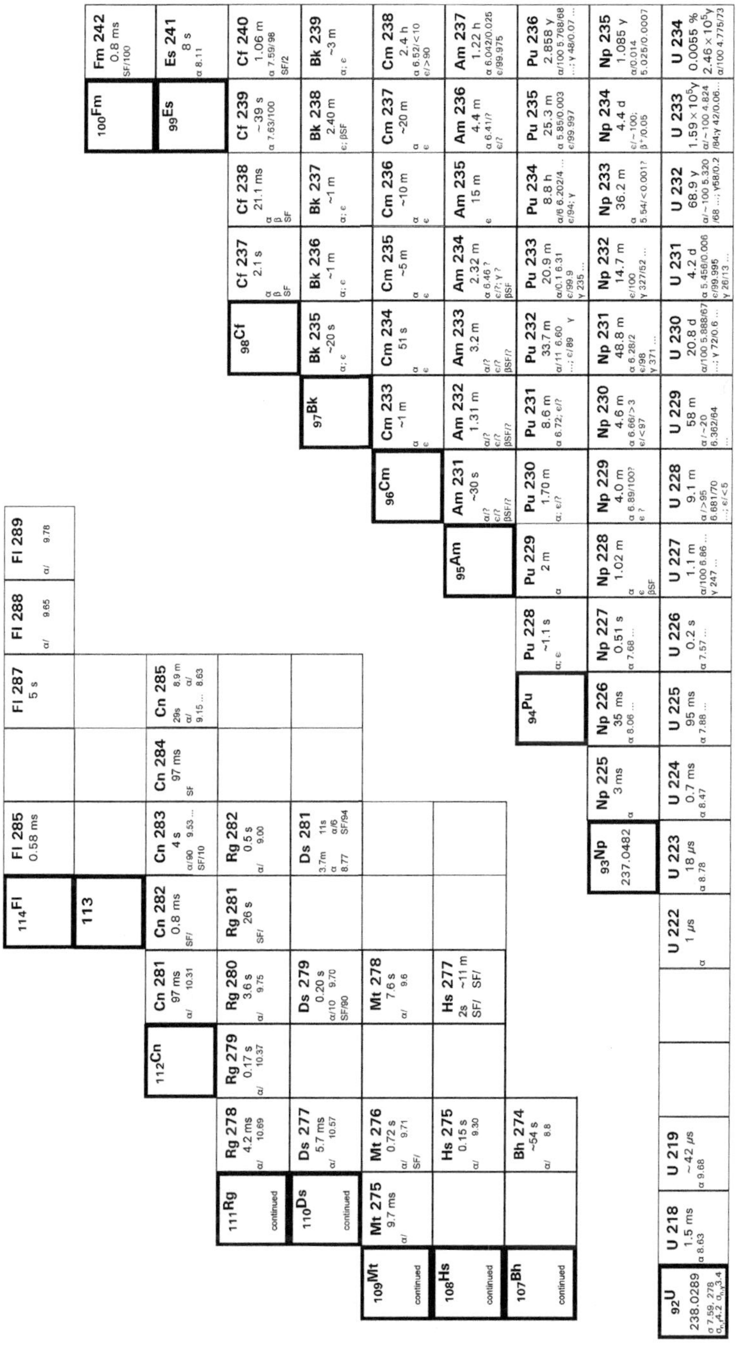

					$_{114}$Fl	Fl 285 0.58 ms		Fl 287 5 s	Fl 288 α/ 9.65	Fl 289 α/ 9.78
					113					
			$_{112}$Cn	Cn 281 97 ms α/ 10.31	Cn 282 0.8 ms SF/	Cn 283 4 s α/90 9.53 ... SF/10	Cn 284 97 ms SF	Cn 285 29s 8.9 m α/ α/ 9.15 ... 8.63		
	$_{111}$Rg continued	Rg 278 4.2 ms α/ 10.69	Rg 279 0.17 s α/ 10.37	Rg 280 3.6 s α/ 9.75	Rg 281 26 s SF/	Rg 282 0.5 s α/ 9.00				
	$_{110}$Ds continued	Ds 277 5.7 ms α/ 10.57		Ds 279 0.20 s α/10 9.70 SF/90		Ds 281 3.7m 11s α α/6 8.77 SF/94				
$_{109}$Mt continued	Mt 275 9.7 ms α/	Mt 276 0.72 s α/ 9.71 SF/		Mt 278 7.6 s α/ 9.6						
$_{108}$Hs continued		Hs 275 0.15 s α/ 9.30		Hs 277 2s ~11 m SF/ SF/						
$_{107}$Bh continued		Bh 274 ~54 s α/ 8.8								

																$_{100}$Fm	Fm 242 0.8 ms SF/100
																$_{99}$Es	Es 241 8 s α 8.11
													$_{98}$Cf	Cf 237 2.1 s α β SF	Cf 238 21.1 ms α β SF	Cf 239 ~39 s α 7.63/100	Cf 240 1.06 m α 7.59/98 SF/2
												$_{97}$Bk	Bk 235 ~20 s α; ε	Bk 236 ~1 m α; ε	Bk 237 ~1 m α; ε	Bk 238 2.40 m ε; βSF	Bk 239 ~3 m α; ε
											$_{96}$Cm	Cm 233 ~1 m α ε	Cm 234 51 s α ε	Cm 235 ~5 m α ε	Cm 236 ~10 m α ε	Cm 237 ~20 m α ε	Cm 238 2.4 h α 6.52/<10 ε/>90
										$_{95}$Am	Am 231 ~30 s α/? ε/? βSF/?	Am 232 1.31 m α/? ε/? βSF/?	Am 233 3.2 m α/? ε/? βSF/?	Am 234 2.32 m α 6.46 ? ε/?; γ ? βSF	Am 235 15 m ε	Am 236 4.4 m α 6.41/? ε/?	Am 237 1.22 h α 6.042/0.025 ε/99.975
								$_{94}$Pu	Pu 228 ~1.1 s α; ε	Pu 229 2 m α	Pu 230 1.70 m α; ε/?	Pu 231 8.6 m α 6.72; ε/?	Pu 232 33.7 m α/11 6.60 ...; ε/89 γ	Pu 233 20.9 m α/0.1 6.31 ε/99.9 γ 235 ...	Pu 234 8.8 h α/6 6.202/4 ... ε/94; γ	Pu 235 25.3 m α 5.85/0.003 ε/99.997	Pu 236 2.858 y α/100 5.768/68 ...; γ 48/0.07 ...
						$_{93}$Np 237.0482	Np 225 3 ms α	Np 226 35 ms α 8.06 ...	Np 227 0.51 s α 7.68 ...	Np 228 1.02 m α ε βSF	Np 229 4.0 m α 6.89/100? ε ?	Np 230 4.6 m α 6.66/>3 ε/<97	Np 231 48.8 m α 6.28/2 ε/98 γ 371 ...	Np 232 14.7 m ε/100 γ 327/52 ...	Np 233 36.2 m α 5.54/<0.001?	Np 234 4.4 d ε/~100; β^+/0.05	Np 235 1.085 y α/0.014 5.025/0.0007
$_{92}$U 238.0289 σ 7.59, 278 $\sigma_{n,f}$4.2 $\sigma_{n,\gamma}$3.4	U 218 1.5 ms α 8.63	U 219 ~42 μs α 9.68			U 222 1 μs α	U 223 18 μs α 8.78	U 224 0.7 ms α 8.47	U 225 95 ms α 7.88 ...	U 226 0.2 s α 7.57 ...	U 227 1.1 m α/100 6.86 ... γ 247 ...	U 228 9.1 m α/>95 6.681/70 ...; ε/<5	U 229 58 m α/~20 6.362/64 ...	U 230 20.8 d α/100 5.888/67 ...; γ 72/0.6 ...	U 231 4.2 d α 5.456/0.006 ε/99.995 γ 26/13 ...	U 232 68.9 y α/~100 5.320 /68 ...; γ58/0.2	U 233 1.59×10^5y α/~100 4.824 /84;γ 42/0.06...	U 234 0.0055 % 2.46×10^5y α/100 4.775/73

Figure 14.1 Isotope chart of uranium, transuranium, and transactinide elements.

												$_{111}$Rg	**Rg 272** 1.6 ms α/ 11.02 ...		**Rg 274** 15 ms α/ 11.23		
								$_{110}$Ds	**Ds 267?** 4 μs α/100		**Ds 269** 0.17 ms α/100 11.11	**Ds 270** 6ms 0.10ms α/100 α/100 11.11 11.03	**Ds 271** 69ms 1.63ms α/100 α/100 10.71 10.74 ...		**Ds 273** 170 ms α/ 11.14		
							$_{109}$Mt		**Mt 266** 1.7 ms α/~100 10.43		**Mt 268** 21 ms α/100		**Mt 270** 1.1s? 5.0ms				**Mt 274** 0.44 s α/100 9.76
						$_{108}$Hs	**Hs 263** 0.74 ms α/ 10.89 ...	**Hs 264** ~0.8 ms α/~100 10.43 SF/<1.5	**Hs 265** 0.75ms 2.0 ms α/? α/ ~100 ...;SF/	**Hs 266** 0.75ms 2.0 ms α/? α/	**Hs 267** 0.8s 52ms α/? α/ 9.83 9.99 ...	**Hs 268** 0.36 s α/ 9.48	**Hs 269** 9.7 s α/100 9.21 ...	**Hs 270** 3.6 s α/100 9.02 ...	**Hs 271** 40 s α/100 9.27 ...		**Hs 273** 0.24 s α/100 9.59
					$_{107}$Bh (Ns)	**Bh 260** ? α/~100	**Bh 261** 11.8 ms α/95 10.40 ... SF/<10	**Bh 262** 8.0ms 102ms α/>70 α/>80 10.37 10.06 ...;SF ...;SF		**Bh 264** 0.44 s α/100 9.48 ...	**Bh 265** 0.44 s α/100 9.48 ...	**Bh 266** 0.44 s α/100 9.48 ...	**Bh 267** 17 s α/ 8.83			**Bh 270** 61 s α/ 8.93	**Bh 271** (5 s) α	**Bh 272** 9.8 s α/ 9.02
				$_{106}$Sg	**Sg 258** 2.9 ms SF/100	**Sg 259** 0.48 s α/90 9.62 ... SF/<20	**Sg 260** 3.6 ms α/26 9.81 ... SF/74	**Sg 261** 0.009ms 0.18s IT/100 α/98 ε/1.3 SF/0.6	**Sg 262** 15 ms SF/100	**Sg 263** 0.9s 0.3s α/87 α/ 9.25 9.06 SF/13	**Sg 264** 68 ms SF/100	**Sg 265** 16.2s 8.9s α/ α/ 8.70 8.90 ...	**Sg 266** 0.36 s SF/100	**Sg 267** 1.4 m α/17 8.30 SF/83		**Sg 269** 2.1 m α/100 8.56		**Sg 271** 1.9 m α/67 8.54 SF/33
			$_{105}$Db	**Db 256** 2.6 s α/~70 9.02 ε/~10; SF/<40	**Db 257** 1.3 s α/82; ε/1 SF/17	**Db 258** 20s 4.4s ε α/67 /100 9.17... ε/33	**Db 259**	**Db 260** 1.52 s α/90.4 9.04/48 ...; ε/<2.5	**Db 261** 1.8 s α/>50 8.93 ... SF/<50	**Db 262** 34 s α/67 8.45/20 SF/33	**Db 263** 27 s α/41 8.36 3 SF/56			**Db 266** 22 m ε/ SF/	**Db 267** 1.2 h SF/	**Db 268** 29 h ε/ SF/		**Db 270** 23.15 h SF/
$_{104}$Rf	**Rf 253** 48μs 1.8s α/~50 SF/ ~50	**Rf 254** 22.3μs 0.5ms α/~0. 3 SF/100	**Rf 255** 1.6ms α/48 8.72 ... γ 203 ... SF/52	**Rf 256** 6.2 ms α/2.2 8.79 SF/98	**Rf 257** 4.7 s α/79.6; ε/18 SF/2.4	**Rf 258** 13 ms α/~13 SF/~87	**Rf 259** 3.1 s α/93 8.77/~60 ...; ε/~0.3	**Rf 260** 20 ms SF/100	**Rf 261** 1.17 m α/>80 8.28 ε/<15; SF/<10	**Rf 262** 47ms 2.1s SF SF /100 /100	**Rf 263** 15 m α/~30 7.90 ? SF/<100		**Rf 265** 2.5 m SF/~100	**Rf 266** 10 h? α/? SF/?	**Rf 267** 1.3 h ε/ SF/<100	**Rf 268** 6 h? α/? SF/?	

Figure 14.1 (*continued*).

					$_{103}$Lr	**Lr 252** 0.36 s α 9.02 ...	**Lr 253** ⟷ 1.49s 0.57s α α 8.72 8.79	**Lr 254** 13 s α/78 8.46 ... ε/22 SF/<0.13	**Lr 255** 21.5 s α/85 8.37/60 ...; ε/<30 SF/<0.1	**Lr 256** 25.9 s α/>80 8.43/37 ...; ε/<20	**Lr 257** 0.646 s α/~85 8.86/82 ... ;	**Lr 258** 3.9 s α/100 8.595/46 ...	**Lr 259** 6.3 s α 8.445/77 ε/<0.5Z	**Lr 260** 3.0 m α 8.03/75 ε/25	**Lr 261** 39 m ε?; SF/100	**Lr 262** 3.6 h ε/100	
				$_{102}$No	**No 250** 0.25 ms SF/~100	**No 251** 0.75 s α / 1 0 0 8.62/80...	**No 252** 2.3 s α/73.1 8.415 /55 ... SF/26.9	**No 253** 1.7 m α/100 8.01 ...	**No 254** 0.26s 55 s IT α /100 8.10 ε;SF	**No 255** 3.1 m α/61.4 8.12/28 ...; ε/38.6	**No 256** 2.91 s α/99.5 8.448 ... SF/0.53	**No 257** 26 s α/100 8.22 ...	**No 258** 1.2 ms SF/100	**No 259** 58 m α/75 7.520 ... ε/25; SF/<10	**No 260** 0.106 s SF/100		**No 262** 5 ms SF/100
$_{101}$Md	**Md 245** 0.35s 0.9ms α SF/100 8.64 ...	**Md 246** 1.0 s α 8.74 ... ε; βSF	**Md 247** 1.12s ~0.2s α SF 8.42 /100	**Md 248** 7 s α/20 8.32/~15... ε/80	**Md 249** 24 s α 8.03/20 ε/80	**Md 250** 52 s α/7 7.75/~4 ... ε/93 βSF	**Md 251** 4.0 m α/<10 7.55/<6 ε/>90	**Md 252** 2.3 m ε/100	**Md 253** ~6 m ε/100	**Md 254** ⟷ 10m 28m ε/100 ε/100	**Md 255** 27 m α 7.326/8 ε/92;γ453/4...	**Md 256** 1.30 h α/9.3 7.221 ... ε/90.7; γ	**Md 257** 5.52 h α/15 7.064/? ε/85; γ 371 ...	**Md 258** 57m 51.5 ε d;α /100 /100 6.71 g	**Md 259** 1.58 h SF/100	**Md 260** 31.8 d SF/100	
Fm 243 0.18 s α 8.55/100	**Fm 244** 3.0 ms SF/100	**Fm 245** 4.2 s α 8.15/100	**Fm 246** 1.1 s α 8.25/92 SF/8	**Fm 247** ⟷ 9.2s 35s α8.18 α/>50 /100 7.87...	**Fm 248** 36 s α/99 7.87/80 ...; ε/~1	**Fm 249** 2.6 m α 7.53/~15 ε/~85	**Fm 250** 1.8s 30m IT α 7.43 />80 />90 ε/<10	**Fm 251** 5.30 h α/1.8 6.832 /1.6 ...; ε/98.2	**Fm 252** 1.058 d α/100 7.040 /~85 ...	**Fm 253** 3.0 d α/12 6.943/5 ...	**Fm 254** 3.24 h α/99.94 7.189 /85...;γ99/0.04	**Fm 255** 20.1 h α/~100 7.022 /93...;γ81/0.8...	**Fm 256** 2.63 h α 6.917/8.1 SF/91.9	**Fm 257** 100.5 d α/99.79 6.520 /94...;γ241/10	**Fm 258** 0.38 ms SF/100	**Fm 259** 1.5 s SF/100	
Es 242 16 s α 7.92	**Es 243** 20 s α 7.899/>30 ε/<70	**Es 244** 37 s α 7.57/~4 ε/96 βSF	**Es 245** 1.1 m α 7.730/40 ε/60	**Es 246** 7.7 m α 7.36/10 ε/90 βSF	**Es 247** 4.55 m α 7.323/~7 ε/~93	**Es 248** 27 m α 6.879/~0.25 ε/~99.75	**Es 249** 1.70 h α 6.776 .../0.57 ε/99.43	**Es 250** 2.22h 8.6h ε/100 ε/100 γ989 γ829 /14... /74...	**Es 251** 1.38 d α/0.5 6.492/0.4 ...; ε/99.5	**Es 252** 1.291 y α/76 6.631/61 ...	**Es 253** 20.47 d α/100 6.633 /90...;γ389/0.0	**Es 254** 1.64d 275.7d α/0.3 α/~10 6.38/ 0 0.2;ε/ 6.43/	**Es 255** 39.8 d α/8 6.301/7 ... β /92; γ	**Es 256** 7.6h 25.4m β⁻/100 β⁻/100 γ135 βSF	**Es 257** 7.8 d β⁻ ?; γ 26 ... SF/100		
Cf 241 3.78 m α 7.342/~25 ε/~75	**Cf 242** 3.68 m α/100 7.392 /~80 ...	**Cf 243** 10.7 m α / ~ 1 4 7.06/~10 ...; ε/~86	**Cf 244** 19.4 m α/100 7.209/75 ...	**Cf 245** 43.6 m α/36 7.137 ... ε/64	**Cf 246** 1.49 d α/~100 6.750 /78...;γ42/0.02	**Cf 247** 3.11 h α 6.296/0.04 ε/99.96	**Cf 248** 333.5 d α/~100 6.258 /83...; γ 388...	**Cf 249** 350.6 y α/~100 5.812 /84...;γ388/66	**Cf 250** 13.08 y α/99.92 6.030 /85...;γ43/0.01	**Cf 251** 898 y α/100 5.679 /35 ...	**Cf 252** 2.645 y α/96.89 6.118 /82...;γ43/0.01	**Cf 253** 17.81 d α/0.3 5.980 /0.3 ...	**Cf 254** 60.5 d α / 0 . 3 5.833/0.3	**Cf 255** 1.4 h β /100	**Cf 256** 12.3 m SF/100		
Bk 240 4.8 m α/10 ε/90 SF/0.002	**Bk 241** 4.6 m α; ε	**Bk 242** 7 m ε	**Bk 243** 4.5 h α/~0.15 6.575/ 0.04...;ε/~99. 85	**Bk 244** 4.4 h α/0.006 6.662/~0.003 ε/~100;γ 892	**Bk 245** 4.90 d α/0.12 5.888/ 0.03...; ε/99.88	**Bk 246** 1.80 d ε/100 γ 799/61 ...	**Bk 247** 1380 y α/100 5.531/45 ...; γ 84/~40 ...	**Bk 248** 23.7h >9 y ε/30 ?? β⁻/70 γ551	**Bk 249** 320 d α / 0 . 0 0 1 5 ; 5.419/	**Bk 250** 3.217 h β 0.7 .../100 γ 989/45 ...	**Bk 251** 55.6 m β ~0.9 .../100 γ 178 ...						
Cm 239 3 h ε/100 γ 188 ...	**Cm 240** 27 d α / 9 9 . 5 6.291/71 ...; γ 45/0.04	**Cm 241** 32.8 d α/1 5.939/0.7 ... ε/99	**Cm 242** 162.94 d α/~100 6.113 /74...;γ44/0.03 ...	**Cm 243** 29.1 y α / 9 9 . 7 5.785/73 ...; ε/0.29	**Cm 244** 18.10 y α/~100 5.805 /76...;γ43/0.02	**Cm 245** 8500 y α/100 5.361/93 ...; γ 175/10 ...	**Cm 246** 4730 y α/100 5.386/79 ...; γ 46/0.03	**Cm 247** 1.56×10^{7} y α/100 4.870/71 ...; γ 404/72 ...	**Cm 248** 3.40×10^{5} y α/91.74 5.078 /75 ...	**Cm 249** 1.069 h β 0.9 .../100 γ 634/2 ...	**Cm 250** ~9700 y SF/~100	**Cm 251** 16.8 m β 1.4 .../100 γ 543/11 ...					
Am 238 1.63 h α 5.940/0.0001 ε/~100; β⁺/0.1	**Am 239** 11.9 h α/0.01 5.774 /0.008 ...	**Am 240** 2.12 d α/0.0002 5.378 /0.0002 ...	**Am 241** 432.2 y α/~100 5.486 /85 ...	**Am 242** 141y 16h α/0.5 ε/17 5.206 β⁻/83 ...;IT γ ...	**Am 243** 7370 y α/~100 5.277 /88 ...;γ75/60	**Am 244** 26m 10.1h β⁻1.5... β⁻0.4 /99.96 /100 ε/0.04 γ744...	**Am 245** 2.05 h β⁻ 0.3 .../100 γ 253/6 ...	**Am 246** 25m 39m β⁻/100 β⁻/100 γ1079/ γ679 28 ... /53 ...	**Am 247** 22 m β⁻/100 γ 286/23 ...								
Pu 237 45.2 d α/0.003 5.334 ...	**Pu 238** 87.74 y α/100 5.499/72 ...; γ 43/0.04 ...	**Pu 239** 24110 y α/~100 5.157 /73 ...; γ ...	**Pu 240** 6563 y α/100 5.168/74 ...; SF/3×10^{-10}	**Pu 241** 14.35 y α/0.002 4.896 /0.002 ...; β⁻	**Pu 242** 3.75×10^{5} y α/~100 4.901 /78...;γ45/0.04	**Pu 243** 4.956 h β 0.6 .../100 γ 84/23 ...	**Pu 244** 8.00×10^{7} y α / 9 9 . 8 7 9 4.589/	**Pu 245** 10.5 h β 0.9 .../100 γ 327/26 ...	**Pu 246** 10.85 h β 0.2 .../100 γ 44/25 ...	**Pu 247** 2.27 d β⁻							
Np 236 22.5h 1.54× ε/52 10^{5}y β⁻/48 α/0.2 γ642 ε/91	**Np 237** 2.14×10^{6} y α/~100 4.790 /47 ...	**Np 238** 2.117 d β⁻ 1.2 .../100 γ 984/28 ...	**Np 239** 2.355 d β⁻ 0.4 .../100 γ 106/24 ...	**Np 240** 7.22m 1.032h β⁻2.2 β⁻0.9 /99.9 /100 IT/0.1 γ 566	**Np 241** 13.9 m β 1.3 .../100 γ 175 ...	**Np 242** 5.5m 2.2m β⁻/100 β⁻2.7 γ 786 .../100	**Np 243** 1.85 m β /100 γ 288/15	**Np 244** 2.29 m β /100 γ 217/39									
U 235 0.7200 % 7.04×10^{8} y α/~100 4.398	**U 236** 2.34×10^{7} y α/100 4.494/74 ...;γ 49/0.08 ...	**U 237** 6.75 d β 0.2 .../100 γ 60/33 ...	**U 238** 99.2745 % 4.47×10^{9} y α/100 4.196/77	**U 239** 23.5 m β 1.2 .../100 γ 75/52 ...	**U 240** 14.1 h β 0.4 .../100 γ 44/2 ...		**U 242** 16.8 m β /100 γ 68/9 ...										

Figure 14.1 (*continued*).

14.2. THE ACTINIDE ELEMENTS

14.2.1. Actinium, element 89

Actinium was discovered in 1899 by Debierne in residues from the work of the Curies. It was named after the Greek word for beam or ray, aktis. Actinium occurs in nature as a member of the long natural decay series, except the one beginning with ^{238}U, see Fig. 1.2. Being the first of them, it has given name to the actinide series of elements. Chemically actinium behaves very similar to the lanthanides and especially as lanthanum.

14.2.1.1. Isotopes

Actinium only occurs in nature as the isotope ^{227}Ac ($t_{½}$ 21.77 y), which is a member of the long decay series starting with ^{235}U. This is also the most stable actinium isotope.

14.2.1.2. Occurrence and production

Due to its very low concentration, actinium is almost never produced from the ^{235}U decay series. However, it has been produced in macro amounts by neuron irradiation of ^{226}Ra:

$$^{226}_{88}Ra + ^{1}_{0}n \rightarrow ^{227}_{88}Ra \rightarrow ^{227}_{89}Ac + \beta^- \quad (14.2)$$

14.2.1.3. Uses

Actinium has been used in radioactive n-sources made by compressing an intimate mixture of ^{227}Ac and beryllium. The neutrons are produced when the α-particles from actinium hit atoms of ^{9}Be. It was earlier suggested that actinium compounds might be used in radiation therapy of cancers, however, better alternatives now exist.

14.2.2. Thorium, element 90

Thorium was discovered by Berzelius in 1828 from a sample of what is now called thorite ($ThSiO_4$) and named by him after the nordic god of thunder, Thor. As a metal it is silvery white but with exposure to air it tarnishes slowly becoming gray and finally black.

14.2.2.1. Isotopes

Thorium occurs in nature as the isotopes $^{227,228,229,230,231,231,232,234}Th$, most of them short lived. Natural thorium consists almost to 100% by weight of the very long-lived isotope ^{232}Th ($t_{½}$ 1.405×10^{10} y), which is the parent nuclide of the thorium decay series. The specific radioactivity for thorium is lower than that of uranium, and it is often treated as a non-radioactive element. For radioactive tracer studies, the nuclide ^{234}Th ($t_{½}$ 24.1 d) is usually separated from natural uranium.

14.2.2.2. Occurrence and production

Thorium is somewhat more common in nature than uranium, with an average content in the earth's crust of 10 ppm (by comparison the average abundance of lead is about 16 ppm in the earth's crust). In minerals thorium mainly exist as oxides but also silicates and phosphates exist. The content of thorium in sea water is $<0.5 \times 10^{-3}$ g/m^3, which is lower than that of uranium because thorium only exist as the low solubility tertavalent form while in the case of uranium oxidation states with higher solubility, e.g. (+VI) also exist and thus increase the total solubility.

The most common thorium mineral is monazite, a golden brown rare earth phosphate containing 1−15% ThO_2 and usually 0.1−1% U_3O_8. It is also found in small amounts in granite and gneiss. The largest deposits of monazite are found in India, Egypt, South Africa, the USA, and Canada, with 200−400 kton ThO_2 in each country. The size of natural resources are defined in terms of ore reserves which can be economically processed at the current market price. Thus, the total reserves at the 2005 market price was estimated by IAEA to ~2.6 Mt Th.

Because thorium often occurs with other valuable metals (in addition to the lanthanides) such as niobium, uranium and zirconium, it can be produced as a byproduct.

The following procedure is used for producing thorium from monazite sand. The sand is digested with hot concentrated alkali which converts the oxide to hydroxide. The filtered hydroxide is dissolved in hydrochloric acid and the pH adjusted between 5 and 6, which precipitates the thorium hydroxide but not the main fraction of lanthanide elements. The thorium hydroxide is dissolved in nitric acid and selectively extracted with methyl isobutyl ketone or tributyl phosphate in kerosene. This gives a rather pure organic solution of $Th(NO_3)_4$. The thorium is stripped from the organic phase by washing with alkali solution.

For the case of production of ^{234}Th it is conveniently done from an ether solution containing uranyl nitrate. Upon contact with some ml of water the thorium is stripped to the water phase. This waster phase is then run through a cation exchange column, washed with 6M HCl and then eluted with 0.5M oxalic acid. The oxalic acid is removed from the solution by wet combustion and the remaining solution is once again passed through a cation exchange column and the procedure is repeated. In the last step the remaining water phase is diluted in the desired acid and the stock solution is ready.

14.2.2.3. Uses

Thorium metal is used as electrode material in gas discharge lamps, and as getter for absorption of rest gases in high vacuum technique. ThO_2 (melting point 3300°C) is highly refractory and used for high temperature furnace linings and in ultrasonic very high temperature thermometers. It is also alloyed with magnesium to improve the high temperature performance Thorium salts are of little practical use. Because Th^{4+} is a stable tetravalent ion with properties very similar to the tetravalent actinides, Th^{4+} is often used as an analog for the An(IV) ions; most commonly used isotopes are ^{228}Th ($t_{½}$ 1.91 y), ^{230}Th ($t_{½}$ 7.54 × 10^4 y) and ^{234}Th ($t_{½}$ 24.5 d), which can be isolated from old ^{232}Th or ^{238}U, see Fig. 1.2. This use of thorium as an analogue for tetravalent actinides is however rather questionable since thorium is considerably larger and do not have any 5f electrons. The size difference matters for so called hard ligands like oxygen while the 5f electrons matter for softer ligands like nitrogen. For hard ligand e.g. zirconium is a better analogue for e.g. plutonium (IV) Thorium may become important as fuel for nuclear reactors due to its relative abundance and the fact that no plutonium is created during reactor operation (Ch. 20). On the other hand, the ^{233}Pa → ^{233}U decay in spent fuel may be used to produce pure ^{233}U, which can replace ^{239}Pu in explosive nuclear devices.

14.2.3. Protactinium, element 91

Protactinium (the parent of actinium) was first discovered in 1913 by Fajans and Göhring and initially named brevium due to the short half-life of the isotope used in the discovery, ^{234}Pa ($t_{½}$ 6.75 h). The elements existence was confirmed in 1918 when two groups independently discovered the longer lives isotope ^{231}Pa. These groups were Otto Hahn and Lise Meitner from Germany and Frederick Soddy and John Cranston from Great Britain. The name was then changed to protactinium as it was recognised as a parent of actinium. It is considered as one of the rarest naturally occurring element.

14.2.3.1. Isotopes

Protactinium occur in nature as the isotopes ^{231}Pa ($t_{½}$ 32760 y), ^{234m}Pa ($t_{½}$ 1.17 m), and ^{234}Pa ($t_{½}$ 6.75 h) — all members of long-lived natural decay chains, see Fig. 1.2. All known isotopes of protactinium are shown in Fig. 14.1.

14.2.3.2. Occurrence and production

Due to its long half-life in comparison with the other isotopes, ^{231}Pa is the dominant isotope in nature. It can be produced in macroscopic amounts from old uranium minerals and sometimes from wastes from U-production and purification. Most of the ^{231}Pa available today stems from a separation project during the early 1960:s where 125 g of ^{231}Pa was recovered from the waste from a defunct U-purification process based on ether extraction. Since then, most of that material has been lost to various wastes.

The shorter lived ^{233}Pa is produced by n-irradiation of ^{232}Th in a nuclear reactor.

$$^{232}_{90}\mathrm{Th} + {}^{1}_{0}\mathrm{n} \rightarrow {}^{233}_{90}\mathrm{Th} \rightarrow {}^{233}_{81}\mathrm{Pa} + \beta^- \qquad (14.3)$$

By dissolution of the irradiated material in nitric acid, dilution to about 1 M and sorption of the ^{233}Pa on a MnO_2-filled column, delay for shorter-lived Pa-isotopes to decay, washing and elution by a dilute fluoride solution, ^{233}Pa can be produced.

14.2.3.3. *Uses*
Long-lived ^{231}Pa has no general use today. The main use of ^{233}Pa is as mother for isotopically pure ^{233}U. It is also an important intermediate in the fuel chain of thermal thorium reactors. Pa is one among the most toxic elements — about 2.5×10^8 times more toxic than HCN.

14.2.4. Uranium, element 92
Uranium was discovered in 1879 by Klaproth, who named it after the newly discovered planet Uranus. The specific radioactivity of natural uranium makes it a weak radiological hazard (see Ch. 15). It is also chemically toxic and precautions should be taken against inhaling uranium dust for which the threshold limit is 0.20 mg/m^3 air (about the same as for lead).

14.2.4.1. *Isotopes*
Natural uranium consists of 3 isotopes, ^{234}U, ^{235}U and ^{238}U, members of the natural decay series discussed in §1.3 and shown in Fig. 1.2. Uranium is an important raw material for nuclear energy production (see Ch. 19, 20 and 21). Fig. 14.1 shows all known uranium isotopes and some of their nuclear properties. A mass of nearly pure ^{235}U above a certain critical amount can undergo a nuclear explosion if suddenly hit by a rather high n-flux. The high initial n-flux requirement is due to the generation time of the fission neutrons and the limited time available until a supercritical amount is dispersed by the heat generated. This is further discussed in Chapter 19.

14.2.4.2. *Occurrence, resources and production capacity*
Uranium appears in a large number of minerals (at least 60 are known). The earth's crust contains 3—4 ppm U, which makes it about as abundant as arsenic or boron. Uranium is found at this relative concentration in the large granitic rock bodies formed by slow cooling of the magma about 1.7—2.5 eons ago (1 eon = 10^9 y = 1 billion years). It is also found in younger rocks at higher concentrations ("ore bodies").

Geochemists now begin to understand how these ore bodies were formed. They are usually located downstream from mountain ranges. As the atmosphere became oxidizing about 1.7 eons ago, rain penetrated into rock fractures and pores, bringing the uranium to its hexavalent state and dissolving it as an anionic complex (likely as a carbonate, $UO_2(CO_3)_n^{2-2n}$, or as a sulphate complex, $UO_2(SO_4)_n^{2-2n}$), often at elevated temperatures. As the water and the dissolved uranium migrated downstream, regions of reducing material were encountered, either inorganic (e.g. pyrite) or organic (e.g. humic) matter, which caused reduction to U(IV). Since many U(IV) compounds are insoluble in the natural environment, the uranium precipitated, possibly as the sulfide or, more likely, as the hydroxide. Many of these original uranium precipitates were later covered by sedimentary material.

In most minerals uranium is in the tetravalent state. The most important one is uraninite (UO_{2+x}, $x = 0.01$ to 0.25), in which the uranium concentration is 50–90%; it is found in Western Europe, Central Africa (e.g. Katanga, Gabon), Canada (e.g. Cigar Lake) and Australia (e.g. Koongara). In the USA and Russia carnotite (a K + U vanadate) is the most important mineral and contains 54% uranium. In the high grade ores the mineral is mixed with other minerals so the average uranium concentration in the crushed ore is much less: e.g. $\leq$0.5% on the Colorado Plateau. The world's largest known uranium deposit is located at the Olympic Dam Mine in South Australia. Uranium is also often found in lower concentration, of the order of 0.01–0.03%, in association with other valuable minerals such as apatite, shale, peat, etc.

At a price of ~$130 per kg of U, the known (total of measured, indicated and inferred) world reserve of uranium which could be recovered economically at that price was estimated to be 4.7 Mtons. Sea water contains some 4 500 Mtons U but it is uncertain if U can be recovered economically from this huge resource.

Uranium production in 2010 was ~50.6 kton. Uranium producers are given in Table 14.1 by country.

Table 14.1 Uranium production by country in 2010

Country	(t U)	%
Kazakhstan	17803	33.18
Canada	9783	18.23
Australia	5900	10.99
Namibia	4496	8.378
Niger	4198	7.823
Russia	3562	6.638
Uzbekistan	2400	4.472
USA	1660	3.093
Ukraine	850	1.584
China	827	1.541
Malawi	670	1.249
South Africa	583	1.086
India	400	0.745
Czech Rep.	254	0.473
Brazil	148	0.276
Romania	77	0.143
Pakistan	45	0.084
France	7	0.013
TOTAL	53663	100

(data from WNA)

14.2.4.3. Production techniques

Uranium ores differ widely in composition, containing a variety of other elements which must be removed. As a result the production methods differ considerably depending on the particular ore to be processed although in every case very selective processes must be used. The following is a common scheme.

The ore is mined in open pits or underground. The ore is crushed and concentrated through flotation. If the uranium is in the tetravalent state it is oxidized in piles by air, sometimes with the aid of bacteria. The material is subsequently treated with sulfuric acid which dissolves the uranium as the sulphate complex, $UO_2(SO_4)_2^{2-}$ (the feed). This complex can be selectively removed from the aqueous solution by means of anion exchange resins or, more commonly, by extraction into an organic solvent (solvent extraction). In the latter case an extractant (i.e. an organic compound with ability to form a U-organic complex soluble in organic solvents) dissolved in kerosene is used; depending on the aqueous feed composition, various extractants have been applied. A typical flow-sheet is illustrated in Figure 14.2, and the basic chemical principles in Table 14.2. The final product is commonly ammonium diuranate, which is referred to as yellow cake; it contains 65–70% U. It is so free of radioactive uranium daughters that it can be handled safely in drums without any radiation protection. Yellow cake is further purified in order to obtain a final pure product of U_3O_8, usually better than 99.98% pure; its content of neutron poisons (nuclides which have high capture cross-sections for neutrons such as B, Cd, Dy, is less than 0.00002%, see Ch. 19).

..A different mining technique is the so called in situ leaching which is used all over the world, e.g. the Beverley mine in Australia. The basic idea is that a leaching solution is injected into the ore body in one place and then the loaded phase is pumped out at a different location (relatively close by). The site is surrounded by monitoring wells to make sure that no leach solution leaves the mining area. After the finishing of the mining activities ground water is pumped back in the holes and the mine is closed. Despite this seemingly simple remediation there are few in situ leaching sites where no active remediation was made leaving the ground heavily polluted with acidic uranium containing water. This problem was clearly shown in Saxonia in the former DDR where large amounts of money now have been spent on remediation since the German unification in 1990.

Heating yellow cake in hydrogen gas produces UO_2 which in sintered form is used as nuclear reactor fuel. If the UO_2 is exposed to F_2, UF_4 is obtained ("green salt"), which in a thermite process with calcium metal yields metallic uranium. The metal is slowly oxidized in air at room temperature; the metal powder is very reactive at elevated temperatures and can then be used to remove practically all impurities from rare gases.

The chemistry of aqueous uranium is discussed in §14.3 together with the chemistry of the other actinides. Production of reactor fuel and reprocessing is described in Chapter 21.

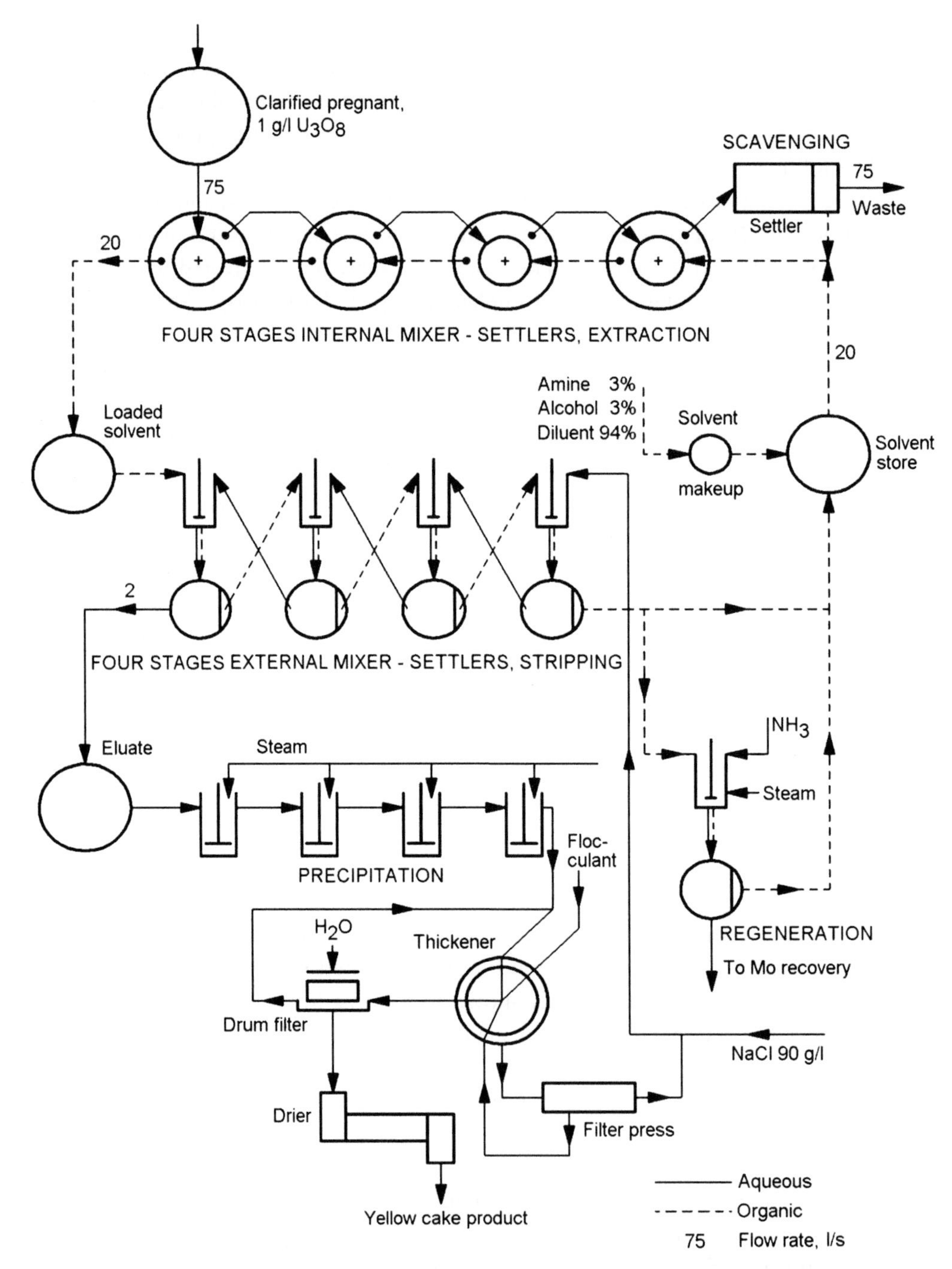

Figure 14.2 Amine extraction circuit, Kerr-McGee Corp., Grants Mill, New Mexico. Each S-X bettery consists of 4 mixer-settlers. Solid lines are aqueous, and dashed lines are organic flow. Numbers are flow rates (l/s).

Table 14.2 Chemical bases for the Ames (amine) and Dapex (HDEHP) processes for uranium recovery from sulfuric acid leach liquors

	AMEX	DAPEX
Extraction	$UO_2^{2+} + SO_4^2 + 2(R_3NH)_2SO_4 \equiv (R_3NH)_4UO_2(SO_4)_3$ R_3N = Alamine 336 or Adogen 364 (trialkylamines with alkyl groups with 8–10 carbons)	$UO_2^{2+} + 2(HDEHP)_2 \equiv UO_2(HDEHP.DEHP)_2 + 2H^+$
Stripping	**Acidic stripping** $(R_3NH)_4UO_2(SO_4)_3 + 4HX \equiv 4R_3NHX + UO_2^{2+} + 3HSO_4 + H^+$ HX = HCl or HNO_3 **Neutral stripping** $(R_3NH)_4UO_2(SO_4)_3 + (NH_4)_2SO_4 + 4\ NH_3 \equiv 4R_3N + UO_2(SO_4)_3^4 + 6NH_4^+ + SO_4^2$ **Alkaline stripping** $(R_3NH)_4UO_2(SO_4)_3 + 7Na_2CO_3 \equiv 4R_3N + UO_2(CO_3)_3^4 + 4HCO_3 + 3SO_4^{2-} + 14Na^+$	**Alkaline stripping** $UO_2(HDEHP.DEHP)_2 + 4Na_2CO_3 \equiv UO_2(CO_3)_3^4 + 4NaDEHP + H_2O + CO_2$ [$+ 4Na^+$ (TBP must be added to avoid the third phase formation owing to low organic phase solubility of NaDEHP.) **Acidic stripping** $UO_2(HDEHP.DEHP)_2 + 2H^+ \equiv 2(HDEHP)_2 + UO_2^{2+}$

(From Musicas and Schulz.)

14.2.4.4. Production wastes

The milling operation produces tailings consisting of fine to coarse particles in a water slurry containing most of the radioactive uranium decay products, of which radium is the most hazardous. As most tailings are not highly radioactive (e.g. from mining of low grade ores) they are dumped on outside the plant; leach water from the dumps will then, combined with mine water, enter local streams. In dry areas, dusts from the tailings may spread by winds.

This is the main picture and explains the high dose commitment values in Table 22.1. However, many health authorities now require conditioning of the tailings to reduce harmful effects to the environment: recycling of waste water, precipitation and removal of radium from solution, and neutralization to precipitate heavy metals. The slurry is usually transported to an impoundment basin where the solid particles settle out, and the effluent is treated for removal of activities before discharge into a settling pond. Treatment ponds may contain low permeability liners where appropriate to control seepage. Barium chloride may be added to the tailings to precipitate dissolved ^{226}Ra as Ra–Ba sulfate. Lime and limestone may be added to the tailings to raise pH.

The final dry waste is stored either on surface or in shallow basins, though underground storage also has been practiced (abandoned mines). In the former case, the tailings are covered by up to 3 meters of earth fill to restrict erosion, and some water tight material to protect it against rain. In some cases (e.g. Sweden) the land produced on top of mine waste and tailings have been reclaimed for farming.

14.2.4.5. Uses

The main use of uranium is as nuclear fuel, see Chapters 19–21. Depleted uranium from isotope separation plants is also used as metallic slugs in special kinds of armorpiercing ammunition. It has also been used as heavy ballast in some extreme racing yachts, but this has sometimes caused political problems. Depleted uranium has also been used as movable balance weights in large commercial aircraft. With the deployment of fast nuclear reactors, depleted uranium will probably find a new use as fertile reactor fuel. Uranium was previously also used as a commodity chemical. Due to its classification as a nuclear material ownership is now restricted in most parts of the world. Previously uranium was also used rather extensively for colouring of glass to a clear green colour which becomes intense when illuminated with UV light.

14.2.5. Neptunium, element 93

Early in 1940 McMillan and Abelson in the United States synthesized and identified a new element with atomic number 93 by bombarding uranium with slow neutrons. They named it neptunium after the planet Neptune. The reaction used in the synthesis was

$$^{238}_{92}\text{U}(\text{n},\gamma)^{239}_{92}\text{U} \xrightarrow[23\text{ min}]{\beta^-} {}^{239}_{93}\text{Np} \xrightarrow[2.3\text{ d}]{\beta^-} \tag{14.4}$$

(cf. Fig. 14.1). The experimental recoil technique for separating the fission and neutron capture products from the target material is described in §17.6.1. Chemical experiments showed that the product had a 2.3 d half-life and could be reduced by SO_2 to a lower valency state (presumably +4), which could be precipitated as a fluoride (carrier: LaF_3); this distinguished the element from uranium. In its oxidized state (using BrO as oxidant) it showed the same chemistry as hexavalent uranium. Because no fission product is expected to have this behavior, the assumption that the element with the 2.3 d half-life was a transuranic element was verified.

14.2.5.1. Isotopes

The longest lived isotope is ^{237}Np with $t_{1/2}$ 2.14×10^6 y. It is produced in nuclear reactors (Ch. 19) through the reactions

$$^{238}\text{U}(\text{n},2\text{n})^{237}\text{U} \xrightarrow[6.75\text{ d}]{\beta^-} {}^{237}\text{Np}(\approx 70\%) \tag{14.5}$$

$$^{235}\text{U}(\text{n},\gamma)^{236}\text{U}(\text{n},\gamma)^{237}\text{U} \xrightarrow[6.75\text{ d}]{\beta^-} {}^{237}\text{Np}(\approx 30\%) \tag{14.6}$$

14.2.5.2. Occurrence and production

About 0.5 kg ^{237}Np is contained in each ton of spent reactor fuel. By separation processes, similar to those described in Chapter 21, hundreds of kg of neptunium has been made available. Though ^{237}Np has too short a half-life to still exist naturally on earth, its optical spectrum has been observed in some stars, from which the star can be deduced to be of relatively young age ($\leq 10^7$ y, cf. §14.2.6.2 on ^{244}Pu).

14.2.5.3. Uses

So far it has only found a practical use as target material for production of ^{238}Pu. Most of the ^{238}Pu produced in this way was earlier used to power pace makers, but the main use is now as a power source (RTG) in space vehicles, used on the Mars surface (Vikings), or travelling far from the sun e.g. the two Mariner spacecraft currently leaving the solar system, or the mission to Pluto and beyond.

14.2.6. Plutonium, element 94

At the end of 1940 an isotope of element 94 (plutonium) was synthesized by Seaborg, McMillan, Kennedy, and Wahl by the bombardment of uranium with deuterons in a cyclotron:

$$^{238}_{92}\mathrm{U(d,2n)}^{238}_{93}\mathrm{Np} \xrightarrow[2.1\ \mathrm{d}]{\beta^-} {}^{238}_{94}\mathrm{Pu} \xrightarrow[88\ \mathrm{y}]{\alpha} \tag{14.7}$$

It was identified as a new element, which was distinctly different from both uranium and neptunium in its redox properties (e.g. the +3 and +4 oxidation states were more stable). A second isotope of element 94, ^{239}Pu, with a half-life of 24 000 y was isolated very shortly afterwards as the α-radioactive daughter of ^{239}Np. Although plutonium was discovered already 1941 its discovery was not revealed until 1946 due to war time issues. It was named after the now planetoid and then planet Pluto.

14.2.6.1. Isotopes

Plutonium has several rather long-lived isotopes: ^{238}Pu ($t_{½}$ 87.74 y), ^{239}Pu ($t_{½}$ 2.41×10^4 y), ^{240}Pu ($t_{½}$ 6.5×10^3 y), ^{241}Pu($t_{½}$ 14 y), ^{242}Pu ($t_{½}$ 3.73×10^5 y) and ^{244}Pu ($t_{½}$ 8.08×10^7 y). Data for all known Pu-isotopes are shown in Fig. 14.1.

14.2.6.2. Occurrence and production

Even though plutonium may have been formed in considerable amounts in the cosmic r-process (§12.5.3), the most stable isotope, ^{244}Pu, has a half-life of only 8.08×10^7 y, which prevents it from having survived on earth to our time. However, traces of ^{244}Pu have been discovered in cerium-rich rare earth minerals. If the amount found is extrapolated, considering the enrichment of plutonium in the mineral, one gram of the earth's crust will contain 3×10^{-25} g ^{244}Pu. This would indicate about 10 g of natural plutonium remaining from the genesis of the earth, at which time it existed in parts per million of matter. ^{244}Pu decays by α-emission to ^{240}U ($t_{½}$ 14.1 h, β^-).

As discussed in §19.10, ^{239}Pu has been formed in natural uranium reactors at a later stage of the earth's evolution. Many thousands of tons of plutonium have been synthesized in commercial and military reactors; the annual global production rate in nuclear power reactors in the year 2000 was $\approx$1000 tons/y, contained in the spent fuel elements. The nuclear reactions and chemical separation processes are presented in Chapters 19 and 21. The build-up of heavier elements and isotopes by n-irradiation of ^{239}Pu in nuclear reactors is illustrated in Figures 14.3 and 14.4.

14.2.6.3. Uses

Experiments with ^{239}Pu confirmed theoretical predictions that it would exhibit high fissibility with both thermal and fast neutrons. This meant that ^{239}Pu in sufficient quantity and suddenly hit by a high n-flux would also experience a nuclear explosion like ^{235}U, see §14.2.4.1 above and Ch. 19. If controlled nuclear fission could be accomplished in a nuclear reactor, it would be possible to produce large amounts of plutonium by neutron bombardment of ^{238}U. The ^{239}Pu could be isolated by chemical methods which were at that time expected to be simpler than the isotopic separation required to obtain pure ^{235}U. As a consequence, the production of ^{239}Pu became a major project of the atomic bomb program of the United States during World War II; see further Ch. 19. Today, the use of Pu in nuclear explosives seems to be abandoned in favour of more easily managed highly enriched ^{235}U, or perhaps even by pure ^{233}U.

Plutonium has also a use as mixed oxide (MOX) fuel for thermal and fast nuclear reactors, see Ch 19—21. Surplus "weapons grade" ^{239}Pu metal is now often converted to PuO_2 powder for use in MOX. The lack of gamma radiation and the 87.7 y half-life of ^{238}Pu has made it a preferred energy source in radioisotope thermal electricity generators (RTG:s), cf §7.9.3. As mentioned in §14.2.5.3 this isotope is produced by n-irradiation of ^{237}Np. Hundreds of kilogram of $^{238}PuO_2$ has been used the space programs, some of which is now even leaving the solar system.

14.2.7. Americium and curium, elements 95 and 96

Americium and Curium were discovered by Seaborg , Ralph A. James, Leon O. Morgan and Albert Ghiorso in the USA in 1944-45. The reaction was irradiation of plutonium with neutrons in a nuclear reactor according to (14.9) and by the irradiation of plutonium with α-particles in a cyclotron according to

$$^{239}_{94}\text{Pu}(\alpha, \text{n})^{242}_{96}\text{Cm} \xrightarrow[163\text{ d}]{\alpha} \tag{14.8}$$

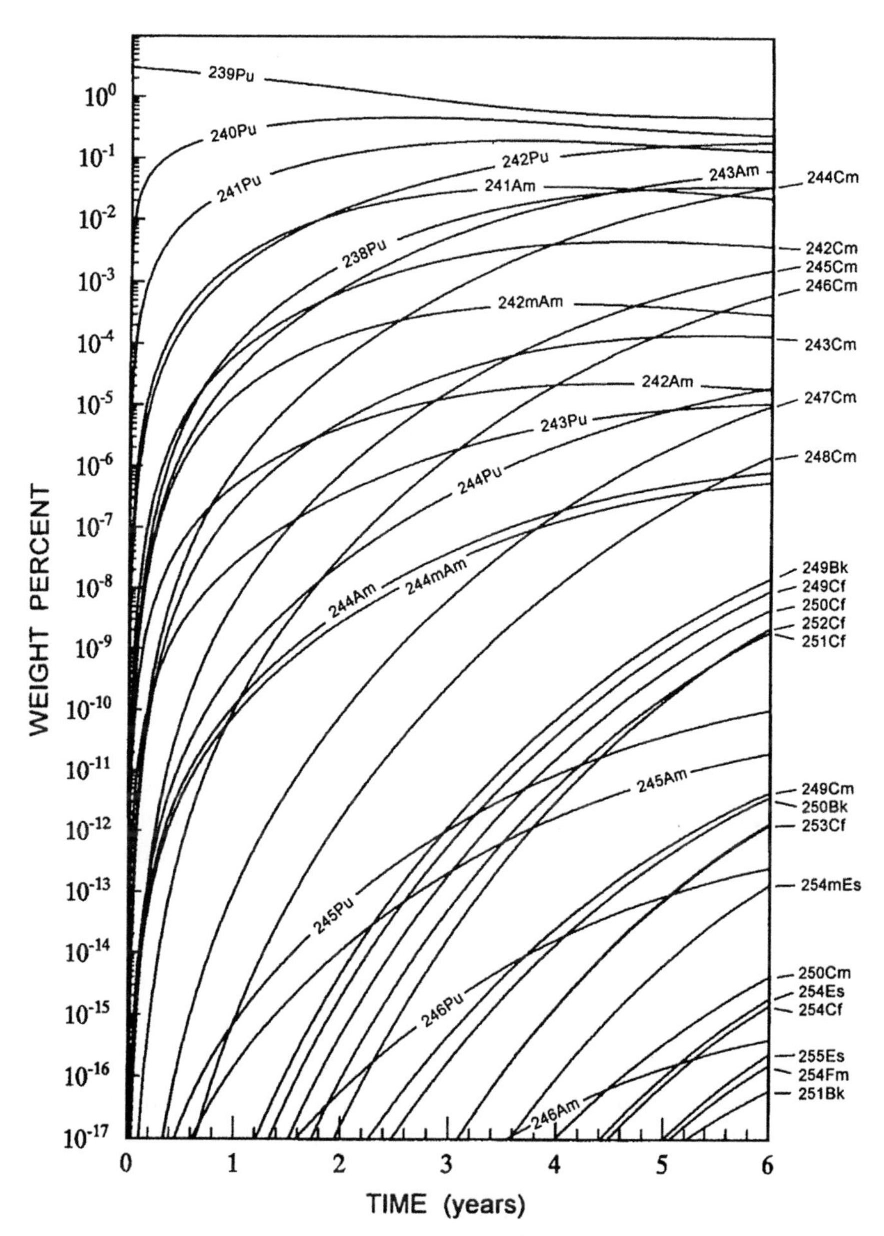

Figure 14.3 Production of higher actinide isotopes by irradiation of mixed oxide fuel (3% ^{239}Pu, rest is depleted U) at constant power in a boiling water reactor.

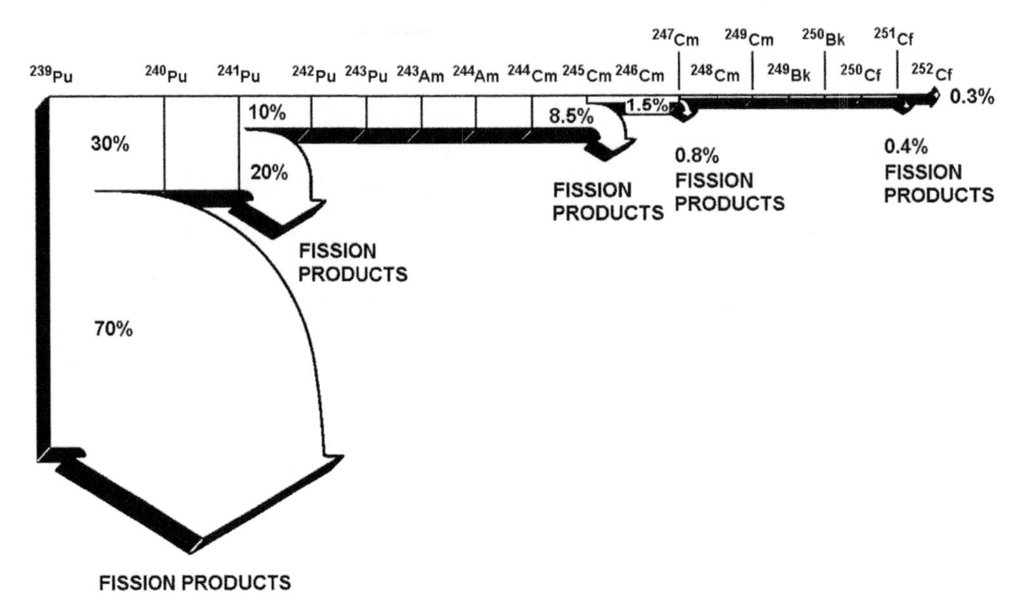

Figure 14.4 The major paths in transuranium element production in a thermal nuclear reactor; fission and n-capture compete for available nuclei *(From Seaborg)*.

14.2.7.1. Isotopes

Successive neutron capture in ^{239}Pu (as occurs in nuclear reactors) produces isotopes of plutonium which undergo β^--decay, forming transplutonium elements with atomic numbers 95 (americium) and 96 (curium) (Fig. 14.2). The reaction sequence is

$$\begin{array}{l} ^{239}_{94}\text{Pu}\,(\text{n},\gamma)\,^{240}_{94}\text{Pu}\,(\text{n},\gamma)\,^{241}_{94}\text{Pu} \\ \qquad\qquad\qquad\qquad\quad \downarrow \beta^-\ 14.4\ \text{y} \\ \qquad\qquad\qquad\qquad\ ^{241}_{95}\text{Am}\,(\text{n},\gamma)\,^{242}_{95}\text{Am} \\ \qquad\qquad\qquad\qquad\qquad\qquad\quad \downarrow \beta^-\ 16.01\ \text{h} \\ \qquad\qquad\qquad\qquad\qquad\qquad\quad ^{242}_{96}\text{Cm} \end{array} \tag{14.9}$$

All of the isotopes so far described have half-lives of sufficient length for ordinary chemistry although the small amounts made special laboratory techniques necessary (§18.2).

^{241}Am and ^{242}Cm also decay through spontaneous fission. The longest lived isotopes of these elements are ^{243}Am, $t_{½}$ 7370 y, and ^{247}Cm, $t_{½}$ 1.56×10^7 y, decaying by α-emission to ^{239}Np and ^{243}Pu (β^-, $t_{½}$ 5.0 h), respectively; see Fig. 14.1.

14.2.7.2. Occurrence and production

The long-lived isotopes of Am and Cm are produced in nuclear power reactors, Figs. 14.3 and 14.4. The much higher stability of the heavier nuclide ^{247}Cm (compared to ^{243}Am) can be explained by its lower Q_α (5.352 as compared to 5.638 MeV) and by the proximity to the next magic neutron number, 152. The current world production of these element amounts to many kilograms per year. When spent fuel is reprocessed both elements are currently routed to the HLW stream. However, reprocessing methods have been developed that are able to isolate both elements in pure form in order to make it possible to try to transmute them into more short-lived or less dangerous nuclides.

14.2.7.3. Uses

The americium isotope ^{241}Am has been used as a radiation source in fire warning devices (about 30 kBq in a typical house hold device) and as a 60 keV gamma source in X-ray fluorescence analytical systems.'

No large scale use of curium exists today. Both these elements are considered among the so called *minor actinides* in used nuclear fuel. Thus, scientific work is ongoing on producing fast nuclear fuels with an appropriate content of at least americium. Thermal neutron irradiation of the mixture of curium isotopes which may be separated from spent reactor fuel initially increses the average halflife of this mixture.

14.2.8. Berkelium and californium, elements 97 and 98

By 1949 Seaborg's group had synthesized a few milligrams of $^{241}_{95}$Am from reactor bombardment of plutonium. This material was used as the target in a cyclotron bombardment. Immediately following irradiation, the target was dissolved and the products separated by passage through a column of ion exchange resin using an eluting solution of ammonium citrate (§18.2.6). An α-emitting species with a half-life of 4.5 h was identified as the isotope of element 97 of mass number 243; the name berkelium was proposed.

Later the same technique was used with a target of a few micrograms of ^{242}Cm and the first isotopes of californium ($Z = 98$) were discovered.

$$^{241}_{95}\mathrm{Am}(\alpha, 2\mathrm{n})^{243}_{97}\mathrm{Bk} \xrightarrow[4.5\ \mathrm{h}]{\mathrm{EC}} \tag{14.10}$$

$$^{242}_{96}\mathrm{Cm}(\alpha, \mathrm{n})^{245}_{98}\mathrm{Cf} \xrightarrow[44\ \mathrm{min}]{\alpha} \tag{14.11}$$

14.2.8.1. Isotopes

The longest-lived isotopes of berkelium and californium are: are ^{247}Bk, $t_{½}$ 1380 y, and ^{251}Cf, $t_{½}$ 898 y, decaying by α-emission to ^{243}Am and ^{247}Cm, respectively, see also Figure 14.1 for other isotopes of these elements.

14.2.8.2. Occurrence and production

Neither berkelium nor californium is found naturally on earth. Both are present in small amounts in fresh spent nuclear fuel. Small amounts have also spread into the environment after atmospheric nuclear weapons tests, from a few severe reactor accidents like Chernobyl, and from releases of waste from the nuclear fuel cycle. However, current concentrations in nature of these elements are very small.

Berkelium and californium have also been produced in small amounts by n-irradiation of targets in high-flux reactors.

14.2.8.3. Uses

$^{252}_{92}Cf$ has been used in small n-sources, for calibration of high energy particle detectors, and as targets for production of so called super heavy elements (SHE), but no practical use of berkelium has been found.

14.2.9. Einsteinium and fermium, elements 99 and 100

In 1952 the United States set off the first test thermonuclear explosion (code name "Mike") at the Eniwetok Island in the Pacific. The early analysis of debris from Mike by Albert Ghiorso and his team showed that a heavy isotope of plutonium, ^{244}Pu, had been made by multi-neutron capture in ^{238}U which had been part of the device. More extensive chemical purification of some of the radioactive coral from the test site proved that isotopes of elements 99 (einsteinium) and 100 (fermium) had been made in the explosion. The neutron flux during the very brief burning time of Mike had been so intense that it resulted in capture by ^{238}U nuclei of as many as 17 neutrons; see Fig. 14.5. This multi-neutron capture ended as the device blew apart and was followed by a sequence of β^--decays.

Within a year, element 99 was again synthesized by cyclotron bombardment of uranium with nitrogen

$$^{238}U + {}^{14}N^{6+} \rightarrow {}_{99}Es \tag{14.12}$$

Several months later, Fm was isolated as a product of reactor irradiation. In this case the neutron capture occurs over a long time and β-decay processes compete with neutron capture depending on the $t_{½}$ (β^-) and the neutron flux. The reaction sequences are shown as the shaded area in Figure 14.4. Symbolically, for example the sequence ^{239}Pu+3n, β^-,+8n, 2β^-, +4n, β^-, +n, β^-, +n produces fermium. See also Figures 14.3 and 14.4 for the production of various nuclides in the sequence as a function of time for a reactor with a predominantly thermal neutron flux.

14.2.10. Mendelevium, nobelium and lawrencium, elements 101 – 103

By 1955 reactor transmutation of ^{239}Pu had produced 10^9 atoms of ^{253}Es. These atoms were deposited on gold and bombarded with helium ions to make the element 101 by the reaction

$$^{253}_{99}Es + {}^{4}_{2}He \rightarrow {}^{256}_{101}Md + {}^{1}_{0}n \tag{14.13}$$

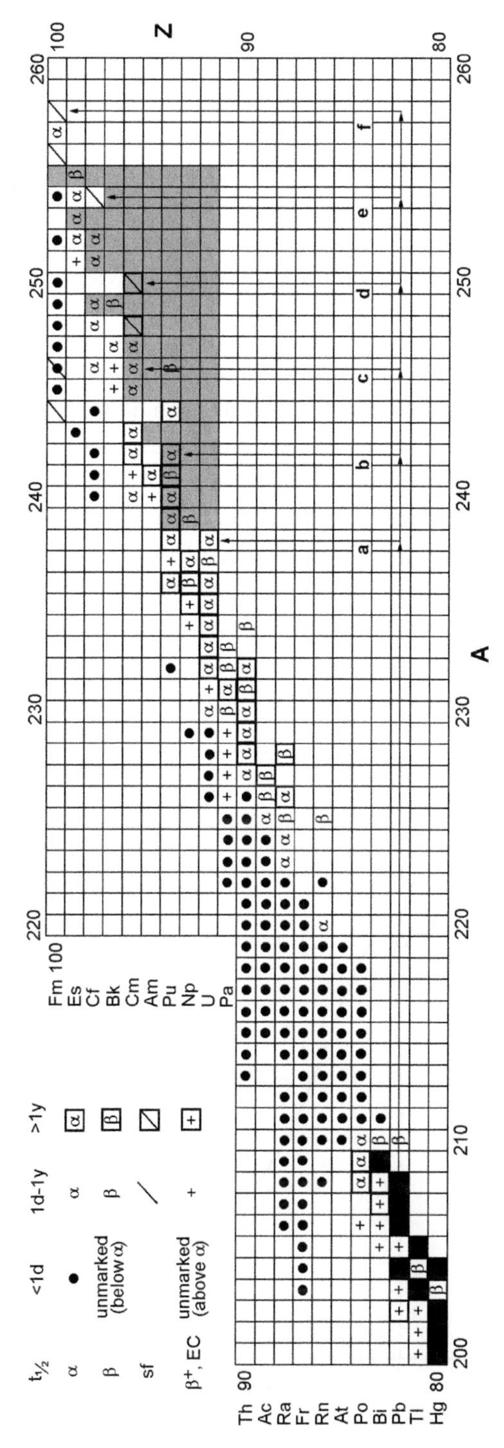

Figure 14.5 The build-up of heavy actinides through rapid multiple n-capture in lead (arrows) and in uranium (shaded area).

A new technique was used that allowed the atoms of ^{256}Md to recoil from the very thin target onto a "catcher" foil. Thirteen atoms of ^{256}Md ($t_{½}$ 1.3 h) were made in 9 h of irradiation and isolated by rapid elution from a column of cation exchange resin using a solution of α-hydroxy isobutyric acid. The elution showed 5 atoms of element 101 (identified by the spontaneous fission of the daughter fermium) and 8 atoms of the fermium daughter (Fig. 14.8; eqn. (14.14)). The recoil – ion exchange technique used in identification of element 101 by Thompson, Choppin, Seaborg, Harvey and Ghiorso has been successfully applied in the discovery of other heavier elements.

$$^{256}_{101}\mathrm{Md} \xrightarrow[1.3\ \mathrm{h}]{\mathrm{EC}} {}^{256}_{100}\mathrm{Fm} \xrightarrow[2.6\ \mathrm{h}]{\mathrm{sf}} \qquad (14.14)$$

So far we have described two methods of production of heavy elements: n-irradiation followed by β^--decay, and accelerator induced reactions according to relations 14.8–14.13. Multiple n-capture has its limitation in the successively decreasing yields, while α-reactions usually yields product nuclides further away from the bottom of the *Z*-versus-*N* stability valley (cf. e.g. Fig. 4.1, 14.1 and 14.6) resulting in shorter half-lives. Heavier isotopes (i.e. isotopes with larger mass numbers, that is more neutron rich) of the product element are expected to be more stable (i.e have longer half-lives). However, by irradiating the heavy target atom by "heavy ions", e.g. cromium ions, three advantages can be achieved in one reaction: the charge of the compound nucleus is increased

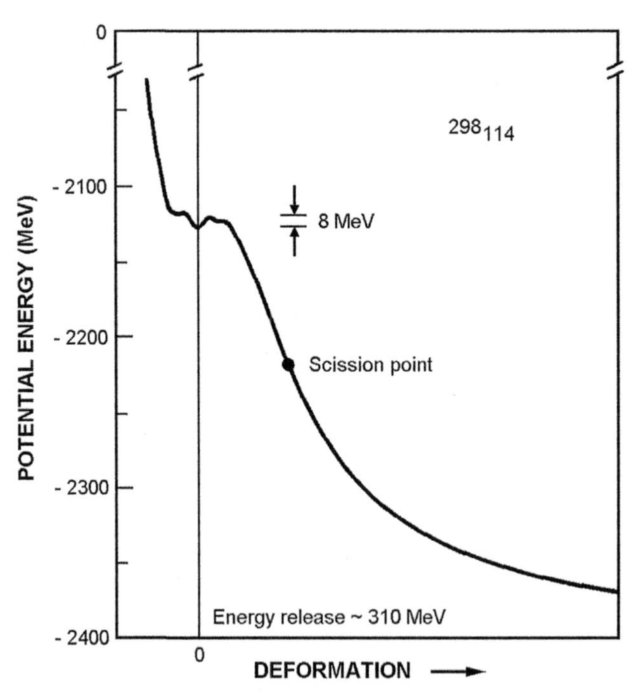

Figure 14.6 Dependence of potential energy on deformation for a nucleus like $^{298}114$. *(From R. Nix.).*

several steps, while at the same time more neutrons ("glue") are added, and, if the reaction energy is negative, giving a less exited compound nucleus, hopefully leading to a much heavier and more stable product. This method is used today for the production of the heaviest actinide elements and of transactinide elements.

In 1957 Russian scientists at the Joint Institute for Nuclear Research (JINR) at Dubna (now Flerov Laboratory for Nuclear Reactions, FLNR) claimed to have synthesized element 102 by irradiating ^{241}Pu with ^{16}O ions, using nuclear emulsions for α-energy determination.

$$^{241}_{94}\mathrm{Pu} + ^{16}_{8}\mathrm{O} \rightarrow ^{252}_{102}\mathrm{No} + 5^{1}_{0}\mathrm{n} \tag{14.15}$$

They assigned an 8.8 ± 0.5 MeV α-decay with a half-life of 2–40 s, to element 102. However, their results have been contested by scientists at the Lawrence Berkeley Laboratory (LBL) in California, who produced and identified element 102 in 1958 by use of a double-recoil technique[1]. The reaction scheme was

$$\begin{array}{ccl} ^{246}_{96}\mathrm{Cm} + ^{12}_{6}\mathrm{C} & \xrightarrow{\text{1 st recoil}} & ^{254}_{102}\mathrm{No} + 4^{1}_{0}\mathrm{n} \\ & & \quad \alpha \downarrow \text{2nd recoil} \\ & & ^{250}_{100}\mathrm{Fm} \end{array} \tag{14.16}$$

Lawrencium, element 103, was synthesized at Berkeley in 1961 by the reactions

$$^{250,1,2}_{98}\mathrm{Cf} + ^{10,11}_{5}\mathrm{B} \rightarrow ^{258}_{103}\mathrm{Lr} + (2-5)^{1}_{0}\mathrm{n} \tag{14.17}$$

using a system where the recoiling products were caught on a moving band, which rapidly transported them to an array of energy sensitive solid state detectors (cf. Fig. 17.3). The half-life of the product (later determined as ~4s) was too short to allow any chemistry. It was the first actinide element to be identified through purely instrumental methods.

14.2.11. The transactinide elements, $Z \geq 104$

The transactinide elements, some of which are often also called superheavy element (SHE), are the elements beyond the end of the actinide series, i.e. $Z > 103$. Many have a nucleus that survives only because of microscopic shell stabilization. When the number of protons increases the repulsive forces between the protons increase, and the number of chemical elements is therefore limited by the increasing probability for spontaneous fission. This macroscopic behavior is governed by shell effects. Nuclei can survive beyond the macroscopic limit, where the necessary balance between the attractive

[1] The name *nobelium* was selected to honor the Swedish industrialist Alfred Nobel, who donated his fortune to be used as prizes to those who have contributed most to progress in the fields of science, the humanities and peace.

nuclear force and the repulsive Coulomb force is achieved only through shell stabilization. The number of protons where this occurs is around Z = 104. Studies of their properties are linked to an understanding of the physics and chemistry at the end of the periodic table. Due to conflicting claims of discovery there have been several names used for some of these heavy elements, see Table 14.7.

According to calculations made e.g. by Nilsson et al., Scharff-Goldhaber, Strutinsky and Swiatecki in the late 1960:ies there may also be an area near the next (predicted) double shell closure above lead where nuclei may have surprisingly long half-lives. The first calculations indicated half-lives of millions of years or even stability. Today the predicted half-lives are a few orders of magnitude shorter but still very long. On the way to this "island of stability" in the "sea of instability" regions with deformed nuclei exhibiting a varying degree of stronger binding have also been predicted.

Some years ago it was generally assumed that the next proton magic number after 82 would be 126 since this was the neutron pattern. However, more elaborate theoretical studies indicate that the proton shell closure occurs for Z = 114 (or possibly 112). The most probable neutron magic number is 184. Taking into account also pairing effects, relativistic theory predicts additional shell closures around Z ~ 108−110 as well as N ~ 162 and possibly N ~ 172.

Experiments show evidence of enhanced nuclear stability around N ~ 162 and possibly in the region around N ~ 172. So far N = 184 has not been reached but experimental data suggest that stability increases with increasing N in the region N = 162−177, with N = 177 as the highest neutron number reached so far.

In 2012 elements up to 118 have been artificially synthesized (only the elements up to and including 112, as well as 114 and 116 are, however, so far acknowledged by IUPAC) and efforts are being made to create Z = 119 and 120 in the laboratory.

14.2.11.1. Synthesis of elements with $Z \geq 104$

The only known way to produce elements with $Z > 103$ is heavy−ion reactions, i.e. fusion of two heavy nuclei. The compound nucleus (CN) is then de-excited through neutron evaporation. The formed nuclide decays by successive alpha decays. For each step in the neutron evaporation process, there is a strong competition between spontaneous fission and alpha decay that makes these heavy systems highly fissile. The survival probability of the compound nucleus is only of the order of 10^{-2} to 10^{-3} per evaporation step. Table 14.8 lists the nuclear reactions used for the first production and identification of the very heavy elements.

The daughter products of the CN are identified by genetic correlations and simultaneous lifetime measurements. Alpha decay half-lives from some seconds and down to microseconds and rapidly decreasing cross-sections for every additional proton limits the study of their chemistry and physics to 'single atoms', which makes experimental studies extremely difficult.

Two types of reaction mechanisms are in use for synthesis of the heaviest elements. "Cold" fusion reactions are performed with lead or bismuth targets that are bombarded with appropriate projectiles. One or two neutrons are evaporated from the CN. The daughter products are typically formed via successive alpha-decays ending by spontaneous fission (SF) in the region of the known elements. They are identified via parent-daughter correlations or the chemical properties of the final decay product. Cold fusion reactions have been used to produce elements 107 to 112 at GSI in Darmstadt. An estimate of the limit of cold fusion experiments comes from the synthesis of a single atom of Element 113 at RIKEN. The production cross-section was on the order of femto-barns, which makes it unlikely that cold fusion can be used to produce even heavier elements.

"Hot" fusion reactions are performed with an actinide target and an appropriate, lighter projectile. The reaction is followed by the evaporation of 3 to 5 neutrons. Hot fusion reactions have been used successfully to produce the more neutron rich nuclides from rutherfordium to element 118. These decay chains often end in unknown regions making an unambiguous identification of the nuclides challenging. The elements 113 to 118 have been produced with "hot" fusion at FLNR in Dubna using actinide targets from uranium to californium with beams of the extremely rare doubly magic isotope ^{48}Ca.

The short half-lives and the small amounts of product did not allow chemical identification of the transactinide elements in the initial experiments. Identification by nuclear properties used to be less conclusive as it usually depends on correlation with predicted energies and half-lives obtained by extrapolation of systematic trends. The controversies resulting from the uncertainties inherent in this approach have been a feature of the history of the light transactinide elements, just as they were for nobelium and lawrencium. In fact, it is still uncertain which claims represent the actual discovery for some of these elements, the Russian group at Dubna or the American group at Berkeley. Both groups undoubtedly have produced isotopes of elements 104—106. Partially the disagreements are due to the different techniques used. At both laboratories nuclear reactions using hot fusion, were used for the first synthesis of elements 104—105, see Table 14.7. This type of reaction was used at Berkeley also for production of element 106, while the Dubna group utilized cold fusion.

During the last 15 years hot fusion reactions with ^{48}Ca as projectile have been used in Dubna to synthesize the elements 113-118. Identification has been made by genetic correlations. A problem in the identification is, however, that all decay chains end with previously unknown spontaneously fissioning nuclides of e.g. dubnium or seaborgium. Thus, there is a need for chemistry experiments to show that these end members of the decay chains follow the expected chemical behavior. To connect the isolated "Dubna island" to the known region of the chart of nuclides is therefore a challenging task. Until now, IUPAC has accepted the discovery of elements 114 and 116 because they have been confirmed in experiments at Berkeley but not 113, 115, 117 and 118.

The predictions that the half-lives of some heavy elements may be very long have led to searches for evidence of their existence in nature. Several reports of tentative evidence for such superheavy elements have appeared but subsequent experimentation has explained the observations as based on known elements with only a few exceptions.

14.3. ACTINIDE PROPERTIES

14.3.1. The actinide series

As early as 1923 N. Bohr suggested that there might exist a group of 15 elements at the end of the Periodic Table that would be analogous in their properties to the 15 lanthanide ("rare earth") elements. This idea, combined with the increasing stability of the +3 oxidation state for the transuranium elements as the atomic number increases from $Z = 93$ to 96, led Seaborg to the conclusion that these new elements constituted a second rare earth series whose initial member was actinium. As the atomic number increases from 90, electrons are added in the 5f subshell similar to the occupation of the 4f subshell in the lanthanides, see Table 14.3. This series would be terminated with element 103 since this would correspond to the addition of 14 electrons for a completed 5f subshell.

Table 14.3 Electronic configuration, radii (in picometres, pm), and oxidation states in (acid, non-complexing) aqueous solution

Atomic number	Element	Metallic radius (pm)	Atomic (g)† config.	Effective ionic radius†		Oxidation* states
				M^{3+}	M^{4+}	
89	Ac	188	$5f^0 6d7s^2$	111.9		**3**
90	Th	180	$5f^0 6d^2 7s^2$	(108)	97.2	(3) **4**
91	Pa	163	$5f^2 6d7s^2$	(105)	93.	(3) 4 **5**
92	U	156	$5f^3 6d7s^2$	104.1	91.8	3 4 5 **6**
93	Np	155	$5f^4 6d7s^2$	101.7	90.3	3 4 **5** 6 (7)
94	Pu	160	$5f^6 7s^2$	99.7	88.7	3 **4** 5 6 (7)
95	Am	174	$5f^7 7s^2$	98.2	87.8	**3** 4 5 6
96	Cm	175	$5f^7 6d7s^2$	97.0	87.1	**3** 4
97	Bk		$5f^9 7s^2$	94.9	86.0	**3** 4
98	Cf		$5f^{10} 7s^2$	93.4	85.1	(2) **3**
99	Es		$5f^{11} 7s^2$	92.5		(2) **3**
100	Fm		$5f^{12} 7s^2$			(2) **3**
101	Md		$(5f^{13} 7s^2)$	89.6		2 **3**
102	No		$(5f^{14} 7s^2)$			**2** 3
103	Lr		$(5f^{14} 6d7s^2)$	88.2		**3**

*The most stable oxidation state in (acid, non-complexing) aqueous solution is given in bold, most unstable (or only found in the solid state) within parenthesis.
†Electronic configuration and ionic radii (coordination number 8) according to Katz, Seaborg and Morss and to Seaborg and Loveland.

Seaborg's actinide hypothesis was initially a subject of considerable objection since the trivalent oxidation state, unlike in the lanthanide family, was not the most stable in aqueous solution for the elements between $Z = 90$ and 94. In aqueous solution the most stable oxidation states are +4 for thorium, +5 for protactinium, +6 for uranium, +5 for neptunium, and +4 for plutonium, see Table 14.1, right column. Only for the elements beginning with americium is the +3 the most stable state in solution. Seaborg, however, had correctly identified Ac as the precursor (analogous to La) and Cm as the midpoint element (analogous to Gd). Recent investigations have shown that mendelevium and nobelium have a divalent state in solution (which probably is the most stable for nobelium). This corresponds to the divalent state observed for ytterbium in the lanthanide elements. For $Z = 90$ to $Z = 94$ the 5f and 6d orbitals are very close in energy and the electronic occupation is variable. However, for these elements one should use Schrödinger orbitals and their energies with considerable caution due to the limited velocity of light in our universe, cf. §2.4.

The recognition of the similarity in chemical properties between the actinide and lanthanide elements was an important contributing factor in the synthesis and isolation of the transcurium elements. Most of the chemical identification was carried out by eluting the elements from columns of cation exchange resin. The pattern of the elution behavior from the resin bed of the lanthanide elements made it possible to predict with good accuracy the expected elution position for a new actinide element (Fig. 14.8). This technique constituted the most definitive chemical evidence in the discovery experiments for the elements from atomic numbers 97 through 101. More recently these conclusions have been confirmed by spectroscopy.

Orbital designations names for solutions to the Schrödinger equation will be used here, although their validity is questionable for these heavy elements. We do this because the more correct descriptions based on Dirac wave functions are seldom used in chemistry and their multiple and multidimensional shapes are difficult to visualize.

The electronic ground state of $_{89}$Ac contains four filled inner electronic shells (the main quantum numbers 1, 2, 3 and 4, designated K, L, M and N, and containing 2, 8, 18 and 32 electrons, respectively); see Ch. 6. The subsequent outer 5th, 6th, and 7th shells (designated "O", "P" and "Q", respectively) are partly empty. In the symbolism used to designate the electronic structure of an atom, the filled shells are usually omitted, as they do not contribute to the chemical properties (and interactions) of the atom. The symbolism $5s^2p^6d^{10}f^0 6s^2p^6d^1 7s^2$ for $_{89}$Ac indicates that in the 5th shell the subshells s, p, d and f contain 2, 6, 10 and 0 electrons, respectively; the 6th subshell contains 2 s-electrons, 6 p-electrons and 1 d-electrons; the outermost 7th shell contains 2 s-electrons. Remembering that the s-, p-, d- and f-subshells can accommodate 2, 6, 10 and 14 electrons respectively, these four outer shells are partly empty. Also leaving out the filled sub-shells, the electronic structure for Ac can be written $5f^0 6d^1 7s^2$; though the 2 7s-electrons could have been left out we keep them for specific purpose. The construction of the electronic configuration for gaseous atomic actinide atoms in Table 14.3 then becomes obvious. These electronic structures have been determined spectroscopically.

From Table 14.3 it is obvious that there can be interaction between the 5f, 6d and 7s orbitals, as a consequence of their very similar binding energies. As these partly empty outer orbitals are the ones which may take part in the chemical bondings we can expect the actinides to show a highly variable chemistry.

The main feature of the electronic structure, when going from $_{89}Ac$ to $_{103}Lr$ is the successive filling of the 5f subshell. This leads to a slight contraction of the atomic and ionic radii with increasing atomic number (the *actinide contraction*), as seen from several columns. However, there are irregularities, which are attributed to the extra stability of the half-filled 5f subshell (at 7 electrons), leading to more labile 6d-electrons.

14.3.2. The solid state

The actinide metals can be produced by heating the tri- or tetra-fluoride with metallic alkaline earth or alkali elements:

$$PuF_4 + 2Ca \rightarrow Pu + 2CaF_2 \tag{14.24}$$

The metals exhibit several allotropic modifications: 3 for uranium and 6 for plutonium between room temperature and the melting point (1130°C for U and 640°C for Pu) (Fig. 14.10). The density of the actinide metals at room temperature shows an unusual variation: Ac ~10 000, Th 11 800; Pa 15 400; U 19 100; Np 20 500; Pu 19 900; Am 13 700; Cm 13 500; Bk 14 800 kg m^{-3}.

All the metals are very electropositive and attacked by water vapor with production of hydrogen. They are slowly oxidized in air and at higher temperatures; in the form of small chips, they are pyrophoric. The hydrides, oxides, nitrides, and halides are produced most easily by heating the metals in the appropriate elemental gas. The fluorides are among the most important solid actinide compounds since they are the starting material for the production of the metals. The volatile hexafluoride of uranium is used in isotopic enrichment (Ch. 3). The preparation and properties of Th and U are described in §§14.2.2.2 , 14.2.4.2, and 14.2.4.3. Actinide chemistry is also discussed in §§21.5 and 22.6.

14.3.3. Actinide oxidation states

The electrons in the 6d and 7s subshells are more loosely bound than the electrons in the filled subshells, and, in general, also than the 5f electrons. In these outer shells the binding energies are in the range of a few eV, i.e. the same order of magnitude as is common in chemical bonding. Thus, from Table 14.3, it is understandable that Ac easily looses its $6d7s^2$ electrons to form Ac^{3+}, and Th its $6d^27s^2$ electrons to form Th^{4+}. For the subsequent elements, from Pa to Am, the situation is more complicated. There are reasons to suppose that the spatial characteristics of the f-subshell orbitals may change abruptly at certain atomic numbers; that is, the f-shell electrons may be shielded more strongly in some elements than in others where the f-orbitals extend close to the surface of the electronic cloud (where chemical interaction occurs), and where the 5f-electrons are in closer contact with the d- and s-shell electrons. There can be little doubt that the 5f-electrons are present in all of the actinide elements after Pa.

The chemical properties of the actinide elements have been intensely studied for the elements available in at least microgram quantities (Z 90 - 99) but much less so for the heaviest members of the family (Z 100 - 103). Table 14.3 lists the valency states. In Figure 14.7 redox diagrams are given for the most important actinides; for comparison, standard potentials are included for some useful redox reagents. Any particular actinide can be obtained in a desired valence state by the use of proper oxidizing or reducing agents (Table 14.4).

The pentavalent state of the actinides (except for Pa and Np) is less stable than the other states and normally undergoes disproportionation in acid solutions. Plutonium is particularly interesting in the variety of oxidation states that can coexist in aqueous solutions (Fig. 14.7, at 1000 mV). For example, a plutonium solution in 0.5 M HCl of 3×10^{-4} M Pu concentration at 25°C, which is initially 50% Pu(IV) and 50% Pu(VI), will equilibrate within a few days via disproportionation reactions to an equilibrium system that is 75% Pu(VI), 20% Pu(IV), and few percent each of Pu(V) and Pu(III), in the absence of complexing anions.

The reactions are:

$$2PuO_2{}^{+} + 4H^{+} \rightleftarrows Pu^{4+} + PuO_2{}^{2+} + 2H_2O \quad (14.25)$$

and

$$PuO_2{}^{+} + Pu^{4+} \rightleftarrows PuO_2{}^{2+} + Pu^{3+} \quad (14.26)$$

Although Pu(V) readily disproportionates at concentrations $\geq 10^{-8}$ M in acidic solutions, it is the state observed in more basic natural and ocean waters, partly due to the stability of the bicarbonate complex, cf. Ch. 22.

14.3.4. Actinide complexes

Since the differences in energy of the electronic levels are similar to chemical bond energies, the most stable oxidation states of the actinides may change from one chemical compound to another, and the solution chemistry will be sensitive to the ligands present. Thus complex formation becomes an important feature of the actinide chemistry (cf. §18.4.3 and §22.7).

The chemical properties are different for the different valency states (Table 14.5), while in the same valency state the actinides closely resemble each other. These

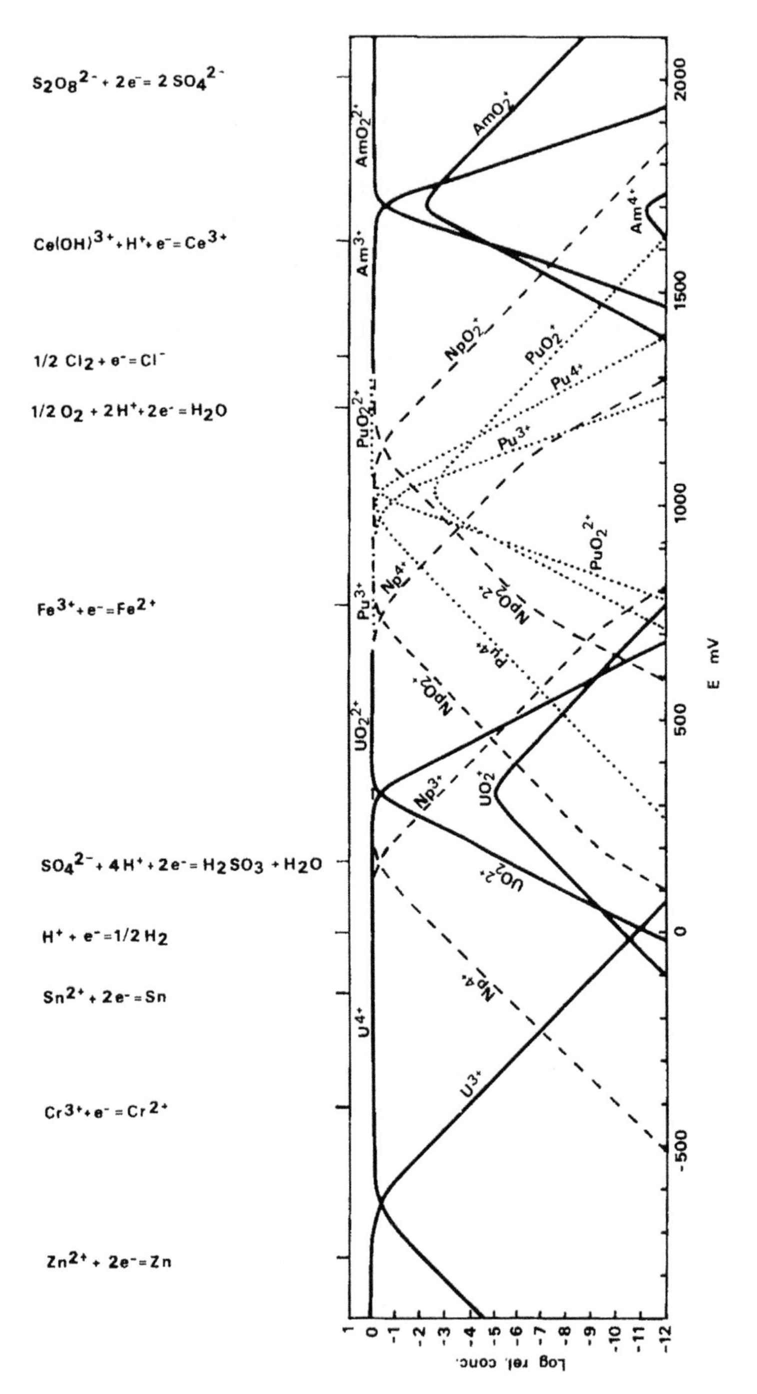

Figure 14.7 Redox diagrams for U, Np, Pu, and Am in 1 M $HClO_4$ at 25 C. Each actinide is assumed to be in equilibrium in its own solution at the indicated redox potential.

Table 14.4 Preparation methods and stability of actinide ions in aqueous solutions

Ion	Stability and method of preparation
U^{3+}	Slow oxidation by water, rapid oxidation by air, to U^{4+}. Prepared by electrolytic reduction (Hg cathode).
Np^{3+}	Stable to water, rapid oxidation by air to Np^{4+}. Prepared by electrolytic reduction (Hg cathode).
Pu^{3+}	Stable to water and air. Oxidizes by action of its own -radiation to Pu^{4+}. Prepared by reduction by SO_2, Zn, U^{4+} or H_2(g) with Pt catalyst.
Md^{3+}	Stable. Can be reduced to Md^{2+}.
No^{3+}	Unstable, reduces to No^{2+}.
Th^{4+}	Stable.
Pa^{4+}	Stable to water. Rapid oxidation by air to Pa(V). Prepared by electrolytic reduction (Hg cathode) and by the action of Zn(Hg), Cr^{2+} or Ti^{3+} in HCl.
U^{4+}	Stable to water. Slow oxidation by air to UO_2^{2+}. Oxidation in nitrate media catalyzed by UV light. Prepared by oxidation of U^{3+} by air, by electrolytic reduction of UO_2^{2+} (Hg cathode) and by reduction of UO_2^{2+} by Zn or H_2(g) with Ni catalyst.
Np^{4+}	Stable to water. Slow oxidation by air to NpO_2^{+}. Prepared by oxidation of Np^{3+} by air or reduction of higher oxidation states by Fe^{2+}, SO_2, I^- or H_2(g) with Pt catalyst.
Pu^{4+}	Stable in concentrated acids, e.g. 6 M HNO_3. Disproportionates to Pu^{3+} and PuO_2^{+} at lower acidities. Prepared by oxidation of Pu^{3+} by BrO_3^-, Ce^{4+}, $Cr_2O_7^{2-}$, HIO_3 or MnO_4^- in acid solution or by reduction of higher oxidation states by HNO_2, NH_3OH^+, I^-, 3 M HI, 3M HNO_3, Fe^{2+}, $C_2O_4^{2-}$ or HCOOH in acid solution.
Am^{4+}	Not stable in water.
Bk^{4+}	Stable to water. Slow reduction to Bk^{3+}. Prepared by oxidation of Bk^{3+} by $Cr_2O_7^{2-}$ or BrO_3^-.
PaO^{3+} or PaO_2^{+}	Stable. Difficult to reduce.
UO_2^{+}	Disproportionates to U^{4+} and UO_2^{2+}. Most stable at pH 2.5. Prepared by electrolytic reduction of UO_2^{2+} (Hg cathode) and by reduction of UO_2^{2+} by Zn(Hg) or H_2(g). pH around 2.5 used.
NpO_2^{+}	Stable. Disproportionates only at high acidities. Prepared by oxidation of lower oxidation states by Cl_2 or ClO_4^- and by reduction of higher oxidation states by NH_2NH_2, NH_2OH, HNO_2, H_2O_2/HNO_3, Sn^{2+} or SO_2.
PuO_2^{+}	Disproportionates to Pu^{4+} and PuO_2^{2+}. Most stable at low acidities. Prepared by reduction of PuO_2^{2+} by I^- or SO_2 at pH 2.
UO_2^{2+}	Stable. Difficult to reduce.
NpO_2^{2+}	Stable. Easy to reduce. Prepared from lower oxidation states by oxidation by Ce^{4+}, MnO_4^-, Ag^{2+}, Cl_2 or BrO_3^-.
PuO_2^{2+}	Stable. Fairly easy to reduce. Reduces slowly under the action of its own -radiation. Prepared by oxidation of lower oxidation states by BiO_3^-, HOCl or Ag^{2+}.
AmO_2^{2+}	Stable. Reduces fairly rapidly under action of its own α-radiation. Prepared by electrolytic oxidation (Pt anode) in 5 M H_3PO_4, by $S_2O_8^{2-}$ in presence of Ag^+ or by Ag^{2+} in K_2CO_3 solution.

Table 14.5 Characteristic reactions of actinide ions of different valency states with some important anions.†

Reagent	Conditions	Precipitated ions	Not precipitated
OH^-	pH ≥ 5	M^{3+}, M^{4+}, MO_2^+, MO_2^{2+}	
F^-	4 M H^+	M^{3+}, M^{4+}	MO_2^+, MO_2^{2+}
IO_3^-	0.1 M H^+	M^{4+} (M^{3+} may oxidize)	MO_2^+, MO_2^{2+}
PO_4^{3-}	0.1 M H^+	M^{4+} (Ac^{3+} partly)	M^{3+} (Pu^{3+} and higher An)
CO_3^{2-}	pH > 10	M^{3+}, M^{4+} (as hydroxide)	MO_2^{2+} (anionic complex)
CH_3COO^-	0.1 M H^+	MO_2^{2+}	M^{3+}, M^{4+}, MO_2^+
$C_2O_4^{2-}$	1 M H^+	M^{3+}, M^{4+}	MO_2^+, MO_2^{2+}

†Cl^-, NO_3^-, and SO_4^{2-} do not precipitate actinide ions. An = actinide.

properties have been extensively exploited for the separation and isolation of the individual elements in pure form; see Ch. 21.

The compounds formed are normally quite ionic. The ionic radii of the actinide elements of the different valency states decrease with increasing atomic number (the actinide contraction, Table 14.3). Consequently the charge density of the actinide ions increases with increasing atomic number and, therefore, the probability of formation of complexes and of hydrolysis increases with atomic number. This is illustrated in Figure 14.8, where the heavier actinides are eluted before the lighter ones because the α-hydroxy-isobutyrate eluant forms stronger complexes as the cation radius decreases.

The pattern of stabilities of complexes in the tetravalent states and hexavalent states with hard ligands is the same and follows the order of decreasing ionic radius, Table 14.3.

$$Th^{4+} < U^{4+} < Np^{4+} < Pu^{4+} \tag{14.27}$$

$$UO_2{}^{2+} < NpO_2{}^{2+} < PuO_2{}^{2+} \tag{14.28}$$

This also explains the extraction behavior (cf. §18.4.3) for the M^{4+} (actinide) ions in Figure 14.9, where Pu^{4+} is better extracted than Th^{4+}; the Figure shows the extraction of some actinide ions from HCl or HNO_3 solutions by the different reagents. In the case of tributyl phosphate ,TBP, dissolved in kerosene, the extracted M(VI), M(IV) and M(III) species are $MO_2(NO_3)_2(TBP)_2$, $M(NO_3)_4(TBP)_2$, and $M(NO_3)_3(TBP)_3$, respectively.

For the same element, the stability of the complexes varies with the oxidation state in the series

$$M^{4+} \geq MO_2{}^{2+} > M^{3+} \geq MO_2{}^{+} \tag{14.29}$$

The reversal between M^{3+} and MO_2^{2+} reflects that the hexavalent metal atom in the linear $[OMO]^{2+}$ is only partially shielded by the two oxygen atoms; thus the metal ion MO_2^{2+} has a higher charge density than M^{3+} (i.e. about 3.2 ± 0.1). Similarly in MO_2^+, the effective charge is ca. 2.2 ± 0.1. This gives a reasonable explanation of the extraction pattern of Figure 14.9, though other factors (molar volume of the complex etc.)

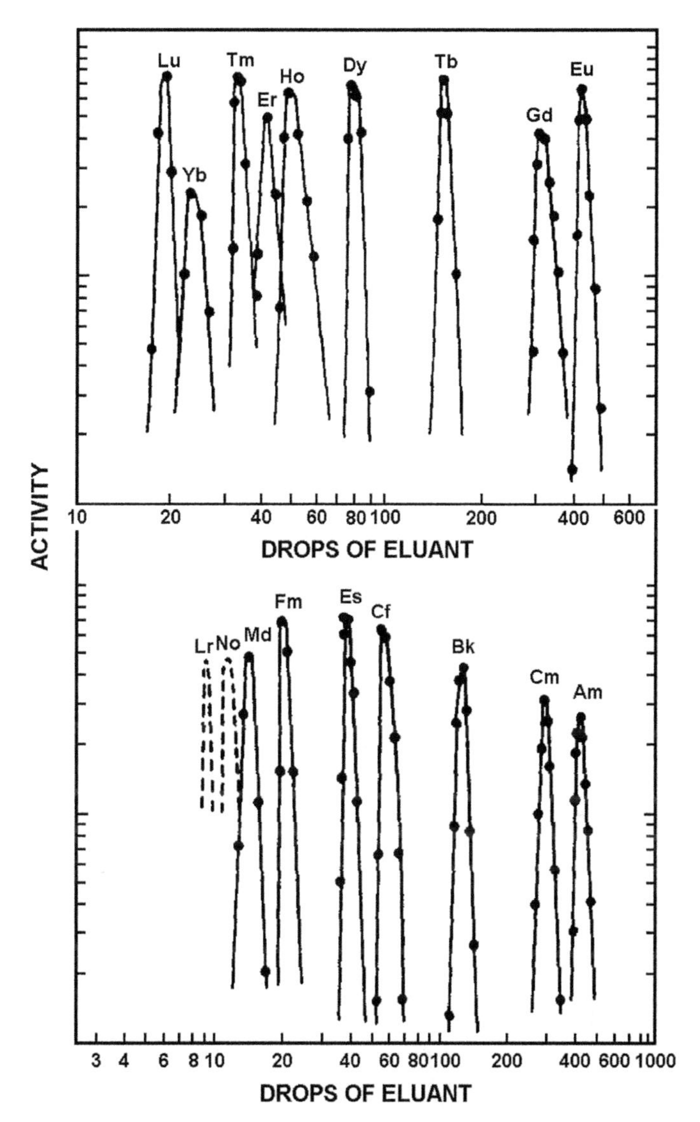

Figure 14.8 Elution curves for 3+ lanthanide and actinide ions from Dowex-50 ion exchange resin with ammonium-α-hydroxy isobutyrate eluant. *(From Katz and Seaborg).*

also contributes. Of importance in reprocessing (§21.6.3) are the low distribution ratios ($D_M \leq 0.01$) of fission products like Cs, Sr, Ru and Zr in the systems in Figure 14.9 (A) and (B).

The extraction of trivalent actinides in general follows the sequence

$$Ac^{3+} < Am^{3+} < Cm^{3+} etc \quad (14.30)$$

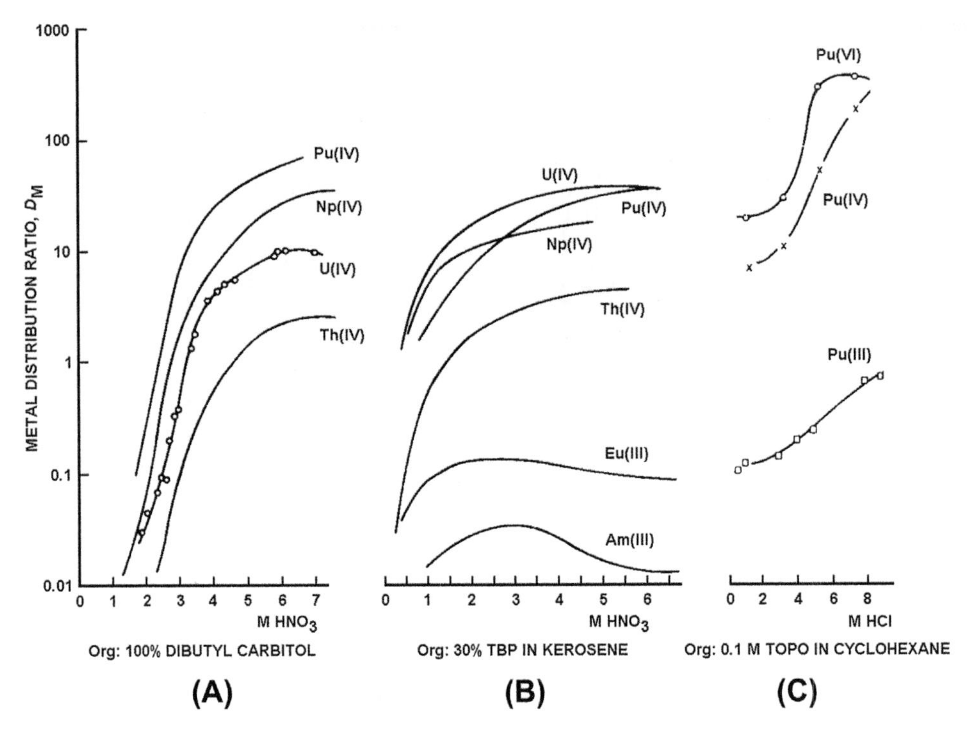

Figure 14.9 Extraction of actinide complexes into various solvents. *(From Ahrland, Liljenzin and Rydberg).*

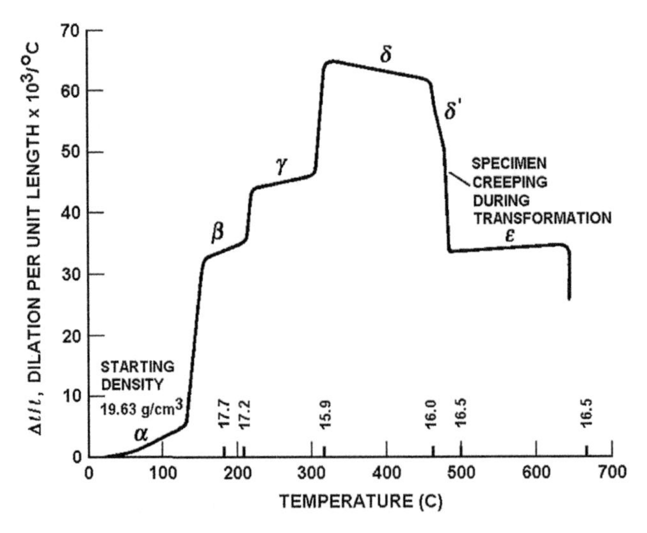

Figure 14.10 Dilation curve and densities of high purity plutonium. *(From Waldron, Garstone, Lee, Mardon, Marples, Poole and Williamson).*

(for example, Cm^{3+} is extracted easier, i.e. at a lower pH, than Am^{3+}, etc.), as the ionic radii decrease in that order.

The chemistry involved in the isolation and purification of the actinide elements from irradiated reactor fuel elements is further discussed in Chapter 21. Actinide chemistry in the ecosphere is discussed in §22.6.

14.4. USES OF ACTINIDES

The use of the actinide elements fall into three categories: (i) for understanding fundamental chemistry and the nature of the periodic system, (ii) as fuels and products in the large scale use of nuclear energy, and (iii) miscellaneous applications, where the particular physical, chemical or nuclear properties are valuable. Only the last aspect is discussed here, the others are treated elsewhere in this book. The availability of transuranium element isotopes suitable for experiments is listed in Table 14.6.

Table 14.6 Availability of transuranium element materials

Nuclide	$t_{1/2}$	Decay mode	Specific amounts† available	Specific activity (Bq/g)
^{237}Np	2.14×10^6 y	α, SF(10^{-10} %)	kg	2.61×10^7
^{238}Pu	87.7 y	α, SF(10^{-7} %)	kg	6.33×10^{11}
^{239}Pu	2.41×10^4 y	α, SF(10^{-4} %)	kg	2.30×10^9
^{240}Pu	6.56×10^3 y	α, SF(10^{-6} %)	kg	8.40×10^{10}
^{241}Pu	14.4 y	β, α(10^{-3} %)	1–10 g	3.82×10^{12}
^{242}Pu	3.76×10^5 y	α, SF(10^{-3} %)	100 g	1.46×10^8
^{244}Pu	8.00×10^7 y	α, SF(10^{-1} %)	10–100 mg	6.52×10^5
^{241}Am	433 y	α , SF(10^{-10} %)	kg	1.27×10^{11}
^{243}Am	7.38×10^3 y	α, SF(10^{-8} %)	10–100 g	7.33×10^9
^{242}Cm	162.9 d	α, SF(10^{-5} %)	100 g	1.23×10^{14}
^{243}Cm	28.5 y	α,ε (0.2 %)	10–100 mg	1.92×10^{12}
^{244}Cm	18.1 y	α, SF(10^{-4} %)	10–100 g	3.00×10^{12}
^{248}Cm	3.40×10^5 y	α, SF(8.3 %)	10–100 mg	1.57×10^8
^{249}Bk	320 d	β,α(10^{-3} %), SF(10^{-8} %)	10–50 mg	6.00×10^{10}
^{249}Cf	350.6 y	α, SF(10^{-7} %)	1–10 mg	1.52×10^{11}
^{250}Cf	13.1 y	α, SF(0.08 %)	10 mg	4.00×10^{12}
^{252}Cf	2.6 y	α, SF(3.1 %)	10–1000 mg	2.00×10^{13}
^{254}Cf	60.5 d	SF, α (0.3 %)	μg	3.17×10^{14}
^{253}Es	20.4 d	α, SF(10^{-5} %)	1–10 mg	9.33×10^{14}
^{254}Es	276 d	α	1–5 μg	6.83×10^{13}
^{257}Fm	100.5 d	α, SF(0.2 %)	1 pg	1.83×10^{14}

†According to Oak Ridge Nat. Lab., USA, late 1990:ies.

Table 14.7 New and historic names of the transactinide elements

Atomic number	IUPAC name	Symbol	Old IUPAC name	Old suggested name
104	rutherfordium	Rf	Db	Rf
105	dubnium	Db	Joliotium (Jl)	Hahnium (Ha)
106	seaborgium	Sg	Rf	Sg
107	bohrium	Bh	Bh	Nielsbohrium (Ns)
108	hassium	Ha	Hahnium (Hn)	Hs
109	meitnerium	Mt	Mt	Mt
110	darmstadtium	Ds		
111	roetgenium	Rg		
112	copernicium	Cn		
113				
114	flerovium	Fl		
115				
116	livermorium	Lv		

Table 14.8 Nuclear reactions used for the first synthesis of elements 104–116

Element	Discovery (year)	Discovery reaction
104	1969	$^{249}_{98}Cf + ^{12}_{6}C \rightarrow ^{257}_{104}Rf + 4n$
105	1970	$^{249}_{98}Cf + ^{15}_{7}N \rightarrow ^{260}_{105}Db + 4n$
106	1974	$^{249}_{98}Cf + ^{18}_{8}O \rightarrow ^{263}_{106}Sg + 4n$
107	1981	$^{209}_{83}Bi + ^{54}_{24}Cr \rightarrow ^{262}_{107}Bh + n$
108	1984	$^{208}_{82}Pb + ^{58}_{26}Fe \rightarrow ^{265}_{108}Hs + n$
109	1982	$^{209}_{83}Bi + ^{58}_{26}Fe \rightarrow ^{266}_{109}Mt + n$
110	1994	$^{208}_{82}Pb + ^{62}_{28}Ni \rightarrow ^{269}_{110}Ds + n$
111	1994	$^{209}_{83}Bi + ^{64}_{28}Ni \rightarrow ^{272}_{111}Rg + n$
112	2001	$^{208}_{82}Pb + ^{70}_{30}Zn \rightarrow ^{277}_{112}Cn + n$
113	2004	$^{243}_{95}Am + ^{48}_{20}Ca \rightarrow ^{288}_{115}Uup + 3n \rightarrow ^{284}_{113}Uut + \alpha$
114	2002	$^{244}_{94}Pu + ^{48}_{20}Ca \rightarrow ^{289}_{114}Fl + 3n$
115	2004	$^{243}_{95}Am + ^{48}_{20}Ca \rightarrow ^{288}_{115}Uup + 3n$
116	1961	$^{248}_{96}Cm + ^{48}_{20}Ca \rightarrow ^{293}_{116}Lv + 3n$

The spontaneous decay of the transuranium elements by α-decay and/or fission results in energy release. Since also very small amounts of some nuclides (e.g. ^{238}Pu, ^{244}Cm, ^{252}Cf) can be sources of appreciable energy, these radionuclides can be used in *small power generators*. This use of actinide elements is discussed in §7.9.3.

^{241}Am emits a 60 keV γ-ray and has been used as a γ-radiation source to measure thickness of metal sheets and of deposited metal coatings, the degree of soil compaction, sediment concentration in flowing streams, and to induce X-ray fluorescence in chemical analysis (§7.8). As an α-particle emitter, ^{241}Am has been mixed with beryllium to make intense neutron sources for oil well logging and for measuring water content in soils and in process streams in industrial plants. It is extensively used for elimination of static electricity and in smoke detectors where its use depends on the ionization of air.

^{252}Cf ($t_{½}$ 2.73 y) has found several uses. It decays to 3.1% by spontaneous fission (the main decay mode is through α emission) leading to a neutron emission rate of 2.3×10^{15} n s^{-1} kg^{-1} (average n-energy 2.35 MeV). The n-dose rate is 22 kSv h^{-1} kg^{-1}. ^{252}Cf is the only nuclide that can provide a useful neutron intensity over a sufficiently long half-life to make it a useful single nuclide neutron source. The low rates of heat emission (38.5 kW/kg), γ-radiation (initially 1.3×10^{16} photons s^{-1} kg^{-1}, yielding an initial dose rate of 1.6 kGy h^{-1} kg^{-1}; γ-ray intensity increases with time due to fission product build-up) and helium evolution (from α decay) allow fabrication of simple, small ^{252}Cf sources that require no external power supply nor any maintenance but that can provide moderately high neutron fluxes. The 2003 production rate of ^{252}Cf from the Oak Ridge facilities (HFIR-TRU) was ~ 250 mg/y and from russian Dimitrograd ~25 mg/y. Among the applications of ^{252}Cf we list the following:

(a) process control by a variety of on-stream non-destructive analytical techniques;
(b) medical diagnosis by activation analysis;
(c) production of short-lived radioisotopes at locations where they will be used, thus avoiding decay during transportation from an accelerator or reactor at another site;
(d) industrial neutron radiography, which images low density materials – especially hydrogenous materials – better than X-rays;
(e) possible medical treatment of tumors with ^{252}Cf sources that can be implanted in the body;
(f) petroleum and mineral exploration in which the compactness and portability of ^{252}Cf neutron sources facilitate testing for valuable deposits, particularly in inaccessible places such as deep wells and the sea floor;
(g) moisture measurements;
(h) hydrology studies to locate sources of water;
(i) nuclear safeguards tests, e.g. for criticality control in reactor fuel storage areas and for nuclear materials accountability (detection and recovery of fissionable material; implementation of nuclear agreements).

14.5. TRANSACTINIDE PROPERTIES

The transactinide elements (sometimes also called Super Heavy Elements) have only been produced a few atoms at a time, When the number of atoms observed in an experiment is very small the usual fundamental thermodynamic descriptions no longer hold. Thus, the phase concept disappears when the medium contains only one or a few radioactive atoms. The law of mass action is also not applicable. In solvent extraction the distribution ratio between two phases loses its significance: a single atom can only exist in one of the two phases at the moment of separation. An average distribution can, however, be obtained by performing a large number of identical experiments. The number of decay events in each phase is recorded. The decay statistics are replaced by the statistical distribution of experimental results.

It is only possible to make compounds that contain one of the rare atoms. Thus, it is possible to make $Rf(NO_3)_4$ but it will probably never be possible to synthesize a compound like Rf_3N_4. Disproportionation is impossible since it involves exchange of electrons between two atoms of the same element. Thus, studies of the heaviest elements involve a lot of statistical considerations.

14.5.1. Chemistry of the transactinide elements

Studies of chemical properties of the transactinide elements are hampered by the short half-lives of most of the isotopes and the few atoms produced. The production cross sections are very small (0.01–0.5 nb) so chemical studies must be done very rapidly on one atom at a time. The detection systems must allow alpha-alpha correlation analysis. However, these elements offer a possibility to investigate the influence of relativistic effects of the electrons which could alter the relative stability of the 7s, 6d and 7p valence electrons. The result would be the existence of the elements in oxidation states other than those predicted from their expected position in the periodic table, see App. I.

Rutherfordium has been studied in the gas phase as well as in aqueous solution. The experiments confirm the tetravalent nature of Rf. Moreover Rf resembles Zr(IV) and Hf(IV) more closely than Th(IV), as expected for a transition element.

Chemical studies in aqueous solution and in the gas phase of dubnium indicated its similarity to the Group 5 transition elements niobium and tantalum, for which +5 is a stable oxidation state, hence a pentavalent nature of element 105 can be assumed. It resembles Nb(V) more than Ta(V). There are also some similarities between Db(V) and Pa(V), which are not expected from a simple extrapolation of Group 5 properties to $Z = 105$.

The limited aqueous chemistry data available on seaborgium imply that Sg has a stable hexavalent oxidation state and that Sg(VI) resembles Mo(VI) and W(VI). This is also indicated by gas phase chemistry data.

The bohrium compound BhO_3Cl has been synthesized in the gas phase using the isotope ^{267}Bh. This compound shows that Bh behaves like its lighter homolog Re, i.e. it belongs to group 7.

The first hassium compound to be synthesized in a gas phase experiment was HsO_4. This places Hs in group 8, below Ru and Os. To verify the group assignment sodium hassate, $Na_2HsO_4(OH)_2$ was successfully synthesized in another experiment.

No chemical studies have so far been performed on meitnerium, darmstadtium or roentgenium.

The chemical behaviour of copernicium has been studied in gas phase experiments where Cn has been allowed to adsorb on surfaces in a system with alpha detectors at different temperature. The temperature gradient goes from room temperature to the temperature of liquid nitrogen (−196 °C). The results imply that Cn is very volatile but it still has the adsorption characteristics of a metal. It is therefore assumed that it belongs to group 12.

No chemical studies have so far been performed on element 113.

The chemical behaviour of element 114 has been studied in gas phase experiments where 114 has been allowed to adsorb in a system with alpha detectors at different temperature. The temperature gradient goes from room temperature to the temperature of liquid nitrogen (−196 °C). The results imply that 114 is very volatile (even more volatile than Cn) but it still has the adsorption characteristics of a metal. It is therefore assumed that it belongs to group 14. Cn and 114 can both be characterized as "gaseous metals". Cn and 114 are the first elements where relativistic effects significantly alter the chemical properties.

No chemical studies have so far been performed on elements 115, 116, 117 and 118.

It should be emphasized that many of the results in this region are based on very few decay events, in some cases only two atoms, due to the low production rates and short half-lives. This gives large uncertainties and there is room for surprises in future experiments. The only trtansactinide elements where a significant amount of data has been collected with different methods in independent experiments are Rf and Db.

14.6. THE END OF THE PERIODIC TABLE

A question that arises in heavy element research is: how far can we reach? Is there a limit on the number of elements that cannot be exceeded? The answer is that theoretically it is possible to find quite a number of new elements above $Z = 118$ but in practice the number of elements will be limited by experimental difficulties. To reach far beyond $Z = 125$ and approach $N = 184$ will require even more neutron-rich actinide targets and more neutron-rich projectiles, e.g. short-lived fission products. The limit of hot fusion may thus be reached short of arriving at the proposed magic region around $N = 184$.

14.7. EXERCISES

14.1. What nuclear reactions would be suitable to make gold?

14.2. What fraction of neptunium is in the +4 state in a 0.1 M $Fe(SO_4)_2$ solution of acidity 1 M H^+, assuming that Np initially is present as Np(V) at a total concentration of 9.4 mM?

14.3. Irradiation of ^{238}U with deuterium yields ^{238}Pu. Will any other plutonium isotope be produced?

14.4. What are the decay products of ^{252}Cf, of ^{253}Cf and of ^{254}Cf?

14.5. ^{244}Pu decays through spontaneous fission with a half-life of 6.6×10^{10} y. Estimate the number of neutrons emitted per fission if the measured n-emission rate is 1890 n s^{-1} g^{-1}.

14.6. Flerov bombarded ^{207}Pb with ^{54}Cr and obtained a product which within 4-10 ms decayed by spontaneous fission. Suggest a product nucleus.

14.7. In Figure 14.9, Eu(III) is more easily extracted than its homolog Am(III). Suggest an explanation.

14.8. A cardiac pacemaker contained 150 mg of ^{238}Pu. What is its heat output? Use data in the isotope chart.

14.9. What electronic configuration would you ascribe to nobelium considering its place in the periodic system and its chemistry?

14.10. Based on the redox diagrams in Figure 14.8, estimate the reduction potentials for the one-electron step in the following series of plutonium species in acid solution: Pu(III), Pu(IV), Pu(V), and Pu(VI).

14.8. LITERATURE

E. K. HYDE and G. T. SEABORG, *The Transuranium Elements*, Handbuch der Physik, Band XLII, Springer, 1957.

J.M. CLEVELAND, *The Chemistry of Plutonium*, Gordon & Breach, 1970.

D. C. HOFFMAN, F. O. LAWRENCE, J. L. MEWHERTER, and F. M. ROURKE, Detection of Plutonium-244 in nature, *Nature* **234** (1971) 132.

Gmelin Handbuch, Ergänzungswerk zur 8. Auflage: Vol. 4, *Transuranium Elements, Part C, The Compounds* (1972); Vol. 7a, *Transuranium Elements, Part A1, II, The Elements* (1974); Vol. 7b, *Transuranium Elements, Part A2, The Elements* (1973); Vol. 20, *Transuranium Elements, Part D1, Solution Chemistry* (1975).

A.J. FREEMAN and J. B. DARBY, JR. (Eds.), *The Actinides: Electronic Structure and Related Properties*, Academic Press, 1974.

S. AHRLAND, J. O. LILJENZIN and J. RYDBERG, in *The Chemistry of the Actinides*, Pergamon Press, 1975.

W. MULLER and R. LINDNER (Eds.), *Transplutonium 1975*, North-Holland/Elsevier, 1976.

W. MULLER and H. BLANK (Eds.), *Heavy Element Properties*, North-Holland/Elsevier, 1976.

H.BLANK and R. LINDNER (Eds.), *Plutonium 1975 and Other Actinides*, North-Holland/Elsevier, 1976.

N. M. EDELSTEIN (Ed.) *Actinides in Perspective*, Pergamon Press, 1982.

W. T. CARNALL and G. R. CHOPPIN (Eds.) Plutonium Chemistry, *ACS Symposium Series 216*, Am. Chem. Soc., 1983.

A. J. FREEMAN and C. KELLER (Eds.) *Handbook of the Physics and Chemistry of the Actinide Elements*, Vol. 1-5. North-Holland, 1984-91.

J. J. KATZ, G.T. SEABORG and L. R. MORSE (Eds.), *The Chemistry of the Actinide Elements*, 2nd ed., Chapman and Hall, London 1986 (2 volumes).

G.T. SEABORG, Recent Research on Transuranium Elements, *J. Nucl. Materials*, **166** (1989) 22-35.

K. KUMAR, *Superheavy Elements*, Adam Hilger, 1989.

G. T. SEABORG and W. D. LOVELAND, *The Elements beyond Uranium*, J. Wiley, 1990.

Radiation Biology and Radiation Protection

Contents

Radiochemistry and Nuclear Chemistry
ISBN 978-0-12-405897-2, http://dx.doi.org/10.1016/B978-0-12-405897-2.00015-X

The biological effects of ionizing radiation were discovered soon after the production of intense radiation sources in the form of X-ray machines and radioactive elements about a century ago. Radiation was found to stop cell division and could therefore be used therapeutically to stop cancer growth. On the other hand, radiation applied locally was found to cause wounds, which were difficult to heal, and to induce cancer. Many serious accidents occurred as a result of the use of radiation before an adequate understanding of its biological effects led to formulation of rules for protection of workers. By 1922 approximately 100 radiologists (not patients) had died as a result of biological radiation damage.

The biological effects of *very large whole-body doses* are radiation sickness and early death, while *large organ doses* lead to local cell destruction and, possibly, organ death. The effects at *lower doses* are cell changes (decreased surviving fraction, decreased rate of division, chromosomal abberations, etc), which usually can be observed by microscope soon after irradiation. The induction of cancer may take years to observe and possible genetic changes may not be discovered until after several generations.

The creation of our world occurred in intense radiation fields. Consequently, we have inherited an Earth drenched in radiation from cosmic sources and the minerals in the ground (Ch. 12 and 13). Life evolved to survive in this normal "eco radiation". Essentially the current natural background levels are the lowest ever experienced by humankind. Though the intensity of these radiation sources is much smaller than produced by human techniques, no human can avoid these natural sources. In some cases the naturally occurring radioactive isotopes have been slightly enriched and taken out of their natural existence, e.g. use of uranium rich material in the building of houses resulting in a too radon rich environment inside the house. These increased levels have become important health issues. Closely related to this problem are the effects of man-made sources of similarly low levels, such as the storage of nuclear waste. Much research is presently devoted to the effects of low-level radiation.

Through extensive research in this field, our knowledge has grown enormously. Initially, the effects of radiation on local organs were studied; then the effects on various specific tissues and cell types were of concern; today, the focus is on the effects at the molecular level. This development is understandable as cancer induction and genetic changes are believed to have their roots in alterations in the DNA molecules in the cells. Table 15.1 lists a number of factors which affect the actual extent of the biological damage caused by radiation; most of these factors are discussed in this chapter. Many groups are actively engaged in research on the biological effects of ionizing radiation: *radiologists* use radiation for diagnosis and treatment of tumors; *health physicists* have the responsibility of controlling the use of irradiation equipment and protecting people from unnecessary exposure to radiation; in collaboration with *oncologists* (tumor researchers) and *geneticists*, *radiobiologists* conduct research to explain the effects of radiation on the cellular and molecular level, *radiation chemists* are interested in the interaction between radiation and the DNA molecule.

Table 15.1 Parameters in biological effects of radiation

Physical parameters
Kind of radiation: α,β, or γ
Internal or external source
Size of total dose
Dose rate
Exposure time: instantaneous, temporary, recurrent, chronic
Microscopic effects
molecular: DNA changes
chromosome: abberations
cellular (single or a few cells, *in vitro* or *in vivo*):
kind of cells (nervous system, bone marrow, liver, etc)
cell cycle (stage, rate of cell division, inactivation, etc)
temperature
oxygen content
cell poisons present (increasing radiation sensitivity)
protective agents present (anti-oxidants)
tissue and organ: cancer growth or death
Macroscopic effects (inactivation, lethal dose (LD), cancer, etc.)
Somatic: rapid (within a month), delayed (up to 20 years)
Genetic: observed in offspring (one or several generations later)

15.1. THE BIOLOGICAL TARGET

The discussion of the effects of radiation on biological systems requires some familiarity with the target composition and common terminology.

The human body contains some 10^{14} cells. Figure 15.1.A shows schematically the cell and its nucleus, which contains thin thread-like *DNA-molecules* (*deoxyribonucleic acid*). The DNA carries the "genetic code" and is the most important molecule of an organism. It consists of two strands of sugar-phosphate chains, attached together by *base pairs* (forming so-called *nucleotides*), in a double helix form (Fig. 15.1.H–J). The human DNA contains 2.9×10^9 nucleotides; the DNA of simpler organisms has fewer nucleotides (down to about 5000). The nucleotides are combined in triplets (called *codons*), each one with the ability to produce (through some intermediate steps) a certain protein. The codons are ordered in long groups; the ordering is referred to as the *genetic code.* Because these long groups carry the information necessary to produce proteins for the different tissues, they are called *genes* ("makers"). In humans, the DNA is distributed over 23 pairs (altogether 46) *chromosomes*. In the chromosome, the DNA is wound around *histone* protein cores (Fig. 15.1.G) and highly twined ("condensed") to facilitate cell division; the DNA+histone unit (containing some 200 base-pairs) is referred to as the *chromatin*; these repeating units are known as *nucleosomes*. Figure 15.1.D shows some of the 23 chromosome pairs, and Figure 15.1.E a single one. The total amount of genetic

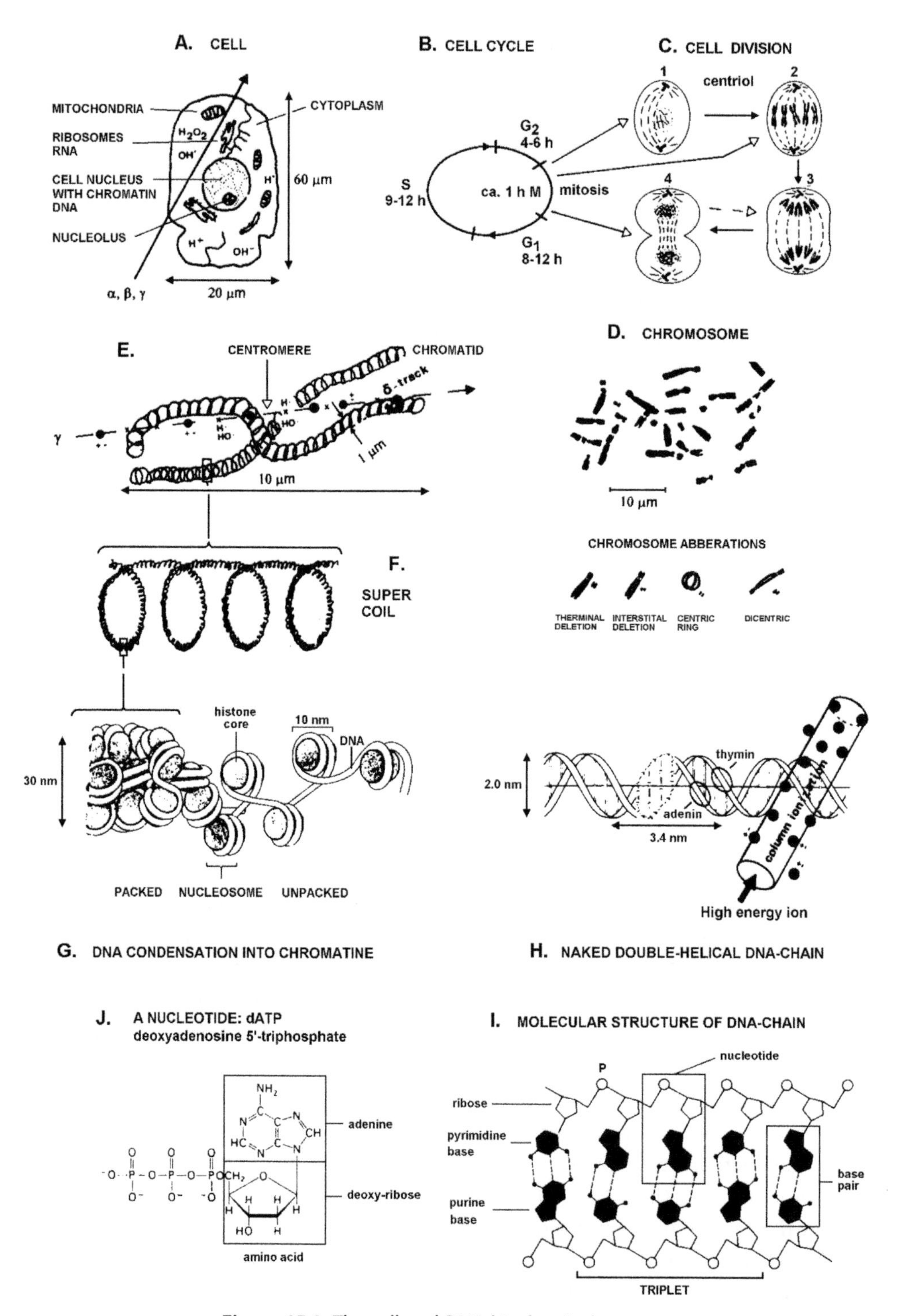

Figure 15.1 The cell and DNA biochemical system.

information in the cell is called the *genom*. Stretched out the total genetic material is about 1.5 m long, but its diameter is only 2 nm; the molecular weight is about 10^{11} per chromosome. While the cells comprise most of the body, the volume of the chromosomes only occupy ~1% of the cell volume and only about 1/10 of the chromosome volume carries a genetic message; thus the genetically significant target is rather small. Thus the probability that radiation should directly "hit" the DNA is small compared to the probability that radiolytically produced radicals will react chemically with the chromosomes.

The nucleotides are held together by hydrogen bonds between the nitrogen bases, which have their nitrogen rings perpendicular to the plane of Figure 15.1.I (i.e. turned 90° compared to the paper surface). The nitrogen rings contain some π-electrons, which interact between the planes and stabilize the chain structure. The unsaturated bonds in the nitrogen bases are sensitive to oxidation. Further, the phosphate oxygens are ionic. Thus, the DNA presents all types of chemical bonding and is sensitive to many types of reactants.

The *cell cycle* plays an important role in radiation damage (Figures 15.1.B and C) (i) during the S-stage proteins are synthesized, (ii) in the "gap 2", G_2, stage, the cell is being prepared for division (the DNA-chains are split up and copied), (iii) in the *cell division* (*mitosis*) the chromatin condenses into chromosomes, which are split and move in opposite directions, while the original cell is divided at mid-level (Figs. 15.1.C.3 and C.4), after which (iv) the cell "rests" or matures in the G_1-state.

15.2. RADIATION EFFECTS ON THE MOLECULAR LEVEL

When a high energy particle (whether α, β or γ) strikes a human cell it produces a narrow track, less than 1 μm thick (cf. Figs. 7.5 and 8.1). The interaction with water, which is the main constituent of the cell, leads to radiolysis products, as described in Chapter 8. To a lesser extent, the radiation may interact directly with the molecules of the cell ("direct hit"). Only the effects (direct or indirect) on the cell nucleus DNA are thought to cause severe biological damage.

15.2.1. Radiation physics

High energy γ-rays lose most of their energy in a few compton scattering events. A "typical" natural radiation background gamma quantum of 1 MeV loses about 75% of its energy in 20 cm of water (a "typical" human thickness). This loss occurs to 50–80% in a single compton scattering event, producing an electron with up to 0.8 MeV kinetic energy, depending on the scattering angle. In the next compton interaction, the $\geq$0.2 MeV γ loses $\geq$50% of its energy; the distance between these initial two interactions exceeds cellular dimensions. Thus, the 1 MeV γ-absorption on the cellular level can be considered as the production of a single high energy electron. However, at very low energies (compton γ's and X-rays), low energy electrons are produced at densities of several electron pairs per μm (cf. Fig. 7.5.d and Table 7.2; 1 mm air is considered equivalent to about 1 μm of water or tissue).

High energy electrons lose most of their energy in low energy collisions (cf. Figures 7.5 and 7.7). A typical LET-value is 200 eV/μm (Table 7.2), corresponding to 5–10 ion pairs/μm; a 10 keV electron produces about 10 ion pairs/μm according to Figure 7.7.

Alpha particles cause dense ionization (Figs. 7.5 and 7.7) with typical LET-values of 200 keV/μm (Table 7.2) and production of several thousand ion pairs per μm (i.e. several ion pairs/nm).

High energy neutrons are absorbed in water mainly by collisions with H-atoms, forming energetic protons which ionize similar to alpha particles (Table 7.2 and Fig. 7.7). In the collision process the H-atoms are knocked from their positions in the molecule; also other atoms may be knocked out. The neutron ends its life by being absorbed in a (n,γ)-process or leaves the system.

15.2.2. Radiation chemistry

The absorbtion of low-LET radiation and high-LET radiation is illustrated in Figure 15.1.E and 15.1.H, respectively. Table 15.2 gives the number of ion pairs formed in the DNA, nucleosome and chromatin by α's and γ's (^{60}Co). Table 15.3 shows the damage caused to the DNA and cell nucleus. The wide range of values in parentheses indicate the uncertainty as well as the dependence on the particle energy (in general, higher ionization at lower energy).

The radiation is largely absorbed in water, as the cell contains some 70% H_2O, and produces ions, free radicals and excited atoms. A momentaneous lethal dose of 20 Sv

Table 15.2 Ionization clusters produced directly in a DNA-related target by a single radiation track (From Goodhead, UNSCEAR)†

Radiation	Average number of ionizations in		
	DNA segment (2 nm long)	Nucleosome (∅ ~ 10 nm)	Chromatin segment (25 nm long)
γ-rays	1 (0 to <8)	2 (0 to <20)	2 (0 to <45)
α-particles	2 (0 to <15)	10 (0 to <90)	50 (0 to <200)

†UNSCEAR = United Nations Scientific Committee on the Effects of Atomic Radiation.

Table 15.3 Damage products in a single-cell nucleus traversed by a single radiation track (From Goodhead, UNSCEAR)

Radiation	Average number of ionizations	Average number of induced breaks	
		DNA single strand	DNA double strand
γ-rays	70 (1–1500)	1 (0–20)	0.04 (0–few)
α-particles	23 000 (1–100 000)	200 (0–400)	35 (0–100)

produces a concentration of 14 μM of reactive products (·OH, e^-_{aq}, H_2O_2, etc). Trace metals in the body (e.g. Cu, Cr, Se) are also poisonous at this concentration level, but essential at lower levels (e.g. a daily intake of 40 μg Se is recommended). The ions produced will probably have little effect as the DNA contains numerous ionizable positions at the phosphate group. Free radicals like HO· and oxidizing products like H_2O_2 are highly reactive and can add to unsaturated bonds, which upsets the sensitive hydrogen-π-bonding and may break the bonding between the two helices. Excited hydrolysis products may transfer the excitation energy to the DNA, leading to a localized break in the sugar-phosphate chain. The damage to the DNA may also lead to a substitution reaction in the nucleotide or a loss of a segment. Actually, hundreds of different DNA-damage products have been identified. If the damage is limited to one of the strands of the helix, it is referred to as a *single strand break*. These damages occur naturally due to what is called oxidative metabolism at a rate of about: 3000 single strand breaks/cell/day, 3000 base changes/cell/day and 8 double strand breaks/cell/day.

The cell is protected by different *DNA repair mechanisms*, which all try to restore the damage. This repair typically result in one of several results. The most common one is an error free repair. However, there are also the possibility of no repair at all which lead to mitotic death of the cell. Even if the repair is made it can be erroneous and this can have two results. Either the cell suffer apoptosis which is the process of programmed cell death, or the evolution of cancerous behaviour, or an inheritable DNA-defect. The details are not known, except when the repair goes wrong (e.g. a replacement of a lost nucleotide by a "wrong" base-pair, etc.). It is believed that most single strand breaks are correctly repaired. The repair system is believed to be more effective in a living organism, where the cells are in continuous exchange with surrounding cells and body fluids, than in the tissue samples often studied in the laboratory. This should be kept in mind below, where effects on whole organisms as well as on cell cultures in "test tubes" (*in vitro* studies) are described.

The cell contains natural *radical scavengers*. A radical scavenger is a molecule which reacts readily with the radicals or other oxidising agents why they are also sometimes referred to as anti-oxidants in biological systems. As long as the radical scavangers are in excess of the radiolysis products, the DNA may be protected. When the products exceed the amount of scavengers, radiation damage and cancer induction may occur. In principle, there could thus be a *threshold* dose for radiation damage, at which the free radicals formed exceed the capacity of scavenging. The scavenging capacity may differ from individual to individual, depending on his/her physical condition.

A low-LET gamma ray "hit" may cause 1 ionization (or <8) in a 2 nm DNA-segment. This corresponds to a deposition of some 30 eV (or <250 eV) in about 6 base-pairs + sugar-phosphates. The radiation energy may distribute over a large number of bonds, so that no bond gets enough energy to dissociate. Thus, the radiation interaction may leave the DNA-segment mainly unchanged.

Low-LET radiation sometimes form clusters of ions along the particle track, i.e. produces high-LET "spots". Such spots mean increased risk of damage because a larger amount of energy is then deposited in the small volume occupied by the DNA-helix. High-LET spots therefore increase the possibility of damage to both strands of the helix, causing a *double strand break*.

It has been found that 1 mGy leads to 1 single strand break in each cell, 1 double strand break in every 25 cells and one aberration in each 10000 cells. Because the DNA-chain is rather rigid, a double strand break does not lead to the two halves snapping away from each other. Instead, they are resting, waiting for the repair mechanism to start. Nevertheless, the repair is more difficult and the chances of "repair errors" (mutations) are much larger than for the single strand break. This is demonstrated by two facts: (i) the repair time for a single break is on average 10 minutes, while that for the double break some hours; (ii) *chromosome abberations* (Fig. 15.1.D) only occur after double strand breaks. The chromosome abberations may be due to that the cell undergoes mitosis before the double strand break has been repaired. Cells with severe chromosome abberations are not viable. Minor repair errors may be carried on to next generation of cells as a change in the genetic code or may stop the cell division process.

In addition to double strand breaks being more difficult to repair than single strand breaks, double strand breaks caused by high-LET radiation are more difficult to repair than those caused by low-LET radiation. The high energy deposited at the hit area causes "havoc" on the DNA (cf. Table 15.3). A few percent of low-LET radiation causes double-strand breaks in contrast to 10–20% for high-LET radiation.

After a low-LET dose of 3 Gy, every cell in the human body contains 3000 single strand and some 100 double strand breaks above normal; if given in a short time the cell (and organism) has great difficulties in repairing this damage (resulting in death in the case of no medical treatment). However, spread out over a week only chromosomal abberations will be noted. In the subsequent sections, we shall discuss macroscopic effects of various kinds and doses of radiation.

One effect of the time needed for repair of damages cells is that radiation sensitivity is larger the faster the cells divide (the Bergonié-Tribondeau law). Hence cancer cells are normally more sensitive to radiation than slower dividing normal cells of the same kind. This is one of the reasons why radiation is used in cancer therapy.

15.2.3. Radiation weighing factors

In order to take into account the biological effects of different kinds of radiation, *radiation weighing factors*, w_R, have been introduced and are given in Table 15.4. Earlier, two similar concepts were used with about the same meaning: the "quality factor", Q, and "relative

Table 15.4 Radiation weighting factors, w_R (ICRP 2007)

LET in water (eV/nm)	Weighting factor w_R	Type and energy of radiation
0.2–35	1	photons (X-rays and γ's)
	1	electrons, muons
	2	protons and charged pions
130	20	α-particles, fission fragments, heavy ions
		continuous function of n-energy neutrons

biological effectiveness", RBE. The ICRP 1990 dose concept, the biologically effective dose, is termed *equivalent dose* and abbreviated H_T, and defined as:

$$H_T\,(\mathrm{Sv}) = \sum_R w_R D_{T,R} \tag{15.1}$$

$D_{T,R}$ is the absorbed dose averaged over the tissue organ T, due to radiation R. H_T is measured in sieverts (Sv); 1 Sv = 1 J kg^{-1}. The earlier, but similar concept was the *rem* (radiation equivalent man), where 1 Sv equaled 100 rem. Until 1990, H_T was called dose quivalent. $D_{T,R}$ is measured in J/kg, which in this context is called gray (Gy). The summation is taken over all sources irradiating the target.

15.3. RADIATION EFFECTS ON DIFFERENT TYPES OF CELLS

We can distinguish between two types of cells: those which are directly involved in the functioning of the organ (e.g. the cells of bone marrow, liver, or the nervous system) and those which are associated with reproduction. Radiation damage gives rise in the former to *somatic effects* (i.e. limited to the organism irradiated) such as cancer induction or cell death, and in the latter to *genetic effects* (i.e. limited to future generations).

Cells which are undergoing frequent division, and organs and tissues in which the cells are replaced slowly, exhibit high radiation sensitivity. Of the some 200 different kinds of cells in our body, some never divide (e.g. in the ovary, some sense organs and part of the central nervous system, except in the embryo state), while others divide frequently (bone marrow, intestinal epithelium, male gonads). The cell cycle time varies from hours to days. Usually, tumor cells divide much faster (3–5 times) than surrounding healthy tissue.

In general, the more differentiated the cells of an organ (i.e. the higher the organ is on the biological evolutionary scale), the greater the sensitivity to radiation. This also holds for different organisms, as reflected in Figure 15.2 and Table 15.5. Figure 15.2 shows the *inactivation dose* (i.e. leading to no further cell division) for organisms with different cell

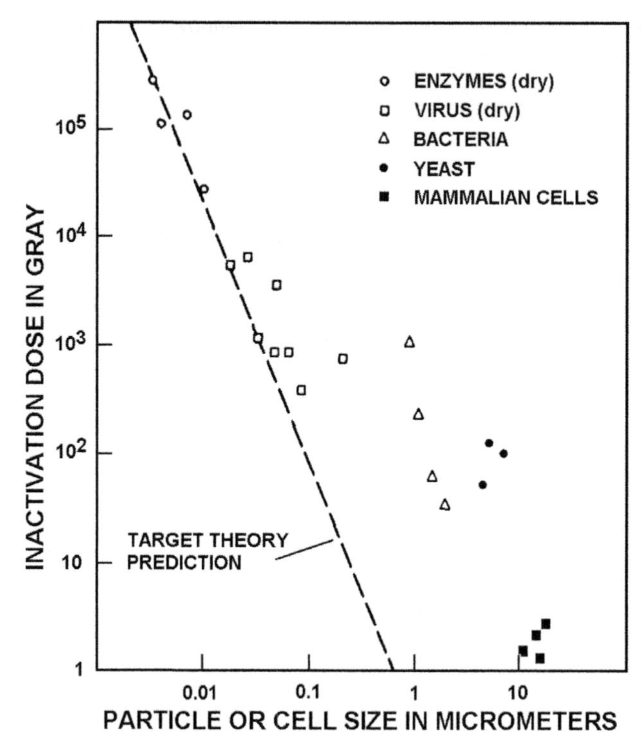

Figure 15.2 Inactivation doses for cells and particles of different sizes.

Table 15.5 Effects of γ-radiation doses on micro organisms, plants and animals

Doses for inactivation (Gy):	enzymes		> 20000
	virus (dry)		300–5000
	bacteria		20–1000
	human cells		≥1
Flowers (Senecio) survive at	10	Gy/d	during the
Trees do not survive at	>1	Gy/d	growing season
Trees normally survive at	≤ 0.02	Gy/d	(normally spring)
$LD_{50/30}$ (Gy) for	amoeba		1000
	fruit fly (Drosophila)		≥600
	shellfish		200
	goldfish		20
	tortoise		15
	song sparrow		8
	rabbit		8
	monkey		6
	man		~4
	dog		3.5

Table 15.6 Tissue weighing factors, w_T (ICRP 2007)

Body part	w_T	$\Sigma\, w_T$
Bone-marrow (red), colon, lung, stomach, breast, remainder	0.12	0.72
Gonads	0.08	0.08
Bladder, oesophagus, liver, thyroid	0.04	0.16
Bone surface, brain, salivary glands, skin	0.01	0.04
Total		1.00

$\Sigma\, w_T$ = sum of w_T-values. Remainder = Adrenals, extrathoracic region, gall, bladder, heart, kidneys, lymphatic nodes, muscle, oral mucosa, pancreas, prostate, small intestine, spleen, thymus, uterus/cervix

sizes. Table 15.5 gives doses which are inactivating or lethal to organisms within 30 days ($LD_{50/30}$) after short-time irradiation. A 10-fold dose is required to kill rather than to inactivate microorganism cells.

Because of the varying radiation sensitivities of the different cell types, a *tissue weighing factor*, w_T, is introduced. It represents the relative contribution of that tissue to the total detriment from uniform irradiation of the body; Table 15.6 gives tissue weighing factors. This leads to another dose concept (cf. eqn. 15.1), the *effective equivalent dose*, H_E, which is the sum of the weighted equivalent doses in all tissues, as given by

$$H_E\,(\mathrm{Sv}) = \sum_T w_T H_T \qquad (15.2)$$

The weighing factors are important in medical radiation therapy and e.g. for evaluating the effects of internal radionuclides, §15.13.5. For low-LET whole-body irradiation w_R and $\Sigma w_T = 1$, i.e. the indici in (15.1) and (15.2) may be dropped.

15.4. SOME CONCEPTS OF RADIATION BIOLOGY

In radiology, the biological effects of radiation are usually discussed along two lines: the matrix effect and target theory. The *matrix effect* considers the particle-water interaction in which ions, radicals, and excited atoms are produced. This is the dominating effect at large radiation doses and dose rates; the lethal dose of 10 Gy generates about 10 000 electron tracks in each human cell. Free radicals and oxidizing products interact directly with cell DNA, causing the DNA-strands to break as described in §15.2.2. One can state that at such high doses the cell is simply poisoned by decomposition products, and the whole organ may be destroyed. It should be noted that in cancer therapy even much higher doses are given.

Inactivation of cell populations (Figure 15.2) and organisms (Table 15.5) requires comparatively large doses and the matrix effect should dominate. However, a 10 Gy dose to cells 1/1000 the size of human cells produces only 10 electron tracks per cell. The cell repair mechanisms are able to overcome the poisonous products. Consequently, doses required for inactivation of enzymes and viruses must be 10^5-10^6 times larger than the ones that are lethal to man.

Target theory depicts the DNA molecule as the site of reaction. Even if matrix effects can never be excluded, target theory is the essential model at low levels of irradiation because a single change in a DNA molecule, however unlikely, may convert it into an *oncogene* (tumor producing gene) with fatal consequence to the organism. Natural background radiation is mainly low-LET. At a common level of 1 mGy/y, each human cell receives on average one track intersection per year. Nevertheless, it is claimed (e.g. UNSCEAR 1993) that this low-level low-LET radiation may be responsible for many of the malignant cancers in the population, though no laboratory experiments have confirmed this, as larger doses are required to produce observable laboratory results.

Figure 15.3 shows typical *dose-response curves*: the effects of X-ray irradiation of hamster cells (*in vitro*) at different stages of the cell cycle. The surviving fraction of cells decreases with increasing dose. Cells in the late G_2 stage and in mitosis are more sensitive than cells at G_1 and the S stage. This is a general observation, which can be explained by assuming

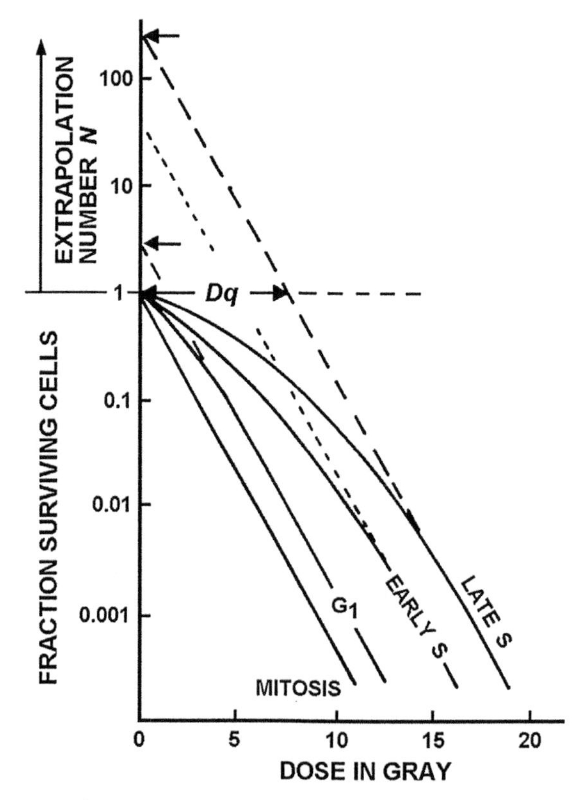

Figure 15.3 Survival curves for Chinese hamster cells at different stages of the cell cycle. *(Sinclair and Morton 1966, from Nias.)*

that the likelihood for double strand breaks of the DNA is greater and the cell repair mechanism less efficient when the DNA is in the condensed chromosome state as compared to when it is in the chromatin state (S-state). In Figure 15.3, the response function for mitosis seems to be linear while the "late S" function is curved. Extrapolation of the straight part of the S-curve extends it to the zero dose line at ~100; this number is referred to as the *extrapolation number, N.* Its significance is not quite clear. The typical values for human cells and X- and γ-radiation are 2–10. The value of the extrapolated line at 100% cell survival (or $N = 1$) is referred to as the "*quasi-threshold dose*", D_q. The "shoulder" in the dose-response curve is typical for X- and γ-irradiation of human cells while neutrons and alpha particles hardly produce any shoulder. Also, a shoulder does not form when very simple organisms, such as viruses, are irradiated by X- and γ-rays. A simple straight dose-response relation (the mitosis line) is taken as support of a *single-cell hit killing*, which is purely random. We have already concluded that n and α radiations are likely to severely damage DNA, particulary during the G_2 and mitosis stages.

The existence of a shoulder is taken as support of a *multi-target model*: the DNA must be damaged at several points by X- or γ-rays in order to cause the cell to die because of the efficient repair mechanism for single-strand breaks.

The slope of the straight part of the line allows calculation of the *mean lethal dose*, D_0, which is the dose required to inactivate the fraction $1 - e^{-1}$ (i.e. 63%) of the cells. Designating the surviving fraction as f_S gives

$$f_S = e^{-D/D_0} \tag{15.3}$$

for the straight line (mitosis). The curve can be described by

$$f_S = 1 - \left(1 - e^{D/D_0}\right)^N \tag{15.4}$$

15.5. FURTHER REGULARITIES AT LARGE DOSES

Figure 15.4 shows the number of dicentric chromosome abberations observed in a cell population as a function of neutron or gamma/X-ray energy. *Dicentric abberations* (Fig. 15.1.D) seem to be the most consistent index of radiation damage and represents about 60% of all observable unstable abberations following acute irradiation. Their background frequency is very low (about 1 in 1000 lymphocyte cells). From Figure 15.4, it is seen that the lower the radiation particle energy is, the larger the damage (of course, until the energy has decreased to such low levels that its ionization effects begin to drop abruptly, see Figure 7.7). This is attributed to the increasing ionization density with decreasing particle energy, as described in chapter 7; see also Tables 15.2 and 15.4.

Figure 15.4 also shows that the damage increases with increasing *dose rate*. This is a general and significant phenomenon which is probably related to the cell (DNA) repair

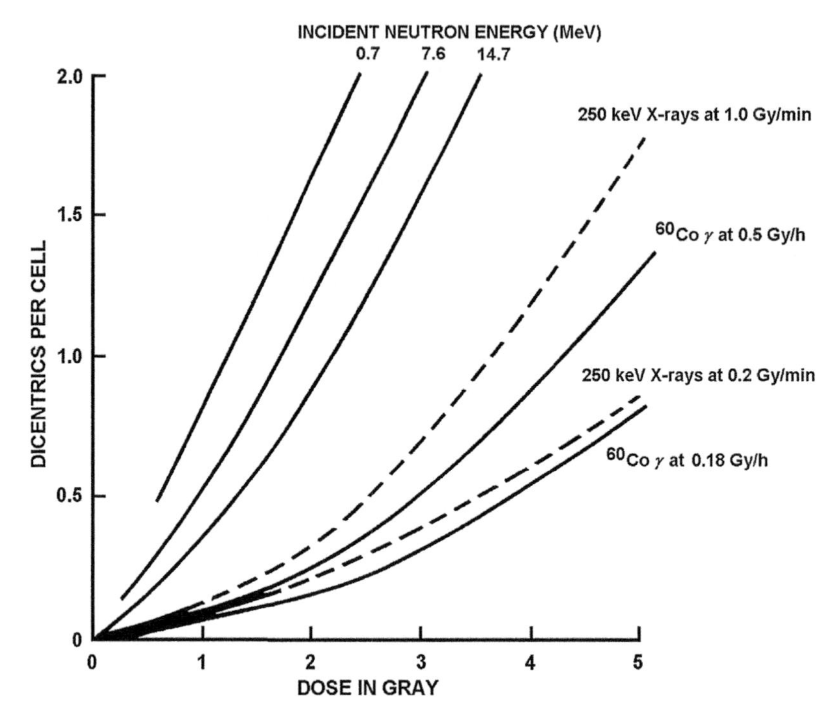

Figure 15.4 Dose response curves of dicentric abberations yields for several qualities of radiation *(Lloyd and Purrot 1981, from Nias.)*

mechanism. At high dose rates, the repair mechanism may become saturated. If a large number of cells are damaged almost simultaneously, the tissue may cease to function. This is seen in the nerve cells at large doses and dose rates.

In animal experiments, it has been found that a higher dose rate also produces earlier cancerogenesis (i.e. shorter latency) and more severe forms of the tumor (higher malignancy).

Figure 15.5.A illustrates the effect of *dose fractionation* on cell survival at high doses and dose rates. After each succeeding dose fraction, the curve exhibits the same shoulder and slope, indicating that within a given period part or all of the damage is repaired.

Figure 15.5.B illustrates a matrix effect during irradiation in air and in an oxygen deficient (*hypoxic*) system. OER, the oxygen enhancement ratio, is the relative dose increase needed to produce the same biological effect in the hypoxic as the oxic case. For X-rays and γ, cell survival increases when oxygen is reduced; the addition of oxidants has the opposite effect. This indicates that it is the oxidative radiolysis products of water which are most damaging to the cell. However, no oxygen effect is seen for α radiation, thereby supporting the model of double strand breaks by direct hits of the α's.

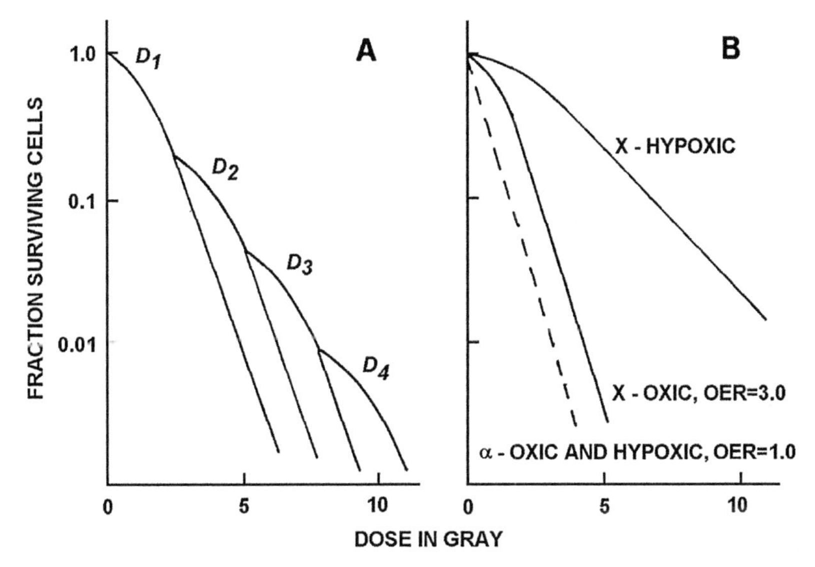

Figure 15.5 Fraction of surviving mammalian cells as a function of dose. A) Effect of dose fractionation. B) Irradiation by X-rays and by α-particles under aerated and hypoxic conditions.

This result suggests that the addition of a reductant would reduce the radiation effect which is also observed. For example, 20 Gy is required to kill 80% of a cell population in the presence of cysteine (or cysteamine), while in the reference system without cysteine the same effect is caused by only 5 Gy. Cysteine is oxidized to cystine.

Several other *radiation protection agents* are known. These compounds are typically amino thiols, similar to the natural amino acid cysteamine. They probably function as scavengers for the products of water radiolysis. Their effectiveness is evaluated by determination of the *dose reduction factor* (DRF), which is the ratio of $LD_{50/30}$ for protected and unprotected animals. Because of their chemical toxicity, many can only be administered in small doses.

No evidence has been found for cells or of higher organisms supporting the development of long-lasting *radiation resistance.* Certain bacteria have been shown to develop a seeming resistance to radiation after receiving small radiation doses over a long period of time. However, this is possibly due to the formation of mutated organisms with a different sensitivity to radiation than the original ones.

15.6. EPIDEMIOLOGICAL OBSERVATIONS OF EFFECTS AT LARGE RADIATION DOSES

We may distinguish between accidental (e.g. Japan and Chernobyl) and deliberate (as given in radiotherapy) exposures to large radiation doses. The former are said to be *stochastic* because the harm caused is statistically distributed over the exposed population; the frequency of tumor induction is assumed to increase linearly with the dose. Deliberate large irradiations are *deterministic* because the damage is caused intentionally to a certain organ or population; such irradiations are considered to have a *threshold value*, below which no effects are observed.

15.6.1. Radiation sickness after accidental exposure

Very large instantaneous doses (>10 Gy) occur in explosions of nuclear weapons, in accidents involving nuclear reactors or criticality accidents, or from carelessness in working with accelerators, X-ray equipment or radioactive installations (e.g. ^{60}Co sources used for technical and therapeutic purposes), criticality accidents, and in handling unshielded strong radiation sources or unshielded radioactive waste. Such doses are very unlikely to be received in work involving amounts of $\leq$1 GBq of radioactivity.

Instantaneous whole body doses (i.e. those received within a few hours) of $\gg$10 Sv lead to death within 24 h through destruction of the neurological system. At 7–8 Sv, the main damage is to the gastrointestinal tract, causing severe bleeding which may lead to death within several days to a month. Doses <0.5 Sv are rarely lethal. For doses between these two levels, intensive hospitalization is required for survival. At the higher end of this range, death usually occurs from 4 to 8 weeks after the accident through infection as because of the destruction of the leukocyte ("white blood cells") forming organs. Those surviving this period usually recover completely. For doses <0.5 Sv, the only proven effect is a decrease in the white blood count (leukopenia). The threshold value for early somatic damage for short irradiation times appears to be about 0.25 Sv.

An example of the effects of high doses was the Tokai-mura criticality accident in Japan in 1999 where 3 workers in a nuclear facility got 4, 8 and 18 Sv respectively. The person receiving the highest dose died within 3 weeks due to a completely destroyed immune system. The man receiving 8 Sv died after 7 months due to important organ failure as a result of the irradiation. The man receiving 4 Sv went back to normal life after 3 months.

The most common type of overexposure in radioactive work involves high instantaneous doses to the hands. Fortunately, the hands, of which the skin is the most sensitive tissue, can stand fairly large doses (Table 15.7). If they do receive extremely high doses ($\gg$10 Sv β or γ), amputation is usually required. Although in some cases, skin transplants have provided temporary relief.

15.6.2. Radiation therapy and deterministic studies

Radiation therapy usually consists of the delivery of large instantaneous or fractionated doses to tissues for which surgical operation is impossible or undesirable. The effects of large organ doses are well described in the literature and is one of the main forms of

Table 15.7 Observed effects of instantaneous radiation doses to the hands

Dose	Effect
<2 Gy	No proven effect
~4 Gy	Erythema, skin scaling, follicle deaths
6–7 Gy	Skin reddening after a few hours, which then decreases, later strongly increases after 12–14 d, then finally disappears within a month; pigmentation
>8.5 Gy	As above, but irreversible degeneration of the skin is visible to the naked eye (the skin becomes hard and cracked); degeneration of the binding tissue with increasing dose
50–80 Gy	Development of non-healing skin cancer; amputation necessary

tumor treatment. Radioactive ^{60}Co (up to 200 TBq), ^{137}Cs (up to 20 TBq) γ-sources, or braking radiation from small linear electron accelerators are used for treatment of deeply located organs; otherwise, X-rays are used more commonly. The organ doses for malignant tumors are usually <100 Sv for mature breast, prostate and blood; typically ≤50 Sv for other internal organs and skin (often given in a series of smaller doses) and 10–20 Sv for the gonads, breast and bone marrow. In comparison, diagnostic radiation doses are ≤ 0.1 Sv. The use of internal radiation sources for therapy is described in §18.5.4.

A special form of radiation therapy uses heavy high energy ions from accelerators. The decreasing velocity of charged particles in matter results in a very high specific ionization near the end of the path (the Bragg peak, Fig. 7.7.b). The energy of the particles is selected according to the depth and type of tissue to be penetrated so that the particles have the proper range to provide a very high local dose in the proper volume of the sick tissue. This technique (the "proton knife") has been particularly important in treating diseases of the pituitary gland, which is located deep inside the brain.

In some cases, local irradiation of some tissues is produced by the use of radioactive nuclides implanted in the tissue by means of needles or small capsules. For example, needles of ^{90}Sr–^{90}Y, pellets of ^{198}Au, etc., have been implanted in the pituitary gland (for acromegalia, Cushing's disease, and cancer), in the breast (for breast cancer), in the prostate, and in the nerves (to reduce pain). The local dose may be 100's of Sv. Also, radioactively labeled tumor seeking compounds are used.

It should be noted that a patient's response to a certain radiation dose is highly individual; thus, all patients cannot be treated alike for the same tumor. Statistically up to 20% of the patients may be "lost" if this individuality is neglected! It is explained by the different "normal levels" of the patient's immune defense, which may have a hereditary cause. Small radiation doses seem to activate the immune response system by stimulating antigen production. This is used in radiation therapy by exposing the patient to a comparably small "pre-treatment" dose (~0.5 Gy) before the large organ doses are delivered in order to hasten the recovery of the patient after the treatment. The immune system then rapidly takes care of the radiation products.

15.6.3. Stochastic cancer induction

The first *radiation induced cancer* in a human was reported in 1902 when skin cancer was observed on the hand of a radiologist working with X-rays. During the following decades, quite a number of such cases were reported and occurred in either persons employed in medical radiology or in patients treated with radiation for benign lesions (e.g. eczema, birth marks or tuberculous lymph nodes). Other types of radiation induced cancers were reported as thyroid cancer and sarcoma in bone or soft tissues. The cumulative doses were usually quite high (several 10ths of Gy). Today this type of radiation induced cancer is rather uncommon due to increased knowledge in radio-physics and clinical radiobiology.

The A-bomb survivors in Japan are the most important source of information on human whole-body irradiation. The population was large and varied as regards age and gender, the time of exposure was short and well defined, and it has been possible to make reliable estimates of the dose to each individual. The most recent evaluation (BEIR V, the so-called DS86 dosimetry) comprises 76 000 persons, who had been exposed to $\leq 7-8$ Gy of instantaneous gamma and neutron irradiation. Their health and their children's health have now been followed in detail for 50 years. Figure 15.6 shows in (A) the chromosomal aberration frequency, and in (B) the number of leukemias observed as a function of dose. The excess leukemia frequency is 3.16 by the original 1965 estimation (T65) and 3.10 by the new 1986 dosimetry (DS86). The excess malignancies are rather low for other types of cancer.

The raw data in Figure 15.6.A show "best" lines for the Hiroshima and the Nagasaki victims; the central line is a common fit. The difference between Hiroshima and Nagasaki is believed to be due to a much higher contribution of neutrons in Hiroshima due to the construction of the bomb; Nagasaki was probably only gamma radiation effects. This is in agreement with the higher damage produced by high-LET radiation in Figures 15.4 and 15.5.B.

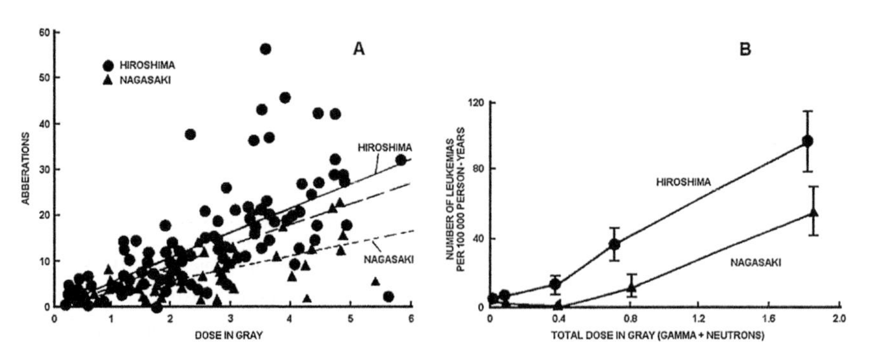

Figure 15.6 Radiation effects on population in Hiroshima and Nagasaki as a function of dose. A. Chromosome abberations. B. Leukemia frequency.

The total leukemia frequency is five times lower for the Japanese bomb victims who received only a slightly larger radiation dose than the average of the whole Japanese population. This may be caused by an early removal of "leukemia sensitive" individuals (i.e. pre-destined victims of leukemia because of inherited leukemia oncogenes) from the group or "protection by radiation" related to the stimulation of the DNA repair mechanism as discussed above, which reduces the susceptibility to the normal incidence of leukemia.

It is important to bear in mind that the number of radiation induced cancers in this cohort (a group of persons with a common statistical characteristic) is relatively small and thus the statistical uncertainty is large: in 1950–1985 a total of 5,936 cases of cancer were reported compared to the statistically expected 5,596 in an unirradiated reference group of similar sex-age composition, i.e. an excess of only 6% (350 cases).

Following large dose exposure, *leukemia* appears first after an approximately short latent period of about 2–3 years. The incidence of leukemia reaches peak frequency around 6–8 years and then declines and almost disappears about 25 years after exposure. This is shown for the population exposed after the Hiroshima and Nagasaki bombings, see Figure 15.7. Cancers others than leukemia, however, tend to exhibit a different behavior. They appear after a latent period of 10 years following the exposure and then show progressive increase with time. The *breast* seems to be one of the most susceptible organs for radiation induced cancer, but the A-bomb cohorts indicate a risk only in the

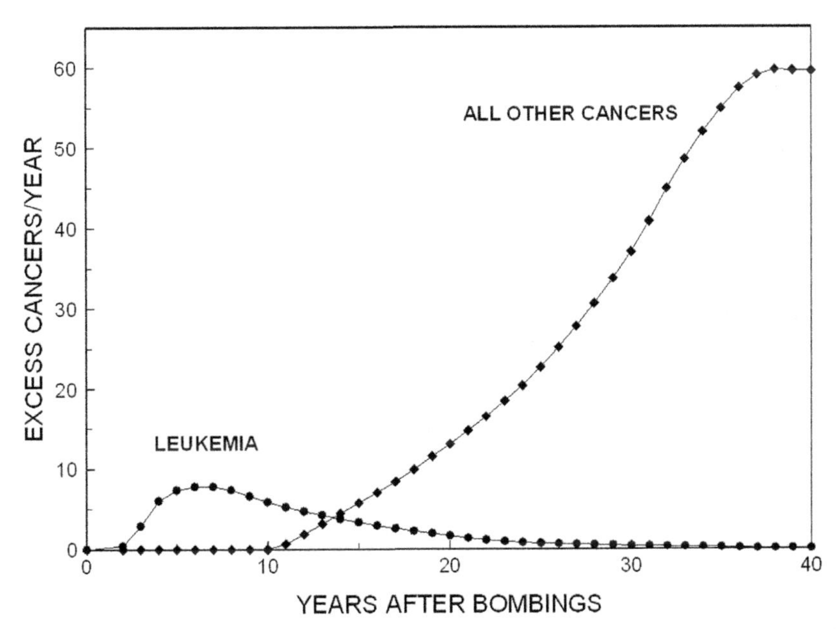

Figure 15.7 Excess cancer incidence rates in the japanese population irradiated by the A-bombs *(After Kato & Shimizi and the RERFoundation).*

younger age group (<40 y at exposure). Except for the breast, the *thyroid* is the most susceptible organ in humans for radiation induced solid tumors. This is the only cancer induction (an 80-fold increase among children) observed so far among the Chernobyl victims. The mortality risk, however, is very low because most of these tumors are benign. *Lung cancer* can be induced by γ-rays, but interest is presently focused on the risk from radon daughters in air. *Bone sarcoma* may develop after local exposure to X-rays or γ-rays. The latency period is 5–10 years for large doses (20–70 Gy).

For the Japanese survivors, increased detection of leukemia (13 times higher than naturally expected), multiple myeloma (6 times), cancer in the colon, urinary tract and breast (about twice) have been reported. In comparing an observed frequency of cancer with an expected one, it is important to correct for differences in sex and age groups.

Information on radiation effects from ingestion or inhalation of large amounts of radioactive substances is mainly limited to four cases: (i) uranium mine workers who inhaled and ingested Rn and Rn-daughters; (ii) the painters of luminous dials in Europe (before 1930, they ingested radium while sharpening their brushes by licking them); (iii) people living where the air has a high Rn concentration; (iv) patients treated by high amounts of ^{131}I for thyroid disorders (a non-stochastic investigation). The first two groups have been followed for more than five decades with the lifetime risks for fatal cancers estimated as 0.05% per Sv for bone sarcoma and 0.2% per Sv for cancer of the bone marrow. Radon is discussed in §§15.10, 15.11.1 and 15.13.5.

Saenger et al. (1968) studied 36 000 patients with hyperthyroidism (a cancer causing the thyroid to grow), of which 22 000 were treated with ^{131}I (only β, no γ; local doses of several hundred Sv) while the rest underwent surgery or chemical therapy. Although the ^{131}I patients received bone marrow doses of about 100 mSv, no difference was observed in relation to the non-irradiated group with respect to incidence of leukemia.

In 1988 Holm et al. reported a similar comprehensive study of 35 000 patients who had been administered 2×10^6 Bq ^{131}I each, which caused a thyroid dose of about 0.5 Sv per person. The mean time of examination was 44 years for patients of all ages. In this group 50 thyroid cancers were discovered (compared to the expected 40 cases for un-treated), indicating a relative risk of 1.27 ± 0.33 (95% confidence interval). The excess of 10 thyroid cancers is so small that it falls within statistical uncertainty. Thus, Holm et al. conclude that "these data provide little proof that ^{131}I is carcinogenic in humans" at these doses and dose rates. The examples illustrate the difficulty in proving carcinogenesis even at relatively high doses.

Information on the effects of large "whole-body internal" radiation doses comes mainly from laboratory experiments on animals. Figure 15.8 shows the excess tumor frequency for mice irradiated with doses up to 120 Gy in a comparison of the effects of different radiation sources: A, ^{90}Sr-induced osteosarcomas in female CBA mice (Nilsson 1975); B, bone tumors in humans from incorporated ^{226}Ra (Rowland 1971); C, kidney tumors in rats by X-radiation (Maldague 1969); D, skin tumors in rats by electrons (Burns

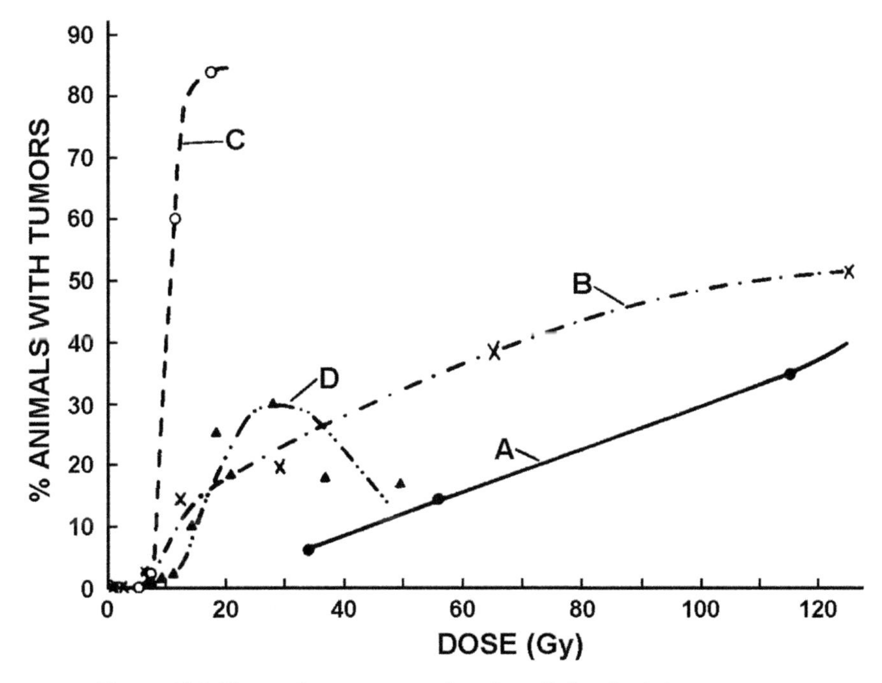

Figure 15.8 Tumor frequency as a function of absorbed dose; see text.

1968). Figure 15.8 seems to indicate a threshold dose at ≤5 Gy below which no effect is observed. However, some animal experiments show what looks like a small beneficial effect at small radiation doses, see Figure 15.12.

15.6.4. Mental retardation

In Hiroshima and Nagasaki, both severe mental retardation and lower intelligence levels (obtained from intelligence tests) occurred following prenatal exposure during the 8–15 weeks after pregnancy. In the following 16–25 weeks after ovulation, lesser vulnerability was observed. The exposure to 1 Gy during the early period increased the frequency of mental retardation 40% (normal frequency about 0.8%), and lowered IQ by 25–30 points. No cases of severe mental retardation has been observed at exposures <0.5 Gy. No radiation effects on the brain have been observed at low dose rates.

15.7. RADIATION STERILIZATION

Figure 15.2 and Table 15.5 illustrate that at appropriate dosages radiation is an effective means of destroying higher organisms with less harm to simpler ones. This has a practical consequence because radiation can be used for the conservation of food in a quite different manner than the conventional methods of heat, canning, and freezing.

Conservation by radiation attempts to achieve the complete destruction of all bacteria with minimum change in the taste of the food (due to the formation of small amounts of decomposition products). Radiation pasteurization (i.e. partial sterilization with lower doses) and irradiation at low temperatures cause correspondingly smaller taste changes.

The safety of food irradiation has been accepted by IAEA, FAO (Food and Agricultural Organization) and WHO. It has been used for several decades without any negative health effects. More than 35 countries have approved some 50 different irradiated food products for unrestricted human consumption. Examples of the foods are (country, maximum dose): potatoes (Canada, 100 Gy), onions (Israel, 100 Gy), shrimp (Netherlands, 1000 Gy), fried meat in plastic bags (Soviet Union, 8000 Gy), and wheat and wheat products (USA, 500 Gy).

Medical supplies, which must be sterile, can be manufactured and packed by conventional techniques, after which the packages may be exposed to high energy penetrating radiation (e.g. ^{60}Co, ^{137}Cs or electrons from small accelerators). The radiation kills all bacteria with little damage to low molecular weight compounds. In this common technique, the irradiation source is quite large (0.1−1 MCi; around 10^{11}Bq). The packages slowly pass through the source, such that the doses are on the order of 10−30 kGy. The same facilities can be used also for other purposes such as food sterilization.

Radiation doses of ~2 Gy produce sterility in the human male for about 12 months while higher doses lead to permanent sterility; libido and potency are unchanged. Permanent sterility occurs in 60% of young women given 8 Gy of fractionated irradiation to the ovary, but occasional pregnancies can occur after doses of up to 5−8 Gy; the children are apparently normal at birth (BEIR V).

Radiation has been used to sterilize (80 Sv) the males of certain insect species which when released (the ratio between the sterilized and untreated males was approximately 4:1), mate with females and prevent further reproduction of the species. This technique has been used in the US, Mexico, Egypt, Libya, etc to eradicate screw worm flies which can cause huge damage to the cattle industry. The technique is now also used against other insect species which are threats to agriculture.

15.8. GENETIC EFFECTS

In 1927, Muller showed how irradiation of fruit flies (Drosophila M.) could produce new species that were defective with regard to the parents (e.g. lacked wings). This defect was carried in the genes to later generations. This dramatic demonstration of the mutagenic effect of radiation has been extended to other primitive species. The vast majority of such changes are recessive, i.e. in order to become dominant both parents must carry the same DNA-damage. From all experience as well as the theoretical considerations described above, the likelihood for two identical DNA impairments to come together in the fertilized cell is extremely small. No radiation induced hereditary effects have been observed in mammals. Irradiation of the sexual organs of 2000 carefully selected mice ($\sim$2 Sv each) for 19 generations (in total 38 000 mice, all inbred) showed no genetic changes (Lüning 1970). A thorough investigation of the 75 000 children born to parents who were exposed to the A-bomb irradiation in 1945 have not shown any increase in the frequency of hereditary diseases (or cancer). The 35 000 children born to parents of which at least one had been exposed to $\leq$3 Sv (average exposure 0.2 Sv) showed no genetic differences (analyzing still births, birth weight, congenital abnormalities, infant mortality, childhood mortality, leukemia or sex ratio) from children born to non-irradiated parents within statistical uncertainty.

Large doses to plants have caused mutations which have either improved the quality of the species or produced effects which are disadvantageous to the species, but desirable to society. Although irradiation of plant seeds results in a ratio of about 1000 to 1 of the harmful to the advantageous mutations, by cultivating those few plants showing improvement in properties, new plant variations have been obtained. This has resulted in species of grains and legumes which have stronger stocks, higher yields, and improved resistance to mold and to fungi. In northern countries such as Sweden, most of the grain that is grown today is radiation-produced species possessing much greater cold resistance.

As a conclusion one might say that genetic effects in higher mammals from radiation has not been observed. After the Chernobyl accident thorough studies of fast reproducing mammals such as rats have shown no genetic effects.

15.9. RADIOMIMETIC SUBSTANCES

Many chemical substances when administrated to the body show the same effects as irradiation. Because cancer caused by radiation has been investigated more than cancer induced by chemicals, such substances are called *radiomimetic* (mimetic = imitative). In order to qualify as a radiomimetic agent, the substance must do the following: stop cell division, stop tumor formation, produce chromosome aberrations, cause mutations, kill lymphocytes, and be carcinogenic. Chemical substances which meet a few, but not all of these requirements are not considered radiomimetic. The effects depend on the concentration of the substance; e.g. cell division is interrupted by many radiomimetic substances at concentrations $\leq 10^{-5}$ mole l^{-1}.

Typical radiomimetic substances are organic peroxides (e.g. ethylene oxide), ethylene di-imine, mustard gas and derivatives, aliphatic dichloro-amines, etc. These compounds or chemical groups occur in many familiar materials such as tobacco smoke. The effect of a certain amount of a radiomimetic substance can be calculated to correspond to a radiation dose. Thus, smoking a pipe of tobacco corresponds to an average radiation dose of about 0.04 mSv.

15.10. RADIATION BACKGROUND

Through life, everyone is constantly exposed to ionizing radiation from a variety of sources. Table 15.8 gives radiation doses which are representative for a large number of countries with the exception of extreme conditions like cosmic radiation at high altitude (e.g La Paz at altitude 3900 m) and locations on monazite sand (e.g. Ramsar, Iran, Kerala, India, and Esperito Santo, Brazil). Table 15.9 gives the relative contribution of sources of radiation to the UK population. The values are representative for most countries in the northern hemisphere. Within the same country, the total background dose commonly varies within a factor of 3–5 between lowest and highest areas. Brick and concrete houses contain varying amounts of uranium and/or radium, depending on the source of the building material. Because of radioactive soil, all foods contain some natural radioactivity. An approximate average value in northern Europe is <0.3 Bq/kg from U-decay products in food or beverage, while vegetables in the monazite areas in India (e.g. roots) have been found to have >10 Bq/kg.

Presently great concern is directed towards people living in dwellings with high radon concentration ($\geq$400 Bq/m^3) because the radon and daughters are assumed to

Table 15.8 Effective dose (normal → elevated value) in mSv/y (UNSCEAR 1993) or average annual dose values (Nias 1990)

Source	Value
Cosmic rays	0.3 → 2.0 (LaPaz) mSv/y
Terrestrial, external	0.3 → 4.3 (Kerala) mSv/y
Intake	0.052 (ingestion), 0.01 (inhalation) mSv/y
Radionuclides in body†	0.23 → 0.6 mSv/y
Rn in water supplies and air:	see §13.6
Rn in body	1.3 → 10 mSv/y
Buildings‡: wood/gypsum	0.18 mSv/y (K 150, Ra 20, Th 5 Bq/kg)
" : typical masonry	0.7 " (K 500, Ra 50, Th 50 Bq/kg)
" : alum shale concrete	5.9 " (K 770, Ra 1300, Th 67 Bq/kg)
Nuclear power plant: to critical group (Sweden)	<0.1 mSv/y
X-ray investigation:	5–8 mGy
barium meal (intestine)	
intravenous urography	4–5 "
abdomen and lung	~1 "
mean gonadal dose (abdomen),	<0.2 "
genetically significant dose	0.1–0.5 "
dental	0.02 "
Total average	2–4 mSv/y

†Except Rn.
‡Staying indoors at 8760 h/y; masonry building in temperate climate.

Table 15.9 Relative contribution of sources of radiation to the population (UK National Radiation Protection Board, 1989); annual average dose is 2.2 mSv

Natural sources	87
radon	47%
food and drink	12%
γ's from ground and buildings	14%
γ's from cosmic rays	10%
Artificial sources	13%
medical	12%
fall-out	0.4%
work	0.2%
occupational†	<0.1%
others	0.4%

†Sellafield reprocessing plant, average 5 mSv/y

make the largest contribution to the present radiation background. Typical indoor values are 50–400 Bq/m^3 air; we discuss this further in §15.13.5.

15.11. SOMATIC EFFECTS OF LOW RADIATION DOSES

The effect of low radiation doses such as the natural radiation background or from the nuclear power fuel cycle (and possible accidents) is controversial because the evidence is inconclusive.

15.11.1. Epidemiological results

Of the large number of epidemiological investigations on the effects of low-LET low-level radiations, we will only mention a few representative ones.

Figure 15.9 shows the malignant mortality rate of the US white population accumulated over 18 years (1950–1967) in 46 states as a function of the natural radiation background. Each point represents an average of about 100 000 deaths and is thus significant. The average background value for the whole US population is 1.3mSv/y and the average annual mortality rate in cancer (horizontal line) corresponds to 0.15%. The only significant trend seems to be a decrease in cancer deaths with increasing background up to about 2.5 mSv/y. Although the background radiation in the 7 highest states is 2.1 mSv/y as compared to 1.3 mSv/y for the US average, the frequency of all malignancies is lower (126 per 100 000) in the former than in the latter (150 per 100 000) case. The frequency of leukemia, breast cancer, thyroid cancer, and all child malignancies (age 0–9 y) is slightly lower or the same. Although attempts have been made to correlate the cancer decrease rate with other factors, such as living

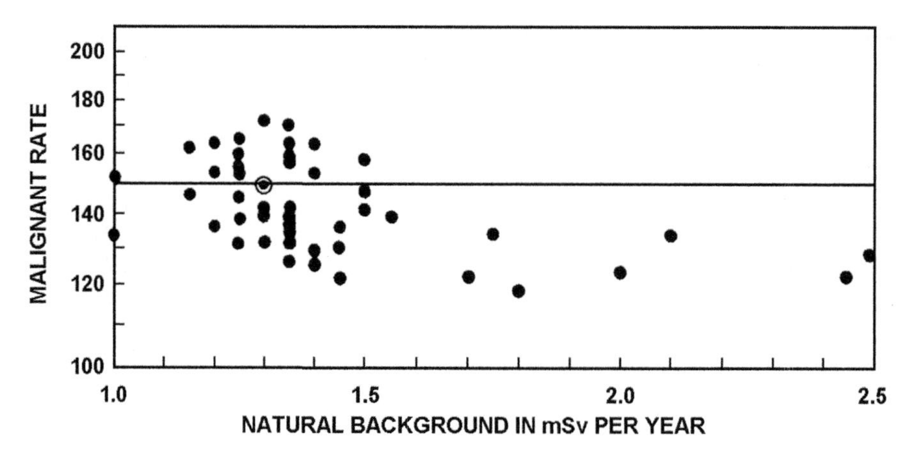

Figure 15.9 Malignant mortality rates per 100 000 by state and natural average radiation background. *(From Frigerio, Eckerman and Stowe, 1973.)*

habits (including sexual frequency and divorce!), no convincing explanation has been given.

This kind of studies are extremely difficult to evaluate since the probability of getting cancer in a country with a high average life span is about 40%. Thus the increased risks due to specific natural radiations will be almost indiscernible.

In the Han province of China, 150 000 peasants with the same genetic background and lifestyle were examined. Half of the group lived in a region where they received almost a threefold higher radiation exposure because of radioactive soil. The investigation of radiation effects such as chromosomal abberations, frequencies of hereditary diseases and deformities, frequency of malignancies, growth and development of children, and status of spontaneous abortions failed to disclose any difference between the two groups.

There are many similar studies from the US, UK, Canada, France, Sweden, Finland, etc, covering millions of people which fail to show a correlation of cancer incidence with small excesses in low-LET radiation. For example, the county in Sweden (Västergötland) with the highest average background radiation has the lowest total cancer frequency.

However, opposite results have also been reported. Kendall et al. (1992; in all 13 authors) have studied the cancer mortality and occupational exposure of 95 000 radiation workers at 24 different sites in England; the mean life-time dose was 34 mSv. They concluded that the frequency of cancers correlated slightly with dose. For multiple myeloma the correlation was "strong" and "for leukemia the trend was significant". 52 leukemia cases were observed while the expected range was 45 $\pm$ 13, excluding chronic lymphatic cancer which is assumed not to be related to radiation. Nevertheless, the observed leukemia frequency was lower than the UK average. The number of all cancers observed was 1363, as compared to an expected 1569 ± 20 in the general population. These "low results" are explained by "the healthy worker syndrome"; i.e. they belonged to a selected group of people with healthy living conditions and good health care and, consequently, the group was assumed not to be representative of the population in general.

A somewhat similar study of Hanford, USA, (see Ch. 19 & 20) workers during 1945−1981 (Gilbert et al. 1989) failed to discover any increase in leukemia, but instead found an increase in multiple myeloma, which also was explained by "the healthy worker syndrome". Multiple myeloma is a skin disease which is common among people who spend much time outdoors and expose themselves excessively to sunshine.

Many studies have tried to relate lung cancer frequency with the Rn-concentrations in ordinary air (≤200 Bq/m^3). Except for the study of uranium miners (§§15.11 and 15.13.5), all these investigations, though carried out with utmost care, have failed to statistically prove such a relation. However, such a relation is still assumed to exist.

15.11.2. Problems in studies of low-level low-LET radiation

In searching for effects of low-level low-LET radiation, the research has focused on the formation of cancer, particulary leukemia, which is a proven effect of large radiation doses, but is a rare form of cancer in the normal population. The normal frequency of leukemia in the US population is 0.012%/year (i.e. 12 cases per 100 000 deaths). The few cases in a large population leads to statistical problems and large uncertainty in the results, as well as difficulties in obtaining a homogenous cohort. Corrections have to be made for sex, aging, changes in the environment (which contain a large number of genotoxic agents), etc. For example, in order to be certain of an increased incidence of breast cancer, which is >10 times more common than leukemia, following an acute exposure of 10 mSv to both breasts at age 35, would require a cohort of 100 million women (Land, Walinder).

Figure 15.10 can be used to illustrate the difficulty with low numbers of incidents. The Figure shows a purely accidental distribution of 165 raindrops on a window. Subdividing the window into 20 squares, A1 to D5 (right figure), one finds a range of 4 (A2) to 13 (D3) drops per square. The difference between the lowest and highest value is over a factor 3! Small excess numbers of cancers, presumably assigned to radiation, carry little weight. Walinder has pointed out that from an epistemological (science theoretical) standpoint, it is impossible to prove that low radiation doses (in the 1 mSv range) are harmful or harmless. As outlined above it is clear that actually trying to prove any effect of ionsing radiation is an epidemiologic challenge. For example in order to ascertain an effect of a dose of 1000 mSv you need a population of 500 persons; 100 mSv need 50000 persons and 10 mSv will require 5 million persons for statistical significance.

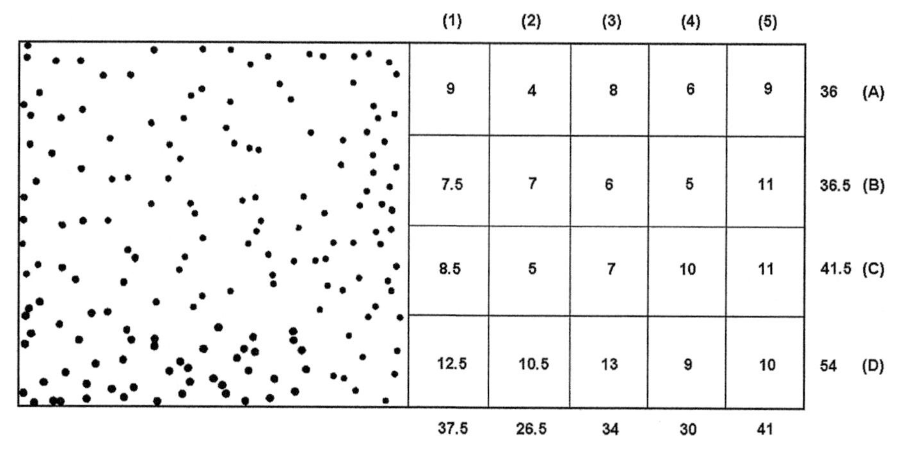

(1) (2) (3) (4) (5)
9 4 8 6 9 36 (A)
7.5 7 6 5 11 36.5 (B)
8.5 5 7 10 11 41.5 (C)
12.5 10.5 13 9 10 54 (D)
37.5 26.5 34 30 41

Figure 15.10 Left, distribution of raindrops on a window. Right, matrix, to be put on top of the left Figure. Counting drops in the squares yields the figures noted; error (1σ) is ± 6.

It has been convincingly proven that several damages to the DNA must occur almost simultaneously before a cancerogenic process starts. In §15.2, it was described that low-LET radiation mainly causes single-strand breaks, which are likely to be repaired by the biological system. It was also mentioned that at low dose rates, the radical scavenging system (more generally, the DNA repair system) is more effective than at high dose rates. Further, it was pointed out that at high doses, each cell in a tissue may be hit several times, while at low doses a "hit" cell usually is surrounded by undamaged cells. Low-level low-LET dose effect studies are further complicated by the fact that damage to the *stem cells* in the M phase is the dominating risk. About 0.01−0.1% of all cells are stem cells, which act as breeding sources (especially in the bone marrow) for other cells (which then mature in the thymus or elsewhere for specialized purposes). Thus, research may have to be concentrated on a very small fraction of cells.

Many researchers claim that there are several arguments for the assumption that responses from high-level and high-LET radiation may not be directly extrapolated to low-dose low-LET radiation. Many are of the opinion that the natural radiation background is harmless, and some even claim that it is beneficial (radiation *hormesis*). Thus, looking for effects at low doses can be a vain study. Walinder has pointed out that it is impossible to arrive to a reliable dose-effect relation at doses <50 mSv by epidemiological studies.

A complicating factor in judging the risk of low-LET low-dose radiation is the different sensitivity of the exposed individuals. Though rats etc. can be inbred to produce a single genetic strand, that is not possible with humans. At the beginning of the 20:th century, dozens of young people, mainly female, were employed in the laboratory of Marie Curie making calibrated radium sources, during this work they exposed themselves to radiation doses of many Sv a year. It is reported – though supressed at that time – that the health of many of these young people rapidly deteriorated so much that they had to leave, some having accured anemia and other radiation related deceases, while others worked for decades without any obvious detrimental health effects. The explanation for this difference is assumed to be due to differences in the individuals immune defence system.

A similar conclusion can be drawn from Figure 15.6.A, where the number of cell abberations – measured a long time after exposure – varies with at least a factor of 5 (i.e. 500% !) for the same dose.

15.12. THE DOSE-EFFECT CURVE

Figure 15.11 shows a number of hypothetical dose-effect relations: The "unmeasurable range" is indicated within the circle. The dashed-dotted line outside the circle indicates the uncertainty in the "measurable range". Line a is based on the by ICRP recommended NTLE (No Threshold Linear Effect) hypothesis and the message is clear: the risk is zero only at zero radiation dose. Curve b indicates a threshold around 50 mSv,

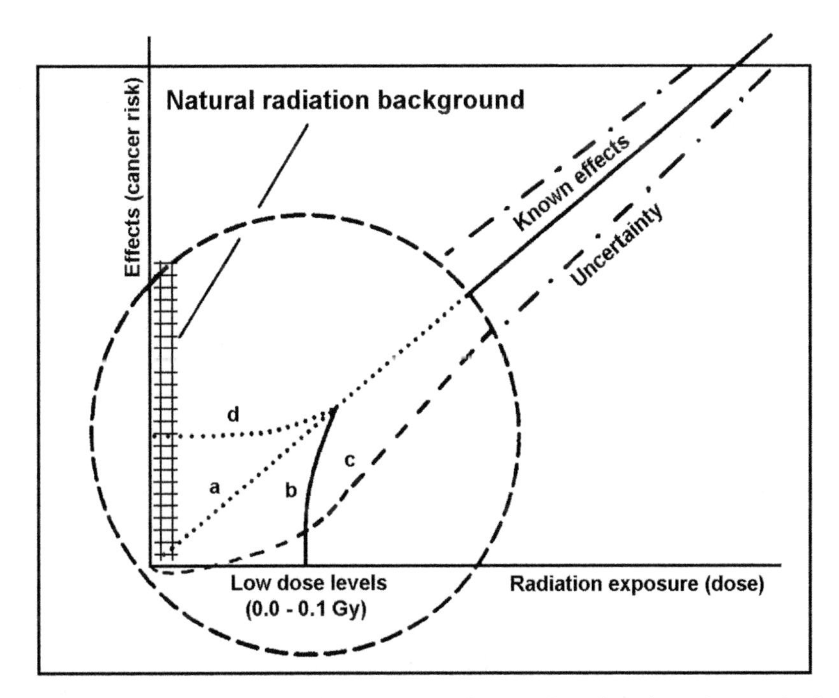

Figure 15.11 Dose effect relations in the known (with spread) and in the unknown (within circle) range, and some hypotheses.

below which there is no increase in cancer (or other radiation induced diseases); many radiologists support this hypothesis. Curve d assumes that there is a constant risk at the lowest doses. Curve c illustrates the "quadratic-linear" model, which presently seems to be favored by several radiation protection agencies (incl. ICRP), who assume that the slope near zero is one half of the slope at higher doses and dose rates. As this slope is unknown, it could as well be less. It is obvious that the NTLE hypothesis is wrong av very high doses because the effects saturate at some high dose level.

Many studies suggest the existence of a dose threshold level around 100 mSv, below which no health effects from radiation are observed. Other studies claim that health effects are observed even in the range 10−50 mSv. This has lead to "schools" of radiation health experts, those who claim that the NTLE thypothesis is scientifically validated, and those who claim that it is not. As an example we may study the effect of radiation on mice as shown by Ullrich and Storer 1979. It is clear from their study that a small radiation dose is not only harmless but actually beneficial for the health of the mice, see Figure 15.12. In conclusion, due to uncertainties the different interpretations of the low dose effects may be drawn as in Figure 15.11.

Figures 15.6, 15.8, and 15.9 show that it is impossible to assess the risk to humans at the 1−5 mSv level, which is the range of the natural radiation background. As epidemiological investigations seem to fall short, it has been suggested (Sondhaus, Yalow, Sagan, Cohen etc) that the only way to resolve this question is through *in vivo* radiation studies at the molecular level.

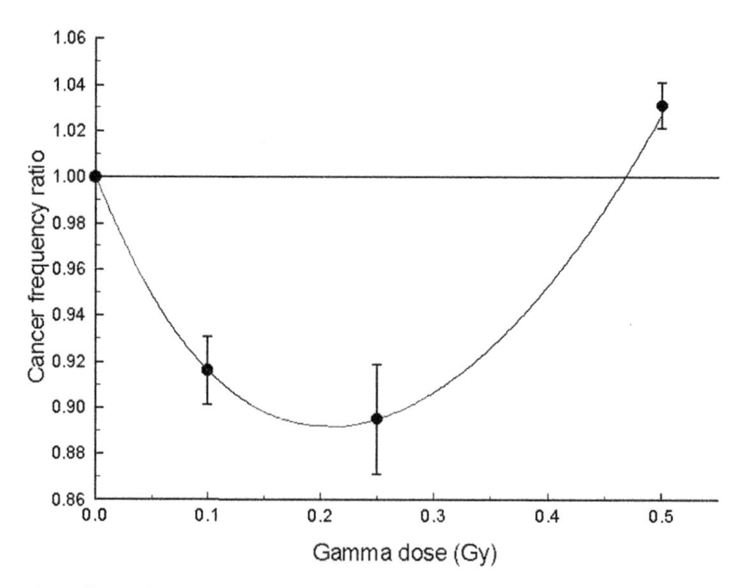

Figure 15.12 The effect of low doses of gamma radiation on the frequency of all cancers in mice, irradiated/unirradiated *(After Ullrich and Storer, 1979)*.

15.13. REGULATORY RECOMMENDATIONS AND PROTECTION STANDARDS

15.13.1. Risk of cancer induction by radiation

The International Commission on Radiation Protection (ICRP) was formed in 1928 and has been the main international organization solely devoted to recommendations to prevent humans from being harmed by ionizing radiation. Their recommendations form the basis for national regulatory decisions.

In 1990, the ICRP concluded (Publ. # 60) that *the risk of cancer induction, L_c, is 5% per Sv* or

$$L_c = 0.05\ H_E \tag{15.5}$$

for low-levels of low-LET radiation (i.e. NTLE is assumed). Thus, for a dose of 20 Sv, the probability of acquiring cancer is set at 100%. Here L_c is the probability for an individual to acquire (and die) of cancer if exposed to the effective equivalent dose H_E (§15.4). This rule is based on the probabilities in Table 15.10.

In §15.6.1, it was stated that the lethal dose for instantaneous low-LET radiation is 10 Sv. The value 5%/Sv is half as large, which is a concession to the fact that biological harm depends on dose-rates as described in §15.5. In ICRP terminology, the *dose reduction factor* is chosen to be 2, although UNSCEAR 1993 (p.688) notes that it is more likely 2−10.

UNSCEAR collaborates closely with the United Nations Environment Program (UNEP), the World Health Organization (WHO), the International Atomic Energy Agency (IAEA), the International Commission on Radiation Units and Measurements (ICRU) and the ICRP.

In the light of previous discussion, one may state that the ICRP rules are "extra-safe". However, it is unlikely that the risk is underestimated, which is an important safety aspect. The ICRP does not recognize a threshold value, below which there is no harm at all. The ICRP notes (see e.g. BEIR V or UNSCEAR 1993) that the straight line a in Figure 15.11 may overestimate the true risks at low radiation doses and low dose rates, but should nevertheless be adhered to for safety reasons. However, it should not be used for prediction of cancer induction in large populations.

Table 15.10 Nominal probability coefficients for stochastic effects in detriment percent per Sv (10^{-2} Sv^{-1}) after exposure to radiation at low dose rate (ICRP 2007)

Exposed population	Fatal cancer	Heritable effects	Total detriment
Adult	4.1	0.1	4.2
Whole population	5.5	0.2	5.7

Table 15.11 Recommended dose limits (ICRP 2007)

Applications	Occupational	Public
Effective dose	20 mSv per year†	1 mSv in a year††
Annual equivalent dose		
in the lens of the eye	150 mSv	15 mSv
in the skin	500 mSv	50 mSv
in the hands and feet	500 mSv	—

†Averaged over defined periods of 5 years, but the effective dose should not exceed 50 mSv in any one year.
††The value may be exceeded as long as the 5-year average dose not exceed 1 mSv.

15.13.2. Recommended dose limits

The ICRP and the IAEA regularly issue recommendations for proper handling of radiation sources. The purposes of the recommended system of *dose limitations* are to ensure that no exposure is unjustified in relation to its benefits, that all necessary exposure is kept as low as is reasonably achievable (the *ALARA principle*[1]), and that the doses received do not exceed the specified limits. The ICRP stresses the values given in Table 15.11; they apply to the sum of the relevant doses from external exposure in the specified period and the 50-year committed dose (for children 70 years; see also next §) in the same period. These values must be respected. They are intended to limit somatic effects in individuals, hereditary effects in the immediate offspring of irradiated individuals, and hereditary and somatic effects in the population as a whole.

The effective equivalent dose limit refers to the sum of the equivalent doses to all tissues from external sources and from radioactivity taken into the body. The limits do not include contributions from any medical procedure or from normal natural radiation.

The dose limits should not be regarded as a dividing line between safety and danger. When limits have been exceeded by a small amount, it is generally more significant that there has been a failure of control than that one or more individuals have exposed themselves to dangerous radiation levels.

15.13.3. The collective dose

The linear relationship between dose and effect, as illustrated by eqn. (15.4) and the line a in Figure 15.11, are based on the assumption that cancer induction is a stochastic single hit process independent of dose rate or dose fractionation. Thus the detriment to the population is the same whether one person receives 20 Sv, or 20 000 receive 1 mSv each. Using the dose-effect relation of (15.4) there will be a 100% chance of cancer in both cases. We can express this by saying that the *collective effective dose* is 20 man sieverts

[1] The ALARA principle is founded on the NTLE hypothesis.

(manSv). The collective effective dose, S_{coll}, is the sum of the effective doses to all n individuals:

$$S_{coll}\,(\text{manSv}) = \overset{n}{\Sigma} H_{E,i} n_i \tag{15.6}$$

For simplicity, we consider only whole body doses although the collective dose concept is equally useful for organ or tissue risk evaluations (mine workers, thyroid patients, etc). In the equation, i refers to a situation where each of the n persons has received different *personal doses*, $H_{E,i}$.

For example, in the Chernobyl accident (1986), it is estimated that the 24 000 people that were evacuated from the Pripyat area received a collective dose of 11 000 manSv. According to (15.4), this would mean an expected increase in cancer incidents of 550 cases spread over the lifetime of these people.

The collective (effective) dose concept is commonly applied to natural radiation background. At an average level of 3 mSv/y; 0.015% (15.4) of the population should die of cancer each year from natural radiation. For a population of 50 M people, the collective dose becomes 150 000 manSv/y and corresponds annually to 7 500 additional cancers (out of an expected 100 000 cancer deaths/y). Consequently, the background radiation may be responsible for about 5–10% of all cancers; a more prudent statement is "≤10% of all cancers". As shown earlier, this claim is not possible to confirm by epidemiological investigations.

Suggestions have been made to also apply the collective dose concept to chemical poisons. We illustrate the consequence of this with an example: copper is a natural and needed constituent of our body (0.00010%) that takes part in enzyme reactions. However, an amount of ~6 g of copper as a salt is lethal (*minimum lethal dose*). Using the collective dose concept, one finds that one of every 6×10^6 persons should die of copper poisoning. This is a condition for which there is no scientific support.

15.13.4. Committed doses

In order to relate the emissions of radioactivity from nuclear power installations or the accumulation of radioactivity in the body from fall-out to a resulting dose to the population, the ICRP has introduced the *committed dose* concept (equivalent or effective), S_{comm}. S_{comm} is the total dose contribution to the population over all future years of a specific release or exposure. It is defined as the infinite time integral of (i) the *per caput dose rate* ($\dot{H} = \mathrm{d}H/\mathrm{d}t$), or (ii) the *man-Sievert dose rate* ($\dot{S} = \mathrm{d}S/\mathrm{d}t$) due to a specific event for a population (such as a critical group or the world population):

$$H_{comm}(t)\,(\text{Sv}) = \int \dot{H}\,\mathrm{d}t \tag{15.7}$$

$$S_{comm}(t)\,(\text{manSv}) = \int \dot{S}\,\mathrm{d}t \tag{15.8}$$

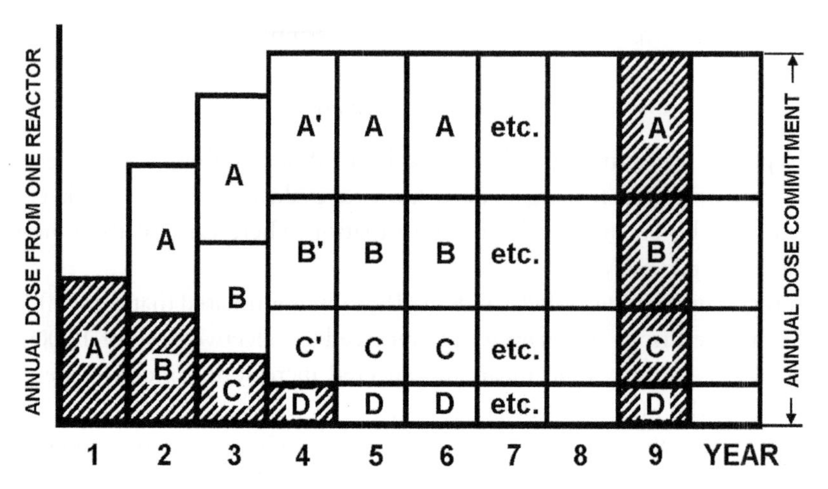

Figure 15.13 Dose commitment for continuous radioactive decay.

The concept is best described by Figure 15.13, where each rectangle represents the dose delivered in one year from an annual release. A is the dose from the first year's release; the next year contains a smaller dose B because of radioactive decay (or other removal processes); the third year yields an even smaller dose C. We assume that the fourth year is the last year that the dose contribution (D) is significant. In this same fourth year, we have a "first year release", A', equal in amount to A, plus what is left from the previous years, B' and C'. Thus, the annual dose at release equilibrium is equal to the dose commitment for one year.

Eqns. (15.7) and (15.8) are integrated between the time of release and infinite time. If infinity is considered unreasonable, one uses "truncated time": e.g. for a person it may be 70 years, while for the human population a common figure of 500 years is often used. It is sometimes claimed in such calculations that improvements in medical science are incorporated (e.g. improved leukemia treatments reduce mortality probabilities).

The collective dose concept allows for extrapolation of the consequences from large scale introduction of nuclear power, which, in turn, establishes the need to ensure that the total annual dose stays within agreed safe limits. If it is assumed that fission power will be used for only about 100 y, the dose commitment integral may be limited to 100 y (sometimes called "incomplete collective dose").

15.13.5. Internal radiation and dose coefficients

When a radioactive substance is taken into the body by ingestion or inhalation in sufficient amounts to be a hazard, the procedure is to attempt to remove it as fast as possible so that it does not have time to become incorporated into tissues that have relatively long *biological lifetimes*, such as the bones. Administration of chemical complexing agents, which form stable complexes with the radioactive substances, provides a mechanism for the removal of radioactivity from the body.

A substance which represents a hazard within the body due to its radioactivity is referred to as *radiotoxic.* The radiotoxicity depends on the properties of the radiation and on a number of physical, chemical, and biological conditions such as mode of intake (via air, in water or food, through wounds, etc.), the size of the ingested or inhaled particles, their chemical properties (e.g. solubility), metabolic affinity, and ecological conditions. Most of these conditions were considered in the ALI concept. ALI (*Annual Limits of Intake*) and DAC (*Derived Air Concentrations*, from the ALI value) and other relevant data are presented in Table 15.12. An annual intake of 1 ALI corresponds to an annual committed dose equivalent of 50 mSv for a "reference human body", or a committed dose equivalent of 500 mSv to any individual organ or tissue, whatever dose is the smaller[2]. ALI and DAC are now obsolete units, although they are still used quite frequently.

In recent years ALI and DAC have been replaced by *dose coefficients* that express the radiotoxicity of a nuclide as (Sv Bq^{-1}) for ingestion or inhalation. The dose coefficients are based on a metabolic model slightly different from that used for ALI and DAC, and they are used to calculate the total effective dose that an internal contamination of a nuclide or mixture of nuclides will cause.

In Table 15.12, t_{eff} is the *effective half-life* of the radionuclide in the body, defined by

$$t_{eff}{}^{-1} = t_{½}(\text{biol})^{-1} + t_{½}(\text{phys})^{-1} \tag{15.9}$$

where $t_{½}$(biol) is the biological half-life, $t_{½}$(phys) is the normal "physical" halflife, and time is counted from the moment when the radionuclide enters the body by ingestion, inhalation, etc. For example, taking the whole body into account $t_{½}$(biol) is 230 d for C, 19 d for Na, 38 d for K, 130 d for I, 190 d for Sr, and 20 000 d for Ra. For strontium incorporated in bone, $t_{½}$(biol) is 4000 d. Radium is assumed to enter into the bone structure. In Table 15.12, a *critical organ* is selected by weighing of two factors. One is the metabolic affinity and the other is the damage by radiation. Some of the commonly used nuclides are listed in Table 15.13 according to their relative radiotoxicities.

Many element shave different paths in the body, e.g. plutonium where an indigestion of Pu(OH)4 results in most of the material passing unaffected through the body while a small part is accumulated in the bone. The biological half life is the calculated in a similar was as for branched decays, cf § 5.14.

Eqn. (15.9) is quite general and may also include terms for other removal processes, such as washing of deposits of ^{90}Sr or ^{137}Cs from the soil, etc.

Beta-ray emitters dissipate their energy over a somewhat larger volume than that of alpha-emitters, but the energy absorbed per unit length is still sufficient to be very damaging to the tissue. Elements such as sodium and potassium represent slight hazards as

2 These general dose limits may be lower in some countries.

Table 15.12 ICRP values (1993) for effective half-lives, dose equivalent conversion factors for ingestion, ALI and DAC values. DAC is Class D (except when W or Y), particle size 1 μm

Nuclide (a)	$t_{1/2}$	t_{eff} (days)	Conversion factors (Sv/Bq) (g, h)	Ingestion ALI (Bq) (b)	Inhalation DAC (Bq/m³) (f)
^{3}H body tissue	12.35 y	12	2×10^{-11} h	10^9	8×10^5
^{14}C fat	5730 y	12	2×10^{-9} g	4×10^7	2×10^4 c
^{24}Na GI(SI)	15.0 h	0.17		5×10^7	2×10^4 c
^{32}P bone	14.3 d	14	2×10^{-9} h	8×10^6	1×10^4
^{35}S testis	87.4 d	76		7×10^7	3×10^4 c
^{42}K GI(S)	12.4 h	0.04		5×10^7	2×10^4 c
^{51}Cr GI(LLI)	27.7 d	0.75	3×10^{-11} h	5×10^8	7×10^5
^{55}Fe spleen	2.7 y	390		1×10^8	4×10^4 c
^{59}Fe GI(LLI)	44.5 d	0.75	4×10^{-9} h	1×10^7	5×10^3
^{60}Co GI(LLI)	5.27 y	0.75	8×10^{-9} g	7×10^6 W	3×10^3 W
^{64}Cu GI(LLI)	12.7 h	0.75		2×10^8	8×10^4 c
^{65}Zn total	243.9 d	190		5×10^6 Y	2×10^3 c
^{89}Sr total	50.5 d		2×10^{-9} h	6×10^6	1×10^4
^{90}Sr bone	29.1 y	6000	4×10^{-8} g	6×10^5	3×10^2
^{95}Zr bone surface	64.0 d	0.75	5×10^{-9} h	2×10^7	3×10^2
^{99}Tc GI(LLI)	2.1×10^5 y	0.75	3×10^{-10} g	3×10^7	1×10^4 c
^{106}Ru GI(LLI)	370 d	0.75	2×10^{-8} h	2×10^6	1×10^3
^{129}I thyroid	1.57×10^7 y	140	6×10^{-8} g	2×10^5	
			5×10^{-8} h		1×10^2
^{131}I thyroid	8.04 d	7.6	9×10^{-9} h	8×10^5	7×10^2
^{137}Cs total	30.0 y	70	1×10^{-8} g	1×10^6	
			9×10^{-9} h		2×10^3
^{140}Ba GI(LLI)	12.7 d	0.75		6×10^6	3×10^3 c
^{144}Ce GI(LLI)	284 d	0.75	5×10^{-8} h	2×10^6	4×10^2
^{198}Au GI(LLI)	2.7 d	0.75		1×10^7	4×10^3 c
^{210}Po spleen	138 d	42	5×10^{-7} g	9×10^4	4×10^1 c
^{222}Rn k lung	3.8	(3.8)	(k)	~70 (k)	
^{226}Ra k bone	1600 y	16000	3×10^{-7} g	9×10^4 W	
			2×10^{-6} h		1×10^1 W
^{232}Th bone	1.4×10^{10} y	73000	4×10^{-4} h	5×10^4 W	2×10^1 W
^{233}U bone,lung	1.58×10^5 y	300	3×10^{-7} g	7×10^5	3×10^2 c
^{238}U lung,kidney	4.5×10^9 y	15	3×10^{-7} g	8×10^5	2×10^1
				3×10^6 W	1×10^1 W
^{238}Pu bone	87.7 y	23000	1×10^{-4} g	4×10^4 W	1×10^1 W
^{239}Pu bone	24065 y	72000	1×10^{-6} g	4×10^4 W	
			1×10^{-4} h		1×10^1 W
^{241}Am kidney	432 y	23000	9×10^{-7} g	3×10^4 W	1×10^1 W

GI gastrointestinal, LLI lower large intestine, S stomach, SI small intestine. (b) From ICRP#61. (f) From ICRP #54, or calculated "c" by the relation DAC = ALI/2400 Bq/m³. (g) For ingestion; ICRP #30 or #56. (h) For inhalation; ICRP #54. (k) Including daughter products;conversion factor for a 1 year exposure is 0.08 mSv/Bq m³; see text.

Table 15.13 Classification of radionuclides according to their radiotoxicity

I.	*Very high*:	^{90}Sr, Ra, Pa, Pu
II.	*High*:	^{45}Ca, ^{55}Fe, ^{91}Y, ^{144}Ce, ^{147}Pm, ^{210}Bi, Po
III.	*Medium*:	^{3}H, ^{14}C, ^{22}Na, ^{32}P, ^{35}S, ^{36}Cl, ^{54}Mn, ^{59}Fe, ^{60}Co, ^{89}Sr, ^{95}Nb, ^{103}Ru, ^{106}Ru, ^{127}Te, ^{129}Te, ^{137}Cs, ^{140}Ba, ^{140}La, ^{141}Ce, ^{143}Pr, ^{147}Nd, ^{198}Au, ^{199}Au, ^{203}Hg, ^{205}Hg
IV.	*Low*:	^{24}Na, ^{42}K, ^{64}Cu, ^{52}Mn, ^{76}As, ^{77}As, ^{85}Kr, ^{197}Hg

their body chemistry does not tend to localize them in any particular organ and their exchange rate is high, leading to rapid elimination. Strontium and iodine, on the other hand, are localized and retained, and therefore are more hazardous.

Heavy elements such as radium and plutonium are often concentrated in the most sensitive areas of the bone where their α-emissions provide essentially lifetime irradiation since the rate of exchange is quite small. The energy is dissipated in the small volume where the element is concentrated and considerably increases the local biological damage.

It is now believed that the largest contributor to population radiation dose comes from low *radon and radon daughter* concentrations in air (see §15.10 and Tables 15.8 and 15.9). These sources were discussed in §13.6. In the UK, the average radon dose is 1.2 mSv/y, with a variation between 0.3 and 100 mSv/y. Though these are large differences, the epidemiological investigations have failed to show any statistically significant correlation between the radon doses and lung cancer. However, at larger doses there is clear evidence of correlation. Of the uranium miners in Erzgebirge from 1875 through 1912, 25–50% (the statistics is somewhat uncertain) died by lung cancers due to inhaled radon. Note that the "Rn-dose" is not delivered by isotopes of radon (when inhaled they are rapidly expelled), but by the Rn-daughters (see Fig. 1.2) which are associated with particles in the air (aerosols) containing the Rn. It has also been found that smokers among uranium miners in the United States have an incidence of lung cancer 10 times higher than nonsmoker miners. Such synergistic effects seem common to cancer. In this particular case, there are several synergistic factors such as the "mine dust" (sharp mineral fragments in the air), but it is not considered in the investigation.

The lifetime risk of lung cancer from radon daughters have been estimated to be 0.2–3 lung cancers per year in a population of one million people continually exposed to 1 Bq/m^3. The probability of obtaining lung cancer from inhaling Rn-daughters is given by ICRP as $1-4 \times 10^{-4}$ WLM^{-1}; 1 Working Level Month is reached after 170 hours of exposure to 3700 Bq (0.1 μCi) Rn-daughters/m^3, corresponding to 72 Bq/m^3 y; in SI-units 1 WLM $= 3.5 \times 10^{-3}$ J h m^{-3}. Authorities apply an expected death rate value for lung cancer of 1.2% for a person living in a space of 100 Bq/m^3 in 80% of the time for 60 years. In several countries, radon "action levels" have been set: e.g. at $\leq$70 Bq/m^3

no action is taken, but at ≥400 Bq/m^3 action must be taken to reduce the level. An international "action limit" is proposed for dwellings exceeding 200–600 Bq/m^3.

It may be noted that the carcinogenic effect of low levels of Rn (i.e. <400 Bq/m^3) is not statistically proven, and therefore contested by some scientists.

15.13.6. Radiotoxicity and risk

A common question in practical work with a particular amount of radioactivity is how "hazardous" it is. Considering both the *intrinsic properties*, *In*, and the *extensive* (*external*) *conditions*, *Ex*, we may designate the hazard (or risk) as a product

$$Ha = In \times Ex \tag{15.10}$$

It is often assumed that a risk, *Ha,* is high if the probability for its occurrence is ≥1%. Authorities usually state that the risk to an individual's health shall be <0.1% to permit an undertaking, such as work with radioactive material. Thus, the probability to induce cancer shall not increase by more than 1/1000.

The intrinsic properties are the amount of radioactivity and factors which give a measure of the risk for the worker (e.g. the radiotoxic properties of the particular nuclide as given by the dose coefficients). It is not possible to draw any definite conclusions about the hazard from a certain amount of a radiotoxic substance. The hazard risk may only be evaluated from its radiotoxicity value. For that purpose, it is also necessary to consider its chemical form and pathways to man, which are considered to be extensive properties. Dose coefficients do take into account if the chemical form is "soluble" or "insoluble", but this is, of course, a rather crude subdivision. They also consider the particle size and time of exposure of airborne contamination (e.g. in a factory). However, it does not consider the particular ways by which the substance is released to the environment (cf. Ch. 21–22).

The pathway of a radioactive substance from its point of release until it reaches a person is the domain of *radio ecology*. Extensive knowledge in this field is essential to the evaluation of the hazards caused by nuclear power. Figure 15.14 pictures the more important pathways for plutonium; dashed lines refer to predominantly liquid flows.

Ex may be >1 if concentration of the radioactive substance occurs (e.g. from grass to cow, or from cow to milk, etc) or <1 if the substance is highly insoluble (precipitates out from the water). This is usually expressed through *transfer* (or "*enrichment*") *factors*, commonly abbreviated k_d. The k_d value is the radionuclide amount per kg product (e.g. milk) divided by the radionuclide amount per kg source (e.g. grass). Typical k_d values for ^{90}Sr and ^{239}Pu from soil to vegetables are 0.2 and 0.0002, respectively, while values for the same nuclides in the water → fish system are typically 1 and 40, respectively. Thus, plutonium is enriched in fish but not in vegetables. Note that there is an alternative pathway: water → sediment → seaweed → fish. The values are "site specific", i.e. measured at two different locations (e.g. Lake Michigan and the Baltic Sea), values are obtained which often differ by more than a factor of ten. See also Ch. 22.

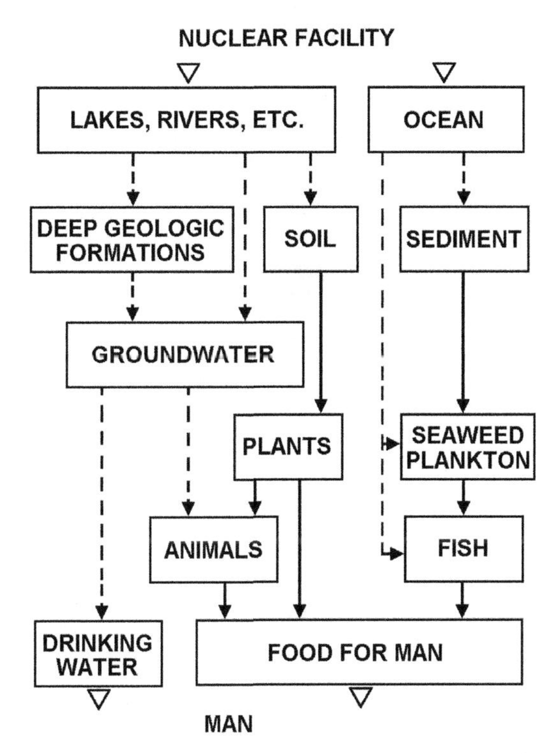

Figure 15.14 Pathways of dissolved plutonium from a nuclear facility to man.

By considering all possible transfer routes, one can estimate what amount of a radionuclide released to the environment may end up in plants, animals, or man. When these figures are combined with the *dose conversion factors* ("committed effective dose equivalent per unit intake", according to ICRP) in Table 15.12, it is possible to calculate the dose received by man from intake of a radionuclide in the environment. The dose conversion factors depend on the mode of intake (usually only inhalation or ingestion). Thus the dose factor of ^{239}Pu is 10^2 larger for inhalation then for ingestion. Both kinds of data are given in Table 15.12, depending on available data (see ICRP publications).

Such calculations are applied in the risk analyses of contemplated nuclear waste repositories and in the risk evaluation of radium or plutonium content in drinking water. They are also used to establish limitations for intake of food from a contaminated area; e.g. in Sweden reindeer meat with >1500 Bq/kg is considered unfit as food (reindeer meat is the main protein intake for the Laps living in Northern Sweden).

15.13.7. Classifications, working rules, etc

Radioactive substances of various activities, concentrations (or specific activities), decay modes, etc., constitute quite different hazards, and must be handled accordingly. Various

Figure 15.15 Label for shipment of radioactive material of class II. "II" in red with a yellow background.

countries classify radioactive material differently and issue different working rules. We give a few classifications and rules, adhering primarily to the recommendations of the IAEA.

Figure 15.15 shows the international transportation label with the symbol for ionizing radiation (with a white or yellow background, wbkg or ybkg). The radiation source and its activity should be given. The category number is shown in red, according to:

I-wbkg, radiation level	$\leq 5\ \mu$Sv/h	at surface of
II-ybkg, "	≤ 0.5 mSv/h	package; for III-
III-ybkg "	≤ 2 mSv/h	≤ 1 mSv/h at 1 m

For high levels, the source must be transported as a special cargo. The transportation carriage must fulfill a number of requirements with regard to resistance against fire, mechanical damage (drop tests), leakage (immersion tests), etc. Special regulations apply for spent reactor fuels and high-level waste (usually >4 000 TBq; see Chapter 20). IAEA has issued rules for transportation of radioactive materials, which are of special importance to large nuclear facilities.

For radio-tracer work in common non-nuclear research laboratories, some general rules can be recommended (see also next section). Spills may result in increase in the radiation background. They may not constitute a hazard to the workers, but may ruin the scientific experiments if not cleaned up immediately. In all work with radionuclides, radioactive waste is produced. It is common practice to collect all such waste in special containers, and to dispose of it according to national rules. For short lived radionuclides of low hazard and low levels of radioactivity (e.g. as in ^{14}C-work), it is common practice to dispose of such waste by normal flushing to the sewer with several liters of tap water if such procedures are permitted by the national radiation protection organizations.

15.14. PROTECTIVE MEASURES FOR RADIOCHEMICAL LABORATORY WORK

Three basic principles are recommended for keeping radiation exposure to a minimum: shielding, control, and distance. If a radiochemical laboratory is designed properly and the work performed in such a manner that the general background contamination is sufficiently low to do valid low level tracer experiments, then the health aspects of radiation control are satisfied. The main principles for work with radioactive substances are indicated, but in each notion, special rules may apply.

15.14.1. Tracer work with moderate β-γ-levels

Ordinary chemical laboratories may be used for radiochemical work at low levels of short lived β-γ radionuclides (e.g. half-lives <14 d and activity levels <10 kBq). However, it is recommended that a special room be used for radioactive work. In the design of such a laboratory it is important that airborne contamination be prevented from spreading to counting rooms and to offices. Therefore, a pressure difference between the laboratories and the other areas is desirable. Thus, the air velocity in the fume hoods should never be below 0.25 m s^{-1}, and 0.5 m s^{-1} is recommended. With such a flow velocity, radioactive dust and fumes are retained in the hood and removed through the vents. The fume hoods should have filters for collecting radioactive particulates.

To minimize the possibility of ingestion as well as the chances of ruining experiments through accidental contamination, limiting the radioactive work to a minimum area is essential. There should be no radioactivity except in the immediate working area, and upon completion of the particular experiment, all activity should be removed and the area cleaned (*decontaminated*) if necessary. For low levels, this means working in a good hood with easily cleaned nonporous surfaces.

One operation which commonly results in contamination involves evaporation of a solution to dryness either on a hot plate or under a heat lamp. Although the percentage of the sample carried away by spray may be very small, it may result in appreciable amounts of activity being spread around the area of the evaporation. Consequently, all evaporation should be performed in a hood and the vicinity should be protected from the active spray by a covering of absorbent paper.

For nuclides which emit only β-particles, the glass walls of the container may provide sufficient shielding. Sheets of glass or plastic (such as lucite) are commonly used to shield exposed solid samples.

For work with higher levels of β-γ-emitters of longer half-lives, special radiochemical laboratories should be used. The working surfaces and floors should be even, nonporous, and with a minimum of seams. Surfaces of plastic material, stainless steel, and artificial stone are acceptable bench materials. It is recommended that fume hood and benches be covered with an absorbing material such as absorbing paper, and that experiments be conducted when possible in trays of stainless steel or plastic. Such arrangements ensure that the radioactive material does not contaminate a larger area if a spill occurs. Radionuclide workers should wear surgical gloves and laboratory coats in the laboratory. If there is any danger of splashing, plastic face-hoods are recommended. It is extremely important that oral contamination be avoided. Beta-radiation from radioactive sources has ranges which rarely exceed 1 g cm^{-2}. Consequently, in a laboratory in which the level of β-emission is less than 1000 MBq, protection from the radiation can be achieved with a 1 cm plexiglass shield.

15.14.2. α-laboratories

For α-emitters the main hazard is internal, not external, to the body. For moderate activity levels a *glove box* under slightly reduced pressure provides a simple and convenient closed chemical laboratory. Samples of the actinide elements usually have high specific radioactivities. Therefore, special care must be exercised in working with them. Because these α- emitters usually are associated with weak γ-radiation or X- rays, chemical work with these elements must be conducted in more advanced gloveboxes (Fig. 15.16), which are kept at a pressure slightly below the surrounding atmosphere by circulation of pure air or inert gas through the box. The boxes should have alarm systems for monitoring hazards such as interrupted water circulation, electrical short circuits, oxygen in the inert gas, heat, etc. In Figure 15.16, the control panel for these protective arrangements is shown above the box. In large alpha-box laboratories, one of the main hazards is radioactive dust. All room surfaces should, therefore, be made with as few seams and sharp corners as possible; particularly the floor must be of high quality. Electric power, water, waste, ventilation, etc., should be connected to piping in the ceiling. The air into the laboratory must pass through filters as must the air exiting the laboratory. The exit air should be monitored for α-activity. Entering and leaving the laboratory should be through airlocks, and the hands and feet must be monitored for activity on exit. All these protective measures make α-laboratories quite expensive, but smaller laboratories working with lower levels of α-activity can be constructed in simpler fashion for correspondingly less cost.

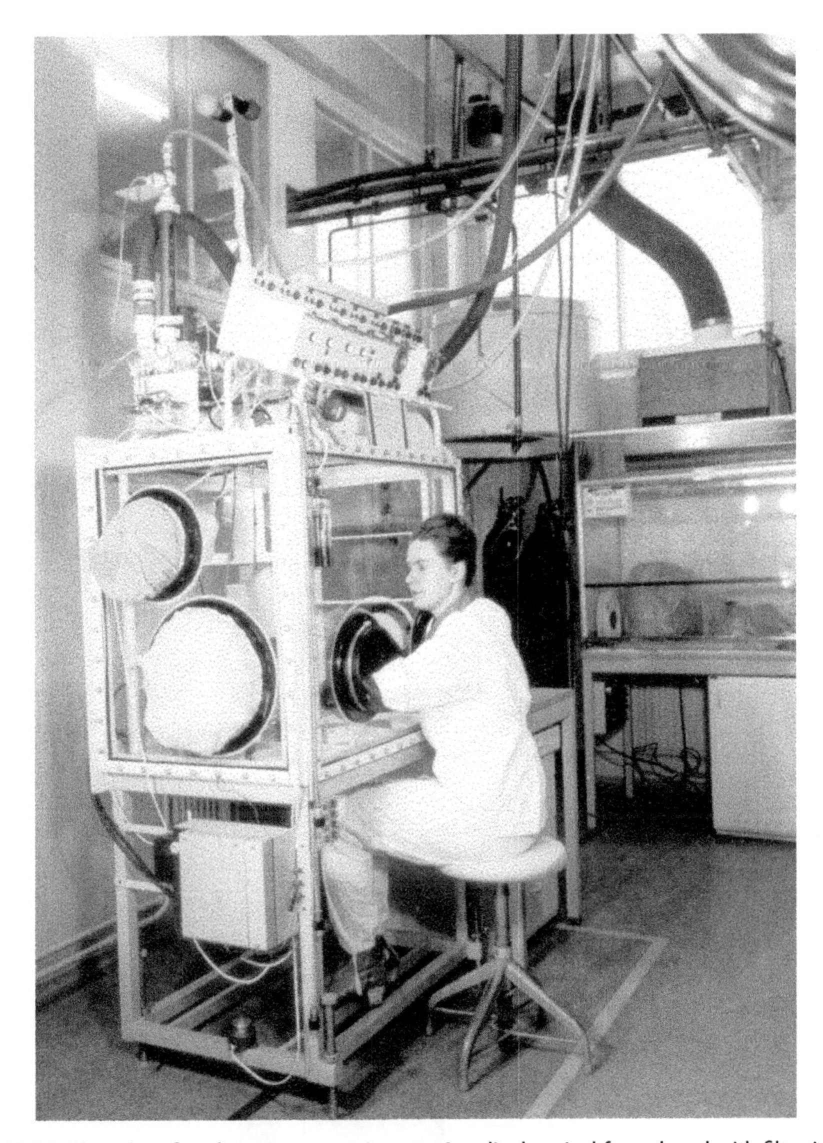

Figure 15.16 Glove-box for plutonium experiments. A radiochemical fume hood with filter is seen in the background. Note that all connections are in the ceiling.

15.14.3. High level β-γ-emitters

Since the intensity decreases as the inverse square of the distance, maintaining maximum distance (by use, when necessary, of remote control apparatus such as tongs) when working with moderate or high levels of activity reduces the exposure appreciably.

High levels of γ, >1000 MBq, require shielding with layers of concrete, water, steel, or lead, and the operations must be carried out by remote control. The eyes can be protected by the use of lead glass windows of high density, periscopes, mirrors, etc., see Figure 15.17.

Commonly, special shielded cells are used. These cells are sealed from the atmosphere and kept at a pressure lower than that for the working personnel. The smallest cells usually have lead walls which sometimes reach to the ceiling; by this arrangement, no scattered radiation reaches the working personnel. Experiments are carried out with the use of tongs passing through the lead walls or, for the thicker cells, with manipulators reaching above or through the walls. In cells for very high activities (≥10 TBq, or >1000 Ci), these manipulators are guided electrically or mechanically (*master-slave manipulators*). All movements of the operator are copied exactly by the slave hands inside the cell. For very complicated work and extremely high radioactivity, robots have been

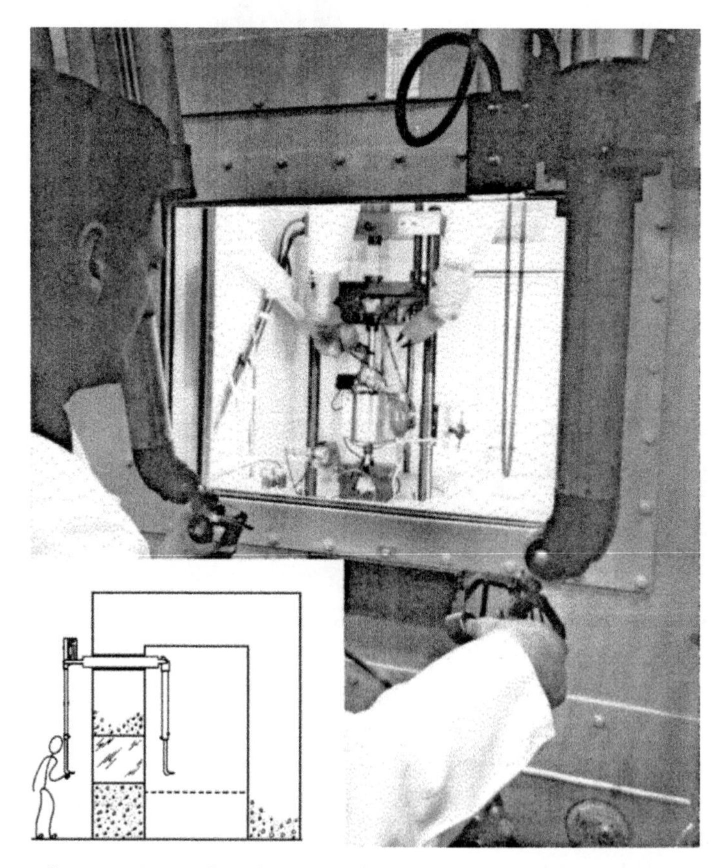

Figure 15.17 High-active β-γ-work with master-slave manipulators. The window is filled with high density $ZnBr_2$ solution.

developed which can be guided to repair such items as unshielded nuclear reactors, radionuclide equipment, heavily contaminated radiochemical apparatus, etc.

15.15. CONTROL OF RADIATION PROTECTION MEASURES

In larger organizations, the control of radiation hazards is the responsibility of specialists known as *health physicists* or, sometimes, *health chemists*. Their main duty is to ensure that work is carried out without hazard to the health of the people involved.

The protection follows three stages: prevention, supervision, and after-control. Preventive measures include use of fume hoods, α boxes, radiation shielding, tongs, etc., as discussed above. The supervision stage involves the use of radiation instruments to monitor the radiation level (see Ch. 9). Small TLD, film or pocket pen dosimeters are used for individual monitoring (§8.9). For spills and contamination of hands, shoes, etc., special contamination instruments (counters) are used which are more sensitive than the monitoring dose instruments.

Contamination in the laboratory must be avoided. This is controlled by smear tests; i.e., a filter paper is wiped over the surface and the paper is checked with a suitable instrument. In a so-called "clean area", the fixed contamination should not exceed 2 Bq for α, and 4 Bq for β-γ on a surface of 100 cm^2. For an "active area" the rules are a maximum of 20 Bq for α per 100 cm^2, and 0.01 $mGy\ h^{-1}$ from β−γ at a distance of 2 cm from the surface. Radioactive aerosols are monitored by air samplers in which a certain amount of air is drawn through a fine filter paper after which the paper activity is measured.

The after-control usually consists of checking personal dosimeters and a medical examination. Depending on the kind of work, the dosimeters are checked from twice a week to once a month. A medical examination is given once or several times a year, depending on the work conditions. In danger of inhalation or ingestion of α- or soft β-emitters, urine samples are analyzed. Such analyses are very sensitive and much less than a kBq in the body is easily detected. For workers who handle hazardous amounts of α-emitters (e.g. plutonium in more than milligram amounts), urine samples should be taken regularly.

If necessary whole-body counts are also taken. *Whole-body counting* is carried out with the subject being surrounded by numerous scintillation or solid state detectors in a heavily shielded room. The natural body content of ^{40}K is easily detected and is a control of the efficiency of this technique.

15.16. EXERCISES

15.1. "Reference man" consists of 18%C, 66% H_2O, 0.2% of K per body weight. He may also have accumulated 10 pCi ^{226}Ra in the body; assume 0.3 decay for each of the following 5 daughters. Calculate for a body weight of 70 kg the number of radioactive decays per unit time from ^{3}H, ^{14}C, ^{40}K and ^{226}Ra. Assume 30 TU in water.

15.2. Using the information above, how many grams of the body's molecules (assume average mole weight of 10^5) will be damaged in a year if the G(damage) value is 3.1×10^{-7} mol/J? Assume $E_{abs}(\beta) = E_{max}/3$.

15.3. Under the same assumption as above, what amount of damage will be caused by cosmic radiation? Assume that the cosmic particles produce 3×10^9 ion pairs s^{-1} m^{-3} of the body.

15.4. With the information in exercise 15.1, calculate annual doses received from (a) ^{40}K, and (b) ^{226}Ra and daughters. Assume $w_r(\alpha) = 20$.

15.5. Ten mg ^{238}U has been collected in the kidneys. Considering the biological half-life of uranium and assuming only one α-emission in ^{238}U decay, calculate the dose (in Sv) received by the organ if the uranium is evenly distributed. The weight of a kidney is 150 g.

15.6. A γ-dose rate of 1 Sv is assumed to inactivate (kill) human cells. The body contains 6×10^{13} cells in a cell weight of 42 kg for a 70 kg man. (a) What average energy (in eV) has to be deposited in a cell to kill it? (b) Calculate the number or kidney cells destroyed for the dose received in exercise 15.5. For simplicity assume the cells to be cubic with a side length of about 11 μm.

15.7. A tumor has the weight of 80 g and we wish to destroy 20% of the cells by irradiating with 180 MeV protons with such penetration that half of the energy is deposited in the tumor. The particle beam is 5 μA. For what time must the irradiation be? A cell of weight 10^{-9} g is assumed to be killed on the absorption of 200 keV and no cell is assumed to be killed twice.

15.17. LITERATURE

E. H. Schultz, *Vorkommnisse und Strahlenunfälle in kerntechnischen Anlagen*, Karl Thiemig, 1965.

K. Z. Morgan and J. E. Turner (Eds.), *Principles of Radiation Protection*, J. Wiley, 1967.

D. J. Rees, *Health Physics*, Butterworths, 1967.

Z. M. Bacq and P. Alexander, *Fundamentals of Radiobiology*, Pergamon Press, 1967.

E. L. Saenger, G. E. Thomas and E. A. Tompkins, Incidence of leukemia following treatment of hyperthyroidism, *J. Amer. Med. Assoc.* **205** (1968) 855.

H. Kiefer and R. Maushart, *Radiation Protection Measurement*, Pergamon Press, 1972.

IAEA, *Population Dose Evaluation and Standards for Man and his Environment*, Vienna, 1974.

K. K. Manocha and R. K. Mohindra, Estimate of natural internal radiation dose to man, *Health Physics* **30** (1976) 485.

G. Walinder (Ed.), *Tumorigenic and Genetic Effects of Radiation*, National Swedish Environmental Protection Board 1976.

B. L. Cohen, The cancer risk from low-level radiation, *Health Physics* **39** (1980) 659.

E. S. Josephson and M. S. Peterson, *Preservation of food by ionizing radiation*, CRC Press, 1983.

C. Ferradini and J. Pucheault, *Biology of ionizing radiation*, Masson, 1983.

J. E. Coggle, *Biological Effects of Radiation*, Taylor & Francis Ltd, 1983.

M. Tubiana, J. Dutreix and A. Wambersie, *Radiobiologie*, Hermann, 1986.
J. Rotblat, A tale of two cities, *New Scientist*, January 7 (1988) 46.
L. Stryer, *Biochemistry*, W H Freeman & Co, 1988.
L.-E. Holm, K. E. Wiklund, G. E. Lundell, N. Å. Bergman, G. Bjelkegren, E. S. Cederquist, U.-B. C. Ericsson, L.-G. Larsson, M. E. Lidberg, R. S. Lindberg, H. V. Wicklund and J. D. Boice Jr, Thyroid cancer after diagnostic doses of iodine-131: A retrospective cohort study. *J. Natl. Cancer Inst.* **80** (1988) 1132.
R. Yalow, *Biological effects of low-level radiation*, Ch. 12 in M. E. Burns (Ed), Low Level Radioactive Waste Regulation: Science, Politics, and Fear, Lewis Publishers, 1988.
E. S. Gilbert, G. R. Petersan and J. A. Buchanan, Mortality of workers at the Hanford site: 1945–1981, *Health Physics* **56** (1989) 11.
National Research Council, *Health effects of exposure to low levels of ionizing radiation. BEIR V*, National Academy Press, 1990.
A. H. W. Nias, *An Introduction to Radiobiology*, J Wiley & Sons, 1990.
C. A. Sondhaus, V. P. Bond and L. E. Feinendegen, Cell-oriented alternatives to dose, quality factor, and dose equivalent for low-level radiation, *Health Physics* **59** (1960) 35.
Annals of the ICRP, Pergamon Press, Oxford: Publ. #54 (1988), *Individual Monitoring for Intakes of Radionuclides by Workers: Design and Interpretation*. #58 (1989), *RBE for Deterministic Effects*. #60 (1991), *1990 Recommendations of the International Commission on Radiological Protection*. #61 (1991), *Annual limits of Intake of Radionuclides by Workers Based on the 1990 Recommendations*.
E. M. Fielden and P. O'Neill (Eds.), *The early effects of radiation on DNA*, NATO Symposium, Springer Verlag, 1991.
R. Mukherjee and J. Mircheva, Radiological effects of low-level radiation and cancer risks, *IAEA Bulletin* **2** (1991) 32.
United Nations Scientific Committee on the Effects of Atomic Radiation (UNSCEAR): *Genetic and somatic effects of ionizing radiation*, United Nations, 1993.
United Nations Scientific Committee on the Effects of Atomic Radiation (UNSCEAR): *Effects of ionizing radiation*, ISBN 92-1-142280-1, United Nations, 2008.
A. D. Wrixon, New ICRP Recommendations, J. Radiol. Prot. **28** (2008) 161.
ICRP 2007 *Recommendations of the International Commission on Radiological Protection*, ICRP Publication #103; Ann. ICRP **37** (2007) (2–4)

M. Tubiana, J. Dutreix and A. Wambersie, Radiobiologie, Hermann, 1986.
[illegible] 1988, [illegible]
[illegible]
[illegible]
[illegible]
[illegible]
[illegible] Kennedy [illegible], M. [illegible] Thomas [illegible]
[illegible]
[illegible]
[illegible]
[illegible] Academy Press, 1990.
[illegible]
[illegible]
[illegible]
[illegible]
[illegible]

CHAPTER 16

Particle Accelerators

Contents

This chapter describes the sources – accelerators – that can be used to obtain energetic charged particles, and secondarily neutrons and photons, that can induce nuclear reactions or chemical changes. Thus we include various accelerators for producing beams of charged particles as well as targets for neutron production. Electrons can be accelerated by the same principles as positive ions but in the following paragraphs we will focus on the acceleration of positive ions because these are more useful for nuclear reactions. Although accelerators can be used to produce neutrons, a more copious neutron source is the nuclear reactor, see Chapter 19. Small neutron sources, consisting of the spontaneous fissioning nuclide ^{252}Cf or of mixtures of elements like radium and beryllium in which nuclear transmutations (mostly (α,n)-reactions) produce neutrons, are discussed in §10.9.

16.1. CHARGED PARTICLE ACCELERATORS

In order to induce nuclear reactions with positively charged projectiles such as protons, deuterons, α-particles, oxygen ions, or uranium ions, it is necessary that the projectile particles have sufficient kinetic energy to overcome the Coulomb barrier created by the repulsion between the positive charges of the projectile and the target nucleus. While there is some probability that a positive projectile can tunnel through the Coulomb barrier at kinetic energies lower than the maximum value of the

Radiochemistry and Nuclear Chemistry
ISBN 978-0-12-405897-2, http://dx.doi.org/10.1016/B978-0-12-405897-2.00016-1

barrier, this probability is quite small until the kinetic energy is close to the barrier maximum.

As mentioned in §10.3, in 1919 Rutherford caused artificial transmutation of one element into another by using the α-particles emitted in the radioactive decay of ^{214}Po as projectiles. These α-particles had sufficient kinetic energy upon emission from the polonium nucleus (7.7 MeV) to overcome the Coulomb barrier and react with nitrogen nuclei. However, the kinetic energy of α-particles emitted in radioactive decay is generally insufficient to overcome the Coulomb barrier to react with nuclei of higher atomic numbers. Consequently, means had to be devised for acceleration of the charged projectiles to kinetic energies sufficient to achieve reaction.

The principle to be used to achieve the higher kinetic energies was obvious. The projectile particles would have to be ionized to obtain positively charged ions. If these ions could be accelerated through a potential difference of 1000 V, they would acquire 1000 eV additional kinetic energy, per unit of charge. If an α-particle of a +2 charge was to be accelerated through a potential difference of 10^6 V, it would acquire an additional kinetic energy or 2 MeV (eqn. (3.6)). The problem of obtaining the desired kinetic energy involved two aspects: first, the production of the charged particles: second, the acceleration through the necessary potential difference.

16.2. ION SOURCE

The problem of producing the positive ions is in principle relatively simple. If a gas is bombarded by energetic electrons, the atoms of the gas are ionized and positive ions produced. Figure 16.1 shows a simple ion source.

As hydrogen gas flows into the region above the filament, the electrons being emitted by the filament are accelerated to an anode (a typical voltage drop over the electrodes B_1-B_2 may be 100 V) and in their passage through the gas cause ionization. The positive ions are extracted by attraction to a negative electrode (the voltage drop over S_1-S_2 may be 1–10 kV) into the accelerator region. The vacuum at the beam extraction is of the order 10^{-4} Pa but in the ionization compartment it is normally 10^{-2} Pa. The basic principle, bombardment with electrons, is similar in the different types of ion sources that have been developed to meet the demands of the variety of ions and accelerator types in use. High ion currents of all elements in a variety of charge states can now be produced.

16.3. SINGLE-STAGE ACCELERATORS

The first successful accelerator for causing nuclear transmutations was developed by Cockcroft and Walton in 1932 and used a high voltage across an acceleration space.

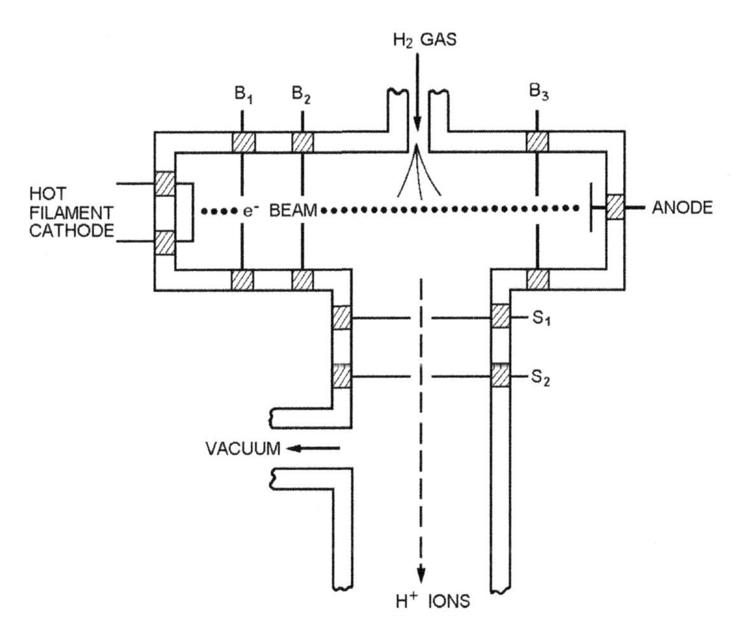

Figure 16.1 A schematic drawing of an ion source producing protons by electron bombardment of hydrogen gas.

Gamow's development of the tunnelling theory for α-decay led Houtermans to suggest an opposite reaction, a high energy proton tunnelling through the Coulomb barrier into a target nucleus. Calculations showed that such tunnelling could occur even at about 100 keV. Cockcroft, who was an electrical engineer, realized that it would be possible to build a high tension generator capable of accelerating protons to such energies (although even Rutherford first thought it technically impossible).

Cockcroft-Walton accelerators are still used for obtaining "low" kinetic energies, up to 4 MeV for protons. Presently, such accelerators are used in many installations as the first stage of acceleration, *injector*, in a more complex machine designed to produce high energy beams (see Figures 16.8 and 16.9).

In a single-stage accelerator, such as a Cockcroft-Walton type, the total potential produced from a high voltage generator is imposed across the accelerator, i.e. between the ion source and the target. The kinetic energy, E_{kin}, of the projectile is:

$$E_{\text{kin}} = nqV \qquad (16.1)$$

where q is the charge of the accelerated ions (Coulomb) and V is the imposed potential across the acceleration gap. This equation is identical to (3.6) except that n is the number of accelerating stages ($n = 1$ for the Cockcroft-Walton machine).

In recent years small and relatively inexpensive accelerators have come into use based on the Cockcroft-Walton principle. These are known as *transformer-rectifier accelerators* and are primarily used for acceleration of electrons or acceleration of deuterons for production of neutrons through the reaction:

$$ {}^{2}_{1}\mathrm{H} + {}^{3}_{1}\mathrm{H} \rightarrow {}^{4}_{2}\mathrm{He} + {}^{1}_{0}\mathrm{n} \tag{16.2}$$

Tritium targets are bombarded by accelerated deuterons. Tunneling of the Coulomb barrier (see Fig. 16.10) results in a good yield for this reaction even for energies of 0.1 MeV.

Figure 16.2 illustrates an inexpensive transformer-rectifier type accelerator. Deuterium molecules leak through a heated palladium foil into the vacuum of the ion source, where a high frequency electric field decomposes the deuterium molecules to form a plasma of D^{+1} ions and electrons. The deuterium ions are extracted from the ion source with a relatively low negative potential to enter the acceleration tube with approximately 2.5 keV kinetic energy. The voltage difference of 100 kV across the acceleration space is obtained from a transformer and rectifier unit coupled to a set of cylindrical electrodes connected by a resistor chain.

The cylindrical electrodes serve the purpose both of acceleration and of focusing the ion beam into the proper path. As the beam particles exit from the last electrode they drift through a short tube and strike the target, which consists of a metal foil covered with titanium in which tritium has been absorbed. Approximately 200 GBq of tritium

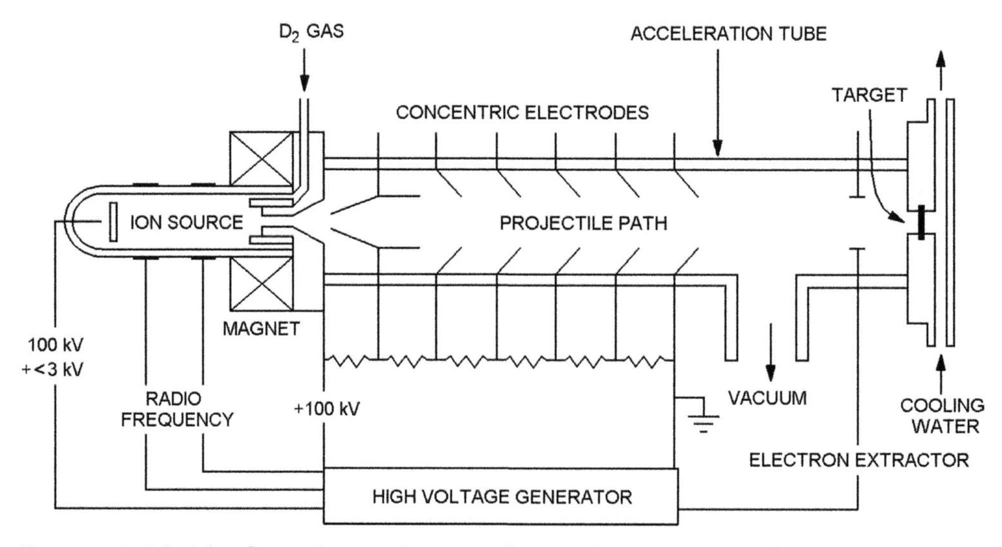

Figure 16.2 Principle of a single-stage linear accelerator. The D_2 gas enters the vacuum through a palladium membrane.

can be absorbed as titanium tritide on an area of 5 cm^2. The target is cooled by water to minimize tritium evaporation. The passage of the beam through the target also results in the emission of electrons which are removed by attraction to an anode to avoid interference with the beam.

With a voltage of 100 kV and an ion current of 0.5 mA, this accelerator can produce approximately 10^{10} n s^{-1} with an energy of 14 MeV. It is often desirable to reduce the kinetic energy of these neutrons to thermal values since thermal neutrons have a much larger probability of interacting with target nuclei. Moderation of the neutrons to thermal energies ($\approx$0.025 eV) can be accomplished by placing water or paraffin around the target (see §§7.6 and 19.5). The flux of the thermal neutrons is relatively low, on the order of 10^8 n cm^{-2} s^{-1}. However, the production rate of neutrons increases as the beam energy and beam current increases.

Small accelerators of similar design are commercially available with such dimensions that they can be lowered into bore-holes for in situ neutron activation analysis during gas or oil prospecting.

More advanced designs of this type of accelerator use coupling of transformers, rectifiers, and condensers to increase the voltage. With the larger single-stage accelerators it is possible to obtain beams of protons and deuterons which have energies of several MeV and currents of about 10 mA.

16.4. VAN DE GRAAFF ACCELERATORS

A type of electrostatic accelerator that has been of great value to nuclear science since its invention is that designed by van de Graaff in 1931. The van de Graaff accelerator (VdG) can provide beams of higher energy than the single-stage Cockcroft-Walton accelerators with a very precisely defined energy. In a modification known as the tandem-VdG beams of 20 MeV protons and 30 MeV α-particles can be achieved. Positive ions of higher Z and electrons can also be accelerated in VdG machines.

The principle of the VdG accelerator is illustrated in Figure 16.3. A rapidly moving belt of a non-conducting material such as rubberized fabric accumulates positive charge as it passes an array of sharp spray points which transfer electrons from the belt to the spray points. The positive charge on the belt is continuously transferred by the movement of the belt away from ground. At the high-voltage terminal, a hollow metal sphere or nearly cylindrical shape, another set of spray points neutralize the charges on the belt by electrons emitted from the spray points. The result is the transfer of the positive charge to the sphere. By a continuous process of transfer of positive charge to the sphere via movement of the belt a high potential relative to ground can be built up on the sphere. The limit of the voltage that can be accumulated on the hollow electrode is determined by its discharge potential to the surrounding housing. If it is insulated by some pressurized non-conducting gas such as N_2, CO_2, or SF_6, potentials of $\sim$15 MV can be

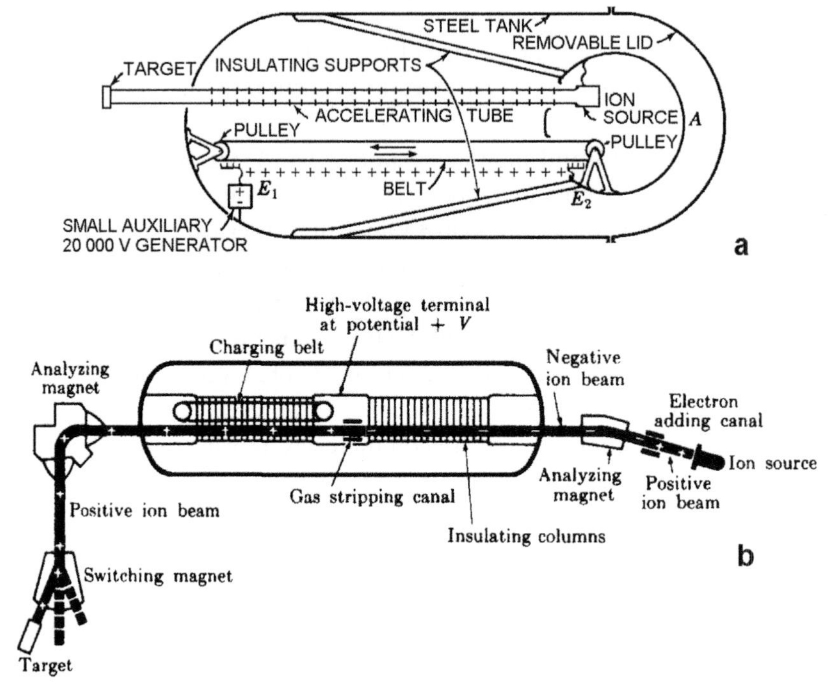

Figure 16.3 (a) The main principle of the van de Graaff accelerator. (According to R. S. Shankland.) (b) The principle of the tandem-VdG accelerator. *(According to R. V. van de Graaff.)*

achieved. This potential can be used to accelerate protons to energies of ~15 MeV in a single stage.

The single-stage VdG has only one accelerating tube. In the tandem VdG modification, atoms are first given a negative charge by bombardment with low energy electrons in the ion source, and then accelerated from ground to the positive potential at the center electrode. At this stage, in the center of the accelerator, the negative ions are stripped of electrons by passage through a thin foil or a gas volume to become positive ions. The average charge, z, of an ion of velocity v (m/s) and nuclear charge Z emerging from a solid material is given by the following empirical relation.

$$z = Z\left[1 + \left\{v/\left(3.6 \times 10^6\left(\mathrm{m/s}\right)\ Z^{0.45}\right)\right\}^{-1.67}\right]^{-0.6} \tag{16.3}$$

In the second stage of acceleration, the *post-stripper*, the positive ions are accelerated as they move from the positive potential on the center electrode back to ground. The emergent beam has a total energy of 20 MeV if the accelerating potential is 10 MV. Beams of hydrogen, helium, nitrogen, oxygen, and heavier ions can be accelerated similarly. The highest energy so for achieved in a tandem VdG accelerator is 60 MeV corresponding to M^- ions in the first stage and M^{5+} in the second ($1 \times 10 + 5 \times 10 =$

60 MeV). By using two separate tanks and alternatingly positively and negatively charged ions, three-stage VdG machines with maximum proton energies up to ~45 MeV have been built.

The energy of a beam produced by a VdG generator is extremely precise. However, the current intensity, 10–100 μA, is somewhat less than that of other accelerators. The beam current i (amperes) is given by the relation

$$i = q\,I_o = \mathbf{e}\,z\,I_0 \tag{16.4}$$

where q is the projectile charge (C), I_0 the incident particle current (particles s^{-1}), $\mathbf{e} = 1.6 \times 10^{-19}$ C, and z the net charge (in electron units) of the beam particle (ion).

16.5. MULTIPLE-STAGE LINEAR ACCELERATORS

The potential obtained from a high voltage generator can be used repeatedly in a multiple-stage accelerator process ($n > 1$ in (16.1)). The linear accelerator operates on this principle. The accelerator tube consists of a series of cylindrical electrodes called drift tubes. In the Wideröe linear accelerator, the electrodes are coupled to a radio frequency generator in the manner shown in Figure 16.4. The high voltage generator gives a maximum voltage V which is applied to the electrodes by the radio frequency so that the electrodes alternate in the sign of the voltage at a constant high frequency. If the particles arrive at the gap between electrodes in proper phase with the radio frequency such that the exit electrode has the same charge sign as the particle and the entrance electrode the opposite, the particles are accelerated across the gap. Each time the particles are accelerated at an electrode gap they receive an increase in energy of qV; for n electrode gaps the total energy acquired is nqV (16.1). Inside the drift tubes no acceleration takes place since the particles are in a region of equal potential.

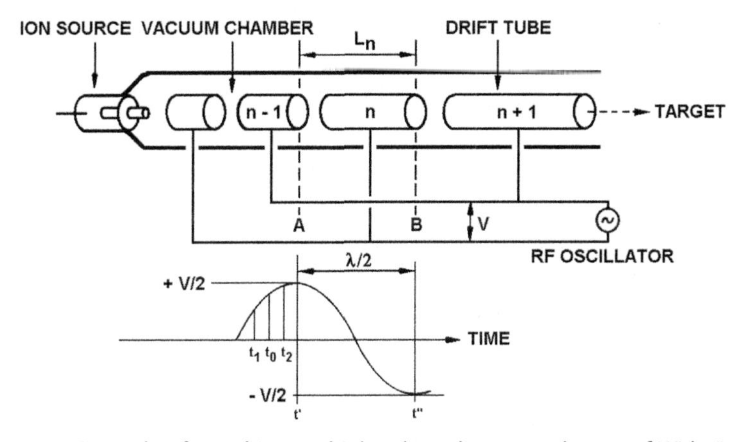

Figure 16.4 Principle of a multi-stage high voltage linear accelerator of Wideröe type.

Veksler and McMillan independently demonstrated the principle of phase stability, which is based on an effect which tends to keep the particles in phase with the radio frequency oscillation of the potential and allows beams of sufficient intensity for use in research. In the lower part of Figure 16.4 the variation of potential at point A with time is shown. In operation, a particle in proper phase should arrive at the acceleration gap at time t_o. If a particle has less energy, it takes longer time to traverse the drift tube and arrives late at the gap, e.g. at time t_2. Since the potential is now higher, it receives a greater acceleration, thereby becoming more in phase. If a particle is too energetic, it arrives too early at the gap t_1 and receives a smaller acceleration, thereby putting it better in phase. This bunching of the ions not only keeps the ions in phase for acceleration but also reduces the energy spread in the beam. When the velocity approaches **c**, phase stability vanishes gradually because changes in energy now mostly becomes changes in mass and not in velocity.

Since the velocity of the beam particles are increasing as they progress down the accelerator but the oscillator frequency remains constant, it is necessary to make the drift tubes progressively longer so as to allow the beam particles to arrive at the exit of each tube in phase with the oscillation of the potential. Acceleration to high energies requires relatively long acceleration tubes. The electric field between the tubes which accelerates the particles also spreads the beam. The defocussing effect can be minimized by the use of electrostatic and magnetic lenses, in the drift tubes, which refocus the beam onto the center line of the acceleration tubes.

For protons and deuterons the increase in energy below 10 MeV is directly related to the increase in the velocity of the particles. However, as the kinetic energy exceeds 10–20 MeV the relativistic mass becomes important since the speed is approaching the velocity of light (Fig. 5.2). At projectile energies above 100 MeV, the relativistic mass increase is greater than the increase in velocity and at very high energies, the GeV range, the energy increase is practically all reflected by an increase in the relativistic mass (at 1.9 GeV, the proton mass is double that of the rest mass).

The phase requirement (Fig. 16.4) is that the time from point A to B (i.e. t''-t') must be

$$L_n/v_n = \lambda/(2\mathbf{c}) \tag{16.5}$$

Taking the relativistic mass increase into consideration it can then be shown that the length of the drift tube at the nth stage is

$$L_n = (\lambda/2)\left[1 - (nk+1)^{-2}\right]^{1/2} \tag{16.6}$$

where

$$k = qV/\left(m^{\circ}\mathbf{c}^2\right) = z\mathbf{e}V/\left(m^{\circ}\mathbf{c}^2\right) \tag{16.7}$$

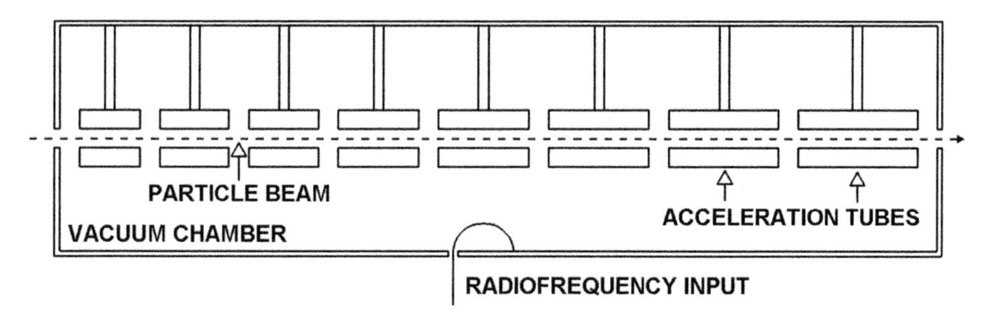

Figure 16.5 Alvarez type of linear accelerator.

When $v \rightarrow \mathbf{c}$, nk becomes $\gg 1$ and thus $L_n \rightarrow \lambda/2$. The length of the drift tubes therefore becomes constant at high particle energies.

In the Alvarez linear accelerator the drift tubes are mounted in a tank acting as a micro-wave resonator, see Figure 16.5. By varying the diameter with drift tube length it is possible to achieve resonance for all drift tubes. A powerful high-frequency radio signal is introduced into the tank by a small antenna. The standing electromagnetic wave introduces oscillating potential differences between the drift tubes in a way similar to the Wideröe accelerator. The drift tubes are grounded in the tank wall through their supports and no insulators are required. This permits higher potential differences between the tubes than in the Wideröe design.

A linear accelerator in use at the Gesellschaft für Schwerionenforschung at Darmstadt, Germany, is the UNILAC (for UNIversal Linear ACcelerator), which initially had a Wideröe section (beam energy 1.4 MeV/u) followed by Alvarez accelerators (beam energy 11.4 MeV/u) and 17 single step linear accelerator cavities. To increase the achievable energy, the partly accelerated ions are further stripped of electrons by passage through gas or nickel foil strippers located after the Wideröe section and after the Alvarez tanks, cf. (16.2). The machine was originally designed for synthesis of transuraniums and super-heavy elements through bombardment of targets of high atomic number elements with heavy ions. The maximum beam current is $\sim 10^{12}$ particles s^{-1} and the energy ≤ 20 MeV/u, which corresponds to ≤ 5 GeV for accelerated uranium ions. The UNILAC has been used to produce a number of transactinide elements (Ns, Hs, Mt). At present the Wideröe section has been replaced by a radio-frequency quadrupole accelerator (see below) followed by two drift tube linear accelerators. This has resulted in a higher ion current than before.

The world's largest linear accelerator is located at Stanford University in California. In this accelerator electrons and positrons are raised to energies of 50 GeV and used to study elementary particles, the distribution of charges within nuclei, etc. The accelerator tube is 3200 m long; the diameter is about 0.1 m. The maximum beam intensity is 8 pA. At 50 GeV the electrons have a velocity of 0.999 999 999 95 $\mathbf{c}$ and the

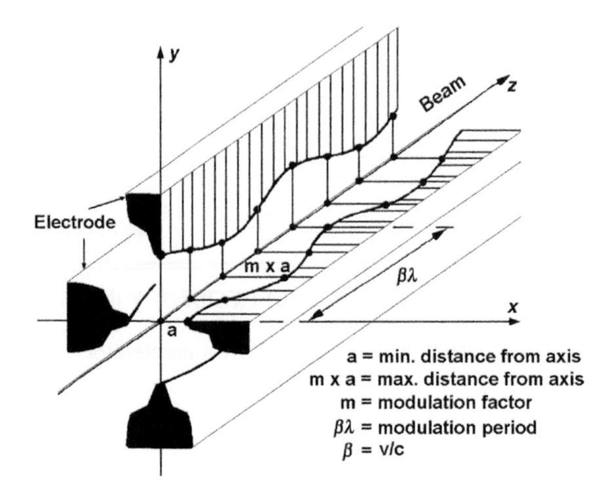

Figure 16.6 RFQ accelerator showing the four shaped electrodes (vanes) of the RF cavity.

electron mass is ~97 800 times greater than its rest mass (a little heavier than a chromium atom).

An important advance in linear accelerator technology is the development of radiofrequency quadrupoles (RFQ:s), permitting either simultaneous acceleration and focusing of high currents of particles or the funnelling of two or more beams into one which is then fed into a linear accelerator operating at a higher frequency. In principle the RFQ is somewhat similar to the Alvarez design but without separate drift tubes, see Figure 16.6. Its main characteristics can be summarized as: i) it is linear accelerator operating in the energy range 20 keV/u to 2 MeV/u, ii) it bunches, focuses, and accelerates particles with RF fields, iii) typical lengths are 1 to 2 m and operating frequencies 50 to 400 MHz, iv) intervane voltage ~100 kV, v) minimum value of *a* (see Figure 16.6) 3–4 mm, Q-factor ~10,000, vi) intervane capacity ~10^{10} F/m. In order to accept an incoming unbunched beam the a-value has to be rather high at the entrance and then progressively decrease to its final constant value. The undulating vane profiles are normally obtained by using numerically controlled milling machines.

16.6. CYCLOTRONS

The difficulties inherent in the length of high energy linear accelerators were ingeniously overcome by Lawrence and Livingston in their construction of the first cyclotron in 1931. The basic operating principles of the cyclotron are shown in Figure 16.7. The particles are accelerated in spiral paths inside two semicircular flat metallic cylinders called dees (D's), placed in a flat vacuum chamber. The two dees are connected to a high frequency alternating voltage. The volume within the dees corresponds to an equipotential condition just as does the volume within the drift tubes of linear accelerators. The

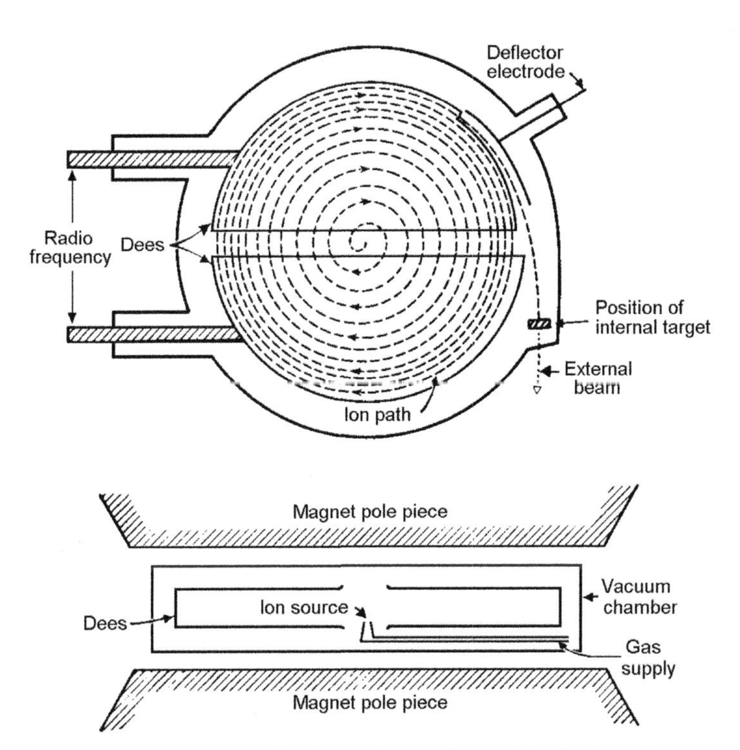

Figure 16.7 Principle of a cyclotron. *(From R.S. Shankland.)*

dees and their vacuum chamber are placed between the two poles of a magnet so that the magnetic field operates upon the ion beam to constrain it to flat circular paths inside the dees. At the gap between the dees the ions experience acceleration due to the potential difference between the dees. This causes an increase in the radius when they enter the next dee. The beam particles originate at the ion source at the center of the cyclotron, and as they spiral outward in the dees they acquire a constant increase in energy for each passage across the gap between the dees. The target is located either inside the vacuum chamber (internal beam) or the beam can be taken out through a port into an evacuated tube or through a thin "window" into the ambient air (external beam) after extraction from the circular path by a deflector.

As the projectiles acquire energy while passing the gaps between the dees, the radius of their path within the dees increases step by step. The longer path-length in the dees is related to the energy of the projectile by the equation (§3.12)

$$E_{\text{kin}} = q^2 r^2 B^2/(2m) \tag{16.8}$$

where r is the radius of the beam path at the energy E_{kin} and B is the value of the magnetic field. This equation is valid for particles of non-relativistic velocities. The successful operation of the cyclotron is based on the principle that the increase in the radius of the path in the dee nearly compensates for the increase in particle velocity. As a result, the particles, as they spiral through the dees, arrive at the gap when the potential difference is of the right polarity to cause acceleration. The principle of phase stability applies also to the cyclotron.

The frequency (in Hz) requirement is

$$\nu = qB/(2\pi m) \tag{16.9}$$

or

$$\nu = v/(2\pi r) \tag{16.10}$$

where v is the particle velocity at radius r. From (16.9) it follows at constant ν that B/m must also be constant.

The maximum energies available in conventional cyclotrons of the constant frequency type are about 25 MeV for protons and deuterons, and 50 MeV for α-particles. The shape of the electrostatic field at the gap between the dees as well as the design of the magnetic field to produce a slight non-uniformity at the outer edges of the dees produce a focusing effect on the beam particles. The ion currents are usually $\leq$0.5 mA for the internal beam, and about a factor 10 less for an external beam.

As the energies increase, so does the relativistic mass. From (16.8) we can see that either the frequency or the magnetic field, or both, must be modified to compensate for the increasing mass in order to have the beam particles arrive at the gap in phase with the voltage. The former is done in synchrocyclotrons (next section) while the magnetic field is varied in sector focused cyclotrons. In the latter, the magnet has ridges (often of spiral form) which provide sufficient modification of the field (retaining azimuthal focussing despite increasing total field strength) to allow the acceleration of particles to much higher energies than is possible with a conventional cyclotron (Fig. 16.8). The energy limit is at about twice the rest mass of the ions. Superconducting magnets are also used to achieve the highest possible magnetic field.

In the *separated-sector cyclotron*, SSC, the magnet is split into two, four, or more, submagnets, each magnet bending the beam only part of the circle. Between the bending magnets the beam follows a linear path. A few (or none) acceleration gaps (corresponding to the dees) are located in each of the linear paths, see Figure 16.8(a). The split of the magnet into several separate sectors permits a larger bending radius to be used, i.e. a higher energy can be achieved with smaller magnet pole pieces although the total magnet weight normally increases compared to a single magnet machine. As an example, in the GANIL accelerator complex (at Caen, France) normal cyclotrons are used as injectors to a SSC with two acceleration cavities, each corresponding to two acceleration steps, and four bending magnets. The beam exiting from the SSC is further stripped by passage through a thin carbon foil and then fed into a second SSC where it achieves its

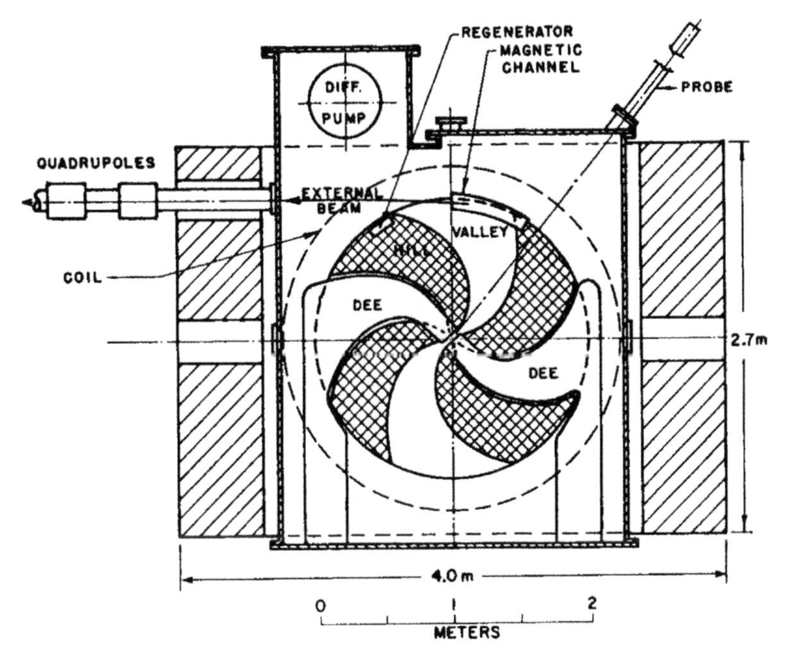

Figure 16.8 A 150 MeV p^+ sector-focused cyclotron showing the magnet with its spiral-ridged pole pieces and ¼-wave D-design.

final energy, see Figure 16.9(b). The beam quality from SSC no 2 is increased by a magnetic spectrometer.

Although there is a limit to the energies achievable with cyclotrons of constant frequency and magnetic field, they have the great advantage of high beam currents. For a heavy ion cyclotron, accelerating particles of charge z ($z \leq Z$), the final energy per mass unit, E/A (MeV/u), is given by

$$E/A = K(z/A)^2 \tag{16.11}$$

and such machines are usually characterized by their K-value. The GANIL SSC:s have e.g. both $K = 400$ but use different z of the ions. The development of ion sources producing higher currents of more highly charged ions is important as it increases the maximum energy achievable at a given K-value.

The magnetic bending of the beam corresponds to a centrifugal acceleration of the charged particles. The acceleration or deceleration of electric charges is always accompanied by the emission of electromagnetic radiation. Although only a small part of the energy input of a cyclotron is lost through this *synchrotron radiation*, it is still sufficient to require considerable radiation shielding (usually concrete walls). Synchrotron radiation is much less of a problem for linear accelerators, except for protons.

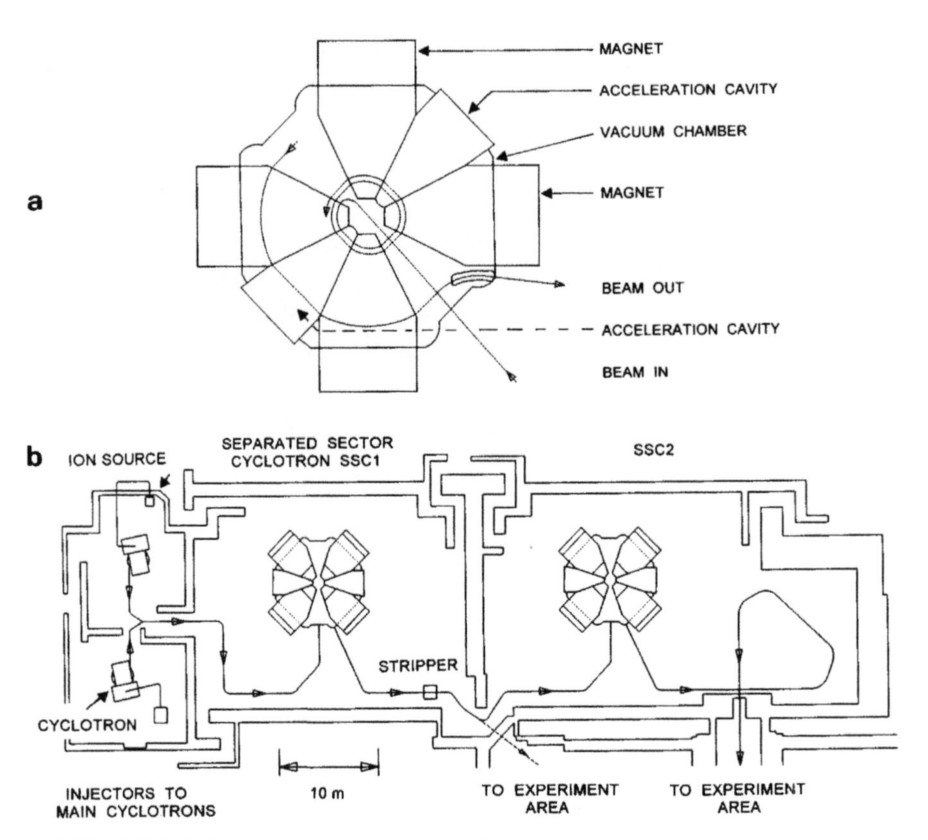

Figure 16.9 (a) Principle of a sector-separated cyclotron and (b) the GANIL accelerator complex (beam-line bending magnets are not shown).

16.7. FREQUENCY MODULATED CYCLOTRONS AND SYNCHROTRONS

The frequency of a cyclotron can be modulated to take into account the variation in velocity and mass as relativistic effects increase in importance. At very high energies the radius becomes very large. This has led to two somewhat different accelerator designs. The *frequency modulated* (FM) or *synchrocyclotron* maintains the original cyclotron principle with a spiral particle path, while in *synchrotrons* the particle path is fixed in a circular orbit; Figure 16.10 is an example of the latter.

In a synchrotron there is a balance between frequency and the magnetic field, one or both being varied during the acceleration (cf. (16.7a)). From (16.6) and (16.7) one obtains

$$E_{\text{kin}} = \nu \pi r^2 B q \tag{16.12}$$

by eliminating the mass from the relation for the kinetic energy of the projectile. In these machines the particles are accelerated in bursts, since the frequency of the accelerating

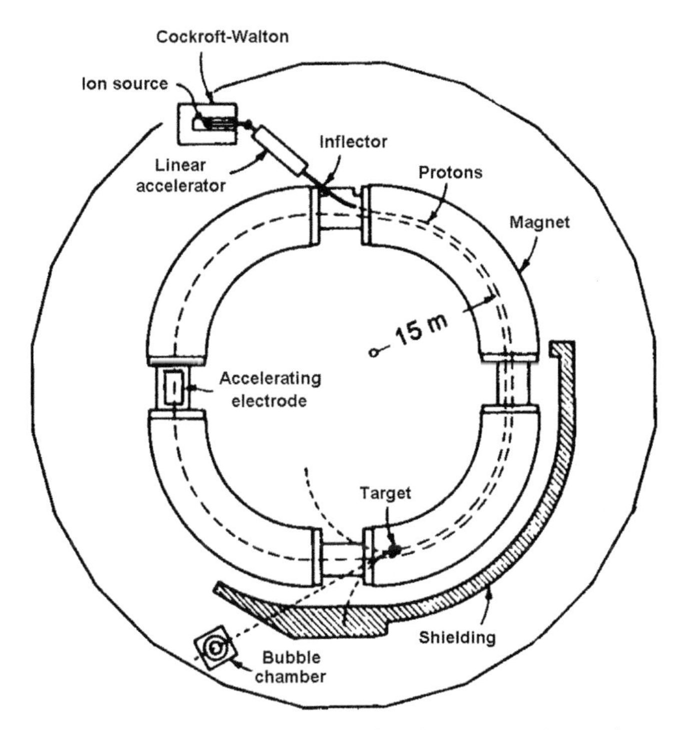

Figure 16.10 Principle of a modulated multistage circular accelerator.

potential must be modulated throughout the traversal of one beam burst in the machine, and only the particles in that burst would be in resonance with the frequency change of the accelerating gap. As a result, average beam currents are much smaller in FM cyclotrons and synchrotrons than in constant frequency cyclotrons.

Synchrotrons which accelerate positive particles use smaller accelerators, often linear, for injecting the beam into the circular synchrotron. Figure 16.10 shows a typical synchrotron design, although it specifically refers to the late Bevatron in Berkeley which provided acceleration of protons to 6.5 GeV (the mass increase of a 6.5 GeV proton is 6.9 u) and heavier ions like carbon or neon to about 2 GeV u^{-1} it illustrates the design of such machines. The Bevatron injectors were a 0.5 MeV Cockcroft-Walton machine followed by a 10 MeV multistage linear accelerator or the Super-HILAC heavy ion linear accelerator. The latter combination was called Bevalac. In the Bevatron, the potential difference over the accelerating step was only 1500 V. As the particles accelerate, the magnetic field continues to increase in such a way that the radius of the particle orbit remains constant. The frequency of the oscillating voltage applied to the accelerating electrode must also increase, since the time required for a complete circuit becomes shorter and shorter until the velocity of the particles approaches the velocity or light, after which the circuit time remains constant. The whole accelerating cycle requires several seconds, during which time the particles make many millions of orbits. Each burst of particles may contain $\sim 10^{12}$ particles, and in the Berkeley machine there was one burst every 6 s. The successful operation of the synchrotron depends on the existence of phase stability.

The Bevatron (later the Bevalac) operated from 1954 to 1993. Larger and more recent (proton) accelerators are the 76 GeV machine of Serpukhov (USSR), and the 500 GeV machine at Batavia in the USA. The Batavia machine contains 1000 magnets along its 6.3 km orbit, each magnet weighing 10 tons.

At CERN, a machine was built that could produce 10^{13} 400 GeV protons s^{-1}; the acceleration diameter was 2.2 km. The injection time (with 10 GeV protons) was 23 s (to fill the acceleration ring), while the acceleration time was 9 s; for each cycle 2.5 MeV was added in energy. The field strength of the 200 bending magnets was 1.8 T and the power consumption 50 MW.

Still higher energies have been achieved in collisions between particles circulating in two beam trains in opposite direction by using intersecting storage rings (ISR). In this system the energy loss due to momentum conservation in the reaction between high speed projectiles and stationary target atoms is eliminated. At CERN proton collision energies of 7 TeV have been obtained and the system is being upgraded to 8 TeV in 2012.

Synchrotrons have been built in which electrons are accelerated in a fixed orbit in a strong magnetic field (≥ 1 T) to energies in the GeV range. The intense electromagnetic (synchrotron) radiation has usually a maximum intensity at ≤ 1 keV and is used for studies of photon induced reactions (e.g. for meson production), for radiation chemistry, as a strong light source, etc.

A special type of synchrotron is the so called storage ring. In a storage ring the particles are kept at a constant velocity generating a constant flux of syncrotron radiation. In order to maximise the synchrotron radiation intensity these rings are often equipped with wrigglers or undulators which give the beam multiple direction changes, see §16.9.1.

16.8. NEUTRON GENERATORS

Many radioactive isotopes are most conveniently produced by irradiation of a target with neutrons. There are three main ways of obtaining useful neutron fluxes:

(a) through accelerator-induced reactions (Table 16.1);
(b) through nuclear reactors (Ch. 19);
(c) through nuclear reactions induced by radioactive isotopes (§10.9).

16.8.1. Reactions

Table 16.1 gives the most common accelerator-induced nuclear reactions leading to the production of neutrons. In general, the neutron yield increases with increasing beam energy up to some maximum value of the latter after which competition of other nuclear

Table 16.1 Neutron yield for some accelerator produced reactions (see also Fig. 16.10)

Reaction	Q(MeV)	Target	Yield(*) ($n\ s^{-1}\ \mu A^{-1}$)	
$^{7}Li(p,n)^{7}Be$	−1.64	Li salt	10^7	for 2.3 MeV H^+
$^{2}D(d,n)^{3}He$	3.27	D_2O ice	7×10^7	for 1 MeV D^+
$^{3}H(d,n)^{4}He$	17.6	^{3}H in Ti	2×10^7	for 0.1 MeV D^+
$^{9}Be(d,n)^{10}B$	3.76	Be metal	2×10^{10}	for 7 MeV D^+
$^{9}Be(\gamma,n)^{8}Be$	−1.67	Be metal	10^9	for 10 MeV e
$U(\gamma,fp)$	−5.1	U metal	10^{11}	for 40 MeV e

(*) 1 $\mu A = 6.24 \times 10^{12}$ single-charged ions per second.

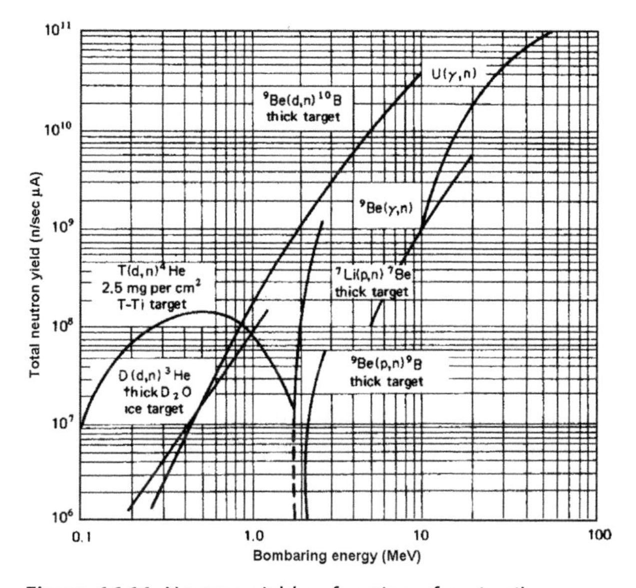

Figure 16.11 Neutron yield as function of projectile energy.

reactions reduces the probability of the initial neutron producing reaction (Fig. 16.11). The least expensive way to get a neutron flux of sufficient intensity for producing radioactive nuclides in easily measurable quantities (e.g. for activation analysis) is through the T + D reaction using a Cockcroft-Walton accelerator. The reaction

$$^{7}Li + {}^{1}H \rightarrow {}^{7}Be + {}^{1}n \tag{16.13}$$

is used for the production of mono-energetic neutrons (cf. §11.3). Higher neutron yields can be obtained by the reaction of ^{9}Be with deuterons. Neutrons can also be obtained from ^{9}Be by irradiation in electron accelerators. The passage of the electrons in the target results in the production of high energy Bremsstrahlung γ-rays (§7.4.2) which react with the beryllium to produce neutrons. An even higher neutron flux is obtained for the same reason when uranium is bombarded by electrons.

Extremely intense neutron sources can be constructed from a high energy (~1 GeV) high current proton linear accelerator and a heavy element (e.g. molten Pb) spallation target. Tentative designs have been described that would be capable of generating ~0.1 g of neutrons (0.1 mole) per second. Such intense neutron sources could be used to transmute longlived radioactive nuclides (into inactive or shorter lived ones) or drive a subcritical assembly for production of energy by fission of actinides.

16.9. AREAS OF APPLICATION FOR ACCELERATORS

Let us consider a few uses of accelerators. In §16.8 we have discussed the use of accelerators for neutron production. Electron as well as positive particle beams from these accelerators can be used directly for the study of radiation effects in materials (see Ch. 8), to produce X-rays, or to generate intense light. Accelerators are used in medicine for producing intense γ- or X-ray beams for cancer treatment, or charged particle beams for similar purposes (the "proton-knife"). Accelerators in the 10 MeV range are very useful in producing radionuclides, which either cannot be made through (n,γ)-reactions in nuclear reactors or are very shortlived. Small cyclotrons of a size suitable for use in hospitals are commercially available. The precise definition of energy available in van de Graaff accelerators makes them particularly useful for the study of the energy levels of nucleons in nuclides and for ^{14}C dating. VdG's, cyclotrons, linear accelerators, and synchrotrons have been extensively used for the study of nuclear reactions.

Very high energy accelerators are used for the production and study of elementary particles. Once the beam energy exceeds a couple of hundred MeV, mesons and other elementary particles are produced in nuclear reactions and can be studied directly or they can be used to cause various types of nuclear reactions.

16.9.1. Syncrotron light sources

High energy electron accelerators, mostly in form of large diameter storage rings, are used to produce high intensity synchrotron radiation for basic studies in chemistry, material science and biology. The synchrotron "light" is usually obtained by placing bending magnets in the path of the circulating electron beam. A series of opposing strong magnets repeatedly bend the beam back and forth many times. These devices are characterized by their K_U-value which is defined as given below.

$$K_U = \frac{eB\lambda}{2\pi v m_e} \tag{16.14}$$

where v is the electron velocity in the beam and λ is the distance beween magnets with the same orientation. When K_U is $\ll 1$ the device is called an undulator and produces x-rays in a narrow energy band, but when K_U is $\gg 1$ the device is called a wiggler and produces a continuum of x-rays within a rather wide energy range. A typical device is

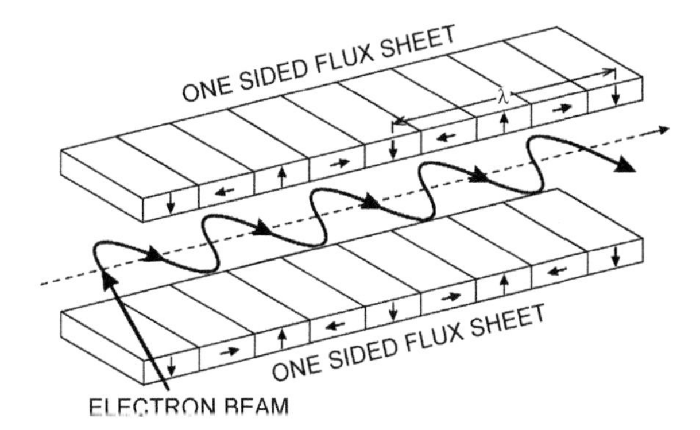

Figure 16.12 Principle of undulators and wigglers for production of intense synchrotron radiation. The X-rays are emitted in the direction shown by the dashed line with an arrow at the right.

shown schematically in Figure 16.12. The direction changes will increase the intensity of the x-ray beam due to the large number of bendings. Syncrotron light sources are used for different kinds of spectroscopy such as, e.g., EXAFS (Extended X-ray Absorption Fine Structure) which has grown considerably in popularity with chemists in the last decade due to the ability to directly study complexation in solution and solids.

16.10. EXERCISES

16.1. In a small linear accelerator containing 30 stages, He^{2+} ions are accelerated by a 150 kV, 100 MHz RF source. The ions are used to bombard a metal target to induce a specific reaction. (a) What is the proper length of the last drift tube? (b) What is the maximum projectile energy achieved? (c) What is the heaviest target in which a nuclear transformation can be induced (no tunnelling)?

16.2. Assume a linear accelerator is built in three sections, each with 30 stages of 100 kV. Between which sections should a thin carbon stripper foil be placed in order to achieve the highest final energy, when accelerating ^{84}Kr ions from an ion source emitting low energy $^{84}Kr^{4+}$ ions? Consider the nearest integer charge after stripping.

16.3. In order to propel a space vehicle to high velocities after its exit from the earth's gravitational pull, "ion rockets" might be used. These can be considered as simple accelerators for charged particles. The electroneutrality of the vehicle is conserved through emission of electrons from a hot filament. The gain in linear velocity Δv is given by the "rocket formula":

$$\Delta v = v_e \ln(m_{Ro}/m_R)$$

where v_e is the exhaust velocity and m_{Ro} and m_R are the initial and final mass of the vehicle respectively. (a) Calculate the propelling power for a rocket, which emits 10 keV protons at a current of 1 A. (b) What is the final velocity gain of such vehicle with 2000 kg initial mass after 1 year's operating time? (c) For the same net available power and operating time, as in (b), calculate the final velocity gain of a 2000 kg vehicle which emits 10 keV Cs^+ ions at a current of 1 A.

16.4. The TD reaction is used to produce 14 MeV neutrons, which are considered to be emitted isotropically from the target. What is the fast neutron "flux" at 5 cm from the target when the ion current is 0.2 mA and the acceleration voltage is 300 kV? Use Figure 16.11.

16.5. Protons are accelerated to 12 GeV in a synchrotron in which the bending magnets have a maximum field strength of 14.3 T. What is the radius of curvature of the proton orbit?

16.6. In a linear accelerator for protons the first drift tube is 1 cm long and the accelerating potential is 25 kV. (a) The speed of light is not approached in the first drift tube. How long should the second and third drift tubes be? (b) What should the full-wave frequency of acceleration be?

16.7. Calculate the maximum energy (a) for protons, deuterons, and helium ions in a cyclotron, whose maximum orbit diameter is 1.25 m and whose frequency is 12 MHz. (b) What magnetic field strength would be required in each case?

16.8. A thin Au target is irradiated with a beam of 800 MeV $^{16}O^{2+}$ ions during 2 hours. After the irradiaten a faraday cup (with current integrator) behind the target shows a total accumulated charge of 92.3 μC. What was the average beam intensity (oxygen ions per second)? Consider the charge-stripping effect of the target.

16.9. In a VdG accelerator a 100 μA beam of He^{2+} ions is accelerated to an energy of 5 MeV before striking a target. How many grams of radium are required to provide the same number of α-particles?

16.10. A cyclotron can accelerate $^{4}He^{2+}$-ions to 35 MeV. (a) What is its K-value? To what energy would it accelerate (b) $^{16}O^{6+}$ and (c) $^{18}O^{8+}$ ions?

16.11. LITERATURE

S. FLÜGGE (ed.) *Handbuch der Physik*, Vol. XLIV: *Instrumentelle Hilfsmittel der Kernphysik*, Springer-Verlag, 1952.

M. S. LIVINGSTON and J. P. BLEWETT, *Particle accelerators*, McGraw-Hill, 1962.

P. H. ROSE and A. B. WITTKOWER, Tandem Van de Graaff accelerators, *Sci. Am.* **223** (Aug. 1970) 24.

CERN/1050, *The 300 GeV Programme*, CERN, Geneva 1972.

R. R. WILSON, The Batavia accelerator, *Sci. Am.* **231** (Febr. 1974) 72.

S. HUMPHRIES, *Principles of Charged Particle Accelerators*, Wiley, 1986.

W. SHARF, *Particle Accelerators and their Uses*, Harwood, 1986.

M. MORTH and M. DIENES (Eds.), *Physics of Particle Accelerators*, AIP Conf. Proc. **184**, 1989.

Proc. Int. Conf. on Cyclotrons and their Applications, since 1975; Vancover 1992, World Scientific Publ, 1993.

CHAPTER 17

Production of Radionuclides

Contents

This chapter discusses production of radionuclides for beneficial use in science, medicine and technology. The nuclear fundamentals for the production processes have been given in Chapters 6, 10, 11 and 16. The formation of radionuclides is discussed in several Chapters: e.g. cosmogenic reactions leading to the formation of short-lived radionuclides in nature (Ch.12 and 13); thermonuclear reactions leading to the formation of long-lived radioactivity in the universe (Ch. 12); the synthesis of trans-uranium elements (Ch. 14 and 19-21). The production and isolation of separated fission products is treated separately (Ch. 19—21). This chapter discusses aspects of fundamental importance to the production of radionuclides by a variety of methods. Initially the principles are reviewed

Radiochemistry and Nuclear Chemistry
ISBN 978-0-12-405897-2, http://dx.doi.org/10.1016/B978-0-12-405897-2.00017-3

and, subsequently, the most advanced techniques for investigating short-lived radionuclides are described.

17.1. GENERAL CONSIDERATIONS

Radioactive nuclides may be prepared by a wide variety of techniques e.g. particle accelerators and nuclear reactors; however, only cyclotrons or reactors of at least moderate particle flux are able to produce sources of sufficiently high specific radioactivity to be of practical interest. These two methods of production complement each other since, in general, they do not produce the same isotopes of an element.

In cyclotrons charged particles such as protons, deuterons, and α-particles bombard the target nuclei, and after emission of one or more particles to remove the excess excitation energy, a radioactive product nuclide may result. In the capture of positively charged particles and the subsequent emission of neutrons the product radionuclides are *neutron deficient* compared to the stable isotopes of the element; see Fig. 4.1 and 5.7. Another important point in cyclotron bombardments is that normally the product is not isotopic with the target. As a result, after chemical separations a product of high specific activity is obtained since it is not diluted by the target material. An example of an important cyclotron-produced radionuclide is ^{22}Na ($t_{½}$ 2.6 y) formed by the reaction:

$$^{24}_{12}\mathrm{Mg} + ^{2}_{1}\mathrm{H} \rightarrow ^{4}_{2}\mathrm{He} + ^{22}_{11}\mathrm{Na} \qquad (17.1)$$

This contrasts with the production of radionuclides by neutron irradiation, e.g. in a nuclear reactor, where the neutron absorption and subsequent de-excitation by emission of γ-rays produces *neutron rich* isotopes, frequently radioactive, which are isotopic with the target. A typical reaction is

$$^{23}\mathrm{Na}\,(\mathrm{n},\gamma)\,^{24}\mathrm{Na} \qquad (17.2)$$

yielding another important isotope, ^{24}Na ($t_{½}$ 15.0 h), which is diluted by the ^{23}Na target atoms. Even if the same radioactive nuclide can be made simpler in a nuclear reactor, cyclotron production may be favored if high specific activity is necessary for a particular use.

A second mode of production in reactors is by the fission process itself as the majority of fission products are radioactive and cover a wide range of atomic numbers of varying abundances. These fission products may be isolated with high specific activity in contrast to the neutron capture products. However, for the isolation of any particular radioactive element from the rest of the fission products can require a fairly lengthy separation; we discuss this further in Ch. 21.

Radioactive nuclides may also be obtained as daughters of parent nuclides which have been previously made either by cyclotron or reactor irradiation or exist naturally (e.g. the production of ^{234}Th from ^{238}U). Several examples of such radioactive isotope generators are discussed in §5.16.

Which mode or production is utilized depends on several factors, e.g. the availability of accelerators, reactors etc., the availability of starting material, the length of time for production, separation and purification processes, the time from production to shipping, and the shipping time determine the half-life with which it is practical to work. The type of experiment and equipment available for handling and measurement determines the preferable decay type and intensity is also important. All these factors must be considered in determining which nuclide is to be used. For example, if in an experiment requiring sodium tracer these aspects indicate the necessity of a time of at least several weeks, the use of reactor-produced ^{24}Na is excluded due to its short half life and cyclotron-produced ^{22}Na is required. If the experiment requires the use of very high specific activities, either cyclotron or fission product activities may be necessary. Reactor activities are usually much cheaper and easier to obtain on a routine basis because many targets can be irradiated simultaneously in the reactor whereas in a cyclotron only one target at a time can be bombarded.

Although these considerations are important in the choice of a radionuclide to use for a particular purpose, the possibilities of production may also be limited by the reaction cross-sections. Figure 5.7 (right side) shows the reaction paths needed to transform a given target nuclide into a product nuclide. If the target nuclide is denoted by $^{A}_{Z}$X, a (n,γ) reaction yields the same product as a (d,p) reaction, i.e. $^{A+1}_{Z}$X. The choice of reaction depends on the particular nuclides available, projectile energy, and the cross-sections (available in charts or tables of nuclear data). As pointed out in Chapter 11, the same projectile may yield several different products depending on e.g. its energy; e.g. Fig. 11.2 shows six different α-induced reactions.

An induced nuclear reaction may be followed by rapid radioactive decay, leading to the desired radionuclide. Thus a (n,2n) reaction on $^{A}_{Z}$X may yield $^{A-1}_{Z}$X, which rapidly decays to $^{A-1}_{Z+1}$X through β^--emission. For example, the formation reaction

$$^{238}\text{U}\ (\text{n}, 2\text{n})\ ^{237}\text{U} \qquad (17.3)$$

followed by

$$^{237}\text{U} \xrightarrow[6.75\ \text{d}]{\beta^-} {}^{237}\text{Np} \qquad (17.4)$$

These reactions occur in nuclear rectors and can used to produce ^{237}Np for investigation of neptunium chemical behaviour. An alternative method for obtaining ^{237}Np is to use ^{241}Am. The americium is deposited on a column and its daughter nuclide, ^{237}Np, is "milked out", by a suitable leachant.

17.2. IRRADIATION YIELDS

The production of a radioactive nuclide by irradiation in a reactor or an accelerator can be written as:

$$X_{target} + X_{proj.} \overset{k}{\rightarrow} X_1 \overset{\lambda_1}{\rightarrow} X_2 \overset{\lambda_2}{\rightarrow} X_3 \quad (17.5)$$

We are mainly concerned with the product X_1 (produced at rate k), which is assumed to be radioactive, decaying in a single step or via a chain of decays to a final stable nuclide. The number of radioactive product atoms N_1 of X_1 present at any time t is the difference between the total number of product atoms formed and the number of radioactive decays that have occurred in time dt. This leads to the differential equation

$$dN_1 = k\,dt - \lambda_1 N_1\,dt \quad (17.6)$$

Integration between the limits of $N_1 = 0$ (this assumes that no radioactive product nuclide X_1 exist at the beginning of the irradiation) and N_1 corresponding to an irradiation time t_{irr}:

$$N_1 = (k/\lambda_1)\left(1 - e^{-\lambda_1 t_{irr}}\right) \quad (17.7)$$

N_1 increases with time towards a maximum value, which is reached when $t_{irr}\lambda_1 \gg 1$ (i.e. $t_{irr} \gg t_{½}$):

$$N_1(\max) = k/\lambda_1 = k\,t_{½}/0.693 \quad (17.8)$$

However, for $t_{irr} \gg t_{½}$, by expanding the exponential into a MacLaurin series, $e^{-\lambda_1 t_{irr}} \approx 1 - \lambda_1\, t_{irr}$, which with (17.7) gives

$$N_1 = k\,t_{irr}\left(\text{for } t_{irr} \ll t_{½}\right) \quad (17.9)$$

If we measure the irradiation time in number of half-lives, $a = t_{irr}/t_{½}$, we obtain

$$N_1 = N_{1,\max}\left(1 - 2^{-a}\right) \quad (17.10)$$

This equation shows that irradiation for one half-life ($a = 1$) produces 50% of the maximum amount, two half-lives give 75%, etc.

After the end of the irradiation time, the radioactive product nuclides continue to decay according to their half-lives. We indicate this decay time after the end of irradiation (termed the cooling time) as t_{cool}. The number of product nuclides, which is a function of the irradiation time t_{irr} and cooling time t_{cool}, is given by

$$N_1 = (k/\lambda_1)\left(1 - e^{-0.693 t_{irr}/t_{½}}\right) e^{-0.693 t_{cool}/t_{½}} \quad (17.11)$$

The effect of the cooling time is shown in Figure 17.1 for two sample nuclides.

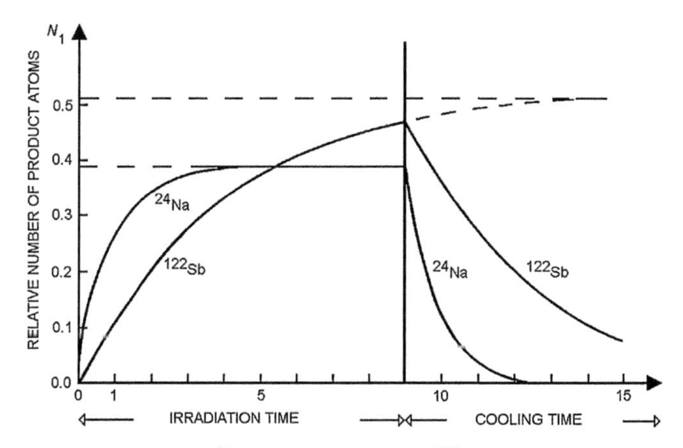

Figure 17.1 Change in number of ^{24}Na ($t_{½}$ 15.0 h) and ^{122}Sb ($t_{½}$ 2.7 d) atoms during and after irradiation.

The production rate k is the number of transformations, N, leading to a number, N_1, of product atoms, divided by the irradiation time, t_{irr}. For a target in a particle flux

$$k = \varphi \, \sigma \, N_t \tag{17.12}$$

where φ is the particle flux (particles of appropriate energy per cm^2 per s), σ the reaction cross section (at the given particle energy, in cm^2) and N_t the number of target atoms with cross section σ; cf. eqn. (11.4). In the case a fission product is produced, $\varphi \, \sigma \, N_t$ is the fission rate, which has to be multiplied by the particular fission yield, y_i, for nuclide i.

If a target is bombarded by a beam of particles the flux is described by a conversion factor and the current I (A) of particles with charge z and the production rate k is

$$k = 6.24 \times 10^{18} I \, \sigma \, N_v \, x \, z^{-1} \tag{17.13}$$

where $N_v \, x$ is the number of target atoms per unit area; cf. eqn. 11.5).

For (17.12) and (17.13) to be valid,

(i) the number of nuclear reactions must remain small compared to the total number of target atoms to justify the assumption of constant N_t or N_v,

(ii) the projectile energy must be maintained constant in order that the reaction cross-section remains constant, and

(iii) the flux must not be decreased in passing through the target. Depending on the conditions of the irradiation only one or two of these requirements may be valid. Corrections must then be introduced for the non-valid component(s).

Usually the radioactive decay rate A_1 is measured, rather than the number of atoms N_1. Recalling $A = \Delta N/\Delta t = \lambda N$, (17.3) and (17.12) yield

$$A_1 = k\left(1 - e^{-0.693 t_{irr}/t_{½}}\right) e^{-0.693 t_{cool}/t_{½}} \tag{17.14}$$

This equation represents the basic relationship for the production of radionuclides; in a reactor k is calculated according to (17.12) but for accelerator irradiation, where the current and not the flux is known, (17.13) is normally used.

Let us again take the production of radioactive sodium by reactor irradiation and cyclotron bombardment as examples of the application of these equations. Na_2CO_3 (M = 106) is irradiated in a reactor to produce ^{24}Na, which has a half-life of 14.66 h and emits β^- and γ-rays. The reaction cross-section of ^{23}Na (100% in natural sodium) is 0.53 b for thermal neutrons. For a 5 g sample, 60 h irradiation time, and a thermal flux of 10^{12} n cm^{-2} s^{-1}, the activity at the end of the irradiation time will be:

$$A = 10^{12} \times 0.53 \times 10^{-24} \times 2 \times 5 \times 6.02 \times 10^{23}\left(1 - e^{-0.693 \times 60/15}\right)/106$$
$$= 2.8 \times 10^{10}\ \text{Bq} \qquad (17.15)$$

^{24}Na may be produced through accelerator irradiation of magnesium by the reaction ^{26}Mg (d,α) ^{24}Na. With 22 MeV D^+ ions the reaction cross-section is about 25 mb. For a 0.1 mm thick magnesium foil (M = 24.3, $\rho = 1.74$ g cm^{-3}) with an area larger than the projectile beam, and a 2 h irradiation at 100 μA, the activity of ^{24}Na produced (natural magnesium contains 11.0% ^{26}Mg) at end of irradiation is:

$$A = 6.24 \times 10^{18} \times 100 \times 10^{-6} \times 25 \times 10^{-27}(0.11 \times (1.74/24.3)$$
$$\times 6.02 \times 10^{23} \times 0.01) \times \left(1 - e^{-0.693 \times 2/15}\right) = 6.53 \times 10^{7}\text{Bq} \qquad (17.16)$$

17.3. SECOND-ORDER REACTIONS

For large cross-sections, short half-lives, and long bombardments, second-order capture products may be formed. If the first-order product is radioactive, then its concentration at any time is dependent on (i) the decay constant, (ii) the cross-section for production of the second-order product, as well as (iii) the cross-section for its own production. These possibilities lead to a scheme such as Figure 17.2. For simplicity, we may assume that the induced nuclear transformations only involve single-neutron capture, and that only β^- decay occurs. However, it should be obvious that the scheme in Figure 17.2, with suitable additions, is applicable to all kinds of nuclear formation and decay reactions.

For the (n,γ) β^- case the upper horizontal row of Figure 17.2 represents the successive formation of higher isotopes of the target element (the constant Z-chain) and the vertical rows the isobaric decay chains of each of these isotopes (the constant A-chains). The first of these two rows is indicated by heavy arrows. Chains which involve both induced transformations and radioactive decay play a central role in theories about the

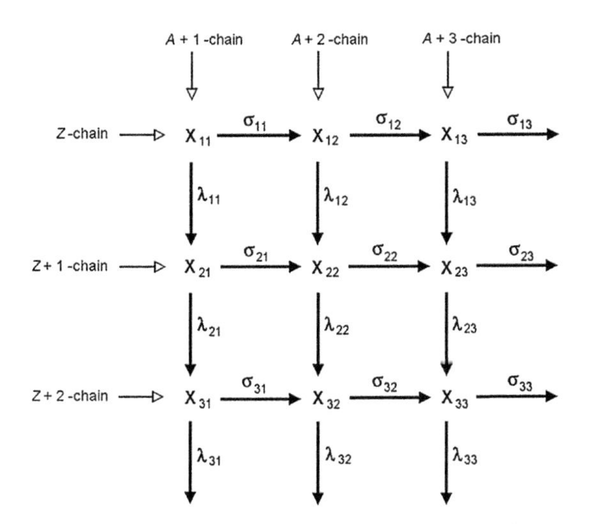

Figure 17.2 Formation of higher nuclides through multiple neutron capture and their associated decay chains.

formation of the elements in the universe, in the thermonuclear reactions in the stars (Ch. 12), and in the synthesis of transuranium elements (Ch. 14).

For each nuclide in Figure 17.2 we can write a differential expression for the change in concentration with time as a function of the irradiation and the radioactive decay:

$$\text{For } X_{11} : dN_{11}/dt = -(\varphi\sigma_{11} + \lambda_{11})N_{11} = \Lambda_{11}N_{11} \tag{17.17}$$

$$X_{21} : dN_{21}/dt = \lambda_{11}N_{11} - (\varphi\sigma_{21} + \lambda_{21})N_{21} = \Lambda^*_{11}N_{11} - \Lambda_{21}N_{21} \tag{17.18}$$

$$X_{31} : dN_{31}/dt = \lambda_{21}N_{21} - (\varphi\sigma_{31} + \lambda_{31})N_{31} = \Lambda^*_{21}N_{21} - \Lambda_{31}N_{31} \tag{17.19}$$

$$\ldots \quad \ldots$$

$$X_{12} : dN_{12}/dt = \varphi\sigma_{11}N_{11} - (\varphi\sigma_{12} + \lambda_{12})N_{12} = \Lambda^*_{11}N_{11} - \Lambda_{12}N_{12} \tag{17.20}$$

$$X_{13} : dN_{13}/dt = \varphi\sigma_{12}N_{12} - (\varphi\sigma_{13} + \lambda_{13})N_{13} = \Lambda^*_{12}N_{12} - \Lambda_{13}N_{13} \tag{17.21}$$

$$\ldots \quad \ldots$$

$$X_{22} : dN_{22}/dt = \lambda_{12}N_{12} + \varphi\sigma_{21}N_{21} - (\varphi\sigma_{22} + \lambda_{22})N_{22} \tag{17.22}$$

Let us now consider three different cases:

Case (i): Successive radioactive decay. This occurs after formation of the first radioactive (parent) member X_{11}. Thus we only consider the left vertical chain, and assume σ_{11}, σ_{21}, etc. $= 0$. In this $A + 1$ chain the second index is constant, and we omit it for case (i). This case is valid for all natural decay chains and for fission product decay chains. Each member of the decay chain can be described by the differential equation

$$dN_i/dt = \lambda_{i-1}N_{i-1} - \lambda_i N_i \ (i > 1) \quad (17.23)$$

We have already solved this relation in §17.2 for the formation of X_1 (the radioactive mother), and in §5.15 we showed how the amount of X_2 (the radioactive daughter) varies with time.

The general solution to the case with many successive decays is usually referred to as the *Bateman equations* (H. Bateman 1910):

$$N_n(t) = \lambda_1\lambda_2 \ldots \lambda_{n-1} \, N_1^0 \sum_1^n C_i e^{-\lambda_i t} \quad (17.24)$$

N_n (t) is the number of atoms for the nth species after time t, N_1^0 is the number of atoms of X_1 at $t = 0$, t is the *decay time*, and

$$C_i = \prod_{j=1}^{j=n} \left(\lambda_j - \lambda_i\right)^{-1} \ \text{for} \ j \neq i \quad (17.25)$$

For $i = 1$, eqn. (5.58) for simple radioactive decay is obtained. It is assumed that $N_i^0 = 0$ for ($i > 1$). If that is not the case, additional chains, starting with each $N_i > 0$, require a new series of Bateman equations which are added together. For $i = 3$ and $N_2^0 = N_3^0 = 0$ at $t = 0$ one obtains

$$N_3(t) = \lambda_1\lambda_2 \, N_1^0 \left(C_1 \, e^{-\lambda_1 t} + C_2 \, e^{-\lambda_2 t} + C_3 \, e^{-\lambda_3 t}\right) \quad (17.26)$$

where

$$C_1 = (\lambda_2 - \lambda_1)^{-1} \, (\lambda_3 - \lambda_1)^{-1} \quad (17.27)$$

$$C_2 = (\lambda_1 - \lambda_2)^{-1} \, (\lambda_3 - \lambda_2)^{-1} \quad (17.28)$$

$$C_3 = (\lambda_1 - \lambda_3)^{-1} \, (\lambda_2 - \lambda_3)^{-1} \quad (17.29)$$

Case (ii): Successive neutron capture. This is the upper horizontal row in Figure 17.2 (the constant Z-chain). Because the first index is constant in this chain, we omit it for case (ii). We assume λ_{11}, λ_{12}, etc. $= 0$. This is a valid approximation often used for the formation of transuranium elements in a nuclear reactor, which are long-lived in relation to the time of irradiation and observation. For each member of the formation chain we may write

$$dN_i/dt = \varphi\sigma_{i-1}N_{i-1} - \varphi\sigma_i N_i \quad (17.30)$$

By replacing λ_i with $\varphi\sigma_i$ this equation becomes identical with (17.23). Thus the solution of case (ii) is obtained by using the Bateman eqn. (17.24) and (17.25). It should be noted that in case (ii) t is the *irradiation time* (t_{irr}).

Case (iii): Combined induced transformation and radioactive decay. These combined effects can be taken into account by using the Bateman equations with some modifications, as developed by Rubinson.

We introduce the abbreviations

$$\Lambda = \varphi\sigma + \lambda \tag{17.31}$$

and

$$\Lambda^* = \varphi\sigma^* + \lambda^* \tag{17.32}$$

Λ_i is referred to as the disappearance constant of nuclide i, while Λ_i^* is the partial formation constant, and σ^* and λ^* the partial reaction cross-sections and decay constants for nuclide $i + 1$; σ, σ^*, λ and λ^* may be equal to zero in some cases. The exact meaning of Λ and Λ^* is explained below with examples. Introducing these abbreviations for the formation of the $X_{i\ 1}$ species of the constant $(A + 1)$-chain, and dropping the second index, which is =1 in the left vertical row in Figure 17.2, we obtain the equation:

$$dN_i/dt = \lambda_{i-1}N_{i-1} - \Lambda_i N_i = \Lambda_{i-1}^* N_{i-1} - \Lambda_i N_i \tag{17.33}$$

Similarly, we find for the formation of the $X_{1\ i}$ species (the constant Z-chain)

$$dN_i/dt = \varphi\sigma_{i-1}N_{i-1} - \Lambda_i N_i = \Lambda_{i-1}^* N_{i-1} - \Lambda_i N_i \tag{17.34}$$

where the first index (=1) has been dropped for simplicity. These two equations turn out to be identical, provided the partial formation constant Λ^* (17.32) for each nuclide (i.e. each index set) is used with the correct σ^* and λ^*; in (17.33) $\sigma^* = 0$, and in (17.34) $\lambda^* = 0$. With this reservation in mind the Bateman equations become

$$N_n(t) = \Lambda_1\Lambda\ldots\Lambda_{n-1}^*\ N_1^0 \sum_1^n C_i e^{-\Lambda_i t} \tag{17.35}$$

where

$$C_i = \prod_{j=1}^{j=n} (\Lambda_j - \Lambda_i)^{-1}(j \neq i) \tag{17.36}$$

The time t is the irradiation time t_{irr}.

The application of this expression and the use of Λ^* are best illustrated with a practical example.

If 1.0 mg of Tb_2O_3 (M = 366) is bombarded for 30 days at a flux of thermal neutrons of 10^{14} cm^{-2} s^{-1}, the following processes occur:

$$\begin{array}{ccccc} {}^{159}\text{Tb} & \xrightarrow[\sigma_{159}\ 25.5\text{ b}]{(n,\gamma)} & {}^{160}\text{Tb} & \xrightarrow[\sigma_{160}\ 525\text{ b}]{(n,\gamma)} & {}^{161}\text{Tb} \\ & & \lambda_{160} \downarrow t_{1/2}\ 72.1\text{ d} & & \lambda_{161} \downarrow t_{1/2}\ 6.90\text{ d} \\ & & {}^{160}\text{Dy (stable)} & & {}^{161}\text{Dy (stable)} \end{array} \tag{17.37}$$

The appropriate equation for the formation of ^{161}Tb (cf. also (17.26)) is:

$$\begin{aligned} N_{161} = {} & \Lambda_{159}\Lambda^*_{160}N^0_{159}\left[e^{-\Lambda_{159}\,t}/\{(\Lambda_{160}-\Lambda_{159})(\Lambda_{161}-\Lambda_{159})\}\right. \\ & + e^{-\Lambda_{160}t}/\{(\Lambda_{159}-\Lambda_{160})(\Lambda_{161}-\Lambda_{160})\} \\ & \left. + e^{-\Lambda_{161}t}/\{(\Lambda_{159}-\Lambda_{161})(\Lambda_{160}-\Lambda_{161})\}\right] \end{aligned} \tag{17.38}$$

By considering the different reaction paths, we obtain

$$(\Lambda_1 =) : \Lambda_{159} = \varphi\sigma_{159} = 10^{14} \times 25.5 \times 10^{-24} = 2.55 \times 10^{-9}\,\text{s}^{-1} \tag{17.39}$$

$$\begin{aligned}(\Lambda_2 =) : \Lambda_{160} = \varphi\sigma_{160} + \lambda_{160} = 10^{14} \times 525 \times 10^{-24} \\ + 0.693/(72.1 \times 24 \times 3600) = 1.64 \times 10^{-7}\,\text{s}^{-1}\end{aligned} \tag{17.40}$$

$$(\Lambda_3 =) : \Lambda_{161} = \lambda_{161} = 0.693/(6.9 \times 24 \times 3600) = 1.162 \times 10^{-6}\,\text{s}^{-1} \tag{17.41}$$

$$(\Lambda_3^* =) : \Lambda^*_{160} = \varphi\sigma_{160} = 10^{14} \times 525 \times 10^{-24} = 5.25 \times 10^{-8}\,\text{s}^{-1} \tag{17.42}$$

$$(N_1^0 =) : N^*_{159} = (2/366) \times 6.02 \times 10^{23} \times 10^{-3} = 3.29 \times 10^{18}\text{ atoms} \tag{17.43}$$

$$(t =) : t = t_{irr} = 30 \times 24 \times 3600 = 2.59 \times 10^6\,\text{s} \tag{17.44}$$

Note that Λ^*_{159} is the rate of formation of ^{160}Tb, and Λ^*_{160} the rate of formation of ^{161}Tb, while Λ_{159}, Λ_{160} and Λ_{161} are the rates of disappearance of ^{159}Tb, ^{160}Tb and ^{161}Tb. Introducing the experimental values yields $N_{161} = 5.68 \times 10^{14}$ atoms. The radioactive decay is $A_{161} = \lambda_{161}\ N_{161} = 1.162 \times 10^{-6} \times 5.68 \times 10^{14} = 6.60 \times 10^8$ Bq or 4×10^{10} dpm.

Alternatively to this approach Laplace transforms can be efficiently used to directly solve the system of differential equations shown in equation 17.17 to 17.22. By using this technique a full flexibility is achieved when it comes to the problem formulation and which events to take into account. Each equation (17.17 to 17.22) is then transformed into Laplace space where a differential equation becomes a linear equation. Thus in Laplace space equation 17.18 becomes:

$$sN_{21} - \widehat{L}N_{21}^{0} = \lambda_{11}\widehat{L}N_{11} - (\varphi\sigma_{21} + \lambda_{21})\,\widehat{L}N_{21} \tag{17.45}$$

The system of now linear equations is solved with respect to the desired nuclide and the solution is transferred back into normal space if the inverse transform is known.

17.4. TARGET CONSIDERATIONS

The success of an irradiation experiment depends to a large extent on the target considerations. In many nuclear experiments as much time has to be devoted to the target preparation as to the rest of the experiment. If the purpose is to produce a radionuclide for a simple tracer experiment, the consideration in this section may be sufficient. If, however, the requirements are a product of extreme purity, very high specific activity, or very short half-life, special techniques must be used, which are discussed in §§17.5-17.7. The production of atoms of the heavy elements also require fast and efficient production and separation methods.

17.4.1. Physical properties

In thin targets the flux and energy of the projectile do not change during passage through the target and calculation of e.g. production rates are straight forward by applying e.g. equation 17.14 using k from 17.12. If, however the projectile energy is considerably changed (in a thick target) it is more complicated to calculate the yield as the variation of energy with distance in the target and the variation of cross section with energy must be taken into account.

The exact specification of when a target is thin depends on the experiment; in most cases thin means a target with a surface weight of not more than a few mg/cm^2 or a thickness of a few μm or less. The *surface weight* (or *linear density*) is $x\rho$, where x is the target thickness and ρ its density. Thin targets are usually used in accelerator irradiations and are made of a solid material, either metal or compound. In order to have sufficient mechanical strength the target is usually supported on a *backing material* such as aluminum. The backing material may be part of a solid target holder which has provisions for cooling to dissipate the heat developed in the bombardment. Thin targets are fixed on the backing material through electrolytic precipitation, vacuum deposition, mass spectrometric collection (for isotopically pure targets), or a number of other techniques.

In reactor irradiations it is more common to use thicker targets in order to obtain higher yields. In traversing the target the effective flux (i.e. $\varphi\sigma$) does not decrease as rapidly for neutrons as for charged particles and, therefore, thicker targets can still be considered "thin" in neutron irradiations. However, if the product σN is large, the neutron flux may decrease appreciably at the center of the target, especially at the neutron energies where cross sections are large.

The target material can be a solid, liquid, or gas. For reactor irradiations it is common to place the target material in a container of polypropylene or of some metal of relatively low neutron capture cross-section. Some other materials should be avoided since they contain elements wich will themselves be activated in the neutron flux and thus complicating the handling of the sample. Such examples are pyrex and soda glass because of the reactions $^{10}B\ (n,\alpha)\ ^{7}Li$ (which has a large σ_{capt}) and $^{23}Na\ (n,\gamma)\ ^{24}Na$ (which yields a long-lived γ-product). In silica some γ-active ^{31}Si ($t_{1/2}$ 2.62 h) is produced ($\sigma_{capt} = 0.11$ b for thermal n). When metal sample holders, usually aluminum and magnesium are used, some γ-activity is produced; however, the cross-sections for the formation of the products ^{28}Al and ^{27}Mg are small, and these nuclides decay rapidly ($t_{1/2} = 2.24$ and 9.46 min, respectively). The primary requirement for the container is that it should be leak-tight and dissipate heat energy at a rate sufficient to avoid melting.

Another factor that affects the yields of an irradiation is the extent of *burn-out* of the target by the irradiation. If the number of target atoms at the beginning of the irradiation is N^0_{target} and at time t_{irr} is N_{target}, then the burn-out can be estimated with the relation

$$N_{target} = N^0_{target}\left(1 - \sigma_{tot}\,\varphi\,t_{irr}\right) \qquad (17.46)$$

where σ_{tot} includes all processes whereby the target atoms are reduced. Burn-out can be neglected when $\sigma_{tot}\,\varphi\,t_{irr} \ll 1$.

In the example of the irradiation of the 0.3 mm thick gold foil in §11.1, from (11.2) we can calculate that the flux reduction is 17%; for a 0.01 mm thick foil it would be 0.62%, but for a 1 mm foil, 44%. The target thickness does not enter into the burn-out, eqn. (17.46). For a 10 min irradiation of 10^{13} n cm^{-2} s^{-1}, the burn-out is negligible. However, a one-month irradiation at the same flux yields 0.26% burn-out, while a month at a flux of 10^{15} gives about 20% burn-out.

Finally, since the production cross sections of the heaviest elements are low, usually in the pb − nb range, synthesis of these require that an actinide target of up to 1 mg/cm^2 thickness on a 1−5 μm backing is irradiated with a heavy ion beam of 1−5 particle μA. Although only a minor part of the beam energy is dissipated when it passes through the target a lot of heat is released. A recent development is therefore rotating target wheels with a number of (usually three) banana shaped targets. When the wheel rotates one target on the wheel is hit by the particle beam while the others are cooled. The recoiling reaction products are then usually swept away from the backside of the targets by a gas-jet transport system.

17.4.2. Chemical properties

There are two ways to produce a pure radionuclide not contaminated with any other radioactivity. An extremely pure target can be used with a reaction path which is unique. Alternatively, the radioactive products can be purified after the end of the bombardment.

For example, a 10 g sample of zinc irradiated for one week with 10^{13} n cm^{-2} s^{-1} yields a sample of ^{65}Zn ($t_{½}$ 244 d) with 7.1×10^9 Bq. If, however, the zinc target is contaminated with 0.1% of copper, in addition to the zinc activity, 3.0×10^9 Bq of ^{64}Cu ($t_{½}$ 12.7 h) is formed. In another example element $_{102}$No was believed to be discovered initially in a bombardment of a target of curium by carbon ions. The observed activity, however, was later found to be due to products formed due to the small amount of lead impurity in the target. Similarly, in neutron activation of samarium the target must be very free of europium contamination because of the larger europium reaction cross-sections. Handbooks of activation analysis often contain information on the formation of interfering activities from impurities.

Even a chemically pure target may yield products of several elements, particularly in cyclotron irradiation, where many reaction paths are often possible. In the bombardment of magnesium with deuterons, the following reactions occur:

$$^{24}\text{Mg}(79\%)(\text{d},\alpha)^{22}\text{Na}(t_{½}\ 2.6\ \text{y}) \tag{17.47}$$

$$^{25}\text{Mg}(10\%)(\text{d,n})^{26}\text{Al}(t_{½}\ 7.2 \times 10^5\text{y}) \tag{17.48}$$

$$^{26}\text{Mg}(11\%)(\text{d},\alpha)^{24}\text{Na}(t_{½}\ 14.66\ \text{h}) \tag{17.49}$$

A short time after the irradiation ^{24}Na is the predominant activity; after longer cooling times the ^{22}Na is always contaminated by ^{26}Al. In such a case chemical purification after the irradiation is required to yield a pure product.

A suitable target material may be unavailable, or very expensive because of a low natural abundance. ^{45}Ca, which is the most useful calcium radioisotope ($t_{½}$ 164 d), is produced through n,γ capture in ^{44}Ca; unfortunately, ^{44}Ca only occurs to 2.1% in nature. While irradiation of natural calcium may yield a specific activity of ~0.3 TBq g^{-1}, isotope enrichment before irradiation to essentially pure ^{44}Ca yields $\leq$20 TBq g^{-1}.

When the target is a chemical compound its radiation stability becomes important. In general organic compounds are not stable towards either neutron or charged particle bombardment. However, with neutrons a very particular and useful kind of reaction, known as the Szilard-Chalmers reaction (see §17.6.3), may take place. Simple inorganic compounds like NaCl are stable, and for some inorganic compounds like Na_2CO_3 the radiation decomposition (to Na_2O and CO_2) are of little importance. However, it may be noted that irradiations producing e.g. alpha active nuclides may eventually cause the inner helium pressure increase so much that crystals etc may rupture.

When chemistry experiments will be performed on the products from the nuclear reaction the products have to be transferred fast and continuously to the chemical separation system. As mentioned above, this is usually done by sweeping the back side of the target with a gas (usually He) containing aerosol particles (often KCl). The formed products have enough energy to recoil out of the target, if it is thin enough, and recoiling ionized atoms attach them selves to the aerosol particles. They can then be swept away several tenths of meters within a few seconds to a low-background area where the chemistry is performed.

When using a gas-filled mass separator between the target and the chemistry there is an additional problem because the products have to pass from the low pressure ($\sim$100 Pa) in the mass separator to the high pressure ($\sim$0.2 MPa) in the gas jet transportation system. This is achieved by using a recoil transfer chamber with a thin window that allows the nuclear reaction products to pass from the low pressure to the high pressure area where they attach to the aerosol particles.

17.5. PRODUCT SPECIFICATIONS

In many countries there are one or more suppliers of *radiochemicals*. To locate suppliers, the simplest way is often to browse the internet but it is also possible to contact the nearest nuclear center, as it may be a producer of radionuclides. Other sources of information may be e.g. the national radiation safety organization, or the annual Byers' Guide of common nuclear journals.

The radionuclides come in the form of calibration/reference sources, and as inorganic compounds and labelled organic compounds. They are packed in various ways, most commonly for compounds in a glass ampoule packed in a small aluminium can. To reduce the surface radiation dose rate (cf. Ch. 15) for high intensity radiation sources (usually >10 MBq) the can is put in a lead block which is packed in a wooden crate for shipment. Opening of the aluminum can and glass ampoule may require remote control and should be done only by experienced, licensed personnel.

17.5.1. Radiochemical processing

In some cases it may be a long process between the removal of the irradiated target from the irradiation facility and the pure radiochemical substance or compound to be used in some experiment. The separation procedures needed to make a pure radiation source are chemically the same and (usually) independent of the "size" (i.e. its total radioactivity), but the experiments become more cumbersome — and often slower — when the radiation source is so large that considerable shielding is needed to perform the work.

Most chemical techniques can be applied in the purification processing: precipitation, ion exchange, solvent extraction, electrodeposition, electrophoresis, distillation, etc. The basic purposes are to eliminate radioactive contaminants and to avoid diluting the radionuclide by isotopic stable atoms. If the radionuclide or radiochemical is to be used in medical and biological work, the sample may have to be sterile. Some publications about radiochemical separation procedures are given in the literature list, §17.9.

17.5.2. Specific activity

The specific activity or concentration of a radionuclide, i.e. radioactivity per gram, mole, or volume, is a very important factor in radionuclide work, e.g. in chemical work with substances of low solubility, in microbiology, etc.

Radionuclides of high specific activity are produced either through accelerator irradiation or through secondary reactions in the target (§17.6) in a reactor. Maximum specific activity is obtained when the radioactive nuclide is the only isotope of the element present in the sample. This is not possible to achieve in regular reactor irradiation through (n,γ) capture processes. For example, reactor-produced ^{24}Na may be obtained in specific activities of $\leq 2 \times 10^{11}$ Bq g^{-1}, while the specific activity of accelerator-produced ^{24}Na may exceed 10^{13} Bq g^{-1}; however, the total activities available are usually the inverse.

A *carrier-free* radioactive sample is usually one in which the radionuclide is not diluted with different isotopes of the same element. In reactor production of ^{24}Na from target ^{23}Na, each ^{24}Na is diluted with a large number of ^{23}Na atoms. ^{24}Na cannot be made carrier-free in a reactor. If a carrier-free radionuclide has been produced, e.g. through accelerator irradiation, which then must be purified, its concentration is so low that it may not follow the normal chemical rules. A macroscopic amount of *carrier*, either isotopic or not, may have to be added to carry the radionuclide through the proper chemical purification steps. We discuss this further in §18.2.

17.5.3. Labelling

A radioactively *labelled* (or "*tagged*") compound is a compound whereat least one of the atoms are radioactive. Preparation of labelled compounds may involve lengthy chemical synthesis starting with the radioactive nuclides in elementary form or in a simple compound.

Most syntheses of ^{14}C-labelled organic or biochemical compounds follow conventional methods with appropriate modifications to contain the radioactivity. $^{14}CO_2$ gas liberated by action of an acid on labelled $Ba^{14}CO_3$ is used commonly as a starting point in the syntheses. Three examples can illustrate how compounds can be labelled at different sites of the molecule:

1. Acetic acid-2-^{14}C:

$$Ba^{14}CO_3 \xrightarrow{HCl} {}^{14}CO_2 \xrightarrow[\text{catalyst}]{H_2} {}^{14}CH_3OH \xrightarrow{PI_3} {}^{14}CH_3I \xrightarrow{KCN}$$

$$^{14}CH_3CN \xrightarrow{H_2O} {}^{14}CH_3CO_2H \tag{17.50}$$

2. Acetic acid-1-^{14}C:

$$Ba^{14}CO_3 \xrightarrow{HCl} {}^{14}CO_2 \xrightarrow{CH_3MgBr} CH_3{}^{14}CO_2H \qquad (17.51)$$

3. Acetic acid-1,2-^{14}C:

$$\text{(a) } Ba^{14}CO_3 \xrightarrow[heat]{Mg} Ba^{14}C_2 \xrightarrow{H_2O} {}^{14}C_2H_2 \xrightarrow[catalyst]{H_2O} {}^{14}CH_3{}^{14}CHO \xrightarrow{[O]}$$

$${}^{14}CH_3\ {}^{14}CO_2H \qquad (17.52)$$

$$\text{(b) } K^{14}CN \xrightarrow{{}^{14}CH_3I} {}^{14}CH_3{}^{14}CN \xrightarrow{H_2O} {}^{14}CH_3{}^{14}CO_2H \qquad (17.53)$$

In these examples labelling is said to be *specific* in that only one specific position in the compound contains a radioactive nuclide. Alternatively, labelling may be *general* when most, or all, positions are labelled, which often is the case for tritium-labelled compounds (cf. §17.6.2).

Complicated organic compounds, e.g. penicillin, can be labelled through *biosynthesis*. For example, if penicillin is grown in a substrate containing a simple ^{35}S compound, ^{35}S is incorporated into the penicillin mold. Since other ^{35}S-containing products may be formed also, the ^{35}S penicillin must be purified, e.g. through solvent extraction or paper chromatography (§18.2.6).

Thousands of labelled organic compounds are commercially available and some organizations offer labelling service on request.

17.5.4. Radiochemical purity

An important consideration in the use of radionuclides is their radiochemical purity since, should several radionuclides of different elements be present in a tracer sample used in an experiment, the result could be ambiguous and misleading. In a radio-chemically pure sample, all radioactivity comes from a single radioactive element. If the radioactivity comes from a single isotope, the sample may be said to be *radio-isotopically pure*.

The radioisotopic purity of a sample may be ascertained by measuring the half-life. This method obviously can be used only with radionuclides that show sufficient decay during the time of observation for a reliable half-life to be calculated. For radioactivity of longer half-lives the radioisotopic purity can be checked by measurement of the type and energy of the emitted radiations (Ch. 9) or by processing the tracers through one or more chemical steps characteristic of the element of which the radioactivity is presumably an isotope. For example, in the use of ^{90}Sr ($t_{½}$ 28.5 y), the radioactivity can be processed through a number of typical reactions characteristic of strontium. If the radioactivity follows the expected chemistry, then the experimenter can be rather certain that it is strontium. This chemical check does not preclude the possibility that there might be several isotopic nuclides in the sample. However, the presence of several isotopes need not be a handicap in tracer studies concerned with the characteristic properties of a particular element.

When a compound labelled with a radioactive nuclide is used to study molecular behavior, it is necessary to ascertain the *radiochemical purity of the compound* and not just the purity of the radioactive isotope. This means it is necessary to be certain that the radioactivity comes only from a specific element and only from a specific compound incorporating that element. This can be determined through selective chemical operations in which it is frequently desirable that the compound should not be destroyed. Examples of such methods involve preparative gas chromatography, thin-layer chromatography, paper chromatography, dialysis, ion exchange, and solvent extraction. Checking that an element or compound is radiochemically pure is important not only for sources of samples prepared in the laboratory but for commercial samples as well. For example, ^{14}C-tagged organic compounds obtained commercially may have been prepared at varying times prior to purchase. Frequently, the decomposition of molecules in the sample due to the radiolysis that accompanies the radioactive decay results in the presence of products that interfere with the course of the experiment for which the labelled compound is to be used.

In some cases the radioactive product after a period of storage will have grown a sufficient amount of radioactive daughter that it is no longer sufficiently pure radiochemically. In these cases, the daughter activity must be separated prior to use. For example, the ^{90}Y normally present in a ^{90}Sr sample must be eliminated if the tracer solution is to be used to study strontium chemistry.

17.6. RECOIL SEPARATIONS

17.6.1. Target recoil products

When a projectile of high energy reacts with a target atom, the reaction products are emitted in the direction of the projectile, measured in laboratory coordinates. We use the term *recoil* to describe this. The products, whose kinetic energy can be calculated by the conservation laws of Ch. 10, are slowed and eventually stopped through collision with other atoms. Consequently, if the target is relatively thin, there is a certain probability that the recoil range of the product exceeds the thickness of the target and the target atoms can therefore escape the target. In such cases the recoiling products be caught on special foils behind the target (*catcher foils*). These catcher foils can be stationary or they can be a moving band or metal or plastic, or even a gas, which sweeps past the target material.

The recoil technique was used in the discovery of the first transuranium element, neptunium, by E. McMillan and P. H. Abelson in 1939. A uranium salt, $(NH_4)_2U_2O_7$, was fixed on a paper and surrounded by more layers of thin paper to form a "sandwich" which was irradiated by thermal neutrons. At the time of fission the fission products ejected with high kinetic energy and were caught in the layers of paper. From the number of layers of paper penetrated by individual fission products, their energy and mass could be calculated. A strong β^- activity was found in the first layers of paper, showing that it had relatively little recoil energy. From chemical experiments it was concluded that this activity was not a fission product but was due to ^{239}U which decayed into ^{239}Np:

$$^{238}U(n,\gamma)^{239}U \xrightarrow{\beta^-} {}^{239}Np \qquad (17.54)$$

In this particular case, the thermal neutron did not carry any kinetic energy, but the emission of the γ-ray caused the ^{239}U to recoil from the uranium sample prior to decay (with very little recoil) to ^{239}Np. This type of reaction is common in neutron irradiation processes, and is referred to as n,γ-recoil.

Recoil techniques have been used extensively in the synthesis of higher transuranium elements. As an example, consider the formation of element 103, lawrencium (Fig. 17.3). The $_{103}Lr$ nuclides, which were formed in the reaction in the target between ^{252}Cf and projectiles of ^{11}B, recoiled out of the thin target into the helium gas where they were stopped by atomic collisions with the He-atoms. These recoil species became electrically charged cations as they lost electrons in the collision with the gas atoms. They could, therefore, be attracted to a moving, negatively charged, metal coated plastic band, which carried the Lr atom on its surface to a number of α-particle detectors where the α-decay of the lawrencium was measured. From the velocity of movement of the plastic band and the measurement of the radioactivity by energy sensitive nuclear detectors, the half-lives and α-energies of the product nuclei could be determined. Comparison of these data with predictions from theory allowed assignment of the mass number of the nuclide.

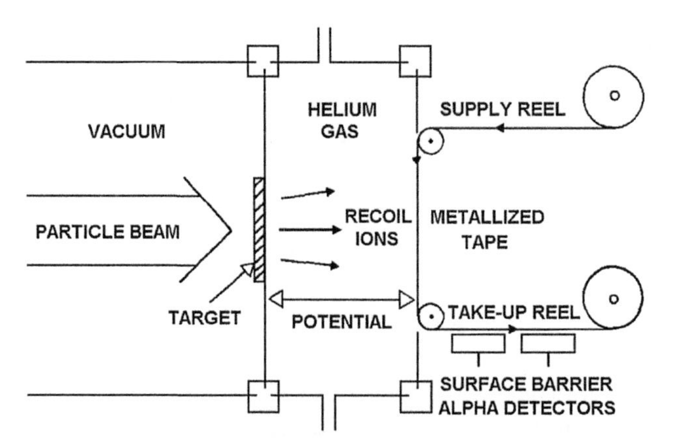

Figure 17.3 Principle of target cage system for α-recoil collection of the products formed.

Target recoil products may have a detrimental effect also. In reactor irradiation of biological samples to determine their trace metal concentration it was found that n,γ products formed in the material of the sample container recoiled out of the container wall into the sample, contaminating it. Such contamination can be eliminated by using a container which forms a minimal amount of such products.

17.6.2. Hot atom reactions

The momentum imparted to a nucleus in a nuclear reaction with a charged particle or a fast neutron almost invariably is sufficient to result in the rupture of chemical bonds holding the atom in the molecule. The same is true for the recoil energy imparted to a nucleus by the emission of an α-particle as well as in β-decay, γ-decay, etc.; the recoil energies in radioactive decay are discussed in §§5.3–5.5.

These recoiling atoms are known as *hot atoms* since their kinetic energy is much in excess of thermal equilibrium values. A hot atom may move as much as several hundred atomic diameters after the rupture of the bond before being stopped even though this takes only about 10^{-10} s. Initially the hot atom is highly ionized (e.g. for $Z = 50$ ionic charges up to +20 have been observed). As the hot atom is slowed down to thermal energies, it collides with a number of other particles in its path producing radicals, ions, and excited molecules and atoms. At thermal energies it may become a neutral atom (particularly if the matrix material is metallic), an ion, possibly in a different oxidation state (a common situation in inorganic material), or it may react to form a compound with molecules in its path (the prevalent situation in matrices of covalent material).

Chemical effects of nuclear transformations (*hot atom chemistry*) have been extensively studied in connection with induced nuclear transformations, both in the gas phase, in solution and in the solid state. In the latter cases the dissipation of the kinetic energy and neutralization of the charge within a small volume produces a high concentration of radicals, ions, and excited molecules in the region where the recoiling atom is slowed to energies where it can form stable bonds. Usually the product molecule is labeled neither very specifically nor completely randomly. The topic of hot atom behavior is also treated in Ch. 8.

Hot atom reactions can be used for synthesis of labelled compounds. So far this technique has been used primarily for ^{3}H and ^{14}C labelling. For example, an organic gaseous compound is mixed with ^{3}He and the sample is irradiated with neutrons. The reaction $^{3}He(n,p)^{3}H^{\rightarrow}$ produces tritium as hot atoms (indicated by the small arrow) with kinetic energies of approximately 0.2 MeV. These hot atoms react with the organic molecules to produce labelled compounds. An example is the formation of labelled butanol

$$C_4H_9OH + {}^{3}H^{\rightarrow} \rightarrow C_4H_9O^{3}H + {}^{1}H^{\rightarrow} \qquad (17.55)$$

The reaction is referred to as a *hot atom displacement* reaction. In an alternate method the solid organic compound may be mixed with Li_2CO_3, and this solid mixture irradiated with neutrons to cause the reaction: $^6Li(n,\alpha)^3H^{\rightarrow}$. Again the hot tritium atom combines with the organic species to cause labeling. In a third and simpler method, the organic compound is simply mixed with tritiated gas. The β-radiation from tritium decay causes excitation and decomposition which leads to capture by the organic radicals of some of the remaining tritium. These reactions are not selective because reactions with decomposition products also occur. Therefore the tritium labeled product must be purified. Gas phase labeling usually results in incorporation of tritium in several sites of the molecule, e.g. toluene has about 9% of the total label in the CH_3 group, 54% in the *ortho* sites of the ring, 24% in the *meta* sites, and the remaining 13% at the *para* site.

^{14}C-labeling can be achieved in a similar manner by using the $^{14}N(n,p)^{14}C^{\rightarrow}$ reaction. The recoiling ^{14}C energy is about 0.6 MeV. A suitable labeling mixture is NH_4NO_3 + the organic compound: cf. ref. Wolf (1960).

17.6.3. The Szilard-Chalmers process

In 1934 L. Szilard and T. A. Chalmers discovered that bond breaking could occur for atoms following nuclear reaction or radioactive decay even though the recoil energy in the initial process is not sufficient to overcome the bonding energy. In the case of thermal neutron capture the processes involved in the emission of the γ-ray, which removes the nuclear excitation energy, impart recoil energy to the atom to break most chemical bonds (n,γ-recoil). If, after rupture of the bonds, the product atoms exist in a chemical state different and separable from that of the target atoms, the former may be isolated from the large mass of inactive target. This provides a means of obtaining high specific activities in reactions where target and product are isotopic.

This process is known as the *Szilard-Chalmers reaction* and was discovered when, following the irradiation of ethyl iodide with thermal neutrons, it was found that radioactive iodide could be extracted from the ethyl iodide with water. Moreover, when iodide carrier and silver ions were added to this aqueous phase, the radioactive iodide precipitated as silver iodide. The obvious interpretation of these results is that the neutron irradiation of the ethyl iodide, which caused the formation of ^{128}I, ruptured the bonding of this atom to the ethyl group. The bond energy of iodine to carbon in C_2H_5I is about 2 eV. Since this exceeds the recoil energies of neutron capture, the bond breakage must have resulted from the γ-emission which followed neutron capture and not the capture process itself. The reaction can be written:

$$C_2H_5\,{}^{127}I + n \rightarrow [C_2H_5\,{}^{128}I]^* \rightarrow C_2H_5 + {}^{128}I^{\rightarrow} + \gamma \qquad (17.56)$$

The $^{128}I^{\rightarrow}$ loses its kinetic energy and is stabilized as an iodine atom or iodide ion; it can also be recaptured by the C_2H_5 radical (*retention* of activity in C_2H_5I). In addition to the necessity that the recoiling species have sufficient energy to rupture the bond, it is also necessary for a successful enrichment of specific activity by the Szilard-Chalmers process that there is no rapid exchange at thermal energies between the active and inactive iodine atoms in ethyl iodide:

$$C_2H_5{}^{127}I + {}^{128}I \overset{\text{slow}}{\rightleftarrows} C_2H_5{}^{128}I + {}^{127}I \tag{17.57}$$

After the rupture of the bond, the hot atoms may also replace iodine or even hydrogen atoms from ethyl iodide molecules, similar to the thermal exchange:

$$^{128}I^{\rightarrow} + C_2H_5{}^{127}I \rightarrow C_2H_5{}^{128}I + {}^{127}I \tag{17.58}$$

Again, the result is retention of the activity within the organic molecule. Retention is decreased by diluting the ethyl iodide with alcohol which reduces the probability of collision and exchange with C_2H_5I molecules as the hot atoms of $^{128}I^{\rightarrow}$ are being slowed.

Isomeric transitions which proceed by emission of γ-rays may not provide sufficient recoil energy to break covalent bonds. However, in these cases or very low energy isomeric transitions, the extent of internal conversion is large. This results in vacancies in the inner electron orbitals (§5.5). When electrons in outer orbitals move to fill the vacancies, the difference in electron binding energies is sufficient to cause some ionization, resulting in relatively high charge states for the atom, leading to bond rupture. An example of such a process is seen in the reaction whereby an alkyl bromide is irradiated with thermal neutrons. During neutron irradiation the reactions are

$$RBr + n \begin{cases} \xrightarrow{\sigma\, 2.9\, b} R + {}^{80m}Br\ (4.4\ h) \\ \qquad\qquad \downarrow IT \\ \xrightarrow[\sigma\, 8.5\, b]{} R + {}^{80}Br\ (18\ min) \end{cases} \tag{17.59}$$

Extraction by water of this bromide produces an aqueous sample containing both ^{80}Br and ^{80m}Br. Some ^{80}Br and ^{80m}Br are retained as alkyl bromide in the organic phase. If, after a period of an hour, a new extraction is carried out on the organic phase, only the ^{80}Br is found in the aqueous sample. This ^{80}Br results from the following reaction

$$R^{80m}Br \xrightarrow{IT} R + {}^{80}Br + \gamma \tag{17.60}$$

In addition to the organic halides a number of inorganic systems undergo similar reactions. From neutron irradiation of acid or neutral solutions of permanganate most of the ^{56}Mn activity is removable as MnO_2. From solid or dissolved chlorates, bromate, iodate, perchlorates, and periodates, active halide samples may be isolated in relatively high yields. Systems with Te, Se, As, Cu, and several Group VIII elements have been shown to be applicable to the Szilard-Chalmers process. The great advantage in all these cases is that a relatively small amount of activity may be isolated from a large mass of isotopic inactive target. Because of retention there is also some labeling of the target molecules.

17.7. FAST RADIOCHEMICAL SEPARATIONS

Short-lived radioactive nuclides are of interest (i) to nuclear science to establish the outer boundary for nuclide formation, which yields information on nuclear structure , (ii) to chemistry to extend the limits of the periodic system, i.e. the actinides and transactinides (or *superheavy elements*); these studies are also of interest to cosmology since these heaviest elements are believed to be formed in supernovae, and (iii) to medicine where they offer a unique tool to identify and treat some diseases (§18.5). Short-lived radionuclides are also used in chemical and industrial research, one of their advantages being that they rapidly disappear, eliminating radioactive waste problems. We arbitrarily define short-lived as meaning half-lives <10 minutes. Because many of the lightest elements of greatest interest in organic chemistry and biology lack long-lived radionuclides, the production of their short-lived isotopes are of particular interest; see Figure 5.7. At the other end of the periodic system there are no stable isotopes at all, and for the trans-uranium elements the half-lives of the most stable isotopes become shorter and shorter for each higher element, e.g. the longest-lived isotope of $Z = 105$ has $t_{1/2} \sim 34$ s. Thus, research on the production of short-lived radionuclides has been particularly intense for the lightest and heaviest elements as well as for the fission products . It is obvious that studies and uses of these short-lived nuclides require special techniques, which often are rather costly. A brief description of some of the techniques used follows.

The sequence of procedures necessary for the production and study of very short-lived nuclei are:

1) target preparation and irradiation,
2) primary (preliminary) target purification (at irradiation site),
3) transport of product (e.g. through pneumatic tubes in which the transport velocity may be up to 100 m s^{-1}) from irradiation site to nuclear chemistry laboratory,
4) isolation of desired product and eventual labeling, and
5) final use in physico-chemical experiments or in applied areas; e.g. hospitals or industry.

A primary requirement for work with short-lived nuclides is that the site or production and the site of use be close together. In 2–30 seconds it is possible to use rapid separation procedures consisting of (a) target dissolution, followed by (b) chemical isolation of the radionuclide through solvent-extraction-stripping or high pressure/temperature ion exchange, and/or (c) precipitation or electro-plating before (d) measurement. In this scheme eventual chemical studies can be included, usually between steps (b) and (c).

There are a number of non-chemical techniques involving time-of-flight and magnetic mass analysis of target recoil products, energy sensitive detectors, etc., with which half-lives down to 10^{-12} s can be measured. However, of more interest to us are the fast chemical systems. Let us review a few representative types.

17.7.1. Production of ^{11}C labeled compounds

In §18.5.3 (c) we describe the use of the PET medical imaging technique, which requires the use of positron emitters. Many interesting positron emitters are very short-lived (see Table 18.5) requiring rapid synthesis of the labeled compound. One of the more frequently used positron emitters is ^{11}C ($t_{½}$ 20.3 m) because it can be introduced into almost any organic compound. Glucose derivatives are of special interest because glucose is the main energy source of the brain; as a consequence, the areas in which glucose accumulates indicate places of highest physiological activity. Also, glucose derivatives are relatively easy to synthesize. To illustrate the technique, consider the synthesis of 2-deoxy-D-[1-^{11}C] glucose (^{11}C-2DG) which is used to study tumor metabolism in the brain (steps 1 and 2 in §18.5.3 (c)):

By irradiating a target consisting of a mixture of 95% N_2 + 5% H_2 in a cylinder, 400 mm long and 60 mm in diameter, at 10 bar pressure by 20 MeV protons, $^{11}CH_4$ and $^{13}N_2$ are produced. After an irradiation of 30-60 minutes, the gas is mixed with ammonia and allowed to react at elevated temperature on a platinum catalyst to produce $H^{11}CN$ (~80% conversion at 850 °C). The yield is about 5 GBq per μA of proton current. The subsequent synthesis is a series of elaborate steps in which $Na^{11}CN$ is digested with preformed reactive glucose derivatives in presence of Ni/Al-alloy in formic acid. The reaction mixture is purified through several ion exchange columns and finally collected on an anion exchanger. The total synthesis, from end of irradiation, takes about 1 hour with an over-all yield of ~80% relative to the $H^{11}CN$ yield.

Many procedures for rapid ^{11}C-labelling are described in the literature. In some cases labeled compounds can be ordered from radiopharmaceutical companies, but for the ^{11}C-compounds the $t_{½}$ mandates production and synthesis at the place of use.

17.7.2. Auto-batch procedures

Figure 17.4 shows an auto-batch procedure for isolation and separation of rare earths obtained from spontaneous fission of ^{252}Cf. The rare earths plus other fission products are washed from the target chamber by nitric acid and extracted by high-pressure liquid chromatography onto a column containing dihexyl diethyl carbamoyl-methylphosphonate (DHDECMP). After removal from this column, the rare earths are sorbed on a cation exchange column, from which they are eluted sequentially by hot (95°C) α-hydroxy isobutyric acid (HIBA). Figure 17.4 shows a typical elution curve. The total separation procedure takes about 10 minutes.

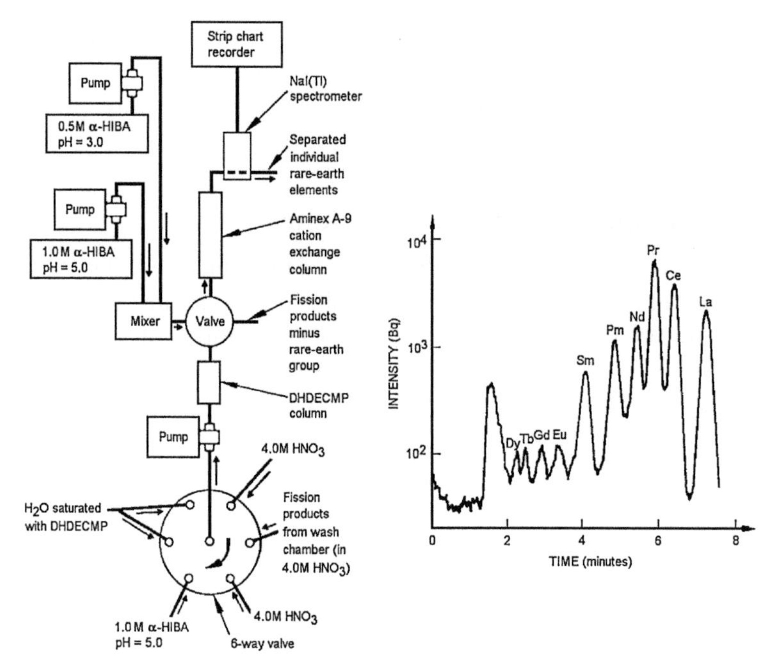

Figure 17.4 Schematic of auto-batch lanthanide chemistry system and elution curve obtained. *(From Baker et al.)*

Figure 17.5 shows a much more rapid procedure to recover zirconium isotopes formed by fission. (a) A solution of ^{235}U sealed in a plastic capsule and irradiated with thermal neutrons is projected pneumatically into the apparatus where the capsule is crushed. (b) The solution is filtered through two layers of preformed, inactive silver chloride, which, by heterogeneous exchange, rapidly removes I and Br. (c) The next step involves solvent extraction with an organic phase containing the reagent tri-butyl phosphate (TBP) adsorbed on the surface of small plastic grains. The filtrate from (b) is mixed with strong $HNO_3 + KBrO_3$ and drawn through the stationary organic phase (marked "movable filter" in the Figure), to remove zirconium. (d) After washing the organic phase with nitric acid, the filter layer containing the zirconium is transferred pneumatically to the detector (to the right in the Figure). (e) Counting is started 4 s after the end of irradiation. The Figure shows the time sequences for the operations in numbers. Stopcocks are operated pneumatically and valves magnetically by signals from electronic timers. The chemical yield of zirconium amounts to about 25%. This experiment led to the identification of ^{99}Zr with a half-life of 1.8 s. The ARCA (Automatic Rapid Chemistry Apparatus) in Darmstadt is a modification of the system in Figure 17.5; because the separations are done by ion exchange resins, it is considerably slower, requiring nuclides with $t_{1/2} > 20$ s.

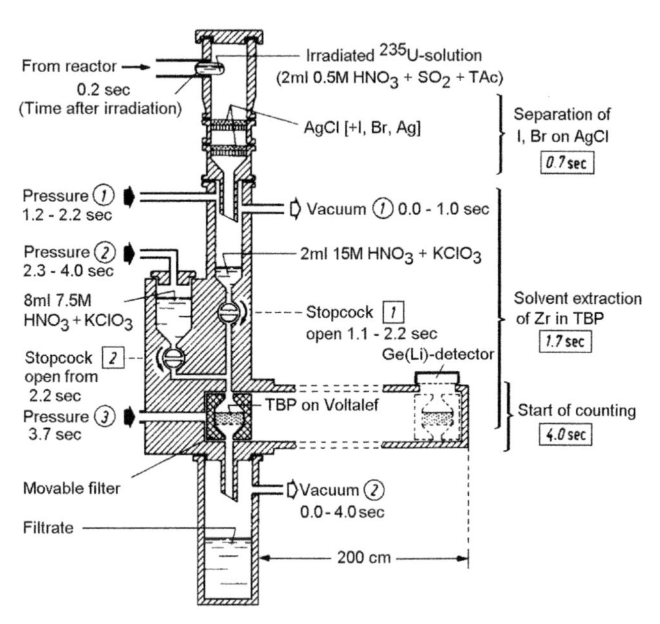

Figure 17.5 Auto-batch system for identification of short-lived zirconium isotopes. *(From Trautmann et al.)*

In these systems proper modification of conventional techniques provided very fast separations. Precipitation (using a preformed precipitate and heterogeneous exchange), solvent extraction (by reversed phase chromatography), and ion exchange (with heating and pressure) have been used in the identification and study of short-lived transuranides (elements 102 and higher).

17.7.3. On-line procedures

For the production of shortlived radioisotopes and for the chemical studies of the heaviest elements it is necessary to have fast chemical systems. In case of the heavy elements this is because the half-lives of most known isotopes of these elements are normally short. In the following we will briefly describe a few such systems and procedures.

17.7.3.1. Gas phase separation

Recoil techniques (§17.6) have been used extensively with *gas-jet transportation systems* for studying short lived radionuclides. As an example, consider the formation of element 103, lawrencium. The $_{103}$Lr nuclides, which were formed in the reaction in the target between ^{252}Cf and projectiles of ^{11}B, recoiled from the thin target into the helium gas where they were stopped by atomic collisions with He-atoms. These recoil species became electrically charged cations as the result of loss of atomic electrons in the collision

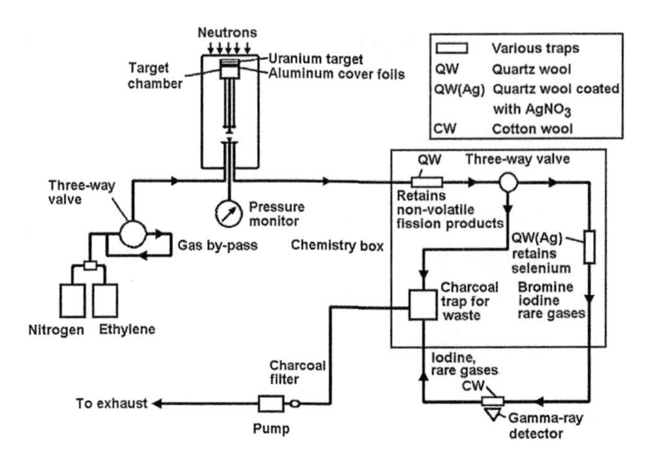

Figure 17.6 On-line gas-phase separation system for isolation of bromine products. *(From Rengan et al.)*

with the gas atoms. They could, therefore, be attracted to a moving, negatively charged, metal coated plastic band, which carried the Lr atoms on its surface to α-particle detectors where the α-decay of the lawrencium was measured; see Fig. 17.3. Instead of collecting the particles on a moving band, they could also be swept by a gas or aerosol stream into an apparatus for chemical processing.

Separation may be achieved by having a target containing a reactive compound or having the recoiling species continue in the gas phase by using either a reactive gas or using gas-solid separation. Figure 17.6 illustrates the separation of bromine from other fission products by using gas-solid reactions on quartz and cotton wool. Figure 17.7 illustrates another technique: a chemically reactive thermo-chromatographic column is placed after the target cage. The target N_2-gas contains a KCl-aerosol to which the fission product nuclides attach. The gas-aerosol stream passes through a tube or column containing a partly selective adsorbent (e.g. quartz powder, KCl, $BaCl_2$, K_2CrO_7). The temperature gradient along the tube leads to deposition at different positions of elements (or compounds) of different vapor pressure. Rather selective deposition can be achieved, although the technique is not totally element specific. Nuclides with half-lives down to 0.1 s have been determined in identification of spallation products, fission products, and transuranium elements. An example is the OLGA system (On-line Gas Chemistry Apparatus) in Darmstadt which was used to identify ^{263}Ha (atomic number 105); see Ch. 14.

17.7.3.2. Gas phase chemistry

Gas phase chemistry has been very successful for the production of SHE and so far this technique has been used to make Rf, Db, Sg, Bh, Hs and Cn. The original version, OLGA (On-Line Gas chemistry Apparatus) consisted of a thermocromatographic column connected to a reaction oven where the products delivered by the gas jet transport from the target were mixed with reactive gases to form compounds that were fed into the

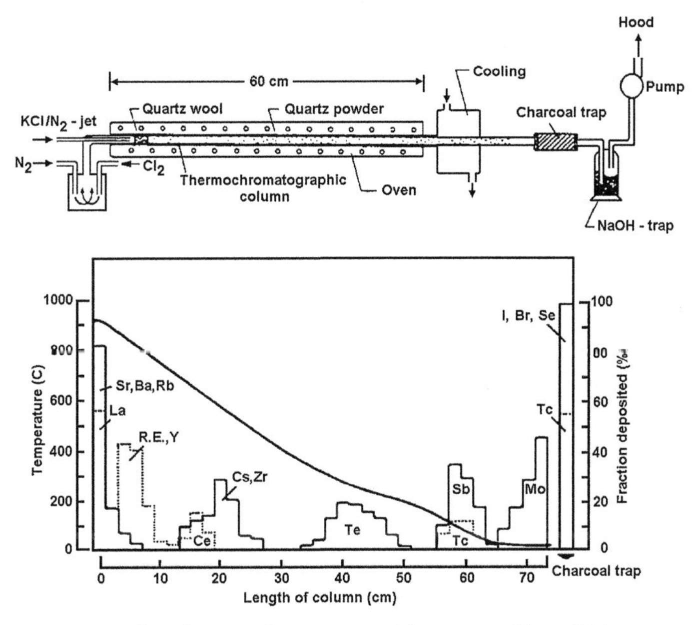

Figure 17.7 On-line thermo-chromatographic system. *(From Hickmann et al.)*

chromatography system. Detection was achieved by alpha-alpha correlations. This technique has been improved and modified for the production of Bh, Hs, and Cn.

17.7.3.3. Solid/liquid extraction and ion exchange

The ARCA (Automated Rapid Chemistry Apparatus) was developed at the Gesellschaft für Schwerionenforschung (GSI), Darmstadt (Germany). ARCA is an HPLC system based on fast repetitive chromatographic separations in micro columns. The columns are 8 mm long with an inner diameter of 1.6 mm and a dead volume of only ~35 μl. Depending on the separation procedure to be performed, the columns can be filled either with an organic extractant adsorbed on an inert supporting material, or with ion exchanger.

The latest ARCA version, ARCA II, consists of three main sections: a section for collecting activity from the gas-jet from the target and for carrying out the chemical separations, an HPLC section consisting of pumps, valves and solvent reservoirs, an electropneumatic valve system and a control unit, see Figure 17.8. All parts in contact with solutions are made of inert materials like Kel-F, PTFE (polytetrafluoro ethylene) or sapphire. The ARCA system is symmetrical and operated in an alternating mode between left and right. At the same time previously collected activity is dissolved and transported to a column for chemical separation on one side, activity from the gas-jet is collected on a polyethylene frit, and the next column is washed and conditioned on the other side. The products of interest are eluted from the columns onto preheated tantalum

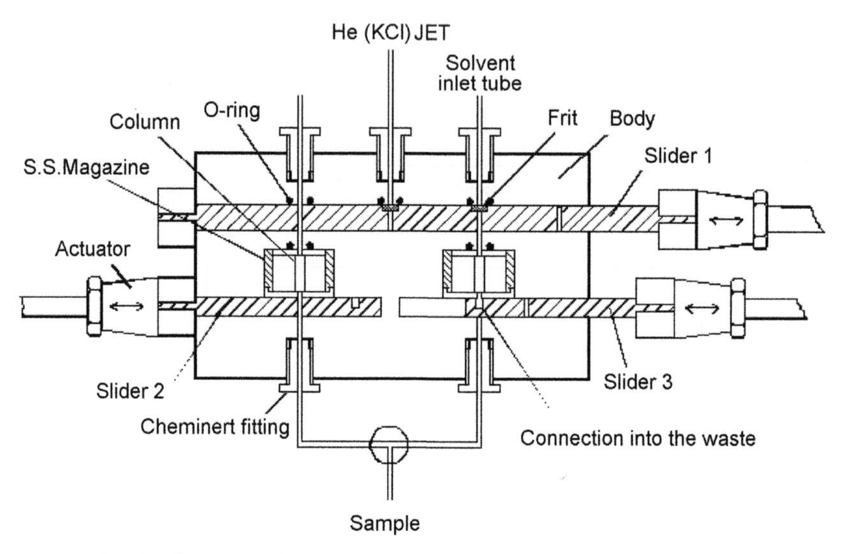

Figure 17.8 A sketch of the general principle of the main chemical separation section of ARCA II.

or titanium disks and the eluent is evaporated. After flaming and cooling the disk is placed in the measuring position. Typical times from the end of irradiation to the start of measurements is 50 s, which makes it possible to study nuclides with a half-life down to 10−15 s.

17.7.3.4. Solvent extraction

The SISAK (Short-lived Isotopes Studied by the AKUFVE) technique is illustrated in Figure 17.9. A target containing an element (e.g. uranium) on a thin foil is bombarded with projectiles, while a gas stream containing "catcher" particles (e.g. KCl) behind the target continually removes the products recoiling into the gas stream. The gas jet is thoroughly mixed with an aqueous solution in the static mixer and then degassed. The aqueous solution containing the radioactive products is passed through a series of fast and selective solvent extraction loops, where continuous flow mixers and centrifugal separators are used (the AKUFVE-technique, §18.4.3). The solution from the final separation passes through a "hold-up" cell, which may contain also some absorbent to capture the element of interest. The cell is viewed by a detector system. The procedure can provide high selectivity for a particular element and takes about 3 s from target to the detection site. Because it is an on-line technique, SISAK has mainly been used mainly to obtain detailed decay schemes of very short lived nuclides ($t_{½}$ ~ 1 to 10 s). It has also been used to study the extraction behaviour of superheavy elements. In this application it is connected to a gas-filled mass separator to reduce interfering background activities. In the newest version, SISAK 3, centrifuges with a hold-up volume of ~ 0.3 ml/phase and a highest flow rate of 3 ml/s are used.

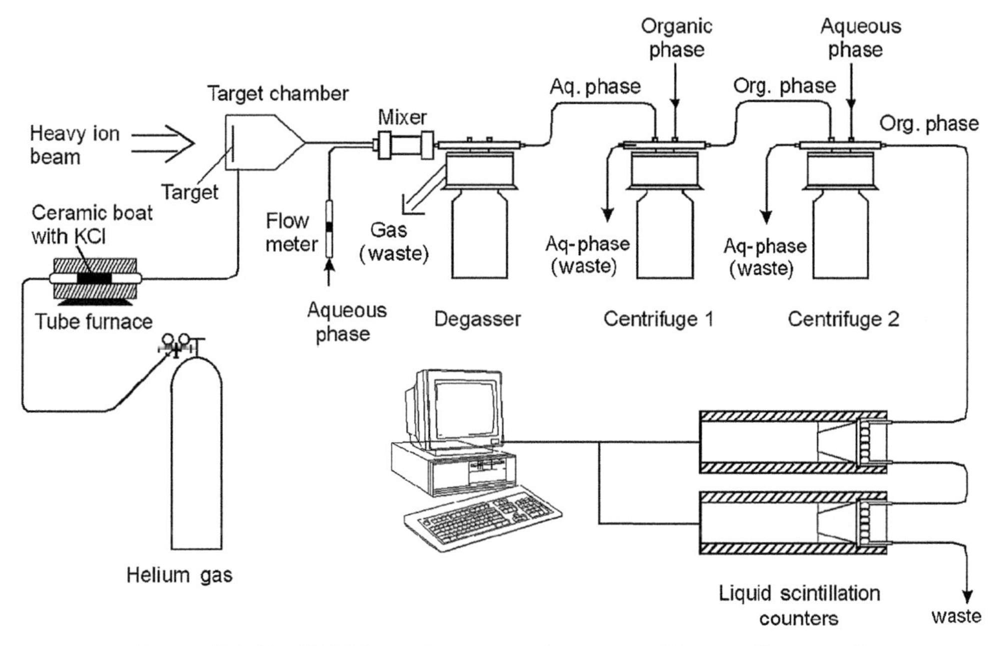

Figure 17.9 The SISAK fast solvent extraction system. *(Courtesy Skarnemark)*

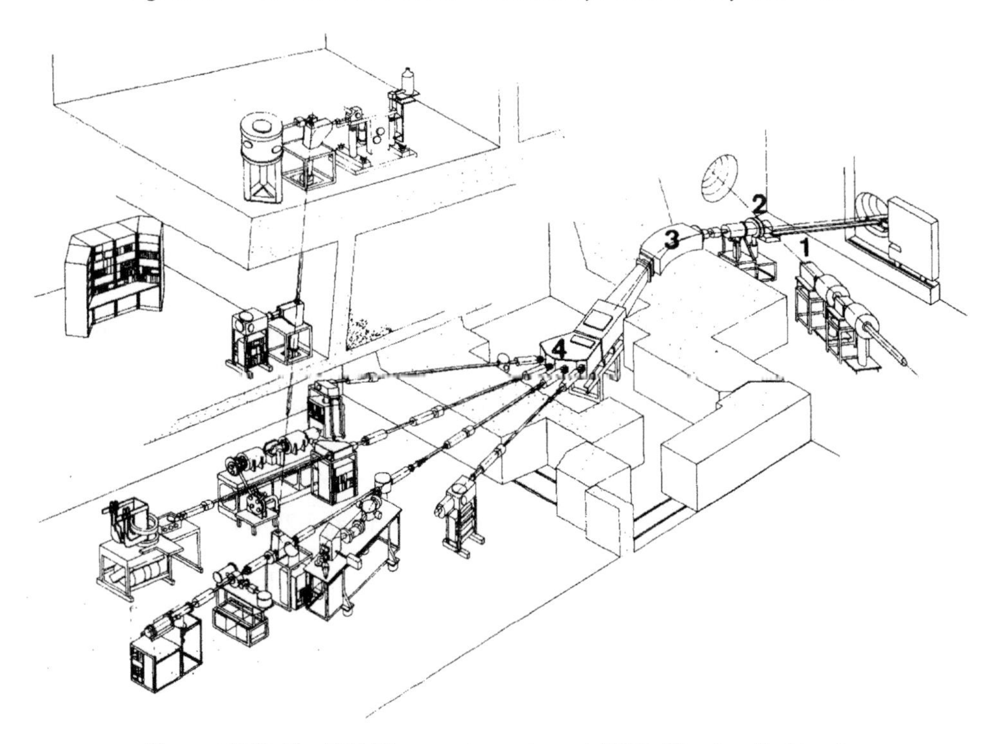

Figure 17.10 The ISOLDE separator system at CERN. *(Courtesy Rudstam)*

17.7.4. Mass separator procedures

In the ISOLDE (Isotope Separation On Line DEvelopment) collaboration at CERN, the target material is fixed in a solid but porous chemical matrix and irradiated by protons or heavy ions. The products react with the target matrix, and those with the "right" chemistry diffuse from the target (which may be heated to increase the diffusion rate) and are swept by a gas stream into the ionization chamber of a mass spectrometer where the products are separated. Figure 17.9 shows the schematic for such a system: At 1, a 600 MeV proton beam is focused on the target 2. The beam of spallation products is accelerated to 60 keV, focused by magnet 3 and separated by electrostatic deflectors at 4 to yield a number of beams which are analyzed by various techniques (nuclear spectroscopy, laser spectroscopy, atomic beam magnetic resonance, range measurements, etc).

17.8. EXERCISES

17.1. ^{24}Na is produced through the reaction ^{26}Mg(d,α)^{24}Na. A 0.2 mm thick magnesium foil is irradiated for 1 h by a current of 130 μA of 22 MeV D^+ ions in a cyclotron. The foil has a much larger area than the cross-section of the beam. What is the specific activity of ^{24}Na if the magnesium foil (3 cm^2) contains 0.003% Na and σ for the reaction is assumed to be 25 mb?

17.2. Oxygen can be determined through the reaction ^{16}O(n,p)^{16}N → (β,γ 7 s); σ_{np} at 14 MeV n is 49 mb. 3.982 g of a fatty acid were irradiated for 20 s in 4×10^{12} n m^{-2} s^{-1}. After the irradiation the sample was rapidly transferred with a rabbit system to a scintillation detector which had an efficiency of 1.1% for the ^{16}N γ-rays (~6 MeV). Exactly 8 s after the end of the irradiation, the sample was counted for 1 min, yielding 13 418 counts above background. What was the oxygen fraction of the sample?

17.3. ^{199}Au can be formed through two successive n,γ-reactions on ^{197}Au (100% in nature). If 1 g ^{197}Au is irradiated with 10^{18} n m^{-2} s^{-1} during 30 h, what will the disintegration rate of ^{199}Au be at the end of the irradiation? The chain of events to be considered is:

$$^{197}\text{Au}\left(\sigma_{n,\gamma}\ 98.8\ \text{b}\right) \rightarrow\ ^{198}\text{Au}\left(t_{1/2}\ 2.694\ \text{d},\ \sigma_{n,\gamma}\ 25\,100\ \text{b}\right) \rightarrow$$
$$^{199}\text{Au}\left(t_{1/2}\ 3.139\ \text{d},\ \sigma_{n,\gamma}\ 30\ \text{b}\right) \rightarrow \text{Neglect self shielding.}$$

17.4. ^{246}Cm has a half-life of 4 730 y. It can be obtained through neutron capture in ^{245}Cm, which has a half-life of 8 500 y; the reaction cross-section is 345 b. Both isotopes are also fissioned by thermal neutrons, σ_{245} 2 020 b and σ_{246} 0.17 b.

^{246}Cm also has a n,γ cross-section of 1.3 b. Because one does not want to loose too much ^{245}Cm, the irradiation is timed to give a maximum yield of ^{246}Cm. If the neutron flux is 2×10^{18} n m^{-2} s^{-1}, (a) when does the ^{246}Cm concentration reach its maximum? (b) What is the ratio between the amount of ^{246}Cm produced and amount of ^{245}Cm consumed at that time?

17.5. A sample containing 50 g ethyl iodide was irradiated with neutrons from a RaBe source of 1×10^{9} n m^{-2} s^{-1} for 2 h. If there is a 20% retention and separation occurs 5 min after end of irradiation, what will be the activity of the AgI sample 10 min after separation? The detector has an 8% efficiency for the emitted γ-radiation. Assume the reaction cross-section for $^{127}I(n,\gamma)^{128}I \rightarrow (\beta^-, 25.0\ min)$ to be 6.2 b.

17.6. ^{59}Fe of high specific activity can be produced through thermal neutron irradiation of a solution of 1 ml 0.1 M potassium hexacyanoferrate. The recoiling (free) iron atoms, which are produced with a 40% yield, are quantitatively extracted into an organic solvent. How long must the irradiation be in a reactor at 3×10^{19} n m^{-2} s^{-1} to obtain 37 MBq ^{59}Fe? The reaction cross-section for $^{58}Fe(n,\gamma)^{59}Fe \rightarrow (\beta,\gamma$ 45.1 d) is 1.15 b and the isotopic abundance of ^{58}Fe is 0.3% in natural iron.

17.9. LITERATURE

H. BATEMAN, The solution of a system of differential equations occurring in the theory of radioactive transformations, *Proc. Cambridge Phil. Soc.* **15** (1910) 423.

W. RUBINSON, The equations of radioactive transformations in a neutron flux, *J. Chem. Phys.* **17** (1949) 542.

I. J. GRUVERMAN and P. KRUGER, Cyclotron-produced carrier-free radioisotopes: thick target yield data and carrier-free separation procedures, *Int. J. Appl. Rad. Isotopes* **5** (1959) 21.

A. P. WOLF, Labeling of organic compounds by recoil methods, *Ann. Rev. Nucl. Sci.* **10** (1960) 259.

IAEA, *Production and Use of Short-lived Radioisotopes from Reactors*, Vienna, 1963.

N. TRAUTMANN, N. KAFFRELL, H. W. BELICH, H. FOLGER, G. HERRMANN, D. HUBSCHER and H. AHRENS, Identification of Short-lived Isotopes of Zirconium, Niobium, Molybdenum, and Technetium in Fission by Rapid Solvent Extraction Techniques, *Radiochim. Acta* **18** (1972) 86.

G. RUDSTAM, The on-line mass separator OSIRIS and the study of short-lived fission products, *Nucl. Instrum. Meth.* **139** (1976) 239.

Radiopharmaceuticals and other compounds labeled with short-lived radionuclides, *Int. J. Applied Rad. Isotopes* **28** (1977) Nos. 1/2.

U. HICKMANN, N. GREULICH, N. TRAUTMANN, H. GÄGGELER, H. GÄGGELER-KOCH, H. EICHLER and G. HERRMANN, Rapid Continuous Radiochemical Separations by Thermochromatography in Connection with a Gas-jet Recoil-transport System, *Nucl. Instr. Methods* **174** (1980) 507. See also I. ZVARA, YU. T. CHUBURKOV, R. CALETKA, and M. R. SHALAEVSKII, *Radiokhimiya* **11** (1969) 163 (= *Sov. Radiochem.* **11** (1969) 161).

T. TOMINAGA and E. TACHIKAWA, Inorganic Chemistry Concepts, Vol. 5, *Modern hot atom chemistry and its applications*, Springer Verlag, 1981.

G. HERRMAN and N. TRAUTMANN, Rapid Chemical Methods for Identification and Study of Short-lived Nuclei, *Ann. Rev. Nucl. and Particle Science* **32** (1982) 117.

J. D. BAKER, R. J. GEHRKE, R. C. GREENWOOD and D. H. MEIKRANTZ, Advanced System for the Separation of Rare Earth Fission Products, *Radioanal. Chem.* **74** (1982) 117.

T. MATSUURA (Ed.), Studies in Physical and Theoretical Chemistry, Vol. 31, *Hot Atom Chemistry: Recent trends and applications in the physical and life sciences and technology*, Elsevier, 1984.

Radiopharmaceuticals and labelled compounds 1984, IAEA, Vienna 1985.
K. RENGAN, J. LIN and R. A. MEYER, Continuous gas-phase separation of fission products with half-lives of 600 ms to 56 s, *Radiochem. Radioanal. Lett.* **50** (1982) 393.
G. SKARNEMARK, M. SKÅLBERG, J. ALSTAD and T. BJÖRNSTAD, Nuclear Studies with the Fast On-line Chemical Separation System SISAK, *Physica Scripta* **34** (1986) 597.
H. PERSSON, G. SKARNEMARK, M. SKÅLBERG, J. ALSTAD, J. O. LILJENZIN, G. BAUER, F. HABERBERGER, N. KAFFRELL, J. ROGOWSKI and N. TRAUTMANN, SISAK-3. An Improved System for Rapid Radiochemical Separations by Solvent Extraction, *Radiochim. Acta* **48** (1989) 177.
K. RENGAN and R. A. MEYER, Ultrafast Chemical Separations, *UCRL report 2C-104963* (1992).
J. P. ADLOFF, P. P. GASPAR, M. IMAMURA, A. G. MADDOCK, T. MATSURA, H. SANO and K. YOSHIHARA, *Handbook of Hot Atom Chemistry*, Kodanska Ltd, Tokyo 1992.

Journals

Radiochemica Acta, Oldenburg Verlag, Munich
Journal of Radioanalytical and Nuclear Chemistry, Elsevier Sequoia/Akademiai Kiado, Budapest
Applied Radiation and Isotopes, Pergamon Press, Oxford
Radioactivity & Radiochemistry, Caretaker Communications, Atlanta

CHAPTER 18

Uses of Radioactive Tracers

Contents

In this chapter some of the ways in which radiochemistry has aided research in various areas of chemistry and related sciences are reviewed.

The first experiments with radioactive tracers were conducted in 1913 by de Hevesy and Paneth who determined the solubility of lead salts by using one of the naturally occurring radioactive isotopes of lead. Later, after discovery of induced radioactivity,

Radiochemistry and Nuclear Chemistry
ISBN 978-0-12-405897-2, http://dx.doi.org/10.1016/B978-0-12-405897-2.00018-5

de Hevesy and Chiewitz in 1935 synthesized ^{32}P (β^- $t_{1/2}$ 14.3 d) and used this tracer in biological studies. In the same year de Hevesy and co-workers also carried out activation analyses on rare earths. Despite the demonstration of the value of the tracer technique by these early studies the technique did not come into common use until after World War II when relatively large amounts of cheap radionuclides became available through the use of nuclear reactors.

While it is not necessary to use radioactive isotopes for tracer studies, in general, the use of radioactivity is simpler and less expensive than the use of stable isotopes. Research with the latter requires rather sophisticated and expensive measuring devices such as mass spectrometers, cf. §3.3.2. We restrict our discussion to the use of radioactive tracers.

Among the advantages of using radiotracers we can list the following: (a) radiotracers are easy to detect and measure with high precision to sensitivities of 10^{-16} to 10^{-6} g; (b) the radioactivity is independent of pressure, temperature, chemical and physical state; (c) radiotracers do not affect the system and can be used in nondestructive techniques; (d) if the tracer is radiochemically pure, interference from other elements is of no concern (common in ordinary chemical analyses); (e) for most radioisotopes the radiation can be measured independently of the matrix, eliminating the need for calibration curves.

18.1. BASIC ASSUMPTIONS FOR TRACER USE

In some experiments answers to scientific questions which require knowledge of the presence and concentration of a specific element or compound at a certain place and at a certain time can be obtained only through the use of a radioactive tracer. For example, self diffusion of metal ions in solutions of their salts cannot easily be studied by any other technique. However, in other cases the use of radioactive tracers is not necessary in principle but is justified by the greater convenience. In either type of investigation there are two assumptions implicit in such uses.

The primary assumption is that radioactive isotopes are chemically identical with stable isotopes of the same element, i.e. the substitution of ^{14}C for ^{12}C in a compound of carbon does not change the type or strength of the chemical bonds nor does it affect the physical properties of the compound. The validity of this assumption depends on the precision of measurement of the chemical and physical properties. The difference in mass between the various isotopes does cause some change in these properties (§3.5) but even in the case of ^{14}C and ^{12}C, with a mass difference of approximately 15%, the isotope effect is rather small and difficult to detect. Normally only for systems involving hydrogen-deuterium-tritium substitution must isotope effects be considered. For heavier elements it can be neglected in almost every situation.

The second assumption in the use of tracer techniques is that the radioactive nature of the isotope does not change the chemical and physical properties. Until the moment of its disintegration the radioactive atom is indistinguishable from its stable isotope except for the isotopic mass difference. When the radioactive disintegration of the atom has been observed ("counted"), the decay product is normally a different element and its subsequent chemical behavior is usually of no interest. If the disintegration rate is very high, the energy released by the radioactive decay can cause observable secondary radiolytic effects (Ch. 8). However, in well-designed tracer experiments the level of radioactivity is high enough to provide accurate data but normally small enough not to produce noticeable chemical effects.

While the radioactivity of the tracers is assumed not to affect the chemical systems, the parent daughter relationship of radioactive nuclides needs special consideration. For example, since strontium and yttrium are not chemically identical, a gross β-count of strontium samples including ^{90}Sr may include an unknown fraction of ^{90}Y activity present from ^{90}Sr decay because of the relationship

$$^{90}\mathrm{Sr}(\beta^-, t_{1/2}\ 28.5\ \mathrm{y})\ ^{90}\mathrm{Y}\ (\beta^-\gamma, t_{1/2}\ 2.671\ \mathrm{d})\ ^{90}\mathrm{Zr(stable)} \qquad (18.1)$$

Beta-absorption, and β- or γ-scintillation techniques which use energy discrimination, are frequently useful in such parent-daughter cases. If equilibrium is rapidly established between the parent and daughter activities it is usually simpler to count the samples after sufficient time for this to occur, the contribution to the observed count rate by the daughter is then proportional to the amount of mother in the sample. In the case of ^{90}Sr—^{90}Y, radioactive equilibrium is established in about 25 d. If ^{137}Cs ($\beta^-\gamma$, $t_{1/2}$ 30.0 y) is being used to study cesium chemistry it is necessary to wait only 15—20 min after sampling until counting as the daughter ^{137m}Ba (IT, $t_{1/2}$ 2.55 min) reaches an equilibrium level within that time. Since the ratio of the ^{137m}Ba and the ^{137}Cs activity is the same in all samples at equilibrium, the total count rate before and after a chemical step is then a true measure of the behavior of cesium alone. If radioactive equilibrium is not re-established in a convenient time, it may be necessary to either discriminate against the activity not involved in the chemical system, to take into account its contributions to the net count rate, or to remove it immediately before counting.

It may be necessary or expedient to use a radioactive nuclide which can undergo significant decay during the chemical investigation. In these cases, in order to compare results at different points in the process, it is necessary to correct all counts to the same time (usually starting time of the experiment).

18.2. CHEMISTRY OF TRACE CONCENTRATIONS

Consider a sample containing a pure radionuclide with a disintegration rate of 10^7 dpm. For a $t_{1/2}$ of 1 h the number of atoms is (§5.11) 8.7×10^8; for a $t_{1/2}$ of 1 y it is 7.6×10^{12}. If such a sample is dissolved in one liter of solution, the respective concentrations would be 1.4×10^{-15} M and 1.3×10^{-11} M. At such concentrations the chemical behaviour may be quite different from what it is at higher concentrations. Addition of macroscopic amounts (e.g. at the gram level) of non-radioactive isotopic) atoms of the element results in concentrations of 10^{-3} to 10^{-1} M. The non-radioactive component is called a *carrier* as it "carries" the radioactive and ensures normal chemical behaviour. Many applications of radiotracers involve mixing the tracer atoms with a much larger amount of nonradioactive isotopic atoms prior to use.

If a radionuclide is to follow the chemical properties of an isotopic carrier it is necessary that the radionuclide and the carrier undergo *isotopic exchange*. If it is not known a priori that such exchange takes place between two compounds with a common element this must be determined by experimentation before it can be assumed that the tracer and the carrier would act similarly in a chemical system. This consideration must be particularly borne in mind if the radioactive tracer and the inert carrier are in different oxidation states when mixed.

In the remainder of this paragraph we discuss the behaviour of trace level concentrations as it is desirable in some applications to use very low concentrations of radiotracers with no carrier.

18.2.1. Adsorption

Solutes in contact with surfaces have a tendency to be adsorbed on the surface. In order to cover the glass surface of a one liter vessel with a monomolecular layer of a hydrated cation only 10^{-7}–10^{-8} moles are required. As indicated in the previous paragraph, the amount of radionuclide in the solution may be less than this and, in principle, all the radioactive atoms could be adsorbed on the walls of the vessel. The *Paneth and Fajans rule* for tracer adsorption states that: "a micro component is adsorbed on a solid macro component or precipitated together with it if it forms an insoluble compound with a counter ion of the macro component".

The amount of radionuclide that is adsorbed on the walls of the container depends on the concentration, on the chemical state of the radionuclide and on the nature of the container material. One of these properties of the material is the so called point of zero charge (pzc). This is the pH where the surface charge is zero. At a pH below this point the surface is positively charged and anios will adsorb. At pH above this point the surface will be negativelay charged and thus only cat ions will be adsorbed. Figure 18.1 shows the variation of the adsorption of thorium on the walls of glass and polyethylene containers as a function of concentration and pH. In case (a) the sorption time is that of pipetting (of an aqueous Th-complex), in case (b) it is that for equilibrium. The variation of adsorption with pH reflects the adsorption of various hydrolytic species formed by thorium as the pH is increased. Curve (a) shows that sorption can be neglected at concentrations $>10^{-4}$ M in this system.

In many cases in nature the pH of the solution in contact with the sold phase is higher than the point of zero charge for that surface. In general adsorption of cations increases

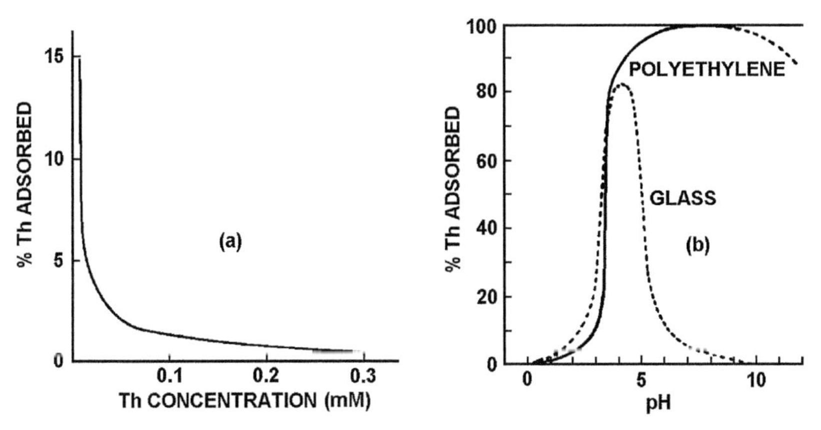

Figure 18.1 Adsorption of Th (a) in a 10 ml pipette at different Th concentrations, (b) from 2×10^{-8} M $Th(ClO_4)_4$ solution on different surfaces and pH's. *(From Rydberg and Rydberg.)*

with ionic charge in the order $M^+ < M^{2+} < M^{3+} < M^{4+}$. The importance of the nature of the surface is obvious in Figure 18.1.b. Adsorption of Pm(III) ions have been shown to increase in the order platinum < silver < stainless steel < polyvinyl chloride. Addition of isotopic carrier dilutes the radiotracer and a smaller fraction of tracer is adsorbed (Fig. 18.1.a). Unfortunately, such isotopic dilution results in a decrease in the specific activity of the trace element, which can be disadvantageous in certain types of experiments. In some cases it is possible to avoid decreasing the specific activity by adding macro amounts of a nonisotopic element which is easily adsorbed and may block the available surfaces from adsorbing the tracer.

In addition to adsorption on the walls of the container, radioactive species frequently adsorb on precipitates present in the system. The nature of the precipitate as well as its mode of precipitation are major factors in the amount of adsorption. If silver iodide is precipitated in an excess of silver ion the precipitate has a positive surface layer due to the excess concentration of silver ions on the surface. By contrast if the precipitation occurs in excess iodide, there is a negative surface charge due to the excess iodide on the surface. When trace amounts of radioactive lead ions are added to a suspension of two such precipitates in water, the precipitate with the negative surface charge adsorbs >70% of the tracer lead ions from the solution, while the precipitate with the positive surface charge adsorbs <5%. The amount of adsorption increases with the ionic charge of the radioactive tracer, e.g. it has been found that with a precipitate of Ag_2S about 7% of Ra^{2+}, 75% of Ac^{3+} and 100% of Th^{4+} is adsorbed.

The adsorption properties of trace elements have been used to advantage for the isolation of the trace elements as well as for the separation of different trace elements with different adsorption properties.

18.2.2. Radiocolloids

Radioactive tracers adsorb not only on solid container surfaces and precipitates but on any kind of solid material suspended or in contact with the solution. Dust, cellulose fibers, glass fragments, organic materials, etc., are examples of substances that readily adsorb radioactive tracers from solution. If the solution contains large molecules as, for example, polymeric metal hydrolysis products, these also tend to adsorb trace elements. In addition to sorption, the presence of such material in the solution can lead to the phenomenon of radiocolloid formation, which is the attachment of radionuclides to semi-colloidal aggregates in solution. If the solution is kept at sufficiently low pH and extremely free from foreign particles, sorption and radiocolloid formation are usually avoided as major problems.

18.2.3. Equilibrium reactions

The low concentration of radioactive tracers can lead to the formation of solute species that are not observed at equilibrium with macro amounts. For example, the hydrolysis of uranyl ions corresponds to the equilibrium

$$m\,UO_2^{2+} + p\,H_2O \rightleftarrows (UO_2)_m(OH)_p^{2m-p} + p\,H^+ \qquad (18.2)$$

With macro concentrations of uranium this equilibrium is shifted to the right with the observation of polymers with properties rather different than that of the uranyl ion. At a uranium concentration of approximately 0.001 M, more than 50% of the uranium is polymerized at pH 6, while for uranium concentrations less than 10^{-6} M the polymerization is negligible. This condition can be used to advantage: trace metal concentrations can be used if one wishes to study the properties of a metal ion at relatively high pH's without interference of polymerization reactions. However, one then must be aware of the adsorption risks.

An additional complication can arise in solution if the radioactive species in trace amounts react with trace concentrations of impurities. For example, in an investigation of the properties of pentavalent protactinium, Pa(V), it was found that the protactinium was extracted into pure xylene from 1 M $HClO_4$ solutions. Further experimentation showed that this extraction was due to the presence in the xylene of organic impurities at concentrations below the detectable limit of 0.01%. Support for this interpretation was provided when the solution was made 10^{-4} M in thorium, which was expected to form complexes with the probable impurity, thereby preventing the reaction with the protactinium. In fact, no protactinium was extracted into xylene from this solution. The thorium in this case acts as a *hold-back carrier.*

18.2.4. Precipitation and crystallization

Due to the low concentration of radioactive tracers in solution the solubility product for an "insoluble" salt is not always exceeded upon the addition of macro concentrations of a *counter ion*. Let us as example take the insoluble lanthanum hydroxides. The solubility product for the reaction $La(OH)_3$ (s) $\rightleftarrows$ La^{3+} + 3 OH^- is K_{s0} $\approx$ 10^{-19}; in 1 mM NaOH the concentration of La^{3+} in the solution is only 10^{-10} M in equilibrium with the $La(OH)_3$ precipitate. If 100 MBq (2.7 mCi) ^{140}La, obtained as a fission product or by milking from ^{140}Ba (Table 5.1), is dissolved in 1 l, the solution will have $[La^{3+}]$ $\approx 3.5 \times 10^{-11}$ M. In this case the solubility product is not exceeded in 1 mM NaOH. To precipitate the $La(OH)_3$ quantitatively the NaOH concentration must be raised to 10–100 mM; however, at these concentrations the precipitate is formed as a colloid in solution and even upon centrifugation the amount of precipitate is so small as to be unweighable by present techniques. With the addition of a *carrier* for ^{140}La, the precipitation can be carried out without any difficulty.

It is possible to remove ions at tracer level concentrations from solutions by precipitation using adsorption or coprecipitation. *Coprecipitation* occurs if the compound of the tracer and the oppositely charged ion of the precipitate is isomorphous with the precipitate. In these cases the active ion may be included in the crystal lattice of the precipitate at a lattice point, particularly if the tracer ion is close in size to the ion which it displaces. However, at trace level concentrations exceptions are found to this requirement of similarity in size as well as to the requirement of isomorphism. When the distribution of the tracer is found to be uniform throughout the precipitate it can be described by the *Berthelot-Nernst homogeneous distribution law* which is expressed as

$$x/y = D'(a - x)/(b - y) \tag{18.3}$$

where x and y are the amounts of A^{z+} and B^{z+} in the precipitate, a and b are the initial amounts of these ions, and D' is the "distribution coefficient". A more "true" distribution constant (D = concentration of tracer in solid/concentration of tracer in solution) can be obtained by using a conversion factor, e.g. C = gram solute per ml of saturated carrier solution divided by the density of the solid

$$D' = D\,C \tag{18.4}$$

The entire precipitate is in equilibrium with solution in this system.

If only the freshly forming surface of the growing crystal is in equilibrium with the solution phase, a nonuniform distribution is observed. In these cases the system is described by a logarithmic (according to Doerner and Hoskins) distribution law, which has the form

$$\ln[a/(a - x)] = \lambda' \ln[b/(b - y)] \tag{18.5}$$

where λ' is the *logarithmic distribution coefficient*, a constant characteristic of the system (Fig. 18.2).

The importance of isomorphism can be illustrated by the coprecipitation of Ra^{2+} in trace quantities with Sr^{2+} in strontium nitrate. If the precipitation is carried out at 34°C, the radium coprecipitates since at this temperature the strontium precipitates as $Sr(NO_3)_2$

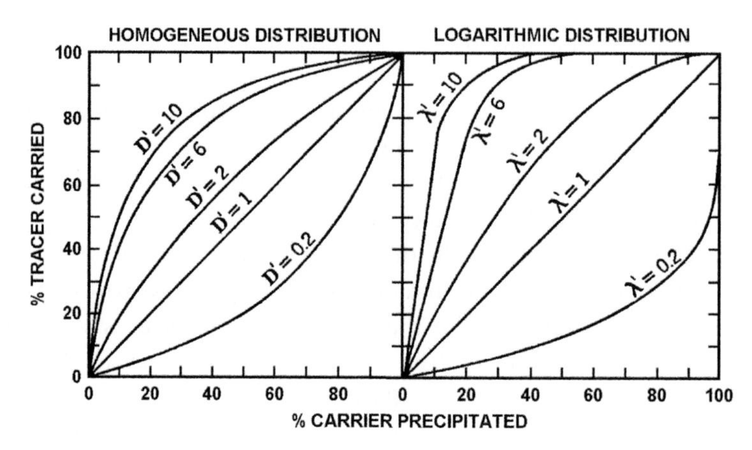

Figure 18.2 Efficiency with which the tracer is carried for various values of the distribution coefficients *D'* and λ'. *(From Wahl and Bonner.)*

with which radium nitrate is isomorphous. However, if the precipitation occurs at 4°C, the strontium crystallizes as $Sr(NO_3)_2 \cdot 4H_2O$ and is no longer isomorphous with $Ra(NO_3)_2$. Due to the lack of isomorphism the radium is not coprecipitated at 4°C.

18.2.5. Electrochemical properties

For the redox equilibrium

$$M^{z+} + ne^- \rightleftarrows M^{z-n} \ (n \geq 1)$$

where M^{z+} and M^{z-n} are the oxidized and reduced states of a chemical species, the *Nernst equation* is valid, i.e.

$$E = E^0 - \{\mathbf{R}T/(n\mathbf{F})\} \ln\left([M^{z-n}]/[M^{z+}]\right) \tag{18.6}$$

In this equation E^0 is the potential for a standard state of 1 M concentration at NTP; the species in brackets relates to the chemical activities in the particular solution phase. This relationship indicates that the redox potential E of a solution is independent of the total concentration of the species and depends only on the ratio of the oxidized and reduced forms. This has been confirmed since concentrations of trace amounts of ions show the same redox behavior as macro concentrations. Reduction and oxidation reactions can, therefore, be carried out in solutions with trace amounts of radioactive species.

Electrolysis of solutions can be used for electrodeposition of a trace metal on an electrode. The selectivity and efficiency which would be present for electrolytic deposition of macro amounts of ions at a controlled potential is not present, however, for trace amounts. The activity of trace amounts of the species is an unknown quantity even if the concentration is known, since the activity coefficient is dependent upon the behavior of the mixed electrolyte system. Moreover, the concentration of the tracer in solution may not be known accurately since there is always the possibility of some loss through adsorption, complex formation with impurities, etc. Nevertheless, despite these uncertainties it has been found that the Nernst equation can be used, with some caution, for calculating the conditions necessary for electrolytic deposition of trace metals.

It is also possible to precipitate insoluble species on electrodes. For example, if a fluorosilicate solution is electrolyzed, thereby freeing a high concentration of fluoride ion at the electrode, a thin uniform layer of UF_4, can be deposited. Similarly, trace amounts of elements which form insoluble hydroxides can be deposited from solutions in which water is being electrolyzed as a region of extremely high pH is present at the cathode.

18.2.6. Tracer separation methods

All the analytical techniques used in conventional chemistry may be used for the separation and isolation of radioactive elements and compounds in macro or trace concentrations. The precipitation method was amply demonstrated by the early radio-chemists M. Curie, Debierne, Rutherford, Hahn, etc., for the separation, concentration and identification of the naturally occurring radioactive elements. However, in §§18.2.1–18.2.4 we have pointed out the many pitfalls in working with tracer concentrations in solutions containing precipitates, etc, as well as in the use of electrochemical methods (§18.2.5).

Normally these separation methods require the addition of a macro amount of isotopic carrier. However, in some cases analytical procedures are available for separation and isolation of carrier free radiotracer concentrations. *Solvent extraction* (see §18.4.3 and App. A), and various forms of *partition chromatography* (§18.4.1), methods have been found to be particularly advantageous in this connection since they are selective, simple, and fast.

Liquid-liquid (or *solvent*) *extraction* is a technique for selectively transferring a species between an aqueous solution and an organic phase (e.g. kerosene, benzene, chloroform, etc) by equilibrating the aqueous phase with an organic solvent. Usually the organic phase contains a reagent A (*extractant*) which forms a neutral compound MA_N with the species M to be transferred between the phases. The number of extractants applied are numerous and the literature must be consulted to determine the most suitable ones for the system of interest; typical extractants are organophosphates, amines and metal chelating agents (usually weak organic acids). The fraction extracted at equal phase volumes is

$$E\,\% = 100\ D/(D+1) \tag{18.7}$$

where D is the *distribution ratio* of the radioactivity between the two phases

$$D = R_{org}/R_{aq} = \psi_{org}A_{org}/\psi_{aq}A_{aq} \approx N_{org}/N_{aq} = [M]_{org}/[M]_{aq} \tag{18.8}$$

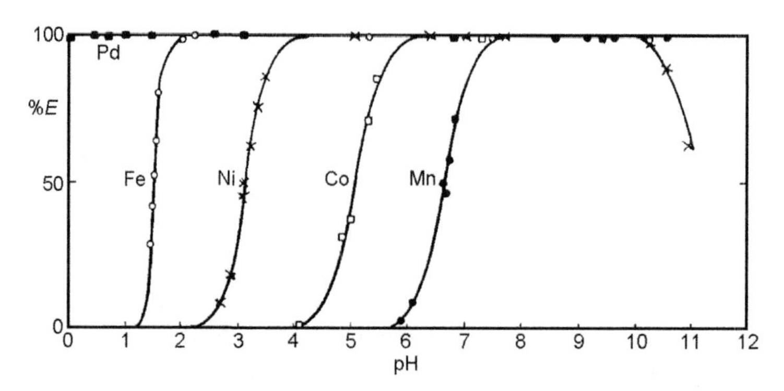

Figure 18.3 Effect of pH on the extraction of some divalent metals from aqueous solution into chloroform by 0.01M 8-hydroxyquinoline (oxine).

The last equality requires that the radioactive measurements R are carried out on equal phase volumes (eventually evaporated to dryness) and that $\psi_{org} = \psi_{aq}$, a requirement easily met by proper choice of radiometric equipment. Thus the D-value directly reflects the concentration ratio of the radioactive species. Figure 18.3 shows as an example how the extraction of a number of metals from an aqueous into an organic solution varies with pH. Such curves are used to select optimal separation conditions: in the Figure at pH <3.5, Pd, Fe and Ni are extracted to 100% into chloroform, while all the Co and Mn stays in the aqueous phase, thus facilitating an easy separation of these two groups of metals.

This technique has a number of applications, i.e.

- on a large scale for the production of valuable metals, such as U, described in §14.2.4,
- for reprocessing spent nuclear fuels as described in Ch. 21,
- at trace metal concentrations for determining equilibrium constants as described in §18.4.3,
- for separation and identification of short lived radionuclides as described in §17.7.

Solid organic resin *ion exchangers* consist of organic polymeric networks containing basic or acidic groups attached to the organic framework. Analogous to (18.8) a distribution ratio D_{iex} is defined as

$$D_{iex} = [M]_{resin}/[M]_{sol} \tag{18.9}$$

for the distribution of a (e.g radioactive) metal between the solid resin and the aqueous solution. D_{iex} depends on resin properties and on solution parameters such as the nature of the metal ion, ionic strength of the solution, temperature, etc. The basic equilibria are discussed in §18.4.3. Because the sorption in the resin phase increases with the valency of the cation, multivalent ions are absorbed more strongly (i.e. have larger D_{iex} values) than divalent or monovalent ions. Most commonly ion exchange columns are used for metal ion separations. In this case, the radioactivity is sorbed in the top layer of a column of the wet ion exchange resin. Following sorption (e.g. of M^+), the metal is eluted from the resin bed by passage of a solution (eluant) through the column. The eluant may contain a complexing anion or another metal ion (e.g. M^{3+}) which displaces M^+ through competition with it for positions on the resin. The metals are eluted by complexing agents in an order depending on their complex formation properties, as e.g. is illustrated for the lanthanides and actinides in Chapter 14 (Fig. 14.7).

In principle, *liquid partition chromatography* (LPC) is a liquid-liquid extraction where one of the liquid phases is stationary and attached to a supporting material, and the other liquid phase is mobile. It can be carried out with either the aqueous or the organic phase stationary; in the latter case the technique is referred to as reversed phase LPC. The aqueous phase can be made stationary by adsorption on silica gel, cellulose powder, etc. In order to make the organic phase stationary, beads (usually 50—200 μm) of PVC, teflon, Kel-F, etc., are being used.

Reversed phase LPC has been useful in radiochemistry for separating individual elements, e.g. lanthanides or actinides. It has also been used for separation of macro amounts of actinides. Instead of using columns in partition chromatography, a sheet of paper may be used to hold the stationary phase (*paper chromatography*) or an adsorbent coated on a glass plate (*thin-layer chromatography*). This technique has an advantage over column separations because the positions of the radioactive species are easily identified on the sheet, either simply by autoradiography (§18.4.4) or by scanning instruments. Paper chromatography is further described in an example in §18.4.1 and illustrated by Figures 18.7 and 18.8.

18.3. ANALYTICAL CHEMISTRY

18.3.1. Radiometric analysis

The term radiometric analysis is often used in a broad sense to include all methods of determination of concentrations using radioactive tracers. In a more restricted sense it refers to a specific analytical method which is based on a two-phase titration in the presence of a radioactive isotope. The endpoint of the titration is indicated by the disappearance of the radioisotope from one of the phases. Figure 18.4 illustrates two cases, (a) the determination of Ag^+ in a solution by titration with NaI solution containing $^{129}I^-$ ($\beta^-\gamma$ $t_{½}$ 1.57×10^7 y), and (b) the determination of Fe^{2+} in an aqueous solution, to which trace amounts of radioactive $^{55}Fe^{2+}$ (EC $t_{½}$ 2.73 y) has been added. In case (a) the AgI precipitate is radioactive but the solution has little radioactivity until all the Ag^+ has been precipitated. The activity of the solution is measured by a liquid flow GM-detector (Ch. 9). In the latter case (b) a two-phase liquid-liquid analytical technique is used (§18.2.6); the titrant contains a substance (oxine) which extracts Fe(II) from the aqueous to the chloroform phase. The radioactivity of the organic phase is followed by liquid scintillation (sampling) to determine the end point of the titration.

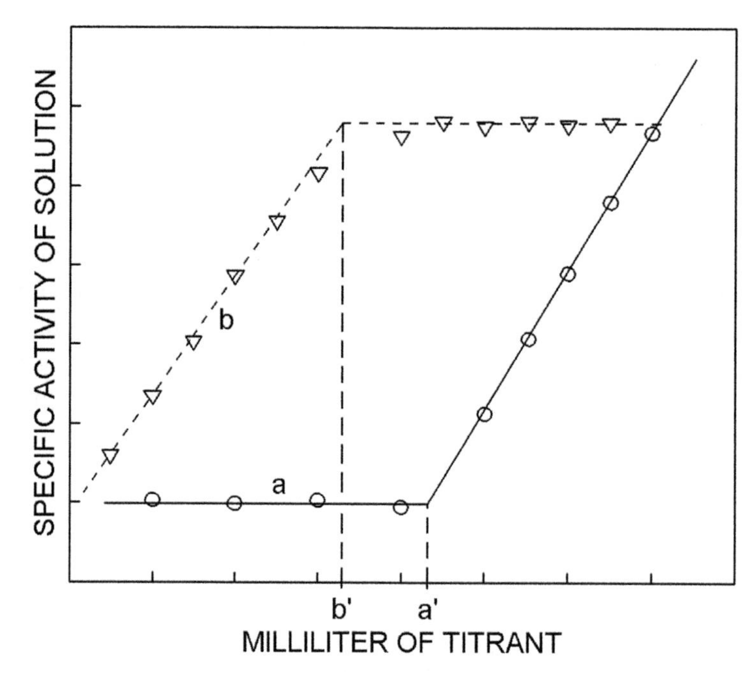

Figure 18.4 Radiometric titration curves.

Radiometric analysis is simple and rapid. Nevertheless, it is rarely used in analytical routine work, as a large number of multiple-element "instrumental" techniques are readily available (though the instruments usually are more expensive). Its most extensive use is for calibration of other techniques, and in analytical comparative techniques (e.g. environmental samples). In some cases results from different laboratories on common samples using various analytical techniques have differed by factors of 10^3 to 10^5! Because the radioactive tracer can be proven to be "atomically pure" (i.e. the radiation given off is unique for the radioisotope and element), it offers an absolute standard.

18.3.2. Isotope dilution analysis using radiotracers

In complex mixtures of compounds (for example, in organic synthesis or biochemical systems) it may be quite difficult to ascertain the exact amount of a specific component. A widely used technique of considerable value is *isotope dilution analysis*. This can be applied either with stable isotopes, in which case the detector is a mass spectrometer, or with radioactive isotopes, using measuring techniques presented in Chapter 9. The use of stable isotopes is usually limited to geologic samples, as described in §3.3, but may be applied to biological samples using highly enriched ^{13}C, ^{15}N or ^{18}O. However, the technique with radiotracers is more common due to its simplicity and lower cost.

A small radiochemically pure amount (w_0 g) of the selected compound ("reference") is added to the complex sample containing the unknown amount (w_u g) of the same compound. The reference may be either an element or a labelled compound, whose

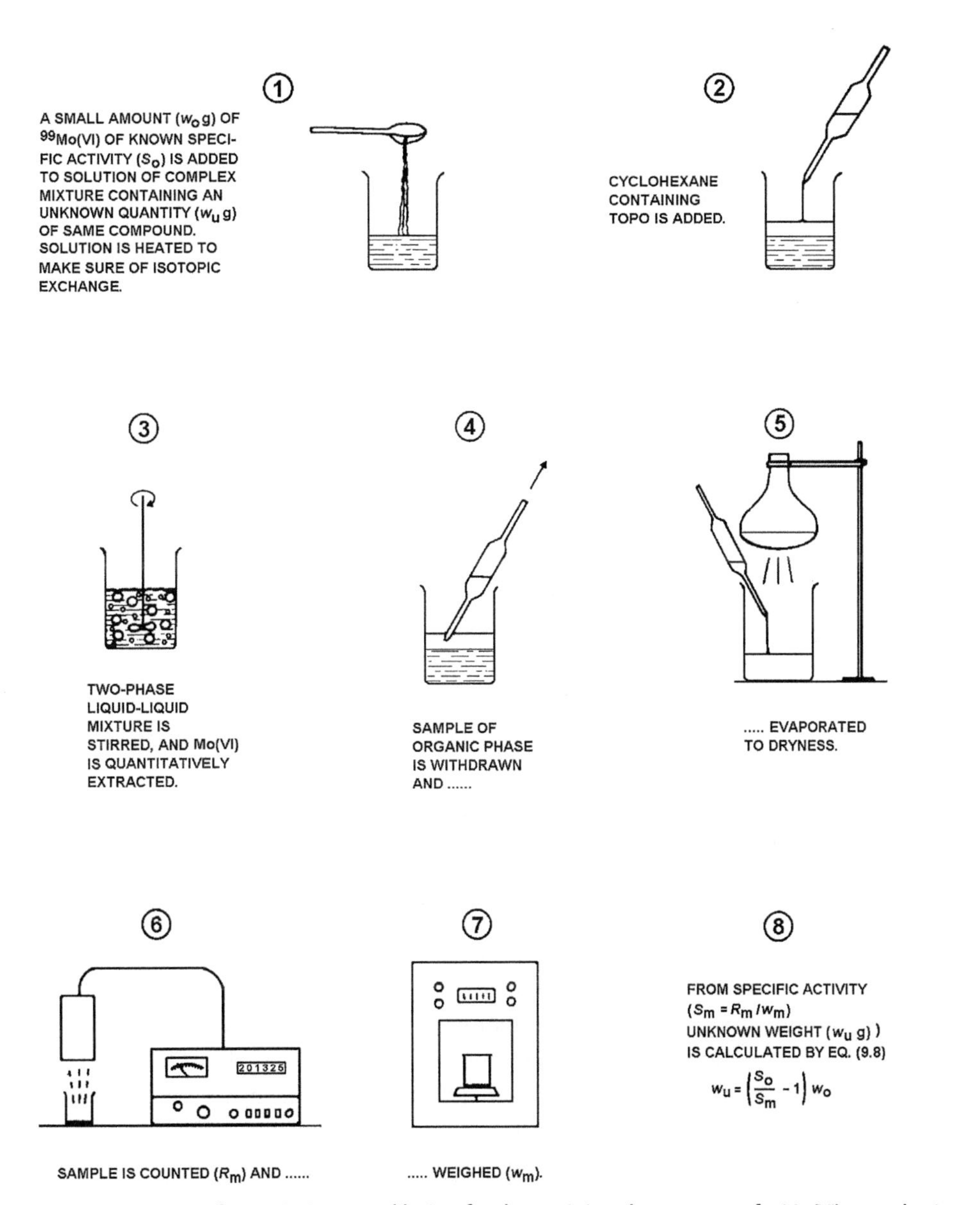

Figure 18.5 Sequence of steps in isotope dilution for determining the amount of a Mo(VI) complex in a composite mixture.

specific activity is known (S_0 Bq/g). After intimate mixing, the selected compound is isolated in high purity but not necessarily in high yield. The separated compound is weighed (w_m, g) and counted (R_m Bq) so that its specific activity (S_m, Bq/g) can be calculated. The method is pictured schematically in Figure 18.5. The weight, w_u, of the selected compound present in the original sample is calculated by

$$w_u = (S_0/S_m - 1)w_0 \tag{18.10}$$

The specific activity is defined by (5.49): $S = A/w\,\mathrm{Bq\,g^{-1}}$; since only the ratio S_0/S_m is used in (18.10), the activity A can be replaced by the measured radioactivity R when the detection efficiency ψ is the same.

This technique is of particular advantage where quantitative separation of the desired compound is not feasible, as illustrated already by de Hevesy in 1932: In determination of micro amounts of lead by anodic precipitation, quite varying results were obtained. By addition of a known amount of "radiolead" and measuring the radioactivity of lead at the anode, the yield of the precipitation could be determined, and — although the electrolytic precipitation was inefficient — an exact analysis was obtained.

In some cases the measurement of the final sample utilizes a technique other than weighing, but the principle remains the same. Isotope dilution is used, for example, in the determination of naphthalene in tar, of fatty acids in mixtures of natural fat, of amino acids in biological material, etc.

18.3.3. Activation analysis

Activation analysis is a highly sensitive nondestructive technique for qualitative and quantitative determination of atomic composition of a sample. It has been particularly useful for determination of elements in complex samples (minerals, environmental samples, biological and archeological objects, etc.), because it provides a simple alternative to much more difficult, tedious and destructive techniques. Its main limitation is the demand for a strong irradiation source.

In activation analysis advantage is taken of the fact that the decay properties such as the half-life and the mode and energy of radioactive decay of a particular nuclide serve to identify uniquely that nuclide. The analysis is achieved by the formation of radioactivity through irradiation of the sample either by neutrons or charged particles. Neutron irradiation is by far the more common technique, and hence this method is often referred to as *neutron activation analysis*, NAA. A major advantage in activation analysis is that it can be used for the simultaneous determination of a number of elements and complex samples. If the counting analysis of the sample is conducted with a Ge-detector and a multichannel analyzer, as many as a dozen or more elements can be measured quantitatively and simultaneously (*instrumental NAA*, or *INAA*).

A sample is irradiated to form an amount R of radioactive nuclide according to the relationship (cf. §17.2):

$$R = \psi\,\varphi\,\sigma\,N\left(1 - e^{-\lambda t_{irr}}\right)e^{-\lambda t_{cool}} \tag{18.11}$$

We assume that irradiation is carried out by a homogeneous particle flux φ, in a neutron reactor. The minimum amount of an element which can be detected increases with the efficiency of the measuring apparatus ψ, the bombarding flux φ, the reaction cross-section σ, the irradiation time t_{irr} (up to saturation activity), the decay constant, λ, of the radioactive nuclide formed, and the time from end of bombardment to start of

Table 18.1 Limits of detection for 71 elements in a thermal neutron flux of 10^{17} n m^{-2} s^{-1} (1 h irradiation)

Limit of detection (μg)	Elements
$1–3 \times 10^{-6}$	Dy
$4–9 \times 10^{-6}$	Mn
$1–3 \times 10^{-5}$	Kr, Rh, In, Eu, Ho, Lu
$4–9 \times 10^{-5}$	V, Ag, Cs, Sm, Hf, Ir, Au
$1–3 \times 10^{-4}$	Sc, Br, Y, Ba, W, Re, Os, U
$4–9 \times 10^{-4}$	Na, Al, Cu, Ga, As, Sr, Pd, I, La, Er
$1–3 \times 10^{-3}$	Co, Ge, Nb, Ru, Cd, Sb, Te, Xe, Nd, Yb, Pt, Hg
$4–9 \times 10^{-3}$	Ar, Mo, Pr, Gd
$1–3 \times 10^{-2}$	Mg, Cl, Ti, Zn, Se, Sn, Ce, Tm, Ta, Th
$4–9 \times 10^{-2}$	K, Ni, Rb
$1–3 \times 10^{-1}$	F, Ne, Ca, Cr, Zr, Tb
10–30	Si, S, Fe

counting, t_{cool}. By proper selection of t_{irr} and t_{cool} the sensitivity for any element can be changed and interferences minimized. Table 18.1 shows the limits of detection in INAA.

Figure 18.6 shows a typical NAA spectrum obtained with a multichannel analyzer equipped with scintillation (upper curve) or semiconductor (lower curve) detectors. Each peak can be ascribed to a certain γ-energy, which in most cases identifies the nuclide. A number of nuclides can be identified simultaneously with semiconductor detectors, but with NaI(TI) scintillation detectors the poor resolution limits simultaneous multi-element analysis.

The area under the peak (shaded area in Fig. 9.5(b)) is proportional to the amount of the radioactive nuclide. If all other factors in (18.11) are known, the number of target nuclide atoms N can be calculated.

When complex mixtures are irradiated, such as geological or biological samples, there may be some difficulties in peak assignment. The energy spectrum is then scanned at repeated time intervals and from the decrease of the peak area with time the half-life of the peak may be established. This is a valuable additional aid in the assignment of the peak to a certain nuclide.

While in principle it is possible to calculate the amount of the desired element through the use of the proper values for the cross-section, flux, irradiation time, and half-life in (18.9), a simpler approach has been developed that avoids errors implicit in the uncertainties of each of these values. The unknown and a known standard of similar composition are irradiated and counted in an identical fashion. A direct comparison can be made according to the following relationship:

$$\frac{\text{Weight of element in unknown}}{\text{Weight of element in standard}} = \frac{\text{Activity of element in unknown}}{\text{Activity of element in standard}}$$

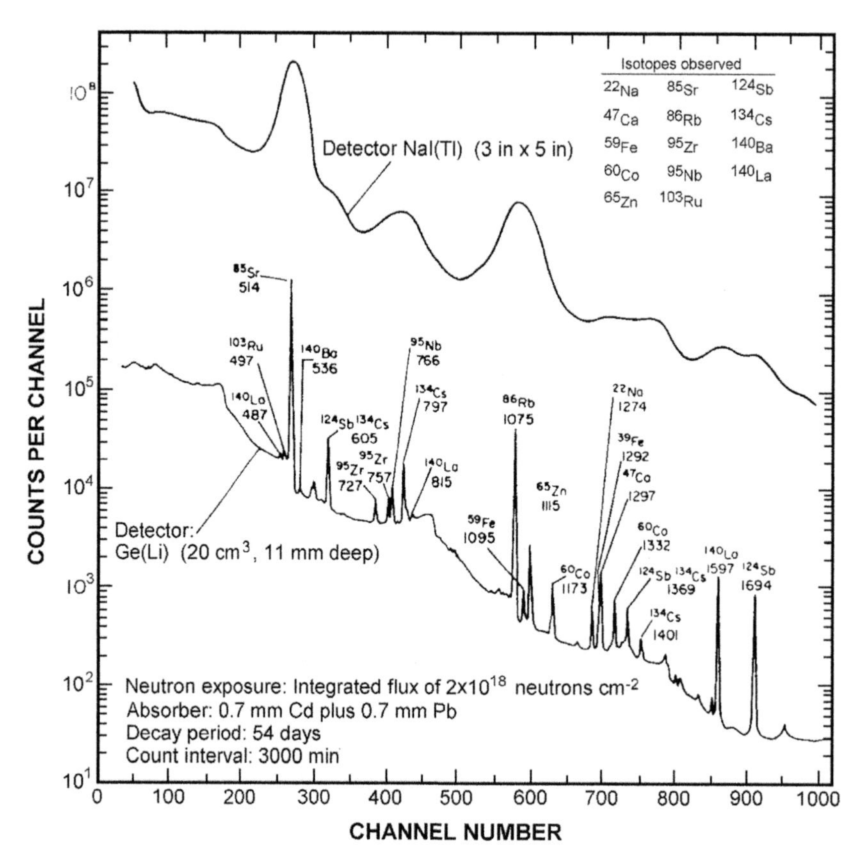

Figure 18.6 γ-spectrum of neutron-activated sea water. *(From Cooper, Wogman, Palmer and Perkins.)*

or

$$w_u/w_0 = R_u/R_0 \tag{18.12}$$

Sometimes such a large number of competing radioactivities are produced in the bombardment process that it may be necessary to conduct some chemical purification (RNAA, radiochemical NAA). This is particularly true if simple counting of β-activity or γ-ray spectrometry using NaI(TI) counting is used. However, with the development of semiconductor detectors and INAA the increased resolution in the spectrum allows simultaneous determination of as many as 15 or 20 competing radioactivities, usually without the necessity of chemical purification. Analysis of trace constituents in air and water, in soil and geological samples, in marine and in biological systems are some of the interesting applications of the NAA technique. Examples of on-line NAA include sorting ore minerals and oil well logging. In forensic science, by using NAA to measure the composition of the material adhering to a hand which has held a gun during firing, it is possible to determine the type of ammunition and even the number of shots fired. Trace metal analysis of plants can be used to determine the location in which that plant has been grown (used, for example, for identification of marijuana growers). The trace constituents of archeological and art objects play an important role in ascertaining their authenticity and the identification of place of origin; the use of nondestructive NAA has been extremely valuable in this field. Activation analysis of the mineral content of pigments has enabled scientists to determine the authenticity of paintings attributed to certain artists since, in times past, each artist prepared his own paints by distinctive and individual formulae. A painting entitled "Christ and Magdalene" done in the Old Dutch style was proved to be a twentieth century forgery when NAA showed <7 ppm silver and <1.3 ppm antimony in the white lead paint. The sixteenth and seventeenth century Dutch paintings have white lead with about 10–1000 ppm silver and 50–230 ppm antimony.

It has been found that hair contains trace metals (e.g., Cu, Au, Ce, Na) in ratios which are typical for a particular individual, and activation analysis can be used to identify hair from a particular person. This application achieved public notice when it was found that hair from Napoleon had a relatively large amount of arsenic, indicating that some time prior to his death he had received large doses of arsenic. Through analysis of the hair of the Swedish king Erik XIV (who died suddenly in 1577 after a meal of pea-soup) it has been found that he must have received lethal amounts of arsenic as well as large amounts of mercury. The latter is assumed to have been taken into his body through the use of a mercury compound for treatment of an old wound.

The high sensitivity of activation analysis has made it very useful in environmental pollution studies. Table 18.2 lists the limits of detection for some elements in sea water under the conditions specified in the table.

Elements with very low sensitivity for thermal neutron bombardment (e.g. the lightest elements) can often be measured through irradiation with either fast neutrons (FNNA) or charged particles (CPAA); in the latter case eqn. (18.11) must be modified, see §17.2. Thus oxygen can be analyzed by bombardment with 14 MeV neutrons ($\sigma =$ 37 mb) yielding ^{16}N, which decays ($t_{½}$ 7.13 s) by emitting energetic β^- and γ (6–7 MeV). In FNAA and CPAA the flux may not be homogeneous, and (17.7)–(17.9) must be used. CPAA is usually employed for elements of atomic number less than 10 and is normally limited to surface analysis because of the short range of charged particles in solid materials. Surface concentrations of 0.01–0.02 $\mu g/cm^2$ can be detected. Irradiation by protons, deuterons, and ^4He have all been used. This technique should not be confused with PIXE described in §7.8.2. In PIXE high energy protons are used to induce X-ray fluorescence in the sample, and the analysis with the multichannel spectrometers is done simultaneously with the bombardment. In activation analysis the sample is counted after the end of the irradiation. Generally, charged particle irradiation techniques are more expensive than neutron activation analysis.

Table 18.2 Estimated minimum detectable concentrations of pollutant elements in sea water by INAA and by RNAA

Trace element	Typical reported concentrations in open ocean (μg l^{-1})	Minimum detectable concentrations (μg l^{-1})	
		INAA[(a)]	RNAA[(b)]
Hg	0.02–0.2	0.05	0.001
Cd	0.06–0.7	16000	0.001
Ag	0.002–0.05	1.0	0.003
As	2–3	Not possible	0.0001
Cu	0.5–2	Not possible	0.002
Cr	0.02–0.6	0.3	0.003
Zn	0.5–10	0.2	0.01
Sn	0.02	Not possible	9
Se	0.08	0.2	0.02
Sb	0.2	0.02	0. 00003

[(a)]25 ml sea water; 1 d irradiation at 10^{17} n m^{-2} s^{-1}; 40 d decay; 1000 min count on 20 cm^3 Ge(Li)-detector; based on 3 × above background-Compton contribution in peak areas.
[(b)]500 ml sea water; elements chemically separated; 1 d irradiation at 10^{17} n m^{-2} s^{-1}; 3 d decay; 500 min count on 20 cm^3 Ge(Li)-detector; based on 2 × above background-Compton contribution in peak areas.

18.3.4. Substoichiometric analysis

Isotope dilution analysis require the determination of either the chemical yield in the separation process or of the specific activity. This can be avoided by applying the *substoichiometric principle*, which may also increase the sensitivity of the analytical method.

The method requires three samples: (1) the test sample containing the unknown weight w_u of the element or compound X of interest; (2) a standard of the same element (preferably in a similar matrix, subscript s); and (3) a nonactive carrier (usually a solution, subscript c), of the same element or compound X.

If a test sample containing the unknown weight w_u of the element or compound X and a standard containing a known weight w_s^0 of the same are irradiated under identical conditions,

$$w_u = w_s^0 A_u / A_s \tag{18.13}$$

where A_u and A_s (or R_u and R_s, if measured under identical conditions) are the total radioactivities contained in sample and standard. If the test sample contains other radioactivities in addition to that in the standard, the preceding relation can be applied only when the measurement technique is isotope specific, for example, when high-resolution gamma spectroscopy is used, in which case interference from undesirable radioactivity is eliminated.

When simple counting equipment must be used or highly complicated test samples are involved, it may be desirable to isolate chemically the element or compound of interest. This can be done through selective chemical procedures using a *non-radioactive carrier* of element X. The carrier is added to the irradiated test sample (containing w_u), which is processed through the different purification steps. Finally a sample of almost (or sufficiently) pure X is isolated, and its activity R_{u+c} and weight w_{u+c} measured. It is also necessary to run the irradiated standard with added carrier through the same chemical separation procedures. With the exception of w_u, the final equation (18.14) has terms that are known or measured:

$$w_u = w_s^0 \{R_{u+c}(w_c^0 + w_u)/w_{u+c}\}/\{R_{s+c}(w_{s+c}^0 + w_s^0)/w_{s+c}\} \tag{18.14}$$

Usually, the amount of carrier is much greater than the unknown and the standard, that is, $w_c^0 \gg w_s^0$, and $\gg w_u^0$. Then if equal amounts (by weight) of the nonactive isotopic carrier are added to the unknown sample and the standard, the equation reduces to

$$w_u = w_s^0 \,(R_{u+c}\, w_{s+c})/(R_{s+c}\, w_{u+c}) \tag{18.15}$$

This equation can be simplified even further. In the chemical separation procedure a *substoichiometric amount* of reagent is added, provided the conditions are such that this reagent quantitatively reacts with X. For example, zinc is extracted quantitatively from aqueous solutions buffered at pH 8 by dithizone in chloroform. If the amount of dithizone is less than that of zinc after adding a carrier of nonradioactive $ZnSO_4$, only part of the zinc (e.g., 25%) is extracted, but the dithizone is quantitatively bound to zinc in the organic phase. Thus varying amounts of zinc in the aqueous phase do not change the amount in the organic phase, which is constant, though the specific activity varies. The two chloroform solutions, from test plus carrier, and from standard plus carrier, contain equal amounts of Zn. Liquid-liquid extraction is commonly used for this technique.

The foregoing is an example of the basic principle of substoichiometric analysis. Under these conditions $w_{s+c} = w_{u+c}$ and

$$w_u = w_s^0 R_{u+c}/R_{s+c} \tag{18.16}$$

Thus by carefully choosing proper experimental conditions, the analytical procedure is reduced to two radioactivity measurements. For precise results the value of w_u must be comparable to w_s^0, which can be ascertained by a few initial survey experiments.

This substoichiometric analysis technique can be applied to most metals with a high degree of accuracy and a sensitivity of 10^{-6} to 10^{-10} g of X.

When the substoichiometric principle is applied to isotope dilution analysis, the relationship becomes

$$w_u = w_s^0 \,(R_s/R_{u+s} - 1) \tag{18.17}$$

where w_s^0 is the weight of the standard added, and R_s and R_{u+s} the radioactivities measured from the substoichiometrically separated samples of the standard and of the mixture of standard and unknown. The specific activities need not be determined as in (18.8) because equal weights of standard (w_s) and sample plus standard (w_{u+s}) are isolated by using the substoichiometric principle.

The usefulness of this technique for routine determination of numerous chemical elements in various matrixes has been demonstrated by Ruzicka, Stary, and others. It is also applicable to organic compounds and known in medicine as radioimmunoassay.

18.4. APPLICATIONS TO GENERAL CHEMISTRY

In no other area have radioactive tracers played such an important role as in the studies of chemical and biological reaction paths. This is, of course, due to that, in principle, in each radioactive decay the atom announces its position. Thus the detection sensitivity can approach the ultimate limit. A radioactive nuclide, as for example ^{14}C (β^- $t_{1/2}$ 5730 y), ^{32}P (β^- $t_{1/2}$ 14.282 d), or ^{198}Au ($\beta^-\gamma$, $t_{1/2}$ 2.6935 d), can be followed through a number of different chemical reaction steps, revealing details of metabolic or process reactions impossible to discover by other techniques.

18.4.1. Determination of chemical reaction paths

The use of radioisotopes in the study of the steps in a chemical reaction system is well established. Let us consider a few examples to illustrate this technique.

If phenol is labelled with deuterium or tritium in the hydroxyl group and warmed to a temperature slightly below decomposition, the labelled hydrogen migrates to other hydrogen positions of the benzene ring either by intra-molecular rearrangement or by inter-molecular reactions. However, using C_6H_5OT and C_6H_4TOH yields C_6H_4TOT, which can be formed only through inter-molecular collisions, thus eliminating intra-molecular rearrangement as the reaction mechanism.

The study of the reaction steps in the photosynthesis of carbohydrates from atmospheric CO_2 in the presence of light and chlorophyll is an outstanding example of the value of the tracer technique. The overall process (which involves many steps) can be written as

$$6\ CO_2 + 12\ H_2O \underset{\text{chlorophyll}}{\overset{\text{light}}{\rightarrow}} C_6H_{12}O_6 + 6\ O_2 + 6\ H_2O \qquad (18.18)$$

Using partition chromatographic technique and tracers of ^{14}C, ^{32}P, and T, Calvin and co-workers were able to identify the intermediate steps involved. The experimental procedure is usually as follows (Fig. 18.7). Plants are placed in atmospheres containing ^{14}C-labelled CO_2 and irradiated with light. After different irradiation times the plants are removed and sections are digested to dissolve the material. A few drops of the solution

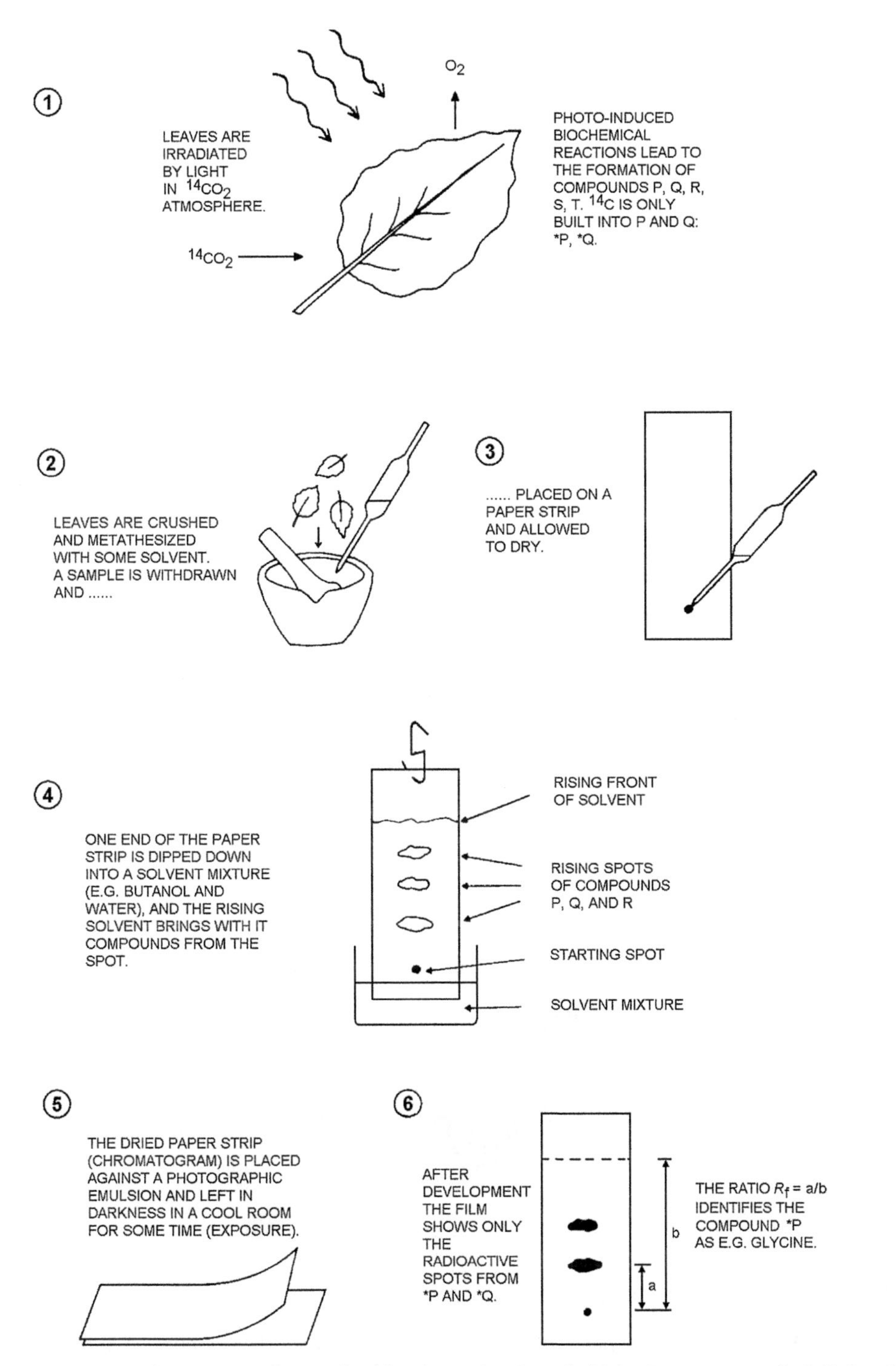

Figure 18.7 Paper chromatographic method for determination of which of components P, Q, R, S and T in a leaf are involved in photosynthesis.

containing the substance to be separated (metal ion, organic molecule, etc.) are placed a few centimeters from the end of a paper strip. The paper strip is hung vertically and dipped into a solution so that the initial point of placement of the substance is near the bottom of the strip above the solution level. Capillary forces draw the solution upwards and bring it into contact with the adsorbed substances at the starting point. As this occurs the substance moves a certain distance up the paper from the starting point with the distance traveled dependent on the kind of paper, the solution used, and the chemical properties of the substance. In such experiments, a certain R_f value for each substance can be defined as

$$R_f = \frac{\text{distance the substance of interest has travelled}}{\text{distance the liquid front has travelled}} \tag{18.19}$$

A typical solvent system for metal ions may be a mixture of acetone, dilute HCl, etc., while for organic substances it is possible to use mixtures of phenol and water, acetylacetonate and water, etc. Figure 18.8 shows a two-dimensional paper chromatogram; in this case it has been run initially with a particular solvent mixture, then (after drying) turned 90° and run with a second solvent mixture, thus increasing the selectivity of the separation. The separated substance can be quantitatively recovered by cutting out the spot and leaching the compound from it. The substance is identified either from its R_f value or further analyzed by standard methods.

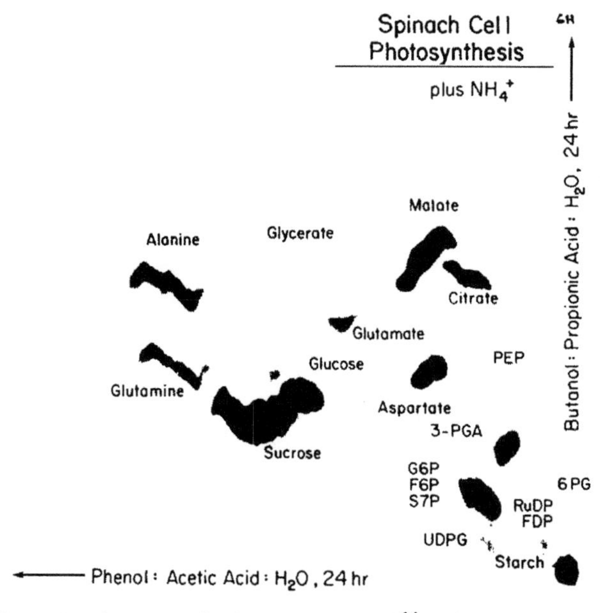

Figure 18.8 Two-dimensional paper radiochromatogram of ^{14}C-labeled products in photo-synthesis after exposure to $^{14}CO_2$ atmosphere. *(Courtesy Calvin and Lemmon.)*

18.4.2. Determination of chemical exchange rates

If two different chemical species with some element in common are mixed in solution, exchange of this common component may occur. The chemical equation would have the form

$$AX + BX^* \rightleftarrows AX^* + BX \quad (18.20)$$

Since the type and concentration of the chemical species remain unchanged, it is impossible to observe the exchange unless the atoms in one reactant are labelled. By using X^*, a radioactive isotope of X, the reaction may be followed, and at equilibrium the activity should be uniformly distributed between the two chemical species, i.e. the specific activity of X^* will be the same for both AX and BX. Of course, if AX and BX are both strong electrolytes, uniform distribution is essentially immediate upon mixing. If at least one of the reactants is an inorganic complex or an organic molecule, the exchange may be measurably slow if it occurs at all.

Since the chemical form of the reactants is not altered by the isotopic exchange, there is no change in heat content. However, the entropy of the total system is increased when uniformity in the distribution of the isotopes of X is achieved throughout the system. This entropy increase provides a decrease in the free energy, making isotopic exchange a spontaneous reaction. Despite this spontaneity, the exchange may be prevented or made very slow by a large energy of activation requirement in the formation of a necessary transition state.

For the exchange reaction represented above, the rate of increase of AX^* is equal to the rate of formation minus the rate of destruction of AX^*. The rate of formation is the product of the rate of reaction k_r, the fraction of reactions which occur with an active BX^*, and the fraction of reactions which occur with an inactive AX. Using the following notation

$$a = [AX] + [AX^*] \quad (18.21)$$

$$b = [BX] + [BX^*] \quad (18.22)$$

$$x = [AX^*] \quad (18.23)$$

$$y = [BX^*] \quad (18.24)$$

the rate of formation k_f is equal to

$$k_f = k_r(y/b)(a - x)/a \quad (18.25)$$

In a similar fashion, the rate of destruction k_d is equal to

$$k_d = k_r(x/a)(b - y)/b \quad (18.26)$$

Therefore

$$dx/dt = k_f - k_d = k_r(ay - bx)/(ab) \quad (18.27)$$

The solution of this equation is

$$\ln(1 - F) = -k_r t\,(a + b)/(ab) \quad (18.28)$$

where $F = x_t / x_\infty$ (x_∞ is the value of x_t at $t = \infty$, i.e. equilibrium). The rate of exchange k_r is evaluated from the slope of a plot log $(1 - F)$ versus t. If more than one rate of exchange is present due to exchange with nonequivalent atoms in a reactant, it may be difficult to resolve this curve sufficiently to obtain values for the reaction rates. Isotopic exchange is a standard tool of the scientist studying the kinetics of chemical reactions whose half-lives are longer than a minute.

One example of isotope exchange can be used to illustrate the value of these studies. Consider the exchange between di- and trivalent chromium in $HClO_4$ solutions. If the total chromium ion concentration is 0.1 M, it takes 14 days for the exchange to reach 50% completion at room temperature. In as much as the di- and trivalent cations are both positively charged, it is unlikely that they can approach each other closely enough to exchange an electron directly to allow a reversal of oxidation state, and a more likely mechanism is that an anion is involved as a bridge between the two cations such that the intrusion of the anion reduces the repulsion between the two cations. If this model of the isotopic exchange mechanism is valid, it could be proposed that the reaction mechanism would be

$$Cr(III)^* + X^- + Cr(II) \rightarrow [Cr^* - X - Cr]^{4+} \rightarrow Cr(II)^* + X^- + Cr(III) \quad (18.29)$$

Such a reaction mechanism would be fostered by the presence of anions that form complexes more readily than perchlorate ion. If HCl solutions are used rather than $HClO_4$ it is found that the exchange takes place more rapidly and the half time of exchange is only 2 min, which agrees with the proposed mechanism since chloride ions are known to be more favorable to complex formation than perchlorate ions. Without the use of radioactive (or isotope separated) chromium to label one of the original oxidation states there would be no means of identifying the exchange.

18.4.3. Determination of equilibrium constants

Determination of cation (M^{z+}) – anion (L^-, for ligand) solution equilibria can advantageously be done using radioactive metal tracers because of the ease with which the metal concentration can be measured, as e.g. in the determination of *solubility products*

$$K_{s0} = [M^{z+}][L^-]^z \quad (18.30)$$

or *complex formation constants*, also named *stability constants*

$$\beta_n = [ML_n^{z-n}]/[M^{z+}][L^-]^n \quad (18.31)$$

At trace concentrations of the metal (i.e. $[M]_{tot} \ll [L^-]$), the complexing anion concentration is unaffected by the metal ion concentration, which allows easy calculation of the values of these two important variables in the system. The experimental techniques used for this purpose are based on two-phase equilibria: solubility (solid/liquid), paper electrophoresis (paper/aqueous solution), solvent extraction (organic solvent/water) and ion exchange (resin/water). The equilibria measured have been shown to be independent of the metal concentration in the range 10^{-3} to $<10^{-13}$ M, as long as no polynuclear complexes (i.e. with several metal atoms per complex molecule) are formed (cf. §18.2.3).

(a) *Solubility*

The sensitivity of tracer detection makes measurement of solubilities relatively simple. This is illustrated by the first radioactive tracer experiment by de Hevesy and Paneth in 1913 in which the solubility of lead chromate was determined. Chromate ions were added to a solution of $PbCl_2$ containing a known amount of ^{210}Pb, precipitating the lead as $PbCrO_4$. The precipitate contained 2030 "radioactive units", and had a weight of 11.35 mg. The specific activity was thus 2030/11.35 = 179 "units" mg^{-1}. Shaking the precipitate with water dissolved 2.14 units per 1000 ml. The solubility was calculated to be 2.14/179 = 0.012 mg l^{-1} or 3.7×10^{-8} M Pb^{2+}. If $[Pb^{2+}] = [CrO_4^{2-}]$ the solubility product would be $K_{s0} = (3.7 \times 10^{-8})^2 = 1.4 \times 10^{-15}$. The modern value is 2×10^{-14}.

(b) *Solvent extraction*

As described in §18.2.6 the solvent extraction technique requires (i) a liquid two-phase system consisting of an organic solvent in contact with an aqueous solution, and (ii) the presence of an extractant (commonly a weak organic acid, abbreviated HA), which reacts with the metal ion to form an uncharged metal-organic complex MA_z, that preferentially dissolves in the organic phase. The distribution of the metal between the organic and aqueous phases can be shown to be fit the relation

$$D_M = K_{DC}\beta_z[A^-]^z \Sigma\beta_n[A^-]^n \tag{18.32}$$

where D_M is the distribution ratio of the metal as defined by eqn. (18.8), and K_{DC} is the *distribution constant* of the uncharged complex MA_z. $[A^-]$ is referred to as the free ligand ion concentration; using trace metal concentration, $[A^-]$ is easily calculated from the amount of acid, HA, added, its dissociation constant K_a, pH and the liquid volumes. From measurements of D_M as a function of $[A^-]$, the formation constants β_n for the complexes MA_n^{z-n} are calculated.

Figure 18.9 shows the distribution of lutetium (using trace concentrations of ^{177}Lu, ($\beta^-\gamma$ $t_{1/2}$ 6.71 d) between an aqueous solution and benzene containing an organic complex former (HA = acetylacetone). Eqn. (18.32) has been fitted to the experimental

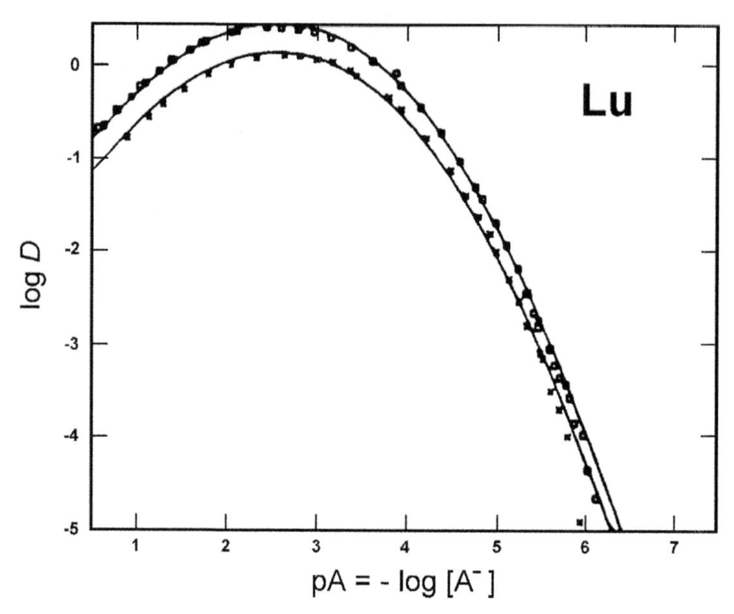

Figure 18.9 Distribution of ^{177}Lu (β^- $t_{½}$ 6.71 d) in 1 M $NaClO_4$/benzene, acetylacetone. Upper curve $[HAa]^0_{org}$ 3.0 M, lower curve 1.0 M. *(From Albinsson.)*

points by regression analysis to yield the equilibrium constants, by which the solid curves have been calculated.

Studies of this kind are often easy to do. However, the Lu-acetylacetone system was chosen as it was not as simple, yet good results were obtained by careful use of the tracer technique: Because the Lu concentration was 10^{-8} M, 10^{-5} M Nd^{3+} was added as *hold-back carrier* to avoid sorption losses (§18.2.3). At the pH's that had to be investigated Lu^{3+} (and also Nd^{3+}) may hydrolyse. Also, the very low *D*-values obtained (1 to 10^{-5}) are difficult to measure exactly with conventional technique. Therefore advanced techniques were developed, see Figure 18.10: (a), AKUFVE, which is a closed flow system, where the two liquid phases are mixed continuously and separated in a special flow-through centrifuge, and, after passing detection devices, returned to the mixing chamber. (b), LISOL, an "on-line" liquid-scintillation detection technique, in which a tiny fraction of the circulating phases is withdrawn, mixed (i) with acid to suppress sorption and hydrolysis, and (ii) with a liquid scintillator, and (iii) pumped to the PMT-detector. The LISOL avoids detector memory effects occurring in conventional on-line detectors and also allows measurements of pure α- and β-emitters (cf. §8.5.2).

As a second example we choose a chemically more complicated system, which, however, in practice is simpler than the one above, and therefore of more common use. The complexation of $^{237}Np^{4+}$ by HSO_4^- in 0.1 M $NaClO_4$ was studied by SX using the

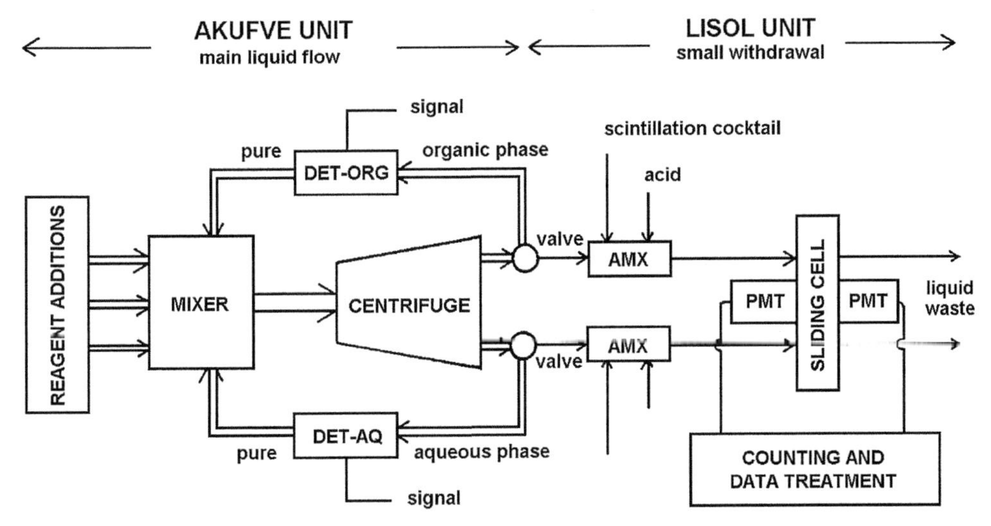

Figure 18.10 The AKUFVE-LISOL system used in investigating the system in Fig. 18.9.

system aqueous solution/$CHCl_3$ containing the organic extractant thenoyltrifluoroacetone (HTTA). In this case the relation

$$D_{Np}/D_p = \Sigma\beta_x\left[HSO_4^-\right]^x \tag{18.33}$$

can be derived, where D_{Np} is the distribution ratio of Np in the presence, and D_p in the absence, of HSO_4^-; in this case pH and [HTTA] must be constant during the experiment, any change requiring correction. From the data in Figure 18.11 the formation constants β_x for Np(IV) sulphate complexes were determined.

The solvent extraction technique has been used extensively for studying complexation of metals for which it may not be possible or desirable to use macroscopic amounts, as e.g. for the heavy actinides or transactinides (Ch. 14).

(c) *Ion exchange*

The technique depends on the distribution of metal cations between a solid cation exchange resin and an aqueous solution. The exchange process for the M^{z+} ion takes place according to the equation

$$M^{z+}(aq) + zRH(resin) \rightleftarrows MR_z(resin) + zH^+(aq) \tag{18.34}$$

$$K_{iex} = ([MR_z][H]^z)/([M][RH]^z) \tag{18.35}$$

omitting ionic charges and phase indices. Introducing D_{iex}, from (18.9), in (18.35) gives

$$D_{iex} = K_{iex}[RH]^z\,[H]^{-z} \tag{18.36}$$

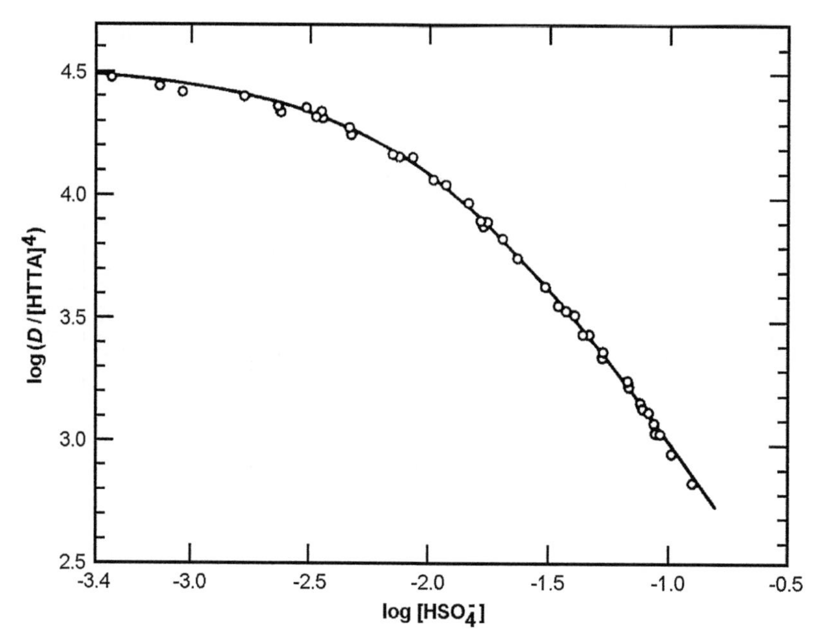

Figure 18.11 Normalized distribution ratio of $^{237}Np(IV)$ TTA complex between chloroform and an aqueous HSO_4^- solution. *(From Sullivan and Hindman.)*

showing that D_{iex} is a constant at constant solution/resin conditions (as in column ion exchange separations). However, when used for determining equilibrium constants in a system with several metal species in solution, each ionic species must be represented by an equation of type (18.35), leading to rather complicated expressions.

18.4.4. Studies of surfaces and reactions in solids

Surface properties of solids have been studied by dipping specimens into a solution containing a suitable radioactive tracer, and, after some "exposure time", removing them, washing their surface carefully and measuring the radiation emitted from them. It has been shown that a very rapid exchange takes place between atoms on a metal surface and the metal ions in solution. While the exchange is a function of the nature of the surface, within minutes it may involve atoms several hundred layers deep. The depth of penetration of sorbed radioactive isotopes can be obtained from a measurement of the absorption of the radiation or by measuring the radioactivity removed by cutting, or grinding, away thin layers. With the same technique the diffusion of atoms in their own solid matrix can be studied. For example, using single crystals of silver suspended in a solution containing silver nitrate labelled with ^{110m}Ag ($\beta^-\gamma$ $t_{½}$ 2418.76 d) it has been possible to demonstrate different rates of diffusion into different faces of the crystal. The surface area of solids can also be determined by measurement of the sorption of radiotracers which do not penetrate into the specimen.

If a radioactive gas is incorporated in a crystalline compound the amount of gas released (the *emanation ability*) can be measured as a function of the temperature. It is found that the emanation increases considerably at certain temperatures, indicating structural changes in the solid at those temperatures. Studies of diffusion and emanation play a valuable role in understanding the mechanism of sintering and in the formation of new solid compounds. This has been of practical importance in the cement and glass industries, in the production of semiconductors, in the paint industry, etc. Studies of surface reactions are of practical importance for flotation, corrosion, metal plating and finishing, and detergent action to name only a few applications.

The distribution of a radioactive element or compound in a composite matrix can made visible either to the naked eye or under a microscope by means of *autoradiography*. The technique is based on the blackening of photographic films when exposed to nuclear radiation (cf. §§8.2 and 8.10). The technique is best illustrated by an example.

Lead is an unwanted impurity in stainless steel even in very small amounts. In order to investigate the mechanism of its incorporation, ^{212}Pb was added to a steel melt. After cooling, the ingot was cut by a saw and the flat surface machine-polished and etched in an electrolytic bath (electro-polishing). This provided a very flat and "virgin" surface. A photographic film was placed firmly with even pressure against the metal surface, and the film was exposed in darkness in a cool room for about a week. After development of the film, darkened spots caused by the radiation from ^{212}Pb showed where on the metal surface lead was present. By taking the results into account in the production process, the negative effect of lead in the raw material could be reduced.

In another technique, a polished surface of the specimen (metal, mineral, etc.) was dipped into a solution containing a radioactive reagent, which selectively reacts with one of the constituents of the surface. A mineral was dipped into potassium ethyl xanthate labelled with ^{35}S (β^- 87.5 d); the xanthate reacted selectively with sphalerite (zinc blende), ZnS, in the sample. The distribution of the xanthate, as shown by the autoradiograph, indicated the ZnS distribution in the mineral. The low β-energy of ^{35}S, E_{max} 0.2 MeV, was an advantage to the technique because the resolution of the autoradiograph increases with decreasing particle range.

18.5. APPLICATIONS TO LIFE SCIENCES

The largest field of application of radionuclides is in the life sciences. A survey is presented in Table 18.3.

In the reactions leading to the desired product (column A) many factors must be considered. They are of such importance that we devote several separate chapters to this: Ch. 10 on nuclear reactions, Ch. 16 on particle accelerators and Ch. 17 on production of

Table 18.3 Survey of radionuclide use in life sciences

A Radionuclide production	**C Technique/application [detection]**
1 Target chemistry,	1 Biochemical analysis
2 irradiation,	**(a)** autoradiography [photographic]
3 isolation	**(b)** immunoassay [counting]
4 and processing, yields	**(c)** DNA-analysis [photographic]
(a) pure radionuclide, or	**(d)** direct tracing [counting]
(b) labelled compound (radiopharmaceuticals)	2 Medical imaging
B Source, position/ administration	**2.1** Transmission Tomography (TCT) photographic, or by counting→computer→display]
1 Internally (α,β or γ)	**2.2** Emission Computed Tomography (ECT) [counting→computer→display]
(a) injection,	**(a)** Single Photon Emission Computed Tomography (SPECT)
(b) inhalation, or	**(b)** Positron Emission Tomography (PET)
(c) oral intake.	3 Irradiation uses
2 Externally as γ-source	**(a)** by internal sources (therapeutically)
	(b) by external sources (therapeutically)
	(c) by external sources (sterilization etc)

radionuclides. The incorporation of the radionuclide in a chemical compound (labelling, §17.5.3) provides it with unique properties, such as specific biological affinity (§18.5.1). When such labelled compounds are taken up by organisms (A4b → B1) they move to specific sites in the organs, signalling normal or abnormal behavior. When used in medicine (primarily C1 and C3) these compounds are referred to as *radiopharmaceuticals* (A4b).

The use of labelled compounds in life sciences is extensive, in fact, the largest single user of radionuclides is medical science. It has been said that radioactive tracers have been of equal importance to medicine as the discovery of the microscope. Presently one out of ten hospitalized patients in the United States is admitted to some nuclear medical procedure. If the intended use of the radionuclide is as an external radiation source (A4a → B2 → C2.1 or C3) its chemical matrix is of minor importance. Such sources are used for radiation treatment of cancer (C3b), radiation sterilization of food (C3c), etc. The *radiation effects* on biological systems are discussed separately in Ch. 15. In this chapter we focus our interest on radionuclides with specific chemical properties, in the order of column C, Table 18.3.

18.5.1. Biological affinity

To study how living species interact with the environment, *ecology*, one can use radioactive tracers to follow the uptake of a trace metal (e.g. cobalt) from the soil by plants, and by animals after having eaten the plant. In agriculture, this is useful in studying the uptake of trace elements necessary for plant growth. For example, it has been found that sheep need plants containing selenium in order to combat white muscle disease. The turnover in nutrients fed to animals can be determined; it was found that 20% of the phosphorus in cow's milk comes directly from the feed, while 80% is taken from the cow's bone.

If a radioactively labelled compound such as an amino acid, a vitamin, or a drug is administered to an animal, the substance is incorporated to varying degrees in different organs (*biological affinity*). The substance undergoes chemical exchange with other substances in the body, is broken down, and, finally, discharged from the body (i.e. *metabolizes*). The radionuclide distribution in samples (cells, tissues, body fluids, etc) removed from living species gives significant information about the (normal or abnormal) physiology of that species. "C1. Biochemical analysis" in Table 18.3 lists these techniques, which are discussed below.

(a) *Autoradiography*

Figure 18.12(a) shows the distribution of ^{14}C-labelled p-aminosalicylic acid, PAS, the first tuberculostatic agent developed, in a 20 μm thick section of a mouse. The autoradiograph was obtained after 10 MBq ^{14}C-PAS had been injected, and the mouse (weight 20 g) had been killed by immersion into a CO_2-acetone (−80°C) mixture and sectioned. It is seen that the PAS is concentrated mainly in the lung, where it is effective against tuberculosis, and the kidney and intestine as it is excreted through these organs.

Figure 18.12(b) shows an autoradiograph of radioactively labelled cells. In the "pulse labelling" (i.e. 10 − 30 min exposure) to ^{3}H-thymidine only cells in the S-phase of the cell cycle reacts with the thymidine and thus become labelled (the very dark spots). Both the cell cycle times and number of cells in the cycle can be measured with this technique, which is important in cancer cell research.

As an alternative to using the blackening of a photographic film for radiation detection "instant imagers" based on semiconductor array detectors are commercially available.

(b) *Radioimmunoassay (RIA)*

Immunoassay is an application of the substoichiometric principle (§18.3.4) developed by Yalow (Nobel laureate in 1977) for protein analysis. In the United States tens of millions radioimmunoassays are made annually in hospitals to measure hormones, enzymes, viruses, serum proteins, drugs, and so forth. Only a drop of the patient's blood is needed, reflecting the versatility and sensitivity of this technique, which can be performed automatically. Commercial RAST-kits (Radio Allergy Sorbent Tests) are used for rapid diagnosis of allergic reactions.

In immunoassay, a known mass, w_s^0, of a labelled protein P* is allowed to react with a much smaller (substoichiometric) mass of an antibody A, so that a complex P*A is formed. The P*A is isolated and its radioactivity, R_s, measured. Under the same

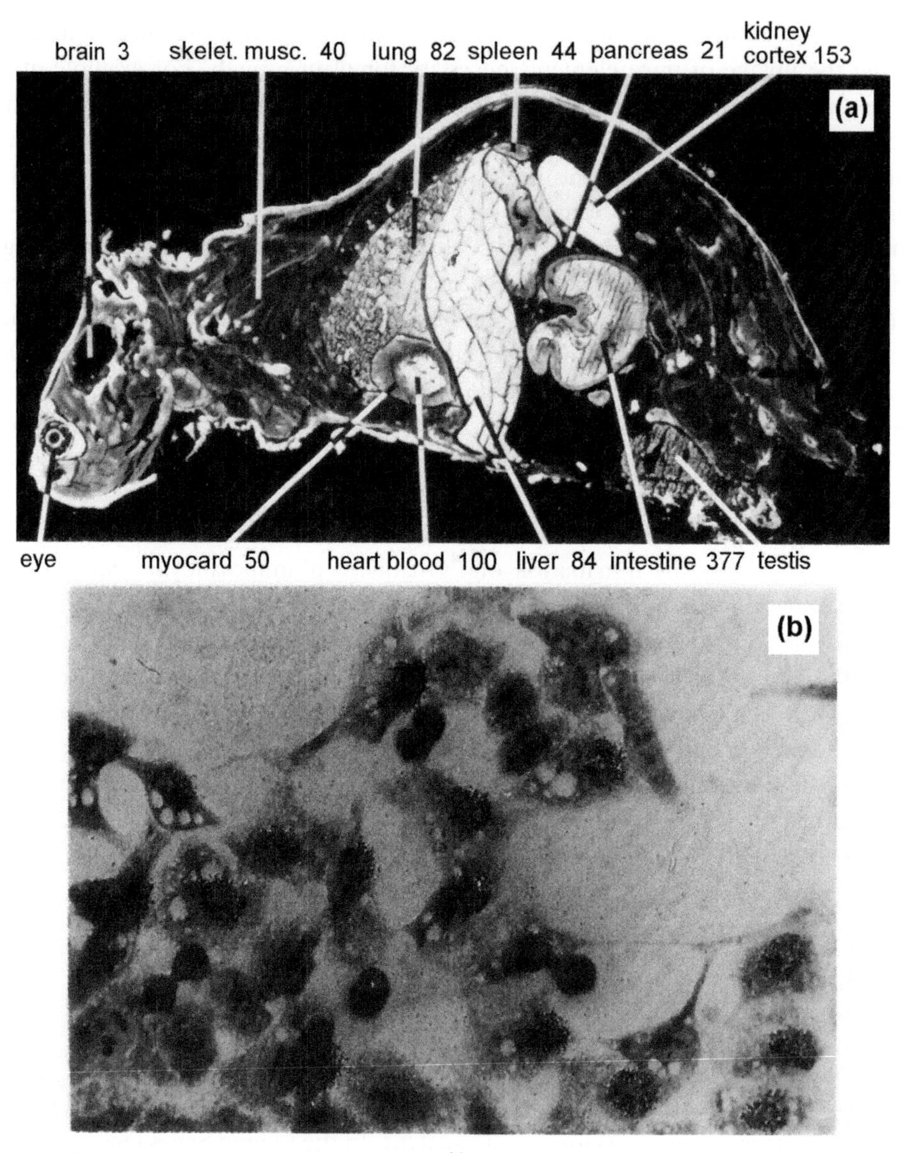

Figure 18.12 Autoradiograms, (a) distribution of ^{14}C-PAS in a mouse 30 minutes after intra-venous injection, (b) of HeLa-cells labeled by ^{3}H-thymidine. *(From Hanngren and Nias.)*

conditions, an identical mass, w_s^0, of labelled protein is mixed with an unknown mass, w_u, of the identical protein to be determined. This sample is also allowed to react with the same amount of antibody A as before; the complex P*A is again isolated (weight w_{u+s}) and its radioactivity, R_{u+s}, measured. The unknown weight, w_u, is then calculated from (18.17).

(c) *DNA-analysis*
The chemical composition (the base sequence) of the DNA molecules, which make up the chromosomes of the cell nucleus, is unique for each species and individual. The detailed analysis of the DNA molecule provides important information about its host. This is used in studies of the evolution of species, in forensic science to identify criminals (e.g. from blood, sperm, etc), in transplantation of organs (kidney, tissue grafting, etc), in detection of genetic diseases, etc. The importance of DNA-analysis is constantly growing.

The sequence of the DNA analysis is as follows. (i) The cell walls are broken up by osmosis etc. and the doublestranded DNA is denatured to pieces of singlestranded DNA. The molecules may be concentrated by centrifugation. (ii) By exposing this DNA to *restriction enzymes* the DNA nucleotide chain is sectioned further into smaller fragments; different restriction enzymes cut the DNA molecule at different positions. The fragments are exposed to radioactively labelled compounds (e.g. containing ^{14}C or ^{32}P) which selectively attach to the different fragments. Alternatively to (i) and (ii), the DNA is directly labelled e.g. by ^{32}P in a cloning process; the clones will then contain ^{32}P at its P- sites also after it has been split by restriction enzymes. (iii) When the DNA samples are exposed to electrophoresis in a suitable gel (agarose, polyacrylamide, etc), the fragments distribute themselves along the potential direction according to their migration velocities. Usually 10−40 samples (each treated by a different restriction enzyme) are run simultaneously. (iv) After the electrophoresis, a photographic plate is placed on the gel and exposed to the radioactive fragments, producing, after development, a pattern of spots or bands on the plate. (v) The band pattern is analyzed to reveal its information about the individual. The technique has many similarities with those demonstrated in Figures 18.7 and 18.8.

18.5.2. Transmission computer tomography (TCT)

Conventional "x-ray pictures" show the morphological structure of the internal organs. The technique is used with x-ray or γ-ray sources; e.g. Ra was extensively used during World War I in operations at the front where electricity was lacking. To improve the contrast in the photographic pictures, dense media like barium sulfate or iodine compounds are administered to the patient; this method is said to be *invasive* and can be painful to the patient.

Though this technique is still widely in use, the 1970'ies saw a large step forward in medical imaging referred to as *computerized tomography* (CT), developed by Cormack and Hounsfield (Nobel laureates in 1979) and others. Though the radiation source is the same, the photographic plate is replaced by one or several radiation sensitive detectors, a digital computer and a display. The radiation source and detector array are moved (*scanned*) in relation to the patient, see Figure 18.13 A. This technique requires computer

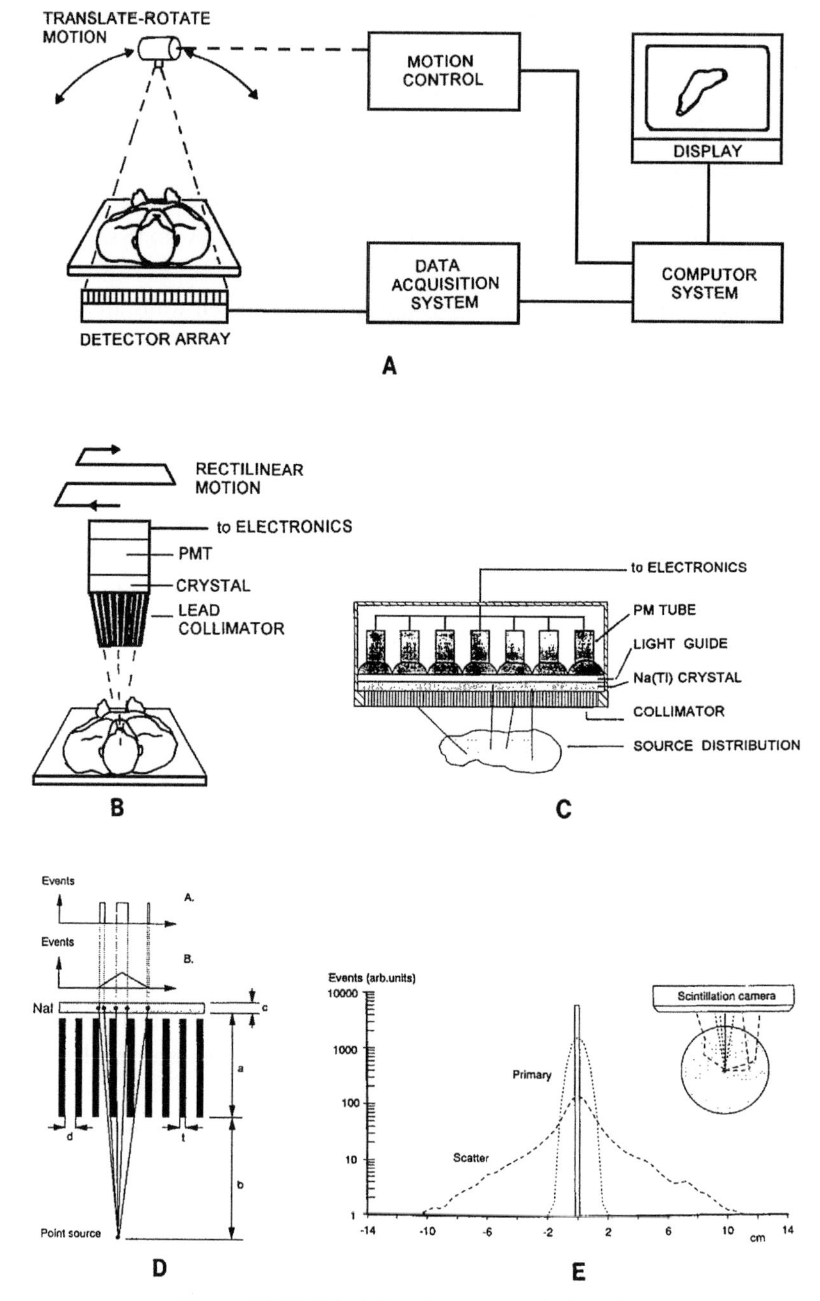

Figure 18.13 Principles of gamma scintigraphy.

software to handle the reconstruction arithmetic needed to provide an image on the screen from the observed changes in count rates by the detectors (due to different absorbancies of the organs) at different geometries. Tomography means "slice"; the technique shows slices through the body on the screen. The resolution of a TCT-scan is usually ~1 mm, allowing quite exact pictures, e.g. of a tumor. Since nowadays only X-ray tubes are used as radiation source, the technique is handled by *radiologists* at the hospital "X-ray department".

Transmission tomography can be compared to a silent movie: one can see what happens physically with the organs. Emission tomography (next section) uses internally administered radioactive isotopes, which provides explanation to why the organs function as they do. Thus, to extend the comparison, computerized emission tomography becomes more like a medical sound movie of the patient.

18.5.3. Emission computer tomography (ECT) and diagnosis

The rate of incorporation and discharge of radioactively labelled substances in the body provides a measure of the metabolism of healthy and of sick tissues. On medical patients this information is obtained by external measurements referred to as *radioisotope scanning* (RIS). Such scanning can yield information about a medical disorder much before it is observed by other means. Since the amount of radioactive tracer is very small, this technique is referred to as *non-invasive*. In hospitals the department of *nuclear medicine* is normally responsible for these investigations.

(a) *Simple scanners*

Simple scanners are designed either with one or several direction sensitive (focusing) detectors, which are moved around or above the patient in a pattern; Figure 18.13.B and C. Figure 18.14 shows the result of a kidney scan, *renography*, of a 38 year old woman who

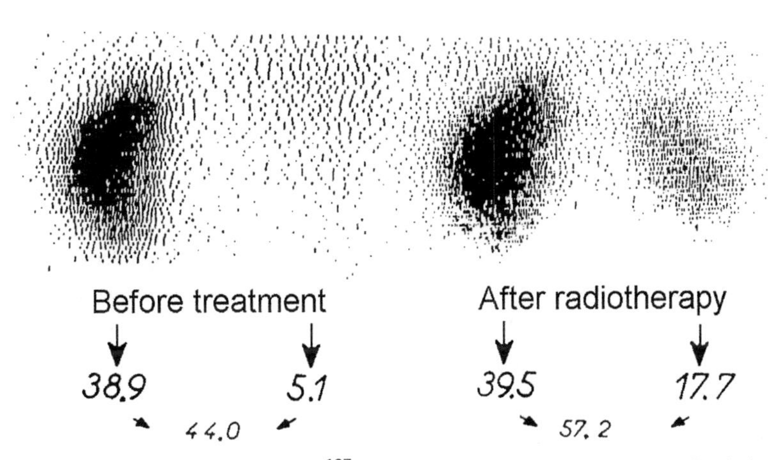

Figure 18.14 Kidney function test by $^{197}HgCl_2$ using scintigraphy. *(From Kellershohn et al.)*

has been administered $^{197}HgCl_2$ (ECγ $t_{1/2}$ 2.672 d); in this case a single NaI(Tl)-PMT with collimator was moved in a recti-linear pattern over the kidneys (Fig. 18.13.B). The left picture shows that 38.9% of the compound has been fixed to the left kidney, and very little to the right kidney. This was caused by a vaginal fibrous sarcoma blocking the urethra from the right kidney. After radiation therapy some improvement is seen (right Figure). Renography is now done with ^{131}I labelled hippuric acid, or a ^{99m}Tc complex.

(b) *Gamma Camera and Single Photon Emission Computer Tomography (SPECT)*

Recent advances in nuclear medical imaging are the gamma camera, SPECT and positron emission tomography (PET, described under (c)).

Figure 18.13.C shows the gamma scintillation camera, *gamma camera*, originally developed by Anger. It consists of a two-dimensional array of 40–100 PMTs (often hexagonally formed for tight stacking) viewing a large flat NaI(Tl) crystal of ~400 mm diameter and 5–10 mm thick, which is located behind a lead collimator containing numerous holes. Typical hole size is 2–3 mm diameter and 40 mm length. Collimator dimensions (Fig. 18.13.D) depend on the E_γ; for ^{99m}Tc (E_γ 0.14 MeV) the wall thickness is 0.2–0.3 mm.

The radionuclide injected into the patient should preferably decay by the emission of a single γ, and the lower the E_γ the better becomes the resolution of the image. Since the detectors are energy sensitive, the camera can be tuned to the primary unperturbed (Ch. 7) γ's, i.e. the signals from the PMTs are accepted only if they are near the photo-peak energy. The energy window of the detector has some width allowing some scattered photons to contribute to the number of events registered from a single point radionuclide source; this is suppressed by narrowing the energy window (Fig. 18.13 E). Figure 18.13 D indicates, in the upper diagram (A), the actual events detected through the pin holes, and how these can be sharpened into a peak if the collimator moves during the exposure time (Fig. 18.13 B); this is of particular importance in SPECT-investigations. The gamma camera is supported by hardware (Fig. 18.13.A) and extensive computer software.

One distinguishes between "scintigraphic" and "SPECT" investigations. Gamma cameras are used in both investigations, but the camera is stationary in the former case, while it usually moves in SPECT. About 95% of all nuclear imaging investigations carried out by the nuclear medicine department of a modern hospital is scintigraphy with stationary gamma camera, usually referred to as "gamma camera (investigation)". In about 2/3 of these investigations a single picture is taken of the patient's organ: heart, kidney, liver, etc. This yields an image of the organ's content of the radionuclide, i.e. of its function with respect to the pharmaceutical administered. About 1/3 of the investigations are studies of the dynamic behavior of an organ. Taking thousands of "pictures" (commonly 10 per second), storing and treating them allows a direct study of the function, e.g. heart beat. Though the resolution in such investigations is rarely better than 10 mm, they nevertheless provide information not available by TCT.

Each picture by the camera is two-dimensional; however, by positioning the camera at different angles around the patient, 3-dimensional pictures of the organs can be constructed. This is commonly done by letting the camera rotate around the patient, usually 360° in increments of a few degrees. Several large cameras at fixed positions or rotating around the patient can be used for the same purpose. The cameras can also be moved in a direction parallel to the patient. By using software similar to that used in TCT-scanning, "radiographic slices" of the patient's organs can be obtained. Such investigations are referred to as SPECT. The resolution of present commercial SPECT equipment is only 12−15 mm.

With the increasing number of radiopharmaceuticals with specific biological affinities, gamma camera and SPECT have become important diagnostic tools with numerous clinical applications, and virtually every organ in the body has been studied. Table 18.4 shows the most frequently performed imaging investigations; Table 18.5 lists data for radionuclides applied in medicine, including the amount of radionuclide needed. The radiation doses received by the patient in diagnostic investigations is usually <10 mGy per investigation.

The SPECT technique is primarily used for cardiovascular and brain imaging. Cardiac stress tests, using $^{201}Tl^{+}$ or ^{99m}Tc labelled radiopharmaceuticals, amount in the U.S. to about 2 million a year. Brain tumors can be located by SPECT after intravenous injection of $Na^{99m}TcO_4$, as brain tumors have a very high affinity for and slow release of Tc. In comparison, the uptake of Tc in brain infarcts is low and the release fast, and from healthy parts of the brain even faster; thus various constrictions to the cerebral blood flow are easily located. A head scan can be made in 10 minutes and virtually instantaneously produces an image of the brain.

Mental disorders are diagnosed by SPECT, gamma-camera or PET using various radiopharma-ceuticals, e.g. after the injection of ^{99m}Tc-HMPAO (Hexa Methyl

Table 18.4 Ten of the most frequently performed imaging investigations

Examination	Radionuclide	Static/ dynamic	Principal application
Bone	^{99m}Tc	S	Secondary spread of malignancy
Liver	^{99m}Tc	S	Secondary spread of malignancy; cirrhotic changes
Brain	^{99m}Tc	S/D	Occult metastases; brain damage; vascular problems
Lung	^{99m}Tc	S	Pulmonary embolism (perfusion)
Lung	$^{133}Xe/^{99m}Tc$	D	Pulmonary emphysema (ventilation)
Kidneys	$^{123}I/^{99m}Tc$	D	Renal function
Hepato-biliary	^{99m}Tc	D	Patency of the biliary tree; liver function system
Thyroid	$^{123}I/^{99m}Tc$	S	Thyroid function
Heart	^{201}Tl	S	Cardiac infarction and ischaemia (perfusion)
Heart	^{99m}Tc	D	Cardiac wall motion (blood pool)

(From Sharp et al.)

Table 18.5 Radionuclides applied in nuclear medicine

Nuclide	Decay		Particle energy (keV)*	Applications**
	Half-life	Mode		
^{3}H	12.33 y	β^-	18.6 β	AN. BA all chem. comp.
^{11}C	20.39 m	EC β^+	960 β, 511 γ	As CO_2, HCN, HCHO, CH_3I. DG PET: brain, glucose
^{14}C	5730 y	β^-	156 β	AN
^{13}N	9.965 m	β^+	1190 β, 511 γ	As NH_3. DG PET
^{15}O	2.037 m	β^+	1723 β, 511 γ	As H_2O. DG PET
^{18}F	1.830 h	β^+ EC	635 β, 511 γ	BA skeleton. As ion or F_2. DG PET
^{24}Na	14.659 h	β^- γ	1389 β, 1369, 2754 γ	AN. BA circulatory system.
^{32}P	14.282 d	β^-	1710 β	AN. BA skeleton, bone marrow. TP: leukemia 200 MBq phosphate.
^{35}S	87.5 d	β^-	1674 β	AN
^{42}K	12.360 h	β^- γ	3523, 1970 β, 1525 γ	AN liquid volume. BA muscles.
^{47}Ca	4.536 d	β^- γ	1981, 684 β, 1297 γ	BA skeleton. DG: bone
^{51}Cr	27.704 d	EC γ	320 γ	BA spleen. DG: kidney funct. (clearance) 4 MBq EDTA.
^{55}Fe	2.73 y	EC		BA: red blood cells
^{59}Fe	44.50 d	β^- γ	475, 273 β, 1099, 1292 γ	DG: ion
^{57}Co	271.77 d	EC γ	122, 137 γ	BA liver.
^{58}Co	70.92 d	β^+	811, 511 γ	DG: vitamin B12
^{60}Co	5.271 y	β^-	315 β, 1333, 1173 γ	TP: cancer. Source $>10^7$ MBq.
^{68}Ga	1.135 h	β^+ EC	1830 β, 511, 1077 γ	BA intestine. DG: ion, EDTA or citrate.
^{75}Se	119.77 d	EC	(864) 136, 265, 280 γ	BA kidney, liver. DG: methionine
^{85}Sr	64.84 d	EC	514 γ	BA skeleton. DG: skeleton 2 MBq ion.
^{99m}Tc	6.006 h	IT	140 γ	DG: Thyroid 100 MBq TcO_4; heart 400 MBq TcO_4; lung (perfusion, emboli) 80 MBq albumin particles; liver (tumor, size, funct.) 100 MBq colloid, or 150 MBq HIDA; kidney 200 MBq DTPA (clearance), DMSA (renography), and other compounds; skeleton (metastasis) 400 MBq phosphate; brain (tumor, cerebral hemorrhage) TcO_4 or DTPA. Etc.
^{111}In	2.807 d	EC	(864) 171, 245 γ	DG: spinal cord ion, oxime or DTPA.
^{113m}In	1.658 h	IT	(392) 392 γ	DG: ion, oxime or DTPA.
^{123}I	13.2 h	EC	(1232) 159 γ	BA thyroid. DG: Kidney funct. 40 MBq hippuran; thyroid.
^{125}I	60.1 d	EC	(178) 35 γ	BA thyroid. DG: Thrombosis 4 MBq fibrinogen. Kidney function hippuran

Table 18.5 Radionuclides applied in nuclear medicine—cont'd

Nuclide	Decay Half-life	Decay Mode	Particle energy (keV)*	Applications**
^{131}I	8.040 d	β^-	606 β, 365 γ	DG thyroid funct. 2 MBq ion. Kidney 1-10 MBq hippuran. TP: 300 MBq ion hyperthyroidism and thyroid cancer.
^{133}Xe	5.24 d	β^-	346 β, 81 γ	DG Lung emboli, 1000 MBq gas (radiospirometry), brain.
^{137}Cs	30.0 y	β^-	514 β, 662 γ	TP: cancer. Source >10^6 MBq.
^{186}Re	3.777 d	β^-	309, 362 β, 137 γ	bone cancer
^{188}Re	16.98 h	β^-	728, 795 β, 155 γ	pain
^{198}Au	2.6935 d	β^-	961 β, 412 γ	TP: 150 MBq spread of ovarian cancer.
^{201}Tl	3.05 d	EC	(480) 167 γ	BA: para-thyroid, kidney. DG Heart (infarct) 50 MBq ion

AN = biochemical analysis, BA = biological affinity, DG = diagnostically (imaging), TP = herapeutically.
* Particle energy (decay energy)
** Common radiochemical species; ion = ionic as chloride, sulfate, etc.

Propylene Amine Oxide) or inhalation of ^{133}Xe. Injecting ~1000 MBq Tc-complex into the blood stream, about 5% of this compound moves to the brain, passes the membrane of the blood vessels and enters into the brain tissue, where it decomposes and decays with its 6.0 h half-life, allowing detailed imaging of the brain. The average absorbed whole body dose in such a SPECT scan is about 0.01 Sv. Figure 18.16 (right) shows three ^{133}Xe gamma-camera pictures of patients with Alzheimers disease (ALZ), frontal lobe dements (FLD) and multi-infarct dements (MID). ^{133}Xe decays by emitting an 80 keV γ, thus the pictures show primarily the ^{133}Xe in the brain cortex (regional-Cerebral Bloods Flow test, r-CBF); in this investigation 256 small scintillation detectors were positioned like a helmet around the patient's head. The pictures in black-and-white shown here do not make justice to the original, more detailed, color pictures, the colors representing the radiation intensity, i.e. blood flow at the point.

^{99m}Tc is a preferred radionuclide due to its convenience of production (from milking ^{99}Mo, §5.16, the ^{99}Mo produced by fission of ^{235}U), and short half-life which reduces radiation risks. There are >20 differently labelled Tc-compounds commercially available for diagnostic purposes. ^{99m}Tc (together with ^{123,125,131}I) is the most frequently used radionuclide for diagnostics; about 7 million such investigations are made per year in U.S. The dominating organs investigated are in order skeleton, kidney, liver and thyroid.

(c) *Positron emission tomography (PET)*

The decay of a positron emitting radionuclide yields two 0.51 MeV γ-rays travelling in opposite directions. If photons with this energy are registered simultaneously by γ-ray

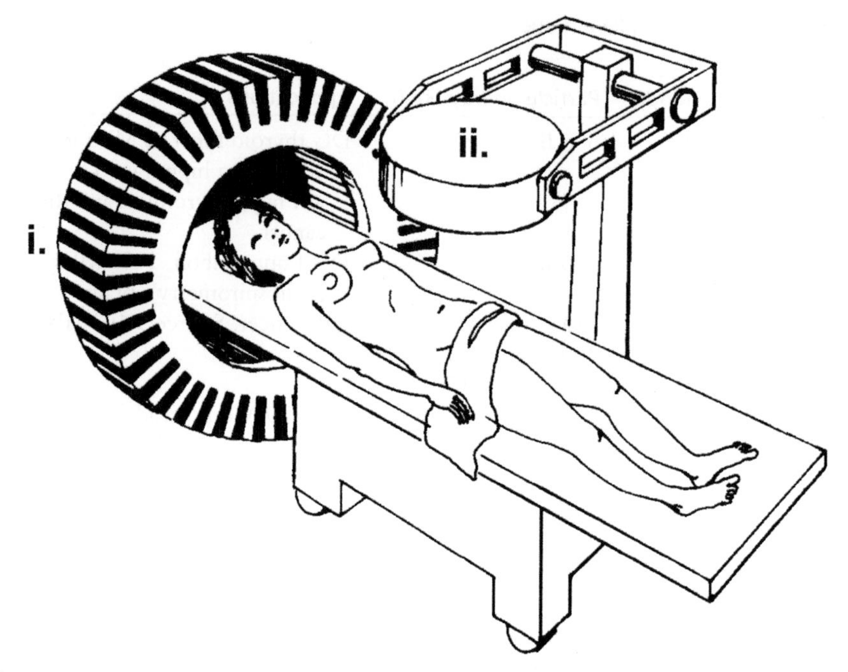

Figure 18.15 Geometry in (i) positron emission tomography (PET) and (ii) gamma-camera.

detectors 180° apart, positron decay must have taken place somewhere along the line between the two detectors. This is used for *positron emission tomography.* A positron emitter is administered to the patient positioned inside a ring or hexagon of 50–100 scintillation-PMT detectors (there is no need for collimators); Figure 18.15. The ring is moved in a translate-rotate pattern. The location of the radioisotope in the body is mapped in a way similar to that described for SPECT. The resolution of this technique is presently of the order of a few mm.

Positron emitters cannot be produced by n-irradiation: from Figure 5.8 it is seen that only charged particle irradiation (using ^{1}H, ^{2}H, ^{4}He, etc) can result in product nuclei on the proton rich side of the stability valley, for which positron emission is the main decay mode (competing with electron capture). Typical positron emitting nuclides used in PET are included in Table 18.5. They can be tagged to a variety of compounds.

For studying brain metabolism, ^{11}C-labelled glucose has been extensively used. The procedure is as follows:

1) $H_3\,^{11}BO_3$ is irradiated by protons in an accelerator, yielding $^{11}CO_2$.
2) Rapid automatic synthesis produces $^{11}C_6H_{12}O_6$ (glucose) or the methyl glucose derivative.

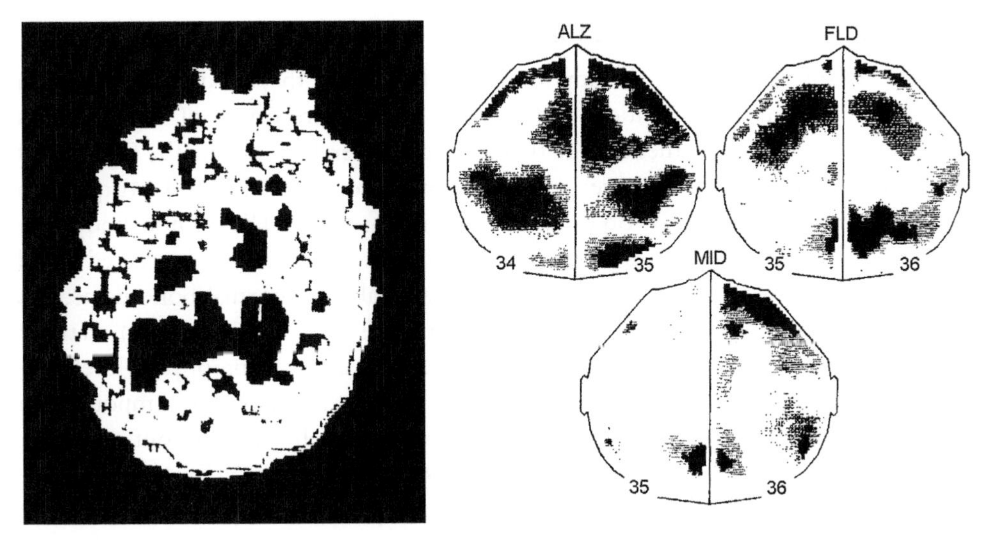

Figure 18.16 (a) A PET scan to identify a blood occlusion, (b) gamma-camera scans of subjects with ALZ, FLD and MID. *(Courtesy Kernforschungsanlage Jülich and Lund Psycho Geriatrics Dept.)*

3) The glucose solution is injected into a patient. Since it is easily metabolized, the glucose goes to the parts of the brain with the highest metabolism, rather than to places with no metabolism.

4) The ^{11}C nuclide decays according to $^{11}\mathrm{C} \rightarrow {}^{11}\mathrm{B} + e^+$ followed by $e^+ + e^- \rightarrow 2\gamma$. The 0.51 MeV $\gamma\gamma$ coincidences are registered by the PET cameras at various positions.

PET is used especially for studies of brain, heart and lungs. Figure 18.16, left, shows a brain investigation. Because glucose is the only source of energy the brain uses, the rate of glucose metabolism can be assessed throughout the brain, which is an indicator of the brain viability. In this case, the patient has been injected with ^{11}C-methyl glucose and the dark spot indicates an occlusion of the left carotid artery. The technique is quite fast: within a few hours the effect of administered drugs can be revealed.

Although TCT has now superseded other techniques in locating brain tumors, SPECT, gamma-camera and PET have provided dramatically new information of various forms of mental illness, such as epilepsy, manic depression, and dementias such as Alzheimer's disease. The development of new, selective radiopharmaceuticals will not only continue to increase the importance of this diagnostic technique but also contribute to our understanding of the functioning of the normal brain.

The use of positron emitters usually requires fast chemical separation and synthesis techniques. Fast chemical separation techniques for producing pure radioisotopes are described in Chapter 17, but for rapid chemical synthesis techniques, specialized texts should be consulted.

18.5.4. Radiation therapy with internal radionuclides

Radiotracers are also used for therapy though to less extent than in diagnosis (C3b in Table 18.3). The main application is the use of ^{131}I for treatment of thyreotoxicos (Graves disease with enlargement of the thyroid gland), thereby reducing the function of the thyroid. Some data for ^{131}I use: amount administered ca. 200–1000 MBq, organ dose to the thyroid ~340 nGy/Bq (total dose to the glands >500 Sv); whole body dose ~0.04 nGy/Bq. ^{32}P compounds are used in the bloodstream to destroy excess red blood cells in polycythemia. Diphosphonate compounds localize in normal bone tissue, but concentrate in cancerous bone at relatively higher levels (5–15 times higher). Patients receiving this type of radiopharmaceutical experience swift lowering of pain levels.

One of the most fascinating aspects of using radionuclide labelled compounds in medical therapy is to develop monoclonal antibodies, which will seek out particular types of cancer cells and bind to them. If a β-, or α-emitting radionuclide is attached to these antibodies, they will deliver a large absorbed dose to the cancer cells without needlessly irradiating the surrounding tissues. Much research is directed towards this goal of cancer therapy.

18.6. INDUSTRIAL USES OF RADIOTRACERS

Industry has applied radiotracers in a very large variety of ways. More than half of the 500 largest manufacturing concerns in the United States use radioisotopes in the production of metals, chemicals, plastics, pharmaceuticals, paper, rubber, clay and glass products, food, tobacco, textiles, and many other products. Radioisotopes are used to study mixing efficiency, effect of chamber geometry, residence time in reactors, flow rates and patterns in columns and towers for fractionation, absorption, racemization, etc. Some of the many uses are listed in Table 18.6 and a few are described below to reflect the scope and value of the industrial applications of radioisotopes. Quite often the radionuclide used is not isotopic with the system studied.

18.6.1. Mixing

Mixing is an important mechanical operation in many industries. Poor mixing may give an unsatisfactory product and low yield of the operation; unnecessary mixing is a waste of time and energy. By adding a radionuclide to the mixing vessel, or by labelling one of the components, the approach to mixing equilibrium can be followed either by external measurement or taking samples at different time intervals. Among examples of this technique are fluxing of cement, gravel, sand, and water to concrete measured by using irradiated pebbles. The homogeneity of glass melts can be determined by adding $^{24}NaHCO_3$ to the melt; organic compounds of ^{95}Zr ($\beta^-\gamma$ $t_{½}$ 64.02 d) have been used to follow homogenization of oil products. Other uses have been the addition of vitamins to flour, coal powder to rubber, water and gas to oil fields, etc.

Table 18.6 Radioisotope technique studies at ICI[†] undertaken in a typical year

Technique	**Number of applications**
Level and interface measurements:	
Gamma-ray absorption	210
Neutron backscatter	480
Gamma-ray backscatter	71
Blockage detection and deposition:	
Gamma-ray absorption	132
Neutron backscatter	129
Entrainment and voidage:	
Gamma-ray absorption	86
Thickness and corrosion measurements	15
Distillation-column scans	108
Flow measurements:	
Pulse velocity	483
Dilution techniques	84
Leak detection	90
Residence-time studies	21
Carryover studies (tracer)	6

The list is not fully comprehensive as less commonly used techniques have not been included.
[†]Imperial Chemical Industries (UK).

18.6.2. Liquid volumes and flows

The liquid volume of a closed system may be difficult to calculate from external container dimensions, particularly if there is a mixing action, either by external circulation or by internal stirring. For example, sulfuric acid volume is desired in an alkylation plant where extensive intermixing between acid and hydrocarbon prevents a well-defined level from forming. ^{137}Cs is added to the sulfuric acid and from the dilution of the added tracer the total volume of acid is calculated.

Another method is applicable to determining volumes of tanks through which there is a known constant flow. A tracer batch is put into the incoming line. Mathematical treatment, assuming complete immediate mixing of the incoming stream with the vessel contents, predicts that the tracer concentration in the tank falls off exponentially with a rate determined by throughput F and volume V according to the equation

$$R = R_0\, e^{-Ft/V} \qquad (18.37)$$

where R_0 is the counting rate at zero time (break through of radioactivity at tank exit) and t is the time when the count rate R is obtained at the same point.

The flow rate F of rivers and streams can be measured by injection of a radionuclide and measurement of the time for its arrival at detectors placed downstream. Because of turbulence the radioactive "cloud" becomes quite diffuse. Therefore a more efficient technique called "total count" is used. A known amount (A_0) of the radionuclide is injected into the river and a downstream detector registers a total count (R_{tot}) as the radioactivity passes; the faster it passes, the lower is the measured radioactivity. Thus

$$F = \psi A_0 / R_{tot} \tag{18.38}$$

ψ is the counting efficiency which has to be determined under known conditions. The technique takes into consideration both longitudinal and transversal mixing.

18.6.3. Wear and corrosion

Wear and materials transfer are easily followed if the material undergoing wear is made radioactive. This has been used for studying wear of parts in automobile engines, cutting tools, ball-bearings, furnace linings, paint abrasions, etc. In this case it is important that the surface undergoing wear has a high specific radioactivity. If the material cannot be tagged by adding (e.g. plating) a radionuclide, the material has to be activated by irradiation (e.g. by accelerator) to produce the radioactive species.

Corrosion in gas and oil pipe-lines on the sea floor is monitored by welding patches containing a series of long-lived radionuclides at different material depth into the pipe wall during construction of the pipe-line. Wall corrosion will remove one isotope after the other as it proceeds. The presence of these isotopes can be monitored from time to time by sending a capsule with a suitable detector and recorder through the pipe.

18.6.4. Chemical processing

We have described some areas of application of radiotracers which are useful to the chemical industry. To be more general, radioactive tracers are used to seek information of flow and mixing patterns for

- parameter evaluations (mass balance, flow rates, kinetics)
- fluid dynamics (trouble shooting, residence time distributions, modelling flow phenomena)
- chemical reaction engineering (effect of flow and mixing on conversions, modelling of reaction systems, process control algorithms).

Each application is rather case specific and described in the chemical engineering literature.

18.7. ENVIRONMENTAL APPLICATIONS

Nuclear weapons tests have released large amounts of radionuclides into the atmosphere, which through their own weight or by rain have been carried to the earth's surface. Geophysics has made use of this weapon "fallout". By measurements on T (as HTO water), ^{90}Sr, ^{137}Cs, and other fission products it has been possible to follow the movements of water from land via lakes and rivers into the sea, as well as to study the water streams of the oceans and the exchange between surface and deep water. As a result the circulation of water on earth has been mapped in quite detail. It has also been possible to analyze how tropical hurricanes are formed by measuring the water taken into the central part (the "eye") of the cyclone, since the HTO concentration in the normal atmosphere is different from that of surface ocean water due to isotopic effects.

Radioactive tracers like T_2O, $^{24}Na^+$, $^{82}Br^-$ ($\beta^-\gamma$ $t_{1/2}$ 1.471 d) and ^{51}Cr-EDTA (ECγ $t_{1/2}$ 27.704 d), are used in hydrology to determine the volume of natural water reserves (even underground) and to map the movement of ground and surface water as well as effluent pathways, Figure 18.17. Also the consumption of water and water flow in industries are readily determined, and leaking dams and pipes checked by radiotracers.

In order to avoid hazardous pollution it is important to discharge communal and industrial waste (silts, liquids, gases) so that the wastes are properly dispersed. The mapping of different dispersal sites is conveniently and commonly carried out by injecting a radioactive tracer at a testing spot, and then following its distribution at various depths, heights, and directions. In such a test it was found that sludge emptied into the River Thames at one point traveled upstream, which led to a repositioning of the sludge pipe exit. The simple technique of identifying hidden pipe leaks is illustrated in Figure 18.17 left. This technique is used not only for liquids, but also for checking the seal of underground electric cable hoses in which case gas tracers like ^{85}Kr ($\beta^-\gamma$ $t_{1/2}$ 10.72 y) or ^{133}Xe ($\beta^-\gamma$ $t_{1/2}$ 5.24 d) are used.

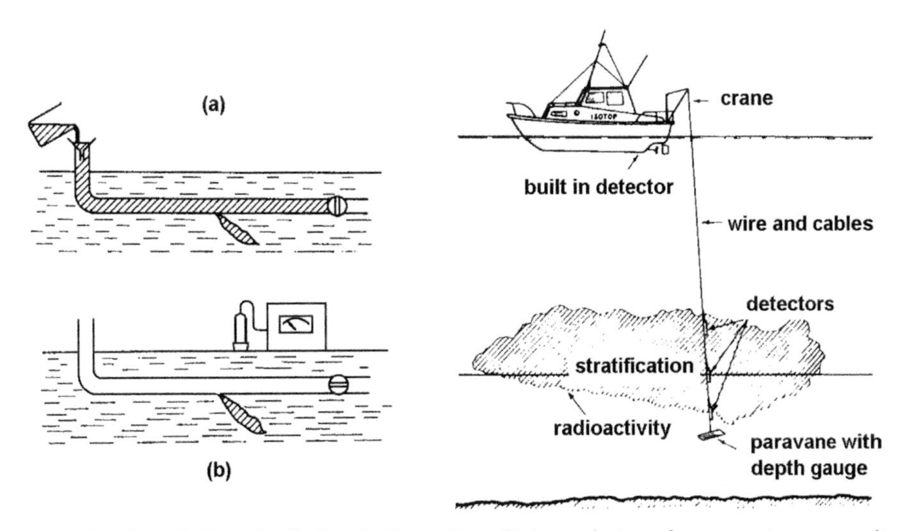

Figure 18.17 Left, technique for finding leaks in pipes. Right, technique for measuring stream lines in the strait of Öresund. *(Courtesy Danish Isotope Center.)*

Westermark and co-workers developed a unique method of revealing environmental history of natural rivers during the last 100–200 years. It is based on the freshwater pearl mussel *Margaritana margaritifera* living in these river. The mussel adds a new outer layer to its shell each year, incorporating into it trace metals (and also to some extent organics) dissolved in the water. The shell is cut and sectioned into small pieces, which are analyzed by α-track autoradiography, INAA and μ-PIXE. The latter is a "microprobe"-PIXE, i.e. the proton beam is focused by lenses to a resolution of a few microns. To get sufficient information the shell sample usually covers 3–15 years. With these techniques concentrations of some 30 elements have been determined retrospectively for >100 years in a number of Swedish rivers. Several unexplained trends have been observed, e.g. decreases in Ag, Au, Fe and Co. This is believed to be related to increasing air concentrations of SO_2. The observation of an increase in Br beginning in the 1940's is believed to be due to Br-additions to gasoline fuel. The technique has great potentials for increasing our knowledge on environmental changes of trace elements in nature. It could also be used as a long term biochemical control system at environmentally strategic places.

18.8. EXERCISES

18.1. The blood volume of a patient is to be determined by means of ^{32}P. For this purpose 15.0 ml of blood is withdrawn from the patient and mixed with a very small volume of $Na_2H^{32}PO_4$ of high specific activity. In 1 h the erythrocytes (red blood cells) take up all the ^{32}P; 1 ml is found to have an activity of 216 000 cpm in the detector system used. Exactly 5.00 ml of this tagged sample is reinjected into the patient, and 30 min later a new sample is withdrawn; 10 ml of this gives 2300 cpm. Calculate the blood volume.

18.2. A mixture of amino acids is to be assayed for cysteine. A 1.0 ml sample (density 1 g ml^{-1}) is withdrawn, and 2.61 mg of ^{35}S labelled cysteine of specific activity 0.862 μCi mg^{-1} is added. From this mixture pure cysteine is isolated by liquid-partition chromatography; 30.6 mg is isolated and measured to give 169 000 cpm in 27% detection efficiency. What is the percentage of cysteine in the original mixture?

18.3. In order to determine the lead content of a color pigment, 8.9871 g was dissolved in conc. HNO_3, and 5.00 ml ^{210}Pb solution added. After excess acid had been removed through evaporation, excess 1 M NaCl was added, the solution heated and filtered. After cooling and crystallization, the $PbCl_2$ was washed and recrystallized, 0.3276 g of the crystals was measured in a scintillation counter, giving 185 160 counts in 5 min. 1.000 ml of the original ^{210}Pb solution gave 57 000 cpm. The background was 362 cpm. Calculate the lead content (%) of the pigment.

18.4. What is the smallest amount of indium which can be determined in a 100 mg aluminum sample using NAA with a neutron flux of 10^{16} n m^{-2} s^{-1}? Consider the neutron capture in both ^{27}Al and ^{115}In: ^{27}Al(n,γ σ 0.230 b)^{28}Al $t_{½}$ 2.25 min; ^{115}In(n,γ σ 45 b)^{116}In $t_{½}$ 14 s. The lowest detectable activity for ^{116}In is assumed to be 10 Bq, and the interference from ^{28}Al not more than 20%.

18.5. In order to determine the amount of gallium in meteorite iron, 373.5 mg meteorite iron (A) and 10.32 mg gallium oxinate (B) were irradiated in a reactor under similar conditions in 30 min. After a short cooling, A was dissolved in concentrated HCl and 4.53 mg inactive Ga^{3+} was added. After a number of chemical separation steps, which were not quantitative, a precipitate of 25.13 mg pure gallium oxinate was isolated (C). Sample B was also dissolved and diluted to 50 ml; 0.50 ml was removed, 4 mg inactive Ga^{3+} added, and gallium oxinate precipitated (D). The radioactive decay curves gave two straight lines: log R_C = 3.401−0.0213 t, and log R_D = 3.445−0.0213 t. What was the gallium content in sample A?

18.6. A 10.0 g sample of iodobenzene is shaken with 100 ml of 1 M KI solution containing 2500 cpm ^{131}I. The activity of the iodobenzene at the end of 2 h is 250 cpm. What percent of the iodine atoms in the iodobenzene have exchanged with the iodide solution?

18.7. A sodium iodide solution contains some radioactive ^{129}I. An ethanol solution was prepared containing 0.135 M of this sodium iodide and 0.910 M inactive C_2H_5I. In the exchange reaction

$$C_2H_5I + {}^{129}I^- \underset{k_r}{\overset{k_f}{\rightleftarrows}} C_2H_5{}^{129}I + I^- \tag{18.39}$$

the reaction rate constant is assumed to be the same in both directions: $k_f = k_r$. One part (A) of the solution was removed and heated to high temperature so that equilibrium was rapidly reached. Another part (B) was kept in a thermostated bath at 30°C. After 50 min ethyl iodide was separated from both solutions. The concentration of radioactive iodine in C_2H_5I in B was found to be only 64.7% of that in A. Calculate k ($k_r = k\ a\ b$ in §18.4.2).

18.8. With Figure 18.9 one can estimate the stability constants (β_1 − β_3) for the lutetium acetylacetonate complexes. Make this estimation using Figure 18.9. A simplified approach to estimate β_n is the use of the approximate relations $k_n = \beta_n/\beta_{n-1} = 1/[\text{Aa}]_{\bar{n}=n-0.5}$ and $\bar{n} = z - \text{d}(\log D)/\text{d}(\log[\text{Aa}])$; for Lu^{3+} $n \leq 4$, but $z = 3$.

18.9. Calculate β_4 and the distribution constant using (18.32) and Figure 18.9.

18.10. One wants to determine the residual liquid volume of a closed sedimentation tank (nominal volume 80 m^3), which has been in use for many years, and in which $CaSO_4$ precipitates. 0.50 ml $^{24}Na_2SO_4$ (specific activity 3.2×10^8 cpm

ml^{-1}) is added to the tank, and 10 ml withdrawn after 2 h of settling; measurements yield a net value (background subtracted) of 500 counts in 10 min. Calculate the free liquid volume in the tank.

18.11. Calculate the critical deposition potential ($E - E^0$) for 10^{-12} M ^{210}Bi on a gold cathode (no over-voltage) from the Nernst equation (18.6), where the chemical activity of the reduced state (Bi^0) is set to unity.

18.12. A mineral ore contains cobalt and small amounts of nickel. In order to determine the nickel concentration it must be separated from cobalt. Solvent extraction using 0.01 M 8-hydroxyquinoline in $CHCl_3$ is chosen. Which metal should be extracted from the other, and at what pH? Consider Figure 18.3 and connected text.

18.13. In a solvent extraction system consisting of uranium and lanthanum in 1 M HNO_3 and 100% TBP, $D_U = 20$ and $D_{La} = 0.07$. If a phase ratio $\theta = V_{org} / V_{aq} = 0.5$ is chosen, how much uranium is removed from the aqueous phase in three repeated extractions? How much of the lanthanum is co-extracted? The fraction extracted with n fresh organic volumes (V_{org}) from one aqueous volume (V_{aq}) is:

$$E_n = 1 - (1 + D\theta)^{-n} \qquad (18.40)$$

18.9. LITERATURE

A. C. Wahl and N. A. Bonner (Eds.), *Radioactivity Applied to Chemistry*, Wiley, 1951 (still useful).

T. Braun and J. Tölgyessy, *Radiometric Titrations*, Pergamon Press, Oxford, 1967.

J. Ruzicka and I. Stary, *Substoichiometry in Radiochemical Analysis*, Pergamon Press, Oxford, 1968.

J. F. Duncan and G. B. Cook, *Isotopes in Chemistry*, Clarendon Press, 1968.

A. K. De, S. M. Khopkar and R. A. Chalmers, *Solvent Extraction of Metals*, van Nostrand Reinhold, 1970.

C. E. Crouthamel, F. Adams and R. Dams, *Applied Gamma-ray Spectrometry*, Pergamon Press, Oxford, 1970.

S. M. Quaim, Nuclear data relevant to cyclotron produced short-lived medical radioisotopes, *Radiochimica Acta* **30** (1982) 147.

J. S. Charlton and M. Polarski, Radioisotope techniques solve chemical process industry problems, *Chem. Eng.*, Jan. 24 (1983) 125.

M. F. L'Annunziata and J. O. Legg (Eds.), *Isotopes and Radiation in Agricultural Sciences*, Academic Press, London, 1984.

IAEA, *Radiopharmaceuticals and Labelled Compounds*, IAEA, Vienna, 1985.

P. F. Sharp, P. P. Dendy and W. I. Keyes, *Radionuclide Imaging Technique*, Academic Press, 1985.

B. Carell, S. Forberg, E. Grundelius, L. Henrikson, A. Johnels, U. Lindh, H. Mutvei, M. Olsson, K. Svärdström and T. Westermark, Can mussel shells reveal environmental history?, *Ambio*, **16** (1987) 2.

G. H. Simmons, *The Scintillation Camera*, The Society of Nuclear Medicine, New York, 1988.

C. E. Swenberg and J. J. Conklin, *Imaging Techniques in Biology and Medicine*, Academic Press, 1988, See especially Ch. 4 by R. R. Eng on *Radiotracer Methodology*.

S. Webb (Ed.), *The Physics of Medical Imaging*, Adam Hilger, 1988.

Y. Albinsson, Solvent extraction studies of lanthanide acetylacetones, *Acta Chem. Scand.*, **43** (1989) 9118.

M. Ljungberg, Diss., University of Lund, 1990.

A. H. W. Nias, *An introduction to radiobiology*, J. Wiley and Sons, 1990.
L. Yuanfang and W. Chuanchu, Radiolabelling of Monoclonal Antibodies with Metal Chelates, *Pure and Applied Chem.* **63** No 2 (1991) 427.
IAEA/UNESCO, *Isotope Techniques in Water and Resources Development*, Vienna, 1992.
C. Yonezawa, Prompt γ-Ray Analysis of Elements Using Cold and Thermal Reactor Neutrons, Analytical Sciences **9** (1993) 185.
J. Rydberg, M. Cox, C. Musikas, G. R. Choppin (Eds), *Solvent Extraction, Principles and Practice, 2:nd Ed.*, Marcel Dekker, New York 2004.

CHAPTER 19

Principles of Nuclear Power

Contents

From earliest times man has become increasingly dependent of a variety of energy sources: heat from burning of wood, animal dung, coal, oil, natural gas, etc., mechanical energy from steam engines, wind mills and water falls, and electricity from the same sources as well as from nuclear reactors. Much effort has been invested in searches for new energy sources, preferably of the "renewable" kind in contrast to the non-renewable sources[1]. Energy is a strategic resource, playing an important role in international and

[1] In a wider context, all energy originates from gravitational collapse, annihilation, fusion reactions or mass loss into black holes and its sources are thus never renewable.

Radiochemistry and Nuclear Chemistry
ISBN 978-0-12-405897-2, http://dx.doi.org/10.1016/B978-0-12-405897-2.00019-7

national politics. This chapter will describe the principles of current nuclear energy generation by fission and also briefly discuss the possibility to use fusion for the same purpose.

After the discovery by Hahn, Strassman, Meitner and Frisch in 1938–1939 that neutrons induced fission in uranium, and that the number of neutrons released in fission was greater than one, many scientists realized that it should be possible to build a chain-reacting system in which large amounts of nuclear energy were released, *a nuclear reactor.* The first such system was constructed in Chicago in the early 1940s under the scientific leadership of Fermi and became critical on December 2, 1942, as part of the World War II Manhattan Project. Since that time many hundreds of nuclear reactors have been built throughout the world, mostly for power production.

In mid 2012, 433 nuclear power reactors (372 GW_e)[2] were in operation; 122 (114 GW_e) in North America including the US, 6 (~4 GW_e) in South America, 2 (~2 GW_e) in Africa, 186 (~162 GW_e) in Europe, including the Russian federation, and 119 (~88 GW_e) in the Far East (mainly Japan). Presently 62 power reactors (~59 GW_e) are under construction, the majority in the Far East and most of the rest in Europe. As small old nuclear power plants are shut down and replaced by bigger new ones, the number of reactors may remain constant in a geographic area, or even decrease slightly, although the total installed nuclear generating capacity often continues to increase.

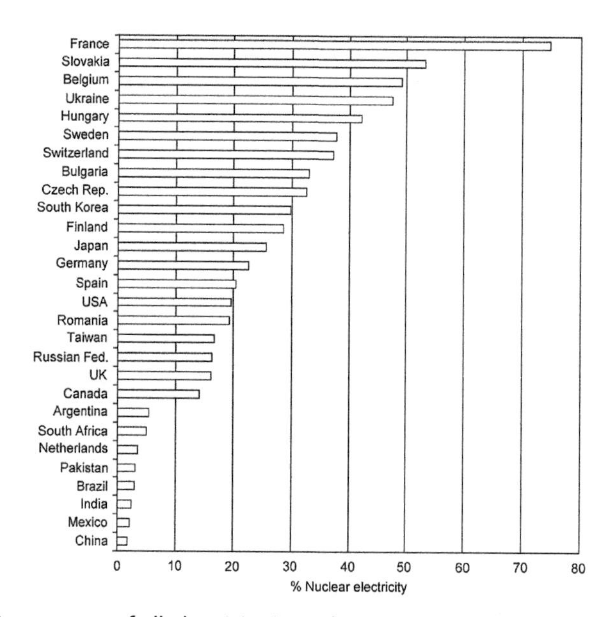

Figure 19.1 Relative amount of all electricity in each country generated by nuclear power stations during 2010.

[2] GW_e = gigawatt electric power installed.

Figure 19.1 shows the fraction of all electricity generated by nuclear power stations in different countries. In France any further large increase in nuclear power production capacity is unlikely as this could lead to a generating capacity in excess of demand. Furthermore, opposition to the peaceful use of nuclear power has led to revised plans for the use of nuclear power in many countries in the West (cf. Ch. 22). In some cases earlier expansion plans have been replaced by moratoria on new plants, by a decision to shut-down some or all operating nuclear power stations (Germany, Italy, and Lithuania), or by plans to force the utilities to shut down all nuclear power stations at some future date (Germany).

Nuclear chemistry plays an essential part in achieving safety and reliability in this source of power. Nuclear chemists and engineers are responsible for much of the nuclear fuel cycle, from uranium ore processing to ultimate disposal of radioactive waste.

In the future fusion power may also contribute to the energy supply. However, not all problems have been overcome to date, see §19.17.

19.1. THE NUCLEAR REACTOR

The fission process with thermal neutrons can be summarized as follows

$$^{235}U + n \rightarrow FP + \nu n \tag{19.1}$$

For each neutron consumed on the average ~2.5 (= ν) new neutrons are released. The new neutrons can be used to fission other ^{235}U nuclei leading to the release of even more neutrons forming a *nuclear chain reacting system*. The kinetic energy of the fission fragments is promptly converted to heat through collisions with other atoms. In a *nuclear reactor* the nuclear chain reactions are controlled so that an equilibrium state is reached, where for each fission exactly one of the new neutrons is used for further fission. Under these conditions the *neutron multiplication factor k* is said to be 1. If the factor is higher the number of neutrons, and consequently the fission rate, increases exponentially. Without any control mechanism the heat evolved would ultimately destroy the chain-reacting system. This is extremely improbable in conventional reactors because of various active and passive control mechanisms. No such control mechanisms are present in *nuclear explosive devices* and, instead, every effort is made to make the reaction as violent as possible.

The neutron transport equation describes the interaction of neutrons with their environment. The accurate design of a chain reacting system requires solution of this equation. The complete solution yields the neutron flux and neutron spectrum as function of the space coordinates and time. As there is no known analytical solution to this equation for a real chain reacting system, numerical methods or various approximations must be used. In this text we do not consider numerical solution methods. We illustrate the general properties of chain reacting systems of interest to the nuclear chemist by using some approximate solutions assuming that the neutron flux and its energy spectrum is known.

There is quite a flexibility in the design of a controlled nuclear chain reacting system. Because each concept has its advantages and drawbacks, more than a dozen different types of nuclear reactors have been developed and tested. However, only a few types are common. About 79% of the currently operating nuclear power plants are of the *light water reactor* type (LWR), i.e. *pressurized water reactors* (PWR)[3] or *boiling water reactors* (BWR); cf. Table 19.1. We therefore begin our discussion of reactors by describing a LWR station with emphasis on the principles and main components. Figure 19.2 shows a schematic diagram of a PWR and a picture of operating LWRs, three PWRs and one BWR (highest chimney).

The principal component in any reactor is the *core* which contains the fissionable *fuel material*. This is usually UO_2 enriched in ^{235}U to 2–4% shaped as short cylinders, *pellets*, which are stacked in zirconium alloy tubes (the *can*, *canning* or *cladding*) forming *fuel rods* or *fuel pins*. The fuel rods are mounted together in clusters forming *fuel elements* or *fuel assemblies*.

When fission occurs in the nuclear fuel practically all of the immediately released fission energy occurs as kinetic energy of fission fragments and neutrons (cf. Table 11.1 and §19.2). The kinetic energy is rapidly converted into heat. In order to maintain a stable operating temperature in the reactor it is necessary to provide a *coolant*. The coolant material, which is water in LWRs, enters the core from below, is heated while passing upwards along the fuel rods and leaves the core at the top as either hot water at high pressure (in the *pressurized water reactor*, PWR, as indicated in Figure 19.2) or as a

Table 19.1 Power reactor types in common use

Reactor type and abbreviation	Number of reactors (%)	Generating capacity (% of nuclear capacity)
Pressurized Water Reactor (PWR)	62.6	67.6
Boiling Water Reactor (BWR)	18.7	19.7
Pressurized Heavy Water Reactor (PHWR)	10.3	6.1
Pressure Tube Boiling Water Reactor (RBMK)	2.5	2.7
Advanced Gas-cooled Reactor (AGR)	3.2	2.3
Advanced Boiling Water Reactor (ABWR)	0.7	1.1
Magnox Reactor (GCR)	0.7	0.3
Fast Breeder Reactor (FBR)	0.5	0.2
Electricity Generating District Heating Reactor (EGP)	0.9	~0.01

Data refer to mars 2012.

[3] Until Ch.20 we will not discuss other existing pressurized water (or boiling water) reactor designs.

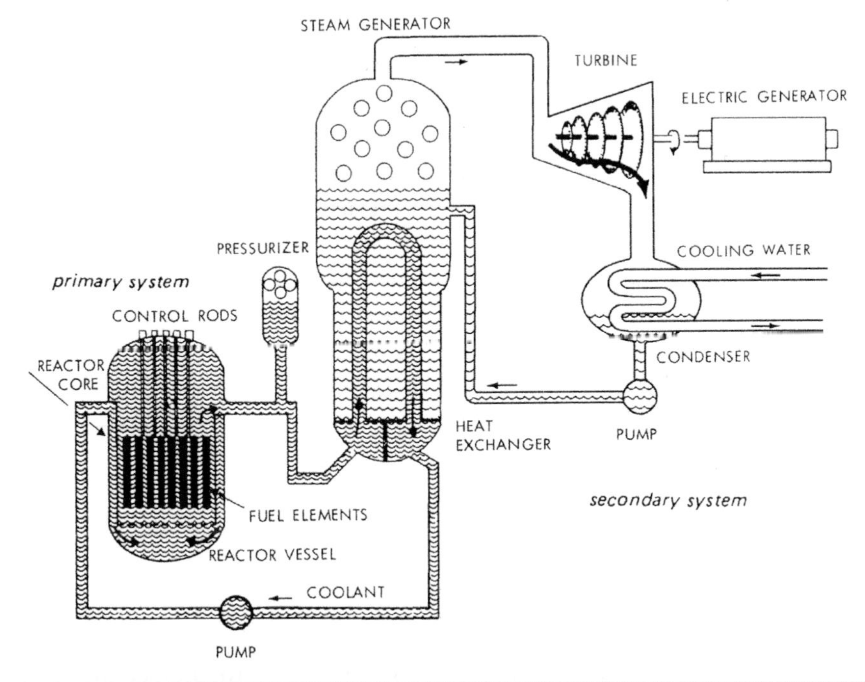

Figure 19.2 Main components of a pressurized light water cooled and -moderated nuclear power reactor (PWR) and a view of the Ringhals plant (Sweden) with 3 PWRs and 1 BWR.

high pressure high temperature steam – water mixture (in the *boiling water reactor*, BWR). The following discussion will concentrate on the operation of a PWR because it is the most common reactor type.

In the PWR steam is generated at lower pressure on the secondary side of separate heat exchangers, *steam generators*. The steam drives one or more turbines connected to electric generators and is condensed to water by an *external cooling system* after the turbine. Except for the external cooling system (sea or river water, cooling towers, etc.), the steam-water flow systems are closed cycles.

The neutron flux, and hence the rate of the fission chain reaction, in the reactor core is controlled by movable *control rods* which contain material with high cross sections for absorption of neutrons and, in a PWR, also by an absorber dissolved in the coolant.

In order to maximize the cross section for fission, which is greatest for low energy neutrons, the neutrons are slowed down or "moderated" by a material (the *moderator*) that elastically scatters neutrons but has a small neutron capture cross section. In LWRs ordinary (but very pure) water serves the purpose of both moderation and cooling (in other reactor types the moderator may be a liquid like D_2O, a solid material like graphite or absent and the cooling medium may be a gas, e.g., helium or carbon dioxide or a metal like lead, mercury or sodium).

Reactors of this kind, in which the fuel is physically separated from the moderator, are said to be of the *heterogeneous type*, while in *homogeneous reactors* the fuel is directly dissolved in the moderator material. Homogeneous reactors have only been built for experimental purposes.

The reactor core system is enclosed in a stainless steel clad pressure vessel of high strength steel, thermally insulated on the outside. In order to protect the operating personnel against hazards from the neutrons and γ-rays emitted in fission, the pressure vessel is surrounded by a thick *biological shield*. The reactor system and steam generators are normally enclosed in a *containment vessel*. In case of a potentially dangerous malfunction in the reactor the containment is sealed off, thus presenting a barrier against escape of radioactive material to the surroundings. Air and water effluents from the reactor station during normal operation are monitored and (when prescribed activity levels are exceeded) purified from radioactive contaminants.

The fuel elements can only be used to much less than 100% consumption of the ^{235}U originally present. Fission leads to the production of fission products, which accumulate in the fuel. Some of the fission products have very high neutron capture cross sections ($\gg$100 b) and compete with the fission chain reaction for the neutrons. Before the reactor becomes "poisoned" by these fission products, part of the fuel elements has to be replaced. In case of LWR:s the reactor is shut down, the top of the pressure vessel removed and some of the older fuel elements replaced by new ones. The replacement is done by means of a fuel *charging* (discharging) *machine* (a few other reactor types permit fuel replacement during operation). Because of the large amounts of radioactive fission products, the used fuel elements are always allowed to "cool" (with respect to both radioactivity and heat) for months, often years, in water-filled *storage pools* located inside the containment (see Fig. 19.16).

The used fuel elements may later be *reprocessed* to recover the remaining amount of fissile material as well as any *fertile material* or regarded as waste; fertile atoms are those which can be transformed into fissile ones, i.e. ^{232}Th and ^{238}U, which through neutron capture and β-decays form fissile ^{233}U and ^{239}Pu, respectively. The chemical reprocessing removes the fission products and actinides other than U and Pu. Some of the removed elements might be valuable enough to be isolated although this is seldom done. The mixed fission products and waste actinides are stored as *radioactive waste.* The recovered fissile materials may be re-fabricated (the U may require re-enrichment) into new elements for reuse. This "back-end" of the nuclear fuel cycle is discussed in Chapter 21.

19.2. ENERGY RELEASE IN FISSION

From Table 11.1 it is seen that in thermal neutron fission of ^{235}U the fission fragments are released with a kinetic energy of ~165 MeV (on the average), the 2.5 *prompt neutrons* have an average kinetic energy of ~5 MeV together, and the *prompt γ-rays* have an average of 7 MeV. This *prompt energy release* of ~177 MeV is absorbed in the surrounding material.

The fission products are radioactive and decay through emission of β^-, γ, and X-rays; their total amount of decay energy is ~23 MeV. About 10 MeV (the value is uncertain) escapes the reactor as radiation, and ~1 MeV of the decay energy remains as undecayed fission products in the spent fuel when unloaded from the reactor; thus ~12 MeV βγ decay energy (divided about equally between $\bar{E}_\beta$ and E_γ) is absorbed in the reactor. The neutrons not consumed by fission are captured in the reactor material with release of binding energy; it is estimated that this amounts to about 10 MeV. Thus the total amount of energy expended per fission in a shielded controlled reactor is about $177 + 12 + 10 \simeq 199$ MeV. The total energy release varies only a few MeV between the different fissile nuclei. As a practical average value 200 MeV per fission can be used regardless of the fissioning nucleus:

$$E_f \approx 200 \text{ MeV per fission} = 3.20 \times 10^{-11} \text{ J per fission} \quad (19.2)$$

Thus $(3.20 \times 10^{-11})^{-1} \approx 3.1 \times 10^{10}$ fissions s^{-1} correspond to the production of 1 W of reactor heat. The heat power of a reactor can be written

$$P = E_f \mathrm{d}N_f/\mathrm{d}t \quad (19.3)$$

where $\mathrm{d}N_f/\mathrm{d}t$ is the number of fissions per second. A nuclear power station producing 3 GW heat (GW$_{th}$; th for thermal) has an electric output of 1 GW$_e$ at a 33% *efficiency* in converting the thermal energy into electric; this corresponds to 8.1×10^{24} fissions d^{-1}. Since the weight of a ^{235}U atom is $M/\mathbf{N}_A = 3.90 \times 10^{-25}$ kg, this would correspond to the fission of $8.1 \times 10^{24} \times 3.9 \times 10^{-25} = 3.16$ kg ^{235}U d^{-1}; the real consumption of ^{235}U is slightly different and varies with time as discussed in §19.9.

19.3. FISSION PROBABILITY

When uranium or plutonium is irradiated by neutrons, in addition to neutron capture followed by fission (n,f), several different processes occur: scattering, (n,γ), (n,2n) reactions, etc. All these reactions are of importance for the reactor designer as well as for the chemists who have to manufacture new fuel pellets and take care of the spent fuel elements. The probability for the various reactions depends on the neutron energy. As is seen in Figure 19.3, three regions are clearly distinguishable:

(1) For thermal neutrons with average kinetic energies ($\bar{E}_n$) $\leq$1 eV, fission of ^{235}U and ^{239}Pu dominates over neutron capture (i.e. $\sigma_f > \sigma_{n,\gamma}$). Though $\sigma_{n,\gamma}$ for ^{238}U is small considerable capture occurs because of the large fraction of ^{238}U normally present.

(2) For epithermal neutrons ($1 \leq \bar{E}_n \leq 10^5$ eV) large radiative capture and fission resonances occur. In this region heavier isotopes are formed by (n,γ) reactions: ^{235}U(n,γ)^{236}U; ^{238}U(n,γ)^{239}U; ^{239}Pu(n,γ)^{240}Pu; ^{240}Pu(n,γ)^{241}Pu; etc (see Fig. 19.5).

(3) For fast neutrons ($\bar{E}_n \geq 0.1$ MeV) the cross sections are relatively small, $\leq$1 b. Fission dominates over radiative capture. Of particular importance is that ^{238}U becomes fissionable at a neutron energy of ~0.6 MeV; its fission cross section increases with neutron energy above the threshold to a constant value of ~0.5 b at $\geq$2 MeV.

It is obvious that the neutron energy spectrum of a reactor plays an essential role. Figure 19.4 shows the prompt (unmoderated) fission neutron spectrum with $\bar{E}_n$ ~2 MeV. In a nuclear explosive device almost all fission is caused by fast neutrons.

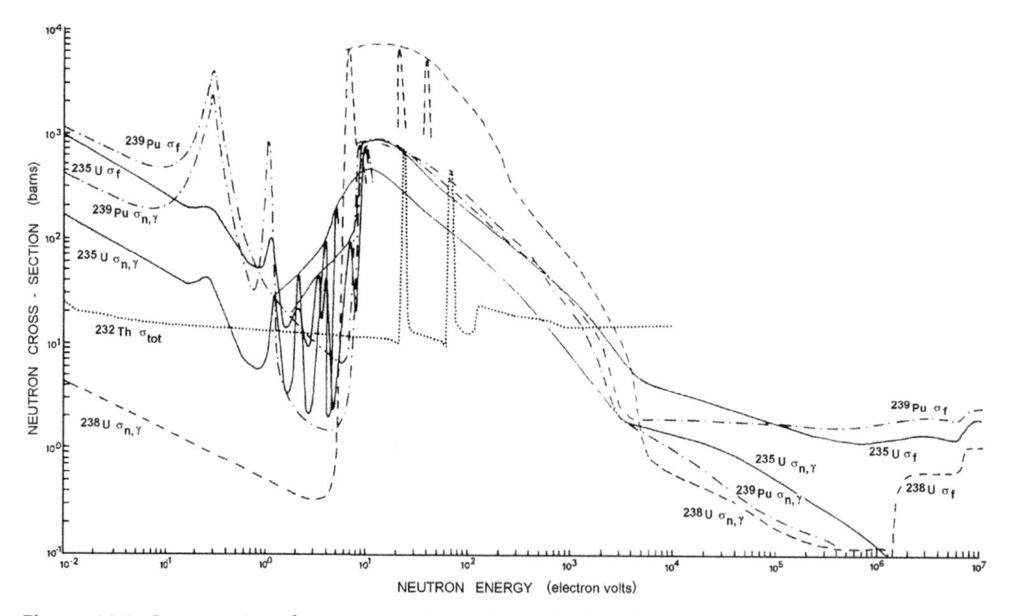

Figure 19.3 Cross sections for n-capture ($\sigma_{n,\gamma}$), fission (σ_f), and total (σ_{tot}) as a function of neutron energy.

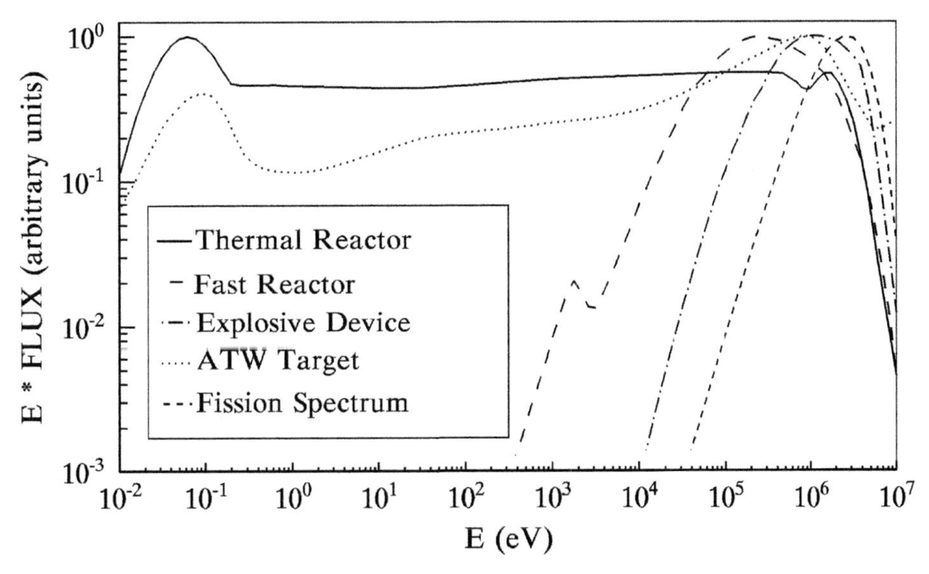

Figure 19.4 Neutron spectra of some chain reacting systems and a proposed accelerator driven system. The abscissa shows neutron flux times energy on an arbitrary scale.

Nuclear reactors can be designed so that fission mainly occurs with fast neutrons or mainly with slow neutrons (by slowing the neutrons to thermal energies before they encounter fuel). This leads to two different reactor concepts — the *fast reactor* and the *thermal reactor*. The approximate neutron spectra for both reactor types are shown in Figure 19.4. Because thermal reactors are more important at present, we discuss this type of reactors first.

In an ideal thermal reactor we may assume that most of the neutrons are in thermal equilibrium with the moderator atoms, though in practice the neutron spectrum in power reactors is much more energetic due to incomplete moderation. From $\bar{E}_n = \mathbf{k}T$ it follows that at about 300°C (a typical moderator temperature in a LWR) $\bar{E}_n$ is 0.05 eV; cf. Fig. 19.4. However, "thermal neutron energy" (E_{th}) cross sections are standardized to mono-energetic neutrons with a velocity of 2200 m s^{-1}, which corresponds to E_n 0.025 eV according to (3.32). Cross sections at this energy are given in Table 19.2. For most (but not all) nuclides the cross section at neutron energies below the resonance region may be estimated by the $1/v$ law (§11.4). Because thermal reactors have neutrons of more than one energy, *effective cross sections*[4] must be used for calculating reactor product yields. Such cross sections are usually calculated as an average over the core and vary with time as well as from reactor to reactor. Typical LWR effective cross sections are given in Table 19.2 and Figure 19.5 for thin targets. Self screening (depression of the n-flux due to a

[4] The effective cross-section is given by $\sigma_{eff} = \int \sigma(E)\ \varphi(E)\ dE / \int \varphi(E)\ dE$, where σ is cross-section, φ flux and E neutron energy.

Table 19.2 Nuclear data for some fissile and fertile nuclides (see also Figs 19.3, 19.5 and 20.5)

	^{232}Th	^{233}U	^{235}U	^{238}U	Nat. U	^{239}Pu	^{240}Pu	^{241}Pu
Radioactive decay	α,sf	α	α	α,sf	—	α,sf	α,sf	α,sf
Half-life (years)	1.405×10^{10}	1.592×10^{5}	7.038×10^{8}	4.466×10^{9}	—	2.411×10^{4}	6.563×10^{3}	14.35
Specific rad. act. (GBq/kg)	4.06×10^{-3}	356	0.0800	0.01244	—	2296	8398	3.83×10^{6}
Specific decay heat (W/kg)	0.003	275	0.056	0.008	—	1876	6950	3250
Thermal neutrons (0.025 eV)								
n,γ capture (σ_γ barns)	7.40	47.7	98.6	2.70	3.39	268.8	289.5	368
fission (σ_f barns)	39 μb	531.1	582.2	<0.5 mb	4.19	742.5	0.030	1009
neutron yield (ν)		2.492	2.418		2.418	2.871	2.90 ?	2.927
fission factor (η)		2.287	2.068		1.34	2.108	0.0003 ?	2.145
Fast neutrons (~0.25; ~1.0 MeV)								
n,γ-capture (σ_γ barns)	0.18; 0.12	0.22; 0.056	0.28; 0.11	0.15; 0.14	0.15; 0.14	0.17; 0.04	0.15; 0.18	2.0; 2.0
fission (σ_f barns)	; 0.067	2.20; 1.90	1.28; 1.19	; 0.018	0.01; 0.03	1.53; 1.73	0.10; 1.06	1.78; 1.56
neutron yield (ν)	; 2.1	2.503; 2.595	2.426; 2.522	; 2.50	2.426; 2.50	2.886; 3.001	2.80; 2.95	3.01; 3.12
fission factor (η)	; 0.75	2.28; 2.52	1.99; 2.31	; 0.28	0.15; 0.44	2.60; 2.93	1.13; 2.52	1.4; 1.4
Reactor spectrum neutrons (~PWR; LMFBR†)								
n,γ-capture (σ_γ barns)	4; 0.30	28; 0.23	107.5; 0.584	1.5; 0.271		559.5; 0.547	2616.8; 0.52	3428.7; 0.568
fission (σ_f barns)	; 0.01	298; 2.52	448.1; 1.951	; 0.040		1048.7; 1.81	; 0.371	1138.6; 2.64
neutron yield (ν)	; 2.4	2.49; 2.6	2.43; 2.445	; 2.900		2.87; 2.922	; 2.976	3.06; 2.968
fission factor (η)	; 0.08	2.27; 2.4	1.96; 1.882	; 0.37		1.86; 2.24	; 1.235	2.22; 2.443

†Creys-Malville central core.

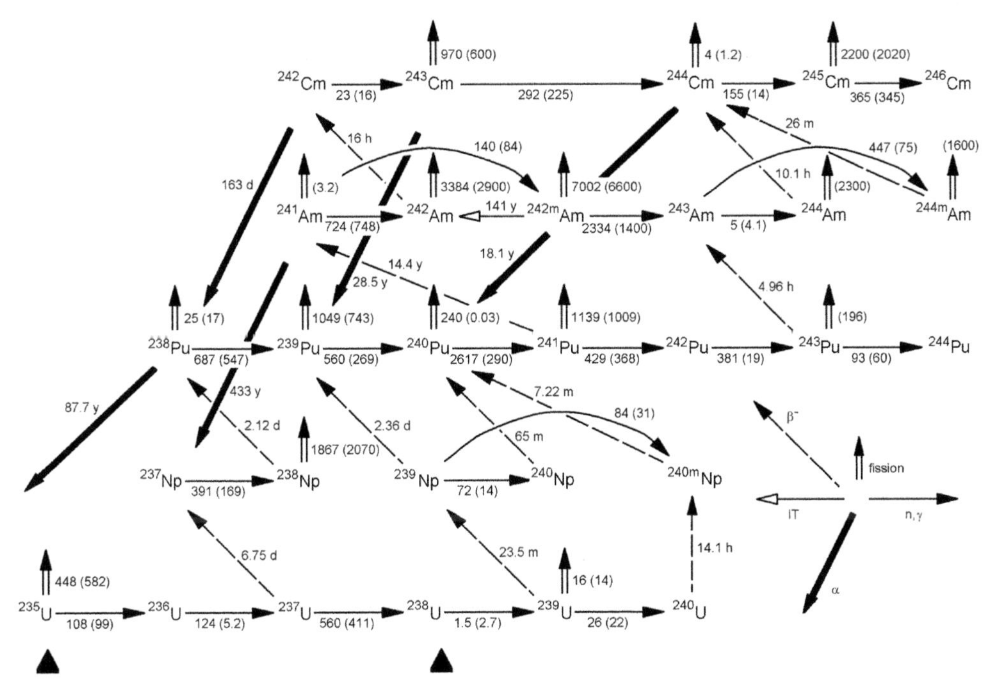

Figure 19.5 Nuclear reactions in irradiation of uranium. Figures along the arrows are half lives or effective cross-sections (b) for a typical power LWR, with thermal (0.025 eV) data within parentheses.

high Σ_a) reduces the effective cross sections below those given in Table 19.2 and Figure 19.5 for isotopes with high concentration in the fuel pins.

19.4. THE FISSION FACTOR

In thermal reactors ^{235}U is consumed mainly by fission and radiative capture

$$^{235}\mathrm{U} + \mathrm{n_{th}} \left[\begin{array}{ll} \xrightarrow[582\ \mathrm{b}]{\sigma_{n,f}} & 2\ \mathrm{FP} + \nu\mathrm{n} \\ \xrightarrow[99\ \mathrm{b}]{\sigma_{n,\gamma}} & {}^{236}\mathrm{U} \\ & \downarrow\alpha \end{array} \right. \tag{19.4}$$

^{236}U decays through α-emission. Due to its long half-life (2.3×10^7 y) it accumulates in the reactor. ^{236}U may capture a neutron, forming ^{237}U, which over a few days time decays to ^{237}Np. The *neutron yield per fission*, ν, is an average value which depends on the neutron energy; it is 2.42 for thermal fission of ^{235}U (see Table 19.2).

If we define the ratio

$$\alpha = \sigma_\gamma/\sigma_f = \Sigma_\gamma/\Sigma_f \tag{19.5}$$

(cf. (11.6)) the probability that the captured neutron gives rise to fission is $\sigma_f/(\sigma_f + \sigma_\gamma) = 1/(1 + \alpha)$. The number of neutrons produced for each neutron captured is

$$\eta = \nu/(1 + \alpha) \tag{19.6}$$

where η is the *neutron yield per absorption*, also called the *fission factor*. A primary requirement for a chain reaction is that $\eta > 1$. Table 19.2 contains η-values for the most important nuclides. At thermal energies η is highest for ^{233}U, while for fast neutrons η is highest for ^{239}Pu.

In mixtures of isotopes the macroscopic cross section Σ must be used in calculating α. Since for natural uranium $\Sigma_f = 0.72 \times 582$ per 100 and $\Sigma_\gamma = (0.72 \times 99 + 99.3 \times 2.70)$ per 100, one obtains $\alpha = 0.81$ and $\eta = 1.34$, which means a chain reaction is possible with thermal neutrons in natural uranium.

19.5. NEUTRON MODERATION

The mode of moderation of the neutrons is one of the crucial design features of a thermal reactor. The fast fission neutrons lose their kinetic energy through elastic scattering with the atoms of the moderator and construction material. In §10.5 the equations are given for the energy change in such collision processes, but only for head-on collisions. Because most collisions involve angular scattering, the number of collisions required to reduce a fast neutron to thermal energy is larger. The *average logarithmic energy decrement* is given by

$$\xi = \overline{\ln(E_n/E'_n)} = 1 + \left[(A-1)^2/(2A)\right]\ln[(A-1)/(A+1)] \tag{19.7}$$

where E'_n is the neutron energy after collision with a moderator atom of mass number A. From Table 19.3 it is seen that "light" water (i.e. H_2O) is most effective in reducing the

Table 19.3 Physical constants of some moderator materials

Property	H_2O	D_2O	Be	C
$N_0 \times 10^{-30}$ (atoms m^{-3})	0.0333	0.0332	0.123	0.0837
ρ (kg m^{-3})	997	1104	1848	1670
σ_a(th) (barns)	0.66	0.0013	0.0076	0.0035
Σ_a(th) (m^{-1})	2.2	0.0044	0.094	0.029
σ_s(epith) (barns)	49	10.6	5.9	4.7
ξ	0.927	0.510	0.206	0.158
$\xi \times \Sigma_s$(epith)/Σ_a(th)	68	4095	160	211
L^2_m(m^2)	0.000713	1.51	0.043	0.287
τ (m^2)	0.0031	0.0125	0.0097	0.0349

neutron velocity. The average number of collisions n required to reduce the neutron energy from E_n^0 to E_n is given by

$$n = \xi^{-1} \ln(E_n^0/E_n) + 1 \quad (19.8)$$

The *slowing down power* (*SDP*) of a moderator depends in addition on the neutron scattering cross section and number of scattering atoms per unit volume (N_0):

$$SDP = \xi \, N_0 \, \sigma_s = \xi \, \Sigma_s \quad (19.9)$$

The *SDP* is an average value over the epithermal neutron energy region. A good neutron moderator should divert few neutrons from the fission process, i.e. the neutron absorption cross section must be small. In this respect both heavy water (i.e. D_2O) and carbon are superior to H_2O, since for H_2O the reaction probability for $^1H(n,\gamma)^2H$ is relatively large (Table 19.3). In order to incorporate this property, the concept *moderating ratio* (*MR*) is used as a criterion of the moderator property in the thermal neutron energy region:

$$MR = SDP/\Sigma_a = \xi \, \Sigma_s/\Sigma_a \quad (19.10)$$

The moderating properties are summarized in Table 19.3. The moderator qualities decrease in the order $D_2O > C > Be > H_2O$. In a commercial reactor, price and other properties must also be considered, so H_2O is favoured over D_2O in most reactors.

19.6. THE NEUTRON CYCLE

In order for a chain reaction to occur at least one of the neutrons released in fission must produce a new fission event. This condition is defined by the *multiplication factor k*:

$$k = \frac{\text{number of neutrons in generation 2}}{\text{number of neutrons in generation 1}} \quad (19.11)$$

when $k > 1$ the number of neutrons in the second generation exceeds the number of neutrons consumed. Under this condition, the neutron flux, and, thus, the number of fission events increase for each successive neutron generation with a resultant increase in the power production of the reactor (§19.8 discusses further the *reactor kinetics*). If $k = 1$ the number of fissions per unit time, and thus the energy production, is constant. For $k < 1$ the chain reaction cannot be maintained. A reactor operating with $k = 1$ is said to be *critical*, $k > 1$ *supercritical*, and $k < 1$ *subcritical*. k is regulated by means of the control rods and in PWRs also by the n-absorber concentration in the water (usually boric acid).

In any neutron generation the neutrons experience a variety of fates (Fig. 19.6). Some neutrons escape the reactor and some are absorbed in the reactor structural materials and shielding (i.e. fuel cans, control rods, moderator, coolant, etc.). To take this loss into

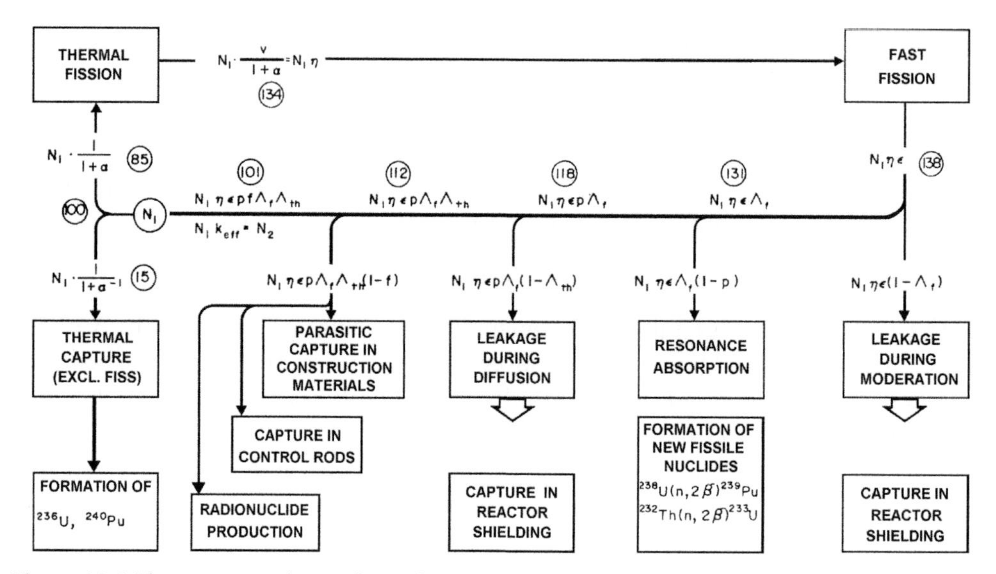

Figure 19.6 The neutron cycle in a thermal reactor. The number of neutrons for 100 in first generation are given in circles. *(data from an old Belgian research reactor, BR1)*

account, two different multiplication factors are used: k_∞ refers to a reactor of infinite dimensions (i.e. no leakage) while k_{eff} refers to a reactor of a finite size:

$$k_{\text{eff}} = k_\infty \Lambda \tag{19.12}$$

Λ is the fraction of neutrons which are *not* lost through leakage to the surroundings (*the non-leakage factor*). In order to minimize the neutron leakage, the reactor core is surrounded with a *neutron reflector* which for thermal neutrons in LWRs is water (graphite or beryllium are sometimes used in other reactor designs); for fast neutron reflection iron is frequently used.

Let us consider the neutron cycle in a reactor, i.e. the fate of the neutrons as they proceed from one generation to the next (Figure 19.6, where leakage, treated in §19.7, also is included). We start by assuming that we have an air cooled reactor containing natural uranium with a graphite moderator in a configuration which allows negligible leakage ($\Lambda = 1$). If 100 neutrons (N_1) are captured by the uranium fuel, the fission releases $N_1\ \eta$ or $100 \times 1.34 = 134$ new neutrons. The fast neutrons may also cause some fission in nearby uranium atoms, thereby releasing more neutrons. This is measured by ε, *the fast fission factor.* Its value in a heterogeneous nuclear reactor depends to a large extent on the moderator and on the dimensions within a fuel element. For natural uranium in graphite ε is about 1.03, so that the 134 neutrons released in fission by the original 100 thermal neutrons are increased to $N_1\ \eta\ \varepsilon$ (or $1.03 \times 134 = 138$) neutrons. In homogeneous reactors $\varepsilon = 1$ because the fission neutrons do not have to pass through any zone with concentrated fuel to reach the moderator.

These energetic neutrons are slowed down through collisions with surrounding atoms and decrease steadily in energy until thermal energies are reached. In the energy range 1 to 10^5 eV, *the epithermal region*, ^{235}U and ^{238}U have large resonance peaks for radiative capture, while fission in ^{235}U dominates at thermal energies (Fig. 19.3). In order to maximize the fission probability, the losses through radiative capture in the epithermal region must be minimized. This can be achieved by physically separating moderator and fuel. The fast neutrons, which easily escape from thin fuel pins, are mainly slowed in the surrounding moderator. Because the moderator is of a material with light atoms, the number of collisions required to slow these neutrons is small. Therefore, it is unlikely that the neutrons encounter a fuel pin (and are captured) until thermal energies have been reached. The probability that the neutrons pass through the energy region of the resonance peaks without capture is called the *resonance escape probability* and denoted by p. From the original N_1 neutrons at this point there would be $N_1\ \eta\ \varepsilon\ p$ neutrons. In a reactor of natural uranium and graphite, p is usually around 0.9. Therefore, from the original 100 neutrons 138 × 0.9 or 124 second generation neutrons reach the thermal energy range.

The cross sections for neutron capture increase for all atoms for thermal energy neutrons. As a result, even though low cross section materials are used some neutrons are captured by the structural and moderator materials. The probability for the non-capture of thermal neutrons in this fashion is signified by f, *the thermal utilization factor*, which in our case can be assumed to be ~0.9. Thus of the original N_1 neutrons 112 thermal neutrons remain in the second generation to cause fission in the nuclear fuel.

These 112 thermal neutrons constitute the second generation of neutrons which according to our definition of neutron multiplication factor is k_∞ so that

$$k_\infty = \eta\ \varepsilon\ p f \tag{19.13}$$

This relation is known as the *four factor formula*. In our case $k_\infty = 1.12$.

The value of η depends on how much fissile material the fuel contains. The other three factors have a more complicated dependence of the reactor design including the fuel/moderator ratio, their amounts and shape; some values are given for a practical case in Table 19.4.

For simplicity, let us for a moment regard a homogeneous reactor. For such a reactor

$$f = \Sigma_a\,(\text{fuel})/[\Sigma_a\,(\text{fuel}) + \Sigma_a\,(\text{mod}) + \Sigma_a\,(\text{other})] \tag{19.14}$$

where the denominator is the macroscopic cross section of all absorption reactions in the reactor; the last term includes absorption by impurities in construction materials and by products formed in the course of the operation, e.g. fission products. When control rods are inserted Σ_a (other) increases. Effective cross sections, valid for the current neutron energy spectrum, must be used for all nuclides.

In order to allow for a decrease in f during operation, commercial reactors are designed with k_∞ 1.2–1.3, rather than the 1.12 given in the example above. Even

Table 19.4 Data for a typical BWR power plant

General: 640 MW_e (1912 MW_{th}) BWR plant at Würgassen, Germany, in commercial operation 1975–1995.

Reactor physics: θ 4.5 × 10^5 (s), η 1.740, f 0.865, p 0.811, ε 1.041, k_∞ 1.270, k_{eff} 1.255, L^2 3.67 (cm^2), τ 34.83 (cm^2), $\boldsymbol{B}^2$ 2.32 × 10^{4} (cm^{-2}), $\phi_{th,average}$ 4.4 × 10^{13} (cm^{-2} s^{-1}), $\phi_{fast,average}$ 1.9 × 10^{14} (cm^{-2} s^{-1}), Δk_{max} 25.5%, $\Delta k/k$ 2.2 × 10^{-3}%/°C

Core: 86 600 kg U, 2.6% ^{235}U; critical mass 38.5 kg ^{235}U; burn-up 27 500 MWd t^{-1}; refueling 1/5 of core annually; storage pool for 125 t fuel; core volume 38 m^3.

Fuel element: UO_2 pellets, 1.43 cm diameter; 49 rods per assembly; 444 assemblies in core; enrichment 2.6% ^{235}U; Gd_2O_3 burnable poison; zircaloy cladding 0.8 mm.

Heat and coolant data: Max. fuel temp. 2580°C, max. can temp. 370°C; coolant inlet temp. 190°C, outlet 285°C, pressure 7 MPa (70 bar); 280 t coolant, 2 × 5200 m^3 h^{-1} internal pump capacity; condenser cooling water 95 000 t h^{-1} (River Weser).

higher k_∞ values occur as is the case for reactors with highly enriched fuel. In reactors with large k_∞, which permits high burnup values, a small amount of suitable n-absorbers (*poisons*) may be included in the fuel at the start to decrease k_∞. Such a poison is gadolinium, especially ^{157}Gd, which occurs to 16% in natural gadolinium and has a thermal neutron cross section of 254 000 b. Even a small amount of gadolinium considerably decreases f (see (19.14)). Although ^{157}Gd is continuously destroyed ("burned") by n,γ-reactions which convert it to ^{158}Gd ($\sigma_{n,\gamma}$ 2.5 b) during operation of the reactor, at the same time fission product poisons are formed and fissile nuclei used-up resulting in a compensating effect, which maintains an acceptable value of k_∞. Another possibility to regulate k, mainly used in PWRs, is the addition of a suitable amount of a water soluble n-absorber to the moderator – coolant, *chemical shim control*. When burnup increases, the concentration of the soluble absorber is gradually reduced in order to compensate the poisoning from the fission products and the consumption of fissile atoms.

Because many elements contain isotopes with high neutron capture cross sections for thermal neutrons all materials in the reactor must be extremely pure in order to keep k_∞ as large as possible.

An important factor is the temperature dependence of k. A temperature increase usually has little effect on $\eta\varepsilon$, but f usually increases (i.e. becomes close to 1) because of decreasing moderator density and average neutron energy increase; this causes more neutrons to be captured in ^{239}Pu, which has a low energy (0.3 eV) fission resonance. On the other hand, p decreases because thermal vibrations of atoms cause a *doppler broadening* of the n,γ resonances in the epithermal region, mostly for ^{238}U. UO_2 fuel has a relatively poor heat conductivity causing the center temperature in the fuel pins to vary strongly with the reactor power level, cf. Ch. 21. This gives the doppler broadening of resonances in the fuel a high fission rate dependence. The doppler broadening also increases capture in the control rods. These latter effects normally dominate over the increase in f, so that k_∞ decreases slightly with increasing reactor power and temperature. This is referred to as a *negative temperature coefficient*. In a BWR a negative temperature coefficient is also caused by the bubbles of steam (*void*) occurring in the moderator, which reduces the thermalization of neutrons while increasing neutron leakage rate. In BWRs the effect of the void normally dominates over doppler broadening in limiting the reactor power. As the reactor tends to keep the void fraction and fuel temperature constant, the power level of a BWR (and to a smaller degree also of a PWR, where mostly the average fuel temperature is affected) can be adjusted by varying the speed of the coolant circulation pumps at a constant control rod setting.

These are important safety features of a reactor. If k increases, so do the fission rate and temperature. If the temperature coefficient was positive, k would increase further, leading to further temperature increase, etc. However, with a negative temperature coefficient the reactor controls itself: an increase in power (and thus temperature) decreases k, which tends to limit the power increase and vice versa. The temperature coefficient is usually given in $\%\Delta k/k$ per °C.

19.7. NEUTRON LEAKAGE AND CRITICAL SIZE

All practical reactors have some leakage of neutrons out of the reactor core. This leakage Λ is approximately described by the fast and thermal leakage factors Λ_f and Λ_{th} respectively:

$$\Lambda = \Lambda_f \Lambda_{th} \tag{19.15}$$

The effect of this leakage is included in Figure 19.6, and results in the number of neutrons in the second generation being reduced, from the 112 calculated previously for k_∞, to $N_2 = 101$. Thus $k_{eff} = 1.01$ and $\Lambda = 0.9$, i.e. 10% of the neutrons are lost through leakage. The example in Figure 19.6 refers to a small air cooled and graphite moderated natural uranium research reactor. In nuclear power stations k_∞ is much larger due to the large η-values (because of the enriched fuel needed to sustain operation for long periods at high power between refuelings), while the leakage is much smaller, $\Lambda \geq 0.97$ (e.g. see Table 19.4).

The leakage is a function of the geometrical arrangement of reactor core and reflector and of the average distance the neutron travels after formation until it causes a new fission:

$$\Lambda_f = e^{-B^2\tau} \tag{19.16}$$

$$\Lambda_{th} = \left(1 + B^2L^2\right)^{-1} \tag{19.17}$$

where B^2 is the *geometrical buckling*, L the *thermal diffusion length*, and τ is referred to as the *neutron* (or *Fermi*) *age*.

Combining these equations gives the *critical equation*:

$$k_{\text{eff}} = k_\infty e^{-B^2\tau}/\left(1 + B^2L^2\right) \approx k_\infty/\left(1 + B^2\left(L^2 + \tau\right)\right) \tag{19.18}$$

The quantity $L^2 + \tau$ is usually denoted by M^2 and is known as the *migration area* and M as the *migration length*:

$$M^2 = L^2 + \tau \tag{19.19}$$

Hence for a large thermal critical reactor one can write

$$k_{\text{eff}} = k_\infty/\left(1 + B^2M^2\right) = 1 \tag{19.20}$$

The thermal diffusion length L is calculated from

$$L^2 = L_{\text{m}}^2(1 - f) \tag{19.21}$$

where L_{m} is the diffusion length in the pure moderator. L_{m}^2 and τ are given in Table 19.3 for various moderator materials.

The geometrical buckling B^2 depends on the neutron flux distribution in the reactor. This distribution in turn depends on the general geometry of the assembly, including boundary conditions. The calculation of the buckling for heterogeneous reactors is quite complicated but for homogeneous and bare (no reflector) reactors the following approximate simple relations hold:

$$B^2 = \pi^2 r^{-2} \quad \text{(sphere)} \tag{19.22}$$

$$= 3\,\pi^2 a^{-2} \quad \text{(cube)} \tag{19.23}$$

$$= (\pi/h)^2 + (2.405/r)^2 \quad \text{(cylinder)} \tag{19.24}$$

$$= 33\,h^{-2} \quad \text{(cylinder; height = diameter)} \tag{19.25}$$

where r is the radius of the sphere and cylinder, a the side of the cube, and h the cylinder height (which is also the diameter of the cylinder in eqn. 19.25).

It is seen that k_{eff} in (19.18) increases with decreasing buckling, and, because $B \propto 1/r$ (19.22), with increasing size of the reactor. This is a result of the neutron production k_∞ being a volume effect (proportional to r^3 for a sphere), whereas leakage is a surface effect (proportional to r^2). For each reactor there is a minimum *critical size* ($k_{\text{eff}} = 1$) below which the surface to volume ratio is so large that neutron leakage is sufficient to prevent the fission chain reaction.

The smallest critical sizes are obtained for *homogeneous systems* of pure fissile nuclides with maximum neutron reflection. For neutrons with the fission energy spectrum, the critical mass of a metallic sphere of pure ^{235}U is 22.8 kg, that of ^{233}U is 7.5 kg, and that of ^{239}Pu is 5.6 kg, assuming a 20 cm uranium metal neutron reflector. For fission by thermal neutrons the smallest critical size of a spherical *homogeneous aqueous solution* of $^{235}UO_2SO_4$ without reflector requires 0.82 kg of ^{235}U in 6.3 l of solution. The corresponding figures for ^{233}U are 0.59 kg in 3.3 l, and of ^{239}Pu, 0.51 kg in 4.5 l.

Homogeneous solutions of fissile nuclides are produced in the reprocessing of spent fuel elements, where care must be exercised that the critical size is not exceeded in any equipment or container in order to prevent an accidental chain reaction. Several such accidents have occurred in the past in fuel production and reprocessing plants in which very high doses were received by nearby personnel even though the duration of the chain reaction usually was very short and a violent explosion never occurred.

For heterogeneous reactors it is more difficult to quote comparable simple values for critical size. These have to be calculated by numerical methods or determined empirically for each particular reactor configuration.

19.8. REACTOR KINETICS

The mean lifetime θ for a neutron in a reactor is the time it takes on the average for the neutrons to complete one loop in the neutron cycle. In thermal reactors θ is 10^{-3}–10^{-4} s due to the comparatively low speed of thermalized neutrons and average distance in the moderator covered by random walk when travelling from and to a fuel pin. For each loop the number of neutrons is multiplied by a factor k_{eff}. Since one neutron is used for maintaining the chain reaction, the neutrons in the reactor change with time according to

$$dN/dt = N(k_{eff} - 1)/\theta + K \quad (19.26)$$

where N is the total number of neutrons in the reactor, and K is the contribution from any constant neutron source present. Solving this equation, we obtain

$$N = N_0\, e^{(k_{eff}-1)t/\theta} + K\,\theta\left(1 - e^{(k_{eff}-1)t/\theta}\right)\Big/\left(1 - k_{eff}\right) \quad (19.27)$$

where N_0 is the number of neutrons at $t = 0$. We can distinguish three different cases:

(i) $k_{eff} < 1$ and $t \gg \theta$: the reactor is subcritical,

$$N(t) = K\,\theta\left(1 - k_{eff}\right)^{-1} \quad (19.28)$$

The initial number of neutrons, N_0, has disappeared and the number of neutrons is directly related to the constant neutron source. The reactor acts as a neutron amplifier, with amplification increasing with k_{eff}.

(ii) $k_{eff} > 1$; the reactor is supercritical,

$$N(t) = [N_0 - K\,\theta/(k_{eff} - 1)]e^{(k_{eff}-1)t/\theta} + K\,\theta/(k_{eff} - 1) \tag{19.29}$$

If K is so small that the terms containing $K\theta$ can be neglected this reduces to

$$N(t) = N_0\, e^{(k_{eff}-1)t/\theta} \tag{19.30}$$

The neutron flux (and consequently also power) increases exponentially. All reactors have $K > 0$, the reason being neutron production through spontaneous fission in ^{238}U and other actinides or through other nuclear reactions. However, the resulting neutron source strength is usually not large enough to give a reliable indication on the control instruments of the power level of a new reactor in the initial start-up procedure. Therefore, extra neutron sources are generally introduced in reactors to facilitate starting. A common type is Sb+Be, where ^{124}Sb (half-life 60 d) has been produced through irradiation in a reactor using the (n,γ) reaction in ^{123}Sb (43% abundance). The neutrons are emitted from (γ,n) reactions in Be. Also ^{241}Am+Be sources are used. When the reactor has reached desired power, control rods are inserted, which decrease k_{eff} to 1. Later the partially spent fuel contains sufficient amounts of nuclides with spontaneous fission, e.g. ^{252}Cf, to make a separate neutron source superfluous.

(iii) $k_{eff} = 1$; the reaction is just critical,

$$N(t) = N_0 + K\,t \tag{19.31}$$

Although this equation indicates that the number of neutrons would increase slightly with time, the $K\,t$ term is usually negligible.

In reactor technology it is common to speak of the *reactivity* ρ and *excess reactivity* (Δk), which are defined by the expression

$$\Delta k \equiv k_{eff} - 1 = \rho\, k_{eff} \tag{19.32}$$

Since k_{eff} is close to 1 in a properly operating reactor, we have $\rho \approx \Delta k$. The *reactor time constant* (or *period*) is defined as

$$t_{per} = \theta/\rho \approx \theta/\Delta k = \theta/(k_{eff} - 1) \tag{19.33}$$

As the neutron flux increases in a supercritical reactor, the second term in (19.29) becomes negligible compared to N_0 and the equation can be simplified to

$$N = N_0\, e^{(t/t_{per})} \tag{19.34}$$

The shorter t_{per} is the faster is the increase in the neutron flux (and reactor power). With $\theta = 10^{-3}$ and $k_{eff} = 1.1$, Δk is 0.1 and t_{per} 10^{-2}. For a longer t_{per} of 1 s, the number of neutrons would increase with a factor of $e^{10} = 10^4$ every 10 s, which still is much too rapid for safe and simple control of a reactor. As described below, nuclear reactors are designed to avoid this problem.

In the equations above, θ is the mean lifetime for the prompt neutrons. As discussed in §11.7.1, some fission products decay by β^- leading to an excited state which emits a neutron, *β^--delayed neutron emission*. For example ^{87}Kr decays partly by β^--delayed neutron emission; it is a daughter of ^{87}Br with a half-life as long as 56 s. A large number of such neutron emitting fission products have been discovered, all with shorter half-lives, see Table 19.5. In reactors where the moderator contains D or Be atoms, γ,n reactions with energetic γ-rays from shortlived fission products and from activation products is also a source of delayed neutrons. The delayed neutrons have lower kinetic energies ($\sim$0.5 MeV) than the prompt ones and amount to <1% of the total number of fission neutrons emitted: the fraction of delayed neutrons, β, is 0.27% for ^{233}U, 0.65% for ^{235}U, 0.21% for ^{239}Pu, and 0.52% for ^{241}Pu. When the delayed neutrons are taken into account, the *effective neutron generation time* is, approximately,

$$\theta_{eff} = \theta + \Sigma(\beta_i/\lambda_i) \tag{19.35}$$

where β_i is the fraction of fission neutrons which are delayed with a decay constant λ_i. The summation gives the mean lifetime of the delayed neutrons; for ^{233}U 0.049 s, ^{235}U 0.084 s, ^{239}Pu 0.033 s, and ^{241}Pu $\sim$0.06 s. Thus for ^{235}U $\theta_{eff} \approx 10^{-3} + 0.084$. Because the generation time determines the period of the reactor the delayed neutrons have lengthened the reactor period by almost a factor of 100, thereby making reactor control much more manageable.

t_{per} depends on Δk in such a way that if $\Delta k > \beta$ the delayed neutrons are not able to make their decisive influence on the reactor period, and the rate of neutron production is

Table 19.5 Some important fission products partly decaying by β-delayed n-emission

Isotope	Half-life (s)	decay by β-n (%)	Cumulative fission yield (%)			
			^{233}U	^{235}U	^{239}Pu	^{241}Pu
^{85}As	2.03	23.0	0.127	0.200	0.061	0.084
^{87}Br	55.7	2.6	2.67	1.82	0.608	0.633
^{88}Br	16.7	6.6	1.90	1.68	0.645	0.617
^{89}Br	4.37	14.2	0.677	1.64	0.346	0.456
^{90}Br	1.92	24.6	0.275	0.620	0.163	0.249
^{93}Rb	5.85	1.4	2.28	2.94	1.71	1.80
^{94}Rb	2.73	10.4	0.849	1.54	0.754	1.22
^{98}Y	2.0	3.4	1.64	2.99	2.59	3.58
^{135}Sb	1.71	17.5	0.031	0.165	0.006	0.490
^{136}Te	17.5	0.7	0.005	2.57	0.005	1.98
^{137}I	24.5	6.4	1.35	2.60	2.46	4.56
^{138}I	6.4	5.5	0.828	1.06	0.436	2.60
^{139}I	2.29	9.5	0.171	0.119	0.376	1.16

dependent only on the prompt neutrons. Such a reactor is said to be *prompt critical*. Nuclear explosives are designed to be prompt critical with a $\Delta k \geq 1$ giving a $t_{per} \leq 10^{-8}$ s, while reactors for power production (thermal as well as fast reactors) should for safety reasons always be operated in the *delayed critical* region. If in a thermal uranium reactor Δk is made $\geq \beta$ ($k_{eff} \geq 1.0065$), t_{per} becomes 13 s for $\theta_{eff} = 0.084$, and (19.34) shows that the neutron amount (and power) doubles in about 10 s. Usually k_{eff} is made smaller than 1.0065 and the power doubling time is correspondingly larger.

Some of the isobar chains from fission contain isotopes with very high n,γ cross sections. One of the most important is $A = 135$ and it's effect on reactor kinetics is used as an example. The genetic relations in the A 135 chain is:

$$\begin{array}{llllllllll}
\text{fission yield\% } 3.5 & & 2.5 & & 0.6 & & & & \\
\quad\downarrow & \scriptstyle 18s & \downarrow & \scriptstyle 6.61\,h & \downarrow & \scriptstyle 9.10\,h & & \scriptstyle 2\,My & \\
{}^{135}Te & \rightarrow & {}^{135}I & \rightarrow & {}^{135}Xe & \rightarrow & {}^{135}Cs & \rightarrow & {}^{135}Ba(\text{stable}) \\
 & & & & n,\gamma \downarrow 2.65\times10^{6}\text{ b} & & & & \\
 & & & & {}^{136}Xe(\text{stable}) & & & & \\
 & & & & n,\gamma \downarrow 0.16\text{ b} & & & &
\end{array} \quad (19.36)$$

During operation at a constant neutron flux the large $\sigma_{n,\gamma}$ continuously converts ^{135}Xe to ^{136}Xe, thereby keeping the amount small and limiting the poisoning effect. However, after a strong reduction in neutron flux ^{135}Xe formation is more rapid by decay of its precursor, ^{135}I, than its destruction, resulting in a transient radioactive equilibrium. As a result the reactivity drops for some hours because of an increasing amount of ^{135}Xe. If the reactor has been shut down for a while, it may be impossible to start it until the amount of ^{135}Xe has been reduced again by decay (*xenon poisoning*). However, if the reactor can be started it burns ^{135}Xe and Δk increases rapidly until a new equilibrium concentration is reached. This is called a *xenon transient* and can temporarily make the reactor difficult to control; if mishandled it may even cause prompt criticality. Once xenon poisoning has become very important after a reactor stop it is practice to delay restart until ^{135}Xe has decreased to a safe level.

19.9. FUEL UTILIZATION

Figure 19.7 shows the consumption of fissile ^{235}U while new fissile ^{239}Pu and ^{241}Pu (as well as some fission products and other actinides) are produced through radiative capture in fertile ^{238}U and ^{240}Pu. The Figure relates to a particular reactor type, and different reactors give somewhat different curves. Figure 19.5 shows the different capture and decay reactions. In §19.2 it is concluded that in a reactor operating at a power of 3 GW_{th} 3.16 kg ^{235}U d^{-1} would be fissioned. Because of the $^{235}U(n,\gamma)^{236}U$ reaction, some additional uranium is consumed (19.4). To account for this, we use

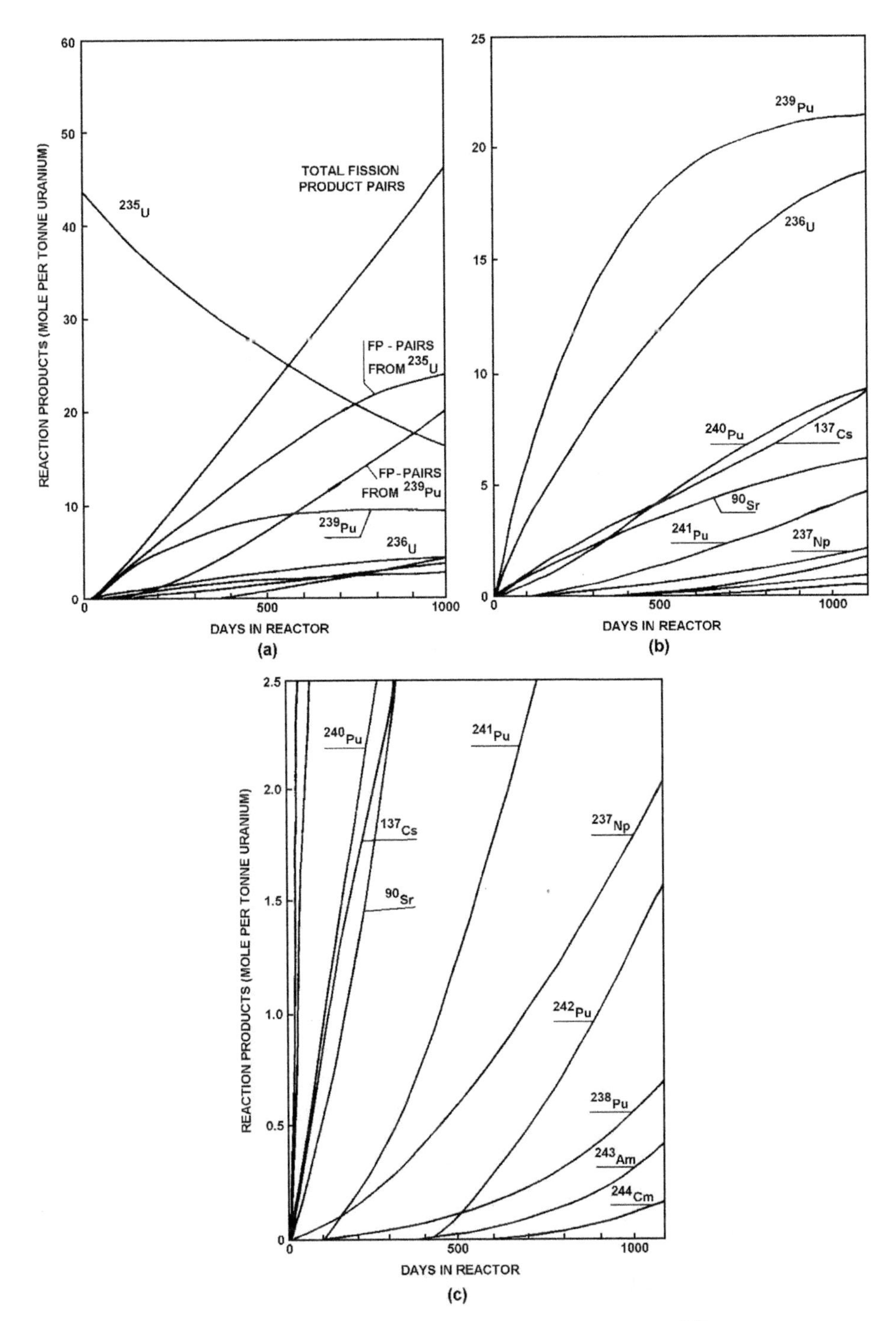

Figure 19.7 Build-up of reaction products in fuel originally enriched in ^{235}U to (a) 1% used in a graphite reactor (~230 MW_{th}), (b) and (c) 3.3% used in a LWR (~1000 MW_e).

$(\sigma_f + \sigma_\gamma)/\sigma_f = 1.17$; thus the ^{235}U consumption is $3.16 \times 1.17 = 3.70$ kg d^{-1}. This value holds for a fresh reactor core; however, after a core has operated for some time, fission in ^{239}Pu formed from neutron capture in ^{238}U begins to contribute to the energy production, and, later, fission from ^{241}Pu also contributes. Therefore the ^{235}U consumption rate successively decreases for the same power production in an ageing core. During the whole lifetime of a 2–3% enriched uranium core about 40% of the total energy production comes from fissioning of plutonium isotopes (Fig. 19.8). On the average a 1000 MW$_e$ (33% efficiency) LWR power station consumes daily 2.2 kg ^{235}U and 2.0 kg ^{238}U. About 3.1 kg fission products are produced each day from the fissioning of uranium and plutonium. To calculate these yields, effective cross sections must be used like those given in Figure 19.5. The more ^{239}Pu and ^{241}Pu that is produced for each atom of ^{235}U consumed, the more efficient can be the utilization of fuel. This is expressed by the conversion factor *C*, which is given by

$$C = \eta - 1 - S \tag{19.37}$$

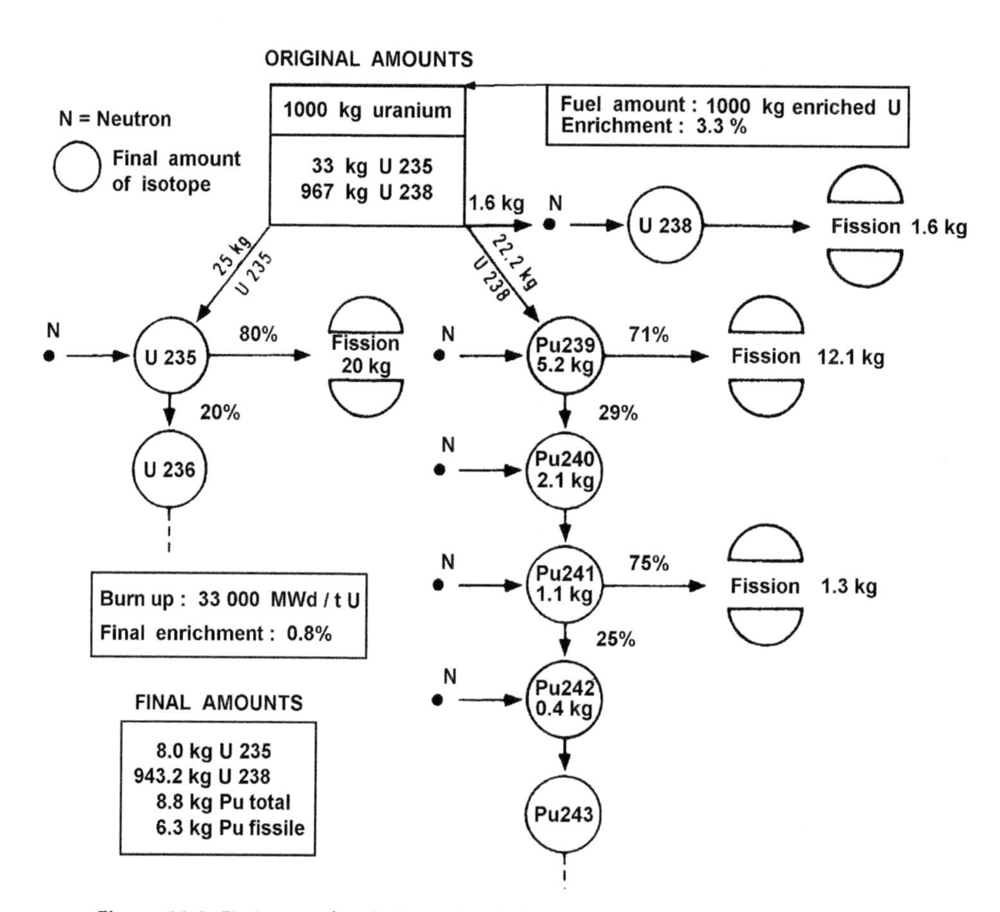

Figure 19.8 Fission product balance in a light water power reactor. *(From F. Hoop.)*

where S is the neutron losses by processes other than fission and capture reactions to produce fissile atoms in the fuel. If we assume $C = 0.8$, which is a valid figure for a heavy water moderated reactor, fission of 100 atoms of ^{235}U leads to the formation of 80 new atoms of ^{239}Pu and ^{241}Pu.

In the second generation one produces C^2 new ^{239}Pu and ^{241}Pu atoms. The fraction of fissile atoms in the fuel which is transformed is $x_i\,(1 + C + C^2 + ...) = x_i\,/(1 - C)$, where x_i is the original atomic fraction of fissile atoms (0.0072 for natural uranium). The ratio x_i / $(1 - C)$ expresses the maximum utilization of the fuel. If a reactor could be operated so that all the fissionable material, both original and produced in operation, were consumed, it would be possible, assuming $C = 0.8$, to obtain five times greater power production than would be provided by the original concentration of ^{235}U only. Because of the high neutron capture cross section of the fission products, unfortunately this is not possible without reprocessing between cycles. Reactors with values of C close to unity are called *converter reactors* while reactors with C considerably less than 1 are referred to as *burners*. Light water reactors usually have $C \leq 0.7$ and are classified as burners, however, designs with a reduced moderator/fuel ratio acting as converters or thermal breeders have been suggested. Figure 19.9 shows the possible fuel utilization in various reactor types.

The lowering of f (eqn. (19.14)) due to the buildup of fission products and decrease in amount of fissile atoms are the main reasons for fuel replacement. It is obvious that if Σ_a (fuel) is very large, as is the case for highly enriched fuel, higher amounts of fission products can be tolerated; i.e. more energy can be produced from the fuel before f becomes too small.

The fuel utilization is referred to as *burnup*. The burnup may be expressed as the percentage of fuel used before it must be replaced. For example, 1% burnup means that for each ton of fuel 10 kg of the fissile plus fertile atoms have been consumed (in fission *and* capture). However, usually the fuel burnup is given in amount of energy obtained per ton of initially present fuel atoms (in case of mixed U – Pu fuels per ton of initial heavy metal, IHM). The production of 33 000 MWd of thermal energy ($MW_{th}d$) from 1000 kg enriched uranium in a LWR consumes 25 kg ^{235}U and 23.8 kg ^{238}U (Fig. 19.8), or 4.88% of the original uranium amount. Then 1% burnup corresponds to 33 000/4.88 = 6762 $MW_{th}d$/t U. However, if credit is made for the 6.3 kg fissile plutonium formed (which "replaces" the ^{235}U), the total consumption is reduced to 48.8 – 6.3 = 42.5 kg (or 4.25%); a 1% burnup is then equal to 7765 $MW_{th}d$/t U.

Considering the amount of total consumed uranium (~7 t natural uranium for each ton enriched to 3.3% in ^{235}U), the utilization of natural uranium is much less; if plutonium is recycled, utilization increases to the overall figures given in Figure 19.9.

The maximum conversion factor at the start of a uranium fueled reactor is given by

$$C = \eta\,\varepsilon(1 - p)\,\mathrm{e}^{-B^2\tau} + \Sigma_a\,(^{238}\mathrm{U})/\Sigma_a\,(^{235}\mathrm{U}) \qquad (19.38)$$

where the two terms give the number of neutrons absorbed by ^{238}U in the resonance absorption region and in the thermal region, respectively, per neutron absorbed in ^{235}U;

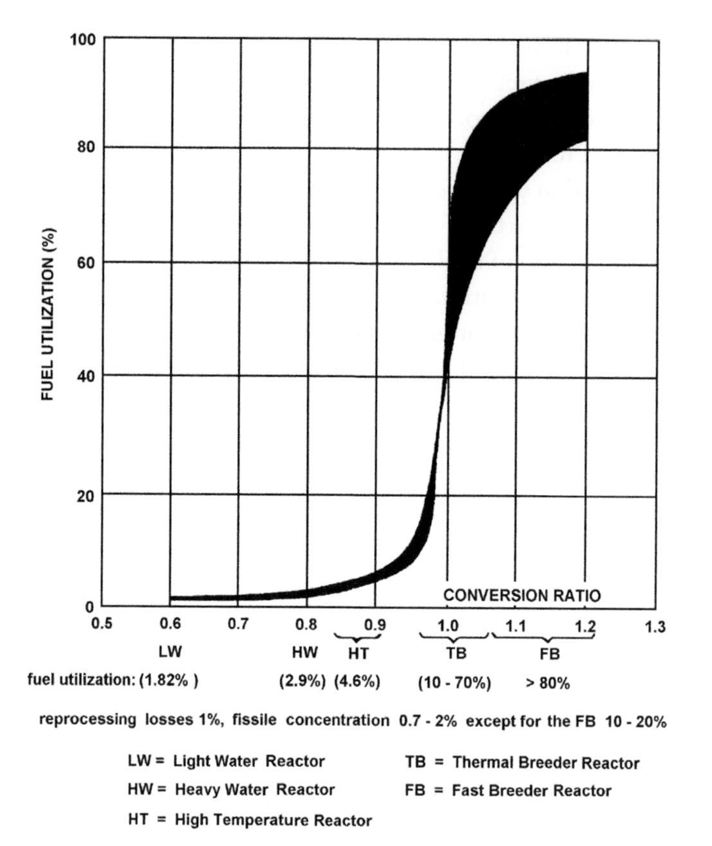

Figure 19.9 Conversion ratios and fuel utilization efficiencies. *(From Thermal Breeder Consultants Group, Salzburg, 1977.)*

the η value refers to ^{235}U. The term $e^{-B^2\tau}$ is the fraction of fast neutrons which does not escape from the reactor (19.16). Using this approximate expression with the reactor example in Table 19.4, we obtain $C = 0.50$. For a reactor core at equilibrium corrections must be introduced for fissions in the plutonium formed; this gives a higher conversion ratio, usually ~0.7 in LWR.

If $C > 1$ the possibility exists of producing more fissile material than is consumed, which is referred to as *breeding*. The breeding ability of a reactor is given by the *breeding gain*, $G = \eta - 2 - S$. Table 19.2 shows that ^{233}U, ^{235}U, ^{239}Pu, and ^{241}Pu all have $\eta > 2$; in practice η is even larger when using effective cross sections (Table 19.2 and Fig. 19.5). In a reactor with these nuclides and fertile material such as ^{232}Th and ^{238}U present it is possible to produce more fissile material than is consumed provided that neutron losses are small; such reactors are called *breeder reactors* or simply *breeders*. While the utilization of the fuel can be increased to perhaps a maximum of 10% with converters, breeding makes it possible to burn ~70% of the fuel material (natural uranium and natural thorium, Fig. 19.9), provided reprocessing and recycling is utilized.

From Table 19.2 it can be seen that the highest η value is exhibited by ^{233}U for thermal neutron energies and by ^{239}Pu for fast neutron energies. This suggests two different types of breeder reactor, the *thermal breeder* based on the reaction $^{232}Th \rightarrow {}^{233}U$ and the *fast breeder* based on the reaction $^{238}U \rightarrow {}^{239}Pu$. Due to the smaller effective cross sections, these breeders must be charged with a fuel having high concentrations (15–30%) of ^{233}U or Pu. These materials can be produced in conventional non-breeder reactors. Because a single inelastic scattering in ^{235}U may reduce neutron energy considerably, a fast breeder should only contain little of this isotope and hence depleted uranium is preferred over natural uranium as fertile material.

For a large fast breeder reactor like the French Creys-Malville (in operation from 1986 to 1997), the fuel utilization is higher than in the LWRs; a 1% burnup corresponds to about 8500 $MW_{th}d\ t^{-1}$ IHM. Burnup figures as high as 15% (130 000 $MW_{th}d\ t^{-1}$ IHM) have been achieved in fast breeders. Breeder reactors are described in Chapter 20.

19.10. THE OKLO PHENOMENON

At the Oklo mine in Gabon uranium has been mined for many years and delivered to the Pierrelatte isotope enrichment plant in France. The ore body, which is estimated to contain 400 000 t of uranium, is quite inhomogeneous, and pockets very rich in uranium are found embedded in sandstone and granite, see Figure 19.10. These pockets are often shaped like a lens, 10 m long and about 1 m in diameter, and contain on the average 10–20% pitchblende, although uranium concentrations up to 85% (pure pitchblende) are found in some spots. In 1972 it was discovered that the isotopic composition of some of the uranium received in France deviated from uranium from other sources. The ^{235}U content was significantly lower than the natural 0.72%. Careful analysis showed that some deliveries contained $<0.5\%$ ^{235}U, a serious disadvantage in materials to be used for ^{235}U enrichment. Analysis at Oklo showed samples even lower in ^{235}U content, as well as other elements whose isotopic composition considerably deviated from the natural one. For example, natural neodymium contains 27% ^{142}Nd, while Oklo neodymium contained $<2\%$, see Fig. 19.10 and 19.11. On the other hand, natural neodymium contains 12% ^{143}Nd, while in Oklo, samples containing up to 24% ^{143}Nd were found. Fission product neodymium contains about 29% ^{143}Nd, while ^{142}Nd is not produced by fission. This should be compared with the lack of ^{142}Nd and the excess of ^{143}Nd in the Oklo samples, a condition which was found to be directly related to a high total concentration of uranium but a deficiency of ^{235}U. The conclusion was obvious: in some ancient time the missing ^{235}U had fissioned, producing ^{143}Nd among other fission products. This conclusion was supported by similar investigations on the isotopic composition of other elements produced by fission.

Because ^{235}U has a shorter half-life than ^{238}U, all uranium ores were richer in ^{235}U in the past. From ^{87}Rb–^{87}Sr analysis the age of the Oklo deposit is known to be 1.74×10^9 y; at

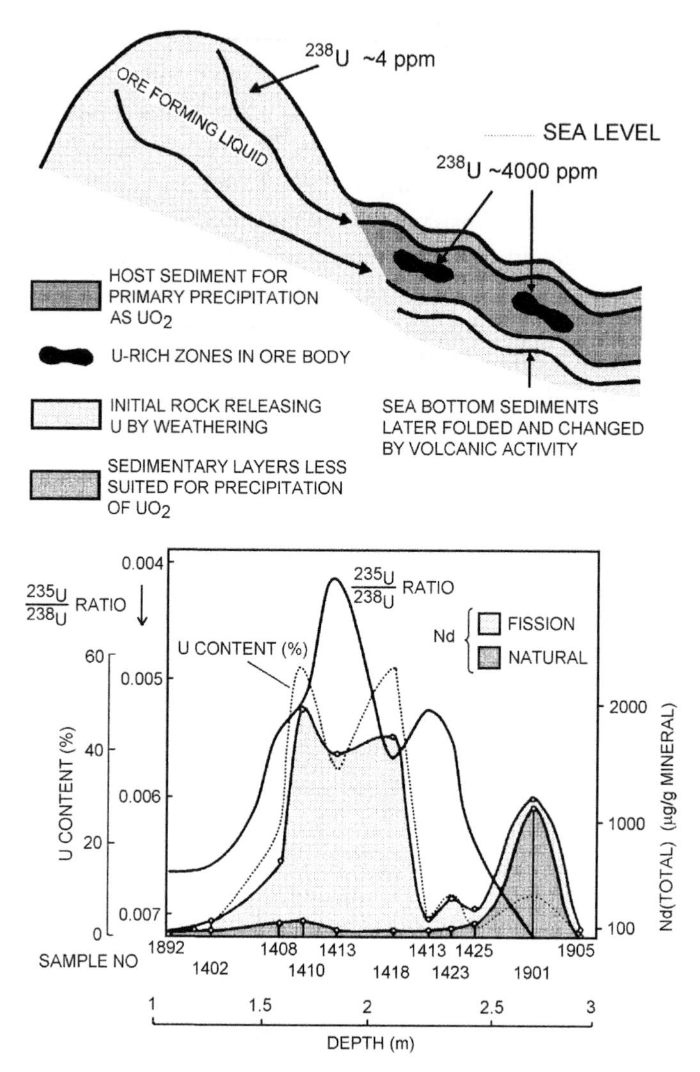

Figure 19.10 Schematic view indicating the formation of the Oklo ore deposit and some data on its current local composition.

that time the ^{235}U content of natural uranium was 3%. Although the fission factor η rapidly increases with ^{235}U content (about 1.8 for 3% ^{235}U), conditions in the natural Oklo deposit were such ($\varepsilon \sim 1.0$, $p \sim 0.4$, $f \sim 1.0$) that $k_\infty < 1$. The deposit is sedimentary and was formed in the presence of water, which greatly increases the resonance escape probability factor p; for an atomic ratio H_2O:U of 3:1, $p \sim 0.8$, and $k_\infty > 1$. Thus conditions existed in the past for a spontaneous, continuing chain reaction to occur in the Oklo deposit.

Further analysis and calculations have shown that these natural Oklo reactors (similar conditions occurred at several places) lasted for $\leq 10^6$ y. Probably, criticality occurred

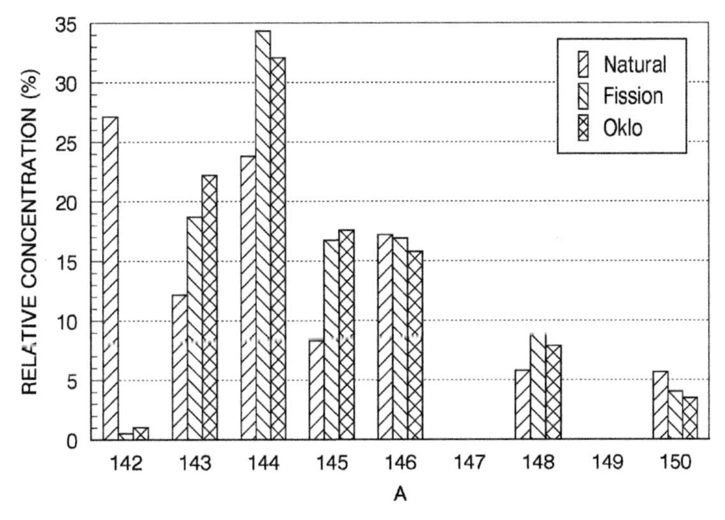

Figure 19.11 Comparison of the isotopic composition of neodymium from normal Nd ore, from ^{235}U fission in a BWR and from Oklo.

periodically as the heat from fission boiled away the water so that the chain reaction ceased after a while. When water returned (as the temperature decreased) the chain reaction would resume. The neutron flux probably never exceeded 10^{13} n m^{-2} s^{-1}.[5] The fluence is estimated to exceed 1.5×10^{25} n m^{-2}. This would have consumed about 6 t ^{235}U, releasing a total energy of 2–3 GWy, at a power level probably $\leq$10 kW. About 1.0–1.5 t of ^{239}Pu was formed by neutron capture in ^{238}U, but the relatively short (on a geological scale) half-life of ^{239}Pu allowed it to decay into ^{235}U. Since a few samples enriched in ^{235}U also have been found, it is believed that in some places in Oklo breeding conditions may temporarily have existed, cf. §22.12.

It has been calculated that many uranium rich ore deposits $2-3 \times 10^9$ y ago must have been supercritical in the presence of moderating water. Therefore natural nuclear chain reactions may have had an important local influence on the early environment of earth.

19.11. REACTOR CONCEPTS

Nuclear reactors are designed for production of heat, mechanical and electric power, radioactive nuclides, weapons material, research in nuclear physics and chemistry, etc. The design depends on the purposes, e.g. in the case of electric power production the design is chosen to provide the cheapest electricity taking long term reliability in consideration. This may be modified by the availability and economy of national resources such as raw material, manpower and skill, safety reasons, etc. Also the risk for proliferation of reactor

[5] n-fluxes and fluencies are given per m^2 in the SI system; however, it is still common to use per cm^2.

materials for weapons use may influence the choice of reactor type. Many dozens of varying reactor concepts have been formulated, so we must limit the discussion in this chapter to a summary of the main variables, and the most common research and power reactors. Fast reactors and some other designs are discussed in Chapter 20.

We have already mentioned three basic principles for reactor design: the neutron energy (thermal or fast reactors), the core configuration (homogeneous or heterogeneous aggregation of fuel and coolant), and the fuel utilization (burner, converter or breeder). In the homogeneous reactor the core can be molten metal, molten salt, an aqueous or an organic solution. In heterogeneous reactors the fuel is mostly rods filled with metal oxide. The fuel material can be almost any combination of fissile and fertile atoms in a mixture or separated as in the core (fissile) and blanket (fertile) concept (Ch. 20). The choice of moderator is great: H_2O, D_2O, Be, graphite, or organic liquid. There is even more choice in the coolant which can be molten metal, molten salt, liquid H_2O, D_2O, or organic liquid, as well as gaseous CO_2, helium or steam.

Some of the more important combinations are summarized in Tables 19.1 and 20.1.

19.12. RESEARCH AND TEST REACTORS

According to the World Nuclear Industry Handbook 2012, 264 research and test reactors were in operation throughout the world at the end of 2011, but some have probably been shut down since then. While the smaller research reactors may have an operating power ≤ 1 kW$_{th}$, and do not need forced cooling, the larger test reactors operate at ≤ 50 MW$_{th}$. Many of these reactors have facilities for commercial radionuclide production.

Research reactors are used in nuclear physics, in nuclear, analytical and structural chemistry, in radiobiology, in medicine, etc. They are usually easy to operate, inherently safe, and of moderate cost. Many of them are of the pool type; the reactor core is located in the center of a stainless steel clad concrete vessel, 6–8 m deep and 2–5 m wide, containing purified water. The water provides the main radiation shielding, moderation, and cooling. The concrete walls are about 2 m thick when the reactor is located above ground; otherwise much thinner walls can be used. The main radiation protection demand comes from the reaction

$$^{16}\mathrm{O(n,p)}^{16}\mathrm{N} \xrightarrow[7.13\ \mathrm{s}]{\beta\gamma} {}^{16}\mathrm{O} \tag{19.39}$$

since the γ's emitted are very energetic (6.1 MeV). At power levels >100 kW$_{th}$ forced cooling may be required. The fuel contains usually highly enriched uranium (20 to 90% ^{235}U) as rods or plates; the amount of ^{235}U is ≤ 3 kg.

Some reactors are designed to produce very high neutron fluxes, either for testing materials (especially for fast reactors) or for isotope production. The high flux reactor at Grenoble, France, has fuel plates of highly enriched (93%) UAl_3 canned in aluminum.

The construction is a swimming pool tank type. Using 8.6 kg ^{235}U and 15 m^3 heavy water as moderator, it provides a maximum thermal neutron flux of 1.5×10^{19} m^{-2} s^{-1} at 60 MW$_{th}$; 10 fold higher fluxes can be achieved during shorter periods. At Oak Ridge, the High Flux Isotope Reactor (HFIR) has a flux of 3×10^{19} m^{-2} s^{-1}, the highest continuous thermal neutron flux so far reported. Some research reactors are built to test new reactor designs which may be candidates for future power reactors. The successful designs lead to prototype power reactors; some of these are described in Chapter 20 together with full-scale power reactors. Other designs never survive the first stage, although the design may be very interesting.

19.13. THERMAL POWER REACTORS

We illustrate the general principles of thermal reactors by a short description of the two most important power reactor types; the pressurized water reactor (PWR) and the boiling water reactor (BWR). They are further discussed in Chapter 20.

19.13.1. The pressurized water reactor

As the name implies a pressurized water reactor (PWR) is cooled by hot high pressure water, H_2O, which also acts as moderator. The main components of a pressurized light water moderated and cooled reactor (PWR) station have been described in §19.1 and Figure 19.2. The Ringhals station, whose first reactor started operation in 1975, is located in Sweden near the sea and seawater is used for cooling; power stations located near large rivers or at the sea shore usually do not need cooling towers.

The important features of the PWR core are shown in Figure 19.12, which, although taken from three different reactors, represents the typical Westinghouse design. A typical core contains ~40 000 fuel rods (a) in 193 assemblies, each with space for 208 fuel pins (b) and (c). The fuel is 88 t of UO_2 with an enrichment of 2.17% (inner region) to 2.67% (outer region) in the initial fuel loading. The reactor vessel (e) is made of stainless steel clad low alloy steel, 13 m high and 4.4 m in diameter, with a wall thickness of 22 cm; (c) shows the distribution of control rods entering from the top of the vessel. The 1060 control rods are made of a silver alloy containing 15 wt% In and 5 wt% Cd; both these elements have high thermal neutron capture cross sections. The main reactivity control is by boric acid, which is fed to the coolant through a special injection system. The boric acid circulates in the primary coolant loop, and acts as a neutron absorber. At the start of a fresh core its concentration is ~1500 ppm (~0.025 M), but it is successively reduced to zero at time of fuel replacement. The concentration is adjusted in a side loop, containing either an anion exchange or an evaporator system. The ion exchange system is so designed that H_3BO_3 is fixed to the ion exchanger (the water is deborated) at low temperature (~30°C) while it is eluted at higher temperature (~80°C).

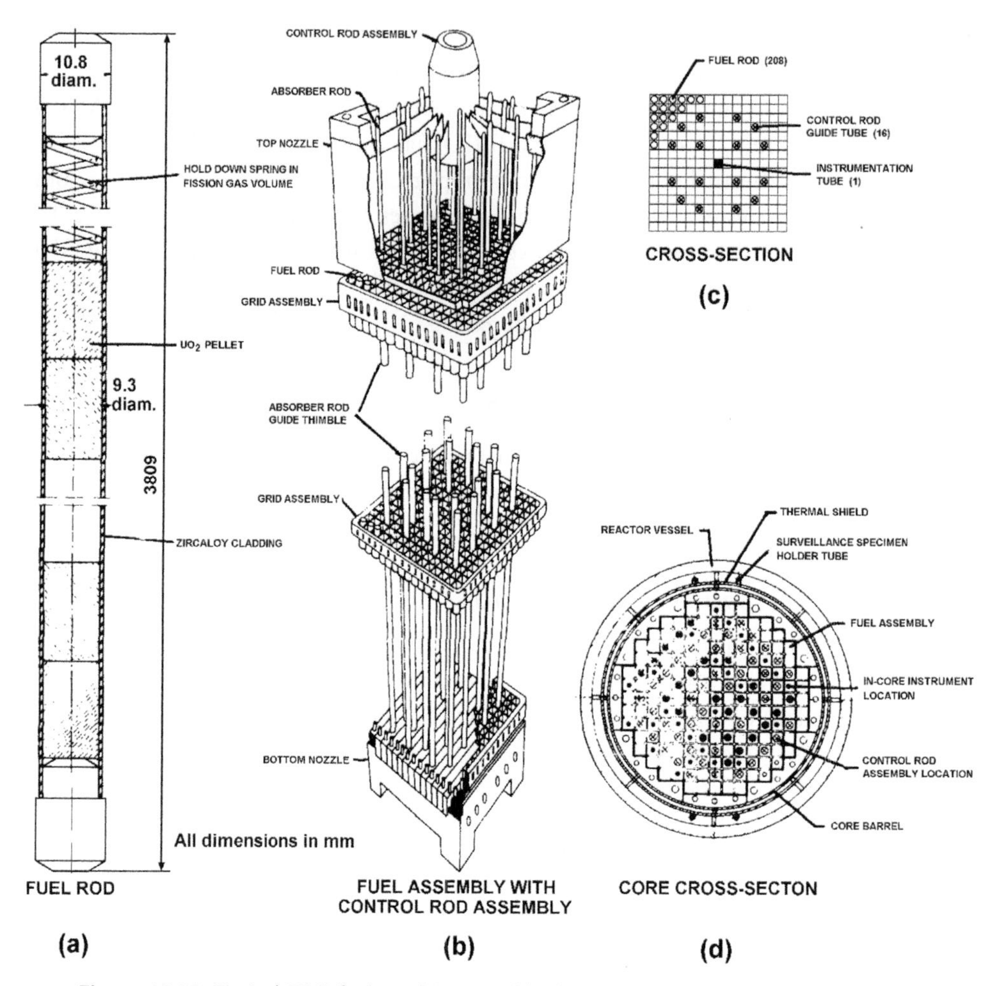

Figure 19.12 Typical PWR fuel pin (a), assembly (b, c), core (d), and vessel (e) designs.

Some of the physical parameters are: k_∞ 1.29 (cold) and 1.18 (operational temperature), temperature coefficient $-1 \times 10^{-3}\%\Delta k/k$ °C, average neutron flux 2.16×10^{17} (thermal), 3×10^{18} (fast) n m^{-2} s^{-1}. Data for heat production are given in Table 19.6.

The Westinghouse-type PWR (Figs. 19.2 and 19.12) reactor is presently the dominating nuclear power reactor in the world.

19.13.2. The boiling water reactor

In a boiling water reactor (BWR) the coolant is brought to boiling while passing along the fuel pins. Thus the need for a separate steam generator and secondary loop is removed. This gives a robust design with less complexity and marginally higher thermal efficiency than a PWR at the expense of some features. To minimize the wear of the

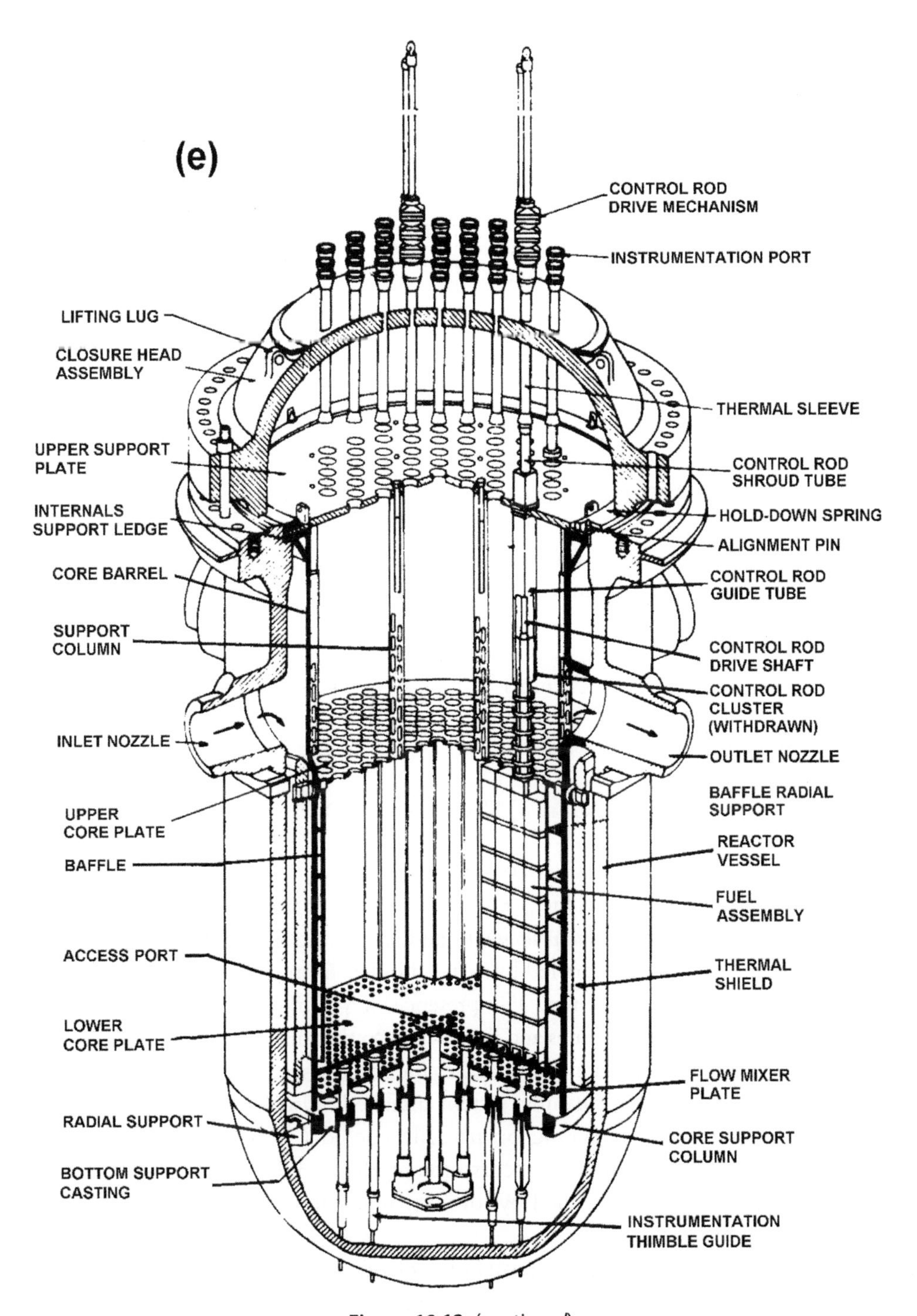
(e)
CONTROL ROD
DRIVE MECHANISM
INSTRUMENTATION PORT
LIFTING LUG
CLOSURE HEAD
ASSEMBLY
THERMAL SLEEVE
UPPER SUPPORT
PLATE
CONTROL ROD
SHROUD TUBE
INTERNALS
SUPPORT LEDGE
HOLD-DOWN SPRING
ALIGNMENT PIN
CORE BARREL
CONTROL ROD
GUIDE TUBE
SUPPORT
COLUMN
CONTROL ROD
DRIVE SHAFT
CONTROL ROD
CLUSTER
(WITHDRAWN)
INLET NOZZLE
OUTLET NOZZLE
BAFFLE RADIAL
SUPPORT
UPPER
CORE PLATE
REACTOR
VESSEL
BAFFLE
FUEL
ASSEMBLY
ACCESS PORT
THERMAL
SHIELD
LOWER
CORE PLATE
FLOW MIXER
PLATE
RADIAL SUPPORT
CORE SUPPORT
COLUMN
BOTTOM SUPPORT
CASTING
INSTRUMENTATION
THIMBLE GUIDE

Figure 19.12 (*continued*).

Table 19.6 The international nuclear event scale (INES) used for the reporting of nuclear events

Type	Definition	Value	Examples	
None	No safety significance	0		
Incident	Anomaly	1		
	Incident	2		
	Serious incident	3	Vandellos	1989
Accident	Accident mainly in installation	4	Saint Laurent	1980
	Accident with off-site risks	5	TMI	1979
	Serious accident	6		
	Major accident	7	Chernobyl	1986

turbine it must be fed with steam that is as dry as possible. For this purpose steam separators and steam dryers are needed. The water not converted to steam is recirculated back into the reactor core by external or internal recirculation pumps.

The single coolant – steam loop permits shortlived activity from neutron activation of oxygen in the coolant to reach the turbine. The main reaction with thermal neutrons is

$$^{18}\mathrm{O}\,(\mathrm{n},\gamma)\,^{19}\mathrm{O} \xrightarrow[27.1\ \mathrm{s}]{\beta,\gamma} {}^{19}\mathrm{F} \tag{19.40}$$

and with high energy neutrons

$$^{16}\mathrm{O}\,(\mathrm{n},\mathrm{p})\,^{16}\mathrm{N} \xrightarrow[7.13\ \mathrm{s}]{\beta,\gamma} {}^{16}\mathrm{O} \tag{19.41}$$

Small amounts of fission products leaking from faulty fuel pins or generated from uranium (leached from damaged fuel or initially present as impurity) deposited on the surfaces in the core (so called *tramp uranium*) and small amounts of activation products (most troublesome is ^{60}Co) are carried by water droplets from the reactor vessel through the piping to turbine and condenser. The presence of γ-emitters in the steam makes it necessary to shield pipes and turbines and limit access to the turbine hall during operation.

The fact that the boiling water never leaves the reactor system makes it necessary to have a special reactor water cleaning system attached. Usually this system contains mixed-bed ion exchangers operating on partially cooled ($\sim$90°C) reactor water. To further reduce the accumulation rate of corrosion products in the reactor, the water from the turbine condenser is also cleaned by ion exchange filters before being returned as feed water to the reactor.

The reactor core and vessel of a typical boiling light water reactor station, a BWR, is shown in Figure 19.13 and supplemental data are given in Table 19.4. The fuel core and elements are rather similar to those of the PWR. The main differences are that boiling

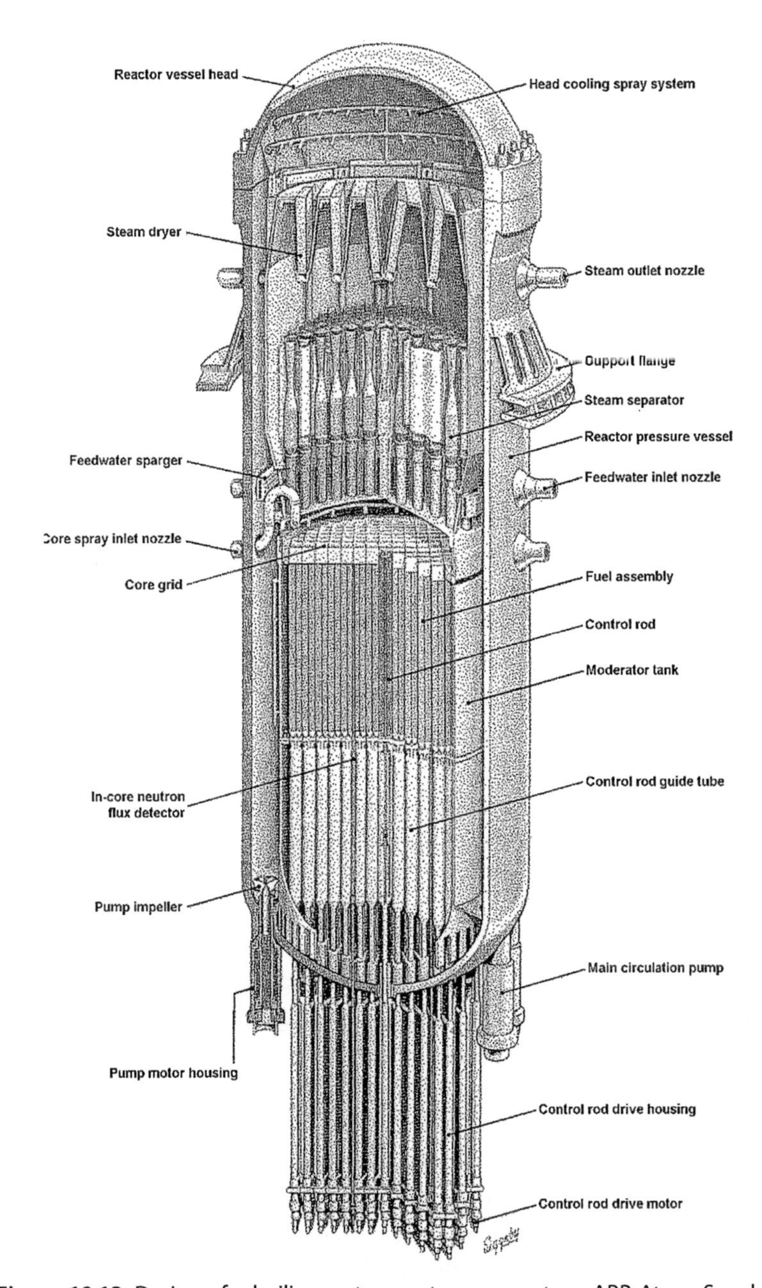

Figure 19.13 Design of a boiling water reactor core system, ABB-Atom, Sweden.

occurs in the reactor vessel, so no external steam generator is required, and that the fuel elements are enclosed in fuel boxes to improve the flow pattern. Although the cores are about the same size for PWRs and BWRs, the reactor vessel is much taller for the BWR; typical values are 22 m in height and 6 m in diameter. Since the pressure in the vessel is lower for BWRs than for PWRs, the vessel wall can be thinner (a typical value is 16 cm). BWRs usually contain more uranium, but with lower enrichment, than PWRs. While only about one-fifth of the core is replaced each year in a BWR, about one-third to one-fourth of the core is replaced annually for a PWR; in both cases this corresponds to about 30 t of used fuel replaced every year.

The BWR is controlled by blades entering from the bottom of the reactor vessel. The blades are usually of cruciform cross section and fit in between the fuel boxes surrounding the fuel elements. The fuel elements often contain burnable poison (e.g. Gd_2O_3). Because of the boiling in the tank boric acid cannot be used for reactivity control; however, it is provided as an extra emergency shutdown feature. The control blades contain boron carbide rods or powder placed in channels in the stainless steel blade-wings. In addition to negative temperature coefficients both for fuel and moderator, BWRs have a negative *void coefficient*. This means that bubble formation (*void*) along the fuel pins reduces the reactivity. As mentioned in §19.6 it is therefore common practice to control the reactor power by the main coolant circulation pumps; increased circulation initially reduces the voids (and fuel temperature) thereby increasing power production until the previous void fraction has been almost restored at a slightly higher average fuel temperature. However, a very large negative void coefficient should also be avoided as it can cause an undesirable amplification of pressure fluctuations in the reactor vessel.

19.14. POWER STATION EFFICIENCY

Not all of the heat produced in a nuclear reactor can be used for work, i.e. for turning the turbine blades connected to the rotor of an electric generator. According to the second law of thermodynamics, the maximum *thermal efficiency* (η) is

$$\eta = (Q_{in} - Q_{out})/Q_{in} \tag{19.42}$$

where Q_{in} is the heat input into a machine and Q_{out} is the heat discharged from that machine. Most reactors operate according to the *Rankine cycle* in which a liquid medium is heated and vaporized at constant pressure, and work is carried out through adiabatic expansion. While the steam entering the turbine should be as dry as possible to reduce turbine blade wear, the steam on the backside is wet due to some condensation. This effect is normally reduced by dividing the turbine into a high pressure and a low pressure part with intermediate reheating by primary steam. Because steam is not an ideal gas (an accurate calculation of Q from heats of vaporization and the ideal gas law is impossible) empirically based steam tables must be used.

The highest thermodynamic efficiency is achieved in the *Carnot cycle* in which energy input (heating the working medium) and work both occur at different but constant temperatures, T_{in} and T_{out}. For a "Carnot engine"

$$\eta_{max} = (T_{in} - T_{out})/T_{in} \tag{19.43}$$

A typical PWR plant may deliver steam of ~275°C to the turbine and have an outlet temperature from the condenser on the turbine backside of ~29°C. Using (19.43) $\eta_{max} = [(275 + 273) - (29 + 273)]/(275 + 273) = 0.45$. A typical BWR usually delivers steam at ~285°C and has a condenser temperature of ~25°C, hence $\eta_{max} = 0.47$. Due to a less efficient energy cycle, friction, heat losses, pumps, etc., the net efficiency (η_{net}) of both reactor types is only about 0.32–0.35 (net electric output delivered to the grid divided by gross thermal output from reactor). In coal-, oil- and gas-fired power plants higher steam temperatures can be achieved, ~500°C with T_{in} 530°C and T_{out} 30°C, $\eta_{max} = 0.65$ leading to an η_{net} ~0.50.

If we compare η_{net} for the two kinds of power plants we find that the fossil fuel power plant has about 40% higher efficiency, i.e. the amount at "*reject heat*" is up to 40% higher for the typical nuclear power plant ((0.50–0.30)/0.50). This lower efficiency is a consequence of the lower steam temperature of the nuclear reactor. However, in the last 10 years considerable development has taken place with respect to coolant, cladding and fuel stability, and inertness, allowing considerably higher fuel temperatures, cf. Ch. 21.

The thermal efficiency of any reactor can be raised by using some of the heat directly for a beneficial purpose, such as for processing in the chemical industry or for district heating. If the diverted heat is at a higher temperature than normal for the condenser coolant effluent such diversion will reduce electricity output correspondingly. If this "lost" electricity had been used to drive heat pumps instead, the total energy efficiency had only increased by a small amount, or not at all, by the direct diversion of heat. In Russia some smaller nuclear power stations produce district heat and electricity for local consumption; EGP reactors in Table 19.1.

The performance characteristics of power stations are commonly described by the availability and load factors, and by the forced outage factor. The *availability factor* is the time the station has been available for operation divided by the length of the desired time period. Thus if it has been desired to operate the station for 6000 h, but it has only been possible to run it for 5500 h (because of repair, etc.) the availability factor is 92%. Typical availability factors are 95–100% for water power, 75–85% for conventional fossil power, and ~70% for nuclear power (usually the reactor has a higher availability, but with a lower value of ~70% for the turbine). Availability factors >80% have been achieved for LWRs.

The *load factor* is the ratio between the electrical net energy delivered by the station and the net electrical energy it would have produced running at maximum power (i.e. design power multiplied by the time). A 1000 MW_e power station which has produced

5.80 TWh during the year has had a load factor of 66% (= 5.80 × 10^6 MWh per 1000 MW × 8760 h y^{-1}). When nuclear power is used for *base load* (i.e. running at full power when operating), its load factor is only limited by how well the nuclear power stations perform. When nuclear power is not used for base load, the load factor depends not so much on the power station performance but rather on the local energy demand, available alternative energy sources and their price. Thus in Sweden during 1999 the annual load factor of hydro power was ~50%, thermal power ~22%, wind power ~19%, and nuclear power ~80%. The low load factors for hydro power and thermal power are due to the use of these in "follow the load" mode, whereas nuclear power was used for base load and wind power operated at its momentary capacity. The rather low value for thermal power is caused by its cost of operation, which is higher than the other alternatives; wind power is subsidized to make it competitive. Typical world wide average yearly load factors for all power reactors weighted with production capacity are (2011 figures): PWR 81%, BWR 73%, PHWR 71%, and RBMK 79%. These figures do not include reactors of less than 150 MWe or data from utilities with less than 4 operating reactors.

A better picture of the power station performance in situations where the demand may be less than the installed capacity is given by the *forced outage rate* which is the ratio between the number of hours a power station has been shut down because of malfunction and the total number of hours during that period. A power station which has been shut off for one month in January for repair of the main feed water pump and three months (June, July, and August) because of no demand for electricity, during which time part or the fuel elements were changed, has an 8% forced outage rate (i.e. one month out of twelve). The forced outage rate is very similar for all power stations of similar size and age. Values for new 1000 MW_e BWR or PWR stations vary from 15% up; a typical average value is 25%, normally decreasing with time.

19.15. REACTOR SAFETY

Greater than 95% of all the operation shutdowns in nuclear power stations are due to failures common to conventional thermal power stations. However, in this text we consider only safety with regard to the nuclear steam supply system which account for the other 5%. We can distinguish between three safety (or risk) levels: (i) small deficiencies due to imperfect technique, which are inevitable in any human enterprise (e.g. small "chronic" releases of radioactivity to the environment), (ii) accidents, which normally should not occur with good equipment, but still are "probable" (e.g. pipe breaks, fires, etc.) and for which protective measures must be included in the reactor design (design basis accident, DBA), and (iii) maximum credible accident (MCA) which is the most serious reactor accident that can reasonably be imagined from any adverse combination of equipment malfunction, operating errors, and other foreseeable causes.

The distinction between DBA and MCA is successively vanishing, as more possible accidents are being included in the DBA. Radioactivity releases from past reactor accidents are discussed in §13.10.2 and Chapter 22.

The broad functions of the safety systems are common to most reactors. In the event of an abnormal condition they should shut down the reactor, ensure a sufficient supply of coolant for the fuel, and contain any fission products which might escape from the fuel elements. Such safety features can be *active* (requiring some action from a control system, involving mechanical devices, and relying on an external power source in order to operate) or *passive* (built-in physical fail-safe features whose operation is not dependent on any control system, mechanical device or external power source).

Active safety systems are normally used and must be designed with a high degree of *redundance* (duplication) and *diversity* (difference in principle), so that if one safety systems fails another shall function. Several redundant active safety systems are required which are logically and physically separated from one another, and from the reactor process systems, as much as possible. They are also often based on different principles in order to reduce the probability of common-mode failures. This confers considerable immunity against events such as external explosions or internal fires. Each safety system alone should be capable of protecting the reactor core and building from further damage.

Passive safety systems are less common than active ones. However, introduction of passive (inherent) safety functions into new reactor designs is much discussed within the nuclear industry. Several ideas have been developed into new design proposals. As an example, the SECURE district heating reactor, developed by ABB-Atom[6], uses a hydro-statically metastable operating condition which causes shut-down of the reactor as soon as the normal coolant flow becomes too low for safe operation or if the core generates too much heat. Figure 19.14 illustrates the operating principle. During normal operation (a in Fig. 19.14) temperature and pressure differences keeps the system in a hydrostatically metastable state with the hot primary coolant separated from the borated shut-down solution. Power is regulated by changing the concentration of boron in the primary coolant. If either temperature in the core increases or flow through the core decreases, the gas bubble above the core will escape and the reactor shuts itself down as cold borated water enters the primary system (b in Fig. 19.14). Cooling is then secured by natural circulation inside the concrete vessel. The same principle is also proposed to be used in an inherently safe pressurized water reactor (PIUS, ABB-Atom), but many other proposed designs exist. In general, reactors with passive safety have to be designed for operation at a lower power level than reactors with active safety features.

The main threat to a reactor would involve the cooling capacity becoming insufficient because of a sudden power excursion of the reactor, a blockage of the cooling

[6] ABB-Atom AB is now owned by Westinghouse and renamed Westinghouse-Atom AB.

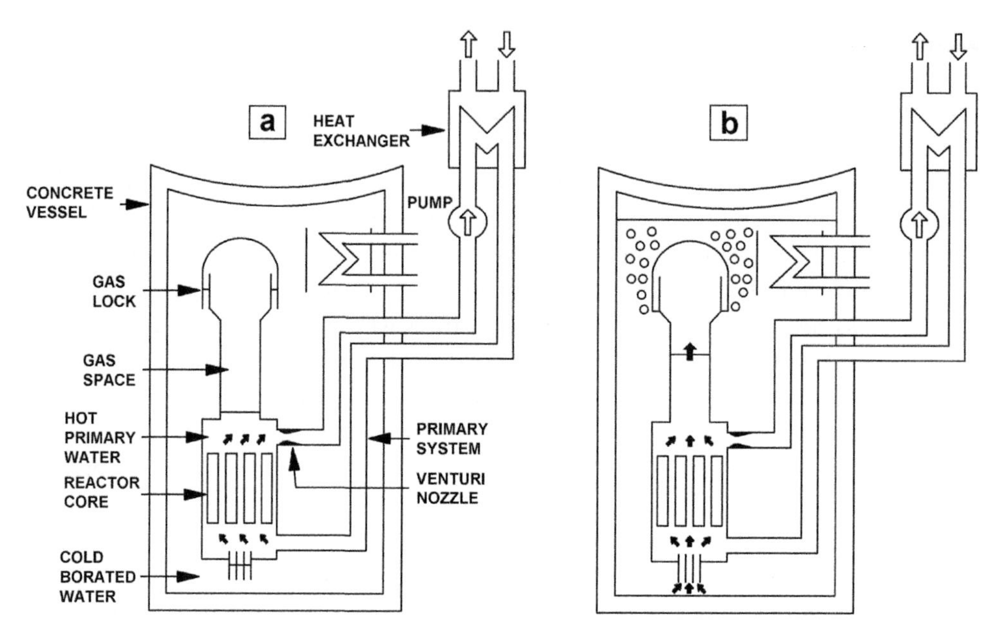

Figure 19.14 Working principle of ABB-Atom SECURE inherently safe district heating reactor.

circuit, or a loss of coolant. In either case the core could become overheated and a core meltdown might begin.

The guarantee against a power excursion is a negative temperature coefficient. For example, if the design parameters of a reactor give a maximum available $k_{eff} = 1.041$, and the temperature coefficient is -4.2×10^{-3} %$\Delta k/k$ °C, a temperature increase of 1000°C would be required to reduce k_{eff} to 0.999, i.e. to automatically stop the fission chain reaction without any of the safety systems operating. Such a reactor would also be prompt critical, as pointed out in §19.8. This is not acceptable, so at starting conditions the maximum available k_{eff} is made <1.0065 by introducing burnable poisons and control rods. After some time fission products with high neutron capture cross sections are formed (Xe, Sm, etc.), reducing k_{eff}. With Δk 0.0065, a temperature increase of 150°C would reduce k_{eff} to <1 for the example given. This can be (and is) used for reactor self-regulation. The main contribution to the negative temperature coefficient in thermal reactors comes from the decreased moderator density (and also from increased void in a BWR) with increasing temperature.

Three types of control rod are used: *(control) rods* for regulating small power fluctuations, *shim rods* for coarse adjustments of k_{eff}, and *scram rods* for suddenly shutting down the reactor. The scram rods can move very fast and are actuated by gravity, spring release, compressed air, etc.; a typical value may be 3 s for complete shutdown. Since the reactor time constant is much larger, a neutron excursion is effectively stopped. The number of control rods is large (≥100) in order to facilitate an even power density in the

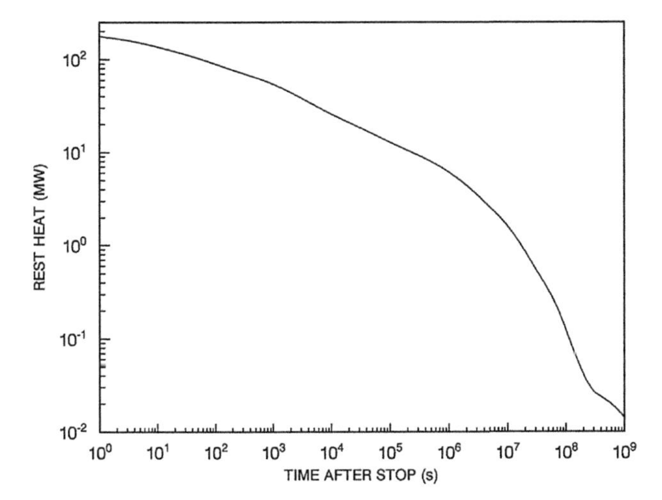

Figure 19.15 Decay heat from a 3000 MW_{th} LWR core, which has been operated continually for 300 days, as function of time after shutdown.

reactor, but the number of scram rods may be <10. Boric acid dissolved in the primary coolant is normally used in PWRs for coarse reactivity adjustment (chemical shim control). In light water reactors (BWRs as well as PWRs) injection of boric acid into the core acts as an additional emergency shutdown feature ("chemical scram").

The power of a reactor does not go to zero when it is shut off. Figure 19.15 shows the heat power due to fission product decay (logarithmic time scale) after a scram. For a 1000 MW_e (3000 MW_{th}) power reactor the rest (or after) power (or heat) is 2% of the full power effect (i.e. ~60 MW_{th}) after 1000 s. If this heat is not removed, the fuel and cladding would reach high temperatures. If nothing else would happen, the rather rapidly declining heat generation (Fig. 19.15), and the considerable heat capacity of the core, would probably limit the fuel temperature to a value below the melting point where the heat lost by thermal radiation and by convection in the remaining steam balances the decay heat. Unfortunately chemical reactions interfere. At normal cladding operation temperatures oxidation of zircaloy is very slow because of a protective oxide layer. However, at temperatures above ~900°C the protective oxide layer fails permitting the rapid exothermic reaction between zirconium and steam according to

$$\mathrm{Zr(s)} + 2\,\mathrm{H_2O(g)} \rightarrow \mathrm{ZrO_2(s)} + 2\,\mathrm{H_2(g)}$$
$$\Delta H = -6.67 + 2.57 \times 10^{-4}\,\mathrm{T}\,(\mathrm{MJ/kg\ Zr}) \tag{19.44}$$

where T is the cladding temperature (°C). If supplied with sufficient steam, this run-away chemical reaction can rapidly heat up the fuel to its melting point resulting in a partial core meltdown. Hot molten fuel would then accumulate on the bottom of the reactor vessel. After some time the melt could penetrate the vessel bottom at the weakest spots; pouring molten core material into the containment.

The *design basis accident* is usually considered to start with a *loss of coolant accident* (LOCA), e.g. as a guillotine break in a main recirculation line. In LWRs this immediately leads to $k_{eff} \ll 1$ because of reduced neutron moderation, independent of the position of the scram rods. However, the rest power would continue to heat the core, so core cooling is required in order to prevent a meltdown by keeping the cladding temperature below the critical range. Initiated by a lack of cooling, the heat from zircaloy oxidation caused a partial core melt-down in one of the Three Mile Island PWR reactors, Harrisburg, in 1979, but all molten material remained inside the pressure vessel. This accident had a large political impact in many countries and led to increased requirements with regard to reactor safety, e.g. improved emergency operating instructions and filtered containment venting systems.

Figures 19.16 show the *emergency cooling systems* in a BWR. There are two such systems for the core – the *core spray* and the *head spray* cooling systems (Fig. 19.16). In a modern BWR these two systems have a fourfold redundancy as Figure 19.16 shows. The systems are located in four separate zones 90° apart around the reactor and at different levels above ground. The cooling water would be taken from the wet well below the reactor. It is assumed that this well would not break; such an accident could only be initiated by an earthquake of unanticipated magnitude. In addition to the high pressure core spray there is a *low pressure coolant injection*. Further protection, especially against fission products leaking out of a damaged reactor vessel, is provided by the *containment spray*. Steam produced in the cooling is blown from the dry well into the wet well

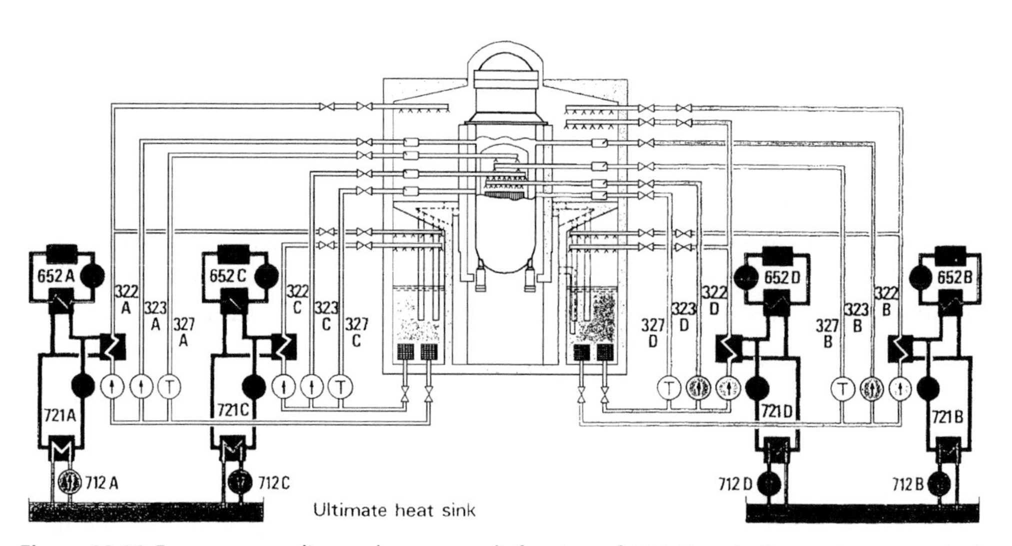

Figure 19.16 Emergency cooling and pressure relief system of ABB-Atom boiling water reactor (only one half of the cooling system is shown).

(the "blow-down" system). The water in the wet well is cooled by independent external cooling systems. The emergency cooling systems are much the same for BWRs and PWRs. In some plants mobile pumps can be connected from outside the containment building or water can be diverted from the fuel storage pool, thus providing an extra source of cooling water to the emergency systems.

A nuclear explosion in a reactor is physically impossible because of the large t_{per}. Rupture of the reactor vessel or pressure tubes due to overpressure is also highly improbable because of a number of safety valves (not shown in any of the Figures). The steam produced can be dumped directly into the condenser or into the containment in case of emergency. All piping into the core has isolation valves, which can be closed.

The reactor vessel is surrounded by several physical barriers. These barriers protect the surroundings from a core accident and protect the core from damage caused by external effects. In a PWR these barriers are the reactor tank (about 15–25 cm steel), the biological shield (1.5–3 m of concrete), and the outer containment. The whole primary circuit, including pressurizer dome, steam generator, and connecting pipes, is surrounded by concrete.

All nuclear power reactors outside the previous USSR and CMEA countries have a *reactor containment building*, the purpose of which is to contain steam and released radioactivities in case of a severe accident, and to protect the reactor from external damage. The containment is designed (and tested) to withstand the internal pressure from a release of the water in the entire primary cooling circuit (and in the case of PWRs of the additional loss of one of the steam generators), corresponding to excess pressures of 0.4 MPa. The containment is provided with a spray, which cools and condenses the steam released and washes out radioactive contaminations. The steam produced in a BWR can also be dumped directly into the condenser in case of emergency. All piping into the core has insulation valves, which can be closed if desired. Modern containments are multilayered pre-stressed concrete and steel and can withstand the impact of a jumbo jet aircraft.

A special hazard, which is taken into account in the emergency systems, is that the cooling would be actuated too late, i.e. after a core meltdown has begun. In practice the cooling system must start $\leq$1 min after a LOCA has occurred. If the core is very hot, a violent steam and hydrogen formation could occur when the cooling water contacts the core possibly causing further heat generation by zircaloy oxidation. In this case the blow-down system, the containment, and the spray systems are still expected to contain the early energy excursion.

Under accident conditions, hydrogen might be produced from steam (or water) and zircaloy; the risk for a fire or an explosion then depends on available oxygen and mixing.

As PWR containments usually contain air, a violent hydrogen explosion could occur which would represent some hazard to a PWR with a large dry containment and even more so in case of smaller dry containments where ice is stored to be used for condensing escaping steam (ice condensers). Such containments therefore contain igniters in order to cause a controlled hydrogen burning.

In case of inerted PS-type containments, a large hydrogen production could still interfere with the blow-down process causing pressures exceeding the design limit. To prevent breaking the containment in case of overpressure all types of containments in many countries now connect by rupture disks or valves to *filtered venting systems* designed to capture most of the vented radioactive material.

In case of any event that clearly deviates form normal behavior of a reactor (ranging from observations with safety implications to catastrophes), the event should be reported to IAEA and the appropriate organizations in other countries. For reports to the press and to the general public of safety related events in nuclear power plants IAEA and OECD/NEA recommend the INES scale, see Table 19.6.

19.16. RADIOACTIVE REACTOR WASTE

In the "ideal" nuclear reactor all fission products and actinides produced are contained in the fuel elements. In all practical reactors there are four processes through which radioactivity leaves the reactor vessel; in all cases the carrier of activity is the coolant:

(i) induced radioactivity in the cooling medium (and moderator if separate);
(ii) corrosion products containing induced activities from construction materials;
(iii) leaked fission products and actinides from faulty fuel elements;
(iv) fission products of actinides deposited on surfaces in the core ("tramp uranium").

In all these respects pressurized and boiling water reactors present some problems of concern.

Although stringent efforts are made to contain all radioactivity formed by the nuclear reactions, it is unavoidable that small releases occur to the environment. Such releases, gaseous as well as liquid, are carefully monitored by measuring stations at the plant and at some locations around the plant. The authorities (nuclear regulatory commission, nuclear power inspectorate, radiation health board, etc.) set limits for such releases. Table 19.7 gives release values for a number of different reactor types and times, as reported by the United Nations. It should be noted that these are actual releases at that time. Usually the measured releases are a small fraction of the allowed limits and tend to decrease with time as more stringent limits are imposed by the regulatory bodies, improved equipment installed or better procedures adopted.

It is important to realize that the legally *permitted release* in fact sets the normal level of releases. Very much lower releases than permitted usually means increased operating costs. Hence, economic and safety considerations sets the practical release level to some chosen fraction of the legal limit, e.g. 1% or 0.1%. A 1% practical limit would give an ample margin to the legal limit, thereby reducing the risk that accidentally increased releases exceed the legal limit (which might have dire economic consequences for the plant operator).

Table 19.7 Average atmospheric and aquatic releases of radionuclides from various reactor types 1990–1997

Release (kBq/kWh)	PWR	BWR	GCR	HWR	RBMK	FBR
Airborne						
Noble gases	2.3	3.0×10^{1}	1.6×10^{2}	1.3×10^{2}	1.2×10^{2}	3.4×10^{1}
^{3}H	2.7×10^{-1}	1.0×10^{-1}	4.9×10^{-1}	5.6×10^{1}	3.0	5.6
^{14}C	1.9×10^{-2}	5.5×10^{-2}	1.1×10^{-1}	3.7×10^{-1}	1.5×10^{-1}	1.4×10^{-2}
^{131}I	2.9×10^{-5}	6.3×10^{-5}	1.0×10^{-4}	2.9×10^{-5}	8.0×10^{-4}	2.9×10^{-5}
Particulates	1.7×10^{-5}	3.0×10^{-2}	2.9×10^{-5}	5.7×10^{-6}	1.3×10^{-3}	7.4×10^{-4}
Liquid						
^{3}H	2.3	1.0×10^{-1}	2.9×10^{1}	4.7×10^{1}	1.3	2.0×10^{-1}
Other	1.5×10^{-3}	3.1×10^{-3}	6.9×10^{-2}	9.9×10^{-3}	6.3×10^{-4}	4.1×10^{-3}

(Data from UNSCEAR 2000)

Legal release limits are usually set so that a critical group of the general public should at most get an insignificant additional dose, typically 0.1 mSv/y. It is also customary to put a legal limit on the collective dose (usually per GW installed generating capacity) arising from operation of a nuclear power plant.

19.17. NUCLEAR FUSION

For more than 50 years scientists have studied methods of copying the fusion energy processes occurring in the sun and other stars in which hydrogen is transformed into helium. The availability of hydrogen and deuterium in the sea is so large that it will outlast any other non-renewable energy sources for any population growth rates: 1 liter of sea water contains deuterium with an energy content equivalent to 300 liter of gasoline (reaction 12.14).

At the first Geneva conference on the peaceful uses of atomic energy in 1955 there was strong sentiment that controlled thermonuclear reactors (CTR) would be in operation within 15 years. The optimism has gradually vanished as research has uncovered numerous technical difficulties. At present, it seems quite unlikely that such reactors will be in operation before the year 2050, although extensive research is being conducted towards CTR's, and great progress is being made. The following sections briefly describe the principal concepts, and some of the problems of controlled thermonuclear (fusion) power generation.

19.17.1. The controlled thermonuclear reactor

Fusion reactions are most easily achieved with hydrogen atoms because of the low coulomb barrier and favorable wave mechanical transmission factor. The threshold reaction energy (E_{CB}(min), eqn. (12.13)) for the $^1H + ^1H$ reaction is 1.11 MeV, which corresponds to an average temperature of 10^{10} K; at 10^8 K the fraction of particles with an energy $\geq$ 1.11 MeV is about 10^{-55}, at 10^9 K about 10^{-5}, and at 10^{10} K about 0.5. The quantum mechanical tunnel effect allows the reaction to proceed at an acceptable rate at lower temperatures: For the D + T reaction the ignition temperature is 3×10^7 K and for the D + D and D + ^{3}He reactions it is 3×10^8 K. These reactions are the prime candidates for controlled fusion; see Figure 12.8. Of the number of designs proposed for CTR's the present discussion is limited to inertial confinement and magnetic confinement systems.

Several factors determine the possibility of achieving a thermonuclear fusion reaction: the particle energy, which is related to the temperature (§3.6.2), the particle density and the reaction rate. These are connected in three design criteria:

(i) the *fusion reaction rate* parameter σv, where σ is the reaction cross section (which depends on the particle energy, cf. Fig. 12.8), and v the relative speed, or temperature, of the ions, averaged over the Maxwellian velocity distribution (the temperature must be $\geq 10^8$ K),

(ii) the *Lawson limit* $n\tau$, where n is the particle density and τ the confinement time, which indicates the ability of the plasma to retain its heat and is also called the *confinement quality* — the product must be $\geq 10^{20}$ particles s m^{-3} for the DT-reaction, and $\geq 10^{21}$ s m^{-3} for the DD-reaction,

(iii) the *fusion product* $n\tau T$, where T is the average ion temperature; this product must be $\geq 5 \times 10^4$ s eV m^{-3}.

The steady state reactor is limited in power density by heat transfer and other considerations to about $n = 10^{20}$–10^{21}. Since each collision involves two particles, the fusion power density varies as the square of the particle density. At 1 Pa (i.e. 3×10^{20} particles m^{-3}) the power density would be tens of MW per m^3. This leads to a required confinement time of about 0.1–1 s.

19.17.1.1. Inertial confinement

Inertial confinement is a pulsed operation system[7]. Small pellets of solid D_2 and T_2 ($\leq$1 mm) are placed into the middle of a chamber where the pellets are irradiated by intense beams of photons (from lasers) or electrons (from accelerators). The surface of the pellet rapidly vaporizes, resulting in a jet-stream of particles away from the pellet and an impulse (temperature-pressure wave) which travels into the pellet, increasing the central temperature to $>10^8$ K. This causes a small fusion explosion, producing energetic ^{4}He and n, eqn. (12.20). Because the particle density is high, the pulse time can be very short and still meet the Lawson criterion. Temperatures of $\sim 10^9$ K have been reached with electron beams, and fusion neutrons produced. With a repetition rate of $\sim$100 pellets

[7] In the uncontrolled case of a hydrogen bomb, it is a single event.

s^{-1}, a power output of 1–10 GW would be achieved. Difficulties which have hampered the development include problems of shielding against the 14 MeV neutrons and extraction of the released kinetic energy.

19.17.1.2. *Magnetic confinement*

At $T = 10^7$ K, hydrogen atoms are completely dissociated into H^+ and free e^- (the plasma state). Because no construction material can withstand a plasma of this energy, it is necessary to keep it away from walls, which can be done by using strong magnetic fields, which is the same principle as used in mass spectrometers or cyclotrons. Of the various designs of "magnetic bottles", the torus (Fig. 19.17) seems to be the most promising and is the design chosen for further development in several national and international programs. The first torus machine was built in Moscow and named Tokamak; these machines have since then been called *tokamaks*.

19.17.2. Experimental fusion reactors

Several large machines based on magnetic confinement have been built. The best results so far obtained, though not simultaneously, are: central ion temperature of 35 keV (4×10^8 K), confinement time of ~2 s, particle density of ~5×10^{19}. Figure 19.18 shows the results obtained with different machines. Table 19.8 lists the results obtained with the JET machine and design parameters for ITER (International Thermonuclear Reactor Experiment). JET (Joint European Torus), the largest tokamak machine in operation, is an international project located at Abingdon, England. ITER is also an international experimental project which is expected to produce >1000 times the power of JET. ITER is being built in Cadarache, France, and it is planned to be operational

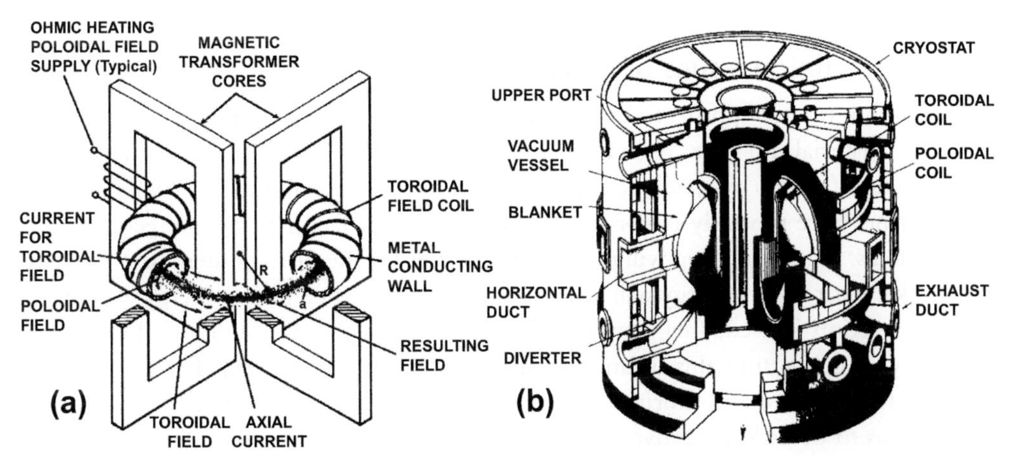

Figure 19.17 (a) The torus magnetic confinement principle. (b) Schematics of the JET (or ITER) tokamak experimental reactor.

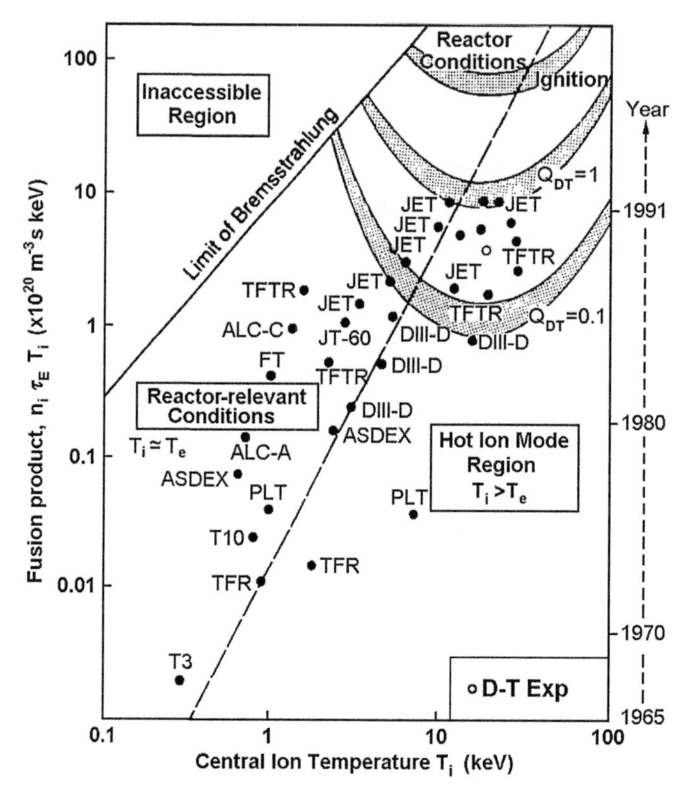

Figure 19.18 Results obtained by different fusion reactor designs.

in 2018. If promising results are obtained with ITER, it is planned to be followed by a power prototype, possibly some time after 2030 A.D.

19.17.3. Technical issues of large fusion reactors

A large fusion reactor appears to be more complicated than a fission reactor. The cost of the raw material for the two energy systems — ^{2}H- and ^{6}Li-enrichment, and T-production (~50% by irradiation in fission reactors), compared to U-production and enrichment — are probably rather similar, and is a minor factor in the net economy of both systems. Waste handling will perhaps be cheaper for the fusion system. On balance, it is unlikely that fusion energy will become much cheaper then fission energy. As long as the DT-reaction is used, fuel supply will be limited by the availability of Li and the efficiency of converting this litium to tritium. Further, considering that with fission breeders the uranium and thorium energy resources are sufficient for several centuries, there seems to be little incentive for the industry to engage in a rapid development of large fusion reactors, although a steady, well planned research is expected to continue on an international basis as fusion energy using the DD-reaction promises to be the long term answer to the world's energy needs.

Table 19.8 Parameters of Tokamak fusion reactors

	JET	ITER
Torus radius, central point (m)	3.0	6.0
Vacuum chamber: width (m)	1.2	2.05
height (m)	3.0	6.0
Total reactor height (m)	11.3	27.6
Total reactor radius (m)	17.5	31.0
Plasma current (MA)	4–7	22
Toroidal field (T)	3.5	4.9
Power in (MW)	36	
Operation:		
ion temperature T (keV)	25	10–20
" (K)	2×10^8	$\sim 10^8$
ion density, n ($\times 10^{19}$ m^{-3})	4	25
pulse length/confinement time τ (s)	1.2 (25)	400
fusion product $n\tau T$ ($\times 10^{20}$ keV s m^{-3})	8–9	50
DT reactions/s	6×10^{17}	10^{21}
T-consumption (kg/d)†	0.43	
DD reactions/s	1×10^{17}	
power out (thermal MW)	$2^\ddagger$	1000

†At continuous operation.
‡For about 2 s in DD reaction.

(a) *Particle injection and withdrawal.* In a steady state machine, whether it operates in pulsed mode or not, fuel (i.e. D and T) must be injected and, after consumption, reaction products (^{3}He, ^{4}He) must be withdrawn. Because of the strong magnetic field, injection of ions is an extremely difficult problem. Consequently, systems with injection of high energy neutral particles are under study. This could be accomplished through particle charge exchange, e.g.

$$D^+(\text{high energy}) + D(\text{low energy}) \rightarrow D(\text{high energy}) + D^+(\text{low energy}) \quad (19.45)$$

If all ions are removed magnetically, a beam of uncharged D-atoms (neutral beam injection, NBI) can enter the fusion volume. Another alternative being tested is high speed injection of small frozen D-T pellets into the hot plasma were they vaporize.

The withdrawal problem is even more difficult. Presently a shut down is necessary when large amounts of helium have been formed, after which the vacuum torus is purged and filled with a fresh D-T mixture. Considering the requirement of short interruptions in the energy production and the need to maintain an extremely high vacuum in the whole torus volume, a successful technical solution to this step is crucial.

(b) *Plasma confinement and heating.* The density must be increased to meet the Lawson criterion. This is achieved by increasing the magnetic field, which also leads to an ohmic heating of the plasma, though possibly not sufficient for ignition. Additional heating is required which possibly can be achieved by high frequency (ion cyclotron resonance) heating. Injection of deuterium/tritium ions of high kinetic energy is another (cf. eqn. (19.45)). After ignition further heating comes from the 3.5 MeV α-particles of reaction, eqn. (12.20).

A magnetic field cannot, of course, increase indefinitely. When it has reached it maximum value, the machine must be shut down to allow the magnets to relax. The cycling time may be 3–10 minutes. The magnetic field goes to extremely high values, requiring superconducting magnets and extremely sturdy construction, as the high current ($\geq$30 000 A) through the magnet coils leads to a pressure of >1 ton cm^{-2}.

Particle injection, withdrawal, and heating lead to the emission of bremsstrahlung and synchrotron radiation of an energy much less than that corresponding to the fusion temperature. It is, therefore, lost by the plasma and absorbed in the walls of the vessel, which must be cooled. An additional difficulty is the heat insulation required between the very hot walls of the vacuum vessel (~1000°C) and the current carrying very cold super-conducting coils (at ~4.5 K).

(c) *Energy extraction and fuel cycle.* In the DT cycle 80% of the energy appears as neutron kinetic energy (14 MeV). Most steady state concepts involve the capture of this energetic neutron in a surrounding blanket containing lithium as a metal or salt (e.g. Li_2BeF_4) in which tritium is produced according to

$$^{6}\mathrm{Li}\,(7.4\%) + \mathrm{n} \rightarrow {}^{3}\mathrm{H} + {}^{4}\mathrm{He} \quad Q = 4.78\ \mathrm{MeV} \tag{19.46}$$

Neutrons reacting with the other Li-isotope yield two ^{4}He:

$$^{7}\mathrm{Li}\,(92.6\%) + \mathrm{n} \rightarrow 2\,{}^{4}\mathrm{He} \quad Q = 18.13\ \mathrm{MeV} \tag{19.47}$$

These reactions develop additional kinetic energy, which is converted into blanket heat. Most of the energy of the fusion process appears therefore in the hot blanket. If the blanket is in the form of molten lithium metal or salt, it can be pumped through a heat exchanger, which, in a secondary circuit, produces steam for a turbine. Although the melting point of Li metal is only 186°C the Li temperature in the system will probably be around 1000°C. The tritium produced must be recovered for recycling as fuel.

The DT reactor needs several kg tritium as starting material (1kg ^{3}H ~ 3.6 × 10^{17}Bq). A likely technique involves the irradiation of a ^{6}Li-Al alloy in a high flux thermal fission reactor which produces both tritium and ^{4}He, see eqns. (19.48) and (19.49)); These can be separated on the basis of their different vapor pressures, different permeability through palladium, or through their different chemical reactivities.

The first fusion reactors probably will use the DT-reaction, as the DD-reaction requires higher temperatures, Figure 12.8. The DT-reaction yields on the average 0.5 neutrons. Provided the blanket consists of only ^{6}Li, this produces 0.5 new T-atoms (per consumed T-atom); if the blanket also contains ^{7}Li, the yield of new T will be less. Therefore, the DT-fusion reactor must be fed continually with new tritium, produced in fission reactors. This demand will not be eliminated until the DD-fusion reactor comes into operation.

(d) *Construction material.* A major area of research is the proper choice of construction material, which should be strong, heat and radiation resistant, and have low neutron capture cross sections. However, materials testing will not be intensely tackled until ITER comes into operation. A preliminary materials choice could be the following, starting from the center of the plasma:

(1) A vacuum chamber of 10 cm stainless steel, on the inside lined with reinforced carbon shields to protect the steel. The chamber walls are water cooled. As the shielding erodes it has to be occasionally replaced by remote control.

(2) A water cooled blanket (and shield), ca. 1.5 m thick, made of vanadium alloy (e.g. V-15Cr-5Ti) and containing solid or liquid lithium. As much heat is produced, the lithium is used both for cooling and breeding and is pumped to heat exchangers for steam production followed by chemical treatment to recover the tritium. Cooling can also be achieved by He-gas, in which case the Li-system must be treated separately.

(3) A water cooled shield and thermal insulator, containing γ- and n-absorbing material.

(4) A liquid helium cooled superconducting magnet coil.

(e) *Health, environmental and economic aspects.* An operating reactor would contain several kilograms of tritium (1 kg of T_2 is about 3.6×10^{17} Bq of radioactivity) which presents a hazard corresponding to the noble gas fission products. However, tritium is more difficult to contain because of its ability to pass through many metals, especially at high temperatures. Moreover, tritium can exchange with hydrogen atoms in water, and thus become an inhalation and ingestion hazard. Tritium is already a problem for fission reactors; for fusion reactors the problem is at least a factor of 1000 greater. Hence, the fusion reactor must be sealed extremely well against tritium leakage.

The structural material will be exposed to high radiation fields, causing radiation damage and induced radioactivity. The preferred material at present seems to be reinforced carbon, special steel and vanadium. Thus, considerable amounts of ^{49}V ($t_{1/2}$, 330 d), and possibly also some very long-lived ^{53}Mn, are formed. This induced activity will be a maintenance hazard, requiring remote control systems. However, compared to a fission reactor of similar size, the fusion reactor will contain less total radioactivity, and (of special importance in waste disposal) be free of long-lived α-activities.

A safety aspect, which has not been thoroughly studied, is the rapidly changing strong magnetic field (1 T s^{-1}), which will put great stress on the coils and structural material. The effects of such high fields on the operators are unknown.

19.18. NUCLEAR EXPLOSIVES

Assume that 50 kg of metallic ^{235}U is rapidly brought together to provide a critical configuration. Using a generation time of 3×10^{-9} s for neutrons, it can be calculated with eqns. (19.33) and (19.34) that it would take 0.2×10^{-6} s to increase the number of neutrons from one to that required for fissioning all ^{235}U atoms. However, long before this time, the energy absorbed in the material would have blown it apart, so 100% efficiency is impossible in a nuclear explosion. If 2% of the 50 kg of uranium had been fissioned, the amount of energy released corresponds to about 20 000 t of TNT (1 t of TNT corresponds to 4.19×10^9 J). The energy production is said to be 20 kt. Only ^{233}U, ^{235}U, and ^{239}Pu (of the more easily available long-lived nuclides) have sufficiently high cross sections for fission by fast neutrons to have a reasonable critical mass for use as nuclear explosives. However, it is a common misconception that when a critical mass is reached, a nuclear explosion will follow. In the cases when a critical mass has been reached by mistake, no energetic nuclear explosion has ever occurred.

A rough estimate of the critical radius of a homogeneous un-reflected reactor may be obtained simply by estimating the neutron mean free path according to (11.6). Assuming ^{235}U metal with a density of 19 g cm^{-3} and a fast fission cross section of 2×10^{-24} cm^2, one obtains $\Sigma_f^{-1} = 10$ cm. A sphere with this radius weighs 80 kg. For an un-reflected metal sphere containing 93.5% ^{235}U the correct value is 52 kg. ^{239}Pu has the smallest un-reflected critical size; for ^{239}Pu (δ-phase, density 15.8 g cm^{-3}) it is 15.7 kg (~6 kg reflected), and for ^{233}U 16.2 kg (~6 kg reflected).

Assume two half-spheres of metallic ^{235}U, each of which is less than critical size but which together exceed criticality. As the distance between the two spheres is diminished the k_{eff} increases for the total system. At a certain distance k_{eff} can be maintained close to 1 but the chain reaction still be prevented. If the distance between the two half-spheres is diminished rapidly to less than this critical distance in the absense of any neutron, a single neutron can then trigger a rapidly multiplying chain reaction. Since stray neutrons are always available, e.g. from spontaneous fission, a chain reaction would be started when the distance between the spheres is less than the critical distance. Hence it is important to achieve the supercritical configuration fast compared to the average time interval between the stray neutrons.

In order to achieve a substantial energy release in the short time available before the supercritical configuration is blown apart, an external or internal neutron source is triggered. By this action the fission chain reaction starts with not a few, but with very many high energy neutrons. This greatly increases the number of neutrons generated in the time available, causes many more fissions, and yields a correspondingly larger energy release.

Nuclear fission weapons contain "normally" either highly enriched ^{235}U (>93%) or fairly pure ^{239}Pu (≥95%). The Pu-metal has to be stabilized by alloying in order to prevent any phase changes that world convert it to powder. Some early nuclear explosives operated by having two subcritical parts of a sphere blown together by the use of chemical explosives (*gun type*), see Figure 19.19. To increase the explosive yield the fissile material is often surrounded by a heavy material (*tamper*) which has the double function of reflecting the neutrons and of acting as a heavy mass (inertia) to increase the time in which the supercritical configuration is held together. If natural uranium is used as the tamper, secondary fission in this shell can contribute to the overall energy release. However, only a fraction of the fission neutrons have energies large enough to fission ^{238}U efficiently.

A critical mass of material can also be arranged as a subcritical spherical shell which can be compressed into a supercritical sphere. This process is called *implosion* and is probably used in most weapons today, see Figure 19.19. The bomb over Hiroshima

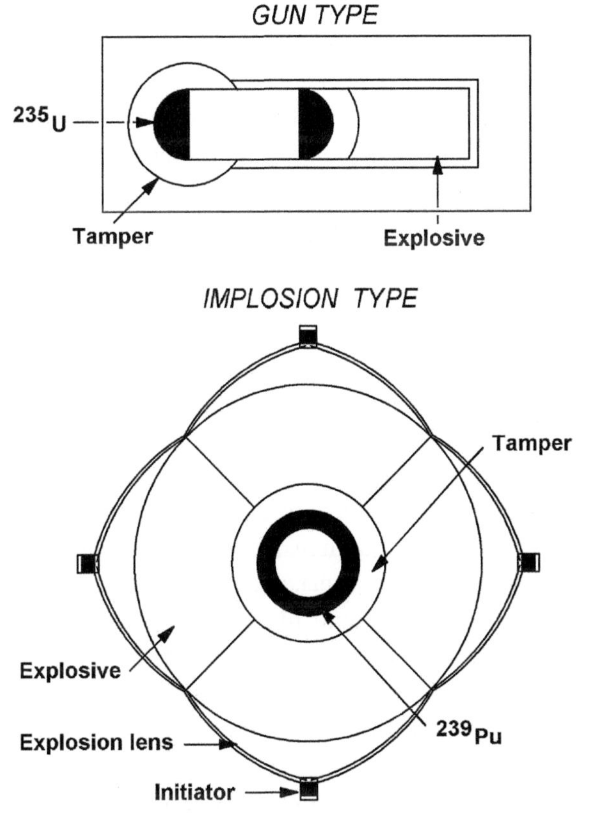

Figure 19.19 Two principles of fission bombs.

($\sim$15 kt) used uranium and was of the gun type, while the Nagasaki bomb ($\sim$22 kt) was of the more efficient implosion type using plutonium. At the end of the implosion the density of the fission charge (U or Pu) may be increased by compression thus reducing the amount needed.

In order for these devices to function efficiently, the neutron multiplication must not begin until the critical size has been well exceeded. Furthermore, the number of neutrons in the initial generation must be sufficiently high in order to reach a very high n-flux in the short lifetime of the supercritical configuration. Therefore, as mentioned above, a few microseconds before maximum criticality a large number of fast neutrons is injected by a special neutron source initiating a rapid neutron multiplication.

The amount of ^{240}Pu in a plutonium charge must be maintained relatively low since this nuclide decays partly through spontaneous fission with the resultant release of neutrons. If the concentration of ^{240}Pu is too high, the neutron multiplication will start too early and with too few neutrons in the first generation, and the energy release is decreased. Also ^{241}Pu, if present in larger percentage, could cause problems because of its rather rapid decay to ^{241}Am. To minimize the ^{240}Pu and ^{241}Pu content, plutonium for weapons used to be produced in special reactors by low burnup (<2000 $MW_{th}d\ t^{-1}$); this is uneconomical to the nuclear power industry where high burnup of the fuel (leading to higher ^{240}Pu concentrations; Fig. 19.7) is a necessity in order to obtain the lowest electricity production costs. It has been stated that also power reactor grade plutonium, even as oxide, can be used in a weapon, though less effective and predictable. The critical mass for such plutonium containing 20% ^{240}Pu has been estimated to be 20 kg for metal and 37 kg for PuO_2. If the Pu contains a measurable concentration ^{241}Pu it must be purified at regular intervals from the ^{241}Am which grows in. In the US this was mainly made at Rock Flats by extraction of the americium from the molten Pu-metal by a chloride melt.

Nuclear weapons of the fission type have been developed in sizes ranging from 0.001 kt to about 500 kt yield. Because the critical amount can never be less than several kilograms of fissile material, a weapon of low yield such as 0.001 kt has very low efficiency ($<5 \times 10^{-6}$). The highest achieved efficiency of a plutonium based charge seems to be an order of magnitude higher than that achieved by a ^{235}U based charge. However, most fission devices are assumed to contain both materials. Eight countries (USA, USSR, UK, France, China, South Africa, Pakistan, India, and North Korea) have carried out >1000 known nuclear weapons tests (fission and fusion devices). Since the Partial Test Ban Treaty in 1963, most explosions have been underground. At the time of writing there is an agreement to ban further tests and South Africa has handed over it nuclear weapons. However, it is uncertain if this has a sufficient effect on all nations aspiring to be nuclear powers.

In the explosion of fission weapons a temperature of approximately 10^8 K is obtained. This temperature is sufficient for producing fusion reactions between deuterium and tritium (see Ch. 17). Although no official information is available on how modern fusion weapons (hydrogen bombs) are constructed, it is generally assumed that they are based on the principle of using a fission charge as the initiator of the fusion reaction. In the debris of hydrogen bombs, lithium has been discovered. It is therefore believed that in hydrogen bombs, deuterium is combined with lithium in the form of solid LiD. ^{6}Li will react with a neutron to form tritium in the reaction

$$^{6}\mathrm{Li} + \mathrm{n} \rightarrow \mathrm{T} + {}^{4}\mathrm{He} \quad Q = 4\,\mathrm{MeV} \tag{19.48}$$

With fast neutrons the reaction

$$^{7}\mathrm{Li} + \mathrm{n(fast)} \rightarrow \mathrm{T} + {}^{4}\mathrm{He} + \mathrm{n} \tag{19.49}$$

also occurs.

The deuterium and the tritium can react as shown by equations (12.13) – (12.24). If the temperature and pressure is high enough the D-D reaction can also contribute. Natural uranium, depleted uranium, or enriched uranium must probably be used to improve the neutron economy when the charge contains LiD. Secondary fission in uranium also leads to an increased energy production in the weapon. The high energy of the neutrons from the fusion reactions make them very effective for fission of ^{238}U. The fission of ^{238}U produces large amounts of radioactive fission products, resulting in a "dirty" weapon. By contrast, if a tamper of a non-fissionable heavy material is used the only fission products released are those obtained from the ignition process and the weapon is relatively "clean".

In order to explode a hydrogen bomb, a small fission device is used. The X- and γ-rays from the fission charge travels much faster than the shock wave from the explosion and also faster than the neutrons. In order to achieve a secondary implosion, the fusion charge is surrounded by styrene foam which is converted to a plasma by the impinging X- and γ-rays. This plasma heats and compresses the thermonuclear charge which is then hit by the neutrons. This principle has been described several times in the open literature and is illustrated in Figure 19.20. By repeating this structure along the bomb axis multiple fusion (two and three stage) designs have been made.

Charges of sizes up to ~60 000 kt TNT have been tested. If it is assumed that half of the energy of a 50 Mt device comes from fusion and the other half from fission, and a 25% efficiency in the explosion, about 8 t of nuclear material is required. It is believed that the 50Mt bomb was a three stage design using ^{238}U cores in the center of the fusion charges.

Very high temperatures are reached in the center of a nuclear explosion (~10^{8} K). The exploded material forms a fireball which rapidly expands and moves upward in the atmosphere. After the first wave has passed, a low pressure area is formed under the fireball which draws material from the ground to produce the typical

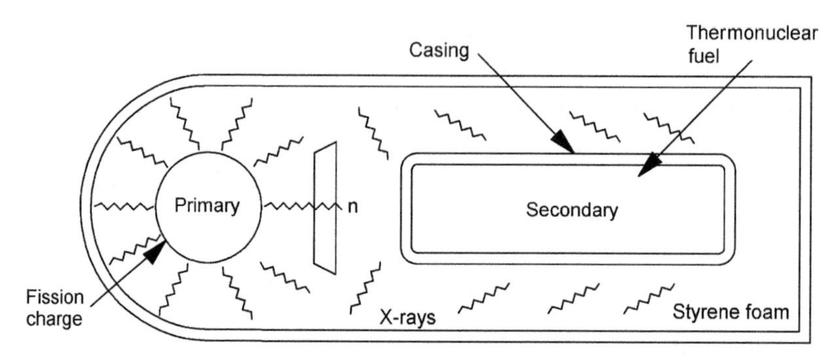

Figure 19.20 Principle of a hydrogen bomb using LiD fuel and styrene foam to produce the compressing and igniting plasma.

mushroom-like cloud associated with nuclear explosions. This cloud contains fission products as well as induced radioactivity from the construction material of the device and from the immediate surrounding ground that has been irradiated. Some of the radioactive particles are carried to great heights and eventually fall from the atmosphere as *radioactive fallout*, see Ch. 22. Much of the fallout returns to the earth close to the explosion site, but some of the material reaches the stratosphere and returns to the earth only slowly over a long period of time and over great distances from the original explosion site. The radioactivity of the fallout decays with time according to the approximate relationship

$$A = A_0 \, e^{-1.2t} \tag{19.50}$$

where A_0 is the radioactivity immediately after the explosion and t is the time in hours after the explosion.

The energy produced in the explosion of a 20 kt weapon is distributed approximately as follows: 50% pressure; 35% heat; 5% instantaneous radiation; 10% radiation from fission products. The radiation dose at 500 m from hypocenter (the vertical point on ground below the explosion center) has been estimated to be ~70 Gy; at about 1.1 km it is ~4 Gy. The pressure wave moves from the explosion site with the velocity of sound. This initial short pressure front is followed by a low pressure wave leading to a rapid change in wind direction. The deaths in Hiroshima and Nagasaki resulted in 20−30% of the cases from primary burns, 50−60% from mechanical injuries and secondary burns, and only ~15% from radiation injuries. The health detriment to the Japanese population is discussed in Chapter 15.

Table 1.1 lists some historical "firsts" with regard to nuclear weapons. The extensive tests in the atmosphere up to 1963 lead to a large global spread of tritium, ^{14}C, fission products and actinides. Scientists have used this to learn more about global wind and water currents. Radiochemists have studied the migration of deposited radionuclides, as discussed in Chapter 22, radioecologists the uptake of radioactive elements by plants and animals, as described in Chapter 15, etc.

A number of ideas have been proposed for the peaceful use of nuclear explosives. It has been estimated that the explosive power of nuclear charges is sufficiently great that their use for large scale explosions would be much cheaper than the use of conventional explosives. About 136 explosions have been used for civil engineering projects in the former USSR; sizes varied from <1 kt to ~50 kt. In the United States crater formation and gas stimulation experiments have been made. Suggested uses include the blasting of harbours on remote coasts to allow greater access to inland mineral deposits, crushing of ore bodies in order to obtain valuable minerals, crushing of underground rocks which may hinder the use of natural gas reservoirs, diverting asteroids threatening the earth, etc.

19.19. EXERCISES

19.1. If the energy developed by a 20 Mt fusion weapon could be used for producing electricity at a value of 2 cents kWh^{-1} what would the "electric value" of the device be? Such a weapon may be expected to cost $ 10 million. One ton TNT releases 1 Gcal of energy in an explosion.

19.2. The bomb over Hiroshima contained ^{235}U. How many grams were fissioned to correspond to 15 kt of TNT?

19.3. Compare two 500 MW_e electric power stations, one burning oil and the other using 3.0% enriched uranium. Both stations operate 6000 h y^{-1} at 35% efficiency (heat to electricity). The oil (43.5 MJ/kg combustion energy) is carried by 100 000 ton d.w. (dead weight, i.e. carrying capacity) oil tankers, and the uranium fuel by train cars of 20 t capacity each. (a) How many oil tankers will be needed every year for the oil-fired station? How many train cars will be needed every year for the nuclear power station for transporting (b) the enriched UO_2 reactor fuel, (c) the corresponding amount of natural uranium as U_3O_8 to the isotope enrichment plant, if the tail is 0.35% in ^{235}U? (See Ch. 3 and §19.9.) Reactor fuel rating 40 $MW_{th}d$/kg U.

19.4. Calculate the number of collisions required to reduce a fast fission neutron ($\bar{E} =$ 2 MeV) to thermal energy (E_n 0.025 eV) in a light-water-moderated reactor, assuming that the data in Table 19.3 are valid.

19.5. Calculate the thermal fission factor for a mixture of 60% ^{239}Pu, 30% ^{240}Pu, and 10% ^{241}Pu.

19.6. The world's first water boiler reactor (LOPO, Los Alamos, 1944) was a homogeneous solution of enriched uranium sulfate as follows: 580 g ^{235}U, 3378 g ^{238}U, 534 g S, 14068 g O, and 1573 g H. From these values, and Tables 19.2 and 19.3, calculate η and f; neglect S. With $p = 0.957$, what will k_∞ be?

19.7. A large homogeneous thermal reactor contains only ^{235}U dispersed in beryllium in the atomic ratio 1:3 × 10^4. The migration area is 0.023 m^2. Assuming $p = \varepsilon = 1$, calculate the size of a cylindrical reactor with height equal to diameter.

19.8. A cubic unreflected graphite moderated natural uranium reactor contains 3% enriched uranium as UC homogeneously dispersed in the graphite matrix; the weight ratio C/U = 10. The resonance passage and thermal utilization factors are both assumed to be 0.9; $\varepsilon = 1.00$. Make an estimate of the critical size of the cube.

19.9. The LOPO reactor in exercise 19.6 has a neutron age $\tau = 31.4$ cm^2, and diffusion area L^2 1.87 cm^2. Calculate (a) the fast neutron leakage factor, and (b) the critical radius for the homogeneous sphere, if $k_\infty = 1.50$.

19.10. Our solar system is considered to be 4.5 billion years old. What was the ^{235}U percentage in natural uranium when the solar system was formed?

19.11. The radiometric sensitivities for discovering ^{59}Fe, ^{131}I, and ^{90}Sr are 75, 25, and 0.74 kBq m^{-3} of water. In the Würgassen plant the total permitted aqueous annual release was 17 Ci β-emitters. Assume an activity ratio in the cooling water of 100:10:1 for the three nuclides above and that none of these activities exceed 1% of the permissible release. How many times must a liquid sample taken each day be concentrated to meet these requirements?

19.12. What amount of tritium (Bq) is produced in the Würgassen nuclear plant assuming that ^3H is only produced through capture in the deuterons of the original cooling water, the amount or which is 50% of the core volume? Data on fluxes, cross sections, and releases are given in Table 19.4. Neglect the tritium decay rate.

19.13. A BWR has operated at full power for a week. At what time after a scram would the Xe poisoning reach its maximum? Use data in Table 19.4.

19.14. The net efficiency of a 1000 MW_e BWR is 34.1%. Estimate the amount of ^{235}U consumed during its first day of full power operation with a completely fresh load of fuel.

19.15. Consider a power reactor in which microspheres ($r = 0.3$ mm) of frozen 1:1 T-D mixture (density 170 kg m^{-3}) are fused by laser irradiation. The laser compresses the spheres to $N_v = N_v^\circ 10^4$, where N_v° is the number of atoms per m^3 at ordinary pressure, and also heats it to a temperature corresponding to almost 20 keV. The energy developed through the T-D fusion reaction leads to expansion of the spheres, which occurs with the velocity of sound ($v_s = 10^8$ m s^{-1}). This leads to no more than 25% of the particles fusing. What power is produced if the fusion micro-explosions occur at a rate of 30 per s?

19.16. Deuterium is to be injected into a fusion reactor at a density of 10^{20} D^+ and 10^{20} e^- m^{-3} and at an energy of 100 keV. How much of the deuterium must fuse to compensate for the ionization and injection energy? The ionization energy of the deuterium atom is 13 eV.

19.20. LITERATURE

H. D. Smyth, *Atomic Energy for Military Purposes*, Princeton University Press, 1946.
S. Glasstone (Ed.), *The Effects of Nuclear Weapons*, USAEC, Washington, 1962.
J. K. Dawson and R. G. Sowden, *Chemical Aspects of Nuclear Reactors*, Butterworths, 1963.
S. Peterson and R. G. Wymer, *Chemistry in Reactor Technology*, Pergamon Press, Oxford, 1963.
ANL-5800, *Reactor Physics Constants*, USAEC, TID, 1963.
Reactor Handbook, 2nd edn., Interscience, 1960–1964.
J. R. Lamarch, *Introduction to Nuclear Reactor Theory*, Addison-Wesley, 1966.
J. G. Wills, *Nuclear Power Plant Technology*, Wiley, 1967.
C. O. Smith, *Nuclear Reactor Materials*, Addison-Wesley, 1967.
E. Teller, W. K . Talley, G. H. Higgins, and G. W. Johnson, *The Constructive Uses of Nuclear Explosives*, McGraw-Hill, 1968.
G. I. Bell and S. Glasstone, *Nuclear Reactor Theory*, van Nostrand, 1970.
W. Fratzscher and H. Felke, *Einführung in die Kerntechnik*, VEB Deutscher Verlag fur Grundstoffindustrie, Leipzig, 1971.
U.N. Conferences on peaceful uses of atomic energy, 1955, 1958, 1965, 1972.
J. A. Richardson, Radioactive waste quantities produced by light water reactors, *Nucl. Eng. Int.* Jan. (1974) 31.
J. R. Lamarch, *Introduction to Nuclear Engineering*, Addison-Wesley, 1975.
IAEA, *The Oklo Phenomenon*, Vienna, 1975.
G. A. Cowan, A natural fission reactor, *Sci. Am.* July 1976.
IAEA, *Management of Radioactive Wastes from the Nuclear Fuel Cycle*, Vols. 1 and 2, Vienna, 1976.
Jahrbuch der Atomwirtschaft 1977, Handelsblatt GmbH, Verlag fur Wirtschaftsinformation, Dusseldorf, Frankfurt, 1977.
Int. Conf. on Nuclear Power and its Fuel Cycle, Salzburg 1977, IAEA, Vienna, 1977.
R. A. Rydin, *Nuclear Reactor Theory and Design*, PBS Publications, 1977.
H. Edelhäuser, H. Bonke, I. Gans, O. Huber, K. J. Vogt, and R. Wolter, *Radioactive Effluents and Present and Future Radiation Exposure to the Population from Nuclear Facilities in the Federal Republic of Germany*, in Int. Conf. on Nuclear Power and its Fuel Cycle, Salzburg 1977, IAEA, Vienna, 1977.
H. W. Graves, Jr., *Nuclear Fuel Management*, J. Wiley & Sons, 1979.
M. Benedict, T. H. Pigford and H. W. Levi, *Nuclear Chemical Engineering*, 2nd Ed., McGraw-Hill, 1981.
The Management of Radioactive Waste, The Uranium Institute, London, 1991.
World Nuclear Industry Handbook – 2012, Vol. 57, No 697, Nuclear Engineering International, ISSN 0029-5507, Aug. 2012.

Nuclear explosives

Nuclear Weapons. Report of the Secretaty-General of the United Nations, Autumn Press, 1981.
S. Glasstone and A. Sesonske, *Nuclear Reactor Engineering*, Van Nostrand Reinhold, 3rd Ed., 1981.
W. Seifritz, *Nucleare Sprengkörper*, Karl Themig, 1984.
U. Ivarsson (Ed.), *FOA orienterar om kärnvapen* (in Swedish), the Swedish National Defense Research Institute, **15** (1990).

Controlled fusion

M. J. Monsler, J. Hovingh, D. L. Cook, T. G. Frank and G. A. Moses, An overview of inertial fusion reactor design, *Nucl. Techn. Fusion*, **1**, July 1981, 302.
J. R. Roth, The Impact of Engineering Constraints on the Feasibility of Advanced Fuel Fusion Reactors, *Nucl. Techn. Fusion*, **2** (1982) 29.
P. A. Ertaud, Fusion thermonucleaire controlee – la grande illusion, *RGN*, Mai-Juin 1985, 242.
R. W. Conn, V. A. Chuyanov, N. Inoue and D. R. Sweetman, The International Thermonuclear Experimental reactor, *Sci. Am.*, April 1992, 75.
J. G. Cordey, R. J. Goldston and R. R. Parker, Progress Toward a Tokamak Fusion Reactor, *Physics Today*, Jan. 1992, 22.
ITER-project link: http://www.f4e.europa.eu/

Nuclear Power Reactors

Contents

We concentrate in this chapter on existing and some future power reactors and discuss the features common to each generic type as well as some peculiarities of special designs. Waste generation and handling at the power plants is also described generically.

Electric power producing reactors have been built to 1500 MW_e but all sizes in the range 400−1400 MW_e are common. With increasing size the cost of the electric power usually decreases. Power reactors of ~200 MW_e are competitive in some areas of high energy costs, such as in many developing countries. These smaller sized plants have some advantages: they are more reliable due to less stress of the components, they can be built more easily in remote areas because the smaller components are easier to transport, the electric grid may be more capable of accepting the moderate power addition, and they can more easily be designed with inherent safety features. Table 20.1 summarizes the characteristics of some operating nuclear power reactors.

In an attempt to systematise the continuous development of nuclear power reactors the concept of reactor generations has been introduced. In 2012 about 4 generations are discussed. The third has actually been divided into two different ones due to the clear trend break for the last generation where the novelty is not only the reactor but a coupled, optimised system of techniques.

Radiochemistry and Nuclear Chemistry
ISBN 978-0-12-405897-2, http://dx.doi.org/10.1016/B978-0-12-405897-2.00020-3

Table 20.1 Characteristics of some nuclear power reactors. Data refer to 2012

Type	Example	Power (MWe)	Fuel (% ^{235}U)	Can	Moderator	Coolant	Coolant temp. (°C)	Coolant press. (MPa)	Net eff. (%)	Burn-up (MWd kg^{-1})	Temp coeff.	Core size (h × d m^2)
PWR	Diablo Canyon 1, USA	1136	93 t UO_2 (4.4)	Za	H_2O	H_2O	320	15.5	32.1	45	Neg.	3.66 × 3.37
VVER*	Novovoronezh 5, Russia	1000	86 t UO_2 (4.4)	Za	H_2O	H_2O	324	15.7	32	<40	Neg.	3.53 × 3.16
BWR	Oskarshamn 3, Sweden	1450	126 t UO_2 (3.1)	Za	H_2O	H_2O	286	6.9	36.4	33	Neg.	3.68 × 4.60
PHWR	Bruce B5, Canada	915	135 t UO_2 (nat)	Za	D_2O	D_2O	305	9.2	30	7.7	Pos.	5.94 × 5.67
RBMK	Kursk 4, Russia	1000	218 t UO_2 (2.4)	ZN	C	H_2O	284	6.9	31	22	Pos.	7.0 × 11.8
GCR	Wylfa A1, UK	550	673 t U (nat)	Mgn	C	CO_2	414	2.7	26	3.5	Pos.	9.1 × 17.4
AGR	Hinkley Point B1, UK	655	119 t UO_2 (3.4)	St	C	CO_2	655	4.2	42	18	Pos.	8.31 × 9.11
LMFBR	Creys-Malville, France**	1240	31.5 t (Pu,U) $O_2 + UO_2$†	St	–	Na	545	0.1	41	70–100	Neg.	1.0 × 7.0

Za = Zircaloy, ZN = Zr + 1% Nb alloy, Mgn = Magnox, St = Stainless steel.
* PWR-like Russian design.
†4.8 t Pu as 19% Pu MOX.
** Creys-Malvills is now permanently shutdown.

The basis for the division has been a development or improvement in certain features like e.g. safety, security and non proliferation, cost effectiveness, and the fuel cycle.

Generation 1: This generation is the early prototype reactors, typically designed in the 1950ies. Some examples are the Calder Hall-1 (1956–2003) in the UK and Shippingport (1957–1982) in the US. Many of these reactors were a "proof of principle" and the power levels were very low. The only remaining Gen 1 reactor operating today is the Wylfa Nuclear Power Station in Wales (UK) which was scheduled to be stopped in 2010, got an extension to late 2012, but is at the moment of writing not shut down.

Generation II. These reactors were typically designed in the 1960ies and were commercial reactors designed to be both reliable and economical. Most commercial power producing reactors operating today is of this generation. It include pressurized water reactors (PWR), CANada Deuterium Uranium reactors (CANDU), boiling water reactors (BWR), advanced gas-cooled reactors (AGR), Reaktor Bolshoy Moshchnosti Kanalniy or High Power Channel-type Reactor (RBMK) and Vodo-Vodyanoi Energetichesky or Water-Water Power Reactors (VVER). The design life time for this generation of reactors were typically 40 years. However, many of them have got a substantial prolongation of this time due to upgrades and replacement of critical parts.

Generation III. These reactors are essentially Gen II reactors which have evolved in some areas like thermal efficiency, fuel technology, modularized construction, and safety systems. In the latter case there has been some development to systems able to handle any event without operators input for several hours. The Gen III reactors comprise e.g. the Advanced Boiling Water Reactor (ABWR) by GE Nuclear Energy (first on line in Japan 1996), the advanced PWR (AP-600) (approved concept 1998) by Westinghouse and the Enhanced CANDU 6 by Atomic Energy of Canada Limited. These reactors have an expected life time of about 60 years.

Generation III+, This a further evolutionary step using essentially the same concepts as previously. The main feature is that more passive safety systems are included. They rely on gravity or natural convection to minimise the impact of abnormal events. Some examples of GEN III+ reactors are: VVER-1200/392M, Advanced CANDU Reactor (ACR-1000), AP1000 (based on the AP600) European Pressurized Reactor (EPR), and Economic Simplified Boiling Water Reactor (ESBWR) based on the ABWR. The GEN III+ reactors are also expected to achieve a higher fuel burnup than their evolutionary predecessors. This will reduce fuel consumption and thus waste production.

Generation IV, This generation is till rather undefined but it should be clear that now it is not only the reactor construction that is the issue but rather a whole concept aiming at superior usage of fuel by recycling the fuel. Thus increasing the energy utilisation and at the same time decreasing the long term radiotoxicity of the used fuel, hence achieving a considerable improvement of both safety and security. In many cases one hears about fast reactors being GEN IV reactors but this is not the case since the first power producing fast reactors (EBR I) went critical in 1951. In the GEN IV system it is assumed that there is a clear integration of both thermal reactors as well as fast reactors, recycling plants for the fuel and fuel fabrication. Regarding the fast reactors some "novel" combinations of fuel and coolant are currently discussed. The proposed recycling plats could also be designed to avoid a pure plutonium steam for proliferation resistance using a separation process similar to the Group Actinide EXtraction (GANEX) process, see Chapter 21. However, with the development of centrifuge based uranium enrichment, the proliferation path using reactor produced plutonium will become less and less attractive for countries desiring nuclear weapons.

20.1. THERMAL REACTORS

The large majority of nuclear power reactors is based on fission by thermal neutrons and the discussion will begin with this class of reactors.

20.1.1. Pressurized water reactors

As mentioned in Chapter 19, the name implies that a pressurized water reactor is cooled by hot high pressure water, either H_2O (PWR, VVER) or D_2O (PHWR). In the PWR and VVER types the coolant is also used as moderator whereas a separate D_2O containing moderator tank is normally used in the PHWR type. These power reactor types have several things in common: primary – secondary coolant circuits separated by heat exchangers (*steam generators*), a *pressurizer* to adjust primary system pressure, and often *chemical shim control* for adjustment of the excess reactivity with fresh fuel.

The pressurized water reactor is generally preferred for propulsion purposes (military surface vessels and submarines), partly because it can react faster to changes in power demand than many other types of thermal reactors.

(a) *PWR*. The main components of a pressurized light water moderated and cooled reactor (PWR) station have been described in §19.1 and Figure 19.2.

The important features of the PWR core are shown in Figure 19.12 and described in §19.13.1. The design with a pressure vessel without facilities for fuel handling under pressure necessitates a yearly shutdown for fuel replacement. However, this is not considered a great inconvenience as a PWR anyhow has to be shutdown on a regular basis for safety inspection of pipes, welds, etc, and for routine maintenance. The need to shut down the reactor and open the pressure vessel in order to replace fuel also makes the PWR (as well as the VVER) resistant to concealed nuclear proliferation.

Leaking tubes in the steam generators have plagued many PWRs. The leaks are usually caused by corrosion. Such leaks are normally fixed by plugging the affected tube. However, too many plugged tubes leads to a reduction in permitted power output; finally requiring either permanent shut down or replacement of the steam generators. In the beginning of 2000 many PWRs have already had their steam generators replaced (52 reactors) and replacements for another 21 reactors was planned.

Another generic problem has been the development of cracks in the control rod guide tubes of inconel which penetrate a PWR pressure vessel head. Such cracks are difficult to locate and repair because of the confined space between the tubes. After some time leaks usually develop when the cracks grow larger, ultimately requiring replacement of the vessel head. Crack formation can probably be prevented by a proper selection of material, its pre-treatment and meticulous adjustment of the water chemistry.

(b) *VVER*. In the former USSR and CMEA a reactor type similar to the PWR is used, referred to as the VVER, cf. Tables 19.1 and 20.1. The older VVER reactors have no containment; except two in Finland (Loovisa 1 & 2) where a Westinghouse type containment was added. The fuel elements have a hexagonal cross section and the steam generators are horizontal rather than vertical. The stainless steel control rods contain boron carbide as neutron absorber. A VVER reactor contains more water in the vessel than a typical PWR. This gives a higher heat capacity allowing a VVER to withstand interrupted coolant flow for a considerably longer time than a PWR. The pressure vessel in the VVER is exposed to rather high neutron fluxes and after many years of operation the wall has to be annealed (heated) in place to remove radiation induced brittleness. The VVER, like the PWR, has to be shutdown for refuelling.

(c) *PHWR*. The advantage of using heavy water for neutron moderation is that its low thermal neutron capture cross section (Σ_a (th), Table 19.3) improves the neutron economy, i.e. $1-p$ becomes much less than in LWRs) so that even natural uranium can be used as fuel. The good neutron economy and harder energy spectrum (cf. Table 19.3) leads to about twice as high plutonium yield in heavy water reactors as compared to LWRs. The first large heavy-water-moderated reactors were built in the United States around 1950 for production of materials for fission weapons. Presently the only commercial PHWR type is the CANDU (CANadian Deuterium Uranium) design. The reactor vessel is a horizontal cylindrical tank filled with D_2O, through which several hundred horizontal aluminum tubes pass. Within each of these *pressure tubes* are bundles of zircaloy clad natural UO_2 pins. The bundles are cooled by pressurized heavy water (~8 MPa), which in the steam generator produces light water steam for turbines. This pressure-tube-tank design is known as a *calandria*. In some designs the heavy water in the moderator tank can be dumped down into an empty tank below the reactor to allow an immediate shutdown, at the sacrifice of shielding in such an event. In other designs (Bruce, Table 19.7) the calandria is surrounded by a vessel full of ordinary water, which provides shutdown shielding; shutdown is then achieved by absorber rods and injection of gadolinium nitrate into the moderator. The calandria design permits on-line refuelling.

Early CANDU-PHWR were built in the 200 MW_e size, but the design can be upgraded to 750 MW_e (Table 20.1). Using natural uranium the burnup is comparatively low, about 9 $MW_{th}d\ kg^{-1}$.

The neutron economy makes it feasible to run PHWRs on the thorium cycle with a conversion ratio of ~0.9. Such a cycle has to be started and operated either with a mixture of $^{233}UO_2$, enriched $^{235}UO_2$ and ThO_2, or — if isotope enrichment is not used — through the following three step fuel schedule: (i) running on natural uranium and extracting Pu; (ii) running on Th + Pu and extracting ^{233}U; (iii) running on Th + ^{233}U.

The investment cost for CANDU reactors is higher than for LWRs because of a more complicated design and the large amount of D_2O required. The running costs are claimed to be lower; the standard loss of D_2O is ~1 kg d^{-1}. The neutron capture by D forms tritium in considerable amounts, so the D_2O becomes contaminated by TDO. Also TD and TDO can leak into the surroundings, which may cause difficulties in biological safety. Two methods have been used to limit the amount of T present: i) In an expanding PHWR system part of the D_2O (contaminated by T) in older reactors can be transferred to new reactors and replaced by fresh heavy water thereby providing a temporary relief by dilution, ii) when a troublesome inventory of T has been built-up, the D_2O is transferred to a special isotope separation plant where tritium is removed.

20.1.2. Boiling water reactors

In boiling water reactors (BWR, ATR and RBMK types) the coolant is brought to boiling while passing along the fuel pins. As mentioned in §19.13.2, no separate steam generator and secondary loop exist and the design has less complexity. In addition to negative temperature coefficients both for fuel and moderator, boiling water reactors have a negative *void coefficient*. This means that bubble formation (*void*) along the fuel pins reduces the reactivity. It is therefore common practice to control the reactor power by the main circulation pumps; increased circulation initially reduces the voids (and fuel temperature) thereby increasing power production until the previous void fraction has been almost restored at a slightly higher average fuel temperature. However, a very large negative void coefficient should also be avoided as it can cause an undesirable amplification of pressure fluctuations in the reactor vessel, generating rapidly increasing power oscillations.

In reactors with *pressure suppression (PS) containments* the steam outlet from the reactor vessel is located in the so-called *upper dry well* (Fig. 19.16). In the case of excess pressure in the tank, safety valves at the steam outlet open, releasing steam into the water in the *wet well* through *blow-down pipes* (the wet well condensation pool typically contains ~2000 m^3 water) thereby condensing the steam. Steam from an inadvertent pipe break inside the containment will also be forced down into the condensation pool. Figure 19.16 shows some emergency systems for the ABB-Atom (Sweden) BWR reactor. In case of multiple failures leading to a too high pressure in the containment, the rupture disk will break and release steam and gas from the wet well through a special filter, thus preventing a failure of the containment.

(a) *BWR*. Most of the typical features of the BWR type is described in §19.13.2. Like the PWR and VVER types, a BWR has to be shut down yearly in order to replace spent fuel. Published data suggests that many BWRs have a marginally higher net efficiency than the typical PWR. The response of a BWR to an increase in power demand is rather slow because of the void effect and it is therefore not used as power plant for propulsion purposes.

(b) *ATR*. The Advanced Thermal Reactor (ATR) was a Japanese D_2O-moderated boiling H_2O cooled MOX-fuelled Pu-burner prototype reactor which is now shut down, see Fig. 20.1. The design has a calandria with vertical pressure tubes, cooled with light water which is allowed to boil in the tubes. This permits on-line refuelling. ATR reactors are seen as a complement to breeder reactors. A large 600 MW_e ATR reactor will require a fuel supply of ~460 kg Pu per year.

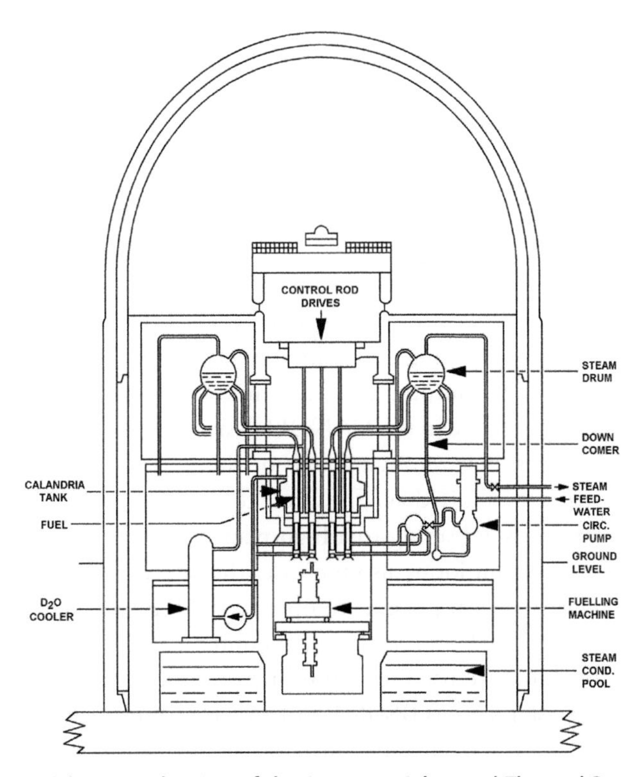

Figure 20.1 Schematic drawing of the Japanese Advanced Thermal Reactor (ATR).

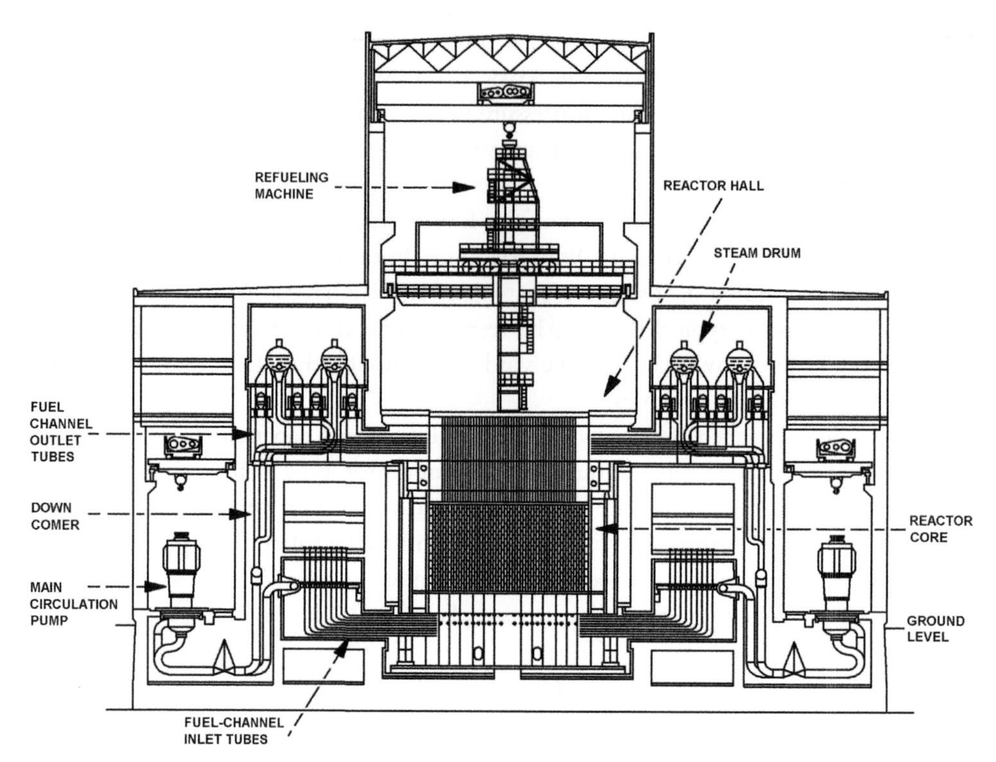

Figure 20.2 Russian graphite-moderated boiling light water cooled power reactor, RBMK.

(c) *RBMK.* The RBMK is a graphite moderated, boiling H_2O cooled, pressure tube reactor of Russian design, see Figure 20.2. The use of many pressure tubes (Zr-Nb alloy), each normally containing two fuel bundles in series, in a large graphite moderator makes it possible to design a boiling water reactor of much larger size and power than possible with a single pressure vessel; up to >2000 MW_e. The typical RBMK has 1000 MW_e output, but two units with 1500 MW_e were built in Lithuania. Furthermore, like in the CANDU design, the use of pressure tubes makes on-line refuelling possible; removing one of the reasons for a yearly shut-down period. Monitoring of individual pressure tubes makes it possible to detect and locate failure rather easily leading to a high plant reliability and availability. Subdivision into several coolant loops also increases the safety of the reactor system. The effective moderator makes it possible to operate the RBMK reactors with fuel only slightly enriched in ^{235}U. Boron carbide containing control rods in special channels through the moderator are used to control neutron flux and power distribution.

The many (1452 in a standard RBMK-1000) pressure tubes makes a pressure tube failure less disastrous than a large pipe break or vessel failure in many other water cooled reactor types. This led to an initial design without any effective containment

(6 RBMK-1000 units). However, the 2nd generation RBMK-1000 (8 units) and the RBMK-1500 (2 units) have a pressure suppression pool for steam condensation in case of pressure tube rupture. Operating statistics shows that the design has been very good with regard to availability and annual load factor.

The disastrous Chernobyl reactor accident highlighted some of the undesirable properties of this reactor type in a dramatic way. The use of a large, very effective, graphite moderator in combination with a light water coolant gives the reactor a large positive void coefficient when operated below ~20% of full power with natural uranium or low enriched fuel; the reactor has a better neutron economy with empty or steam-filled pressure tubes than with water-filled pressure tubes. This feature allowed one of the Chernobyl reactors to increase its power from below 20% to ~100 times full power in about 4 seconds, causing a catastrophe. Furthermore, the large core makes the neutron fluxes in different parts of the core only weakly coupled, e.g. one or more regions of the large core may still be critical when one section is made subcritical by control rods. After the Chernobyl disaster, safety margins in the RBMK reactors have been increased by using higher enrichment in the fuel, special absorbers, an improved control system and better operating procedures. The consequences of this accident are discussed in §13.10.2 and Chapter 22.

It may be noted that the RBMK reactor design was chosen by the USSR (despite warnings from the Soviet Academy of Sciences) because it was better suited to available production facilities than the VVER types (which required the manufacture of large pressure vessels). The rapid introduction of RBMK reactors in the USSR made previously used energy resources (oil and gas) available for export to the West, giving a needed hard currency income.

20.1.3. Special reactor systems

So far we have described some reactors designed for conventional power production. In §19.12, reactors for research or isotope production were discussed. We have purposely not included reactor designs which presently are considered obsolete (e.g. the sodium cooled graphite moderated reactor such as Hallam USA, 75 MW_e), designs which presently seems to have little promise for the future (e.g. gas cooled graphite moderated reactors, GCR), or are unsuited for large scale power production (e.g. the organic moderated and cooled thermal heterogeneous reactor, Piqua, USA, 11 MW_e). Most reactors of these old types have been, or are planned to be, shut down.

Nuclear power reactors may be used for other purposes than large scale production of electricity (or heat) for community consumption. A new type of helium cooled and graphite moderated prototype reactor, HTTR, using a special thorium based high-temperature fuel, has been built at Orai (Japan). The design goal was to heat helium to ≥1000°C by the 30 MW_{th} reactor. Presently operation is permitted with a gas outlet temperature of about 950°C. The hot helium produced is intended to drive a new,

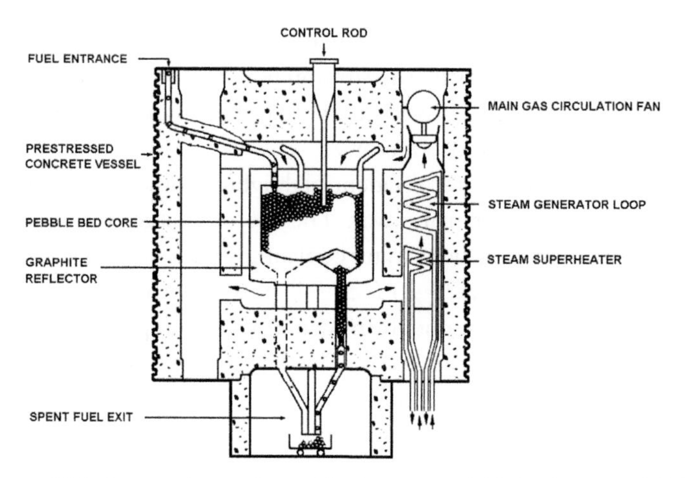

Figure 20.3 Schematic drawing of the German pebble bed reactor (THTR).

continuous, chemical two-cycle process for production of hydrogen from water or from reforming of methane, thus converting nuclear energy into a possible source of hydrogen fuel for combustion engines. A possible alternative is to use the hot helium directly in a He-gas turbine connected to an electrical generator. For safety reasons the HTTR reactor at Orai is located under ground.

A rather unique design is the pebble bed reactor, see Figure 20.3. It is a helium cooled graphite moderated reactor with a core consisting of a bed of spheres; each about 6 cm in diameter. The fuel is initially based on a mixture of $^{235}U/^{232}Th$ carbides as microspheres in a graphite matrix, see §21.7. The fuel spheres are covered with a layer of silica or silicon carbide. This type of reactor can operate as a thermal breeder or converter with its own ^{233}U replacing the ^{235}U in the initial fuel. However, in order to recover new ^{233}U, the spent fuel has to be reprocessed. Spent fuel is withdrawn at the bottom and new fuel added at the top of the core. The possibility to use slightly different sizes of fuel spheres would permit a mechanical sorting of the spent fuel according to initial type before reprocessing. Also this reactor can produce helium at a rather high temperature. A 15 MW_e prototype was operated at Jülich (Germany) from 1967 to 1988 and a larger 300 MW_e reactor was built and operated for about one year (THTR 300). A small prototype HTR-10 is in operation in China.

Both the Japanese and German designs described above have a fuel that can withstand extreme temperatures without melting. Hence a core melt accident in such reactors is practically impossible. Even at high temperatures the fuel integrity is not likely to be compromised, thus strongly reducing the possibility of radioactive releases to the environment in case of a severe accident. However, after air ingress into an overheated core, burning graphite might cause a release of radioactive material. The THTR 300 was permanently shutdown after a minor accident which released some radioactive material into the environment.

The large amount of energy stored in the small volume of fuel makes nuclear reactors suitable as energy sources when repeated refuelling is a disadvantage. For example, a nuclear reactor was used for a long time as a power source in the Arctic. Several nuclear powered merchant ships have been built in order to determine their feasibility, but none is presently in use. The now scrapped USSR icebreaker *Lenin* (44 000 shaft horse power, launched in 1959) had three reactors of the pressurized water type, of which one was a spare. Nuclear power is used in many naval vessels in the USA, Russia, UK, China and France. The first nuclear powered vessel was the US submarine *Nautilus*, launched in 1955; in 1958 it made a famous northern voyage from the Pacific to the Atlantic, passing under the ice at the North Pole. It could travel 250 000 km ($\sim$6 times around the earth) without refuelling. At the end of the year 2000, about 250 nuclear powered vessels (8 icebreakers and the rest naval vessels - mostly submarines) were in operation. These vessels correspond to a total of about 400 reactors.

The present nuclear fleets are dominated by advanced submarines which now number several hundred. Their reactors (small PWR:s) are fuelled by enriched uranium metal (20–50% ^{235}U) alloyed with zirconium. The high initial enrichment in combination with burnable poisons and the use of extra neutron shielding to protect the pressure vessel gives a long service life of both the fuel and the reactor; from $\sim$4 years between refuellings for the Russian icebreakers to $\sim$30 years for the 48 MW reactors used in the French Rubis-class submarines. Hence, many submarines today have reactors that can operate without fuel replacement or pressure vessel annealing during the expected life of the submarine itself. Because of their long endurance submerged, high underwater speed, and the difficulty of detecting and destroying these submarines they are considered to act as deterrents against a nuclear war.

The world's first large nuclear powered naval vessel was the US aircraft carrier *Enterprise*, (CVN-65) which has 8 PWR reactors, producing 300 000 shp. With a displacement of $\sim$80 000 t (and $\sim$100 airplanes) it can cruise at 35 knots for 60 days without refuelling.

Finally, 35 small nuclear reactors have been employed by the former USSR and 1 by USA as high power energy sources in satellites. Two naval-type reactors on barges have also been designed in Russia to be used for heat and electricity supply at remote locations in the arctic. Reactors have also been developed and tested in the US for use as aircraft and rocket engines (e.g. NERVA, which was intended as a drive for a manned Mars visit).

20.2. CHEMISTRY OF WATER COOLED REACTORS

The conditions in water cooled reactors are approximately: temperatures $\leq 350°C$, pressures ≤ 16 MPa, and intense γ- and n-irradiation. This causes potentially severe corrosive conditions.

The radiation decomposition in water cooled and/or water moderated reactors is considerable. About 2% of the total energy of the γ- and n-radiation is deposited in the water. We have seen in §8.6 that this produces H_2, O_2, and reactive radicals. In a 1 GW_{th} BWR the oxygen production is about 1 l $min^{?1}$, but it is considerably less in a PWR. Because of the explosion risk from $H_2 + O_2$, the two are recombined catalytically to H_2O in all water reactors.

Both the radicals and the O_2 formed increase the corrosion rate of reactor materials. In a BWR the corrosion rate of stainless steel (18% Cr, 8% Ni) is about 10^{-4} mm y^{-1}, leading to the release or small amounts of Fe, Cr, Co, and Ni into the cooling circuit. Many of the activation products of these are highly radioactive, e.g. ^{58}Co and ^{60}Co. The corrosion products form insoluble voluminous colloid-like products referred to as *crud* (said to be the acronym for Chalk River Unidentified Deposits). In addition to polymeric metal hydroxides, the crud contains small amounts of other materials in contact with the wet circuit: zirconium from the cladding material, possibly copper from the BWR condenser system, silicon and organic material from the water purification systems, boron from the boric acid control system for PWR, together with fuel and fission products from leaking fuel, etc. Deposition of crud on the fuel element surfaces may block cooling channels and be detrimental for the cooling of the fuel rods because of the poor thermal conductivity of crud. This can lead to a local overheating of the cladding, and a corresponding accelerated corrosion, and may cause a penetration of the cladding and a subsequent release of fuel and fission products into the reactor water. For high duty PWR-plants, local sub-nucleate boiling may pose an additional problem for high concentrations of boron in the reactor water. In this operating regime, boron accumulates in the crud, and because of the high absorption cross section for thermal neutrons in ^{10}B, an unwanted reactivity altering mechanism (parasitic absorption) is introduced. This may cause deviations in the axial power profile for individual fuel assemblies, so called Crud Induced Power Shift (CIPS), and hence affect thermal margins in other parts of the core in a negative way. In extreme cases, the axial power profile of the total core is affected, and further operation of the reactor may violate safety criteria as defined in the safety analyses of the plant. An immediate remedy, if CIPS is detected, is to decrease the power of the reactor, and replacement of fuel assemblies may be necessary during outage. These are both costly options, and to ensure that this scenario is avoided, extreme care is taken to minimise the risk for CIPS. Important parameters for optimising PWR primary water chemistry, in combination with core physics parameters, regarding the risk for CIPS are mainly pH, boron concentration and power peaking factors (ratios between local and global power).

The solubility of various corrosion products depends mainly on the temperature and pH-value, and because these properties vary throughout the primary system, different corrosion products deposit on various places in the system. In order to determine a favourable pH-value for core conditions, one should utilize the axial temperature gradient through the core. From the core inlet, the cooling water is heated when passing through the core, and therefore, the solubility limits for corrosion products vary with core elevation. Since iron is a major corrosion product from stainless steel system surfaces, a good idea is to optimise the pH value so that iron remains in solution for the entire temperature range expected when the coolant is transported through the core, and thereby try to avoid iron depositing on the cladding surfaces (the hottest place in the core). This is a good starting strategy for minimising crud formation, and hence to prevent the problem with CIPS for PWRs.

Important corrosion products are Ni and Co, and in case of maintenance and modernisation of power plants, fresh unpassivated system surfaces are introduced (for example when steam generators are replaced in a PWR). For such cases, the source term of nickel increases immediately after introduction of high Ni-content alloys, and becomes an important source for crud formation and radioactivity distribution in the plant. A special program for "start up chemistry" may be necessary to employ after major projects, especially if power up-rates are planned in connection to modernisation of a nuclear power plant. The releases of Ni and Co are especially important to suppress, since these elements constitute the highest sources for dose rates in the plant.

Since corrosion products constantly pass the core, elements with sufficiently high neutron absorption cross sections will become activated. The radioactive corrosion products carried through the cooling loop create a serious radiation problem for the reactor personnel. Considerable effort is put into the development and selection of corrosion-resistant materials. Equally important in this respect is the selection of water conditions which minimize corrosion and the deposition of such products within the reactor core, and effective water cleaning systems. In a PWR, the crud is transported through the whole primary cooling circuit, but removed in the purification circuit. In a BWR, the crud accumulates in the reactor vessel; therefore water boiling reactors have a special cleaning circuit attached to the reactor vessel. The cleaning circuits are discussed in §§19.13.2 and 20.4. Because of the radiation problem in the plant caused by activated corrosion products, the main corrosion rates should be as low as possible, and the residence time for corrosion products in the core (activation time) should be kept to minimum by optimising the water chemistry parameters.

The amount of crud deposited in the reactor core can be partly removed during shutdown by mechanical cleaning or by washing with chemical decontamination solutions. These solutions contain mildly oxidizing agents such as alkaline permanganate and/or organic complexing agents like ammonium citrate (APAC treatment), oxalic acid or EDTA. Another way to remove crud from fuel assemblies is by ultrasonic fuel cleaning, where ultrasonic waves are used to create small oscillating air bubbles which collapse (cavitation) and case small shock waves, which in turn shatters the crud on the fuel cladding surfaces.

There are a number of ways to reduce corrosion. One is to increase the pH of the water to ~8 by adding alkali, e.g. LiOH or NH_3 (≤10 ppm). While reactors of US type use ^7LiOH in order to reduce the formation of tritium from n,γ-capture in ^6Li, the VVER reactors normally use KOH. There are benefits and drawbacks from choosing either KOH or LiOH as buffer solutions, and the latter is discussed in some detail. Supposed corrosion of zirconium alloys (cladding material) for high Li concentrations has been an issue for some time. The problem arises when large amounts of boron are needed in the coolant for controlling the access reactivity at the beginning of long operating cycles on high power. In order to maintain the desired pH-value, high LiOH concentration is needed, and the corresponding risk for accumulation of Li in the cladding oxide increases. Experiments have suggested that these regimes (~10 ppm Li) may cause accelerated corrosion of the cladding, but recent findings suggest that the problem has been exaggerated. When KOH is used instead, the corrosion of the cladding is not an issue. However, activation of potassium causes an additional radiation problem in the plant (VVER).

When ammonia is used, the radiolysis yields HNO_2 and HNO_3; it is necessary to add H_2 gas to shift the equilibrium from the acidic products. At an H_2 concentration of ~2 ppm, the concentration of dissolved O_2 is greatly reduced. Instead of NH_3, hydrazine or N_2 may be added to the water to increase the pH via the reactions

$$\tfrac{1}{2}N_2 + 3H \rightarrow NH_3; \quad NH_3 + H_2O \rightarrow NH_4^+ + OH^- \qquad (20.1)$$

The hydrogen content of the water must be kept fairly low to avoid brittleness of zircaloy by hydride formation. Hydride formation may also lead to mechanical deformation of zircaloy parts. Surface pre-treatment is important in order to reduce these effects.

Another measure to suppress the amount of corrosion products in the plant is to add small amounts of zink to the reactor coolant, so called Zn-chemistry. The aim for doing this is to minimise the primary corrosion of components. The principle works by substituting the chemically similar Ni (and some amounts of Co) for Zn in the surface layers of e.g. stainless steel piping and steam generator tubes. This process forms a thinner and more stable oxide surface layer which makes migration of Ni, Co and Cr through the surface more difficult. After the initial replacement of Ni in the surface matrix (passivation of the surface), the concentration of corrosion products decrease in the primary system, and a general decrease of dose rates in the plant can be achieved.

The purity of the reactor water is checked by measurement of pH, conductivity, turbidity, oxygen concentration, radioactivity, etc., either by sampling or – more usual – by on-line analyses. If the operating conditions changes, e.g. the thermal power, a chemistry transient may occur, and in such situations, the manual sampling rate of the reactor coolant should increase in order to more accurately follow any changes in the water chemistry.

20.3. BREEDER REACTORS

The possibility to breed excess fissile nuclei by consumption of fissile + fertile nuclei increases with increasing fission factor η according to §19.9. Since η depends on the fission cross section, which is highly energy dependent (Fig. 19.3), the neutron spectra in reactors must be considered. In a reactor with large amounts of highly moderating atoms, the original un-moderated spectrum is strongly shifted towards lower energies. However, even under optimal moderating conditions, a thermal Maxwell neutron distribution is never achieved in a reactor, but can be closely approached in thermal columns of research reactors. Because breeding occurs through neutron capture in fertile ^{232}Th and ^{238}U (the *only* fertile atoms in nature), with maximum probability between 600 and 700 eV, it is desirable for breeding that the neutron flux not be too low in this region. Just as breeding cannot be achieved with a thermal Maxwell neutron distribution it cannot be achieved with a fission neutron energy spectrum. Even if attempts are made to produce very "hard" neutron spectra (i.e. high energy), some moderation always occurs. Thus in a fast reactor, in which as few as possible light atoms are used, the neutron spectrum is still shifted strongly towards a lower energy region (Fig. 19.4) so that typically the neutron flux at 0.25 MeV is higher than that at 1 MeV. The curves shown for thermal and fast reactors in Figure 19.4 are only indicative, depend strongly on the particular reactor and fuel design, change with time, and vary over the core.

Because breeding is only possible for $\eta > 2$, ^{233}U is the favored fissile material for breeding in thermal reactors, and ^{239}Pu for breeding in fast reactors. Considering the shift towards lower neutron energies in fast reactors it is seen that η for ^{239}Pu is always greater than for ^{233}U. Thus ^{239}Pu can be more rapidly obtained by breeding than ^{233}U. Consequently much more effort has been put into the development of fast plutonium breeders than into thermal ^{233}U breeders.

20.3.1. The thermal breeder

In the thermal breeder reactor, ^{233}U is produced from ^{232}Th. The reactor may be designed either with a core containing a mixture of ^{232}Th and ^{233}U, or with a central zone (*core*) of ^{233}U surrounded by an outer layer (*blanket*) of ^{232}Th. However, it is necessary to minimize parasitic neutron capture in structural materials including monitoring systems, control rods, etc. Calculations have shown that a conversion ratio of 1.06 should be possible.

As an alternative to using a solid fuel matrix, both core and blanket can be liquid. In the HRE-2 (homogeneous reactor experiment), USA, the core consisted of a ^{233}U sulfate solution in D_2O, while the blanket was a suspension of ThO_2 in D_2O. The project was terminated because of various technical difficulties.

The molten salt reactor (MSR) was a somewhat similar design tested at Oak Ridge, USA, for some years. Here UF_4 and ThF_4 were dissolved in an eutectic melt of BeF_2, ZrF_4, and LiF (or NaF). The reactor may be homogeneous with respect to ^{233}U and Th, or of the core and blanket type in which case graphite may be used as moderator. The molten salt system has several advantages: (i) radiation effects are less in molten salts than in solid fuels, (ii) the expense of fabricating complicated fuel elements is avoided, (iii) the fission products can be removed continuously, (iv) fuel recharging can be made continuous, (v) the core contains little absorbing structural materials, and (vi) the system has a high negative temperature coefficient and is therefore easy to control. Points (iii) and (v) makes it possible to obtain a conversion ratio of 1.07, i.e. a 7% breeding gain; the fissile doubling time has been estimated to be 20 y. In the MSR high salt temperatures ($\leq$700°C) were used. Nevertheless, the system could be operated near atmospheric pressure because of the low vapour pressure of the melt. The salt is also inert to air and water. By using a dump tank for the core, the risk of criticality accidents was minimized. The on-line fission product removal required novel chemistry, and these problems were never fully solved.

Figure 20.4 shows the production of heavy atoms in the ^{232}Th system. Because ^{233}Th is produced until the fuel is removed from the core (or blanket), the long half-life of ^{233}Pa (27 d) requires long cooling times before reprocessing if a maximum yield of ^{233}U is to be obtained. Furthermore, when continuous on-line reprocessing with removal of ^{233}Pa is not used, the decay of ^{233}Pa will lead to a slow increase in reactivity with time after

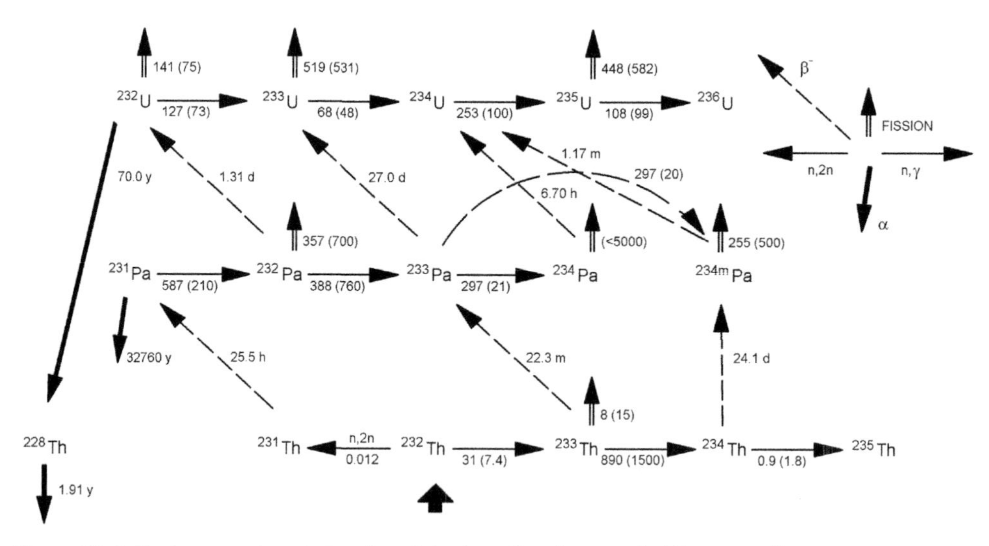

Figure 20.4 Nuclear reactions in irradiated thorium. Numbers are half-lives or effective reaction cross-sections for a power LWR; 0.025 eV data are given within parentheses.

shutdown due to a slightly increasing amount of ^{233}U from decay of ^{233}Pa. The thorium would become contaminated by the isotope ^{234}Th and 240 d would be required to reduce this activity to 0.1%. The ^{233}U would be contaminated by ^{234}U and ^{232}U; the latter constitutes a problem because of its relatively rapid decay into the $^{228}Th \rightarrow {}^{224}Ra$ – chain. These isotopic contaminations of the fertile and fissile products in the ^{232}Th–^{233}U fuel cycle would cause handling difficulties, which are more severe than for the ^{238}U–^{239}Pu fuel cycle. At present no ^{233}U breeder exists.

Several research groups have argued that a subcritical thorium breeder, driven by an accelerator based high intensity neutron source, might be a good and safe future energy producer. Using a molten salt/thorium fuel and coolant combination in continuous loops through a D_2O moderator tank or a graphite moderator it could generate a relatively high power from a small inventory of fissile material. It is also possible to design a fast, lead/bismuth cooled, Th-based subcritical breeder by using an intense external n-source. However, no such devices have yet been built and tested.

20.3.2. The fast breeder

The principal design of a fast breeder reactor consists of a central core of plutonium in which fission occurs, surrounded by an outer blanket of ^{238}U in which neutrons are captured to form new ^{239}Pu. This blanket is surrounded by a reflector, usually of iron. The fission yield curve of ^{239}Pu is similar to that of ^{235}U (see Fig. 14.9) but its heavy mass peak is shifted up by a few atomic mass units. Some neutron capture occurs in ^{239}Pu yielding fertile ^{240}Pu, which through another neutron capture produces fissile ^{241}Pu; similarly, successively higher elements are formed in time. In a strong neutron flux their fate is destruction by fission. Thus, in the long run, by conversion of ^{238}U into ^{239}Pu, etc., fission energy is always released and, in principle, all the ^{238}U can be used for producing fission energy. In practice a value of ~70% is considered more realistic. Still this would mean ~100 times more energy than is available from fission of ^{235}U only. Since the fast breeder reactor concept extends the fission energy resources by about a factor of 100, using a proven technology, it makes uranium the largest fossil energy resource presently available on earth. In addition fast reactors may also be used to fission long lived minor actinides such as americium and thereby not only increase the energy usage of the fuel but also shorten the longlevity of the final nuclear waste. Thus, if a separation process is designed to recover both uranium and plutonium together with the minor actinides a new fuel could be made directly using this composition. This is often called homogenous recycling.

The conversion efficiency of a reactor of this design can be estimated from

$$C = \eta + \varepsilon' - p' - \Lambda' - 1 \tag{20.2}$$

where ε' is the contribution to the neutron production from fertile nuclides, p′ is the neutron loss due to parasitic absorption, and Λ' due to leakage. Some representative values are

	η	ε'	p'	Λ'	C
Core	2.56	0.27	0.049	1.15	0.63
Blanket	–	0.10	0.020	0.032	–
Total	2.56	0.37	0.069	0.032	1.83

It should be observed that the neutrons lost from the core are caught by the blanket. C is higher than in any other known chain-reacting system.

The longest experience with fast research reactors has been provided by the BN-350 reactor in the former USSR, now Kazakhstan. It was operated at a power level up to 1000 MW_{th} from 1972 to 1994. Most of the heat was used for water desalination (120 000 m^3/day) and only a small amount for electricity production. Larger (100–1200 MW_e) prototype fast breeder power reactors have been designed in the former USSR, France, the UK, and Japan. The largest prototype FBR built and operated so far is the Creys-Malville plant (located in France), which could produce 1 200 MW_e.

Figure 20.5 shows the Creys-Malville reactor; in operation from 1986 to 1997. The thermal rating was 3000 MW_{th} (gross 1240 MW_e); the neutron flux was very intense, $\leq 6 \times 10^{19}$ n $m^{-2}s^{-1}$, which puts a severe strain on construction materials. The core consisted of an inner and an outer fuel zone surrounded by a blanket, a steel reflector and a neutron shield, all immersed in liquid sodium (3 300 t) contained in a 25 cm thick austenitic steel vessel at atmospheric pressure. The reactor vessel was enclosed in a safety vessel and surrounded by a cooled concrete (biological) shield. The fuel was made of pins placed in hexagonal shrouds. The fuel zones contained 15 vol.% PuO_2 and 85 vol.% depleted UO_2, while the breeding blanket only contained depleted UO_2 (0.2% ^{235}U). The fuel pins were clad in 0.5 mm stainless steel. The total plutonium amount was 4 800 kg. The neutron shield consisted of hollow steel tubes. Because the sodium becomes extremely radioactive due to ^{24}Na formation, the whole primary cooling system was contained in the reactor vessel. By means of a sodium/sodium intermediate heat exchanger within the reactor vessel (but outside the neutron flux) heat from the primary circuit (542°C) was transferred to a secondary sodium stream (525°C) and transported to a sodium/water steam generator which produced 487°C steam at 17.7 MPa. Heat exchanger and piping were made of steel.

The prompt neutron lifetime in a fast reactor is about 1000 times smaller than in a thermal reactor. A reactor which can go critical on the prompt neutrons only would be exceedingly difficult to control. Therefore, fast reactors are designed to depend on the delayed neutrons (like thermal reactors). The time period (t_{per}) is large enough to allow reactor control through the use of neutron absorbing rods. Since the neutron spectrum is

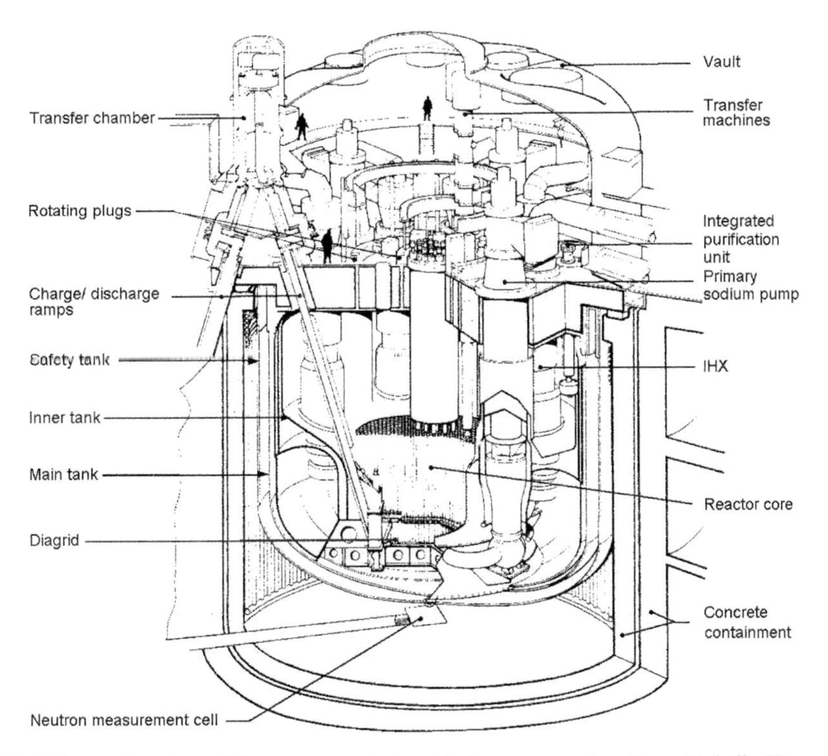

Figure 20.5 Vertical section of European prototype fast power reactor Creys-Malville (Superphenix). The vessel was filled with 3 300 t of liquid sodium.

such that several percent of the flux is in the resonance region, control rods with boron can be used; in practice boron carbide (the carbon atom reduces the neutron energy further) and/or tantalum (which has large absorption peaks $\geq 10^4$ b at 3–100 eV) are used.

Although the temperature coefficient of the Creys-Malville reactor was negative, this is not so for fast Pu-fuelled reactors with harder neutron spectra, e.g. the Russian BN-350. If the fast reactor becomes overheated, the core could be deformed, making it prompt critical. The power would then increase rapidly with a doubling time in the microsecond range, and a severe accident would be unavoidable. To prevent this the core is designed to achieve negative reactivity upon sudden power transients by using $^{238}UO_2$ in the core. When the temperature rises, doppler broadening occurs in the ^{238}U resonance capture region, and consequently more neutrons are consumed by ^{238}U, which limits the power increase. The shorter the neutron lifetime, the less would be the power excursion. In addition the thermal expansion of the fuel reduces the reactivity, which makes the temperature coefficient smaller or more negative.

The total breeding ratio in the Creys-Malville reactor was 1.24. In BN-350 a breeding ratio of 1.5 was obtained because it had less neutron absorbing materials. Still higher breeding gain can be achieved in fast reactors at some sacrifice of safety control.

Considerable radioactivity is induced in the sodium in the primary cooling circuit of a liquid metal cooled fast breeder reactor (LMFBR): $^{23}Na(n,\gamma)\,^{24}Na$, $^{23}Na(n,p)\,^{23}Ne$, etc. ^{24}Na has a 15 h $t_{1/2}$ and emits energetic γ's. The primary cooling loop must therefore be well shielded. Its activity is a nuisance only in case of repair work in the primary system, requiring considerable waiting time before the loop can be approached. The sodium dissolves many of the corrosion and fission products eventually released. To remove these, the primary loop is provided with cold trap purification systems.

Other coolants currently discussed, e.g. in Europe, is molten lead and helium. For all coolants there are advantages and disadvantages. For example lead is excellent for accident management since a core melt will dissolve in the lead and thus disarm the situation leaving a large, heavily contaminated lead block. Also some of the more troublesome volatile fission products such as iodine will react with the hot lead and be trapped preventing their release to the environment. A considerable disadvantage with using lead as a coolant is the increased corrosion in pumps and pipes. Sometimes bismuth is added to the lead in order to lower the melting point and make the environment less aggressive. This will, however, produce polonium during reactor operation. Any release of gas from a gas-cooled reactor will contain some radioactive iodine formed from tramp uranium and leaking fuel. Thus special venting systems and filters will be needed.

20.4. REACTOR WASTE

In the "ideal" nuclear reactor all fission products and actinides produced are contained in the fuel elements. As mentioned in §19.17, there are four processes in a water cooled reactor through which radioactivity leaves the reactor vessel; in all cases the carrier of activity is the coolant:

(i) induced radioactivity in the cooling medium (and moderator if separate);
(ii) corrosion products containing induced activities (mainly from materials in the feed water);
(iii) leaked fission products and actinides from faulty fuel elements;
(iv) fission products of actinides deposited on surfaces in the core ("tramp uranium").

We order the discussion of reactor wastes according to the phase which contains the radioactivity (gas, liquid or solid) and end by a brief discussion of waste from decommissioning of nuclear reactors.

20.4.1. Gaseous wastes

The intense flux of fast and thermal neutrons induces several radioactivities in H_2O: $^{2}H(n,\gamma)^{3}H$, $^{16}O(n,p)^{16}N$, $^{18}O(n,\gamma)^{19}O$, and $^{18}O(n,p)^{18}F$. Considerable activities of ^{13}N are produced through the reactions $^{13}C(n,p)^{13}N$ and $^{14}N(n,2n)^{13}N$ with carbon in the steel, and CO_2 and nitrogen in the water. ^{14}C is produced through reactions with ^{13}C, ^{14}N (mainly dissolved to 10–60 ppm in the fuel), and ^{16}O; 400–800 GBq ^{14}C is produced per GW_e-year, mainly contained in the fuel. Activation products from B, Li and Ar occur when the cooling water contains these atoms due to the reactions $^{10}B(n,2\alpha)^{3}H$, $^{6}Li(n,\alpha)^{3}H$, and $^{40}Ar(n,\gamma)^{41}Ar$. In a PWR the tritium content may be 40–400 GBq m^{-3} because of the boric acid and lithium content, while 0.04 GBq m^{-3} is more typical for a BWR. Though most of these activities are short lived many emit energetic γ-rays in their decay. In a BWR the active gases are transported with the steam to the turbines, where a considerable fraction of the decays occur, thus raising the background in this part of the power station to such a level that personnel cannot be allowed permanently in the turbine hall during operation. Also some radioactivity may be released through leaking gland seals; however, this can be very much reduced by adding a little inactive steam from a small electrically heated boiler to the high pressure side of the seals.

Other activities are also found in the steam: ^{24}Na, activated corrosion products like ^{60}Co, and fission products like ^{91}Sr, ^{99m}Tc, and ^{137}Cs. These activities are many powers of ten lower than for the gases.

To take care of gaseous products transported with the steam, efficient drainage and suction (air ejector) systems are installed at the condenser side of the turbines. Here most of the activity is swept out and passes through a hydrogen-oxygen recombiner and a delay system (delay times of ~30 min are used in older and ~30 h in modern BWRs) before the rest of the gases are caught in filters, consisting of absorbing material (e.g. charcoal). Because some of the activity may leak out into the reactor and turbine buildings, the air flow is from areas with low risk to areas with higher risk for contamination. The building ventilation system also contains filters before the air is vented through the stack. Figure 20.6 shows the principle of the off gas system of a modern BWR. By operating the system at reduced pressure leakage to the environment can be avoided. When the charcoal absorbers become saturated they can be heated and the adsorbed gas driven off back into the turbine condenser, thus causing a recirculation which multiplies the available decay time.

The primary release of gaseous activities in a PWR is at the de-aerator after the heat exchanger in the primary loop. The hydrogen-oxygen recombiner is also located here. The gases are compressed and stored for up to 2 weeks before they (via filters) are vented through the stack; these volumes are much smaller than for a BWR, whose off-gas also contains air leaked into the turbine on the condenser side.

If fission products are released by faulty fuel elements, the gaseous ones appear most rapidly in the cooling water: ^{133}Xe, ^{135}Xe, ^{85}Kr, ^{133}I, etc. The noble gases can be delayed in sand-filled hold-up tanks (permitting shortlived activity to decay) and the remaining activity retained in charcoal filters. Iodine can be removed by various special techniques. As mentioned above and indicated in Figure 20.6, many BWR plants cyclically

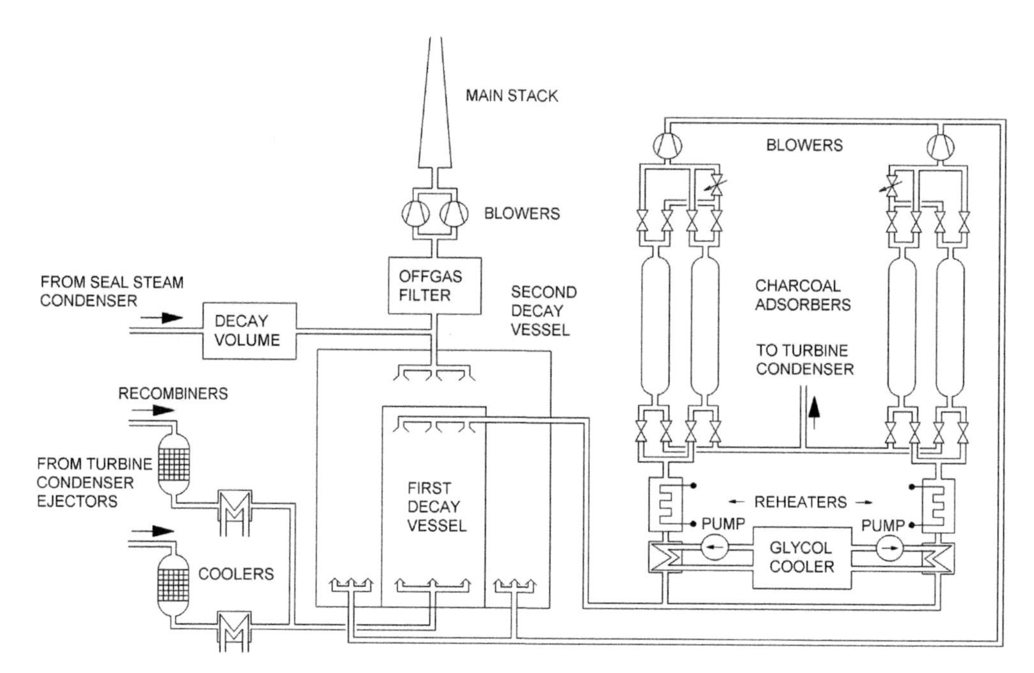

Figure 20.6 BWR ventilation and off gas treatment system (ABB Atom).

regenerate the charcoal filters feeding the released activity back into the coolant system. Some actual release values are given in Table 19.7, see also Ch. 22.

The presence of radioactive noble gases in the off-gas is monitored continuously. Abrupt increases occur when fuel pins are penetrated. By considering the different half-lives and activities it is often possible to make a rough differentiation between tramp uranium, old leaks, pinholes and large new leaks in the fuel.

20.4.2. Liquid wastes

Corrosion and fission products appear in dissolved ionic form and "precipitated" in the crud, depending on the chemistry and water conditions. Most of the corrosion products giving rise to induced activity enter with the feed water. The dominating activated corrosion products are ^{51}Cr, ^{54}Mn, ^{59}Fe, ^{58}Co, ^{60}Co, ^{65}Zn, and ^{124}Sb, and the dominating fission products are ^{3}H, ^{131}I, ^{134}Cs, and ^{137}Cs. Other fission products and actinides are released in minor amounts depending on the kind and size of fuel element leak. These products are continually removed by cleaning circuits. The aqueous chemistry, and related corrosion and waste problems, of operating reactors is one of the main areas for (nuclear) chemists working in the nuclear power industry.

In a PWR a continuous liquid stream is withdrawn from the coolant in the primary circuit, on the back side of a main heat exchanger. After cooling in another heat

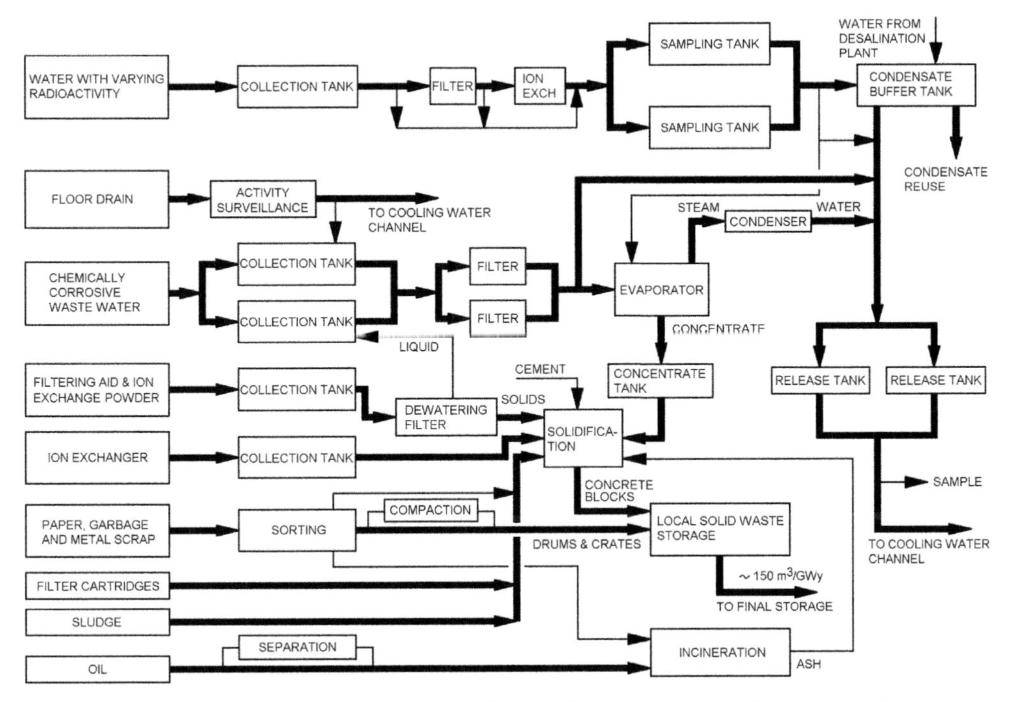

Figure 20.7 Typical waste management system for a nuclear power station, but many variations exist.

exchanger, the water is allowed to pass a filter (e.g. of the pre-coat type, 25 μm particle size) and a demineralizer which removes ionic species as well as all particulates. The liquid waste streams are detailed in Figure 20.7, which is typical for both PWRs and BWRs. The system is designed to permit segregation of waste streams according to characteristics and type of treatment. The largest liquid waste source from a PWR is the steam generator blow-down, while the largest activities originate at the clean-up systems (condensate demineralizer and filter back-washes); the "demineralizer" may be an inorganic substance (e.g. kieselguhr) or an organic ion exchanger (among the most efficient ones is the Powdex system, which consists of very finely ground, ~400 mesh, organic cation and anion exchangers). The boric acid adjustment system of some PWRs may be considered as part of the cleaning circuit; it is discussed in §19.13.1.

A BWR has three liquid purification circuits: one after the condenser for the feed water, one with withdrawal of a small stream from the reactor vessel, and one for the water in the fuel storage pool. The condensate purification system contains similar demineralizers as for the PWR. The reactor water purification system consists of a cooler followed by a mixed bed ion exchanger; also other substances have been used (magnetite, kieselguhr, etc.). A typical capacity figure for the purification system is ~100 m^3 h^{-1}.

The radioactivity becomes very high in the purification systems. The organic ion-exchange resins used are limited to absorbed doses of ~10^6 Gy. The filters are back-washed and resins changed through remotely controlled systems, when measurements show that too much activity is leaking through, the pressure drop becomes too large, or water quality drops. At present resins are not regenerated.

As an additional safety, when filters and demineralizers show insufficient purification of circulating water and of water for release to the environment, an evaporator system may be incorporated. Such a system is normally a part of the PWR boric acid control circuit.

A nuclear power plant contains numerous liquid streams, not all indicated in Figure 20.7, and small controlled amounts of radioactivity are released from the plant. Typical releases are given in Table 19.7 and Chapter 22; these are actual measured values, as pointed out in §19.16 normally constituting a small fraction of the legally permitted releases.

20.4.3. Solid wastes

Solid radioactive wastes are produced from many systems and purification circuits of the reactor station, see Figure 20.7. Many contaminated items can be decontaminated by proper treatment, saving both money and waste storage space. Low level waste can also sometimes be "declassified" and disposed of as normal, inactive, waste or scrap. Combustible solids may be reduced in volume through incineration. Several techniques are used for fixation of wastes in such a way that they can be safely stored with a minimum of surveillance. These techniques all have in common an enclosure of the activity in an insoluble material (normally concrete or bitumen) and in blocks of sizes and activities small enough to be handled with a fork lift.

The waste amounts are typically ~500 m^3 ILW and ~5000 m^3 LLW annually from a 1000 MW_e power station, but vary considerably between different stations depending on purification and concentration techniques used. The following annual averages are thought to be fairly representative for most LWRs: spent ion-exchange resins ~30 m^3 (~50% water), evaporator residues 0–20 m^3 (~80% water), filter (condenser) concentrates ~10 m^3, other wastes from primary reactor loop ~5 t. The activity in the resins is typically 2–7 TBq ^{60}Co, ~0.5 TBq ^{137}Cs, and ~0.05 TBq ^{90}Sr, and in evaporator and filter concentrates it is <1/10 of the resin amount.

Figure 20.8 shows waste solidification and fixation with cement, using prefabricated concrete boxes. This technique is useful both for contaminated or activated equipment and for resins, sludges, etc. The technique, however, increases the waste volume considerably, by a factor 4 to 40 times the untreated waste.

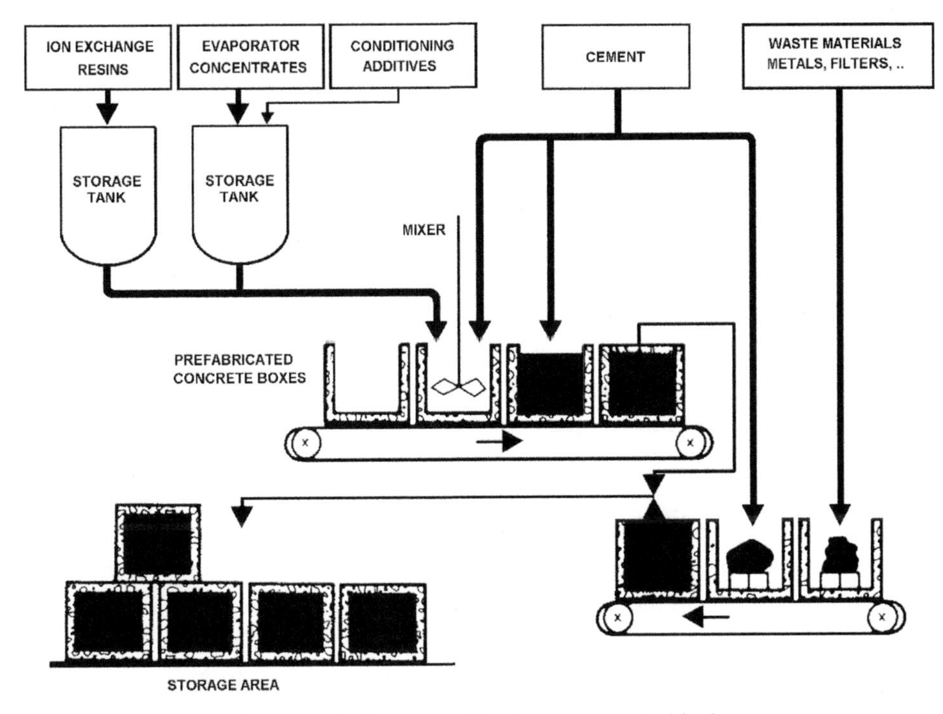

Figure 20.8 Waste solidification in concrete blocks.

Concrete is a cheap fire and corrosion resistant material. However, active species, especially easily soluble ions such as caesium, can be leached from it by water. Addition of plastic binders to the concrete in order to improve its properties has been suggested. In the continuous bitumen extruder process for semi-liquid wastes all water is directly eliminated, considerably reducing the waste volume. The bitumen mixture is placed in steel drums of standard size (150−200 l). When additional shielding is desired, the filled drums are placed into disposable or reusable sleeves of concrete, iron, or lead. Such a sleeve of 12 cm lead weighs 7 t, and reduces the surface dose rate by a factor of $\sim 10^3$. A typical unshielded bitumen drum may have a surface dose rate of 1 Sv h^{-1}, necessitating remote handling.

Comparing cement and bitumen, bitumen produces smaller waste volumes (often 5 times less than concrete) with lower initial leach rates. The disadvantage of bitumen is its in-flammability and lower radiation resistance; a fire could release large activities of fairly long-lived nuclides. Further, bitumen in wet areas has poor ageing properties, slowly deteriorating. However, for low and medium active solid wastes, bitumen is presently the preferred fixation material at many power stations. About 300 bitumen drums (200 l) with concentrated waste are produced annually per 1300 MW_e reactor in Germany;

about 50 of these contain medium active waste and have outer concrete shielding (outer volume ~400 l).

If the waste does not contain any fission products, a storage time of 50 y would be sufficient to allow all activity (mainly ^{60}Co) to decay to negligible amounts; storage at plant site would be feasible. If the amount of ^{90}Sr, ^{137}Cs, or actinides is high, much longer storage times are required, and a special storage site will be required. If possible, it is desirable to separate these products from other waste; as an alternative the isolated ^{90}Sr, ^{137}Cs, and actinides could then be added to the high level waste of reprocessing plants.

The cementation and bituminisation techniques are also used for intermediate level wastes arising in other parts of the nuclear fuel cycle, e.g. reprocessing. The long-term storage of the solidified waste is described in Chapter 21 and some of its possible future impact on the environment in Chapter 22.

20.4.4. Decommissioning nuclear reactors

Most of the waste from the dismantling of decommissioned nuclear reactors is non-radioactive. Also, large volumes can be decontaminated and "declassified" which permits them to be treated as normal industrial waste. Of the remaining waste, most contains low levels of relatively short-lived nuclides similar to wastes from normal reactor operation. Hence, it can be treated and disposed of in the same manner, e.g. put in bitumen or concrete, and stored in repositories for LLW and ILW (§21.9). The volumes may be rather large, however. Some internal parts of the reactor have relatively high activity. The induced activity is limited to part of the pressure vessel (closest to the core) and some core components.

The inner surface of the pressure vessel is normally covered by highly radioactive deposits. These deposits can be dissolved and removed by a repeated treatment with suitable chemicals. As an example, after more than 20 years of operation the pressure vessel of the BWR reactor Oskarshamn 1, Sweden, was decontaminated to such an extent that workers could enter the empty pressure vessel to perform necessary modifications and tests with only a cylindrical radiation shield placed inside the central part of the vessel wall to protect against induced radioactivity in the wall. Upper and lower parts of the pressure vessel are normally fairly well shielded from the neutron flux and only contain a low level of induced radioactivity. For chemical decontamination, i.e. dissolution of deposits, weak acids will dissolve hydroxides and complex formers (such as chelating agents) bind metal ions in soluble complexes. However, in some countries (e.g. Sweden) authorities forbid the inclusion of strong complex formers in the reactor waste. Therefore, in the case of Oskarshamn 1, oxalic acid was used; after dissolution of the deposits and fixing the metal ions on ion-exchange resins oxalic acid was destroyed by ultraviolet light yielding CO_2 and H_2O. Thus, a pressure vessel of a decommissioned reactor could, after chemical decontamination, be cut into smaller sections, some of which would only be slightly radioactive. The resulting scrap can either be encapsulated in concrete and put in a repository for normal reactor waste (Swedish plans), or melted to be used as raw material for new reactor parts where the remaining level of activity is acceptable (France).

20.5. SAFE OPERATION OF NUCLEAR REACTORS

The major risks from reactor operation stem from operator mistakes ($<10\%$) or failure of critical hardware ($>90\%$). The risk from human errors is reduced by recurrent training of personnel, especially persons working in the control room, using mockups and simulators. Thereby personnel in critical functions train their behaviour in normal situations, but also in situations which seldom or never occur during routine operation of the reactor. Also exchange of experience and the results from analysis of actions violating or threatening to violate safety instructions help to reduce the risk for human errors. Risks from hardware failures or malfunctions are minimized by a continuous work aiming to identify risk dominating sequences and critical hardware. Once identified, such sequences should be modified, procedures changed, equipment replaced or duplicated in order to reduce the overall risk until it reaches an acceptable level. It is important to remember that simple duplication will not reduce the risk from common mode failures very much.

Figure 20.9 shows an example of an *event tree* starting with an *initiating event* to the left and branching each time an action takes place. The upward branch corresponds to a success of the action specified at the top, whereas the downward branch represents a failure. Each branch is normally the compounded effect of actions in a sub-tree describing the function and interaction of affected hardware. The figures at each downward branch are estimated probabilities for a failure of the action at the top. The line representing the outcome of an event is then followed to the right; either to a final state or to another branch. To the extreme right is the final state with its estimated probability. For example, the most important risk in Fig. 20.9 arises from failure to restore AC power within 20 minutes; probability $(1/25) \times 10^{-6} = 4 \times 10^{-8}$ per reactor year. This was the main reason for the Fukoshima accident where the auxiliary diesel generators were drowned by the tsunami. For independent outcomes of events in a branch (e.g. probability of failure of too many of the available safety valves together with failing operation of the blow-down pipes), the probability is the product of the probabilities at the branches taken in the sub-tree. In case of common mode failures, the compounded probability is the same as the probability of a single event of that type. The probabilistic event tree is an example of the type of tools used to identify the current risk dominating event chain. When carried out meticulously, operator training, experience feedback and risk analysis help to continually improve the safety of reactor operation.

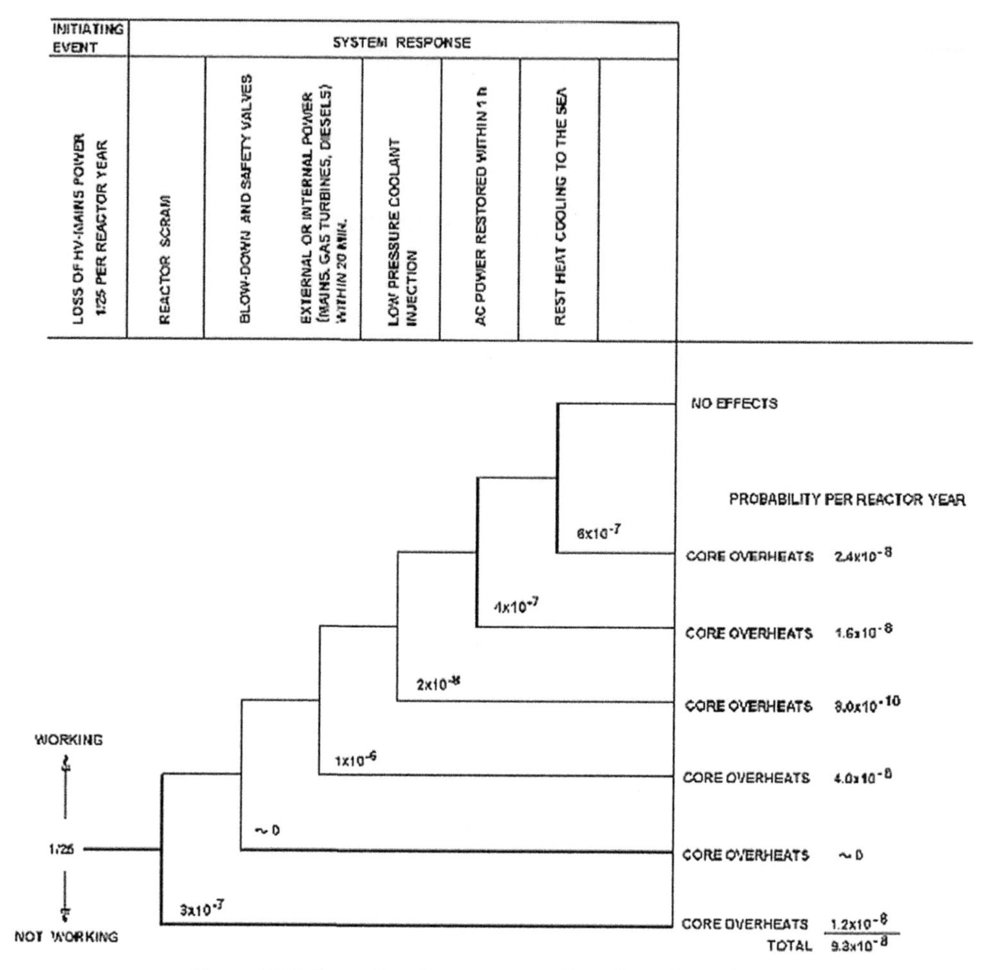

Figure 20.9 Example of an event tree for a hypothetical BWR.

20.6. EXERCISES

20.1. The gas space (volume 6000 m^3) in the PS system of a large BWR is filled with N_2 at 97 kPa and 25°C. Assume the same temperature in the dry and wet wells and that the heat of evaporation of water and its heat capacitivity are ~2.26 MJ/kg and ~4.18 kJ kg^{-1} $°C^{-1}$, respectively, up to ~100°C. The 2000 m^3 water in the condensation pool is at 25°C. The primary system contains steam at 287°C and 7 MPa (energy content ~2.8 MJ/kg steam). How many kg of such steam could be dumped into the condensation pool before the pressure in the PS system exceeds 0.1 MPa?

20.2. The amount of steam found in example 20.1 is released into an air-filled large dry containment where the pressure may not exceed 0.7 MPa. Assume steam and air in the containment can be treated as ideal gases. Estimate the necessary containment volume.

20.3. In the Chernobyl accident, reactor output increased from ~20% to ~100 times full power in about 4 seconds. Assume a constant reactivity excess during this time. a) What was the average reactor period? b) How much energy was released during this time?

20.4. KOH is normally used for pH control in VVER reactors. Which radioisotopes will be formed from potassium in the coolant?

20.5. ^{232}U is formed in a thermal ^{232}Th based breeder. a) By which chain of reactions is it formed? b) Will the choice between a continuous or discontinuous (once a year) reprocessing of the Th blanket affect the isotopic composition of the isolated uranium (assume that all other actinides remain in the blanket)?

20.6. Use eqn. (19.43) to estimate the maximum thermal efficiency for the Creys-Malville plant.

20.7. Assume that the probability of not restoring AC power within 20 minutes is reduced by a factor of 5 in Fig. 20.9. a) Which is now the risk dominating event chain? b) Is it better to reduce the total risk in Fig. 20.9 by this change or by improving the stability of the HV mains with a factor of 2 by e.g. building more power stations?

20.8. What is the fate of the gas released into the turbine condensers during regeneration of the charcoal absorbers in Fig. 20.6?

20.7. LITERATURE

J. K. DAWSON and R. G. SOWDEN, *Chemical Aspects of Nuclear Reactors*, Butterworths, 1963.

S. PETERSON and R. G. WYMER, *Chemistry in Reactor Technology*, Pergamon Press, Oxford, 1963.

Reactor Handbook, 2nd edn., Interscience, 1960?1964.

J. R. LAMARCH, *Introduction to Nuclear Reactor Theory*, Addison-Wesley, 1966.

J. G. WILLS, *Nuclear Power Plant Technology*, Wiley, 1967.

C. O. SMITH, *Nuclear Reactor Materials*, Addison-Wesley, 1967.

G. I. BELL and S. GLASSTONE, *Nuclear Reactor Theory*, van Nostrand, 1970.

W. FRATZSCHER and H. FELKE, *Einführung in die Kerntechnik*, VEB Deutscher Verlag für Grundstoffindustrie, Leipzig, 1971.

U.N. Conferences on peaceful uses of atomic energy, 1955, 1958, 1965, 1972.

J. R. LAMARCH, *Introduction to Nuclear Engineering*, Addison-Wesley, 1975.

IAEA, *Management of Radioactive Wastes from the Nuclear Fuel Cycle*, Vols. 1 and 2, Vienna, 1976.

Jahrbuch der Atomwirtschaft 1977, Handelsblatt GmbH, Verlag für Wirtschaftsinformation, Düsseldorf, Frankfurt, 1977.

Int. Conf. on Nuclear Power and its Fuel Cycle, Salzburg 1977, IAEA, Vienna, 1977.

R. A. RYDIN, *Nuclear Reactor Theory and Design*, PBS Publications, 1977.

H. EDELHÄUSER, H. BONKE, I. GANS, O. HUBER, K. J. VOGT, and R. WOLTER, *Radioactive Effluents and Present and Future Radiation Exposure to the Population from Nuclear Facilities in the Federal Republic of Germany*, in Int. Conf. on Nuclear Power and its Fuel Cycle, Salzburg 1977, IAEA, Vienna, 1977.

H. W. Graves, Jr., *Nuclear Fuel Management*, J. Wiley & Sons, 1979.
M. Benedict, T. H. Pigford and H. W. Levi, *Nuclear Chemical Engineering*, 2nd Ed., McGraw-Hill, 1981.
S. Glasstone and A. Sesonske, *Nuclear Reactor Engineering*, Van Nostrand Reinhold, 3rd Ed., 1981.
R. F. Mould, *Chernobyl. The Real Story*, Pergamon Press, 1988.
The Management of Radioactive Waste, The Uranium Institute, London, 1991.

CHAPTER 21

The Nuclear Fuel Cycle

Contents

Radiochemistry and Nuclear Chemistry
ISBN 978-0-12-405897-2, http://dx.doi.org/10.1016/B978-0-12-405897-2.00021-5

The nuclear fuel cycle comprises the handling of all fissile and fertile material necessary for nuclear power production and of the radioactive products formed in this process (Fig. 21.1). The fuel cycle is suitably divided into a front end and a back end part, where the nuclear power station is the dividing line. The front end comprises uranium exploration, mining, and refining (§14.2.4), isotope enrichment (§3.8), and fuel element fabrication (§21.1). Reactor operation involves fuel behaviour during operation, canning corrosion etc., while the back end involves reprocessing and radioactive waste ("radwaste") handling. Health and environmental aspects are important in all these steps, but being of a more general nature (see Ch. 15 and 22), they are not considered as "steps" in the nuclear fuel cycle.

The nuclear fuel used in almost all commercial reactors is based on uranium in the form of UO_2, either enriched so that the ^{235}U content has been increased to a few percent, or – less commonly – with the natural 0.7% abundance of the fissile isotope ^{235}U. Some power reactors also use fuel containing depleted uranium (~0.3% ^{235}U) in which plutonium is bred and/or Pu mixed with ^{238}U as a replacement for ^{235}U ("mixed oxide fuel"). ^{232}Th, in which fissile ^{233}U is bred, has also been used in a few cases.

The fuels for the next generation of reactor systems are typically either, oxides, nitrides, carbides, inert matrix or metallic. All these fuel types have distinct advantages and disadvantages raging from ease of manufacturing to direct safety issues by themselves or in combination with the selected coolants in the reactor. The coolants often discussed for future fast reactors are: molten sodium, molten lead or a gas such as helium or carbon dioxide. They also have their distinct pros and cons. Currently the majority of fast reactor hours have been run with liquid sodium as coolant even if its interactions with water is one of the major draw backs. In e.g. Russia a considerable experience with liquid lead, or liquid lead–bismuth, as coolant exist, but it has probably caused at least one accident in a nuclear submarine. In this case violent reactions with water are avoided but instead significant erosion and corrosion of e.g. coolant pumps exist.

Whether based on uranium, thorium, or plutonium, a fuel must be capable of resisting temperatures considerably above 1000°C without physical or chemical deterioration due to heat or to radiation. Nitride based fuels are of interest for future fast reactors due to their good heat conductivity. However, due to the internal N_2-pressure they need another canning material than zircaloy. Metallic fuels have the high heat conduction necessary to minimize temperature gradients. Uranium melts at 1130°C and plutonium at 640°C. However, metallic uranium has three and plutonium six allotropic forms between room temperature and their melting points. As a consequence, either the separate or combined effects of the radiation field, the high pressure during operation,

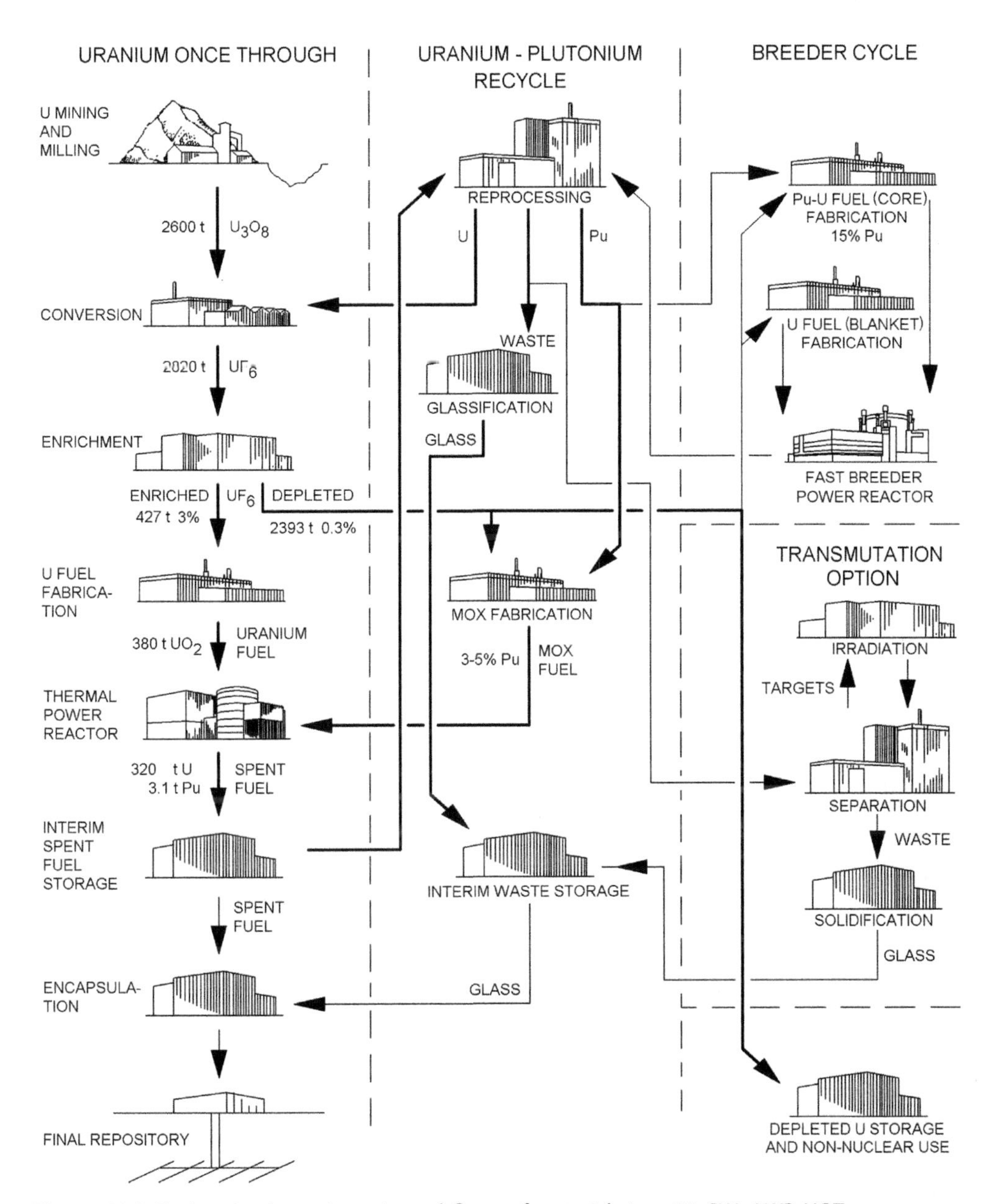

Figure 21.1 Fuel cycle alternatives. Annual flows of materials in a 10 GW_e LWR UOT program are indicated.

and the high temperature can cause recrystallization into different allotropic forms with significantly different volume. Volume changes within the fuel element during operation cause mechanical deformations, reduce the mechanical strength, and increase the problem of corrosion even if the elements are clad in another corrosion resistant metal.

With the exception of some older gas cooled reactors, power reactors use ceramic pellets of UO_2, PuO_2, and ThO_2, or a mixture of these oxides, as fuels. UC has also been tested in some reactors. The size of the cylindrical pellets is $\sim 1 \times 1$ cm (diameter $\times$ height). To improve the possible burnup and to increase the temperature margin in the pin centers, the pellets are now often thinner in diameter, e.g. 0.8 cm. Fuel rods consisting of ceramic fuel pellets stacked in metallic tubes of zircaloy or stainless steel are quite temperature resistant, do not have the phase transformations of the metals, and have greater resistance to radiation effects. Unfortunately, the heat conduction is not as good as in the metallic fuel elements, and as a result rather high temperature gradients (up to 300°C mm^{-1}) often exist in ceramic UO_2 elements.

21.1. PRODUCTION OF FUEL ELEMENTS

The normal raw material for production of UO_2 based fuels is enriched uranium in the form of UF_6, which is delivered in special containers. By heating the container to $\sim$100°C it is possible to transfer the hexafluoride as gas to the conversion plant, where UO_2 powder is produced, see Fig. 21.2. Several possible reactions can be used, e.g. hydrolysis of UF_6 by dissolution in water

$$UF_6(g) + 2H_2O \rightarrow UO_2F_2(aq) + 4HF(aq) \quad (21.1)$$

followed by precipitation with ammonia

$$2UO_2F_2(aq) + 6NH_4OH(aq) \rightarrow (NH_4)_2U_2O_7(s) + 4NH_4F(aq) + 3H_2O \quad (21.2)$$

after filtration, washing and drying the diuranate is converted to UO_2 by reduction with hydrogen at 820°C.

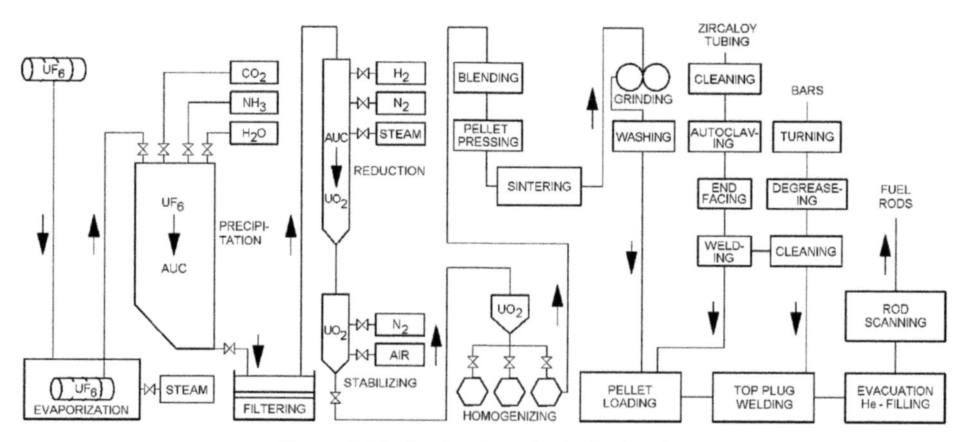

Figure 21.2 Production of UO_2-fuel rods.

Another often used reaction sequence begins with the formation and precipitation of ammonium uranyl carbonate (AUC) by reaction of UF_6 with water, CO_2 and NH_3, see Figure 21.2. The AUC is reduced to UO_2 by reduction in a fluidized bed using a mixture of hydrogen, nitrogen and steam and cooled to room temperature in a mixture of air and nitrogen.

The UO_2 powder is pressed into "green" pellets of slightly larger dimensions than the final product. These have ~50% of the theoretical density. The green pellets are sintered at ~1700°C in a dry H_2 atmosphere which gives a small controlled oxygen excess in the product; UO_{2+x}, where $x \sim 0.05$ is normally desired for best fuel performance. As exact dimensions of the pellets are needed in order to fit into the cladding tubes, the sintered pellets are ground to final shape.

Typically a density of ~10 400 kg m^{-3} is desired, which corresponds to a porosity of ~5%. If the density is too high the pellets swell excessively during irradiation due to the volume of the fission products gaseous at the operating temperature. Strongly swelling pellets may cause deformation, and failure, of the can. On the other hand a too low density can cause an initial shrinking of the pellet leading to an increased pellet – can gap, thereby reducing the heat transfer coefficient and increasing pellet temperature.

The finished pellets are put into a zircaloy tube with a welded bottom plug. Thereafter the pellet column hold-down spring and top plug are fitted and the resulting fuel pin sealed by electron beam welding, see Figures 19.12 and 21.2.

In order to improve heat transfer between pellets and can and to reduce pressure induced strain during operation, the space between pellets and tube is usually first evacuated and then pressurized with helium through a hole in the top end plug. It important to remove all traces of water as remaining water may form hydrogen by reaction with hot UO_2 during operation. The hydrogen reacts with the inside of the can forming zirconium hydride which may cause can failure.

Heat conduction between fuel and can is further improved by using *bonding materials* such as molten sodium, graphite powder, etc. The bonding material should also provide some lubrication between pellets and can. The canning material itself must not only be corrosion resistant to the coolant at all temperatures but should also not react with the fuel or the bonding material.

The can should be as thin as possible, consistent with satisfactory mechanical strength and corrosion resistance (Fig. 19.12(a)). To reduce the danger of hydride formation a protective oxide layer can be produced by autoclaving the tube before filling it with pellets. In case of UO_2 pellets in zircaloy, the bonding material, e.g. graphite, is put onto the inner surface of the zircaloy tubes before the pellets are introduced. In case of stainless steel clad fast reactor fuel the production and assembly is similar, but the bonding is usually by sodium metal.

The purpose of cladding the fuel is to protect it against corrosion and to protect the coolant from radioactive contamination by the fission products from the fuel element.

Aluminum has been used in water-cooled reactors, but at temperatures >300°C zirconium alloys show superior strength. At steam temperatures >400°C zirconium absorbs hydrogen, which increases brittleness, so stainless steel becomes preferable. In sodium cooled fast reactors, stainless steel is normally used. The most common alloys are zircaloy-2 (containing 1.58 Sn and 0.3% Cr, Ni, and Fe) and stainless steel type 302B (containing 10% Cr and 8% Ni). Stainless steel is not used at lower temperatures because of its larger neutron capture cross-section: σ 0.23 b for Al, 0.18 b for zircaloy-2, and about 3 b for 302B steel.

Metallic uranium is usually produced by conversion of UF_6 to UF_4 followed by metallothermic reduction of UF_4 by magnesium or calcium metal; however, several other methods exist. Metallic fuel is encased in a canning (cladding) of aluminum, magnesium, or their alloys. Fuel for high flux research reactors based on highly enriched uranium (>10% ^{235}U) is often made in the form of uranium metal alloys (or compounds like USi_2) canned in aluminum to improve mechanical and thermal stability.

The fuel elements for use in high temperature gas cooled reactors consist of graphite rods or balls filled with oxide or carbide kernels produced by the sol-gel process. The kernels are covered by several layers of graphite and silicon carbide achieved by pyrolyzing methane or acetylene in a fluidized bed of the kernels.

Fuel cost and performance is an important part of the economy of power reactors. Approximately 20% of the expense of the electrical production in a power reactor can be attributed to the cost of the fuel. This is due about equally to the expense of the consumption of fissile material and to the production and, when applicable, reprocessing costs or intermediate storage costs. In the fast breeder reactors, it is anticipated that fuel costs would be substantially lower because of a higher burn-up.

When mixed uranium-plutonium oxide (MOX) fuel elements are used as, for example, in plutonium recycling (<5% PuO_2) in LWRs or in fast breeders ($\leq$15% PuO_2), the UO_2–PuO_2 mixture must be very intimate. This can be achieved by coprecipitation of the tetravalent actinides, normally as oxalates, followed by calcination. However, in industrial scale production of MOX fuel a rich mixture containing 15–20% PuO_2 is first very finely ground and then diluted with coarse grained pure UO_2. The final mixture is pressed into pellets and sintered similarly to UO_2-fuel pellets. This yields a fuel which can be dissolved to 99% (or better) in nitric acid. MOX fuel elements have been regularly added to the cores of many European LWRs for many years without any technical difficulties. Mixed uranium and thorium oxide fuels have also been used in a few heavy water reactors, see §20.1.1.

Instead of ground powders, spherical fuel particles can be used as a starting material. This has advantages with respect to fabrication, reactor utilization, and fuel reprocessing. These oxide or carbide particles are very small, <1 mm in diameter. The particles are produced by the *sol–gel process*, which in principle consists of the following steps:

(i) An aqueous colloidal solution or the actinide or actinide mixture is prepared. The actinide(s) may be in the form of a hydrated complex of high concentration (3–4 M).

(ii) The solution is added to an inert solvent, which dehydrates the complex and causes the droplet to gelate. In one technique, hexamethylenetetramine, $(CH_2)_6N_4$, is added to the aqueous solution, which is added dropwise to a hot (~95°C) organic solvent. The heat causes the amine to decompose, forming NH_3, which leads to hydroxide precipitation in the droplet. The droplet dehydrates and solidifies rapidly, forming a "kernel".

(iii) The kernels are washed, air dried at 150–200°C, and – in the case of uranium – reduced by hydrogen gas at higher temperature to form UO_2.

(iv) The kernels are sintered at high temperature in an inert atmosphere.

Kernels of actinide carbides can be made in a similar manner. In many cases the kernels are covered by the addition of protective layers of graphite, silica or silicon carbide. The kernels are placed in fuel rod cans, pressed into pellets, or incorporated in a graphite matrix for use in high temperature reactors.

21.2. POWER GENERATION

During operation a large amount of heat is generated inside the fuel and has to be transferred to the coolant. Assuming a constant power per unit volume in the fuel, and fuel pins so long that longitudinal conduction can be neglected, the temperature profile can be estimated from the *specific power* (p, W m^{-3}) in the fuel from the following equations using the notation for radii from Figure 21.3.

Inside the fuel

$$T(r) = T(r_f) + p\left(r_f^2 - r^2\right)(4\,k_f)^{-1} \tag{21.3}$$

where $T(r)$ (°C) is the temperature at radius r (m) inside the fuel pellet, $T(r_f)$ (°C) the temperature at the pellet surface, r_f the pellet outer radius (m) and k_f the heat conductance in the fuel (W m^{-1} K^{-1}).

Across the gap between fuel and can

$$\Delta T_{gap} = p\,r_f\left(r_i - r_f\right)(2\,k_g)^{-1} = p\,r_f(2\,\alpha_g)^{-1} \tag{21.4}$$

where ΔT_{gap} is the temperature difference across the fuel – can gap (°C), r_i the inner radius of the can (m), k_g the heat conductivity across the gap (W m^{-1} K^{-1}) and α_g is the heat transfer coefficient across the gap (W m^{-2} K^{-1}).

For a thin can

$$\Delta T_{can} = p\,r_f^2\,(r_c - r_i)(2\,r_i\,k_c)^{-1} \tag{21.5}$$

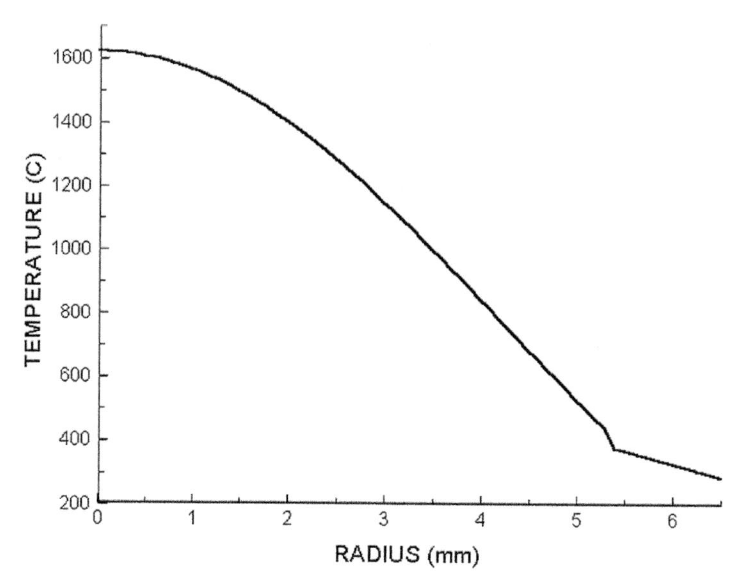

Figure 21.3 Calculated temperatures in a BWR UO_2-fuel pin during operation at high load.

where ΔT_{can} is the temperature difference across the can (°C), r_c the outer radius of the can (m) and k_c the heat conductivity in the can (W m^{-1} K^{-1}).

Between the can surface and coolant,

$$\Delta T_o = p\, r_f^2\, (2\, r_c\, h)^{-1} \tag{21.6}$$

where ΔT_o is the temperature difference between the outer surface of the can and the coolant (°C) and h the film coefficient for heat transfer between can and coolant (W m^{-2} K^{-1}). The film coefficient is affected by the coolant velocity along the surface, the temperature gradient and also by the onset of boiling in a BWR.

Temperatures are best calculated by starting with eqn. (21.6) and proceeding inward to the center. A typical temperature profile is shown in Figure 21.3. However, in a more accurate calculation we must also consider the variation in specific power with fuel radius (because of self-screening effects etc.), the variation of heat conductivity (and film coefficient) with temperature, the change in heat conductivity of the fuel caused by accumulation of fission products, by pellet breakup, and by the densification of the fuel caused by high operating temperatures; the heat conductivity of UO_2 as function of temperature is shown in Figure 21.4 and some typical data at room temperature are given in Table 21.1.

As a result of these heat gradients it has been found that ceramic fuel elements may melt in the center (2865°C mp for pure UO_2) at high loading even though the surface temperature is much below the melting point. High center temperatures, especially in breeder fuel, may cause so much densification that a central hole is formed during operation.

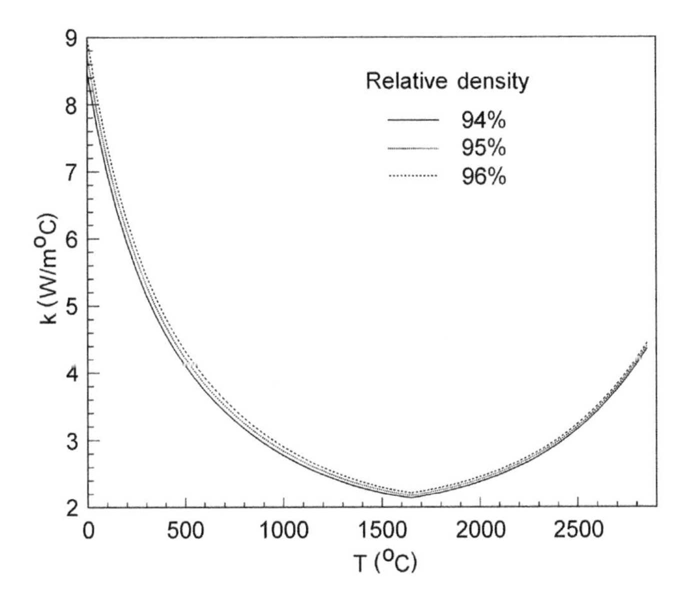

Figure 21.4 The heat conductivity of UO_2 as function of temperature.

Table 21.1 Data (~25°C) on some materials which may be used in making nuclear fuel

Material	Density (kg m^{-3})	Melting point (K)	Thermal conductivity (W m^{-1} K^{-1})	Heat capacitivity (J kg^{-1} K^{-1})	Thermal linear expansion coeff. (K^{-1})
Th (metal, α-)	11720	2023	41	118	11.2×10^{-6}
ThO_2	10001	3663	0.56	256	
U (metal, α-)	19070	1405	25	116	13.5×10^{-6}
UC	13630	2525	25.1	202	10.4×10^{-6}
UN	14320	2850	16.9	189	7.5×10^{-6}
UO_2	10970	3138	Fig. 21.4	360	14×10^{-6}
NpN	14190	3103			7.8×10^{-6}
NpO_2	11138	2820	6.2	238	8.4×10^{-6}
Pu (metal, α-)	19860	912.5	8	137	57×10^{-6}
PuC	13520	d 1927	9.7 (500 K)		10.7×10^{-6}
PuN	14150	d 2862	15.0 (740 K)	196	10.0×10^{-6}
PuO_2	11510	2663	~6	258	11.0×10^{-6}
AmN	13620	d≈2700			9.4×10^{-6}
AmO_2	11751	2773		241	8.3×10^{-6}
CmN	13380				
CmO_2	12044		8.2	221	1.02×10^{-6}
Aluminum	2700	933.4	238	903	23.2×10^{-6}
Magnox A12‡	1740	~650	~167	~1024	26×10^{-6}
SS (type 304)	8030	1673	19	500	18×10^{-6}
Zircaloy 2†	6550	~2090			5.2×10^{-6}
Zircaloy 4†‡	6440	~2120	16	330	4.4×10^{-6}

†Zr + 12-17‰ Sn, 0.7-2‰ Fe, 0.5-1.5‰ Cr, 0.3-0.8‰ Ni.
†‡Zr + 12-17‰ Sn, 1.8-2.4‰ Fe, 0.7-1.3‰ Cr.
‡0.8% Al, 0.01% Be; cf. Magnox ZA with 0.5-0.65% Zr, d = decomposes.

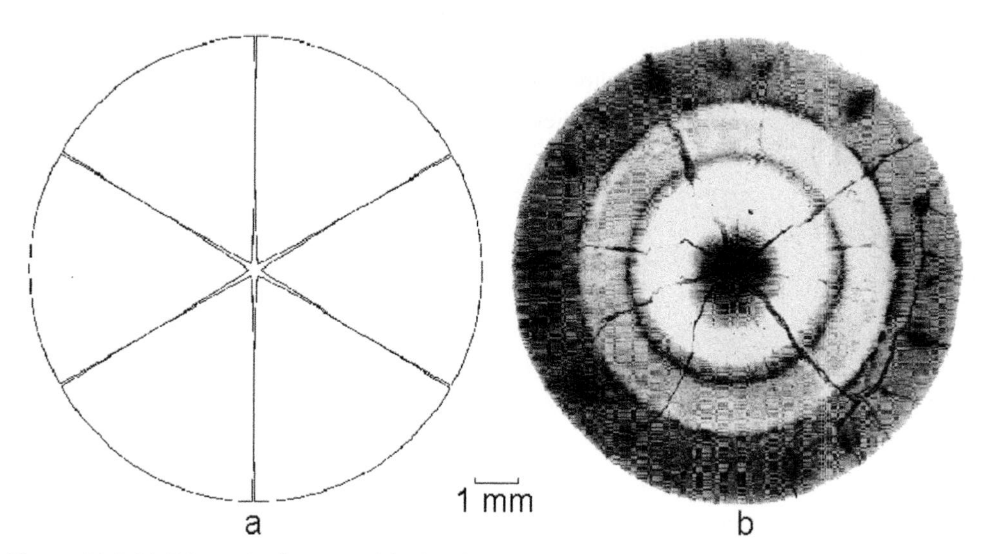

Figure 21.5 (a) Schematic drawing of fuel pellet deformation during operation and (b) autoradiograph of a cut spent fuel pellet showing the typical pattern of cracks.

The steep temperature gradients in the fuel during operation can cause a gradient of thermal expansion in the pellet; the expansion increasing from the surface to the center. The induced stresses lead to formation of a series of radial and annular cracks in the pellet and a deformation during operation, see Figure 21.5(a). Upon cooling the cracks close, but the characteristic pattern of cracks can be seen when a used fuel pin is cut and inspected, see Figure 21.5(b).

During operation, a slow corrosion of the can is unavoidable. As long as the corrosion products stick to the surface, corrosion rates drop with time. For zircaloy clad fuel in water cooled reactors the corrosion rate follows a parabolic equation (in the normal operating temperature range)

$$ds/dt = (k/s)e^{-u} \tag{21.7}$$

where s is the thickness of the zirconium dioxide layer (m), t is the exposure time (s), k the rate constant (3.937×10^{-5} m^2 s^{-1}) and u is given by

$$u = \Delta E (\mathbf{R}\, T_c)^{-1} \tag{21.8}$$

where $\mathbf{R}$ is the gas constant (8.318 J mole^{-1} K^{-1}), ΔE the activation energy (1.905×10^5 J mole^{-1}) and T_c the surface temperature of the can (K). The zirconium – steam reaction becomes very violent above ~1200°C, see §19.15. At about the same temperature a less well studied reaction between UO_2 and Zr begins in the fuel – cladding gap leading to the formation of a metallic U+Zr melt and ZrO_2.

Accumulation of the various fission products (and other impurities) occur where their chemical potential is at minimum. In contrast to an isothermal system where concentration gradients tend to disappear, the large temperature differences in operating nuclear fuel can result in a lower chemical potential at a higher concentration leading to an increase in the concentration gradient. Hence, some fission products, e.g. the noble gases, Cs and I, migrate to the fuel – cladding gap (lowest temperature), whereas others, e.g. Zr and Nb, migrate to the center line (highest temperature), cf. Fig. 21.5(b). A penetration of the can during reactor operation thus leads to an initial rapid release of those fission products which have accumulated in the gap followed by a slow release of fission products present in the cracks and at grain boundaries and finally by a much slower dissolution and release of fuel (U, other actinides and other fission products).

The high concentration of uranium (and/or Pu) atoms in the fuel in combination with strong resonance peaks at certain neutron energies also leads to self-screening effects (n-flux depression at energies with large reaction cross sections), especially for $^{238}U(n,\gamma)$, $^{239}Pu(n,\gamma)$ and $^{239}Pu(n,f)$ (see Fig. 19.3). Thus, most of the Pu formed is located near the fuel surface and relatively little at the fuel center. This is easily seen in α-autoradiographs of spent fuel pins where most of the α:s are found in a ring near the fuel surface. Hence, thinner fuel pins yield a better fuel utilization than thicker pins, but thinner pins means higher fuel fabrication costs for the same amount of uranium. In practice an economic compromise usually results in pin diameters slightly less than 1 cm.

Radiation effects and oxidation causes changes in the tensile properties of the canning material. Fresh fuel has a can that is very ductile whereas the can of spent high burn-up fuel normally is hard and brittle.

In an operating power reactor, only part of the fuel is replaced annually; e.g. 1/6 to 1/4 of the total number of fuel elements. The most burnt up fuel elements are removed from the core as spent fuel and replaced by fresh fuel. In order to achieve as even as possible heat generation in the core (permitting the highest power output), the fresh fuel elements are mostly loaded in the outer core regions whereas partly spent fuel is moved in towards the center. This results in a checker board pattern of fuel of varying age in the reactor. However, several variants to these movements are in use.

21.3. COMPOSITION AND PROPERTIES OF SPENT FUEL ELEMENTS

The composition of spent reactor fuels varies as a function of input composition (kinds and amounts of fissile and fertile atoms), neutron spectrum, flux, fluency (or burnup), design of the pins and fuel elements, positions occupied in the reactor during operation, and the cooling time after removal from the reactor. A harder neutron spectrum increases fertile to fissile conversion (Fig. 19.3). Hence, after refuelling, some BWRs are initially operated at the highest possible void fraction in order to maximize conversion of ^{238}U to plutonium. This permits a higher final burnup of the fuel. Increased burnup increases the concentration of fission products and larger amounts of higher actinides are formed (Fig. 19.7). Thinner fuel pins increases conversion to and burning of plutonium due to less self screening. A high neutron flux results in more high order reactions (§17.3), while a long irradiation time produces relatively larger amounts of longlived products. With increased cooling time the fraction of shortlived products is reduced.

Because of such effects, spent uranium fuel elements from PWR, BWR, HWR, GCR and FBR differ in composition both from each other and between fuel batches from the same reactor. Furthermore, the composition differs between pins in the same fuel element and for each pin also along its length, especially when initial burnable poison concentration and enrichment is graded along pins. The difference is not so large that very different fuel cycles (e.g. other reprocessing schemes) are required as long as the fuel is based on uranium metal or uranium dioxide. In the following subsections we mainly discuss uranium dioxide fuel elements, and, more specifically, LWR elements. Fission product and actinide yields for a typical BWR fuel are given in Table 21.2.[1] The uranium once through part of Figure 21.1 shows the annual materials flow in a mixed BWR–PWR conglomerate with a total average power of 10 GW_e (i.e. about twelve 1000 MW_e plants) running at full power for 7000 hours per year (load factor ~80%).

Table 21.2 Calculated average composition after 1 y cooling of 1 t U as 3.6% enriched UO_2 fuel with 60 MWd/kg U burnup in a typical BWR where the n-flux changes yearly after fuel rearrangement. Straight figures are weight in kg/t IHM and italics are radioactivity in TBq/t IHM. Total weight is 63 kg FPs and 14 kg transuranium elements

H																	He
Li	Be											B	C 0.14 *0.04*	N	O 134.5 *0*	F	Ne
Na	Mg											Al	Si	P	S	Cl	Ar
K	Ca	Sc	Ti	V	Cr	Mn	Fe	Co	Ni	Cu	Zn	Ga	Ge $8\cdot10^{-4}$ *0*	As $2\cdot10^{-4}$ *0*	Se 0.096 *0.025*	Br 0.036 *0*	Kr 0.68 *1930*
Rb 0.44 *$2\cdot10^{-6}$*	Sr 1.3 *3900*	Y 0.72 *3900*	Zr 5.7 *54*	Nb $9\cdot10^{-5}$ *130*	Mo 6.2 *0*	Tc 1.1 *0.71*	Ru 4.8 *5900*	Rh 0.53 *5900*	Pd 3.6 *0.010*	Ag 0.12 *7.3*	Cd 0.30 *$2\cdot10^{-5}$*	In 0.001 *$4\cdot10^{-13}$*	Sn 0.12 *11*	Sb 0.028 *270*	Te 0.99 *0.004*	I 0.44 *0.002*	Xe 9.8 *$4\cdot10^{-13}$*
Cs 4.6 *15000*	Ba 3.3 *$6\cdot10^{-8}$*	*Ln	Hf	Ta	W	Re	Os	Ir	Pt	Au	Hg	Tl $3\cdot10^{-12}$ *0.041*	Pb $6\cdot10^{-7}$ *$5\cdot10^{-4}$*	Bi $3\cdot10^{-12}$ *$5\cdot10^{-4}$*	Po $2\cdot10^{-15}$ *$8\cdot10^{-4}$*	At $5\cdot10^{-22}$ *$3\cdot10^{-8}$*	Rn $1\cdot10^{-14}$ *$5\cdot10^{-4}$*
Fr $6\cdot10^{-18}$ *$3\cdot10^{-6}$*	Ra $2\cdot10^{-10}$ *$5\cdot10^{-4}$*	**An	Rf	Db	Sg	Bh	Hs	Mt	Ds	Rg	Cn	113	Fl	115	Lv	117	118

*Lanthanides	La 2.2 *$8\cdot10^{-7}$*	Ce 4.3 *4000*	Pr 2.0 *4000*	Nd 7.2 *$5\cdot10^{-10}$*	Pm 0.061 *2100*	Sm 1.3 *9.3*	Eu 0.33 *1600*	Gd 0.52 *0.32*	Tb 0.006 *0.36*	Dy 0.003 *$2\cdot10^{-13}$*	Ho 0.006 *0*	Er	Tm	Yb	Lu
**Actinides	Ac $6\cdot10^{-11}$ *$2\cdot10^{-7}$*	Th $3\cdot10^{-6}$ *0.012*	Pa $7\cdot10^{-7}$ *0.042*	U 923 *0.19*	Np 1.2 *3.6*	Pu 12 *6500*	Am 0.74 *37*	Cm 0.44 *1500*	Bk $4\cdot10^{-7}$ *0.027*	Cf $2\cdot10^{-8}$ *0.005*	Es $8\cdot10^{-12}$ *$6\cdot10^{-7}$*	Fm $2\cdot10^{-19}$ *$5\cdot10^{-12}$*	Md	No	Lr

[1] As the exact composition of spent fuel varies considerably and depend on many factors, the reader will find slightly varying figures in this text.

21.3.1. Fission products

About 34 kg fission products (FP, including gaseous) are formed in each initial ton of uranium irradiated to 33 MWd/kg. To accommodate mixed oxide fuels (i.e. fuels initially containing both U and Pu), burnup, composition, activities, etc, are usually normalized to the amount of *initially present heavy metal, IHM.* The composition of spent fuel varies with burnup, power history and reactor. The formation rates of the various primary fission products depend on the fission rate, chain yields of the fissioning nuclide and on the corresponding charge distribution (Ch. 11). Increased fission of heavier actinides displaces the lower mass peak in the yield curve towards higher mass numbers while the heavier mass peak remains about the same (Fig. 11.9). An increased contribution from fast fission increases the yield in the valley between the peaks. Because of the continued n-irradiation, secondary n,γ-reactions occur with the primary fission products and their daughters (Ch. 17). As an example, ^{134}Cs is not formed to any appreciable extent in fission because it is shielded by the stable ^{134}Xe. Hence, no ^{134}Cs is normally observed in the remains after a nuclear explosion in the atmosphere. However, primary fission products in the $A = 133$ isobar chain have time to decay to stable ^{133}Cs during reactor operation and ^{134}Cs is produced by the reaction ^{133}Cs (n,γ) ^{134}Cs. Given the cooling time, the ratio between the decay rates of ^{134}Cs and ^{137}Cs can be used to estimate the burnup of fuel from a given reactor, see Figure 21.6.

Using effective cross-sections and yield values the amounts and radioactivities in Figure 21.7 and Table 21.2 were calculated. It is seen that Xe, Zr, Mo, Nd, Cs, and Ru, which are the elements formed in largest amounts in thermal fission (both by mole percent and by weight), constitute about 70% of the fission product weight after a cooling time of 1 y.

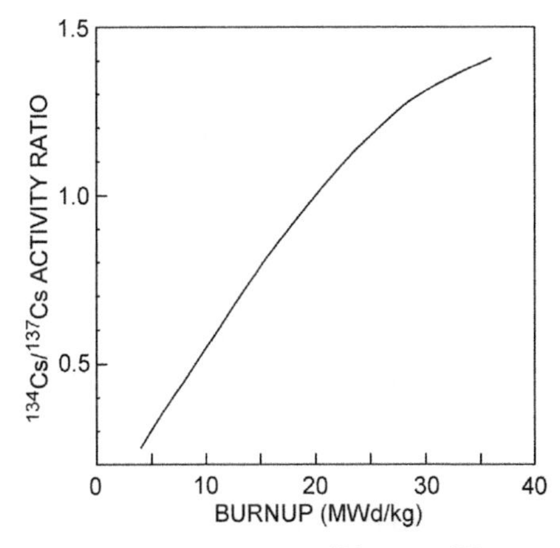

Figure 21.6 Typical variation of the ratio between the ^{134}Cs and ^{137}Cs radioactivities with fuel burnup.

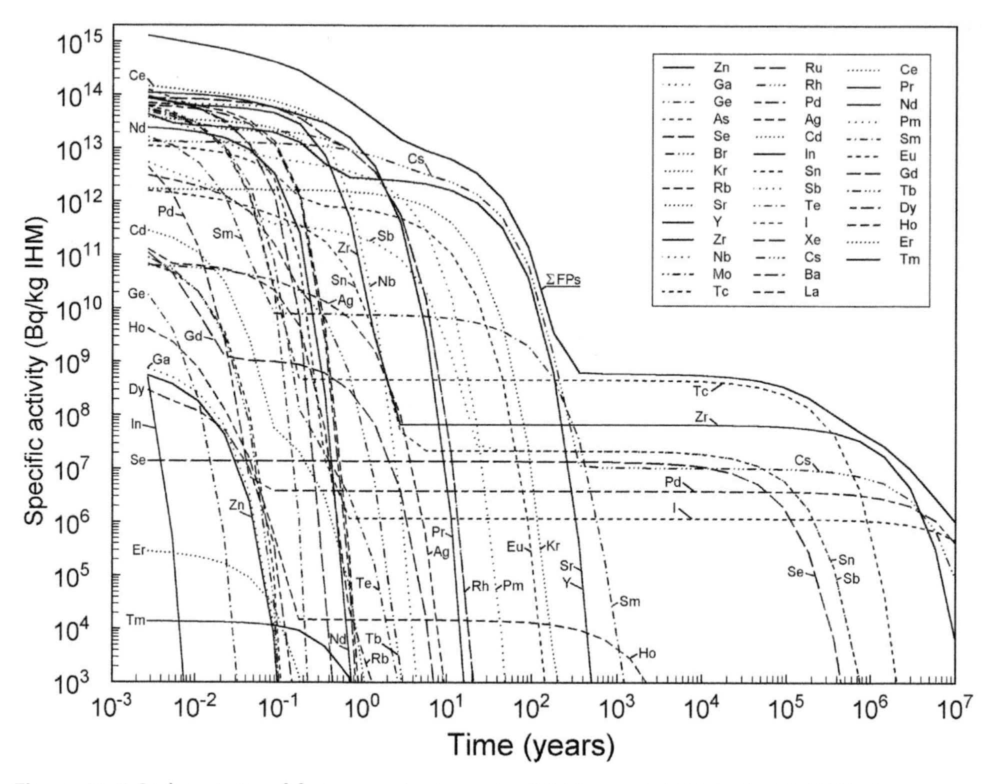

Figure 21.7 Radioactivity of fission products per kg IHM in spent PWR fuel at 33 MWd/kg burnup. Inflexion points indicate the existence of several radioisotopes.

At cooling times 10–1000 y the activities of ^{90}Sr and ^{137}Cs (with daughters) dominate among the fission products. Later the fission product activity is due to very longlived nuclides of low activity (Figure 21.7, Table 21.3). ^{131}I, which for very short cooling times is one of the most hazardous FPs because of its affinity to the thyroid gland, has practically disappeared after a cooling time of 6 months. The $A = 95$ isobar chain, which is formed in highest yield (6.52%), leads to ^{95}Zr ($t_{½}$ 64 d) → ^{95}Nb($t_{½}$ 35 d) → ^{95}Mo (stable).

From Table 21.2 it is seen that only stable isotopes remain for some fission elements at $t_{cool} \geq 1$ y, while some others are of low activity. Some of the more active ones after 1 y have disappeared almost completely at 100 y (T, Sb, Ce, Eu, Pm, Ru, Rh, and Kr), leaving essentially only the ^{90}Sr–^{90}Y and ^{137}Cs–^{137m}Ba pairs and ^{151}Sm.

21.3.2. Actinides

Neutron capture and β-decay lead to the formation of higher actinides. This is illustrated in Figs. 14.3, 14.4, 19.5 and 19.7. ^{239}Pu and ^{241}Pu also fission, contributing significantly

Table 21.3 Some long-lived radionuclides produced in fission; t_{cool} 1 y, data as in Table 21.02

Nuclide	Half-life (years)†	Decay; β^- (energy MeV)‡	Thermal fission yield(%)††	Activity (TBq/t U)
^{79}Se	$<6.5 \times 10^4$	0.1509	0.0443	251
^{85}Kr	10.72	0.6874	1.318	1930
^{87}Rb	4.8×10^{10}	0.2823	2.558	1.2×10^{-6}
^{90}Sr→^{90}Y	28.5	0.5462 + 2.2815	5.772	3890
^{93}Zr→^{93m}Nb	1.5×10^6	0.0905	6.375	0.112
^{99}Tc	2.13×10^5	0.2936	6.074	0.710
^{107}Pd	6.5×10^6	0.033	0.147	0.0104
^{126}Sn→^{126m2}Sb	$\sim 1 \times 10^5$	0.368 + 3.670	0.0536	0.0493
↓ ^{126m1}Sb				
↓ ^{126}Sb				
^{129}I	1.57×10^7	0.192	0.757	0.0023
^{134}Cs	2.062	2.0585	0	8760
^{135}Cs	3.0×10^6	0.205	6.536	0.0261
^{137}Cs→^{137}mBa	30.0	0.5134	6.183	6619
^{151}Sm	90	0.0763	0.4196	9.34
^{154}Eu	8.8	1.9689	0	984
^{155}Eu	4.96	0.2527	0.0320	574

†Only for the longer lived mother nuclide.
‡Decay energy, not particle energy (see decay schemes).
††Thermal fission of ^{235}U (fission of ^{238}U, fission of Pu isotopes and n,γ-reactions are important effects in a nuclear reactor).

to the energy production (Fig. 19.8). Truly, all plutonium isotopes lead to fission, since the n-capture products and daughters, ^{241}Am, ^{242}Am, ^{243}Cm, ^{245}Cm, and most other actinides are either fertile, fissible or fissile.

Fast neutrons in the reactor induce (n,2n) reactions, e.g. ^{238}U (n,2n) ^{237}U (β^-) ^{237}Np (n,γ) ^{238}Np (β^-) ^{238}Pu (n,γ) ^{239}Pu (see Fig. 19.5), as well as fast fission. ^{237}Np is also formed through the reaction ^{235}U (n,γ) ^{236}U (n,γ) ^{237}U (β^-) ^{237}Np. These reactions are the main sources of neptunium and of ^{238}Pu. Many actinides formed are fissile but have a short half-life. The ratio, x, of the amount fissioned to the amount decayed or reacted with neutrons at constant flux and steady state is given by

$$x = \varphi\,\sigma_f / \left[\lambda + \varphi\left(\sigma_f + \sigma_{n,\gamma}\right)\right] \tag{21.9}$$

where φ is the neutron flux, σ_f the effective fission cross-section, $\sigma_{n,\gamma}$ the effective cross-section for radiative capture and λ the decay constant. As can be seen from (21.9) a higher fraction of the nuclide will fission in a very high neutron flux ($x \rightarrow \sigma_f/(\sigma_f + \sigma_{n,\gamma})$ when $\varphi \rightarrow \infty$), whereas most of it will disappear by decay in a low flux ($x \rightarrow 0$ when $\varphi \rightarrow 0$). As an example, practically all ^{238}Np formed by ^{237}Np(n,γ) ^{238}Np will fission when

$\varphi\sigma_f \gg \lambda$ ($\sigma_f = 2070$ b, $\sigma_{n,\gamma} \sim 0$ b and $\lambda = 3.8 \times 10^{-6}$ s^{-1}). Hence, the buildup of many higher actinides is less efficient in a high flux reactor than in a reactor with a more moderate neutron flux although the total buildup rate might still be higher in the higher flux. This is accentuated in fast breeder reactors where the combination of a very high neutron flux and a hard neutron spectrum (increased effective σ_f for fissible nuclides) strongly reduces the amount of higher actinides formed (at a given burnup) compared to a thermal reactor.

From the fission and capture cross-sections, and half-lives the radioactivity of each actinide element in one kg spent fuel (and radium) has been calculated and shown in Figure 21.8 for a PWR UO_2 fuel with a burnup of 33 MWd per kg IHM. Due to the use of a log-log scale in Figure 21.8, the decay curve of any single nuclides is s smooth curve bending downwards. Inflexion points indicate the existence of several radioactive isotopes of the same element with different half-lives.

The total amount of plutonium formed in various reactors is given in Table 21.4. The old gas-graphite reactors and heavy water reactors are the best thermal plutonium

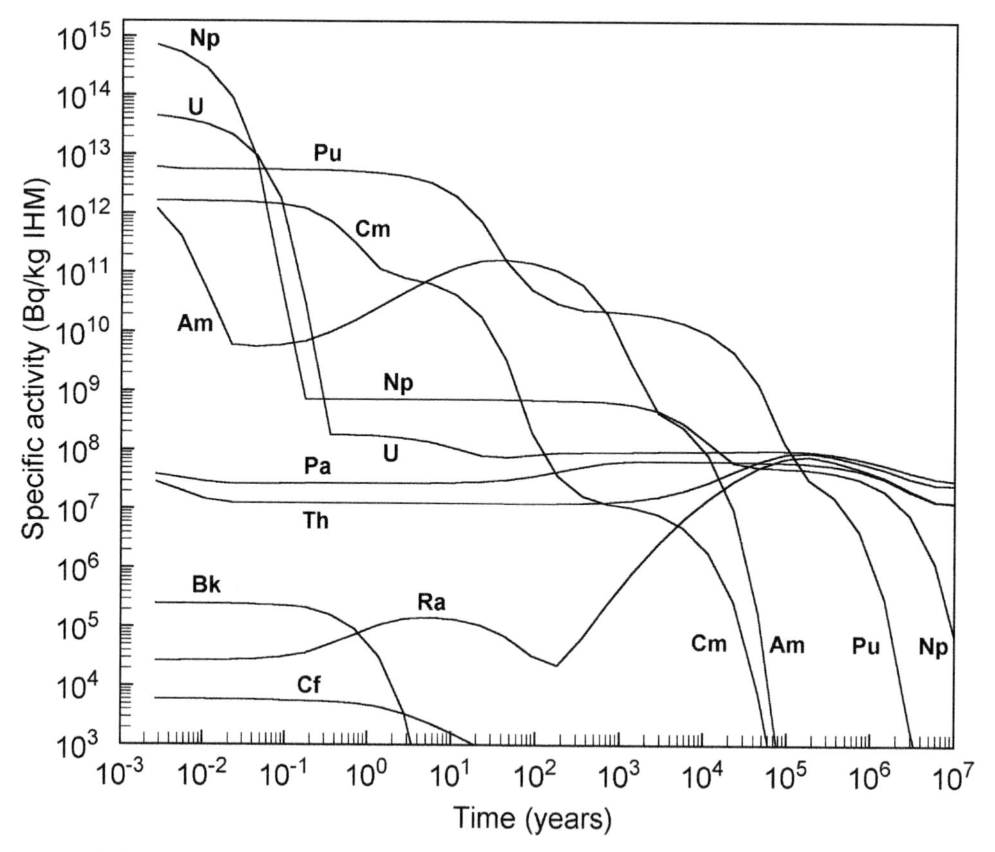

Figure 21.8 Radioactivities of actinides and radium per kg IHM in spent PWR fuel after 33 MWd/kg burnup. Inflection points indicate the presence of several radioisotopes of the element.

Table 21.4 Production of plutonium (kg/MW_ey) in various reactor types

Reactor type	Total Pu	Fissile Pu
Light water reactors	0.26	0.18
Heavy water reactors	0.51	0.25
Gas-graphite reactors	0.58	0.43
Advanced gas-cooled reactors	0.22	0.13
Liquid metal fast breeder reactors	1.35	0.7–1.0

producers. They have therefore been used in weapons fabrication. The fast breeder reactor is also an efficient Pu producer. Whereas thermal reactors (except at very low burnup) produce a mixture of odd and even *A* Pu isotopes, a fast breeder loaded with such a mixture, by a combination of fission and n-capture increases the relative concentration of Pu isotopes with odd *A* in the core and produces fairly pure ^{239}Pu in the blanket. Hence the combined Pu product from core and blanket elements has a much higher concentration of fissile Pu isotopes than plutonium from a thermal reactor (the fast breeder not only produces more Pu than it consumes but also improves Pu quality, i.e. increases the concentration of fissionable isotopes). The LWR and AGR are the poorest plutonium producers.

Composition of spent fuel varies somewhat with cooling time, e.g. the amount of ^{241}Am increases because it is the daughter of 14.4 y ^{241}Pu; 116, 280 and 579 g/t IHM of Am after 0, 3 and 10 years, respectively.

21.3.3. Decay heat and physical properties

As the radioactivity of the FP and actinides decreases by time, so does the energy absorbed in the shielding material (and by self-absorption) which is seen as *decay heat.* Table 21.5 gives data on the decay heat with contributions separately for the fission products and the actinides. Figure 21.9 shows the variation of the average energy per decay with time. The maxima and minima in the total and actinide curves in Figure 21.9 are caused by the presence of both α- and β,γ emitters with different activities and half-lives and also by the evolution in some decay chains. For cooling times $>10^3$ y the decay heat from the actinides and their daughters dominates.

The decay heat is considerable at short cooling times due to the very high decay rate (see Fig. 19.15). Before unloading spent fuel from a reactor, the used fuel elements are first allowed to cool in the reactor by forced circulation. Within a few weeks they are then transferred under water to the cooling basin at the reactor site for an additional cooling time, usually 6–12 months, after which they may be transferred to a central spent fuel storage facility. In the absence of such facilities, spent fuel elements can be stored in the reactor pools for many years. During this time the radiation level and heat production decrease considerably.

Table 21.5 Decay heat from unpartitioned fuel, fission products and heavy elements

Cooling time	Decay heat (W/kg initial U)		
	Total	**Fission products**	**Heavy elements**
1 d	54.3	38.7	15.6
90 d	16.2	13.0	3.27
180 d	11.6	9.20	2.42
1 y	7.67	5.93	1.73
5 y	2.98	1.70	1.28
10 y	2.17	1.12	1.05
100 y	0.28	0.016	0.27
10^3 y	0.13	0.000036	0.13
10^4 y	0.017	0.000034	0.017
10^5 y	0.0080	0.000020	0.0080
10^6 y	0.0021	0.0000007	0.0021

Limits: Fission products 28<Z<68, heavy element Z>79.
(Basic data as in Table 21.2)

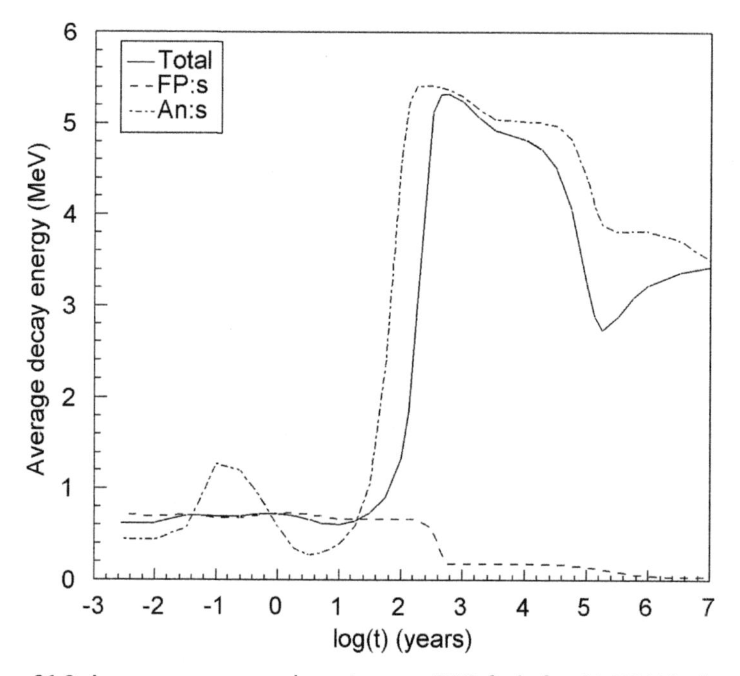

Figure 21.9 Average energy per decay in spent PWR fuel after 33 MWd/kg burnup.

21.4. MANAGEMENT OF SPENT FUEL

There are in principle several options for the handling of spent fuel. It can be:

a) Stored temporarily in pools, or dry storage, awaiting a decision on its future treatment

b) Reprocessed on order to recover fissionable material like U and Pu

c) Placed in a deep underground geologic repository
d) Ejected into space (not very likely at present)
Figure 21.11 shows the known and predicted total amounts of spent fuel, reprocessed fuel, and stored fuel as a function of time.

21.4.1. Transport of spent reactor fuel

The storage capacity of reactor pools is normally several years' production but can be increased by adding neutron absorbers to the storage racks. Eventually the fuel assemblies must be transferred in special transport flasks to (interim) storage sites, sent for reprocessing, or sent to final disposal.

The loading of used fuel the assemblies in the transport flask requires shielding and remote handling, and the heat continual cooling. Therefore, almost all operations are carried out under water. Because each transport is expensive the transport flasks are designed to carry several assemblies of different types. A 30 t (gross weight) flask may carry 4 PWR or 9 BWR assemblies ($\sim$1 t U), a 100 t flask $\sim$12 PWR or $\sim$30 BWR assemblies ($\sim$6 t U). The inner cavity of such flasks is normally surrounded by a neutron absorbing shield; see Figure 21.10. They have shock absorbers, and sometimes cooling fins on the outside. A filled 100 t flask with 1 year old fuel develops $\sim$60 kW heat; the

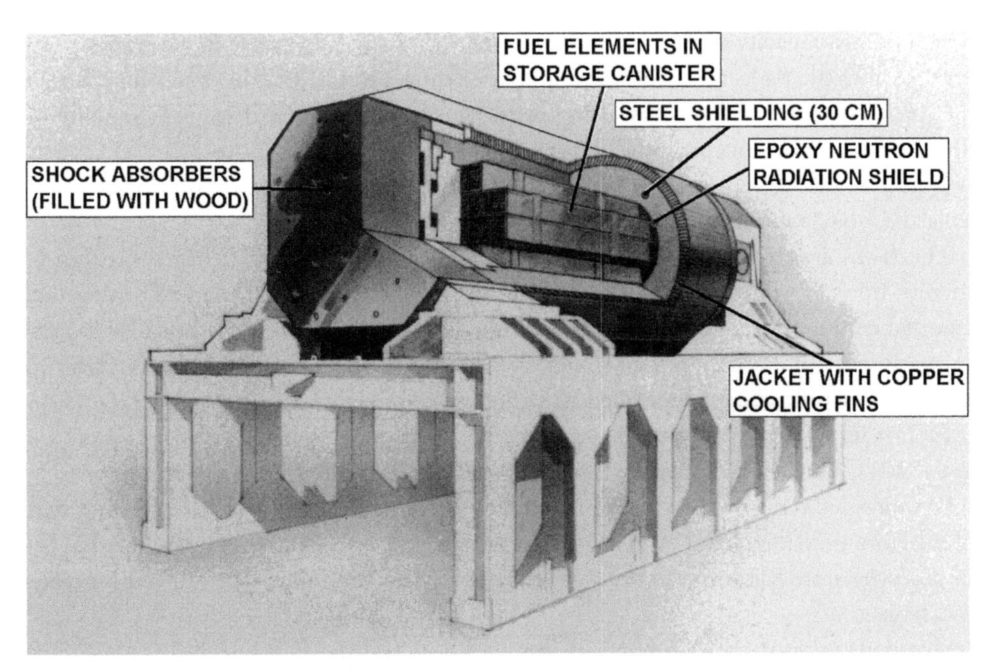

Figure 21.10 Transport cask TN-17 for 7 PWR or 17 BWR fuel elements; 1.96 m, length 6.15 m, empty weight 76 tons (SKB, Sweden).

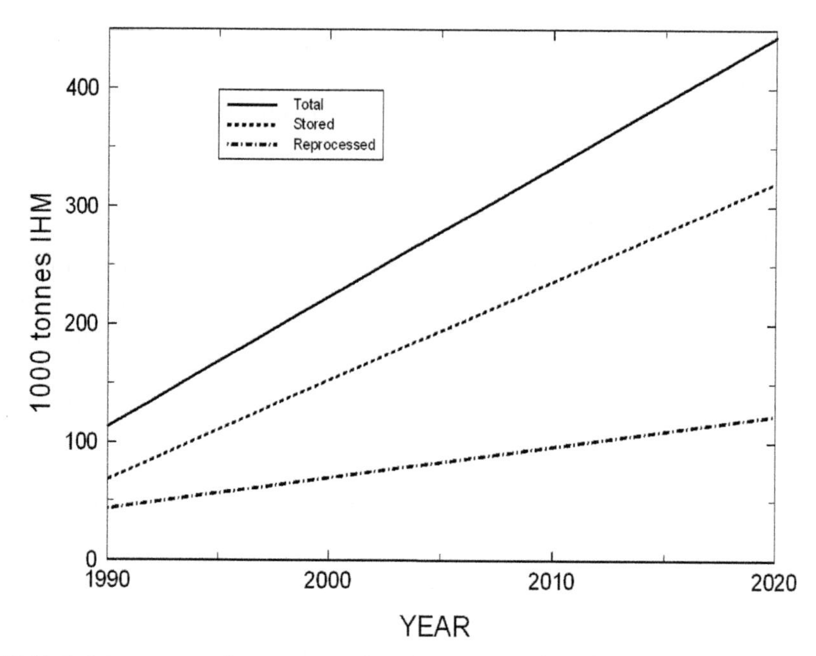

Figure 21.11 Total amounts of spent, stored, and reprocessed nuclear fuels as predicted by IAEA 2010.

design cooling capacity of the flask is ~100 kW. The cavity in some flasks is filled with water because the flasks are loaded and unloaded under water, and the water functions as heat conducting and neutron absorbing material (several actinides decay by spontaneous fission; α,n-reactions occur with light target atoms). Figure 21.10 shows a fuel cask for dry transport of 7 PWR or 17 BWR assemblies by special trucks, by rail, or by sea in special RO-RO type ships.

The flasks are designed for exceedingly severe treatment: free fall from 9 m onto a concrete floor, 30 min gasoline fire (~800°C), submersion into 15 m of water, etc., without being damaged. After loading a flask the outside of the flask must be decontaminated; often it is quite difficult to remove all activity, and a removable plastic covering is used. Before unloading, water filled flasks are flushed to remove any activity leaked from the fuel or suspended crud, which is present on the used fuel elements from BWRs and most PWRs. Further decontamination is carried out after the flask has been emptied.

Dry air filled flasks are used in many countries. They are either cooled by filling with water before unloading (the escaping steam is collected and condensed) or unloaded hot and dry by remote handling. In the former case, further handling is the same as for water filled flasks.

Shielded transport is also required for solidified high level waste, hulls, plutonium containing material, and for some intermediate and α-active waste. Special containers are used for each type. As an example, in the United Kingdom plutonium containers made of wood and cadmium are limited to carrying 10 kg Pu; the container weighs 175 kg, is 1.3 m high, and 0.8 m in diameter.

21.4.2. Interim storage facilities

In the once-through fuel cycle, or with a limited capacity for reprocessing, separate spent fuel element interim storage facilities become compulsory. These facilities are either of the wet pool type (Czech Republic, Finland, Germany, Japan, Sweden, United Kingdom) or of the dry vault type (France, Germany, United Kingdom). The pool types consist of large water filled basins containing a geometrically safe (with regard to criticality) lattice of stainless steel racks for the spent fuel elements. The water is circulated for cooling. In general such pools are either located in a building at ground level or underground at 30–50 m depth. To increase pool capacity, boron steel plates are sometimes inserted thereby permitting denser packing of fuel elements without any criticality problem.

Zircaloy clad oxide fuel elements can be stored for decades in storage pools with very little risk of leakage. Metal fuels, especially those canned in magnesium or aluminum alloys, are less resistant and should not be stored as such in this manner for a prolonged time. The corrosion resistance of aluminum or magnesium clad fuel can be improved by electrolytic treatment yielding a protective oxide layer.

Some of the stringent requirements on the storage pools are: $k_{eff} < 0.95$ (even if unused fuel elements are introduced), earthquake safety, no possible water loss, water level automatically kept constant, adequate leakage and radiation monitoring systems, water temperature $<65°C$, acceptably low radiation level in working areas, etc.

Dry vault storage may also be used. The fuel elements are stacked horizontally or vertically in concrete pipes which allows cooling of the fuel elements by air convection (forced or natural).

For long term storage ($\geq$10 y) the fuel elements must usually be recanned. For this purpose single fuel elements or bundles are placed in cylindrical containers, and the void is filled with some suitable material like lead, which has good heat conductivity and also provides some radiation protection. Depending on the external condition at the final storing place (humidity, temperature, etc.) the canisters are surrounded by an additional container to improve the lifetime of the fuel elements, which – preferably – should not be exposed to the biosphere until all radioactivity has disappeared.

21.5. ALTERNATIVE FUEL CYCLES

Fission energy can be obtained from uranium, using the *uranium once-through* option and the *uranium-plutonium fuel cycle*, and from thorium, by the *thorium-uranium fuel cycle*. Each fuel cycle offers a number of alternative routes with respect to reactor type, reprocessing, and waste handling. Although the uranium based cycles are described with special reference to light water reactors, the cycles also apply to the old uranium fuelled gas cooled reactors.

In recent years several special fuel cycles have also been discussed which aim at the destruction of the minor actinides Np, Am and Cm in fast reactors. However, so far none of these have been implemented.

21.5.1. The uranium once-through (UOT) option

The heavy arrows in Figure 21.1 indicate the steps in the nuclear fuel cycle presently used on a large commercial scale. The cycle stops at the spent fuel interim storage facility; from here two alternative routes are available, one leading to the uranium-plutonium fuel cycle (reprocessing of the spent fuel elements, as described in the next section) and another leading to final storage of the unreprocessed spent fuel elements. The latter is referred to as the *once-through fuel cycle* (UOT) option.

In the UOT option, the energy content of unused ^{235}U, fertile ^{238}U, and fissile and fertile plutonium is not retrieved for future use, and the "waste" contains large amounts of these and other α-emitting nuclides. At present this is the cheapest option and also withholds plutonium from possible diversion to weapons use.

So far only four countries (Canada, Finland, Sweden, USA) have decided to use the UOT strategy, while all other nuclear power countries either are undecided or have elected to use the reprocessing strategy. The technical aspects of the UOT option are further discussed in §21.13.

21.5.2. The uranium-plutonium (U-Pu) fuel cycle

"Cycle" infers to some mode of recirculation of material. The term "fuel cycle" was originally used for the steps in which fissile and fertile material was isolated from used fuel elements (*reprocessing*) and returned to the front end of the process for use in new fuel elements, see Figure 21.1. The proper time to start reprocessing is a balance between loss of fissile material (^{241}Pu $t_{1/2}$ 14.4 y), economic interest loss on unused fissile and fertile material, and storage costs of the unprocessed fuels, on one hand, and, on the other hand, savings due to simplified reprocessing and waste handling. Originally a 180 d cooling time was considered appropriate (this time has been said to have been used in military programs). Presently the average cooling time for commercial fuel elements is 7−10 y because of lack of reprocessing capacity.

The fissile fraction in spent LWR fuel elements amounts to ~0.9% ^{235}U and 0.5−0.7% $^{239+241}$Pu. By recovering these and returning them to the LWR fuel cycle the demand for new uranium and enrichment services is reduced by ~30%. The uranium recovered may either be re-enriched and used in normal uranium oxide fuel or blended with the plutonium recovered to form *mixed oxide* (MOX) fuel elements (§21.1). MOX fuel can also be made from recovered plutonium and depleted uranium. MOX fuel elements for LWRs contain up to 5% ^{239}Pu+^{241}Pu. Many tons of plutonium has already been used as MOX fuel in LWRs.

The re-enrichment of recovered uranium leads to a small contamination of the enrichment plant by ^{232}U, ^{233}U and ^{236}U, which are introduced as part of recovered uranium. Today, most batches of enriched uranium contain small amounts of these uranium isotopes. This phenomenon is usually called "rainbowing".

In this cycle "old" plutonium from earlier production is successively exposed to increasing neutron fluency, which changes its isotopic composition. This is shown in Table 21.6 for both virgin plutonium and virgin + recycled plutonium. The fissile fraction (^{239}Pu+^{241}Pu) decreases, requiring successively increased plutonium fractions in the MOX elements. The concentration of the most toxic isotope, ^{238}Pu, increases substantially. Its short half-life (88 y) results in a high specific radioactivity which increases the heat evolution and radiolysis of the reprocessing solutions. ^{238}Pu, ^{240}Pu, and ^{242}Pu all decay partly through spontaneous fission, with the emission of neutrons ($\sim 2 \times 10^6$ n kg^{-1} s^{-1}). This makes the MOX elements more difficult to handle. (All these changes also make the plutonium less suitable for weapons use.) ^{241}Pu has a critical mass of about half that of ^{239}Pu, so as ^{241}Pu builds up, criticality risks increase. The buildup of other actinides by recycling high mass isotopes in LWR fuel further increases reprocessing, handling, fuel manufacturing and waste problems.

Another possibility is to recycle only uranium, while the plutonium is left with the waste. This would produce a waste with a very high concentration of plutonium, except for highly enriched uranium fuel; the weight ratio of plutonium to fission products would be 10:35 (kg/t U, Table 21.2).

Table 21.6 The isotopic composition of plutonium (weight %) as function of number of recycles as MOX fuel in a PWR with a burnup of 33 000 MW$_{th}$d/t IHM per cycle

Recycles	^{236}Pu	^{238}Pu	^{239}Pu	^{240}Pu	^{241}Pu	^{242}Pu	^{244}Pu	Fraction left
0	7.2×10^{-7}	1.5	56.6	26.0	10.8	5.2	0.0004	1.00
1	4.6×10^{-8}	4.7	32.2	33.7	10.0	19.5	0.0013	0.47
2	6.6×10^{-8}	5.9	22.7	25.5	7.9	38.1	0.010	0.23
3	9.2×10^{-8}	4.4	19.8	20.2	5.0	50.6	0.020	0.12
4	1.1×10^{-7}	2.9	18.2	19.9	3.9	55.1	0.040	0.066
5	1.2×10^{-7}	2.3	17.6	20.3	3.6	56.1	0.080	0.036
6	1.2×10^{-7}	2.1	17.4	20.4	3.6	56.3	0.16	0.020
7	1.2×10^{-7}	2.1	17.4	20.5	3.5	56.3	0.29	0.011
8	1.2×10^{-7}	2.1	17.4	20.4	3.5	56.1	0.53	0.006
9	1.2×10^{-7}	2.0	17.3	20.4	3.5	55.8	0.96	0.003

The recycling is assumed to be in the form of 5% Pu and 95% depleted U.

The reuse of plutonium in LWRs would only be temporary if fast breeder reactors would become common (Fig. 21.1). FBRs are designed with a core, containing ~15% fissile Pu and ~85% ^{238}U (as depleted uranium) in the form of mixed oxides or carbides surrounded by a blanket of depleted uranium. In such a fast reactor, the presence of ^{235}U is undesirable as it reduces the neutron energy considerably by inelastic scattering. The actinide and fission product contents in discharged FBR fuel (core and blanket) and discharged LWR fuel are roughly the same on a GW_ey basis. Also the masses of total discharged fuel per GW_ey are comparable because the high burnup in the core is balanced by a very low burnup in the blanket. Breeding occurs only in the blanket (cf. §20.3). The burnup in the core elements is ~3 times higher than in LWRs, and the fraction of fission products is also 3 times larger. Since only a small part of the plutonium is burnt and the remainder has a high content of the fissile Pu isotopes, the used core fuel elements retain a high economic value, making it desirable to reprocess them after a short cooling time. Used FBR core elements may have a tenfold greater specific radioactivity than spent LWR fuel elements at the time of reprocessing and a much higher content of plutonium. Hence, criticality risks require a special reprocessing plant for core elements. The FBR blanket elements are simpler to handle because of a lower content of fission products and plutonium and they may be reprocessed in plants for LWR fuel elements. However, if the core elements are sufficiently diluted with blanket elements such a mix may be reprocessed in a "conventional" plant for LWR fuels.

In the FBR, ^{238}U is consumed both by fission (i.e. energy production) and by ^{239}Pu formation. Because the α-value of ^{239}Pu is 0.42, at least 70% $(100(1+\alpha)^{-1})$ of all ^{238}U is useful for energy production in the U–Pu cycle. This value should be compared with the fairly small fraction, $\leq 0.7\%$, of the natural uranium which is used in the UOT cycle (taking enrichment also into account), or $\leq 1\%$ in the LWR MOX fuel recycle. The FBRs not only increase the useful energy of natural uranium by a factor of ~100, they also make it possible to burn current stockpiles of depleted uranium and also make it economic to mine low grade uranium ore, vastly extending the available uranium resources.

Currently, some 13 nuclear energy countries have decided to use reprocessing as part of their strategy for spent fuel management (Argentina, Belgium, Brazil, Bulgaria, China, France, Hungary, India, Japan, the Netherlands, Switzerland, Russia and the United Kingdom).

21.5.3. The thorium-uranium (Th-U) fuel cycle

Nuclear energy cannot be produced by a self-sustained chain reaction in thorium alone because natural thorium contains no fissile isotopes. Hence the thorium-uranium cycle must be started by using enriched uranium, by irradiation of thorium in a uranium- or plutonium-fuelled reactor or by using a strong external neutron source, e.g. an accelerator driven spallation source.

Fertile ^{232}Th can be transformed into fissile ^{233}U in any thermal reactor. The reactions in ^{232}Th irradiated by neutrons are given in Fig. 20.3. Of the thermally fissile atoms ^{233}U has the highest $\sigma_f/\sigma_{n,\gamma}$ ratio, i.e. highest fission efficiency. The η-value is high enough to permit breeding in the thermal region. Capture of neutrons in ^{233}U is not a serious drawback as a second capture (in ^{234}U) yields fissile ^{235}U, but reduces the breeding gain because two neutrons are consumed without a net increase in fissile material. Since σ_{tot} increases with neutron energy (from ~7 b at 0.025 eV to ~26 b for reactor conditions; cf. also Fig. 19.3) the slightly harder neutron spectrum in PHWR and GCR make such reactors the prime candidates for a thorium-uranium fuel cycle together with the molten salt reactor (§20.3). The fuel may be arranged in a core (^{233}U) and blanket (^{232}Th) fashion, or mixed fissile and fertile material as, for example, in the HTGR prototype (High Temperature Gas-cooled Reactor) graphite matrix fuels, or as a metal fluoride melt. The initial ^{233}U must be produced from thorium in reactors fuelled with ^{235}U (or ^{239}Pu) or in special accelerator driven devices. After sufficient amounts of ^{233}U have been produced, the Th-U fuel cycle may become self sustaining, i.e. thermal breeding is established.

The advantage of the Th—U fuel cycle is that it increases nuclear energy resources considerably because thorium is about three times more abundant on earth than uranium and almost as widely distributed. In combination with the uranium fuel cycle it could more than double the lifetime of the uranium resources by running the reactors at a high conversion rate (~1.0) and recycling the fuel. Very rich thorium minerals are more common than rich uranium minerals. The presence of extensive thorium ores has motivated some countries (e.g. India) to develop the Th—U fuel cycle.

No full-scale Th-U fuel cycle has yet been demonstrated and reprocessing has only been demonstrated on an experimental scale. The fuel cycle has to overcome the high activity problems due to the presence of ^{228}Th formed in the thorium fraction and ^{232}U formed in the uranium fraction (Fig. 20.3). The Th—U fuel cycle has a rather specific advantage over the U—Pu cycle in that its high active waste from reprocessing contains a much smaller amount of long-lived heavy actinides, although it probably contain some ^{231}Pa. With regard to nuclear weapons proliferation ^{233}U is almost as good a weapons material as ^{239}Pu and easier to produce as a single isotope by continuous withdrawal of protactinium, since it is the decay product of ^{233}Pa ($t_{½}$ 27 d), see also §21.7.

21.6. REPROCESSING OF URANIUM AND MIXED OXIDE FUELS

The main purpose of commercial reprocessing is:

(1) to increase the available energy from fissile and fertile atoms;

(2) to reduce hazards and costs for handling the high level wastes.

Two other reasons are sometimes mentioned:

(3) to reduce the cost of the thermal reactor fuel cycle;

(4) to extract valuable byproducts from the high active waste.

The 30% savings in natural uranium for LWR and similar reactors and the hundredfold energy resource expansion for FBRs when reprocessing spent fuel, has already been discussed in the previous section. The economic advantage of reprocessing depends on the cost and availability of natural ("yellow cake") uranium, on enrichment and other front end activities, and on the prevailing energy price (mainly based on fossil fuels). At present, cheap uranium is abundant.

However, as mentioned before, older aluminium canned fuel needs either to be treated in order to cover it by a thick and resistant oxide film, or to be reprocessed, in order to avoid severe corrosion of the can and release of fuel constituents into the storage facility.

The reprocessing plants at La Hague, France, have a total capacity of ~1600 tons IHM/y, the Magnox reprocessing plant and the THORP (THermal Oxide Reprocessing Plant) at Sellafield, UK, have capacities of ~1500 and ~850 tons IHM/y, respectively. A commercial reprocessing plant under construction in Japan is designed to have a capacity of 800 tons IHM/y. In Siberia, a large russian reprocessing plant is under construction. A number of smaller, older plants are also in operation in several countries.

Figure 21.12 is a schematic representation of reprocessing of spent LWR fuel. The main steps are: (i) the *head end* section, in which the fuel is prepared for chemical separation; (ii) the main fractionation (partitioning) of U, Pu, and FP; (iii) purification of uranium; (iv) purification of plutonium; (v) waste treatment; and (vi) recovery of chemicals. These steps are described in the following sections.

21.6.1. Head end plant

Figure 21.13 is a simplified drawing of one of the French oxide fuel head end plants at La Hague. The flasks with the used fuel assemblies are lifted by a crane into water-filled pits, where the flasks are unloaded and decontaminated. The assemblies are stored for a desired time and then transferred to a shielded dismantling and chopping section. Some BWR fuels have end parts, which can be mechanically dismantled; this is not the case for PWR fuels, in which the end parts have to be cut off. The fuel pins are cut into pieces 3–5 cm long, either under water or in air. At THORP, UK, and in the newest part of the La Hague plant, dry charging and chopping is used.

The chopping is usually achieved with a shearing knife (cutter), but other techniques for removing or opening up the zircaloy (or stainless steel) cans have been tried. Previously, chemical decanning was used at some plants, e.g. Hanford and Eurochemic, but such techniques increase the amounts of active waste considerably.

The chopped pieces are transferred to the dissolver unit, where the oxide fuel is leached by boiling in 6–11 M HNO_3 (the cladding hulls do not dissolve) in thick stainless steel vessels provided with recirculation tubes and condenser. The hulls are measured for residual uranium or plutonium, and, if sufficiently clean of fissile material,

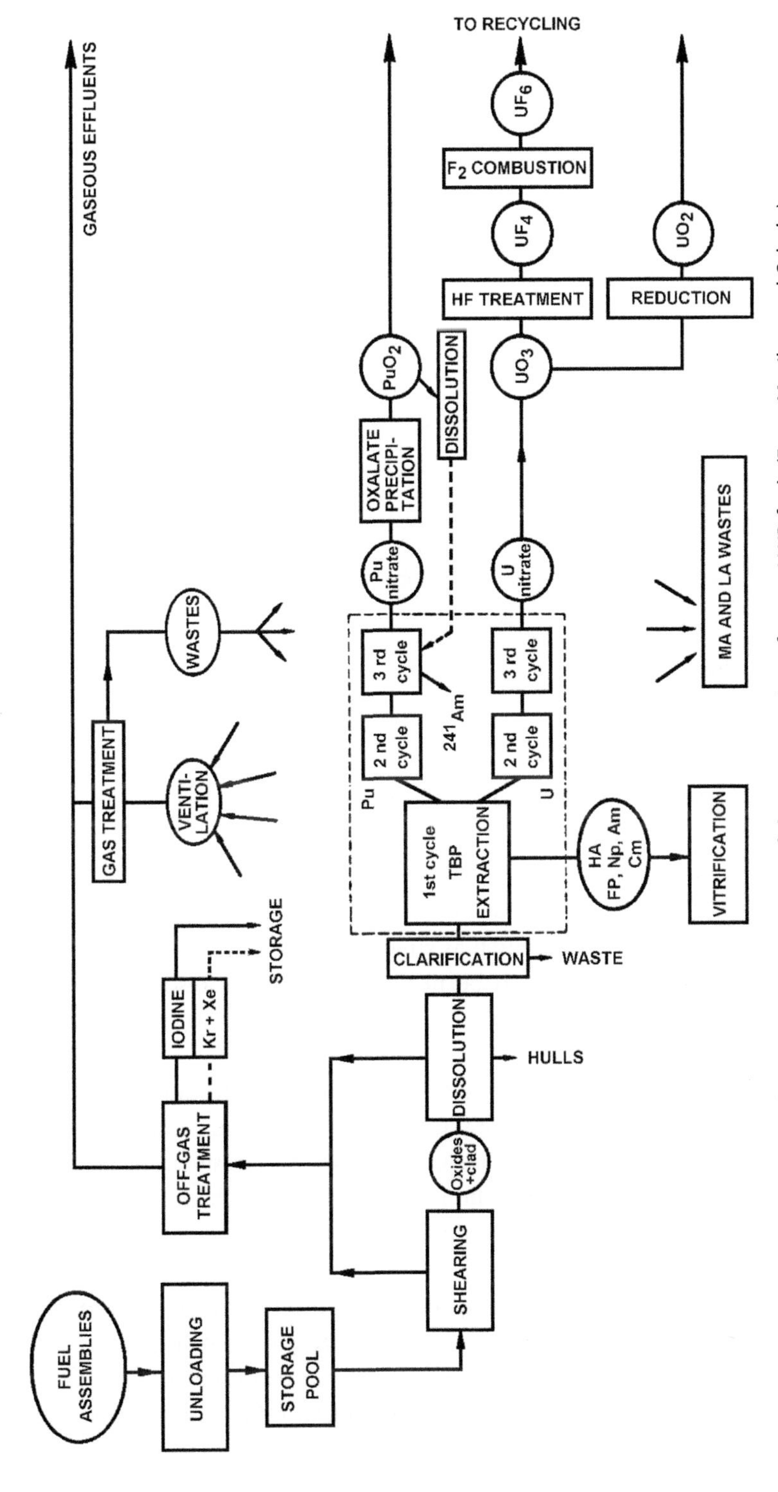

Figure 21.12 Schematic representation of the reprocessing of spent LWR fuel. *(From Musikas and Schulz.)*

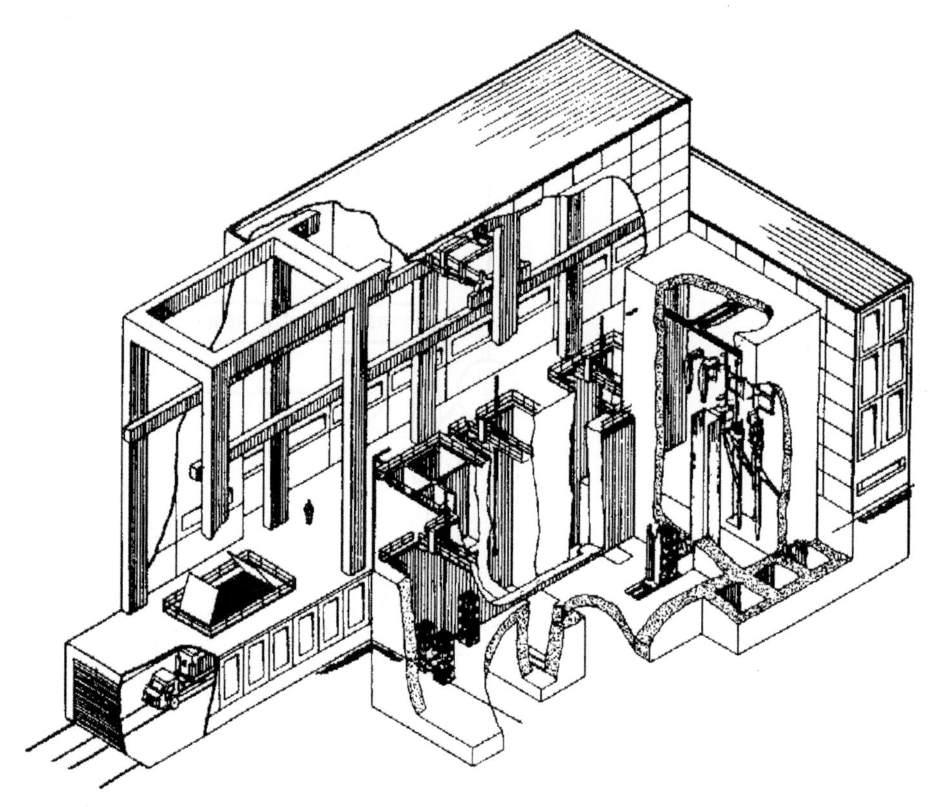

Figure 21.13 Head end oxide fuel building at La Hague, France.

are discharged to the waste treatment section of the plant. To improve the dissolution, some fluoride ($\leq$0.05 M AlF_3) may be added to the HNO_3. The F^- forms strong complexes with some metal ions such as zirconium, while its corrosion of the stainless steel equipment has been found to be negligible. Soluble poisons, such as cadmium nitrate or gadolinium nitrate, are often added to the nitric acid to assure the criticality control of the dissolution operation.

High burnup of fissile material leads to a high fission product content in the fuel elements resulting in the formation of seminoble metal fission product alloys, which are insoluble in boiling nitric acid. The insoluble material consists of mm sized metal particles of Ru, Rh, Tc, Mo, and Pd, the larger ones often called "ingots". These metal particles usually contain negligible amounts of uranium and plutonium and can be filtered as high level solid waste, HLSW.

When the fuel pins are cut, the volatile fission products contained in the gas space between the fuel oxide pellets and the canning is released (mainly ^{85}Kr, ^{3}H, ^{129}I, and ^{131}I). These gases are ducted to the dissolver off-gas treatment system. Gases released during dissolution are Kr, Xe, I_2, T_2, THO, RuO_4, CO_2, minor amounts of fission product aerosols, and large amounts of H_2O, HNO_3, and nitrogen oxides. Oxygen or air is fed into the off-gas stream to allow recovery of part of the nitrogen oxides. The overall dissolution stoichiometry is

$$UO_2 + 2HNO_3 + \frac{1}{2}O_2 \rightarrow UO_2(NO_3)_2 + H_2O \qquad (21.10)$$

The gas streams pass to a condenser which reclaims and returns some nitric acid to the dissolver. The noncondensibles are discharged to the off-gas treatment system.

When the dissolution is completed, the product solution is cooled and transferred to the input measurement—clarification (filter and/or centrifuge) feed adjustment unit. At this point the uranium is in the hexavalent state, and plutonium in the tetravalent.

21.6.2. Separation methods

The specifications for purified uranium and plutonium to be recycled are summarized in Table 21.7. Comparing these data with those presented before shows that at $t_{cool} \sim 1$ y the fission product activity must be reduced by a factor of $\sim 10^7$ and the uranium content in plutonium by a factor of $\sim 2 \times 10^4$. The large number of chemical elements involved (FPs, actinides *and* corrosion products) make the separation a difficult task. Additional complications arise from radiation decomposition and criticality risks and from the necessity to conduct all processes remotely in heavily shielded enclosures under extensive

Table 21.7 Specifications for reprocessed uranium and plutonium

Uranyl nitrate:
- Uranium concentration 1–2 M
- Free HNO_3 $\geq$1 M
- Impurities:
  - Fe, Cr, Ni $\leq$500 ppm
  - Boron equivalents $\leq$8 ppm†
- Fission products $\leq$19 MBq/kg U
- α-activity (excluding uranium) $\leq$250 kBq/kg U

Plutonium nitrate:
- Plutonium concentration $\sim$1 M
- Free HNO_3 2–10 M
- Impurities:
  - Metallic $\leq$5000 ppm
  - Uranium $\leq$5000 ppm
  - Boron equivalents $\leq$10 ppm
  - Sulfate $\leq$1000 ppm
- Fission products ($t_{1/2} > 30$ d) $\leq$1.5 GBq/kg Pu‡
- ^{241}Am content (9 months after delivery to MOX-plant) $\leq$5000 ppm

†The equivalent values are B 1.0, Cd 0.4, Gd 4.4, Fe 0.0007, etc. The amount measured for each of these elements multiplied by the factor indicated must not be more than 8 ppm.
‡^{95}Zr – Nb $\leq$ 185 MBq/kg Pu.
(From IAEA 1977)

health protection measures. As a result reprocessing is one of the most complicated chemical processes ever operated on an industrial scale.

The problem encountered by the chemists of the Manhattan Project in the 1940's was the selection of satisfactory separation techniques. Advantage was taken of the relative stability of the oxidation state of uranium (+6) and most fission products, and the redox lability of plutonium (+3, +4, and +6). In the earliest process, only plutonium was isolated by precipitating plutonium in the reduced state as PuF_3 or PuF_4 together with all insoluble FP fluorides. This was followed by a second stage dissolving the precipitate, oxidizing plutonium to the +6 state, and a new fluoride precipitation, leaving relatively pure plutonium in the supernatant. In a final step plutonium was again reduced and precipitated as fluoride. This principle was used for the first isolation of hundreds of kilograms of plutonium at the Hanford Engineering Works, USA, with phosphate precipitation of Pu(+3) and Pu(+4), but not of Pu(+6) which does not form an insoluble phosphate. Since Bi^{3+} was used as a carrier for the precipitate, it is referred to as the *bismuth phosphate process*.

This principle of oxidizing and reducing plutonium at various stages of the purification scheme has been retained in all subsequent processes. No other element has the same set of redox *and* chemical properties as plutonium, though some elements behave as Pu^{3+} (e.g. the lanthanides), some like Pu^{4+} (e.g. zirconium) and some like PuO^+ (e.g. uranium). Numerous redox agents have been used, e.g. $K_2Cr_2O_7$ (to PuO^+), $NaNO_2$ (to Pu^{4+}), hydrazine, ferrous sulfamate, and U^{4+} (to Pu^{3+}), cf. Table 14.4

The precipitation technique is not suitable for large-scale, continuous remote operations in which both uranium and plutonium have to be isolated in a very pure state from the fission products. It was therefore replaced in the late 1940's by *solvent extraction* in which the fuels were dissolved in nitric acid and contacted with an organic solvent which selectively extracted the desired elements. The technique has been mentioned in §§14.2.4.3, 17.7.4, and 18.4.3 but is described in more detail in Appendix A.

The first solvent to be adopted at Hanford was methyl isobutyl ketone ("MIBK" or "Hexone"). This solvent forms *adduct compounds* with coordinatively unsaturated compounds like the actinide nitrates, e.g. $Pu(NO_3)_4S_2$, where S represents the adduct molecule

$$Pu^{4+} + 4NO_3^- + 2S(org) \rightarrow Pu(NO_3)_4S_2(org) \qquad (21.11)$$

The corresponding adduct compounds for 3- and 6-valent actinides are $An(NO_3)_3S_3$ and $AnO_2(NO_3)_2S_2$. These chemically saturated neutral compounds are soluble to different extent in organic solvents like kerosene, and – in the case of hexone – by hexone itself. The process using hexone is referred to as the *Redox process*.

In the United Kingdom, β,β'-dibutoxydiethylether ("dibutyl carbitol" or "Butex") was selected as organic solvent; it forms the same kind of adduct compounds as hexone. Though more expensive, it was more stable, less flammable and gave better separations.

Many other similar solvent systems, as well as *chelating agents*, have been tested. Thenoyltrifluoroacetone (HTTA) was found to form strong complexes with the actinides (e.g. $Pu(TTA)_4$, $UO_2(TTA)_2$), which show very high distribution ratios in favour of organic solvents. Though useful in the laboratory they were not found suitable for large scale commercial nuclear fuel reprocessing. One of the most useful recent extraction agents for actinide separation is di-2-ethylhexylphosphoric acid (HDEHP) which has found several industrial uses outside the nuclear energy industry, e.g. separation of rare earth elements.

A drawback of hexone and butex is the need to use salting-out agents (salts like $Al(NO_3)_3$ added to the aqueous phase) in order to obtain sufficiently high extraction factors. Such salt additions increase the liquid waste volumes. Further, hexone was unstable at high nitric acid concentrations. All this led to the search for a better extractant.

Presently, tributyl phosphate (TBP) is the extractant used in all reprocessing plants. It acts as an adduct former and is normally used as a 30% solution in kerosene. It forms the basis for the *PUREX process* (Plutonium Uranium Redox EXtraction). TBP is cheaper than Butex, more stable, less flammable, and gives better separations. The limitation to 30% in TBP is needed in order to prevent such an increase in density of the organic phase, due to extracted uranium and plutonium, that it becomes equally dense as the aqueous phase or even denser[2].

Other extractants, especially tertiary amines, have been tested for some steps in reprocessing. The amines form organic soluble complexes with negatively charged metal complexes (used in uranium purification, §14.2.4.2). The use in the basic PUREX cycle of a secondary extractive reagent can improve the decontamination factor[3].

As alternatives to the aqueous separation processes, "dry" techniques have also been studied, but none has been used on an industrial scale. Examples are the following:

(a) *Halide volatility.* Many FP and the high valency actinides have appreciable vapor pressures; this is particularly true for the fluorides. In *fluoride volatilization* the fuel elements are dissolved in a molten fluoride salt eutectic ($NaF + LiF + ZrF_4$, 450°C) in the presence of HF. The salt melt is heated in F_2, leading to the formation of UF_6, which is distilled; it may be possible also to distill PuF_6, though it is much less stable. The process has encountered several technical difficulties.

(b) *Molten salt extraction.* The fuel is dissolved as above or in another salt melt. With a heat resistant solvent of low volatility (e.g. 100% TBP), actinides and FP distribute themselves between the two phases analogous to solvent extraction. This technique is of interest for continuous reprocessing of molten salt reactor fuel or partitioning in an accelerator driven transmuter (§20.3.1). An advantage is the higher radiation resistance of the nonaqueous system.

2 In one process developed for the steel industry, 75% TBP was used.

3 The decontamination factor is defined as (concentration before separation)/(concentration after separation).

(c) *Molten salt transport.* The fuel is dissolved in a metallic melt, e.g. a molten Cu-Mg alloy, which is in contact through a stirred molten chloride salt at about 800°C with another metallic melt containing a reductant, e.g. a molten Zn-Mg alloy. Noble metal FPs are retained in the Cu-containing melt, whereas U and/or Pu is collected in the Zn-containing melt. The less noble FPs concentrate in the molten salt.

(d) *Molten salt electrorefining.* The spent fuel acts anode in a molten salt also containg a pure metal cathode. By applying an electric field between the anode and cathode, material is dissolved at the anode and deposited at the cathode. Careful control of the applied potential makes it possible to obtain an extremely pure cathode product. The process was initially developed for purification of plutonium metal alloys proposed as fuel for the Los Alamos Molten Plutonium Reactor Experiment (LAMPRE).

(e) *Molten metal purification.* Metallic fuel elements can be molten and/or dissolved in molten metals (e.g. a zinc alloy). In the presence of (deficient amounts of) oxygen, strongly electropositive fission elements form oxides, which float to the surface of the melt as slag and can thus be removed, while volatile FPs distill. The residual melt would mainly contain U, Pu, Zr, Nb, Mo, and Ru ("fissium alloy") and can be reused in new metallic fuel elements. This melt refining technique has been tested on metallic breeder reactor fuel elements (in EBR II). Molten chlorides have also been used to remove americium from molten plutonium metal scrap.

21.6.3. PUREX separation scheme

The distribution of uranium, plutonium, and some FPs between 30% TBP (in kerosene) and aqueous solutions of varying HNO_3 concentration is shown in Figure 14.9. D_{Sr} and D_{Cs} are $\ll 0.01$ for all HNO_3 concentrations. The distribution of the trivalent lanthanides and actinides fall within the Eu-Am area. Many fission products (most I, II, III, V, and VII-valent species) are not extracted, i.e. $D \leq 0.01$; see Appendix A. Thus at high HNO_3 concentration Pu(IV), Pu(VI), and U(VI) are extracted but very little of the FPs. At low HNO_3 concentration the *D*-value for actinides of all valency states is $\ll 1$, and consequently the tetra- and hexavalent actinides are stripped from the organic phase by dilute HNO_3. This chemistry is the basis for the PUREX process.

The PUREX process is presented schematically in Figure 21.12, where the solvent extraction steps are within the dotted frame. Three purification cycles for both uranium and plutonium are shown. High levels of beta and gamma activity is present only in the first cycle, in which >99% of the fission products are separated. The principle of the first cycle is shown in Fig. 21.14. The two other cycles are based upon the same chemical reactions as in the first cycle; the purpose is to obtain additional decontamination and

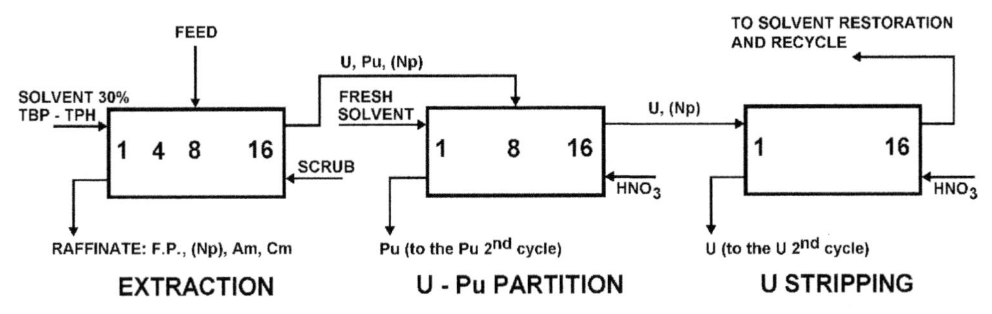

Figure 21.14 Flowsheet of the first purification cycle in the PUREX process; TPH, tetrapropylene, is a commercial dodecane. *(From Musikas and Schulz.)*

overall purity of the uranium and plutonium products. Each square in Figure 21.14 indicates a number of solvent extraction stages of the particular equipment used: pulsed columns, mixer-settlers, etc. (see Appendix A).

In the first cycle > 99.8% U and Pu (in VI and IV state, respectively) are co-extracted from 3–4 M HNO_3 into the kerosene-TBP phase, leaving >99% of the FP in the aqueous raffinate. In the *partitioning* stage, plutonium is reduced to the III state by a solution containing a suitable reductant, e.g. U(IV) nitrate or Fe(II) sulfamate; the plutonium is stripped to a new aqueous phase and transferred to the plutonium purification section. Uranium, which as U(VI) (and U(IV)) stays in the organic phase, is stripped by dilute HNO_3 in a third stage. After concentration by evaporation it is sent to the uranium purification section.

The uranium purification contains two extraction-stripping stages. Plutonium tracers are removed by a reducing agent, e.g. U(IV):

$$U^{4+} + 2Pu^{4+} + 2H_2O \rightarrow UO^{+} + 2Pu^{3+} + 4H^{+} \qquad (21.12)$$

U(IV) is preferred over Fe(II)-sulfamate in some plants as it avoids the introduction of foreign substances. The final, concentrated uranium solution may be percolated through a column filled with silica gel, which removes residual FP, particularly Zr–Nb.

The plutonium purification may be achieved by additional TBP extraction cycles. U(IV) cannot be used as reductant in this part of the process. The final uranium and plutonium products are nitrate solutions whose conversion to oxides, fluorides, etc., have partly been described earlier (§21.1).

The chemical problems encountered in the solvent extraction are:

(i) The choice of diluent: A mixture of aliphatic hydrocarbons, and occasionally pure dodecane, are most common; improper choice of diluent may lead to formation of a third liquid phase, slow extraction kinetics, difficult phase disengagement (i.e. separation of the organic and aqueous phases in the extraction equipment), etc.

(ii) The choice of reductant for the reaction Pu(IV) → Pu(III): The use of U(IV) as reductant for Pu(IV) introduces new uranium in the streams; Fe(II) sulfamate ($Fe(SO_3NH_2)_2$) adds objectionable inorganic salts to the aqueous high level waste; hydroxylamine (NH_2OH) is a kinetically slow-reducing agent; electrolytic reduction of Pu(IV) to Pu(III) is lacking experience.

(iii) Solvent degradation: Radiation decomposes TBP into lower phosphates and butyl alcohol. The main products are dibutyl phosphate (DBP, $(BuO)_2POOH$) and monobutyl phosphate (MBP, $BuOPO(OH)_2$) which form strong complexes with many of the fission products as well as plutonium. As these radiolysis products are formed the decontamination efficiency decreases and losses of fissile material to the aqueous waste streams increase. The solvent is treated to remove the degradation products prior to recycle in the process, e.g. by washing the TBP solution successively with Na_2CO_3, NaOH, and dilute acid solutions.

To reduce some of these difficulties and to generally improve the efficiency of the U and Pu extraction (leaving a waste almost free of these elements) as well as the decontamination factors, other extractants than TBP have been suggested, for example dialkylamides, R-CON-R'$_2$ where R may be C_3H_7 and R' is $CH_2CHC_2H_5C_4H_9$, "DOBA". The advantages of this type and other proposed alternatives to TBP are that they are completely incinerable (while TBP leaves a phosphate waste), the radiolytic products are not deleterious for the process performance, and that no reducing agent is necessary to partition uranium and plutonium. Finally, the ash from combustion could be leached and its content of actinides recovered and returned to the process – further reducing the content of these elements in intermediate level waste.

During the decontamination steps, acid streams containing small amounts of actinides and fission products are produced. These streams are evaporated to concentrate the metal ions and recycle them. Nitric acid is recovered from the condensates and recycled. Excess HNO_3 may be destroyed by formaldehyde. Fission product concentrates are routed to the aqueous raffinate from the first extractor of the partitioning cycle which contains >99% of the FP. This constitutes the *high level liquid waste* (HLLW, or alternatively called HAW, *high active waste*). All other liquid wastes can be subdivided into *intermediate level waste.*

(ILW or MLW) or *low level waste* (LLW). It is an important goal in all reprocessing operations to reduce the amount of intermediate and low level waste streams as far as possible and to route most of what remains to the HLLW stream. In order to reduce volume the HLLW is often concentrated further by evaporation. The concentrate is usually designed as HLC. Waste treatment is discussed in §21.8 .

As we have seen above the completely dominating recycling of nuclear fuel today is based on the recovery of uranium and plutonium through the PUREX process. However, in a more sustainable future nuclear system recycling of essentially all the actinides will be desirable. Due to their abundance in the used nuclear fuel the main focus has been on recovering neptunium, americium and curium from the streams in, or out from, the PUREX process. Neptunium can be recovered by adjustment of the redox properties of the initial extraction stages. Then it will follow uranium to the uranium purification cycle, where it can be separated from U and recovered separately. The main problem encountered, however, is that both the heavy actinides as well as the lanthanides are trivalent and of similar size making an Am-Cm separation from the PUREX waste stream difficult.

Two main strategies exist to perform this separation of the trivalent actinides from the lanthanides. Both techniques start by extracting both the actinides and the lanthanides from the PUREX raffinate but then the routes differ. In the technique utilised by e.g. the reversed TALSPEAK (Trivalent Actinide – Lanthanide Separation by Phosphorous reagent Extraction from Aqueous Komplexes) process in the US the actinides are selectively stripped from the mixture extracted by HDEHP in while in the European techniques exemplified by the SANEX (Selective Actinide Extraction) processes the actinides are selectively extracted from the mixture. Both these techniques have been tested successfully using genuine spent nuclear fuel.

In recent times, however, questions have been raised about proliferation issues using the PUREX process considering the fact that one of the results is a pure plutonium stream. This makes the safe guard issues in relation to PUREX based processes complicated. Instead it has been suggested to extract all the actinides together from the dissolved used nuclear fuel. Such a process is called a GANEX (Group Actinide Extraction) process. This extraction is performed in two steps. First the bulk uranium is extracted and in the next step all the other actinides (and remaining uranium) is extracted. Such a process opens for the possibility of a homogeneous recycling where essentially the new nuclear[1] fuel is produced directly from the waste of the used one utilising a similar composition where as much uranium as needed can be added. Naturally the use of this fuel types will require the use of fast neutron spectrum reactors.

21.6.4. Engineering aspects and operation safety

All operations in a reprocessing plant have to cope with the necessity of preventing nuclear criticality and of protecting operations personnel and the environment from exposure to or contamination by radioactivity. Thus all equipment has dimensions which are safe against criticality, as, for example, annular or pipe shaped tanks for liquid storage instead of conventional tanks, or are provided with neutron absorbers. All equipment is made of stainless steel and is installed in concrete cells with wall thicknesses up to 2 m at the head end, partitioning, and waste treatment sections. All operations are carried out in airtight enclosures at reduced pressure relative to working areas.

In the event of failure of equipment within a radioactive area, three courses of action may be taken: (i) switching to duplicate equipment, (ii) replacing or repairing equipment

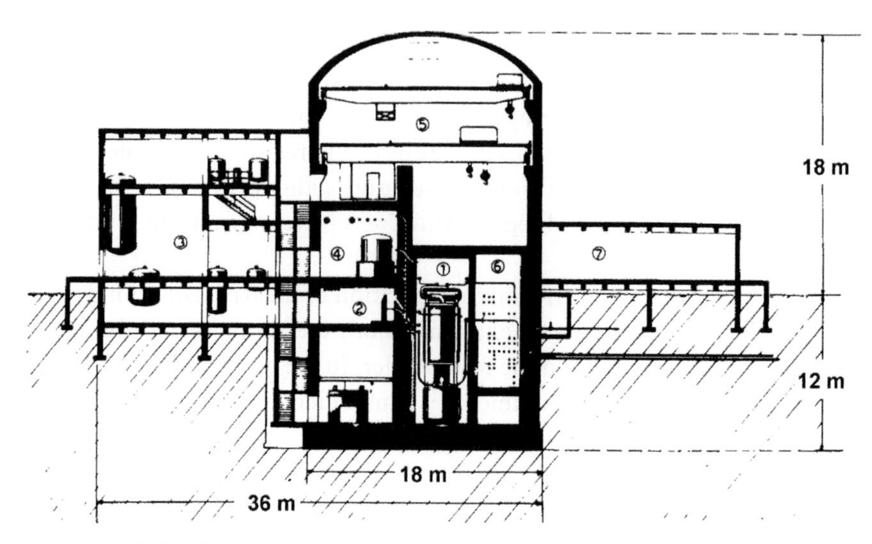

Figure 21.15 Principle of a reprocessing plant with remote maintenance. 1, canyon with process equipment; 2, control room; 3, feed preparations; 4, feed input; 5, cranes; 6, piping; 7, analytical section.

by remote methods; *remote maintenance*, or (iii) repairing by *direct maintenance* after decontamination.

Plants were originally constructed for either completely remote maintenance or completely direct maintenance. Figure 21.16 illustrates the principle of a plant for remote maintenance. All equipment is installed in a large canyon with piping in a parallel corridor. The equipment can be replaced by an overhead crane operated from a shielded room; malfunctioning equipment could be transferred to a decontamination and repair unit or to an equipment "grave". With this philosophy conventional type chemical plant equipment can be used, though redesigned for the remote replacement. Of particular importance in such plants are good joints for piping, electricity, etc., for remote connection.

The British Nuclear Fuels Plc. (BNFL) Magnox and THORP plants at Sellafield, UK, are designed for no maintenance, which in practice might mean either direct maintenance or remote maintenance; Fig. 21.16. All kinds of common reasons for failure in chemical plants have been eliminated or minimized by using welded joints and no moving parts. Thus there are no leaking fittings, frozen valves, or stuck pumps. All welding is carefully controlled by ultrasound or radiography. Active liquids are transported by gas (usually steam) jets or lifts, or by vacuum. Liquid levels and volumes are measured by differential pressure gages and by weighing. Samples for analysis are remotely withdrawn and analyzed in shielded boxes or glove boxes depending on their activity outside the enclosure.

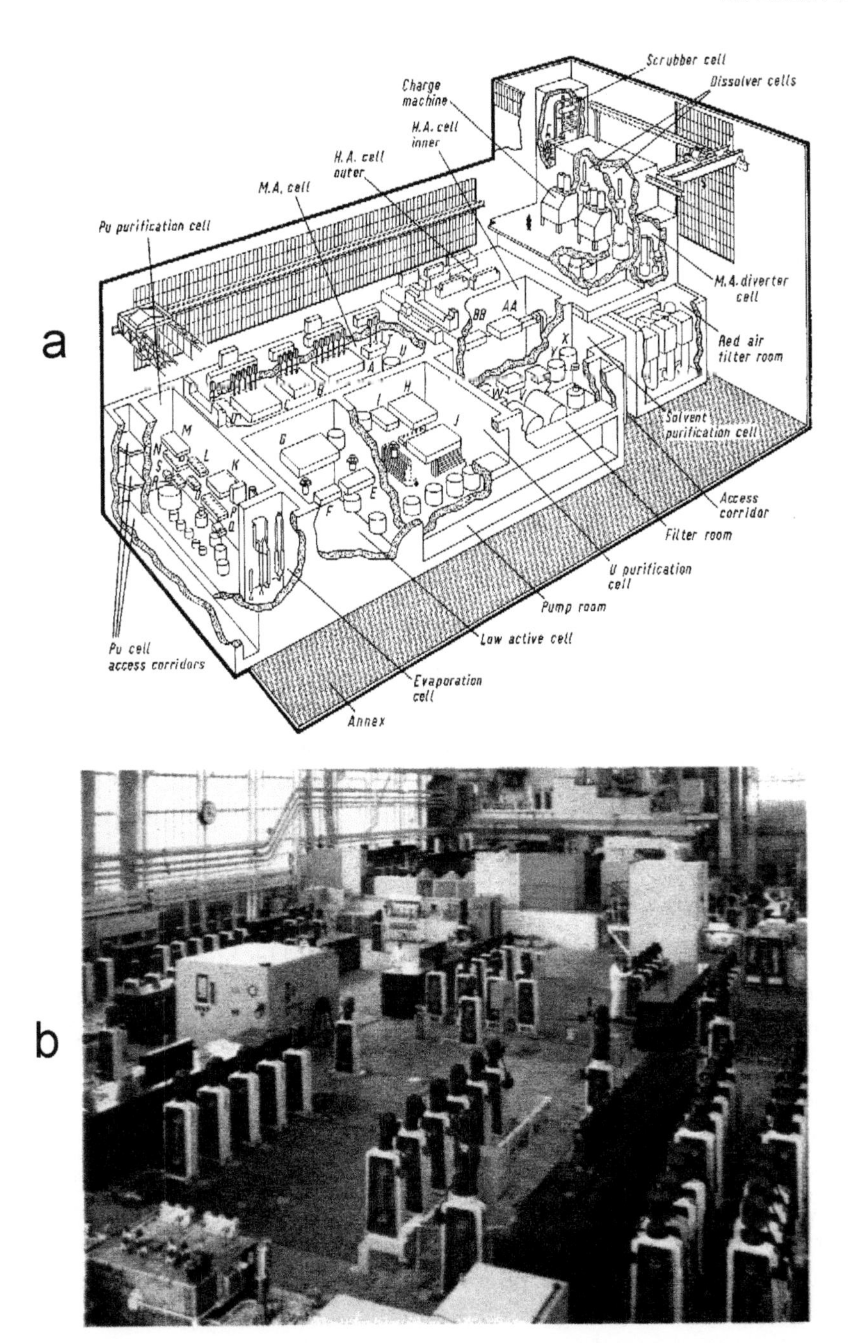

Figure 21.16 (a) General layout of BNFL Magnox reprocessing plant at Sellafield, UK. (b) Cell top of primary separation plant showing stirrer motors for mixer-settlers below.

Even such systems may fail and dual equipment is therefore sometimes installed in parallel cells. With the process running on the spare equipment, the failing equipment must be repaired. This requires efficient decontamination both on inside and outside of the equipment, provisions for which must be incorporated in the original design. This design requires dividing the plant in a large number of shielded cells so that the repair workers are protected from the radiation of the functioning plant.

Remote maintenance is more expensive, but may be safer for the personnel and desirable from the standpoint of continuity of operations because equipment replacements can be carried out quickly and interruption of operation is relatively brief. Modern robot technology simplifies such remote maintenance.

21.7. REPROCESSING OF THORIUM FUELS

The thorium-containing fuels of present interest are only those of the kind used in HTGR and HWR; in the future fuel from MSR-like[4] (or other) transmutation devices may become important. In the HTGR fuel elements the fertile ThO_2 and fissile $^{235}UO_2$ (or $^{233}UO_2$) particles are coated differently and embedded in a graphite matrix.

In case of the HTGR, the spent graphite fuel elements are mechanically crushed and burned to eliminate the graphite matrix and pyrolytic carbon coating from the fuel particles. Leaching permits separation of the fissile and the fertile particles because the fissile particles have a silicon carbide coating which remains intact during burning and leaching, while the all-pyrolytic carbon coatings on the fertile particles are burned away, allowing the oxide ash to be dissolved by a leach solution, consisting of HNO_3 and F^-. The solution is clarified and adjusted to proper acidity for solvent extraction (paragraph (i) below). The undissolved residue resulting from clarification is dried and classified for further treatment (paragraph (ii) below). The burner off-gas streams are passed through several stages of filtration, scrubbing, and chemical reaction to remove the entrained and volatile fission products, as well as ^{14}C containing CO_2, prior to atmospheric discharge.

In case of oxide fuel, it has to be cut and dissolved in a way similar to that described for uranium fuels.

(i) The *acid Thorex solvent extraction process* is used to purify and to separate the ^{233}U and the thorium. Three solvent extraction cycles are used. In the first, the uranium and thorium are coextracted by 30% tributyl phosphate (TBP) from $\sim$5 M HNO_3 and then stripped into an aqueous phase. In the second cycle, the uranium and thorium are separated by controlling the extraction conditions using $\leq$1 M HNO_3. The uranium is extracted and processed by an additional solvent extraction cycle for final purification, while the thorium remains in the aqueous raffinate stream. Following

[4] MSR = Molten Salt Reactor.

concentration and assay, the uranium is ready for fuel refabrication. The partially decontaminated thorium is concentrated and stored.

(ii) The separated silicon-carbide coated fissile particles from the head end process for HTGR fuel are mechanically crushed to expose the fuel and are burned to remove carbon and oxidize the fuel material: the ash is leached to separate the fuel and fission products from the coating waste. The ^{235}U is separated from the fission products by solvent extraction using a PUREX flowsheet. An organic solvent containing only 3–5% TBP is used in order to avoid criticality problems. A reductant such as Fe(II) sulfamate is added to the feed in order to force the small amount of plutonium present into the aqueous raffinate stream. The waste streams are treated as in the PUREX process.

The process is plagued by both chemical and nuclear difficulties. The decay chain ^{233}Th → ^{233}U forms 27 d half-life ^{233}Pa. For a complete decay of all ^{233}Pa to ^{233}U, the spent fuel elements must be cooled for about a year. A still considerable amount of longlived ^{231}Pa ($t_{½}$ 32 760 y) is present in the spent fuel (about 1/2000 of the amount of ^{233}U); protactinium complicates the reprocessing chemistry and constitutes an important long-lived waste hazard.

The ^{233}U isolated contains some ^{232}U formed through reactions indicated in Figure 20.3. Since the half-life of ^{232}U is rather short and its decay products even shorter lived, a considerable γ-activity will grow in with the ^{233}U stock, complicating its handling. Since the first decay product is ^{228}Th, some ^{228}Th forms in the fuel elements, making the ^{232}Th contaminated by this isotope. Thus neither the ^{233}U nor the ^{232}Th produced can be free from γ-activity.

21.8. WASTES STREAMS FROM REPROCESSING

Fuel reprocessing generates a large variety of wastes which can be classified in categories according to activity (low, medium and high), physical state (gas, liquid or solid package), or decay characteristics (shortlived, longlived), each treated separately. The amounts of different categories are given in Table 21.8; these wastes are treated internally at the plant and not released to the environment. Because reprocessing plants vary considerably the amounts of wastes produced also differs, especially for the liquid medium and low level categories. In this section we briefly discuss the various waste streams appearing at the plant and the treatment methods.

21.8.1. Gaseous wastes

Gaseous wastes originate mainly from the chopping and dissolution operations. In current practice the volatile radionuclides are discharged to the stack after scrubbing with sodium hydroxide and filtration through a special zeolite or charcoal filter. The hydroxide scrubbing removes the acidic nitrous oxides which pass through the

Table 21.8 Typical annual amounts of waste arising in a reprocessing plant (per t IHM and 3 years cooling time)

Type	Form	Volume (m^3)	β,γ-activity (TBq)	Pu-content (kg)
HLLW	liquid	5	10^5	~0.05
MLLW	liquid	5		
LLLW	liquid	100		
HLLW	conc. liquid	0.5	10^5	0.05
Fission products	glass	0.15	10^5	0.05
Hulls	concrete	0.8	400	0.01
Ion/exch., precip.	Bitumen	0.8	~100	0.008
Iodine	concrete	0.03	<0.01	
ML and LL α	solid	0.1		
Technical waste	concrete	3.2	~0.5	~0.01

recombination unit above the dissolver. The silver impregnated zeolite or charcoal removes remaining traces of iodine.

The most hazardous volatile constituents are the iodine and ruthenium fission products. Though more than 95% of the iodine is volatilized in the dissolver (as I_2, HI and HIO mainly) most of it is caught in the off-gas scrubber and most of what remains is removed by the filters. With these techniques the retention of iodine in the plant is >99.5%.

Ruthenium forms volatile RuO_4 in the dissolver. Almost all RuO_4 is retained in the gas purification system. As an additional feature some plants use a steel wool filter (after the nitric acid recombination) to catch the RuO_4.

Of the noble gases, radioactive xenon has completely decayed after 1 y cooling, but krypton contains ^{85}Kr with 10.7 y half-life. This isotope is produced in appreciable amounts, and though commonly it has been released to the atmosphere, this is no longer acceptable. Many processes have been devised for krypton removal. Krypton in dry, clean air is effectively trapped on a charcoal filter at cryogenic temperature; however, because of explosion risk (due to reaction between radiolytically formed ozone and carbon), the favored process is condensation by liquid N_2 (krypton boils at −153°C) followed by fractional distillation. This removes >99% of ^{85}Kr. The krypton can be stored in pressurized cylinders until ^{85}Kr has decayed (>100 years).

The amount of tritium released in reprocessing is considerable. In the chopping section it is released as T_2, but as HTO in the dissolver, where >90% of all tritium formed is present. While T_2 can be caught (particularly if chopping is done in air), a recovery of the HTO formed in the dissolver would be expensive. Where it cannot be released to the environment the tritium can be trapped. In one process, called Voloxidation, the chopped fuel elements are treated with oxygen at 450−700°C before dissolution. Tritiated water is generated which should be relatively free of ordinary water and consequently occupy a much smaller volume than tritiated wastes do in present plants. Voloxidation may collect ~99% of the tritium present in unprocessed fuel. Even when all tritium is released to the environment, some precautions against its spreading in the plant are usually necessary in order to reduce the dose to operators. By proper design of the first extraction cycle most of the tritium can be confined to this part of the plant.

^{14}C is formed through the $^{14}N(n,p)^{14}C$ reaction in the nitrogen contained in the fuel elements. It is released mainly as CO_2 at the dissolution. Though only a small amount is formed (~40 GBq/GW$_e$y for each ppm nitrogen in the fuel), its release to the environment, makes it the dominating dose commitment of the fuel cycle back end (from reprocessing in 1989, 97 manSv from ^{14}C, and 98 manSv from all other nuclides). Presently $^{14}CO_2$ is released to the atmosphere, but techniques for its retention are available. It will probably finally be caught as $CaCO_3$, with a retention of ~80%. However, the future use of nitride fuels will increase this proportional to the ^{14}N content (as mentioned above, ~40 GBq/GW$_e$y for each ppm of ^{14}N in the nitride fuel, i.e. ≈2 PBq/ GW$_e$y for UN made from natural nitrogen).

21.8.2. Liquid wastes

The high level liquid PUREX waste (HLLW) contains typically >99.5% of the FPs, <0.5% of the U and <0.2% of the Pu present in the fuel as ~1 M HNO_3 solution, see Table 21.9. It is pumped to storage tanks as discussed further in §21.10. Total actual Pu losses during reprocessing is presently ≤0.2% of the feed.

The medium level liquid waste (MLLW) results essentially from evaporating various streams from the chemical process, such as solvent clean-up, off-gas scrubbers, product concentration, etc; Table 21.8. It may contain up to 0.5% of the uranium and up to 0.2% of the plutonium processed. The radioactivity is usually <40 GBq/l (an average value is 4 GBq/l). The solutions also contain appreciable amounts of solids (e.g. $NaNO_3$, iron, etc.). The waste may be neutralized and is stored in steel tanks at the reprocessing site.

Liquid waste is generated in numerous places with activities <0.1 GBq/m^3. Such waste is classified as low level. Some of these liquids may be clean enough to be released directly into the environment. Others are cleaned by flocculation, ion exchange, sorption, and similar processes. The general philosophy for liquid wastes is to concentrate all radioactivity to the next higher level because the waste volumes decrease in the order LLLW > MLLW > HLLW. Thus, in principle, the three kinds of wastes are reduced to two (HLLW and MLLW) and cleaned aqueous effluent. The MLLW and residues from LLLW cleaning are treated as the wastes of the nuclear power stations, i.e. concentrated and put into a disposal matrix such as concrete or bitumen (see §20.4.3). At some coastal sites it has been the practice to release the LLLW to the sea, with official permission. The nuclides of main concern are ^3H, ^{90}Sr, ^{137}Cs, ^{106}Ru, and the actinides.

Table 21.9 Composition of HLLW waste from PUREX reprocessing of 1 t IHM LWR fuel with a burnup of 33 000 $MW_{th}d$/t IHM and a cooling time of 3 years

Component		Weight (kg) in original waste volume (~5 m^3)	Approx. molarity in 0.5 m^3 concentrate
H^+		1.4	~1.0
NO_3^-		900	~2.4
Fission products:	Group I (Rb, Cs)	2.94	0.046
	Group II (Sr, Ba)	2.37	0.041
	Group III (Y, Ln)	10.31	0.15
	Zr	3.54	0.076
	Mo	3.32	0.068
	Tc	0.77	0.016
	Group VIII (Ru, Rh, Pd)	4.02	0.078
	Te	0.48	0.0075
	Others	0.35	0.004
	Total fission products	28.1	0.487
Corrosion products:	Fe	1.1	0.04
	Cr	0.2	0.008
	Ni	0.1	0.003
	Total corrosion products	1.4	0.051
Phosphate (from TBP)		0.9	0.02
Actinides:	U (~0.5%)	4.8	0.040
	Np (~100%)	~0.44	0.0037
	Pu (~0.2%)	~0.018	0.00015
	Am (100%)	~0.28	0.0023
	Cm (100%)	~0.017	0.00014
	Total actinides	5.5	0.047
Neutron poison (e.g. Gd)		12	0.15

21.8.3. Organic wastes

The liquid organic waste consists of spent TBP diluent mixtures originating from the organic solvent clean-up circuits and from the diluent (kerosene) washings of the aqueous streams (to remove entrained solvent); in addition to degradation products of TBP and the diluent it contains small amounts of actinides (mainly U and Pu) and FP (mainly Ru, Zr, and Nb).

This type of waste is disposed of by incineration or decomposed by hydrolysis and pyrolysis leading to the formation of inactive hydrocarbons, which are distilled, and active phosphoric acid, which is treated together with other aqueous wastes.

21.8.4. Solid wastes

High level solid waste (HLSW) originates at the dissolver. The hulls from the dissolution contain activation products and small amounts of undissolved fuel (≤0.1%). The dissolver solution contains finely divided particles of undissolved seminoble metal alloys (Ru, Rh, Mo, Pd, etc.). This suspension is treated by filtering or centrifugation prior to the solvent extraction. Past practice in the USA and the former USSR was to put the HLSW in shielded containers which are transported to and stored at a dry disposal site. In the future the same disposal is expected to be used for the HLSW as for the solidified HLLW (§21.12).

Medium and low level solid wastes are produced at numerous places. They are divided in various ways: combustible, noncombustible, α-bearing, non-α-bearing, etc., and treated independently, when possible, to reduce volume. The wastes are then fixed in the disposal matrix (§20.4.3). Table 21.8 gives the relative amounts of solid waste produced in reprocessing. The final deposition of these wastes is further discussed in §21.13.

21.8.5. Environmental releases from reprocessing plants

As described in §21.8.1 relatively large amounts of radioactive gases with a low hazard index (^{3}H, ^{14}C and ^{85}Kr) are released to the environment, see Table 21.10. Other large

Table 21.10 Annual releases to the environment from some reprocessing plants

Source		Half-life (y)	Sellafield† (TBq)	PUREX‡ (TBq/GW$_e$y)
^{3}H	Airborne	12.33	222	41
	Liquid		1050	643
^{14}C	Airborne	5730	4.1	2.0
	Liquid			0.54
^{85}Kr	Airborne	10.72	26000	12300
^{90}Sr		28.5	600	11
^{95}Nb		0.0958	150	
^{95}Zr		0.175	150	
^{99}Tc		2.13×10^{5}	180	
^{106}Ru		1.02	810	39
^{129}I	Airborne	1.57×10^{7}		0.006
	Liquid		0.074	39
^{131}I	Airborne	0.022		0.0007
	Liquid			0.032
^{137}Cs	Airborne	30.0		0.002
	Liquid		4090	13
^{144}Ce		0.781	100	
^{241}Pu		14.4	1800	
U			11000 (kg)	

†Releases in 1978 only; 1.8 GW$_e$ produced from fuel reprocessed.
‡Average for 3 plants normalized to 1985-1989; La Hague, France, Sellafield, UK, and Toka-Mura, Japan.
(From UNSCEAR 1993.)

non-gaseous activities are contained within the plant, but small releases occur to the environment according to limits set by the regulating authorities. These releases were rather large in earlier days, but have been considerably reduced as improved procedures and equipment are coming into operation, see Table 21.10. The authorized releases in the UK in 1978 were 10^4 TBq β-emitters (excl. T) per year. The ^{137}Cs release produced ~1000 Bq/l water in the middle of the Irish Sea (~150 km SW Windscale), and could be traced as far away as in the Baltic Sea. The release into the Atlantic in 1978 was ≤10 000 TBq. Not counting releases from plants in the US, former USSR and India, it is assumed to be <1000 TBq in 1992. It should be noted that the release of radionuclides into the sea from nuclear power operations is dominated by the effluents from a few reprocessing plants. It is, however, only a small fraction of the natural radioactivity in the oceans.

21.9. TREATMENT AND DEPOSITION OF LOW AND MEDIUM LEVEL WASTES

It has become a common practice at reprocessing plants to store low level solid waste (and sometimes higher level wastes) in trenches dug from the soil. These trenches, which commonly are 5–8 m deep, are sometimes lined with concrete or simply have a gravel bottom. For this purpose dry areas are selected (deserts, when available) or isolated areas with controlled groundwater conditions with respect to water table depth, flow rate, and direction. The disposed material normally should not exceed a dose rate of 1 mGy h^{-1} at 0.3 m distance, or contain a specific activity exceeding 1 MBq kg^{-1}. However, this varies, and may be a factor of 10 higher or lower in some places. When the trench is full, it is backfilled with earth, after which the surface dose rate usually is <0.01 mGy h^{-1}. Trenches of this kind are used in the USA, the UK, France, Russia, etc., where tens of thousands of cubic meters have been disposed annually; Fig. 21.17(A).

Since some of the waste products are longlived, and the physical protection of the waste in surface trenches is poor, radioactivity may ultimately leak into the groundwater, see Ch. 22. Therefore, in many countries repositories for final deposition of solid intermediate and low level wastes have been or are being built. The Swedish repository at Forsmark for ILW and LLW from reactors, hospitals, research facilities, etc. has been in operation since 1988; Fig. 21.17(B). It is located in solid granite rock below 50 m of sea water at the Baltic Sea coast. This underground site contains large concrete silos for the ILW and storage tunnels for ILW and LLW containers. The space between the concrete walls and the rock is filled with clay (bentonite). The repository is partly operated by remote control. Fig. 21.17(C) shows Belgian plans for a mixed ILW and HLW repository located in a large clay deposit. Also in this repository, concrete and clay are the main protective barriers. Geologic barriers are discussed in §21.13.

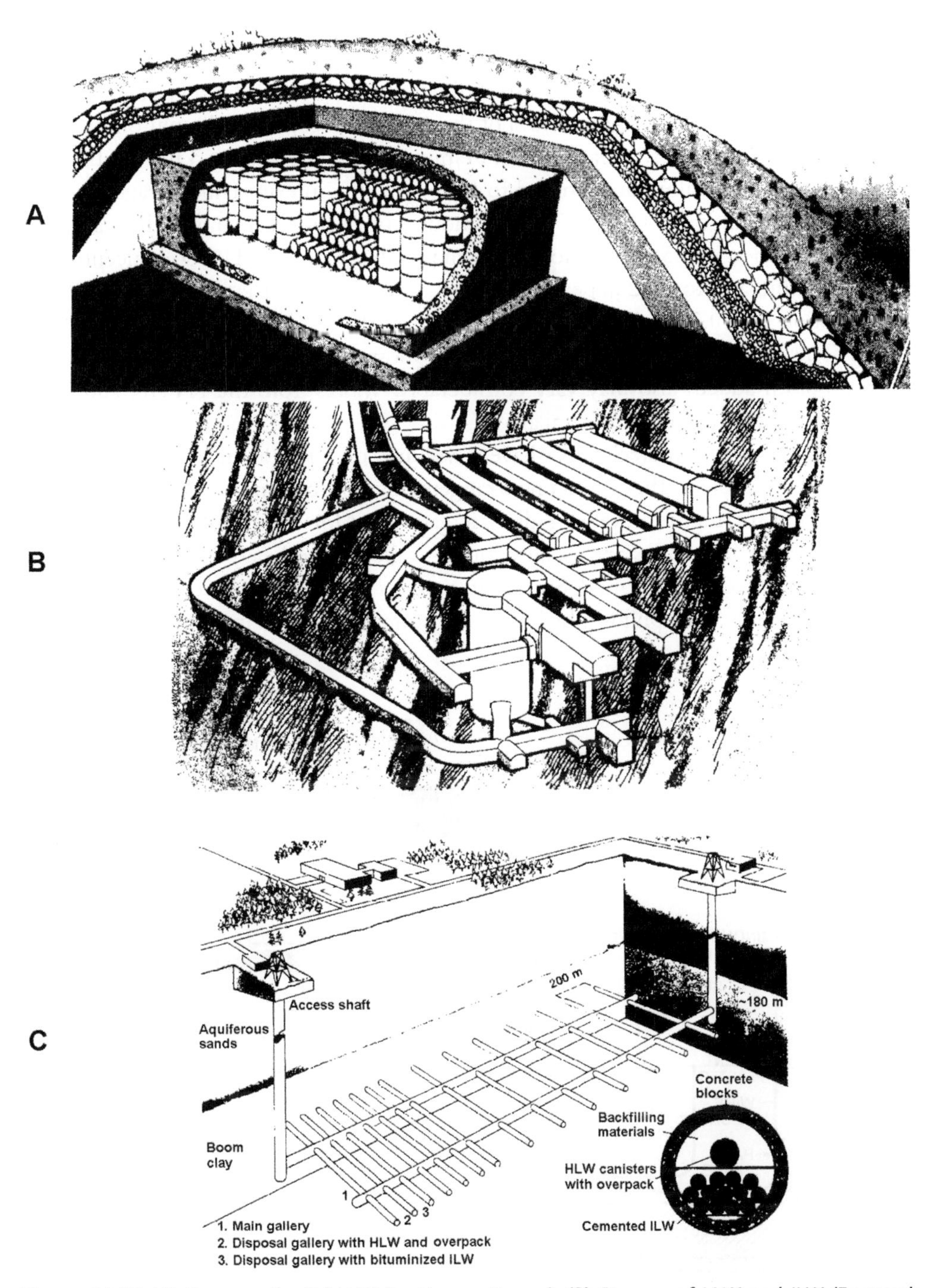

Figure 21.17 (A) Storage of solid LLW (La Hague, France); (B) Storage of LLW and ILW (Forsmark, Sweden); (C) Suggested repository for solid ILW and HLW (Belgium).

21.10. TANK STORAGE OF HIGH LEVEL LIQUID WASTES

The main part of the HLLW is aqueous raffinate from the PUREX cycle. It contains ~99.9% of the nonvolatile FPs, <0.5% of the uranium, <0.2% of the plutonium, and some corrosion products. For each ton of uranium reprocessed about 5 m^3 of HLLW is produced. This is usually concentrated to 0.5−1 m^3 for interim tank storage; specific activity is in the range 10^7 GBq m^{-3}. The amounts of various elements in the waste and their concentration in 0.5 m^3 solution is shown in Table 21.9. The HNO_3 concentration may vary within a factor of 2 depending on the concentration procedure. The metal salt concentration is ~0.5 M; it is not possible to keep the salt in solution except at high acidity. The amounts of corrosion products, phosphate, and gadolinium (or other neutron poison added) also may vary considerably. Wastes from the HTGR and FBR cycles are expected to be rather similar.

The HLLW stainless steel tanks have a volume of 50−500 m^3. They are rather elaborate (Fig. 21.18); they contain arrangements for cooling (≤65°C to reduce corrosion) and stirring, removal of radiolytic gases, and for control of liquid level, pH, and radioactivity. The tanks are usually double-walled, have heavy concrete shielding, and are often placed underground. Storage in stainless steel tanks has been used in the last 40 y without failure. The philosophy is that tank storage is only an interim procedure and

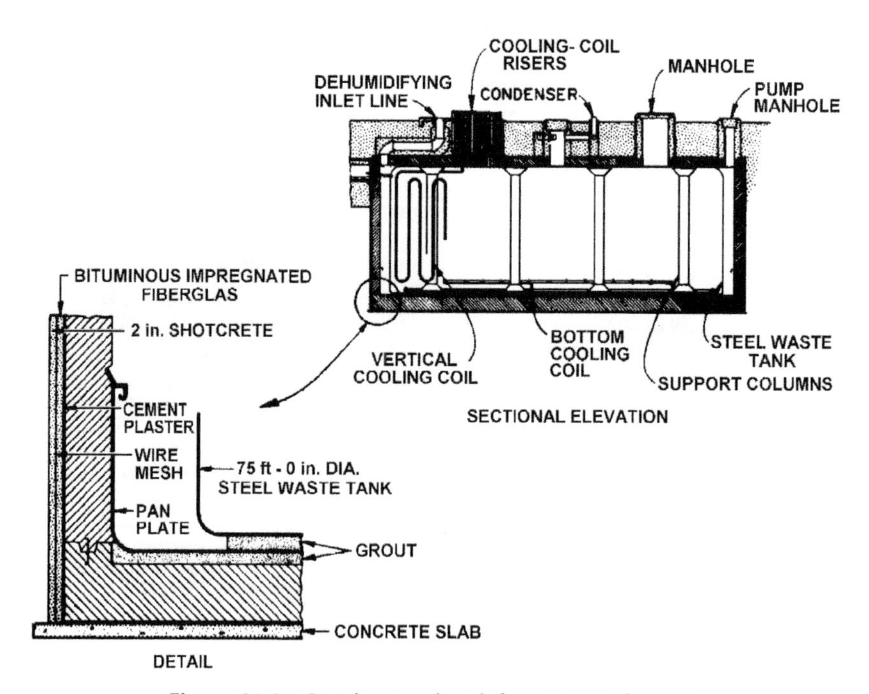

Figure 21.18 Stainless steel tank for storage of HLLW.

will not last for more than a few years, but the capacity for solidification has often been insufficient, so in practice tank storage has been of much longer duration.

Tank storage has not been without failure. The mild steel tanks built in the 1940's at Hanford, USA, were later (in the 1960's) found to leak due to corrosion. The waste nuclides seeped into the soil, some ultimately reaching ground water; NO_3^-, Ru and Cs were found to have moved hundreds of meters, while Sr, Rare Earths and Pu were less mobile in decreasing order.

In 1957 (possibly also in 1967) a serious accident occurred near Chelyabinsk (*Kyshtym*), south of Ekaterinburg, in the former USSR, probably a chemical explosion between organic wastes and nitric acid, in a high-level waste storage facility (tank or underground repository), leading to the contamination of approx. 1600 km^2 by 8×10^{16} Bq fission products. Local values as high as 10^{10} Bq/m^2 and values of 2×10^8 Bq/m^2 for ^{90}Sr and ^{137}Cs have been reported. A large area is now excluded to the public.

21.11. OPTIONS FOR FINAL TREATMENT OF HIGH LEVEL WASTES

Many concepts are being studied to treat the high level liquid waste from reprocessing so that the environment is protected against short and long term radiation damage. The final treatment concepts are:

(a) dispersion to achieve environmentally acceptable concentrations;
(b) partitioning followed by
  (i) nuclear transmutation;
  (ii) disposal into space;
  (iii) burial in a nonaccessible place;
(c) solidification followed by geologic deposition.

In principle, dispersion is only applicable for the gaseous and liquid wastes, which would need negligible pretreatment. The limitations are practical (how efficient is the dispersion into air and sea?), radiological (what radioactive concentrations are acceptable in air and sea?) and political/legal (can it be permitted?).

The options (b) are not feasible for all high level waste products, and therefore has to be limited to the most hazardous ones. It would require an isolation of these products, usually referred to as partitioning, or fractionation, of the HLW.

Option (c) is the main route considered for high level wastes from reprocessing. Because of its importance it is discussed in greater detail in a separate subsection (§21.12). Geologic deposition is also the main choice for un-reprocessed spent fuel elements.

21.11.1. Dispersion into sea and air

The main danger in release of radioactive waste is the risk of ingestion or inhalation. As the hazard differs considerably between the various waste products, depending on their activity, half-life, and biochemical properties, each radionuclide can be assigned a radiotoxicity value (*In*), §15.13.6, defined by

$$In_w = A/\mathrm{ALI}\,(\text{man-years/kg spent fuel}) \tag{21.13}$$

and

$$In_a = A/\mathrm{DAC}\left(\mathrm{m}^3\,\text{air/kg spent fuel}\right) \tag{21.14}$$

where A is the radioactivity (Bq) of a particular nuclide (cf. Fig. 21.7 and 21.8) and ALI and DAC values are taken from Tables like 15.12; see also §15.13.6. Figure 21.19 shows the radiotoxicity values for ingestion (In_w) and inhalation (In_a) according to the ALI and DAC values recommended by the ICRP for the most hazardous fission products and actinides in 1 kg of unreprocessed spent PWR fuel at a burnup of 33 MWd/kg as a function of time. It is seen that ^{90}Sr and ^{137}Cs dominate for the first 10 y, then followed by various actinides. The values for spent BWR fuel are approximately the same.

The ocean water volume needed to contain all water soluble radionuclides and the volumes of atmosphere needed to contain the gaseous waste products from all nuclear power in the world at a level below the DAC and ALI values recommended by the ICRP for safe breathing and drinking can be estimated. The basic data are the total toxicity values (In_w and In_a), which have to be compared with the global (free ocean) water volume (1.4×10^{18} m^3) and the atmospheric volume (the troposphere volume up to 12 km is 6×10^{18} m^3).

Taking all nuclides into account and multiplying by 15 000 000 (assumed to be the amount in kg/year of spent fuel removed from nuclear power plants around the turn of the century) one finds that the water volume needed is <0.1% of the ocean water volume. Thus in principle the ocean capacity much exceeds the dispersion demand for quite a long time. Similarly, one finds that the air volume needed is only a small fraction of the global air volume, assuming all gaseous products to disperse evenly, i.e. the air volume can accommodate all nuclear power gaseous waste products without approaching the DAC value; however, this is not the case if all nuclides including the actinides were dispersed in the air. There is, however, no process for such a dispersal.

Considering that uniform dispersion is impossible, and that biological processes may enrich some radionuclides, local concentrations would be expected leading to unacceptable doses by exceeding the ALI and DAC values. Therefore only limited amounts are allowed to be disposed of into air and the sea, and within strict rules on the kind and amount of nuclide and the packaging prescribed by the London Convention. The London Convention on the prevention of marine pollution by dumping of wastes (1972) limits the amounts to 100 Ci y^{-1} for ^{226}Ra, 10 Ci t^{-1} for other α-active waste of $t_{1/2} > 50$ y, 10^6 Ci t^{-1} for tritium, 100 Ci t^{-1} for ^{90}Sr + ^{137}Cs, and 1000 Ci t^{-1} for other β,γ-waste

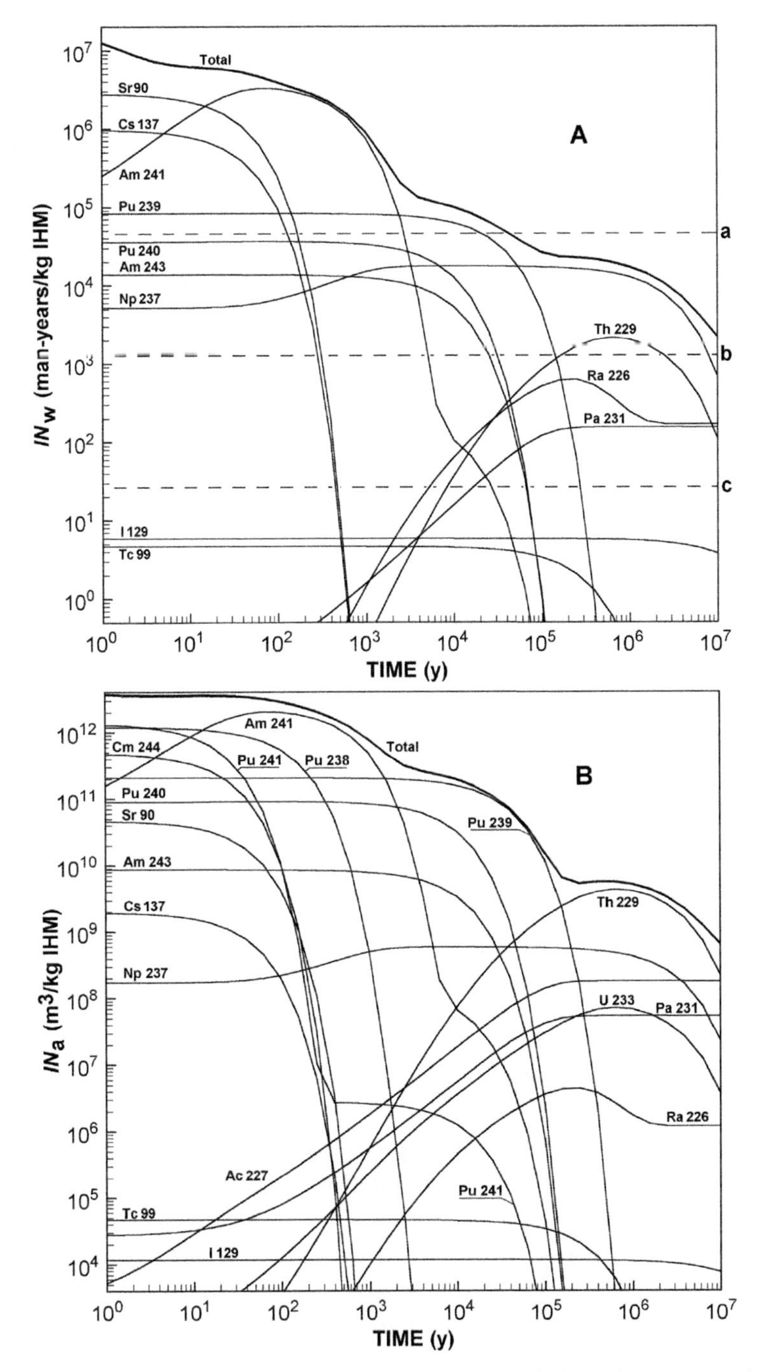

Figure 21.19 Radiotoxicity of the dominating nuclides in spent fuel; (A) for ingestion based on ALI and (B) for inhalation based on DAC. PWR fuel after 33 MWd/kg burnup.

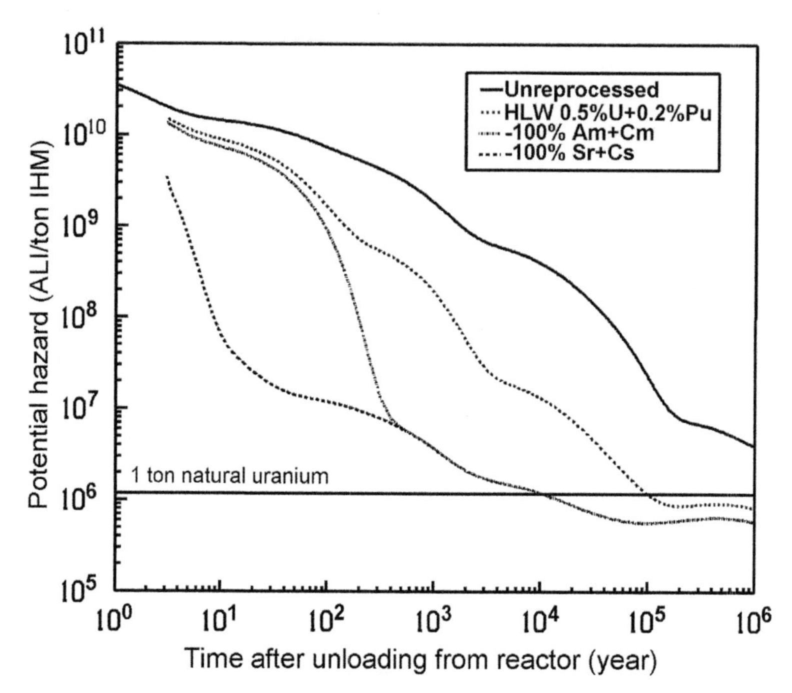

Figure 21.20 Relative radiotoxicity after 3 y cooling of PWR fuel at 33 Mwd/kg U burnup. From top: spent fuel, HLW, HLW after removal of all Am and Cm, and after additional removal of all Sr and Cs.

at any site. (N.B., 1 Ci is 37 GBq.) These figures are based on a dumping rate of not more than 100 000 t y^{-1} at each site. Dumping is controlled by the IAEA.

In the diagram (Fig. 21.19(A)), the horizontal dashed lines indicate the In_W values for the uranium involved in the fuel cycle. The three lines refer (a) to 7 kg natural U (i.e. the potential hazard from the amount of ore that must processed in order to produce 1 kg U enriched to ~3.3% in ^{235}U), (b) 1 kg enriched U, and (c) ~0.041 kg U (the amount $^{235+238}$U consumed in 1 kg IHM at 33 MWd/kg); in the calculation of the *In* values all uranium daughters were taken into account. It may be argued that when the radiotoxicity value for the waste goes below such a line, the potential hazard is not greater than the natural material provided the external conditions (*Ex*, eqn. (15.10)) can be made the same. The In_W line for HAW from reprocessing, shown in Figure 21.20, and for 1 t natural U cross each other at a decay time about 10^5 y. Thus if the HAW is evenly dispersed and equally well fixed as the original uranium and daughters in the mine, it does not constitute any greater hazard than the uranium ore itself. Consequently, it has been proposed to mix the HAW with cement and/or mine refuse, or, in a more elaborate scheme, to mix it with rock-forming materials and convert it into a "synthetic rock" (SYNROCK), and dispose of it in empty mines or natural underground cavities.

21.11.2. Partitioning

The term partitioning is used in two senses: in §21.6.3, to indicate steps in the PUREX cycle where U and Pu are separated; here, to indicate a separation of the most hazardous products from the high level waste. Figure 21.20 shows that the hazard of the HAW is dominated by a few fission products (mainly ^{90}Sr and ^{137}Cs) and the remaining actinides; the actinides in the HLW consists of $\leq$0.5% U, $\leq$0.2% Pu and all Np, Am and Cm. In a successful partitioning cycle, the U and Pu recovered will be recycled, leaving "the minor actinides" Np, Am and Cm, which amount to about 0.8 g kg^{-1} y^{-1} spent fuel from a 1000 MW$_e$ LWR reactor; the corresponding amount of Sr + Cs is about 3.7 g kg^{-1} y^{-1}.

The amount of the minor actinides is small enough to make special disposal of them interesting. If they could be removed from the HLW, the waste hazard would be considerably reduced in time, see Fig. 21.20. Presently, research is directed towards the complete removal of all long-lived actinides from the reprocessing waste, followed by their "elimination" through the alternatives described later.

These separation projects go under different names: in The US the approach is referred to as the CURE (Clean Use of Reactor Energy), in which the TRUEX (TRansUranium EXtraction; also Truex) process is used; the Japanese have the OMEGA (Option for Making Extra Gain from Actinides) program; the French the SPIN and ACTINEX (for Separation Project Incineration Nucleair, and ACTINide EXtraction), etc. Many of these projects are expected to run for decades before they can be realized.

It is a difficult task to isolate the higher actinides in the HLW, particularly to separate them from the lanthanides, because these elements all are present in solution as trivalent ions of similar size and therefore have very similar chemical properties. The separation methods utilize their slightly different complex forming abilities in techniques such as solvent extraction, ion exchange, and reversed phase partition chromatography. Three solvent extraction processes have been run on a larger experimental scale:

(a) In the *Reversed Talspeak process* the extractant is di-2-ethylhexylphosphoric acid (HDEHP) in a suitable aliphatic diluent. By adding lactic acid to the aqueous phase and adjusting pH to 2.5–3.0 the actinide and lanthanides are selectively co-extracted. The actinides can then be stripped by an aqueous phase containing diethylenetriaminopentaacetic acid (DTPA) as complex former and lactic acid as kinetic promotor, leaving the lanthanides in the organic phase.

(b) The *Truex process* is based on a carbamoyl organophosphorus extractant, abbreviated CMPO (for octylphenyl-N,N-diisobutylcarbamoylmethylphosphine oxide, chemical formula $C_8H_7(C_6H_5)POCH_2CON(CH_2CH(CH_3)_2))$. CMPO is used together with TBP in a mixture of aliphatic hydrocarbons. The Truex flow sheet is shown in Fig 21.21. This process (like the other ones) is based on a common principle: combining complexation of the species in the aqueous phase with a selective extraction process or reagent.

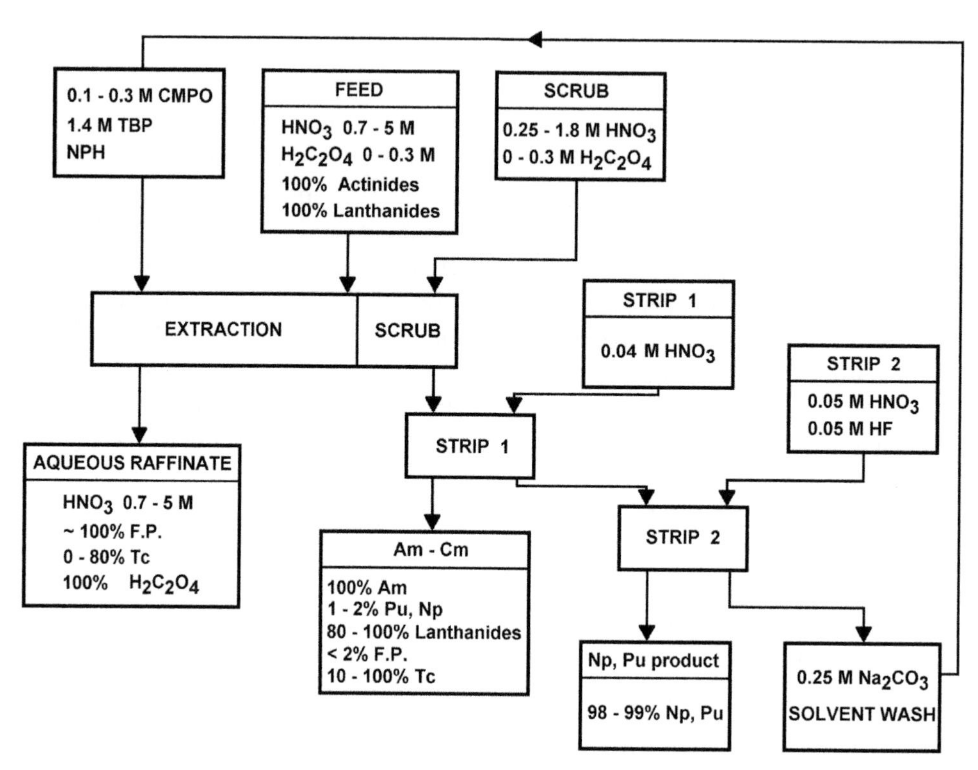

Figure 21.21 Truex flow sheet for removal of minor actinides from HLLW. *(From Musikas and Schulz.)*

A "complete partitioning" requires the removal of strontium and cesium from the HLW, as ^{90}Sr and ^{137}Cs dominate the waste hazard during the first 400 years. If these are removed, the waste becomes almost "harmless" (i.e. it will be quite easy to handle), see Figure 21.20.

Solvent extraction processes have been developed which effectively remove Sr and Cs. Such a process have to be integrated with the actinide removal processes for an efficient back end cleaning procedure. Sr can also be removed from HLLW by the *Srex process*, in which it is very selectively extracted by a macrocyclic ether (ditertiarybutyldicyclo-hexanone-18-crown-6 dissolved in n-octanol, also ^{99}Tc is extracted). Recently another group of crown ethers, calixarenes, have shown a high selectivity for Cs: e.g. in 1 M HNO_3, D_{Cs} is ~20, while D_{Sr} is only 0.01. Thus improved methods are being developed for the efficient removal of Sr and Cs from high level waste.

The main point of partitioning the high level waste is that it shall lead to a safer waste (more acceptable to the public) as well as a cheaper back end fuel cycle (to the advantage of the nuclear energy industry). Koyama has analyzed the different waste handling options, Fig. 21.22, and concluded that full partitioning (actinides as well as Sr+Cs, Tc and I) will lead to the cheapest fuel cycle.

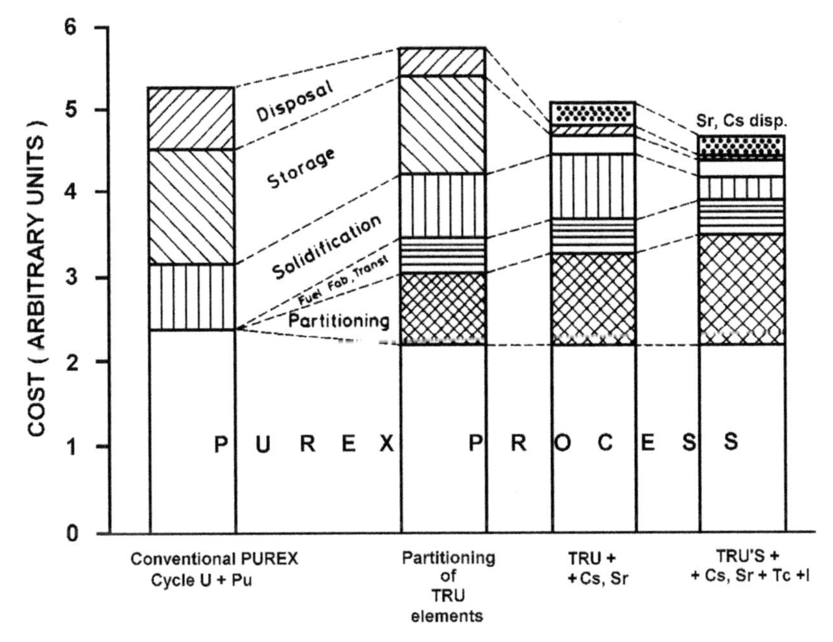

Figure 21.22 Comparison of costs of various options for managing and disposing of PUREX high-level waste. *(From Koyama.)*

21.11.3. Disposal into space

Assuming 20 t IHM spent fuel are removed from a 1000 MW_e LWR annually, the partitioned waste would amount to ~20 kg minor actinides and ~80 kg Sr+Cs annually. The payload of modern rockets exceed several tons. Thus, even including the weight of packaging material, it does not seem unreasonable to assume that the minor actinides from a considerable number of LWRs (>10) could be transported by a rocket or other means to a location off the Earth. Several different space trajectories have been considered. These include:

(i) a high Earth orbit (altitude 150 000 km); Δv 4.15 km s^{-1};
(ii) transport to the sun; Δv ~22 km s^{-1};
(iii) an inner solar orbit; Δv ~15 km s^{-1};
(iv) solar system escape; Δv ~52 km s^{-1}.

Δv is the incremental velocity required to leave a circular Earth orbit at 200 km and is a direct indication of the size and propulsion energy of the rockets required. Vehicles that could be used include existing rockets and future substitutes for the space shuttle.

A high Earth orbit has the advantage of low Δv and possible later retrieval of the waste, but it requires long term container integrity and very long orbit lifetime (not yet proven). Transport to the sun or solar escape has the advantages that the waste is permanently eliminated from the Earth, but a very high Δv is required. The inner solar orbit has the next lowest Δv but requires a very stable orbit, so that it does not return to Earth.

Space disposal has an economic disadvantage since the cost of transportation is likely to be very high, as the weight of shielding could create significant economic penalty. There are also requirements for capsule integrity to provide a reasonable degree of assurance of survival in the case of an abort. At present, this is not considered to be a serious option. However, many kilograms of ^{238}Pu produced from ^{237}Np have already been used as a power source in deep space probes or unmanned landers, c.f. §7.9.3. At present some of this is leaving the solar system aboard Pioneer 10 and 11 and Voyager 1 and 2.

21.11.4. Remote deposition

It has been suggested that the most hazardous actinides be removed from the reprocessing waste and stored separately. The advantage is (i) to eliminate all actinides from the HAW, so that its hazardous potential follows that of ^{90}Sr, ^{137}Cs and ^{99}Tc, by which the main hazard is gone in about 400 years, and completely in about 100 000 y; (ii) to reduce the waste actinides to a small volume, 1/100 to 1/1000 of that of the HAW (and, of course, even much less when compared to the volume of the spent fuel elements), which will simplify the storage problem. For example, such actinide waste could be stored uniquely in very deep bore holes in the ground, eventually in the earth's molten interior. A similar procedure seems possible also for the Sr+Cs fraction.

21.11.5. Transmutation

The fission products ^{90}Sr and ^{137}Cs can be transformed into shorter lived or stable products by charged particle or neutron irradiation. Charged particle irradiation would be very expensive, and irradiation by reactor neutrons would produce almost as much fission products as are destroyed. Therefore the use of intense accelerator driven spallation neutron sources for transmutation by n-irradiation has been suggested. If controlled thermonuclear reactors (CTR) are developed, their excess neutrons could be used for ^{90}Sr transformation, but less efficient for ^{137}Cs.

In the long term ($\geq$600 y) the actinides dominate the risk picture. Continuous neutron irradiation of the actinides finally destroys all of them by fission (c.f. Figs. 14.3 and 14.4). The annual production of americium and curium is ~5 kg in a 1000 MW_e LWR, but considerably less in a FBR. Thus if pins of these elements are inserted in a FBR, more americium and curium is destroyed than formed; it is estimated that 90% will have been transformed into fission products after 5−10 y. In the future, CTRs could be used for the same purpose. As an alternative it has been suggested to leave the americium and curium in the uranium returned in the LWR cycle. Wastes from transmutation processes will contain some amount of longlived nuclides, thus a safe final repository is still needed.

21.12. SOLIDIFICATION OF HIGH LEVEL LIQUID WASTES

The solidification of HLLW, followed by geologic deposition, is presently considered as the only realistic technique to create conditions for a safe long term disposal of HAW. The objectives of solidification is to immobilize the radioactive elements and to reduce the volume to be stored. The solidified product must be nondispersable (i.e. not finely divided as a powder), insoluble, and chemically inert to the storage environment, be thermally stable, have good heat conductance (this determines the maximum radioactivity and volume of the final product), be stable against radiation (up to 10^{10} Gy), and have mechanical and structural stability.

Figure 21.23 shows the options for solidification and encapsulation of HLLW. The first step is usually a calcination in which nitrates are destroyed and all metals converted to oxides. Thousands of cubic meters of HLLW have been solidified by fluidized bed calcination. However, calcine has low leach resistance, low heat conductivity, and can be dispersed in air. It is, therefore, only considered as an interim product.

Most countries have focused development work on the fixation of the active waste in borosilicate or phosphate glass. Large continuous vitrification plants producing borosilicate waste glass are located at La Hague in France and at Sellafield in the UK.

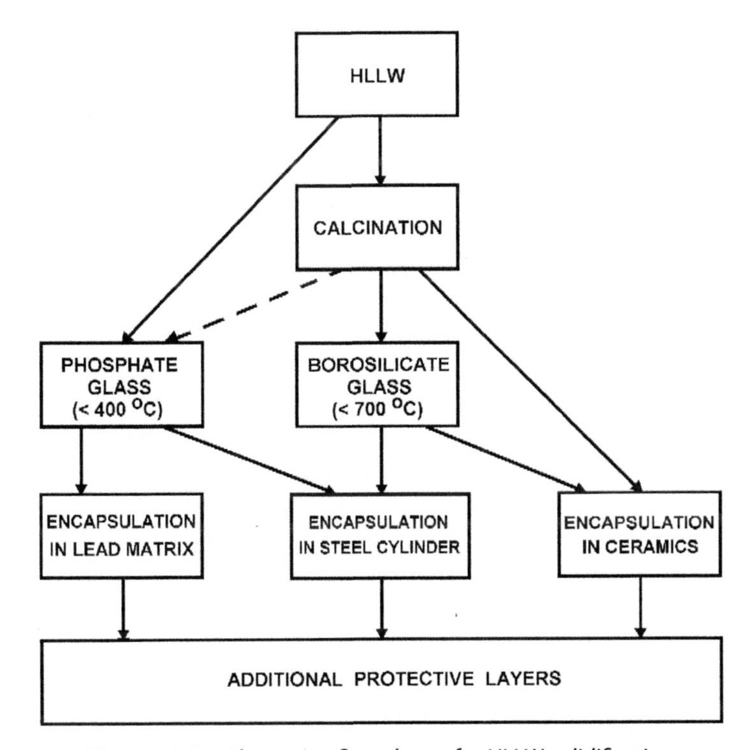

Figure 21.23 Alternative flow sheets for HLLW solidification.

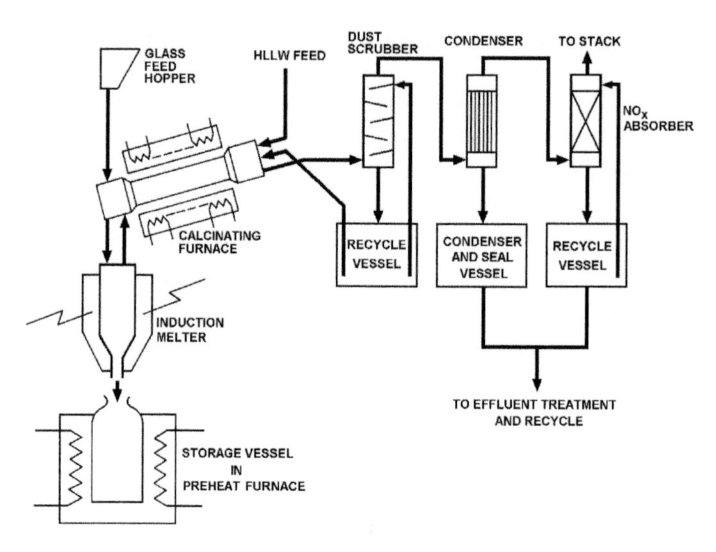

Figure 21.24 Continuous HLLW calcination and borosilicate vitrification process.

Figure 21.24 shows a rotating calciner which feeds a continuous borosilicate glass melter. Preformed and crushed borosilicate glass is continuously added near the end of calcination process. In the melter the calcines and glass mixture is heated to 1000–1200°C, leading to the formation of a homogeneous glass, which is poured into stainless steel cylinders. From a chemical standpoint (stability) the glass can contain up to 30 wt% fission product oxides, but 20% is a more normal value. Thus the waste nuclides in 1 t of LWR fuel can be contained in about 200 kg glass; the exact value depends on how long time the waste has cooled. For short cooling times a high waste content in the glass could cause it to melt and possibly crystallize (at ~700°C), which may reduce corrosion resistance because of the larger internal surface formed. The properties of some glasses are given in Table 21.11. The incorporation of the solidified waste into phosphate glass is an alternative to the use of borosilicate glass. In this case the calcination step can be bypassed. The HLLW, together with phosphoric acid, is evaporated and denitrated, and then fed to a continuous melter operating at 1000–1200°C, from where the molten glass flows into the storage pot. This process is more corrosive and produces a glass somewhat inferior to the borosilicate. Further disadvantages are that the glass recrystallizes at relatively low temperature (~400°C), and that the devitrified (crystallized) phosphate glass exhibits a rather high leach rate, 0.01–0.1 kg m^{-2} d^{-1}.

In the PAMELA process, earlier considered for use in Germany, granulated phosphate glass is incorporated into a metal alloy matrix in a steel cylinder. This offers high chemical and mechanical stability, as well as good heat conductivity (~10 W m^{-1} K^{-1} for a lead matrix; cf. 1.2 W m^{-1} K^{-1} for borosilicate glass). This decreases the central temperature and allows the incorporation of larger amounts of radioactivity (up to 35%

Table 21.11 Characteristics of some solidified high level waste products

Property	Fluidized bed calcine	Phosphate glass	Borosilicate glass
Physical form	Granular, 0.3–0.7 mm	Monolithic	Monolithic
Bulk density (kg m^{-3})	1.0–1.7	2.7–3.0	3.0–3.5
Maximum weight % FP	50	35	30
Thermal conductivity (W m^{-1} K^{-1})	0.2–0.4	1.0	1.2
Leachability (20°C, kg m^{-2} d^{-1})	1–10	10^{-5}–10^{-3}†	10^{-6}-10^{-4}‡
Maximum center temperature (°C)	550††	400*	700*
Maximum allowable heat density (kW m^{-3}) #	45	53	127
Maximum allowable heat density (kW m^{-3}) @	38	36	92

†Devitrified (crystalline) glass has leach rates 10^{-2}–10^{-1} kg m^{-2} d^{-1}.
‡At 100°C the leach rate is ~10^{-2} kg m^{-2} d^{-1}.
††Because of risk for FP volatilization.
* Devitrifies at higher temperature.
#Forced water cooling of a cylinder; Ø 0.3 m, surface temperature ≤100°C.
@Natural air cooling of a cylinder; Ø 0.3 m.

FPs) in a single cylinder. The high heat conductivity makes it feasible to use short fuel cooling times (down to 0.5 y) and diminishes the demand for interim fuel element storage basins or HLLW storage tanks. The solidified waste glass is collected in stainless steel cylinders.

A typical waste canister may contain ~9% FPs, emit ~10 kW after 1 y, ~1 kW after 10 y, and ~0.5 kW after 40 y. They can be stored in air-cooled vaults or water-cooled pools. Before final deposition the HAW containers are reconditioned, i.e. enclosed in an additional canister of type described below. At the higher power level they are stored in water-filled pools. After >10 y air-cooled vaults may be satisfactory; forced cooling is usually considered, but with precaution for sufficient convection cooling in case of ventilation failure.

21.13. DEPOSITION IN GEOLOGIC FORMATIONS

A general consensus has developed that disposal of radioactive waste should be in the country in which the nuclear energy is produced. Thus, even if the spent fuel elements are sent from one state to another for reprocessing (like spent fuel from Japanese reactors being sent to the Sellafield reprocessing plant in the United Kingdom), the producer state (Japan) must guarantee its readiness to accept the processed waste. This condition has led to intensive national investigations of the safest way to dispose of the high level waste, either in the form of solidified HLLW, or spent fuel elements in the once-through option. In these studies geologic repositories have been the prime consideration, e.g. crystalline silicate rock, clay layers, salt domes, the sea bed, etc.

21.13.1. Properties of geologic formations

Many geologic formations are being studied for final storage of the waste canisters: rock salt, crystalline rock (e.g. granite), volcanic tuff, clay, etc. The main requirements are:

(i) geologic stability, i.e. regions of very low seismic, volcanic, or other geologic activity;
(ii) absence of large fractures, holes, etc.;
(iii) impermeability to surface water;
(iv) negligible groundwater circulation with no flow-lines leading to nearby potential intake sources;
(v) good heat conductivity;
(vi) of little interest to and as remote as possible from human activities.

There are large geologic formations, which are very old (many hundred million years) and geologically stable, e.g. the Scandinavian and Canadian shields. These are usually fractured, even if considerably large volumes can be found (10^6 m^3) which are free of fractures. A nuclear waste storage facility will, however, for reasons given below, cover rather large areas, several km^2; thus for crystalline rock fracture zones, which carry ground water, must be taken into account. Rock salt and clay formations, on the contrary, are usually free of fractures: rock salts are dry, but clay formations are usually wet, though the water migration through the formation is extremely small (see below).

A limiting factor for geologic disposal is heat conductivity. If the thermal conductivity of the geological deposit is too poor, the material in the waste canisters may melt and react with the encapsulation, possibly destroying it. Because a safe disposal concept must rely on conduction cooling, the relatively low heat conductivity of the geologic deposit makes it necessary to disperse the waste material over a large volume. The thermal conductivity of $\sim$3 W m^{-1} K^{-1} for granitic rock (compare rock salt $\sim$9 and wet clay $\sim$1) limits the waste heat density of the rock wall to $\leq$20 W m^{-2}. Therefore storage repositories would consist of long tunnels or deep holes at large interspace. Three concepts are under investigation: (i) very deep holes (VDH, up to 3000 m) in which waste canisters would be stacked, (ii) deep repositories (300–500 m) with parallel horizontal tunnels of moderate length ($\leq$100 m), 20–100 m apart, upper part in Figure 21.25, and (iii) deep repositories with a few very long tunnels (>1 km, VLH).

With a typical heat production of $\sim$500 W per canister containing vitrified HAW (at 40 years cooling) the wall heat density would be 6–7 W m^{-2}. The internal temperature of a glass canister would then never exceed 90°C, and the surface temperature (above the storage holes, Fig. 21.25) not more than 65°C. Slightly higher temperatures can be accepted for spent fuel elements.

Such deep (geologically) repositories are safe against surface activities (including nuclear explosions) except drilling or mining. In order to avoid such an occurrence after the repository has been forgotten, it should be located in a formation of no interest to

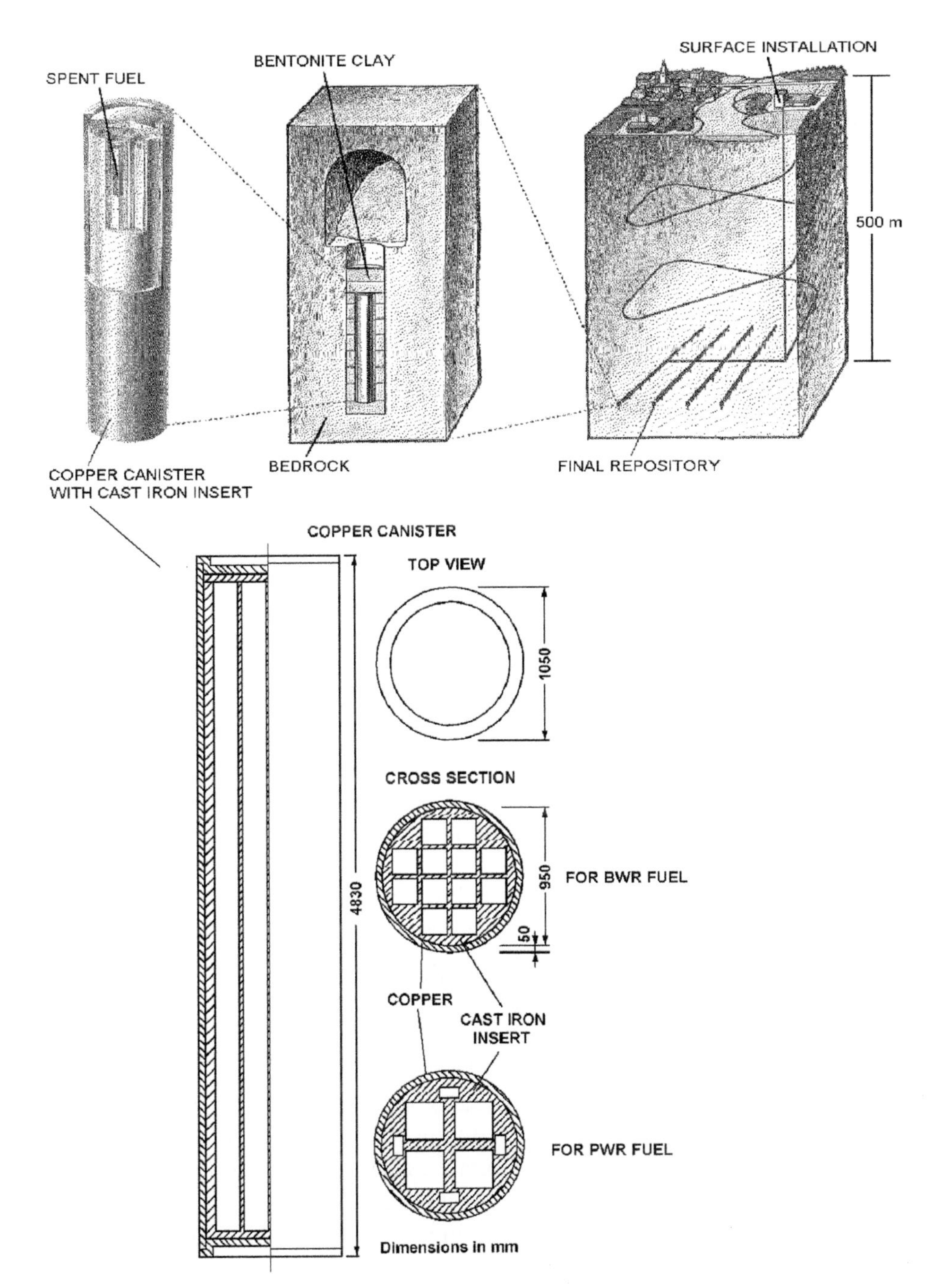

Figure 21.25 Principles of storage of HLW: upper part, Swedish concept; lower part, canister with inserts for final storage of spent BWR fuel and spent PWR fuel. *(From SKB.)*

society; e.g., the formation should not contain any valuable minerals. From this point of view, location in deep seabeds has been suggested. Such a location can be chosen either deep in the bottom silt or in a bedrock under the sea floor, or, possibly, in a tectonically active trench, where with time the waste would be pulled down into the interior of the earth.

The risk for a land-based geologic repository is that a combination of more or less unpredictable circumstances could lead to rupture of the protective barriers around the waste matrix, followed by dissolution and transport of the most hazardous products by water to a place where it can enter into the food chain. Therefore a clay layer (e.g. bentonite) between the canister and the rock wall is suggested, partly to act as a mechanical buffer, so that even considerable slippage caused by earthquakes would have little effect on the mechanical integrity of the canister and partly to reduce water flow around the canister.

Most rocksalt deposits are very old. In many cases they rise in the form of a dome, the top of which may be at a few hundred meters below surface level, and with a depth up to more than 1000 m. The dome is usually protected from groundwater by a calcite ($CaCO_3$) cap. Salt has been mined in such formations for centuries. The formation contains only microcrystalline water and is extremely dry (otherwise it would have dissolved during the geologic ages). Therefore, canisters emplaced in such formations are not believed to dissolve (as long as no water enters through the hole mined through the calcite layer). The good heat conductance allows emplacement of waste of higher power density than for deposits in clay or granite. Since the salt is plastic, the holes and corridors, which are backfilled with crushed salt, selfseal and no clay buffer is needed. The canisters would probably be crushed in the plastic salt, and therefore could be irretrievable after a hundred years.

Granite and clay formations are percolated by groundwater except at depths which presently are considered impractical. The water flow rate is given by *Darcy's law*

$$F_w = k_p i\, S\, (m^3 s^{-1}) \tag{21.15}$$

where k_p is the permeability (m s^{-1}), i the hydrostatic gradient (m m^{-1}), and S the flow cross section (m^2). A typical value for Scandinavian granite formations at 500 m depth is $F_w = 2 \times 10^{-4}$ m^3 m^{-2} y^{-1}, which gives a ground-water velocity of 0.1 m y^{-1}, and a time of 5000 y for the water to move 500 m. In this case the rock permeability is taken as 10^{-9} m s^{-1}; rock formations closer to the surface usually have higher permeabilities, but many formations with much lower values are also known ($\sim 10^{-13}$).

Clays consist of small particles, usually <2 μm and with an average size of <0.1 μm, of various minerals like quartz, feldspar, montmorillonite (a hydrated aluminum silicate with high ion-exchange capacity), mica, etc. The overall chemical composition is mainly a mixture of silica, alumina, and water. The small particles in the clay give rise to a very large surface area (1 cm^3 of particles of 0.1 μm diameter have a total surface area or about

60 m^2) and correspondingly high sorption capacity for (eventually) dissolved waste products. Although the clay can take up large amounts of water (up to 70%) without losing its plasticity, the water permeability is extremely low. For the sodium bentonite clay (~90% montmorillonite, plenty available) considered in many projects, the permeability is 2×10^{-14} m s^{-1} (10% water in clay compacted to a density of 2100 kg m^{-3}). Such clay can be considered impermeable to groundwater. Natural clays of this type occur in many countries (the Netherlands, Italy, etc.).

21.13.2. Waste conditioning before final storage

The basic philosophy in storing high levels of radioactivity in geologic formations is the use of a multi barrier system to protect it against dissolution. These barriers are:

(i) The waste matrix itself, which is made highly insoluble; this is achieved by vitrification of the reprocessing HLLW as described above, while the spent unreprocessed fuel elements keep the radioactive products in a highly insoluble UO_2 matrix.

(ii) The waste matrix (glass or UO_2) is encapsulated in an "insoluble" canister.

(iii) The canister is surrounded by a clay buffer in the case of a crystalline rock deposit.

Here we shall only discuss the encapsulation, for which numerous concepts have been suggested.

(a) *Vitrified waste.* The stainless steel cylinders containing the vitrified radioactive waste are not considered to provide sufficient protection against long term corrosion in the final repository. Additional encapsulation in various corrosion resistant materials is therefore suggested: lead, titanium, copper, gold, graphite, ceramics, etc. For example, in one proposal canisters containing 170 l vitrified waste in a 3 mm thick stainless steel cylinder will be surrounded by 100 mm lead in a casing of 6 mm titanium; the overall dimensions of the cylinder will be 1.6 m long with a diameter 0.6 m. One such cylinder contains solidified HLLW from reprocessing 1 t spent fuel from a 1000 MW_e light water reactor; this corresponds to about 20 cylinders per year.

(b) *Spent fuel.* Although UO_2 dissolves more slowly than glass in groundwater, spent fuel elements must be recanned before entering the final storage facility. For this purpose single fuel elements or bundles are placed in cylindrical canisters. The lower part of Figure 21.25 shows an example. The outer part of the canister may consist of iron, stainless steel, copper, etc. selected to resist the repository environment (e.g. dry or wet). For storage in very long or deep dry holes, rounded caps are likely to be used. Waste canisters for unchopped spent fuel elements will be rather long, but it has also been suggested that the fuel is chopped as in the German Pollux concept, Fig. 21.26, which is designed for multi-purpose use, both to fit all kinds of spent fuel elements (from BWRs, PWRs, HTGRs etc) as well as vitrified high level waste. It is a double-shell concept, consisting of leak-tight welded steel; between the two steel

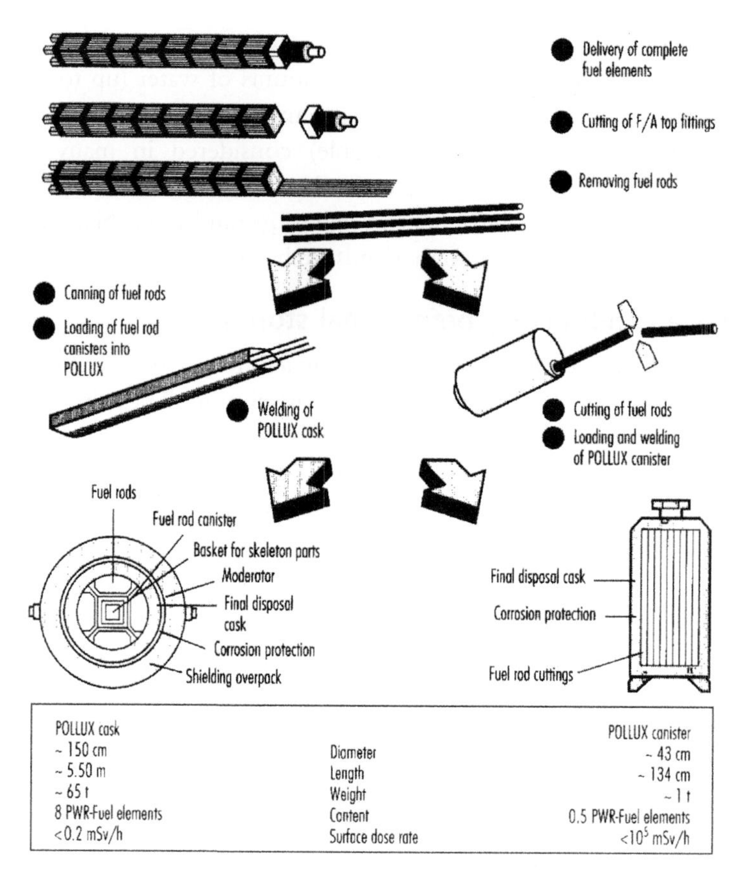

POLLUX cask		POLLUX canister
~ 150 cm	Diameter	~ 43 cm
~ 5.50 m	Length	~ 134 cm
~ 65 t	Weight	~ 1 t
8 PWR-Fuel elements	Content	0.5 PWR-Fuel elements
<0.2 mSv/h	Surface dose rate	$<10^5$ mSv/h

Figure 21.26 The proposed German POLLUX cask for final disposal of spent fuel and/or vitrified high level waste in salt.

walls is a corrosion resistent sheet of Hastelloy C4. The void is either gas filled (He) or filled with some suitable material like lead, which has good heat conductivity and also provides some radiation protection. Figure 21.25, shows the design of a spent fuel element canister where the fuel elements are inserted in a cast iron cylinder which is surrounded by 50 mm of copper metal. The thick metal layer has several purposes: (i) to simplify handling during deposition operations; (ii) to protect its content from mechanical deformation; (iii) to reduce the surface radiation to so low values that the radiolysis of groundwater will not contribute to canister corrosion; (iv) to protect, as long as possible, the waste-containing matrix from corrosion. For example, by using an outer copper wall, the lifetime of the canister is expected to greatly exceed 1000 y for oxygen containing groundwater, and 10^4 y for reducing groundwater (the lifetime of a 5 cm thick copper layer is now estimated to be $>10^6$ y). The dissolution of waste canisters and leaching of waste is discussed in §22.10.

21.13.3. Repository projects

In one concept the canisters are stored in dry areas (e.g. a desert floor) after having been surrounded by concrete. A space may be left between the inner cylinder and the surrounding concrete wall to allow air convection cooling. As long as the climate stays dry the pillars will erode very slowly and last for tens of thousands of years. This procedure also allows for easy retrievability. An alternative encapsulation is achieved by using hot (~1500°C) isostatic (100–300 MPa) compression to surround the fuel bundles by a homogeneous, dense ceramic material, like corundum (microcrystalline Al_2O_3) or graphite. Since corundum and graphite are natural minerals, the long term resistance should be very high, even against water.

The main concept is to store the vitrified HLW and the spent fuel elements irretrievably in deep underground geologic formations, according to designs described above. Many such formations are now being evaluated: rock salt (mainly Germany, the Netherlands and USA), crystalline rock (Canada, Finland, France, Japan, Sweden, USA), volcanic tuff (USA), clay (Belgium, Italy), etc. Cross sections of geologic repositories for HLW are shown in Figures 21.17(C) and 21.25. Also disposal into polar ice sheets (USA) and the seabed (UK, USA) have been investigated, but are no longer considered feasible.

The most advanced projects are (i) at the abandoned Asse salt mine in Germany, where drums with low and intermediate level activities have been deposited at 300 m depth in 1967–78; the repository is now used for research; a final repository in salt at Gorleben has been planned, (ii) the waste isolation pilot plant (WIPP) at a bedded salt deposit in New Mexico at 600 m depth; the planned repository is currently examined by the authorities for full scale operation; it will primarily be used for α-bearing military waste; in Yucca Mountain extensive studies have been made for a repository 300 m deep in dry tuff for civilian and defense nuclear waste (total capacity ~7000 t IHM) – however, according to current political decisions it will not be used as a repository, (iii) the Swedish KBS project for disposal in granite; the Stripa mine has been studied for more than a decade by an international team and a new underground laboratory has been built at Äspö; other hard rock laboratories are in operation in Canada, Finland and Switzerland, and (iv) a laboratory at 250 m depth in clay at Mol, Belgium, has been in operation since 1983, etc. These sites are all located in geologic formations, which have been unaltered for more than 100 million years. From the tectonic plate theory continual tectonic stability is expected for at least the next 10 million years. Although ice ages may alter surface conditions at northerly located sites, they are expected to have a negligible effect on repositories at great depth (>500 m). In 2001 the Finnish parliament gave its permission to begin construction of a KBS-like final repository for spent fuel in granite. It will be the first of its kind in the world.

21.14. BENEFICIAL UTILIZATION OF NUCLEAR WASTES

The amount of spent fuel generated during 2003 amounts to about 8000 t. It contains large and potentially valuable sources of metals and radioactive nuclides. The known and predicted cumulative amounts are indicated in Figure 21.11. Though today considered a liability it may in the future become a needed asset. Since the extraction and utilization of some of the fission products or actinides will probably not be economic after the waste has been vitrified and placed in permanent geologic storage, the nuclear fuel reprocessing scheme should therefore be designed for byproduct extraction. Presently the most interesting products in the waste are the platinum group metals (due to their metal values) and ^{90}Sr, ^{137}Cs, ^{85}Kr, ^{237}Np/^{238}Pu, and ^{241}Am (due to their radiation properties, cf. §§8.11, 18.5, etc).

The waste contains considerable amounts of Ru, Rh, and Pd, all metals in scarce abundance on earth. These elements are used as catalysts in the chemical industry, for catalytic exhaust cleaning in cars, and as corrosion resistant materials. The United States demand exceeds the domestic production by about a factor of 100. The United States would be independent of import from year 2000 if these elements were recovered from the yearly generated spent fuel. This is particularly true if technetium is recovered, since it can often replace platinum. Some of the recovered elements would be radioactive, but the activities would be small enough to make the elements easy to handle.

Beta radiation from ^{85}Kr on phosphors causes visible light. Radiokrypton light sources have widespread applications where reliable lights are required as, for example, at airports, railroads, hospitals, etc., or where sources of electricity could cause dangerous explosions, as in coal mines, natural gas plants, etc. However, fission product krypton contains only 4% of ^{85}Kr, which makes it unsuitable for high intensity lightning applications. ^{85}Kr must therefore be enriched about a factor of 10, which presently can be done e.g. by thermal diffusion.

By the year 2000, over 100 MW heat will be produced by radiostrontium. Strontium fuelled thermoelectric generators are used in several countries to power unmanned weather data acquisition systems, lighthouses, and other navigation aids, etc. Their reliability surpasses any other remote power source. The current thermoelectric generators have a thermal-to-electrical efficiency of about 5%, while recently developed thermal-to-mechanical systems show efficiencies of 25–30%. The use of such systems could be expanded vastly by increased recovery of suitable fission products and actinides. However, due to environmental concerns many of the generators deployed so far at sea or on shores are now being recovered and scrapped.

Food sterilization by radiation is potentially of global importance (§8.11). Though this presently is done by ^{60}Co or accelerator radiation, the advent of large quantities of radiocesium recovered from nuclear wastes may have a considerable positive impact on the economics and scale of food irradiation. An almost equally important use of ^{137}Cs would be for sewage sludge treatment, which may become increasingly important as the requirements for sterilization and secondary treatment of the sludge increases. The thermoradiation of sludge may also make it useful as a sterile fertilizer.

The low penetrating radiation, long half-life, and high power density of ^{238}Pu makes it ideal for special purpose power supplies (§7.9.3). The main present use is in space research, and ~30 kW_{th} power sources have been launched into space, ^{238}Pu has been used in heart pacemakers and is still a candidate as a power source for completely artificial hearts. Production of ^{238}Pu requires the isolation of ^{237}Np, which is then irradiated and reprocessed to produce pure ^{238}Pu.

The demand for ^{241}Am is larger than present production capacity because of its use in logging oil wells, in smoke detectors, and for various gauging and metering devices. However, the potential source is large. At the end of 2003, the amount of americium produced as a result of civil nuclear power reactor operation was about 87 tonnes. About 68 tonnes of this americium resulted from the decay of ^{241}Pu subsequent to its discharge from the reactors. Currently, the americium inventory increases by about 4 tonnes per year.

If all the mentioned waste products are recovered, this would mean (a) that the waste is turned into an essential asset with benefits in food production, health, and safety, and (b) that the hazard of the remaining waste would be much lower, considerably simplifying final waste storage.

21.15. EXERCISES

21.1. In a BWR the minimum and maximum heat fluxes at the fuel rod surface are 0.46 and 1.08 MW m^{-2} at a coolant temperature of 283°C. The rods have an outer diameter of 12.7 mm with a cladding of 0.8 mm thick Zircaloy-4; assume a negligible fuel cladding gap and neglect the temperature drop between coolant and cladding and across the fuel-cladding gap. Assume that data in Table 21.1 are valid at all temperatures and $k_{UO2} = 4$ W/m°C. What are the highest and lowest fuel-center temperatures?

21.2. Using the data for the Würgassen reactor and the thermal neutron capture cross-section (Table 19.4) it can be calculated how many kg Pu should be formed per t U at a burn-up of 27 500 MWd/t. (a) Make this calculation assuming that plutonium disappears only through fission in ^{239}Pu. (b) According to Table 21.2 each t U from a PWR contains 8.69 kg Pu; why is your result much lower?

21.3. It is desired that 98% of all ^{233}Th formed by neutron capture in ^{232}Th decays to ^{233}U. How long a time must elapse between end of irradiation and start of reprocessing?

21.4. In a ^{233}U fuelled reactor, some ^{233}U is converted into ^{235}U. Calculate the amount of ^{235}U formed in 1 t ^{233}U from neutron capture in ^{233}U and ^{234}U ($\sigma_{n,\gamma}$

97 b) for a fluence of 10^{25} n m^{-2} (a) assuming no consumption of ^{235}U formed, (b) taking ^{235}U fission and capture into account.

21.5. Explain why ^{228}Th and ^{232}U is a nuisance in the thorium-uranium fuel cycle.

21.6. A reactor starting with 3% ^{235}U produces 6000 MWd energy/t U fuel each year. Neglecting fission in ^{238}U, (a) how much fission products have been produced after 5 years? (b) What is the ^{235}U concentration if plutonium fission also is taken into account?

21.7. In the example above, 1 t U as fuel elements is removed from the reactor after 2 years. Using Fig. 21.7, (a) what is the total radioactivity from the fission products after 1 y cooling time? (b) Which FP elements are the most radioactive ones at this time?

21.8. Calculate the decontamination factor required for (a) fission product activity, and (b) for gadolinium in commercial plutonium nitrate produced from PWR fuel (Tables 21.2 and 21.7) at t_{cool} 10 y.

21.9. 0.0015 Ci ^{239}Pu is released annually from a reprocessing plant. What will be the corresponding release of ^{238}Pu and ^{240}Pu for typical isotopic plutonium composition of LWR fuel?

21.10. Calculate the natural radiotoxicity value In_W of 1 km^3 of land (density 2 600 kg m^{-3}) containing 3 weight ppm ^{238}U with daughter products. Only ^{226}Ra has to be considered.

21.11. A tank contains 100 m^3 5 y old HLLW. Analyses show that a 1 ml sample contains 1.09 GBq of ^{90}Sr, which is the only Sr activity. (a) Calculate the heat production for a waste of composition in Table 21.9 left column. (b) How many 500 kg glass cylinders would be needed (assume the glass contains 10% FPs) to contain all the solidified waste? (c) How many 1000 MW_e PWR reactor years does this waste correspond to?

21.12. Assume that Ru, Rh, and Pd are recovered from the waste from a 10 GW_e program. What will the annual amounts and specific radioactivities be at t_{cool} = 10 y for each of them?

21.16. LITERATURE

J. PRAWITZ and J. RYDBERG, Composition of products formed by thermal neutron fission of ^{235}U, *Acta Chem. Scand.* **12** (1958) 369, 377.

R. STEPHENSON, *Introduction to Nuclear Engineering*, McGraw-Hill, 1958.

J. FLAGG (Ed.), *Chemical Processing of Reactor Fuels*, Academic Press, 1961.

J. T. LONG, *Engineering for Nuclear Fuel Reprocessing*, Gordon & Breach, 1967.

K. J. SCHNEIDER and A. M. PLATT (Eds.), *High-level Radioactive Waste Management Alternatives*, Battelle Pacific Northwest Laboratories, Richland, Wash. 99352, Report BNWL-1900, NTIS, Springfield, 1974.

S. AHRLAND. J. O. LILJENZIN and J. RYDBERG, (Actinide) Solution Chemistry, in *Comprehensive Inorganic Chemistry*, Vol. 5, Pergamon Press, Oxford, 1975.

Convention on the Prevention of Marine Pollution by Dumping of Wastes and other Matter ? "The Definition Required by Annex I, paragraph 6 to the Convention and the Recommendations Required by Annex II, Section D", IAEA, INFCIRC/205/Add. 1, 10 January 1975.
IAEA, *Transuranium Nuclides in the Environment*, Vienna, 1976.
IAEA, *Management of Radioactive Wastes from the Nuclear Fuel Cycle*, Vienna, 1976.
ERDA, *Alternatives for Managing Wastes from Reactors and Postfission Operations in the LWR Fuel Cycle*, ERDA-76-43. NTIS, Springfield, 1976.
IAEA, *Regional Nuclear Fuel Centres*, Vienna, 1977.
NEA-OECD, *Objectives, Concepts and Strategies for the Management of Radioactive Waste Arising from Nuclear Power Programmes*, OECD, 1977.
H. A. C. McKay and M. G. Sowerby, *The Separation and Recycling of Actinides*, EEC, EUR 5801e, 1977.
KBS Project, *Nuclear Fuel Cycle Back-End*, 1978, Kärnbränslesäkerhet, Fack, 10240 Stockholm.
R. G. Wymer and B. L. Vondra, *Technology of the Light Water Reactor Fuel Cycle*, CRC-Press, 1980.
M. Benedict, T. H. Pigford and H. W. Levi, *Nuclear Chemical Engineering*, 2nd Ed., McGraw-Hill, 1981.
C. Musikas, W. W. Schulz, J. O. Liljenzin, Solvent Extraction in Nuclear Science and Technology, p. 507-557, in J. Rydberg, M. Cox, C. Musikas, G. R. Choppin (Eds), Solvent Extraction, Principles and Practice, 2:nd Ed., ISBN 0-8247-5063-2, Marcel Dekker Inc., New York 2004.
SKB, Swedish Nuclear Fuel and Waste Management Company, Annual and other reports, Stockholm, Sweden.
Sources and Effects of Ionizing Radiation (UNSCEAR), United Nations, New York, 1993.
IAEA Bulletin **36**, No 1, 1994. (Radiation Technologies).
Sources and Effects of Ionizing Radiation (UNSCEAR 2008), United Nations, New York, 2011.

CHAPTER 22

Behavior of Radionuclides in the Environment

Contents

The main objection against nuclear power is the risk of spread of "radioactivity" (radioactive elements) to the environment where it may cause health effects in humans. Such effects have already been discussed in chapter 15. As mentioned previously there are several suggested methods for the treatment of used nuclear fuel. However, all of them will require a final repository for the waste. The main difference is the total storage time, the different barriers, and the radiotoxicity of the final waste. In any case, if the repository is breached and ground water comes in contact with the fuel the readionuclides will be dissolved to a smaller or greater extent, depending on element and form. In this chapter, the chemical aspects of the sources of releases, and of the migration of the radionuclides in the environment, will be discussed. Their chemical properties, together with hydrology, determine how fast they will move from their point of entry into the

Radiochemistry and Nuclear Chemistry
ISBN 978-0-12-405897-2, http://dx.doi.org/10.1016/B978-0-12-405897-2.00022-7

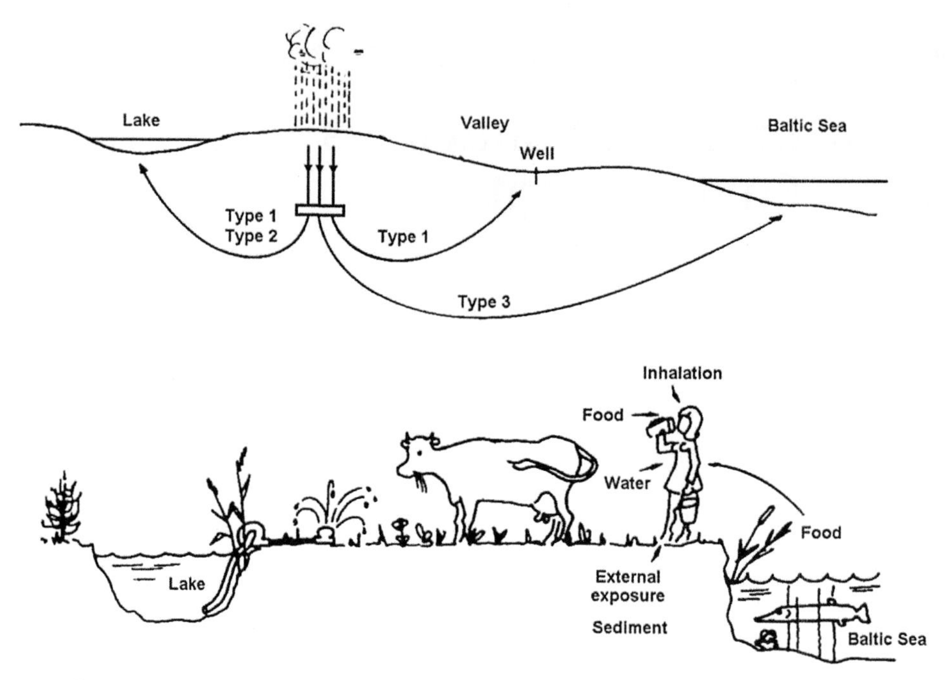

Figure 22.1 Migration path of radioactive nuclides from a waste repository to man.

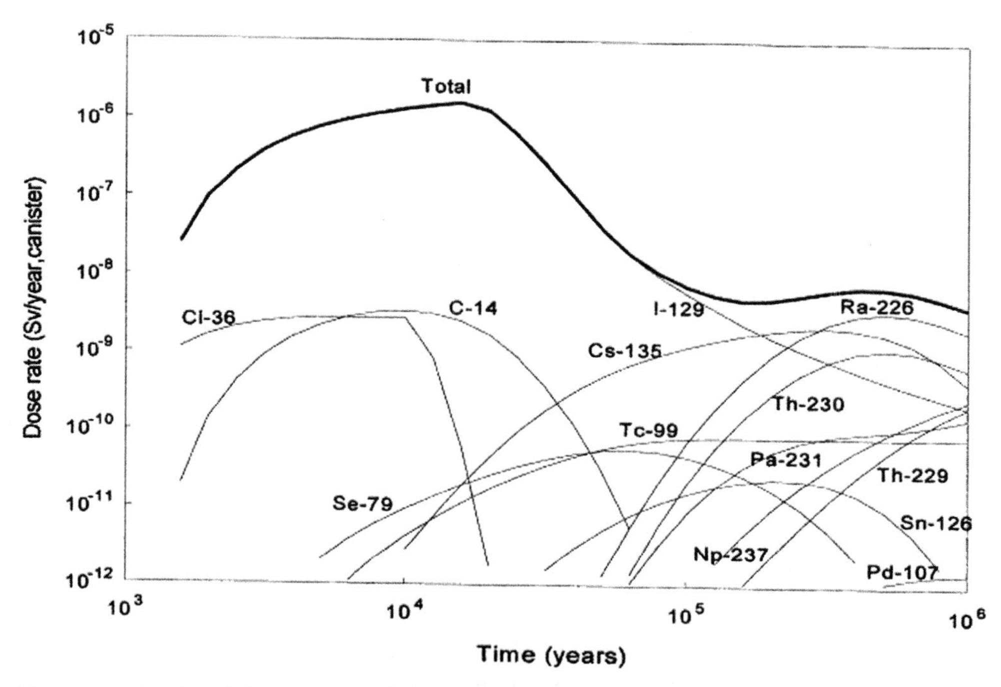

Figure 22.2 Predicted dose rates to individuals from the release of a defective canister, assuming the release occurs directly to the biosphere, i.e. without retention in the ground (*From SKI Site97*).

groundwater to water resources used by man; this is schematically illustrated in Figure 22.1. In particular we discuss actinide behavior as these elements have the most hazardous radionuclides which may be released in the different steps of the nuclear fuel cycle, and, especially, from nuclear waste repositories. However, despite this it may be noted here that the actinides contribution to the dose from a leaking cannister to the recipient is negligible. This is exemplified by the dose to a recipient from one leaking cannsiter in the planned Swedish repository shown in Figure 22.2.

22.1. RADIOACTIVE RELEASES AND POSSIBLE EFFECTS

In earlier chapters, there have been brief discussions of the release of radionuclides to the environment. Such releases occur from mining and milling operations, particularly of uranium ores (§14.2.4.4), from the nuclear fuel fabrication processes, from normal operation of nuclear reactors (§19.16), from reprocessing of spent nuclear fuel (§21.8), from nuclear weapons production and recovery, from transportation of nuclear material, from testing of nuclear weapons and accidents (§13.10), and from storage of nuclear wastes (which is discussed in subsequent sections). Table 22.1 compares the estimated collective dose to the public from the various activities of the nuclear fuel cycle, from

Table 22.1 Collective effective dose to the public from radionuclides released in effluents from the nuclear fuel cycle 1995-97 (UNSCEAR 2000)

Source	Normalized collective effective dose (manSv/GW$_e$y)
	Local and regional effects
Mining	0.19
Milling	0.008
Mine and mill tailings (releases over 5 years)	0.04
Fuel fabrication	0.003
Reactor operation	
Atmospheric	0.4
Aquatic	0.04
Reprocessing	
Atmospheric	0.04
Aquatic	0.09
Transportation	<0.1
	Solid waste disposal and global effects
Mine and mill tailings	
Releases of radon over 10 000 y	7.5
Reactor operation	
Low-level waste disposal	0.00005
Intermediate-level waste disposal	0.5
Reprocessing solid waste disposal	0.05
Globally dispersed radionuclides (truncated to 10 000 years)	40

Table 22.2 Radionuclides released in atmospheric nuclear testing (UNSCEAR2000)

Radio-nuclide	Half-life	Estimated release Total (EBq)	Radio-nuclide	Half life	Estimated release Total (EBq)
^{3}H	12.33 y	186	^{125}Sb	2.73 y	0.741
$^{14}C^{\dagger}$	5730. y	0.213	^{131}I	8.02 d	675
^{54}Mn	312.3 d	3.98	^{137}Cs	30.07 y	0.948
^{55}Fe	2.73 y	1.53	^{140}Ba	12.75 d	759
^{89}Sr	50.53 d	117	^{141}Ce	32.50 d	263
^{90}Sr	28.78 y	0.622	^{144}Ce	284.9 d	30.7
^{91}Y	58.51 d	120	^{239}Pu	24110. y	0.00652
^{95}Zr	64.02 d	148	^{240}Pu	6560. y	0.00435
^{103}Ru	39.26 d	247	^{241}Pu	14.36 y	0.142
^{106}Ru	1.023 y	12.2			

For the non-gaseous fission products a total non-local fission explosion yield of 160.5 Mt, obtained from measured ^{90}Sr deposition, was assumed in deriving the total amounts released.
†For simplicity, all ^{14}C is assumed to be due to fusion.

mining and milling through nuclear waste disposal; the significant point is the recognition that the mining operation for uranium is the major contributor in the dose to the global public by quite a large fraction.

An estimate of the total radioactivity, by nuclide, released into the atmosphere by above-ground nuclear tests is given in Table 22.2; the total release is ~2 600 EBq. To provide comparison, the estimated global releases from reactors and reprocessing plants are listed in Table 22.3; globally this gives ~3.9 EBq from reactors and 3.4 EBq from reprocessing plants up to 1998. The major releases are the noble gases, ^{3}H and ^{14}C. However, the total ^{14}C release over 30 years is less than one percent of the normal ^{14}C level from cosmic ray production (§ 13.4.3). This is a small fraction of the radioactivity released in nuclear weapons testing. To further place the releases in proper relation, an estimate of the total atmospheric release in the Chernobyl accident was 2 EBq (Table 22.4), in the Three Mile Island Reactor (TMI) accident, 10^{-6} EBq, and from the Fukushima accident ~0.9 EBq. From such figures the United Nations Scientific Committee on the Effects of Atomic Radiation (UNSCEAR) estimates that the average annual dose of radiation per person during year 2000 corresponded to:

Natural background	2.4 mSv
Medical diagnostics	0.4 mSv
Nuclear weapons tests	0.005 mSv
Chernobyl accident	0.002 mSv
Nuclear fuel cycle	0.0002 mSv

Table 22.3 Global release of radionuclides by reactors and reprocessing plants up to 1998 (UNSCEAR2000)

Radionuclides	Reactor releases (PBq)	Reprocessing releases (PBq)
Noble gases	3631	3190
^{3}H	269	144
^{14}C	1.97	0.44
^{90}Sr	—	6.6
^{106}Ru	—	19
^{129}I	—	0.014
^{131}I	0.046	0.004
^{137}Cs	—	40
Particulates	0.121	—
Others	0.839	—
Total	~3900	~3400

Table 22.4 Radionuclides released into the atmosphere in the Chernobyl accident and local contamination

		Release		Core	Contamination	
Nuclide	Half-life	Tot Ebq	%	inventory tot EBq	Air (Bq m ^{3}) Stockholm †	Ground (kBq m ^{2}) Gävle area ‡
^{85}Kr	10.73 y	0.033	100	0.033		
^{89}Sr	50.5 d	0.094	4.0	2.4		
^{90}Sr	28.6 y	0.0081	4.0	0.20		
^{95}Zr	64.0 d	0.16	3.2	5.0	0.6	5.9
^{103}Ru	39.4 d	0.14	2.9	4.8	6.4	14.6
^{106}Ru	368 d	0.059	2.9	2.0	1.8	4.0
^{131}I	8.04 d	0.67	20	3.3	15	179
^{133}Xe	5.24 d	1.7	≈100	1.7		
^{134}Cs	2.07 y	0.019	10	0.19	2.4	14.4
^{136}Cs	13.2 d				0.7	6.0
^{137}Cs	30.2 y	0.037	13	0.28	4.5	24.7
^{140}Ba	12.8 d	0.28	5.6	5.0	25.6	2.1
^{141}Ce	32.5 d	0.13	2.3	5.6	0.5	5.9
^{144}Ce	284 d	0.088	2.8	3.1	0.3	3.7
^{239}Np	2.36 d	0.97	3.2	30		2.7
^{238}Pu	87.7 y	3.0×10^{-5}	3	0.001		
^{239}Pu	24100 y	2.6×10^{-5}	3	0.0009		
^{240}Pu	6570 y	3.7×10^{-5}	3	0.0012		
^{241}Pu	14.4 y			0.170		

†First days, April 28-29, 1986.
‡Ullbolsta, outside of Gävle, among highest depositions outside Russia; corrected to April 28, 1986. References: USSR State Comm. on the Utilization of Atomic Energy 1986. The accident at the Chernobyl nuclear power plant and its consequences, IAEA expert meeting 25-29 Aug. 1986, Vienna.

22.2. RADIONUCLIDES OF ENVIRONMENTAL CONCERN

Most of the radionuclides produced in nuclear tests, accidents and in the normal fuel cycle are short lived. In Table 21.3, longer lived fission products and activation products from these systems are listed as these represent the major concern to the general public if they are allowed to enter the environment; we exclude nuclides with insignificant contribution to the total activity after 10 years. In addition, several heavy radionuclides are formed by neutron capture reactions such as ^{236}U, ^{237}Np, $^{238-242}$Pu, ^{241}Am, etc.

To assess the potential for these radionuclides to cause harm to humans, their geochemical and biological behavior must be evaluated. For example, since Kr is a chemically inert gaseous element, it would have little effect on a person who inhaled and immediately exhaled the small amount which might be present in the air. By contrast, other nuclides with high activity, ^{90}Sr–^{90}Y and ^{137}Cs–^{137m}Ba, have active geological and biological behavior and can present much more significant radiation concerns to humans. In normal operations, these nuclides would of course be released in quite insignificant amounts, as represented by the value for aquatic releases in reactor operation and reprocessing, Table 19.7 and 21.10. Similarly, the heavy elements (Np, Pu, Am) would not be released in normal operation at levels that would be of concern. We have pointed out that some low level releases as a result of normal operations are allowed by the health authorities, who also monitor these levels. In the effluents from reprocessing plants (e.g. Sellafield in the United Kingdom and La Hague in France), the relatively long-lived nuclides such as ^{3}H, ^{14}C, ^{85}Kr, ^{99}Tc and ^{129}I are of major concern. The liquid effluents from nuclear power plants and from reprocessing plants are about equally responsible for the global collective dose commitment of nuclear power generation (i.e., ~0.8 man Sv per GW_ey of the total ~2.5 man Sv).

22.3. RELEASES FROM LARGE REACTOR ACCIDENTS

So far there have only been two major nuclear reactor accidents, Chernobyl and Fukushima. In both cases large amounts of radioactive material was released into the environment. However, in these events there was only an insignificant addition of ^{14}C to the atmosphere relative to the average atmospheric amount of ^{14}C at the corresponding times.

22.3.1. Chernobyl

On April 26, 1986, a low power engineering test was being conducted at one of the reactors of the Chernobyl nuclear power station in the Ukraine (then the USSR). The reactor became unstable, resulting in thermal explosions and fires that caused

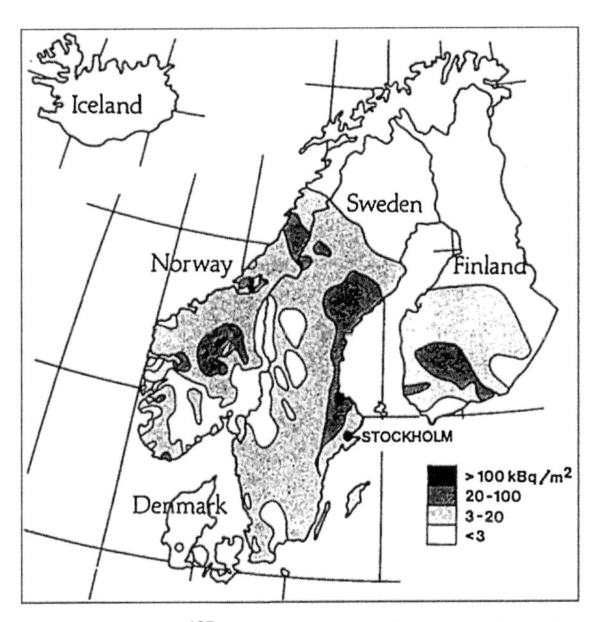

Figure 22.3 Deposition of ^{137}Cs in Scandinavia from the Chernobyl accident.

severe damage to the reactor and its building (§§13.10.2 and 20.1.2.c). Radioactivity was released over the next ten days until the fires were extinguished and the reactor entombed in concrete. The radioactivity was released as gas and dust particles and initially blown by winds in a northerly direction. Outside Russia, the accident was first detected by increased radioactivity levels at the Forsmark nuclear power plant, about 110 km north of Stockholm, Sweden, where it caused a full alarm as the radioactivity was believed to come from the Swedish plant. Subsequently, the radioactivity released at Chernobyl was spread more to the west and southwest (Figures 13.10, 22.3, and 22.4).

For the exposed population in the Byelorussia region near Chernobyl the estimated average increased dose in the first year after the accident was approximately the same as the annual background radiation. In northern and eastern Europe in general, the increased exposure during the first year was 25–75% above background levels. The highest dose will be delivered in southeastern Europe and is estimated to be 1.2 mSv up to year 2020, which can be compared to ~70 mSv from natural background radiation during the same period. Figure 22.3 shows the levels of ^{137}Cs deposited in Scandinavia in the first days after the accident. Table 22.4 gives the fraction of the core activity released and the air and ground contamination of various nuclides at two Swedish locations. Figure 22.4 shows the fallout in part of the northern hemisphere.

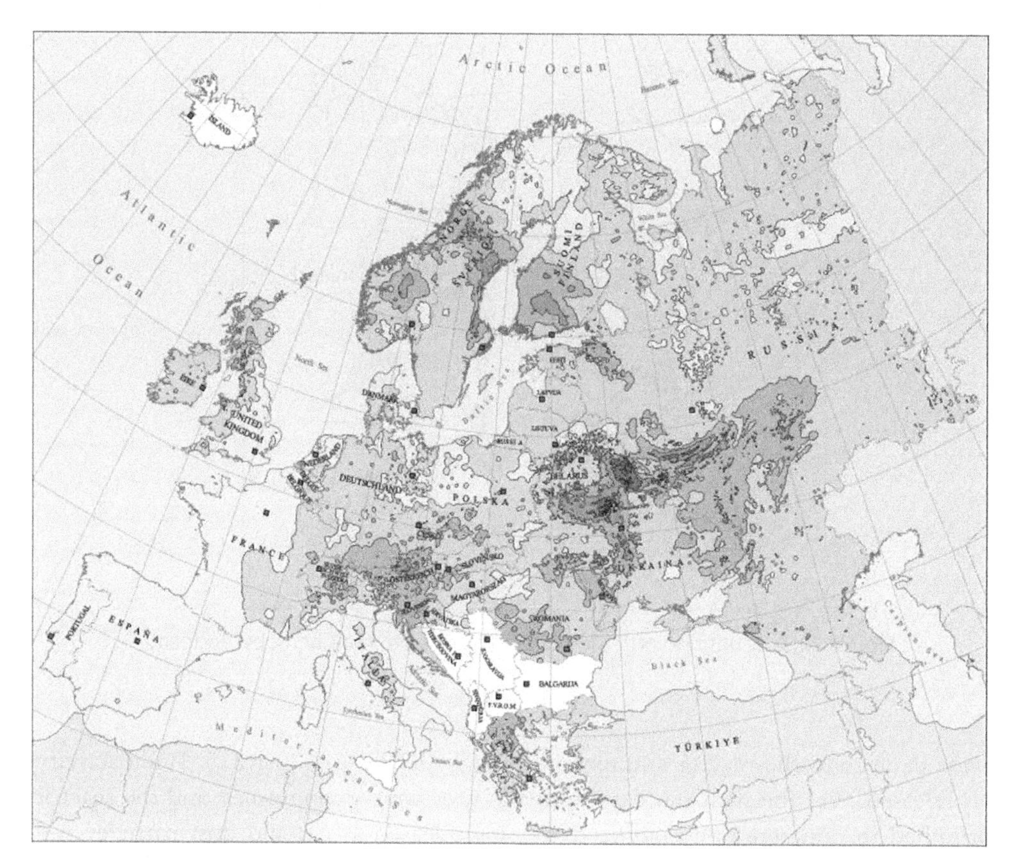

Figure 22.4 Deposition from Chernobyl in much of the northern hemisphere.

It was shown that the larger airborne particulates from the Chernobyl accident had a composition which was quite similar to that of the reactor fuel. A comparison of the composition of these "hot particles" with that of the reactor fuel is given in Table 22.5. About a tenth of these hot particles had a high concentration of ^{103}Ru and ^{106}Ru while others were depleted in the ruthenium fission products. The Ru-rich particles may have originated from a part of the reactor where burning graphite produced CO which reduced the ruthenium to non-volatile metallic Ru. In other sections, oxidation occurred, forming volatile RuO_3 and/or RuO_4 which vaporized from the particles.

In contrast to the larger particles, the composition of the smaller ones varied considerably and they were distributed over much greater distances. Other measurements reflect this variability in the fall-out from Chernobyl. For example, 70% of the ^{137}Cs from Chernobyl measured in Great Britain was water soluble. By contrast, the ^{137}Cs measured in Prague, much closer to the accident site, was only 30% water

Table 22.5 Nuclide composition (%) of a "hot particle" from Chernobyl compared to reactor fuel after 3 years of burning

Nuclide	"Hot Particle"	Reactor fuel
^{95}Zr	17.9	17.6
^{95}Nb	20.7	19.3
^{103}Ru	14.2	15.0
$^{106}Ru(Rh)$	3.5	2.4
$^{140}Ba(La)$	12.6	11.1
^{141}Ce	15.8	14.3
$^{144}Ce(Pr)$	15.8	14.4

soluble. Further insight into the variety of species present in the Chernobyl dust is found in data on the deposition of some radionuclides during periods of rain, and dry weather. Rainy periods accounted for 70–80% of the total deposition of ^{134}Cs, ^{137}Cs, ^{103}Ru, ^{106}Ru, and ^{132}Te while deposition during dry weather was more important for ^{131}I.

These observations indicate that the speciation of radionuclides in the atmosphere is dependent on their source, their mechanisms of production and the nature of the particular environment. While some species are gaseous, others are associated to particles with properties and suspension times that are strongly dependent on the particle size and density.

22.3.2. Fukushima

On the 11th of March 2011 an earthquake of the magnitude 9.0 on the Richter scale occurred outside the eastern coast of Japan. All affected nuclear reactors shut down as planned as a response to the earthquake. However, an unusually large tsunami hit the Fukushima Daiichi nuclear power station afterwards causing a total blackout. As the fuel heated up the zircaloy cladding underwent a highly exothermic oxidation reaction according to eqn. (19.44).

$$Zr + 2H_2O \rightarrow ZrO_2 + 2H_2 \quad \Delta H = -6.67 + 2.57 \times 10^{-4}\,T\,\left(MJ/kg\,Zr\right) \quad (19.44)$$

The hydrogen produced escaped up into the secondary reactor building where it accumulated and later exploded. Such explosions occurred in two reactors, but at different times. This nuclear accident and its release was more fully described earlier, see §13.10.2.

In comparison, the total releases of ~900 PBq total into the air from Fukushima are modest when compared to the ~5200 PBq total from Chernobyl. The fallout in Japan resulted in evacuation of many of the most afflicted areas. The effect of rain on the fallout was rather similar in the two cases.

22.4. INJECTION OF TRU INTO THE ENVIRONMENT

Of the artificial radionuclides released to the environment by nuclear activities, the transuranium (TRU) species are a major concern. This concern arises from the very long half-life of a number of the nuclides as well as their high radiotoxicity values. Although reactor operation and spent fuel reprocessing activities have released small amounts of TRU's to the environment, testing of nuclear weapons has released rather large quantities. Since the first nuclear test detonation in New Mexico in 1945, approximately 3 500 kg of plutonium has been released in atmospheric explosions and another 100 kg in underground tests. This corresponds to about 11 PBq of $^{239+240}$Pu ejected into the atmosphere. In addition 0.6 PBq of ^{238}Pu were released over the south Pacific in the high altitude destruction of the SNAP-9 satellite power source in 1964. By contrast, a total of ~0.58 PBq of $^{239+240}$Pu has been released into the Irish Sea from the Sellafield (UK) reprocessing plants between 1971 and 1999; most of this before 1985. About 37 kg of ^{241}Am is present in the environment from the decay of ^{241}Pu from the nuclear testing. Table 22.6 compares the TRU's released in nuclear tests in the atmosphere, in the Irish Sea from the Sellafield plants, and into the English Channel from the LaHague plants. As the amount of spent nuclear fuel increases, the contribution to the total plutonium in the environment could become more significant over a longer time, especially if nuclear waste disposal sites release actinide elements slowly to the environment. Whatever the sources of plutonium and other actinides, their presence represents a contamination of the environment by highly toxic material. An understanding of the factors involved in their retention and/or migration in the ecosphere is therefore highly desirable. Studies of the environmental behavior of releases from tests provide data needed to understand and predict the behavior of smaller releases from the nuclear power industry. Therefore, in the next few paragraphs we concentrate our discussion to the behavior of the actinide elements in the environment, particulary the mobility of plutonium.

Table 22.6 TRU released from nuclear tests and from reprocessing (UNSCEAR2000, BNFL and COGEMA)

	Tests (TBq)	Sellafield (TBq)		LaHague (TBq)	
Nuclide	**1945–1980**	**1971–84**	**1985–94**	**1995–99**	**2000**
^{237}Np	–	–	0	0.33	
^{239}Pu	6520 }	}	}		
^{240}Pu	4350 }	559 }	15 }	0.92	
^{241}Pu	142000	–	–	21.8	0.039†
^{241}Am	–	442	9	0.31	
^{242}Cm	–	–	–	0.052	
$^{243+244}$Cm	–	–	–	0.024	

†Total α.

The majority of the plutonium from weapons testing was injected initially into the stratosphere. The plutonium originally in the weapon which survived the explosion would have been formed into high-fired oxide which would be expected to remain insoluble as it returned to earth. Such insoluble particles would have sunk in a rather short time into the bottom sediments of lakes, rivers, and oceans or would become incorporated in soils below the surface layer. However, in most nuclear weapon explosions a considerable amount of plutonium is generated in the explosion via ^{238}U (n,γ) reactions and subsequent α-decay of the product ^{239}U, ^{240}U, ^{241}U, etc. In total, about two thirds of the plutonium released was generated in this way. The nuclides from the (n,γ) reactions would exist as single atoms, and, hence, were never formed into high-fired oxides. The plutonium from this formation path would have been soluble and, as a result, more reactive and its behavior would be more similar to that of plutonium released from nuclear reactors, reprocessing plants and from nuclear waste repository sites.

22.5. PRESENT LEVELS OF TRU IN THE ECOSPHERE

The United Nations Scientific Committee on the Effects of Atomic Radiation (UNSCEAR), in 2000 reports data that corresponds to the following total average global nuclide depositions from atmospheric nuclear weapon tests: ^{89}Sr 11, ^{90}Sr 1.2, ^{95}Zr 19, ^{106}Ru 12, ^{137}Cs 1.8, ^{239}Pu 0.013, ^{240}Pu 0.009, and ^{241}Pu 0.278 kBq m^{-2}. Some of these radionuclides have now decayed because of their short half-lives and the end of atmospheric testing in 1980 and almost all of the ^{241}Pu has decayed to ^{241}Am.

Near test sites, reprocessing facilities, etc., the concentration of plutonium in the soil and water is much higher than in more distant locations. Generally, the great majority of plutonium is associated with sub-surface soils or sediments or with suspended particulates in water. For example, when vegetation, animals, litter and soils are compared, $\geq$99% of the plutonium is present in the soil. Similarly, in shallow bodies of water, more than 96% of the plutonium is found associated with the sediments. However, it is via the species that are soluble or attached to suspended colloids and/or particulate matter in water that plutonium is transported in the environment. Analysis of vertical plutonium migration in soils near Chernobyl and in eastern Europe from the Chernobyl accident has shown that most of the plutonium is still in the first 0.5 cm from the surface for soils with significant humic acid content. In these soils, the plutonium is mostly associated with the insoluble calcium-humate fraction. In non-humic, carbonate rich soils, the plutonium has moved several centimeters downward. Migration rates of $\leq$0.1 cm y^{-1} is associated with the humic soils and of 1–10 cm y^{-1} with the carbonate rich ones. Presumably, migration is retarded by the interaction with the immobilized humic material in soils.

Table 22.7 Concentration of plutonium in filtered samples of natural waters

Water[†]	Concentration of Pu (M)
Lake Michigan	2.0×10^{-17}
Great Slave Lake, Canada	1.5×10^{-17}
Okefenokee River, Florida	1.5×10^{-16}
Hudson River, New York	1.0×10^{-17}
Irish Sea:	
1 km from Windscale	1.6×10^{-14}
110 km from Windscale	1.1×10^{-15}
Mediterranean	2.6×10^{-18}
North Pacific (surface)	3.0×10^{-17}
South Pacific (surface)	1.0×10^{-17}

[†]Samples were passed through 0.45 μm filters.

In subsurface oxic soil near Los Alamos National Laboratory, USA, plutonium is relatively mobile and has been transported primarily by colloids in the 25−450 μm size range. Moreover, the association with these colloids is strong and removal of Pu from them is very slow. By contrast, near Sellafield in wet anoxic soil, most of the Pu is quickly immobilized in the sediments although a small fraction remain mobile. Differences in oxidation state (Pu(V) vs. Pu(IV)) as well as in humic content of the soils may explain these differences in mobility.

Table 22.7 lists the concentration of plutonium, after filtration (0.45 μm), in the surface layers of some natural waters. The higher concentration in the Okefenokee River is assumed to reflect the effect of complexing by humic materials. This agrees with the observation that adding humic material to seawater samples containing plutonium increases the solubility by more than a factor of five over a period of one month.

It is difficult to obtain reliable values of plutonium concentration in natural aquatic systems as it is very low, approximately 0.001 dpm per liter sea water. Moreover, the plutonium associated with suspended particles may be more than an order of magnitude greater than that in true solution. In tests of water from the Mediterranean Sea, filtration (0.45 μm) reduced the concentration of plutonium by a factor of 25. In laboratory tests with filtered seawater to which plutonium was added, after one month the total concentration of Pu was 1.3×10^{-11} M, but only 40% (5×10^{-12} M) was in solution as ionic species and the other 60% was probably in colloidal form. The mean residence time of Pu in the water column is proportional to the concentration of particulate matter. As a consequence, >90% of the Pu is rapidly removed from coastal waters whereas, in mid-ocean waters where the particulate concentrations are lower, the residence time for Pu is much longer. It also means that Pu is accumulated in clays and silt near reprocessing plant effluent release points. Known cases are near Sellafield, and in the river downstream from the old norvegian reprocessing plant at Lilleström.

22.6. ACTINIDE CHEMISTRY IN THE ECOSPHERE

22.6.1. Redox properties

Before proceeding to more detailed discussion of the behavior of actinides in the environment, it is useful to review some of their chemical properties. A general discussion of the actinide solution chemistry is given in §14.3; here, the focus is on their behavior in aqueous solutions and primarily in solutions of pH 5–9 which is the pH range of natural waters (e.g. the oceans have pH = 8.2). The actinide ions have an unusually broad range of oxidation states in aqueous solution, from II to VII; The II and VII States are not discussed further as they do not form in ecosystems. Following the normal pattern for polyvalent cations, lower oxidation states are stabilized by more acidic conditions while higher oxidation states are more stable in basic solutions. Of course, this generalization can be negated by other factors, such as complexing, which may cause a reversal of the relative stability of different oxidation states. The greater strength of complexing of An(IV) cations relative to that of An(III) can significantly increase the apparent redox stability of the An(IV) species compared to An(III). The greater tendency to hydrolysis of Pu(IV) causes Pu(III), which is stable in acid solution, to be oxidized to Pu(IV) in neutral media. The disproportionation of Pu(V) is discussed in §14.3.3, where it is pointed out that in the higher pH and very low concentrations of Pu in natural waters, disproportionation is not a factor in the redox behavior of plutonium.

The actinide elements in a particular oxidation state (e.g. Th(IV), U(IV), Pu(IV), Np(IV), Am(IV)) have similar behavior. However, their redox behavior is quite different, as mentioned in Chapter 14. The pH affects this redox behavior significantly as reflected in Figure 22.5 which compares the redox potentials of U, Np and Pu at pH = 0, 8 and 14.

Am (III) is the most stable oxidation state in aqueous solutions while Pu(III) and Np(III) are present under reducing conditions (e.g. anoxic waters). Th(IV) is the common and stable state for that element. U(IV) and Np(IV) do not react with water but are oxidized by O_2 in oxic systems. Pu(IV) is stable at low concentrations in acidic solution, but $Pu(OH)_4$ has a very low solubility product.

NpO_2^+ is stable except at high acidities and high concentrations under which conditions it disproportionates. UO_2^+ and PuO_2^+ increase in stability as the pH is increased. U, Np and Pu form AnO_2^{2+} ions in solution with the stability decreasing in the order U > Pu > Np. UO_2^{2+} is the most stable uranium species in natural waters.

Redox properties are often described by the aid of potential-pH (or Pourbaix) diagrams, Figure 22.6. The shaded area represents typical groundwaters in granitic rock, containing iron minerals; such waters are usually reducing, i.e. have *Eh*-values below 0. Natural groundwaters (including oceans, lakes, rivers, etc) fall within the area enclosed by the dashed curve; they can have rather high *Eh* values due to atmospheric oxygen, and also be rather alkaline in contact with carbonate rocks. The sloping lines follow the Nernst equation (defined by eqn. (18.6)). We discuss this Figure further in §22.7.1.

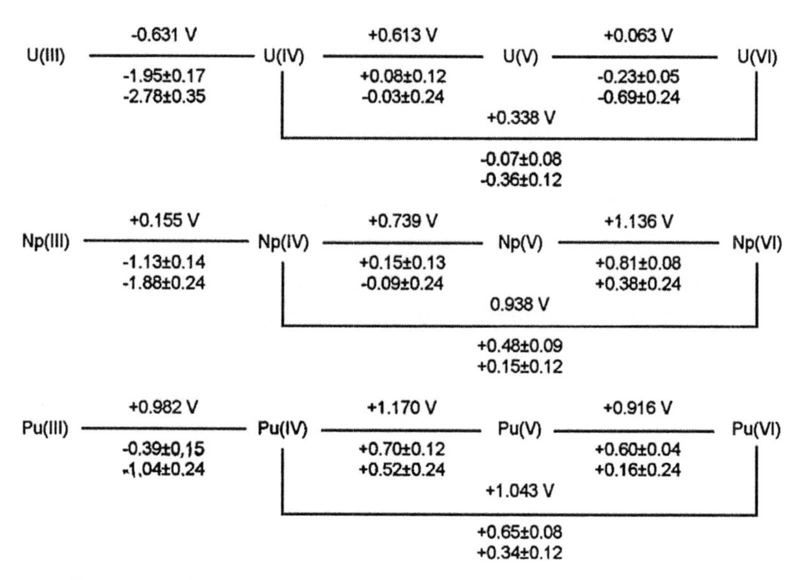

Figure 22.5 Redox potential diagrams of U, Np and Pu; the reduction potentials as listed are for pH values: pH = 0; pH = 8; pH = 14.

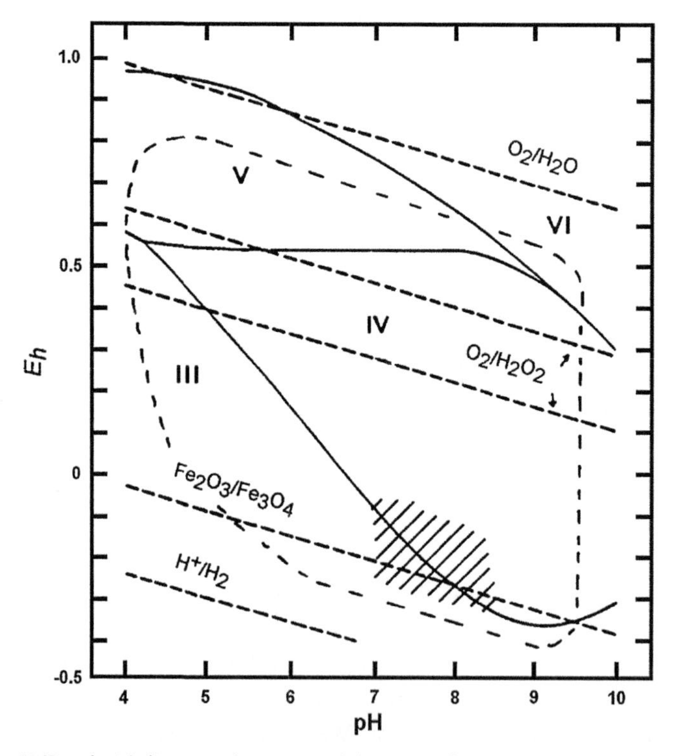

Figure 22.6 Eh-pH (Pourbaix) diagram showing stability areas for Pu(III), Pu(IV), Pu(V) and Pu(VI); at the division line there is an equal concentration of the two oxidation states.

22.6.2. Hydrolysis

Hydrolysis is an important factor in actinide behavior in natural waters as the pH is high enough to result in such reactions as:

$$An^{+n} + mH_2O = An(OH)_m^{+(n-m)} + mH^+ \tag{22.1}$$

Since hydrolysis is higly ionic in its nature the main governing factor for the strength of the hydroxide complex is the charge/size ratio. The larger this ratio the stronger the hydrolysis. Therefore, the order of increasing pH for onset of hydrolysis follows the sequence (indeed, in principle all complexation with hard ligands will follow this trend):

$$An^{4+} > AnO_2^{2+} > An^{3+} > AnO_2^{+} \tag{22.2}$$

The variation of the concentration of the free (non- hydrolyzed) cations with pH is shown for the oxidation states of III to VI of Pu in Figure 22.7. These curves are based on estimated values of the hydrolysis constants, but are of sufficient accuracy to indicate the pH values at which hydrolysis becomes significant (e.g. ~6–8 for Pu^{3+}, ≤ 0 for Pu^{4+}, 9–10 for PuO_2^+ and 4–5 for PuO_2^{2+}.

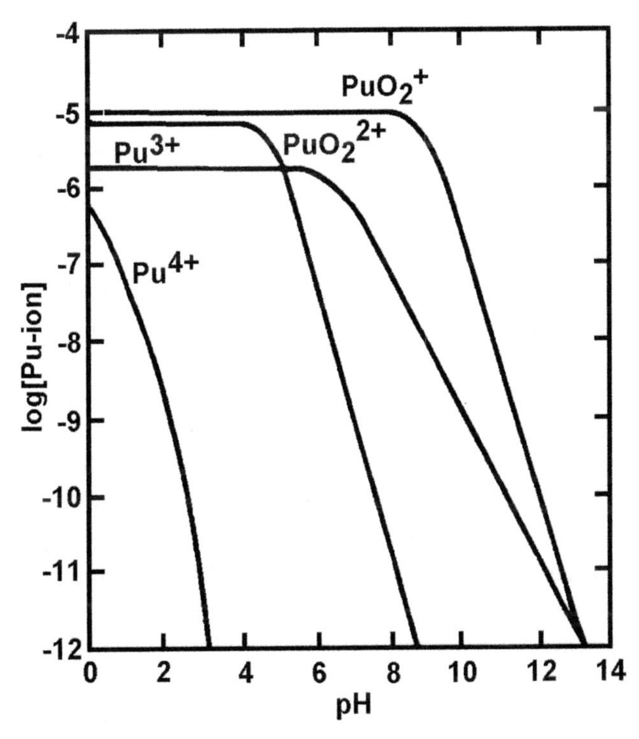

Figure 22.7 Concentration of free plutonium ions at different oxidation states in solutions of different pH, showing the effect of hydrolysis.

The study of plutonium hydrolysis is complicated by the formation of oligomers and polymers once the simple mononuclear hydrolytic species start forming. The relative mono/oligomer concentrations are dependent on the plutonium concentration: e.g. the ratio of Pu present as $(PuO_2)_2(OH)_2^{2+}$ to that as $PuO_2(OH)^+$ is 200 for total $[Pu]_T$ 0.1 M, decreases to 5.6 for 10^{-4} M and is only 0.05 for 10^{-8} M. The hydrolysis of Pu^{4+} can result in the formation of polymers which are very difficult to convert back to simpler species. Generally, such polymerization requires $[Pu]_T > 10^{-6}$ M. However, the irreversibility of polymer formation prevents the destruction of the polymers by dilution of more concentrated hydrolysed solutions to concentrations below 10^{-6} M. Soon after formation, such polymers in solution can be decomposed to simpler species by acidification or by oxidation to Pu(VI). However, as the polymers age, the depolymerization process requires an increasingly vigorous treatment. A reasonable model of the aging involves initial formation of aggregates with hydroxo bridging with conversion over time to structures with oxygen bridging:

The relative percentage of oxygen bridges presumably determines the relative inertness of the polymers. The polymers apparently increase in aggregate size as the pH increases. At pH 4, the polymers are small enough that essentially all of the Pu remains suspended in solution after a week while at pH 5 less than 10% remains in solution and at pH 6, only 0.1% remains.

22.6.3. Solubilities

In marine and natural waters, the limiting solubility is usually associated with either the carbonate or the hydroxide depending on the oxidation state, pH and carbonate concentration. For example for Am^{3+}, the reported value of the solubility product ($\log K_{s0}$, eqn. 18.30) is −26.6 for crystalline $Am(OH)_3(c)$ at very low ionic strength and −22.6 for $Am(OH)(CO_3)(c)$. At pH 6, if $[CO_3^{2-}]$ free $>10^{-12}$ M and at pH 8, if $[CO_3^{2-}]$ free $>10^{-8}$ M , the solubility of Am^{+3} would be expected to be limited by the formation of $Am(OH)(CO_3)$.

Plutonium solubility in marine and natural waters is limited by the formation of $Pu(OH)_4$(am) (for amorphous) or PuO_2(c) (for crystalline). The K_{s0} of these species is difficult to measure, in part due to the problems of the polymer formation. A measured value for $Pu(OH)_4$(am) is $\log K_{s0} = -56$. This value puts a limit on the amount of plutonium present, even if Pu(V) or Pu(IV) are the more stable states in the solution phase. Moreover, hydrolyzed Pu(IV) sorbs on colloidal and suspended material, both inorganic and biological.

The strong preference for neptunium to form the NpO_2^+, which has relatively weak complexing and hydrolysis tendency, lead to solubilities as large as 10^{-4} M under many geochemical conditions. However, the reducing environment found at large depths in some granites would make Np(IV) the dominant oxidation state. As with plutonium, the solubility of neptunium in all oxidation states seems to be limited by the low solubility of $Np(OH)_4$ (am) or NpO_2(c).

In oxic waters, uranium is present as U(VI) and strongly complexes with carbonate; e.g. in sea water, the uranium is present at 10^{-8} M concentration as $UO_2(CO_3)_3^{4-}$. The solubility of uranium in some waters may be limited by the formation of an uranyl silicate species.

22.7. SPECIATION CALCULATIONS

An essential step in the safety analysis of potential waste repositories is the prediction of what chemical species are formed in the actual water. For example, the relatively high solubility of uranium in sea water is due to this strong carbonate complexation which forms $UO_2(CO_3)_3^{4-}$. Figure 22.8 shows the variation of uranyl species in a surface water under normal atmospheric pressure of CO_2 ($p_{CO_2} \sim 3.2 \times 10^{-4}$; $\log[CO_3^{2-}] = 2pH - 18.1 + \log(p_{CO_2})$). These speciation diagrams are calculated from the equilibrium constant for formation of each species plus mass balance equations. In this section we describe the use of equilibrium constants in modeling the speciation in a natural water.

Assume the reactions $M + nX = MX_n$ where X is OH^-, and CO_3^{2-} and $n = 1$ to 3. The equilibrium constants (assuming for simplicity that the chemical activity coefficients all are unity), expressed as β are given by (cf. eqn. (18.31)):

$$\beta_n = [MX_n]/[M]\,[X]^n \tag{22.3}$$

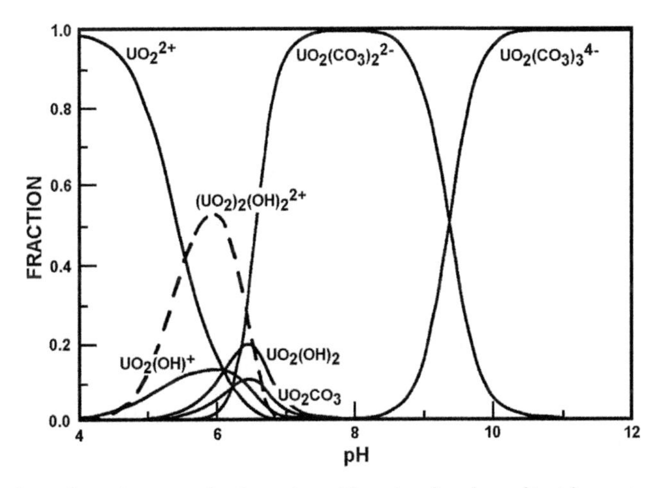

Figure 22.8 Fraction of uranium species in water with natural carbon dioxide content and at different pH, showing hydrolysis and CO_3^{2-} complexation.

The equilibrium constants for Am(III) are listed in Table 22.9 for hydrolysis and CO_3^{2-} complexation. Let us consider the case of Am(III) should it be released into the environment.

The first step in modeling the speciation of Am(III) is to rearrange the above equation to express the ratio of complexed to free metal, e.g. (omitting ionic charges)

$$[AmX]/[Am] = \beta_1[X] \tag{22.4}$$

$$[AmX_2]/[Am] = \beta_2[X]^2 \tag{22.5}$$

and so forth. The mass balance equation is:

$$[Am]_T = [Am] + [AmX] + [AmX_2] + \ldots \tag{22.6}$$

where $[Am]_T$ is the total analytical concentration of americium. Dividing by [Am] gives:

$$[Am]_T/[Am] = [Am]/[Am] + [AmX]/[Am] + [AmX_2]/[Am] + \ldots \tag{22.7}$$

$$[Am]_T/[Am] = 1 + \beta_1[X] + \beta_2[X]^2 + \ldots \tag{22.8}$$

and

$$\alpha_n = \beta_n[X]_n([Am]/[Am]_T) \tag{22.9}$$

where α_n is the fraction of all americium in the form of AmX_n. With these equations, β_n's, and a given value of $[Am]_T$, we can calculate the concentration of each species for any value of [X]. In Figure 22.9 the speciation of Am(III) as hydroxide and carbonate complexes is shown as a function of pH. The concentration of CO_3^{2-} is based on the

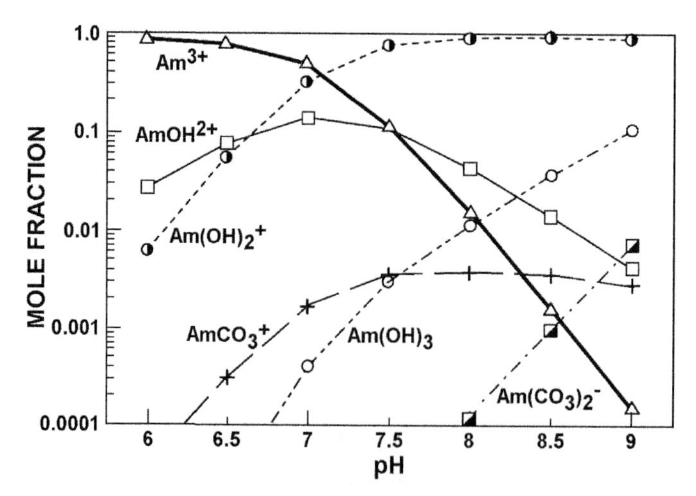

Figure 22.9 Results of speciation calculations for Am(III) in natural water.

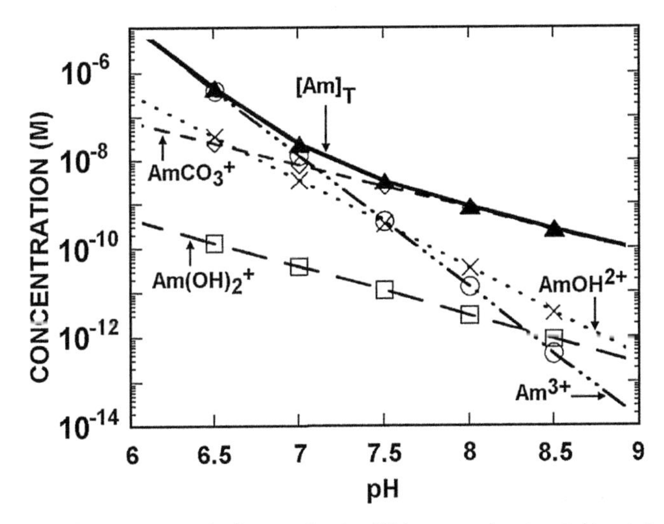

Figure 22.10 Results of speciation calculations for Am(III) in natural water, taking into account the low solubility of $Am(OH)_3$.

atmospheric partial pressure (3.2×10^{-4} atm) and the resultant concentration of CO_3^{2-} at different pH values in a natural water.

If the solubility is limited by a solid phase for which the K_{s0} is known, the actual value to be expected for each species can be calculated by assuming no over-saturation. These can be included in the mass balance equation to predict the total minimum solubility. If $\log K_{s0} = -28.9$ is assumed for $Am(OH)_3$, Figure 22.10 shows the results of such a calculation for the solubility of americium in the same system as in Figure 22.9, but now including the effect of the limited solubility of $Am(OH)_3$.

We have stated that Pu(V), as PuO_2^+, is the dominant dissolved species while $Pu(OH)_4$ is the solubility limiting precipitate. The redox reaction can be written as:

$$PuO_2^+ + 4H^+ + e^- = Pu^{4+} + 2H_2O \tag{22.10}$$

The E^0 value for the IV/V pair is 1.17 V. Using the equilibrium expression with the Nernst equation, assuming the redox potential in a fresh water lake to be 0.4 V, we obtain at pH 7:

$$[PuO_2^+]/[Pu^{4+}] \approx 10^{15} \tag{22.11}$$

From the value of $\log K_{s0} = -56$ for $Pu(OH)_4$(am) we can calculate that $[Pu^{4+}]$ in a solution of pH = 7 in contact with solid $Pu(OH)_4$ is 10^{-28} M. This gives us a value of $10^{15} \times 10^{-28} \approx 10^{-13}$ M for the expected concentration of PuO_2^+ in this solution.

A number of geochemical modeling codes have been developed which use such speciation and solubility equilibrium equations to calculate the concentration of different species of a metal ion as well as its net solubility in various waters (common codes are PHREEQC and EQ 3/6). The results from such calculations are as good as the equilibrium constants or the thermodynamic values used in the calculations. However, most important, the calculations must include the equilibrium equations for all species present which may contribute significantly to the solution phase concentrations and all solids which can provide the limiting solubility to the solution species. Furthermore, the degree of oversaturation with regard to each solid phase must be prescribed in order to calculate a realistic speciation. These modeling codes are presently based on the assumption that the natural systems are all at equilibrium whereas in nature this may not be true. Many systems are kinetically controlled and are often in a steady state, but not in true equilibrium. In these cases, perhaps the majority of the systems, the equilibrium modelling codes cannot accurately describe the actual conditions, but may provide a set of limiting species, approximate relative concentrations and baseline net solubilities. A further complication arises in assessing the role of colloids, and of sorption which may reduce the concentration of soluble species below that estimated for the least soluble solid phase. On the other hand, sorption on suspended colloids may also increase the total concentration (dissolved plus amount in colloids, e.g. Pu in sea water §22.5). In general, the equilibrium code calculations can easily give lower limit values of maximum solubilities by assuming no oversaturation. Such calculations are valuable in waste management risk assessment since, if the lower limit solubilities from the equilibrium calculations fall well below the accepted safety limits even when assuming reasonable degrees of oversaturation, it is very likely that the actual total concentrations will also be below the acceptable limits, cf. Fig. 22.11, Table 22.10 and §22.10.

22.7.1. Calculated species in solution

The diversity of reactions which actinides can undergo in natural waters is presented schematically in Figure 22.11. Complexation by anions such as hydroxide, carbonate, phosphate, humates, etc. determine the species in solution. Sorption to colloids and suspended material increases the actinide concentration in the water while precipitation of hydroxides, phosphates, carbonates, and/or sorption to mineral and biological material limit the amount in the solution phase.

In natural oxic waters, americium is present in the trivalent and thorium in the tetravalent state while uranium is hexavalent, UO_2^{2+}. The total concentrations of uranium and thorium in surface sea water are $1.1-1.5 \times 10^{-8}$ and $\sim 2.5 \times 10^{-12}$ M, respectively in both the Atlantic and Pacific oceans. The amount of uranium associated with particulate matter in the water is small. By contrast, for tetravalent ^{232}Th (originating from thorium minerals), about 50% is bound in aluminosilicate particles

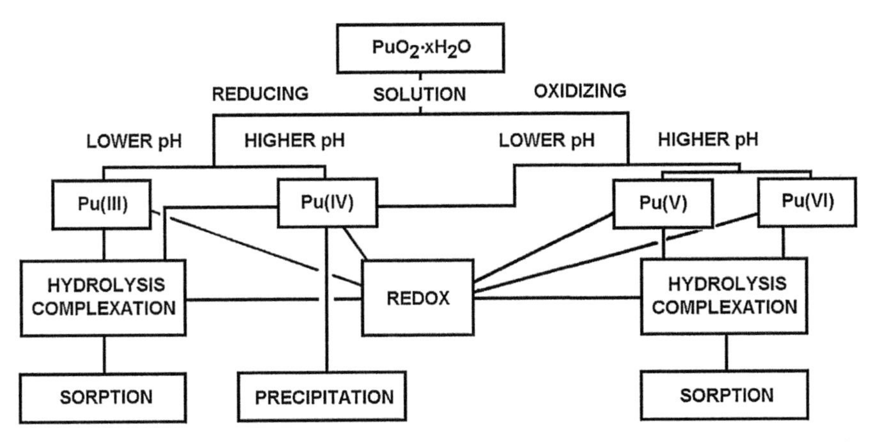

Figure 22.11 Speciation diagram for the range of reactions to be considered in studying the environmental behavior of plutonium.

and 50% is dissolved (e.g. passes through a 1 μm filter). For ^{228}Th and ^{230}Th (radiogenic decay products), about 90% is found in solution. Sorption of ^{234}Th tracer is a reversible process, possibly due to an organic material coating on the surface with increased sorption as the particles age. In such studies, solubility is defined as passage through a 0.45 μm filter. Speciation calculations show typically that thorium occurs as a hydrolyzed species whereas uranium is present in surface waters, at low in Ca^{2+}, normally as the $UO_2(CO_3)_2^{2-}$ and $UO_2(CO_3)_3^{4-}$ species. In neutral waters of very low carbonate content, speciation calculations indicate that uranyl cations hydrolyze to form oligomers for total uranyl concentrations as low as 10^{-7} M.

Neptunium, in oxic waters, is present in the pentavalent state. The hydrated cation is calculated to be the dominant species for pH≤8 unless the free carbonate concentration exceeds approximately 10^{-4} M in which case $NpO_2(CO_3)_n^{1-2n}$ is more common.

The rather similar values for the plutonium reduction half reactions at pH 8 indicate that plutonium may exist in oxic waters in more than one oxidation state. The reduction potential at pH 8 of the Pu(III)/Pu(IV) couple indicates that Pu(III) is unlikely to exist in oxic waters in the absence of a reductant, but may be present in anoxic waters. Each oxidation state of plutonium differs in chemical behavior from that of the other states so modelling the geochemical behavior of plutonium must include the correct oxidation state, or states, of plutonium which are likely to be present in a particular system.

In principle, the calculation of the oxidation states of plutonium requires knowledge of the redox potential, *E*h, of the aqueous phase. However, the *E*h measured with a certain type of electrode may not be the potential for the particular redox couple with which the plutonium reacts. One of the reasons for this is that the

Table 22.8 Plutonium oxidation states (in % of total Pu) in natural waters

State	Pu(III+IV)	Pu(V)	Pu(VI)
Rain	34	66	
Mediterranean	42	58	
Irish Sea	23	77	0
Pacific (I)	39	52	9
Pacific (II)	40	46	14
Lake Michigan	13	87	0

Table 22.9 Equilibrium Constants for Am(III)

Species	$\log\beta_n$	Species	$\log\beta_n$
$Am(OH)^{2+}$	6.44	$Am(CO_3)^+$	5.08
$Am(OH)_2^+$	13.80	$Am(CO_3)_2^-$	9.27
$Am(OH)_3$	17.86	$Am(CO_3)_3^{3-}$	12.12

Eh-electrode usually catalyses the reaction rate for its specific redox couple. For example, in surface sea water, the measured *E*h is about 0.8 V and is due to the O_2/H_2O couple. In the log *E*h versus pH diagram, Figure 22.6, the area of existence of plutonium in different oxidation states, including the effects of hydrolysis and carbonate complexation, is marked by roman numerals. From that diagram, it can be seen that Pu(VI) would be the predominant state in solution in the ocean. In fact, the predominant species is Pu(V). Table 22.8 summarizes the reported oxidation state distribution in several natural systems.

Like NpO_2^+, PuO_2^+ has a low tendency to hydrolysis and complexation and is much less likely to be sorbed to solid surfaces and on colloidal particles than the Pu species in other oxidation states. As a consequence, plutonium can be expected to migrate most rapidly if it is in the pentavalent oxidation state. The total solubility is limited by the formation of the highly insoluble $Pu(OH)_4$. The sorption of hydrolyzed Pu(IV) in neutral water on mineral and organic-coated surfaces is accountable for the very low concentrations of dissolved Pu even in the absence of $Pu(OH)_4$(am) or PuO_2(c). Desorption is accomplished only by strong complexing and/or redox reagents. For example, citrate extracts little plutonium from soil, but a combination of citrate and the redox agent dithionite provides good extraction. The intractable nature of $Pu(OH)_4$ and its strong tendency to sorb on surfaces is a dominant and often controlling feature of plutonium geochemistry.

Silicates and humic substances present in natural waters form colloids and pseudocolloids with which actinides can react. The pseudocolloids formed by humic substances in ground waters have been shown to be efficient scavengers of americium and plutonium.

22.8. NATURAL ANALOGUES

The geochemical modeling calculations based on measured equilibrium data are of primary importance in the safety assessments of proposed nuclear waste repositories (and should be important for other waste repositories as well). Another useful tool in such assessments is the data from studies of natural analogue sites.

The modeling calculations use data from laboratory and field studies which have been obtained over the last few decades. These data are used in the codes to predict the solubilities, and nuclide migration of material which might be released from nuclear repositories over thousands and, even, hundreds of thousands of years. It is not possible to demonstrate rigorously that the models used are accurate as they may simplify the natural system, use incorrect data or misrepresent (possibly, ignore) important processes which occur over very long time periods. However, some validation of the models and data used in the modeling calculations can be obtained from careful comparison of calculated values with those measured at appropriate geologic sites, known as a natural analogues.

The natural analogue sites are areas in which uranium ores have been present for geologic time periods. In most cases, these sites have not been affected by human activities, so the record of geologic, long term effects are well preserved. A number of such sites are being studied around the world; studies from a few sites of different characteristics are reviewed here.

From studies of granitic sites, the dominant role of fractures and fissures in the transport of fluids has been convincingly demonstrated. Thus any model of water in a granitic site must include both advection which is dominant in the fractures and fissures, and diffusion which is important in regions of highly altered rocks ("alteration rims"). In clays, mass transport seems to proceed primarily by ionic and molecular diffusion although some fluid mass transport occurs at discontinuities in the formation.

Often, the mobilization of many elements (e.g. U, but not Th, in crystalline rocks) is correlated with the flow of oxidizing water. The mobilization and fixation of uranium involves complexation, redox and retention on minerals via adsorption, and ion exchange. In clay media, the redox potential is strongly buffered if significant amounts of organic substances are present.

In a section of the Pocos de Caldas (Brazil) formation, most of the thorium and the rare earths, and, to a lesser extent, the uranium, is associated with goethite (FeOOH) particles and transport by organic colloids is much less important. This region is reducing, as shown by the presence of Fe(II). However, in another region of this formation, the thorium and rare earths have a much higher mobility and are associated with organic (humic) colloids. At an analogue site in Scotland, the flow from the ore has passed into a peat bog in which the uranium is associated predominantly with humic material while the Th is found on Fe/Al oxyhydroxide colloids and particles. In many clay deposits, the organic material is most significant in maintaining a reducing potential which restricts actinide migration and provides a sorption source of the mobilized fraction.

22.9. THE OKLO REACTOR

Analysis of the Oklo natural reactors (§19.10) indicates that they must have lasted for 100 000 to 500 000 years with criticality occurring periodically. They probably would have consumed about twelve metric tons of ^{235}U, releasing a total energy of 2–3 GWy at a probable power level of ~10 kW. About 1.0–1.5 t of ^{239}Pu were formed by neutron capture in ^{238}U, but the half-life of ^{239}Pu has resulted in its total decay to ^{235}U. Since a few samples enriched in ^{235}U (presumably due to this decay) have been found, it is believed that in some places in Oklo breeding conditions may have temporarily existed. The decaying fission products are estimated to have had 10^{28} β,γ disintegrations over the operating time of the reactors. The average energy release in the reactor zone was ~50 W m^{-2} which is several times greater than that planned for geological nuclear waste repositories. As a consequence, it is estimated that the fluids in the inclusions of the mineral grains in the Oklo reactors had temperatures of 450 to 600°C, well above those anticipated in deep waste repositories. There is evidence of connectively driven circulation of the fluids for distances of 30 m from the main reactor zones as well as significant dissolution and modification of minerals due to the thermal and radiation conditions. Redistribution of some elements resulted from the convective flow of the hot liquids.

The uranium minerals appear to have remained relatively stable despite this heating process. The uranium and the lanthanide elements show evidence of some small degree of localized redistribution, but were mostly retained within the reactor zone. By contrast, fission product rare gases, halogens, molybdenum and the alkali and alkaline earth elements migrated significant distances from the reactor zones. In general, it seems that elemental redistribution took place over a period of 0.5 to million years while the area was thermally hot (during and after nuclear criticality). It has been estimated that as much as a total of 10^{12} liter of hot water flowed through the reactor zones. The water leaving the reactor zones had 5×10^{-3} g U m^{-3} and ~10^{-10} M concentrations of Tc, Ru, and Nd. The Tc and Ru were oxidized to TcO_4^- and RuO_4^-, and these soluble oxyanions moved with the water as it flowed from the area, creating significant (25–35%) deficiencies of these elements in the reactor zones. The Nd was less soluble and, apparently, migrated much less during this hot period. For Tc and Ru, the migration rate seems to have been ~10^{-5} m y^{-1} in water moving at a flow of 5 m y^{-1}.

A very important observation is the evidence in Oklo that the plutonium produced by the reactors did not move during its lifetime from the site of its formation.

In summary, essentially 100% of the Pu, 85–100% of the Nd, 75–90% of the Ru and 60–85% of the Tc were retained within the reactor zones. The migrating fission products were held within a few tens of meters of these zones. Thermodynamic calculations of the temperature dependent solubilities indicate that the loss of fissiogenic elements is diffusion controlled, whereas, retention in the surrounding rocks is due to temperature dependent deposition from an aqueous solution.

While the conditions at Oklo differ in a number of aspects from those expected in nuclear repository sites, they frequently were much less favorable to retention of the radionuclides. The lack of migration of the actinides and the much slower release of Tc agree with the predictions of laboratory studies and indicate their value in validating the safety of nuclear repositories.

22.10. PERFORMANCE ASSESSMENTS OF WASTE REPOSITORIES

In this Chapter we have presented some information on studies of the environmental behavior of some fission products and actinides. We have shown how laboratory data can be used in speciation calculations to predict solubilities, etc. The knowledge gained from studies of natural analog sites and, very importantly, from the Oklo natural reactor site, has been reviewed. All of these types of studies and data are of use in assessing the probable ability of nuclear waste repositories to protect the public over the time needed required for the radioactivity to decay. This time is largely dependent on used nuclear fuel handling as discussed previously and thsu the time required may vary between some thousends of years for a full transmutation cycle to more than 100 000 years for the once through cycle. The most common design goal for these repositories is that no radioactivity shall be released within the next 1 000 years, and that the leakage after that time would be so small that it presents no danger to living species. The Environmental Protection Agency (EPA), USA, standard requires that no more than 1 000 cancers shall be caused in 10 000 years by radionuclides released from the waste, using the ICRP predictions of radiation dose effects (Ch. 15).

In the case of direct disposal of used nuclear fuel, the initiated radioactive inventory consists of actinides, fission products and activation products. As noted previously, (Ch. 21) the shorter lived fission products, such as ^{90}Sr and ^{137}Cs, and transuranic elements, such as ^{238}Pu, ^{241}Pu, ^{244}Cm, are the main contributors to the radioactivity. However, performance assessments strongly indicate that the waste form matrix and the near field engineered barriers (e.g. clay backfill, etc.), can successfully retain and prevent any migration to the far field environment for one thousand years and probably much longer (>10^4 years). After the first thousand years the long lived nuclides such as ^{129}I, ^{135}Cs, ^{126}Sn, ^{99}Tc and ^{79}Se among the fission products, ^{226}Ra, and the actinides ^{234}U, ^{236}U, ^{237}Np, ^{239}Pu, ^{240}Pu, and ^{241}Am become the major concern.

22.10.1. Release scenarios

Two major types of scenarios are considered in the performance assessments for release of radionuclides from the repository. One scenario evaluates all the processes which are expected to occur normally in the region of the repository which could effect the rate of release and migration. In this scenario, ground water penetrates the waste packages and leaches the radionuclides which, then, can migrate through and out of the repository, Figure 22.12. R_1, R_2, etc represent mathematical models describing the migration of one particular radionuclide in that specific region. The R_i's are then added into the overall equation.

To assess the release by this scenario, it is necessary to evaluate the rate of release from the package, the flow rate of the underground fluids, the speciation and solubility of the different radionuclides and their diffusion (migration) rates.

The second scenario includes "disturbed" conditions which could be the result of geologic events such as earthquakes, volcanic activity and changes in hydrological conditions. This scenario also includes the effect of "human intrusion," in which people in future generations unknowingly penetrate a repository and release a portion of its radioactive contents to the earth's surface via the groundwater system. For example, this release could be the result of a drilling operation. Such events are assumed to occur over the history of the repository including some at very early times (less than 1 000 years). In this case, sorption and half-life and, for the most part, solubility are not major factors in determining the potential release. The dominant contribution would be from transuranic elements (mainly ^{239}Pu, ^{240}Pu, ^{241}Am) with some contributions from fission products.

22.10.2. Canister dissolution

The most common canister materials are copper and iron. In rocks, where the ground water is reducing (as e.g. in the Canadian and Scandinavian shields), copper is practically insoluble as shown by the existence of native copper, million years old, found in this environment. Detailed studies have shown that the radiation from the waste nuclides have a very small effect on the dissolution. It is therefore predicted that a 50 mm thick copper canister will be intact for at least one million years. Iron, or steel, can be expected to dissolve more rapidly, especially in oxidizing groundwater. This anoxic oxidation will produce hydrogen gas which will contribute to keep the general redox potential reducing which would be important in limiting the waste nuclide migration.

As the metal encapsulation is dissolved or cracks (Figure 22.12.B), the radionuclides in the waste matrix (UO_2 or glass) will be released mainly with the dissolution rate of the matrix (congruent dissolution). A reasonable figure for the glass dissolution rate is $2 \times 10^{-3}\ g\ m^{-2}\ d^{-1}$ exposed glass surface, assuming no limit with regard to the solubility

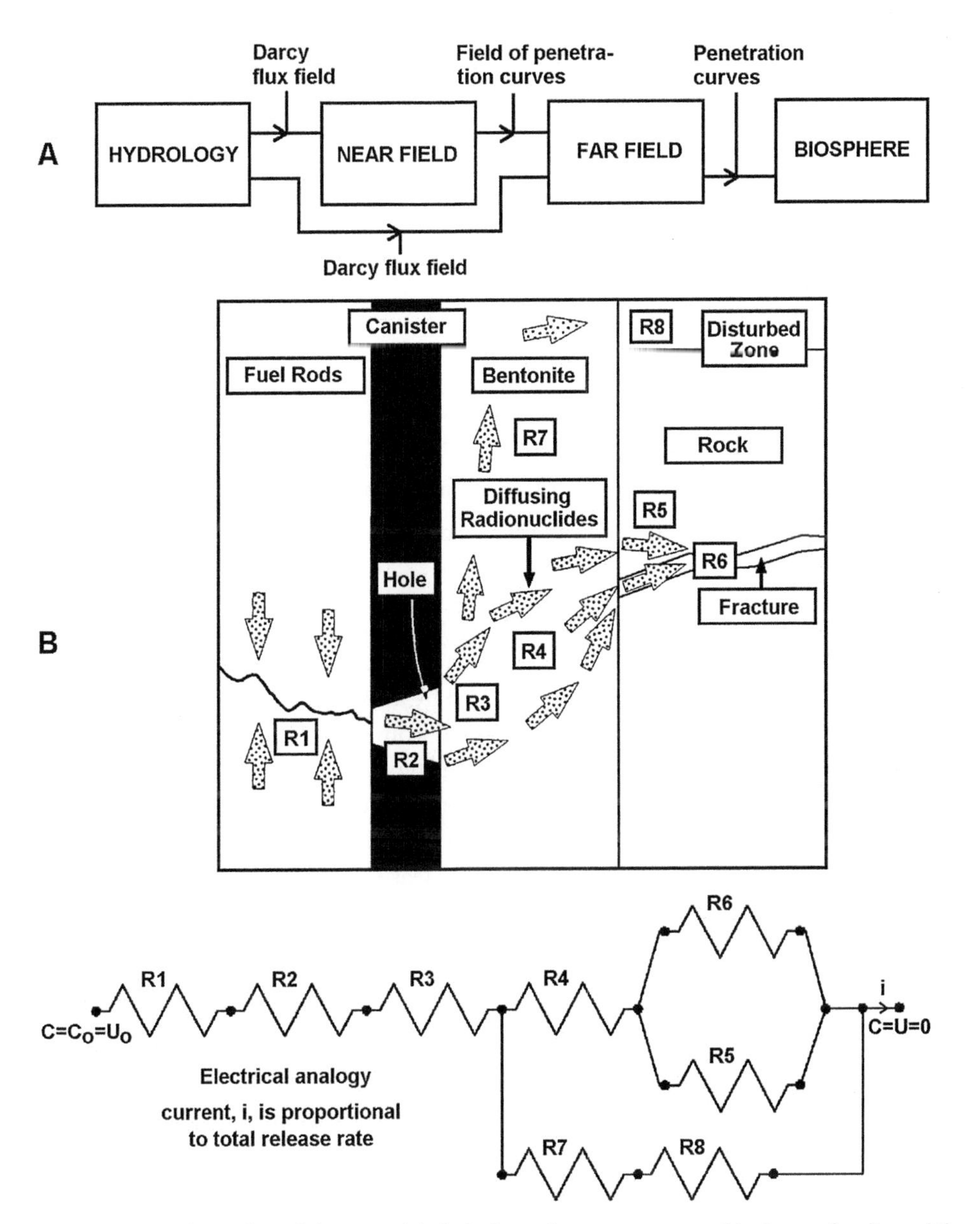

Figure 22.12 Coupling of models in model chain for safety assessment of leakage of radionuclides from fractured fuel rod and canister in a bentonite filled rock hole. (*From SKB91*).

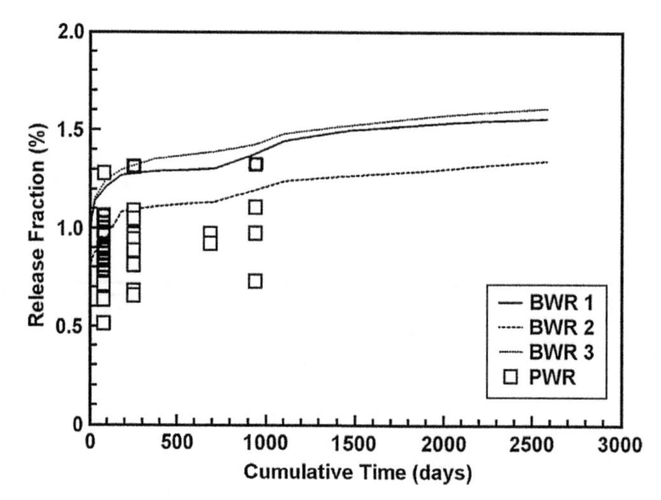

Figure 22.13 Fraction of released cesium from BWR and PWR fuels in different waters as a function of contact time (*From SKB91*).

product (i.e. unlimited amount of water); this corresponds to a corrosion rate of the glass surface of ~2.7×10^{-4} mm y^{-1}. Experience shows that this rate rapidly decreases with time. Thus strontium release rates were reduced by a factor of 10^6 in 15 y. Figure 22.13 shows measurements of the fraction of Cs released from spent fuel as a function of contact time with simulated ground water. By the time, the fuel matrix will be converted into hydrous oxide. The fraction altered has been measured under various conditions and are extrapolated to very long times in Figure 22.14. These results indicate that even if the barrier surrounding the canisters break down, the leakage of waste products into the near field would be quite small.

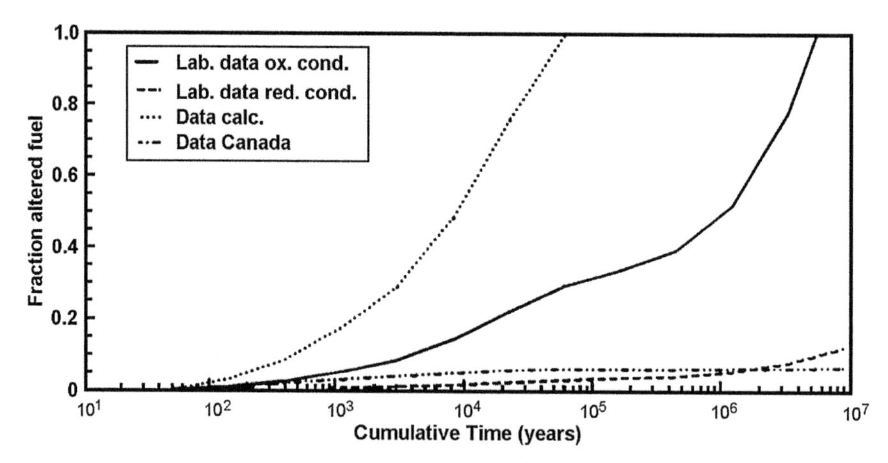

Figure 22.14 Fraction of fuel altered in ground water as a function of time; the reaction is assumed to start 40 years after fuel discharge from the reactor (*From SKB91*).

Table 22.10 Radionuclide solubilities and limiting phase in reducing and oxidizing Finnsjö fresh water (SKB91)

	Reducing conditions		Oxidizing conditions	
	Solubility (M)	**Limiting phase**	**Solubility (M)**	**Limiting phase**
Se	very low	M_xSe_y	high	—
Sr	1×10^{-3}	Strontianite	1×10^{-3}	Strontianite
Zr	2×10^{-11}	ZrO_2	2×10^{-11}	ZrO_2
Tc	2×10^{-8}	TcO_2	high	—
Pd	2×10^{-6}	$Pd(OH)_2$	2×10^{-6}	$Pd(OH)_2$
Sn	3×10^{-8}	SnO_2	3×10^{-8}	SnO_2
I	high	—	high	—
Cs	high	—	high	—
Sm	2×10^{-4}	$Sm_2(CO_3)_3$	2×10^{-4}	$Sm_2(CO_3)_3$
Am	2×10^{-8}	$AmOHCO_3$	2×10^{-8}	$AmOHCO_3$
Pu	2×10^{-8}	$Pu(OH)_4$	3×10^{-9}	$Pu(OH)_4$
Pa	4×10^{-7}	Pa_2O_5	4×10^{-7}	Pa_2O_5

The dissolution rate of canister encapsulation and waste matrix is limited by water solubility. As mentioned above, of the actinides the most hazardous, plutonium, would most likely be in the +4 state and form a very insoluble hydroxide. This is particulary true in an H_2 andFe(II) dominated environment caused by dissolution of an iron canister. Table 22.10 gives limiting solubilities and phases for the more important waste elements in reducing and oxidizing groundwater. Figure 22.15 shows the leakage rate from the near field of an initially defective canister, extrapolated to long times; all nuclides that are expected to leak out at a rate of more than 1 Bq y^{-1} are included in the Figure. In this prediction, the stipulations mentioned in the beginning of §22.10 are met.

22.10.3. Releases from bitumen and concrete encapsulations

The low-level and intermediate-level wastes are to be encapsulated in bitumen, concrete or glass. The bitumen would be highly water resistant, but it ages with time and begins to lose strength in 10–20 years. Special bitumen materials have to be developed for wastes which must be contained in >50 years. The bitumen drums are normally stored in a containment building, e.g. of concrete.

The groundwater surrounding concrete rapidly becomes very basic, pH>13, as a consequence of leakage of Na^+ and K^+ ions. At a later stage, Ca^{2+} ions begin to leak, reducing the pH to about 10.5. Thus nuclides migrating through the concrete encounter very basic media, leading to the formation of insoluble hydroxides for most polyvalent ions, including all actinides. The leaching of Ca^{2+} from the concrete causes it to lose its mechanical strength and to begin to deteriorate. By suitable additives, the onset of deterioration can be delayed, and at pH<10 the concrete is stable for very long times.

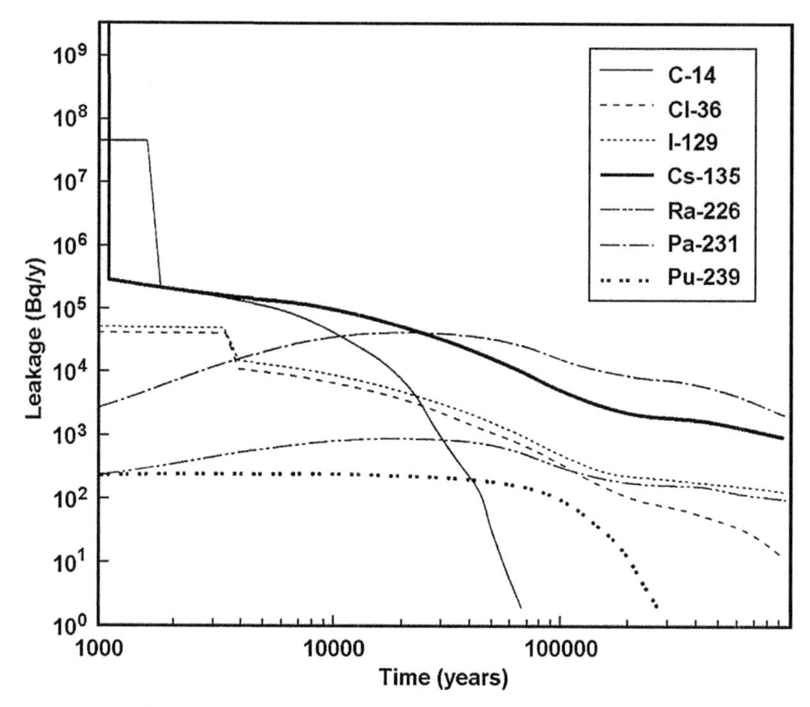

Figure 22.15 Leakage from the near field of radionuclides from an initially defect canister; all nuclides that leak out at a rate of >1 Bq/y are included (*From SKB91*).

However, concrete containers for LLW and MLW can not be considered to have an infinite lifetime, and therefore further containment is necessary.

22.10.4. Migration from the repository

After release into the near field, the radionuclides can only migrate via a water transport path (Figure 22.12). Migration in the far field may occur for radionuclides with long lifetimes, high solubility in ground water, and low sorption along the transport pathway. In repositories where the water is confined to interstitial fracture and pore areas, the actinides would have low solubility and high sorption compared with the fission products. All nuclides which adsorb would move slower than the free groundwater due to the sorption-desorption equilibria. As a consequence ^{129}I and ^{99}Tc would be expected to be the principle contributors to a radioactivity release to the environment, as they are poorly adsorbed. This is also true for ^{14}C, which contributes because it is transported as dissolved CO_2 (Fig. 22.15).

The effect of elevated temperatures resulting from the energy released by the radioactive decay must be included in the evaluation of release and near field migration (i.e. within the repository volume). Elevated temperatures could alter the geology as well as the chemical speciation and solubility of the released nuclides. If the temperature exceeds the boiling point of water in the fluids, it would result in a drier repository with reduced or no release and migration.

When granite or clay is contacted with water containing dissolved cations, sorption or exchange of these ions with ions of the solid phase are observed. For example, montmorillonite has such a high exchange capacity that it is used as a natural ion exchanger, e.g. for water purification. The ion exchange elution curves in Figure 14.9, however, do not depend on sorption capacity, but on sorption strength and on aqueous complexation. The sorption strength depends on ionic charge (higher charged species sorb more strongly), ionic size (smaller ions sorb more strongly), etc., while aqueous complexation depends on the nature of the complexant (ligand), as well as on the cation properties.

All ground/soil/rock materials sorb ions. This sorption has traditionally been defined as the sorption distribution, k_d, eqn. (22.12). The k_d depends on so many factors that they may be considered site specific; i.e. dependent on the ion released, the near field matrix composition (buffer material, dissolved canister, etc), the ground water composition and the rock mineral composition. However, clay and granite have about the same k_d values in similar groundwaters. A more recent redefinition claim that instead use the term k_d only for one species while the complete sorption is described by the term R_d analogous to D and k_d in solvent extraction. However, most of the collected literature data use the older definition why we for simplicity also do so.

$$k_d = \frac{\text{conc. of radionuclide per kg soil/rock/etc}}{\text{conc. of radionuclide per m}^3\text{ water}} \qquad (22.12)$$

Typical k_d values are listed in Table 22.11. From the k_d values and the soil and groundwater properties, the retention time for a radionuclide may be calculated. The radionuclide retention factor (RF) is defined as

$$RF = v_w/v_n = 1 + k_d\,\delta(1-\varepsilon)/\varepsilon \qquad (22.13)$$

Table 22.11 Far field radionuclide distribution values for granitic rock (SKB91)

Element	k_d value ($m^3\ kg^{-1}$)	Element	k_d value ($m^3\ kg^{-1}$)
Zirconium	1	Carbon	0.001
Radium	0.15	Chlorine	0
Protactinium	1	Palladium	0.001
Thorium	2	Selenium	0.001
Uranium	2	Strontium	0.015
Neptunium	2	Cesium	0.15
Plutonium	0.2	Iodine	0
Technetium	1		

Conditions: specific surface area 0.1 m^2 per m^3 of rock, (equal to) 100 m^2 per m^3 of water; matrix diffusion coefficient $3.2 \times 10^{-6}\ m^2\ y^{-1}$; diffusion porosity in the rock matrix 0.005.

where ν_w is the groundwater velocity (in granite typically 0.1 m y^{-1}), ν_n the nuclide transport velocity (which must be $<\nu_w$), δ the soil density (typically 1 500–2 500 kg m^{-3}), ε the soil porosity (or void fraction, typically 0.01–0.05), The groundwater velocity can be obtained from eqn. (21.15), if geologic parameters are known. Observed k_d values for Swedish granite are given in Table 22.11. Typical retention times in granite are about 400 y for Cs, 1 500 y for Sr, 200 000 y for Ln(III) and 40 000 y for An(IV). Rather similar (within a factor of ten) values have been found for other geologic formations such as tuff from New Mexico, basalt from Idaho, and limestone from Illinois. The high retention values combined with a water transport time of a few years lead to retention of ^{90}Sr, ^{137}Cs, and most Pu, Am, and Cm isotopes over their lifetimes. With the negligible water flow rate through clay, diffusion is the dominating transport process. It is estimated that it will take 700 y for Cs, 1 800 y for Sr, 22 000 y for Am and 8 000 y for Pu to penetrate 0.4 m bentonite. During those times most of the radioactivity of these nuclides have decreased to negligible values. Studies for a salt repository (WIPP) in New Mexico indicate that, of all waste nuclides released in the repository, only ^{14}C, ^{99}Tc, and ^{129}I may reach the surface before they have decayed. Thus, ground retention of the radionuclides plays an essential role in the risk evaluation.

Both dissolution rate and transport rate of nuclides depends on the existence of complex formers in the groundwater. Such complex formers are Cl^-, F^-, SO_2^{4-}, HPO_4^{3-}, CO_3^{2-}, and organic anions (e.g. humic acid). Complexes with these anions, in most cases, increase solubility, and – through formation of less positively charged metal species – reduce the retention factors. The groundwater conditions, therefore, play a central role in evaluation of the risks of a waste repository. Since these conditions vary, they have to be evaluated for each site. This point is illustrated by the Maxey Flats facility in the USA, where radionuclides were found to move from a storage basin much more rapidly through soil than would be expected from the equations above. This was found to be caused by the addition of strong metal chelating complex formers like diethylenetriaminopentaacetic acid (DTPA) to the basin liquid. The DTPA complexes have negative charge, resulting in lower k_d values (usually $\sim$1). This reflects the necessity to know in adequate detail the chemistry of the waste and of the storage site for proper evaluation of the safety.

Figure 22.2 shows the result of a safety analysis, indicating the estimated amounts of released radioactivities carried by ground water from a spent fuel repository to a recipient, and the expected doses delivered, without consideration of retention. Retention will cause the ^{226}Ra and ^{231}Pa to reach zero value before 10 000 years.

Calculations for individual doses and the risks for intrusion into a granite repository of LWR spent fuel at 100 and at 100 000 years after burial are listed in Table 22.12. These calculations are based on very extensive modelling (cf. Fig. 22.12) of the releases and uptakes from wastes produced from a 20 GWe reactor park operating in Europe for

Table 22.12 Individual doses and risks from intrusion scenarios in a granite repository of unreprocessed LWR spent fuel (Mobbs et al.)

	100 years		100 000 years	
Radionuclide	**Dose Rate (Sv y^{-1})**	**Risk† (y^{-1})**	**Dose Rate (Sv y^{-1})**	**Risk‡ (y^{-1})**
^{99}Tc	5.0×10^{-6}	1.5×10^{-14}	1.9×10^{-9}	2.5×10^{-14}
^{129}I	1.8×10^{-6}	5.4×10^{-16}	3.2×10^{-10}	4.2×10^{-15}
^{135}Cs	2.4×10^{-6}	7.0×10^{-15}	1.3×10^{-7}	1.7×10^{-12}
^{137}Cs	1.5	1.8×10^{-7}	—	—
^{234}U	7.3×10^{-4}	2.1×10^{-12}	5.0×10^{-7}	6.5×10^{-11}
^{235}U	1.8×10^{-5}	5.1×10^{-14}	1.4×10^{-6}	1.9×10^{-11}
^{236}U	1.9×10^{-4}	5.4×10^{-13}	2.8×10^{-7}	3.7×10^{-12}
^{238}U	2.0×10^{-4}	5.9×10^{-13}	5.9×10^{-7}	7.7×10^{-12}
^{237}Np	1.1×10^{-3}	3.1×10^{-12}	3.6×10^{-6}	4.7×10^{-11}
^{238}Pu	3.6	1.8×10^{-7}	5.1×10^{-6}	6.7×10^{-11}
^{239}Pu	1.0	1.8×10^{-7}	1.2×10^{-4}	1.5×10^{-9}
^{240}Pu	1.5	1.8×10^{-7}	3.1×10^{-7}	4.0×10^{-12}
^{241}Pu	1.0×10^{1}	1.8×10^{-7}	9.2×10^{-6}	1.2×10^{-9}
^{242}Pu	5.4×10^{-3}	1.6×10^{-11}	8.9×10^{-6}	1.2×10^{-10}
^{241}Am	2.0	1.6×10^{-7}	1.8×10^{-6}	2.4×10^{-11}
^{243}Am	6.4×10^{-2}	1.9×10^{-10}	3.4×10^{-6}	4.4×10^{-11}
^{244}Cm	8.7×10^{-2}	2.5×10^{-10}	4.7×10^{-9}	6.2×10^{-14}
^{245}Cm	4.8×10^{-4}	1.4×10^{-12}	7.3×10^{-9}	9.6×10^{-14}
Total	2.1×10^{1}	1.8×10^{-7}	1.6×10^{-4}	2.1×10^{-9}

†From S. F. Mobbs et al
‡Dose-risk conversion factor = 1.65×10^{-2} Sv^{-1} recommended by ICRP in 1977 [ICRP, 1977]. However, ICRP recommended a higher value, 5.0×10^{-2} Sv^{-1}, in 1990 [ICRP #60, 1990].

30 years, including reprocessing and storage of LLW and ILW in surface facilities, and HLW, either in the form of spent fuel elements or vitrified HLLW in an underground granite repository. To model the dissolution of radionuclides and their migration in the ground existing facilities in their actual environments have been chosen. Similarly, for the uptake actual food chains (crops, fish , etc and eating habits) have been used. The quantity of spent fuel is assumed to be 18 000 t IHM. The LLW and ILW repositories are assumed to begin to leak at time 0, the dissolved radionuclides migrating into the actual soils, reaching streams and wells, etc, while it is assumed that the spent fuel elements or high level waste glass begins to leak after 1 000 years. The radionuclides migrate by groundwater under various retention conditions, depending on hydrogeologic conditions and chemical properties. The exposed population is divided into four groups: the critical group around the installations, the population of the country, the population in Europe and in the world. As expected, the population dose is the highest for the critical and national groups for aqueous releases, but about the same for all groups for gaseous

releases. The model also includes an intrusion scenario using a well drilling probability, beginning after 300 years (in the main case), as it is expected that control will last that long. It is assumed that as soon as a drill core from a waste repository is taken, it will immediately be recognized; this means that high doses from the core will only be received by drilling and laboratory personal. The risk[1] from intrusion after 100 years is calculated to be 8.1×10^{-7} y^{-1} and after 100 000 years 2.1×10^{-9} y^{-1}. By comparison, the calculated risk from the "normal" scenario of releases from spent fuel elements and migration in the environment over a long period is 5×10^{-5} y^{-1}. With this exception all the maximum calculated individual risks from disposal of solid wastes are below the limit of 10^{-5} y^{-1} recommended by the ICRP. For vitrified high level waste and spent fuel disposal, the doses calculated for intrusion scenarios are very high but the probability of occurrence are low, so the risks from intrusion are lower than those from radionuclide migration with groundwater. The maximum individual doses from the migration scenarios for solid waste disposal are predicted to arise beyond 10 000 years after closure of the repository.

These values indicate that release and migration is the more significant concern, although even for this scenario, geologic disposal is calculated to present very small risks to future generations.

22.11. CONCLUSIONS

A rather large amount of nuclear fission products and actinide elements have been released to the environment from nuclear weapons testing and from accidental and intentional discharges from nuclear reactor operations and fuel reprocessing. The research on the fate of these released radionuclides suggest that the long lived actinides form quite insoluble or strongly sorbed species while ^{129}I and ^{99}Tc have relatively high dissemination in natural systems. The most active shorter-lived species (^{90}Sr, ^{137}Cs) also have more mobility in ecosystems.

These conclusions are confirmed by the results of the investigations of radionuclide behavior in natural analog sites. Even more relevant are the results of the natural reactor region at Oklo.

The data from field studies is largely confirmed by the performance assessments of proposed nuclear waste repositories. The "intrusion" scenario is calculated to have a lower risk than the undisturbed natural leach and migration scenario. The latter qualitatively agrees with the Oklo data and indicates no unacceptable risks result from a carefully chosen and designed geologic repository in which the nuclear wastes are emplaced with appropriate packaging.

[1] Risk is here assumed to be equal to expected frequency regardless of the severity of the consequence.

22.12. EXERCISES

22.1. What is the purpose of the clay buffer of a waste repository in granitic rock?

22.2. Which of the following ions move slower than the groundwater: K^+, Cs^+, La^{3+}, TcO_4^-, HCO_3^-?

22.3. Why is Np assumed to move faster than Pu in most groundwaters?

22.4. (a) What types of geologic formations are considered for waste repositories? (b) Which types were used/planned for Asse and WIPP?

22.5. What will be the concentration of $Am(CO_3)_2^-$ at pH 7 under the conditions given in §22.7.

22.6. What will be the ratio of [Pu(III)]/Pu[(IV)] in a groundwater containing small concentrations of iron in the relation Fe(II) 99% and Fe(III) 1%? E^0 for Fe(II)/Fe(III) = 0.743 V. Neglect hydrolysis.

22.7. (a) Calculate the Pourbaix-line for Np(IV)/Np(V) in Figure 22.6. (b) What value must the *E*h value of water exceed at pH 6 for Np(V) to dominate? Neglect hydrolysis.

22.8. A miner has a deep well 160 km away from a waste repository. During an earthquake, the rock fractures and a groundwater stream opens between the repository and the well so that 10 Ci ^{90}Sr momentarily dissolves and moves toward the well at v_w 160 km y^{-1}. (a) What will be the amount of ^{90}Sr (in Bq) reaching the well (assume plug flow)? (b) Will that water be harmful to the miner?

22.13. LITERATURE

B. ALLARD, H. KIPATSI AND J. O. LILJENZIN, Expected species of U, Np and Pu in neutral aqueous solutions, J. Inorg. Nucl. Chem. 42 (1980) 1015.

D. L. PARKHURST, D. C. THORSTENSON AND L. N. PLUMMER, PHREEQE – A Computer Program for Geochemical Calculations, U. S. Geologic Survey, Water Resources Investigations, Report 90-86, Denver, 1980.

E. R. SHOLKOVITZ, The Geochemistry of Plutonium in Fresh and Marine Environments, Earth-Science Rev. 19 (1983) 95.

P. J. COUGHTREY, D. JACKSON, C. H. JONES AND M. C. THORNE, Radionuclide Distribution and Transport in Terrestrial and Aquatic Ecosystems: A Critical Review of Data, CEC and U.K. Ministry Agriculture, Fish and Food, A. A. Balkema, Rotterdam, 1984.

B. ALLARD, U. OLOFSSON AND B. TORSTENFELT, Environmental actinide chemistry, Inorg. Chimica Acta 94 (1984) 205.

D. RAI, Solubility Product of Pu(IV) Hydrous Oxide and Equilibrium Constants of Pu(IV)/Pu(V), Pu(IV)/Pu(VI), and Pu(V)/Pu(VI) Couples, Radiochim. Acta, 35 (1984) 97.

D. C. HOFFMAN AND G. R. CHOPPIN, Chemistry Related to Isolation of High-Level Waste, J. Chem. Educ. 63 (1986) 1059.

J. W. MORSE AND G. R. CHOPPIN, The Chemistry of Transuranic Elements in Natural Waters, Rev. in Aquatic Sci. 4 (1991) 1.

S. F. MOBBS, M. P. HARVEY, J. S. MARTIN, A. MAYALL AND M. E. JONES, Comparison of the Waste Management Aspects of Spent Fuel Disposal and Reprocessing; Post Disposal Radiological Impact, NRPB report EUR 13561 EN, UK, 1991.

Swedish Nuclear Fuel and Waste Management Co., SKB 91: Final Disposal of Spent Nuclear Fuel, Importance of the Bedrock for Safety, SKB Tech. Report 92-20, Stockholm, 1992.

J.-C. Petit, Migration of Radionuclides in the Geosphere: What Can We Learn from Natural Analogues?, in Chemistry and Migration Behavior of Actinides and Fission Products in the Geosphere, Radiochim. Acta, special issue 58/59 (1992).

F. Branberg, B. Grundfelt, L. O. Höglund, F. Karlson, K. Shaquis and J. Smellie, Studies of Natural Analogues and Geologic Systems, SKB Tech. Rept., 93-05, Stockholm, 1993.

H. R. von Gunten and P. Benes, Speciation of Radionuclides in the Environment, Paul Scherrer Institute Report, No. 94-03, Würenlingen, 1994.

United Nations Scientific Committee on the Effects of Atomic Radiation, Sources and Effects of Ionizing Radiation, UNSCEAR 2000 Report to the General Assembly, ISBN 92-1-142238-8, United Nations, 2000.

United Nations Scientific Committee on the Effects of Atomic Radiation, Sources and Effects of Ionizing Radiation, UNSCEAR 2008 Report to the General Assembly, ISBN 92-1-142280-1, United Nations, 2011.

The ExterneE report, http://www.externe.info/externe_d7/?q=node/4

APPENDIX A

Solvent Extraction Separations

Contents

Solvent extraction is a technique which has been highly developed within many national nuclear energy programs because of its suitability as a selective separation process for fission products, actinides, and other radioactive substances. The technique is briefly described in several sections of this book (Sections 14.2.4.3, 14.3.4, 17.7.3.4, 18.2.6, 18.4.3, 21.7, and especially in 21.6 on the Purex process). It is based on the formation of uncharged organic metal complexes which are preferentially soluble in organic solvents. The three main types of such compounds are:

1. Organic chelate (i.e. binding through two positions in the organic molecule, like a claw) complexes such as plutonium tetraacetylacetonate, $PuAa_4$ (general formula MA_n) (Section 18.4.3, cf. Figs 18.3, 18.9 and 18.11);
2. Inorganic metal complexes forming adducts with solvating organic compounds like TBP and hexone, e.g. $UO_2(NO_3)_2)(TBP)_2$ (or generally ML_nS_s, where L is an inorganic anion and S is the organic adduct former (Section 21.6, cf. Fig. 14.9);
3. Ion pair complexes between large organic cations (like alkylammonium ions, R_3NH^+) and negatively charge inorganic complexes like $UO_2(SO_4)^-$ (generally RB^+ML^-, Section 14.2.4.2).

The physical and chemical principles governing the formation of extractable metal complexes, the conclusions which can be drawn about the chemical system from solvent extraction studies, and analytical and industrial applications are described in several monographs, e.g. 'Principles and Practices of Solvent Extraction' (2004).

A.1. SINGLE STAGE BATCH EXTRACTIONS

Let us consider a solute (e.g. uranium) which is distributed between an organic and an aqueous phase, independent of the kind of compounds the solute forms in the two phases. After equilibrium the weight of solute in the organic phase (*extract*) is w_{org} and in the aqueous phase (*raffinate*) w_{aq}. Thus

$$w_{tot} = w_{org} + w_{aq} \tag{A.1}$$

The weight fraction in the organic phase is

$$\psi_1 = P/(P+1) \tag{A.2}$$

where the index 1 refers to the conditions after one extraction. P is called the *extraction* (or *partition*) *factor* and defined by

$$P = D\theta \tag{A.3}$$

where D is the distribution factor, defined by

$$D = \frac{\text{Concentration of all M-species in the organic phase}}{\text{Concentration of all M-species in the aqueous phase}} \tag{A.4}$$

and θ is the phase volume ratio v_{org}/v_{aq}. The percentage solute extracted, denoted $E(\%)$, is equal to 100 ψ_1. The fraction of nonextracted solute left in the organic phase is φ_1, and $\psi_1 + \varphi_1 = 1$,

$$\varphi_1 = 1/(P+1) \tag{A.5}$$

Assume that one wants to separate uranium from the fission product lanthanum using the two-phase system: aqueous 1 M HNO_3, the organic adduct former tributyl phosphate, TBP (100%, i.e. undiluted). In this system $D_U = 20$ for U(VI), while for lanthanum $D_{La} = 0.07$. If one extracts with a phase ratio of 0.5, then $\psi_U = 0.909$ ($E_U = 100\ \psi_U = 90.9\%$) and $\psi_{La} = 0.034$ (E_{La} 3.4%). This may be unsatisfactory with respect to both uranium yield and purity.

The yield can be increased by repeated extractions of the same aqueous phase (multiple extraction with one stationary phase, or 'crosscurrent extraction'). For n such extractions one finds that

$$\varphi_n = (P+1)^{-n} \tag{A.6}$$

Suppose $n = 3$ for our example, then $\varphi_{3,U} = 0.00075$, i.e. for the three organic phase volumes taken together $E_U = 100(1 - \varphi_{3,U}) = 99.92\%$. However, for lanthanum $\varphi_{3,La} = 0.902$, i.e. $E_{La} = 9.8\%$. Although the uranium yield is high, the lanthanum impurity may be intolerable.

A more elaborate technique must be employed in order to obtain both high yield and high purity under such conditions. Many such batch laboratory techniques have been described using alternatively fresh organic (*extraction*) and aqueous (*washing*) solutions. The extractions are carried out in special multistage equipment (Section A.4).

A.2. MULTIPLE STAGE CONTINUOUS PROCESSES

A single partitioning of a compound between an organic solvent and water may not be sufficient for isolating it in acceptably pure form and good yield. Various multiple extraction techniques may therefore be required. Such techniques have been described in Section 21.6.4 and their technical application for uranium production (Section 14.2.4.3) and spent fuel reprocessing (Section 21.6.3).

Continuous processes are preferred in industry, where the most common and simple solvent extraction equipment is the mixer-settler (Fig. A.1(a); cf also Fig. 14.2). This type of equipment is also becoming standard in laboratories engaged in process development. In the uranium industry a single mixer-settler may hold as much as 1000 m^3. The mixer-settlers, each closely corresponding to a single ideal extraction stage, are arranged in

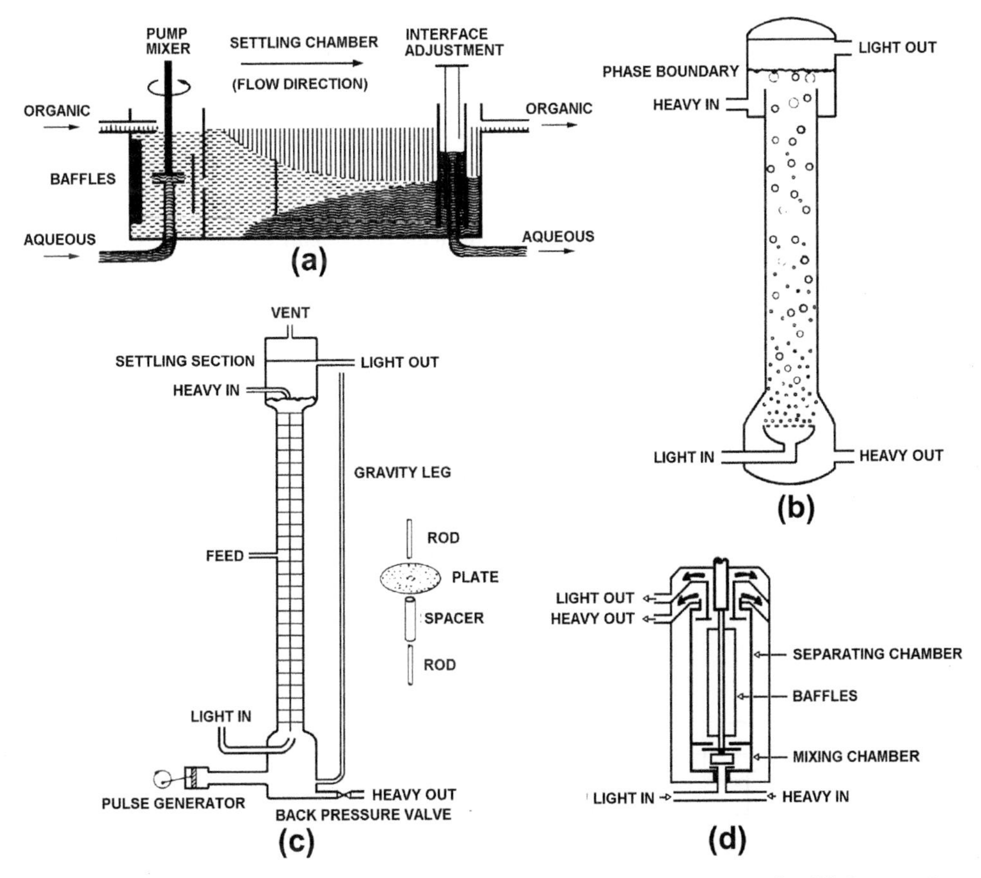

Figure A.1 Different types of continuous extraction equipment. (a) Mixer-settler (b) Spray column (c) Pulsed column (d) Tubular centrifugal contactor.

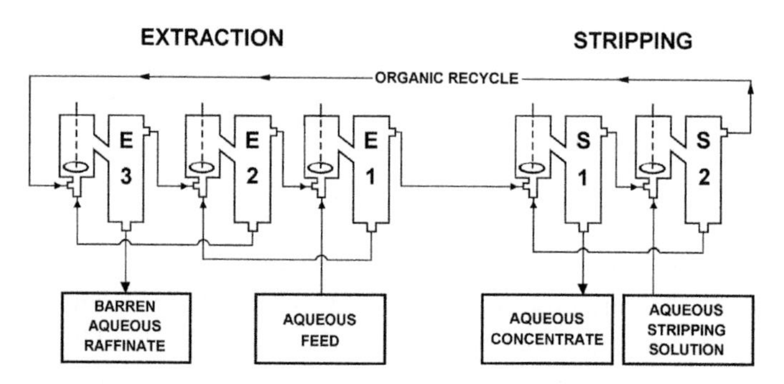

Figure A.2 Mixer-settler counter current solvent extraction battery.

batteries containing any number of stages. In these batteries the aqueous and organic phases flow counter current to each other (Fig. A.2).

For counter current solvent extraction, either batch or continuously, one finds that for the stationary state and for n stages

$$\varphi_n = (P-1)/(P^{n+1}-1) \tag{A.7}$$

provided P is constant through all stages. In our example one obtains for $\theta = 0.5$ (flow rate ratio organic: aqueous) and $n = 3$ that $E_U = 99.91\%$ and $E_{La} = 3.5\%$. Thus about the same yield of uranium, but with a somewhat lower lanthanum impurity as compared to the crosscurrent extraction procedure. This impurity figure can be lowered substantially by modification of the extraction process according to Fig. A.3 so that the extraction battery contains n extraction stages with extraction factor P_1, and $m-1$ *washing* stages with the *extraction* factor P_2. Then

$$\varphi_{m,n} = \frac{(P_1-1)(P_2^m-1)}{(P_1^{n+1}-1)(P_2-1)P_2^{m-1}+(P_2^{m-1}-1)(P_1-1)} \tag{A.8}$$

which in the case that $P_1 = P_2$ reduces to

$$\varphi_{m,n} = (P^m-1)/(P^{m+n}-1) \tag{A.9}$$

In the latter case, using three extraction stages and one washing stage, we find for our example that E_U is 99.9% and E_{La} is 0.12%. Thus both high yield and high purity are

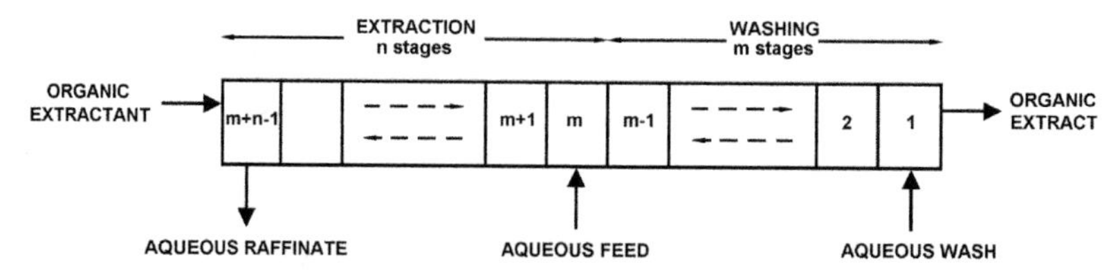

Figure A.3 Counter current solvent extraction with n extraction and m washing stages.

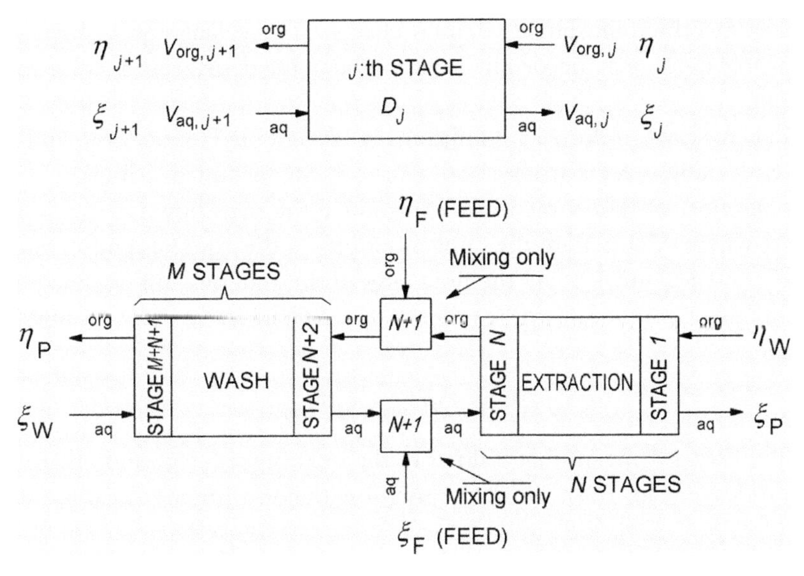

Figure A.4 Notations for counter current extraction with wash stages.

achieved with the counter current, central or intermediate feed solvent extraction technique. This is extensively used in uranium production, nuclear fuel reprocessing, and transuranium element separations. When the conditions are selected so that a high ($P \gg 1$) extraction factor is obtained for the desired product and low ones ($P \ll 1$) for the impurities, high purity and good yield can be obtained with relatively few stages. However, θ is always lower in the extraction stages than in the washing stages because the central feed flow rate adds to the wash flow rate in the extraction stages and the organic flow rate is unchanged. Hence, cases with the same P-value in extraction and washing stages seldom occur.

The purpose of the *washing* (or *scrubbing*) stages is to clean the desired product from impurities, and should not be confused with the *stripping* stages, the purpose of which is to transfer the product from the organic phase to a new aqueous phase. However, stripping can also be so arranged that further purification is achieved, sometimes even using a central organic feed, cf. Fig. A.4. If no additional purification is desired after extraction and the extractant is volatile, the organic phase may be distilled, leaving a pure solid product.

A.3. HIGH LOADINGS

A further industrial requirement on solvent extraction processes is high capacity, which means high concentrations of several solutes in the aqueous and organic solvents ('high loading'). The extraction factor P (or distribution factor D_M) may then vary from stage to stage because D-values and phase volumes change. The calculation of the number of stages, flow rate ratios, etc., needed in order to obtain the desired product has to take this into account.

In order to treat industrial conditions with coupled extraction and washing batteries with central feed where both phase volumes and distribution ratios may change as an effect of the extraction we introduce the mass flow rates: $\xi = X\, V_{aq}$ and $\eta = Y\, V_{org}$, where X and Y are the concentrations in the aqueous and organic phases respectively, and V_{aq} and V_{org} are the flow rates of the aqueous and organic phases. We also redefine the extraction factor for the jth stage, P_j, in terms of actual flow rates out of that stage as

$$P_j = D_j V_{org,j+1}/V_{aq,j} \tag{A.10}$$

where the index $j+1$ means that it flows to the next higher stage and j that it flows to the previous stage (Fig. A.4). Now five useful extraction functions can be defined as follows.

$$e_1 = \left(1 + \sum_{i=2}^{N} \prod_{k=i}^{N} P_k\right) \tag{A.11}$$

$$e_2 = \left(1 + \sum_{i=2}^{N} \prod_{k=i}^{N} P_k\right) \tag{A.12}$$

$$s_1 = \left(1 + \sum_{i=N+2}^{M+N+1} \prod_{k=i}^{M+N+1} P_k\right) \tag{A.13}$$

$$s_2 = \left(1 + \sum_{i=N+3}^{M+N+1} \prod_{k=i}^{M+N+1} P_k\right) \tag{A.14}$$

$$\zeta = \prod_{k=N+2}^{M+N+1} P_k \tag{A.15}$$

Through a series of mass balances we can obtain the following equation for mass flow rate with the outgoing aqueous phase, ξ_P

$$\xi_P = [\xi_w + \eta_w(e_2\zeta + s_2) + \xi_F s_1 + \eta_F s_2]/(e_1\zeta + s_2) \tag{A.16}$$

and the outgoing organic phase, η_P

$$\eta_P = \xi_F + \xi_W - \xi_P + \eta_F + \eta_W \tag{A.17}$$

The symbols refer to those given above and in Fig. A.4. Mass flow rates and conditions inside each battery can in principle be obtained recursively from the relations

$$\xi_{n+1} = \xi_n(1 + P_n) - \eta_n \quad \text{and} \quad \eta_{n+1} = \xi_n P_n \qquad \text{(A.18)}$$

However, numerical problems tend to make the use of these equations difficult for batteries with many stages. Special cases occur when P is either constant or constantly equal 1.

The flow rates out from any stage can usually be calculated with sufficient accuracy from the apparent molar volumes, Φ_i, of the i compounds according to

$$V_{aq} = V_{0,aq}\sum(1 + \Phi_{i,aq}X_i) \quad \text{and} \quad V_{org} = V_{0,org}\sum(1 + \Phi_{i,org}\,Y_i) \qquad \text{(A.19)}$$

where $V_{0,aq}$ and $V_{0,org}$ are the flow rates of the pure unloaded phases in each stage. In many cases one of the central feeds is missing (usually the organic feed) and the corresponding mass and volume flow is zero. The wash stream (index W) is then the real feed for that phase.

When the flow rates vary considerably in the batteries due to loading effects, the whole calculational procedure becomes recursive and a computer code based on these equations may be needed. However, in most cases it is only necessary to treat one, or perhaps two, of the extracted components recursively as it dominates volume changes and defines the P-values for all minor components, e.g. in the high active first stage in the Purex process nitric acid and uranium define the flow rates and extraction factors for all other elements.

In industrial processes it is not only important to extract the desired compound with high yield and purity, but also to be able to strip it again with good efficiency using a minimum amount of chemicals. Contrary to single stage batch extraction, the possibility to vary the P-value by changes in the flow rate ratio (remember that $P = D \times V_{org}/V_{aq}$) makes the use of conditions with D-values for the product not too far from 1 is desirable. In such a case the product can be stripped by a decrease in the flow rate ratio (in more dilute form than in the feed) and a moderate decrease in the D-value without the need to introduce new chemicals for stripping. A typical example is the uranium purification part of the Purex process. Here uranium (and nitric acid) is extracted at high concentration in the presence of 1–6 M HNO_3 and stripped at a low flow rate ratio by more dilute nitric acid. If needed, the separation can be repeated after reconcentration by evaporation.

A.4. SOLVENT EXTRACTION EQUIPMENT

The less conventional part – from a chemical engineering viewpoint – of a reprocessing plant is the solvent extraction equipment, even though the technique is becoming increasingly common in the chemical industry. The principle of all such equipment is illustrated by Fig. A.1(a); it contains a mixing section for efficient transfer of materials between the phases and a settling section for efficient phase separation. The inputs and outputs provided for connecting stages in the counter current extraction scheme (Fig. A.2). *Mixer-settlers* (Fig. A.1(a)) provide good mixing and reasonably good phase separation performance but rather large hold-ups.

For reprocessing the equipment must be highly reliable, have high stage efficiency, short contact times, small liquid inventory (hold-up), be easy to decontaminate and to service, and not least be safe against criticality. High reliability usually means simple design and few (if any) moving parts. The *packed columns* meet this requirement. They are simply long columns (often 10–20 m with a diameter of 0.3–3 m) filled with small pieces of material obstructing a straight flow through the column, which is by gravity (Fig. A.1(b)). However, high stage efficiency requires mechanical agitation of the two phases and clean phase separation, which cannot be met by packed columns.

In *pulsed columns* (Fig. A.1(c)) the mechanical agitation provides good mixing but poor phase separation. Each plate is perforated (a sieve). The organic and aqueous phases separate between the plates ('settling chambers'). In the down movement (<¼ of the interplate distance) the aqueous phase is forced through the sieves, forming droplets, which by gravity fall through the lighter organic phase and coalesce when reaching the interface boundary. In the upward stroke, organic droplets form and rise through the aqueous phase until they meet the organic phase boundary.

The phase separation in a mixer-settler battery or in a column is usually not better than ~99%, i.e. each outgoing phase contains some percent entrained droplets of the other phase. This separation efficiency can be improved to almost 100% by using *centrifugal extractors* (Fig. A.1(d)). Centrifugal extractors affect good mixing, good phase separation, and have very small hold-ups. The organic-aqueous phase contact time in centrifugal extractors can be made much shorter than in mixer-settlers or columns. The small hold-up volume and short residence time cooperate to reduce radiation decomposition.

Packed columns were used in the first Windscale plant (Sellafield, UK). Pulsed columns were used at Hanford (USA), in the old Eurochemic plant at Mol (Belgium), and are currently in use in the newer La Hague and THORP plants. Mixer-settlers have been used at Savannah River (USA), in the Magnox plant at Sellafield (UK), and at La Hague (France). Centrifugal extractors have been installed at Savannah River and at La Hague.

A.5. EXERCISES

A.1. In a solvent system the distribution ratio, D_U, is 2 for uranium and D_{Cs} is 0.003 for caesium. If 99.5% U is to be extracted in a repeated batch fashion (Eqn (A.5)), (a) how much Cs is coextracted? If instead a counter current process is used with

10 extraction and 2 wash stages, what percentage of (b) uranium and (c) caesium is extracted? In an extraction equipment D_{Cs} 0.003 cannot be maintained, because droplets are carried over between stages; the practical value will be D'_{Cs} 0.02. (d) How much caesium is extracted in this latter case with the counter current equipment? Assume equal phase volumes.

A.2. After scrubbing the solvent (1M TBP in kerosene) with a carbonate solution it should be reacidified to equilibrium with 6 M nitric acid before reuse in reprocessing operations. Assume that the partial molar volume of nitric acid can be neglected for both the organic phase and the aqueous phase. Thus, in this example extraction of nitric acid will not increase the volume of the organic phase nor reduce the volume of the aqueous phase. Extraction of nitric acid is assumed to only occur as the complex $TBP \cdot HNO_3$. For simplicity we will assume that the Eqn $D_{HNO_3} = k_{ex} \times [HNO_3]_{aq} \times [TBP]_{free,org}$, with $k_{ex} = 0.1$ can be used to calculate the *D*-value for nitric acid. The reacidification will be performed in a simple counter current mixer-settler battery. (a) How many stages and (b) what flow rate ratio (org/aq) would be needed in order to produce the required nitric acid concentration in the organic phase using the smallest possible feed of 6.4 M nitric acid? (c) What would be the concentration of the spent nitric acid?

A.6. LITERATURE

G. H. Morrison and H. Freiser, *Solvent Extraction in Analytical Chemistry*, Wiley, 1957.

Y. Marcus and A. S. Kertes, *Ion Exchange and Solvent Extraction of Metal Complexes*, Wiley, 1969.

A. K. De, S. M. Khopkar and R. A. Chalmers, *Solvent Extraction of Metals*, Van Nostrand Reinhold, 1970.

Yu. A. Zolotov, *Extraction of Chelate Complexes*, Humprey Sci. Publ., Ann Arbor, 1970.

T. Sekine and Y. Hasegawa, *Solvent Extraction Chemistry*, Marcel Dekker, 1977.

T. C. Lo, M. H. I. Baird and C. Hanson, *Handbook of Solvent Extraction*, Wiley, 1983.

G. Ritcey and A. W. Ashbrook, Solvet Extraction. Part I and II, Elsevier, 1984. *J. Solvent Extr. Ion Exch.*, Marcel Dekker **11** (1993).

J. Rydberg, M. Cox, C. Musikas, G. R. Choppin (Eds.), *Solvent Extraction, Principles and Practice*, 2nd Ed., Marcel Dekker, New York 2004.

APPENDIX B

Answers to Exercises

Chapter 2: (1) π-mesons, conversion of excess energy into mass. **(2a)** gravitation, the electromagnetic force, the weak interaction force, and the strong interaction force. **(2b)** See e.g. Tables 2.2 and 2.3. **(3a)** Bosons are a group of 'particles' to which the photon and mesons belong. The bosons are the carriers of forces. When two fermions interact they continually emit and absorb bosons. The bosons have an even spin (0, 1, etc.), they do not obey the Pauli principle, and they do not require the formation of antiparticles in their reactions. Particles of nonintegral spin are called fermions because they obey the statistical rules devised by Fermi and Dirac, which state that two such particles cannot exist in the same closed system (nucleus or electron shell) having all quantum numbers the same (referred to as the Pauli principle). Fermions are the building blocks of nature. Fermions can be created and destroyed only in conjunction with an antiparticle of the same class. For example if an electron is emitted in β-decay it must be accompanied by the creation of an antineutrino. Conversely, if a positron – which is an antielectron – is emitted in the β-decay, it is accompanied by the creation of a neutrino. **(3b)** Answer is included in, and follows from, answer to (a). **(4)** Radiation pressure, recoil in photon and γ-emission, energy increase and decrease when falling into or moving out against a gravity field. **(5)** Several possibilities, e.g.: Reaction with ^{1}H to produce a fast neutron and a positron. Then detection of annihilation γ and neutron, $^{37}Cl + n = {}^{37}Ar + e^-$ followed by collection of Ar and detection of its decay, $^{71}Ga + n = {}^{71}Ge + e^-$, conversion to GeH_4, collection and detection of its decay.

Chapter 3: (1) 1.553×10^{22}. **(2)** 0.100. **(3a)** 2111 m s^{-1}. **(3b)** 0.0462 V **(4)** 472 V **(5)** 11.00934 u. **(6)** 0.129 nm **(7a)** 3.032×10^{17}. **(7b)** 3322 K **(8)** 2.080. **(9)** 4. **(10)** 0.030 m **(11a)** By series expansion $e^{\delta} = 1 + \delta + \delta^2/2 + \ldots$but $\alpha = 1 + \delta$ hence $e^{\delta} \simeq \alpha$. **(11b)** 1.097. **(11c)** 22. **(12)** 5.742 SWU.

Chapter 4: (1) 8.261 MeV **(2)** 1.581×10^6. **(3)** 47371 km **(4)** Fusion gives 7 times more per gram. **(5)** −1.02 MeV, −2.76 MeV, and −1.04 MeV **(6)** 6.262 MeV u^{-1}, 8.366 MeV u^{-1}, 8.734 MeV u^{-1}, and 7.489 MeV u^{-1}. **(7)** 4.8, 12.6, 26.4, and 91.8.

Chapter 5: (1) 0.088 MeV, ^{235}U. **(2a)** 0.87 MeV **(2b)** 0.46 MeV (measured 0.51 MeV). **(3)** 99 eV **(4)** 1.26 keV **(5)** 547 keV (measured 662 keV). **(6)**. 1.28×10^9 years **(7)** 5.9×10^9 years **(8)** 2.21×10^{22} J year^{-1}. **(9a)** 16.9 GBq. **(9b)** 3.0×10^{-10} m^3 **(10a)** Multiply (5.25) by $\mathbf{c}^2$. **(10b)** $m^0_\gamma = 0 \rightarrow m_\gamma = E_\gamma/\mathbf{c}^2$ hence $E_\gamma(E_\gamma/\mathbf{c}^2) = E_d m_d$. **(11a)** 1.46×10^4 Bq. **(11b)** 8.09×10^8 atoms. **(12)** $t_{½} = 104$ h and 99 Bq, $t_{½}$ 4.4 h and 1030 Bq. **(13)** R_0 507 cpm for $t_{½}$ 107 min, R_0 440 cpm for $t_{½}$ 24 min (cf. ^{298}At $t_{½}$ 98 min, ^{206}At $t_{½}$ 31 min) **(14)** 2.3×10^5 years.

Chapter 6: **(1a)** 10. **(1b)** 10. **(2a)** 2.19×10^6 m s^{-1}. **(2b)** 0.003%; no. **(3)** 21 MHz **(4)** $2.689 \times 2/3$. **(5)** 1i, 2g, 3d, 4s; 56 nucleons. **(6)** 283 MeV, **(7a)** 7/2, $3.81 \times (1/5)$. **(7b)** 5, $4.50 \times (2/5)$. **(7c)** 5/2. **(8a)** 0.108. **(8b)** 7/2 (also observed). **(9)** 1p½. **(10a)** 43.8 and 307 keV **(10b)** 3.47 fm (8.68 fm from (4.14), meas. 5.48 fm). **(11)** 0.460 T **(12)** 3.2×10^9 years **(13)** 8.86×10^{10} years, hindrance factor 1.2.

Chapter 7: **(1)** From (7.12) R_{Al} 5.8 mg cm^{-2}, and from Fig. 7.6 ~6 mg cm^{-2}, diff. ~3%. **(2)** E_α 2.0 MeV **(3)** 0.71 mm **(4)** e^- 3.8 m, H^+ 23 cm, α 3.3 mm **(5)** Al 0.030, Ni 0.012, Pt 0.0074 mm **(6)** 6.54×10^7 γ m^{-2} s^{-1}. **(7)** ^3H 5 μm, ^{14}C 270 μm, ^{32}P 5.3 mm, ^{90}Sr 1.1 mm **(8)** 2.92×10^8 m s^{-1}. **(9)** Lead $105 000, concrete $3730. **(10)** Na_K 57.53, Na_L 58.54, I_K 25.43, I_L 53.80 keV **(11)** 72% and 52%. **(12)** 28, 39, and 49 g cm^{-2}. **(13a)** Emax 0.62 MeV **(13b)** ~0.8 MeV **(14a)** 9.7 days. **(14b)** 19 min **(15)** 4.9 m **(16a)** 5.5 cm **(16b)** 11.9. **(16c)** 2.5.

Chapter 8: **(1a)** **1.47** $\times 10^5$. **(1b)** 2.94×10^4. **(1c)** 972. **(2)** ~1.2%. **(3)** 3.8 pGy, **(4)** 0.34%. **(5)** ~11%. **(6)** ^{90}Sr 0.10, ^{90}Y 0.068, ^3H 2.5 keV μm^{-1} **(7)** 6.3 mGy s^{-1} **(8)** 38 pF.

Chapter 9: **(1a)** 1.33 GBq. **(1b)** 2.5%. **(2a)** 25 μm **(2b)** 1.05×10^5 μm^{-1} **(2c)** 4.6 nm **(2d)** 3.8 eV $atom^{-1}$. **(2e)** 29 400 K **(3)** 0.11 μM **(4)** 0.88 Bq. **(5)** 5.8 V **(6a)** 509 μs. **(6b)** 12.53. **(6c)** 17.84. **(7)** 3.2×10^{10} counts. **(8)** 45 ns. **(9a)** 6.6 μg cm^{-2} **(9b)** 1.59. **(9c)** 71%. **(10a)** 9.33 ± 0.11 cps. **(10b)** ± 0.072 cps. **(11)** 0.5%.

Chapter 10: **(1a)** 37.3 MeV **(1b)** 1305 m s^{-1}, 0.035 eV **(2)** 47.4 fm **(3a)** 32.9 b sr^{-1}. **(3b)** 0.423 m **(4)** 0.75 MeV **(5a)** 39.5 fm **(5b)** 7.56 fm **(6a)** 8.03 MeV **(6b)** 1.64 MeV **(7)** 3.09×10^6 m s^{-1}. **(8)** 3.57×10^{28} kg **(9)** 77.3 MeV **(10)** Q (0.87 MeV) > 0, He thus unstable. **(11a)** −5.85 MeV **(11b)** 5.85 MeV **(11c)** 2.96 MeV **(11d)** 8.81 MeV.

Chapter 11: **(1)** 9.9×10^{-8} s^{-1}. **(2)** 6.0%. **(3a)** 3.17×10^{14}. **(3b)** 1 in 19,900. **(3c)** max 29 kW **(4)** 36.7 m^{-1} **(5)** yes. **(6)** light 103.4 MeV, heavy 74.3 MeV **(7)** 1.78 MeV **(8)** $E_\gamma \geq 11.46$ MeV **(9)** 2.7×10^8 K.

Chapter 12: **(1a)** 1.9 keV **(1b)** $\sim 1.8 \times 10^{-30}$. **(2a)** 12.3 × Lawson crit. **(2b)** 49 MW **(3)** 283 1 **(4)** 3.35×10^{10} kg s^{-1}. **(5)** 1.68%. **(6)** Water power $\sim 9.8 \times 10^7$ W, fusion $\sim 1.2 \times 10^{15}$ W.

Chapter 13: **(1)** Isotropic flux above 1 GeV. No variation with solar activity. **(2a)** High energy particles; ~70% p, ~20%, ~0.7% Li + Be + B ions, ~1.7% C + N + O ions, rest $Z > 10$. **(2b)** Mostly not. **(3)** 200−300 cps. **(4)** See Chapter 2, energy to mass conversion. **(5)** 0.131 Bq. **(6)** 9.76×10^6 years **(7)** average 266 Bq kg^{-1}. **(8)** 1.65×10^9 years **(9)** 5.0×10^8 years **(10)** 3.70×10^9 years **(11)** 1.17×10^{-3}, 1.41×10^{-3}, and 2.33×10^{-4} W m^{-2}. **(12a)** 193 t $year^{-1}$. **(12b)** 4.6×10^5 t $year^{-1}$. **(13)** ~18, ~53, and ~211 years.

Chapter 14: **(1)** n, on ^{196}Hg, or n,2n on ^{198}Hg. **(2)** 93%. **(3)** ^{239}Pu from d,n in ^{238}U. **(4)** ^{248}Cm and FP; ^{253}Es and ^{249}Cm; ^{250}Cm and FP. **(5)** 2.30 n per fiss. **(6)** possibly ^{259}Sg. **(7)** Am(III) has larger ionic radius than Eu(III) and is thus less strongly solvated by water. **(8)** 0.085 W **(9)** 5 f^{14} 7 s^2. **(10)** Pu(III)/Pu(IV) ~970 mV, Pu(IV)/Pu(V) ~1050 mV, Pu(V)/Pu(VI) ~930 mV; correct values in Fig. 14.7.

Chapter 15: (1) ^{3}H 163.5 Bq, ^{14}C 2.85 kBq, ^{40}K 4.40 kBq, Ra and daughters 0.93 Bq, sum 7.41 kBq/70 kg **(2)** 4.2×10^{-7}%. **(3)** 1.8×10^{-6}%. **(4a)** 124 μSv year^{-1}. **(4b)** 39 μSv year^{-1}. **(5)** 21 mSv. **(6a)** 4.37 MeV **(6b)** 698 million. **(7)** 1.14 μs.

Chapter 16: (1a) 0.329 cm **(1b)** 9.0 MeV **(1c)** ~^{38}Ar. **(2)** Between 1st and 2nd. **(3a)** 10 kW **(3b)** 228 m s^{-1}. **(3c)** 2646 m s^{-1}. **(4)** 7.6×10^{11} n m^{-2} s^{-1} **(5)** 4.105 m **(6a)** $L_2 = 1.41$ cm, $L_3 = 1.73$ cm **(6b)** 111.4 MHz **(7)** H^+ 11.8 MeV, 0.80 T; D^+ 23.6 MeV, 1.598 T: He^{2+} 46.9 MeV, 1.588 T **(8)** 1.01×10^{10} s^{-1}. **(9)** 8.53×10^{3} g **(10a)** 35. **(10b)** 78.8 MeV **(10c)** 124 MeV.

Chapter 17: (1) 28 TBq g^{-1} Na. **(2)** 10.6%. **(3)** 0.85 TBq. **(4a)** 1.56×10^{7} s **(4b)** 0.145. **(5)** 4873 cps. **(6)** 10.5 d.

Chapter 18: (1) 4.7 l **(2)** 24%. **(3)** 21.0%. **(4)** 1.1 mg **(5)** 45 ppm Ga. **(6)** 20.4%. **(7)** 3.32×10^{-4} s^{-1} M^{-1} **(8)** Calculated log(β_n) values are: 6.1, 10.6, 13.8. **(9)** log $\beta_4 = 15 \pm 1$ and log $K_{DC} = 0 \pm 0.3$. **(10)** 29 m^3 **(11)** −0.24 V **(12)** $D = 1$ for V at pH = 0.5 and for U at pH = 2.6; optimal separation at pH = 1.55. **(13)** 99.9% U with 9.8% La.

Chapter 19: (1) $465 million. **(2)** 0.765 kg ^{235}U. **(3a)** 7 oil tankers (7.1). **(3b)** One train car (8.9 t). **(3c)** 4 train cars (75.4 t). **(4)** 21 collisions (20.63). **(5)** 1.898. **(6)** $\eta = 2.022, f = 0.750, k_\infty = 1.451$. **(7)** 1.257 m **(8)** 2.131 m side. **(9a)** $\Lambda = 0.607$. **(9b)** $r = 0.256$ m **(10)** 23.1%. **(11)** ~1100 times. **(12)** 3.71×10^{12} atoms s^{-1}. **(13)** 9.0 h **(14)** 3.07 kg **(15)** 49 MW (16) 8.4×10^{17} fusions or ~1.7% consumed of all D^+.

Chapter 20: (1) ~34000 kg **(2)** 14600 m^3 **(3a)** 0.64 s **(3b)** 193 GJ **(4)** ^{36}Cl, ^{37}Cl, ^{40}K, and ^{42}K. **(5a)** ^{232}Th(n,2n)^{231}Th(β$^-$)^{231}Pa(n,γ)^{232}Pa(β$^-$)^{232}U. **(5b)** Yes. **(6)** 60%. **(7a)** Rest heat cooling to the sea. **(7b)** HV-stability more important. **(8)** Recirculated back to off-gas system (multiplies decay time).

Chapter 21: (1) 674 and 1200 °C. **(2a)** 1.40 kg **(2b)** Effective cross-sections (Fig. 19.5), self-screening, and capture to ^{240}Pu. **(3)** 152 d **(4a)** 2.333 kg ^{235}U. **(4b)** 1.141 kg ^{235}U. **(5)** See Fig. 19.17. **(6a)** 31.6 kg FP. **(6b)** 0.91%. **(7a)** ~202 TBq. **(7b)** Ce, Pr, Nb, Zr, Y, Sr, and Ru. **(8a)** 8.7×10^{5}. **(8b)** 3.9×10^{2}. **(9)** 404 MBq ^{238}Pu, 94 MBq ^{240}Pu. **(10)** 9.04×10^{15}. **(11a)** 95 kW **(11b)** 34 cylinders. **(11c)** 1.5 reactor years. **(12)** Ru 710 kg, 10.2 TBq kg^{-1}; Rh 153 kg, 47 TBq kg^{-1}; Pd 456 kg, 2.92 MBq kg^{-1}.

Chapter 22: (1) See Section 21.13.1. **(2)** None. **(3)** NpO, Pu^{4+}. **(4a)** See Section 21.13.3. **(4b)** ASSE salt dome, WIPP bedded salt. **(5)** $10^{13.8}$. **(6)** 3.7×10^{5}. **(7a)** $E = 0.739$ 0.236 pH. **(7b)** $Eh > 0.677$. **(8a)** 2.92 kBq. **(8b)** No.

Appendix A: (1a) 1.5%. **(1b)** 99.9%. **(2)** Use smaller solution for P; $\theta_{extr} = 8.60$, $\theta_{wash} = 94.6$.

APPENDIX C

APPENDIX I. PERIODIC TABLE OF THE ELEMENTS

1	2	3	4	5	6	7	8	9	10	11	12	13	14	15	16	17	18
1 H 1.008																	2 He 4.003
3 Li 6.94	4 Be 9.01											5 B 10.81	6 C 12.01	7 N 14.01	8 O 16.00	9 F 19.00	10 Ne 20.18
11 Na 22.99	12 Mg 24.31											13 Al 26.98	14 Si 28.09	15 P 30.97	16 S 32.06	17 Cl 35.45	18 Ar 39.95
19 K 39.10	20 Ca 40.08	21 Sc 44.96	22 Ti 47.90	23 V 50.94	24 Cr 52.01	25 Mn 54.94	26 Fe 55.85	27 Co 58.93	28 Ni 58.71	29 Cu 63.54	30 Zn 65.37	31 Ga 69.72	32 Ge 72.59	33 As 74.92	34 Se 78.96	35 Br 79.91	36 Kr 83.80
37 Rb 85.47	38 Sr 87.62	39 Y 88.91	40 Zr 91.22	41 Nb 92.91	42 Mo 95.94	43 Tc (97.91)	44 Ru 101.07	45 Rh 102.91	46 Pd 106.4	47 Ag 107.87	48 Cd 112.40	49 In 114.82	50 Sn 118.69	51 Sb 121.75	52 Te 127.60	53 I 126.90	54 Xe 131.30
55 Cs 132.91	56 Ba 137.34	57 - 71 *Ln	72 Hf 178.49	73 Ta 180.95	74 W 183.85	75 Re 186.2	76 Os 190.2	77 Ir 192.2	78 Pt 195.09	79 Au 196.97	80 Hg 200.59	81 Tl 204.37	82 Pb 207.19	83 Bi 208.98	84 Po (208.98)	85 At (209.99)	86 Rn (222.02)
87 Fr (223.02)	88 Ra (226.03)	88 - 103 **An	104 Rf (267)	105 Db (268)	106 Sg (271)	107 Bh (270)	108 Hs (269)	109 Mt (278)	110 Ds (281)	111 Rg (281)	112 Cn (285)	113	114 Fl (289)	115	116 Lv (293)	117	118

*Lanthanides	57 La 138.91	58 Ce 140.12	59 Pr 140.91	60 Nd 144.24	61 Pm (144.91)	62 Sm 150.35	63 Eu 151.96	64 Gd 157.25	65 Tb 158.92	66 Dy 162.50	67 Ho 164.93	68 Er 167.26	69 Tm 168.93	70 Yb 173.04	71 Lu 174.97
**Actinides	89 Ac (227.03)	90 Th 232.04	91 Pa (231.04)	92 U 238.03	93 Np (237.05)	94 Pu (244.06)	95 Am (243.06)	96 Cm (247.07)	97 Bk (247.07)	98 Cf (251.08)	99 Es (252.08)	100 Fm (257.10)	101 Md (258.10)	102 No (259.10)	103 Lr (262)

APPENDIX II. QUANTITIES AND UNITS

Quantity	Symbol	Unit	Symbol	Dimensions (within brackets), derived units etc.
Length	l	meter	m	1 fermi = 10^{-15} m = 1 fm; 1 μ = 10^{-6} m; 1 Å = 10^{-10} m
Mass	m	kilogram	kg	Basic SI unit; 1 ton (t) = 10^3 kg; 1 kg = 10^3 g (gram)
Time	t	second	s	" ; 1 (ephem.) year (y or a) = 365.24 days (d) =
Electric current	I	ampere	A	" ; = 8765.8 hours (h) = 3.1557 × 10^7 s
Thermodynamic temperature	T	kelvin	K	" ; $t_C = t_K - 273.15$, t_C = temp. in degree Celsius, °C
Amount of substance	n, ν	mole	mol	" ; molarity (mol/l) = $m_a M^{-1} V^{-1}$, m_a = mass of pure
Luminous intensity	I^*	candela	cd	" ; substance;
Atomic (molecular) weight	M	atomic mass unit	u	$M = m_i \times 10^3 \times \mathbf{N}_A$, m_i mass of atom (molecule); 1 mole = M g
Volume	V	cubic meter	m^3	1 m^3 = 10^3 liter (l, dm^3, L), 1 l = 10^3 milliliters (ml), 1 ml = 1 cm^3
Density (mass per unit volume)	ρ		kg m^{-3}	1 g cm^{-3} = 10^3 kg m^{-3}
Pressure	P	pascal	Pa	[Pa = Nm^{-2} = kg s^{-2} m^{-1}]; 1 atm = 1.013×10^5 Pa; 1 bar = 10^5 Pa
Energy	E	joule	J	[J = Ws = N m = kg m^2 s^{-2}] 1 torr = 133.3 Pa
Frequency	f, ν	hertz	Hz	[Hz = s^{-1}]
Force	F	newton	N	[N = kg m s^{-2}]; 1 dyne = 10^{-5} N; 1 kp = 9.8067 N
Angle, flat		radian	rad	1 rad = 57.30°; full circle 360° = 2π rad
" , space		steradian	sr	full space angle = 4π sr
Power	P	watt	W	[W = J s^{-1} = N m s^{-1} = kg m^2 s^{-3}]
Angular frequency	ω		rad s^{-1}	
Electric potential (voltage)	U	volt	V	[V = W A^{-1} = kg m^2 s^{-3} A^{-1}]
resistance	R	ohm	Ω	[Ω = V A^{-1} = W A^{-2} = kg m^2 s^{-3} A^{-2}]
charge	q	coulomb	C	[C = A s]; 1 C = 0.1 **c** statcoulomb (esu)
capacitance	C	farad	F	[F = C V^{-1} = s A^2 W^{-1} = s^4 A^2 kg^{-1} m^{-2}]
Magnetic inductance	L^*	henry	H	[H = Wb A^{-1} = V s A^{-1} = kg m^2 s^{-2} A^{-2}]
induction (flux density)	B	tesla	T	[T = Wb m^{-2} = V s m^{-2} = kg s^{-2} A^{-1}]; 1 gauss = 10^{-4} T
flux	Φ_B^*	weber	Wb	[Wb = V s = W s A^{-1} = kg m^2 s^{-2} A^{-1}]
Radioactivity	A	becquerel	Bq	[Bq = (radioactive events) s^{-1}]; 1 curie (Ci) = 3.7 × 10^{10} Bq
Radiation exposure		kerma		[C kg^{-1} = A s kg^{-1}]
Radiation dose, absorbed	D	gray	Gy	[Gy = J kg^{-1} = m^2 s^{-2}]; 1 Gy = 100 rad **
Radiation dose, equivalent	H	sievert	Sv	[Sv = J kg^{-1} = m^2 s^{-2}]; 1 Sv = 100 rem **

* Not used in this text. ** Old radiation dose unit (see §7.3).

APPENDIX III. FUNDAMENTAL CONSTANTS

Quantity	Symbol	Value	SI-unit	Auxiliary value
Speed of light in vacuum	$\mathbf{c}$	299 792 458	m/s	Basic SI unit (exact)
Elementary charge	$\mathbf{e}$	$1.602\,177\,3 \times 10^{-19}$	C	
Planck constant	$\mathbf{h}$	$6.626\,076 \times 10^{-34}$	J s	$= 4.135\,669 \times 10^{-15}$ eV s; $\hbar = \mathbf{h}/2\pi = 1.054\,572\,7 \times 10^{-34}$ J s
Avogadro constant	$\mathbf{N_A}$	$6.022\,137 \times 10^{23}$	mol^{-1}	
Atomic mass unit	u	$1.660\,540 \times 10^{-27}$	kg	$= 931.494\,3$ MeV; mass of $^{12}C = 12$ u
Electron rest mass	$\mathbf{m_e}$	$9.109\,390 \times 10^{-31}$	kg	$\mathbf{M_e} = \mathbf{N_A} \times \mathbf{m_e} = 5.485\,799\,0 \times 10^{-4}$ u $= 0.510\,999\,1$ MeV
Proton rest mass	$\mathbf{m_p}$	$1.672\,623 \times 10^{-27}$	kg	$\mathbf{M_p} = \mathbf{N_A} \times \mathbf{m_p} = 1.007\,276\,47$ u $= 938.272\,3$ MeV
Neutron rest mass	$\mathbf{m_n}$	$1.674\,929 \times 10^{-27}$	kg	$\mathbf{M_n} = \mathbf{N_A} \times \mathbf{m_n} = 1.008\,664\,90$ u $= 939.565\,6$ MeV
Faraday constant	$\mathbf{F}$	96 485.31	$C\,mol^{-1}$	$= \mathbf{N_A} \times \mathbf{e}$
Rydberg constant	$\mathbf{R_\infty}$	10 973 731.53	m^{-1}	$\mathbf{R_\infty} \times \mathbf{h} \times \mathbf{c} = 13.605\,698$ eV
Inverse fine structure constant	α^{-1}	137.035 990		$= \mu^0 \times \mathbf{c} \times \mathbf{e}^2/2 \times \mathbf{h}$; μ^0 (permeability of vacuum) $= 4\pi \times 10^{-7}\,H\,m^{-1}$
Bohr radius	a_0	$0.529\,177\,25 \times 10^{-10}$	m	$= \alpha / 4\pi \mathbf{R_\infty}$
Electron magnetic moment	μ_e	$-9.284\,770 \times 10^{-24}$	$J\,T^{-1}$	
Proton magnetic moment	μ_p	$1.410\,607\,6 \times 10^{-26}$	$J\,T^{-1}$	
Neutron magnetic moment	μ_n	$-0.966\,237\,0 \times 10^{-26}$	$J\,T^{-1}$	
Bohr magneton	μ_B	$9.274\,015 \times 10^{-24}$	$J\,T^{-1}$	$= \mathbf{e} \times \hbar/2 \times \mathbf{m_e}$; ($1\,J\,T^{-1} = 10^3$ erg $gauss^{-1}$)
Nuclear magneton	μ_N	$5.050\,787 \times 10^{-27}$	$J\,T^{-1}$	$= \mathbf{e} \times \hbar/2 \times \mathbf{m_p}$
Molar gas constant	$\mathbf{R}$	8.314 51	$J\,mol^{-1}\,K^{-1}$	$= 0.082\,06$ l atm $mol^{-1}\,K^{-1}$
Molar volume of ideal gas at STP	$\mathbf{V_0}$	$22.414\,1 \times 10^{-3}$	$m^3\,mol^{-1}$	$= \mathbf{R}\,T_0/p_0$; $T_0 = 273.15$ K, $p_0 = 1$ atm $= 101\,325$ Pa
Boltzmann constant	$\mathbf{k}$	$1.380\,66 \times 10^{-23}$	$J\,K^{-1}$	$= \mathbf{R}/\mathbf{N_A} = 8.617\,39 \times 10^{-5}\,eV\,K^{-1}$; $1/\mathbf{k} = 11\,604.4\,K\,eV^{-1}$
Gravitational constant	$\mathbf{G}$	$6.672\,0 \times 10^{-11}$	$N\,m^2\,kg^{-2}$	
Acceleration of gravity at sea level	$\mathbf{g}$	9.806 65	$m\,s^{-2}$	(at 45° latitude)

APPENDIX IV. ENERGY CONVERSION FACTORS (CODATA 1986)

Joule (J)	Kilowatthour (kWh)	Atomic mass unit (u)	Kilocalorie (kcal)	Electronvolt (eV)
1	$2.777\,78 \times 10^{-7}$	$6.700\,53 \times 10^{9}$	$2.388\,46 \times 10^{-4}$	$6.241\,46 \times 10^{18}$
$3.600\,00 \times 10^{6}$	1	$2.412\,19 \times 10^{16}$	$8.598\,46 \times 10^{2}$	$2.246\,93 \times 10^{25}$
$1.492\,419 \times 10^{-10}$	$4.145\,61 \times 10^{-17}$	1	$3.564\,58 \times 10^{-14}$	$9.314\,943 \times 10^{8}$
$4.186\,80 \times 10^{3}$	$1.163\,00 \times 10^{-3}$	$2.805\,38 \times 10^{13}$	1	$2.613\,20 \times 10^{22}$
$1.602\,177 \times 10^{-19}$	$4.450\,53 \times 10^{-26}$	$1.073\,544 \times 10^{-9}$	$3.826\,77 \times 10^{-23}$	1

1 eV $= 1.602\,177 \times 10^{-19}$ J; 1 eV $atom^{-1} = 23.045\,0$ kcal $mol^{-1} = 96.485$ kJ mol^{-1}
1 Q $= 10^{18}$ Btu; 1 Btu (British thermal unit) $= 1.055\,06$ kJ; 1 hp (horse power) $= 0.746$ kW
1 toe (ton oil equivalent) = 10 Gcal = 11.63 MWh = 41.87 GJ
1 ton hard coal (tce) = 0.65 toe = 27.2 GJ; 1000 m^3 natural gas = 0.80 toe
1 g ^{235}U fissioned at 200 MeV/fission = 82.11 GJ = 0.95 MWd (heat)
1 erg (dyne cm) $= 1.000 \times 10^{-7}$ J
Energy-wavelength product ($\Delta E\, \lambda$) = 12 398.5 eV Å

Prefixes for powers of ten		
E	exa	10^{18}
P	peta	10^{15}
T	tera	10^{12}
G	giga	10^{9}
M	mega	10^{6}
k	kilo	10^{3}
h	hecto	10^{2}
d	deci	10^{-1}
c	centi	10^{-2}
m	milli	10^{-3}
μ	micro	10^{-6}
n	nano	10^{-9}
p	pico	10^{-12}
f	femto	10^{-15}
a	atto	10^{-18}

Some numerical values	
e	2.718 28
log e	0.434 29
ln 2	0.693 15
ln 10	2.302 59
ln 2/ln 10	0.301 03
π	3.141 59
ln a = ln 10 × log a	

Some English measures	
1 inch	0.025 4 m
1 (statue) mile	1 609.34 m
1 (int.) nautical mile	1 852 m
1 (US liq.) gallon	0.003785 m^3
1 barrel	0.159 0 m^3
1 cubic foot	0.028 32 m^3
1 pound (mass)	0.4536 kg

ELEMENT AND NUCLIDE INDEX

0-9

INDEX

Note: Page numbers with "*f*" denote figures; "*t*" tables; and "*b*" boxes.

C

D

G

H

I

J

K

L

M

N

U

Printed and bound by CPI Group (UK) Ltd, Croydon, CR0 4YY
08/05/2025
01864863-0001